Wartime Jeeps
Model GPW
Ultimate Military Technical
Manual Collection

TM 9-803
Operating and Maintenance Instructions

TM 10-513
Maintenance Manual May 1942 Change 1

TM 9-1803A
Engine and Engine Accessories Maintenance Manual

TM 9-1803B
Power Train, Body and Frame Maintenance Manual

SNL G-503
Ordinance Catalog

AR-850
Army Regulations - Marking of Equipment, Property and Vehicles

Use the black index marks to quickly find each manual in this book.

edited by
Brian Greul

The GPW Model MB, commonly referred to as the Jeep is probably the most ubiquitous American military vehicle ever produced. Simple, rugged, and capable of hard work it began as a World War II vehicle and descendant vehicles are still produced today as passenger vehicles.

This book is intended to support enthusiasts and their restoration efforts by providing a professionally printed, 8.5x11 compilation of the key manuals for this vehicle.

Every effort has been made to faithfully reproduce the document while cleaning up the pages to make them usable to you the reader. However, we are dealing with original works that have been electronically preserved from nearly 80 years ago. There are a number of artifacts in the source documents. Your understanding is appreciated.

An 8.5x11 3 hole punched loose leaf copy may be purchased for your 3 ring binder. Email books@ocotillopress.com for current information.

Should you have suggestions or feedback on ways to improve this book please send email to Books@OcotilloPress.com

---

Edited 2021 Ocotillo Press
ISBN 978-1-954285-10-1

No rights reserved. This content of this book is in the public domain as it is a work of the US Government. It is reproduced by the publisher as a convenience to enthusiasts and others who may wish to own a quality copy of it. It has been adjusted to accomodate the printing and binding process.

Printed in the United States of America

Ocotillo Press
Houston, TX 77017
Books@OcotilloPress.com

Disclaimer: The user of this book is responsible for following safe and lawful practices at all times. The publisher assumes no responsibility for the use of the content of this book. The publisher has made an effort to ensure that the text is complete and properly typeset, however omissions, errors, and other issues may exist that the publisher is unaware of.

WAR DEPARTMENT TECHNICAL MANUAL

TM 9-803

*This manual supersedes TB 9-803-4, 5 January 1944. For supersession of Quartermaster Corps 10-series technical manuals, see paragraph 1

1/4-Ton 4x4 TRUCK (WILLYS-OVERLAND MODEL MB & FORD MODEL GPW)

WAR DEPARTMENT .   FEBRUARY 1944

United States Government Printing Office
Washington : 1947

TM 9-803

¼-TON 4 x 4 TRUCK (WILLYS-OVERLAND MODEL MB and FORD MODEL GPW)

CONTENTS

PART ONE—OPERATING INSTRUCTIONS

			Paragraphs	Pages
SECTION	I	Introduction	1	5–9
	II	Description and tabulated data	2–3	10–12
	III	Driving controls and operation	4–6	13–20
	IV	Operation under unusual conditions	7–11	21–27
	V	First echelon preventive maintenance service	12–16	28–36
	VI	Lubrication	17–18	37–48
	VII	Tools and equipment stowage on the vehicle	19–21	49–51

PART TWO—VEHICLE MAINTENANCE INSTRUCTIONS

SECTION	VIII	Record of modifications	22	52
	IX	Second echelon preventive maintenance	23	53–67
	X	New vehicle run-in test	24–26	68–72
	XI	Organization tools and equipment	27–28	73
	XII	Trouble shooting	29–49	74–103
	XIII	Engine—description, data, maintenance, and adjustment in vehicle	50–59	104–115
	XIV	Engine—removal and installation	60–61	116–117
	XV	Ignition system	62–69	118–125
	XVI	Fuel and air intake and exhaust systems	70–78	126–136
	XVII	Cooling system	79–86	137–143
	XVIII	Starting system	87–90	144–145
	XIX	Generating system	91–94	146–149

3

¼-TON 4 x 4 TRUCK (WILLYS-OVERLAND MODEL MB and FORD MODEL GPW)

			Paragraphs	Pages
Section	XX	Battery and lighting system	95–106	150–166
	XXI	Clutch	107–112	167–171
	XXII	Transmission	113–116	172–175
	XXIII	Transfer case	117–121	176–178
	XXIV	Propeller shafts and universal joints	122–125	179–180
	XXV	Front axle	126–137	181–188
	XXVI	Rear axle	138–145	189–192
	XXVII	Brakes	146–152	193–204
	XXVIII	Springs and shock absorbers	153–157	205–210
	XXIX	Steering gear	158–163	211–215
	XXX	Body and frame	164–175	216–220
	XXXI	Radio interference suppression system	176–179	221–227
	XXXII	Shipment and temporary storage	180–182	228–232
References				233–234
Index				235

TM 9-803
1

PART ONE—OPERATING INSTRUCTIONS

Section I

INTRODUCTION

Paragraph

Scope 1

1. SCOPE.

a. This technical manual* is published for the information and guidance of the using arm personnel charged with the operation and maintenance of this materiel.

b. In addition to a description of the ¼-ton 4 x 4 Truck (Willys-Overland model MB and Ford GPW), this manual contains technical information required for the identification, use, and care of the materiel. The manual is divided into two parts. Part One, sections I through VII, contains vehicle operating instructions. Part Two, sections VIII through XXXII, contains vehicle maintenance instructions to using arm personnel charged with the responsibility of doing maintenance work within their jurisdiction, including radio suppression and shipment and temporary storage information.

c. In all cases where the nature of the repair, modification, or adjustment is beyond the scope of facilities of the unit, the responsible ordnance service should be informed so that trained personnel with suitable tools and equipment may be provided, or proper instructions issued.

d. This manual includes operating and organizational maintenance instructions from the following Quartermaster Corps 10-series technical manuals. Together with TM 9-1803A and TM 9-1803B, this manual supersedes them:

(1) TM 10-1103, 20 August 1941.
(2) TM 10-1207, 20 August 1941.
(3) TM 10-1349, 3 January 1942.
(4) TM 10-1513, Change 1, 15 January 1943.

---

*To provide operating instructions with the materiel, this technical manual has been published in advance of complete technical review. Any errors or omissions will be corrected by changes or, if extensive, by an early revision.

TM 9-803
1
¼-TON 4 x 4 TRUCK (WILLYS-OVERLAND MODEL MB and FORD MODEL GPW)

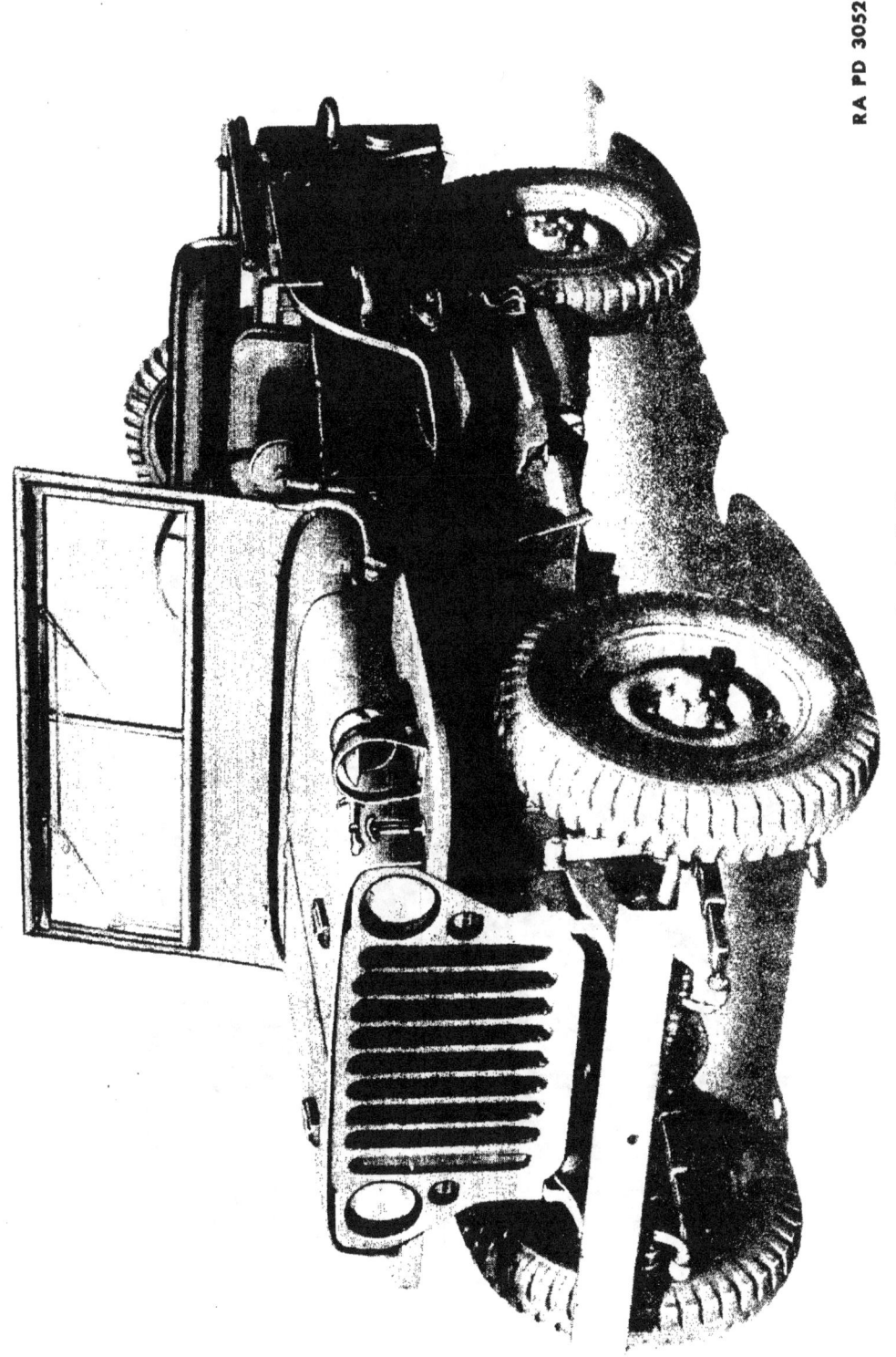

Figure 1—¼-Ton 4 x 4 Truck—Left Front

RA PD 305251

INTRODUCTION

Figure 2—1/4-Ton 4 x 4 Truck—Right Rear

TM 9-803
1

¼-TON 4 x 4 TRUCK (WILLYS-OVERLAND MODEL MB and FORD MODEL GPW)

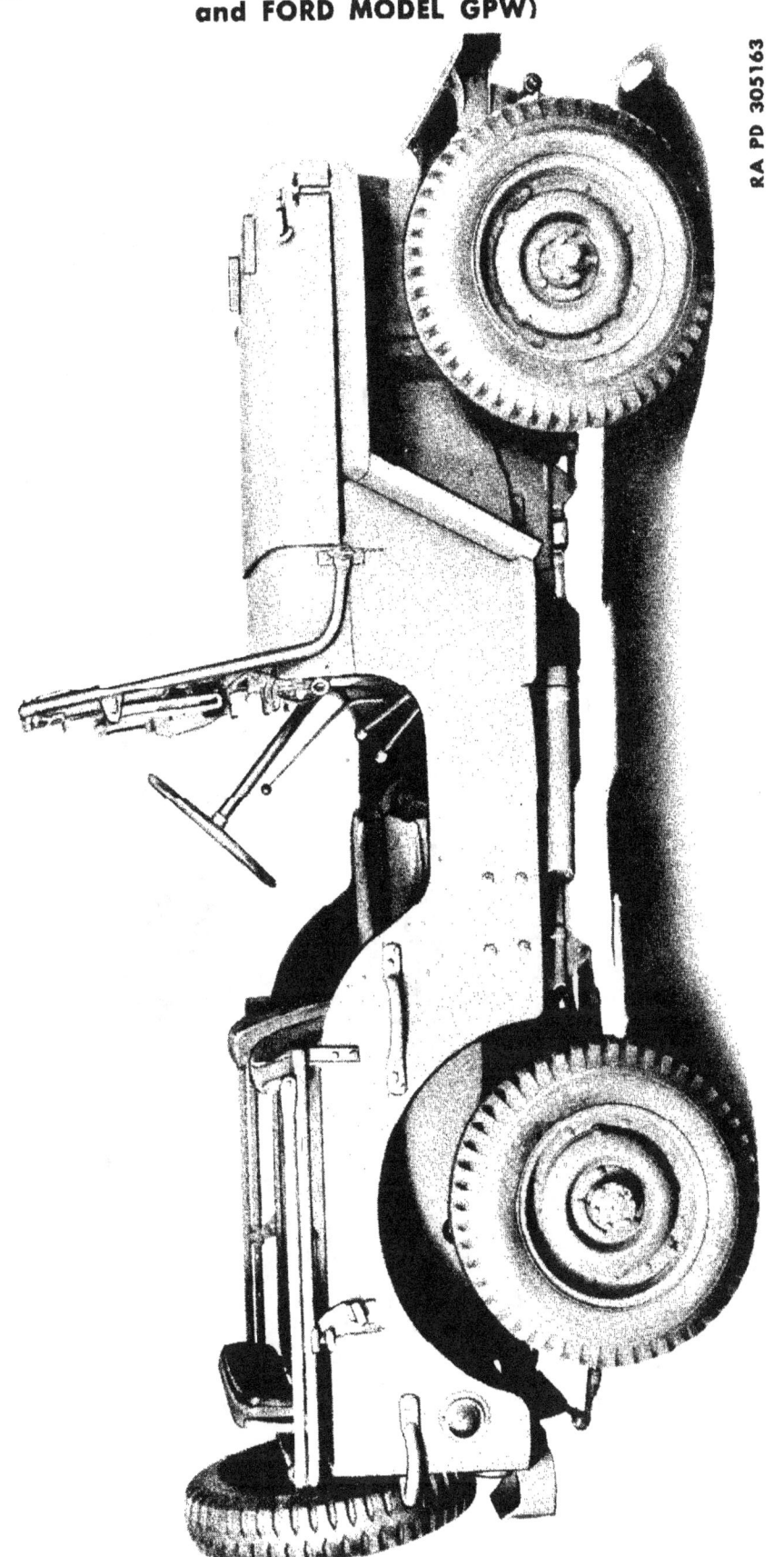

Figure 3 — ¼-Ton 4 x 4 Truck — Right Side

RA PD 305163

TM 9-803
1

INTRODUCTION

Figure 4 — 1/4-Ton 4 x 4 Truck — Right Front

TM 9-803

¼-TON 4 x 4 TRUCK (WILLYS-OVERLAND MODEL MB and FORD MODEL GPW)

Section II

DESCRIPTION AND TABULATED DATA

	Paragraph
Description	2
Data	3

2. DESCRIPTION.

a. Type. This vehicle is a general purpose, personnel, or cargo carrier especially adaptable for reconnaisance or command, and designated as ¼-ton 4 x 4 Truck. It is a four-wheel vehicle with four-wheel drive. The engine is a 4-cylinder gasoline unit located in the conventional place, under the hood at the front of the vehicle. A conventional three-speed transmission equipped with a transfer case provides additional speeds for traversing difficult terrain. The body is of the open type with an open driver's compartment. The folding top can be removed and stowed; and, the windshield tilted forward on top of the hood, or opened upward and outward. A spare wheel equipped with a tire is mounted on the rear of the body, and a pintle hook is provided to haul trailed loads. Specifications of the vehicle are given under "Data" (par. 3). General physical characteristics are shown in figures 1 through 4.

b. Identification. The manufacturer's chassis serial number is stamped on a plate inside the left frame side member at the front end, and on the name plate (fig. 6). The engine serial number is stamped on the right side of the cylinder block, front upper corner. The U.S.A. registration number is painted on both sides of the hood.

3. DATA.

a. Vehicle Specifications.

Wheelbase	80 in.
Length, over-all	132¼ in.
Width, over-all	62 in.
Height, over-all—top up	69¾ in.
—top down	52 in.
Wheel size	combat 16 x 4.50 E
Tire size	16 x 6.00 in.
Tire pressure (front and rear)	35 lb
Tire type	mud and snow
Tire plies	6
Tread (center-to-center)—front	49 in.
—rear	49 in.
Crew, operating	2
Passenger capacity including crew	5

TM 9-803

DESCRIPTION AND TABULATED DATA

Weights:
Road, including gas and water 2,453 lb
Gross (loaded) 3,253 lb
Shipping (less water and fuel) 2,337 lb
Boxed gross 3,062 lb
Maximum pay load 800 lb
Maximum trailed load 1,000 lb
Ground clearance 8¾ in.
Pintle height (loaded) 21 in.
Kind and grade of fuel (octane rating) Gasoline (68 mm)
Approach angle 45 deg Departure angle 35 deg Shipping dimensions—cubic feet 331 —square feet 57

b. Performance.

Maximum allowable speeds (mph) with transfer case in "HIGH" range:
  High gear (3rd) 65
  Intermediate gear (2nd) 41
Low gear (1st) 24

Reverse gear 18
Maximum allowable speeds (mph) with transfer case in "LOW" range:
High gear (3rd) 33
Intermediate gear (2nd) 21
Low gear (1st) 12
Reverse gear 9
Maximum grade ability 60 pct
Minimum turning radius—right 17½ ft
  —left 17 1/2 ft Maximum fording depth. 21 in.
Towing facilities—front none
  —rear pintle hook
Maximum draw-bar pull 1,930 lb Engine idle speed 600 rpm Miles per gallon—(high gear—high range)
average conditions 20
Cruising range—(miles) average conditions 20

c. Capacities.

Engine crankcase capacity—dry 5 qt
    —refill 4 qt
Transmission capacity ¾ qt Transfer case capacity 1 1/2 qt

¼-TON 4 x 4 TRUCK (WILLYS-OVERLAND MODEL MB and FORD MODEL GPW)

Front axle capacity (differential) .... 1¼ qt
Rear axle capacity (differential) .... 1¼ qt
Front axle steering knuckle universal joint .... ¼ qt
Steering gear housing .... ¼ qt
Air cleaner (oil bath) .... ⅝ qt
Fuel tank capacity .... 15 gal
Cooling system capacity .... 11 qt
Brake system (hydraulic brake fluid) .... ¼ qt
Shock absorbers—front .... 5 oz
          —rear .... 5¾ oz

d. Communications.

(1) RADIO OUTLET BOX. A radio outlet box is provided on the later vehicles to use the vehicle battery (6-volt current supply). This outlet is located against the body side panel at the right front seat.

(2) AUXILIARY GENERATOR. A 12-volt, 55-ampere auxiliary generator is furnished on some vehicles. The generator is driven by a V-belt from a power take-off unit on the rear of the transfer case. Instructions for operation and care accompany those vehicles.

Section III

DRIVING CONTROLS AND OPERATION

	Paragraph
Instruments and controls	4
Use of instruments and controls in vehicular operation	5
Towing the vehicle	6

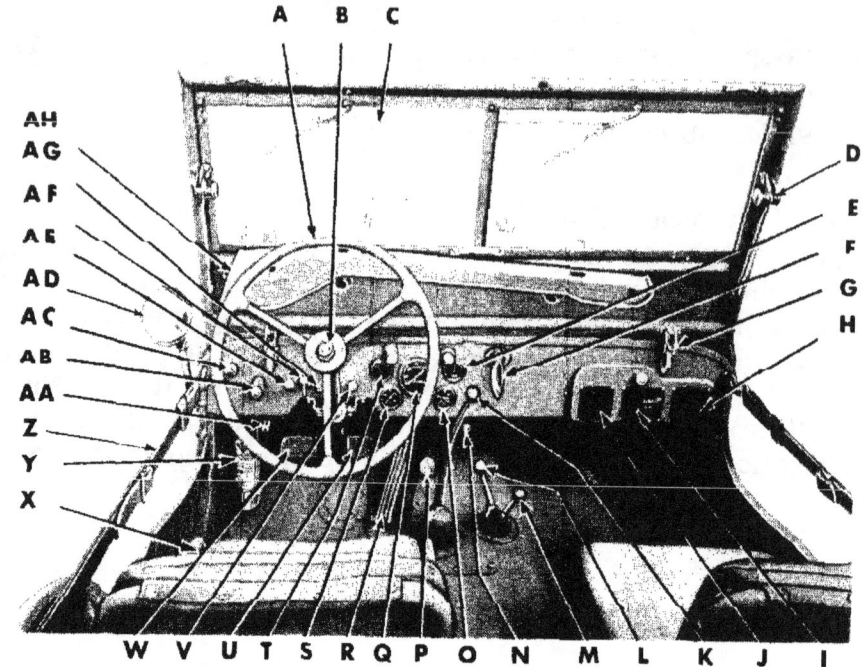

A STEERING WHEEL	R ACCELERATOR (FOOT THROTTLE)
B HORN BUTTON	S OIL PRESSURE GAGE
C WINDSHIELD WIPERS	T FUEL GAGE
D WINDSHIELD ADJUSTING ARMS	U BRAKE PEDAL
E AMMETER	V INSTRUMENT PANEL LIGHT SWITCH
F HAND BRAKE	W CLUTCH PEDAL
G WINDSHIELD CLAMPS	X FUEL TANK
H CAUTION PLATE	Y FIRE EXTINGUISHER
I NAME PLATE	Z SAFETY STRAP
J SHIFT PLATE	AA HEADLIGHT FOOT SWITCH (BEAM CONTROL)
K TRANSMISSION GEAR SHIFT LEVER	AB BLACKOUT LIGHT SWITCH
L TRANSFER CASE SHIFT LEVER—FRONT AXLE DRIVE	AC BLACKOUT DRIVING LIGHT SWITCH
M TRANSFER CASE SHIFT LEVER—AUXILIARY RANGE	AD REAR VISION MIRROR
N STARTING SWITCH	AE CHOKE CONTROL
O TEMPERATURE GAGE	AF IGNITION SWITCH
P ACCELERATOR FOOT REST	AG HAND THROTTLE
Q SPEEDOMETER	AH RIFLE HOLDER

RA PD 334753

Figure 5—Instruments and Controls

4. INSTRUMENTS AND CONTROLS.

a. Instruments.

(1) AMMETER (fig. 5). The ammeter on the instrument panel indicates the rate of current flow when the generator is charging the battery, and also indicates the amount of current being consumed when the engine is idle.

(2) FUEL GAGE (fig. 5). The fuel gage on the instrument panel

TM 9-803
4
¼-TON 4 x 4 TRUCK (WILLYS-OVERLAND MODEL MB and FORD MODEL GPW)

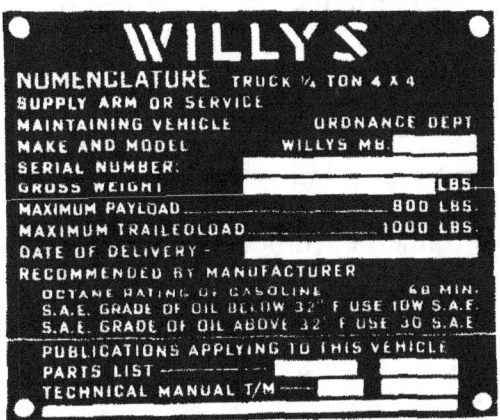

Figure 6—Name Plate

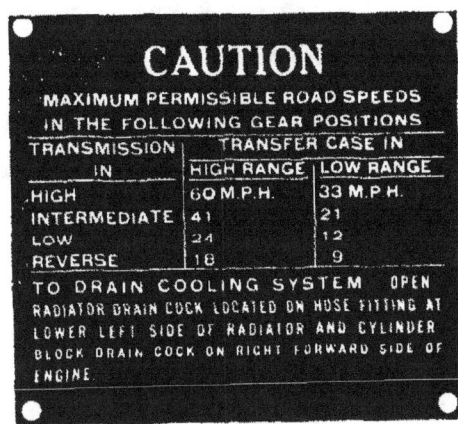

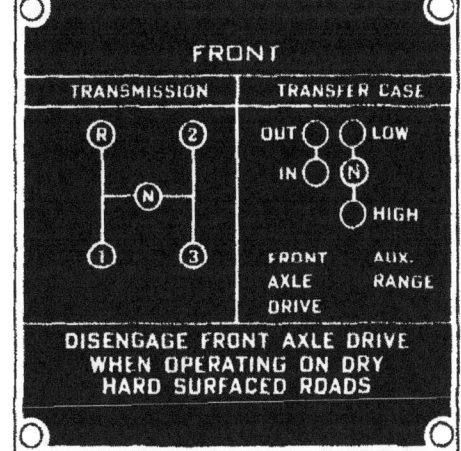

Figure 7—Caution Plate   Figure 8—Shift Plate

DRIVING CONTROLS AND OPERATION

is an electrical unit which indicates the fuel level in the tank, and only registers while the ignition switch is turned on.

(3) OIL PRESSURE GAGE (fig. 5). The oil pressure gage located on the instrument panel indicates the oil pressure when the engine is running.

(4) SPEEDOMETER (fig. 5). The speedometer on the instrument panel indicates in miles per hour the speed at which the vehicle is being driven. The odometer (in upper part of speedometer face) registers the total number of miles the vehicle has been driven. A trip indicator (in lower part of speedometer face) gives distance covered on any trip. Set trip indicator by turning the knurled control shaft extending through back of the speedometer.

(5) TEMPERATURE GAGE (fig. 5). The temperature gage registers the temperature of the solution in the cooling system.

b. Controls.

(1) BLACKOUT DRIVING LIGHT SWITCH (fig. 5). The blackout driving light switch (B.O. DRIVE) on the instrument panel controls the blackout driving light located on the left front fender, to furnish additional light during blackout periods. To operate light, first pull the blackout *light* switch button to the first position, then pull blackout *driving* light switch knob. To switch off the light, push in blackout *driving* light switch knob.

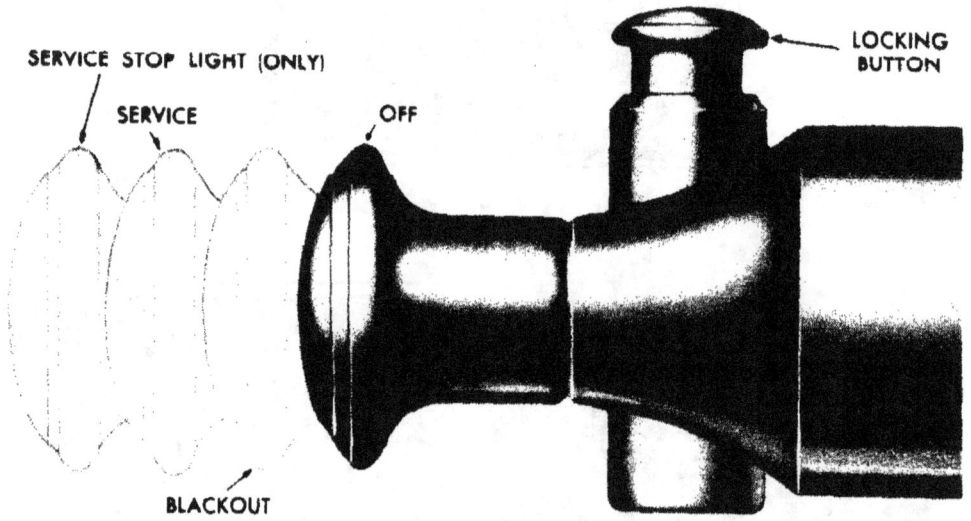

Figure 9—Blackout Light Switch Operating Positions

(2) BLACKOUT LIGHT SWITCH (fig. 5). The knob on the instrument panel (LIGHTS) controls the entire lighting system, including the instrument panel lights, blackout driving light, and stop lights. A circuit-breaker type fuse, on the back of the switch, opens when a short circuit occurs, and closes when the thermostatic element cools. The light switch is a four-position push-pull type with a safety lock (fig. 9). When the control knob is pulled out to the first position, the blackout headlights and blackout stop and taillights are turned on.

15

TM 9-803
¼-TON 4 x 4 TRUCK (WILLYS-OVERLAND MODEL MB and FORD MODEL GPW)

The switch control knob travel is automatically locked in this position by the lock-out button to prevent accidentally turning on of the service (bright) lights in a blackout area. To obtain service lights, push in on lock-out control button on the left side of the switch, and pull out control knob to second position. When switch is in this position service headlights, service stop and taillights are turned on, and the panel lights can be turned on by pulling out on the knob (PANEL LIGHTS). CAUTION: *When driving during the day, press in lock-out control button, and pull control knob out to the last or stop light position to cause only the regular stop light to function.*

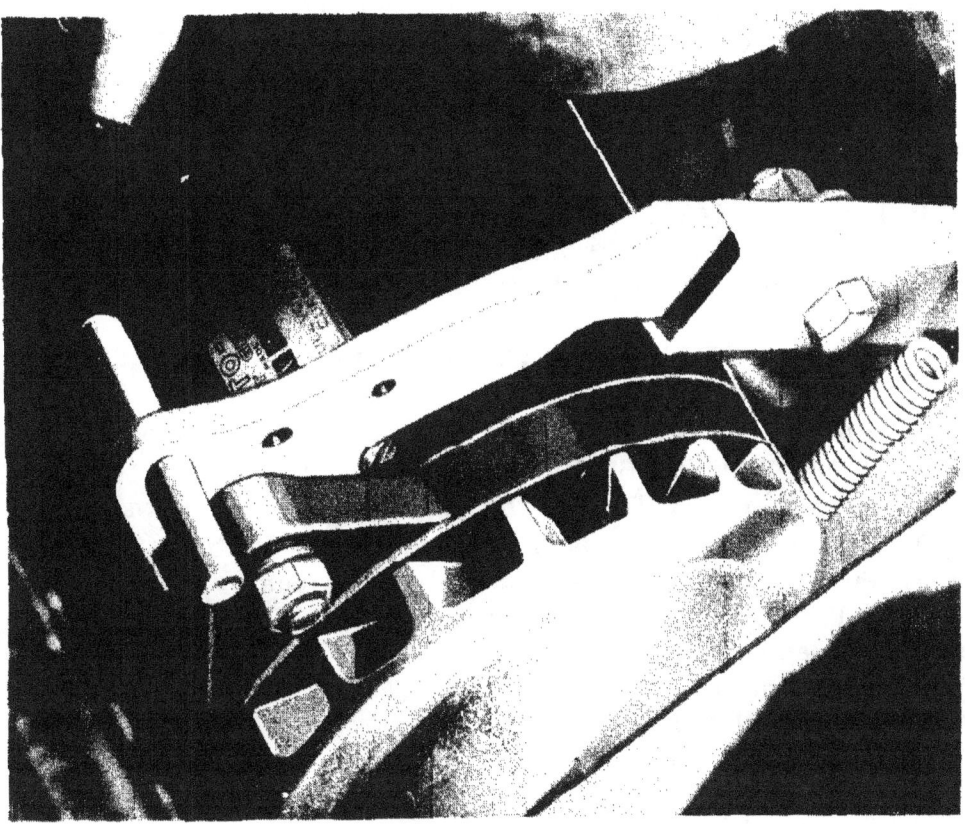

RA PD 305165

Figure 10—Generator Brace

(3) PANEL LIGHT SWITCH (fig. 5). The panel light switch knob (PANEL LIGHTS), located on the instrument panel, controls the lights to illuminate the panel instruments and controls. The blackout light switch (subpar. b (2) above) must be in service (bright light) position for this switch to control the panel lights.

(4) FIRE EXTINGUISHER (fig. 5). The fire extinguisher is mounted inside the left cowl panel. To remove, pull outward on the clamp release lever. To operate extinguisher, hold body in one hand and with the other, turn handle to left one-quarter turn, which releases plunger lock. Use pumping action to force liquid on base of fire. Read instructions on fire extinguisher plate.

TM 9-803
4-5

DRIVING CONTROLS AND OPERATION

(5) HAND BRAKE (fig. 5). The hand brake is applied by pulling out on the handle at the center of the instrument panel. Pull the handle out in a vertical position when the vehicle is parked. The brake is released by turning the handle one-quarter turn.

(6) WINDSHIELD ADJUSTING ARMS (fig. 5). The windshield adjustment arms are mounted on each end of the windshield frame. To open windshield, loosen knobs and push forward on lower part, then set by tightening the knobs.

(7) WINDSHIELD CLAMPS (fig. 5). The windshield clamps are located on the lower part of the windshield. Pull up on both clamps and unhook them, after which the windshield can be lowered on top of the hood. Be sure to hook down the windshield, using the hold-down catches on both sides of the hood.

(8) GENERATOR BRACE (fig. 10). The generator brace can be pulled up to release tension on the fan belt and stop the fan from throwing water over the engine when crossing a stream. Pull generator out to running position as soon as possible thereafter, and it will lock in place. CAUTION: *Be sure fan belt is on pulleys.*

(9) OTHER INSTRUMENTS AND CONTROLS. Other instruments and controls are of the conventional type, and are shown in figure 5.

5. USE OF INSTRUMENTS AND CONTROLS IN VEHICULAR OPERATION.

a. Before-operation Service. Perform the services in paragraph 13 before attempting to start the engine.

b. Starting Engine. To start the engine proceed as follows:

(1) Put transmission gearshift lever in neutral position (fig. 8).

(2) Pull out hand throttle button about ¾ inch to 1 inch.

(3) Pull out choke button all the way. NOTE: *Choking is not necessary when engine is warm.*

(4) Turn ignition to "ON" position.

(5) Depress clutch pedal to disengage clutch, and hold pedal down while engine is started.

(6) Step on starting switch to crank again. Release switch as soon as engine starts.

(7) Adjust choke and throttle control buttons to obtain proper idling speed. As engine warms up, push choke button all the way in.

(8) Check oil pressure gage reading: at idle speed the indicator hand should show at least 10 on the gage.

(9) Check ammeter for charge reading. Check fuel gage for indication of fuel supply.

(10) After engine has operated a few minutes, check temperature gage reading. Normal operating temperature is between 160°F and 185°F.

(11) In extremely cold weather refer to paragraph 7.

c. Placing Vehicle in Motion.

(1) For daytime driving turn on service stop light (par. 4 b (2)).

(2) Place transfer case right-hand shift lever in rear position to

¼-TON 4 x 4 TRUCK (WILLYS-OVERLAND MODEL MB and FORD MODEL GPW)

engage "HIGH" range, then place center shift lever in forward position to disengage front axle (fig. 8).

(3) Depress clutch pedal, and move transmission shift lever toward driver and backward to engage low (1st) gear (fig. 8).

(4) Release parking (hand) brake.

(5) Slightly depress accelerator to increase engine speed, and at the same time slowly release clutch pedal, increasing pressure on accelerator as clutch engages and vehicle starts to move. NOTE: *During the following operations perform procedures outlined in paragraph 14.*

(6) Increase speed to approximately 10 miles per hour, depress clutch pedal, and at the same time release pressure on accelerator. Move transmission shift lever out of low gear into neutral, and then into second gear. No double clutching is required. Release clutch pedal and accelerate engine.

(7) After vehicle has attained a speed of approximately 20 miles per hour, follow the same procedure as outlined above in order to shift into high (3rd) gear, moving the gearshift lever straight back.

d. Shifting to Lower Gears in Transmission. Shift to a lower gear before engine begins to labor, as follows: Depress clutch pedal quickly, shift to next lower gear, increase engine speed, release clutch pedal slowly, and accelerate. When shifting to a lower gear at any rate of vehicle speed, make sure that the engine speed is synchronized with vehicle speed before clutch is engaged.

e. Shifting Gears in Transfer Case (fig. 8). The transfer case is the means by which power is applied to the front and rear axles. In addition, the low gear provided by the transfer case further increases the number of speeds provided by the transmission. The selection of gear ratios depends upon the road and load conditions. Shift gears in the transfer case in accordance with the shift plate (fig. 8), and observe the instructions on the caution plate (fig. 7). The transmission gearshift does not in any way affect the selection or shifting of the transfer case gears. Vehicle may be driven by rear axle, or by both front and rear axles. The front axle cannot be driven independently.

(1) FRONT AXLE ENGAGEMENT. Front axle should be engaged only in off-the-road operation, slippery roads, steep grades, or during hard pulling. Disengage front axle when operating on average roads under normal conditions.

(a) Engaging Front Axle with Transfer Case in "HIGH" Range. With transfer case in "HIGH" range, move front axle drive shift lever to "IN" position. Depressing the clutch pedal will facilitate shifting.

(b) Disengaging Front Axle with Transfer Case in "HIGH" Range. Move front axle drive shift lever to "OUT" position. Depress the clutch pedal to facilitate shifting.

(c) Disengaging Front Axle when Transfer Case is in "LOW."
1. Depress clutch pedal, then shift transfer case lever into "HIGH."
2. Shift front axle drive lever into "OUT" position.

DRIVING CONTROLS AND OPERATION

3. Release clutch pedal and accelerate engine to desired speed.

(2) ENGAGING TRANSFER CASE LOW RANGE. Transfer case LOW range cannot be engaged until front axle drive is engaged.

(a) Engage front axle drive (subpar. e (1) above).

(b) Depress clutch pedal and move transfer case shift lever into "N" (neutral) position.

(c) Release clutch pedal and accelerate engine.

(d) Depress clutch pedal again and move transfer case shift lever forward into "LOW" position.

(e) Release clutch pedal, and accelerate engine to desired speed.

(3) ENGAGING TRANSFER CASE—"LOW" to "HIGH." This shift can be made regardless of vehicle speed.

(a) Depress clutch pedal and move transfer case shift lever into "HIGH" position.

(b) Release clutch pedal, and accelerate engine to desired speed.

f. Stopping the Vehicle. Remove foot from accelerator, and apply brakes by depressing brake pedal.

(1) When vehicle speed has been reduced to engine idle speed, depress clutch pedal and move transmission shift lever to "N" (neutral) position (fig. 8).

(2) When vehicle has come to a complete stop, apply parking (hand) brake, and release clutch and brake pedals.

g. Reversing the Vehicle. To shift into reverse speed, first bring the vehicle to a complete stop.

(1) Depress clutch pedal.

(2) Move transmission shift lever to the left and forward into "R" (reverse) position.

(3) Release clutch pedal slowly, and accelerate as load is picked up.

h. Stopping the Engine. To stop the engine turn the ignition switch to "OFF" position. NOTE: *Before a new or reconditioned vehicle is first put into service, make run-in tests as outlined in section 10.*

6. TOWING THE VEHICLE.

a. Attaching Tow Line. To tow vehicle attach the chain, rope or cable to the front bumper bar at the frame side rail gusset (fig. 11). Do not tow from the middle of the bumper. To attach tow line, loop chain, rope, or cable over top of bumper, bring tow line up across front of bumper, and back on opposite side of frame, then hook or tie.

b. Towing to Start Vehicle. Place transfer case (aux. RANGE) shift lever of towed vehicle to the rear ("HIGH"). Place front axle drive shift lever in "OUT" (forward) position. Depress clutch pedal and engage transmission in high (3rd) speed. Switch ignition "ON," pull out choke control knob (if engine is cold), pull out throttle knob about 1 inch, release parking (hand) brake, and tow vehicle. After

¼-TON 4 x 4 TRUCK (WILLYS-OVERLAND MODEL MB and FORD MODEL GPW)

vehicle is under way, release clutch pedal slowly. As engine starts, regulate choke and throttle controls and disengage clutch, being careful to avoid overrunning towing vehicle or tow line.

c. Towing Disabled Vehicle. When towing a disabled vehicle exercise care so that no additional damage will occur.

(1) ALL WHEELS ON GROUND.

(a) If transfer case is *not* damaged, shift transmission and transfer case into neutral position and follow steps *(c)* and *(d)* below.

(b) If transfer case *is* damaged, disconnect both propeller shafts at the front and rear axles by removing the universal joint U-bolts, being careful not to lose the bearing races and rollers. Securely fasten the shafts to the frame with wire or remove dust cap and pull apart at the universal joint splines. Place bolts, nuts, rollers, and races in the glove compartment.

Figure 11—Chain Tow

(c) If the front axle differential or propeller shaft is damaged, remove front axle shaft driving flanges. Place front axle drive shift lever in "OUT" (forward) position and drive vehicle under own power.

(d) If the rear axle differential is damaged, remove the rear axle shafts; remove rear propeller shaft at rear universal joint U-bolts and front universal joint snap rings in forward flange, then drive out bearing cups. Place front axle drive shift lever in "IN" (rear) position and this will allow front axle drive to propel vehicle under own power.

(e) If rear propeller shaft only is damaged, remove as described in step *(d)* above.

(2) TOWING VEHICLE WITH FRONT OR REAR WHEELS OFF GROUND. If vehicle is to be towed in this manner be sure that transfer case shift lever is placed in "N" (neutral) position and front axle drive shift lever is placed in "OUT" (disengaged) position.

TM 9-803
7

Section IV

OPERATION UNDER UNUSUAL CONDITIONS

	Paragraph
Operation in cold weather	7
Operation in hot weather	8
Operation in sand	9
Operation in landing	10
Decontamination	11

7. OPERATION IN COLD WEATHER.

a. Purpose. Operation of automotive equipment at subzero temperatures presents problems that demand special precautions and extra careful servicing from both operation and maintenance personnel, if poor performance and total functional failure are to be avoided.

b. Gasoline. Winter grade of gasoline is designed to reduce cold weather starting difficulties; therefore, the winter grade motor fuel should be used in cold weather operation.

c. Storage and Handling of Gasoline. Due to condensation of moisture from the air, water will accumulate in tanks, drums, and containers. At low temperatures, this water will form ice crystals that will clog fuel lines and carburetor jets, unless the following precautions are taken:

(1) Strain the fuel through filter paper, or any other type of strainer that will prevent the passage of water. CAUTION: *Gasoline flowing over a surface generates static electricity that will result in a spark, unless means are provided to ground the electricity. Always provide a metallic contact between the container and the tank, to assure an effective ground.*

(2) Keep tank full, if possible. The more fuel there is in the tank, the smaller will be the volume of air from which moisture can be condensed.

(3) Add ½ pint of denatured alcohol, Grade 3, to the fuel tank each time it is filled. This will reduce the hazard of ice formation in the fuel.

(4) Be sure that all containers are thoroughly clean and free from rust before storing fuel in them.

(5) If possible, after filling or moving a container, allow the fuel to settle before filling fuel tank from it.

(6) Keep all closures of containers tight to prevent snow, ice, dirt, and other foreign matter from entering.

(7) Wipe all snow or ice from dispensing equipment and from around fuel tank filler cap before removing cap to refuel vehicle.

d. Lubrication.

(1) TRANSMISSION AND DIFFERENTIAL.

¼-TON 4 x 4 TRUCK (WILLYS-OVERLAND MODEL MB and FORD MODEL GPW)

(a) Universal gear lubricant, SAE 80, where specified on figure 14, is suitable for use at temperatures as low as −20°F. If consistent temperature below 0°F is anticipated, drain the gear cases while warm, and refill with Grade 75 universal gear lubricant, which is suitable for operation at all temperatures below +32°F. If Grade 75 universal gear lubricant is not available, SAE 80 universal gear lubricant diluted with the fuel used by the engine, in the proportion of one part fuel to six parts universal gear lubricant, may be used. Dilute make-up oil in the same proportion before it is added to gear cases.

(b) After engine has been warmed up, engage clutch, and maintain engine speed at fast idle for 5 minutes, or until gears can be engaged. Put transmission in low (first) gear, and drive vehicle for 100 yards, being careful not to stall engine. This will heat gear lubricants to the point where normal operation can be expected.

(2) CHASSIS POINTS. Lubricate chassis points with general purpose grease, No. 0.

(3) STEERING GEAR HOUSING. Drain housing, if possible, or use suction gun to remove as much lubricant as possible. Refill with universal gear lubricant, Grade 75, or, if not available, SAE 80 universal gear lubricant diluted with fuel used in the engine, in the proportion of one part fuel to six parts SAE 80 universal gear lubricant. Dilute make-up oil in the same proportion before it is added to the housing.

(4) OILCAN POINTS. For oilcan points where engine oil is prescribed for above 0°F, use light lubricating, preservative oil.

e. Protection of Cooling Systems.

(1) USE ANTIFREEZE COMPOUND. Protect the system with antifreeze compound (ethylene-glycol type) for operation below +32°F. The following instructions apply to use of new antifreeze compound.

(2) CLEAN COOLING SYSTEM. Before adding antifreeze compound, clean the cooling system, and completely free it from rust. If the cooling system has been cleaned recently, it may be necessary only to drain, refill with clean water, and again drain. Otherwise the system should be cleaned with cleaning compound.

(3) REPAIR LEAKS. Inspect all hoses, and replace if deteriorated. Inspect all hose clamps, plugs, and pet cocks and tighten if necessary. Repair all radiator leaks before adding antifreeze compound. Correct all leakage of exhaust gas or air into the cooling system.

(4) ADD ANTIFREEZE COMPOUND. When the cooling system is clean and tight, fill the system with water to about one-third capacity. Then add antifreeze compound, using the proportion of antifreeze compound to the cooling system capacity indicated below. Protect the system to at least 10°F below the lowest temperature expected to be experienced during the winter season.

OPERATION UNDER UNUSUAL CONDITIONS

ANTIFREEZE COMPOUND CHART
(for 11-quart capacity cooling system)

Temperature	Antifreeze Compound (ethylene-glycol type)
+10°F	3 qt
0°F	3¾ qt
−10°F	4½ qt
−20°F	4¾ qt
−30°F	5½ qt
−40°F	6 qt

(5) WARM THE ENGINE. After adding antifreeze compound, fill with water to slightly below the filler neck; then start and warm the engine to normal operating temperature.

(6) TEST STRENGTH OF SOLUTION. Stop the engine and check the solution with a hydrometer, adding antifreeze compound if required.

(7) INSPECT WEEKLY. In service, inspect the coolant weekly for strength and color. If rusty, drain and clean cooling system thoroughly, and add new solution of the required strength.

(8) CAUTIONS.

(a) Antifreeze compound is the only antifreeze material authorized for ordnance materiel.

(b) It is essential that antifreeze solutions be kept clean. Use only containers and water that are free from dirt, rust, and oil.

(c) Use an accurate hydrometer. To test a hydrometer, use one part antifreeze compound to two parts water. This solution will produce a hydrometer reading of 0°F.

(d) Do not spill antifreeze compound on painted surfaces.

f. **Electrical Systems.**

(1) GENERATOR AND CRANKING MOTOR. Check the brushes, commutators, and bearings. See that the commutators are clean. The large surges of current which occur when starting a cold engine require good contact between brushes and commutators.

(2) WIRING. Check, clean, and tighten all connections, especially the battery terminals. Care should be taken that no short circuits are present.

(3) COIL. Check coil for proper functioning by noting quality of spark.

(4) DISTRIBUTOR. Clean thoroughly, and clean or replace points. Check the points frequently. In cold weather, slightly pitted points may prevent engine from starting.

(5) SPARK PLUGS. Clean and adjust or replace, if necessary. If it is difficult to make the engine fire, reduce the gap to 0.005 inch less than that recommended for normal operation (par. 67 b). This will make ignition more effective at reduced voltages likely to prevail.

¼-TON 4 x 4 TRUCK (WILLYS-OVERLAND MODEL MB and FORD MODEL GPW)

(6) TIMING. Check carefully. Care should be taken that the spark is not unduly advanced nor retarded.

(7) BATTERY.

(a) The efficiency of batteries decreases sharply with decreasing temperatures, and becomes practically nil at −40°F. Do not try to start the engine with the battery when it has been chilled to temperatures below −30°F until battery has been heated, unless a warm slave battery is available. See that the battery is always fully charged, with the hydrometer reading between 1.275 and 1.300. A fully charged battery will not freeze at temperatures likely to be encountered even in arctic climates, but a fully discharged battery will freeze and rupture at +5°F.

(b) Do not add water to a battery when it has been exposed to subzero temperatures unless the battery is to be charged immediately. If water is added and the battery not put on charge, the layer of water will stay at the top and freeze before it has a chance to mix with the acid.

(8) LIGHTS. Inspect the lights carefully. Check for short circuits and presence of moisture around sockets.

(9) ICE. Before every start, see that the spark plugs, wiring, or other electrical equipment is free from ice.

g. Starting and Operating Engine.

(1) INSPECT CRANKING MOTOR MECHANISM. Be sure that no heavy grease or dirt has been left on the cranking motor throwout mechanism. Heavy grease or dirt is liable to keep the gears from being meshed, or cause them to remain in mesh after the engine starts running. The latter will ruin the cranking motor and necessitate repairs.

(2) USE OF CHOKE. A full choke is necessary to secure the rich air-fuel mixture required for cold weather starting. Check the butterfly valve to see that it closes all the way, and otherwise functions properly.

(3) CARBURETOR AND FUEL PUMP. The carburetor, which will give no appreciable trouble at normal temperatures, is liable not to operate satisfactorily at low temperatures. Be sure the fuel pump has no leaky valves or diaphragm, as this will prevent the fuel pump from delivering the amount of fuel required to start the engine at low temperatures, when turning speeds are reduced to 30 to 60 revolutions per minute.

(4) AIR CLEANERS. At temperatures below 0°F do not use oil in air cleaners. The oil will congeal and prevent the easy flow of air. Wash screens in dry-cleaning solvent, dry, and replace. Ice and frost formations on the air cleaner screens can cause an abnormally high intake vacuum in the carburetor air horn hose, resulting in collapse.

(5) FUEL SYSTEM. Remove and clean sediment bulb, strainers, etc., daily. Also drain fuel tank sump daily to remove water and dirt.

OPERATION UNDER UNUSUAL CONDITIONS

(6) STARTING THE ENGINE. Observe the following precautions in addition to the normal starting procedure (par. 5 a and b).

(a) Clean ignition wires and outside of spark plugs of dirt and frost.

(b) Free distributor point arm on post and clean points.

(c) Be sure carburetor choke closes fully.

(d) Operate fuel pump hand lever to fill carburetor bowl (fig. 12).

(e) Free up engine with hand crank or use slave battery.

(f) Stop engine if no oil pressure shows on gage.

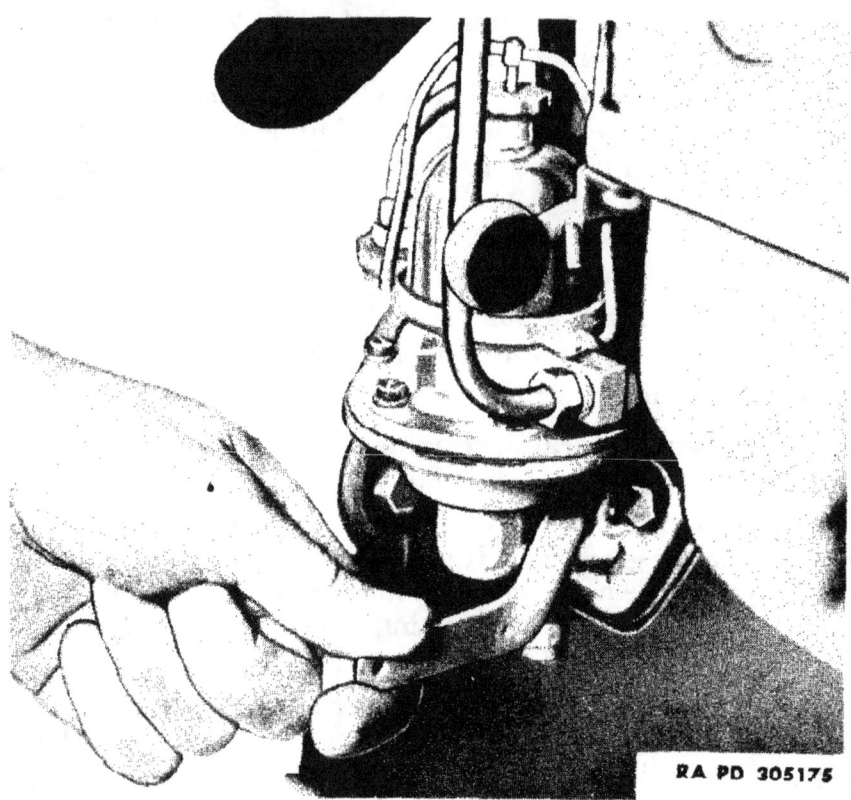

Figure 12—Fuel Pump, Hand Operation

(g) Engage clutch to warm up transmission oil before attempting to move vehicle.

(h) Check engine operation for proper condition (par. 13 b (22)).

h. **Chassis.**

(1) BRAKE BANDS. Brake bands, particularly on new vehicles, have a tendency to bind when they are very cold. Always have a blowtorch handy to warm up these parts, if they bind prior to moving, or attempting to move, the vehicle. Parking the vehicle with the brake released will eliminate most of the binding. Precaution must be taken, under these circumstances, to block the wheels or otherwise prevent movement of the vehicle.

TM 9-803
7-9
¼-TON 4 x 4 TRUCK (WILLYS-OVERLAND MODEL MB and FORD MODEL GPW)

(2) EFFECT OF LOW TEMPERATURES ON METALS. Inspect the vehicle frequently. Shock resistance of metals, or resistance against breaking, is greatly reduced at extremely low temperatures. Operation of vehicles on hard, frozen ground causes strain and jolting which will result in screws breaking, or nuts jarring loose.

(3) SPEEDOMETER CABLE. Disconnect the oil-lubricated speedometer cable at the drive end when operating the vehicle at temperatures of —30°F and below. The cable will often fail to work properly at these temperatures, and sometimes will break, due to the excessive drag caused by the high viscosity of the oil with which it is lubricated.

8. OPERATION IN HOT WEATHER.

a. Protection of Vehicle. In extremely hot weather avoid the continuous use of low gear ratios whenever possible. Check and replenish oil and water frequently. If a flooded condition of the engine is experienced in starting, pull the throttle control out, push choke control in, and use the cranking motor. When engine starts, adjust throttle control.

(1) COOLING SYSTEM. Rust formation occurs more rapidly during high temperatures; therefore, add rust preventive solution to the cooling system, or clean and flush the system at frequent intervals.

(2) LUBRICATION. Lubricate the vehicle for hot weather operation (par. 8).

(3) ELECTRICAL SYSTEM. Check the battery solution level frequently during hot weather operation, and add water as required to keep it above the top of the plates. If hard starting is experienced in hot, damp weather or quick changes in temperature, dry the spark plugs, wires, and both inside and outside of distributor cap.

9. OPERATION IN SAND.

a. Operation. Reduce tire pressures in desert terrain if character of sand demands this precaution. When operating in sand deep enough to cause the use of a lower gear, do not exceed the speed specified on the caution plate for the particular gear ratio (fig. 7).

b. Starting the Vehicle. When starting the vehicle in sand, gravel, or soft terrain, engage the front wheel drive (par. 5 e (1)). Release clutch pedal slowly so the wheels will not spin and "dig in," necessitating a tow or "winch-out."

c. Clutch. Do not attempt to "jump" or "rock" the vehicle out with a quick engagement of the clutch, particularly if a tow or winch is available. Racing the engine usually causes the wheels to "dig in" farther.

d. Air Cleaner. In sandy territory clean the carburetor air cleaner more often. The frequency of cleaning depends upon the severity of the sandy condition.

OPERATION UNDER UNUSUAL CONDITIONS

e. Radiator. In desert operation check the radiator coolant supply frequently, and see that the air passages of the core do not become clogged.

f. For additional information on technique of operating the vehicle in sand, refer to FM 31-25.

10. OPERATION IN LANDING.

a. Inspection. As soon as possible after completing a landing or operation in water, inspect the vehicle for water in the various units.

(1) ENGINE. Drain the engine crankcase oil. If water or sludge is found, flush the engine, using a mixture of half engine oil SAE 10 and half kerosene. Before putting in new oil, clean the valve chamber, drain and clean the oil filter, and install a new filter element.

(2) FUEL SYSTEM. Inspect the carburetor bowl, fuel strainers, fuel pump, filter, fuel tank, and lines. Clean the air cleaner and change the oil.

(3) POWER TRAIN. Inspect the front and rear axle housings, wheel bearings, transmission, and transfer case lubricant for presence of sludge. If sludge is found, renew the lubricant after cleaning the units with a mixture of half engine oil SAE 10 and half kerosene. Lubricate the propeller shaft universal joints and spring shackles to force out any water which might damage parts.

11. DECONTAMINATION.

a. Protection. For protective measures against chemical attacks and decontamination refer to FM 17-59.

¼-TON 4 x 4 TRUCK (WILLYS-OVERLAND MODEL MB and FORD MODEL GPW)

Section V

FIRST ECHELON PREVENTIVE MAINTENANCE SERVICE

	Paragraph
Purpose	12
Before-operation service	13
During-operation service	14
At-halt service	15
After-operation and weekly service	16

12. PURPOSE.

a. To ensure mechanical efficiency it is necessary that the vehicle be systematically inspected at intervals each day it is operated, also weekly, so that defects may be discovered and corrected before they result in serious damage or failure. Certain scheduled maintenance services will be performed at these designated intervals. The services set forth in this section are those performed by driver or crew before operation, during operation, at halt, and after operation and weekly.

b. Driver preventive maintenance services are listed on the back of "Driver's Trip Ticket and Preventive Maintenance Service Record," W.D. Form No. 48, to cover vehicles of all types and models. Items peculiar to specific vehicles, but not listed on W.D. Form No. 48, are covered in manual procedures under the items to which they are related. Certain items listed on the form that do not pertain to the vehicle involved are eliminated from the procedures as written into the manual. Every organization must thoroughly school each driver in performing the maintenance procedures set forth in manuals, whether they are listed specifically on W.D. Form No. 48 or not.

c. The items listed on W.D. Form No. 48 that apply to this vehicle are expanded in this manual to provide specific procedures for accomplishment of the inspections and services. These services are arranged to facilitate inspection and conserve the time of the driver, and are not necessarily in the same numerical order as shown on W.D. Form No. 48. The item numbers, however, are identical with those shown on that form.

d. The general inspection of each item applies also to any supporting member or connection, and generally includes a check to see whether the item is in good condition, correctly assembled, secure, or excessively worn.

(1) The inspection for "good condition" is usually an external visual inspection to determine whether the unit is damaged beyond safe or serviceable limits. The term "good condition" is explained further by the following: not bent or twisted, not chafed or burned, not broken or cracked, not bare or frayed, not dented or collapsed, not torn or cut.

FIRST ECHELON PREVENTIVE MAINTENANCE SERVICE

(2) The inspection of a unit to see that it is "correctly assembled" is usually an external visual inspection to see whether or not it is in its normal assembled position in the vehicle.

(3) The inspection of a unit to determine if it is "secure" is usually an external visual examination, a hand-feel, wrench, or prybar check for looseness. Such an inspection should include any brackets, lock washers, lock nuts, locking wires, or cotter pins used in assembly.

(4) "Excessively worn" will be understood to mean worn, close to or beyond, serviceable limits, and likely to result in failure if not replaced before the next scheduled inspection.

e. Any defects or unsatisfactory operating characteristics beyond the scope of the first echelon to correct must be reported at the earliest opportunity to the designated individual in authority.

13. BEFORE-OPERATION SERVICE.

a. This inspection schedule is designed primarily as a check to see that the vehicle has not been tampered with or sabotaged since the After-operation Service was performed. Various combat conditions may have rendered the vehicle unsafe for operation, and it is the duty of the driver to determine whether or not the vehicle is in condition to carry out any mission to which it is assigned. This operation will not be entirely omitted, even in extreme tactical situations.

b. **Procedures.** Before-operation Service consists of inspecting items listed below according to the procedure described, and correcting or reporting any deficiencies. Upon completion of the service, results should be reported promptly to the designated individual in authority.

(1) ITEM 1, TAMPERING AND DAMAGE. Examine exterior of vehicle, engine, wheels, brakes, and steering control for damage by falling debris, shell fire, sabotage, or collision. If wet, dry the ignition parts to ensure easy starting.

(2) ITEM 2, FIRE EXTINGUISHER. Be sure fire extinguisher is full, nozzle is clean, and mountings secure.

(3) ITEM 3, FUEL, OIL, AND WATER. Check fuel tank, crankcase, and radiator for leaks or tampering. Add fuel, oil, or water as needed. Have value of antifreeze checked. If, during period when antifreeze is used, it becomes necessary to replenish a considerable amount of water, report unusual losses.

(4) ITEM 4, ACCESSORIES AND DRIVES. Inspect carburetor, generator, regulator, cranking motor, and water pump for loose connections and security of mountings. Inspect carburetor and water pump for leaks.

(5) ITEM 6, LEAKS, GENERAL. Look on ground under vehicle for indications of fuel, oil, water, brake fluid, or gear oil leaks. Trace leaks to source, and correct or report to higher authority.

¼-TON 4 x 4 TRUCK (WILLYS-OVERLAND MODEL MB and FORD MODEL GPW)

(6) ITEM 7, ENGINE WARM-UP. Start engine, observe cranking motor action, listen for unusual noise, and note cranking speed. Idle engine only fast enough to run smoothly. Proceed immediately with following services while engine is warming up.

(7) ITEM 8, CHOKE. As engine warms, push in choke as required for smooth operation, and to prevent oil dilution.

(8) ITEM 9, INSTRUMENTS.

(a) Fuel Gage. Fuel gage should indicate approximate amount of fuel in tank.

(b) Oil Pressure Gage. Normal oil pressure should not be below 10 with engine idling, and should range from 40 to 50 at running speeds (at normal operating temperature). If gage fails to register within 30 seconds, stop engine, and correct or report to higher authority.

(c) Temperature Indicator. Temperature should rise slowly during warm-up. Normal operating temperature range is 160°F to 185°F.

(d) Ammeter. Ammeter should show high charge for short period after starting and positive (plus) reading above 12 to 15 miles per hour with lights and accessories off. Zero reading is normal with lights and accessories on.

(9) ITEM 10, HORN AND WINDSHIELD WIPERS. Sound horn, tactical situation permitting, for proper operation and tone. Check both wipers for secure attachment and normal full contact operation through full stroke.

(10) ITEM 11, GLASS AND REAR VIEW MIRROR. Clean windshield and rear view mirror and inspect for cracked, discolored, or broken glass. Adjust mirror.

(11) ITEM 12, LIGHTS AND REFLECTORS. Try switches in each position and see if lights respond. Lights and warning reflectors must be securely mounted, clean, and in good condition. Test foot control of headlight beams.

(12) ITEM 13, WHEEL AND FLANGE NUTS. Observe whether or not all wheel and flange nuts are present and tight.

(13) ITEM 14, TIRES. If time permits, test tires with gage, including spare; normal pressure is 35 pounds with tires cold. Inspect tread and carcass for cuts and bruises. Remove imbedded objects from treads.

(14) ITEM 15, SPRINGS AND SUSPENSION. Inspect springs for sagged or broken leaves, shifted leaves, and loose or missing rebound clips.

(15) ITEM 16, STEERING LINKAGE. Examine steering gear case, connecting links, and Pitman arm for security and good condition. Test steering adjustment, and free motion of steering wheel.

(16) ITEM 17, FENDERS AND BUMPERS. Examine fenders and bumpers for secure mounting and serviceable condition.

FIRST ECHELON PREVENTIVE MAINTENANCE SERVICE

(17) ITEM 18, TOWING CONNECTIONS. Examine pintle hook for secure mounting and serviceable condition. Be sure pintle latches properly and locks securely.

(18) ITEM 19, BODY AND LOAD. Examine body and load (if any) for damage. Be sure there is a cap on front drain hole under fuel tank. See that rear drain hole cap is available in glove compartment. CAUTION: *Rear drain hole cap should be installed when about to pass through deep water.*

(19) ITEM 20, DECONTAMINATOR. Examine decontaminator for full charge and secure mountings.

(20) ITEM 21, TOOLS AND EQUIPMENT. See that tools and equipment are all present, properly stowed, and serviceable.

(21) ITEM 23, DRIVER'S PERMIT AND FORM 26. Driver must have his operator's permit on his person. See that vehicle manuals, Lubrication Guide, Form No. 26 (accident report) and W.D. AGO Form No. 478 (MWO and Major Unit Assembly Replacement Record) are present, legible, and properly stowed.

(22) ITEM 22, ENGINE OPERATION. Accelerate engine and observe for unusual noises indicating compression or exhaust leaks; worn, damaged, loose, and inadequately lubricated parts or misfiring.

(23) ITEM 25, DURING-OPERATION SERVICE. Begin the During-operation Service immediately after the vehicle is put in motion.

14. DURING-OPERATION SERVICE.

a. While vehicle is in motion, listen for any sounds such as rattles, knocks, squeals, or hums that may indicate trouble. Look for indications of trouble in cooling system, and smoke from any part of the vehicle. Be on the alert to detect any odor of overheated components or units such as generator, brakes, or clutch; check for fuel vapor from a leak in fuel system, exhaust gas, or other signs of trouble. Any time the brakes are used, gears shifted, or vehicle turned, consider this a test and notice any unsatisfactory or unusual performance. Watch the instruments frequently. Notice promptly any unusual instrument indication that may signify possible trouble in system to which the instrument applies.

b. **Procedures.** During-operation Service consists of observing items listed below according to the procedures following each item, and investigating any indications of serious trouble. Notice minor deficiencies to be corrected or reported at earliest opportunity, usually at next scheduled halt.

(1) ITEM 27, FOOT AND HAND BRAKES. Foot brakes must stop vehicle smoothly without side pull and within reasonable distance. There should be at least $\frac{1}{3}$ reserve brake pedal travel and $\frac{1}{2}$-inch free travel. Hand brake must securely hold vehicle on reasonable incline with $\frac{1}{3}$ reserve ratchet travel. There must be $\frac{1}{2}$-inch clearance (on cable) between relay crank and lower end of hand brake conduit.

¼-TON 4 x 4 TRUCK (WILLYS-OVERLAND MODEL MB and FORD MODEL GPW)

(2) ITEM 28, CLUTCH. Clutch must operate smoothly without chatter, grabbing, or slipping. Free clutch pedal travel of three-quarter inch is normal.

(3) ITEM 29, TRANSMISSION. Gearshift mechanism must operate smoothly, and not creep out of mesh.

(4) ITEM 29, TRANSFER CASE. Gearshift mechanism must operate smoothly and not creep out of mesh.

(5) ITEM 31, ENGINE AND CONTROLS. Observe whether or not engine responds to controls, and has maximum pulling power without unusual noises, stalling, misfiring, overheating or unusual exhaust smoke. If radio noise is reported during operation of the vehicle, the driver will cooperate with the radio operator in locating the interference. See paragraph 178.

(6) ITEM 32, INSTRUMENTS. During operation observe the readings of all instruments frequently to see if they are indicating properly.

(a) Fuel Gage. Fuel gage must register approximate amount of fuel in tank.

(b) Oil Pressure Gage. Oil pressure gage should register 10 with engine running idle, and 40 to 50 at operating speeds.

(c) Temperature Indicator. Temperature indicator should show a temperature of 160°F to 185°F after warm-up under normal conditions.

(d) Speedometer. Speedometer should show speed of vehicle without noise or fluctuation of indicator needle. Odometer should register accumulating trip and total mileage.

(e) Ammeter. Ammeter should show zero reading with lights on, zero or positive (plus) charge with lights off, and slightly higher positive (plus) charge for short time immediately after starting.

(7) ITEM 33, STEERING GEAR. Observe steering for excessive pulling of vehicle to either side, wandering, or shimmy.

(8) ITEM 34, CHASSIS. Listen for unusual noises from wheel or axles.

(9) ITEM 35, BODY. Observe body for sagging springs, loose or torn top or windshield cover, if in use.

15. AT-HALT SERVICE.

a. At-halt Service may be regarded as the minimum maintenance procedure, and should be performed under all tactical conditions, even though more extensive maintenance services must be slighted or omitted altogether.

b. Procedures. At-halt Service consists of investigating any deficiencies noted during operation, inspecting items listed below according to the procedures following the items, and correcting any deficiencies found. Deficiencies not corrected should be reported promptly to the designated individual in authority.

FIRST ECHELON PREVENTIVE MAINTENANCE SERVICE

(1) ITEM 38, FUEL, OIL AND WATER. Check fuel supply, oil, and coolant; add, as required, for complete operation of vehicle to the next refueling point. If, during period when antifreeze is used, an abnormal amount of water is required to refill radiator, have coolant tested with hydrometer, and add antifreeze if required.

(2) ITEM 39, TEMPERATURES. Feel each brake drum and wheel hub, transmission, transfer case, and front and rear axles for overheating. Examine gear cases for excessive oil leaks.

(3) ITEM 40, AXLE AND TRANSFER CASE VENTS. Observe whether axle and transfer case vents are present, and see that they are not damaged or clogged.

(4) ITEM 41, PROPELLER SHAFT. Inspect propeller shaft for looseness, damage, or oil leaks.

(5) ITEM 42, SPRINGS. Look for broken spring leaves or loose clips and U-bolts.

(6) ITEM 43, STEERING LINKAGE. Examine steering control mechanism and linkage for damage or looseness. Investigate any irregularities noted during operation.

(7) ITEM 44, WHEEL AND FLANGE NUTS. Observe whether or not all wheel and axle flange nuts are present and tight.

(8) ITEM 45, TIRES. Inspect tires, including spare, for flats or damage, and for cuts or foreign material imbedded in tread.

(9) ITEM 46, LEAKS, GENERAL. Check around engine and on ground beneath the vehicle for excessive leaks. Trace to source, and correct cause or report to higher authority.

(10) ITEM 47, ACCESSORIES AND BELTS. See that fan, water pump and generator are securely mounted, that fan belt is adjusted to 1-inch deflection, and is not badly frayed. If radio noise during operation of the engine was observed, examine all radio noise suppression capacitors, at coil, ignition and starting switches, generator, regulator, and radio terminal box; suppressors at spark plugs and distributor, and all bond straps for damage, and loose mountings or connections.

(11) ITEM 48, AIR CLEANER. If dusty or sandy conditions have been encountered, examine oil sump for excessive dirt. Service if required. CAUTION: *Do not apply oil to element after cleaning.*

(12) ITEM 49, FENDERS AND BUMPERS. Inspect fenders and bumpers for looseness or damage.

(13) ITEM 50, TOWING CONNECTIONS. Inspect pintle hook and trailer light socket for serviceability.

(14) ITEM 51, BODY LOAD AND TARPAULIN. Inspect vehicle and trailed vehicle loads for shifting; see that tarpaulins are properly secured and not damaged.

(15) ITEM 52, APPEARANCE AND GLASS. Clean windshield, mirror, light lenses, and inspect vehicle for damage.

¼-TON 4 x 4 TRUCK (WILLYS-OVERLAND MODEL MB and FORD MODEL GPW)

16. AFTER-OPERATION AND WEEKLY SERVICE.

a. After-operation Service is particularly important because at this time the driver inspects his vehicle to detect any deficiencies that may have developed, and corrects those he is permitted to handle. He should report promptly, to the designated individual in authority, the results of his inspection. If this schedule is performed thoroughly, the vehicle should be ready to roll again on short notice. The Before-operation Service, with a few exceptions, is then necessary only to ascertain whether the vehicle is in the same condition in which it was left upon completion of the After-operation Service. The After-operation Service should never be entirely omitted, even in extreme tactical situations, but may be reduced, if necessary, to the bare fundamental services outlined for the At-halt Service.

b. Procedures. When performing the After-operation Service the driver must remember and consider any irregularities noticed during the day in the Before-operation, During-operation, and At-halt Services. The After-operation Service consists of inspecting and servicing the following items. Those items of the After-operation Service that are marked by an asterisk (*) require additional Weekly Service, the procedures for which are indicated in step (b) of each applicable item.

(1) ITEM 54, FUEL, OIL, AND WATER. Check coolant and oil levels, and add as needed. Fill fuel tank. Refill spare cans. During period when antifreeze is used, have hydrometer test made of coolant if loss from boiling or other cause has been considerable. Add antifreeze with water if required.

(2) ITEM 55, ENGINE OPERATION. Listen for miss, backfire, noise, or vibration that might indicate worn parts, loose mountings, faulty fuel mixture, or faulty ignition.

(3) ITEM 56, INSTRUMENTS. Inspect all instruments to see that they are securely connected, and not damaged.

(4) ITEM 57, HORN AND WINDSHIELD WIPERS. Test horn for sound, if tactical situation permits. See that horn is securely mounted and properly connected. Operate both windshield wipers. See that blades contact the glass effectively throughout full stroke.

(5) ITEM 58, GLASS AND REAR VIEW MIRROR. Clean glass of windshield and rear view mirror. Examine for secure mounting and damage.

(6) ITEM 59, LIGHTS AND REFLECTORS. Observe whether or not lights operate properly with the switch in "ON" positions, and go out when switch is off. See that stop light operates properly. Clean lenses and warning reflectors.

(7) ITEM 60, FIRE EXTINGUISHER. Be sure fire extinguisher is full, nozzle is clean, and that extinguisher is mounted securely.

(8) ITEM 61, DECONTAMINATOR. Examine decontaminator for good condition and secure mounting.

FIRST ECHELON PREVENTIVE MAINTENANCE SERVICE

(9) ITEM 62, *BATTERY.

(a) See that battery is clean, securely mounted, and not leaking. Inspect electrolyte level, which should be ½ inch above plates with caps in place and vents open. Clean cables as required.

(b) Weekly. Clean top of battery. Remove battery caps, and add water to ½ inch above plates. (Use distilled water if available; if not use clean, drinkable water.) CAUTION: *Do not overfill.* Clean posts and terminals if corroded, and apply light coat of grease. Tighten terminals as needed. Tighten hold-down assembly. Clean battery carrier if corroded.

(10) ITEM 63, *ACCESSORIES AND BELTS.

(a) Test fan belt for deflection of 1 inch. Examine belt for good condition; it must not be frayed. Timing hole cover must be closed and tightened.

(b) Weekly. Tighten all accessories such as carburetor, generator, regulator, cranking motor, fan, water pump, and hose connections; examine fan belt for fraying, wear, cracking, or presence of oil.

(11) ITEM 64, *ELECTRICAL WIRING.

(a) See that all ignition wiring and accessible low voltage wiring is in good condition, clean, correctly and securely assembled and mounted.

(b) Weekly. Tighten all loose wiring connections or electrical unit mountings. Pay particular attention to radio noise suppression units such as: capacitors, bond straps, and spark plug and distributor suppressors.

(12) ITEM 65, *AIR CLEANER.

(a) Examine oil in air cleaner oil cup to see that it is at proper level, and not excessively dirty. Clean element and refill oil cup as required. CAUTION: *Do not apply oil to element after cleaning.*

(b) Weekly. Remove, clean, and dry air cleaner element and oil cup. Fill cup to indicated oil level (approximately ⅝ qt). Do not apply oil to element after cleaning.

(13) ITEM 66, *FUEL FILTERS.

(a) Examine fuel filter for leaks.

(b) Weekly. Remove plug from bottom of dash-mounted fuel filter. Allow water and sediment to drain out. Be sure plug is replaced tightly, and does not leak.

(14) ITEM 67, ENGINE CONTROLS. Examine engine controls for wear or disconnected linkage.

(15) ITEM 68, *TIRES.

(a) Inspect tires for cuts or abnormal tread wear; remove foreign bodies from tread; inflate to 35 pounds when tires are cold.

(b) Weekly. Replace badly worn or otherwise unserviceable tires.

(16) ITEM 69, *SPRINGS.

(a) Examine springs for sag, broken or shifted leaves, loose or missing rebound clips, or shackles.

(b) Weekly. Aline springs, and tighten U-bolts and shackles as required.

¼-TON 4 x 4 TRUCK (WILLYS-OVERLAND MODEL MB and FORD MODEL GPW)

(17) ITEM 70, STEERING LINKAGE. Examine steering wheel column, gear case, Pitman arm, drag link, tie rod, and steering arm to see if they are bent, loose, or inadequately lubricated.

(18) ITEM 71, PROPELLER SHAFT. Inspect propeller shaft and universal joints for loose connections, lubrication leaks, or damage.

(19) ITEM 72, *AXLE AND TRANSFER VENTS.

(a) See that axle and transfer case vents are in good condition, clean, and secure.

(b) *Weekly.* Remove, clean, and replace vents.

(20) ITEM 73, LEAKS, GENERAL. Check under hood and beneath the vehicle for indications of fuel, oil, water, or brake fluid leaks.

(21) ITEM 74, GEAR OIL LEVELS. After units have cooled, inspect differential transmission and transfer unit lubricant levels. Lubricant should be level with bottom of filler hole. Observe gear cases for leaks.

(22) ITEM 76, FENDERS AND BUMPERS. Fenders and bumpers must be in good condition and secure.

(23) ITEM 77, *TOWING CONNECTIONS.

(a) Inspect pintle hook and towed-load connections for looseness or damage.

(b) *Weekly.* Tighten pintle hook mounting bolts, and lubricate pintle hook as required.

(24) ITEM 78, BODY AND TARPAULINS. Inspect body, top, and windshield cover for damage and proper stowage. Make sure rear drain below fuel tank is open, and that cap is in glove compartment.

(25) ITEM 82, *TIGHTEN.

(a) Tighten any loose wheel, axle drive flange, and spring U-bolt nuts.

(b) *Weekly.* Tighten all vehicle assembly or mounting nuts or screws that inspection indicates require tightening.

(26) ITEM 83, *LUBRICATE AS NEEDED.

(a) Lubricate spring shackles and steering linkage, if lubrication is needed.

(b) *Weekly.* Lubricate points indicated on current vehicle Lubrication Guide as requiring weekly attention, also points that experience and operating conditions indicate need lubrication. Observe latest lubrication directives.

(27) ITEM 84, *CLEAN ENGINE AND VEHICLE.

(a) Clean dirt and trash from inside of body. Keep sump under fuel tank cleaned of dirt and water. Remove excessive dirt or grease from exterior of the engine.

(b) *Weekly.* Wash vehicle if possible. If not possible, wipe off thoroughly; clean engine.

(28) ITEM 85, TOOLS AND EQUIPMENT. Check to see that all tools and equipment assigned to vehicle are present and secure.

Section VI

LUBRICATION

	Paragraph
Lubrication Guide	17
Detailed lubrication instructions	18

17. LUBRICATION GUIDE.

a. War Department Lubrication Guide No. 501 (figs. 13 and 14) prescribes lubrication maintenance for the ¼-ton 4 x 4 truck.

b. A Lubrication Guide is placed on or is issued with each vehicle and is to be carried with it at all times. In the event the vehicle is received without a Guide, the using arm shall immediately requisition a replacement from the Commanding Officer, Fort Wayne Ordnance Depot, Detroit 32, Mich.

c. Lubrication instructions on the Guide are binding on all echelons of maintenance and there shall be no deviations from these instructions.

d. Service intervals specified on the Guide are for normal operation conditions. Reduce these intervals under extreme conditions such as excessively high or low temperatures, prolonged periods of high speed, continued operation in sand or dust, immersion in water, or exposure to moisture, any one of which may quickly destroy the protective qualities of the lubricant and require servicing in order to prevent malfunctioning or damage to the materiel.

e. Lubricants are prescribed in the "Key" in accordance with three temperature ranges; above +32°F, +32°F to 0°F, and below 0°F. Determine the time to change grades of lubricants by maintaining a close check on operation of the vehicle during the approach to change-over periods. Be particularly observant when starting the engine. Sluggish starting is an indication of thickened lubricants and the signal to change to grades prescribed for the next lower temperature range. Ordinarily it will be necessary to change grades of lubricants *only when air temperatures are consistently in the next higher or lower range*, unless malfunctioning occurs sooner due to lubricants being too thin or too heavy.

18. DETAILED LUBRICATION INSTRUCTIONS.

a. Lubrication Equipment. Each piece of materiel is supplied with lubrication equipment adequate to maintain the materiel. Be sure to clean this equipment both before and after use. Operate lubricating guns carefully and in such manner as to insure a proper distribution of the lubricant.

b. Points of Application.

(1) Red circles surrounding lubrication fittings, grease cups, oilers and oil holes make them readily identifiable on the vehicle. Wipe clean such lubricators and the surrounding surface before lubricant is applied.

TM 9-803
18

¼-TON 4 x 4 TRUCK (WILLYS-OVERLAND MODEL MB and FORD MODEL GPW)

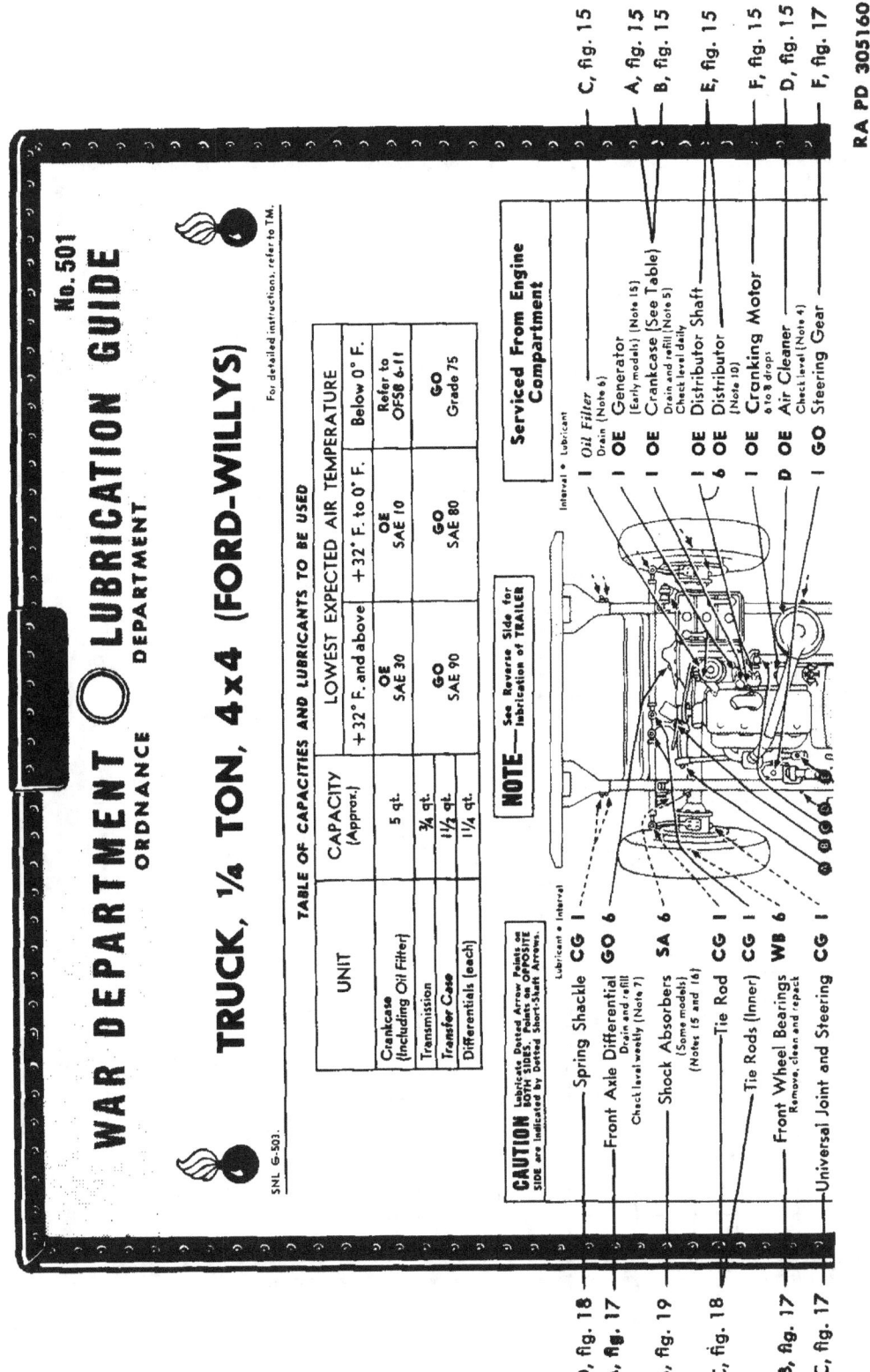

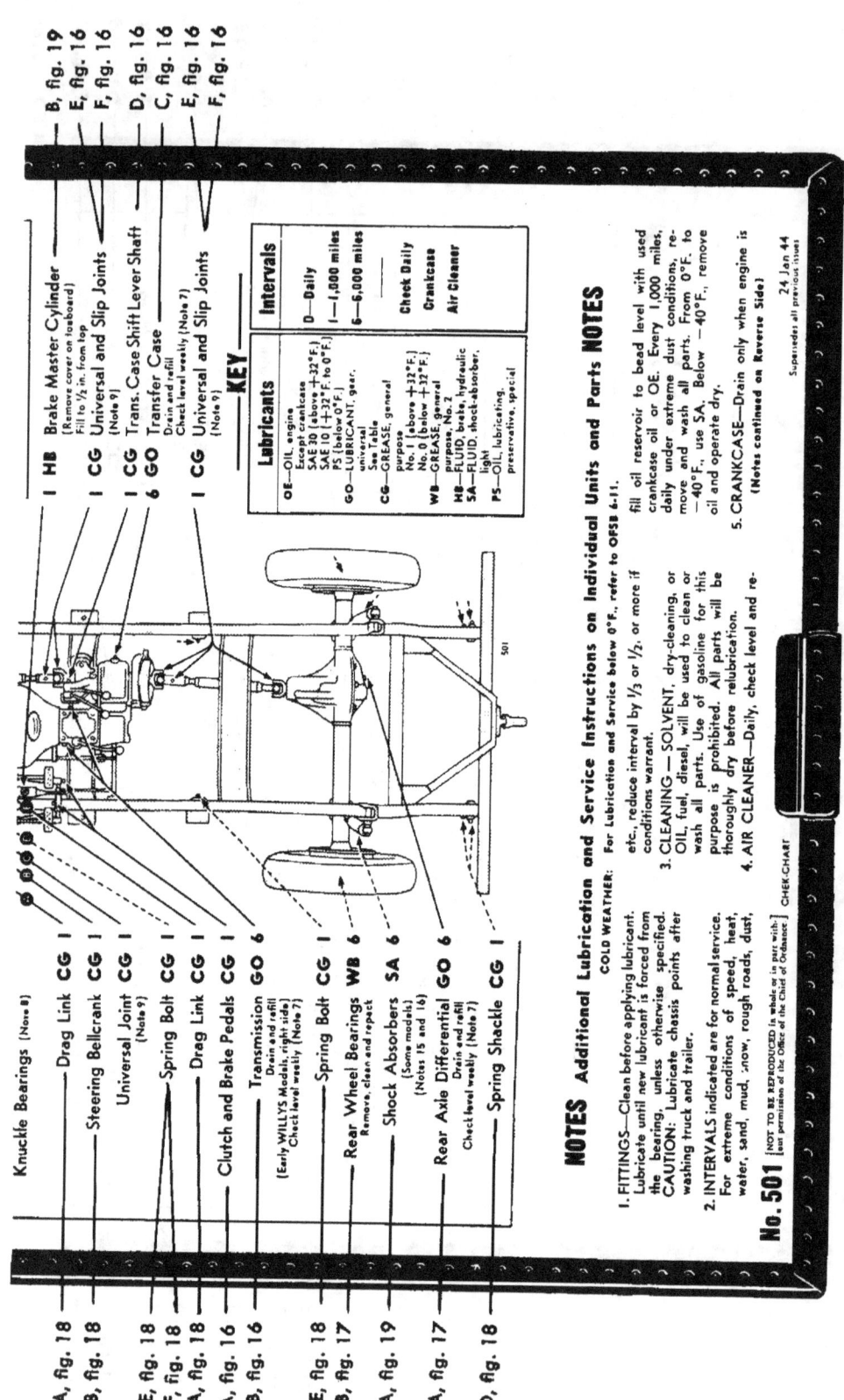

Figure 13 — Lubrication Guide—Truck, 1/4-Ton, 4 x 4 (Ford-Willys)

TM 9-803
18

¼-TON 4 x 4 TRUCK (WILLYS-OVERLAND MODEL MB and FORD MODEL GPW)

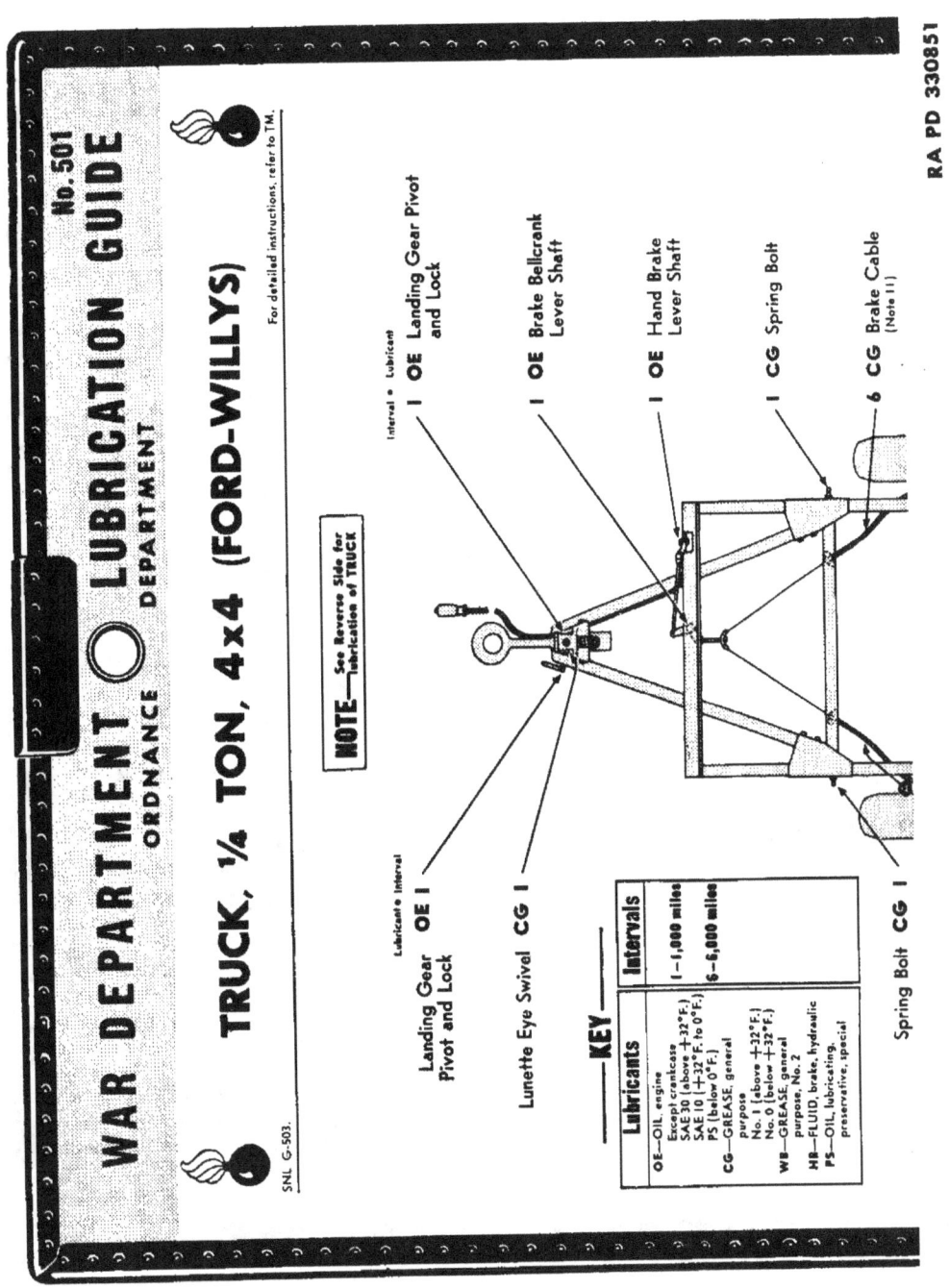

LUBRICATION

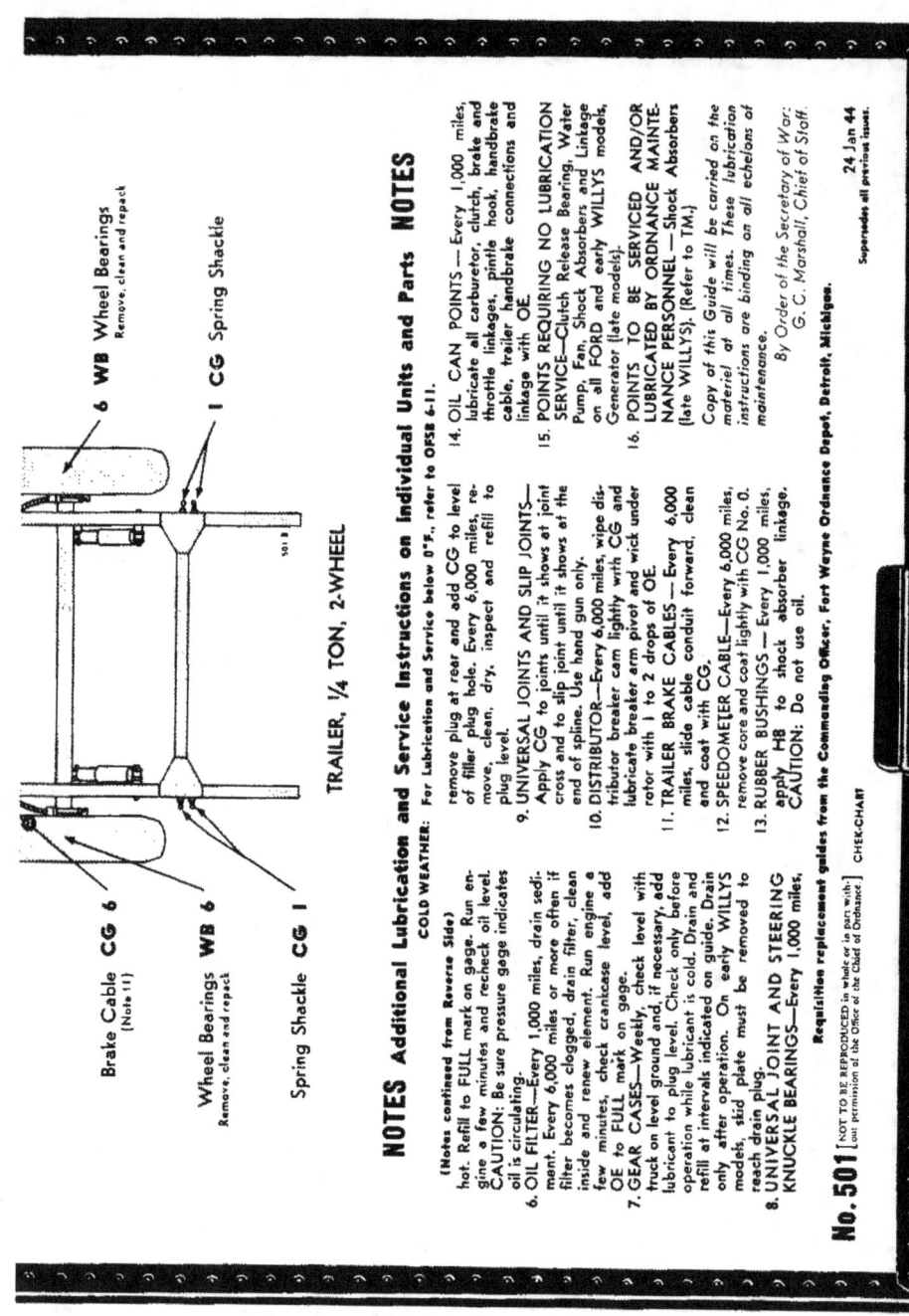

Figure 14—Lubrication Guide—Trailer, 1/4-Ton, 2-wheel

¼-TON 4 x 4 TRUCK (WILLYS-OVERLAND MODEL MB and FORD MODEL GPW)

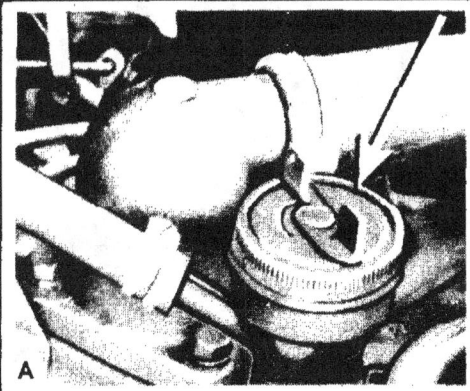

A

ENGINE CRANKCASE—OE

Oil level indicator in oil filler pipe. Check level at least daily. Keep oil up to FULL mark. Capacity five quarts; refill four quarts.

B

ENGINE CRANKCASE DRAINING—

Remove drain plug to drain. At least once a year, remove the oil pan and clean floating oil intake screen.

C

OIL FILTER—OE

One filter—Remove drain plug to drain. To replace element, remove drain plug, filter cover then element. After completing installation run engine a few minutes and refill crankcase to FULL mark on oil level indicator.

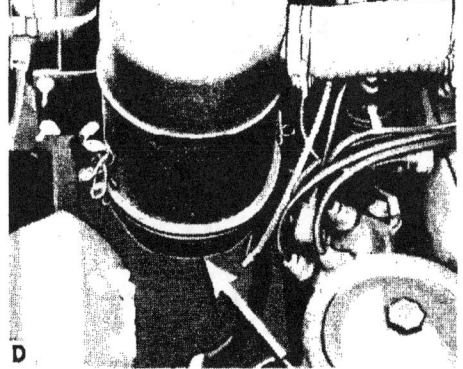

D

AIR CLEANER—OE

One air cleaner. Clean cleaner and refill reservoir to indicated level. Capacity ⅝ quart.

E

DISTRIBUTOR—OE

One distributor. Total places—four. Use oil can for oiler and lubricate sparingly wick and post; grease cam lightly.

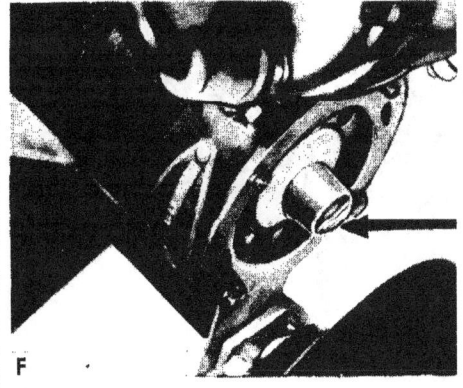

F

CRANKING MOTOR—OE

Total oilers—one. Use oil can, push aside oil hole cover. Oil and replace cover.

RA PD 305166

Figure 15—Engine Lubrication Points

LUBRICATION

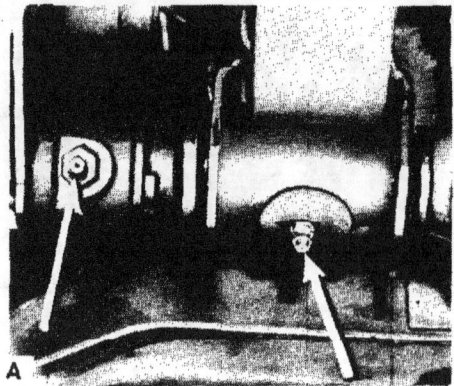

A
CLUTCH AND BRAKE PEDAL SHAFT—CG
One pedal shaft. Total fittings—two. Use pressure gun on fittings until grease shows.

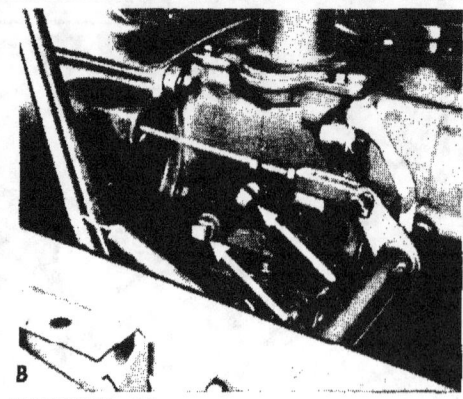

B
TRANSMISSION—GO
One transmission. Total plugs—two (filler and drain). Use gear oil pump. Drain and refill to bottom of filler plug hole. Capacity ¾ quart.

C
TRANSFER CASE—GO
One transfer case. Total plugs—two (filler and drain). Use gear oil pump. Drain and refill to bottom of filler plug hole. Capacity 1½ quarts.

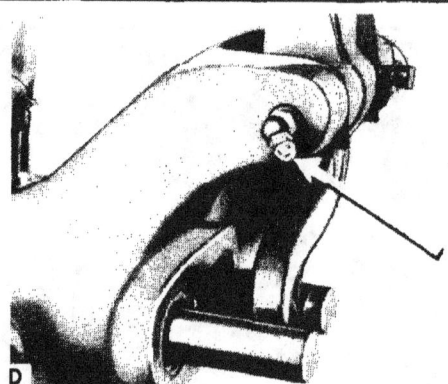

D
TRANSFER CASE SHIFT LEVER SHAFT—CG
One shift lever shaft. Total fittings—one. Use pressure gun on fitting until grease shows.

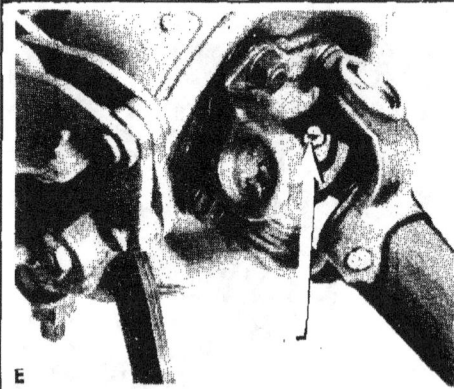

E
PROPELLER SHAFT UNIVERSAL JOINTS—CG
Four universal joints. Total fittings—four. Use pressure gun (hand) with adaptor. CAUTION: Do not use high pressure grease gun because of damage to seals.

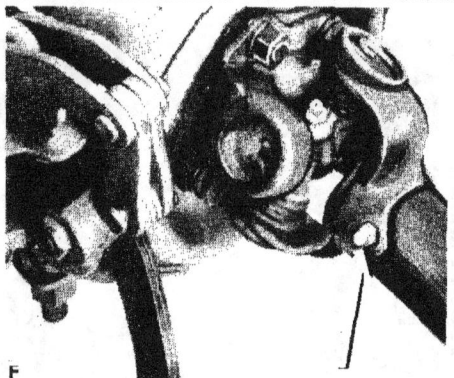

F
PROPELLER SHAFT SLIP JOINT—CG
Two slip joints. Total fittings—two. Use pressure gun on fittings until grease shows.

RA PD 305167

Figure 16—Pedal Shafts and Power Train Lubrication Points

TM 9-803
18
¼-TON 4 x 4 TRUCK (WILLYS-OVERLAND MODEL MB and FORD MODEL GPW)

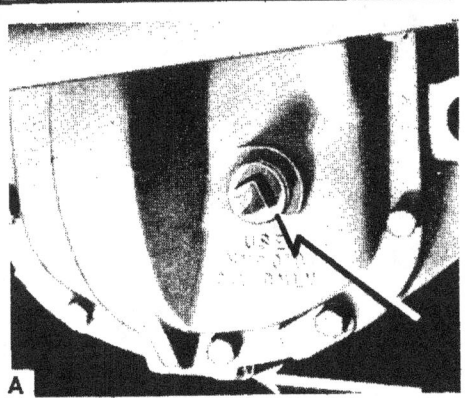

A

AXLE HOUSINGS—GO

Two axle housings. Total plugs—four (filler and drain). Use gear oil pump. Drain and refill to bottom of filler plug hole. Capacity 1¼ quarts.

B

WHEEL BEARINGS—WB

Four wheels. Total bearings—eight. Use bearing lubricator or hand pack thoroughly. Apply grease also around outside of cage and rollers. Clean out wheel hub, inspect bearing races, put three ounces of grease in each hub.

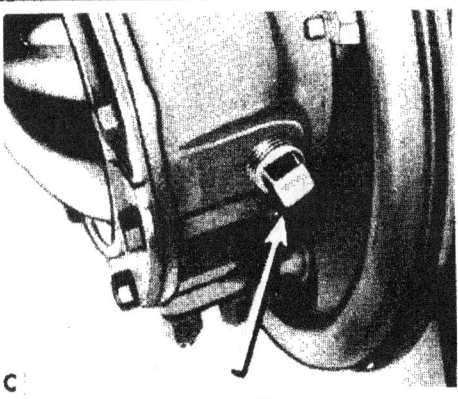

C

FRONT AXLE UNIVERSAL JOINTS—CO

Two universal joints. Total plugs—two. Use pressure gun (hand) and fill housing slowly to level of filler plug hole.

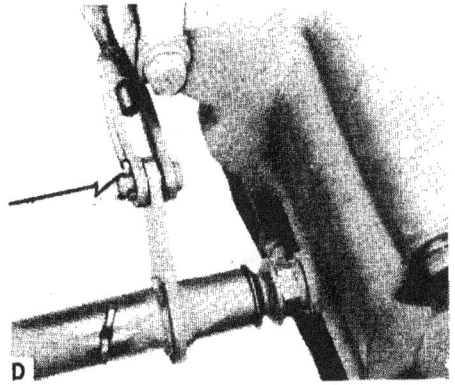

D

LINKAGE CLEVIS PINS—OE

All clevis pins and hood and windshield catches. Use oil can and apply in proper quantity.

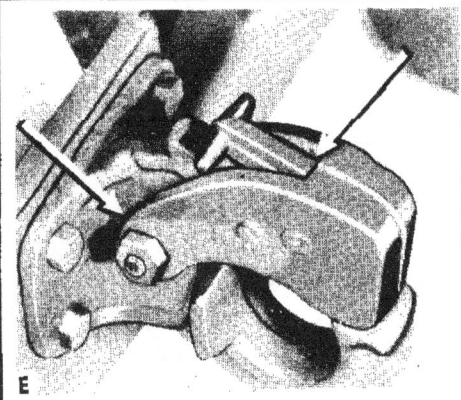

E

PINTLE HOOK—OE

One hook—With an oil can lubricate pins, connections and sliding surfaces.

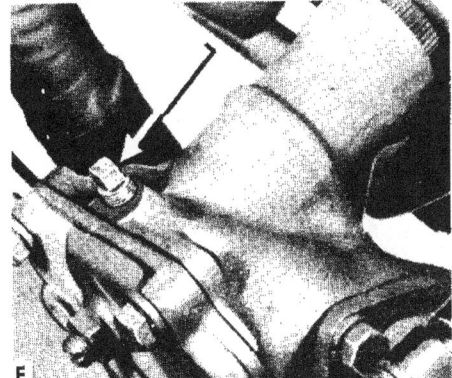

F

STEERING GEAR HOUSING—GO

One housing. Total plugs—one. Use pressure gun (hand) and fill housing slowly until full.

RA PD 305168

Figure 17—Axle, Wheel, Pintle, and Steering Gear Housing Lubrication Points

LUBRICATION

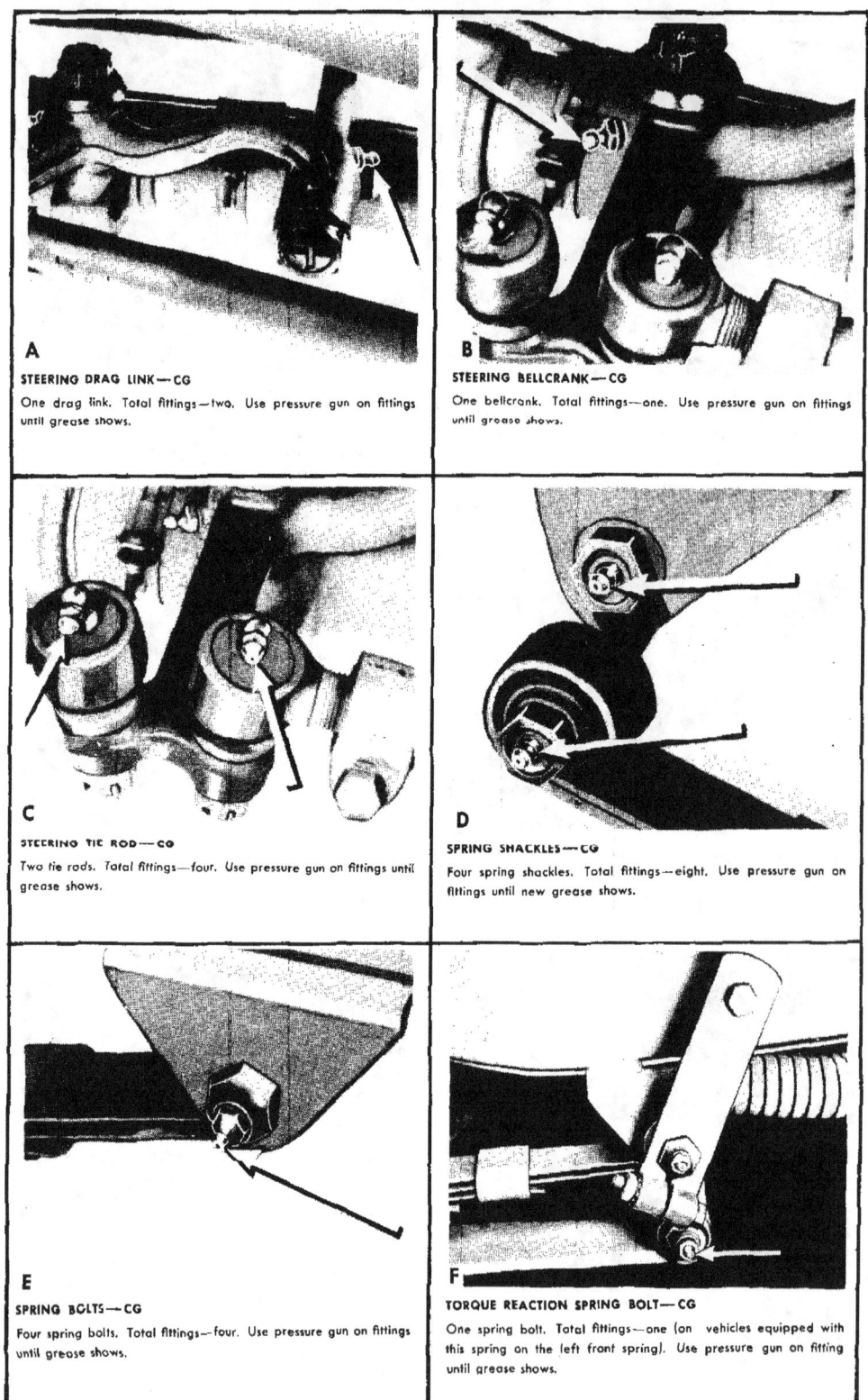

Figure 18—Steering Gear and Spring Lubrication Points

¼-TON 4 x 4 TRUCK (WILLYS-OVERLAND MODEL MB and FORD MODEL GPW)

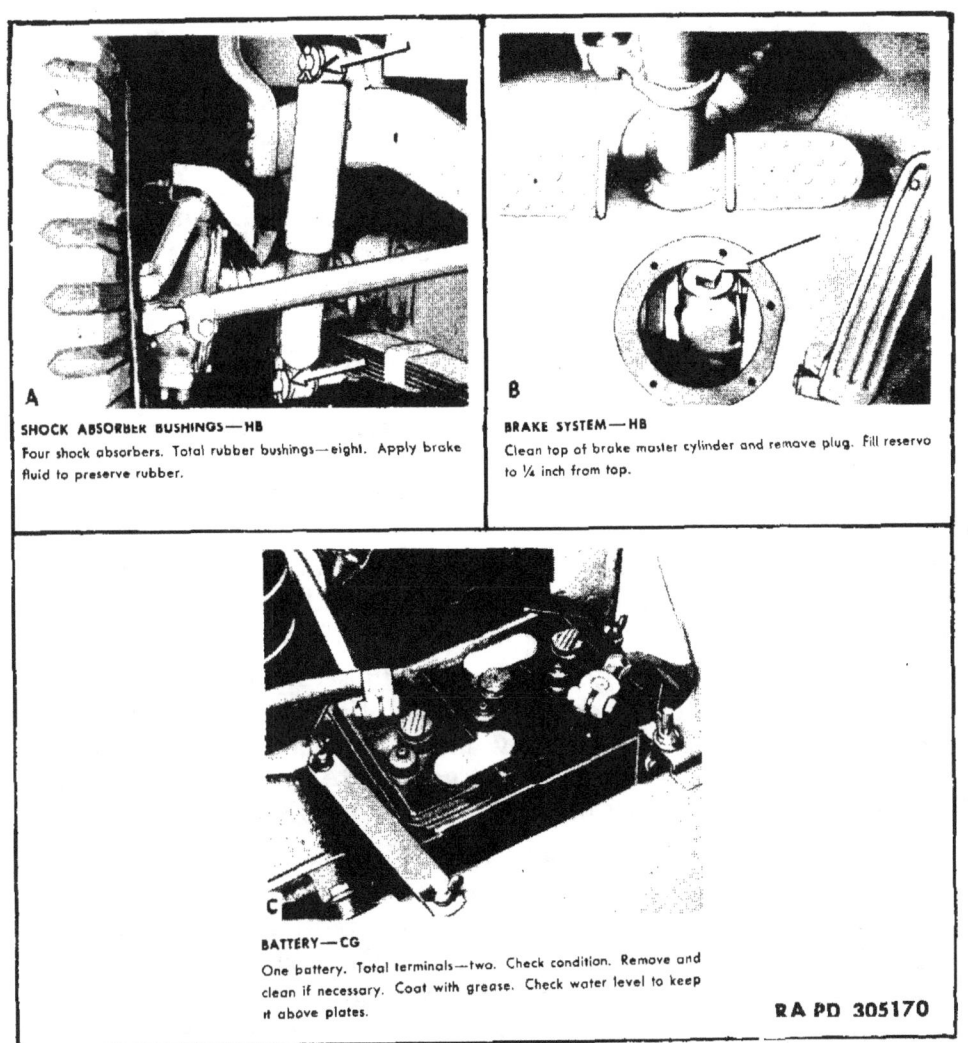

Figure 19—Shock Absorber, Master Cylinder, and Battery Lubrication Points

(2) Where relief valves are provided, apply new lubricant until the old lubricant is forced from the vent. Exceptions are specified in notes on the Lubrication Guide.

c. **Cleaning.** Use SOLVENT, dry-cleaning, or OIL, fuel, Diesel, to clean or wash all parts. Use of gasoline for this purpose is prohibited. After washing, dry all parts thoroughly before applying lubricant.

d. **Lubrication Notes on Individual Units and Parts.** The following instructions supplement those notes on the Lubrication Guide which pertain to lubrication and service of individual units and parts. All note references in the Guide itself are to the paragraph below having the corresponding number.

LUBRICATION

(1) FITTINGS. Clean before applying lubricant. Lubricate until new lubricant is forced from the bearing, unless otherwise specified. CAUTION: *Lubricate chassis points after washing truck and trailer.*

(2) INTERVALS. Intervals indicated are for normal service. For extreme conditions of speed, heat, water, sand, mud, snow, rough roads, dust, etc., reduce interval by one-third or one-half, or more if conditions warrant.

(3) CLEANING. SOLVENT, dry-cleaning, or OIL, fuel, Diesel, will be used to clean or wash all parts. Use of gasoline for this purpose is prohibited. All parts will be thoroughly dry before relubrication.

(4) AIR CLEANER. Daily, check level and refill oil reservoir to bead level with used crankcase oil or OIL, engine, SAE 30 above +32°F or SAE 10 from +32°F to 0°F. Every 1,000 miles, daily under extreme dust conditions, remove and wash all parts. From 0°F to −40°F, use FLUID, shock-absorber, light. Below −40°F, remove oil and operate dry.

(5) CRANKCASE. Drain only when engine is hot. Refill to "FULL" mark on gage. Run engine a few minutes and recheck oil level. CAUTION: *Be sure pressure gage indicates oil is circulating.*

(6) OIL FILTER. Every 1,000 miles, drain sediment. Every 6,000 miles or more often if filter becomes clogged, drain filter, clean inside and renew element. Run engine a few minutes, check crankcase level, add OIL, engine, to "FULL" mark on gage. (SAE 30 above +32°F; SAE 10 from +32°F to 0°F; below 0°F, refer to OFSB 6-11.)

(7) GEAR CASES. Weekly, check level with truck on level ground and, if necessary, add lubricant to plug level. Check only before operation while lubricant is cold. Drain and refill at intervals indicated on Guide. Drain only after operation. On early Willys models, skid plate must be removed to reach drain plug.

(8) UNIVERSAL JOINT AND STEERING KNUCKLE BEARINGS. Every 1,000 miles, remove plug at rear and add GREASE, general purpose, No. 1 above +32°F or No. 0 below +32°F, to level of filler plug hole. Every 6,000 miles, remove, clean, dry, inspect and refill to plug level.

(a) Remove brake tube and brake backing plate screws. This permits the removal of the axle spindle, the complete axle shaft, and the universal joint assembly. Care should be taken not to injure the outer oil seal assembly in the housing.

(b) Wash the axle shaft and universal joint thoroughly in SOLVENT, dry-cleaning, and dry.

(c) Clean and repack upper and lower steering spindle bearings within the universal housing and reassemble entire unit.

(9) UNIVERSAL JOINTS AND SLIP JOINTS. Apply GREASE, general purpose, No. 1, above +32°F, or No. 0 below +32°F, to joints until it shows at joint cross, and to slip joint until it shows at the end of spline. Use hand gun only.

¼-TON 4 x 4 TRUCK (WILLYS-OVERLAND MODEL MB and FORD MODEL GPW)

(10) DISTRIBUTOR. Every 6,000 miles, wipe distributor breaker cam lightly with GREASE, general purpose, No. 1, above +32°F or No. 0, below +32°F, and lubricate breaker arm pivot and wick under rotor with 1 to 2 drops of OIL, engine, SAE 30 above +32°F; SAE 10 from +32°F to 0°F; OIL, lubricating, preservative, special, below 0°F.

(11) TRAILER BRAKE CABLES. Every 6,000 miles, slide cable conduit forward, clean and coat with GREASE, general purpose, No. 1 above +32°F and No. 0 below +32°F.

(12) SPEEDOMETER CABLE. Every 6,000 miles, remove core and coat lightly with GREASE, general purpose, No. 0.

(13) RUBBER BUSHINGS. Every 1,000 miles, apply FLUID, brake, hydraulic, to shock absorber linkage. CAUTION: *Do not use oil.*

(14) OILCAN POINTS. Every 1,000 miles, lubricate all carburetor, clutch, brake and throttle linkages, pintle hook and hand brake cable with OIL, engine, SAE 30, above +32°F; SAE 10, +32°F to 0°F; OIL, lubricating, preservative, special, below 0°F.

(15) POINTS REQUIRING NO LUBRICATION SERVICE. These are the clutch release bearing, water pump, fan, shock absorbers and linkage on all Ford and early Willys models, generator (late models), speedometer cable.

(16) POINTS TO BE SERVICED AND/OR LUBRICATED BY ORDNANCE MAINTENANCE PERSONNEL ONLY. These are the shock absorbers (late Willys). Every 6,000 miles, remove and disassemble the shock absorbers. Unscrew linkage eye and refill with FLUID, shock-absorber, light.

(17) WHEEL BEARINGS. Remove bearing cone assemblies from hub and wash spindle and inside of hub. Inspect bearing races and replace if necessary. Wet the spindle and inside of hub and hub cap with GREASE, general purpose, No. 2, to a maximum thickness of $\frac{1}{16}$ inch only to retard rust. Wash bearing cones and grease seals. Inspect and replace if necessary. Lubricate bearings with GREASE, general purpose, No. 2, with a packer or by hand, kneading lubricant into all spaces in the bearing. Use extreme care to protect bearings from dirt and immediately reassemble and replace wheel. The lubricant in the bearings is sufficient to provide lubrication until the next service period. Do not fill hub or hub cap. Any excess might result in leakage into the brake drum.

e. **Reports and Records.** If lubrication instructions are closely followed, proper lubricants used, and satisfactory results are not obtained, make a report to the ordnance officer responsible for the maintenance of the materiel. A complete record of lubrication servicing may be kept in the Duty Roster (W.D., A.G.O Form No. 6).

f. **Localized Views.** The localized views of lubrication points (figs. 15, 16, 17, 18, and 19) supplement the instructions on the Guide and in the notes.

TM 9-803
19

Section VII

TOOLS AND EQUIPMENT STOWAGE ON THE VEHICLE

	Paragraph
Vehicle tools	19
Vehicle equipment	20
Vehicle spare parts	21

19. VEHICLE TOOLS.

a. Unless the vehicle is equipped with extra tool equipment, the following are supplied (one of each unless otherwise specified):

Tool	Federal Stock No.	Where Carried
HAMMER, machinist's, ball peen, 16 oz	41-H-523	Tool bag
JACK, screw type, 1½-ton, w/handle	41-J-66	Tool compartment
PLIERS, combination, slip joint, 6-in.	41-P-1650	Tool bag
PULLER, wheel hub	41-P-2962-700	Tool compartment
WRENCH, drain plug	41-W-1962-50	Tool bag
WRENCH, engineer's open-end, ⅜- x ⁷⁄₁₆-in.	41-W-991	Tool bag
WRENCH, engineer's open-end, ½- x ¹⁹⁄₃₂-in.	41-W-1003	Tool bag
WRENCH, engineer's open-end, ⁹⁄₁₆- x ¹¹⁄₁₆-in.	41-W-1005-5	Tool bag
WRENCH, engineer's open-end, ⅝- x ²⁵⁄₃₂-in.	41-W-1008-10	Tool bag
WRENCH, engineer's open-end, ¾- x ⅞-in.	41-W-1012-5	Tool bag
WRENCH, hydraulic brake, bleeder screw	41-W-1596-125	Tool bag
WRENCH, adjustable, auto type, 11-in.	41-W-449	Tool bag
WRENCH, socket, screw fluted	41-W-2459-500	Tool bag
WRENCH, socket, spark plug, w/handle	41-W-3335-50	Tool bag
WRENCH, wheel bearing nut, 2⅛-in. hex	41-W-3825-200	Tool compartment
WRENCH, wheel stud nut, ⁴⁹⁄₆₄-in. hex	41-W-3837-55	Tool compartment

802011 O—48——4

TM 9-803
20-21

¼-TON 4 x 4 TRUCK (WILLYS-OVERLAND MODEL MB and FORD MODEL GPW)

20. VEHICLE EQUIPMENT.

a. Unless vehicle is equipped with special equipment, the following are supplied (one of each unless otherwise specified):

Tool	Federal Stock No.	Where Carried
ADAPTER, lubr. gun		Tool bag
APPARATUS, decontaminating, 1½ qt		Driver's compartment
AX, chopping, single-bit	41-A-1277	Body left side
BAG, tool	41-B-15	Tool compartment
CATALOG, ord. std. nom. list	SNL-G-503	Glove compartment
CHAINS, tire, 6.00 x 16	8-C-2358	Tool compartment (4)
CONTAINER, 5-gallon		Bracket on rear
COVER, headlight		Under right seat (2)
COVER, windshield		Under right seat
CRANK, starting		Under rear seat
EXTINGUISHER, fire	58-E-202	Inside cowl, left
GAGE, tire pressure	8-G-615	Tool compartment
GUN, lubr., hand-type	41-G-1330-60	Tool compartment
MANUAL, technical	TM 9-803	Glove compartment
NOZZLE, flexible tube		
OILER, straight spout, ½-pt	13-O-1530	Front of dash
PUMP, tire, w/chuck	8-P-5000	Behind rear seat
RIFLE		On dash
SHOVEL, D-handle, rd. pt.	41-S-3170	Body, left side
TAPE, friction, roll	17-T-805	Parts bag
WIRE, iron, roll	22-W-650	Parts bag

21. VEHICLE SPARE PARTS.

a. Unless the vehicle is equipped with a special assortment of parts, the following are supplied (one of each unless otherwise specified):

Name of Spare Part	Federal Stock No.	Where Carried
BAG, spare parts	8-B-11	Glove compartment
BELT, fan	33-B-76	Parts bag
CAPS, tire valve (boxed)	8-C-650	Parts bag (5)
CORES, tire valve (boxed)	8-C-6750	Parts bag (5)

TOOLS AND EQUIPMENT STOWAGE ON THE VEHICLE

Name of Spare Part	Federal Stock No.	Where Carried
Lamp, elec. incand. 6-8V sing-tung-fil., 3 cp (MZ63)	17-L-5215	Parts bag
Lamp-unit, blackout, stop, sealed, one opng., 6-8V, 3 cp	8-L-421	Parts bag
Lamp-unit, blackout, tail, sealed, 4 opngs., 6-8V, 3 cp	8-L-415	Parts bag
Lamp-unit, service tail and stop, sealed, 6-8V, 21-3 cp	8-L-419	Parts bag
Pin, cotter, split, s. type B boxed ass't.	42-P-5347	Parts bag
Plug, spark, with gasket	17-P-5365	Parts bag

TM 9-803
22

¼-TON 4 x 4 TRUCK (WILLYS-OVERLAND MODEL MB and FORD MODEL GPW)

PART TWO
VEHICLE MAINTENANCE INSTRUCTIONS

Section VIII
RECORD OF MODIFICATIONS

Paragraph

MWO and major unit assembly replacement record........ 22

22. MWO AND MAJOR UNIT ASSEMBLY REPLACEMENT RECORD.

  a. **Description.** Every vehicle is supplied with a copy of A.G.O. Form No. 478 which provides a means of keeping a record of each MWO completed or major unit assembly replaced. This form includes spaces for the vehicle name and U.S.A. registration number, instructions for use, and information pertinent to the work accomplished. It is very important that the form be used as directed, and that it remain with the vehicle until the vehicle is removed from service.

  b. **Instructions for Use.** Personnel performing modifications or major unit assembly replacements must record clearly on the form a description of the work completed, and must initial the form in the columns provided. When each modification is completed, record the date, hours and/or mileage, and MWO number. When major unit assemblies, such as engines, transmissions, transfer cases, are replaced, record the date, hours and/or mileage, and nomenclature of the unit assembly. Minor repairs and minor parts and accessory replacements need not be recorded.

  c. **Early Modifications.** Upon receipt by a third or fourth echelon repair facility of a vehicle for modification or repair, maintenance personnel will record the MWO numbers of modifications applied prior to the date of A.G.O. Form No. 478.

TM 9-803
23

Section IX

SECOND ECHELON PREVENTIVE MAINTENANCE

Paragraph

Second echelon preventive maintenance services............ 23

23. SECOND ECHELON PREVENTIVE MAINTENANCE SERVICES.

a. Regular scheduled maintenance inspections and services are a preventive maintenance function of the using arms and are the responsibility of commanders of operating organizations.

(1) FREQUENCY. The frequency of the preventive maintenance services outlined herein is considered a minimum requirement for normal operation of vehicles. Under unusual operating conditions such as extreme temperatures, and dusty or sandy terrain, it may be necessary to perform certain maintenance services more frequently.

(2) FIRST ECHELON PARTICIPATION. The drivers should accompany their vehicles and assist the mechanics while periodic second echelon preventive maintenance services are performed. Ordinarily the driver should present the vehicle for a scheduled preventive maintenance service in a reasonably clean condition: that is, it should be dry and not caked with mud or grease to such an extent that inspection and servicing will be seriously hampered; however, the vehicle should not be washed or wiped thoroughly clean, since certain types of defects, such as cracks, leaks, and loose or shifted parts or assemblies are more evident if the surfaces are slightly soiled or dusty.

(3) INSTRUCTIONS. If instructions other than those which are contained in the general procedures in step (4), or in the specific procedures in step (5) which follow, are required for the correct performance of a preventive maintenance service or for correction of a deficiency, other sections of the vehicle operators' manual pertaining to the item involved, or a designated individual in authority should be consulted.

(4) GENERAL PROCEDURES. These general procedures are basic instructions which are to be followed when performing the services on the items listed in the specific procedures. NOTE: *The second echelon personnel must be thoroughly trained in these procedures so that they will apply them automatically.*

(a) When new or overhauled subassemblies are installed to correct deficiencies, care should be taken to see that they are clean, correctly installed, and properly lubricated and adjusted.

(b) When installing new lubricant retainer seals, a coating of the lubricant should be wiped over the sealing surface of the lip of the seal. When the new seal is a leather seal, it should be soaked in engine oil SAE 10 (warm if practicable) for at least 30 minutes, then, the leather lip should be worked carefully by hand before installing the seal. The lip must not be scratched or marred.

(c) The general inspection of each item applies also to any supporting member or connection, and usually includes a check to see

¼-TON 4 x 4 TRUCK (WILLYS-OVERLAND MODEL MB and FORD MODEL GPW)

whether or not the item is in good condition, correctly assembled, secure, or excessively worn. The mechanics must be thoroughly trained in the following explanations of these terms.

1. The inspection for "good condition" is usually an external visual inspection to determine if the unit is damaged beyond safe or serviceable limits. The term "good condition" is explained further by the following: not bent or twisted, not chafed or burned, not broken or cracked, not bare or frayed, not dented or collapsed, not torn or cut.

2. The inspection of a unit to see that it is "correctly assembled" is usually an external visual inspection to see if it is in its normal assembled position in the vehicle.

3. The inspection of a unit to determine if it is "secure" is usually an external visual examination, a hand-feel, wrench, or a pry-bar check for looseness. Such an inspection should include any brackets, lock washers, lock nuts, locking wires, or cotter pins used in assembly.

4. "Excessively worn" will be understood to mean worn, close to or beyond serviceable limits, and likely to result in a failure if not replaced before the next scheduled inspection.

(d) Special Services. These are indicated by repeating the item numbers in the columns which show the interval at which the services are to be performed, and show that the parts or assemblies are to receive certain mandatory services. For example, an item number in one or both columns opposite a *Tighten* procedure means that the actual tightening of the object must be performed. The special services include:

1. Adjust. Make all necessary adjustments in accordance with the pertinent section of the vehicle operator's manual, special bulletins, or other current directives.

2. Clean. Clean units of the vehicle with dry-cleaning solvent to remove excess lubricant, dirt, and other foreign material. After the parts are cleaned, rinse them in clean fluid and dry them thoroughly. Take care to keep the parts clean until reassembled, and be certain to keep cleaning fluid away from rubber or other material which it will damage. Clean the protective grease coating from new parts, since this material is not a good lubricant.

3. Special lubrication. This applies both to lubrication operations that do not appear on the vehicle Lubrication Guide and to items that do appear on such charts, but which should be performed in connection with the maintenance operations if parts have to be disassembled for inspection or service.

4. Serve. This usually consists of performing special operations, such as replenishing battery water, draining and refilling units with oil, and changing the oil filter cartridge.

5. Tighten. All tightening operations should be performed with sufficient wrench-torque (force on the wrench handle) to tighten the unit according to good mechanical practice. Use torque-indicating wrench where specified. Do not overtighten, as this may strip

SECOND ECHELON PREVENTIVE MAINTENANCE

threads or cause distortion. Tightening will always be understood to include the correct installation of lock washers, lock nuts, and cotter pins provided to secure the tightening.

(e) Conditions. When conditions make it difficult to perform the complete preventive maintenance procedures at one time, they can sometimes be handled in sections, planning to complete all operations within the week, if possible. All available time at halts and in bivouac areas must be utilized, if necessary, to assure that maintenance operations are completed. When limited by the tactical situation, items with special services in the columns should be given first consideration.

(f) The numbers of the preventive maintenance procedures that follow are identical with those outlined on W.D., A.G.O. Form No. 461, which is the Preventive Maintenance Service Work Sheet for Wheeled and Half-track Vehicles. Certain items on the work sheet that do not apply to this vehicle are not included in the procedures in this manual. In general, the numerical sequence of items on the work sheet is followed in the manual procedures, but in some instances there is deviation for conservation of the mechanic's time and effort.

(5) SPECIFIC PROCEDURES. The procedures for performing each item in the 1,000-mile (monthly) and 6,000-mile (6-month) maintenance procedures are described in the following chart. Each page of the chart has two columns at the left edge corresponding to the 6,000-mile and the 1,000-mile maintenance respectively. Very often it will be found that a particular procedure does not apply to both scheduled maintenances. In order to determine which procedure to follow, look down the column corresponding to the maintenance due, and wherever an item appears, perform the operations indicated opposite the number.

ROAD TEST

MAINTENANCE	
6000 Mile	1000 Mile

NOTE: *When the tactical situation does not permit a full road test, perform those items which require little or no movement of the vehicle, namely, items 3, 4, 5, 6, 9, 10, and 14. Make a full road test of 5, but not more than 10 miles, over varied terrain if possible.*

6000	1000	
1	1	**Before-operation Service.** Perform Before-operation Service as outlined in paragraph 13.
3	3	**Dash Instruments and Gages.** Observe instruments frequently during road test.

AMMETER. Ammeter should show high charge for short time after starting, then zero or slight positive (plus) reading above speeds of 12 to 15 miles per hour with lights and accessories off. Zero reading is normal with lights and accessories on.

SPEEDOMETER. See that speedometer indicates vehicle speed, operates without excessive fluctuation or noise, and that odometer registers accumulating trip and total mileage correctly.

¼-TON 4 x 4 TRUCK (WILLYS-OVERLAND MODEL MB and FORD MODEL GPW)

MAINTENANCE	
6000 Mile	1000 Mile
4	4
5	5
6	6
7	7
8	8
9	9
10	10
13	13

TEMPERATURE INDICATOR. Temperature indicator should gradually increase to normal operating range of 160°F to 180°F.

FUEL GAGE. Fuel gage must indicate the approximate amount of fuel in tank.

Horn, Mirror, and Windshield Wiper. Test horn for proper operation and tone, tactical situation permitting. Adjust mirror, and inspect for broken or discolored glass. Wiper should have sufficient arm tension to stay in "UP" position. Examine blade for good condition and full contact with glass throughout entire stroke.

Brakes. Test brakes for smooth, even stop, excessive pedal travel before application, "spongy" pedal, or loss of pedal pressure when brakes are held on. Brakes must not squeak or require excessive pedal pressure. Test pedal free travel, which should be ½ inch. Hand brake must hold vehicle on a reasonable grade, must have positive ratchet action and ⅓ reserve handle travel. There should be ½-inch reserve clearance between hand brake relay crank and lower end of hand brake cable conduit.

Clutch. Clutch must have free pedal travel of three-quarter inch. Test clutch for slip, grab, gear clash, or rattle. Listen for noises that would indicate dry or defective release bearing or pilot bushing.

Transmission and Transfer Case. Shift through entire range of transmission and transfer, noting whether the levers move easily and snap into each position. With shifting levers in each position, accelerate and decelerate engine, noting any unusual noises or tendency of levers to slip into neutral. Inspect for loose mountings.

Steering. Steering gear must not bind. There should be no excessive free play with wheels in straightahead position. Test for existence of front-end shimmy, wander, or side pull.

Engine. Engine must idle smoothly without stalling. Test acceleration and pulling power in each transmission speed. Listen for detonation and "ping," misses, popping, spitting, or other noises that might indicate need for engine repair.

Unusual Noises. Listen for noises that might indicate loose, damaged, or faulty parts.

Temperatures. Feel brake drums and wheel hubs for abnormally high temperatures. Overheated brake drum or wheel hub may indicate dragging brake or defective, dry, or improperly adjusted wheel bearing. Examine

SECOND ECHELON PREVENTIVE MAINTENANCE

MAINTENANCE	
6000 Mile	1000 Mile
14	14
16	16
17	17
22	22
22	22
18	18

differentials, transmission, and transfer case for too-high running temperature. NOTE: *Transfer case operates at a higher temperature than other cases.*

Leaks. Look on ground under vehicle for indications of coolant, fuel, oil, or hydraulic fluid leaks.

Gear Oil Level and Leaks. Examine lubricant levels of transmission, transfer case, and differentials. Inspect cases for leaks. Safe level when cold is even with filler plug. If an oil change is due, drain and refill, according to Lubrication Guide (par. 18). Capacities: transmission, ¾ quart; transfer case, 1½ quarts; front differential, 1¼ quarts; rear differential, 1¼ quarts.

MAINTENANCE OPERATIONS

Unusual Noises. With engine running, proceed as follows: Accelerate and decelerate engine slightly, and listen for unusual engine noises. With transmission in third gear, front wheel drive engaged, and engine at fast idle, listen for unusual noises in operating units. Observe propeller shaft and universal joints, wheels, and axles for excessive vibration and run-out.

Battery. Inspect battery case for cracks and leaks. Inspect cables, terminals, bolts, posts, straps, and hold-downs for good condition and secure mounting. Clean top of battery. Test specific gravity and voltage, and record on W.D., A.G.O. Form No. 461. Specific gravity readings below 1.225 indicate battery should be recharged or replaced. Electrolyte level should be above top of plates, and may extend ½ inch above plates.

SERVE. Perform high-rate discharge test according to instructions for "condition" test which accompany test instrument, and record voltage on W.D., A.G.O. Form No. 461. Cell variation should not be more than 30 percent. NOTE: *Specific gravity must be above 1.225 to make this test.*

CLEAN. Clean entire battery and carrier, and repaint carrier if corroded. Clean battery cable terminals, terminal bolts and nuts, and battery posts; grease lightly; inspect bolts for serviceability. Tighten terminals and hold-downs carefully to avoid damage to battery. Add clean water to ½ inch above plates.

Cylinder Head and Gasket. Look for cracks, and indications of water or compression leaks. Tighten cylinder head (only if leaks are indicated and after performing item 21) with torque wrench; tighten headscrews to from 65 to 75 foot-pounds; head stud nuts

¼-TON 4 x 4 TRUCK (WILLYS-OVERLAND MODEL MB and FORD MODEL GPW)

MAINTENANCE	
6000 Mile	1000 Mile
	19
19	
	20
20	
21	21
23	23
23	
24	24
24	
25	25

to from 60 to 65 foot-pounds. Tighten in correct order (fig. 25). Be sure cylinder head to dash bond strap is in good condition and securely connected.

Valve Mechanism. Adjust valves only if noisy.

ADJUST. Check clearance and adjust valves. Proper clearances are: intake valve, 0.014 inch when hot or cold; exhaust valve, 0.014 inch when hot or cold.

Spark Plugs. Wipe off plugs without removing; inspect for insulator cracks and leakage through insulators and gaskets. Service if required.

SERVE. Clean and adjust plugs to gap of 0.030 inch, using round gage. Plugs with broken insulators, excessive carbon deposits, electrodes burned thin or otherwise unserviceable, must be replaced. Correct plug (AN-7). NOTE: *If sand blast cleaner is not available install new or reconditioned plugs.*

Compression. Test compression with all plugs removed, and with throttle and choke wide open. Standard pressure is approximately 110 pounds at cranking speed; minimum pressure is 70 pounds. Maximum variation between cylinders must not be more than 10 pounds. If variation is greater than 10 pounds, recheck weak cylinders, using oil test, and report to higher authority. Record all readings.

Crankcase. Observe vehicle for crankcase, valve cover, timing case, or flywheel housing oil leaks. Check oil level. Drain and refill crankcase if change is due. See Lubrication Guide (par. 18).

CAUTION: *Do not start engine until completion of item 24.*

Oil Filters and Lines. Inspect filters, lines, and connections for good condition or leaks.

SERVE: Remove filter cartridge, clean filter case and install new cartridge and gaskets. Refill crankcase (5 quarts with new filter cartridge). Again inspect for leaks with engine running and check oil level after engine is stopped.

Radiator. Observe radiator core, hose, cap and gaskets for good condition and inspect for leaks. CAUTION: *System operates under 3¼ to 4¼ pounds pressure (be careful in removing cap).* Examine air passages and guards for obstructions and clean out any dirt, insects, or trash. Test and record antifreeze value (as climate demands). Examine coolant for oil, rust, or foreign

SECOND ECHELON PREVENTIVE MAINTENANCE

MAINTENANCE	
6000 Mile	1000 Mile
25	
26	26
27	27
27	
29	29
31	31
31	
32	32

material. Clean and flush radiator as needed. CAUTION: *Save and filter coolant if antifreeze is present.* Add inhibitor and antifreeze if needed.

TIGHTEN. Tighten hose clamps. Inspect radiator cap and gasket for tight seal.

Water Pump and Fan. Loosen fan belt; test water pump shaft and bearing for play. Inspect pump for secure attachment, good condition, and for leaks. Inspect fan for alinement and secure mounting.

Generator, Cranking Motor, and Switch. Inspect these units to see if they are in good condition, clean and securely connected or mounted; particularly radio noise suppression capacitor on generator and starting switch terminal, and bond straps from generator and cranking motor.

SERVE. Inspect commutators and brushes for good condition and wear. Brushes should be free in holders, and have full contact with commutator. Clean commutators with 2/0 flint paper if needed. Blow out with compressed air. Replace generator or cranking motor when commutator is scored, rough, worn, or brushes are less than half their original length.

Drive Belt and Pulleys. Inspect fan belt for fraying, wear, and deterioration. Inspect pulleys for cracks and misalinement. Replace or adjust belt as needed. Adjust to deflection of 1 inch between pulleys.

Distributor. Clean and remove distributor cap. Examine cap and rotor arm for cracks, corrosion and burned conductors. Clean breaker plate assembly, if dirty. Inspect breaker points for burning, pitting, alinement, and adjustment. Replace and aline burned or badly pitted points. Feel to determine excessive distributor shaft play. Turn distributor shaft (with rotor), and release to test centrifugal advance for binding.

SPECIAL LUBRICATION. Sparingly lubricate cam surfaces, movable breaker arm pin, wick and camshaft according to Lubrication Guide (par. 18). Adjust breaker point gap to 0.020 inch.

Coil and Wiring. Examine coil, high tension, and exposed low voltage wiring for cleanliness, and secure connections and attachment. Clean and tighten as required. Pay particular attention to see that spark plug

¼-TON 4 x 4 TRUCK (WILLYS-OVERLAND MODEL MB and FORD MODEL GPW)

MAINTENANCE		
6000 Mile	1000 Mile	
		and coil to distributor wire, radio noise suppressors, and coil terminal capacitor are in good condition, and securely mounted or connected.
33	33	**Manifolds and Heat Control.** Tighten manifold stud nuts as required to from 31 to 35 foot-pounds. Inspect for gasket leaks. Heat control valve must be free and bimetal spring must be in good condition.
34	34	**Air Cleaner.** Examine air cleaner for good condition and secure mounting. Examine oil cup. If dirty, remove and clean filter element; do not apply oil to element after cleaning. Clean oil cup and refill (⅝ qt).
36	36	**Carburetor.** Make certain that the choke and throttle open and close fully. Lubricate linkage, and inspect for worn parts.
37	37	**Fuel Filter, Screens, and Lines.** Clean fuel pump screen, renew gaskets, inspect unit for leaks. Remove disk filter element from fuel filter mounted on dash; clean element and bowl. Reinstall with new gasket. Inspect for leaks after unit has been refilled.
38	38	**Fuel Pump.** Observe fuel pump for leaks, secure mounting, and pressure reading. Pressure should be 1½ to 2½ pounds with engine running at approximately 30 miles per hour vehicle road speed.
39	39	**Cranking Motor.** Start engine and observe cranking motor for positive action, normal speed, and unusual noise. Make sure oil pressure gage and ammeter readings are satisfactory.
40	40	**Leaks.** Look around engine and on ground under engine for oil, fuel, coolant, or hydraulic fluid leaks.
41	41	**Ignition Timing.** With neon light, check ignition timing. Observe if spark advances automatically. Adjust timing as required (par. 65). CAUTION: *Close timing hole cover and tighten screw.*
42	42	**Engine Idle and Vacuum Test.** Adjust engine to smooth idle, using vacuum gage; obtain highest possible steady vacuum reading.
	43	**Regulator Unit.** See that regulator and radio noise capacitors are in good condition, and that all connections and mounting are secure.
43		TEST. Connect low voltage circuit tester and test voltage regulator, current regulator, and cut-out for output control.
47	47	**Tires and Rims.** Inspect valve stems for correct position and missing caps. Inspect tires for cuts, bruises, blisters, irregular and excessive tread wear. Remove imbedded glass, nails, or stones. Directional and non-

SECOND ECHELON PREVENTIVE MAINTENANCE

MAINTENANCE	
6000 Mile	1000 Mile

directional tires should not be installed on same vehicle. If equipped with directional tires, open end of chevron should meet ground first on front tires, and last on rear tires. Tires should match on all wheels within ¾-inch over-all circumference, and as to type of tread. Take measurements with all tires equally inflated. Inspect tire carrier for looseness and damage. Tighten all lug nuts securely. Inflate tires to 35 pounds (cold).

48 | 48 — **Rear Brakes.** Remove grease and dirt from brake drums and backing plates, and inspect for excessive wear or scoring and loose mounting bolts. Inspect brake hose for proper fit and for deterioration. Inspect wheel cylinders (exterior) for good condition, secure mounting, and for leaks. Tighten brake support and drum mounting bolts securely.

49 | 49 — **Rear Brake Shoes.** Remove right rear wheel and inspect linings for wear, oil, and dirt, and possibility of rivets scoring drum before next 1,000-mile inspection. If lining on right rear wheel requires replacement, remove all wheels for lining inspection.

49 — SERVE. Remove all wheels and drums. Observe linings for wear, oil, and dirt, and determine if shoes are secure and guided by anchor pins. Inspect return springs for good action. Lightly lubricate anchor pins. Adjust brake shoes to 0.005 inch at heel, and 0.008 inch at toe.

52 — **Rear Wheels.** Inspect wheel for good condition and, without removal, test for evidence of looseness of wheel bearing adjustment, and dry or damaged bearings. Inspect around drive flanges, brake supports, and drums for lubricant or brake fluid leaks. Tighten drive flange and wheel nuts. CAUTION: *If it is known that vehicle has operated in deep water which may have entered wheel bearings, inspect right wheel bearing for contamination. Remove, clean, repack, and adjust as for 6,000-mile service. If contamination of lubricant has occurred, service other wheel bearings likewise.*

52 — CLEAN. Disassemble wheel bearings and seals, clean, and inspect for damage.
SPECIAL LUBRICATION. Pack wheel bearings, install new seals, and adjust bearings.

53 | 53 — **Front Brakes.** Examine brake hose for chafing, leakage, and deterioration. Inspect wheel cylinders (exterior) for good condition, secure mounting, and leaks.

53 — DRUMS AND SUPPORTS. Clean drums and backing plates thoroughly, and tighten backing plate bolts. Inspect drums for damage, looseness, excessive wear, and scoring. Lightly lubricate anchor pins.

54 — **Front Brake Shoes.** Inspect brake shoes, linings, and anchors for damage or looseness. Replace worn parts

¼-TON 4 x 4 TRUCK (WILLYS-OVERLAND MODEL MB and FORD MODEL GPW)

MAINTENANCE	
6000 Mile	1000 Mile

and worn linings. Clean dust from linings. Adjust brake shoes to 0.005-inch clearance at heel, and 0.008-inch clearance at toe.

6000	1000	
55	55	**Steering Knuckles.** Inspect steering knuckle housings and oil seals for serviceable condition. Check lubricant for contamination. Refill to bottom of filler hole.
56	56	**Front Springs.** Inspect front springs for good condition, correct alinement, and excessive deflection. Inspect springs for excessive wear of spring bushing and clips. Tighten U-bolts securely and uniformly. Examine U-shackles and pivot bolts for wear.
57	57	**Steering.** Observe steering gear, Pitman arm, drag link, tie rod, and steering connecting rods for good condition, correct assembly, and secure mounting.
57		TIGHTEN. Tighten and adjust assembly mounting nuts and screws, arms, tie rods, drag link, Pitman arm, and gear, and steering wheel nuts. Replace broken seals or worn parts.
58	58	**Front Shock Absorbers.** Inspect shock absorbers to see if they are in good condition and secure, if bodies are leaking fluid, and if rubber bushings have deteriorated. If rubber bushings are hard or cracked, apply a film of brake fluid. NOTE: *If fluid is leaking or bodies are defective, shock absorber must be replaced.*
60	60	**Front Wheels.** Inspect for good condition, security, end play, and lubricant leaks. Rotate wheels and observe for loose, broken, or dry bearings.
60		CLEAN AND LUBRICATE. Remove, clean, inspect, lubricate, and replace bearings. Adjust bearings and test for wheel shake before removing jack.
61	61	**Front Axle.** Examine front axle housing for good condition and lubricant leaks. Inspect pinion shaft for end play and grease leaks. Inspect axle for apparent alinement, and see that vent is open.
62	62	**Front Propeller Shaft.** Inspect propeller shaft for damage and incorrect assembly, excessive wear, and lubricant leaks. Inspect universal and slip joints for alinement, wear, and leakage.
62		TIGHTEN. Tighten flange yoke bolts.
63	63	**Engine Mountings and Braces.** See that engine mountings and bond straps are in good condition and secure, and that rubber mountings are not separated from metal backing. Tighten front mountings if loose. Adjust rear

TM 9-803
23
SECOND ECHELON PREVENTIVE MAINTENANCE

MAINTENANCE	
6000 Mile	1000 Mile

6000	1000	
		mounting bolts to from 38 to 42 foot-pounds with torque wrench. Tighten radio noise suppression bond strap mountings securely.
64	64	**Parking (Hand) Brake.** See that drum is not scored or oily; that lining is not oil-soaked nor worn thin. Inspect ratchet for positive holding action. Lubricate upper end of conduit tube at cable with engine oil.
64		ADJUST. Adjust clearance between drum and lining to from 0.005 inch to 0.010 inch. Reserve lever travel should be one-third the ratchet range. There must be ½-inch reserve clearance (on cable) between relay crank and lower end of hand brake conduit.
65	65	**Clutch Pedal.** Clutch pedal linkage must be secure and not worn; return spring must be operative; clutch should have free pedal travel of ¾ inch.
65		ADJUST. Adjust clutch pedal free travel to ¾ inch.
66	66	**Brake Pedal.** Test brake pedal operation; brake linkage must be secure and not worn excessively; return spring must be operative; brake should have ⅓ reserve travel.
66		ADJUST. Adjust brake pedal free travel to ½ inch.
67	67	**Brake Master Cylinder.** Inspect master cylinder for good condition and secure mounting; check master cylinder boot for good condition and correct installation; inspect stop light switch for terminal attachment and correct operation. Look for brake fluid leaks; clean out filler plug vent. Fill master cylinder reservoir to ¼ inch below plug.
71	71	**Transmission.** Inspect oil seals and gaskets for leakage. Test control for looseness, excessive wear, and improper operation. Inspect mounting and assembly bolts and cap screws for looseness.
71		TIGHTEN. Tighten mounting and assembly bolts and cap screws.
72	72	**Transfer Case.** Inspect oil seals and gaskets for leakage. Test controls for looseness, excessive wear, and improper operation. Inspect mounting and assembly bolts and cap screws for looseness. Clean vent.
72		TIGHTEN. Tighten mounting and assembly bolts, nuts, and cap screws.
73	73	**Rear Propeller Shaft.** Remove any trash that may be wrapped around shaft or universal joints. Inspect

63

¼-TON 4 x 4 TRUCK (WILLYS-OVERLAND MODEL MB and FORD MODEL GPW)

MAINTENANCE	
6000 Mile	1000 Mile
		mounting of universal and slip joints for misalinement, wear, and grease leaks.
73		TIGHTEN. Tighten flange yoke cap screws.
75	75	**Rear Axle.** Inspect rear axle housing for leaks; feel for excessive play in pinion shaft; clean vent. Make sure differential carrier mounting cap screws are tight.
77	77	**Rear Springs.** Check springs for shifted leaves due to broken center bolt, loose spring clips, or U-bolts. If found loose, tighten U-bolts to from 50 to 55 foot-pounds. Tighten spring pivot bolt nut to from 29 to 30 foot-pounds.
78	78	**Rear Shock Absorbers.** Inspect in the same manner as for item 58.
80	80	**Frame.** Examine frame for loose side rails and cross members. Tighten loose bolts. If frame appears to be bent, or out of alinement, report condition to higher authority.
81	81	**Wiring, Conduits and Grommets.** Inspect all wiring for looseness and broken insulation; check conduits and grommets for proper position and good condition.
82	82	**Fuel Tank and Lines.** Inspect tank and lines for good condition, secure mounting, and leaks; check cap for defective gasket or clogged vent.
82		SERVE. Remove fuel tank drain plug briefly, and drain off accumulated water and sediment.
83	83	**Brake Lines and Connections.** Inspect brake lines for proper mounting, cracks, worn spots in lines, leaks, deteriorated or damaged hose and connections.
84	84	**Exhaust Pipe and Muffler.** Inspect exhaust pipe and muffler for secure mounting, rusted condition, damage or leaks. Inspect tail pipe for stoppage.
85	85	**Vehicle Lubrication.** Lubricate according to Lubrication Guide (par. 18) in this manual. Observe latest issued lubrication directives.

LOWER VEHICLE TO GROUND

86	86	**Toe-in and Turning Stops.** With front wheels on ground, straight-ahead position, use wheel alining gage, and check toe-in. Normal toe-in range is 3/64-inch to 3/32-inch. Turn front wheels fully in both right and left

SECOND ECHELON PREVENTIVE MAINTENANCE

MAINTENANCE	
6000 Mile	1000 Mile

6000 Mile	1000 Mile	
		directions, and determine if turning stops hold tires clear of all parts of vehicle in these positions. Examine axle for loose turn stops.
91	91	**Lights.** Determine that switches for head, tail, instrument, and blackout lights operate properly. Operate stop light by depressing brake pedal. Test foot switch, noting whether beam is controlled for high and low positions. Inspect all lights; these must be clean, securely mounted, and in good condition; lenses must not be broken, cracked, or discolored; reflectors must not be discolored; blackout lights must be in good condition with shield in proper position.
91		ADJUST. Adjust and aim headlight beams.
92	92	**Safety Reflectors.** Safety reflectors must be present, clean, and secure. Replace if cracked or broken.
93	93	**Front Bumper and Grille.** Front bumper and grille must be present, in good condition, and securely mounted.
94	94	**Hood, Hinges and Fasteners.** Examine hood for alinement and secure mounting when fastened; see that fasteners are present, secure, undamaged, and not excessively worn or bent. Lubricate hinges and fasteners lightly. See that radio noise bond straps from hood to dash and grille are secure.
95	95	**Front Fenders.** Inspect front fenders for good condition and secure mounting.
96	96	**Body Hardware.** Inspect body of vehicle according to following standards: Hardware should operate properly and be adequately lubricated; top should be clean, having no holes or tears, and all grommets must be present and in good condition. Windshield should be free from cracks or discoloration; windshield frame and hold-down hooks at hood should be in good condition. Seats and upholstery should be clean and undamaged; safety straps should be present and in place; body handles should be present, secure, and undamaged; floor drain plugs (2) should be present, and in good condition.
98	98	**Circuit Breaker, Terminal Blocks, or Boxes.** Inspect points of thermal circuit breaker (30 amperes, located on main light switch) for pitting or corrosion. Be sure all radio noise suppression bond straps and capacitor on radio terminal box (if so equipped) are in good condition and secure.
101	101	**Rear Bumpers and Pintle Hook, Latch and Lock Pin.** Inspect rear bumpers and pintle hook to see if they are

TM 9-803
23

¼-TON 4 x 4 TRUCK (WILLYS-OVERLAND MODEL MB and FORD MODEL GPW)

MAINTENANCE	
6000 Mile	1000 Mile
103	103
104	104
105	105
131	131
132	132
133	133
143	143
135	135

present, in good condition, and secure. Pintle hook safety latch should be free, and lock securely.

Paint and Markings. Inspect paint of entire vehicle for good condition and bright spots that might cause glare or reflection. Vehicle markings and identification must be legible. Inspect identification plates and their mountings (if furnished) for good condition, secure mounting, and legibility.

Radio Bonding (Suppressors, Filters, Condensers, and Shielding). See that all units not covered in the foregoing specific procedures are in good condition, and securely mounted and connected. Be sure all additional noise suppression bond straps and toothed lock washers listed in paragraph 177, are inspected for looseness or damage, and see that contact surfaces are clean. NOTE: *If objectionable radio noise from vehicle has been reported, make tests in accordance with paragraph 178. If cleaning and tightening of mountings and connections, and replacement of defective radio noise suppression units does not eliminate the trouble, the radio operator will report the condition to the designated individual in authority.*

Armament. Examine gun mounts and covers (if present) for good condition, cleanliness, and secure attachment. NOTE: *Guns, parts, and covers are to be referred to armorer or gun commanders for all inspections or service.*

TOOLS AND EQUIPMENT

Tools and Equipment. Standard vehicle tools, Pioneer tools, and equipment must be present, clean, serviceable, and securely mounted. Sharpen cutting tools and darken bright parts of exposed tools in combat areas. Check against stowage list (par. 19).

Fire Extinguisher. Inspect fire extinguisher for full charge and secure mounting. See that nozzle is clean.

Decontaminator. Inspect decontaminator for damage, secure mounting, and full charge. Make latter check by removing filler plug. Drain and refill with fresh solution every 90 days. See date of last filling on attached tag.

First Aid Kit. Examine contents of first aid kit for good condition, completeness, and satisfactory packing. Report any deficiency.

Publications and Form No. 26. See that the vehicle manuals and Lubrication Guide, Form No. 26 (Acci-

TM 9-803

SECOND ECHELON PREVENTIVE MAINTENANCE

MAINTENANCE	
6000 Mile	1000 Mile
136	136
139	139
140	140
141	141
142	142

dent Report) and W.D., A.G.O. Form No. 478 (MWO and Major Unit Assembly Replacement Record), are present, legible, and properly stowed.

Traction Devices. Inspect tire chains for broken or worn links, missing cross chains, or damaged fasteners.

Fuel Can and Bracket. Inspect fuel can and bracket for damage, leaks, loose mounting, and presence of cap on chain.

Fuel Can Nozzle and Bucket. See that fuel can nozzle and bucket are not damaged, are clean, and properly stowed.

Modifications (Completed). Inspect entire vehicle to be sure all Modification Work Orders have been completed, and enter any modifications or major unit replacements made at time of this service, on Form No. 478.

Final Road Test. Road test, rechecking items 2 to 16. Recheck transmission, transfer case, and differentials for lubricant level and for leaks. Confine this test to minimum distance necessary to satisfactory observations. NOTE: *Correct or report all defects found during final road test to higher authority.*

¼-TON 4 x 4 TRUCK (WILLYS-OVERLAND MODEL MB and FORD MODEL GPW)

Section X

NEW VEHICLE RUN-IN TEST

	Paragraph
Purpose	24
Correction of deficiencies	25
Run-in test procedures	26

24. PURPOSE.

a. When a new or reconditioned vehicle is first received at the using organization, it is necessary for second echelon personnel to determine whether or not the vehicle will operate satisfactorily when placed in service. For this purpose, inspect all accessories, subassemblies, assemblies, tools, and equipment to see that they are in place and correctly adjusted. In addition, they will perform a run-in test of at least 50 miles as directed in AR 850-15, paragraph 25, table III, according to procedures in paragraph 26 below.

25. CORRECTION OF DEFICIENCIES.

a. Deficiencies disclosed during the course of the run-in test will be treated as follows:

(1) Correct any deficiencies within the scope of the maintenance echelon of the using organization before the vehicle is placed in service.

(2) Refer deficiencies beyond the scope of the maintenance echelon of the using organization to a higher echelon for correction.

(3) Bring deficiencies of serious nature to the attention of the supplying organization.

26. RUN-IN TEST PROCEDURES.

a. Preliminary Service.

(1) FIRE EXTINGUISHER. See that portable extinguisher is present and in good condition. Test it momentarily for proper operation, and mount it securely.

(2) FUEL, OIL, AND WATER. Fill fuel tank. Check crankcase oil and coolant supply; add oil and coolant as necessary to bring to correct levels. Allow room for expansion in fuel tank and radiator. During freezing weather, test value of antifreeze, and add as necessary to protect cooling system against freezing. CAUTION: *If there is a tag attached to filler cap or steering wheel concerning engine oil in crankcase, follow instructions on tag before driving the vehicle.*

(3) FUEL FILTER. Inspect main fuel filter for leaks, damage, and secure mountings and connections. Drain sediment bowl. Clean fuel pump filter screen and bowl. If any appreciable amount of dirt or water is present, remove main filter bowl and clean bowl and element

NEW VEHICLE RUN-IN TEST

in dry-cleaning solvent. Also, drain accumulated dirt and water from bottom of fuel tank. Drain only until fuel runs clean.

(4) BATTERY. Make hydrometer and voltage test of battery, and add clean water to bring electrolyte $3/8$ inch above plate.

(5) AIR CLEANER. Examine carburetor air cleaner to see if it is in good condition and secure. Remove element and wash thoroughly in dry-cleaning solvent. Fill oil cup to indicated level with fresh oil, and reinstall securely. Be sure oil cup and body gaskets are in good condition, and that air horn connection is tight.

(6) ACCESSORIES AND BELT. See that accessories such as carburetor, generator, regulator, cranking motor, distributor, water pump, fan, and oil filter, are securely mounted. Make sure that fan and generator drive belt is in good condition, and adjusted to have 1-inch finger-pressure deflection.

(7) ELECTRICAL WIRING. Examine all accessible wiring and conduits to see if they are in good condition, securely connected, and properly supported.

(8) TIRES. See that all tires, including spare, are properly inflated to 35 pounds, cool; that stems are in correct position; all valve caps present and finger-tight. Inspect for damage, and remove objects lodged in treads and carcasses.

(9) WHEEL AND FLANGE NUTS. See that all wheel mounting and axle flange nuts are present and secure.

(10) FENDERS AND BUMPER. Examine fenders and front bumper for looseness and damage.

(11) TOWING CONNECTIONS. Examine towing shackles and pintle hook for looseness and damage, and see that pintle latch operates properly and locks securely.

(12) BODY. See that all body mountings are secure. Inspect attachments, hardware, glass, seats, grab rails and safety straps, top and frame, curtains and hood, to see if they are in good condition, correctly assembled, and securely mounted or fastened. Examine body paint or camouflage pattern for rust, or shiny surfaces that might cause glare. See that vehicle markings are legible.

(13) LUBRICATE. Perform a complete lubrication service of the vehicle, covering all intervals, according to instructions on Lubrication Guide (par. 18), except gear cases, wheel bearings, and other units already lubricated or serviced in items (1) to (12). Check all gear case oil levels, and add as necessary to bring to proper levels. Change only if condition of oil indicates the necessity, or if gear oil is not of proper grade for existing atmospheric temperatures. NOTE: *Perform following items (14) through (17) during lubrication.*

(14) SPRINGS AND SUSPENSIONS. Inspect front and rear springs and shocks to see that they are in good condition, correctly assembled, secure, and that bushings and shackle pins are not excessively loose, or damaged.

¼-TON 4 x 4 TRUCK (WILLYS-OVERLAND MODEL MB and FORD MODEL GPW)

(15) STEERING LINKAGE. See that all steering arms, rods, and connections are in good condition and secure; and that gear case is securely mounted and not leaking excessively.

(16) PROPELLER SHAFTS. Inspect all shafts and universal joints to see if they are in good condition, correctly assembled, alined, secure, and not leaking excessively.

(17) AXLE AND TRANSFER VENTS. See that axle housing and transfer case vents are present, in good condition, and not clogged.

(18) CHOKE. Examine choke to be sure it opens and closes fully in response to operation of choke button.

(19) ENGINE WARM-UP. Start engine and note if cranking motor action is satisfactory, and if engine has any tendency toward hard starting. Set hand throttle to run engine at fast idle during warm-up. During warm-up, reset choke button so that engine will run smoothly, and to prevent overchoking and oil dilution.

(20) INSTRUMENTS.

(a) Oil Pressure Gage. Immediately after engine starts, observe if oil pressure is satisfactory. (Normal operating pressure, hot, at running speeds is 40 to 50 pounds; at idle, 10 pounds). Stop engine if pressure is not indicated in 30 seconds.

(b) Ammeter. Ammeter should show slight positive (+) charge. High charge may be indicated until generator restores to battery, current used in starting.

(c) Temperature Gage. Engine temperature should rise gradually during warm-up period to normal operating range, 160°F to 185°F

(d) Fuel Gage. Fuel gage should register "FULL" if tank has been filled.

(21) ENGINE CONTROLS. Observe if engine responds properly to controls, and if controls operate without excessive looseness or binding.

(22) HORN AND WINDSHIELD WIPERS. See that these items are in good condition and secure. If tactical situation permits, test horn for proper operation and tone. See if wiper arms will operate through their full range, and that blade contacts glass evenly and firmly.

(23) GLASS AND REAR VIEW MIRROR. Clean all body glass, curtain windows, and mirror, and inspect for looseness and damage. Adjust mirror for correct vision.

(24) LAMPS (LIGHTS) AND REFLECTORS. Clean lenses and inspect all units for looseness and damage. If tactical situation permits, open and close all light switches to see if lamps respond properly.

(25) LEAKS, GENERAL. Look under vehicle, and within engine compartment, for indications of fuel, oil, coolant, and brake fluid leaks. Trace to source any leaks found, and correct or report them to designated authority.

(26) TOOLS AND EQUIPMENT. Check tools and On Vehicle Stowage Lists, paragraphs 19 and 20, to be sure all items are present, and see that they are serviceable, and properly mounted or stowed.

NEW VEHICLE RUN-IN TEST

b. Run-in Test. Perform the following procedures, steps (1) to (11) inclusive, during the road test of the vehicle. On vehicles which have been driven 50 miles or more in the course of delivery from the supplying to the using organization, reduce the length of the road test to the least mileage necessary to make observations listed below. CAUTION: *Continuous operation of the vehicle at speeds approaching the maximum indicated on the caution plate should be avoided during the test.*

(1) DASH INSTRUMENTS AND GAGES. Do not move vehicle until engine temperature reaches 135°F. Maximum safe operating temperature is 200°F. Observe readings of ammeter, oil temperature, and fuel gages to be sure they are indicating the proper function of the units to which they apply. Also see that speedometer registers the vehicle speed, and that odometer registers accumulating mileage.

(2) BRAKES: FOOT AND HAND. Test service brakes to see if they stop vehicle effectively, without side pull, chatter, or squealing; and observe if pedal has at least ½-inch free travel before meeting push rod-to-piston resistance. Parking brake should hold vehicle on reasonable incline, leaving one-third lever ratchet travel in reverse. CAUTION: *Avoid long application of brakes until shoes become evenly seated to drums.*

(3) CLUTCH. Observe if clutch operates smoothly without grab, chatter, or squeal on engagement, or slippage (under load) when fully engaged. See that pedal has ¾-inch free travel before meeting resistance. CAUTION: *Do not ride clutch pedal at any time, and do not engage and disengage new clutch severely or unnecessarily.*

(4) TRANSMISSION AND TRANSFER. Gearshift mechanism should operate easily and smoothly, and gears should operate without excessive noise, and not slip out of mesh. Test front axle declutching for proper operation.

(5) STEERING. Observe steering action for binding or looseness, and note any excessive pull to one side, wander, shimmy, or wheel tramp. See that column, bracket, and wheel are secure.

(6) ENGINE. Be on the alert for any abnormal engine operating characteristics or unusual noise, such as lack of pulling power or acceleration, backfiring, misfiring, stalling, overheating, or excessive exhaust smoke. Observe if engine responds properly to all controls.

(7) UNUSUAL NOISE. Be on the alert throughout road test for any unusual noise from body and attachments, running gear, suspension, or wheels, that might indicate looseness, damage, wear, inadequate lubrication, or underinflated tires.

(8) HALT VEHICLE AT 10-MILE INTERVALS FOR SERVICES (steps (9) and (10) below).

(9) TEMPERATURES. Cautiously hand-feel each brake drum and wheel hub for abnormal temperatures. Examine the transmission, transfer case, and differential housing for indications of overheating

¼-TON 4 x 4 TRUCK (WILLYS-OVERLAND MODEL MB and FORD MODEL GPW)

and excessive lubricant leaks at seals, gaskets, or vents. NOTE: *Transfer case temperatures are normally higher than other gear cases.*

(10) LEAKS. With engine running, and fuel, engine oil, and cooling systems under pressure, look within engine compartment and under vehicle for indications of leaks.

c. Upon completion of run-in test, correct or report any deficiencies noted. Report general condition of vehicle to designated individual in authority.

TM 9-803
27-28

Section XI

ORGANIZATION TOOLS AND EQUIPMENT

	Paragraph
Standard tools and equipment	27
Special tools	28

27. STANDARD TOOLS AND EQUIPMENT.

a. All standard tools and equipment available to second echelon are listed in SNL N-19, and their availability is determined by the table of equipment for any particular organization.

28. SPECIAL TOOLS.

a. The special tools available to second echelon for repair of this vehicle are listed in the Organizational Spare Parts and Equipment List of SNL G-503. The special tools required for the operations described in this manual are listed below:

Tool	Federal Stock No.
COMPRESSOR, shock absorber grommet	41-C-2554-400
WRENCH, tappet, double-end, $11/32$- x $17/32$-in.	41-W-3575

73

TM 9-803
29-30

¼-TON 4 x 4 TRUCK (WILLYS-OVERLAND MODEL MB and FORD MODEL GPW)

Section XII

TROUBLE SHOOTING

	Paragraph
General	29
Engine	30
Clutch	31
Fuel system	32
Intake and exhaust systems	33
Cooling system	34
Ignition system	35
Starting and generating systems	36
Transmission	37
Transfer case	38
Propeller shafts	39
Front axle	40
Rear axle	41
Brake system	42
Wheels, wheel bearings, and related parts	43
Springs and shock absorbers	44
Steering system	45
Body and frame	46
Battery and lighting system	47
Radio suppression	48
Instruments	49

29. GENERAL.

a. The following listed possible vehicle troubles and remedies will assist in determining the cause of unsatisfactory operation. A separate list is provided for each unit. If the remedy is not given, reference is made to a paragraph where more complete information will be found.

b. The information in this section applies to operation of the vehicle under normal conditions. If extreme conditions are encountered, it is assumed the vehicle has received the attention outlined in section IV.

30. ENGINE.

a. Diagnosing Troubles. Determine troubles in a general way first as follows:

(1) CHECK MECHANICAL CONDITION. Check for mechanical trouble such as broken or deficient parts in engine or cylinder compression.

TROUBLE SHOOTING

(2) CHECK IGNITION SYSTEM. Remove spark plug wire at a plug. Hold terminal end of wire about ¼ inch from a metal part of engine, and check for a good spark by having someone turn ignition switch on and operate cranking motor. If no spark is obtained, check ammeter operation to determine condition of ignition primary circuit. Ammeter must show slight deflection from zero to discharge side (with lights off) when cranking motor is operated and ignition switch is on. If ammeter drops to zero when starting switch is pressed, starting system is defective, or battery is discharged.

(3) CHECK FUEL SYSTEM. Operate priming lever on rear side of fuel pump; to determine if fuel is reaching carburetor. Resistance to operation indicates carburetor is empty or no fuel; no resistance indicates carburetor is full. A flooded carburetor and engine may prevail so the spark plugs are shorted.

b. Cranking Motor Will Not Crank Engine.

(1) AMMETER DROPS TOWARD ZERO WHEN STARTING SWITCH IS PRESSED.

Possible Cause	Possible Remedy
Battery discharged.	Replace or charge battery (par. 97).
Battery terminals or ground cables loose or corroded.	Remove and clean.
Cranking motor drive gear jammed in flywheel teeth.	Rock vehicle backwards or loosen cranking motor (par. 89).
Excessive engine friction due to seizure or improper oil.	Change oil to proper grade (par. 18); if seizure has occurred, report to higher authority.

(2) AMMETER REMAINS UNCHANGED WHEN STARTING SWITCH IS PRESSED.

Possible Cause	Possible Remedy
Battery cable terminal corroded or broken.	Clean or replace.
Poor starting switch contacts.	Replace switch (par. 90).

(3) CRANKING MOTOR RUNS BUT FAILS TO CRANK ENGINE WHEN SWITCH IS PRESSED.

Possible Cause	Possible Remedy
Cranking motor gear does not engage flywheel.	Remove cranking motor and clean gear (par. 89).
Cranking motor or drive gear faulty.	Replace cranking motor (par. 89).

c. Engine Will Not Start.

(1) NO SPARK.

(a) *Ammeter Shows No Discharge (Zero Reading) with Ignition Switch "ON."*

Possible Cause	Possible Remedy
Ignition switch partly on.	Turn on fully.
Ignition switch faulty.	Replace switch (par. 68).

¼-TON 4 x 4 TRUCK (WILLYS-OVERLAND MODEL MB and FORD MODEL GPW)

Possible Cause	Possible Remedy
Ignition primary wires, or cranking motor cables broken, or connections loose.	Repair or replace and tighten
Ignition coil primary winding open.	Replace coil (par. 66).
Distributor points burned, pitted, or dirty.	Clean or replace and adjust (par. 64).
Distributor points not closing.	Clean and adjust; put one drop of oil on arm post (par. 63).
Loose or corroded ground or battery cable connections.	Clean or replace and tighten.
Open circuit in suppression filter.	Test for trouble by removing ignition switch and coil wires, and connect together; if filter is faulty, report to higher authority.

(b) Ammeter Reading Normal.

Possible Cause	Possible Remedy
High tension wire from coil to distributor broken, grounded, or out of terminals.	Repair or replace (par. 69).
Short-circuited secondary circuit in coil.	Replace coil (par. 66).
Short-circuited condenser.	Replace condenser (par. 64).
Short-circuited or burned distributor cap or rotor.	Replace part (par. 64).
Spark plugs, distributor cap, or wires wet (shorted).	Dry and clean thoroughly.
Spark plug gaps wrong.	Reset gaps (par. 67).
Ignition timing incorrect.	Set timing (par. 65).
Ignition wires installed wrong in distributor cap.	Put in proper places (par. 69).

(c) Ammeter Indicates Abnormal Discharge.

Possible Cause	Possible Remedy
Short-circuited wire between ammeter and ignition switch or coil.	Repair or replace wire.
Short-circuited primary winding in ignition coil.	Install new coil (par. 66).
Radio filter short-circuited.	Disconnect temporarily, and report to higher authority.
Short-circuited condenser or broken lead.	Repair lead or replace condenser (par. 64).

TROUBLE SHOOTING

Possible Cause	Possible Remedy
Distributor points not opening.	Clean or replace and adjust (par. 63).
Distributor does not operate cam to open points.	Report to higher authority.

(2) Weak Spark.

Possible Cause	Possible Remedy
Distributor points pitted or burned.	Clean or replace and adjust (par. 64).
Distributor condenser weak.	Replace (par. 64).
Ignition coil weak.	Replace (par. 66).
Primary wire connections loose.	Tighten.
High tension or spark plug wires or distributor cap wet.	Dry thoroughly.
High tension or spark plug wires or distributor cap damaged.	Replace (par. 69).
Distributor rotor burned or broken.	Replace (par. 64).

(3) Good Spark.

Possible Cause	Possible Remedy
Fuel tank empty.	Refill tank (par. 75).
Dirt or water in carburetor or float stuck.	Report to higher authority.
Carburetor and engine flooded by excessive use of choke.	Pull out throttle; crank engine with motor; when engine starts, regulate throttle; leave choke control "IN."
Choke control not operating properly.	Adjust (par. 72).
Fuel does not reach carburetor.	Check for damaged or leaky lines; air leak into line between tank and fuel pump.
Dirt in fuel lines or tank.	Disconnect drain tank and blow out lines.
Fuel line pinched.	Repair or replace.
Fuel strainer clogged.	Dismantle and clean (par. 76).
Fuel pump does not pump.	Clean screen; replace pump if inoperative (par. 74).
Lack of compression.	Report to higher authority.

(4) Backfiring.

Possible Cause	Possible Remedy
Ignition out of time.	Retime (par. 65).
Spark plug wires in wrong places in distributor cap or at spark plugs.	Install in proper places (par. 69).

¼-TON 4 x 4 TRUCK (WILLYS-OVERLAND MODEL MB and FORD MODEL GPW)

Possible Cause	Possible Remedy
Distributor cap cracked or shorted.	Replace (par. 64).
Valve holding open—due to lack of compression.	Report to higher authority.

d. Engine Runs but Backfires and Spits.

Possible Cause	Possible Remedy
Overheated engine.	Check (subpar. l below).
Improper ignition timing.	Reset (par. 65).
Spark plug wires in wrong place in distributor cap.	Install in proper places (par. 69).
Dirt or water in carburetor.	Clean and adjust (par. 72).
Carburetor improperly adjusted.	Check idle adjustment (par. 72).
Carburetor float level low.	Report to higher authority.
Valve sticking or not seating properly, burned, or pitted.	Report to higher authority.
Excessive carbon in cylinders.	Remove carbon (par. 54).
Valve springs weak.	Report to higher authority.
Heat control valve not operating.	Free-up and check thermostat spring position (par. 53).
Fuel pump pressure low.	Clean screen; replace pump, if faulty (par. 74).
Fuel strainer clogged.	Dismantle and clean (par. 76).
Partly clogged or pinched fuel line.	Clean or repair.
Intake manifold leak.	Check gaskets (par. 52).
Distributor cap cracked or shorted.	Replace (par. 64).

e. Engine Stalls on Idle.

Possible Cause	Possible Remedy
Carburetor throttle valve closes too far, or idle mixture incorrect.	Adjust (par. 72).
Carburetor choke valve sticks closed.	Free-up and lubricate.
Dirt or water in idle passages of carburetor.	Replace carburetor (par. 72).
Air leak at intake manifold.	Tighten manifold stud nuts or replace gaskets (par. 52).
Heat control valve faulty.	Free-up and adjust (par 53).
Spark plugs faulty, gaps incorrect.	Clean or replace, set gaps (par. 67).
Ignition timing too early.	Reset (par. 65).
Low compression.	Report to higher authority.

TROUBLE SHOOTING

Possible Cause	Possible Remedy
Water leak in cylinder head or gasket.	Replace gasket, or report cylinder head leak to higher authority.
Crankcase ventilator valve stuck open.	Clean (par. 59).

f. Engine Misfires on One or More Cylinders.

Possible Cause	Possible Remedy
Dirty spark plugs.	Clean and adjust or replace (par. 67).
Wrong type spark plugs.	Replace with correct type (par. 67).
Spark plug gap incorrect.	Reset gap (par. 67).
Cracked spark plug porcelain.	Replace spark plug (par. 67).
Spark plug or distributor suppressors faulty.	Replace (par. 67).
Spark plug wires grounded.	Replace.
Spark plug wires in wrong places in cap or at spark plugs.	Install correctly (par. 69).
Distributor cap or rotor burned or broken.	Replace (par. 64).
Valve tappet holding valve open.	Service (par. 56).
Compression poor—valve trouble.	Report to higher authority.
Leaky cylinder head gasket.	Replace gasket (par. 54).
Cracked cylinder block or broken valve tappet or tappet screw.	Report to higher authority.

g. Engine Does Not Idle Properly—(Erratic).

Possible Cause	Possible Remedy
Ignition timed too early.	Reset (par. 65).
Dirty spark plugs or gaps too close.	Clean and adjust (par. 67).
Ignition coil or condenser weak.	Replace (par. 66).
Distributor points sticking, dirty or improperly adjusted.	Adjust or replace (par. 64).
Distributor rotor or cap cracked or burned.	Replace (par. 64).
Weak or broken valve spring.	Report to higher authority.
Leaky cylinder head gasket.	Replace (par. 54).
Uneven cylinder compression.	Report to higher authority.
High tension or spark plug wires leaky—cracked insulation.	Replace.
Dirt or water in carburetor, or float level incorrect.	Report to higher authority.
Carburetor adjustment or choke not set right.	Adjust (par. 72).

¼-TON 4 x 4 TRUCK (WILLYS-OVERLAND MODEL MB and FORD MODEL GPW)

Possible Cause	Possible Remedy
Fuel pump pressure low.	Clean screen; replace pump (par. 74).
Crankcase ventilator valve leaks.	Clean (par. 59).
Leaky intake manifold.	Tighten manifold stud nuts or replace gaskets (par. 52).

h. Engine Misses On Acceleration.

Possible Cause	Possible Remedy
Dirty spark plugs or gaps too wide.	Clean and adjust (par. 67).
Wrong type spark plug.	Replace (par. 67).
Ignition coil or condenser weak.	Replace (par. 66).
Distributor breaker points sticking, dirty or improperly adjusted.	Adjust or replace (par. 64).
Distributor cap or rotor cracked or burned.	Replace (par. 64).
Distributor cap, spark plugs or wire wet or dirty.	Clean and dry thoroughly.
High tension or spark plug wires leaky—cracked insulation.	Replace (par. 69).
Carburetor choke not adjusted.	Adjust (par. 72).
Carburetor accelerating pump system faulty, dirt in metering jets or float level incorrect.	Report to higher authority.
Fuel pump faulty—lack of fuel.	Clean screen; replace faulty pump (par. 74).
Air cleaner dirty.	Clean and reoil (par. 73).
Heat control valve faulty.	Check and adjust (par. 53).
Valves sticking—weak or broken valve springs.	Report to higher authority.
Overheated engine.	Check (subpar. l below).
Fuel strainer clogged.	Dismantle and clean (par. 76).

i. Engine Misses at High Speeds.

Possible Cause	Possible Remedy
Distributor points sticking, adjusted too wide or burned.	Clean and adjust (par. 64).
Weak distributor arm spring.	Replace (par. 64).
Incorrect type of spark plugs.	Replace (par. 67).
Excessive play in distributor shaft bearing.	Replace distributor (par. 64).
Spark plugs faulty, dirty or incorrect gap.	Clean, adjust or replace (par. 67).
Weak ignition coil or condenser.	Replace (par. 66).

TROUBLE SHOOTING

Possible Cause	Possible Remedy
Valves sticking—weak or broken springs.	Report to higher authority.
Fuel supply lacking at carburetor.	Check fuel system (par. 71 a).
Heat control valve faulty.	Free-up and adjust (par. 53).
Air cleaner dirty.	Clean and reoil (par. 73).
Carburetor metering rod incorrectly set.	Report to higher authority.

j. Engine Pings (Spark Knock).

Possible Cause	Possible Remedy
Ignition timing early.	Reset (par. 65).
Distributor automatic spark advance stuck in advance position or spring broken.	Replace distributor (par. 64).
Overheated engine.	Check (subpar. l below).
Excessive carbon deposit in cylinders.	Remove cylinder head and clean (par. 54).
Heat control valve faulty.	Free-up and adjust (par. 53).
Wrong type spark plug.	Replace (par. 67).
Old or incorrect fuel.	Drain and use correct fuel (par. 3).

k. Engine Lacks Power.

Possible Cause	Possible Remedy
Ignition timing late.	Reset (par. 65).
Ignition system faulty.	Check (subpar. c above).
Old or incorrect fuel.	Use correct gasoline.
Leaky gaskets.	Replace.
Engine overheated.	Check (subpar. l below).
Excessive carbon formation.	*Remove cylinder head and clean* (par. 54).
Engine too cold.	Test thermostat (par. 85); in cold weather, cover radiator.
Insufficient oil or improper grade.	Use correct grade (par. 18).
Oil system failure.	Report to higher authority.
Air cleaner dirty.	Clean; change oil in reservoir (par. 73).
Spark plug gaps too wide.	Reset (par. 67).
Choke valve partially closed or throttle does not open fully.	Adjust (par. 72).
Manifold heat control inoperative.	Check valve operation; see that spring is in proper position (par. 53).

¼-TON 4 x 4 TRUCK (WILLYS-OVERLAND MODEL MB and FORD MODEL GPW)

Possible Cause	Possible Remedy
Exhaust pipe, muffler or tail pipe damaged or clogged.	Service or replace (par. 78).
Low compression—broken valve springs or sticking valves or improper tappet adjustment.	Report to higher authority.
Lack of fuel.	Clean filter (par. 76) check fuel pump (par. 74) check carburetor for water or dirt (par. 72).

l. Engine Overheats.

Possible Cause	Possible Remedy
Cooling system deficient.	Water low; air flow through radiator core restricted, clean from engine side; clogged core, clean or replace radiator (par. 81).
Radiator or water pump leaky.	Replace (par. 82).
Leaky cylinder head gasket.	Tighten or replace gasket (par. 54).
Damaged or deteriorated hose or fan belt.	Replace (par. 83).
Loose fan belt.	Adjust, or generator brace not hooked (par. 83).
Cylinder block, head or core hole plugs leaky.	Report to higher authority.
Ignition timing incorrect.	Reset (par. 65).
Damaged muffler; bent or clogged exhaust pipe.	Service or replace (par. 78).
Excessive carbon in cylinders.	Remove cylinder head and clean (par. 54).
Insufficient oil or improper grade.	Use correct grade (par. 18).
Air cleaner restricted.	Clean and renew oil (par. 73).
Inoperative thermostat or radiator cap.	Replace (par. 85).
Ignition system faulty.	Check (subpar. c above).
Water pump impeller broken.	Replace pump (par. 82).
Poor compression or valve timing wrong.	Report to higher authority.
Oil system failure (clogged screen).	Check (subpar. p below).

m. Low Fuel Mileage.

Possible Cause	Possible Remedy
High engine speeds (unnecessary and excessive driving in lower gear range).	Correct driving practice.

TROUBLE SHOOTING

Possible Cause	Possible Remedy
Air cleaner clogged.	Clean and renew oil (par. 73).
Carburetor float level too high. Metering rod, accelerating pump not properly adjusted.	Report to higher authority.
Fuel line leaks.	Tighten or replace.
Overheated engine.	Check (subpar. l above).
Carburetor parts worn or broken.	Replace carburetor (par. 72).
Fuel pump pressure too high or leaky diaphragm.	Replace fuel pump (par. 74).
Engine running cold.	Check thermostat (par. 85); cover radiator.
Heat control valve inoperative.	Free-up and put spring on bracket (par. 53).
Choke partially closed.	Adjust (par. 72).
Ignition timed wrong.	Reset (par. 65).
Spark advance stuck.	Replace distributor (par. 64).
Leaky fuel pump bowl gasket.	Replace gasket (par. 74).
Low compression.	Report to higher authority.
Carburetor controls sticking.	Free-up and lubricate.
Engine idles too fast.	Adjust carburetor throttle stop screw (par. 72).
Spark plugs dirty.	Clean or replace (par. 67).
Weak coil or condenser.	Replace (par. 64).
Clogged muffler or bent exhaust pipe.	Service or replace (par. 78).
Loose engine mountings permitting engine to shake and raise fuel level in carburetor.	Tighten; if damaged replace.

n. Low Oil Mileage.

Possible Cause	Possible Remedy
High engine speeds or unnecessary and excessive driving in low gear ranges.	Correct driving practice.
Oil leaks.	Replace leaky gaskets.
Improper grade or diluted oil.	Use new oil of proper grade (par. 18).
Overheating of engine causing excessive temperature and thinning of oil.	Check (subpar. l above).
Oil filter clogged.	Clean; replace element (par. 58).

¼-TON 4 x 4 TRUCK (WILLYS-OVERLAND MODEL MB and FORD MODEL GPW)

Possible Cause	Possible Remedy
Faulty pistons, or rings or rear bearing oil return clogged; excessive clearance of intake valves in guides; cylinder bores worn (scored, out-of-round, tapered); excessive bearing clearance; misalined connecting rods.	Report to higher authority.

o. Poor Compression.

Possible Cause	Possible Remedy
Incorrect tappet adjustment.	Adjust (par. 56).
Leaky, sticking or burned valves; sticking tappets; valve springs weak or broken; valve stems and guides worn; piston ring grooves worn or rings worn, broken or stuck; cylinders scored or worn excessively.	Report to higher authority.

p. Low Oil Pressure.

Possible Cause	Possible Remedy
Insufficient oil supply.	Check oil level.
Improper grade of oil or diluted oil foaming at high speeds.	Change oil; check crankcase ventilator (par. 59); check for water in oil by inspecting dip stick.
High oil temperature causing oil to be thin.	Check (subpar. 1 above).
Oil too heavy (funneling in cold weather).	Dilute engine oil (par. 18).
Floating oil intake loose or gasket leaky.	Renew gasket, tighten (par. 57).
Oil screen clogged.	Remove oil pan and clean screen (par. 57).
Oil leak causing lack of oil.	Inspect and service.
Faulty oil pump or pressure regulator valve stuck or spring broken.	Report to higher authority.
Oil filter restriction hole too large.	Replace oil filter (par. 58).
Oil pressure too high.	Faulty oil pump regulator valve stuck closed or improperly adjusted, report to higher authority.

q. Faulty Valves.

Possible Cause	Possible Remedy
Incorrect tappet adjustment.	Adjust tappets (par. 56).
Other valve troubles.	Report to higher authority.

TROUBLE SHOOTING

r. Abnormal Engine Noises.

Possible Cause	Possible Remedy
Loose fan, fan pulley or belt, heat control valve, or noisy generator brush.	Tighten or service.
Leaky intake or exhaust manifold or gaskets, cylinder head gasket or spark plug.	Replace or tighten (pars. 52 and 54).
Overheated engine; clogged exhaust system.	Remove obstruction from muffler tail pipe. Check (subpar. 1 above).
Other abnormal engine noises.	Report to higher authority.

31. CLUTCH.

a. Clutch Slips.

Possible Cause	Possible Remedy
Improper pedal adjustment.	Adjust pedal free travel (par. 109).
Release linkage binding.	Free-up and lubricate.
Clutch facings burned or worn, torn loose from plate, or oil-soaked.	Replace clutch driven plate (pars. 110 and 111).
Weak pressure spring.	Report to higher authority.
Sticking pressure plate.	Report to higher authority.

b. Clutch Grabs or Chatters.

Possible Cause	Possible Remedy
Control linkage binding.	Free-up and lubricate.
Loose engine mountings.	Tighten.
Engine stay cable not adjusted.	Adjust; just taut.
Facings burned, worn, or loose on driven plate; driven plate crimped or cushion flattened out, worn, or binding on splined shaft.	Replace clutch driven plate (pars. 110 and 111).
Pressure plate or flywheel face scored or rough; pressure plate broken; improper clutch lever (finger) adjustment; excessive looseness in power train.	Report to higher authority.

c. Clutch Drags.

Possible Cause	Possible Remedy
Too much pedal play.	Adjust pedal free play (par. 109).
Driven plate warped; facings torn or loose.	Replace clutch driven plate (pars. 110 and 111).
Pressure plate warped or binds in bracket; improper finger adjustment; excessive friction in flywheel bushing.	Report to higher authority.

¼-TON 4 x 4 TRUCK (WILLYS-OVERLAND MODEL MB and FORD MODEL GPW)

d. Clutch Rattles.

Possible Cause	Possible Remedy
Clutch pedal return spring is broken or disconnected.	Replace or connect.
Release fork loose on ball stud.	Adjust clutch pedal free travel to ¾ inch (par. 109).
Driven plate springs broken. Worn release bearing.	Replace (pars. 110 and 111).
Worn pressure plate or broken return springs at driving lugs; worn driven plate hub on splined shaft; worn release bearing; fingers improperly adjusted; pilot bushing worn in flywheel.	Report to higher authority.

32. FUEL SYSTEM.

a. Fuel Does Not Reach Carburetor.

No fuel in tank.	Fill tank.
Fuel filter clogged.	Service fuel filter (par. 76).
Fuel pump inoperative.	Replace.
Fuel line air leak between tank and fuel pump.	Locate and correct.
Fuel line clogged.	Disconnect and blow out lines.
Fuel tank cap not functioning.	Replace cap.

b. Fuel Reaches Carburetor but Does Not Enter Cylinders.

Choke does not close.	Free-up and lubricate; inspect for proper operation.
Fuel passages in carburetor clogged.	Replace carburetor (par. 72).
Carburetor float valve stuck closed.	Report to higher authority.

c. Low Fuel Mileage.

Engine at fault.	Check (par. 30 m above).
Lubricant in power train too heavy.	Use correct lubricant (par. 18).
Tires improperly inflated.	Inflate (par. 3).
Vehicle overloaded.	Reduce to 500 pounds if possible.

d. Low Fuel Pressure.

Air leak in fuel lines.	Tighten connections; repair if damaged; hand-tighten fuel pump dome nut.

TROUBLE SHOOTING

Possible Cause	Possible Remedy
Fuel pump faulty; diaphragm broken; valves leaky; linkage worn.	Replace fuel pump (par. 74).
Fuel lines clogged.	Clean or replace lines.

e. Engine Idles Too Fast.

Improper carburetor throttle adjustment.	Adjust throttle stop screw (par. 72).
Carburetor control sticking.	Free-up and lubricate.
Control return spring weak.	Replace.

f. Fuel Gage Does Not Register.

Loose wire connection at instrument panel or tank units.	Tighten connection.
Instrument panel unit or tank unit inoperative.	Replace (pars. 75 and 77).

33. INTAKE AND EXHAUST SYSTEMS.

a. Intake System.

Leaky gaskets, sand hole or crack in manifold.	Replace (par. 52).
Leaky crankcase ventilator valve.	Replace (par. 59).

b. Exhaust System.

Leaky gaskets, sand hole or crack in manifold.	Replace (par. 52).
Exhaust pipe and connections loose or leaking.	Service and/or replace (par. 78).
Muffler leaks or rattles.	Replace (par. 78).
Exhaust system or muffler restricted; exhaust pipe kinked or tail pipe plugged.	Service or replace parts.
Heat control valve inoperative, causing miss on acceleration or slow warm-up.	Free-up; install spring in place on bracket (par. 53).

34. COOLING SYSTEM.

a. Overheating.

Abnormal conditions.	Check (par. 30 l).

b. Loss of Cooling Solution.

Loose hose connection.	Tighten.
Damaged or deteriorated hose.	Replace.

¼-TON 4 x 4 TRUCK (WILLYS-OVERLAND MODEL MB and FORD MODEL GPW)

Possible Cause	Possible Remedy
Leaky radiator.	Replace (par. 81).
Radiator cap inoperative.	Replace.

c. Engine Running Too Cool.

Possible Cause	Possible Remedy
Thermostat stuck open.	Replace (par. 85).
Low air temperatures.	Cover radiator; refer to operation under unusual conditions (par. 7).

d. Noises.

Possible Cause	Possible Remedy
Frayed or loose fan belt.	Replace or adjust (par. 83).
Water pump faulty.	Replace (par. 82).
Fan blades striking.	Aline blades.

35. IGNITION SYSTEM.

a. Ignition System Troubles.

Possible Cause	Possible Remedy
No spark.	Refer to paragraph 30 c (1).
Weak spark.	Refer to paragraph 30 c (2).
Timing incorrect.	Retime ignition (par. 65); refer to paragraph 30 j for other causes.
Moisture on distributor wires, coil, or spark plugs.	Dry and clean thoroughly with cloth dampened with carbon tetrachloride.
Ignition switch "OFF."	Turn "ON" fully.
Ignition switch does not make contact.	Replace switch (par. 68).
Primary or secondary wiring loose, broken, or grounded.	Service.
Primary or secondary wiring wrong.	Check against wiring diagram (par. 62 and fig. 30); install secondary wires correctly in distributor cap and on spark plugs.
Ground strap connections (engine to frame) loose or dirty.	Clean and tighten.
Coil faulty.	Refer to subparagraph b below.
Distributor faulty.	Refer to subparagraph c below.
Spark plug or distributor suppressors faulty.	Replace (par. 67).
Filter unit open or grounded.	Replace filter (par. 69).

TROUBLE SHOOTING

b. Ignition Coil Troubles.

Possible Cause	Possible Remedy
Connections loose; dirty or broken external wire; wet.	Clean and tighten or repair; dry thoroughly.
Coil internal fault.	Replace coil (par. 66).

c. Distributor Troubles.

Possible Cause	Possible Remedy
Distributor breaker points dirty or pitted; gap incorrect.	Clean or replace and adjust (par. 64).
Distributor breaker point arm spring weak.	Replace breaker point arm (par. 64).
Distributor breaker points stuck open.	Free-up and lubricate arm on post.
Distributor automatic advance faulty.	Lubricate and free up; if "frozen" replace distributor (par. 64).
Distributor cap or rotor shorted, cracked, or broken.	Replace.
Distributor rotor does not turn.	Report to higher authority.
Distributor cap cracked or shorted.	Replace cap (par. 64).
Condenser or lead wire faulty.	Replace condenser (par. 64).

d. Spark Plug Troubles.

Possible Cause	Possible Remedy
Cracked, broken, leaky, or improper type.	Replace spark plug (par. 67).
Spark plug wires installed on wrong plugs, or in distributor cap.	Install in correct place (par. 69).
Spark plugs dirty; gaps incorrect.	Clean or replace; set gaps (par. 67).
Spark plug porcelain cracked or broken.	Replace plug.
Spark plugs wrong type.	Replace with correct type (par. 67).

36. STARTING AND GENERATING SYSTEMS.

a. Cranking Motor Troubles.

(1) CRANKING MOTOR CRANKS ENGINE SLOWLY.

Possible Cause	Possible Remedy
Engine oil too heavy.	Change to proper seasonal grade (par. 18).
Battery low.	Replace or recharge (par. 97).
Battery cell shorted.	Replace battery (par. 97).

TM 9-803
36

¼-TON 4 x 4 TRUCK (WILLYS-OVERLAND MODEL MB and FORD MODEL GPW)

Possible Cause	Possible Remedy
Battery connections corroded, broken, or loose; or engine ground strap to frame connections dirty or loose.	Clean and tighten or replace (par. 97).
Dirty commutator.	Clean (par. 89).
Poor brush contact.	Free-up brush or replace cranking motor (par. 89).
Cranking motor internal fault.	Replace cranking motor.
Starting switch faulty.	Replace switch (par. 90).

(2) CRANKING MOTOR DOES NOT CRANK ENGINE.

Possible Cause	Possible Remedy
Engine oil too heavy.	Change to proper seasonal grade (par. 18).
Cranking motor, starting switch or cables faulty; loose connections.	Replace; tighten loose connections.

b. Generator Troubles.

(1) NO OUTPUT.

Possible Cause	Possible Remedy
Generator faulty.	Replace generator (par. 93).
Filter unit or suppressors faulty.	Replace (par. 93).
Regulator faulty.	Replace (par. 94).

(2) LOW OR FLUCTUATING OUTPUT.

Possible Cause	Possible Remedy
Loose fan belt.	Adjust (par. 83); generator brace not hooked (par. 4 b (8)).
Poor brush contact, weak brush springs; worn commutator; broken or loose connections.	Replace generator (par. 93).
Dirty commutator.	Clean (par. 93).
Regulator faulty.	Replace (par. 94)
Loose or dirty connections in charging circuit.	Clean and tighten.
Ground strap (engine to frame) broken.	Replace.
Filter unit faulty.	Replace (par. 93).

(3) EXCESSIVE OUTPUT.

Possible Cause	Possible Remedy
Short circuit between field coil and armature leads.	Replace generator (par. 93).
Regulator faulty.	Replace regulator (par. 94).

(4) NOISY.

Possible Cause	Possible Remedy
Loose pulley or generator mounting.	Tighten.

TROUBLE SHOOTING

Possible Cause	Possible Remedy
Faulty bearings, improperly seated brushes, or armature rubbing on field poles.	Replace generator (par. 93).

c. Generator Regulator Troubles.

Possible Cause	Possible Remedy
Loose connections or mounting.	Clean and tighten.
Regulator internal defect.	Replace regulator (par. 94).

37. TRANSMISSION.

a. Excessive Noise.

Possible Cause	Possible Remedy
Incorrect driving practice.	Correct practice (par. 5).
Insufficient lubricant.	Add lubricant (par. 18).
Incorrect lubricant.	Use correct lubricant (par. 18).
Gears or bearings broken or worn; shift fork bent; gears worn on splines.	Replace transmission (pars. 115 and 116).
Overheated transmission.	Check lubricant grade and supply (par. 18).

b. Hard Shifting.

Possible Cause	Possible Remedy
Clutch fails to release.	Adjust clutch pedal free travel (par. 109).
Clutch driven plate binds on splines, or pressure plate faulty.	Report to higher authority.
Gearshift binding in housing.	Lubricate and free-up.
Shift rods binding in case.	Report to higher authority.
Transmission loose on bell housing.	Tighten.
Clutch shaft pilot binding in bushing case or shift housing damaged.	Report to higher authority.

c. Slips Out of Gear.

Possible Cause	Possible Remedy
Weak or broken poppet spring.	Report to higher authority.
Interlock plunger not in place.	Install plunger (par. 116).
Transmission gears or bearings worn.	Replace transmission (pars. 115 and 116).
Shift fork bent, causing partial gear engagement.	Report to higher authority.
Transmission loose on bell housing.	Tighten.
Damaged bell housing.	Report to higher authority.

d. Loss of Lubricant.

Possible Cause	Possible Remedy
Worn or damaged seals or gaskets.	Report to higher authority.

¼-TON 4 x 4 TRUCK (WILLYS-OVERLAND MODEL MB and FORD MODEL GPW)

Possible Cause	Possible Remedy
Overfilled with lubricant.	Drain to proper level.
Loose bolts and screws.	Tighten.

38. TRANSFER CASE.

a. Slips Out of Gear.

Possible Cause	Possible Remedy
Shift rod poppet spring weak or broken; gears not fully engaged; shift fork bent; end play in sliding gear shaft.	Report to higher authority.
Parts damaged or worn.	Replace transfer case (pars. 119 and 120).

b. Hard Shifting.

Possible Cause	Possible Remedy
Improper driving practice.	Use correct procedure (par. 5).
Lack of lubrication.	Replenish supply.
Shift lever seizing on shaft.	Lubricate and free-up.
Shift rod tight in case; poppet scored or stuck; shift fork bent, or parts worn or damaged.	Report to higher authority.
Low or uneven tire pressures; odd tires on (front and rear) wheels.	Service.

c. Oil Leaks.

Possible Cause	Possible Remedy
Leaks at gaskets or seals.	Report to higher authority.
Lubricant level too high.	Reduce to correct level.
Vent on top of unit clogged.	Clean.

d. Excessive Noise.

Possible Cause	Possible Remedy
Insufficient lubricant.	Replenish supply.
Incorrect lubricant.	Drain and refill with correct lubricant (par. 18).
Gears or bearings worn, improperly adjusted, or damaged.	Replace transfer case (pars. 119 and 120).

e. Overheats.

Possible Cause	Possible Remedy
Insufficient lubricant.	Replenish supply.
Vent on top of unit clogged.	Clean.
Bearings adjusted too tight.	Report to higher authority.

f. Backlash.

Possible Cause	Possible Remedy
Universal joint yoke loose on output shaft.	Report to higher authority.

TROUBLE SHOOTING

Possible Cause	Possible Remedy
Transfer case loose on transmission or snubbing rubber.	Tighten.
Parts worn or damaged.	Report to higher authority

39. PROPELLER SHAFTS.

a. Excessive Vibration or Noise.

Possible Cause	Possible Remedy
Foreign material around shaft.	Clean out.
Universal joints not in same plane.	Match arrows on joint and propeller shaft (par. 125).
Lack of lubricant.	Lubricate (par. 18).
Universal joint parts worn, or propeller shaft sprung.	Replace shaft.

b. Universal Joint Leaks.

Possible Cause	Possible Remedy
Overfilled.	Lubricate correctly (par. 18).
Oil seals leak.	Report to higher authority.
Lubricant fitting leaks.	Replace fitting.

40. FRONT AXLE.

a.	Steering trouble.	Refer to paragraph 45.
b.	Noisy gears or backlash.	Report to higher authority.
c.	Damaged axle.	Replace axle (pars. 136 and 137).
d.	Abnormal tire wear.	Inflate tires (par. 13 b (13)) (do not use front wheel drive except where needed); correct toe-in; report to higher authority incorrect caster or camber.
e.	Lubrication leaks.	Replace steering knuckle oil seals; for other remedies refer to paragraph 41 c.

41. REAR AXLE.

a.	Noisy gears or backlash.	Report to higher authority.
b.	Damaged axle.	Replace axle (par. 145).
c.	Lubrication leaks.	Drain excessive lubricant; clean housing vent; replace wheel bearing grease seals; remove excessive grease in wheel hubs; tighten or replace housing cover gasket.

¼-TON 4 x 4 TRUCK (WILLYS-OVERLAND MODEL MB and FORD MODEL GPW)

42. BRAKE SYSTEM.

a. All Brakes Drag.

Possible Cause	Possible Remedy
Improper pedal adjustment.	Adjust brake pedal free travel (par. 148).
Clogged master cylinder port.	Replace (par. 150).
Brake pedal return spring broken or weak.	Replace.
Brakes improperly adjusted.	Adjust (par. 148).
Rubber parts swollen from use of mineral oil in brake fluid.	Report to higher authority.

b. One Brake Drags.

Possible Cause	Possible Remedy
Brake shoe adjustment faulty.	Adjust (par. 148).
Brake shoe anchor pin tight in shoes.	Free-up and lubricate lightly.
Brake shoe return spring broken or weak.	Replace.
Brake hose clogged or pinched.	Replace.
Loose or damaged wheel bearings.	Adjust or replace (pars. 128 and 141).
Wheel cylinder pistons or cups faulty.	Replace wheel cylinder (par. 150).

c. One Brake Grabs (Vehicle Pulls to One Side).

Possible Cause	Possible Remedy
Tires underinflated.	Inflate tires (par. 13).
Tires worn unequally.	Replace.
Insufficient brake shoe clearance or brake anchor pin adjustment faulty.	Adjust (par. 148).
Axle spring clips or brake backing plate loose.	Tighten.
Brake shoes binding on anchor pin.	Free-up and lubricate lightly.
Weak or broken shoe return spring.	Replace spring.
Grease or brake fluid on linings.	Correct leakage; clean up and install new shoes and lining assemblies.
Dirt imbedded in linings or rivet holes.	Clean with wire brush.
Drums scored or rough.	Replace drums and brake shoe and lining assemblies.

TROUBLE SHOOTING

Possible Cause	Possible Remedy
Primary and secondary brake shoes reversed in one wheel.	Change shoes to proper place and adjust brakes (par. 148).
Odd kinds of brake lining on opposite wheels.	Replace shoe and lining assemblies in both wheels.
Loose or broken wheel bearings.	Adjust or replace (pars. 128 and 141).
Obstruction in brake line.	Clean or replace tube (par. 152).

d. Severe Brake Action on Light Pedal Pressure.

Possible Cause	Possible Remedy
Brake shoes improperly adjusted.	Adjust (par. 148).
Grease or brake fluid on linings.	Correct leakage, clean up and replace shoe and lining assemblies.
Loose brake shoe anchor.	Adjust and tighten (par. 148).
Improper linings.	Replace shoe and lining assemblies (par. 148).

e. Brakes Locked.

Possible Cause	Possible Remedy
Brake pedal lacks free travel.	Adjust pedal free travel (par. 148).
Bleed hole in master cylinder clogged.	Replace master cylinder (par. 150).
Dirt in brake fluid.	Flush system (par. 151).
Wheel cylinder stuck.	Replace cylinder (par. 150).
Brakes frozen to drums (cold weather).	Break loose by driving vehicle.

f. Brakes Noisy or Chatter.

Possible Cause	Possible Remedy
Brake lining worn out.	Replace shoe and lining assemblies (par. 148).
Grease or brake fluid on linings.	Correct leakage, clean up and replace shoe and lining assemblies (par. 148).
Improper adjustment of anchor bolts.	Adjust (par. 148).
Dirt imbedded in linings and rivet holes.	Clean with wire brush.
Improper or loose linings.	Replace shoe and lining assemblies (par. 148).
Brake shoes, drums, or backing plate distorted.	Straighten or replace.
Loose spring clips or shackles.	Tighten.

¼-TON 4 x 4 TRUCK (WILLYS-OVERLAND MODEL MB and FORD MODEL GPW)

g. Excessive Pedal Travel.

Possible Cause	Possible Remedy
Normal lining wear.	Adjust brake eccentrics only (par. 148).
Lining worn out.	Replace shoe and lining assemblies (par. 148).
Brake not properly adjusted.	Adjust (par. 148).
Improper pedal adjustment.	Adjust (par. 148).
Brake line leaky or broken.	Locate and tighten or repair.
Low fluid level in master cylinder or air in brake system.	Fill master cylinder and bleed lines (par. 151).
Scored brake drums.	Replace (pars. 131 and 144).
Incorrect brake lining.	Replace with correct shoe and lining assemblies.
Pedal goes to floorboard (disconnected from master cylinder).	Connect or replace faulty part (par. 149).
Leaky piston cup in master or wheel cylinders.	Replace cylinder.

h. Excessive Pedal Pressure.

Possible Cause	Possible Remedy
Grease or brake fluid on linings; worn or glazed lining.	Correct cause, clean up and replace shoe and lining assemblies (par. 148).
Warped shoes or improper brake linings.	Replace shoe and lining assemblies (par. 148).
Shoes improperly adjusted.	Adjust (par. 148).
Brake drums scored or distorted.	Replace damaged parts.
Improper brake fluid.	Clean system and fill with correct fluid.
Obstructed main brake line.	Locate and correct.

i. Spongy Brake Pedal Action.

Possible Cause	Possible Remedy
Air or insufficient fluid in brake system.	Fill master cylinder and bleed lines (par. 149).
Brake anchor adjustment faulty.	Adjust (par. 148).

j. No Brakes—Pedal Will Pump Up.

Possible Cause	Possible Remedy
Brake shoe clearance excessive.	Adjust brake eccentrics (par. 148).
Leaky master or wheel cylinder piston cup.	Replace cylinder.
Leaky brake line or hose.	Locate and tighten or replace.

TROUBLE SHOOTING

k. Pedal Goes to Floor Slowly When Brakes Are Applied.

Possible Cause	Possible Remedy
Leaky master cylinder piston cup.	Replace master cylinder (par. 149).
Leaky brake line or hose.	Tighten or replace part.

43. WHEELS, WHEEL BEARINGS, AND RELATED PARTS.

a. Wheel Troubles.

Wheel wobbles: bent.	Check mounting on hub; replace bent wheel.
Wheel loose on hub.	Tighten.
Wheel out of balance.	Remount tire correctly.
Wheel bearings run hot (pull vehicle to one side).	Adjust (pars. 128 and 141).
Wheels misalined.	Refer to paragraph 135.
Excessive or uneven tire wear.	Refer to paragraph 45.
Diameter of front tires not the same in size or wear.	Replace or match up.

44. SPRINGS AND SHOCK ABSORBERS.

a. Broken Springs.

Improper handling of vehicle on rough terrain.	Use correct practice when possible (par. 5).
Overloaded vehicle.	Reduce load (par. 3).
Overlubricated springs.	Do not lubricate unless rusty.
Rebound clips off or out of place.	Service.
Shackles or pivot bolts too tight.	Free-up and lubricate.
Main leaf broken at end.	Replace spring (par. 156).
Axle clips loose (spring broken at center).	Keep clips tight.
Shock absorbers not adjusted correctly, lack fluid, or damaged.	Adjust or replace shock absorbers (par. 157).
Clutch or brakes grab.	Service.

b. Noisy Springs.

Worn shackles, pivot pins, or bushings.	Replace worn parts (par. 155).
Spring clips loose on axle or leaves.	Tighten.
Spring hangers loose on frame.	Report to higher authority.
Spring shackle bushing loose; inner spring eye opened up.	Replace spring (par. 156).

TM 9-803
44-45

¼-TON 4 x 4 TRUCK (WILLYS-OVERLAND MODEL MB and FORD MODEL GPW)

Possible Cause	Possible Remedy
No fluid in shock absorbers, or bushings worn out.	Replace (par. 157).

c. Bottomed Springs.

Possible Cause	Possible Remedy
Overloaded vehicle.	Reduce load (par. 3).
Overlubricated springs.	Do not lubricate unless rusty.
Broken spring leaves.	Replace spring (par. 156).
Shock absorbers broken, lack fluid or proper adjustment.	Replace shock absorbers (par. 159).

d. Overflexible Springs.

Possible Cause	Possible Remedy
Overlubrication causes springs to bottom.	Do not lubricate springs.
Shock absorbers not adjusted right, lack fluid, or are broken.	Service, adjust, or replace shock absorbers (par. 157).
Rebound clips damaged or lost.	Replace.
Broken spring.	Replace spring (par. 156).

e. Stiff Springs.

Possible Cause	Possible Remedy
Rusted spring leaves.	Lubricate.
Shackle or pivot bolts too tight.	Free-up and lubricate.
Shock absorber adjustment not right.	Adjust (par. 157).

f. Noisy Shock Absorbers.

Possible Cause	Possible Remedy
Rubber bushings worn out.	Replace bushing (par. 157).
Mounting bracket loose.	Report to higher authority.
Shock absorber faulty.	Replace (par. 157).

g. Shock Absorber Control Too Stiff or Too Soft.

Possible Cause	Possible Remedy
Shock absorber adjustment wrong.	Adjust (par. 157).
Shock absorber damaged or lacks fluid.	Replace shock absorber (par. 157).

45. STEERING SYSTEM.

a. Steering Difficult.

Possible Cause	Possible Remedy
Lack of lubrication.	Lubricate (par. 18).
Tire pressures low.	Inflate (par. 13).
Tight steering system connections.	Lubricate and adjust (par. 159).
Tight steering gear; misalined front wheels (caster or camber); or bent frame.	Report to higher authority.

TM 9-803

TROUBLE SHOOT—

Possible Cause

Improper front wheel toe-in. Bent steering connecting parts. Misalined steering gear mounting.
  b. Wander or Weaving. Improper toe-in
Improper camber or caster (axle twisted).
Front springs settled or broken.
Axle shifted (spring center bolt broken).

Loose or lost spring clips.

Loose or worn spring shackles or bolts.

Tire pressures uneven.

Steering system connections or king pin bearings not properly adjusted.

Loose wheel bearings. Faulty shock absorbers. Steering gear worn or out of adjustment.
Steering gear mounting loose.
Steering Pitman arm loose.

Possible Remedy

Adjust (par. 135).

Loose steering connections.

Spring clips or shackles loose.

Front axle loose on spring (broken spring center bolt).

Insufficient toe-in.

Improper caster or twisted axle. Steering gear worn, or adjust-ments too loose.

Loose wheel or king pin bearings.

d. High Speed Shimmy or listed in subparagraph c above).

Tire pressures low or uneven.
  Straighten or replace.

Adjust mounting.

Adjust (par. 135).
Report to higher authority.

Replace spring (par. 156). Replace part.

Tighten or replace.
Replace or tighten.

Inflate (par. 13).
Lubricate and adjust (par. 159).

Adjust (pars. 128 and 141).
Replace (par. 157).
Report to higher authority.

Tighten.
Tighten.
e. Low Speed Shimmy or Wobble.

Wheels and tires out of balance.
Adjust.
Adjust or replace.
Replace bolt.

Adjust (par. 135).
Report to higher authority.
Report to higher authority.

Adjust (par. 128).

Tight Wheel (Refer to remedies

Inflate (par. 13).
Check tire mounting; report other trouble to higher authority.

TM 9-803
45

¼-TON 4 x 4 TRUCK (WILLYS-OVERLAND MODEL MB and FORD MODEL GPW)

Possible Cause	Possible Remedy
Wheel run-out; tire radial run-out or wheel camber incorrect.	Report to higher authority.
Front springs settled or broken.	Replace spring (par. 156).
Bent steering knuckle arm.	Report to higher authority.
Shock absorbers not effective.	Adjust or replace.
Steering gear loose in frame.	Tighten.
Front springs too flexible.	Do not lubricate.
Worn spring bolts, shackles, or bushings.	Replace (par. 155).
Axle housing or frame damaged.	Report to higher authority.

e. Wheel Tramp (High Speed).

Possible Cause	Possible Remedy
Wheels and tires out of balance.	Check tire mounting; report other trouble to higher authority.
Uneven tire wear.	Shift tires.
Shock absorbers ineffective.	Replace or adjust (par. 157).

f. Vehicle Pulls to One Side.

Possible Cause	Possible Remedy
Tires not inflated evenly.	Inflate (par. 13).
Unequal caster or camber (bent axle).	Report to higher authority.
Odd size, or new and old tires on opposite front wheels.	Switch tires.
Tight wheel bearing.	Adjust (pars. 128 and 141).
Bent steering arm or connection.	Straighten or replace.
Brake drag.	Adjust brakes (par. 148).

g. Road Shock.

Possible Cause	Possible Remedy
Tightness in steering connecting parts.	Adjust (par. 159).
Excessive spring flexibility.	Do not lubricate.
Loose wheel bearings.	Adjust.
Loose Pitman arm or mounting.	Tighten.
Looseness in steering gear.	Report to higher authority.
Shock absorbers out of adjustment or faulty.	Adjust or replace (par. 157).

h. Steering Dive.

Possible Cause	Possible Remedy
Steering gear loose on frame.	Tighten.
Broken front spring leaves.	Replace.
Worn spring shackles, bushings or bolts.	Replace.

TROUBLE SHOOTING

Possible Cause	Possible Remedy
Spring hangers loose on frame.	Report to higher authority.
Spring clips loose, broken, or lost.	Tighten or replace.
Spring center bolt broken and/or clips loose.	Replace.
Axle housing on frame damaged.	Report to higher authority.

i. Unequal Steering (Right and Left).

Possible Cause	Possible Remedy
Pitman arm not installed in proper position on steering gear.	Remove and install in correct position (par. 162).
Drag link bent.	Straighten or replace.

46. BODY AND FRAME.

a. Body.

Possible Cause	Possible Remedy
Worn or damaged seat cushion.	Replace.
Badly damaged fender, radiator guard, hood, fuel can rack, seats, top, or windshield.	Replace; report minor damage to higher authority.
Windshield wiper faulty.	Service or replace.

b. Frame.

Possible Cause	Possible Remedy
Badly damaged bumpers and pintle hook.	Replace; report minor damage to higher authority.
Damaged frame.	Report to higher authority.

47. BATTERY AND LIGHTING SYSTEM.

a. Battery.

(1) BATTERY DISCHARGED.

Possible Cause	Possible Remedy
Battery solution level low.	Add distilled water to bring level above plates; check for cracked case.
Short in battery cell.	Replace battery (par. 97).
Generator not charging.	Check generator, fan belt and regulator (par. 92).
Loose or dirty connections; broken cables.	Clean and tighten connections; replace cables.
Excessive use of cranking motor.	Tune up engine; charge battery.
Idle battery, or excessive use of lights.	Replace or charge battery.

(2) BATTERY (OTHER TROUBLES.)

Possible Cause	Possible Remedy
Overheated battery.	Check for short circuit or excessive generator charge.

¼-TON 4 x 4 TRUCK (WILLYS-OVERLAND MODEL MB and FORD MODEL GPW)

Possible Cause	Possible Remedy
Case bulged or out of shape.	Check for overcharging and too tight hold-down screws.

b. Switch.

Possible Cause	Possible Remedy
Loose or dirty connections or broken wire.	Clean and tighten; replace broken wire.
Internal fault.	Replace switch.

c. Fuse (Circuit Breaker).

Possible Cause	Possible Remedy
Points dirty.	Clean.
Other troubles.	Replace fuse assembly (par. 104).

d. Wiring.

Possible Cause	Possible Remedy
Loose or dirty connections, broken wire or terminal.	Clean, tighten or replace.

e. Lights do Not Light.

Possible Cause	Possible Remedy
Switch not fully on.	Turn switch on fully.
Loose or dirty connection, or broken wire or terminal.	Clean and tighten; replace or repair wire or terminal.
Wiring circuit shorted or open.	Localize and repair.
Headlight, blackout driving light, tail or stop light burned out.	Replace lamp-unit (par. 96).
Blackout headlight burned out.	Replace lamp (par. 100).

f. Lights Dim.

Possible Cause	Possible Remedy
Loose or dirty connection or poor ground connection.	Clean and tighten.
Wire grounding.	Localize and replace (par. 98).
Poor switch contact.	Replace switch.
Headlight aim not right.	Adjust lights (par. 99).

g. Trailer Connection Trouble.

Possible Cause	Possible Remedy
No current supply.	Tighten loose wires; connect to correct terminals.

h. Horn Troubles.

Possible Cause	Possible Remedy
Loose or dirty connections.	Clean and tighten.
Sounds continuously (short circuit in wiring between horn and horn button).	Replace wire.
Improper tone.	Adjust points; tighten cover or bracket screws; clean and tighten loose or dirty wiring connections.

TROUBLE SHOOTING

Possible Cause	Possible Remedy
Internal defect.	Replace horn.
Battery low.	Charge or replace battery.

48. RADIO SUPPRESSION.

a. Radio Interference.

Possible Cause	Possible Remedy
Faulty ignition.	Check distributor, spark plugs, and suppressors. Tighten braided bonding straps. Tighten radiator and fender supporting bolts. Check high-tension insulation. Tighten loose wiring connections, or replace corroded distributor cap towers. Replace defective switches or gages (par. 178).
Faulty generator.	Tighten generator to regulator bond. Check for faulty commutator, brushes, or holders. If defective, replace generator. Check for discharged battery causing high charging rate (par. 178).
Erratic noises.	Tighten or clean loose or dirty lock washer ground. Install lock washers in correct position (par. 178).

49. INSTRUMENTS.

a. Faulty Instruments.

Possible Cause	Possible Remedy
Dirty or loose connections.	Clean and tighten.
Internal defects.	Replace instrument.
Broken speedometer cable.	Replace (par. 166).

¼-TON 4 x 4 TRUCK (WILLYS-OVERLAND MODEL MB and FORD MODEL GPW)

Section XIII

ENGINE—DESCRIPTION, DATA, MAINTENANCE, AND ADJUSTMENT IN VEHICLE

	Paragraph
Description and tabulated data	50
Engine tune-up	51
Intake and exhaust manifolds	52
Manifold heat control valve	53
Cylinder head gasket	54
Valve cover gasket	55
Valve tappet adjustment	56
Oil pan gasket	57
Oil filter	58
Crankcase ventilator valve	59

50. DESCRIPTION AND TABULATED DATA.

a. Description. The engine (figs. 20 and 21) is of the conventional 4-cylinder, L-head, internal-combustion type. The engine with the clutch, transmission, and transfer case is built into a unit power plant which is mounted at four points in the chassis. For identification refer to paragraph 2 b.

b. Tabulated Data.

Type	L-head
Number of cylinders	4
Bore	3⅛ in.
Stroke	4⅜ in.
Piston displacement	134.2 cu in.
Compression ratio	6.48 to 1
Net horsepower	54 at 4,000 rpm
Compression	110 lb per sq in. at 185 rpm
SAE horsepower	15.63
Maximum torque	95 ft-lb
Firing order	1-3-4-2
Tappet clearance—intake and exhaust (hot or cold)	0.014 in.

51. ENGINE TUNE-UP.

a. Procedure.

(1) Perform preventive maintenance and corrective operations listed in paragraph 16.

(2) Remove spark plugs and clean. Adjust gaps (par. 67).

ENGINE—DESCRIPTION, DATA, MAINTENANCE, AND ADJUSTMENT IN VEHICLE

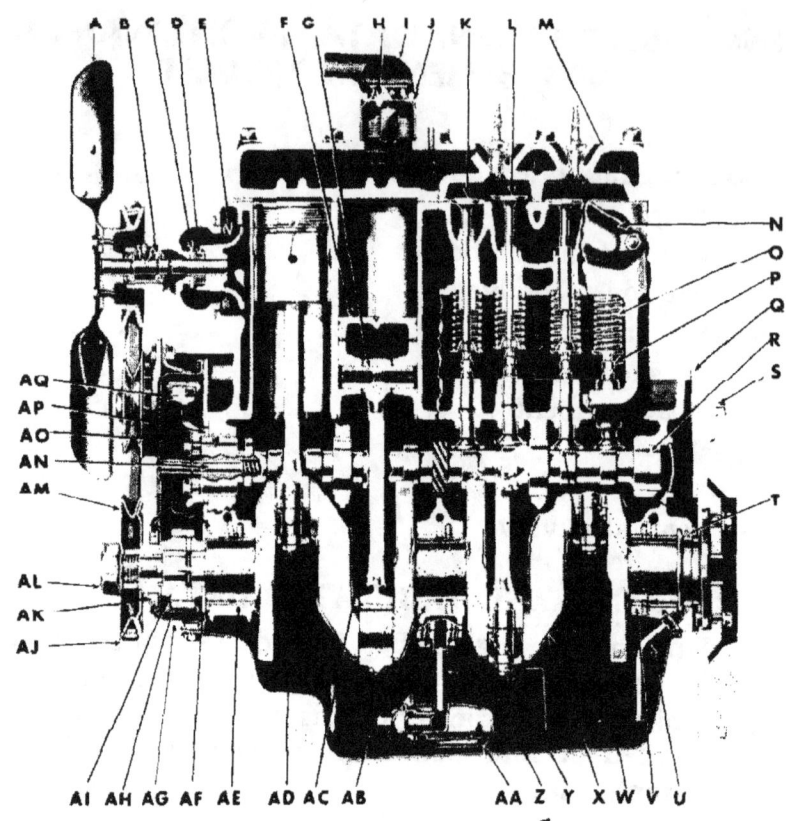

A	FAN	V	CRANKSHAFT BEARING—REAR LOWER
B	WATER PUMP BEARING AND SHAFT	W	VALVE TAPPET
C	WATER PUMP SEAL WASHER	X	CRANKSHAFT
D	WATER PUMP SEAL	Y	CONNECTING ROD CAP BOLT
E	WATER PUMP IMPELLER	Z	OIL FLOAT SUPPORT
F	PISTON	AA	OIL FLOAT
G	PISTON PIN	AB	CRANKSHAFT BEARING—CENTER LOWER
H	THERMOSTAT	AC	CONNECTING ROD
I	WATER OUTLET ELBOW	AD	CONNECTING ROD BOLT NUT LOCK
J	THERMOSTAT RETAINER	AE	CRANKSHAFT BEARING—FRONT LOWER
K	EXHAUST VALVE	AF	CRANKSHAFT THRUST WASHER
L	INLET VALVE	AG	TIMING CHAIN COVER
M	CYLINDER HEAD	AH	TIMING CHAIN
N	EXHAUST MANIFOLD	AI	CRANKSHAFT SPROCKET
O	VALVE SPRING	AJ	FAN BELT
P	VALVE TAPPET ADJUSTING SCREW	AK	CRANKSHAFT PACKING—FRONT END
Q	ENGINE PLATE—REAR	AL	STARTING CRANK NUT
R	CAMSHAFT	AM	FAN AND GENERATOR DRIVE PULLEY
S	FLYWHEEL RING GEAR	AN	CAMSHAFT THRUST PLUNGER
T	CRANKSHAFT PACKING—REAR	AO	CAMSHAFT BUSHING—FRONT
U	CRANKSHAFT REAR BEARING DRAIN PIPE	AP	CAMSHAFT THRUST WASHER
	AQ CAMSHAFT SPROCKET		

RA PD 305283

Figure 20—Sectional View of Engine

¼-TON 4 x 4 TRUCK (WILLYS-OVERLAND MODEL MB and FORD MODEL GPW)

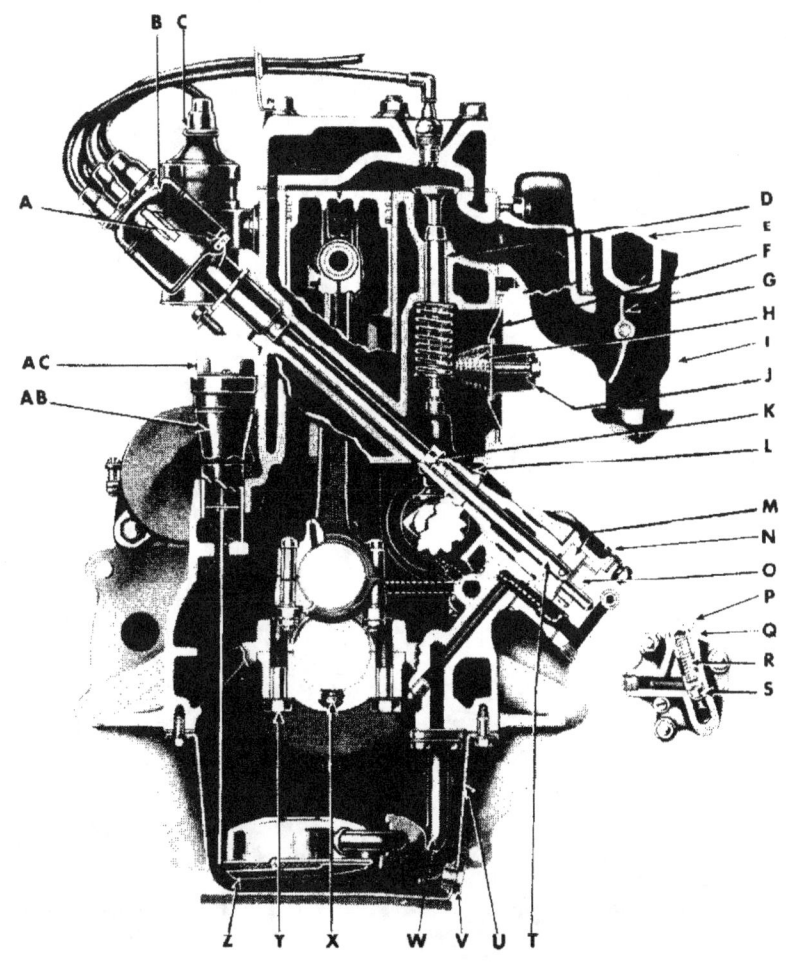

A DISTRIBUTOR OILER
B IGNITION DISTRIBUTOR
C IGNITION COIL
D EXHAUST VALVE GUIDE
E INTAKE MANIFOLD
F VALVE SPRING COVER
G HEAT CONTROL VALVE
H CRANKCASE VENTILATOR BAFFLE
I EXHAUST MANIFOLD
J CRANKCASE VENTILATOR
K DISTRIBUTOR SHAFT FRICTION SPRING
L OIL PUMP DRIVEN GEAR
M OIL PUMP ROTOR DISK
N OIL PUMP
O OIL PUMP PINION
P OIL RELIEF PLUNGER SPRING RETAINER
Q OIL RELIEF PLUNGER SPRING SHIMS
R OIL RELIEF PLUNGER SPRING
S OIL RELIEF PLUNGER
T OIL PUMP SHAFT AND ROTOR
U OIL PAN
V OIL PAN DRAIN PLUG
W OIL FLOAT SUPPORT
X CRANKSHAFT BEARING DOWEL
Y CRANKSHAFT BEARING CAP TO CRANKCASE SCREW
Z OIL FLOAT
AB OIL FILLER TUBE
AC OIL FILLER CAP AND LEVEL INDICATOR

RA PD 305284

Figure 21—Sectional View of Engine

TM 9-803
51

ENGINE—DESCRIPTION, DATA, MAINTENANCE, AND ADJUSTMENT IN VEHICLE

Figure 22—Manifolds

(3) Test cylinder compression with gage. The gage must read more than 70 pounds, and the variation between cylinders remain less than 10 pounds. Normal compression is approximately 110 pounds per square inch at cranking speed. Report lack of compression to higher authority.

(4) Make sure that ground strap at engine left front support is in good condition and tight.

(5) Remove distributor cap and rotor. Check for cracks and leaks. Clean or replace breaker points and adjust (par. 64).

TM 9-803
¼-TON 4 x 4 TRUCK (WILLYS-OVERLAND MODEL MB and FORD MODEL GPW)

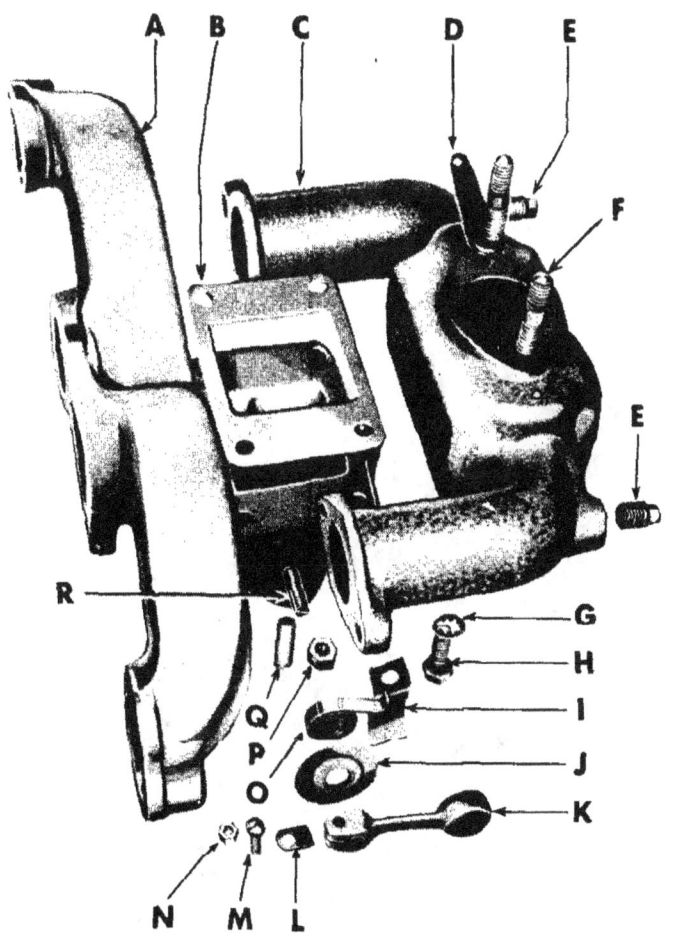

A EXHAUST MANIFOLD
B INTAKE TO EXHAUST MANIFOLD GASKET
C INTAKE MANIFOLD
D ACCELERATOR SPRING CLIP
E INTAKE MANIFOLD PLUG
F INTAKE MANIFOLD TO CARBURETOR STUD
G INTAKE TO EXHAUST MANIFOLD SCREW LOCKWASHER
H INTAKE TO EXHAUST MANIFOLD SCREW
I HEAT CONTROL VALVE BI-METAL SPRING STOP
J HEAT CONTROL VALVE BI-METAL SPRING WASHER
K HEAT CONTROL VALVE COUNTERWEIGHT LEVER
L HEAT CONTROL VALVE LEVER KEY
M HEAT CONTROL VALVE LEVER CLAMP SCREW
N HEAT CONTROL VALVE LEVER CLAMP SCREW NUT
O HEAT CONTROL VALVE BI-METAL SPRING
P EXHAUST PIPE TO EXHAUST MANIFOLD STUD NUT
Q EXHAUST PIPE TO EXHAUST MANIFOLD STUD
R HEAT CONTROL VALVE SHAFT

RA PD 334754

Figure 23—Manifolds, Disassembled

ENGINE—DESCRIPTION, DATA, MAINTENANCE, AND ADJUSTMENT IN VEHICLE

(6) Check ignition timing (par. 65).

(7) Check valve tappet clearance (par. 56).

(8) Install spark plugs, assemble distributor, start engine, and allow to run until normal temperature is reached; then set throttle valve stop screw so that engine will idle at 600 revolutions per minute (vehicle speed 8 mph).

(9) Adjust idle adjustment screw until engine idles smoothly. If carburetor float level, accelerating pump, or metering rod require adjustment, report to higher authority.

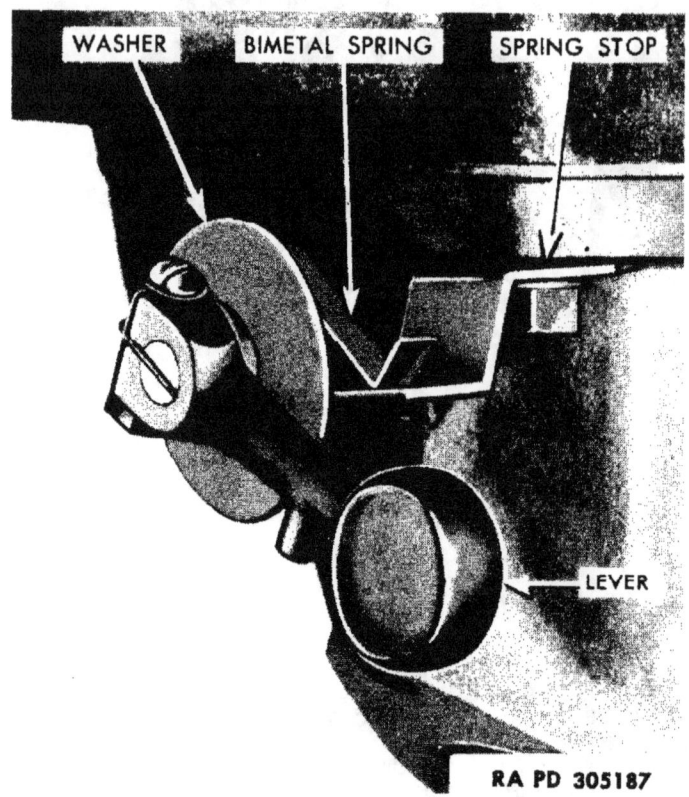

Figure 24—Heat Control Valve

(10) Tighten cylinder head screws and nuts, using torque wrench (par. 54).

(11) Check operation of manifold heat control (par. 53).

52. INTAKE AND EXHAUST MANIFOLDS.

a. Description. The intake and exhaust manifolds (figs. 22 and 23) are attached to each other with four screws, making a unit in which a heat control valve is used to regulate the intake manifold temperature (par. 53).

b. Remove Intake and Exhaust Manifolds. Remove carburetor air horn at top of carburetor, and disconnect hand throttle, choke,

TM 9-803
54
¼-TON 4 x 4 TRUCK (WILLYS-OVERLAND MODEL MB and FORD MODEL GPW)

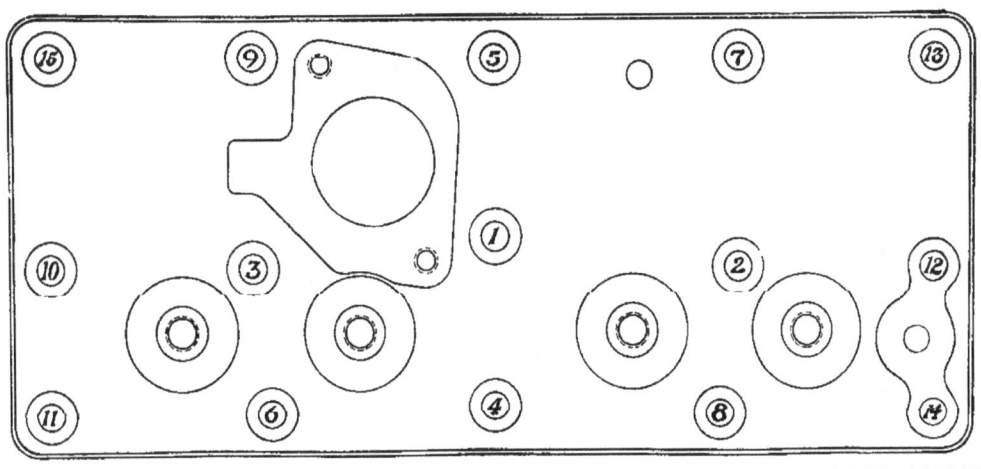

Figure 25—Cylinder Head Tightening Chart

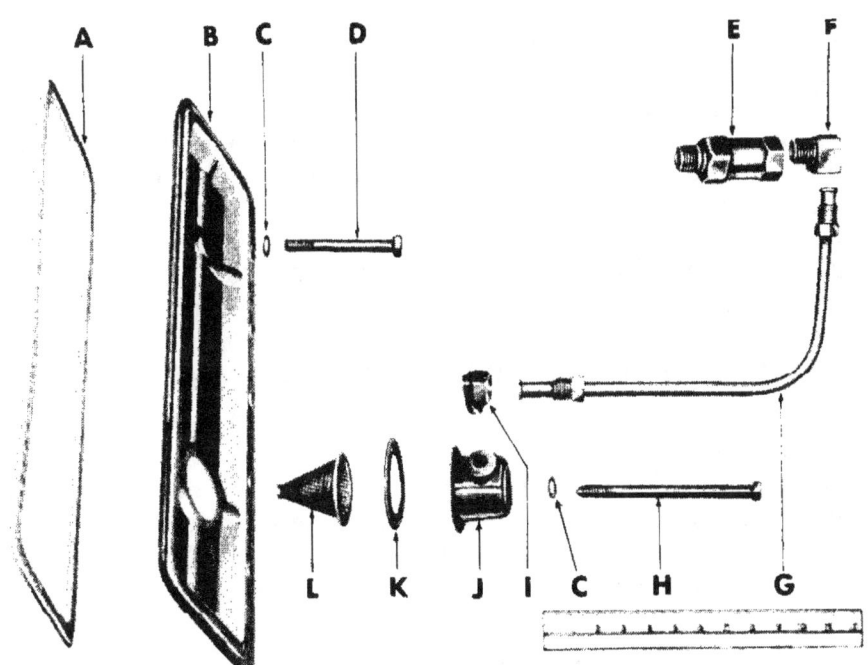

A VALVE SPRING COVER GASKET	G CRANKCASE VENTILATOR TUBE
B VALVE SPRING COVER	H VALVE SPRING COVER SCREW—FRONT
C VALVE SPRING COVER SCREW GASKET	I CRANKCASE VENTILATOR BODY ELBOW
D VALVE SPRING COVER SCREW—REAR	J CRANKCASE VENTILATOR BODY
E CRANKCASE VENTILATOR VALVE	K CRANKCASE VENTILATOR BODY GASKET
F CRANKCASE VENTILATOR VALVE ELBOW	L CRANKCASE VENTILATOR BAFFLE

Figure 26—Valve Spring Cover, Disassembled

ENGINE—DESCRIPTION, DATA, MAINTENANCE, AND ADJUSTMENT IN VEHICLE

and accelerator at carburetor. Loosen fuel line at fuel pump, and disconnect at carburetor. Remove two nuts attaching carburetor to intake manifold, and remove carburetor with accelerator spring clip. Loosen the valve spring cover front screw to relieve any pull on crankcase ventilator tube, and then remove the tube. Disconnect exhaust pipe at the manifold. Remove all nuts and washers from manifold studs in cylinder block, remove manifolds as an assembly, and remove ventilator valve.

c. Separate Intake Manifold from Exhaust Manifold. Remove four screws holding intake and exhaust manifolds together, and remove intake to exhaust manifold gasket.

d. Assemble Intake Manifold to Exhaust Manifold. Attach intake manifold to exhaust manifold loosely, using a new gasket. Tighten screws only slightly until manifolds are installed on cylinder block. Install ventilator valve.

e. Install Intake and Exhaust Manifolds. Clean contact surfaces of manifolds and cylinder block. Place new gasket on studs in cylinder block, and install manifold. Install washers and nuts with convex side of washers against manifolds, and tighten evenly (torque wrench reading 31 to 35 ft-lb). Tighten the four screws attaching intake manifold to exhaust manifold. Attach exhaust pipe to manifold, using new gasket, and tighten in place with nut and screw. Install ventilator tube, and tighten valve spring cover front screw. Install carburetor, accelerator clip, and spring. Attach fuel line at carburetor, and tighten at fuel pump. Connect accelerator rod, hand throttle, and choke at carburetor. Push controls in on instrument panel (throttle closed and choke fully open). Install carburetor air horn and secure in place. Operate fuel pump priming lever to put fuel in carburetor, then start engine, and check for leaky gaskets.

53. MANIFOLD HEAT CONTROL VALVE.

a. Description. The heat control valve (figs. 23 and 24) is controlled thermostatically by a bimetal spring. This valve diverts exhaust gases around the central portion of the intake manifold during the warm-up period of the engine. NOTE: *The manifold heat control valve is an integral part of the exhaust manifold. For replacement follow procedure outlined in paragraph 52.*

54. CYLINDER HEAD GASKET.

a. Removal. Drain the cooling system by opening the drain cock under the radiator at the left front. If there is antifreeze in the cooling system, drain into a pan so it can be used again. Disconnect spark plug wires at the plugs, and remove distributor cap from distributor. Remove two nuts on cylinder studs holding air cleaner tube bracket, and remove bracket with wires and distributor cap. Remove radiator upper tube with hoses attached. Disconnect oil filter upper tube, remove two nuts holding filter to engine, and remove filter. Remove all cylinder head screws and nuts. Remove cylinder head and bonding strap, taking care not to damage oil filler tube, and discard gasket.

b. Installation. Clean cylinder head, tops of pistons, and cylinder block thoroughly. Place cylinder head gasket in position on cylinder block. NOTE: *The front and rear center studs are pilot studs to correctly position the gasket.* Install cylinder head. CAUTION: *Do not damage oil filler pipe.* Install rear bonding strap, oil filter, and air cleaner tube bracket. Install cylinder head bolts and nuts. Tighten cylinder screws and nuts evenly and in sequence (fig. 25), using a

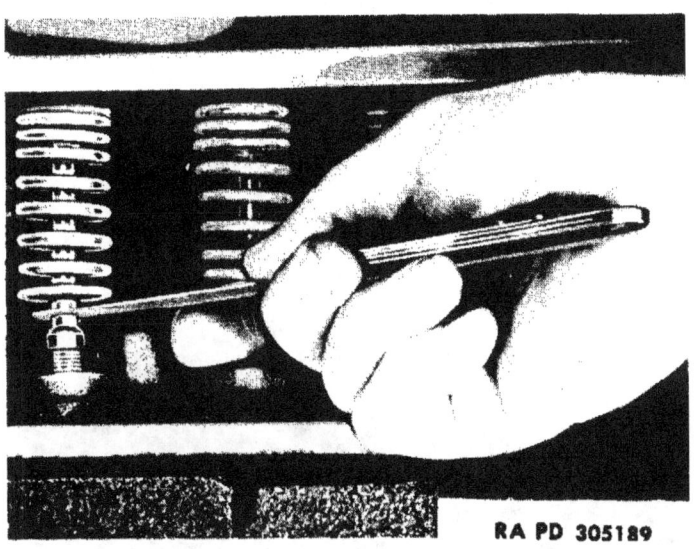

RA PD 305189

Figure 27—Valve Tappet Adjustment

torque-type wrench (screws, 65 to 70 ft-lb; nuts, 60 to 65 ft-lb). Connect oil filter tube, install distributor cap, and attach spark plug wires to correct plugs. Install radiator upper tube, tighten hose clamps, and close radiator drain cock. Fill the cooling system, giving due attention to antifreeze, if required (par. 7). Start engine, and check cooling system for leaks. See that cooling solution level has not gone down; replenish if necessary.

55. VALVE COVER GASKET.

a. Removal. Remove front valve spring cover bolt (fig. 26). Remove crankcase ventilator tube at ventilator valve, and remove tube and cap. Remove rear valve spring cover bolt, and slide valve spring cover forward, up, and out over the fuel pump. Discard gasket.

b. Installation. Clean cover and gasket seat on cylinder block. Cement cork gasket to cover. Position cover on cylinder block by sliding it to the rear over fuel pump. Install cover rear screw and copper gasket, but do not tighten. Install cover front screw and copper

ENGINE—DESCRIPTION, DATA, MAINTENANCE, AND ADJUSTMENT IN VEHICLE

gasket, with ventilator cap, baffle, and gasket. Connect ventilator tube to valve, and tighten both cover screws evenly. Start engine and check for oil leaks.

56. VALVE TAPPET ADJUSTMENT.

a. Adjustment. Remove the valve spring cover (par. 55). Adjust the self-locking tappet screws while they are cold (or warm) to 0.014 inch (fig. 27). Set tappet screws, starting with No. 1 cylinder on compression stroke at top center, then adjust valves in cylinder firing order (par. 62 b), turning the crankshaft one-half turn for each cylinder. NOTE: *The valve tappets will then be on the heel of the cam.* After adjusting, replace valve spring cover (par. 55).

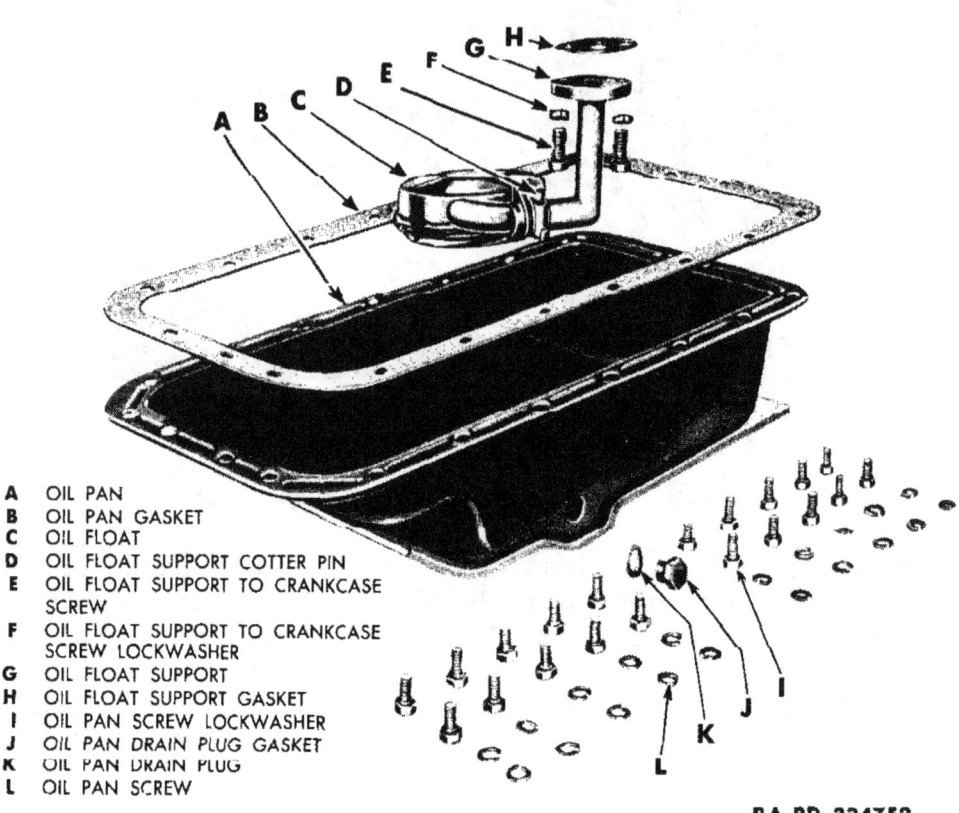

A OIL PAN
B OIL PAN GASKET
C OIL FLOAT
D OIL FLOAT SUPPORT COTTER PIN
E OIL FLOAT SUPPORT TO CRANKCASE SCREW
F OIL FLOAT SUPPORT TO CRANKCASE SCREW LOCKWASHER
G OIL FLOAT SUPPORT
H OIL FLOAT SUPPORT GASKET
I OIL PAN SCREW LOCKWASHER
J OIL PAN DRAIN PLUG GASKET
K OIL PAN DRAIN PLUG
L OIL PAN SCREW

RA PD 334752

Figure 28—Floating Oil Intake and Oil Pan

57. OIL PAN GASKET.

a. Removal. Drain oil by removing drain plug in lower left side of oil pan (fig. 28). Remove oil pan screws, exercising care not to lose spacers under fan belt guard. Remove oil pan, then remove gasket.

TM 9-803
57-58
¼-TON 4 x 4 TRUCK (WILLYS-OVERLAND MODEL MB and FORD MODEL GPW)

b. Installation. First clean oil pan thoroughly. Check condition of floating oil intake screen and if dirty, clean in dry-cleaning solution. Clean face of oil pan and crankcase where gasket is installed, and cement gasket to oil pan. Put oil pan in position, and install screws. Be sure that spacers under belt guard are in position, and tighten all screws evenly. Torque wrench reading must be 10 to 14 foot-pounds.

58. OIL FILTER.

a. Description. The oil filter is the military standard type located on the right front side of the engine (fig. 36). Part of the oil circulated through the oiling system is sent through the filter. The filter-

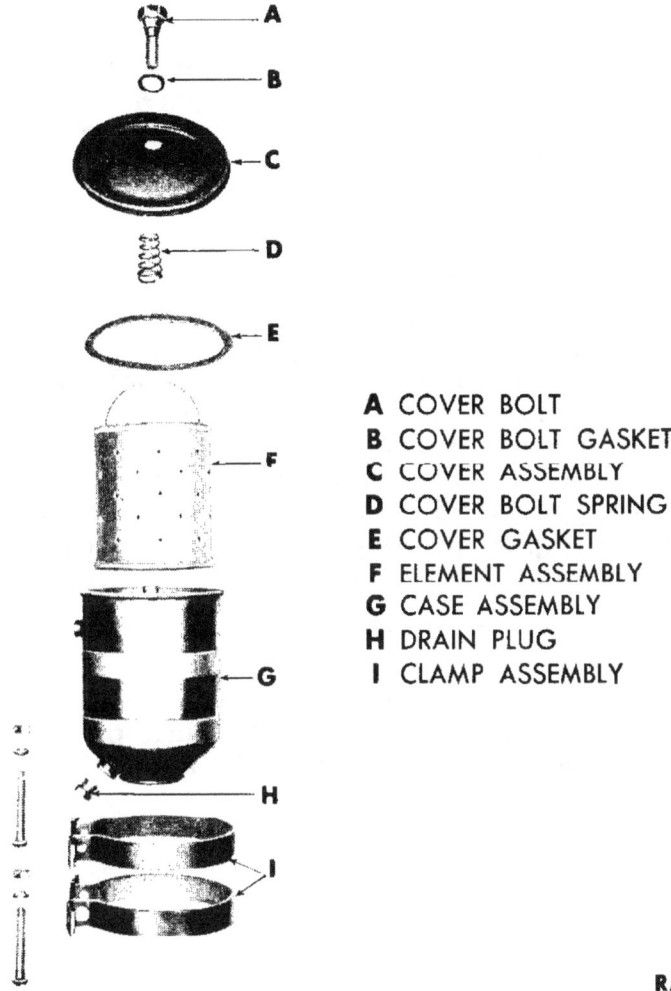

A COVER BOLT
B COVER BOLT GASKET
C COVER ASSEMBLY
D COVER BOLT SPRING
E COVER GASKET
F ELEMENT ASSEMBLY
G CASE ASSEMBLY
H DRAIN PLUG
I CLAMP ASSEMBLY

RA PD 305275

Figure 29 — Oil Filter, Disassembled

ing element is a cylindrical replaceable unit which should be changed each 6,000 miles, more often if the oil gets dirty quickly. The inlet line at the top of the filter connects to the oil distribution line at the

ENGINE—DESCRIPTION, DATA, MAINTENANCE, AND ADJUSTMENT IN VEHICLE

plug in the left front side of the engine. The outlet or oil return line connects to the timing chain cover.

b. Remove Element. Unscrew cover bolt in top of unit and remove cover, exercising care not to damage gasket (fig. 29). Remove drain plug in lower side of filter to drain filter, then lift out element.

c. Installing Element. Clean filter thoroughly. Install drain plug, and put new element in filter. Inspect cover gasket, and replace if necessary. Install cover and tighten in place with cover bolt. Start engine, and check filter for oil leaks. Add enough oil to crankcase to bring level up to "FULL" mark on gage.

d. Remove Filter. Drain filter by removing drain plug in lower side. Disconnect upper and lower tubes at filter. Remove four bolts holding filter in bracket, and remove filter. Remove tube fittings, using care not to distort them; then install drain plug.

e. Install Filter. Install tube fittings, exercising care not to damage them. Mount filter in bracket, and tighten in place. Connect tubes, being careful not to cross threads. Start the engine and check for oil leaks after which check engine oil level in crankcase and replenish supply to "FULL" mark on gage.

59. CRANKCASE VENTILATOR VALVE.

a. Description. The crankcase ventilator valve (fig. 22) is located at the center of the intake manifold. This valve is spring-loaded, and is operated by the intake manifold vacuum. The valve is closed when the engine is idling (manifold vacuum high). When the engine speed is increased the manifold vacuum is lowered, and the valve opens to allow clean air to be drawn from the air cleaner tube through the engine oil filler pipe to ventilate the crankcase. If this valve fails to seat properly, an engine operating condition will occur similar to a leaky intake manifold.

b. Remove Valve. Loosen valve spring cover front screw. Remove ventilator tube at valve, and remove ventilator valve (fig. 22).

c. Installing Valve. Place ventilator valve in a vise and remove top. Clean the valve and seat. Be sure that spring operates freely, and reassemble valve. Install valve in manifold and attach tube. Tighten valve spring cover front screw.

TM 9-803
60

¼-TON 4 x 4 TRUCK (WILLYS-OVERLAND MODEL MB and FORD MODEL GPW)

Section XIV

ENGINE—REMOVAL AND INSTALLATION

Paragraph

Removal 60
Installation 61

60. REMOVAL.

a. Open Hood. Unhook hood by pulling up catches at forward sides of hood. Raise hood and lay back against windshield. Hook or tie hood to windshield to avoid accidental closing.

b. Drain Cooling System. Open drain cocks at lower left-hand corner of radiator, and at right-front lower corner of cylinder block.

c. Remove Battery. Disconnect battery cables. Remove two wing nuts and washers from hold-down bolts. Remove battery hold-down frame, and lift out battery.

d. Remove Radiator. Remove upper and lower radiator hoses. Remove radiator stay rod nuts and remove rod. Remove two radiator stud nuts on bottom of radiator, and lift off radiator. Do not lose radiator pads.

e. Remove Air Cleaner. Disconnect air cleaner flexible hose at cleaner. Loosen wing nuts on side toward engine. Remove nuts on the opposite side, and remove cleaner. CAUTION: *Do not tip and spill oil from reservoir.*

f. Remove Cranking Motor. Disconnect cranking motor wires. Remove two bolts in engine rear plate, and one screw in side of crankcase. Remove motor.

g. On Right Side of Vehicle. Disconnect wires on generator. Disconnect ignition switch wire at ignition coil. Loosen fuel tank cap to relieve any pressure, and disconnect fuel line at flexible connection on right side of engine. Unscrew heat indicator unit from cylinder head. Remove two bolts holding engine front support insulator to frame, and disconnect bond strap.

h. On Left Side of Vehicle. Remove horn from bracket by removing two screws. Remove rear center cylinder head stud nut, and detach bond strap. Disconnect throttle and choke controls at carburetor. Remove fuel line (fuel pump to carburetor). Disconnect oil gage line at upper end of flexible tube on front of dash. Disconnect accelerator rod at lower end of bell crank on back of engine. Remove one bolt and one screw to separate exhaust pipe from manifold. Remove two engine front support bolts in frame, and disconnect engine ground strap from frame.

i. Disconnect Bell Housing. Remove bell housing upper bolts. Wrap a rope or cable around front and rear end of engine, attach chain hoist, and pick up weight of engine. Underneath vehicle, remove

ENGINE—REMOVAL AND INSTALLATION

the remaining bell housing bolts, and disconnect engine stay cable at the frame crossmember. Drive out two bell housing bolts at side of engine. Raise engine and guide out of frame.

61. INSTALLATION.

a. Installing Engine. Wrap rope or cable around front and rear end of engine, and attach chain hoist. Raise engine and lower into position. Insert transmission shaft in clutch driven plate hub, and work engine back into place. Install dowel bolts from engine side. Install bell housing bolts and tighten. Install engine front support insulator bolts in frame. Attach engine ground strap at left support and bond strap at right support. Install engine stay cable. Run rear adjusting nut up to the bracket, then with cable just taut, tighten lock nut on the front side of bracket.

b. On Right Side of Vehicle. Install cranking motor and attach wires. Attach ignition coil wire. Install heat indicator unit in cylinder head. Attach generator wires and ground strap. Connect flexible fuel line. Check air cleaner oil in reservoir, and install cleaner on dash with wing nuts. Tighten air flexible connection. Install battery, and secure in place with hold-down frame and wing nuts. Clean cable connections, grease, and attach to battery posts.

c. On Left Side of Vehicle. Install accelerator rod with cotter pin. Connect oil gage tube. Attach exhaust pipe to manifold with bolt and screw. Gasket must be in good condition. Install bond strap on cylinder head rear stud, and tighten nut to from 60 to 65 foot-pounds. Install carburetor choke and throttle wires. NOTE: *Controls on instrument panel must be all the way in, the throttle in the carburetor in the closed position, and the choke fully open.* Install fuel line between pump and carburetor. Attach horn to bracket.

d. Installing Radiator. See that radiator pads are in place, and install radiator on frame. Install upper and lower radiator hoses. Install radiator stay rod. Fill radiator, giving due attention to antifreeze, if required.

e. Inspection. Tighten fuel tank cap. Check engine oil (par. 18 e). Start engine, check for leaks, tune-up, and finally check level of solution in radiator. Close hood and hook properly.

TM 9-803
¼-TON 4 x 4 TRUCK (WILLYS-OVERLAND MODEL MB and FORD MODEL GPW)

Section XV
IGNITION SYSTEM

	Paragraph
Description and data	62
Maintenance	63
Distributor	64
Ignition timing	65
Coil	66
Spark plugs	67
Ignition switch	68
Ignition wiring	69

62. DESCRIPTION AND DATA.

a. Description. The ignition system (fig. 30) is a 6-volt system and consists of the spark plugs, high- and low-tension ignition wires,

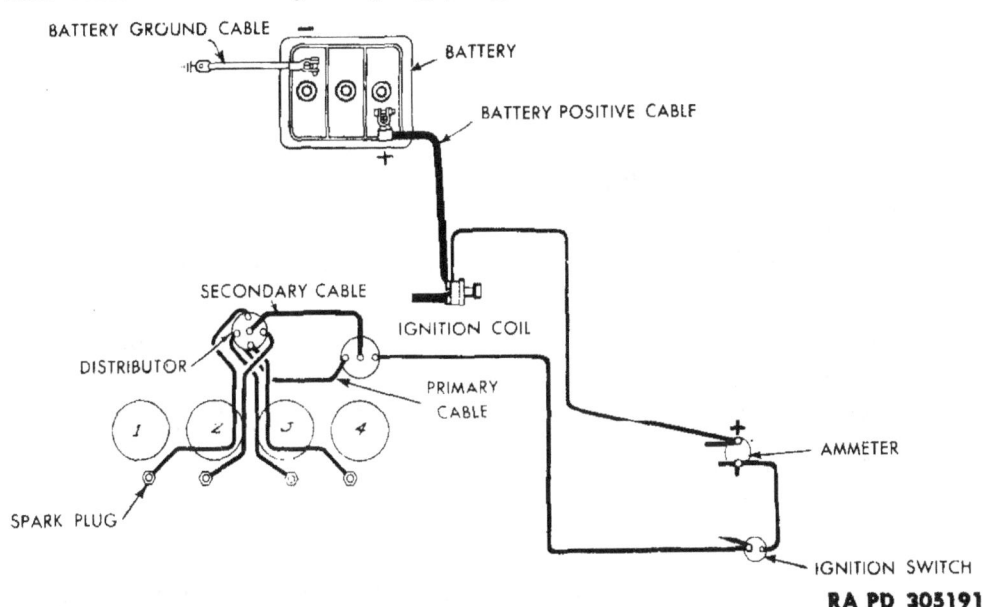

Figure 30—Ignition System Circuit

distributor, coil, and an ignition switch through which it is connected to the electrical system of the vehicle. There are two separate circuits in the ignition system (primary and secondary) which combine to develop the high-voltage current necessary to make a spark jump the plug gaps in the engine combustion chambers, and ignite the fuel mixture. In operation, with the ignition switch turned on, and the distributor points closed, current flows through the primary winding of the ignition coil and builds up a strong magnetic field. When the distributor points open, the magnetic field collapses and induces a high-voltage current in the secondary winding of the coil. This

IGNITION SYSTEM

happens each time the points open. This high-voltage current is delivered to the spark plugs at the correct time by the distributor rotor, cap, and secondary wires. To prevent burning of the distributor points by current arcing across the open points, a condenser is connected across the points (in parallel). This provides a place (capacity) for current to go until the points open enough to prevent arcing. Discharge of this condenser current back through the primary winding of the coil causes the magnetic field to collapse much faster in developing the high-voltage current for the spark plugs.

b. Data.

Distributor
- Make and model .... Auto-Lite IAD-4008
- Type advance .... Centrifugal
- Rotation .... Counterclockwise
- Firing order 1-3-4-2
- Point gap .... 0.020 in.
- Breaker arm spring tension .... 17 to 20 oz
- Condenser capacity .... 0.18 to 0.26 mfd

Ignition coil
- Make and model .... Auto-Lite IG-4070L
- Voltage 6-8
- Draw (engine stopped) .... 5 amps at 6.4 volts
- Draw (engine idling) .... 2.5 amps

Spark plugs
- Make and model .... Auto-Lite AN-7
- Thread size 14-mm
- Gap .... 0.030 in.

Ignition switch
- Make and model .... Douglas No. 6282

Ignition wires
- Primary (gage) .... No. 14
- Secondary (gage) .... No. 16

c. Tests. The following procedure will assist in localizing trouble in the ignition system without the use of instruments:

(1) First check the brilliancy of the headlights and operate cranking motor, to judge the condition of the battery and connections as far as the ammeter.

(2) Remove an ignition wire at a spark plug and hold about three-eighths inch away from a bare metal part of the engine. A good spark should result when the cranking motor is operated with the ignition switch on. If spark is weak or absent, proceed as follows:

(3) Pull coil wire out of center of distributor cap; remove cap and crank engine until distributor points are fully closed. Turn on ignition switch, hold cap end of coil secondary wire about three-eighths inch away from cylinder block, and open breaker points with

TM 9-803
62-64

¼-TON 4 x 4 TRUCK (WILLYS-OVERLAND MODEL MB and FORD MODEL GPW)

the fingers, or rock the distributor cam. If a good spark occurs, fault is located in distributor cap, rotor, or wires; inspect cap and rotor for cracks or carbon runners and ignition wires for short circuits. To test distributor cap place wire back in cap and operate cranking motor. Short circuit will be evidenced by a spark within or outside the cap. To check rotor, remove from distributor; put end of coil secondary wire in rotor, and hold top against cylinder block; operate distributor points (spark will show where "short" occurs). If no spark occurs in preceding test, proceed as follows:

(4) Open distributor points and notice if a slight spark is obtained. If spark occurs, current is reaching points. If no spark occurs, detach condenser, and repeat above operation. If spark is obtained, condenser is at fault, and must be replaced. If no spark is obtained, determine if coil primary wires are faulty, as follows:

(5) Remove switch wire at ignition coil, and strike wire terminal against cylinder block. If no spark is obtained, check wiring up to ammeter for loose connection or open circuit. If spark is obtained, current is reaching coil, and this indicates coil is faulty. Replace coil.

63. MAINTENANCE.

a. The distributor requires periodic lubrication at various points (par. 18). Keep coil and distributor wires pushed down in towers. Keep distributor cap and spark plug porcelains free from dirt and grease. All wire terminals must be clean and tight. Replace any wires that are frayed or have cracked insulation. Clean and adjust spark plugs and distributor points (par. 67).

64. DISTRIBUTOR.

a. **Description.** The distributor (fig. 32) is mounted on the right side of the engine. A full automatic spark advance is mechanically governed by two counterweights which advance the spark as the engine speed increases. The distributor is driven by a shaft extending into the oil pump driven gear, which is driven by a gear on the camshaft. The lower end of the distributor shaft has an offset tongue which must be in correct relation to the oil pump shaft before the two can be assembled. A friction spring on the distributor shaft (fig. 21) engages in the oil pump gear to prevent backlash at this point, and uneven engine performance.

b. **Removal.** After raising hood, remove wires from distributor cap and primary wire from terminal on side of distributor. Remove screw holding advance arm to crankcase, and pull out distributor assembly.

c. **Installation.** Remove distributor cap. Insert distributor in crankcase, pushing it down into place, turn rotor until offset tongue on lower end of distributor shaft fits into oil pump shaft, then push farther down into position. Some resistance will be experienced, caused by friction of the spring on the lower end of the distributor shaft fitting into the oil pump gear (fig. 21). Install screw in advance arm loosely. Attach primary wire to distributor. Set timing (par. 65). Install distributor cap and wires (fig. 32).

IGNITION SYSTEM

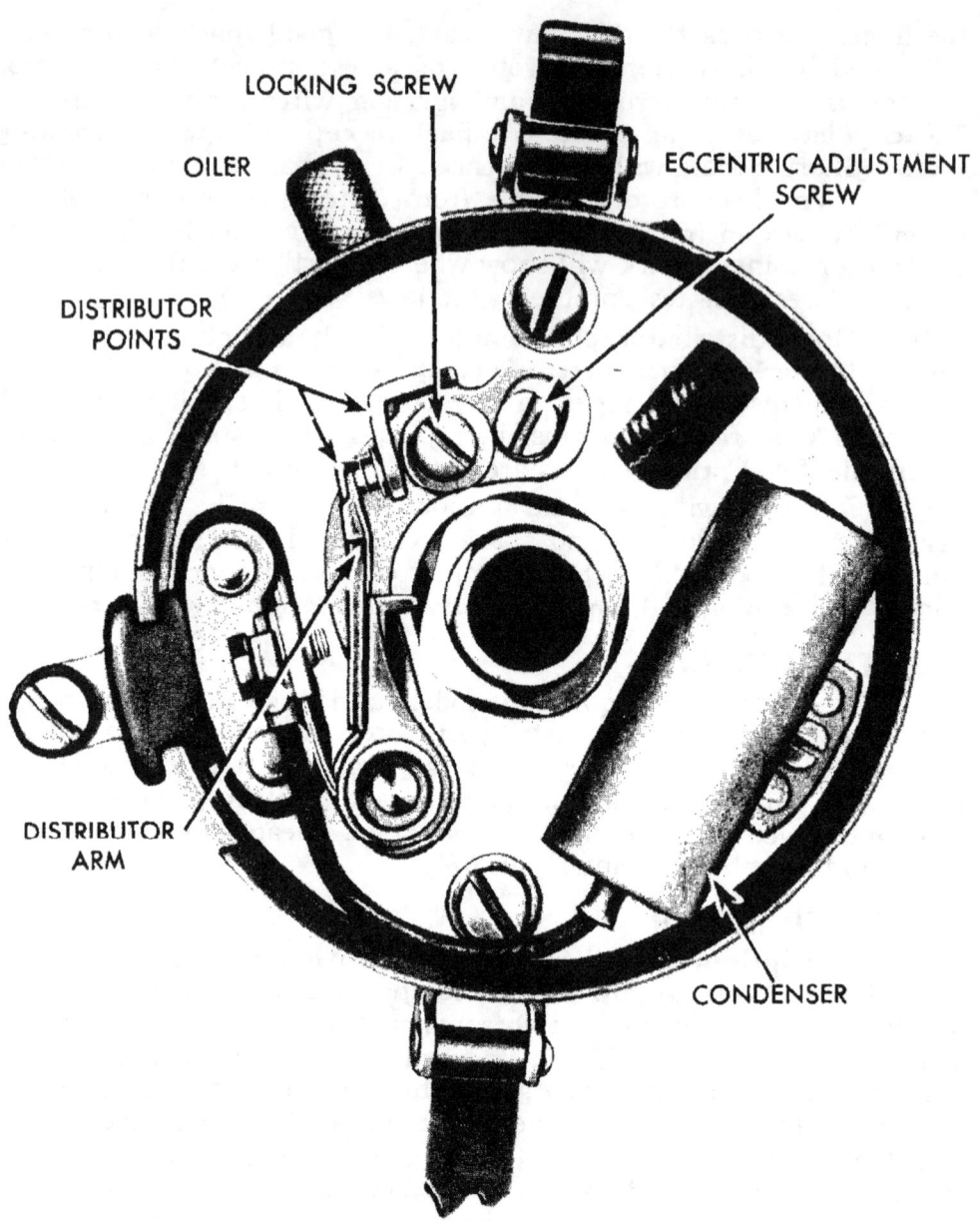

Figure 31—Distributor Points and Condenser

d. Distributor Points.

(1) ADJUSTMENT. Slip off the two clips holding the distributor cap in place and remove cap. Lift off rotor. Crank engine until point arm rubbing block is on top of a cam. Loosen lock screw (fig. 31), and turn eccentric screw until point gap is 0.020 inch measured with a thickness gage. Tighten lock screw and recheck gap. Install rotor and cap. Push wires well down into cap.

(2) REMOVAL. Slip off the two clips holding the distributor cap in place and lift off cap, then remove rotor (fig. 31). Using a small screwdriver, unscrew condenser lead which will release breaker arm

¼-TON 4 x 4 TRUCK (WILLYS-OVERLAND MODEL MB and FORD MODEL GPW)

Figure 32—Distributor Wires

spring, then lift off breaker arm. Remove screw in stationary breaker point, and lift out point.

(3) INSTALLATION. Place stationary breaker point in distributor, and install locking screw loosely. Lightly lubricate breaker arm pivot pin, and install breaker arm. Place spring in position with condenser lead, insert screw, and tighten securely in place. Aline points if necessary. To adjust points refer to step (1) above.

e. **Distributor Condenser.**

(1) DESCRIPTION. The condenser is attached by one screw to the support plate in the distributor, and connected by a short flexible wire across the distributor points. The condenser functions to absorb momentarily any current which has a tendency to arc across the points when they open. The condenser must be firmly attached to the

IGNITION SYSTEM

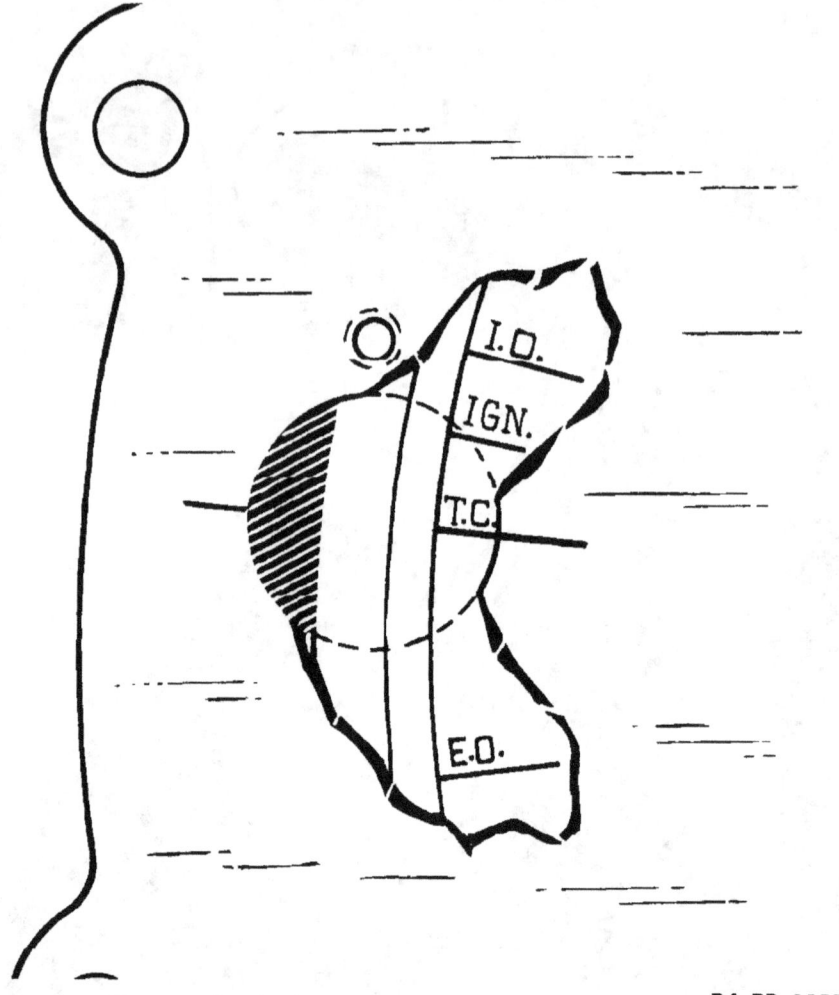

Figure 33—Timing Marks (Flywheel)

support plate, and the cable in good condition. Refer to paragraph 62 c for tests. Test condenser on a condenser tester, if available.

(2) REMOVAL. Slip off distributor cap clips and remove cap. Lift off rotor. Remove screw holding condenser to support plate. Remove screw in lead and remove condenser.

(3) INSTALLATION. Position condenser on plate and attach with screw. Attach lead. NOTE: *Determine if distributor points have been disturbed.* Install rotor and cap.

65. IGNITION TIMING.

a. To Set Timing without Timing Light. Remove timing hole cover on engine rear plate at right side under cranking motor. Remove distributor cap. Crank engine to No. 1 cylinder firing stroke (distributor rotor toward lower front corner of distributor). Set flywheel with ignition mark (fig. 33) in center of timing hole. Turn distributor housing so that breaker points are just opening, and tighten screw holding distributor to engine. Install timing hole cover.

¼-TON 4 x 4 TRUCK (WILLYS-OVERLAND MODEL MB and FORD MODEL GPW)

b. To Set Timing with Timing Light. Remove timing hole cover on engine rear plate at right side under cranking motor. Attach one lead of timing light to No. 1 spark plug, without removing the high-tension wire, and ground the other lead to the engine. Start engine and run slowly. Hold timing light in position to illuminate timing hole. Observe timing mark on flywheel, as illuminated by light, in relation to timing mark on engine plate. If marks do not coincide, loosen distributor clamp screw, and turn distributor housing in proper direction until marks do coincide. NOTE: *Use mirror for a better view.* Tighten clamp screw. Accelerate engine speed, and with light observe timing marks. They should separate to indicate that centrifugal spark advance is functioning.

66. IGNITION COIL.

a. Description. The ignition coil is mounted on the right side of the engine at the rear (fig. 21). A terminal is provided for the primary wire from the switch, a terminal for the primary wire to the distributor, and a terminal for the high-tension wire to the distributor cap, and a ground strap connection. The coil steps up the primary current, furnished by the battery and generator, to high-tension current of sufficient voltage to cause a spark to jump the gaps of the spark plugs.

b. Removal. Remove air cleaner by loosening clamp on flexible tube to carburetor; loosen wing nuts on air cleaner bracket at center of dash, and remove those on the right side. Pull secondary wire out of top of coil. Disconnect primary wires, then remove coil and bracket from engine, after which remove the bond strap.

c. Installation. Attach bond strap to coil, and mount coil on engine with bond strap on coil bracket front stud, and tighten securely. Attach primary wires, and push secondary wire into top of coil. Install air cleaner, and tighten hose clamp.

67. SPARK PLUGS.

a. Description. Spark plugs (fig. 22) are located in the top of the cylinder head at the left side. The plugs are of the one-piece type. A copper-silver alloy gasket is used on each plug for heat transfer to the cylinder head, and to prevent leakage of compression. Push-on type wire terminals are used with radio filters at each plug. A spark plug insulator cap is fitted on each plug to protect the plugs from water and dirt. No shielding is used on the plugs.

b. Adjustment. To adjust the gap, bend the side electrode only, and gage the plug with a round thickness gage to a gap of 0.030 inch.

c. Removal. To avoid breakage of the porcelain, remove the plugs with the socket wrench and handle furnished in the vehicle tool equipment.

d. Installation. Install new plug gaskets if available. Tighten spark plugs snug so that gasket will compress.

IGNITION SYSTEM

68. IGNITION SWITCH.

a. Description. On the earlier production, a key-type ignition switch was used which has been superseded by an interchangeable lever-type switch (fig. 5). Turn the switch lever clockwise for "ON" position.

b. Removal. Disconnect battery positive cable at battery. Unscrew retaining nut against face of instrument panel, and remove switch from panel. Disconnect wires and remove switch.

c. Installation. Install the wires on proper terminals, then fit the switch into the hole in the instrument panel, screw on retaining nut, and tighten securely.

69. IGNITION WIRING.

a. Description. The ignition wiring (fig. 30) consists of low-tension and high-tension wires. The low-tension, or primary wires, carry the current from the ignition switch, which is connected to the electrical system of the vehicle, to the ignition coil, and from the coil to the distributor. The high-tension or secondary wires carry the high-voltage current generated in the secondary wires of the coil, to the distributor, where it is distributed to the spark plugs. No shielding harness is used.

b. Removal. Before removing ignition switch wire, disconnect the battery negative (ground) cable. It is not necessary to disconnect the battery cable to replace the coil-to-distributor primary wire or the secondary wires. Disconnect wires at terminals, and remove wire or harness, opening wire clips on the harness in which the switch wire is a part. When removing secondary wires mark terminal tower in distributor cap for No. 1 cylinder spark plug. Pull wires out of distributor cap and off spark plug terminals.

c. Installation. To install primary wires, run wire or harness through clips, and attach terminals securely. Install secondary wires, for spark plugs, through support bracket. Push terminals down well into the proper distributor towers (fig. 32). Push rubber secondary wire tip down in place on distributor towers. Push terminals down on proper spark plugs. Refer to firing order (par. 62 **b**).

TM 9-803
70

¼-TON 4 x 4 TRUCK (WILLYS-OVERLAND MODEL MB and FORD MODEL GPW)

Section XVI

FUEL AND AIR INTAKE AND EXHAUST SYSTEMS

	Paragraph
Description and data	70
Maintenance	71
Carburetor	72
Air cleaner	73
Fuel pump	74
Fuel tank	75
Fuel strainer	76
Fuel gage	77
Exhaust system	78

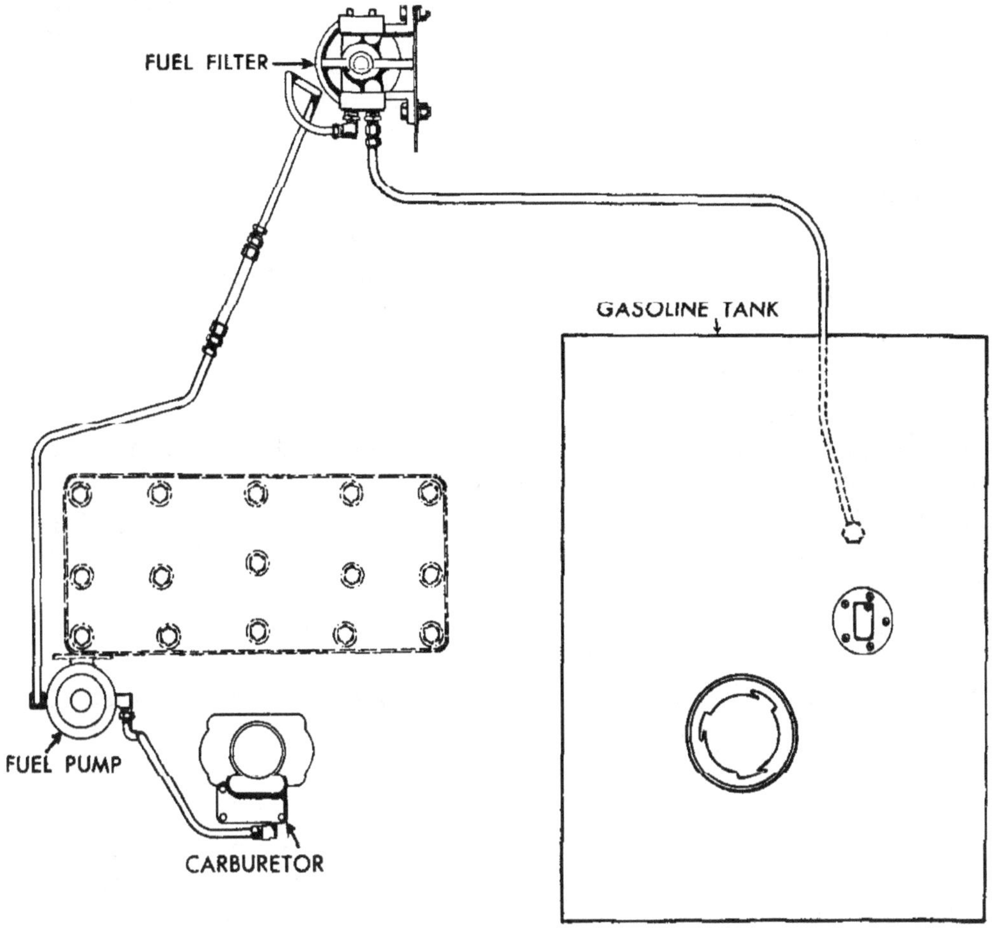

Figure 34—Fuel System   RA PD 305196

70. DESCRIPTION AND DATA.

a. Description. The fuel system (fig. 34) consists of the fuel tank, fuel lines, fuel strainer, fuel pump, carburetor, and air cleaner. In

126

FUEL AND AIR INTAKE AND EXHAUST SYSTEMS

addition to these units an electric-type fuel gage is mounted on the instrument panel, and is connected by one wire to a fuel tank unit.

b. Data.

Carburetor	Carter WO-539S
Air cleaner	Oakes 613300
Fuel pump	AC 1538312
Fuel pump static pressure	4.5 lb at 1,800 rpm
Fuel tank capacity	15 gal
Fuel strainer	Type T-2; AC1595848
Fuel gage	Electric actuation

c. Operation. Fuel in the tank is drawn through the fuel strainer by the action of a pump mounted on the forward left side of the engine. The pump also forces the fuel into the carburetor bowl until the flow is shut off by the carburetor float valve. In the carburetor, the fuel is proportioned and mixed with air drawn through the carburetor, from the oil-bath type air cleaner, by the action of the engine pistons. The fuel system must be inspected and cleaned periodically (par. 16). Tighten connections which show signs of leakage, and replace kinked or damaged lines.

71. MAINTENANCE.

a. The carburetor requires attention only to the idle adjustment. Fuel lines and vacuum connections must be tight. All mounting screws must be tight. Choke and throttle control clamp screws must be tight. Exterior of carburetor must be kept clean. All linkage must be lubricated at regular intervals, and be free to operate.

b. The air cleaner requires periodic check of correct oil level and condition of oil. Element must be kept clean and free from dirt. Mounting screws and clamps must be tight.

c. The fuel pump requires periodic cleaning of screen. Mounting screws and fuel line connections must be tight. Failure to function properly requires replacement of unit.

d. Drain fuel tank periodically to remove dirt and water. All connections and mounting bolts must be tight. Filler cap must be kept clean, and gasket checked for seat. Filler neck screen must be cleaned at regular intervals.

e. Clean fuel strainer at regular intervals. Gasket must be replaced if damaged. All connections and mounting bolts must be tight.

f. Fuel gage requires no attention other than that mounting screws must be tight. Replace damaged or frayed wires. Electrical connections must be clean and tight. Failure to operate requires replacement of unit.

72. CARBURETOR.

a. Description. The carburetor (fig. 35) is of the conventional downdraft, plain-tube type, with a throttle operated accelerator pump and economizer device. The carburetor is a precision instru-

¼-TON 4 x 4 TRUCK (WILLYS-OVERLAND MODEL MB and FORD MODEL GPW)

Figure 35—Carburetor Idle Adjustment

ment which delivers the proper fuel and air mixture for all speeds and operating requirements of the engine.

b. Adjustment. The idle adjustment screw indicated in figure 35 is the only service adjustment provided on the carburetor. To obtain the approximate correct setting, turn the adjustment screw to the right and all the way in, but do not jam the screw against the seat, then, back out adjustment screw between one and two turns. To make the final adjustment, warm up the engine, and adjust the screw until the engine runs smoothly. Set the throttle stop screw so the engine will idle at 600 revolutions per minute (vehicle speed, 8 mph). Replace carburetor if it requires other attention.

c. Removal. Loosen clamp on air horn and flexible air hose, and remove air horn. Remove throttle and choke control wires. Disconnect throttle control rod at throttle lever. Disconnect fuel line at carburetor. Remove carburetor flange nuts and retracting spring clip, and lift off carburetor.

d. Installation. Inspect condition of gaskets between carburetor and manifold, and install new gaskets, if required. Install carburetor retracting spring clip, and carburetor flange nuts. Tighten nuts evenly. Connect throttle control rod, and throttle and choke wires. Install carburetor air horn, and tighten clamp screws. Adjust carburetor (subpar. **b** above).

FUEL AND AIR INTAKE AND EXHAUST SYSTEMS

Figure 36—Air Cleaner and Oil Filter

73. AIR CLEANER.

a. Description. The air cleaner (fig. 36) is of the oil-bath type, and mounted on the right-front side of the dash. Air enters through louvers in the dash side of the unit, passes down and across the surface of the oil, up through the filtering element, then to the carburetor. For servicing of air cleaner refer to subparagraph d below.

b. Removal of Oil Cup. Hold one hand under cup and spring loose the two retaining clamps. Remove cup, clean, and refill to indicated oil level.

c. Installation. Place cup in position on bottom of cleaner, push up into place, and lock with the retaining clamps.

d. Removal and Servicing of Air Cleaner. Loosen the hose clamp and two wing nuts at the center of the dash. Remove two wing nuts on right side of dash. and lift out cleaner assembly. Unfasten the two clamps holding oil cup in place, and remove cup (fig. 37). Unscrew element wing bolt in bottom center of cleaner, and pull out cleaner element. Plunge the element up and down in dry-cleaning solvent to wash out any dirt, then dry with compressed air.

¼-TON 4 x 4 TRUCK (WILLYS-OVERLAND MODEL MB and FORD MODEL GPW)

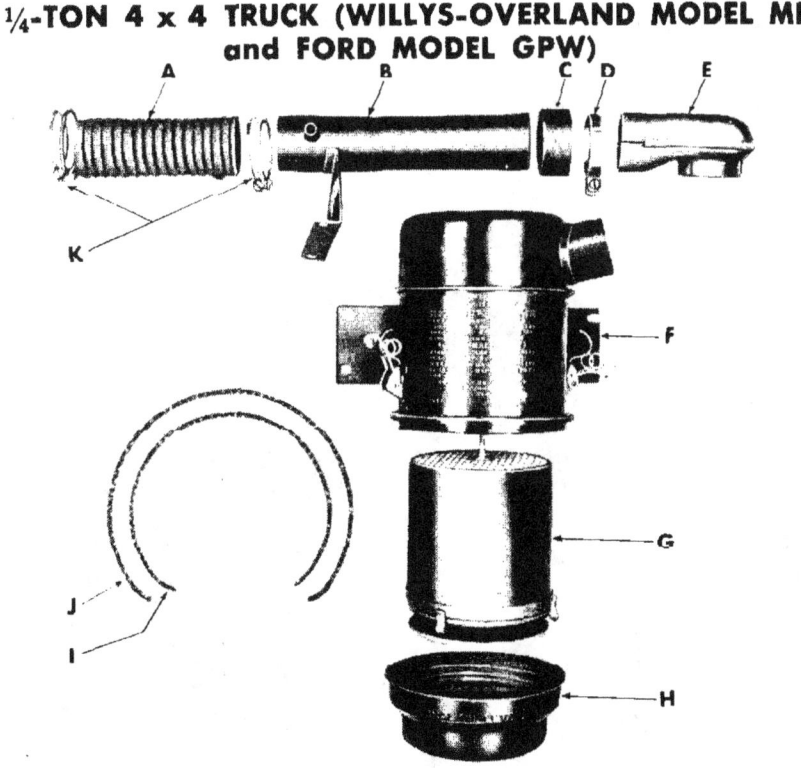

A FLEXIBLE HOSE CONNECTION
B TUBE AND BRACKET ASSEMBLY
C BUSHING
D HOSE CLAMP
E CARBURETOR AIR CLEANER HORN
F BODY ASSEMBLY
G ELEMENT AND WING BOLT ASSEMBLY
H CUP
I LOWER GASKET
J BODY GASKET
K HOSE CLAMP

RA PD 305281

Figure 37—Air Cleaner, Disassembled

e. **Installation.** Install the element in cleaner housing, and secure in place with retaining bolt. Clean and refill oil cup to indicated level, and clamp into place on cleaner. Mount cleaner on dash, install and tighten mounting nuts, and tighten air hose clamp in place.

74. FUEL PUMP.

a. **Description.** The fuel pump (fig. 38) located on the forward, left side of the engine is a diaphragm type, operated by a lever against an eccentric on the engine camshaft. The pump is equipped with a hand lever on the rear side which can be used to operate the pump for priming the carburetor bowl. Lever must be placed down for camshaft to operate pump. A filter screen in incorporated in the fuel bowl, and should be cleaned periodically (subpar. **b** below).

b. **To Clean Screen.** Unscrew knurled nut on bowl clamp, swing clamp aside and lift off bowl. Clean screen with dry-cleaning solvent and small brush. Dry with compressed air. Blow out fuel chamber lightly with compressed air and install screen in place. Install new bowl gasket or turn over old one if serviceable. Install bowl, place bail clamp in position and tighten securely.

FUEL AND AIR INTAKE AND EXHAUST SYSTEMS

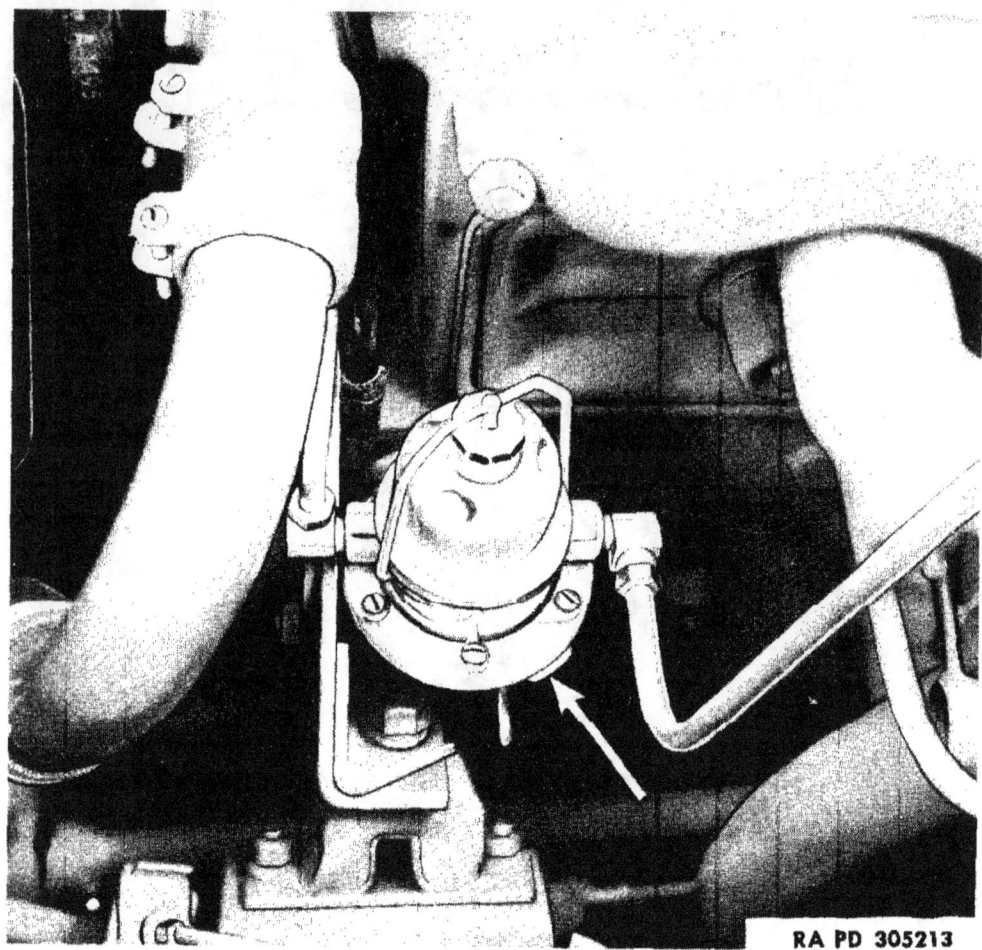

Figure 38 — Fuel Pump

c. Removal. Remove the inlet and outlet lines, remove two screws holding pump to side of engine and remove pump.

d. Installation. Install pump in place on crankcase and tighten securely with two screws. Inspect gasket and replace if unserviceable. Attach inlet and outlet lines and tighten. Prime pump by operating priming lever, start engine, and check connections for leaks.

75. FUEL TANK.

a. Description. The fuel tank (fig. 39) is located under the driver's seat. An extension filler neck can be pulled up to facilitate filling the tank from a container. After removing the filler cap, pull up on the filler extension, and turn to the right to lock it in place. To remove the filler extension from the tank turn it to the left and pull up.

b. Removal. Drain fuel by removing drain plug in left side of tank. Remove bolts in seat rear flange and front legs, then lift out seat. Remove filler cap. Disconnect fuel gage wire and remove fuel

¼-TON 4 x 4 TRUCK (WILLYS-OVERLAND MODEL MB and FORD MODEL GPW)

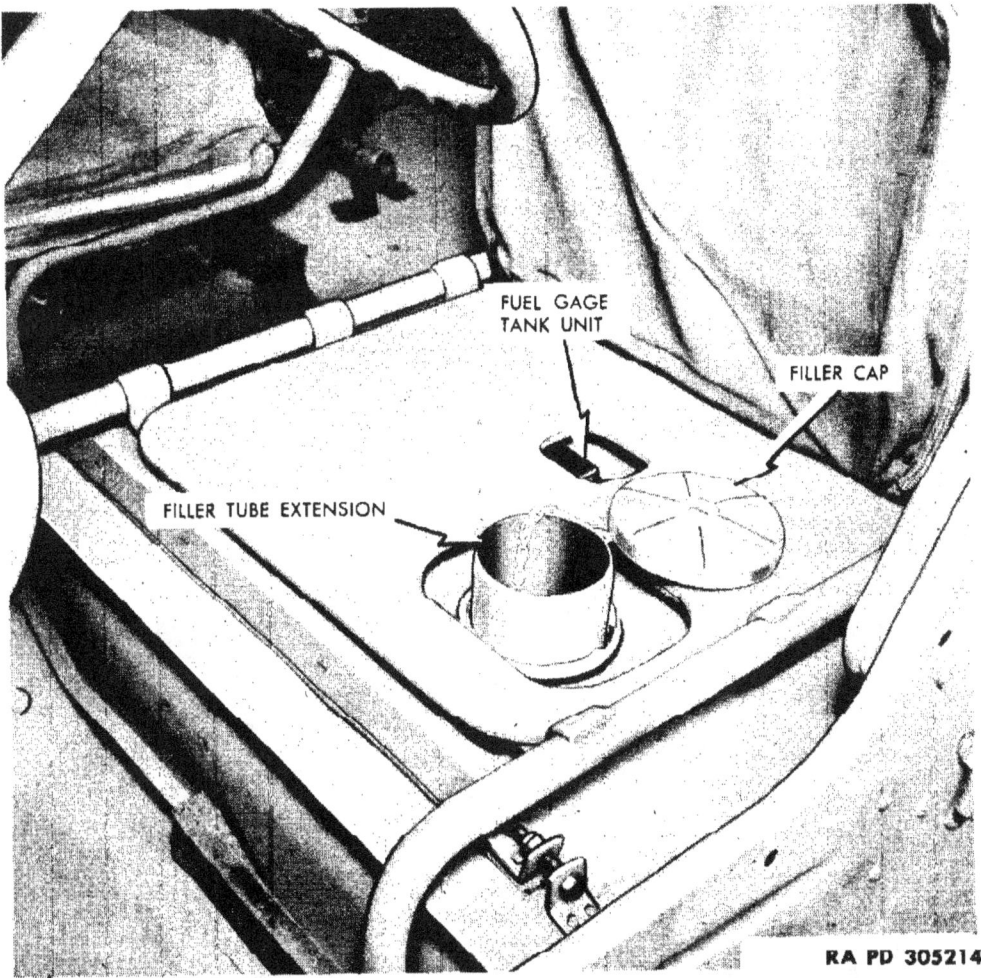

Figure 39—Fuel Tank

gage unit by taking out five screws. Disconnect fuel line from tank, under vehicle. Remove bolt holding tank to wheel housing, remove bolts in tank straps, and lift out tank.

c. Installation. Clean out fuel tank sump in body, and install tank in place. Attach tank straps and bolts, also bolt holding tank to wheel housing. Connect fuel line to tank. Install fuel gage in tank, and attach wire to gage. Set seat in position and secure in place with bolts in front legs and seat flange. Fill tank, install cap and check tank and connections for leaks.

76. FUEL STRAINER.

a. Description. The fuel strainer (fig. 41) has a disk-type (laminated) element, and a settling bowl for dirt and water. It is located between the fuel tank and the pump, and is mounted on the right-front side of the dash. Service this unit in accordance with preventive maintenance procedure (par. 16).

FUEL AND AIR INTAKE AND EXHAUST SYSTEMS

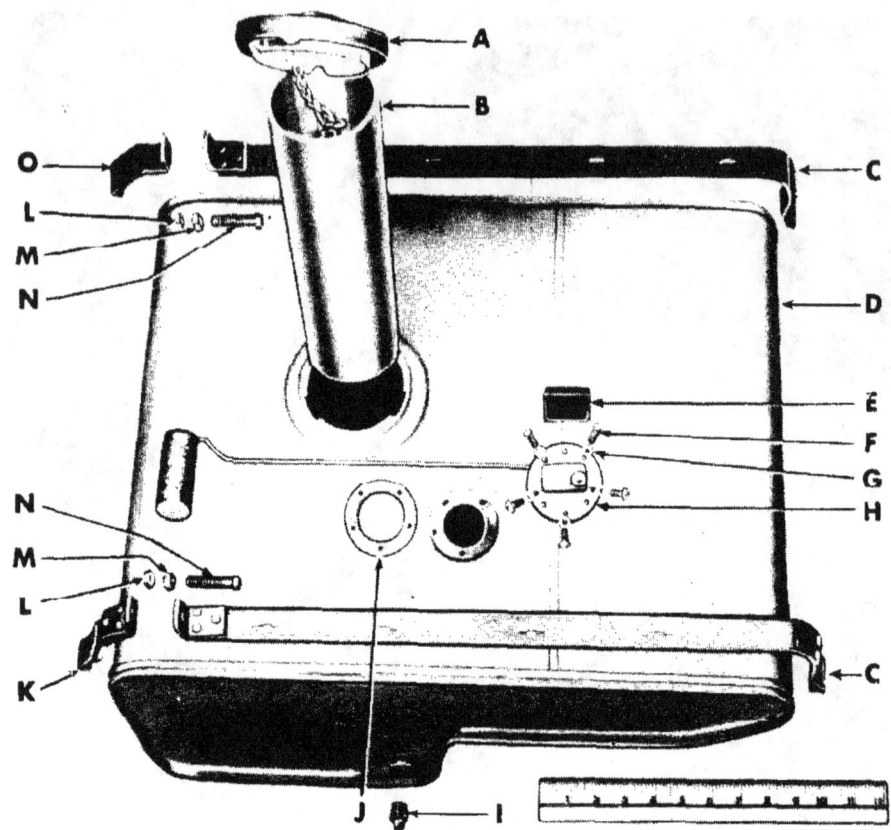

A	CAP	H	GAGE TANK UNIT ASSEMBLY
B	FILLER TUBE EXTENSION	I	DRAIN PLUG
C	HOLD DOWN STRAP	J	GAGE GASKET
D	FUEL TANK ASSEMBLY	K	HOLD DOWN CLAMP
E	GAGE TANK UNIT PROTECTOR	L	HOLD DOWN STRAP LOCK NUT
F	GAGE TO TANK SCREW	M	HOLD DOWN STRAP NUT
G	GAGE TO TANK SCREW LOCKWASHER	N	STRAP CLAMP SCREW
		O	STRAP CLAMP

RA PD 334756

Figure 40—Fuel Tank, Disassembled

b. To Clean Strainer. Remove drain plug and allow strainer to drain into a container. NOTE: *Do not allow fuel to drain onto cranking motor.* Unscrew cover bolt and remove strainer bowl and element. Do not damage bowl gasket. Remove element from bowl, and clean thoroughly in dry-cleaning solvent. Be sure all dirt particles are removed from between disks of the element. Blow element dry with compressed air, but do not use extreme pressure. Wash strainer bowl, and dry with clean cloth. Install element spring and element in bowl. Check condition of element gasket, bowl gasket and cover bolt gasket, and replace, if damaged. Install gaskets in position, and assemble bowl to cover. Install cover bolt and gasket, and tighten securely. Install drain plug. Start engine, run a few minutes, stop engine, and check for leaks.

¼-TON 4 x 4 TRUCK (WILLYS-OVERLAND MODEL MB and FORD MODEL GPW)

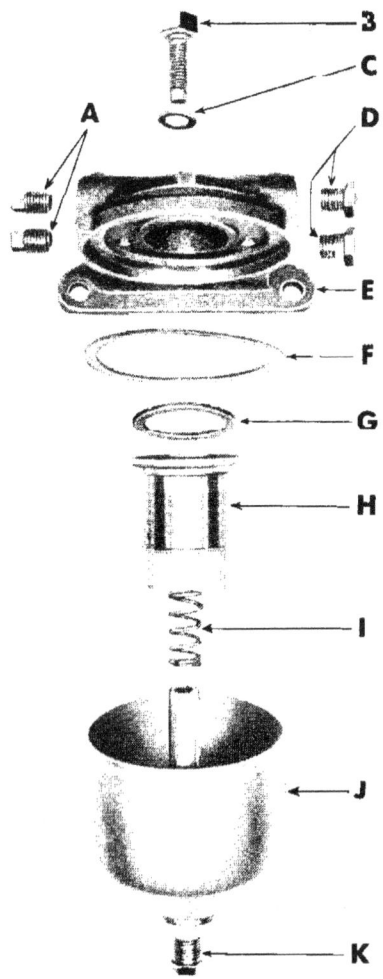

A PIPE PLUG
B COVER CAP SCREW
C COVER CAP SCREW GASKET
D PIPE REDUCING BUSHING
E COVER
F STRAINER BOWL GASKET
G STRAINER UNIT GASKET
H STRAINER UNIT ASSEMBLY
I STRAINER UNIT SPRING
J STRAINER BOWL AND CENTER STUD ASSEMBLY
K STRAINER DRAIN PLUG

RA PD 305282

Figure 41—Fuel Strainer, Disassembled

c. Removal. Disconnect inlet and outlet lines from strainer (fig. 36). Open glove compartment door, and remove mounting bolt nuts, then remove strainer.

d. Installation. Place mounting bolts in strainer bracket, and reinstall on dash. Install nuts on bolts through glove compartment. Attach inlet and outlet lines to strainer. Start engine, run a few minutes, stop engine, and check for leaks.

77. FUEL GAGE.

a. Description. The fuel gage consists of an electrical indicating unit located in the instrument panel, and a tank unit in the fuel tank, for actuating the instrument panel unit. A float in the tank operates a rheostat, which governs the current flowing through the

FUEL AND AIR INTAKE AND EXHAUST SYSTEMS

instrument panel gage. The gage registers only while the ignition switch is on. The fuel gage wiring circuit is shown in figure 42.

b. Removal of Tank Unit. Remove bolts in seat rear flange and front legs. Take out seat. Disconnect fuel gage wire, and remove tank unit, by unscrewing five screws.

c. Installation of Tank Unit. Inspect gasket, and replace if damaged. Install tank unit, and secure with five screws. Attach wire to fuel gage. Check operation of gage by turning ignition switch to "ON" position; gage must register amount of fuel in tank. Turn off ignition

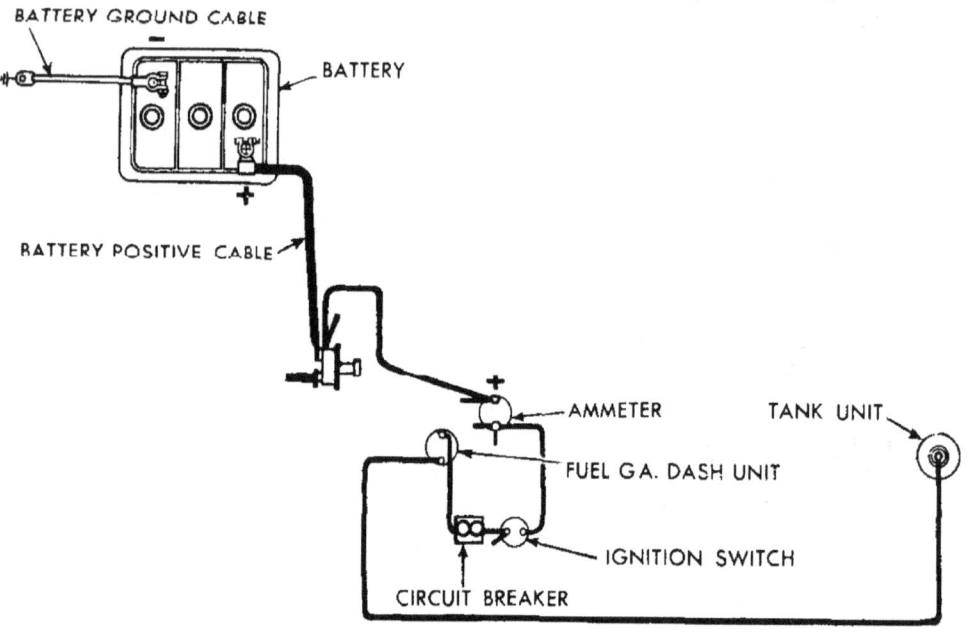

Figure 42—Fuel Gage Wiring Circuit

switch. Install seat, and secure in place with bolts in front legs and seat rear flange.

d. Removal of Fuel Gage (Instrument Panel). Disconnect positive terminal cable at battery. Remove wires on back of gage, and two retaining nuts on the retaining bracket. Remove bracket and take gage out through front of panel.

e. Installation of Fuel Gage (Instrument Panel). Place gage in panel, and install retaining clamp and nuts. Position gage correctly, and tighten retaining clamp nuts. Attach wires (wire to tank unit is attached to left terminal; wire from circuit breaker attaches to right-hand terminal). Clean terminal, and install positive cable on battery.

78. EXHAUST SYSTEM.

a. Description. The exhaust system (fig. 43) consists of the exhaust pipe which, for maximum road clearance, passes under the vehicle to the muffler located under the right side of the body. A

¼-TON 4 x 4 TRUCK (WILLYS-OVERLAND MODEL MB and FORD MODEL GPW)

flexible section of the exhaust pipe permits movement of the engine in the frame. The muffler is mounted on brackets with fabric inserts.

b. Removal of Exhaust Pipe. Remove three bolts in exhaust pipe shield at right frame member and remove shield. Remove two exhaust pipe clamp bolts in skid plate under transmission. Loosen clamp on exhaust pipe, at front end of muffler. Remove bolt and

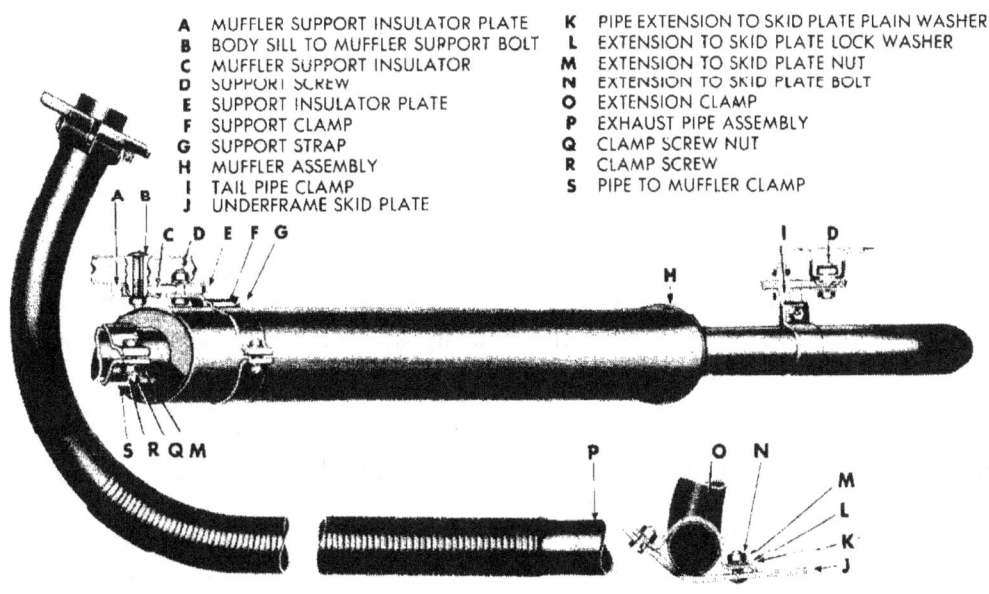

RA PD 305257

Figure 43—Exhaust System

A MUFFLER SUPPORT INSULATOR PLATE
B BODY SILL TO MUFFLER SUPPORT BOLT
C MUFFLER SUPPORT INSULATOR
D SUPPORT SCREW
E SUPPORT INSULATOR PLATE
F SUPPORT CLAMP
G SUPPORT STRAP
H MUFFLER ASSEMBLY
I TAIL PIPE CLAMP
J UNDERFRAME SKID PLATE
K PIPE EXTENSION TO SKID PLATE PLAIN WASHER
L EXTENSION TO SKID PLATE LOCK WASHER
M EXTENSION TO SKID PLATE NUT
N EXTENSION TO SKID PLATE BOLT
O EXTENSION CLAMP
P EXHAUST PIPE ASSEMBLY
Q CLAMP SCREW NUT
R CLAMP SCREW
S PIPE TO MUFFLER CLAMP

screw in exhaust pipe flange at exhaust manifold. Drop front end of exhaust pipe, and remove from the left side of vehicle. Discard gasket.

c. Installation of Exhaust Pipe. Install pipe in position, and insert rear end in muffler. Install exhaust pipe manifold flange gasket, and attach pipe to manifold with screw and nut, and tighten evenly. Attach clamp at muffler, and attach exhaust pipe shield with three bolts.

d. Removal of Muffler. To remove the muffler, loosen exhaust pipe clamp. Remove muffler front support bolt and tail pipe support bolt. Remove muffler assembly, after which, remove muffler strap and tail pipe clamp.

e. Installation of Muffler. Loosely install tail pipe clamp and support clamp on muffler. Place muffler on exhaust pipe. Install bolt through tail pipe clamp and flexible mounting. Tighten tail pipe clamp bolt. Install bolt in muffler support and flexible mounting. Tighten muffler strap bolt and exhaust pipe clamp.

Section XVII

COOLING SYSTEM

	Paragraph
Description and data	79
Maintenance	80
Radiator	81
Water pump	82
Fan belt	83
Fan	84
Thermostat	85
Temperature gage	86

79. DESCRIPTION AND DATA.

a. Description. The cooling system (fig. 44) consists of the radiator, pressure-type filler cap, fan, fan belt, water pump, thermostat, and temperature gage. The system is of the sealed type, operating under pressure when the engine is warmed up. When in proper condition, the units of the cooling system automatically maintain the engine at the proper operating temperature. The filler pipe is in the top of the radiator at the right side. There are two drain cocks, one in the radiator outlet at the lower left corner, and the other in the right side of the cylinder block at the forward end. In operation the water pump draws the coolant from the bottom of the radiator through the hose connection, and forces it through the cylinder block, past the thermostat, and back to the radiator, where it is cooled by the action of the fan drawing air through the radiator core. The cooling system capacity is 11 quarts.

b. Data.

Cooling system
    Capacity .... 11 qt
Radiator
    Type .... Fin and tube
    Filler cap .... Pressure type
Water pump
    Type .... Centrifugal
    Drive .... Fan belt
    Bearings .... Prelubricated ball
Fan belt
    Type .... Vee
    Length .... 44 1/8 in.
    Width .... 11/16 in.
    Angle of Vee .... 42 deg
Fan
    Blades 4
    Diameter .... 15 in.

TM 9-803
79-80

¼-TON 4 x 4 TRUCK (WILLYS-OVERLAND MODEL MB and FORD MODEL GPW)

Thermostat
    Opens .... 145° to 155°F
    Fully open .... 170°F
Temperature gage type .... Capillary

80. MAINTENANCE.

a. The cooling system must be inspected in accordance with preventive maintenance procedures (pars. 13 and 16). When draining the cooling system refer to caution plate on the instrument panel (fig. 7). General maintenance of the cooling system consists of the following procedures.

(1) Keep sufficient coolant in the system. Use clean water to which must be added the specified rust inhibitor; at temperatures below 32°F, add proper quantity of antifreeze solution (par. 7).

(2) Drain, flush, and refill system whenever inspection reveals any accumulation of rust or scale. Clean system seasonally as well as before and after using antifreeze solution.

(3) If engine overheats due to lack of coolant in the system, do not add cold water immediately. Let engine cool so that radiator does not boil, start engine, and add water slowly to prevent damage to cylinder block and head.

(4) Do not overfill radiator. Fill radiator to bottom of baffle visible through the radiator filler hole.

(5) Keep cylinder head, water pump, hose clamps, and connections leakproof. Replace deteriorated or leaky hose.

(6) Adjust fan belt and replace as required.

(7) Test periodically for air suction and exhaust gas leaking into system (subpars. c and d below).

b. Draining and Refilling System.

(1) Drain the cooling system, when required, by opening the drain cocks at the lower left corner of the radiator, and at the right front corner of the cylinder block. Loosen the pressure-type radiator cap to break any vacuum which might prevent proper draining. If solution is to be saved, catch it in a clean container. If system is not to be refilled immediately, attach a tag to the steering wheel, warning personnel about the system being drained.

(2) Refill the cooling system by first closing the two drain cocks tightly. Use clean water available, preferably soft water (water with low alkali content or other substances that promote rust and scale). Fill system through radiator filler pipe until level is up to lower edge of baffle visible through filler hole. Install radiator cap, and turn clockwise to tighten. Start engine and warm up. Check coolant level in radiator, and add more if required.

c. Air Suction Test. The air suction test is used to determine if air is entering the coolant, possibly due to low coolant level in the radiator, leaky water pump, or loose hose connections. To make test,

COOLING SYSTEM

Figure 44—Cooling System

fill system to bottom edge of baffle in top of radiator. Replace pressure type cap with plain cap, and tighten securely (airtight). Attach length of rubber tubing to lower end of overflow pipe (this connection must be airtight). Run engine, with transmission in neutral, at a moderate speed until warmed up. Place tubing in glass container of water, and without changing engine speed, watch for bubbles in water. Continuous appearance of bubbles indicates that air is entering coolant. The cause will be one of the above, and must be corrected.

d. Exhaust Gas Leakage Test. The exhaust gas leakage test is used to determine if gas is entering the coolant, possibly due to leaky cylinder block, cylinder head, or gasket. NOTE: *Make this test with engine cold.* Remove fan belt. Open radiator drain cock until coolant is below cylinder head water outlet. NOTE: *Determine by loosening three screws holding outlet to head.* Remove water outlet, and fill cylinder head with coolant until level is up to top of head. With

TM 9-803
80-82

¼-TON 4 x 4 TRUCK (WILLYS-OVERLAND MODEL MB and FORD MODEL GPW)

transmission in neutral, start engine, "gun" it several times, and watch for bubbles in water. Appearance of bubbles indicates leakage from one of the above conditions, which must be corrected. Replace leaky gasket; report other causes to higher authority.

e. Cleaning and Flushing Procedure. This procedure is used to clean out loose rust. Run engine at moderate speed to stir up loose rust. Drain cooling system. Close drain cocks, and fill system with specified cleaning compound. Install radiator cap. Operate engine as directed for prescribed solution. Stop engine and completely drain system by opening both drain cocks. To flush system close drain cocks, fill system with water, run engine until warmed up again, or run water through system, then completely drain. Close drain cocks, refill system, and add inhibitor corrosion compound to prevent formation of rust and scale. Inhibitor compound must be renewed periodically to be effective.

81. RADIATOR.

a. Description. The radiator assembly (fig. 44) consists of the fin-and-tube type core with a coolant tank at the top, and a sediment tank at the bottom. It is located in the conventional place at the front of the vehicle. A pressure-type filler cap maintains up to 4¼-pound pressure to give better engine efficiency, and prevent evaporation of the coolant. When the engine is warm, release pressure by turning cap slightly before removal.

b. Removal. Remove the radiator filler cap. Open the drain cock on bottom of radiator outlet and right front corner of cylinder block to drain system. Remove radiator stay rod nut at front end. Loosen hose clamp on upper hose at front end. Loosen hose connection at water pump. Remove two radiator hold-down nuts and lift off radiator, then remove drain cock and pads.

c. Installation. Install drain cock and place pads on bracket; set radiator in place. NOTE: *A light coating of grease in hose connections will facilitate assembly.* Install stay rod and tighten lock nut. Tighten upper and lower hose clamps. Install hold-down nuts and bond strap. Close radiator drain cock and cylinder block drain cock. Fill radiator, install cap, and check system for leaks. Start engine, check coolant level after engine is warmed up, and fill, if needed, to proper level.

82. WATER PUMP.

a. Description. The water pump (fig. 44) is of the centrifugal-impeller type and is located in the front end of the cylinder block. A double-row ball bearing is integral with the shaft, and is packed with lubricant when it is made. No lubrication attention is required. A packless self-sealing gland is used to prevent water leakage, and requires no attention. The water pump with the generator and fan is driven by a belt from the engine crankshaft fan pulley.

COOLING SYSTEM

Figure 45—Fan Belt Deflection

b. Removal. Open the radiator drain cock in outlet pipe at lower left corner of radiator, and also drain cock in right side of cylinder block at forward end. Remove radiator filler cap. Pull up on handle of generator brace to loosen fan belt, and slip off belt. Loosen pump hose clamp and remove hose. Remove fan blade screws. Remove screws holding water pump in cylinder block, and remove water pump.

c. Installation. Check condition of pump to cylinder block gasket. Replace if damaged. Install pump in cylinder block, and tighten with screws. Install fan and screws. Install fan belt, and pull out generator until brace drops into position. Attach hose connection. Fill radiator, install cap, and check system for leaks. Start engine and check radiator coolant level after engine is warmed up.

83. FAN BELT.

a. Description. The fan belt (fig. 45) is of the V-type and drives the fan, water pump, and generator. Proper adjustment is necessary

¼-TON 4 x 4 TRUCK (WILLYS-OVERLAND MODEL MB and FORD MODEL GPW)

for efficient operation and maximum life of belt. Do not adjust the belt extremely tight, causing excessive wear on water pump and generator bearings.

b. Removal. Pull up on generator brace handle, and move generator toward engine as far as it will go. Remove belt from generator, pump, and crankshaft pulleys, and lift over fan blades.

c. Installation and Adjustment. Place belt over fan and crankshaft pulleys, then over the generator pulley. Pull generator out until

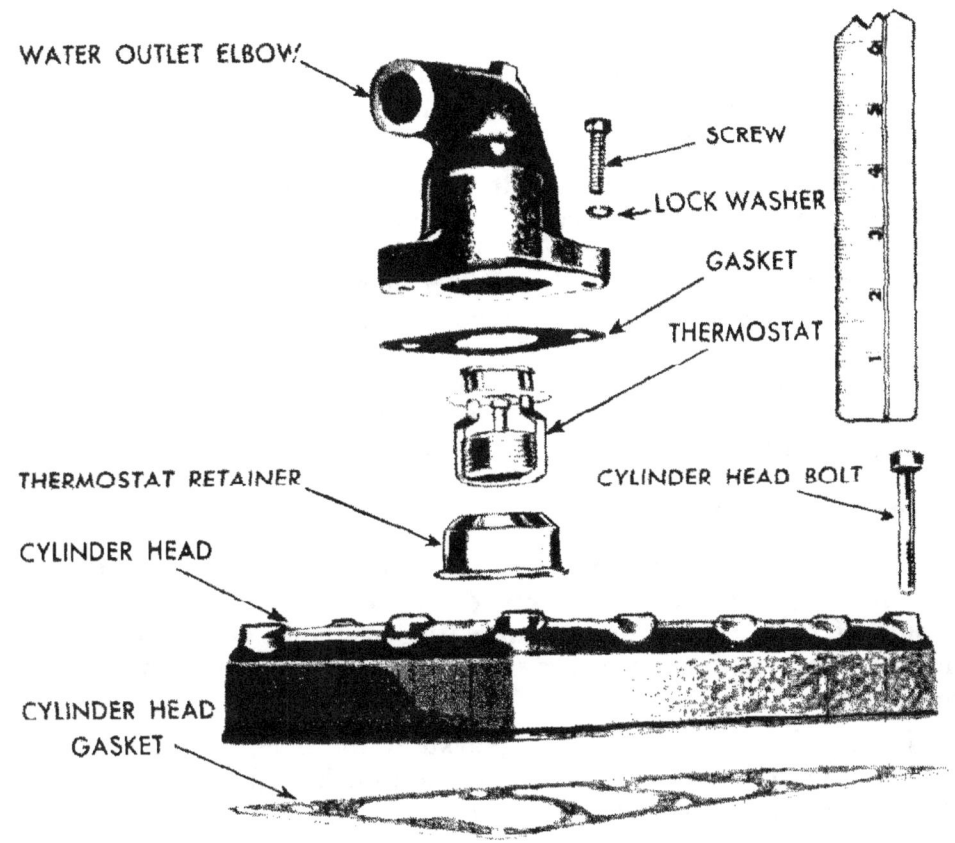

Figure 46—Thermostat, Disassembled

generator brace locks in position. To adjust fan belt tension, loosen generator brace nut, and move generator until fan belt has about one-inch deflection midway between fan and generator pulleys (fig. 45), then tighten nut.

84. FAN.

a. Description. A four-blade, 15-inch fan (fig. 44) draws air through the radiator core. It is mounted on the front end of the water pump shaft and is driven from the crankshaft by the same belt which drives the generator.

COOLING SYSTEM

b. Removal. Remove four screws holding fan on fan pulley, and lift fan out of shroud.

c. Installation. Place fan in position on fan pulley, install four screws, and tighten securely.

85. THERMOSTAT.

a. Description. The thermostat (fig. 44) is of the bellows-type, and located in the water outlet elbow on top of the cylinder head. It is designed to open between 140°F and 155°F, and is fully open at 170°F.

b. Removal. Drain the cooling system by opening drain cock at lower left side of radiator. Loosen hose clamp at cylinder head water outlet elbow; remove three screws, and lift off elbow. Pull out thermostat retaining ring in elbow, and take out thermostat (fig. 46).

c. Installation. Place thermostat in water outlet elbow with bellows down, so that coolant can reach bellows, to cause valve to operate. Install retaining ring with flanged edge against thermostat. Check condition of gasket and cylinder head surface. Install new gasket if necessary. Insert outlet elbow in hose connection. Install outlet elbow on cylinder head, and tighten in place with three screws. Tighten hose connection. Close radiator drain cock, and fill cooling system, giving due attention to antifreeze, if required. Start engine, and check for leaky connections after engine is warmed up.

86. TEMPERATURE GAGE.

a. Description. The temperature gage (fig. 5) is of the Bourdon-type with a capillary tube connecting it to an expansion bulb in the right side of the cylinder head. The entire system is sealed into one assembly; if any difficulty is experienced, the whole assembly must be replaced.

b. Removal. Drain the cooling system by opening drain cock under left side of radiator. Remove engine unit in right side of cylinder head by unscrewing retaining nut in reducing bushing. Remove two nuts holding retaining bracket on back of gage. Remove grommet around tube through dash and draw out gage assembly, tube, and engine unit through hole in panel.

c. Installation. Insert engine unit through hole in panel, mounting bracket, and dash. Position gage correctly, install mounting bracket and nuts; tighten securely. Insert grommet in dash. Install engine unit, and tighten retaining nut. Fill cooling system, and check for leaks.

¼-TON 4 x 4 TRUCK (WILLYS-OVERLAND MODEL MB and FORD MODEL GPW)

Section XVIII

STARTING SYSTEM

	Paragraph
Description and data	87
Maintenance	88
Cranking motor	89
Starting switch	90

87. DESCRIPTION AND DATA.

a. Description. The starting system (fig. 47) is a 6-volt system. It consists of the starting switch, cranking motor, and the cables through which they are connected to the battery of the electrical system. When the starting switch is pressed with the right foot, the battery current energizes the cranking motor, causing the armature to turn. This results in the inertia-type cranking motor gear engaging the teeth on the outer circumference of the flywheel. When the engine starts, the gear is automatically thrown out of engagement with the flywheel teeth.

b. Data.

System voltage	6
Cranking motor	
Normal engine cranking speed	185 rpm
Make and model	Auto-Lite MZ-4113
Bearings	3
Brushes	4
Brush spring tension	42 to 53 oz
Drive (direct)	R.H. outboard Bendix
Starting switch	
Make and model	Auto-Lite SW-4015

88. MAINTENANCE.

a. The cranking motor requires lubrication only at the forward end. Check tightness of mounting screws. Wire terminals must be in good condition, clean, and tight. As part of the starting system, check the battery condition periodically. Clean the cranking motor drive periodically (par. 89).

89. CRANKING MOTOR.

a. Description. The cranking motor is a 6-volt, four-brush type, located at the right-rear side of the engine. The drive is transmitted to the engine from an inertia-type gear to the teeth on a removable ring gear on the circumference of the flywheel. Rotation of the cranking motor shaft causes the pinion of the cranking motor drive

STARTING SYSTEM

to advance and mesh with the gear on the flywheel. After the engine starts, and the speed of the flywheel exceeds that of the cranking motor, the flywheel disengages the pinion automatically. A removable cover at the front end of the cranking motor permits inspection of the brushes and commutator.

b. Removal. Remove cable from motor. Remove screw from front support bracket, and remove two attaching screws at flywheel. Remove cranking motor. Remove front bracket.

c. Installation. Clean cranking motor gear. *Do not oil drive.* Install front bracket and place cranking motor in position on engine. Insert attaching screws, also bracket screw, and tighten securely. Attach cable.

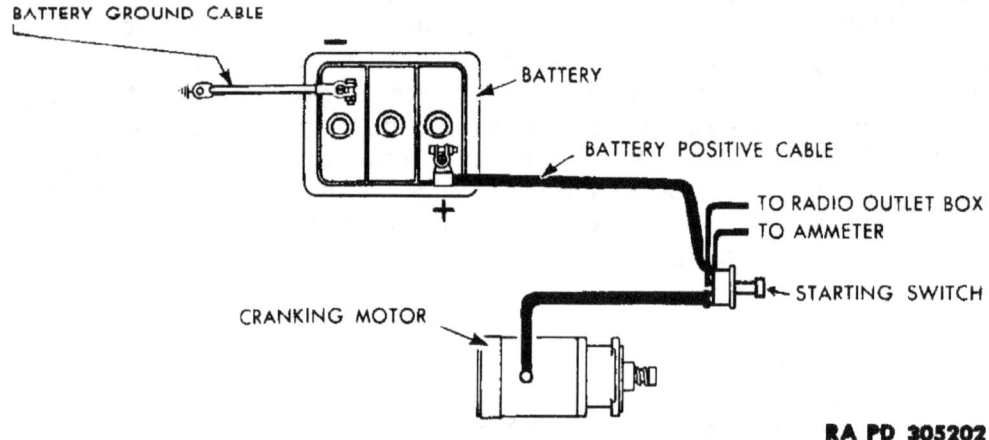

Figure 47 — Starting System Circuit

90. STARTING SWITCH.

a. Description. The starting switch is a push-button type operated with the right foot, and located to the right of the accelerator treadle on the toeboard (fig. 5). Press the button to close the switch and cause the cranking motor to operate.

b. Removal. Remove battery cable at negative post of battery. Remove air cleaner (par. 73). Remove cable terminal nuts on switch, and remove cables and wires. Remove two screws holding switch to toeboard, and remove switch from underneath.

c. Installation. Remove cable terminal nuts, place switch in position under toeboard, and tighten in place with two screws. Attach cables and wires. NOTE: *Battery positive cable and wires to radio outlet box, filter, and ammeter are attached to top terminal.* Install air cleaner (par. 73), and connect battery cable to negative post on battery.

TM 9-803
91-92

¼-TON 4 x 4 TRUCK (WILLYS-OVERLAND MODEL MB and FORD MODEL GPW)

Section XIX

GENERATING SYSTEM

	Paragraph
Description and data	91
Maintenance	92
Generator	93
Regulator	94

91. DESCRIPTION AND DATA.

a. Description. The generating system (fig. 49) is a 6-volt system, single-wire, ground-return type. This system consists of the generator, regulator, and wires connecting it to the ammeter. For information concerning the battery and lighting system refer to paragraph 95. The system develops current to keep the battery charged, and furnishes current for ignition, lighting, and other electrical accessories if the engine operation is sufficient. The regulator governs the generator output in accordance with the condition of the battery, and the requirements of the other electrical units used in the operation of the vehicle.

b. Data.

System voltage	6 to 8
Generator:	
Make and model	Auto-Lite GEG-5101D
Ground polarity	Negative
Controlled output	40 amps
Rotation (drive end)	Clockwise
Control	Current voltage regulator
Brushes	2
Output	8.0 amps; 7.6 volts; 955 rpm
	40.0 amps; 7.6 volts; 1460 rpm
	40.0 amps; 8.0 volts; 1465 rpm
Regulator:	
Make and model	Auto-Lite VRY-4203 A
Type	Current voltage
Volts	6
Amperes	40
Ground polarity	Negative

92. MAINTENANCE.

a. The generator requires attention to lubrication of those generators provided with oilers. Properly adjust the drive belt. Check mounting screws and bracket attached to engine. Generator brace must be in position. Wire terminals must be in good condition, clean,

GENERATING SYSTEM

and tight. External connecting wires with damaged or cracked insulation must be replaced. Bond straps must be cleaned and securely tightened. If trouble is experienced with the generator, report to higher authority.

93. GENERATOR.

a. Description. The generator is located on the right side of the engine at the forward end (fig. 49). It is a 6-8 volt, shunt-wound, two-brush unit rotating clockwise as viewed from the drive pulley end. It has a controlled output of 40 amperes, and is air-cooled by a fan built into the drive pulley. Air is drawn in at the back of the generator, and discharged at the fan.

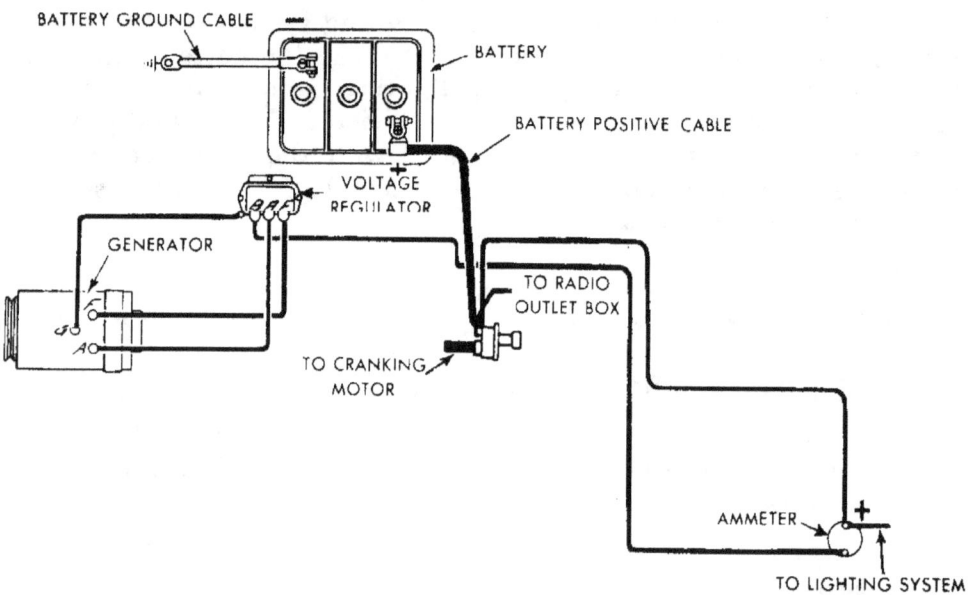

RA PD 305203

Figure 48—Generating System Circuit

b. Removal. Remove generator brace spring (fig. 10) and slip fan belt off pulley. Disconnect wires at generator; remove radio filter and ground strap. Remove two generator support bolts and remove generator.

c. Installation. Place generator in position, and install a drift or pin through front hole as a pilot. Install flat washer and bushing at rear support, and install bolt through hole. Remove drift, and install washer and bolt in front hole. Install generator brace spring. Install fan belt, and adjust if necessary (par. 83). Attach wires, filter, and ground strap. Start engine and check charging rate on ammeter in instrument panel (fig. 5).

¼-TON 4 x 4 TRUCK (WILLYS-OVERLAND MODEL MB and FORD MODEL GPW)

94. REGULATOR.

a. Description. The regulator (fig. 49) automatically governs the output of the generator in accordance with the condition of the battery and the current requirements in the operation of the vehicle, thus, when the battery is low the output is increased, and as the

Figure 49—Generator and Regulator

battery becomes fully charged the generator develops less current to avoid overcharging. The regulator is mounted inside the splasher of the right front fender. It is a precision instrument, sealed at the factory, and no attempt should be made to adjust it. The regulator consists of three separate units: the cut-out or circuit breaker, the voltage regulator, and the current regulator. The circuit breaker automatically closes the circuit between the generator and the battery when the generator voltage rises above that of the battery, and opens the circuit when the generator current falls below that of the battery.

GENERATING SYSTEM

The voltage regulator governs the generator so that it will not develop more voltage than the value for which the voltage regulator is set. The current regulator controls the generator current (amperage) output so that it will develop enough to keep the battery charged (providing the engine is run sufficiently), and also prevent damage to the generator due to an overload.

b. Removal. Remove battery cable at battery negative post. Remove wires from regulator terminals. Mark wires to assure correct installation. Remove bond strap, and remove four bolts holding regulator to fender.

c. Installation. Place regulator in position on fender, and tighten securely with four bolts. Install bond strap. Connect wires to proper terminals. Install battery cable on post, and tighten securely. Start engine and check charging rate of ammeter in instrument panel.

¼-TON 4 x 4 TRUCK (WILLYS-OVERLAND MODEL MB and FORD MODEL GPW)

Section XX

BATTERY AND LIGHTING SYSTEM

	Paragraph
Description and data	95
Maintenance	96
Battery	97
Wiring system	98
Headlights	99
Blackout headlights	100
Blackout driving light	101
Taillights and stop lights	102
Instrument panel lights	103
Blackout (main) light switch	104
Panel and blackout driving light switches	105
Trailer connection	106

95. DESCRIPTION AND DATA.

a. Description. The lighting system (fig. 50) functions on 6 volts supplied by a three-cell storage battery. The system consists of two service headlights, two blackout headlights, one blackout driving light, service and blackout stop and taillights, two instrument panel lights, and operating switches and battery. The entire lighting system is controlled by the blackout (main) light switch on the instrument panel. When the blackout light switch is set in the proper operating position, other light switches such as blackout driving light, instrument panel, stop lights and dimmers, are controlled by the respective switches. The lighting system is protected by a thermal-type fuse on the back of the blackout switch.

b. Data.

Battery:
- Make and model Auto-Lite-TS-2-15
- Willard SW-2-119
- Volts 6
- Plates per cell 15
- Capacity (ampere hours) 116
- Length .... 10 in.
- Width .... 7 in.
- Height .... $8\frac{5}{16}$ in.
- Ground terminal .... Negative

Wiring system:
- Volts 6
- Wiring identification reference .... Fig. 50

BATTERY AND LIGHTING SYSTEM

Figure 50 — Wiring System — Phantom View

TM 9-803
95

¼-TON 4 x 4 TRUCK (WILLYS-OVERLAND MODEL MB and FORD MODEL GPW)

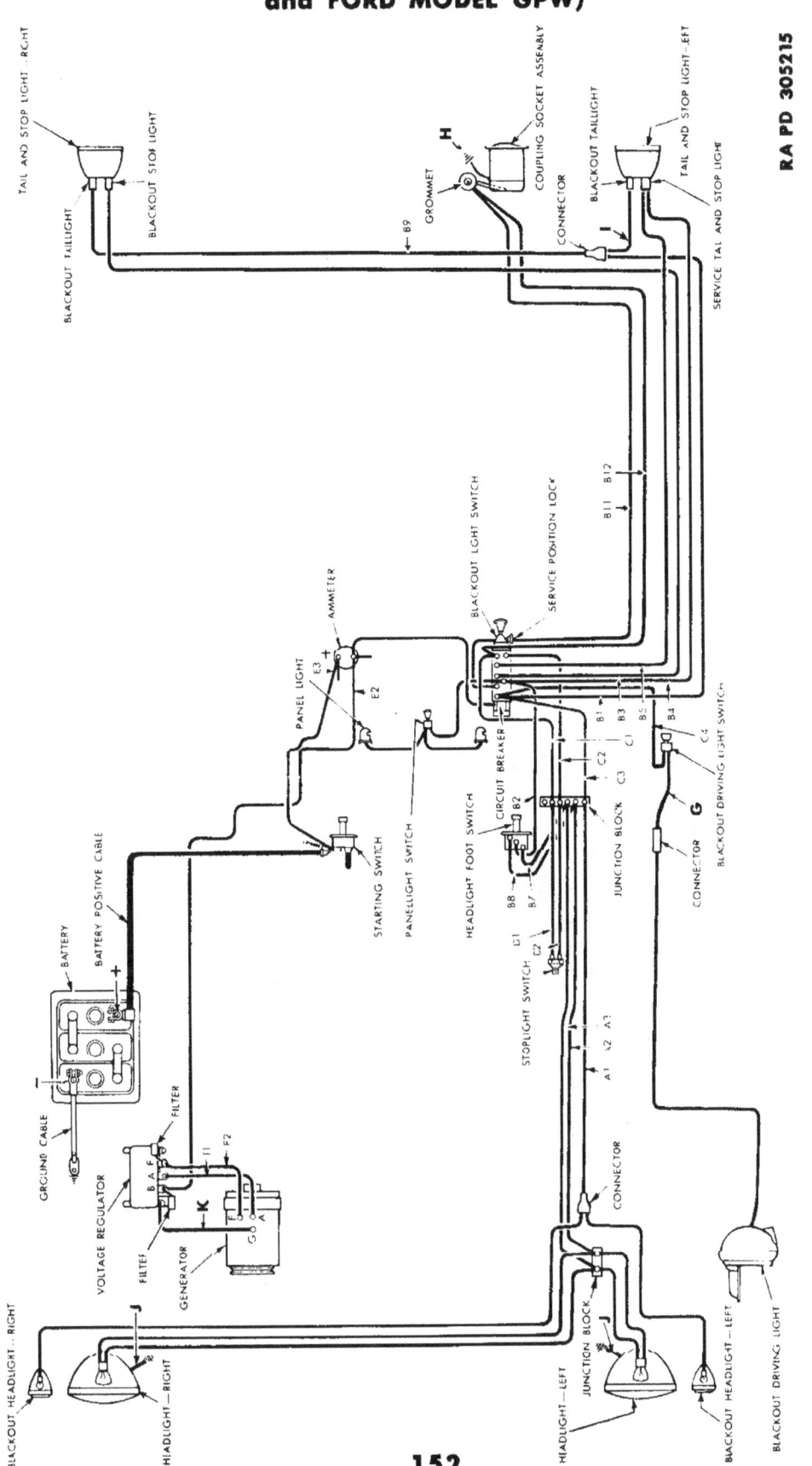

Figure 51—Lighting Circuits

BATTERY AND LIGHTING SYSTEM

		GAGE	COLOR
A—HEADLIGHT WIRING HARNESS			
A-1	BLACKOUT HEADLIGHT JUNCTION BLOCK TO JUNCTION BLOCK	14	YELLOW—2 BLACK TR.
A-2	HEADLIGHT JUNCTION BLOCK TO JUNCTION BLOCK (UPPER BEAM)	12	RED—3 WHITE TR.
A-3	HEADLIGHT JUNCTION BLOCK TO JUNCTION BLOCK (LOWER BEAM)	14	BLACK—2 WHITE TR.
B—BODY WIRING HARNESS—LONG			
B-1	LIGHT SWITCH TERMINAL "B T" TO BLACKOUT TAIL CONNECTION	14	YELLOW—2 BLACK TR.
B-2	LIGHT SWITCH TERMINAL "H T" TO FOOT DIMMER SWITCH CENTER TERMINAL	12	BLUE—3 WHITE TR.
B-3	LIGHT SWITCH TERMINAL "B S" TO BLACKOUT STOPLIGHT	14	WHITE—2 BLACK TR.
B-4	LIGHT SWITCH TERMINAL "H T" TO SERVICE TAILLIGHT AND INSTRUMENT LIGHT SWITCH	14	BLUE—2 WHITE TR.
B-5	LIGHT SWITCH TERMINAL "S" TO SERVICE STOPLIGHT	14	RED—2 WHITE TR.
B-6	HORN CIRCUIT BREAKER TO HORN	14	BLACK—2 RED TR.
B-7	JUNCTION BLOCK TO FOOT DIMMER SWITCH (LOWER BEAM)	14	BLACK—2 WHITE TR.
B-8	JUNCTION BLOCK TO FOOT DIMMER SWITCH (UPPER BEAM)	12	RED—3 WHITE TR.
B-9	CONNECTOR TO BLACKOUT TAILLIGHT	14	YELLOW—2 BLACK TR.
B-11	LIGHT SWITCH TERMINAL "T" TO COUPLING SOCKET TERMINAL "T L"	14	GREEN—2 BLACK TR.
B-12	LIGHT SWITCH TERMINAL "S S" TO COUPLING SOCKET TERMINAL "S L"	14	RED—2 BLACK TR.
C—BODY WIRING HARNESS—LEFT SIDE—SHORT			
C-1	JUNCTION BLOCK TO LIGHT SWITCH TERMINAL "S S"	14	RED—2 WHITE TR.
C-2	JUNCTION BLOCK TO LIGHT SWITCH TERMINAL "S W"	14	GREEN—2 BLACK TR.
C-3	JUNCTION BLOCK TO LIGHT SWITCH TERMINAL "B H T"	14	YELLOW—2 BLACK TR.
C-4	BLACKOUT LIGHT SWITCH TO LIGHT SWITCH TERMINAL "B H T"	14	BLACK—2 WHITE TR.
D—CHASSIS WIRING HARNESS—LEFT			
D-1	STOPLIGHT SWITCH TO JUNCTION BLOCK	14	RED—2 WHITE TR.
D-2	STOPLIGHT SWITCH TO JUNCTION BLOCK	14	GREEN—2 BLACK TR.
E—BODY WIRING HARNESS—RIGHT			
E-1	COIL TO IGNITION SWITCH TO GASOLINE GAGE CIRCUIT BREAKER	14	BLACK—2 WHITE TR.
E-2	VOLTAGE REGULATOR TERMINAL "B H T" TO AMMETER	12	RED—3 WHITE TR.
E-3	STARTING SWITCH TO AMMETER	12	BLACK—3 WHITE TR.
E-4	AMMETER TO HORN CIRCUIT BREAKER	14	BLACK—2 RED TR.
F—GENERATOR TO VOLTAGE REGULATOR AND FILTER HARNESS			
F-1	GENERATOR TO REGULATOR ARMATURE	12	RED—3 WHITE TR.
F-2	GENERATOR TO REGULATOR FIELD	14	GREEN—2 BLACK TR.
G—BLACKOUT DRIVING LIGHT CONNECTOR TO SWITCH		14	BLACK—2 WHITE TR.
H—COUPLING SOCKET TO GROUND		14	YELLOW—2 WHITE TR.
I—CONNECTOR TO BLACKOUT TAILLIGHT		14	BLACK—1 WHITE TR.
J—HEADLIGHT GROUND		12	FLEXIBLE BRAIDED STRAP
K—GENERATOR TO VOLTAGE REGULATOR—GROUND			

RA PD 305215B

Legend for Figure 51—Lighting Circuits

¼-TON 4 x 4 TRUCK (WILLYS-OVERLAND MODEL MB and FORD MODEL GPW)

Lamps:
   Headlights .... Sealed unit
   Blackout headlights .... Mazda No. 1245
   Blackout driving light .... Sealed unit
   Taillights and stop lights .... Sealed unit
   Instrument lights .... Mazda No. 51

Trailer connections:
   Make .... Wagner
   Socket model .... No. 3604
   Plug model .... No. 3544

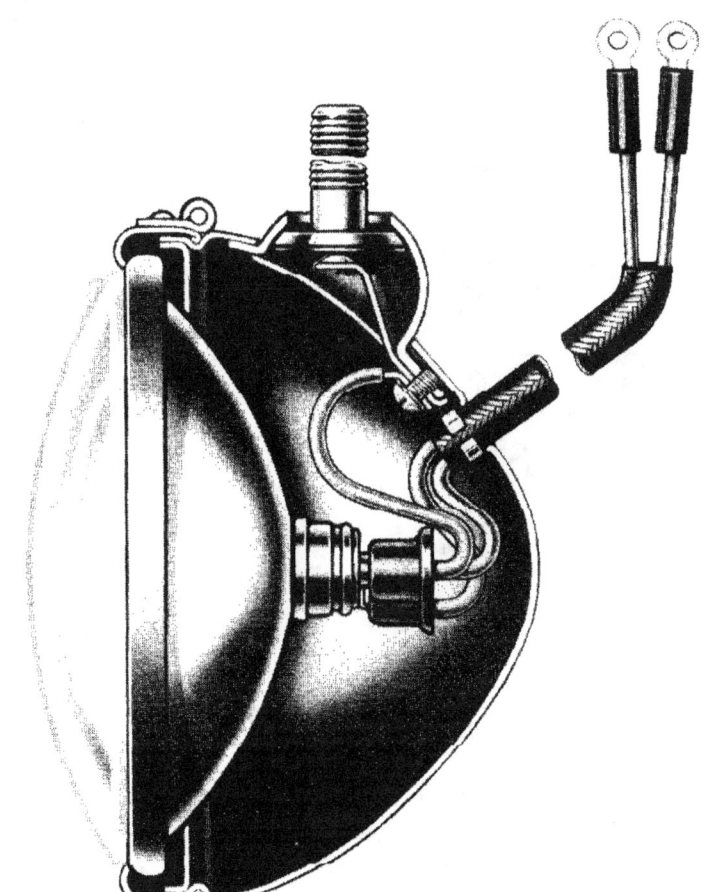

Figure 52— Headlight

RA PD 305216

96. MAINTENANCE.

a. The battery requires periodic checking of the proper level of electrolyte. Battery case must be kept clean. All vent plugs must be tight and breather holes kept open. Keep battery terminals and posts clean and securely tightened. Clean battery carrier when corroded. Tighten battery hold-down clamps. All light mounting screws and

BATTERY AND LIGHTING SYSTEM

nuts must be kept clean and tight. Light lenses and reflectors must be kept clean and securely fastened. Headlights must be aimed properly. All loose and dirty electrical connections must be cleaned and tightened. All damaged or frayed wires must be replaced. Bond straps must be clean and securely tightened.

97. BATTERY.

a. Description. The battery (fig. 36) is a 6-volt, 116-ampere-hour storage battery consisting of three side-by-side cells of 15 plates each. The battery is located under the hood at the right side,

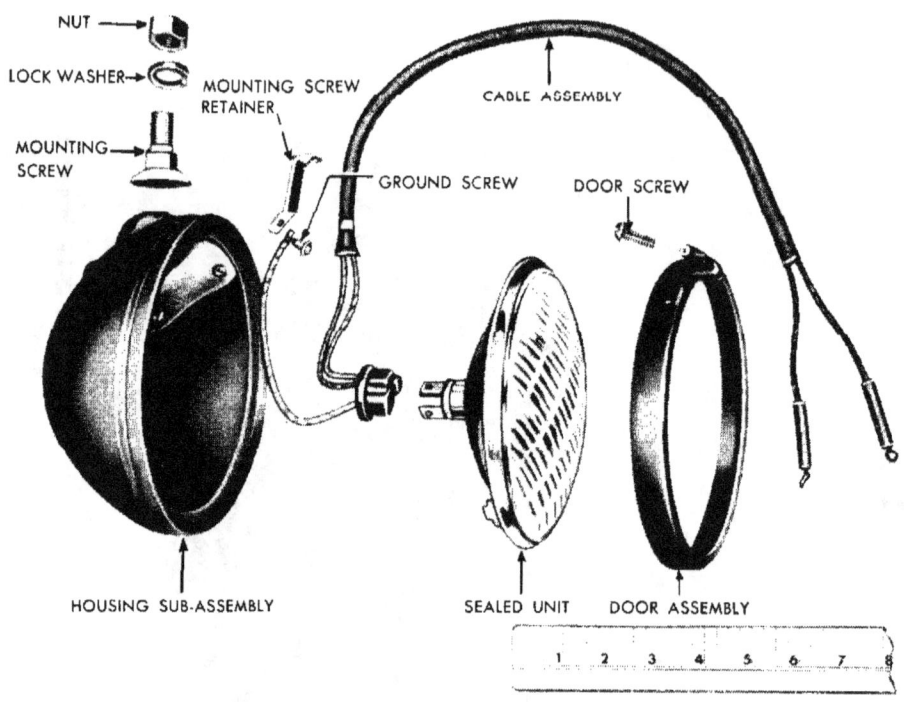

Figure 53—Headlight, Disassembled

and has the negative (small) post grounded. In normal temperatures the battery should be recharged when the specific gravity reads 1.175 or lower, and is fully charged when the specific gravity is 1.275 to 1.285.

b. Removal. Loosen bolts in cable terminals, and remove terminals from battery. Loosen two nuts holding frame on battery, and remove hold-down frame. Lift out battery.

c. Installation. Place battery in position with positive post to rear. Place hold-down frame in position, and tighten securely with nuts. Clean cable terminals; if necessary; grease and install on bat-

TM 9-803
97-98
¼-TON 4 x 4 TRUCK (WILLYS-OVERLAND MODEL MB and FORD MODEL GPW)

tery posts. Check level and specific gravity of electrolyte in battery. Electrolyte level must be ½ inch above plates, and specific gravity must be 1.275 to 1.285 at 80°F.

98. WIRING SYSTEM.

a. Description. The entire vehicle wiring system is shown in figure 50, and the circuits of the lighting system are shown in figure 51. A single-wire system is used, and the negative terminal of the battery is grounded. A single cable connects the battery positive post to the upper terminal of the starting switch, to which are also attached wires running to the radio outlet box, filter, and right-hand terminal of the ammeter, where the horn wire is connected. From the left-hand ammeter terminal wires run to the voltage regulator, ignition switch, and blackout (main) light switch, which controls

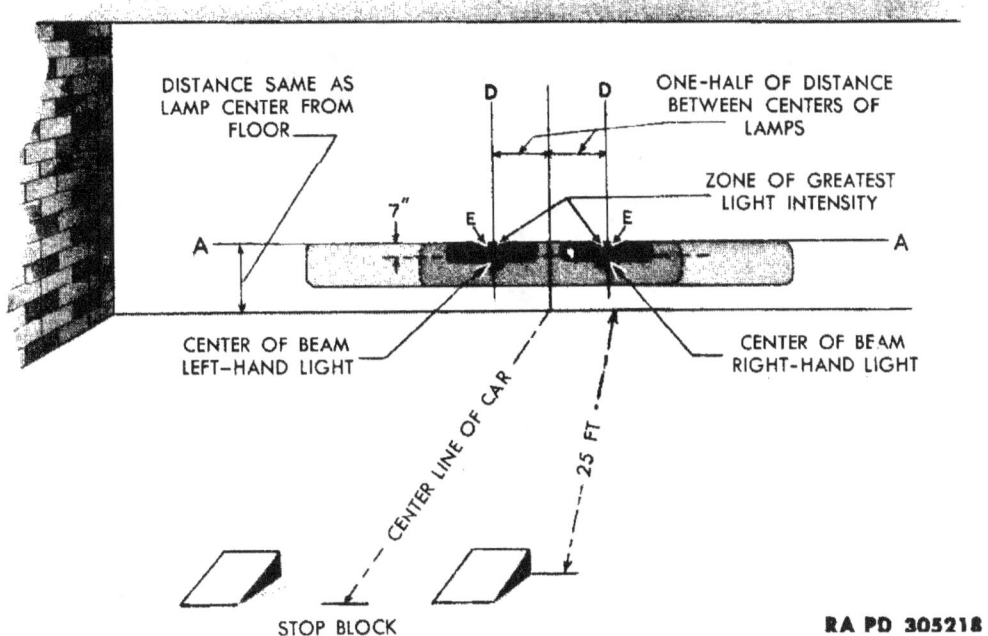

Figure 54—Headlight Aiming Chart

all lights in the lighting system. The wiring system consists of the harnesses and wires which are identified by colors and tracers as shown in figure 50, and described in the accompanying legend.

b. Removal. When necessary to renew wires in the wiring system, remove grounded (negative) cable at battery. Note location of wires; disconnect wires at terminals; loosen clips along harness, and remove wires.

c. Installation. When installing wires and harnesses see that all connections are correctly located and secure. Double-check location and tightness of connections; install retaining clips along wires.

BATTERY AND LIGHTING SYSTEM

99. HEADLIGHTS.

a. Description. Two headlights (fig. 52) are mounted at the front of the vehicle and are protected by the radiator guard. Each headlight is mounted on a hinged bracket. By loosening the wing nut, the light can be swung over and used as a trouble light in the engine compartment. The headlights are of the double filament sealed beam-unit type consisting of reflector, lamp, and lens, which can be

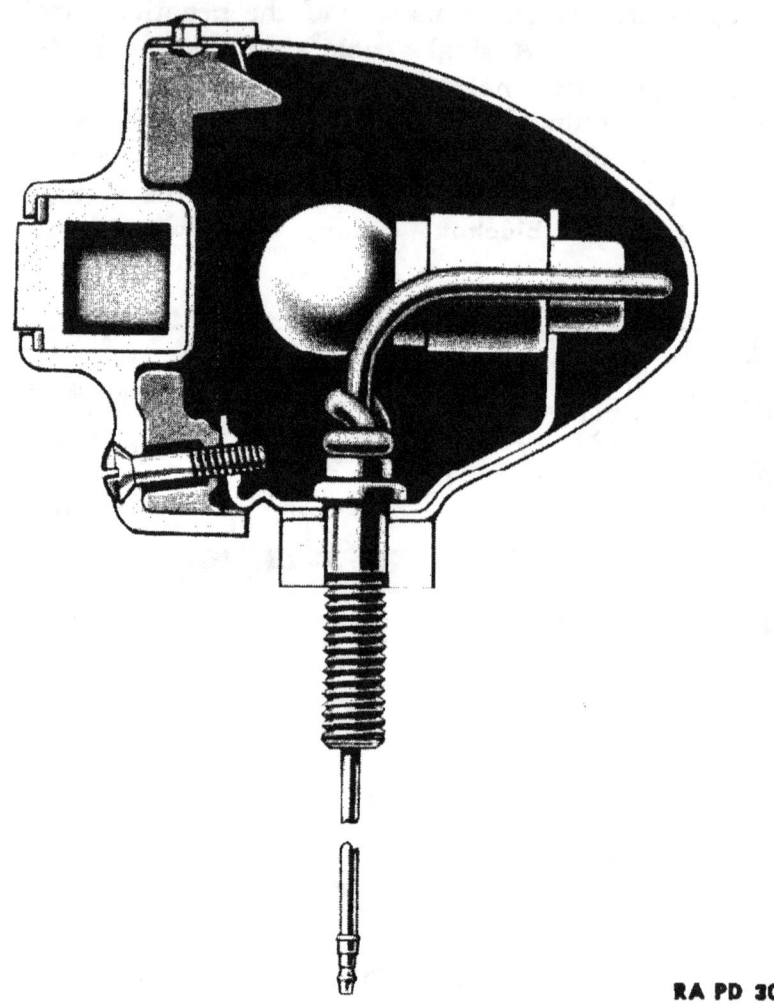

Figure 55—Blackout Headlight

replaced only as a unit (Mazda No. 2400). The headlights light when the blackout lighting switch is in service position. The upper and lower beams are controlled by the foot-operated dimmer switch. The lower beam filament is positioned slightly to one side of the focal point deflecting the beam slightly to illuminate the right side of the road.

b. Removal of Sealed Beam-unit. Remove door clamp screw and remove door (fig. 53). Remove sealed beam assembly, and pull connector from back of unit.

TM 9-803
99
¼-TON 4 x 4 TRUCK (WILLYS-OVERLAND MODEL MB and FORD MODEL GPW)

c. Installation of Sealed Beam-unit. Attach connector to back of unit. Place unit in position in headlight housing. Install door, and secure in place with screw.

d. Removal of Headlight. Disconnect headlight wires at junction block on left fender splash under hood. Remove clip on radiator grille (two on right headlight wire). Remove nut on headlight. Loosen headlight support wing nut, raise support, and remove headlight.

e. Installation of Headlight. Mount the headlight on the support. Install wire clip on grille (two on right headlight wire). Connect wires to junction block (red wire on bottom terminal; black wire on top terminal). Lower support, and fasten with wing nut. Aim headlight (subpar. **f** below).

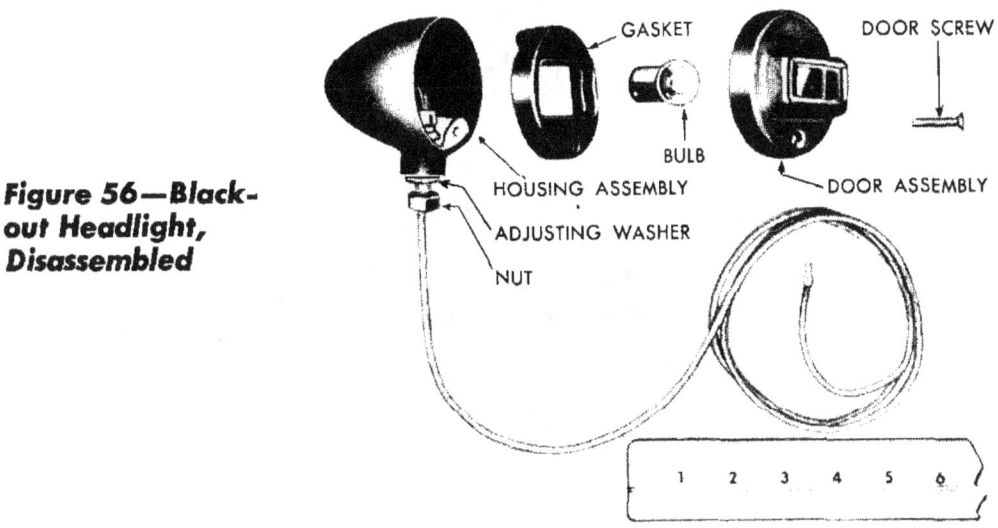

Figure 56—Blackout Headlight, Disassembled

RA PD 305220

f. Aiming Headlights. Aim headlights by using headlight aimer, aiming screen, or wall shown in figure 54. NOTE: *Use a light background with black center line for centering vehicle on screen. Mark two vertical lines on each side of center line equal to the distance between headlight centers. Mark screen 7 inches less than the height of centers of headlights.* Inflate all tires to recommended pressure (par. 3). Set vehicle so headlights are 25 feet away from screen and center line of vehicle is in line with center line of screen. To determine center line of vehicle, stand at rear and sight through windshield down across cowl and hood. Turn on headlight upper beam, cover one headlight, and observe location of upper beam on screen. Adjust headlight so that center of bright light area is on intersection of vertical and horizontal lines. Tighten headlight mounting nut. Cover headlight just aimed, and adjust other in same manner.

BATTERY AND LIGHTING SYSTEM

100. BLACKOUT HEADLIGHTS.

a. Description. There are two blackout headlights (fig. 55) with lens which permit only horizontal rays to pass through. These headlights are illuminated only when the blackout (main) light switch is in blackout position. Mazda No. 1245 lamps are used.

b. Removal of Lamp. Remove door screw in lower side of rim and remove door by pulling out on lower side (fig. 56). NOTE: *The door and lens are in one unit.* Push in on lamp, turn lamp to left and remove.

c. Installation of Lamp. Insert lamp in socket, push in and turn lamp to right. Replace door gasket, if damaged. Replace door and tighten door screw.

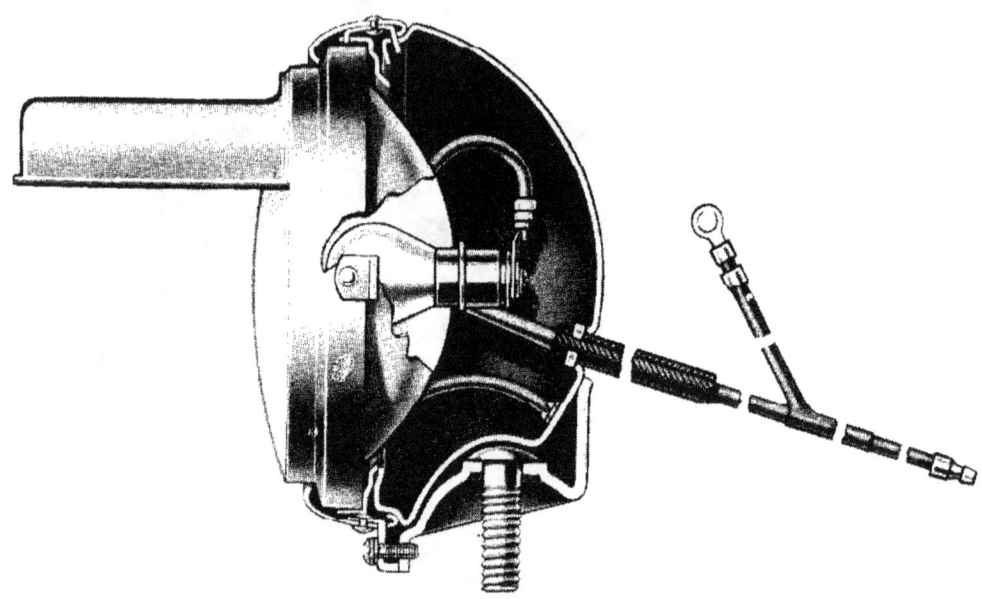

RA PD 305221

Figure 57—Blackout Driving Light

d. Removal of Blackout Headlight. Pull wire out of connection just behind left headlight. (For right headlight remove three clips across bottom of grille.) Remove mounting nut by reaching in from the rear and lift out headlight.

e. Installation of Blackout Headlight. Put light in position and tighten in place with mounting nut. (Attach three clips for right headlight.) Attach wire in connector.

101. BLACKOUT DRIVING LIGHT.

a. Description. The blackout driving light (fig. 57) is a hooded light having a sealed unit which throws a horizontal diffused beam to illuminate any vertical object. This light is mounted on the left front fender, and is controlled by a push-pull type switch (marked "B.O. DRIVE") in the instrument panel. The blackout driving light

¼-TON 4 x 4 TRUCK (WILLYS-OVERLAND MODEL MB and FORD MODEL GPW)

can be illuminated only when the blackout (main) light switch is in blackout position.

b. Removal of Sealed Unit. Remove screw at bottom of light door (fig. 58). Pull bottom of door forward and up to remove. Disconnect socket wire, and remove door by releasing mounting ring.

c. Installation of Sealed Unit. Place sealed unit in door, and install unit mounting ring. Attach wire connection. Install door on housing, and tighten retaining screw.

d. Removal of Blackout Driving Light. Pull wire out of connector at dash. Remove three wire clips, and pull wire through fender. Remove mounting nut on bottom of light, and lift off light.

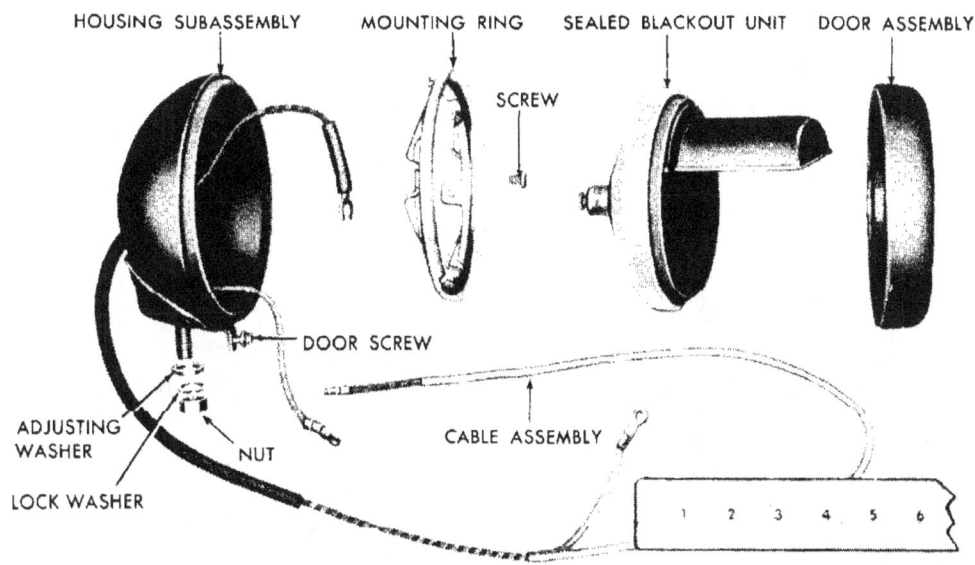

Figure 58 — Blackout Driving Light, Disassembled

e. Installation of Blackout Driving Light. Install light on bracket and tighten nut; run wire through fender and splasher; attach three clips, and push wire into connector at dash.

f. Adjustment. The light is aimed with vehicle on a level surface and loaded. Hold a 4-foot wood stick on floor close to the light, and mark stick where bright spot appears. Move stick 10 feet ahead of light; bright spot should be 2.1 inches lower than mark.

102. TAILLIGHTS AND STOP LIGHTS.

a. Description. Two combination taillights and stop lights (fig. 59) are located in the rear panel of the body. Each light consists of two separate units in a housing. The left-hand light contains a combination service tail and stop light unit in the upper part, and a blackout taillight in the lower part. The upper unit consists of

BATTERY AND LIGHTING SYSTEM

the taillight lens, gasket, reflector, and a 21-3-candlepower lamp. The lower unit consists of blackout lens, gasket, reflector, and 3-candlepower lamp. The right-hand light contains a blackout stop light unit in the upper part, and a blackout taillight in the lower part. The upper unit consists of the blackout stop light lens, gasket, reflector, and 3-candlepower lamp. The lower unit is the same as the lower unit in the left light. When a lamp burns out, the unit must be replaced. These lights are controlled by the blackout (main) light switch.

b. Removal of Light Unit. Remove two screws in light door, and remove door (fig. 60). Pull each unit straight out of socket.

c. Installation of Light Unit. Be sure that unit is correct type, and push into socket. Install light door and screws.

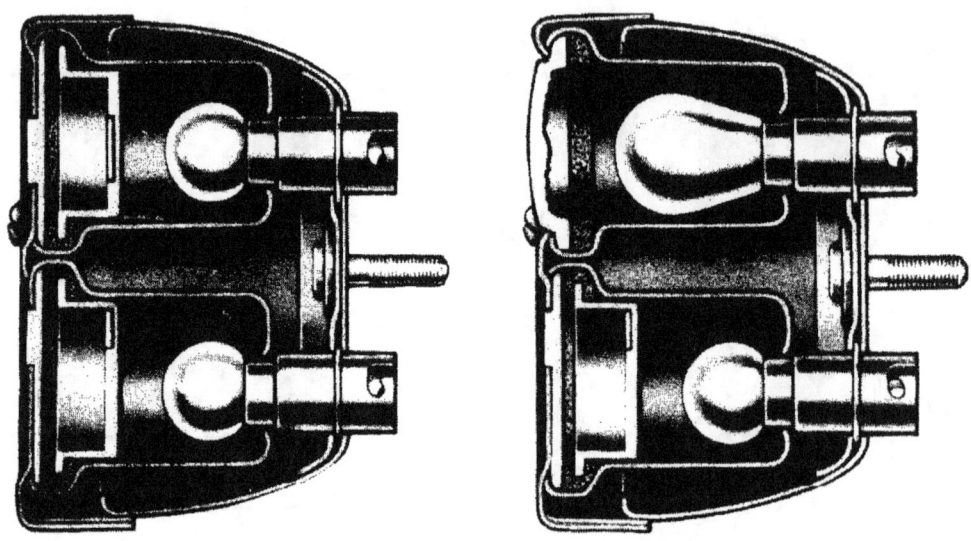

RA PD 305223

Figure 59 — Taillights and Stop Lights

d. Removal of Light. Reach up under body and disconnect wire connector; push in on connector, turn counterclockwise and pull connector out of socket. Remove two nuts holding light to bracket, and remove light.

e. Installation of Light. Place light in position and secure with nuts. Attach connectors. Double contact connector goes in upper socket of left light. Turn on blackout (main) light switch to blackout position to see if right blackout taillight (lower unit) lights; if not, interchange connectors in sockets.

103. INSTRUMENT PANEL LIGHTS.

a. Description. There are two instrument panel lights (fig. 5) located above instruments and on the outside of the instrument panel. They are controlled by the panel light switch when the blackout (main) light switch is in service position.

TM 9-803
103
¼-TON 4 x 4 TRUCK (WILLYS-OVERLAND MODEL MB and FORD MODEL GPW)

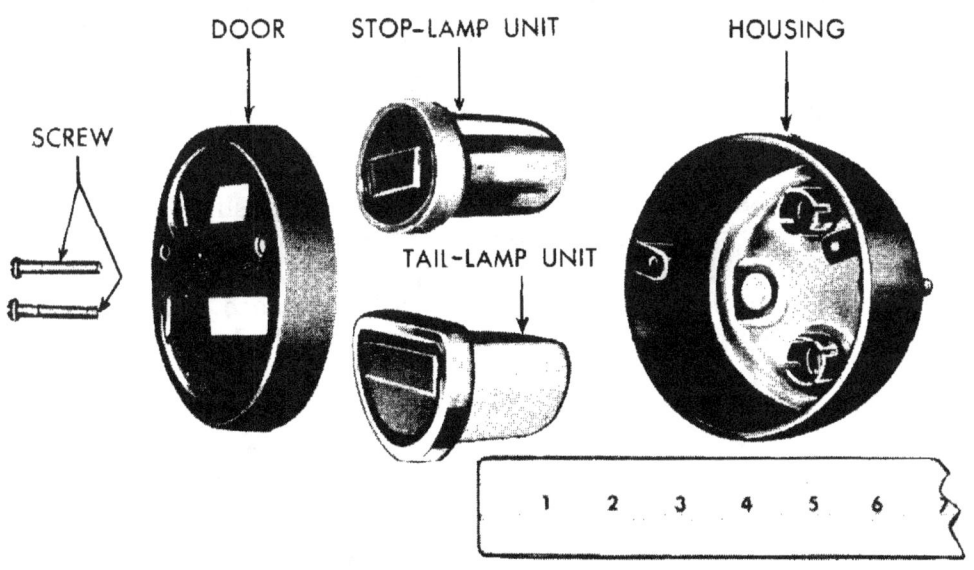

Figure 60 — Taillights and Stop Lights, Disassembled

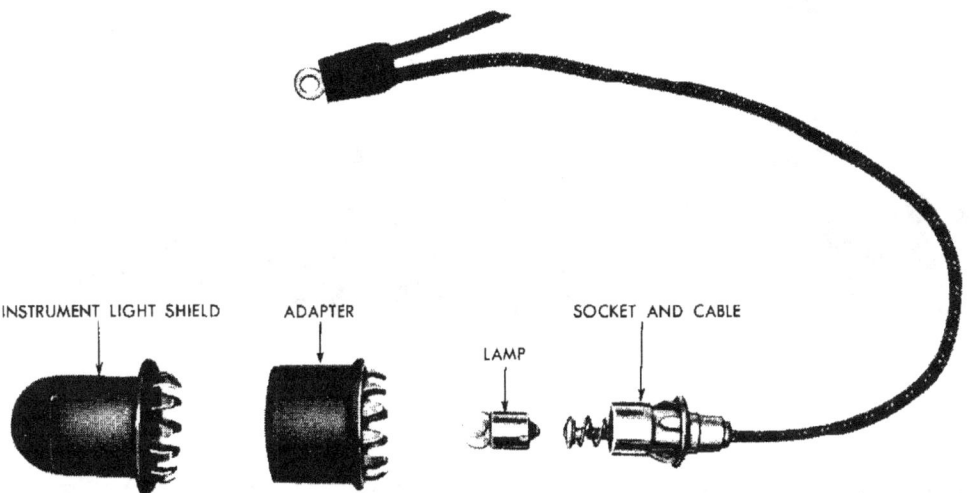

Figure 61 — Instrument Light, Disassembled

b. Removal of Lamp. Pry off shield by using a sharp tool behind flange (fig. 61). Pull light socket out of shield; press in on lamp, turn lamp counterclockwise, and remove.

c. Installation of Lamp. Put lamp in socket, push lamp in, and turn clockwise. Push socket into shield, and push shield into instrument lamp adapter.

BATTERY AND LIGHTING SYSTEM

d. Removal of Instrument Light. Disconnect wire on back of light switch, and dismantle as outlined in subparagraph **b** above. Remove wire and sockets for both lamps under instrument panel.

e. Installation of Instrument Light. Attach wire terminal to panel light switch, and place sockets through holes in instrument panel. Assemble light as outlined in subparagraph **c** above.

104. BLACKOUT (MAIN) LIGHT SWITCH.

a. Description. The blackout light switch is a push-pull type (fig. 62), mounted in the instrument panel to the left of the steering gear (fig. 5). The switch controls all the lights, and has four positions (fig. 9). With the knob all the way in, all lights are off. Pull knob out to first (blackout) position to illuminate blackout headlights,

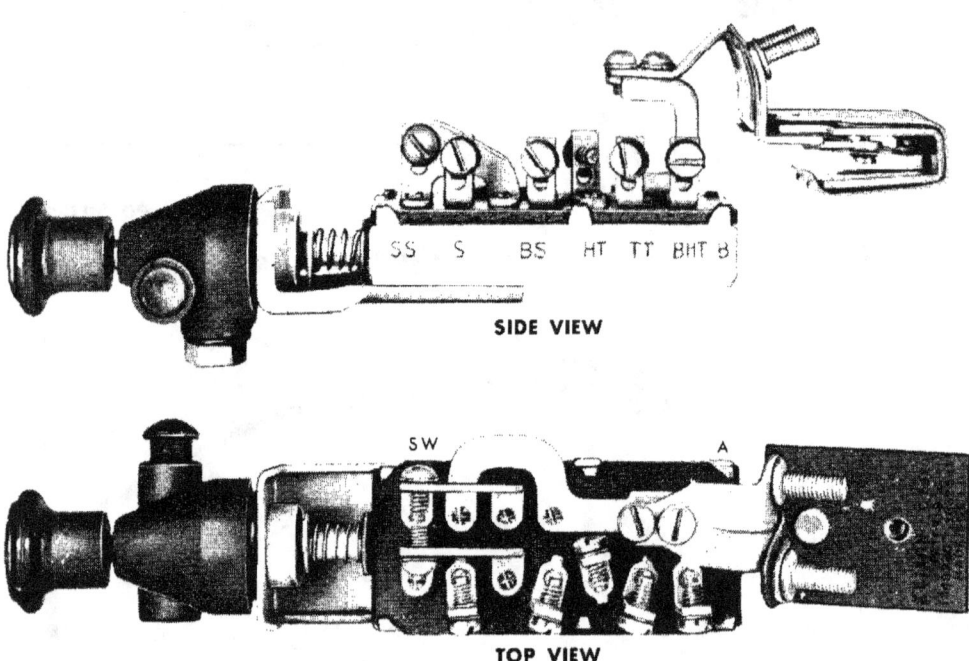

Figure 62—Blackout (Main) Light Switch

blackout taillights, and establish connection so blackout stop light will function when foot brake is applied. Press lockout control button, and pull switch knob out to second (service) position to illuminate service headlights and taillight, and to establish connections so service stop light will function when foot brake is applied. Pull knob out to third (service stop light) position, when vehicle is operated during daylight, to cause stop light only to function when foot brake is applied. A thermal-type fuse is mounted on the back of the switch having a bimetal spring, which causes a set of contact points to open and close, if a short circuit occurs in the lighting system.

b. Removal. Disconnect ground cable at negative post of battery. Loosen set screw in switch knob and unscrew knob. Loosen

TM 9-803
104-105

¼-TON 4 x 4 TRUCK (WILLYS-OVERLAND MODEL MB and FORD MODEL GPW)

hexagon head screw at side of switch bushing on front of panel: press lockout control button, and pull off bushing. Remove mounting nut, and take switch out from under panel. Remove wire terminal screws. As wires are removed, mark them for identification.

c. Installation. Attach wires to proper terminals. NOTE: *Switch terminals are marked for easy identification.* Wires are to be attached as outlined below:

Figure 63—Light Switch

RA PD 305227

Terminal	Circuit	Wire Color
A	Extra	Not used
B	Ammeter	Red-white tr.
BHT	Blackout headlights	Yellow-black tr.
	Blackout taillights	Yellow-black tr.
	Blackout driving light	Black-white tr.
BS	Blackout stop light	White-black tr.
HT	Service headlight	Blue-white tr.
	Service taillight	Blue-white tr.
	Panel lights	Blue-white tr.
S	Service stop light	Red-white tr.
SS	Blackout driving light	Red-white tr.
	Trailer coupling socket	Red-black tr.
SW	Stop light	Green-black tr.
TT	Trailer coupling socket	Green-black tr.

Install switch in panel and secure in place with mounting nut. Install bushings and tighten screw. Install knob and tighten set screw. Attach ground cable to battery.

105. PANEL AND BLACKOUT DRIVING LIGHT SWITCHES.

a. Description. The panel and blackout driving light switches

TM 9-803
105
BATTERY AND LIGHTING SYSTEM

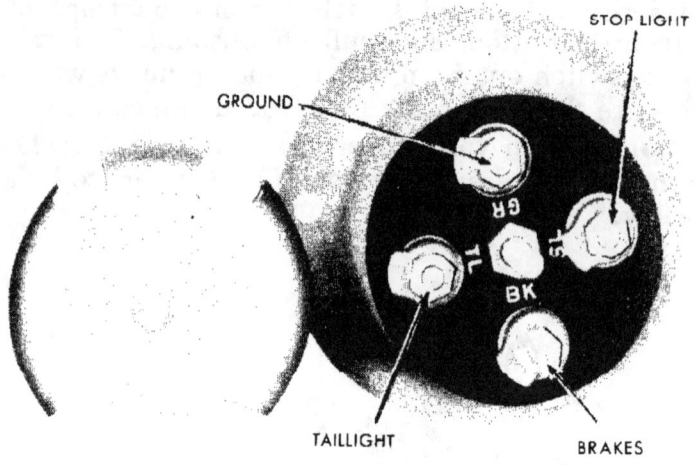

Figure 64—Trailer Socket Terminals

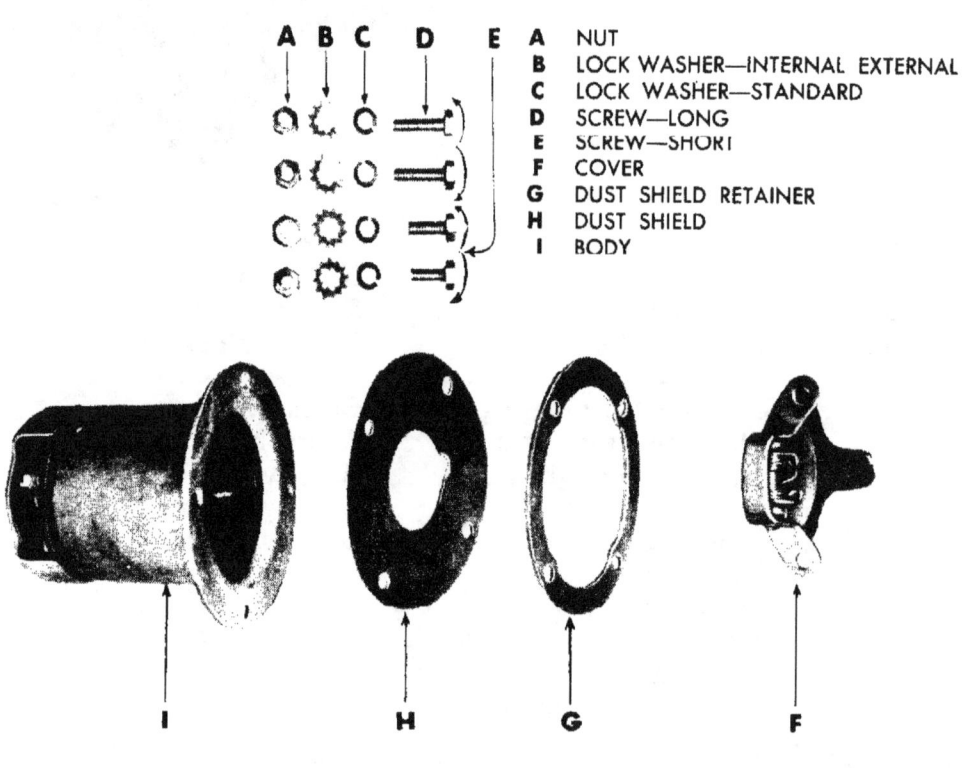

Figure 65—Trailer Socket, Disassembled

TM 9-803
105-106

¼-TON 4 x 4 TRUCK (WILLYS-OVERLAND MODEL MB and FORD MODEL GPW)

are push-pull (off-and-on) type (fig. 63). The panel light switch knob is marked "PANEL LIGHTS"; the blackout driving light switch is marked "B.O. DRIVE." The panel light switch controls these lights only when the blackout (main) light switch is in the service position. The blackout driving light switch controls that light only when the blackout (main) light switch is in blackout position.

b. Removal. As a precaution remove ground cable from battery. Loosen set screw in knob and unscrew knob. Remove mounting nut, and remove switch from in back of panel. Remove screws in wire terminals, and disconnect wires.

c. Installation. Attach wires to switch, install in place in panel, and secure with mounting nut. Install knob and secure with set screw so that marking on knob is in correct position.

106. TRAILER CONNECTION.

a. Description. The trailer connection is a socket located in the body rear panel at the left side (fig. 2). The blackout (main) light switch controls the current to socket and trailer lights.

b. Removal. Remove equipment from left rear tool compartment. Remove screw at top edge of protecting cover, and remove cover. Remove four bolts holding socket in body panel. Pull out socket, and remove cover over the terminals. Remove wires, noting their proper place.

c. Installation. Attach wires to terminals. NOTE: *Attach green wire to terminal "TL"; attach red wire to terminal "SL"; attach small terminal of black wire to terminal "GR."* Install cover over terminals. To install socket in body, install the two long mounting screws in socket cover hinge. Place dust shield retainer ring over screws, followed by the dust shield with slot in center opening, opposite the hinge. Place cover against outside of body panel with the two screws through upper mounting holes so that cover opens upward. Place internal-external lock washers on each bolt. Install socket with drain hole down, and install lock washers and nuts loosely. NOTE: *Toothed lock washers must be installed as directed for a good ground connection.* Install the lower bolts with the lock washers in correct position. Install ground wire on a lower bolt, and tighten all four nuts holding socket in body panel. Install protecting cover, and secure in place with screw at top edge to complete the installation.

Section XXI

CLUTCH

	Paragraph
Description and data	107
Maintenance	108
Pedal adjustment	109
Removal	110
Installation	111
Clutch release bearing	112

107. DESCRIPTION AND DATA.

a. Description. The clutch (fig. 66) located between the engine and the transmission is a single plate, dry-disk type. The clutch consists principally of two units, the clutch driven plate which has a spring-center vibration neutralizer, and the clutch pressure plate unit, which is bolted to the flywheel. The controlled pressure of the driven plate against the flywheel provides a means of engaging and disengaging the engine power to the transmission. A ball-type release bearing operates three clutch release levers to control the clutch. The release bearing is controlled by a rod and cable to the clutch pedal. This type clutch has only one service adjustment. This adjustment is for the foot pedal, and regulates the amount of free pedal travel. As the clutch facings wear, adjust the free pedal travel to three-quarters of an inch (fig. 68).

b. Data.

Type	Dry single plate
Torque capacity driven plate	132 foot-pounds
Make	Borg and Beck No. 11123
Size	7⅞ in.
Facings	1 woven and 1 molded
Diameter	5⅛ in. inside, 7⅞ in. outside
Thickness	⅛ in. (0.125)
Pressure plate:	
Make	Atwood No. TP-2B-7-1
Number of springs	3
Clutch release bearing type	Prelubricated ball
Clutch shaft bushing (in flywheel) size	I.D. 0.628 in.
Clutch pedal adjustment (free play)	¾ in.

108. MAINTENANCE.

a. The clutch requires attention to pedal adjustment. Pedal adjustment must be periodically checked, due to the natural wear of lining. Report grabbing or slipping condition of clutch to higher authority.

109. PEDAL ADJUSTMENT.

a. Adjust Pedal. Loosen clutch control cable adjusting yoke lock nut (figs. 67 and 68). Using a wrench, unscrew cable until clutch pedal has ¾ inch free play. Tighten lock nut.

TM 9-803
¼-TON 4 x 4 TRUCK (WILLYS-OVERLAND MODEL MB and FORD MODEL GPW)

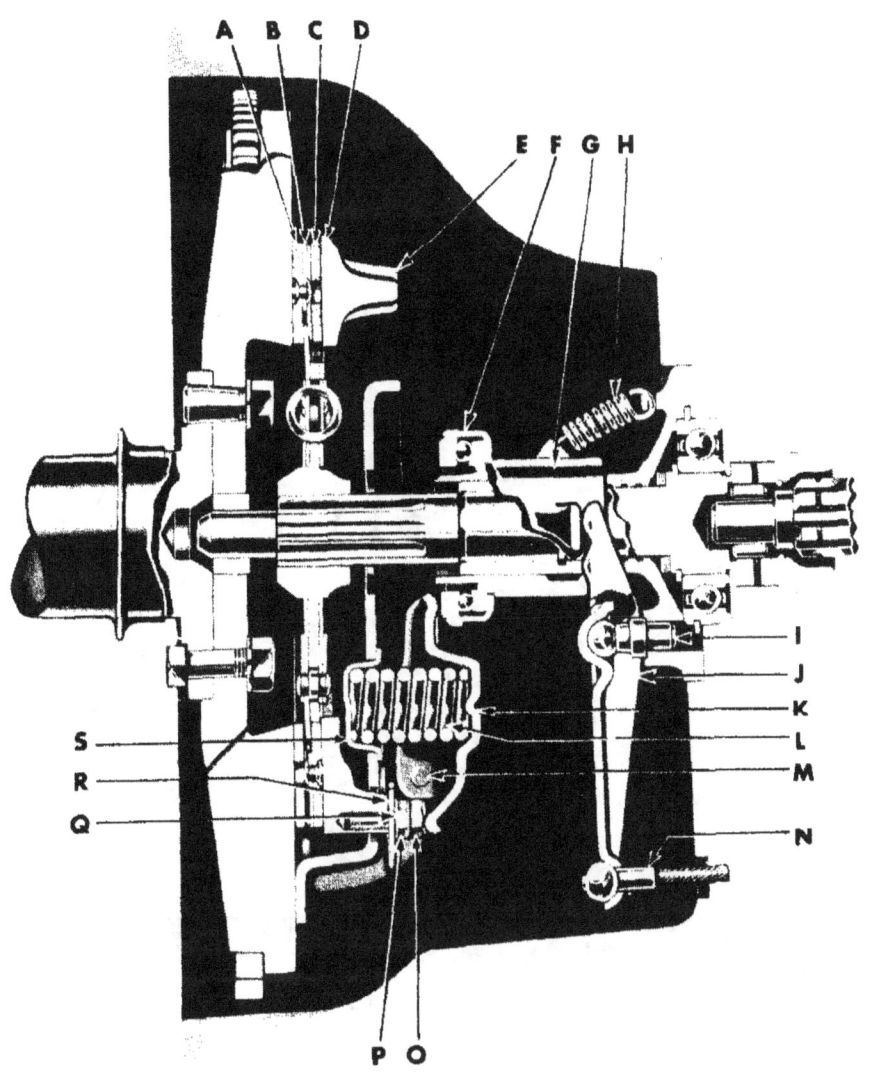

- **A** FACING—FRONT
- **B** DRIVEN PLATE ASSEMBLY
- **C** FACING—REAR
- **D** PRESSURE PLATE
- **E** CLUTCH PRESSURE PLATE ASSEMBLY
- **F** RELEASE BEARING
- **G** RELEASE BEARING CARRIER
- **H** RELEASE BEARING CARRIER SPRING
- **I** CLUTCH CONTROL LEVER FULCRUM
- **J** CLUTCH CONTROL LEVER
- **K** CLUTCH LEVER
- **L** PRESSURE SPRING
- **M** CLUTCH LEVER PIVOT PIN
- **N** CONTROL LEVER CABLE
- **O** ADJUSTING SCREW
- **P** ADJUSTING SCREW LOCK NUT
- **Q** ADJUSTING SCREW WASHER
- **R** PRESSURE PLATE RETURN SPRING
- **S** PRESSURE SPRING CUP

RA PD 305274

Figure 66—Clutch—Sectional View

TM 9-803
109

CLUTCH

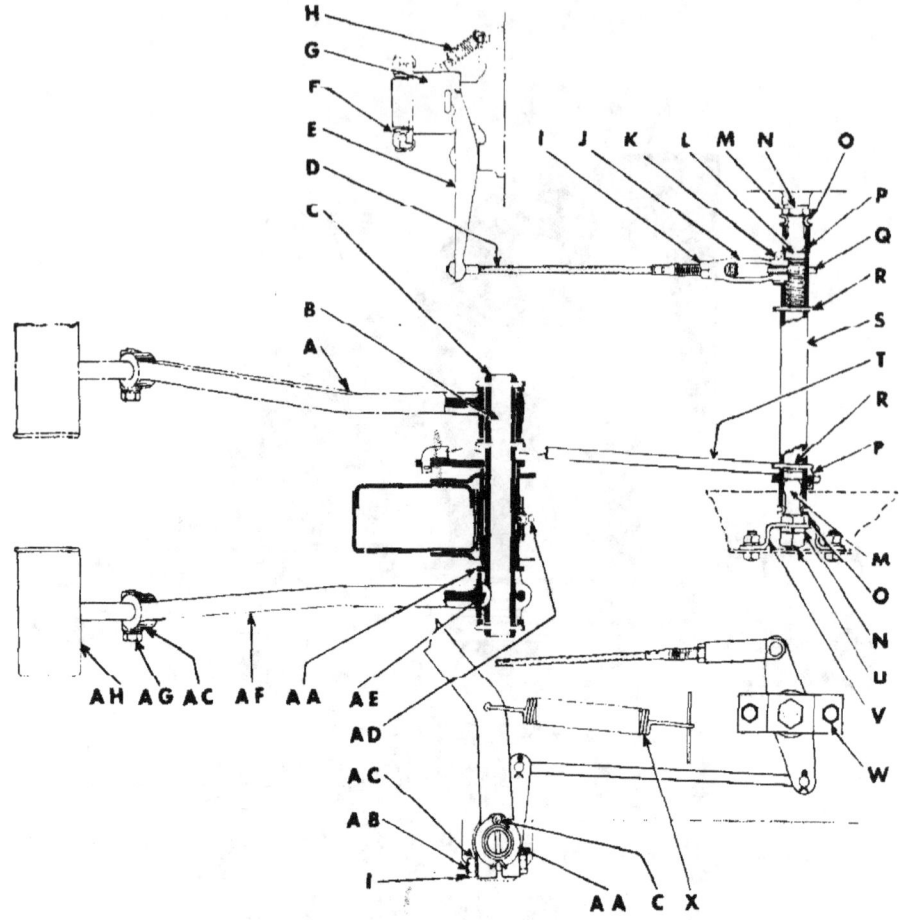

A	BRAKE PEDAL	Q	SPRING
B	PEDAL SHAFT ASSEMBLY	R	TUBE SPRING COTTER PIN
C	PEDAL SHAFT COTTER PIN	S	LEVER AND TUBE ASSEMBLY
D	LEVER CABLE	T	PEDAL ROD
E	LEVER	U	BALL STUD NUT
F	RELEASE BEARING	V	FRAME BRACKET
G	RELEASE BEARING CARRIER	W	FRAME BRACKET SCREW
H	RELEASE BEARING CARRIER SPRING	X	PEDAL RETRACTING SPRING
I	LEVER CABLE YOKE END LOCK NUT	AA	PEDAL SHAFT WASHER
J	LEVER CABLE YOKE END	AB	PEDAL CLAMP BOLT
K	LEVER CABLE CLEVIS PIN	AC	PEDAL CLAMP BOLT LOCK WASHER
L	TUBE WASHER	AD	PEDAL SHAFT HYDRAULIC GREASE FITTING
M	BALL STUD	AE	PEDAL TO SHAFT KEY
N	BALL STUD LOCK WASHER	AF	PEDAL
O	TUBE DUST WASHER	AG	PEDAL CLAMP BOLT
P	TUBE FELT WASHER	AH	PEDAL PAD

RA PD 305286

Figure 67—Clutch Control

¼-TON 4 x 4 TRUCK (WILLYS-OVERLAND MODEL MB and FORD MODEL GPW)

110. REMOVAL.

a. Remove engine assembly (par. 60) or transmission and transfer case assembly (par. 115). NOTE: *The easiest method is to remove the engine assembly.*

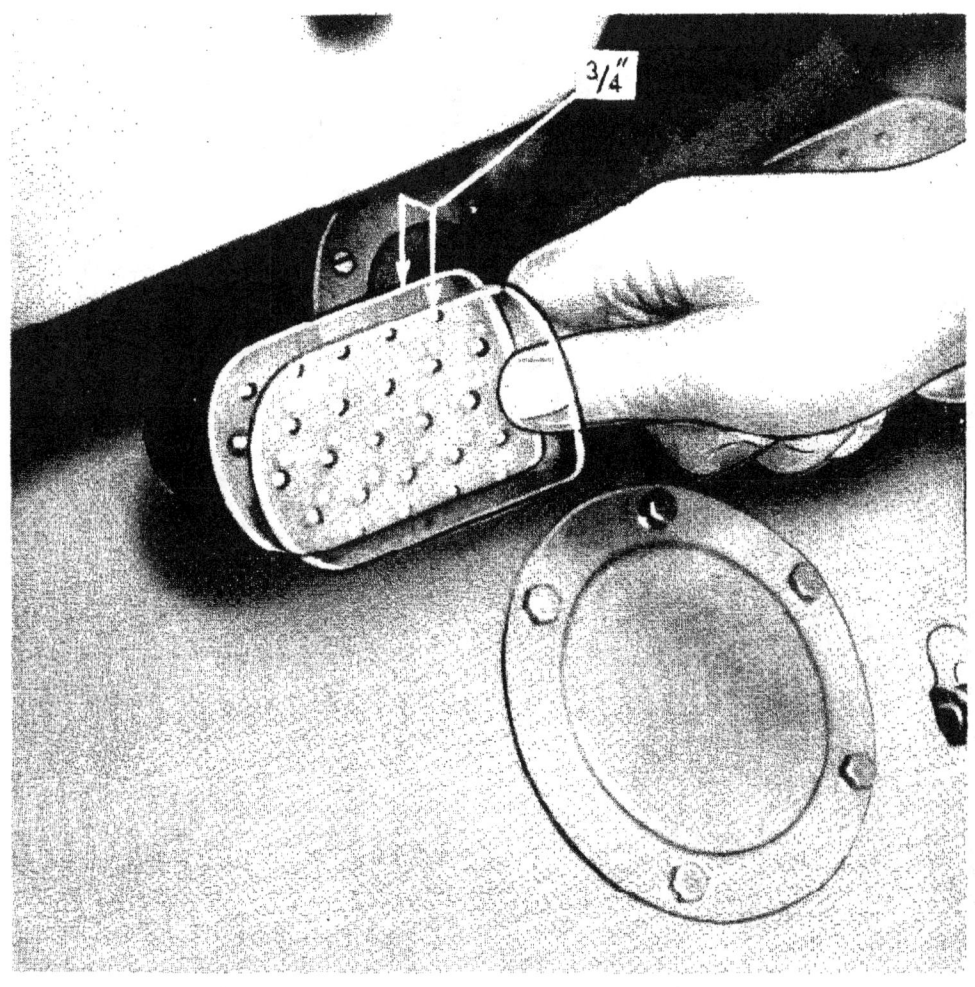

Figure 68—Clutch Pedal Free Travel

b. Mark clutch pressure plate and flywheel to assure correct position when installing. Loosen evenly and remove screws holding pressure plate to flywheel. Remove pressure plate; remove driven plate.

111. INSTALLATION.

a. Clean flywheel and clutch. Install small amount of light grease in clutch shaft flywheel bushing. Install driven plate against flywheel with short end of hub toward flywheel. Install clutch pressure plate loosely with screws. Use a clutch shaft or clutch pilot arbor to aline driven plate, and tighten clutch pressure plate screws evenly. Remove pilot and check adjustment of clutch fingers (fig.

CLUTCH

69), which should be $27/32$ inch. To adjust clutch fingers, loosen lock nut on adjusting screws, and turn screws until measurements from face of fingers (release bearing contacts) to face of clutch bracket measures $27/32$ inch; set lock nuts. Install engine (par. 61), or transmission and transfer case assembly (par. 116), as required.

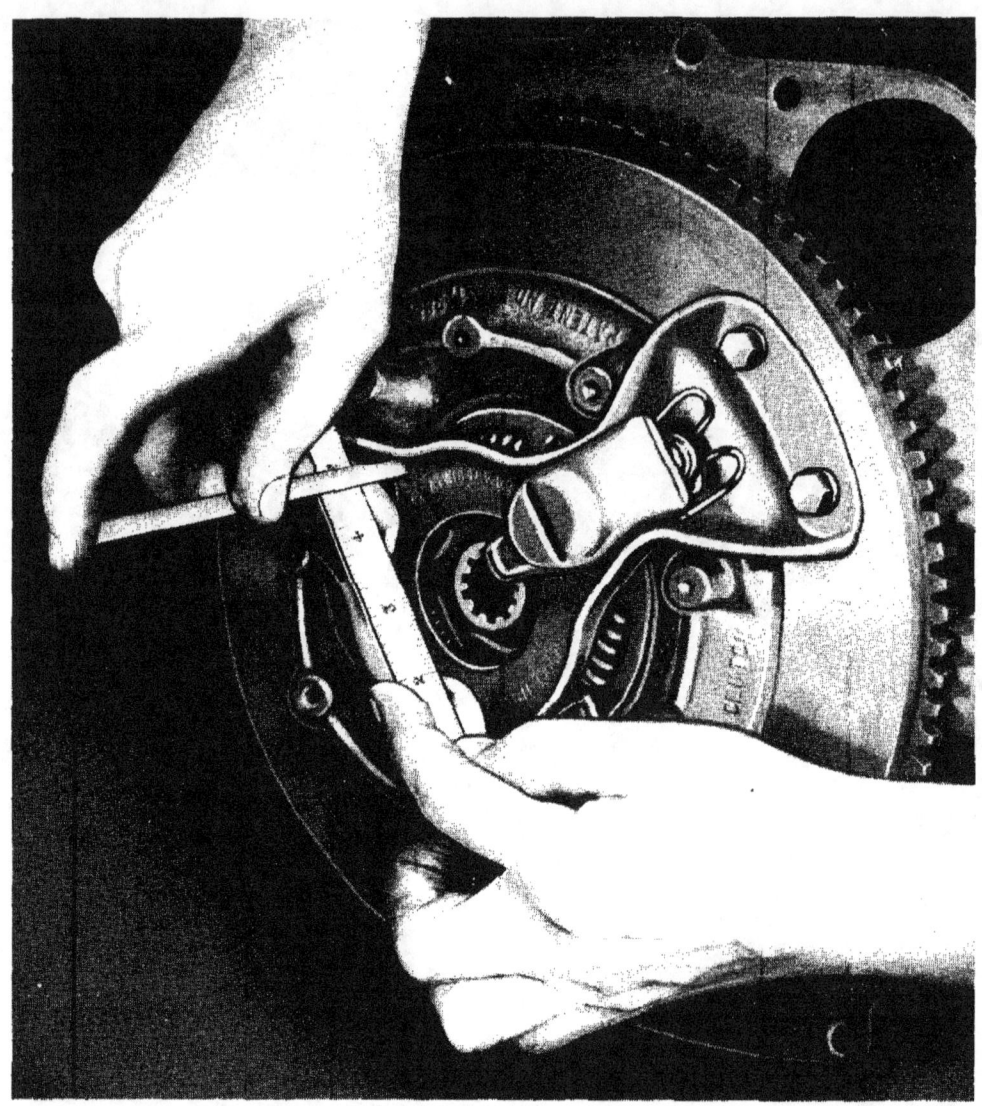

Figure 69—Clutch Finger Adjustment

112. CLUTCH RELEASE BEARING.

a. Removal. Follow procedure outlined in paragraph 60 for removing engine. After release bearing can be reached, unhook release bearing carrier spring, and pull off bearing and carrier (fig. 66). Press carrier out of bearing.

b. Installation. Press carrier into bearing. Slip bearing and carrier onto transmission bearing retainer, install release lever on fulcrum, and hook spring to bearing carrier. Follow the procedure outlined in paragraph 61 for completion of assembly.

¼-TON 4 x 4 TRUCK (WILLYS-OVERLAND MODEL MB and FORD MODEL GPW)

Section XXII

TRANSMISSION

	Paragraph
Description and data	113
Maintenance	114
Removal	115
Installation	116

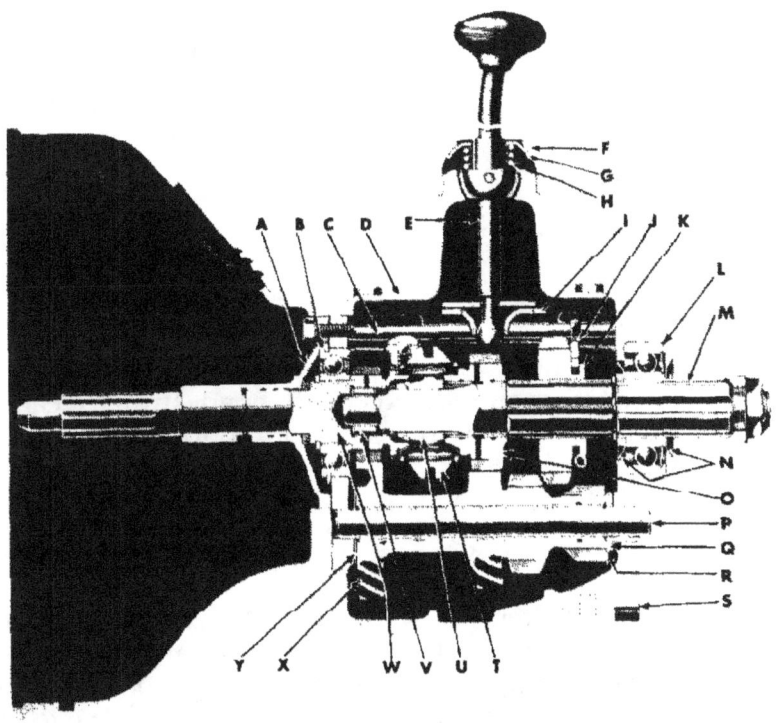

A — MAIN DRIVE GEAR BEARING RETAINER
B — MAIN DRIVE GEAR BEARING
C — SHIFT RAIL—LOW AND REVERSE
D — CONTROL HOUSING ASSEMBLY
E — CONTROL LEVER ASSEMBLY
F — CONTROL HOUSING CAP
G — CONTROL HOUSING CAP WASHER
H — CONTROL LEVER SUPPORT SPRING
I — SHIFT PLATE
J — LOW AND REVERSE SHIFT FORK
K — LOW AND REVERSE SLIDING GEAR
L — MAIN SHAFT BEARING
M — MAIN SHAFT
N — OIL RETAINING WASHER
O — MAIN SHAFT SECOND SPEED GEAR ASSEMBLY
P — COUNTERSHAFT
Q — REAR COUNTERSHAFT THRUST WASHER (STEEL)
R — REAR COUNTERSHAFT THRUST WASHER (BRONZE)
S — CASE
T — SECOND AND DIRECT SPEED CLUTCH SLEEVE
U — HIGH AND INTERMEDIATE CLUTCH HUB
V — MAIN SHAFT PILOT ROLLER BEARING
W — MAIN DRIVE GEAR
X — COUNTERSHAFT GEAR ASSEMBLY
Y — COUNTERSHAFT THRUST WASHER-FRONT

RA PD 305285

Figure 70—Transmission—Sectional View

113. DESCRIPTION AND DATA.

a. Description. The transmission (fig. 70) is a selective, 3-speed, synchromesh type with synchronized second and high speed gears.

TRANSMISSION

It is located in the power plant unit with the engine, clutch, and transfer case. Third speed is a direct drive through the transmission; all other speeds are through gears of various sizes to obtain the necessary gear reduction. The gears are shifted by a lever extending out of the top of the transmission, and through the floor at the right of the driver. For shifting instructions, refer to paragraph 5 e.

b. Data.

Make and model	Warner T-84-J
Type	Synchromesh
Speeds	3 forward—1 reverse
Ratios:	
Low (1st)	2.665 to 1
Intermediate (2nd)	1.564 to 1
High (3rd)	1 to 1
Reverse	3.554
Lubricant capacity	¾ qt
Lubricant grade reference	par. 18

114. MAINTENANCE.

a. The transmission requires periodic checking of lubrication level. Mounting screws must be tight. Gearshift ball cap must be removed and cleaned as required. Keep all bond straps clean and securely tightened. Vent hole in transmission control housing must be kept clean at all times. Report transmission gear noise to higher authority.

115. REMOVAL.

a. Raise hood and fasten to windshield to prevent accidental closing. Open drain cock at bottom of radiator and drain cooling system. Remove radiator upper hose. Unscrew balls on shift levers. Remove bolts around transmission cover on floor, and remove cover. Remove transmission shift lever by unscrewing retainer collar at top of shift housing. Remove transfer case shift lever pin set screw. Remove lubricator in right end of shift shaft. Drive out shaft and remove levers. Place jack under engine oil pan. Remove exhaust pipe guard. Remove exhaust pipe clamp on skid plate. Remove skid plate. Remove bolts in front and rear propeller shaft universal joint flanges at transfer case end, and tie propeller shafts up to frame. Disconnect speedometer cable at transfer case. Remove transfer case support (snubber) bolt at cross member (fig. 73). Remove clevis pin in lower end of hand brake cable, and remove hand brake retracting spring. Remove clevis pin in clutch release cable yoke at cross tube lever. Disconnect engine stay cable at cross member. Remove bonding strap on transfer case and transmission. Unhook clutch pedal pull-back spring. Remove nuts on engine rear support insulator studs in cross member. Place a second jack under the transmission. Remove frame to cross member bolts at each end, and remove cross member. Place rope around the transmission. Push transmission to

¼-TON 4 x 4 TRUCK (WILLYS-OVERLAND MODEL MB and FORD MODEL GPW)

right, and remove clutch release lever tube from ball joint on transfer case. Remove four bolts holding transmission to bell housing. Remove two screws in inspection cover. Remove core and clutch release fork. Hold weight of assembly with rope, and remove jack from under transmission. Slide transmission assembly back until clutch shaft clears bell housing, lowering jack under engine just enough so that transmission will clear floor pan. Remove assembly from under vehicle.

b. Remove Transmission from Transfer Case. Clean outside of units. Drain lubricant from transmission and transfer case. Remove screws holding rear cover on transfer case, and remove cover and gasket. Remove cotter pin, nut, and washer on rear end of transmission main shaft. Pull off main shaft gear and oil slinger.

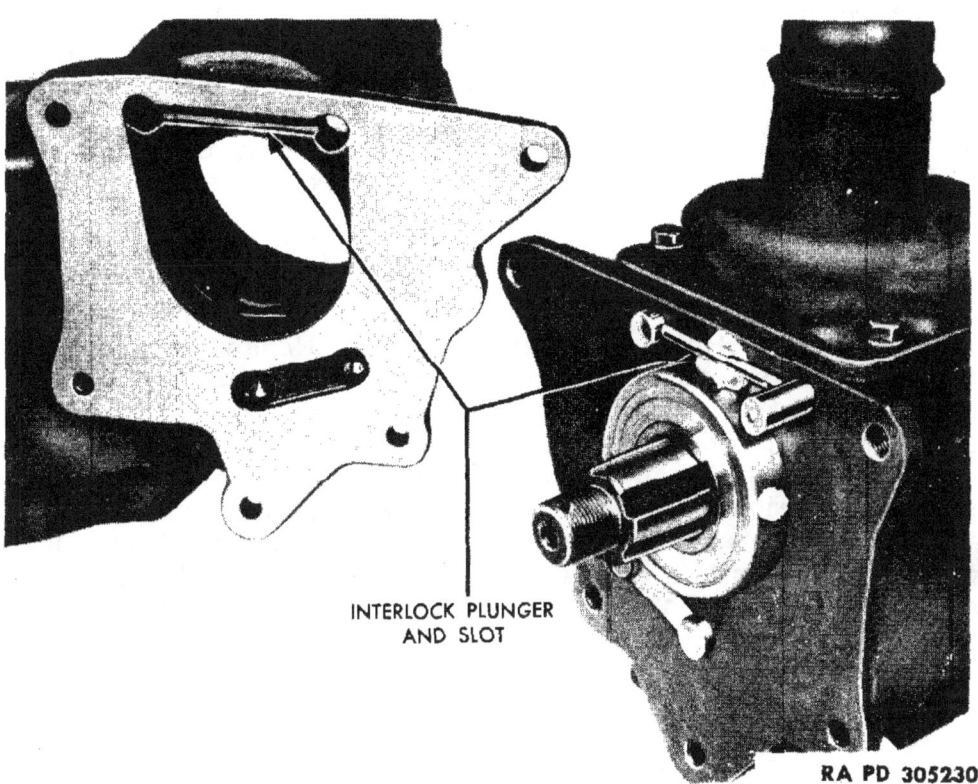

Figure 71—Gearshift Interlock Plunger

Remove four screws holding control housing on top of transmission, and remove housing. Remove shifter plate spring, and remove shifter plate. Loop piece of wire around main shaft to the rear of second speed gear, and attach wire tightly to front of transmission. Remove five screws holding transfer case to rear of transmission. Support transfer case, and tap lightly on end of transmission main shaft; at the same time draw transfer case away from transmission. NOTE: *Do not lose transmission gearshift interlock plunger* (fig. 71).

TRANSMISSION

116. INSTALLATION.

a. Assemble Transmission to Transfer Case. Install transmission shift interlock plunger in position, using grease to hold it temporarily (fig. 71). Attach transfer case to transmission with five screws. Install shifter plate and spring. Install control housing on top of transmission with four screws. Using new gasket, install oil retaining washer on transmission main shaft in transfer case, with open face to rear. Install main drive gear with small power take-off gear to rear. Install washer, nut, and cotter pin. Install rear cover with screws and lock washers, using new gasket. Put lubricant in transmission and transfer case (par. 18).

b. Install in Vehicle. Place unit under vehicle. Place rope around transmission. Raise transmission, and insert transmission main drive gear shaft (clutch shaft) in hub of clutch plate and flywheel. Push transmission forward into position. Place jack under transmission to take the weight. Install clutch release fork on pivot ball, and install clutch release cable in end. Install transmission to bell housing bolts. Remove rope. Raise transmission with jacks. Push transmission to the right, and install clutch release tube on ball on transfer case. Install frame cross member. Remove jack under transmission. Install engine support insulator stud nuts at cross member. Place transfer case support rubber in place, and install bolt through cross member. Attach bond straps. Connect clutch release cable to lever. Connect hand brake cable and spring. Install engine stay cable and adjust so it is just taut. Install pedal pull-back spring. Adjust clutch release cable for pedal play (par. 109). Remove jack under engine. Connect speedometer cable at transfer case. Install transfer case shift levers and springs, driving shaft in from right side. Install shaft lock screw, and wire in place. Install lubricator in right end of shaft. Attach front and rear propeller shafts at transfer case. Install transmission skid plate. Install exhaust pipe clamp on skid plate. Install exhaust pipe guard. Install transmission shift lever in top of housing. Install transmission floor cover. Install gearshift lever balls. Install radiator upper hose. Fill cooling system, giving due attention to antifreeze, if required. Run engine until warm, and check coolant supply. Check cooling system for leaks. Lower hood and hook into place.

¼-TON 4 x 4 TRUCK (WILLYS-OVERLAND MODEL MB and FORD MODEL GPW)

Section XXIII

TRANSFER CASE

	Paragraph
Description and data	117
Maintenance	118
Removal	119
Installation	120
Shifting levers	121

117. DESCRIPTION AND DATA.

a. Description. The transfer case (fig. 72) is an auxiliary gear unit attached to the rear of the transmission. The transfer case is essentially a two-speed transmission which provides an additional

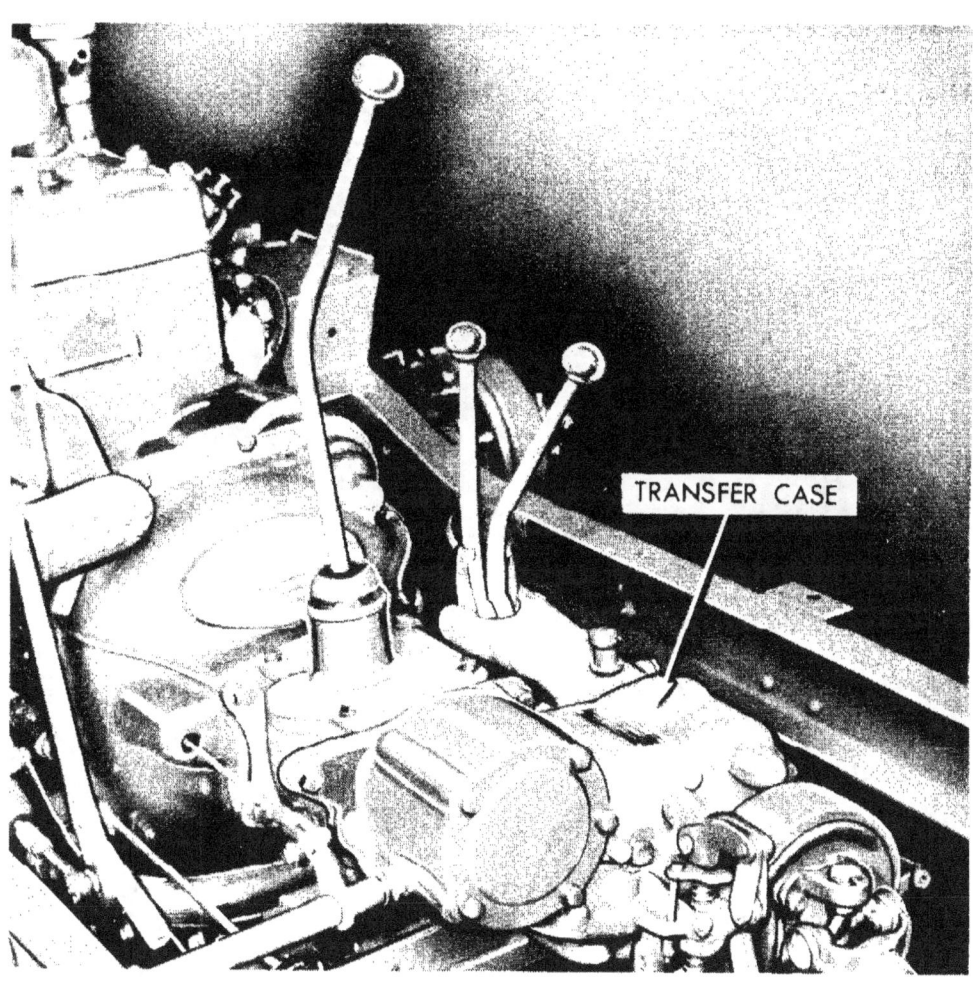

Figure 72—Transfer Case—Rear View

TRANSFER CASE

gear reduction for any selection of the transmission gears, also a means of engaging and disengaging power to drive the front axle. The shifting mechanism (fig. 73) is operated by two levers on the top of the case. A power take-off aperture is located at the rear just behind the transmission main shaft. The speedometer drive gear is located in the output shaft housing for the drive to the rear axle, and a hand brake is located on the same housing.

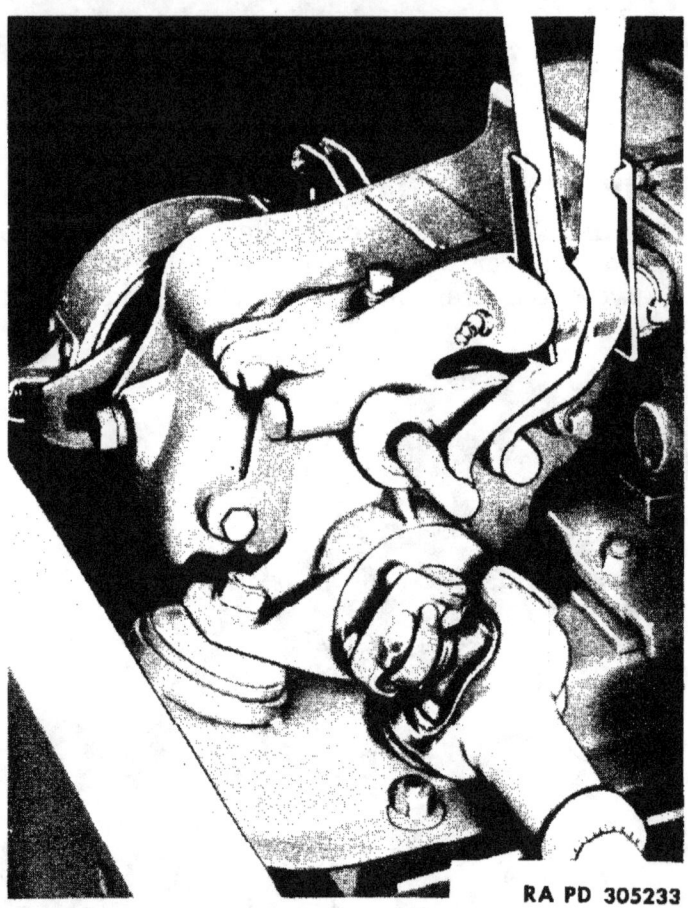

Figure 73—Transfer Case Shifting Mechanism

b. Data.

Make and model	Spicer—18
Ratio—high	1 to 1
—low	1.97 to 1
Speedometer teeth—drive gear	4
—driven gear	14
Lubricant capacity	1½ qt

118. MAINTENANCE.

a. The transfer case requires lubrication at regular intervals. All mounting bolts must be kept tight. Universal joint yoke and com-

¼-TON 4 x 4 TRUCK (WILLYS-OVERLAND MODEL MB
and FORD MODEL GPW)

panion flange nuts must be tight. Bond strap must be clean and tightened securely. Tighten drain and filler plugs. Report failure to stay in gear and noisy gears to higher authority.

119. REMOVAL.

a. Remove transmission and transfer case as an assembly as outlined in paragraph 115.

120. INSTALLATION.

a. Attach transfer case to transmission, and install in vehicle as outlined in paragraph 116.

121. SHIFTING LEVERS.

a. **Removal.** Unscrew accelerator footrest and remove 10 screws holding transmission floor cover, also remove two screws holding transfer case shift lever housing cover. Remove covers. Raise hood and secure to windshield. Remove transfer case shift lever set screw and lock wire. Unscrew hydraulic fitting from right end of shaft. Drive out shaft, and remove levers (fig. 73).

b. **Installation.** Place shift levers and springs in position. Drive in shaft. Install shaft set screw, and wire in place. Install hydraulic fitting in right end of shaft. Install transmission floor cover and accelerator footrest.

Section XXIV

PROPELLER SHAFTS AND UNIVERSAL JOINTS

	Paragraph
Description and data	122
Maintenance	123
Removal	124
Installation	125

122. DESCRIPTION AND DATA.

a. Description. Two propeller shafts are used, one to drive each axle. Two universal joints are used on each shaft (fig. 74). A splined slip joint is used at the rear of the front shaft, and, at the front of the rear shaft. NOTE: *The slip joint is marked with an arrow on the spline shaft and on the sleeve yoke, and these arrows must aline*

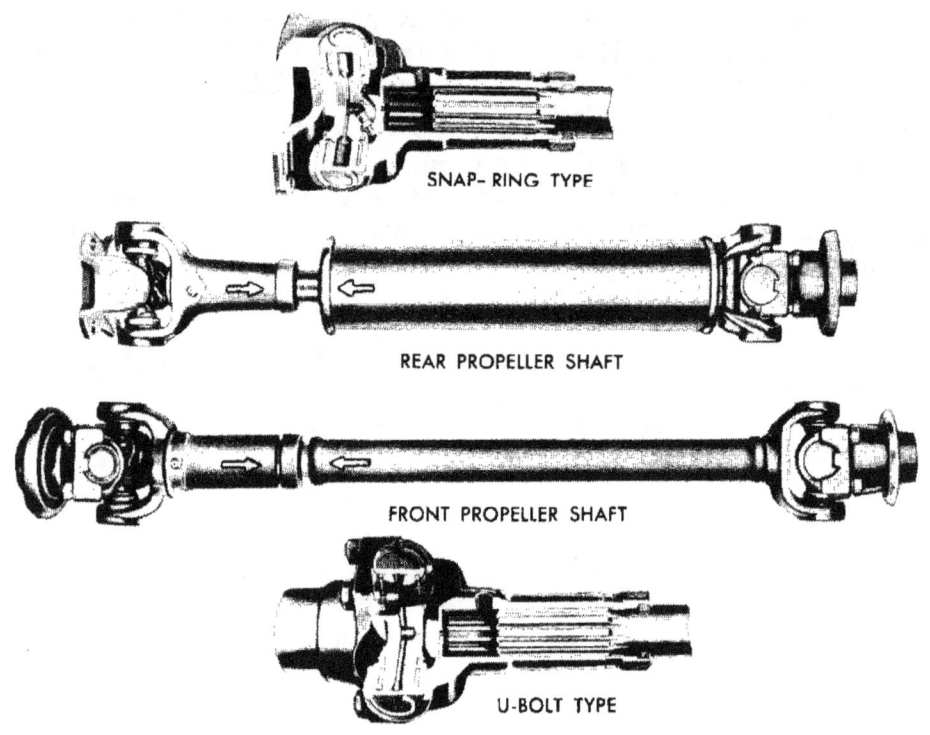

Figure 74—Propeller Shafts and Universal Joints

for correct assembly (fig. 74). The propeller shaft connecting the transfer case to the front axle has U-bolt type joints at both ends. The rear propeller shaft has a U-bolt type joint at the rear where it attaches to the rear axle, and a snap-ring type joint at the front end. The trunnion bearings are of the needle bearing type which are lubricated through a hydraulic fitting and an X-channel in the trunnion. Refer to lubrication instructions in paragraph 18.

¼-TON 4 x 4 TRUCK (WILLYS-OVERLAND MODEL MB and FORD MODEL GPW)

b. Data.

Propeller shafts
  Make .... Spicer
  Installed length normal load-joint center to center
    Front .... $21\tfrac{25}{32}$ in.
    Rear .... $21\tfrac{5}{8}$ in.
  Front Universal Joint (front shaft)
    Type .... Snap-ring and U-bolt
    Model 1268
  Rear Universal Joint (front shaft)
    Type .... Snap-ring and U-bolt
    Model 1261
  Front Universal Joint (rear shaft)
    Type .... Snap-ring
    Model 1261
  Rear Universal Joint (rear shaft)
    Type .... Snap-ring and U-bolt
    Model 1268

123. MAINTENANCE.

a. Propeller and universal joints require proper lubrication at regular intervals. Attaching bolts and nuts must be securely tightened. The yokes of front and rear universal joints must be assembled in the same plane. Snap rings must be securely locked in recess. Trunnion gaskets must be grease-tight. Tighten U-bolts evenly.

124. REMOVAL.

a. Front Propeller Shaft. Remove exhaust pipe shield. Remove U-bolts from universal joint yoke on axle. Remove U-bolts from universal joint yoke on transfer case. Remove shaft and universal joints as a unit.

b. Rear Propeller Shaft. Remove four bolts in front universal joint yoke flange. Remove two U-bolts from rear universal joint yoke on axle. Remove propeller shaft and universal joints as a unit.

125. INSTALLATION.

a. Front Propeller Shaft. Install propeller shaft and universal joint assembly in position on vehicle. Install U-bolts at axle. Install U-bolts at transfer case. NOTE: *Tighten U-bolts evenly.* Install exhaust pipe shield. Lubricate universal joints (par. 18 e).

b. Rear Propeller Shaft. Install propeller shaft and universal joint assembly in position. Install U-bolts in rear universal joint, and tighten evenly. Attach front universal joint flange at transfer case with four bolts. Lubricate universal joints (par. 18 e).

Section XXV

FRONT AXLE

	Paragraph
Description and data	126
Maintenance	127
Wheel bearings	128
Wheel grease retainer	129
Wheel hub	130
Brake drums	131
Steering knuckle housing oil seal	132
Steering tie rod	133
Drag link bell crank	134
Wheel alinement (toe-in)	135
Removal	135
Installation	137

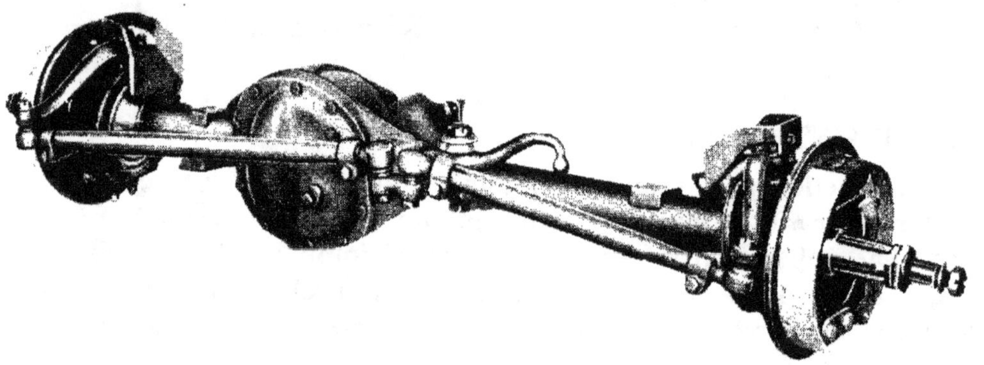

RA PD 305253

Figure 75—Front Axle

126. DESCRIPTION AND DATA.

a. Description. The front axle (fig. 75) is a full-floating type enclosing a front wheel driving unit having a single-reduction, two-pinion differential, and hypoid drive gears. The differential carrier housing is offset to the right so that the propeller shaft is located to the right of the engine for maximum ground clearance. A cover provides easy access to the differential unit. The front wheels are driven by axle shafts, each equipped with a constant-velocity type universal joint enclosed within a steering knuckle at the outer end of the axle housing. The differential assembly is the same as used in the rear axle. Power is transmitted by a propeller shaft from the transfer case, where a shift lever permits the vehicle operator to engage or disengage the drive.

b. Data.

Make and model	Spicer–25
Drive gear ratio	4.88 to 1

¼-TON 4 x 4 TRUCK (WILLYS-OVERLAND MODEL MB and FORD MODEL GPW)

Drive .... Hotchkiss (through springs)
Type .... Full floating
Road clearance .... 8 7/16 in.
Differential type .... 2 pinion
Differential drive gears .... Hypoid
Differential bearings .... Tapered roller
Turning angle .... 26 deg

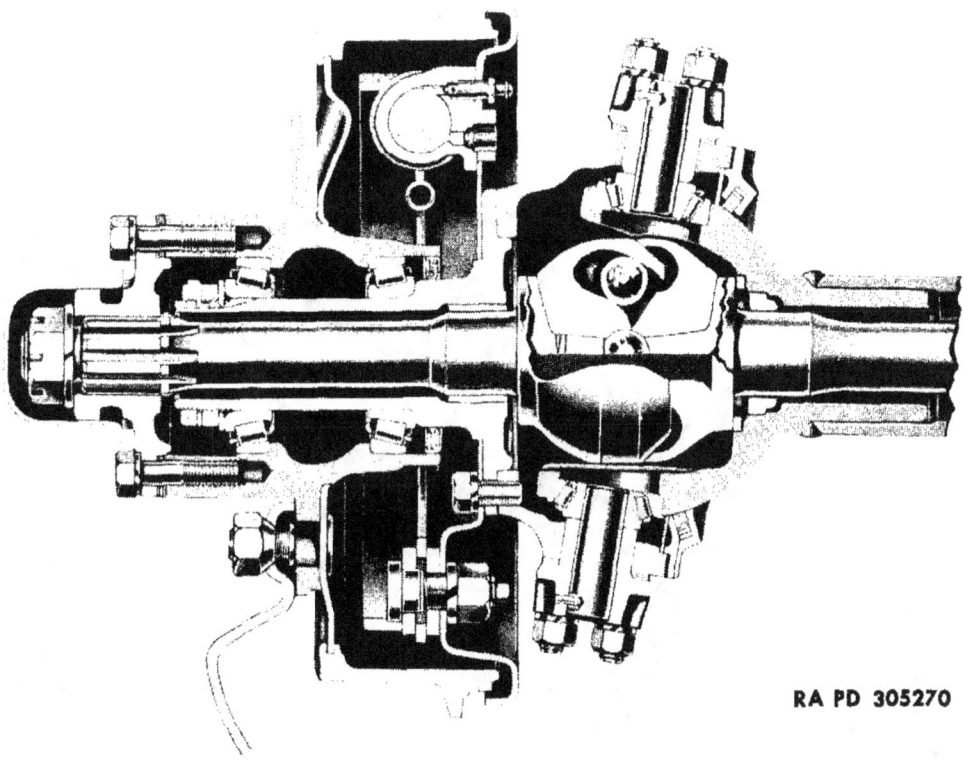

RA PD 305270

Figure 76—Front Wheel Hub

Tie rods:
  Number 2
  Right-hand length, center-to-center .... 24¼ in.
  Left-hand length, center-to-center .... 17 11/32 in.
Steering geometry:
  King pin inclination .... 7½ deg
  Wheel camber .... 1½ deg
  Wheel caster .... 3 deg
  Wheel toe-in .... 3/64 in. to 3/32 in.
Bearings:
  Differential side .... Tapered roller
  Pinion shaft .... Tapered roller
  Wheel hub .... Tapered roller

FRONT AXLE

Steering knuckle . . . . Tapered roller
Steering bell crank . . . . Tapered roller
Lubricant capacity . . . . 1¼ qt

127. MAINTENANCE.

a. Correct any lubricant leakage. Lubricate differential and steering housings, wheel bearings, and steering control as required (par.

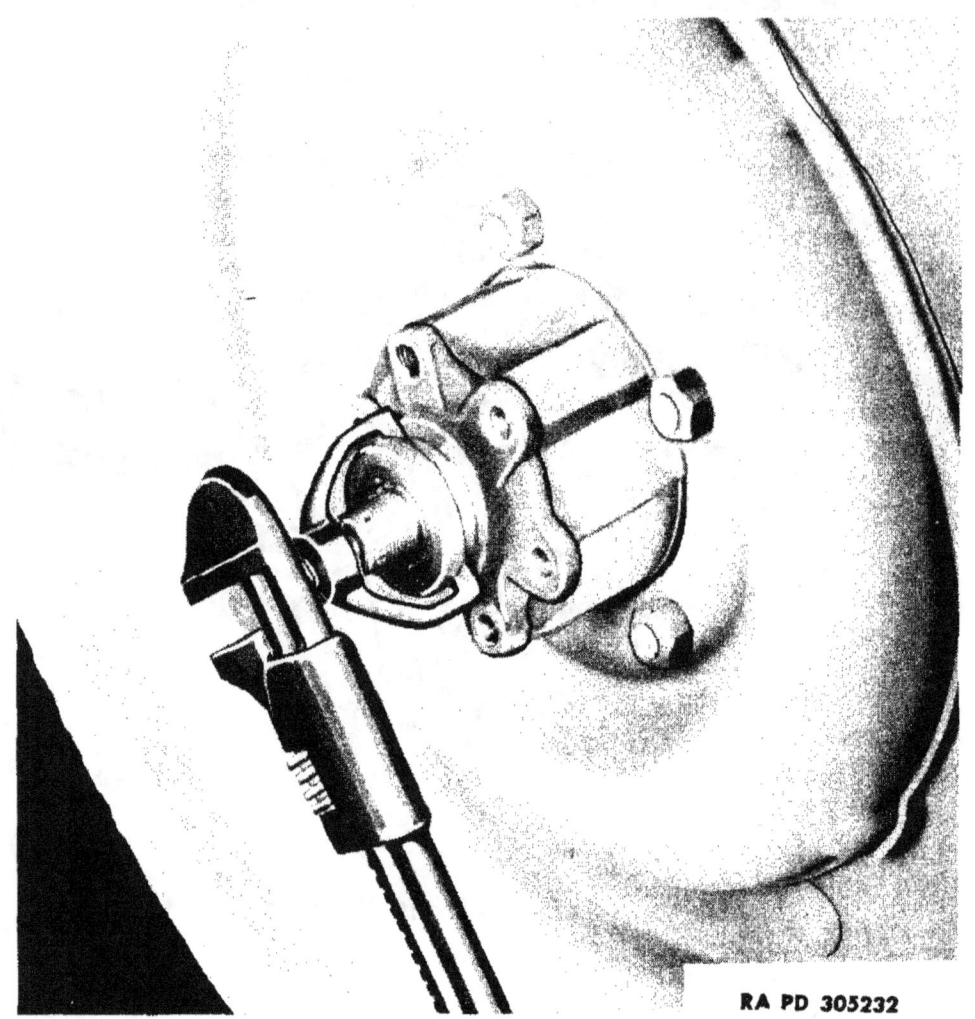

Figure 77—Removing Driving Flange, Using Puller (41-P-2905-60)

18). Keep vent cleared of dirt. Wheel bearings must be properly adjusted (par. 128). Replace damaged brake drums and hubs, and steering control rods (par. 133). Check wheel toe-in periodically, and correct if necessary (par. 135). Keep all mounting bolts tight. Report to higher authority any caster and camber trouble, or unusual noises.

128. WHEEL BEARINGS.

a. Adjustment. Raise front of vehicle so that tire clears floor. Pry off hub cap (fig. 76). Remove axle shaft nut cotter pin, nut, and

¼-TON 4 x 4 TRUCK (WILLYS-OVERLAND MODEL MB and FORD MODEL GPW)

washer. Remove driving flange screws. With puller, pull off flange (fig. 77). NOTE: *Do not lose flange shims.* Bend lip of nut lock washer away from nut. Remove lock nut with box-type socket wrench (fig. 78). Remove lock washer. Spin wheel and tighten wheel bearing nut until wheel binds. Back off nut about one-sixth turn, or more if necessary, until wheel turns freely. Install lock washer and lock nut. NOTE: *Bend over lip of lock washer against lock nut.* Check adjustment of bearings by gripping front and rear side of tire and moving it from side to side. A slight perceptible shake should be felt in the bearings. Install flange shims and flange. Check axle shaft end play by tightening flange nut without the lock washer. Swing wheel to maximum left or right; have punch mark on end of axle shaft up or down. Back off nut until 0.050-inch thickness gage will go between hub and nut. Shaft will move in by the amount of end play when end is tapped with soft hammer. Measure clearance between nut and flange, and deduct amount from 0.050 inch to determine end play. If less than 0.015 inch or more than 0.035 inch, correct thickness of shim pack (with Rzeppa joint disregard these instructions and use 0.060-inch shim pack). Install axle shaft lock washer, nut, and cotter pin. Install new hub cap.

b. Removal. Loosen wheel stud nuts. NOTE: *Wheel studs have left-hand threads on left side of vehicle.* Raise front of vehicle so that tire clears floor. Remove wheel stud nuts and remove wheels. Pry off hub cap. Remove axle shaft nut cotter pin, nut, and washer. Remove driving flange screws and using puller, pull flange. NOTE: *Do not lose flange shims.* Bend lip of nut lock washer away from lock nut and remove nut. Remove lock washer. Remove wheel bearing nut and bearing lock washer. Shake wheel until outer bearing comes free of hub, and lift off wheel. Drive or press out inner bearing along with oil seal. Turn wheel over, and drive or press out outer bearing cup. Clean lubricant out of hub, and wash all parts in dry-cleaning solvent.

c. Installation. Press bearing cups solidly into place in hub. Spread $\frac{1}{16}$-inch layer of lubricant inside hub to prevent rust. Thoroughly lubricate inner bearing cone and roller assembly. NOTE: *Pack lubricant thoroughly into bearing rollers and cage.* Install bearing in hub. Press oil seal into hub (with lip of seal toward bearing), until seal is even with end of hub. NOTE: *Before installation, soak seal in oil to soften leather.* Lubricate outer bearing cone and roller assembly. Install wheel on axle. Install outer bearing, lock washer, and nut. Adjust wheel bearings, and complete installation of parts (subpar. **a** above).

129. WHEEL GREASE RETAINER.

a. Removal. Remove retainer as outlined in paragraph 128 **b**.

b. Installation. Install retainer as outlined in paragraph 128 **c**.

FRONT AXLE

130. WHEEL HUB.

a. Removal. Remove wheel and hub (par. 128 b). Support brake drum inside at hub, and drive out studs. Remove brake drum.

b. Installation. Assemble brake drum on hub. Install new wheel studs. NOTE: *Left-hand thread studs are used in wheels for left side of vehicle.* Support studs and swedge shoulder over against tapered hole in hub. Install hub on axle and mount wheel (par. 128 c).

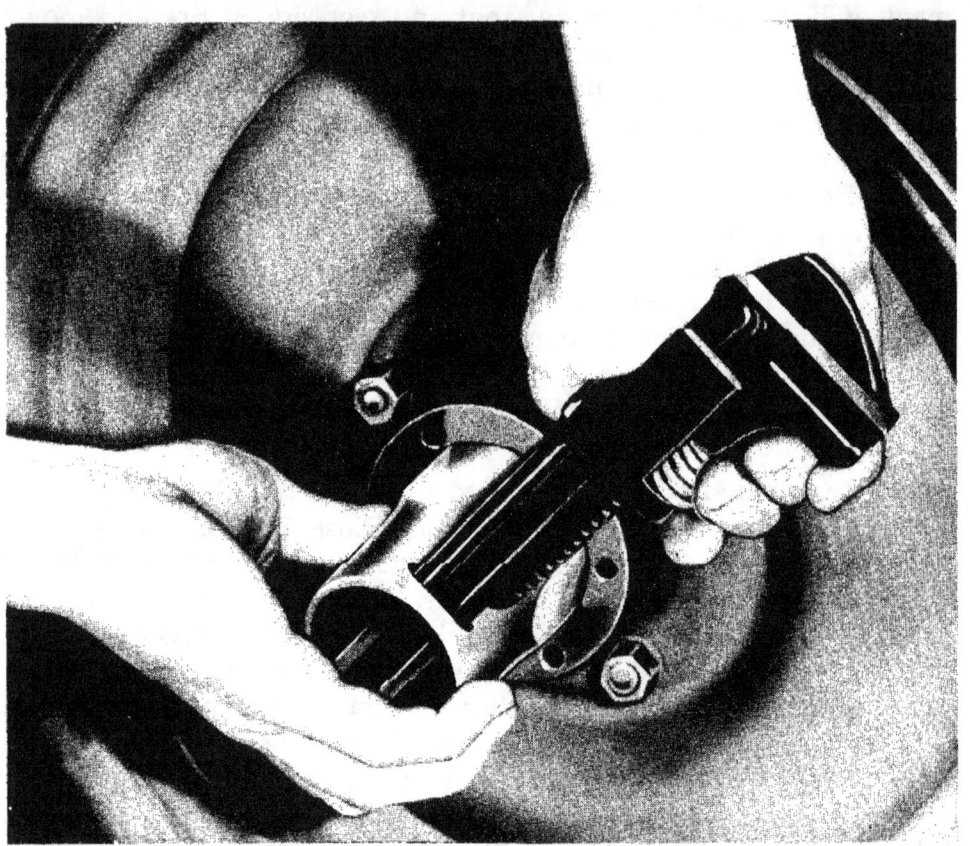

RA PD 305235

Figure 78—Removing Lock Nut, Using Wrench (41-W-3825-200)

131. BRAKE DRUMS.

a. Removal. Remove wheel hub and drum as outlined in paragraph 130 a.

b. Installation. Install drum and wheel hub as outlined in paragraph 130 b. Adjust brakes (par. 148).

132. STEERING KNUCKLE HOUSING OIL SEAL.

a. Removal. Raise front of vehicle. Remove screws holding oil seal in place, and remove both halves of oil seal assembly (fig. 79).

b. Installation. NOTE: *Before installing new oil seal smooth spherical surface of axle with aluminum oxide abrasive cloth.* Grease

TM 9-803
132-134
¼-TON 4 x 4 TRUCK (WILLYS-OVERLAND MODEL MB and FORD MODEL GPW)

spherical surface of axle, also oil seal. Install seal in place so that ends fit snugly together, and tighten in place. Check lubricant level in steering knuckle housing, and replenish if necessary (par. 18).

133. STEERING TIE ROD.

a. Removal. Remove tie rod cotter pins and nuts from tie rod ends (fig. 75). Drive out tie rod ends from steering arms, and remove dust washers and springs.

b. Installation. Install dust washers and springs on tie rod ends. Install tie rod ends in steering knuckle arm and bell crank, and secure with nuts and cotter pins. Check wheel alinement, and adjust if necessary (par. 135).

RA PD 305240

Figure 79 — Steering Knuckle Oil Seal

134. DRAG LINK BELL CRANK.

a. Removal. Remove cotter pin in front end of steering connecting rod. Remove slotted adjusting plug (ball seat). Lift rod off bell crank ball. Remove cotter pins and nuts on tie rod ends at bell crank. Drive tie rod ends out of bell crank. NOTE: *Do not lose dust washers and springs.* Remove cotter pin in bell crank stud and remove nut, dust washer, and thrust washer. Remove bell crank. Clean all parts in dry-cleaning solvent. To remove bell crank stud, remove thrust washer, and drive out tapered lock pin toward left front wheel. Drive stud up out of axle.

b. Installation. If bell crank has been removed, drive stud into axle so that slot will line up with tapered pin hole. Drive tapered pin into position, and stake edge of hole at large end. Install thrust washer on stud. Lubricate roller bearings, and install on stud. Install

FRONT AXLE

thrust washer, dust washer, nut, and cotter pin. Install steering connecting rod. Install tie rod ends in bell crank arm, and secure each with nut and cotter pin. Check front wheel toe-in, and adjust if necessary (par. 135).

135. WHEEL ALINEMENT (TOE-IN).

a. Caster and Camber. Caster is the backward tilt of the axle. Camber is the outward tilt of the wheels at the top. If these conditions require attention, notify higher authority.

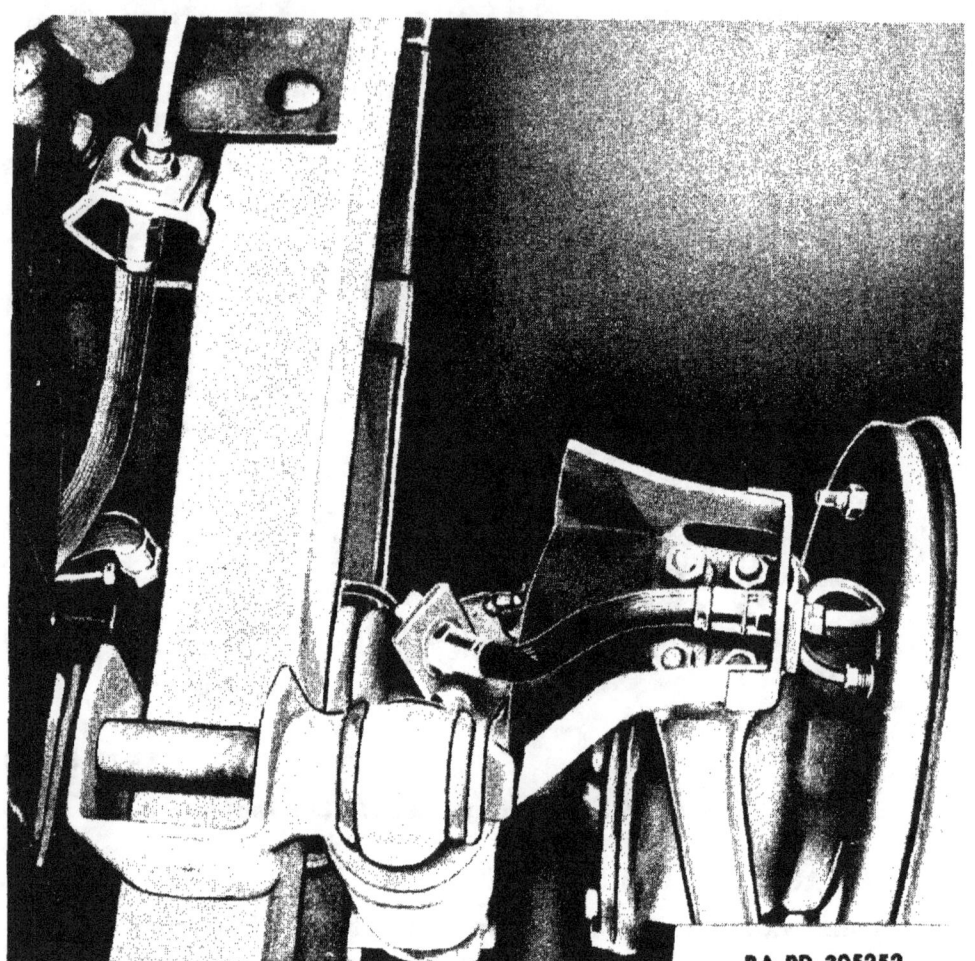

Figure 80—Brake Hose

b. Toe-in. Wheel toe-in is the difference in distance between the front wheels at the front and at the rear of the axle. To adjust toe-in, set tie rod arm of steering bell crank at right angles to front axle. Use straightedge or line against outside of left wheels, as a guide. Adjust left tie rod so that left wheel is straight ahead. While bell crank remains at right angle to axle, check right front wheel, and adjust tie rod if necessary. Set toe-in of front wheels at $3/64$ inch to $3/32$ inch by shortening right tie rod approximately one turn.

¼-TON 4 x 4 TRUCK (WILLYS-OVERLAND MODEL MB and FORD MODEL GPW)

136. REMOVAL.

a. Loosen wheel stud nuts. Raise front of vehicle, and support underframe side members at rear of spring pivot brackets. Remove wheels. Disconnect brake line at front cross member (fig. 80). Remove universal joint U-bolts at front axle. Jack up front springs. Remove axle spring clip nuts and clips. Remove spring pivot bolt at rear end of right spring. Remove jacks from under springs. Disconnect steering connecting rod at bell crank. Install a jack between left spring and frame. Spread spring until axle assembly will clear. Move axle assembly to the right, and remove. Remove brake hose from axle.

137. INSTALLATION.

a. Attach brake hose at axle. Install axle assembly on springs. Remove jack from between left spring and frame. Install right spring pivot bolt. Position axle on springs. Jack up springs, and install spring clips, plates, and nuts. Remove jacks from under springs. Connect brake hose at cross member. Install dust cover on bell crank, and attach steering connecting rod. Attach propeller shaft. Draw universal joint U-bolts up evenly. Lubricate front axle universal joints, and check axle lubricant (par. 18). Adjust brakes if necessary (par. 147). Remove master cylinder inspection cover on toeboard between foot pedals. Fill master cylinder, and bleed brakes (par. 151). Replace master cylinder inspection cover. Install wheels and adjust (par. 128 a). Lower vehicle to floor.

Section XXVI

REAR AXLE

	Paragraph
Description and data	138
Maintenance	139
Axle shaft	140
Wheel bearings	141
Wheel bearing grease retainer	142
Wheel hub	143
Brake drum	144
Rear axle replacement	145

RA PD 305254

Figure 81—Rear Axle

138. DESCRIPTION AND DATA.

a. Description. The rear axle (fig. 81) is a full-floating type enclosing a single-reduction driving unit, two-pinion differential, and hypoid-drive gears. The differential carrier housing is offset to the right so that the propeller shaft will have a straight drive from the transfer case. A cover provides easy access to the differential unit. The axle shafts are splined to fit into the differential side gears, and flanged at the outer end where they attach to the wheel hub. The wheel bearings are adjusted by two nuts threaded onto the axle tube.

b. Data.

Make and model	Spicer 23-2
Drive gear ratio	4.88 to 1
Drive	Hotchkiss (through springs)
Type	Full floating
Road clearance	$8\frac{7}{16}$ in.
Differential type	Two-pinion
Differential bearings	Tapered roller

¼-TON 4 x 4 TRUCK (WILLYS-OVERLAND MODEL MB and FORD MODEL GPW)

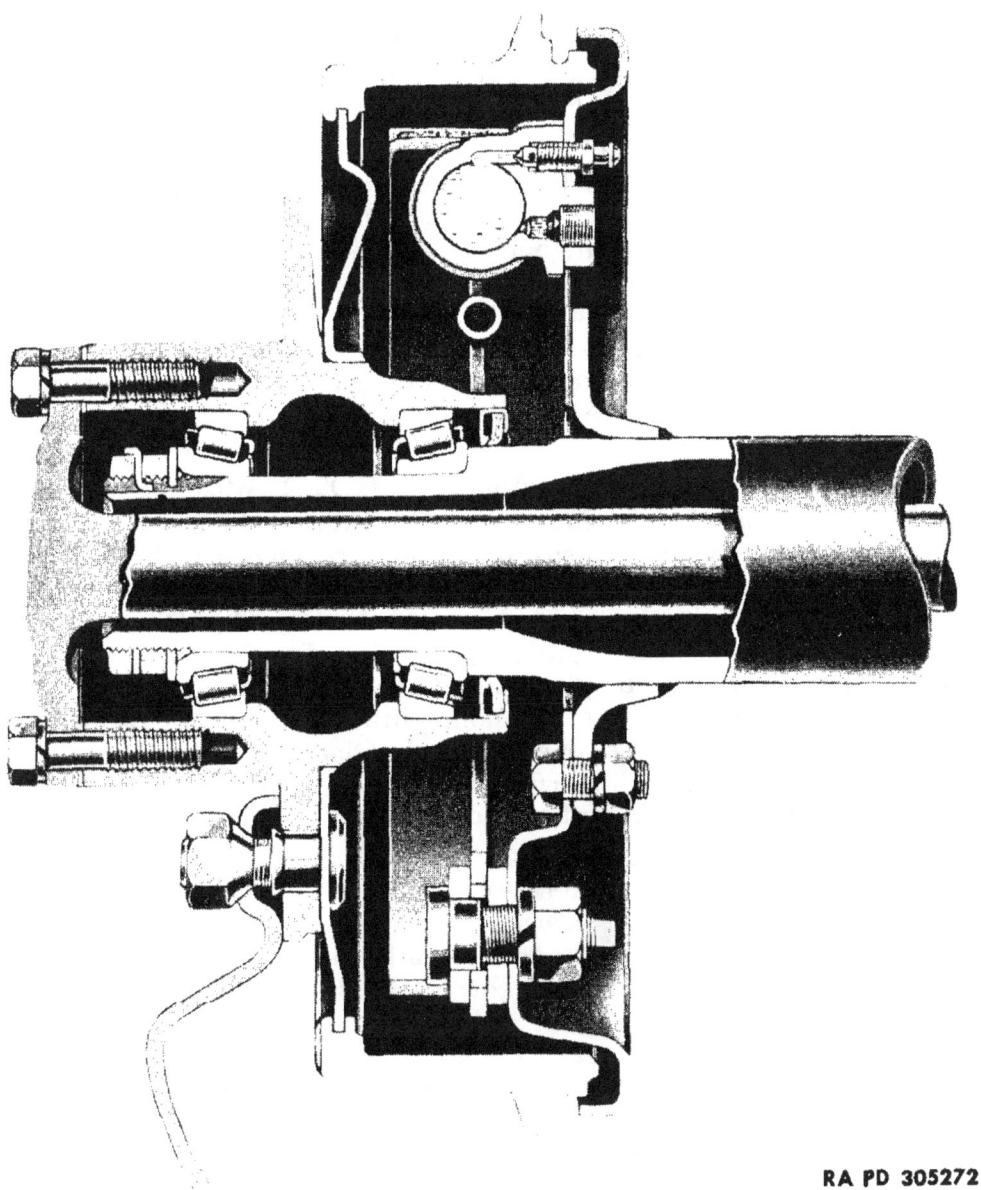

Figure 82 — Rear Wheel Hub

139. MAINTENANCE.

a. Correct any lubricant leakage. Lubricate differential and wheel bearings as required (par. 18). Keep vent cleared of dirt. Wheel bearings must be properly adjusted (par. 141). Replace damaged drums and hubs (par. 143). Keep all mounting bolts tight. Report unusual noise to higher authority.

140. AXLE SHAFT.

a. Removal. Remove six axle shaft flange screws and lock washers. Pull out axle shaft (fig. 82), and remove flange gasket.

REAR AXLE

b. Installation. Install new axle shaft flange gasket. Install axle shaft in axle housing, rotating shaft so that shaft will enter differential side gear. Take care not to damage inner oil seal in axle housing. Install axle flange screws and lock washers, and tighten securely.

141. WHEEL BEARINGS.

a. Adjustment. Place jack under axle housing, and raise wheel so that tire clears floor. Remove axle shaft (par. 140 a). Bend lip of lock washer away from lock nut, and remove nut with box-type socket wrench (fig. 78). Remove lock washer. Spin wheel, and tighten wheel bearing nut until wheel just binds. Back off nut one-sixth turn or more, if necessary, until wheel turns freely. Install lock washer and lock nut. NOTE: *Bend over lip of lock washer against lock nut.* Check adjustment by shake of wheel. Install axle shaft (par. 140 b). Lower vehicle to floor.

b. Removal. Loosen wheel stud nuts. NOTE: *Wheel studs have left-hand threads on left side of vehicle.* Raise vehicle so that tire clears floor. Remove wheel stud nuts, and remove wheels. Remove axle shaft (par. 140 a). Bend lip of lock washer away from lock nut, and remove nut with box-type socket wrench (fig. 78). Remove lock washer. Remove bearing adjusting nut and bearing lock washer. Shake wheel until outer bearing comes free of hub, and lift off wheel. Drive or press out inner bearing along with oil seal from wheel hub. Drive or press out bearing cups from hub. Clean old lubricant out of hub, and wash all parts in dry-cleaning solvent. Examine parts for excessive wear or damage, and replace if unserviceable.

c. Installation. Press bearing cups solidly into place in hub. Spread $\frac{1}{16}$-inch layer of lubricant inside of hub to prevent rust. Thoroughly lubricate inner bearing cone and roller assembly. NOTE: *Pack lubricant into bearing cage.* Install bearing in hub. Press oil seal into hub (with lip of seal toward bearing) until seal is even with end of hub. NOTE: *Before installation, soak seal in oil to soften leather.* Lubricate outer bearing cone and roller assembly. Install wheel on axle. Install outer bearing lock washer and nut. Adjust wheel bearings, and complete installation of parts (par. 141 a).

142. WHEEL BEARING GREASE RETAINER.

a. Removal. Remove retainer as outlined in paragraph 141 b.

b. Installation. Install retainer as outlined in paragraph 141 c.

143. WHEEL HUB.

a. Removal. Remove wheel and hub as outlined in paragraph 141 b. To remove brake drum from hub, support brake drum at hub, and drive out studs.

b. Installation. Place brake drum on hub. Install new wheel studs. NOTE: *Left-hand thread studs are used in wheels on left side of vehicle.* Support studs and swedge shoulder over against tapered hole in hub. Install hub on axle and mount wheel (par. 141 c). Tighten wheel stud nuts securely. Check brake action.

¼-TON 4 x 4 TRUCK (WILLYS-OVERLAND MODEL MB and FORD MODEL GPW)

144. BRAKE DRUM.

a. Removal. Remove wheel hub and drum as outlined in paragraph 143 a.

b. Installation. Install drum and wheel hub as outlined in paragraph 143 b.

145. REAR AXLE REPLACEMENT.

a. Loosen wheel stud nuts. Raise rear of vehicle and support underframe side member just ahead of spring pivot brackets. Remove wheels. Remove universal joint U-bolts at rear axle. Disconnect brake hose at frame cross member. Remove brake hose at axle. Place jack under each rear spring. Remove spring clip nuts, clips, and plates. Remove jacks from under springs, place between frame and spring, and spread spring. Remove axle, sliding it to left until right end clears spring, then slide to right and remove.

b. Installation. Install axle assembly on springs. Remove jacks from between springs, and place under each spring. Position axle on springs. Install spring clips, plates, lock washers, and nuts. Tighten nuts securely. Remove jacks from under springs. Attach propeller shaft. Draw universal joint U-bolts up evenly. Attach brake hose at axle, then at frame cross member. Check axle lubricant. Remove master cylinder inspection cover on toeboard between foot pedals. Fill master cylinder, and bleed brakes (par. 151). Replace master cylinder inspection cover. Adjust brakes if necessary (par. 148). Install wheels. Lower vehicle to floor.

Section XXVII

BRAKES

	Paragraph
Description and data	146
Maintenance and adjustment	147
Service (foot) brakes	148
Master cylinder	149
Wheel cylinder	150
Flexible lines, hoses and connections	151
Parking (hand) brake	152

146. DESCRIPTION AND DATA.

a. Description. The service, or foot brake, system is of the hydraulic type with brakes in all four wheels (fig. 83). The parking, or hand brake, is cable-controlled and mounted on the rear side of the transfer case (fig. 84). The service, or foot brakes, are of the two-shoe, double-anchor type. The brake pedal, through a connection, operates a piston in the master cylinder to force brake fluid through the lines to the brake cylinders in the wheels. The fluid enters the wheel cylinders between two pistons of equal diameter, forcing them apart to apply the brake shoes against the drums. Releasing the brake pedal permits the brake fluid to flow back through the lines to the master cylinder. Adjustments are provided to compensate for wear of the brake linings. The hand brake is designed for parking the vehicle, or as an emergency brake. The hand brake lever is located at the center of the instrument panel. Pulling out on the lever draws a flexible cable through a conduit to actuate an external contracting brake band at the rear of the transfer case. The brake cable is of a predetermined length, and cannot be adjusted. When adjustment is required, the brake band lining will be worn to the point where replacement is necessary. Adjustments are provided on the brake to set the band correctly, and to limit the release action.

b. Data.

Service brakes:
- Type .... Four-wheel, hydraulic
- Size .... 9 in. x 1¾ in.
- Fluid capacity .... ¼ qt

Master cylinder:
- Type .... Combination reservoir and cylinder
- Size .... 1 inch

Wheel cylinders:
- Type .... Straight bore
- Size .... Front, 1 in.; rear, ¾ in.

TM 9-803
146

¼-TON 4 x 4 TRUCK (WILLYS-OVERLAND MODEL MB and FORD MODEL GPW)

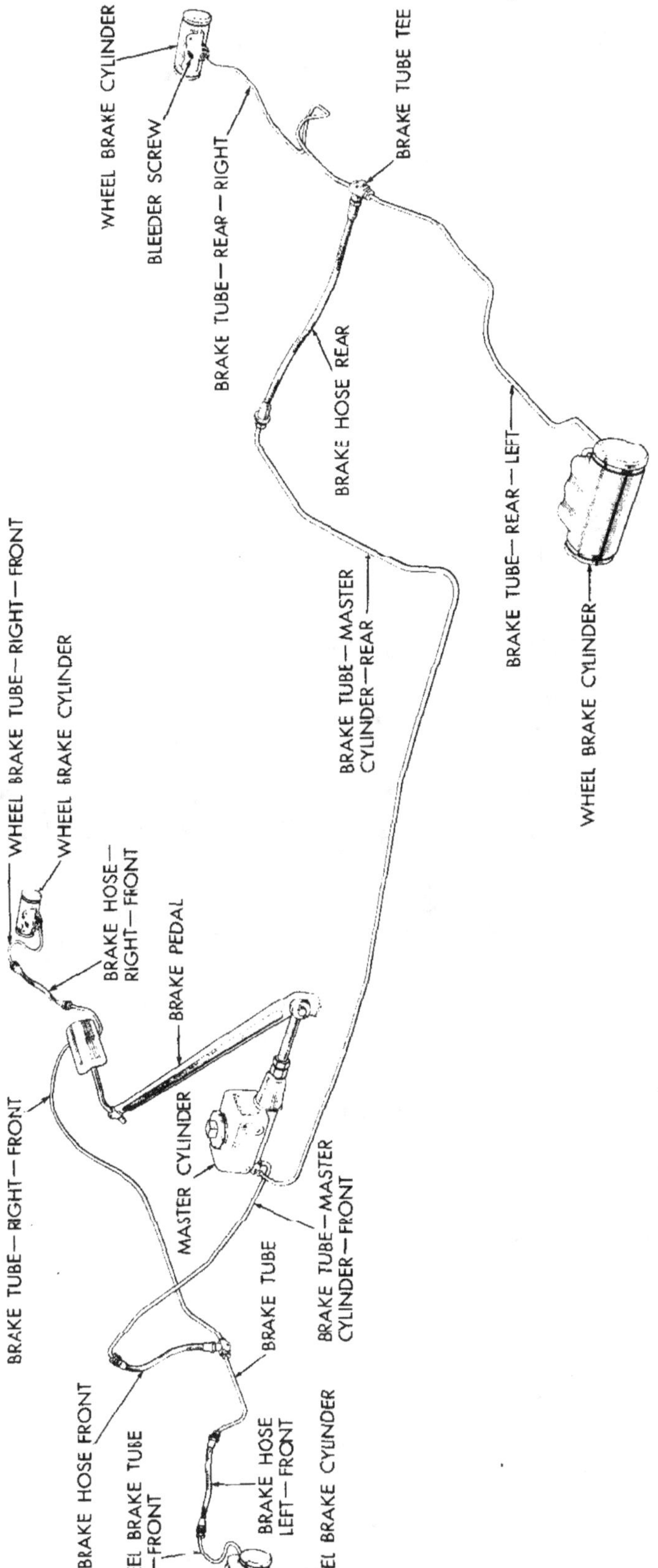

RA PD 305255

Figure 83—Service (Foot) Brake System

194

BRAKES

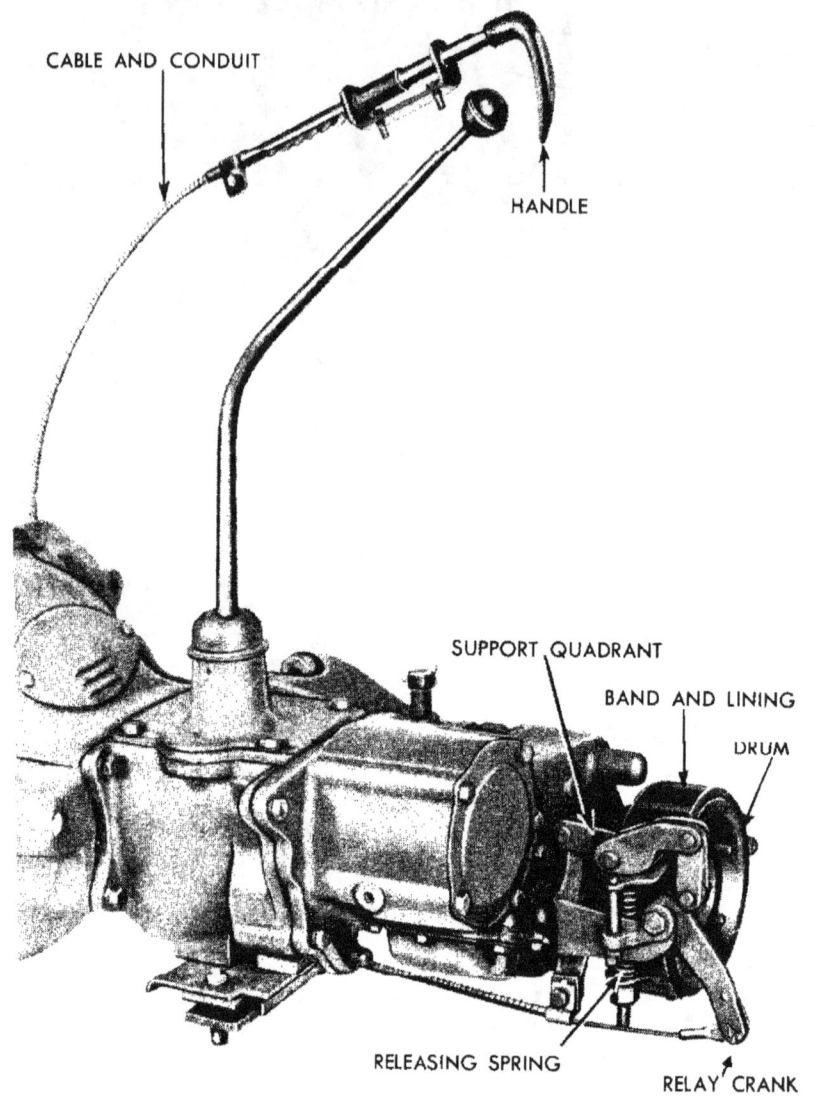

Figure 84 — Parking (Hand) Brake System

Brake shoes:
    Lining length—forward shoe (moulded) .... $10\tfrac{7}{32}$ in.
    Lining length—reverse shoe (moulded) .... $6\tfrac{39}{64}$ in.
    Width .... $1\tfrac{3}{4}$ in.
    Thickness .... $\tfrac{3}{16}$ in.
Hand brake:
    Type .... Mechanical
    Lining length (woven) .... $18\tfrac{9}{16}$ in.
    Width .... 2 in.
    Thickness .... $\tfrac{5}{32}$ in.

147. MAINTENANCE AND ADJUSTMENT.

a. The service, or foot, brakes require periodic checking of the brake fluid supply in the master cylinder. Keep master cylinder sup-

¼-TON 4 x 4 TRUCK (WILLYS-OVERLAND MODEL MB and FORD MODEL GPW)

plied with fluid to avoid air entering the lines. Wheel bearings and brakes must be properly adjusted to provide emergency stops. All brake lines, hoses, and connections must be tight and leakproof. Scored brake drums or saturated brake linings must be replaced. Clean brake drums when wheels are removed. Brake anchor bolt and eccentric adjustment bolt lock nuts must be kept tight. Brake backing plate screws and axle spring clips must be kept tight. Brake pedal must have ½-inch free travel to assure full release of brakes. Brake control linkage must be free to operate, and should be inspected periodically for condition.

RA PD 305260

Figure 85 — Wheel Brake

b. Adjustment. Adjust brake pedal free travel by lengthening or shortening brake master cylinder eyebolt so that pedal has ½-inch free play (par. 148). Follow procedure outlined in paragraph 148 to adjust brakes when lining has worn so that brake pedal goes almost to the toeboard. Three adjustments are provided on the hand brake (par. 152).

148. SERVICE (FOOT) BRAKES.

a. Adjustment (minor). Adjust brake pedal free play to one-half inch by lengthening or shortening brake master cylinder eyebolt. Set lock nut securely. Raise vehicle until tires clear floor. NOTE: *Do not adjust brakes when drums are hot.* Loosen eccentric lock nut on forward shoe of one brake (fig. 86). Place wrench on eccentric so

BRAKES

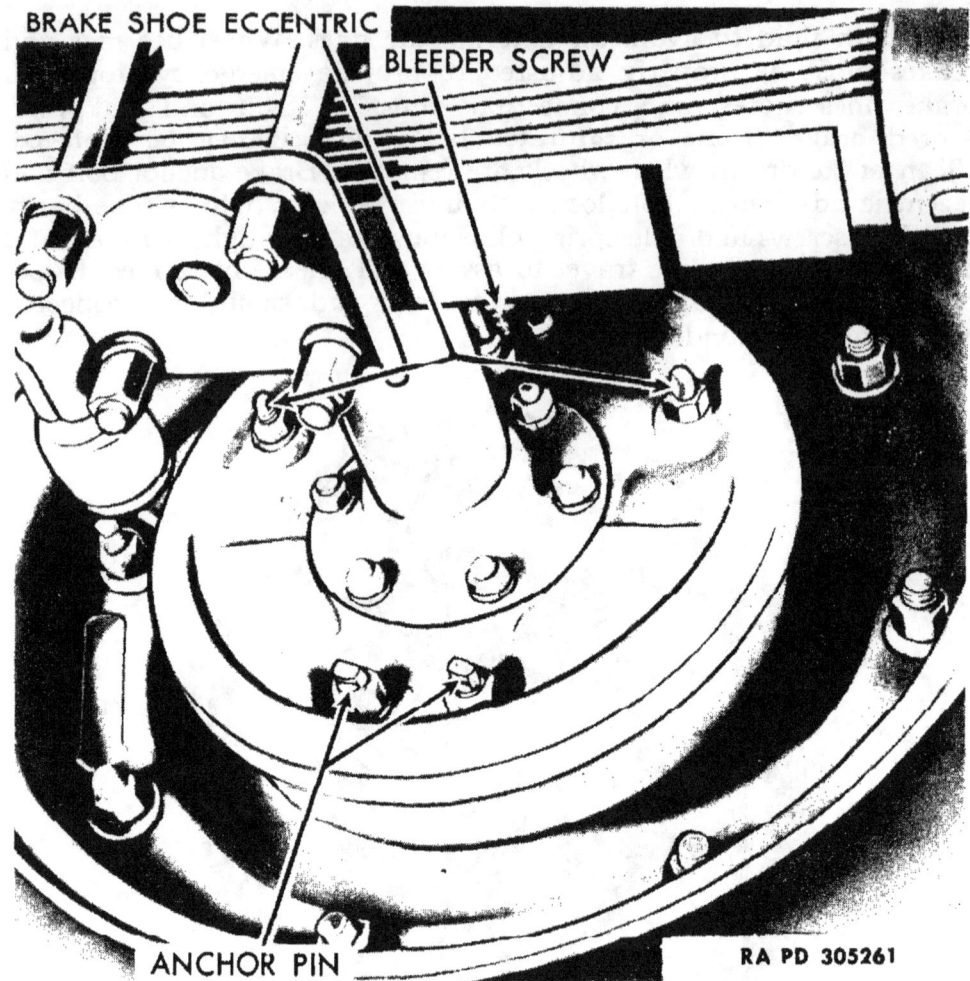

Figure 86 — Wheel Brake Adjustment Points

that handle extends up. Rotate wheel, and turn wrench handle toward wheel rim, or forward, until brake drags. Turn wrench in opposite direction until wheel turns freely. Hold wrench on eccentric, and tighten lock nut. Loosen eccentric lock nut on reverse shoe. Place wrench on eccentric with handle up. Rotate wheel, and turn wrench toward wheel rim, or to the rear, until brake drags. Turn wrench in opposite direction until wheel turns freely. Hold wrench on eccentric, and tighten lock nut. Make the same adjustment on the other wheel brakes. Replenish brake fluid in master cylinder (par. 149). Lower vehicle to floor. Apply brake pedal to test brakes.

b. Adjustment (major). Adjust brake pedal free play to one-half inch by lengthening or shortening brake master cylinder eyebolt. Set lock nut securely. Raise vehicle until tires clear floor. NOTE: *Do not adjust brakes when drums are hot.* Remove wheel stud nuts, and remove wheels from hubs. Insert 0.008-inch thickness gage through slot in brake drum, and turn drum so that gage is at upper (toe) end of forward brake lining. NOTE: *Check clearance 1 inch*

¼-TON 4 x 4 TRUCK (WILLYS-OVERLAND MODEL MB and FORD MODEL GPW)

from end of lining. Loosen eccentric lock nut on forward brake shoe. Place wrench on eccentric so that handle is up, and turn wrench handle toward wheel rim, or forward, until 0.008-inch clearance is obtained by feel of gage. Hold wrench on eccentric, and tighten lock nut. Turn brake drum so that gage is at upper end of reverse brake shoe lining. Loosen eccentric lock nut on reverse shoe. Place wrench on eccentric so that handle is up, and turn wrench handle toward wheel rim, or to the rear, until 0.008-inch clearance is obtained by feel of gage. Hold wrench on eccentric, and tighten lock nut. Remove 0.008-inch thickness gage, and insert 0.005-inch gage in slot. Turn brake drum so that gage is at lower (heel) end of forward brake shoe

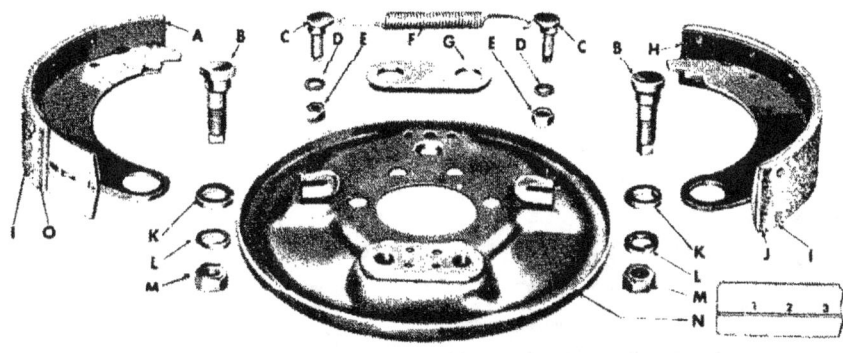

A SHOE AND LINING ASSEMBLY—REVERSE	H SHOE AND LINING ASSEMBLY—FORWARD
B ANCHOR PIN	I LINING TUBULAR BRASS RIVET
C ECCENTRIC	J LINING—FORWARD
D ECCENTRIC LOCK WASHER	K ANCHOR PIN CAM
E ECCENTRIC NUT	L ANCHOR PIN LOCK WASHER
F RETURN SPRING	M ANCHOR PIN NUT
G ANCHOR PIN PLATE	N BACKING PLATE ASSEMBLY
	O LINING—REVERSE

RA PD 305277

Figure 87—Wheel Brake Shoes, Disassembled

lining. Loosen lock nut on anchor pin of forward shoe. Place wrench on anchor pin with handle down, and punch marks on ends of anchor pins toward each other; turn wrench toward rim, or forward, until 0.005-inch clearance is obtained by feel of gage. Hold anchor pin and tighten lock nut. Turn brake drum so gage is at lower end of reverse brake shoe lining. Loosen anchor pin lock nut on reverse shoe. Place wrench on anchor pin with handle down, and punch mark on end of anchor pin toward other anchor pin; turn wrench handle toward rim, or to the rear, until 0.005-inch clearance is obtained by feel of gage. Hold anchor pin and tighten lock nut. Follow same procedure on the other three brakes. Check amount of fluid in master cylinder (par.

BRAKES

149), and apply foot brake pedal to test brakes. Bleed brakes if *soft* pedal is experienced (par. 151). Install wheel. Lower vehicle to floor.

c. Removal of Brake Shoes and Linings. Raise vehicle. Remove wheel hubs (pars. 128 and 141). Loosen eccentric lock nuts (fig. 87). Turn eccentric so that low side is against the shoes. Install brake cylinder clamp to hold pistons in place. Remove brake shoe return spring. Remove anchor pin nuts, lock washers, anchor pins, and anchor pin plate from backing plate. Remove brake shoes. Remove brake shoe anchor pin cam. Inspect exterior of wheel brake cylinder for leakage of brake fluid. If leakage is apparent, replace cylinder assembly (par. 150).

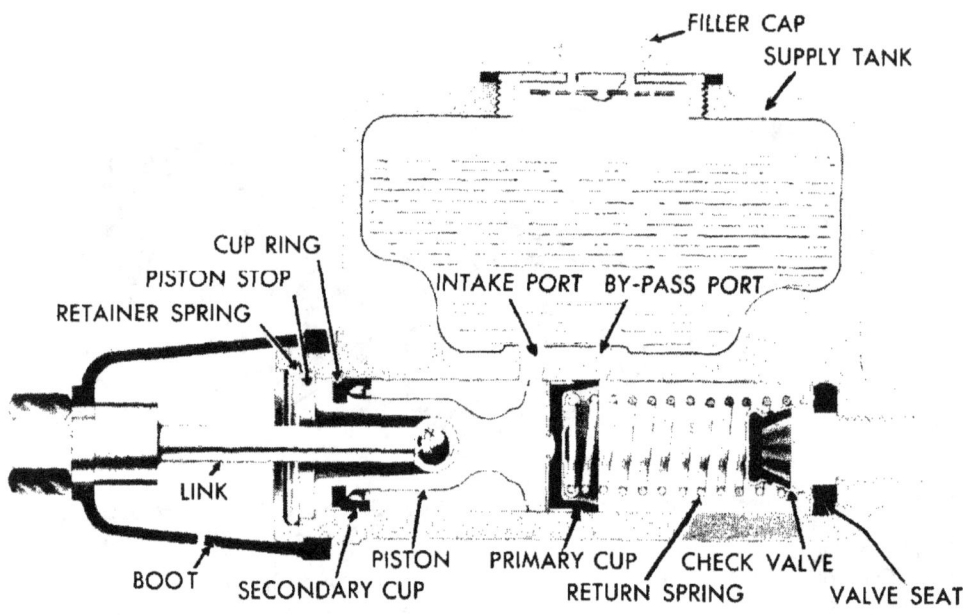

Figure 88—Master Cylinder

d. Installation of Brake Shoes and Linings. Install cam in brake shoes. Install anchor pin plate on anchor pins; install pins in brake shoes, and mount assembly on brake backing plate. NOTE: *Forward shoe has longest lining.* Install brake return spring, and remove brake cylinder clamp. Install brake anchor pin lock washers and nuts. NOTE: *Turn brake anchor pins so that punch marks on ends are toward each other. Do not tighten anchor nuts.* Install hubs (pars. 128 and 141). Make major brake adjustment (par. 149 b).

149. MASTER CYLINDER.

a. Removal. Raise hood and disconnect battery ground at battery terminal. Remove two bolts holding master cylinder shield and remove shield. Pull stop light switch wires out of terminal on switch. Remove stop light switch. Remove outlet fitting screw. Remove master cylinder front screw attaching cylinder to frame. Remove master cylinder rear bolt nut. Remove cotter pin holding master cylinder tie bar on pedal cross shaft. Remove master cylinder boot (fig. 88). Remove master cylinder and tie bar. Remove tie bar from master cylinder.

b. Installation. Fill master cylinder with brake fluid. Install tie bar and rear bolt on master cylinder, and install master cylinder in frame with tie bar on pedal shaft. Install cotter pin in pedal shaft. Install eyebolt link in cylinder. Install master cylinder front screw,

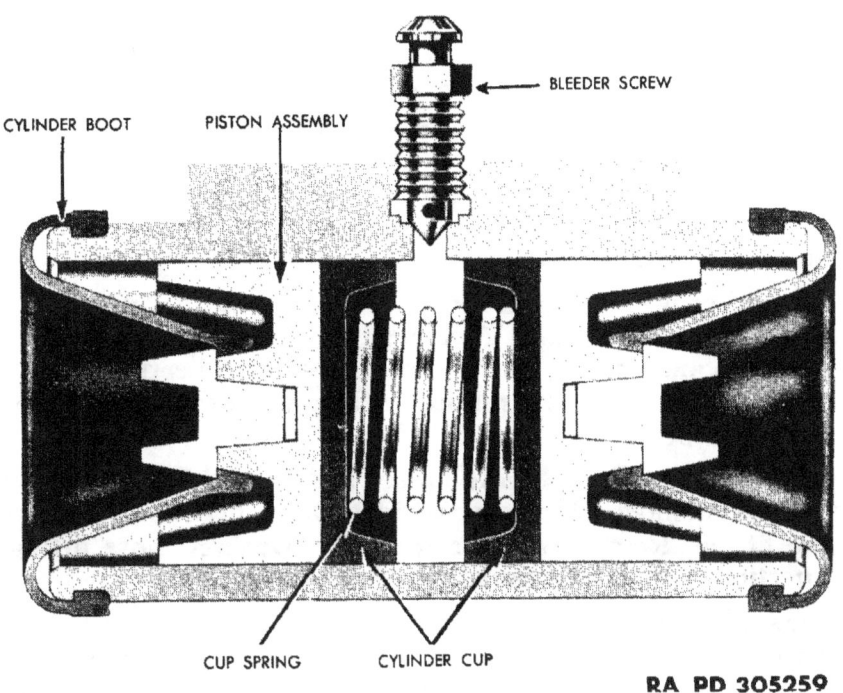

Figure 89—Wheel Cylinder

and tighten rear bolt. Install master cylinder boot with drain hole down. Install outlet fitting bolt. Install stop light switch. Insert stop light wires in terminals. Install master cylinder shield with two bolts. Bleed brakes (par. 151). Attach battery ground cable. Lower hood and hook.

150. WHEEL CYLINDER.

a. Removal. Raise vehicle so that tire clears floor. Remove wheel and hub (pars. 128 and 141). Remove brake shoe return spring. Spread shoes until clear of brake cylinder. Disconnect brake tube at

BRAKES

backing plate. Remove two screws holding cylinder to backing plate, and remove cylinder.

b. Installation. Place cylinder in position on backing plate, and attach with two screws and lock washers. Attach brake tube. Enter brake shoes in slots of cylinder pistons (fig. 89). Install brake shoe return spring. Replace wheel and hub (pars. 128 and 141). Bleed brake (par. 151). Apply foot brake pedal to test brakes. If soft pedal is experienced, bleed all brakes. Lower vehicle to floor.

151. FLEXIBLE LINES, HOSES, AND CONNECTIONS.

a. Removal of Brake Hose at Front Wheels. Remove brake tube connections at each end. With screwdriver slip hose lock off ends of hose fitting, and remove hose.

b. Installation of Brake Hose at Front Wheels. Place hose in brackets and drive locks into place in the fittings. Attach brake tube connections. Bleed brake. Press brake pedal; if soft pedal is experienced, bleed all brakes (subpar. s below).

c. Removal of Brake Hose at Frame and Front Axle. Remove brake tube connection at frame bracket, upper end of hose. With screwdriver, remove hose spring lock from fitting at bracket. Remove fitting from bracket. Unscrew brake hose lower fitting from T-connection on axle and remove.

d. Installation of Brake Hose at Frame and Front Axle. Screw brake hose lower fitting into T-connection on axle. Insert upper fitting into bracket, and install spring lock. Attach brake tube connection. Bleed both front brakes (subpar. s below). Press brake pedal; if soft pedal is experienced, bleed all brakes.

e. Removal of Rear Brake Hose. Remove brake tube connection frame cross member. With screwdriver, drive brake hose spring lock off hose fitting. Remove hose from frame. Unscrew hose fitting from T-connection on rear axle housing.

f. Installation of Rear Brake Hose. Screw brake hose into T-connection on rear axle housing. Insert hose fitting into frame, and drive spring lock into fitting. Attach tube connection. Bleed both rear brakes (subpar. s below). Press brake pedal; if soft pedal is experienced, bleed all brakes.

g. Removal of Master Cylinder to Front Hose Brake Tube. Remove clip from frame. Disconnect tube from brake hose fitting (frame to axle). Disconnect tube from master cylinder connection, and remove tube.

h. Installation of Master Cylinder to Front Hose Brake Tube. Connect tube at master cylinder. Connect tube at brake hose (frame to front axle). Install tube clip at frame. Bleed front brakes (subpar. s below). Press brake pedal; if soft pedal is experienced, bleed all brakes.

i. Removal of Master Cylinder to Rear Hose Brake Tube. Remove clip on underside of frame rear cross member. Remove clip

¼-TON 4 x 4 TRUCK (WILLYS-OVERLAND MODEL MB and FORD MODEL GPW)

on frame side member. Disconnect tube at rear brake hose. Remove master cylinder shield. Disconnect tube at master cylinder. Withdraw tube to rear of vehicle.

j. Installation of Master Cylinder to Rear Hose Brake Tube. Install tube in frame side member. Connect tube to master cylinder, and install master cylinder shield. Install tube in frame rear cross member, and attach to hose fitting. Install tube clips on frame side member and rear cross member. Bleed rear brakes (subpar. s below). Press brake pedal; if soft pedal is experienced, bleed all brakes.

k. Removal of Tee to Front Hose Brake Tube—Left. Disconnect brake tube at tee connection. Disconnect tube at brake hose fitting and remove tube.

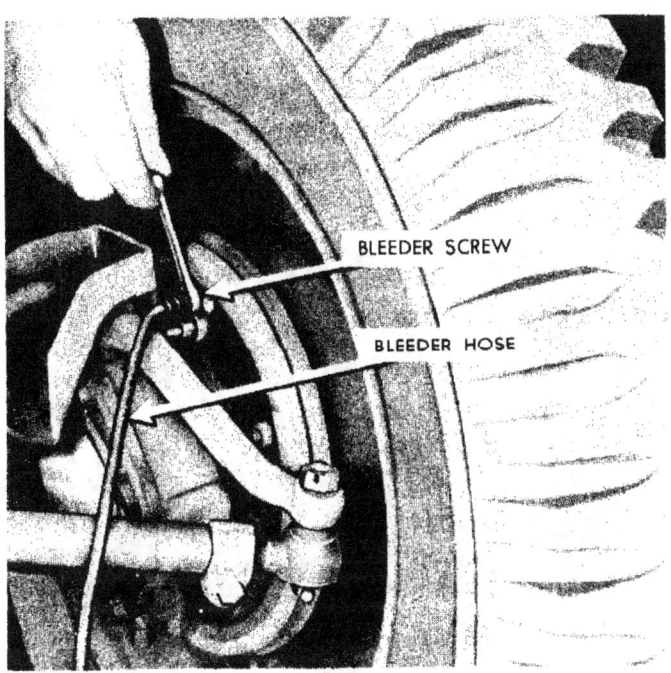

RA PD 305263

Figure 90—Bleeding Brakes

l. Installation of Tee to Front Hose Brake Tube—Left. Connect brake tube at tee connection. Connect tube at hose fitting. Bleed left brake (subpar. s below). Press brake pedal; if soft pedal is experienced, bleed all brakes.

m. Removal of Tee to Front Hose Brake Tube—Right. Remove clips and clamps on axle. Disconnect tube at tee connection. Disconnect tube at hose fitting and remove tube.

n. Installation of Tee to Front Hose Brake Tube—Right. Connect brake tube at tee connection. Connect tube at brake hose fitting. Install clips and clamps on axle. Bleed right front brake (sub-

BRAKES

par. s below). Press brake pedal; if soft pedal is experienced, bleed all brakes.

o. Removal of Front Wheel Cylinder to Hose Brake Tube. Disconnect tube at brake hose. Disconnect tube at wheel cylinder and remove tube.

p. Installation of Front Wheel Cylinder to Hose Brake Tube. Attach brake to wheel cylinder. Attach tube to hose fitting. Bleed brake (subpar. s below). Press brake pedal; if soft pedal is experienced, bleed all brakes.

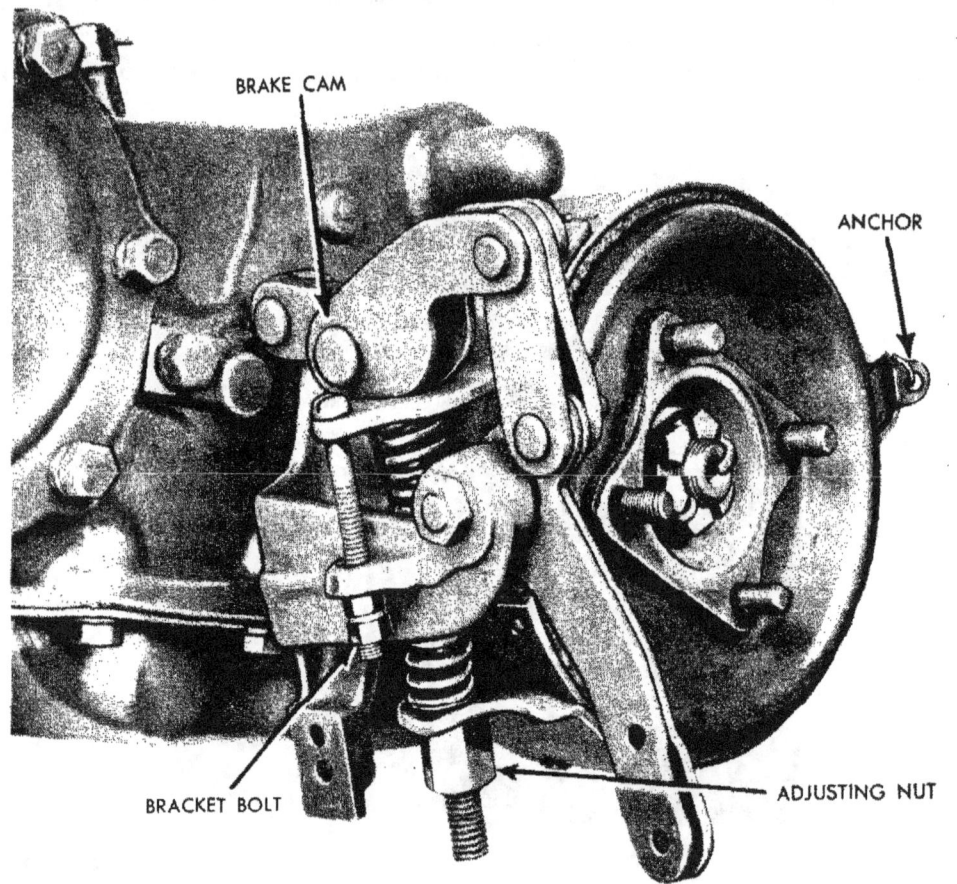

Figure 91—Parking (Hand) Brake

q. Removal of Tee to Rear Brake Tube–Right. Disconnect brake tube at tee connection. Disconnect tube at wheel cylinder. Remove tube clamp on axle. Bend tube slightly and remove.

r. Installation of Tee to Rear Brake Tube. Attach brake tube to wheel cylinder. Attach tube to tee connection. Install clamp on axle. Bleed brake (subpar. s below). Press brake pedal; if soft pedal is experienced, bleed all brakes.

s. Bleeding Brakes. Remove screws holding brake master cylinder inspection cover to toeboard between foot pedals and remove cover. Reach through hole, and clean around master cylinder filler cap. Remove cap and fill master cylinder with brake fluid. Replace cap temporarily. Clean all bleeder connections at wheel cylinders (fig. 90). Attach bleeder hose to *right rear* wheel cylinder bleeder screw, and place end in a glass jar or bottle so that the end is submerged in brake fluid. Open bleeder screw a three-quarter turn. Press brake pedal by hand, allowing it to return slowly. Continue action until air bubbles cease to appear at end of bleeder hose. Tighten bleeder screw and remove hose. Follow the same procedure on the *right front* brake, then the *left rear*, and finally, the *left front* brake. Replenish master cylinder brake fluid supply. Install filler cap and inspection cover.

152. PARKING (HAND) BRAKE.

a. Adjustment. Place hand brake grip in released position. Check brake levers to see that cable is free and released. Remove lock wire from anchor adjusting screw (fig. 91). Place 0.005-inch thickness gage between band and drum at anchor screw, and adjust screw to secure clearance. Install lock wire. Tighten adjusting nut until brake band is tight around drum. Loosen bracket bolt lock nut and rear nut. Back nut off two turns and set lock nut. Loosen adjusting nut so that brake band has approximately 0.010-inch clearance on drum.

b. Removal of Parking (Hand) Brake Band. Remove anchor bolt lock wire. Remove anchor bolt. Remove bracket bolt. Remove cotter pin from brake cam clevis pin and remove pin. Remove brake band adjusting nut. Remove brake adjusting bolt and spring. Remove retracting spring, and remove brake band assembly.

c. Installation of Parking (Hand) Brake Band. Install brake band on drum. Install brake release spring and adjusting bolt. Install clevis pin in brake cam and head of adjusting bolt. Install cotter pin in clevis pin. Install brake band adjusting nut. Install bracket bolt and nuts. Install anchor bolt. Adjust brake (subpar. a above). Install retracting spring.

Section XXVIII

SPRINGS AND SHOCK ABSORBERS

	Paragraph
Description and data	153
Maintenance	154
Spring shackles and bolts	155
Springs	156
Shock absorbers	157

153. DESCRIPTION AND DATA.

a. Description. The springs (figs. 92 and 93) are of the semi-elliptic type with the second leaf wrapped around the spring eye of the first (main) leaf. The front springs appear to be identical, but

RA PD 305265

Figure 92—Left Front Spring

have different load carrying ability. The left spring has an "L" painted on the underside of the second leaf at the front end. Four spring leaf clips keep the leaves in alinement, and hold the leaves together to take the rebound. The front spring is shackled at the front end; the rear spring is shackled at the rear end. A spring pivot bolt attaches the opposite end of the spring to the frame. The shackles are of the threaded U-bolt type with threaded bushings having right- and left-hand threads. The left-hand threaded shackle ends and bushings are used in the spring eye of the left front spring and right rear spring. Left-hand threaded shackles have a small forged boss on the lower shank of the shackle. The left-hand threaded bushings have a groove cut around the hexagon head. The left front spring is equipped with a torque reaction spring to stabilize the front axle in extremely rough service. The shock absorbers are of the hydraulic-cylinder type, direct-acting, two-way control, adjustable, and refillable.

¼-TON 4 x 4 TRUCK (WILLYS-OVERLAND MODEL MB and FORD MODEL GPW)

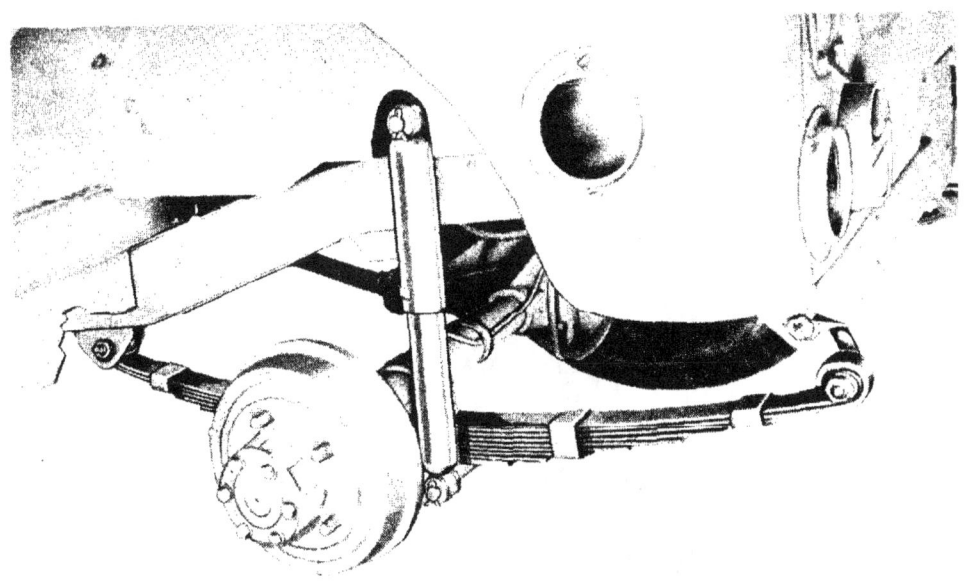

RA PD 305266

Figure 93—Left Rear Spring

b. Data.

Front Springs
- Length—center line of eyebolts.............. 36¼ in.
- Width .... 1¾ in.
- Number of leaves............................. 8
- Spring center bolt........................... At center
- Spring eye bushed............................ Rear

Rear Springs
- Length—center line of eyes................... 42 in.
- Width .... 1¾ in.
- Number of leaves............................. 9
- Spring center bolt........................... At center
- Spring eye bushed............................ Front

SPRINGS AND SHOCK ABSORBERS

Shock Absorbers

Type	Hydraulic
Action	Double
Length—compressed—front	$10\frac{9}{16}$ in.
Length—compressed—rear	$11\frac{9}{16}$ in.
Length—extended—front	$16\frac{1}{8}$ in.
Length—extended—rear	$18\frac{1}{8}$ in.
Adjustable	Yes
Refillable	Yes
Mountings	Rubber bushings

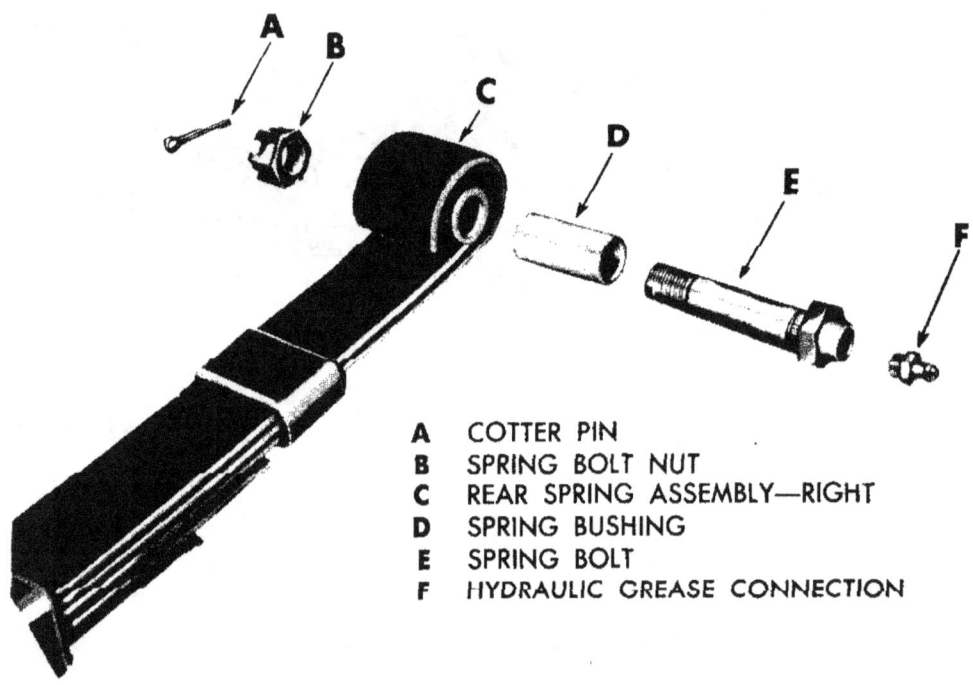

A COTTER PIN
B SPRING BOLT NUT
C REAR SPRING ASSEMBLY—RIGHT
D SPRING BUSHING
E SPRING BOLT
F HYDRAULIC GREASE CONNECTION

RA PD 305280

Figure 94—Right Rear Spring Bolt

154. MAINTENANCE.

a. The springs and shock absorbers should be inspected periodically in accordance with preventive maintenance (par. 23). Lubricate springs to prevent breakage, and excessive wear of spring pivot bolts and shackles (par. 18). Spring bushings and shackles must be free to move. Adjust shock absorbers correctly (par. 157). Replace worn or damaged shock absorber mounting bushings (par. 157).

155. SPRING SHACKLES AND BOLTS.

a. **Removal of Spring Bolt.** Raise vehicle frame until tires just rest on floor. Pull cotter pin in spring pivot bolt nut. Remove nut and drive out bolt (fig. 94).

TM 9-803
155
¼-TON 4 x 4 TRUCK (WILLYS-OVERLAND MODEL MB and FORD MODEL GPW)

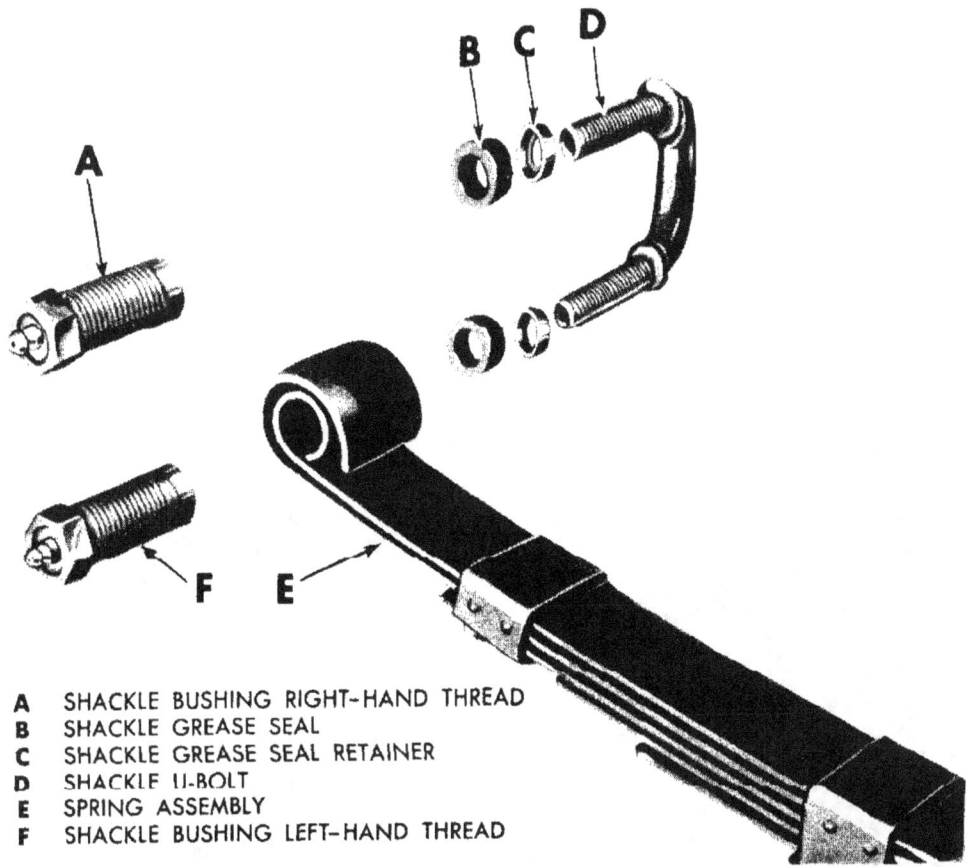

A SHACKLE BUSHING RIGHT-HAND THREAD
B SHACKLE GREASE SEAL
C SHACKLE GREASE SEAL RETAINER
D SHACKLE U-BOLT
E SPRING ASSEMBLY
F SHACKLE BUSHING LEFT-HAND THREAD

Figure 95, Left Front Spring Shackle

RA PD 305279

Figure 95—Left Front Spring Shackle

b. Installation of Spring Bolt. Line up holes in spring bracket and spring. Drive spring pivot bolt into place with oil groove *up*. Install nut and cotter pin. Lubricate with high pressure grease gun. Lower vehicle to floor.

c. Removal of Spring Shackle. Raise vehicle frame until tires just rest on floor. Remove shackle bushings (fig. 95). NOTE: *Left-hand threaded bushings are used in spring end of shackles on left front spring and right rear spring.*

d. Installation of Spring Shackle. Install shackle grease seal and retainer over threaded end, and up to the shoulder. Insert shackle through frame bracket and eye of spring, giving due attention to right- and left-hand threads. Hold shackle tightly against frame, and start upper bushing on shackle. Run in about half-way, then start lower bushing, holding shackle tightly against spring eye. Run bushing in about half-way. Then alternately tighten bushings until upper bushing is tight against frame bracket, and lower bushing hexagon

SPRINGS AND SHOCK ABSORBERS

head is about $\frac{1}{32}$ inch away from spring eye. Lubricate bushings, and try flex of shackle, which must be free. If tight, remove bushings and reinstall.

156. SPRINGS.

a. Removal. Remove spring shackle and pivot bolt (par. 155). Remove four axle spring clip bolt, nuts, and lock washers. Remove spring plate, or torque spring, and pivot bolt lock. Remove spring.

Figure 96—Shock Absorber

b. Installation. Install spring pivot bolt (par. 155 **b**). Install shackle (par. 155 **d**). Raise vehicle, and place center bolt in spring saddle on axle. Install axle spring clips and nuts. NOTE: *Axle spring clip nut torque wrench reading should be 50 to 55 foot-pounds; spring pivot bolt nut, 27 to 30 foot-pounds; torque reaction spring bolt, 60 to 65 foot-pounds.*

¼-TON 4 x 4 TRUCK (WILLYS-OVERLAND MODEL MB and FORD MODEL GPW)

157. SHOCK ABSORBERS.

a. Removal. Pull cotter pins holding upper and lower washers against rubber bushing on mounting brackets. Remove washers, and pull off shock absorbers and rubber bushings (fig. 96).

b. Installation. Check shock absorber adjustment; compress shock absorber, and turn one end to engage adjusting keys in slots. Turn end in clockwise direction until limit of adjustment is reached, then turn end counterclockwise two turns for average adjustment. NOTE: *Turn end clockwise for firmer control, and counterclockwise for softer control, allowing faster spring action.* Install inner mounting rubber bushing on upper and lower bracket pins. Install shock absorber. Install outer bushing and flat washer. Use bushing compressor (41-C-2554-400) to compress bushing, and install cotter pin. Spread both ends of cotter pin to hold washer evenly in proper position.

Section XXIX

STEERING GEAR

	Paragraph
Description and data	158
Maintenance	159
Steering connecting rod	160
Steering wheel	161
Steering Pitman arm	162
Steering gear	163

RA PD 305269

Figure 97—Steering Gear—Phantom View

158. DESCRIPTION AND DATA.

a. Description. The steering gear (figs. 97 and 99) is of the conventional type, mounted on the left frame side member, and connected to the front axle steering ball crank by a Pitman arm and steering connecting rod (fig. 98). The steering gear is of the cam and lever type with a variable-ratio cam. The steering wheel is of the

TM 9-803
158-159
¼-TON 4 x 4 TRUCK (WILLYS-OVERLAND MODEL MB and FORD MODEL GPW)

3-spoke, safety type, with 17¼-inch diameter. The steering connecting rod is of the adjustable, ball-and-socket type.

b. Data.

Make and model .... Ross T-12
Type .... Cam and twin pin lever
Ratio .... Variable; 14-12-14 to 1
Wheel .... 3-spoke; safety type; 17¼ in.

159. MAINTENANCE.

a. Maintenance consists primarily of proper lubrication (par. 18) and periodic inspection in accordance with preventive maintenance

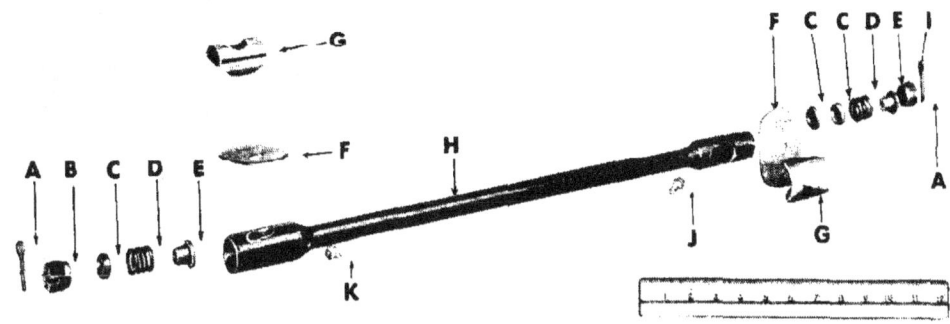

A COTTER PIN
B ADJUSTING PLUG—LARGE
C BALL SEAT
D SPRING
E SAFETY PLUG
F DUST COVER
G DUST COVER SHIELD
H CONNECTING ROD ASSEMBLY
I ADJUSTING PLUG—SMALL
J HYDRAULIC GREASE FITTING
K HYDRAULIC GREASE FITTING

RA PD 305278

Figure 98—Steering Connecting Rod, Disassembled

procedures (par. 23) to include the Pitman arm and steering connecting rod. A systematic inspection for steering troubles is as follows:

(1) Equalize tire pressures and set car on level floor.
(2) Inspect king pin and wheel bearing for looseness.
(3) Check wheel run-out.
(4) Check for spring sag.
(5) Inspect brakes and shock absorbers.
(6) Check steering assembly and connecting rod.
(7) Check toe-in.
(8) Check toe-out on turns.

212

STEERING GEAR

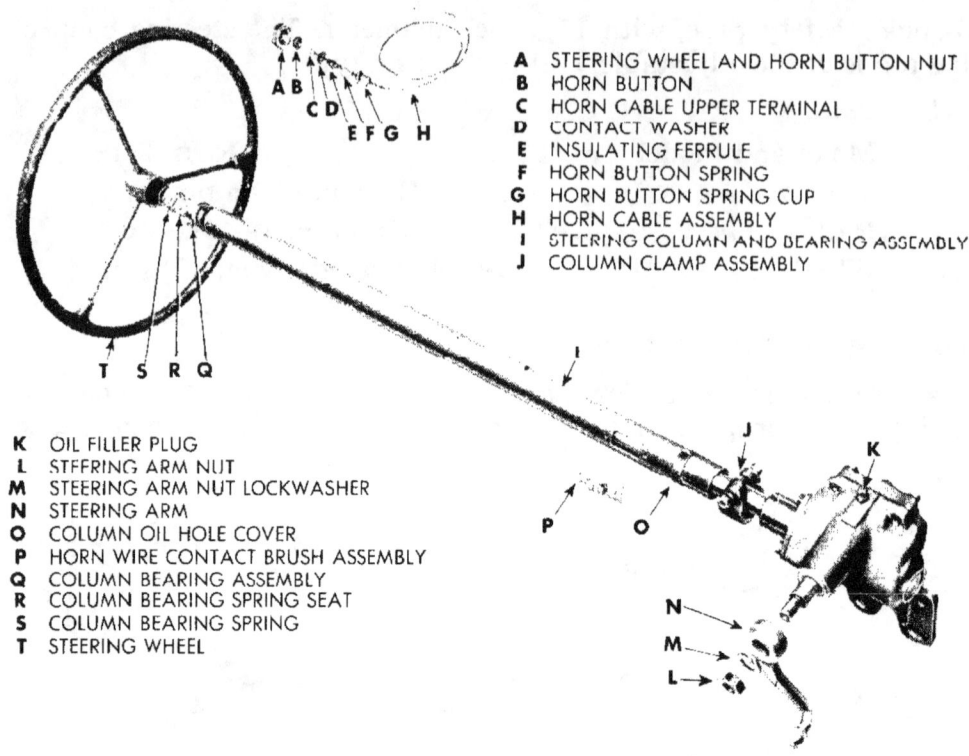

A STEERING WHEEL AND HORN BUTTON NUT
B HORN BUTTON
C HORN CABLE UPPER TERMINAL
D CONTACT WASHER
E INSULATING FERRULE
F HORN BUTTON SPRING
G HORN BUTTON SPRING CUP
H HORN CABLE ASSEMBLY
I STEERING COLUMN AND BEARING ASSEMBLY
J COLUMN CLAMP ASSEMBLY

K OIL FILLER PLUG
L STEERING ARM NUT
M STEERING ARM NUT LOCKWASHER
N STEERING ARM
O COLUMN OIL HOLE COVER
P HORN WIRE CONTACT BRUSH ASSEMBLY
Q COLUMN BEARING ASSEMBLY
R COLUMN BEARING SPRING SEAT
S COLUMN BEARING SPRING
T STEERING WHEEL

RA PD 305268

Figure 99—Steering Gear, Disassembled

(9) Check tracking of front and rear axle.

(10) Check frame alinement.

b. If steering difficulty is experienced after checking and correcting the above items, report to higher authority, because the trouble may be due to wheel balance, caster, camber, or king pin inclination.

160. STEERING CONNECTING ROD.

a. **Removal.** Pull cotter pin at each end of rod. Unscrew plugs and remove rod.

b. **Installation.** Correct end of connecting rod to be attached to front axle bell crank will have the lubrication hydraulic fitting to the right. Install safety plug, spring, and ball seat in this end of rod. Install rod on bell crank ball. Install adjusting plug. Screw plug in firmly against ball, back off one-half turn, and lock with cotter pin. Insert ball seat in other end of rod. Install rod on steering Pitman arm. Install second ball seat, spring, safety plug, and adjusting plug in order. Screw plug in firmly against ball, back off one-half turn, and lock with cotter pin. Lubricate with high pressure gun.

161. STEERING WHEEL.

a. **Removal.** Raise hood, remove horn wire at steering post terminal, and tape end so it will not ground. Remove steering wheel

TM 9-803
161-163
¼-TON 4 x 4 TRUCK (WILLYS-OVERLAND MODEL MB and FORD MODEL GPW)

nut, and lift off horn button. Pull steering wheel off with steering wheel puller.

b. Installation. Set front wheels straight ahead. Install steering wheel so that one spoke of wheel is in vertical position above steering post. Drive wheel down on post. Install horn button and steering wheel nut. Untape horn wire, attach to steering post terminal, try horn, and lower hood.

162. STEERING PITMAN ARM.

a. Removal. Remove Pitman arm nut and lock washer. Remove Pitman arm by using wedge type Pitman arm remover, or as follows: Drive a chisel between the arm and the steering gear case at the front side, and using a bar, strike rear side of arm to loosen it on the tapered serrations. Pull cotter pin in rear end of steering connecting rod, and remove adjusting plug. Take steering connecting rod off Pitman arm ball.

b. Installation. Turn steering wheel maximum distance to right; turn wheel to left, and count turns. Turn wheel to right exactly one-half of the turns. Install steering connecting rod on Pitman arm (par. 160 b). Set front wheels in straight-ahead position, and install Pitman arm on steering gear. Install lock washer and nut. Tighten nut securely. Lubricate connecting rod hydraulic fitting.

163. STEERING GEAR.

a. Removal. Raise hood and tie to windshield. Remove battery ground cable at post on battery. Release headlight bracket wing nut and tilt headlight away from fender. Remove headlight wires from junction block on fender splasher. Remove blackout headlight wire clip on fender. Remove horn from bracket. Remove headlight wires from junction block on dash. Disconnect blackout driving light wire from slip connector at dash. Remove two screws attaching horn wire contact brush assembly to steering column. Loosen steering column clamp bolt. Remove bolts attaching fender to body, frame, and radiator grille. Remove fender support bolts in frame and remove fender. Remove cotter pin from rear end of steering connecting rod, unscrew adjusting plug, and remove from Pitman arm ball. Remove steering wheel nut and horn button. Pull steering wheel with steering wheel puller. Remove steering column bracket. Remove bolts in steering column floor seal, and remove seal and retainer. Pull steering column up off tube. Remove steering column to frame bolts. Lower upper end of steering column, and lift lower end out over frame side member.

b. Installation. Check steering gear lubricant; replenish if necessary. Insert upper end of steering gear through toeboard, and position in chassis. Install steering gear to frame bolts, but do not tighten. Install steering column floor seal, retainer, and screws. Install steering column over tube, with horn contact brush opening up. Tighten steering column clamp. Install horn wire contact brush, and tighten

STEERING GEAR

screws. Attach steering column bracket, and tighten steering gear to frame bolts. Install steering wheel (par. 161 b). Install steering connecting rod (par. 160 b). Install fender and support bolts, also screws and bolts to frame, body, and radiator grille. Install blackout headlight wire clip on fender. Attach headlight wires to junction block on fender splasher. Connect blackout headlight wire at slip connector. Attach headlight wire to junction block on dash. Connect blackout driving light wire in slip connector at dash. Install horn on bracket. Tilt headlight, and tighten wing nut. Attach battery cable. Check operation of horn and lights. Lower hood and lock.

c. **Adjustment.** Loosen lock nut on side adjusting screw. With front wheels straight ahead, adjust screw for minimum backlash of studs in cam groove. Tighten lock nut. For other adjustments, report to higher authority.

¼-TON 4 x 4 TRUCK (WILLYS-OVERLAND MODEL MB and FORD MODEL GPW)

Section XXX

BODY AND FRAME

	Paragraph
Description and data	164
Maintenance	165
Instruments	166
Seats and cushions	167
Windshield wiper	168
Windshield	169
Top	170
Rifle holder	171
Shovel and ax	172
Hood	173
Radiator guard	174
Fenders	175

164. DESCRIPTION AND DATA.

a. Description. The body (figs. 1 to 4) is of the open type, identified by a name plate located on the instrument panel (figs. 5 and 6). There are two individual tubular frame front seats and a rear seat. The left front seat cushion can be raised to fill the fuel tank; the right front seat can be raised forward for stowage of the vehicle removable top, curtains, and windshield and light covers. The rear seat can be raised to reach the tire pump. Tools and accessories are stowed in two compartments in the rear corners of the body. The windshield is equipped with dual, hand-operated wipers, and can be opened forward or folded down on top of the hood. A fire extinguisher and an adjustable rear vision mirror are mounted on the left side of the cowl. Safety straps are provided in the entrance ways. A rifle holder is mounted on the lower panel of the windshield over the instrument panel. A strap and sheath carry a shovel and ax on the left side of the body. Hand grips on the side of the body facilitate lifting. The fuel tank sets in a sump in the floor pan under the driver's seat. The vehicle top is supported by top bows which can be folded down along the body sides to form a hand rail. A fuel can rack, trailer connection, and spare tire and wheel are mounted on the body rear panel. The chassis has five cross members; the rear intermediate cross member having a gun platform. Box-type, reinforced frame side members are used for maximum strength. Bumpers at the front and rear, and a radiator guard provide protection against damage. A pintle hook at the rear provides a means of hauling a trailed load.

b. Data.

Body type	Open	Windshield type	Folding
Driver's position	Left side	Cross members	5

Chassis frame type .... Double drop

BODY AND FRAME

165. MAINTENANCE.

a. General maintenance of the body requires periodic tightening of all loose parts, and lubrication of wearing parts. Keep the body clean and touch up bare spots to prevent rust. Keep the sump under the fuel tank free of dirt, stones, and water. Keep the sump front drain hole cover on so that dirt and water thrown by the front wheel will not enter. Keep the rear drain hole cover in the glove compartment in the instrument panel except when crossing water, when it must be installed. Water in the body can be drained by removing drain plugs in the floor at the side of the cowl. Chassis maintenance concerns primarily, proper lubrication of connecting parts (par. 18).

166. INSTRUMENTS.

a. Procedure for the removal and installation of the various panel instruments is identical, and as follows:

b. Removal. Remove battery ground cable at battery post as a safety precaution. Remove connecting wires or tubes. Remove two nuts holding retaining clamp in place, and remove instrument through face of instrument panel.

c. Installation. Install instrument in place in panel. Install retaining clamp and nuts. Attach tubes or wires.

167. SEATS AND CUSHIONS.

a. Removal of Seat Cushions and Backs. Remove five screws holding front seat cushion to frame at rear side, and remove cushion. Remove 10 screws holding seat back to frame, and remove seat back. Lift up back edge of rear seat cushion, remove five screws holding front edge of cushion to frame, and remove cushion. Remove five screws in top edge of seat back and two in lower edge, and remove seat back.

b. Installation of Seat Cushions and Backs. Place rear seat back in position, and install two lower screws. Pull edge of seat back up in place, and install five screws in top side. Place rear seat cushion in position, top side down. Install screws, and turn cushion over into place. Place front seat back in position, and install screws. Place seat cushions in position, and install screws.

c. Removal of Front Seats. Remove three screws holding back of driver's seat to floor. Remove screw in wheel housing holding seat. Remove two bolts holding front of seat frame to floor, and lift out seat. Remove two bolts holding right front seat bracket to floor, and lift out seat.

d. Installation of Front Seats. Place seat in position. Install bolts in place at front of seat. On driver's seat install screws and bolts holding seat back to floor and wheel housing.

e. Removal of Rear Seat. Pull up front edge of seat to fold seat. Remove bolt in tool compartment holding retainer bracket at seat bracket. On same side remove two bolts holding seat back bracket. Raise end of seat and lift out.

f. Installation of Rear Seat. Place seat in position in brackets. Install retainer bracket. Install seat back bracket and bolts.

168. WINDSHIELD WIPER.

a. Removal. Remove nuts holding wiper handle, and remove handles. Remove plain washer, and remove wiper blades.

b. Installation. Install blades and arms in place through windshield frame. Install plain washers, handles, and nuts.

169. WINDSHIELD.

a. Removal. Unhook windshield clamps on instrument panel (fig. 5). Remove wing screws at sides of cowl, and lift off windshield.

b. Installation. Place windshield in position, and install wing screws at sides of cowl. Clamp windshield to instrument panel.

170. TOP.

a. Installation. Loosen the two wing screws at the pivot brackets (fig. 4). Slide tubular bows back out of front bracket. Install front ends in rear brackets, and tighten winged screws. Allow front bow to drop down over seats. Remove top from under right front seat. Attach top to fasteners at top of windshield. Stretch top over bow and down to body back panel. Place straps in metal loops, and attach to body panel; stretch top, and buckle straps. Raise front bow into position at bow flaps, and snap flaps around bow. The curtains are attached in the conventional way with snap fasteners.

b. Removal. Remove curtains by releasing snap fasteners. Unsnap bow flaps, and lower front bow on front seat. Unbuckle top straps at body rear panel. Unsnap top at top of windshield, and remove. Fold top and stow under right front seat. Loosen wing screws in top rear brackets. Fold front bow against rear bow. Raise bows out of rear brackets, and insert lower ends in front brackets. Tighten rear bracket screws.

171. RIFLE HOLDER.

a. Removal. Swing the rifle bumper to the right, at the right end of holder, and remove rifle. Remove two bolts holding rifle holder to windshield lower panel, and remove holder (fig. 5).

b. Installation. Place rifle holder in position on windshield panel with butt end to the left, insert bolts, and tighten securely. Swing rifle bumper to the right. With barrel up, insert butt end of rifle in holder at the left. Push rifle up against spring pressure, and turn bumper to left under rifle.

BODY AND FRAME

172. SHOVEL AND AX.

a. Removal. Release straps and remove shovel or ax individually.

b. Installation. Turn bit, or blade, of ax up. Insert handle in front clamp. Insert blade in sheath. Pull up clamp under ax head, and strap in place. Turn face of shovel against cowl and place in strap on cowl side. Wrap fabric strap, through handle, over grip, between grip and side of body, through loop, over outside of grip, and buckle. NOTE: *This will hold the shovel forward in the strap on the cowl side (fig. 100).*

173. HOOD.

a. Removal. Unhook hood and raise against windshield. Remove screws in hinge at cowl, and disconnect bond strap.

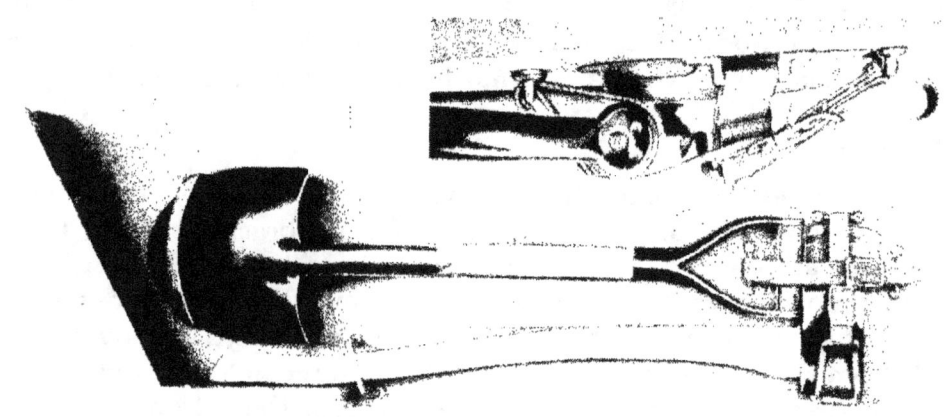

RA PD 305271

Figure 100—Shovel and Ax Mounting

b. Installation. Place hood in position and install hinge screws in cowl, but do not tighten. NOTE: *Install bonded screws last as follows:* Install flat washer on screw. Install screw through bond strap. Install flat washer. Install washer between hinge and hood. Install screw through hinge, and tighten to cowl. Lower hood for alinement. Raise hood and tighten screws. Lower hood and hook down both sides.

174. RADIATOR GUARD.

a. Removal. Raise hood. Remove headlight hinge bolts. Remove headlight wire clips on guard. Remove blackout headlight wire clip on left front fender. Remove wires from slip connector at left front fender. Remove fender to guard bolts. Remove guard from chassis. Remove blackout headlight wires, clips, and loom. Remove rubber shield at headlight. Remove blackout headlight nut and washer, and remove both light and wire assemblies.

¼-TON 4 x 4 TRUCK (WILLYS-OVERLAND MODEL MB and FORD MODEL GPW)

b. Installation. Install both blackout headlight, spacer, and wire assemblies on guard. Install washer and nut. Install shield over wire. Install loom on wire. Install wire clips on loom and clips on guard. Install frame bolts in guard. Install guard on chassis. Install fender to guard bolts loosely. Install guard to frame bolt nuts. Tighten fender to guard bolts. Install headlight hinge bolts. Install headlight wire clips on guard. Install blackout headlight wire clip on fender. Connect wires to slip connector. Check operation of lights. Lower hood and lock down.

175. FENDERS.

a. Removal of Right Front Fender. Raise hood. Loosen wing nut on headlight bracket, and tilt light up. Remove battery to front fender strap. Remove battery ground cable at battery post. Remove voltage regulator bolts. Remove fuel line clip to fender. Remove hood catch assembly. Remove fender to radiator guard bolts. Remove fender bolts in support, body, and frame. Remove fender.

b. Installation of Right Front Fender. Place fender on chassis. Install one fender to body bolt. Install one fender to guard bolt. Install other fender bolts and tighten all. Install fuel line clip. Install voltage regulator. Install battery to fender strap. Install battery cable. Position headlight, and tighten wing nut. Install hood catch. Lower hood and lock.

c. Removal of Left Front Fender. Raise hood. Remove headlight bracket wing nut, and tilt lamp up. Remove two wire clips on splasher. Remove wires from junction block, and slip connector at dash. Remove blackout driving light clip on top of fender. Remove three bolts in blackout driving light bracket. Remove blackout driving light wire grommet and clips from fender. Remove junction block on fender. Remove blackout headlight wire clip. Remove hood catch. Remove fender shield to frame bolt. Remove bolts between fender and guard support, body, and frame. Remove fender.

d. Installation of Left Front Fender. Place fender on chassis. Install guard upper bolt. Install all fender bolts loosely. Install blackout headlight wire clip (front) on fender. Install junction block to fender. Install blackout driving light wire through fender and splasher. Install three bolts in blackout driving light bracket and fender. Install blackout headlight wire clip on fender. Install wire grommet in fender. Install two wire clips to fender splasher. Connect wires to junction block at dash and slip connector. Install hood catch. Place headlight in position, and secure with wing nut. Check operation of lights. Lower hood and lock down.

Section XXXI

RADIO INTERFERENCE SUPPRESSION SYSTEM

	Paragraph
Description	176
Data	177
Tests	178
Maintenance, removal, and replacement	179

176. DESCRIPTION.

a. Description. Radio noise suppression is the elimination, or minimizing, of electrical disturbances which interfere with radio reception, or disclose the location of the vehicle to sensitive electrical detectors. Electrical disturbances or radio frequency waves may originate as static discharges between adjoining parts of the vehicle, or may be given off by the electrical systems during operation of the vehicle. These waves are actually radiated as disturbing signals that interfere with any radio receiving apparatus that may be operating in the vehicle or immediate vicinity. Each disturbance (at plugs, breaker points, generator brushes, or elsewhere) creates a surge of electricity, which produces interfering radio waves. Their origin can generally be determined by the nature of the noise heard in the receiver. Radio interference suppression, therefore, involves the suppression of these waves at their sources, or confining them within an area where they cannot be picked up by the antenna of a radio-equipped vehicle. Suppression is accomplished by the use of resistor-suppressors, and condensers. In addition, the hood and other metal parts in the vicinity of the engine are made to form a shield by the use of internal-external toothed lock washers and bond straps; thus, the hood and side panels form a box within which radio frequency waves are confined to prevent their acting on the antenna of receiving equipment. Wiring that may carry interfering surges to a point where interference will affect radio reception, is shielded. In attaching condensers and bond straps, the lock washers must be placed between the parts to be grounded, and tinned spots must be cleaned, but not painted. This is necessary to obtain good connections between the component parts, and to permit electrical energy to dissipate without causing electrical disturbances. The suppression components have no effect on engine performance as long as they are maintained in good condition. The sources of electrical noise interference may be basically divided into three groups: the ignition system, including coil, distributor, and spark plugs; the generator system, including generator and regulator; and the wiring.

177. DATA.

a. Ignition (both high-tension and primary-circuit suppression).

(1) High-tension suppression is of the resistor-suppressor type and consists of:

(a) Coil to distributor high-tension wire at distributor, resistance 10,000 ohms.

¼-TON 4 x 4 TRUCK (WILLYS-OVERLAND MODEL MB and FORD MODEL GPW)

(b) Spark plug high-tension wire at spark plugs, resistance 10,000 ohms.

(2) Primary circuit suppression is of the capacitive-filter type and consists of:

(a) Ignition coil terminal (+) to cylinder block, capacity 0.10 microfarad.

(b) Ignition switch terminal (lower) to instrument panel, capacity 0.01 microfarad.

b. Charging Circuits (includes generator, regulator, ammeter, and battery).

(1) Generator suppression is of the capacitive-filter type and consists of generator armature terminal (A) to ground on generator, capacity 0.10 microfarad.

(2) Regulator suppression is of the capacitive-filter type and consists of:

(a) Regulator field terminal (F) to ground, capacity 0.01 microfarad.

(b) Regulator field terminal (B) to ground, capacity 0.25 microfarad.

c. Miscellaneous Circuits (including radio box and starting circuits).

(1) Radio terminal box suppression is of the capacitive-filter type and consists of:

(a) Radio box terminal to ground, capacity 0.50 microfarad.

(b) Starting switch battery terminal to floor, capacity 0.50 microfarad.

d. Bonding.

(1) BOND STRAPS (ground straps) (figs. 101 and 102). Bond straps are installed from:

(a) Hood to dash, right side (D).

(b) Hood to dash, left side (I).

(c) Cylinder head stud to dash (H).

(d) Cables (hand brake, speedometer, heat indicator) to dash (E).

(e) Generator mounting bolt to cranking motor bracket to engine support bracket (A).

(f) Generator to regulator wire shield to ground on generator and regulator (B).

(g) Front engine bracket to frame, left side (J).

(h) Hood ground to grille, left side (L).

(i) Hood ground to grille, right side (N).

(j) Radio terminal box to ground wire (F).

(2) TOOTHED LOCK WASHERS (figs. 101 and 102). Toothed lock washers are supplied from:

RADIO INTERFERENCE SUPPRESSION SYSTEM

(a) Radiator to frame, right side (O).
(b) Radiator to frame, left side (K).
(c) Body bracket ground to frame, right side (S).
(d) Body bracket ground to frame, left side (R).
(e) Fender splasher ground to frame, right side (U).
(f) Fender splasher ground to frame, left side (T).
(g) Air cleaner mounting (C).
(h) Body hold-down bolts (G).
(i) Radiator grille to cross member (M).
(j) Fender to cowl, left side (P).
(k) Fender to cowl, right side (Q).

178. TESTS.

a. General. Electrical disturbances which cause radio interference are loose bonds, loose toothed lock washers, broken or cracked resistor-suppressors, loose connections, or faulty filters. Following are tests which can be made to determine the cause of interference. The radio equipment in the vehicle may be used as a test instrument to localize troubles, and to determine when faulty parts or conditions have been eliminated or corrected. If the vehicle has no radio equipment, utilize a radio-equipped vehicle placed about 10 feet from the vehicle under test. Here the cooperation of the radio operator is required. Determine the circuits causing the noise by checking as follows:

(1) Operate engine while listening to radio. A regular clicking which varies with engine speed, and ceases the instant the ignition is shut off, is caused by the ignition circuit.

(2) An irregular clicking which continues a few seconds after the ignition is shut off, is caused by the regulator.

(3) A whining noise which varies with engine speed, and continues a few seconds after the ignition is shut off, is caused by the generator.

b. Noise Caused by Ignition Circuit.

(1) Make certain ignition system is functioning properly (section XV). Improper plug gaps, late timing, poor adjustment of breaker points, and damaged or worn distributor, will affect the suppression system.

(2) Inspect resistor-suppressors in spark plug leads. Replace any that are scorched, cracked, or otherwise faulty. Be sure wires are screwed in tightly.

(3) Inspect resistor-suppressor at distributor. Replace if necessary.

(4) Inspect and tighten all bonds in engine compartment.

(5) Inspect capacitive-type filters in primary circuit at ignition coil and ignition switch. Make certain mounting bolts are tight. Replace filter and test for noise.

¼-TON 4 x 4 TRUCK (WILLYS-OVERLAND MODEL MB and FORD MODEL GPW)

c. **Noise Caused by Regulator.**

(1) Check all connections to regulator.

(2) Check capacitive-type filter mounting bolts for tightness and correct placement of lock washers.

(3) Check regulator mounting bolts for tightness and correct placement of lock washers.

(4) Test for noise.

(5) If noise is still present, replace battery circuit filter attached to regulator (B) terminal. Test for noise.

(6) Replace field circuit filter attached to regulator (F) terminal. Test for noise.

(7) Replace armature circuit filter attached to generator (A) terminal.

(8) Test for noise.

d. **Noise Caused by Generator.**

(1) Check to make certain there is no excessive sparking at brushes. Correct if necessary.

(2) Inspect filter mounting, and check placement of lock washers. Tighten.

(3) Inspect ground strap.

(4) Replace filter.

(5) Test for noise.

e. **Noise Caused by Miscellaneous Circuits** (radio box and starting switch).

(1) Inspect mounting of filter attached to circuit. Tighten and test.

(2) Replace filter.

(3) Test for noise.

f. **Noise Observed While Vehicle Is in Motion, but Not When Stopped.**

(1) Inspect and tighten all body bonds (par. 177 d (1) above and figs. 101 and 102).

(2) Inspect and tighten all points where toothed lock washers are used (par. 177 d (2) above and figs. 101 and 102).

(3) Test for noise.

179. MAINTENANCE.

a. **General.** General maintenance of the radio suppression system (par. 23 a (5), item 104) must be made in connection with preventive maintenance items, particularly in regard to spark plugs, distributor and wires, late ignition timing, generator brushes, loose switch contacts, and discharged battery causing high generator charging rate.

TM 9-803
RADIO INTERFERENCE SUPPRESSION SYSTEM

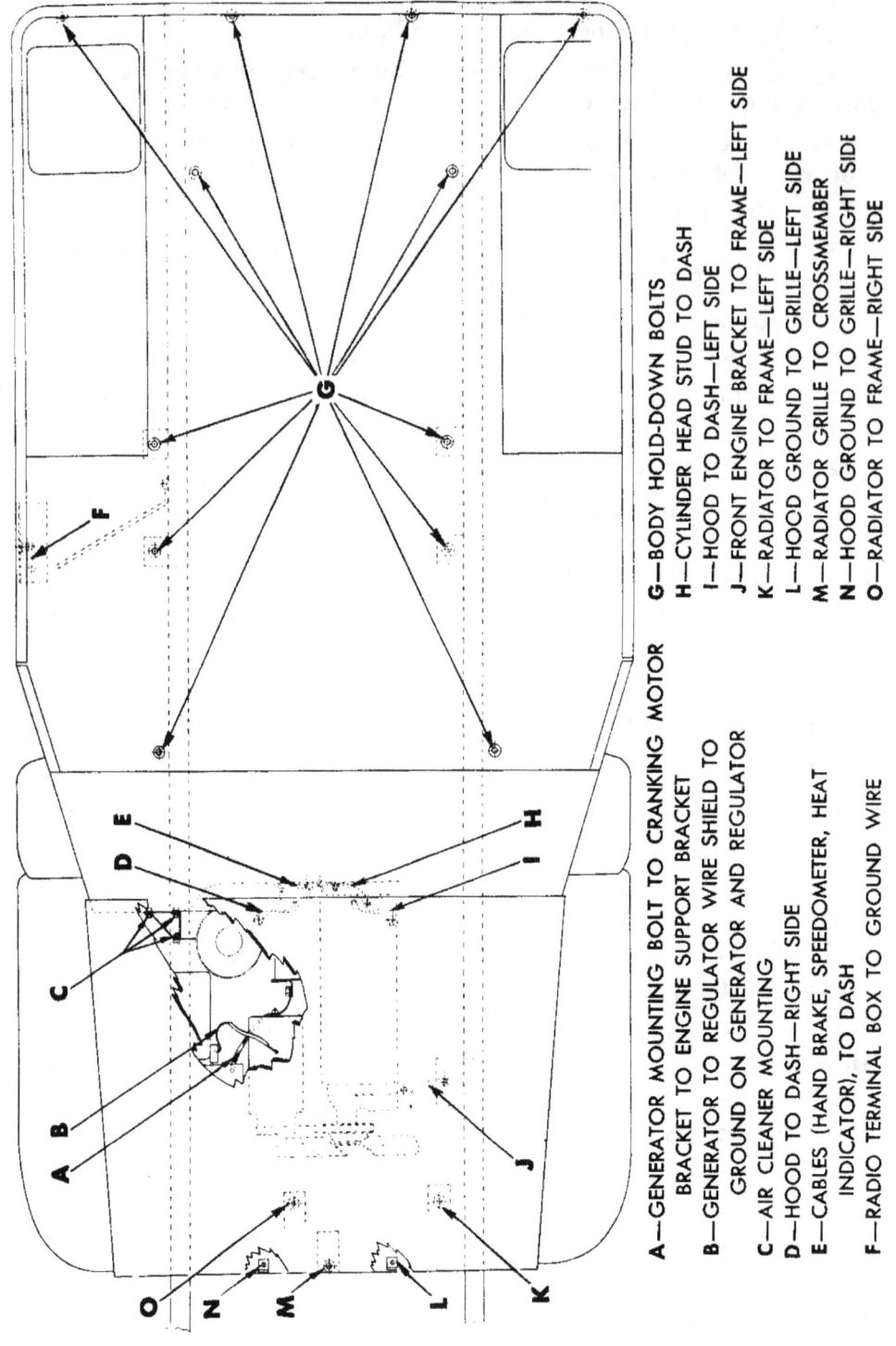

A—GENERATOR MOUNTING BOLT TO CRANKING MOTOR BRACKET TO ENGINE SUPPORT BRACKET
B—GENERATOR TO REGULATOR WIRE SHIELD TO GROUND ON GENERATOR AND REGULATOR
C—AIR CLEANER MOUNTING
D—HOOD TO DASH—RIGHT SIDE
E—CABLES (HAND BRAKE, SPEEDOMETER, HEAT INDICATOR), TO DASH
F—RADIO TERMINAL BOX TO GROUND WIRE
G—BODY HOLD-DOWN BOLTS
H—CYLINDER HEAD STUD TO DASH
I—HOOD TO DASH—LEFT SIDE
J—FRONT ENGINE BRACKET TO FRAME—LEFT SIDE
K—RADIATOR TO FRAME—LEFT SIDE
L—HOOD GROUND TO GRILLE—LEFT SIDE
M—RADIATOR GRILLE TO CROSSMEMBER
N—HOOD GROUND TO GRILLE—RIGHT SIDE
O—RADIATOR TO FRAME—RIGHT SIDE

RA PD 305295

Figure 101—Location of Bond Straps and Fastenings (1)

¼-TON 4 x 4 TRUCK (WILLYS-OVERLAND MODEL MB and FORD MODEL GPW)

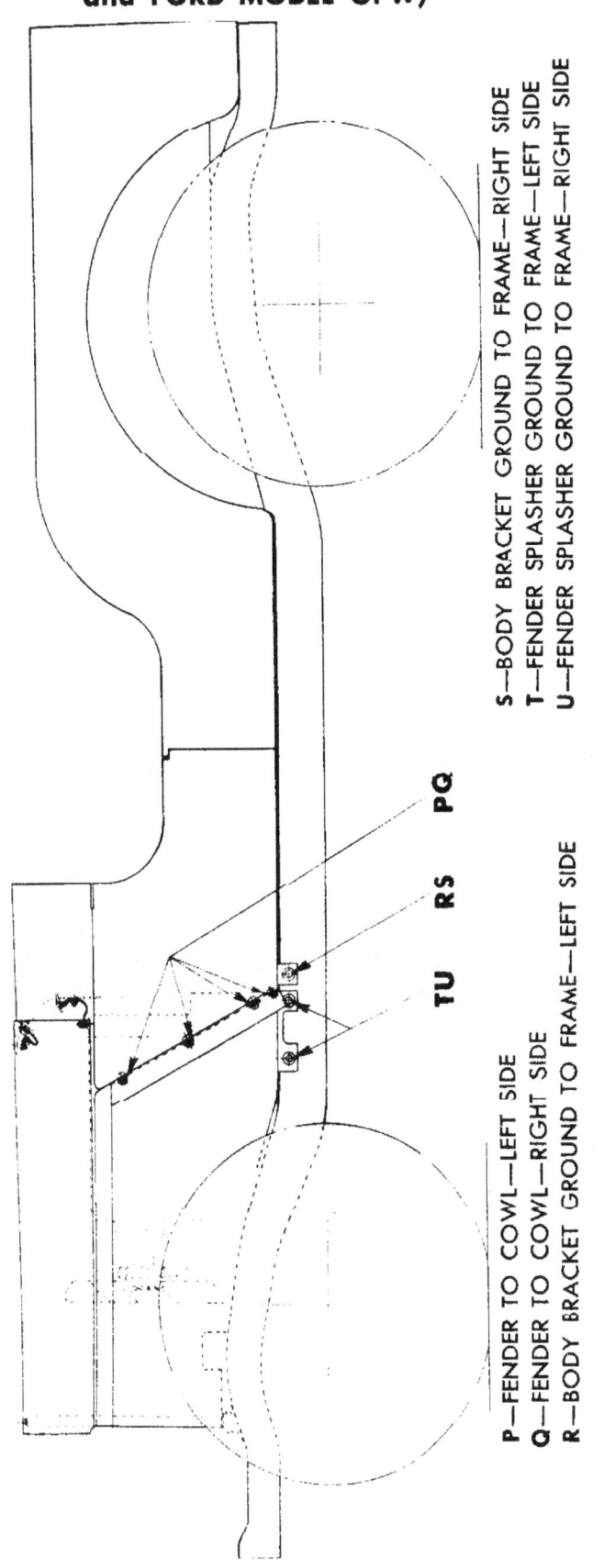

P—FENDER TO COWL—LEFT SIDE
Q—FENDER TO COWL—RIGHT SIDE
R—BODY BRACKET GROUND TO FRAME—LEFT SIDE
S—BODY BRACKET GROUND TO FRAME—RIGHT SIDE
T—FENDER SPLASHER GROUND TO FRAME—LEFT SIDE
U—FENDER SPLASHER GROUND TO FRAME—RIGHT SIDE

RA PD 334758

Figure 102—Location of Bond Straps and Fastenings (2)

RADIO INTERFERENCE SUPPRESSION SYSTEM

b. Ignition Circuits.

(1) Resistor-suppressors, of which there are five (one at each spark plug and one at the distributor), consist of a high resistance element in an insulated housing. Inspect each suppressor for cracked or broken housing. Each suppressor must be threaded tightly into end of spark plug wire so that screw enters strands of cable. Wire terminals must be tight, well pushed down into place, and free of corrosion or dirt.

(2) Capacitive-filters, of which there are two in the ignition circuit, are located at the ignition coil and ignition switch. Replace if faulty by disconnecting wire at terminal, removing mounting nut or screw, and removing filter. Install coil filter with toothed lock washer between filter bracket and mounting nut. Install ignition switch filter with toothed lock washer between filter bracket and panel; also between bracket and mounting nut.

c. Charging Circuit.

(1) A capacitive-filter is mounted on the generator, and attached to the armature (A) terminal. The mounting screw must be tight with the internal-external toothed lock washer between the filter bracket and the generator housing, and the external toothed lock washer under the screw head.

(2) Two capacitive-filters are used on the regulator; one between the battery terminal and ground, the other between the field coil terminal and ground. The mounting screw must be tight with the internal-external toothed lock washer between the filter bracket and the regulator base.

(3) For battery removal and replacement, refer to paragraph 97.

d. Miscellaneous Circuits (radio box and starting circuits). The capacitive-filter used in the radio box between the terminal and the ground wire must be tight with the lock washer under the head of the mounting screw. The capacitive-filter in the starting circuit is mounted on the bottom of the starting switch with the connection on the "live" terminal. The filter bracket is located under the left mounting bolt with an internal-external toothed lock washer between the bolt head and the bracket, and an internal-external toothed lock washer between the nut and the rear side of the dash.

e. Bonding. Bonding points indicated in paragraph 117 d must be clean and tight. Tinned spots must be clean but not painted. Where bonding is obtained by use of an internal-external toothed lock washer, the lock washer must be between the parts to be grounded.

TM 9-803
180-181

¼-TON 4 x 4 TRUCK (WILLYS-OVERLAND MODEL MB and FORD MODEL GPW)

Section XXXII

SHIPMENT AND TEMPORARY STORAGE

	Paragraph
General instructions	180
Preparation for temporary storage or domestic shipment	181
Loading and blocking for rail shipment	182

180. GENERAL INSTRUCTIONS.

a. Preparation for domestic shipment of the vehicle is the same as preparation for temporary storage or bivouac. Preparation for shipment by rail includes instructions for loading and unloading the vehicle, blocking necessary to secure the vehicle on freight cars, number of vehicles per freight car, clearance, weight, and other information necessary to properly prepare the vehicle for rail shipment. For more detailed information, and for preparation for indefinite storage refer to AR 850-18.

181. PREPARATION FOR TEMPORARY STORAGE OR DOMESTIC SHIPMENT.

a. Vehicles to be prepared for temporary storage or domestic shipment are those ready for immediate service but not used for less than 30 days. If vehicles are to be indefinitely stored after shipment by rail, they will be prepared for such storage at their destination.

b. If the vehicles are to be temporarily stored or bivouacked, take the following precautions:

(1) LUBRICATION. Lubricate the vehicle completely (par. 18).

(2) COOLING SYSTEM. If freezing temperature may normally be expected during the limited storage or shipment period, test the coolant with a hydrometer, and add the proper quantity of antifreeze compound to afford protection from freezing at the lowest temperature anticipated during the storage or shipping period. Completely inspect the cooling system for leaks.

(3) BATTERY. Check battery and terminals for corrosion and if necessary, clean and thoroughly service battery (par. 97).

(4) TIRES. Clean, inspect, and properly inflate all tires. Replace with serviceable tires, tires requiring retreading or repairing. Do not store vehicles on floors, cinders, or other surfaces which are soaked with oil or grease. Wash off immediately any oil, grease, gasoline, or kerosene which comes in contact with the tires under any circumstances.

(5) ROAD TEST. The preparation for limited storage will include a road test of at least 5 miles, after the battery, cooling system, lubrication, and tire service, to check on general condition of the vehicle. Correct any defects noted in the vehicle operation, before the vehicle

SHIPMENT AND TEMPORARY STORAGE

is stored, or note on a tag attached to the steering wheel, stating the repairs needed or describing the condition present. A written report of these items will then be made to the officer in charge.

(6) FUEL IN TANKS. It is not necessary to remove the fuel from the tanks for shipment within the United States, nor to label the tanks under Interstate Commerce Commission Regulations. Leave fuel in the tanks except when storing in locations where fire ordinances or other local regulations require removal of all gasoline before storage.

(7) EXTERIOR OF VEHICLES. Remove rust appearing on any part of the vehicle with flint paper. Repaint painted surfaces whenever necessary to protect wood or metal. Coat exposed polished metal surfaces susceptible to rust, such as winch cables, chains, and in the case of track-laying vehicles, metal tracks, with medium grade rust-preventive lubricating oil. Close firmly all cab doors, windows, and windshields. Vehicles equipped with open-type cabs with collapsible tops will have the tops raised, all curtains in place, and the windshield closed. Make sure paulins and window curtains are in place and firmly secured. Leave rubber mats, such as floor mats, where provided, in an unrolled position on the floor; not rolled or curled up. Equipment such as Pioneer and truck tools, tire chains, and fire extinguishers will remain in place in the vehicle.

(8) INSPECTION. Make a systematic inspection just before shipment or temporary storage to insure all above steps have been covered, and that the vehicle is ready for operation on call. Make a list of all missing or damaged items, and attach it to the steering wheel. Refer to "Before-operation Service" (par. 13).

(9) ENGINE. To prepare the engine for storage, remove the air cleaner from the carburetor. Start the engine, and set the throttle to run the engine at a fast idle. Pour 1 pint of medium grade, preservative lubricating oil, Ordnance Department Specification AXS-674, of the latest issue in effect, into the carburetor throat, being careful not to choke the engine. Turn off the ignition switch as quickly as possible after the oil has been poured into the carburetor. With the engine switch off, open the throttle wide, and turn the engine five complete revolutions by means of the cranking motor. If the engine cannot be turned by the cranking motor with the switch off, turn it by hand, or disconnect the high-tension lead and ground it before turning the engine by means of the cranking motor. Then reinstall the air cleaner.

(10) BRAKES. Release brakes and chock the wheels.

c. Inspections in Limited Storage. Vehicles in limited storage will be inspected weekly for conditions of tires and battery. If water is added when freezing weather is anticipated, recharge the battery with a portable charger, or remove the battery for charging. Do not attempt to charge the battery by running the engine.

¼-TON 4 x 4 TRUCK (WILLYS-OVERLAND MODEL MB and FORD MODEL GPW)

182. LOADING AND BLOCKING FOR RAIL SHIPMENT.

a. **Preparation.** In addition to the preparation described in paragraph 181, when ordnance vehicles are prepared for domestic shipment, the following preparations and precautions will be taken:

(1) EXTERIOR. Cover the body of the vehicle with a canvas cover supplied as an accessory.

(2) TIRES. Inflate pneumatic tires from 5 to 10 pounds above normal pressure.

(3) BATTERY. Disconnect the battery to prevent its discharge by vandalism or accident. This may be accomplished by disconnecting the positive lead, taping the end of the lead, and tying it back away from the battery.

(4) BRAKES. The brakes must be applied and the transmission placed in low gear after the vehicle has been placed in position, with a brake wheel clearance of at least 6 inches (fig. 101 "A"). The vehicles will be located on the car in such a manner as to prevent the car from carrying an unbalanced load.

(5) All cars containing ordnance vehicles must be placarded "DO NOT HUMP."

(6) Ordnance vehicles may be shipped on flat cars, end-door box cars, side-door box cars, or drop-end gondola cars, whichever type car is the most convenient.

b. **Facilities for Loading.** Whenever possible, load and unload vehicles from open cars under their own power, using permanent end ramps and spanning platforms. Movement from one flat car to another along the length of the train is made possible by cross-over plates or spanning platforms. If no permanent end ramp is available, an improvised ramp can be made from railroad ties. Vehicles may be loaded in gondola cars without drop ends by using a crane. In case of shipment in side-door cars, use a dolly-type jack to warp the vehicles into position within the car.

c. **Securing Vehicles.** In securing or blocking a vehicle, three motions, lengthwise, sidewise, and bouncing, must be prevented. There are two approved methods of blocking the vehicles on freight cars, as described below. When blocking dual wheels, all blocking will be located against the outside wheel of the dual.

(1) METHOD 1 (fig. 101). Locate eight blocks "B", one to the front and one to the rear of each wheel. Nail the heel of each block to the car floor, using five 40-penny nails to each block. That portion of the block under the tread will be toenailed to the car floor with two 40-penny nails to each block. Locate two blocks "D" against the outside face of each wheel. Nail the lower block to the car floor with three 40-penny nails, and the top block to the lower block with three 40-penny nails. Pass four strands, two wrappings, of No. 8 gage, black annealed wire "C" around the bumper support bracket at the front of the vehicle, and then through a stake pocket on the railroad car. Perform the same operation at the rear of the vehicle, passing the

SHIPMENT AND TEMPORARY STORAGE

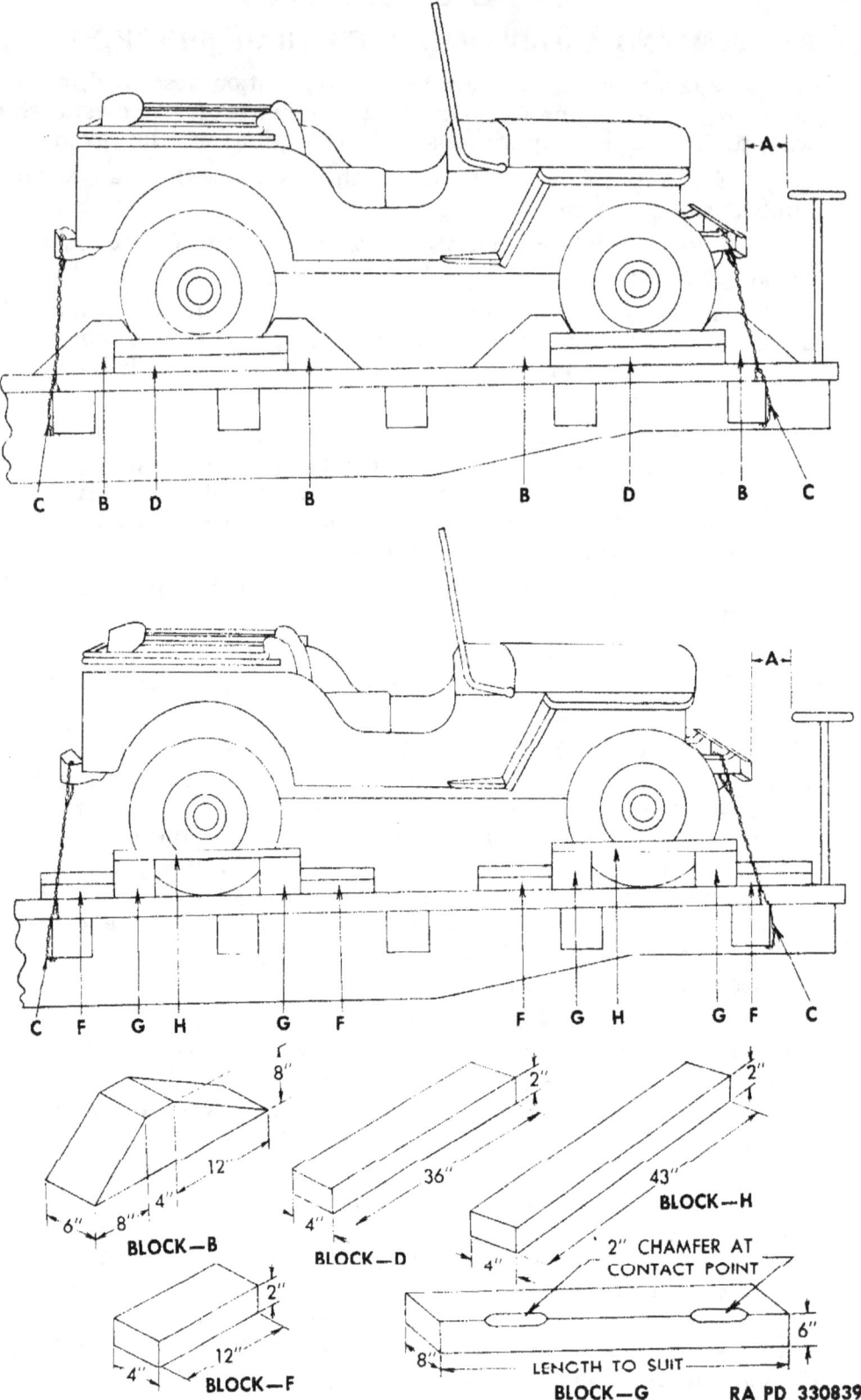

Figure 103—Blocking Requirements for Securing Wheeled Vehicles on Railroad Cars

¼-TON 4 x 4 TRUCK (WILLYS-OVERLAND MODEL MB and FORD MODEL GPW)

wire through the opening in the rear bumper. Duplicate these two operations on the opposite side of the vehicle. Tighten the wires enough to remove slack. When a box car is used, this strapping must be applied in a similar fashion, and attached to the floor by the use of blocking or anchor plates. This strapping is not required when gondola cars are used.

(2) METHOD 2 (fig. 101). Place four blocks "G", one to the front and one to the rear of each set of wheels. These blocks are to be at least 8 inches wider than the over-all width of the vehicle at the car floor. Using sixteen blocks "F", locate two against blocks "G" to the front of each wheel, and two against blocks "G" to the rear of each wheel. Nail the lower cleat to the floor with three 40-penny nails, and the top cleat to the cleat below with three 40-penny nails. Locate four cleats "H" on the outside of each wheel to the top of each block "G" with two 40-penny nails. Pass four strands, two wrappings, of No. 8 gage, black annealed wire "C" around the bumper support bracket (front), and also opening in the rear bumper (rear), as described in Method 1 above.

d. **Shipping Data.**

Length, over-all	132.25 in.
Width, over-all	62 in.
Height, top down	32 in.
Shipping weight	2453 lb
Approximate floor area	57 sq ft
Approximate volume	152 cu ft
Bearing pressure (lb per sq ft)	4½

TM 9-803

REFERENCES

PUBLICATIONS INDEXES.

The following publications indexes should be consulted frequently for latest changes to, or revisions of the publications given in this list of references and for new publications relating to materiel covered in this Manual:

Introduction to Ordnance Catalog (explains SNL system)	ASF Cat. ORD-1 IOC
Ordnance publications for supply index (index to SNL's)	ASF Cat. ORD-2 OPSI
Index to Ordnance publications (lists FM's, TM's, TC's, and TB's of interest to Ordnance personnel, MWO's, OPSR's, RSD, S of SR's, OSSC's, and OFSB's. Includes alphabetical listing of Ordnance major items with publications pertaining thereto)	OFSB 1-1
List of publications for training (lists MR's, MTP's, T/BA's, T/A's, FM's, TM's, and TR's, concerning training)	FM 21-6
List of training films, film strips, and film bulletins (lists TF's, FS's, and FB's by serial number and subject)	FM 21-7
Military training aids (lists graphic training aids, models, devices, and displays)	FM 21-8

STANDARD NOMENCLATURE LISTS.

Truck ¼-ton, 4 x 4, command reconnaissance (Ford and Willys)	SNL G-503
Cleaning, preserving and lubrication materials, recoil fluids, special oils, and miscellaneous related items	SNL K-1
Soldering, brazing and welding materials, gases and related items	SNL K-2
Tool sets—motor transport	SNL N-19

EXPLANATORY PUBLICATIONS.

Fundamental Principles.

Automotive electricity	TM 10-580
Automotive lubrication	TM 10-540
Basic maintenance manual	TM 38-250
Driver's manual	TM 10-460
Driver selection and training	TM 21-300
Electrical fundamentals	TM 1-455

233

¼-TON 4 x 4 TRUCK (WILLYS-OVERLAND MODEL MB and FORD MODEL GPW)

Military motor vehicles	AR 850-15
Motor vehicle inspections and preventive maintenance service	TM 9-2810
Precautions in handling gasoline	AR 850-20
Standard Military Motor Vehicles	TM 9-2800
The internal combustion engine	TM 10-570

Maintenance and Repair.

Cleaning, preserving, lubricating and welding materials and similar items issued by the Ordnance Department	TM 9-850
Cold weather lubrication and service of combat vehicles and automotive materiel	OFSB 6-11
Maintenance and care of pneumatic tires and rubber treads	TM 31-200
Ordnance Maintenance: Engine and engine accessories for ¼-ton 4 x 4 truck (Ford and Willys)	TM 9-1803A
Ordnance Maintenance: Power train, chassis, and body for ¼-ton 4 x 4 truck (Ford and Willys)	TM 9-1803B
Ordnance Maintenance: Electrical equipment (Auto-Lite)	TM 9-1825B
Ordnance Maintenance: Hydraulic brake system (Wagner)	TM 9-1827C
Ordnance Maintenance: Carburetors (Carter)	TM 9-1826A
Ordnance Maintenance: Fuel pumps	TM 9-1828A
Tune-up and adjustment	TM 10-530

Protection of Materiel.

Camouflage	FM 5-20
Chemical decontamination, materials and equipment	TM 3-220
Decontamination of armored force vehicles	FM 17-59
Defense against chemical attack	FM 21-40
Desert operations	FM 31-25
Explosives and demolitions	FM 5-25

Storage and Shipment.

Ordnance storage and shipment chart, group G—Major items	OSSC-G
Registration of motor vehicles	AR 850-10
Rules governing the loading of mechanized and motorized army equipment, also major caliber guns, for the United States Army and Navy, on open top equipment published by Operations and Maintenance Department of Association of American Railroads.	
Storage of motor vehicle equipment	AR 850-18

INDEX

A

Adjustment
- blackout driving light 160
- brakes 196
- carburetor 128
- clutch pedal 167
- steering gear 215
- valve tappet 113
- wheel bearings (rear axle) 191

Air cleaner
- after-operation and weekly service ... 35
- at halt service 33
- description 129
- installation 129
- lubrication 47
- operation in cold climates 24
- operation in sand 26
- removal and servicing 129
- removal from engine 116
- run-in test preliminary service ... 69
- second echelon maintenance 60

Ammeter
- description 13
- run-in test preliminary service ... 70
- second echelon service 55

Assembly
- intake to exhaust manifold 111
- transmission to transfer case ... 175

Axle, front
- brake drums 185
- description and data 181
- drag link bell crank 186
- installation 188
- maintenance 183
- removal 188
- second echelon maintenance 62
- steering knuckle housing oil seal ... 185
- steering tie rod 186
- trouble shooting 93
- wheel alinement 187
- wheel bearings 183
- wheel grease retainer 184
- wheel hub 185

Axle, rear
- axle shaft 190
- brake drum 192
- description and data 189
- maintenance 190
- replacement 192
- trouble shooting 93
- wheel bearings 191
- wheel hub 191

Axle shaft
- installation 191
- removal 190

B

Battery
- after-operation and weekly service ... 35
- data 150
- description 155
- installation 155
- maintenance 57
- operation in cold weather 24
- preparation for shipment 230
- removal 155
- from engine 116
- run-in test preliminary service ... 69
- trouble shooting 101

Battery and lighting system
- battery 155
- blackout driving light 159
- blackout headlights 159
- blackout (main) light switch .. 163
- description and data 150
- headlights 157
- instrument panel lights 161
- maintenance 154
- panel and blackout driving light switches ... 164
- taillights and stop lights 160
- trailer connection 166
- wiring system 156

Before-operation service 29

Blackout driving light
- adjustment 160
- description 159

Blackout driving light switch,
- description 15

Blackout headlights 159

Blackout (main) light switch
- description 15-16, 163
- installation 164
- removal 163

Brake bands, care in cold climates ... 25

Brake master cylinder, second echelon maintenance 63

Brake shoes, front, second echelon maintenance 61

Brake shoes, rear, second echelon maintenance 61

TM 9-803

¼-TON 4 x 4 TRUCK (WILLYS-OVERLAND MODEL MB and FORD MODEL GPW)

B—Cont'd

	Page No.
Brakes	
description and data	193
flexible lines, hoses, and connections	201
maintenance and adjustment	195
master cylinder	199
parking (hand) brake	204
service (foot) brakes	196
trouble shooting	94
wheel cylinder	200

C

	Page No.
Carburetor	
adjustment	128
description	127
operation under unusual conditions	24
second echelon maintenance	60
removal and installation	128
Circuit breakers, second echelon maintenance	65
Clutch	
clutch release bearing	171
description and data	167
during-operation	31
installation	170
maintenance	167
operation in sand	26
pedal adjustment	167
removal	170
run-in test	71
second echelon service	56
trouble shooting	85
Cold weather operation	21
Communications	12
Cooling system	
description and data	137
fan	142
fan belt	141
maintenance	138
operation in hot weather	26
preparation for storage or shipment	228
protection in cold weather	22
radiator	140
temperature gage	143
thermostat	143
trouble shooting	87
water pump	140

	Page No.
Crankcase	
lubrication	47
maintenance	58
Crankcase ventilator valve, description, removal, and installation	115
Cranking motor	
data	144
description	144
protection in cold weather	23
removal and installation	145
removal from engine	116
second echelon maintenance	59, 60
trouble shooting	75, 89
Cranking motor mechanism, inspection before starting engine	24
Cylinder head gasket	
installation	112
maintenance	57
removal	111

D

	Page No.
Data, tabulated	
battery and lighting system	150
body and frame	216
brakes	193
clutch	167
cooling system	137
engine	104
front axle	181
fuel and air intake and exhaust systems	127
generating system	146
ignition system	119
propeller shafts and universal joints	180
radio interference suppression system	221
rear axle	189
springs and shock absorbers	206
starting system	144
steering gear	212
transfer case	177
transmission	173
Description	
battery and lighting system	150
body and frame	216
brakes	193
clutch	167
cooling system	137
distributor	120

236

INDEX

D—Cont'd

Description—Cont'd
- engine.................... 104
- front axle................. 181
- fuel and air intake and exhaust systems................ 126
- generating system........... 146
- ignition system............. 118
- propeller shafts and universal joints................... 179
- radio interference suppression system................... 221
- rear axle................... 189
- springs and shock absorbers..... 205
- starting system............. 144
- steering gear............... 211
- transfer case............... 176
- transmission................ 172
- truck...................... 10

Differential, lubrication in cold weather...................... 21

Distributor
- care in cold weather........ 23
- data........................ 119
- description................. 120
- distributor condenser....... 122
- distributor points.......... 121
- lubrication................. 48
- removal and installation.... 120
- second echelon maintenance... 59
- trouble shooting............ 89

Drag link bell crank
- installation................ 186
- removal..................... 186

Dragging clutch, trouble shooting.. 85

Drive belt and pulleys, second echelon maintenance............ 59

Driving controls and operation
- instruments and controls.... 3
- towing vehicle.............. 19
- use of instruments and controls in vehicular operation.......... 17

E

Electrical system
- operation in hot weather.... 26
- protection in cold weather.. 23

Engine
- crankcase ventilator valve... 115
- cylinder head gasket........ 111
- description and tabulated data... 104
- installation................ 117
- intake and exhaust manifolds.... 109
- manifold heat control valve..... 111
- oil filter.................. 114
- oil pan gasket.............. 113
- preparation for storage or shipment................. 229
- removal..................... 116
- trouble shooting............ 74
- tune-up..................... 104
- valve cover gasket.......... 112
- valve tappet adjustment..... 113

Exhaust system
- description................. 135
- exhaust pipe................ 136
- muffler..................... 136

F

Fan
- data........................ 137
- description................. 142
- removal and installation.... 143

Fan belt
- adjustment and installation..... 142
- data........................ 137
- description................. 141
- removal..................... 142

Fenders, removal and installation.. 220

Fire extinguisher
- after-operation and weekly service................... 34
- run-in test preliminary service... 68

First echelon preventive maintenance service
- after-operation and weekly service................... 34
- at-halt service............. 32
- before-operation service.... 29
- during-operation service.... 31

Frame and body
- description and data........ 216
- fenders..................... 220
- hood........................ 219
- instruments................. 217
- maintenance................. 217
- radiator guard.............. 219
- rifle holder................ 218
- seats and cushions.......... 217
- shovel and axe.............. 219
- top......................... 218
- trouble shooting............ 101
- windshield.................. 218
- windshield wiper............ 218

TM 9-803
¼-TON 4 x 4 TRUCK (WILLYS-OVERLAND MODEL MB and FORD MODEL GPW)

F—Cont'd

Fuel and air intake and exhaust systems
- air cleaner 129
- carburetor 127
- description and data 126
- exhaust system 135
- fuel gage 134
- fuel pump 130
- fuel strainer 132
- fuel tank 131
- maintenance 127

Fuel filter
- run-in test preliminary service ... 68
- second echelon maintenance 60

Fuel gage
- description 13, 134
- removal and installation 135

Fuel pump
- description 130
- operation under unusual conditions 24
- removal and installation 131
- second echelon maintenance 60

Fuel strainer
- description 132
- removal and installation 134

Fuel system
- operation in cold climates 24
- operation in landing 27
- trouble shooting 86

Fuel tank
- description 131
- removal and installation ... 131-132

G

Gear cases, lubrication 47

Generating system
- description and data 146
- generator 147
- maintenance 146
- regulator 148

Generator
- data 146
- description 147
- protection in cold weather 23
- removal and installation 147
- second echelon maintenance 59
- trouble shooting 90

H

Hand brake
- description 17
- during-operation service 31
- second echelon service 56

Headlights
- description 157
- installation 158
  - sealed beam unit 158
- removal 158
  - sealed beam unit 157

Hood, removal and installation 219

I

Ignition coil
- data 119
- description 124
- removal and installation 124

Ignition switch
- data 119
- description 125
- removal and installation 125

Ignition system
- description and data 118
- distributor 120
- ignition coil 124
- ignition switch 125
- ignition timing 123
- ignition wiring 125
- maintenance 120
- spark plugs 124
- trouble shooting 88

Ignition wiring
- data 119
- description, removal, and installation 125

Inspection of vehicle in limited storage 229

Intake and exhaust manifolds
- description 109
- installation 111
- removal 109
- trouble shooting 87

Intervals of lubrication 47

Instrument panel lights
- description 161
- removal and installation 163

Instruments and controls
- controls 15
- instruments 13

238

INDEX

L

	Page No.
Light switches	
blackout	15
blackout driving	15
panel light	16
Lighting system, trouble shooting	102
Lubrication	
cold weather	21
vehicle for run-in test	69
Lubrication guide	37

M

Maintenance	
battery and lighting system	154
body and frame	217
brakes	195
clutch	167
cooling system	138
front axle	183
fuel and air intake and exhaust systems	127
generating system	146
ignition system	120
propeller shafts and universal joints	180
radio interference suppression system	224
rear axle	190
springs and shock absorbers	207
starting system	144
steering gear	212
transfer case	177
transmission	173
Manifold heat control valve, description	111
Manifolds, second echelon maintenance	60
Master cylinder	
data	193
installation	200
removal	199
Modifications, record of	52
Muffler	
installation	136
removal	136
MWO and major unit assembly replacement record	52

N

New vehicle run-in test	68

O

	Page No.
Oil filter (and lines)	
description	114
installation	115
element	115
lubrication	47
maintenance	58
removal	115
element	115
Oil pan gasket, removal and installation	113
Oil pressure gage	
before-operation service	30
description	15
Oilcan points	22, 48
Operation	
fuel and air intake and exhaust systems	127
under unusual conditions	
cold weather	21
decontamination	27
hot weather	26
in sand	26
in landing	27
Organization tools and equipment	73

P

Parking (hand) brake	
adjustment	204
data	195
Panel and blackout driving light switches	
description	164
removal and installation	166
Panel light switch, description	16
Pitman arm, steering, removal and installation	214
Power train, operation in landing	27
Propeller shafts (and universal joints)	
description and data	179
installation	180
maintenance	180
removal	180
run-in test preliminary service	70
second echelon maintenance	62
trouble shooting	93

R

Radiator	
data	137
description	140

R—Cont'd

Radiator—Cont'd
- installation on engine 117
- maintenance 58
- operation in sand 27
- removal and installation 140
- removal from engine 116

Radiator guard
- installation 220
- removal 219

Radio bonding, second echelon maintenance 66

Radio interference suppression system
- data 221
- description 221
- maintenance 224
- tests 223
- trouble shooting 103

Regulator
- data 146
- description 148-149
- removal and installation 149

Reports and records of lubrication .. 48

Rifle holder, removal and installation 218

Road test chart 55

Run-in test, new vehicle
- correction of deficiencies 68
- procedures 68

S

Second echelon preventive maintenance services
- maintenance operations 57
- road test chart 55
- tools and equipment 66

Service (foot) brakes
- adjustment 196
- data 193
- removal and installation of brake shoes and linings 199

Shipment and temporary storage
- loading and blocking for rail shipment 230
- preparation for temporary storage or domestic shipment 228

Shock absorbers and springs
- description and data 205
- maintenance 207
- second echelon maintenance 62
- shock absorbers 210
- spring shackles and bolts 207
- springs 209
- trouble shooting 97

Shovel and ax, removal and installation 219

Spare parts, vehicle 50

Spark plugs
- care in cold weather 23
- data 119
- description and adjustment 124
- maintenance 58
- trouble shooting 89

Special tools 73

Speedometer
- description 15
- second echelon service 55

Springs and suspension
- before-operation service 30
- run-in test preliminary service .. 69

Starting switch
- data 144
- removal and installation 145

Starting system
- cranking motor 144
- description and data 144
- maintenance 144
- starting switch 145
- trouble shooting 89

Steering connecting rod, removal and installation 213

Steering gear
- adjustment 215
- description and data 211
- during-operation service 32
- installation 214
- maintenance 212
- steering connecting rod 213
- steering Pitman arm 214
- steering rod 213
- removal 214

Steering gear housing, lubrication in cold weather 22

Steering knuckle bearings, lubrication 47

Steering knuckle housing oil seal
- installation 185
- removal 185

INDEX

S—Cont'd

	Page No.
Steering knuckles, second echelon maintenance	62
Steering linkage	
at-halt service	33
after-operation and weekly service	36
before-operation service	30
run-in test preliminary service	70
Steering system, trouble shooting	98, 101
Steering tie rod, removal and installation	186
Steering wheel	
installation	214
removal	213

T

	Page No.
Taillights and stop lights	
description	160
removal and installation	161
Temperature gage, removal and installation	143
Terminal blocks, second echelon maintenance	65
Thermostat	
data	138
description, removal, and installation	143
Tires, before-operation service	30
Tools and equipment stowage on vehicle	
equipment	50
spare parts	50
tools	49
Trailer connections	
data	154
removal and installation	166
Transfer case	
description	176
during-operation service	32
installation	174
maintenance	177
removal	173
run-in test	71
second echelon maintenance	56, 63
shifting levers	178
trouble shooting	92
Transmission	
description and data	172-173
during-operation service	32
installation	175

	Page No.
lubrication in cold weather	21
maintenance	173
removal	173
run-in test	71
second echelon maintenance	56, 63
trouble shooting	91
Trouble shooting	
battery and lighting system	101
body and frame	101
brake system	94
clutch	85
cooling system	87
engine	74
front axle	93
fuel system	86
ignition system	88
instruments	103
intake and exhaust systems	87
propeller shafts	93
radio suppression	103
rear axle	93
springs and shock absorbers	97
starting and generating system	89
steering system	98
transfer case	92
transmission	91
wheels, wheel bearings, and related parts	97
Tune-up of engine	104

U

	Page No.
Universal joint lubrication	47

V

	Page No.
Valve cover gasket, installation and removal	112
Valve mechanism, maintenance	58
Valve tappet adjustment	113
Vehicle specifications	10
Vehicular operation, use of instruments and controls in	
before-operation service	17
placing vehicle in motion	17
reversing vehicle	19
shifting gears in transfer case	18
shifting to lower gears in transmission	18
starting engine	17
stopping engine	19
stopping the vehicle	19

¼-TON 4 x 4 TRUCK (WILLYS-OVERLAND MODEL MB and FORD MODEL GPW)

W

	Page No.
Water pump (and fan)	
data	137
description	140
maintenance	59
removal and installation	141
Wheel alinement	187
Wheel cylinder	
data	193
installation	201
removal	200
Wheel bearings	
adjustment	183
lubrication	48
removal and installation	184
trouble shooting	97
Wheel bearings (rear axle), adjustment, removal, and installation	191
Wheel hub, removal and installation	185
Windshield adjusting arms, description	17
Windshield wipers	
before-operation service	30
second echelon service	56
Wiring system	
data	150
description	156
removal and installation	156

MAINTENANCE MANUAL

FOR
WILLYS TRUCK
¼ TON 4 x 4

BUILT FOR
U. S. GOVERNMENT

MODEL MB

Contract Number
W-398-qm-11423

U. S. A. Reg. Numbers
2073506 to 2078606

★ ★ ★

Parts are designated in this book under both Ford and Willys part numbers since all parts are interchangeable for vehicles produced by Ford Motor Company.

Contract W-398-qm-11424 Model GPW

U. S. A. Registration Numbers
20100000S to 20163145S

TM-10-1349

WILLYS-OVERLAND MOTORS, INC.
TOLEDO, OHIO, U. S. A.

TM-10-1513
CHANGE NO. 1
MAY 15, 1942

TM-10-1513
CHANGE NO. 1
MAY 15, 1942

INDEX

	*Section	Page
Unloading	0-1	4
Drivers Instructions	0-2	5
Lubrication and Inspection	0-3	9
Engine	0100	17
Clutch	0200	37
Fuel System	0300	42
Exhaust System	0400	52
Cooling	0500	53
Electrical	0600	57
Transmission	0700	73
Transfer Case	0800	79
Propeller Shaft and Universal Joints	0900	83
Front Axle	1000	86
Rear Axle	1100	93
Brakes	1200	101
Wheels, Hubs and Drums	1300	108
Steering	1400	111
Frame	1500	119
Springs and Shock Absorbers	1600	121
Body	1800	125
Tools	2300	126

* These numbers refer to Parts Group Classifications in Parts List.

FOREWORD

This Motor Vehicle has been thoroughly tested and inspected. Like any other piece of machinery, to maintain it in proper operating condition, it should be lubricated at the time specified using the proper grades of oil and grease. All working parts as well as oil holes should be kept clean and free from dirt and grit. This vehicle should periodically have a systematic inspection.

All parts in this vehicle are completely interchangeable with those manufactured by Ford Motor Company under the contract listed on the preceding page. Both Ford and Willys part numbers are therefore listed under the illustrations showing views of the various assemblies. These part numbers should be used only for the purpose of identifying parts as they are mentioned in the text, and, the accuracy of the part number should be verified by referring to the parts book when placing orders for parts.

In the following pages we have described how to take care of this unit and handle it in such a way that it will give maximum service and dependable performance.

In the forepart of this Manual will be found complete instructions relative to conditioning the unit for Service, Driver's Instructions, Lubrication and Inspection.

In the Maintenance and Repair Section will be found instructions which will enable one to make proper adjustments and repairs.

See Index on preceding page; bend back edge of pages to find Section desired.

Read and follow these instructions carefully.

WILLYS-OVERLAND MOTORS, INC.

GENERAL DATA

ENGINE

Type	Gasoline
Number of Cylinders	4
Bore	3 1/8"
Stroke	4 3/8"
Piston Displacement	134.2 cu. in.
Compression Ratio	6.48-1
Horsepower—S.A.E.	15.6
Horsepower { Actual	60
{ Revolutions per minute	4000
Torque { Maximum Lbs.-Ft.	105
{ Revolutions per Minute	2000
Wheelbase	80"
Tread	48 1/4"—with combat wheels 49"
Overall Width	62"
Overall Length	132 3/4"
Overall Height—Normal Load	
To top of cowl	40"
To top of steering wheel	51 1/4"
Top up	69 3/4"
Weight—Maximum Pay Load	800 lbs.
Maximum Trailed Load	1000 lbs.
Shipping (Less water, fuel and chains)	2125 lbs.
Road	2315 lbs.
Gross	3125 lbs.

CAPACITIES

	U.S.	Imperial	Metric
Fuel Tank (Gals.)	15	12 1/2	56.78 liters
Engine Crankcase-Refill (Qts.)	4	3 1/2	3.78 "
Cooling System (Qts.)	11	9 1/4	10.41 "
Transmission (Pts.)	2	1 3/4	.95 "
Transfer Case (Pts.)	3	2 1/2	1.42 "
Front Axle Differential (Pts.)	2 1/2	2	1.18 "
Rear Axle Differential (Pts.)	2 1/2	2	1.18 "
Oil Bath Air Cleaner (Pts.)	1 1/2	1	.71 "
Brake System Brake Fluid (Pts.)	3/4	3/4	.36 "

See Lubrication Chart, Page 12

LAMP BULBS

	Mazda
Head Lamp (Sealed Beam type)	2400
Upper Beam	45 Watts
Lower Beam	45 Watts
Blackout Lamp Bulb (1)	3 Cp. SC 63
Left Tail Lamp Bulb (1)	21-3 Cp. DC 1154
Left Tail Lamp Bulb (1)	3 Cp. SC 63
Right Tail Lamp Bulbs (2)	3 Cp. SC 63
Instrument Lamp Bulb (2)	1.5 Cp. SC 51

Fuse (Thermal Type)—On Light Switch-30 Amperes

IDENTIFICATION

Chassis Serial Number located on inside of frame at left front end.

Engine Number located on right side of cylinder block front upper corner.

UNLOADING INSTRUCTIONS

Spot freight car along side of the unloading platform. Open freight car door and make visual inspection of vehicles for damage, loose blocking and shortages, due to rough handling or pilferage while vehicles were in transit. If any evidence of carrier's responsibility, the railroad representative should inspect shipment and note it on Bill of Lading.

Vehicles are shipped from one to six in a freight car, therefore, the manner varies in which the vehicles are anchored in the car. Where shipment does not exceed two vehicles per freight car, the regular 36 foot box car is used. Where three or more vehicles are shipped an "Evans" or "Channel" automobile freight car is used. These freight cars are equipped with upper deck platforms operated by chain falls and have anchor chains in flooring; to operate follow printed instructions on inside wall at controls.

One or Two Vehicles per Car

The vehicles are anchored to floor with grooved blocks spiked to the floor at front and rear of each wheel. Spring rebound straps are anchored to front end of front springs and rear end of rear springs and spiked to the floor.

To remove vehicles from car, use a crow bar to pry loose wheel blocks and straps from floor. Remove bolt in spring rebound strap at springs and remove straps.

Roll one vehicle to end of car, then jack or lift the other vehicle so it can be removed through door to platform, then remove second vehicle, and check all items listed in Tool and Accessory list.

Three Vehicles per Car

Where three vehicles are shipped, the two end vehicles are fastened at front end with car equipment anchor chains. The rear wheels have grooved blocks spiked to the floor. Spring rebound straps at end of rear springs are also spiked to the floor.

The center vehicle is anchored at the ends of front and rear axles with car equipment chains. Spring rebound straps at end of front and rear springs are spiked to the floor.

To remove vehicles first remove all wooden blocks, spring rebound straps and anchor chains from the three vehicles. Run end vehicles to extreme ends of freight car; jack or lift center vehicle so it can be rolled through door to platform. Repeat this operation to remove other two vehicles.

Four Vehicles per Car

Where four vehicles are shipped, one is decked and three anchored to the floor the same as in three vehicle shipment.

To remove vehicles, first remove anchor chains and wooden blocks from the three vehicles on floor and remove vehicles to platform. Follow instructions printed on inside of freight car at controls in ends of car for lowering Deck platform. Lower platform and remove anchor chains, then remove vehicle.

Five Vehicles per Car

Two vehicles are decked and three anchored to flooring in same manner as four to a car.

The removal of vehicles should be in the same sequence as outlined under three and four car shipment.

Six Vehicles per Car

Where six vehicles are shipped, two are decked and four are anchored to the floor.

The two end vehicles are fastened at front ends with anchor chains, the rear end of vehicles are anchored with grooved blocks and spring rebound straps spiked to the floor.

The two center vehicles are fastened in the opposite manner, rear ends with anchor chains and front ends with wheel grooved blocks and spring rebound straps spiked to the floor.

To remove vehicles remove wheel blocks, spring rebound straps and anchor chains. Roll end cars and one center car to end of freight car, jack or lift other center vehicle so it can be removed to platform, then remove other three.

Lower one decked vehicle by chain falls, following instructions printed on wall. Then remove second decked vehicle in same manner.

PRE-OPERATION INSTRUCTIONS

All vehicles are carefully tested and inspected before leaving the factory, however, while in transit and unloading some things may happen which will require attention before putting vehicle into Service. We therefore suggest checking the following items before operating vehicle.

1. Fill radiator and check all connections for water leaks.
2. Check oil in engine, transmission, transfer case, front and rear differential housings.
3. Fill gasoline tank and check full system for leaks.
4. Check battery fluid level.
5. Check terminal connections at battery, generator, voltage control, starter, distributor and spark plugs.
6. Check operation of lights and horn.
7. Check brake fluid level in master cylinder and check connections for leaks or damage.
8. Check steering connections and front wheel alignment.
9. Check tire pressure, inflate to 30 lbs.
10. Check hand brake operation.
11. Check cylinder head screws and nuts.

DRIVER'S INSTRUCTIONS

This vehicle should not be driven faster than 40 miles an hour for the first 100 miles nor more than 50 miles an hour from 100 to 500 miles. If the vehicle is operated at excessive speeds while new, the closely fitted parts may possibly become overheated, resulting in serious damage to mechanical units. Never race the Engine while making adjustments or when vehicle is standing idle.

FIG. 1—CONTROLS

It is very important that the driver of this vehicle be thoroughly familiar with the various Controls and their proper use. The most experienced driver should study the Controls because there are a number which are not ordinarily found on standard vehicles.

Illustrations show the controls, instruments and instruction plates; in the following paragraphs we refer to these illustrations by the key numbers so the reader may easily follow the instructions.

Ignition Switch—No. 9, Fig. 1

Is operated by a key, turning key to right (clockwise) closes the ignition circuit. Turning key to left (counter clockwise) opens the ignition circuit and shuts off the engine.

Light Switch—No. 6, Fig. 1

The light switch is the push-pull type with safety lock.

This switch controls the entire lighting system including the instrument panel lights and stop lights.

When the control knob is pulled out to the first position, the blackout lamp circuit is closed—which consists of two blackout lamps, stop and tail lamps.

To obtain bright lights, push in lockout control button on left of switch and pull out control knob to second position. This closes entire bright light circuit, which consists of two head lamps—instrument panel lamps, stop and tail lamps.

CAUTION: When driving during the day press in lockout control button and pull Control Knob out to the last or Stop Light position to cause regular Stop Light to operate.

Panel Light Switch—No. 12, Fig. 1.

The Panel Light switch controls the Panel Lights when the main Light Switch is in Service (bright light) position, otherwise the Panel Lights do not operate.

Head Lamp Beam Control Switch—No. 8, Fig. 1

Pressing and releasing the button of the selector foot switch with the left foot alternately changes the headlight beam from high to low.

Starter Switch—No. 23, Fig. 1

Toe board mounted to the right of the accelerator; pushing button down closes starter circuit and causes starter to crank engine—release the button as soon as the engine starts.

Hand Throttle—No. 10, Fig. 1

Pulling control button out opens carburetor throttle valve and increases engine speed.

Carburetor Choke Control—No. 7, Fig. 1

Pulling control button out closes choke valve in carburetor to enrich gas mixture for starting the engine when cold, and opens throttle valve slightly for faster idle speed.

Oil Gauge—No. 15, Fig. 1

The instrument panel oil gauge indicates oil pressure delivered to camshaft, crankshaft, timing chain and connecting rod bearings when engine is running.

Proper registration should be not below 10 on idle nor more than 80 at speeds above 10 miles per hour.

This gauge does not indicate the amount of oil in crankcase.

Ammeter—No. 20, Fig. 1

The ammeter is used to indicate when the generator is charging the battery. It also indicates the amount of current being consumed.

If the ammeter shows discharge at all times, the cause should be immediately investigated and corrected, otherwise the wiring may be damaged and battery discharged.

Fuel Gauge—No. 13, Fig. 1

The fuel gauge registers the amount of fuel in the fuel tank when ignition switch is turned on. The dial graduations are for—empty, ¼, ½, ¾, and full.

Temperature Indicator—No. 19, Fig. 1

This is a thermal type gauge and registers the temperature of the liquid in the cooling system. The operator should watch this instrument closely.

The normal operating temperature is indicated when hand stands between 160 and 185. The driver should immediately investigate the cause if temperature becomes excessive. Continuous operation of an overheated engine will cause serious damage.

Never fill cooling system with cold water when engine is overheated.

Speedometer—No. 17, Fig. 1

The Speedometer indicates the speed at which vehicle is being driven. The Odometer (in upper part of speedometer face) registers the total number of miles the vehicle has been driven.

A trip mileage indicator (in lower part of speedometer face) gives miles covered on any trip. It can be reset by turning a knurled control shaft extending through the rear of the speedometer.

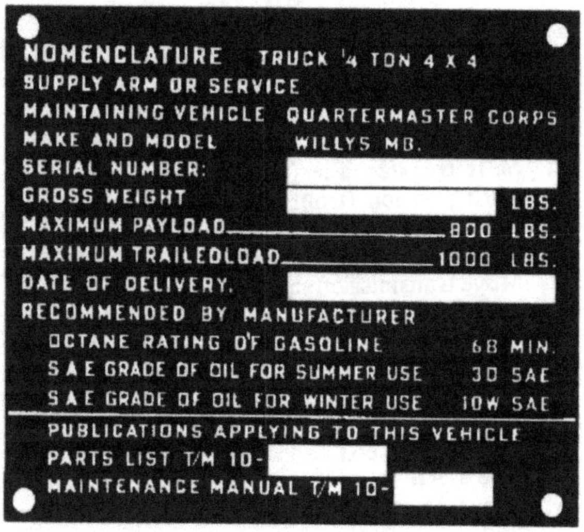

Nomenclature Plate (Name Plate)—Fig. 2

The nomenclature plate identifies the vehicle and gives the manufacturer's model and serial number, date of delivery, recommended fuel and lubricating oil. Service publication numbers are also given for reference. (When ordering parts be sure to give serial number). See No. 26, Fig. 1.

FIG. 3—CAUTION PLATE

Caution Plate—Fig. 3 & No. 29, Fig. 1

Covers maximum permissible road speeds in different gear positions and gives instructions relative to complete draining of the cooling system.

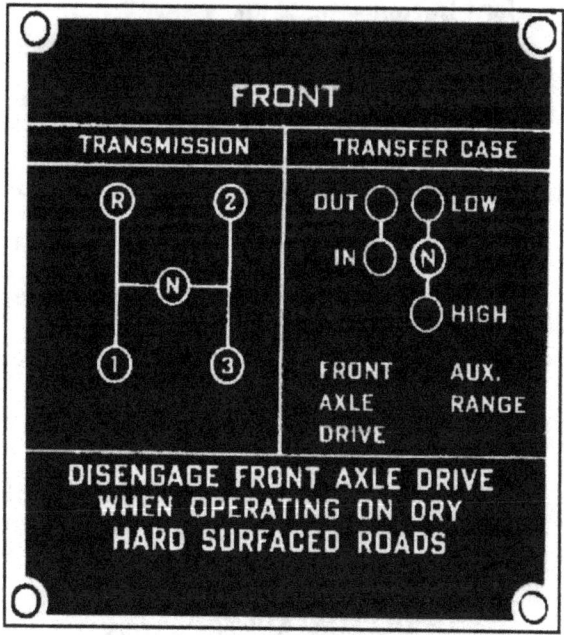

FIG. 4—SHIFT PLATE

Transfer Case Shifting Instructions—Fig. 4

This diagram gives relative position of shifting levers for front axle drive, low and high gear ratios.

On hard surface and flat roads disengage front axle drive by placing center shift lever, (front axle drive) in forward position. No. 21 & No. 25, Fig. 1.

FIG. 2—NAME PLATE

The right hand lever (third from driver) controls transfer case gear ratio—low or high. No. 22, Fig. 1. The low gear ratio can only be used when front axle drive lever is in the rear position to engage front wheel drive.

Proper position for disengaging axles to use power take-off unit is shown as "N" in Fig. 4.

Clutch Pedal—No. 11, Fig. 1

The clutch pedal is used to disengage the engine power from transmission when shifting gears. Driving with the foot on pedal will cause excessive wear of clutch facings and release bearing. There should be 3/4" free pedal travel before clutch starts to disengage.

Brake Pedal—No. 14, Fig. 1

Depressing the pedal applies the hydraulic brakes at all four wheels. Avoid driving with foot on brake pedal, as brakes will be partially applied and cause unnecessary wear of brake linings requiring early adjustment.

Hand Brake Lever—No. 24, Fig. 1

By pulling out on brake handle the external contracting brake at the transmission on rear propeller shaft is applied mechanically. Whenever vehicle is parked, the lever should be pulled out as far as possible. Before moving vehicle be sure lever is released.

Accelerator—No. 16, Fig. 1

The accelerator is foot operated and is used to govern the engine speed under ordinary driving conditions.

Transmission Gearshift Lever—No. 18, Fig. 1

This lever is used to shift the gears in the transmission. There are four positions in the movement of the lever in changing the gears in transmission. See diagram for lever location in different gears. Fig. 4.

Horn Button—No. 2, Fig. 1

Pressing on button closes circuit in horn wiring and causes horn to sound.

Steering Knuckle Oil Seal—Fig. 11, Page 90

When parking during cold, wet weather swing the front wheels from right to left to wipe away moisture adhering to the front axle universal joint housings and oil seals. This will prevent freezing with resulting damage to the oil seal felts. When the car is stored for any period the front axle universal joint housings should be coated with light grease to prevent rusting.

Fuel Tank

The fuel tank is located under Drivers seat. To fill tank raise seat cushion and remove filler cap. See Capacity Chart, Page 3.

Draining Radiator

When draining radiator be sure to open the drain cock on the right forward side of cylinder block as well as at the bottom of the radiator outlet. Remove the radiator cap to break any vacuum and thoroughly drain system. See Caution Plate, Fig. 3.

OPERATING INSTRUCTIONS

Before any trip the following inspections should be made before starting engine.

1. Check the oil level in crankcase, See "Lubrication" Section. Remove oil level indicator located in oil filler pipe on right side of engine, wipe off clean. Insert indicator in filler pipe to full depth. Remove indicator and note position of oil film, if below the full level add sufficient oil to bring to full mark.
2. Remove radiator filler cap and note water level. Should be up within 1/2" below filler neck. Check all hose connections for leaks also fan belt tension.
3. Check all lights and signal devices. Note condition of tires and see that they are properly inflated to 30 lbs.
4. DRIVING THROUGH WATER. See that Cap is on front drain hole under fuel tank so as to keep out stones and dirt. An extra cap is provided for the rear drain hole and this cap should be kept in the glove compartment so it will be readily available for use when fording small streams. Before driving through streams or deep puddles of water, INSTALL THE CAP ON THE REAR DRAIN HOLE. Remove this cap and return it to the glove compartment after passing through the water.
5. When there is a possibility of water being thrown over the engine by fan action in crossing streams, pull up on handle of the generator brace, then remove the fan belt. This will stop the fan. As soon as possible the belt should be replaced, then pull out on the generator. The generator will lock in place by spring action of the brace.

STARTING THE ENGINE

1. Transmission gearshift lever must be in neutral position. See Fig. 4.
2. Pull out hand throttle about 3/4" to 1".
3. Pull out choke button all the way to obtain proper fuel and air mixture for starting, No. 7, Fig. 1. Choking is not necessary when engine is warm.
4. Insert key in ignition switch and turn to right.
5. Disengage clutch by depressing pedal, holding down till engine starts, No. 11, Fig. 1.
6. Step on starter button No. 23, Fig. 1, to crank engine. Release button when engine starts.
7. Push in on choke button and adjust hand throttle to obtain proper idling speed. When engine is cold, it is advisable to leave choke button pulled out about 1". As engine warms up, push choke button all the way in.

STARTING VEHICLE

(For day time driving, turn on Stop Light; See Light Switch, Page 02-5.)

1. Push clutch pedal down to disengage clutch No. 11, Fig. 1.
2. Shift transfer case (center hand lever) in forward position (front axle disengaged) No. 21, Fig. 1, right hand lever No. 22, Fig. 1 in rear position (high gear ratio).
3. Move transmission shift lever toward driver and back for first gear. Fig. 4.
4. Release hand brake, No. 24, Fig. 1, increase engine speed with accelerator by gradually pressing down on accelerator treadle No. 16, and slowly release clutch pedal, No. 11, increasing engine speed as load is picked up and vehicle starts to move.
5. As vehicle speed increases, release accelerator pedal and depress clutch pedal, move gearshift lever to neutral then in to second gear.

Press on accelerator and release clutch pedal slowly. Repeat these operations until transmission is in high gear.

SHIFTING GEARS IN TRANSFER CASE

Instructions for shifting gears in transfer case and engagement of the front axle drive are as follows: No. 21, Fig. 1 and Fig. 4.

1. Transfer case may be operated in either high or low speed range when front axle drive is ENGAGED.
2. The transfer case can be operated only in "High" (direct drive) when front axle drive is DISENGAGED.
3. To engage front axle drive, depress clutch pedal, release accelerator and move center shift lever to rear position, No. 21, Fig. 1.
4. To disengage front axle drive, release accelerator and shift lever to forward position.
5. Shifting from high to low gear should not be attempted except when the vehicle is being

FIG. 5—CHAIN TOW

operated at low speeds or at a standstill. The front axle drive must be engaged for this shift. Release accelerator and depress clutch pedal—move center shift lever to rear position, engaging front wheel drive, No. 21, Fig. 1, then move right hand shift lever, No. 22, to forward position.
6. Shifting from low to high gear may be accomplished at any time, regardless of vehicle speed. Release accelerator and depress clutch pedal—shift right hand lever into rear position.

SHIFTING TO LOWER SPEED IN TRANSMISSION

The transmission gears should always be shifted to the next lower speed before engine begins to labor or before vehicle speed is reduced appreciably. Shifting to lower speed is accomplished as follows:

1. Depress clutch pedal quickly, increase engine speed and shift to next lower gear, release clutch slowly and accelerate.

It is advisable to use the same transmission gear going down a long hill as would be required to climb the same hill.

STOPPING THE VEHICLE

1. Remove foot from accelerator pedal and apply brakes by pressing down on brake pedal, No. 14, Fig. 1.
2. When vehicle speed has been reduced to idling engine speed, disengage clutch and move transmission shift lever, No. 18, Fig. 1 to neutral position. See Fig. 4.
3. When vehicle has come to a complete stop apply hand brake, No. 24, Fig. 1, and release clutch and brake pedals.

SHIFTING INTO REVERSE

Before attempting to shift into reverse the vehicle must be brought to a complete stop.

1. Push clutch pedal down to disengage clutch.
2. Shift transmission lever to the left and forward toward instrument board. Fig. 4.
3. Release clutch pedal slowly and accelerate as load is picked up.

FIG. 6—ROPE TOW

TOWING VEHICLE

When necessary to tow vehicle the tow chain, rope or cable, should be attached to the front bumper bar and frame side rail gusset, Fig. 5 and 6.

Loop chain or rope over top of bumper and frame gusset bringing it up across face of bumper and back on opposite side of frame, then hook or tie. Do not tow from the middle of the bumper.

FIRE EXTINGUISHER

The fire extinguisher is mounted on left side cowl panel Fig. 1. To remove pull outward on clamp release lever.

To operate extinguisher, hold body in one hand and with the other turn handle to left ¼ turn which releases plunger lock. Use pumping action to force fluid on fire.

Read instructions on fire extinguisher plate.

GENERAL LUBRICATION

Lubrication of any vehicle is important to prevent damage to moving parts. Because all moving parts are not subjected to the same operating conditions, the lubricants specified are those which most nearly meet the requirements of the parts involved. In some places excessive heat or cold is a problem to overcome, in others it is extreme pressure, water, sand or grit. The type of operating surfaces must also be taken into consideration as parts rotate or oscillate on various types of bearings. Each of the above conditions in construction make necessary the application of the specified lubricant.

Lubricants should be applied regularly to secure maximum useful service from the vehicle. It is of equal importance that not only the proper grade of lubricant be used but that it be applied in accordance with a definite schedule.

The chart in this section should be referred to for instructions on mileage of application, grade and quantity of lubricant required for all parts of the vehicle. A more detailed account of certain phases of lubrication is given in the following paragraphs.

ENGINE

Lubrication of the engine is accomplished by means of a force-feed continuous circulating system. This is effected by means of a planetary gear type pump located externally on the left side of the engine, and is driven by a spiral gear on the camshaft.

The oil is drawn into the circulating system through a floating oil intake. The floating intake does not permit water or dirt to circulate, which may have accumulated in the bottom of the oil pan, because the oil is drawn horizontally from the top surface. Oil pressure is maintained under all driving and climatic conditions.

Oil is forced to the crankshaft and camshaft bearings through drilled passages in the cylinder block and then to the connecting rod bearings through drilled passages in the crankshaft. A drilled passage in the crankshaft, from the front bearing to holes in the crankshaft sprocket provides positive lubrication for the timing chain. Direct spray from connecting rod bearings lubricates the cylinder walls, pistons, piston pins and the valve mechanism.

The pressure under which the oil is forced to the bearings is controlled by a pressure regulator or relief valve, located in the cover of the oil pump. The valve is set to relieve at an indicated pressure of 75 lbs. at a car speed of approximately 30 miles per hour, with warm oil, assuring ample lubrication at all speeds. An oil pressure gauge is mounted in the instrument panel, and indicates the pressure being supplied. Failure of the gauge to register may indicate absence of oil or leakage and the engine should be stopped immediately.

If there is plenty of oil in the reservoir, the oiling system should be carefully checked before starting the engine.

For capacity of the oiling system see Page 3. Care should be taken to replenish the supply when the oil level indicator, which is combined with the oil filler cap located in the oil filler pipe, shows the oil below the full mark, No. 6, Fig. 1. Fresh oil should be poured into the reservoir through the filler pipe sufficiently to bring level to full mark.

WHEN TO CHANGE CRANKCASE OIL

When the vehicle leaves the factory the crankcase is filled to the correct level with oil of the proper viscosity for the "break-in" period. (Decked vehicles in freight cars have engine oil drained and five quarts of oil in cans in freight car for each vehicle.)

At 500 miles and 1500 miles, then every 2500 miles thereafter completely drain the oil by removing the drain plug, in the lower left side of the oil pan and refill with 4 quarts of fresh lubricant, in accordance with specifications.

To insure continuation of best performance and long engine life, it is necessary to change the crankcase oil whenever it becomes diluted or contaminated with harmful foreign materials. Under the adverse driving conditions described in the following paragraphs, it may become necessary to drain the crankcase oil more frequently.

Vehicles operated in extremely dusty country, should have the oil drained both winter and summer, at 1,000 mile intervals or oftener, and extra precaution should also be taken to keep the carburetor airfilter clean and supplied with oil. The frequency of cleaning the Carburetor Oil Bath Air Cleaner depends upon severity of dust conditions and no definite refill periods can be recommended.

Thinning of the oil by unburned fuel leaking by the piston rings and mixing with the oil, is known as crankcase dilution.

Leakage of fuel into the oil pan mostly occurs during the "Warming-Up" period, when the fuel is not thoroughly vaporized and burned.

Short runs in cold weather do not permit thorough warming up of the engine and water may accumulate in the crankcase from condensation of moisture from piston blow-by.

Practically all present-day engine fuels contain a small amount of sulphur which, in the state in which it is found, is harmless; but this sulphur on burning, forms certain gases, a small portion of which is likely to leak past the pistons and rings and reacting with water when present in the crankcase forms sulphurous acid. As long as the gases and the internal walls of the crankcase are hot enough to keep water vapor from condensing no harm will result, but when an engine is run in low temperatures moisture will collect and unite with the gases formed by combustion, thus acid will be formed and is likely to cause serious etching or pitting. This manifests itself by broken valve springs and excessively rapid wear on piston pins, crankshaft bearings and other parts of the engine.

In view of these conditions it is necessary to drain the crankcase oil at regular intervals. It is always advisable to drain the oil when the engine is warm. The benefit of draining is, to a large extent, lost if the crankcase is drained when the engine is cold because some of the foreign material will remain in the bottom of the oil pan and will not drain out readily with the oil.

At least once a year, preferably in the Spring, the oil pan and floating oil intake should be removed from the engine and thoroughly washed with cleaning solution.

Lubrication Chart Index

See Page 13 for Details

1—Spring Shackle (4)
2 hydraulic fittings
Pressure gun
Chassis grease #1

2—Spring Bolt (5)
1 hydraulic fitting
Pressure gun
Chassis grease #1

3—Tie Rod (2)
2 hydraulic fittings
Pressure gun
Chassis grease #1

4—Drag Link
2 hydraulic fittings
Pressure gun
Chassis grease #1

6—Universal Joint Needle Bearings (4)
1 hydraulic fitting in center
Pressure gun (hand) & adapter
Mineral oil gear lubricant

7—Slip Joint (2)
1 hydraulic fitting
Pressure gun
Chassis grease #1

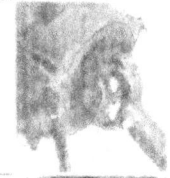

10—Lever Shaft
Transfer case shift
1 hydraulic fitting
Pressure gun
Chassis grease #1

10—Lever Shaft
Clutch and brake pedal
2 hydraulic fittings
Pressure gun
Chassis grease #1

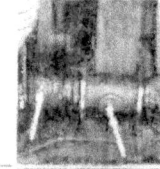

10—Lever Shaft
Steering belcrank
1 hydraulic fitting
Pressure gun
Chassis grease #1

19—Wheel Bearings (4)
Remove and repack
Chassis grease #1

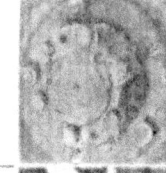

21—Linkage
All clevis pins
Oil can
Engine oil

22—Steering Gear Housing
1 plug
Pressure gun (hand)
Chassis grease #1

27—Front Axle Universal (2)
1 plug—fill to level
Pressure gun
Chassis grease #1

28—Transmission
1 fill and 1 drain plug
Pump
Mineral oil gear lubricant

29—Transfer Case
1 fill and 1 drain plug
Pump
Mineral oil gear lubricant

30—Axle Housing (2)
1 fill and 1 drain plug
Pump
Hypoid gear lubricant

34—Distributor
1 oiler—1 wick—1 post
Oil can
Engine oil
Grease cam

36—Starter
1 oil hole
Oil can
Engine oil

39—Crankcase
1 fill pipe and 1 drain plug
Oil can
Engine oil

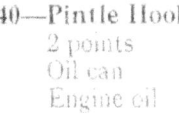

40—Pintle Hook
2 points
Oil can
Engine oil

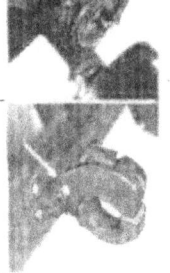

LUBRICATION CHART
¼ Ton 4 x 4 Chassis
Hydraulic Brakes

Make Willys
Model MB

1—Spring Shackle
2—Spring Bolt
3—Tie Rod
4—Drag Link
6—Universal Joint Needle Bearings
7—Slip Joint
10—Lever Shaft—Transfer Case Shift
10—Lever Shaft—Clutch Shaft and Brake Pedal
10—Lever Shaft—Steering Belcrank
19—Wheel Bearings
21—Linkage
22—Steering Gear Housing
27—Front Axle Universal
28—Transmission
29—Transfer Case
30—Axle Housing
34—Distributor
36—Starter
39—Crankcase
40—Pintle Hook

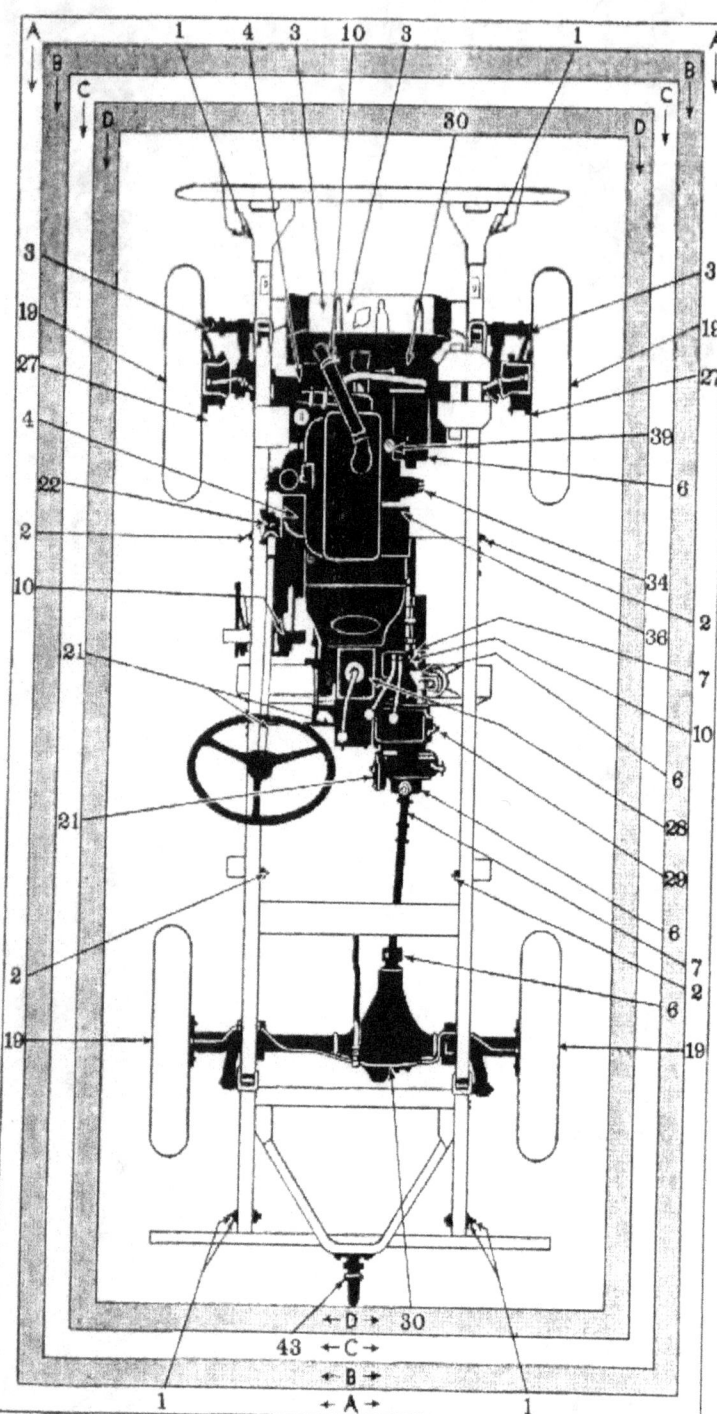

FIG. 1—CHASSIS

TOOLS

Cleaning Rag
Adjustable Wrench
Square Shank Wrench, ⅜"
Wheel Bearing Nut Wrench
Screw Driver

INSTRUCTIONS

Clean and lubricate all points in the order indicated, except those which require disassembly. Clean all vents. Check and adjust level in housings. Disassemble as separately instructed. Drain as separately instructed. See page 13.

Frame A—Chassis Lubricant
Frame B—(Mineral Oil) Gear Lubricant.
Frame C—Engine Oil
Frame D—(Hypoid) Gear Lubricant.

Predominating Temperature

	Above 32° F.	Between 32° F. and 0° F.	Below 0° F.
Chassis Grease	#1	#1	#1
Gear Lubricant	90	90	80
Engine Oil	30	10W	10W plus 10% Kerosene

LUBRICATION CHART — U. S. GOVERNMENT

For Canadian and British Lubrication Specifications see reverse side of Index Page in front of Manual.

ITEM TO BE LUBRICATED See Page 10 & 11	HOW APPLIED	Capacity	GRADE RECOMMENDED						MILES
			Winter			Summer			
			SAE	Navy	Army	SAE	Navy	Army	
Engine Crankcase (39)	Filler pipe R. side check level daily	Refill 4 qts	10W	1042		30	1065		2500
*Transmission Case (28)	Filler plug R. side—add Oil to level of plug	2 pts	90	1100		90	1100		6000
Transfer Case (29)	Filler plug—add oil to level of plug	3 pts	90	1100		90	1100		6000
Differential F. & R. (30)	Filler plug in cover—add Hypoid oil to level of plug	2½ pts	90EP		Fed. Spec. VVL 761 Class 2	90EP		Fed. Spec. VVL 761 Class 2	6000
Propeller Shaft Universal Joints F & R (6)	Fitting		140	1120		140	1120		1000
Air Cleaner	Remove Cover	1¼ pt	10W	1042		30	1065		2000
Front Axle Shaft Universal Joint & Steering Knuckle Bearings (27)	Filler plug outer casing	½ lb	NLGI No. 1		NLGI No. 1			NLGI No. 1	1000
F & R Wheel Bearings (19)	Remove and Repack		NLGI No. 1		NLGI No. 1			NLGI No. 1	6000
Steering Gear Housing (22)	Remove Plug	6½ oz	NLGI No. 1		NLGI No. 1			NLGI No. 1	1000
Steering Bell Crank (10)	Fitting								
Steering Tie Rods (3)	Fitting each end								
Steering Connecting Rod (4)	Fitting each end		Lubricate with NLGI No. 1 Summer & Winter						1000
Spring Shackles F & R (1)	Fittings 8								
Spring Pivot Bolts F & R (2)	Fittings 2								
Clutch & Brake Pedal Shaft (10)	Fittings 5								
Propeller Shaft Slip Joints (7)	Fitting 1 each								
Starter Front (36)	Oil Hole	5 Drops	10W	1042		30	1065		1000
Distributor (34)	Oil Cup on side	5 Drops	10W	1042		30	1065		1000
Distributor Shaft Wick (34)	Oil Can	1 Drop	10W	1042		30	1065		2500
Distributor Arm Pivot (34)	Oil Can	1 Drop	10W	1042		30	1065		2500
Distributor Cam (34)	Wipe with grease		NLGI No. 1			NLGI No. 1		NLGI No. 1	2500
Clevis Pins, Yokes & Cables (21)	Oil Can	5 Drops	10W	1042		30	1065		1000
Pintle Hook (40)	Oil Can	5 Drops	10W	1042		30	1065		1000
Hydraulic Brake System	Oil Can	¾ Pts.	Lockheed No. 21 Brake Fluid						

*Remove Skid Plate to drain Transmission Shock Absorbers Non-Refillable

See Page 3 for Imperial and Metric qualities

The following table shows at a glance the specification of oil to use in the engine according to temperature conditions:

	GRADE
Above 90° Fahr.	30
Not lower than 32° above zero Fahr.	20 or 20W
As low as 10° above zero Fahr.	20W
As low as 10° below zero Fahr.	10W
Lower than 10° below zero Fahr.	10W plus 10% Kerosene

Always select an oil with a temperature range which agrees as closely as possible with the outdoor temperature range likely to be encountered. When the crankcase is drained and refilled, the oil should be chosen, not on the basis of the existing temperature at the time of change, but on the minimum temperatures that might reasonably be expected until time to change the oil again.

In warm weather, light oil tends to be used up a little faster than heavier oil; accordingly heavier oil is recommended for Summer use. In cold weather, however, it is important to use a light oil so that the engine can be started easily and to assure an adequate, early flow of oil to every part of the engine when first started and cold.

CHASSIS LUBRICATION

All hydraulic lubrication fittings indicated by No. 1-2-3-4 & 10 in Fig. 1 should be wiped clean and gone over with a compressor every 1,000 miles.

Make certain that each bearing surface is properly lubricated. All clevis pins, yokes and upper end of hand brake conduit should be oiled.

STEERING GEAR

Check level of lubricant in steering gear housing No. 22, Fig. 1, every 1,000 miles, keeping it filled at all times with NLGI No. 1 lubricant. Avoid the use of cup grease, graphite, white lead or heavy solidified oil.

Remove plug in steering housing and with a hand gun fill the housing slowly. When housing is full, replace the filler plug.

FAN AND WATER PUMP

The fan and water pump bearings are prelubricated and the lubricant lasts for the life of the bearings.

IGNITION DISTRIBUTOR

The oiler, on the distributor indicated by No. 34 in Fig. 1, should be lubricated every 1,000 miles with several drops of engine oil.

Every 2,500 miles when engine oil is changed apply a drop of light engine oil on the wick located in the top of the shaft which is accessible by removing the rotor arm. Also put a wipe of soft grease on the breaker arm cam, and a drop of oil on the breaker arm pivot.

GENERATOR

The Generator Bearings are prelubricated and require no attention.

STARTER

The oil hole cover on Commutator (Front) End slips to one side; 3 to 5 drops of medium engine oil is recommended every 1,000 miles, No. 36, Fig. 1. Be sure to slip cover back in place.

UNIVERSAL JOINTS—(Propeller Shaft)

Every 1,000 miles lubricate the propeller shaft universal joints with a hand gun and adaptor using S.A.E. 140 No. 6, Fig. 1. Use NLGI No. 1 in the slip joint, No. 7, Fig. 1.

UNIVERSAL JOINTS—(Front Axle Shaft)

Front axle shaft universal joints should be checked every 1,000 miles thru plug hole in rear of housing and add NLGI No. 1 lubricant to level of filler plug. Every 12,000 miles remove, clean, inspect and refill with ½ lb. No. 27, Fig. 1.

WHEEL BEARINGS

Wheel Bearings should be removed, thoroughly cleaned and repacked every 6,000 miles, No. 19, Fig. 1, using NLGI No. 1 lubricant

TRANSMISSION AND TRANSFER CASE

Transmission and Transfer cases are filled with S.A.E. 90 mineral oil lubricant at the factory, this being satisfactory for "year round" use except where extremely cold temperatures are experienced in which case use S.A.E. 80 or the oil should be diluted 10% to 20% with Kerosene. It should be checked each 1,000 miles when vehicle is lubricated and renewed each 6,000 miles. See Fig. 1, No. 28 and Fig. 29. To drain transmission remove skid plate. Lubricate transfer case shift lever shaft hydraulic fitting with NLGI No. 1 lubricant.

FRONT AND REAR AXLE

Hypoid gears require extreme pressure lubricant, therefore the lubricant manufacturers have developed a special lubricant, which is suitable for hypoid axles.

The level of the lubricant in these units should be checked every 1,000 miles. Do not mix different types of Hypoid Lubricants.

Seasonal changes of lubricant are not required except where extremely cold temperatures are experienced in which case use S.A.E. 80 or the oil should be diluted 10% to 20% with Kerosene. It is recommended that the housings be drained and refilled with 2½ pts. of S.A.E. 90 EP lubricant at least twice a year or every 6,000 miles. Use a light engine oil or flushing oil to clean out housings. See Fig. 1, No. 30.

Note—Do not use water, steam, kerosene, or gasoline for flushing.

AIR CLEANER

Each engine oil change or oftener, when vehicle is operated in sandy or dusty areas, remove, clean and refill oil cup to indicated oil level. See Lubrication Chart on Page 12. To clean element, see Page 50.

PERIODIC INSPECTION

OPERATION	Daily	Each 1000 Miles	Each 6000 Miles	12,000
Front Axle				
Check Wheel Alignment		X		
Inspect Tie Rod Ends for Wear		X		
Inspect Steering Arms for Tightness		X		
Inspect Steering Knuckle Bearings and Oil Seals for Looseness and Wear		X		
Check Axle Shaft Universal Joints for Wear			X	
(Make same Inspections as for Rear Axle)				
Rear Axle				
Check Axle Shaft Flange Bolts for Tightness		X		
Check Wheel Bearings for Looseness and Wear		X		
Inspect for Oil Leaks at Pinion Shaft Oil Seals		X		
Check Axle Housing Cover Bolts		X		
Inspect Axle Shaft Oil Seals for Grease Leak		X		
Check for End Play in Pinion Shaft		X		
Check Universal Joint Flanges for Looseness		X		
Body				
Check Body Bolts		X		
Brakes				
Inspect Fluid Supply in Master Cylinder		X		
Make Visual Inspection of Brake Lines and Hoses	X			
Test Service Brakes; adjust if necessary	X			
Remove Wheels; inspect brake lining			X	
Flush entire system with new fluid				X
Check Brake Pull Back Springs		X		
Test Hand Brake	X			
Clutch				
Check Free Pedal Travel; adjust when necessary	X			
Check Adjustment on Clutch Cable to determine when Driven Plate is required—(Last adjustment)		X		
Cooling				
Check Water in Radiator	X			
Test Anti-Freeze Solution (During Winter)	X			
Check Fan Belt for Tension	X			
Inspect Radiator Hoses and Connections	X			
Check Water Pump for Leaks	X			
Check Temperature Gauge	X			
Flush System Twice a year. Before and after using Anti-Freeze; test Thermostat; replace Radiator Hoses				X
Battery				
Check Gravity and add Distilled Water every 2 weeks				
Check Terminals		X		
Check and Tighten Ground Straps		X		
Check Hold Clamp Bolts		X		
Wiring				
Inspect all Connections		X		
Inspect for Chaffed or Broken Wires		X		
Inspect Retaining Clips and Grommets		X		

PERIODIC INSPECTION—Continued

OPERATION	Daily	Each 1000 Miles	Each 6000 Miles	12,000
Starting Motor				
Clean Commutator with .00 Sand Paper			X	
Check Brushes for Wear and Tension			X	
Check Mounting Bolts for Tightness			X	
Overhaul				X
Clean Bendix Drive				X
Check Cable Connections			X	
Generator				
Clean Commutator with .00 Sand Paper			X	
Check all Terminals			X	
Check Brushes for Wear and Tension			X	
Overhaul				X
Check Voltage Regulator			X	
Lights and Switches				
Check Operation of Lights	X			
Horn				
Check for operation	X			
Distributor and Spark Plugs				
Clean and Adjust Distributor Points and Spark Plug Gaps			X	
Replace Plugs				X
Check and Test Condenser			X	
Overhaul Distributor				X
Check Timing			X	
Check Automatic Spark Advance			X	
Clean Distributor Cap—Terminal Towers			X	
Test Coil (Heat Test)				X
Engine				
Check Cylinder Head Nuts and Bolts			X	
Check Manifold Nuts and Gaskets		X		
Check Oil Pan Bolts; Check for Oil Leaks		X		
Check Compression			X	
Check Tappet Clearance			X	
Check Engine Mounting Bolts and Nuts			X	
Check Oil Pressure Gauge Reading	X			
Check Oil Level. (Add Necessary)	X			
Remove and Clean Oil Pan and Floating Intake				X
Change Oil Filter				X
Definite Periods for Major operations or overhauling cannot be predetermined. They are dependent upon service to which Engine has been subjected.				
Fuel System				
Remove, clean and refill air cleaner		X		
(Vehicles operated in Dusty or Sandy regions, the air cleaner should be inspected and cleaned if necessary)	X			
Clean Fuel Pump Sediment Bowl and Strainer		X		
Clean Fuel Filter		X		
Tighten Carburetor Flange to Manifold Nuts		X		
Check Carburetor Adjustments		X		
Inspect all Fuel Lines and Connections for leaks		X		

PERIODIC INSPECTION—Continued

OPERATION	Daily	Each 1000 Miles	Each 6000 Miles	12,000
Fuel System				
Test Fuel Pump Pressure			X	
Overhaul Carburetor and adjust			X	
Drain Fuel Tank and Flush out Sediment				X
Lubrication				
Refer to Lubrication Chart	X	X	X	X
Springs				
Inspect Spring Clips to Axle for Tightness		X		
Inspect Spring Shackles and Bushings		X		
Check condition Front and Rear Springs		X		
Shock Absorbers				
Inspect Mounting Bushings, replace when necessary		X		
Inspect Mounting Brackets		X		
Check for Control (off car) adjust or replace				X
Steering System				
Check Steering System (Loss Motion)	X			
Inspect Steering Connecting Rod, Ball and Sockets		X		
Check Steering Post Bolts to frame		X		
Check Steering Pitman Arm Nut		X		
Transmission and Transfer Case				
Inspect for Oil Leaks		X		
Check Transmission to Bell Housing Bolts			X	
Check Oil Seals at Propeller Shafts		X		
Universal Joints				
Check Flange Nuts		X		
Inspect Joint Bearings for wear		X		
Wheels and Tires				
Check Tire Pressures	X			
Tighten Wheel Hub Bolt Nuts		X		
Remove Wheel Bearings, inspect, replace worn or chipped cups or cone repack and adjust			X	
Check Tire Wear, check toe-in, caster and camber		X		

Inspection

The importance of regular inspection cannot be over-emphasized. Making adjustments, tightening bolts, nuts and wiring connections when needed, will go far towards avoiding trouble and delay on the road and uphold the high standards of reliability and performance built into the vehicle by the Manufacturer.

After maneuvers involving operations in swamps and streams inspect for water and sludge in engine, transmission, transfer case, front and rear axles, wheel bearings and front universal joints.

ENGINE

(NEVER RUN ENGINE IN CLOSED GARAGE)

Due to the presence of carbon monoxide (a poisonous gas in the exhaust of the engine) never run the engine for any length of time while this vehicle is in a small closed garage. Opening the doors and windows lessens the danger considerably, but it is safest, if adjustments are being made that require operation of engine, to run the vehicle out of doors.

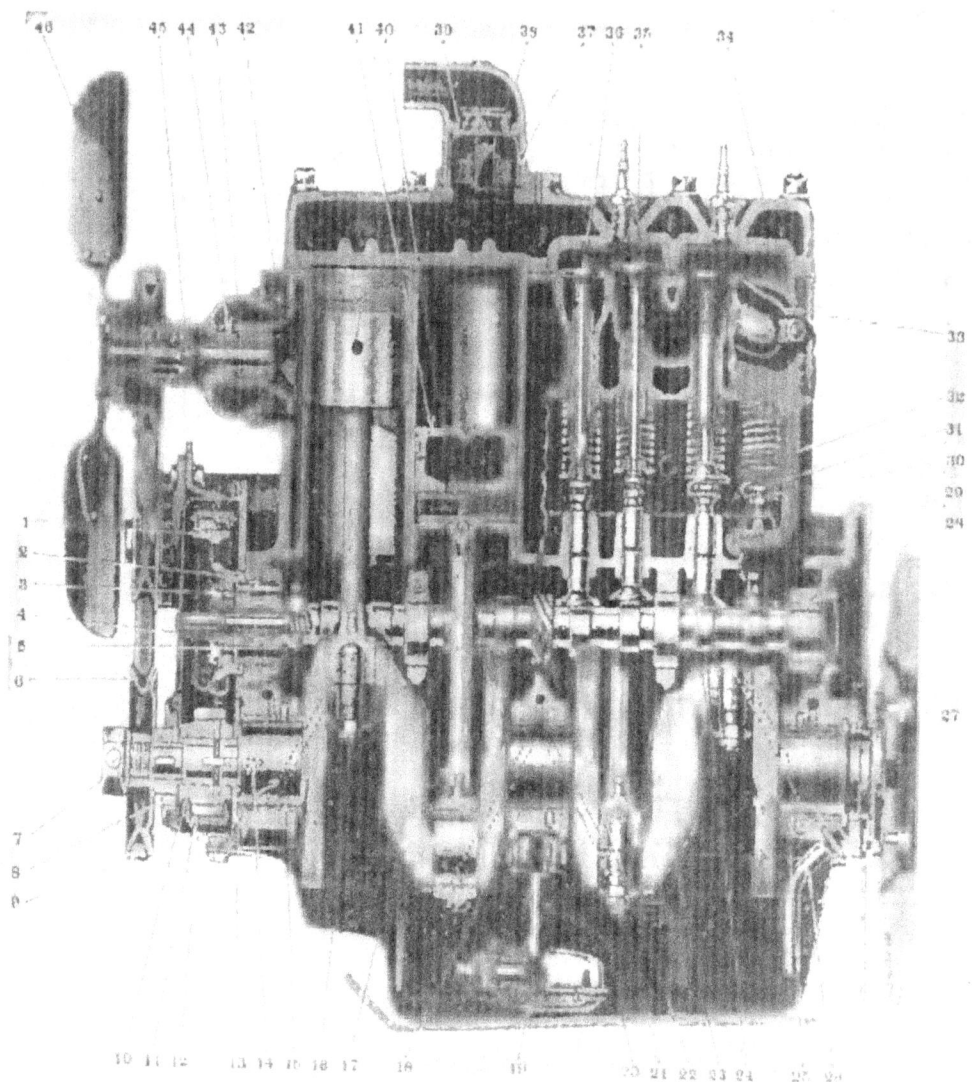

FIG. 1—SIDE SECTIONAL OF ENGINE

The engine is a Four-Cylinder L-head type unit equipped with a counter-balanced crankshaft.

At end of this section will be found the Engine specifications. When adjustments are necessary we recommend that reference be made to these specifications for proper running tolerances and clearances of all component parts. On Page 34 headed "Engine Troubles and Causes" are listed many reasons for engine failure or poor performance. For correction of these difficulties you will find the procedure to follow under separate headings in this section.

FIG. 1—SIDE SECTIONAL VIEW OF ENGINE

No.	Willys Part No.	Ford Part No.	Name
1	638458	GPW-6256	Camshaft Sprocket
2	375900	GPW-6245	Camshaft Thrust Washer
3	639051	GPW-6262	Camshaft Front Bushing
4	375907	GPW-6243	Camshaft Thrust Plunger
5			Fuel Pump Eccentric
6	638113	GPW-6312	Fan and Generator Drive Pulley
7	387633	GPW-6319	Starting Crank Nut Assembly
8	637096	GPW-6700	Crankshaft Packing—Front End
9	A-1495	GPW-8620	Fan and Generator Drive Belt
10	638459	GPW-6306	Crankshaft Sprocket
11	638457	GPW-6260	Camshaft Drive Chain
12	A-1190	GPW-6016	Chain Cover Assembly
13	634796	GPW-6308	Crankshaft Thrust Washer
14			Crankshaft Oil Passages
15	637008	GPW-6338-A	Crankshaft Bearing Front—Lower
16	52825	356028-S	Connecting Rod Bolt Nut Lock
17	639859		Connecting Rod Assembly No. 2
18	638731	GPW-6341-A	Crankshaft Bearing Center—Lower
19	630396	GPW-6615	Oil Float Assembly
20	630397	GPW-6617	Oil Float Support
21	637020		Connecting Rod Cap Bolt
22			Oil Pump and Distributor Drive Gear
23	638121	GPW-6303-A1	Crankshaft
24	637047	GPW-6500	Valve Tappet
25	638733	GPW-6337-A	Crankshaft Bearing Rear—Lower
26	630294	GPW-6326	Crankshaft Bearing Rear Drain Pipe
27	637237	GPW-6702	Crankshaft Packing—Rear End
28	635394	GPW-6384	Flywheel Ring Gear
29	637065	GPW-6250	Camshaft
30	637704	GPW-6550	Valve Tappet Clearance Spring
31	637048	GPW-6549	Valve Tappet Adjusting Screw
32	638636	GPW-6513	Valve Spring
33	A-912		Exhaust Manifold Assembly
34	A-1534	GPW-6050	Cylinder Head
35	637182	GPW-6507	Inlet Valve
36	637183	GPW-6505	Exhaust Valve
37	639651	GPW-8578	Thermostat Retainer
38	A-1192	GPW-8260	Water Outlet Elbow
39	637646	GPW-8575	Thermostat Assembly
40	636961	GPW-6135-A	Piston Pin
41	636957	GPW-6110-A	Piston
42	639993	GPW-8512	Water Pump Impeller
43	639663	GPW-8524	Water Pump Seal Assembly
44	639994	GPW-8557	Water Pump Seal Washer
45	636297	GPW-8530	Water Pump Bearing and Shaft Assembly
46	A-447	GPW-8600	Fan Assembly

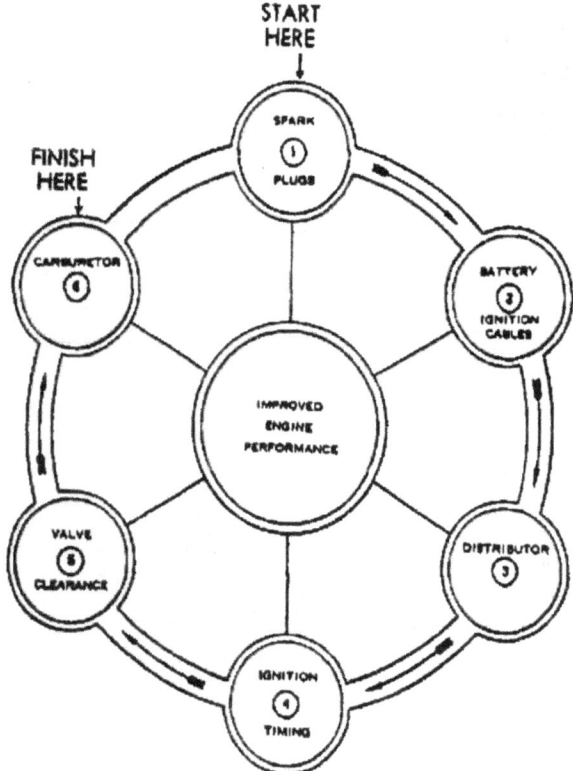

FIG. 2 — THE CYCLE OF ENGINE TUNE-UP

Engine Tune-Up

For best performance and dependability, the engine should have a periodic tune-up every 6,000 miles, see Fig. 2. The following procedure is recommended when performing this operation:

1. Remove spark plugs and clean. Adjust the Electrodes to .030″ gap.
2. Check Battery Terminals, ground cable and ground straps on left side of engine at front engine support and cylinder head for clean and tight connections.
3. Remove distributor cap and inspect points.
4. Check ignition timing.
5. Check valve tappet clearance-set .014″ cold.
6. Set carburetor float level, accelerator pump travel, and metering rod as covered under heading "Carburetor".
7. Start engine and allow to run until thoroughly warmed up, then set carburetor idle screw so the engine will idle at 600 R.P.M. (vehicle speed approximately 8 miles per hour).
8. Adjust low speed idling screw so that engine will idle smoothly.

Carburetor

Complete information regarding dismantling, cleaning and adjusting will be found in the Fuel section, under heading "Carburetor".

Distributor

For complete information regarding cleaning, adjusting and setting ignition timing refer to Electrical section under heading "Distributor".

Grinding Valves

Lack of power in an engine is sometimes caused by poor seating of the valves in the valve seats which allows the gases in the compression chamber to escape into the intake or exhaust manifold.

Through the use of a cylinder compression gauge one can readily determine which valves are not properly seating.

Compression gauge readings should all be within 10 pounds of each other and not less than 70 lbs.

If no gauge is available, remove all spark plugs, hand crank engine and have mechanic place thumb over one spark plug hole at a time. When no compression is experienced that particular cylinder is at fault.

The valve grinding operation from the standpoint of engine power and performance is very important. Extreme care should be used whenever valves are ground to maintain factory limits and clearances, as only by maintaining these can one expect to get good engine performance.

When it is necessary to perform a valve job, it will be best to follow the procedure as outlined in the next few paragraphs.

1. Drain radiator by opening drain cock at the bottom left corner of the radiator.
2. Remove oil filter and bracket by removing the nuts on the cylinder studs and lay filter on generator.
3. Remove fuel line from fuel pump to carburetor.
4. Remove carburetor air cleaner and accelerator rod.
5. Remove choke and throttle control wires.

WILLYS MODEL "MB" ¼-TON 4 x 4 GOVERNMENT TRUCK

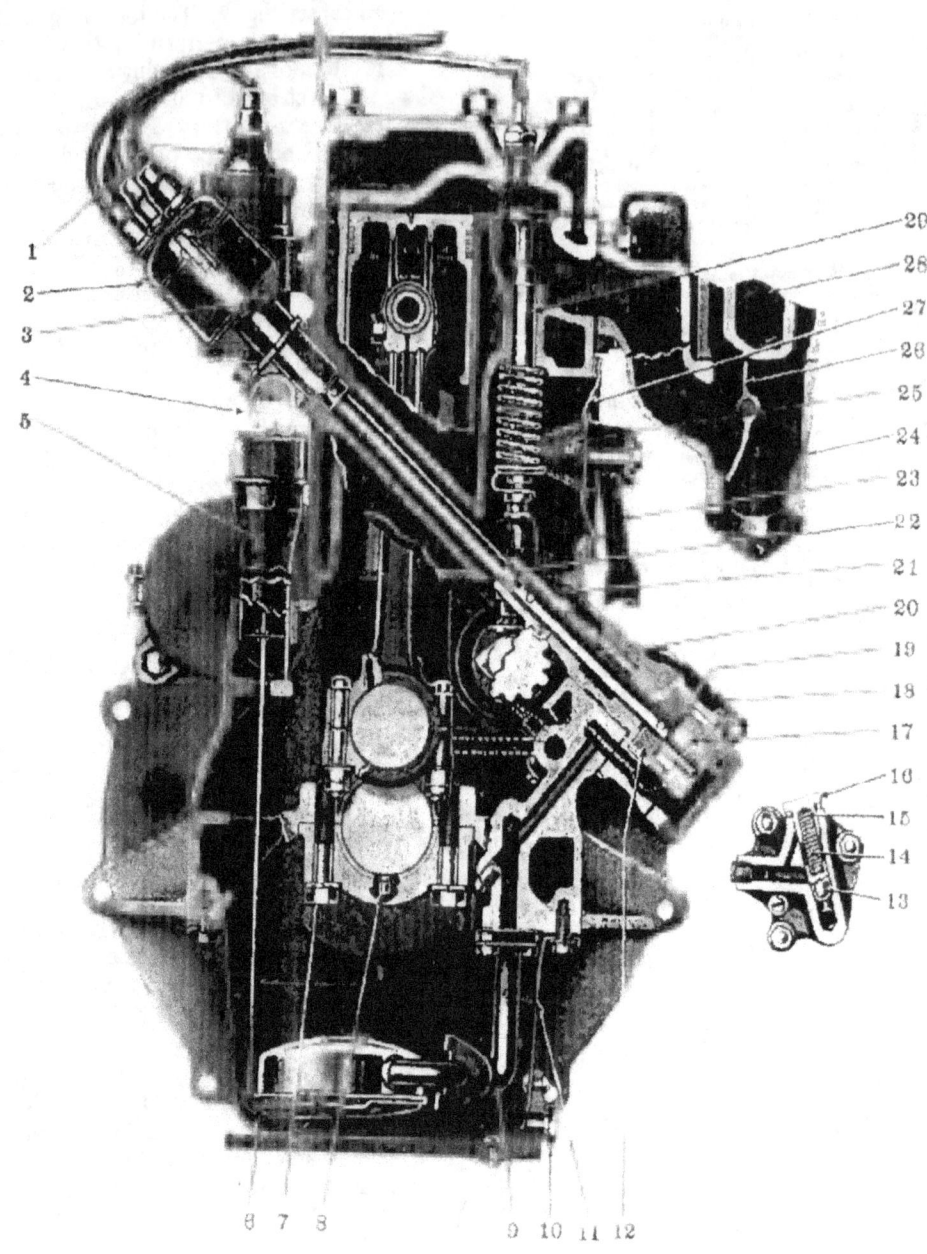

FIG. 3—FRONT SECTIONAL OF ENGINE

No.	Willys Part No.	Ford Part No.	Name
1	A-1527	GPW-12000	Ignition Coil
2	A-1244	GPW-12100	Ignition Distributor
3	107128	B-10141	Distributor Oiler
4	A-5168	GPW-6766-B	Oil Filler Cap and Level Indicator
5	A-5165	GPW-6763-B	Oil Filler Tube
6	630396	GPW-6815	Oil Float Assembly
7	381519	GPW-6345	Crankshaft Bearing Cap to Crankcase Screw
8	635377	GPW-6369	Crankshaft Bearing Dowel
9	630397	GPW-6617	Oil Float Support
10	639979	GPW-6727	Oil Pan Drain Plug
11	A-1167	GPW-6675	Oil Pan Assembly
12	636599	GPW-6608	Oil Pump Shaft and Rotor Assembly
13	630518	GPW-6663	Oil Relief Plunger
14	356155	GPW-6654	Oil Relief Plunger Spring
15	630389	GPW-6628	Oil Relief Plunger Spring Shims
16	630390	GPW-6644	Oil Relief Plunger Spring Retainer
17	343306	GPW-6614	Oil Pump Pinion
18	637636	GPW-6600	Oil Pump Assembly
19	636600	GPW-6673	Oil Pump Rotor Disc
20	630386	GPW-6609	Oil Pump Shaft
21	630394	GPW-6630	Oil Pump Body to Cylinder Block Gasket
22	637615	GPW-12083	Distributor Shaft Friction Spring
23	A-1061	GPW-6758	Crankcase Ventilator Assembly
24	A-912		Exhaust Manifold Assembly
25	630298	GPW-6762	Crankcase Ventilator Baffle
26	636439	GPW-9460	Heat Control Valve
27	636554	GPW-6519	Valve Spring Cover Assembly
28	A-1166		Intake Manifold Assembly
29	375811	GPW-6510	Exhaust Valve Guide

6. Remove nuts holding carburetor to manifold and remove carburetor.
7. Remove bolt and nuts holding exhaust pipe to manifold.
8. Remove manifold stud nuts and manifolds.
9. Remove the upper radiator hose. Remove all spark plugs by using the socket wrench furnished in the tool kit. Remove the cylinder head cap screws, stud nuts and the temperature gauge bulb, then lift head from engine block. Removal is made easy by using lifting hooks screwed in No. 1 and 4 spark plug holes. CAUTION—Do not use screw driver or any other sharp instrument to drive in between the cylinder head and the block to break the head loose from the gasket.
10. Remove the valve cover plate screws and the valve cover. Care should be taken when removing the valve cover breather tube not to lose the copper gasket on each screw as well as the screen and gasket. With a piece of cloth or cotton waste cover the three holes in the valve chambers to prevent the valve keys dropping into crankcase upon removal.
11. Remove valve tappet clearance springs by placing screw driver on top edge and snapping out. With valve spring compressor inserted between valve tappet and spring retainer raise springs on those valves which are in closed position and remove valve locks. Turn crankshaft with crank or by fan belt until those valves which are open become closed and repeat the operation.
12. Remove valves and place them in a valve carrying board, so that they can be identified as to cylinders from which they were removed. Remove valve springs. The valve springs should be tested for pressure which should show 116 lbs. when valves are open (Length 1¾") or 50 lbs. pressure when closed (Length 2⁵⁄₆₄"). The free length of the valve spring is 2½" inches. Any springs which are distorted or do not fall within these specifications should be replaced with new springs.
13. Clean carbon from cylinder head, top of pistons, valve seats and cylinder block, clean valve guides with guide brush. Clean valves on a wire wheel brush making sure that all carbon is removed from the top and bottom of the heads, as well as the gum which might have accumulated on the stems. The valve heads should then be refaced at an angle of 45°. If valve seats in block show signs of excessive pitting it is advisable to reface the seats and check with dial gauge—Fig. 4. Then by hand, touch up the valves to the valve seats with fine valve grinding compound.

The clearance between the intake valve stem and the valve guide is .0015" to .00325", the exhaust valve stem clearance to guide is .002" to .00375". Excessive clearance between the valve stem and the valve guide will cause improperly seating and burned valves. If there is too much clearance between the inlet valve stem and the valve guide, on the suction stroke there will be a tendency to draw oil vapors up the guide into the combustion chamber causing excessive oil consumption, fouled spark plugs and poor low speed performance.

To check the clearance of the valve stem to the valve guide, take new valve and place in each valve guide and feel the clearance by moving the valve stem back and forth. If this tolerance is excessive it will be necessary to replace the valve guide, otherwise the valve stem is worn.

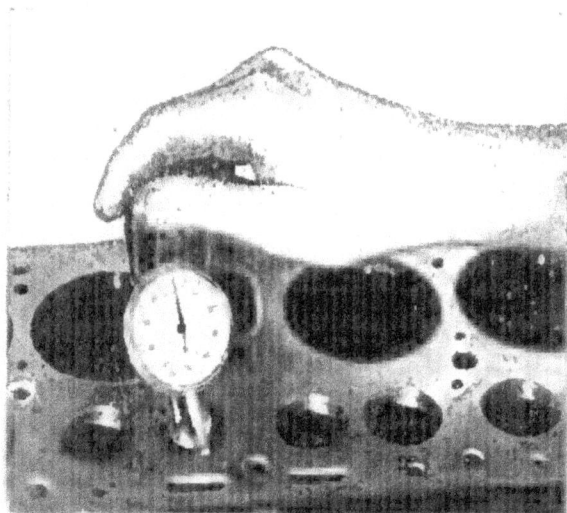

FIG. 4—GAUGING VALVE SEATS

FIG. 5—PULLING VALVE GUIDE

Removing and Replacing of Valve Guide

When removing the valve guides use a valve guide puller such as shown in Fig. 5 to prevent damage to cylinder block. If a regular puller is not available, a suitable tool can be made from a 2" pipe, 6" long and a 3/8" bolt 10" to 12" long with a long threaded end, a small hexagon nut which will pass through the hole in the cylinder block and a 2" washer with a 3/8" hole in it.

The valve guides are installed with a replacer or a driver as shown in Fig. 6. Taking a piece of half inch round stock 6" long and turning down one end to 3/8" diameter 2" long will make a suitable driver.

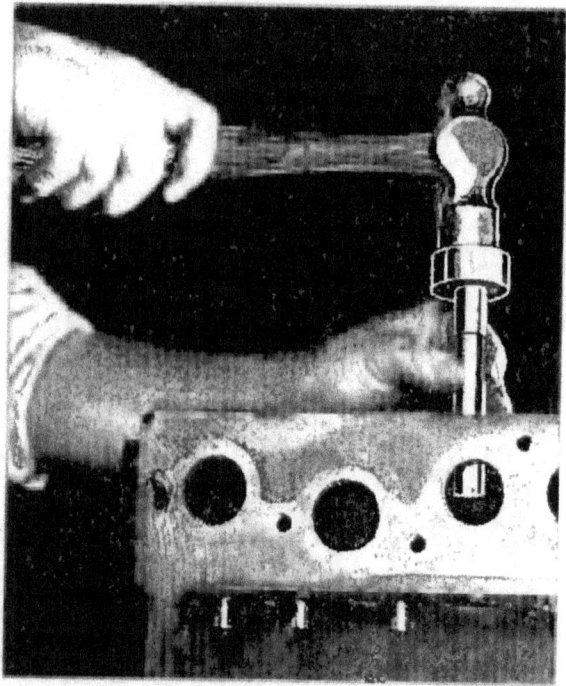

FIG. 6—INSTALLING VALVE GUIDE

The exhaust valve guide is installed in the cylinder block so that there will be a distance of 1" from the top of the guide to the top of the block. The intake valve guide is set at 1 5/16" from top of valve guide to the top of block. Fig. 7.

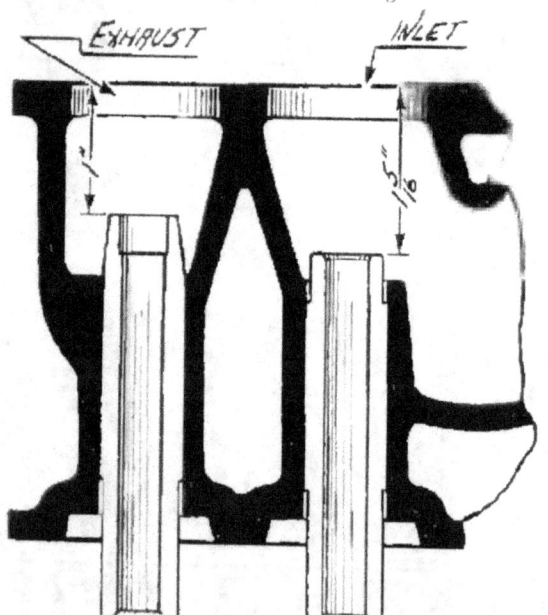

FIG. 7—POSITION OF VALVE GUIDES

The valve tappet clearance in the guide should be .0005" to .002". It is advisable to check the clearance of the valve tappet by moving it back and forth in the guide. If the clearance seems to be excessive it might be necessary to install a new valve tappet. This operation is covered in this section under "Camshaft and Valve Tappet".

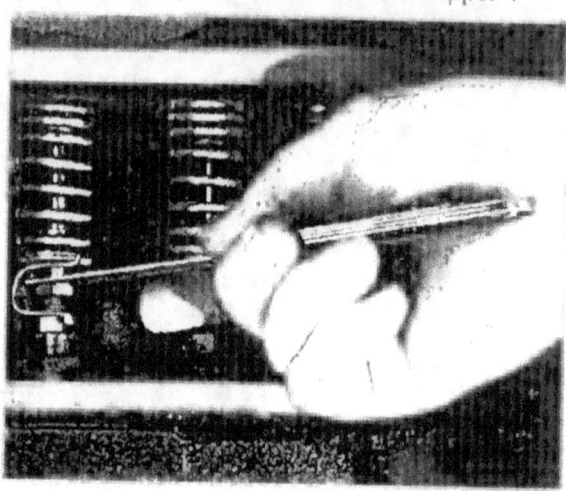

FIG. 8—VALVE TAPPETS AND SPRINGS

When assembling valve springs and retainers in engine make sure that the closed coils are up against cylinder block. See Fig. 8. Then make installation of the valves in their respective positions as they were disassembled. Through the use of a valve spring compressor raise valve springs on those valves which are in the closed position, and with valve key inserting tool, insert the valve spring locks. If no tool is available, hold keys in place by sticking them to valve stem with grease.

Adjust the valve tappet to valve stem clearance to .014". Fig. 8. Remove cloth or waste from valve chamber.

Clean top of block and pistons of all foreign matter and install cylinder head gasket. Clean carbon from cylinder head and wipe off all foreign matter then install over studs on cylinder block.

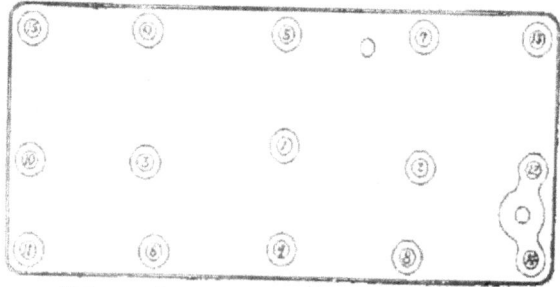

FIG. 9—CYLINDER HEAD TIGHTENING

Install oil filter and air cleaner bracket and tube assembly. Install cylinder head cap screws and nuts bringing them down finger tight, then with a tension wrench tighten cylinder head screws and nuts in sequence as shown on Fig. 9, tightening screws to 65 to 75 foot pounds or 780 to 920 inch pounds and the nuts to 60 to 65 foot pounds or 720 to 780 inch pounds.

Clean and adjust spark plugs, setting the electrode gaps at .030", Fig. 10. Install spark plugs in cylinder head to prevent any foreign matter from entering the combustion chamber during the remaining operations.

Install manifold with new gaskets. Install manifold clamp washers with convex surface toward manifold. Install manifold nuts drawing them up tight. Install exhaust pipe to manifold with new gasket.

Overhaul and recondition carburetor as per instructions given under the heading "Carburetor". Install carburetor to manifold and attach controls.

Recondition distributor and set ignition timing in accordance with instructions given under "Distributor".

NOTE—Make sure when installing distributor assembly in crankcase that it fits down in the crankcase properly.

Install upper radiator hose, and all line connections and fill radiator with water. Start engine and allow it to idle for a period of five or ten minutes, then recheck tappet clearance.

If necessary, install new valve cover plate gasket (shellac to cover). Install cover plate to engine block. Clean valve chamber ventilator tube and screen and reinstall with gaskets.

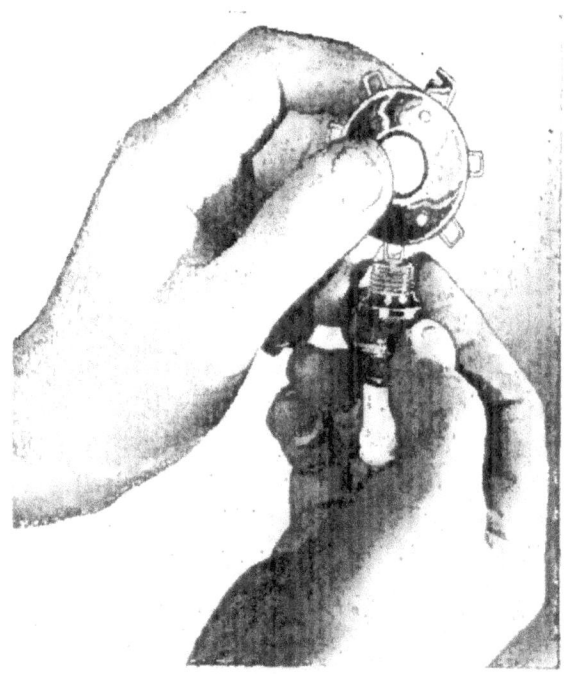

FIG. 10—SETTING SPARK PLUG

FIG. 11—VALVE MECHANISM

No.	Willys Part No.	Ford Part No.	Name
1	637193	GPW-6505	Exhaust Valve
2	638636	GPW-6513	Valve Spring
3	637044	GPW-6514	Valve Spring Retainer Lower
4	376994	GPW-6546	Valve Spring Retainer Lower Lock
5	637704	GPW-6550	Valve Tappet Clearance Spring
6	637048	GPW-6549	Valve Tappet Adjusting Screw
7	375910	355671-S	Valve Tappet Adjusting Screw Lock Nut
8	637047	GPW-6500	Valve Tappet
9	639051	GPW-6262	Camshaft Bushing—Front
10	637065	GPW-6250	Camshaft
11	375900	GPW-6265	Camshaft Thrust Washer
12	638457	GPW-6260	Camshaft Drive Chain
13	638458	GPW-6256	Camshaft Sprocket
14	638459	GPW-6306	Crankshaft Sprocket
15	315932	GPW-6269	Camshaft Sprocket Cap Screw Lockwasher
16	634850	355499-S	Camshaft Sprocket Cap Screw
17	375908	GPW-6244	Camshaft Thrust Plunger Spring
18	375907	GPW-6243	Camshaft Thrust Plunger

Camshaft and Valve Tappets

The alloy cast steel camshaft Fig. 11 rotates on four bearings which are lubricated under oil pressure through drilled passages in the crankcase. The front bearing carries the thrust and is a steel back babbitt-lined shell. This bearing is staked in place to prevent rotation and endwise movement. See Fig. 12.

The valve tappets are lubricated through oil troughs cast in crankcase and drilled passages to valve tappet guides. The oil troughs are filled from oil spray holes at connecting rod bearing ends. A groove cut in center of valve tappet shank carries the oil up and down in guides.

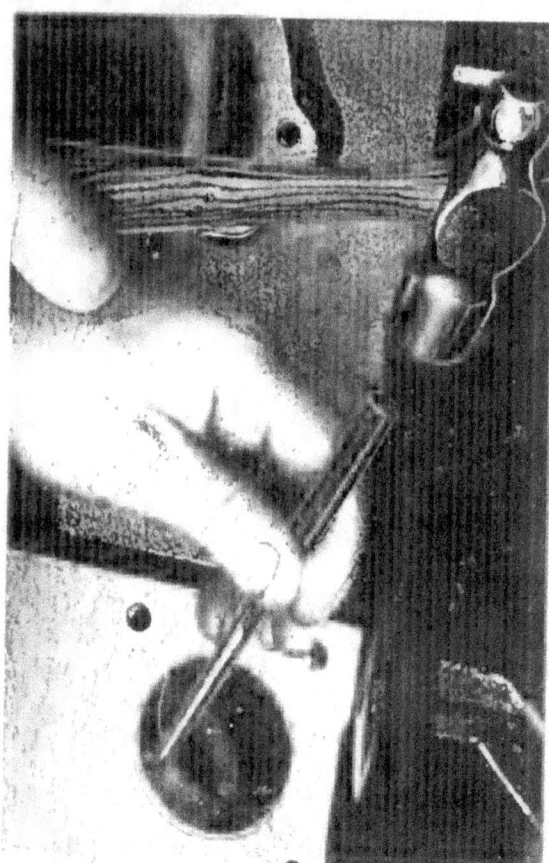

FIG. 12—STAKING CAMSHAFT BEARING

Removal of Camshaft or Valve Tappet

Drain water from radiator, remove radiator and grille, cylinder head, manifold, valves and valve springs. Follow instructions given under sub-heading "Grinding Valves" Page 18.

Remove oil pump and fuel pump assemblies.

Remove oil pan, crankshaft pulley, fan belt and fan assembly.

Remove nuts holding front engine supports to rubber insulators.

Remove timing chain cover, camshaft sprocket screws and timing chain.

Tie all valve tappets up with a string wrapped around heads of screws and attach to manifold studs.

Place jack with block under crankcase and raise front end of engine until camshaft will clear front frame cross member. Remove camshaft and valve tappets.

Carefully inspect camshaft for scores, roughness of cams and bearings. Examine valve tappet faces where they contact cams and replace if found to be scored, rough or cracked. Check clearance of tappets to guides, renewing those which have worn excessively. Oversize available .004".

Replacing Camshaft or Valve Tappets

Install valve tappets and tie up in place with string. Install camshaft. Install camshaft thrust washer.

To set the valve timing, see instructions given under heading "Valve Timing" Page 24.

For installation of oil pump see section under heading "Oil Pump" Page 29.

Install the plunger and spring in the front end of camshaft with round end out. Inspect pin in timing chain cover to see that it stands perpendicular to the cover face. Put a light smear of cup grease on end of pin and on the end of plunger, then assemble cover to the engine.

Balance of the assembly is the same operations used in removal of camshaft only in the reverse order.

NOTE—On earlier engines, when replacing oil pan, the four short cap screws are used across front end.

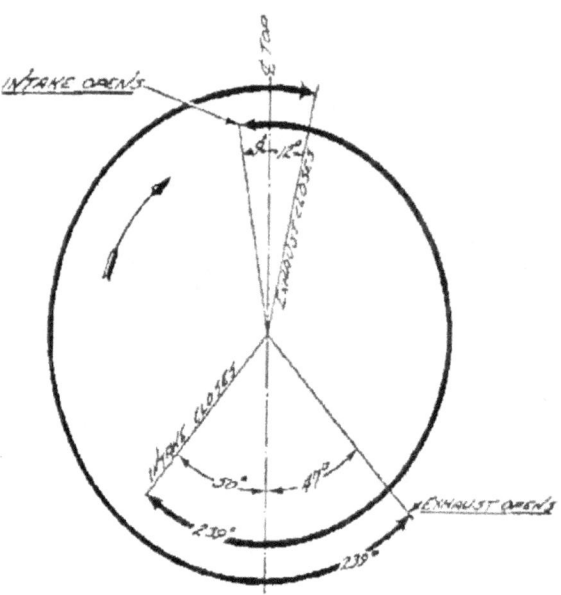

FIG. 13—VALVE TIMING

TIMING CHAIN AND SPROCKETS

The silent type timing chain is non-adjustable. The lubrication is positive through drilled passages in the crankshaft and sprocket from the front main bearing and oil filter return. These should be checked whenever the chain or sprockets are replaced.

To replace timing chain, it is necessary to remove fan blades, radiator, fan belt, crankshaft pulley and timing case cover. Remove screws holding camshaft sprocket to camshaft and remove chain.

When chain has been removed, for any purpose, it will be necessary to set the valve timing when chain is replaced.

VALVE TIMING

To set the valve timing turn the crankshaft so No. 1 and No. 4 pistons are at top dead center.

Place camshaft sprocket on camshaft, turn camshaft so punch mark faces punch mark on crankshaft sprocket.

Remove sprocket and place the timing chain on, then place timing chain over crankshaft sprocket, changing position of camshaft sprocket within chain until the cap screw holes in sprocket and camshaft are in line.

Timing is correct when a line drawn between sprocket centers cuts through timing marks on both sprockets. See Fig. 14.

Inlet opens 9° before top center measured on flywheel or .039" piston travel from top center.

To check valve timing, Fig. 13 adjust inlet valve tappet No. 1 cylinder to .020". Rotate crankshaft clockwise until piston in No. 1 cylinder is ready for the intake stroke, at which time the tappet should just be tight against end of valve stem and mark on flywheel "I O" is in center of timing hole in flywheel housing, see Fig. 15.

TIMING CHAIN COVER AND SEAL

The timing chain cover is a pressed steel stamping heavily ribbed for strength.

The stationary pin in cover is so located as to bear against the plunger in the end of camshaft which controls the end play of camshaft.

The crankshaft oil seal is woven asbestos impregnated with graphite and oil. When necessary to install new oil seal, the steel retainer should also be renewed.

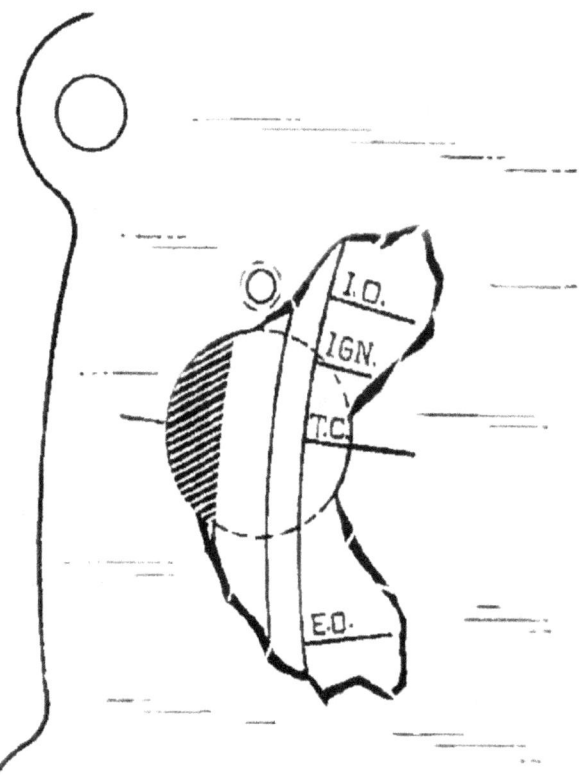

FIG. 15—TIMING MARKS (FLYWHEEL)

Crankshaft

The crankshaft Fig. 16 is of drop forged steel, grain direction following shape of crankshaft with four integral counterweights. This short shaft is rugged and reduces torsional vibration, weight 41½ lbs. After machining, the crankshaft is balanced statically and dynamically, and then dynamically balanced with clutch and flywheel as an assembly. It rotates on three steel back babbitt lined bearings, the front bearing taking the thrust. The packing at the rear bearing prevents oil getting into clutch and loss when parked on steep inclines.

The main bearing journal diameter and length dimensions are as follows:

Front—2.3340"—1.920"
Center—2.3340"—1¹³⁄₁₆"
Rear—2.3340"—1¾"

The steel back babbitt lined bearings are made to size and are interchangeable without line reaming. The running tolerance of the bearing is established at .001". No adjustment is provided on the main bearings. Should they require attention they should be replaced to maintain proper control of oil. Main bearing cap screw torque wrench reading 65-70 ft. lbs.

The end play of the crankshaft is .004" to .006" and adjusted by shims between the crankshaft sprocket thrust washer and end of Main Bearing, Fig. 17.

To adjust end play the crankshaft sprocket must be removed with gear puller, Fig. 18.

Whenever it is necessary to remove the crankshaft or install new crankshaft bearings, the engine has to be removed from the frame. See Note under "Flywheel" Page 32.

Undersize main bearings are available in .010"; .020" and .030".

FIG. 14—TIMING SPROCKETS

WILLYS MODEL "MB" ¼ TON 4 x 4 GOVERNMENT TRUCK

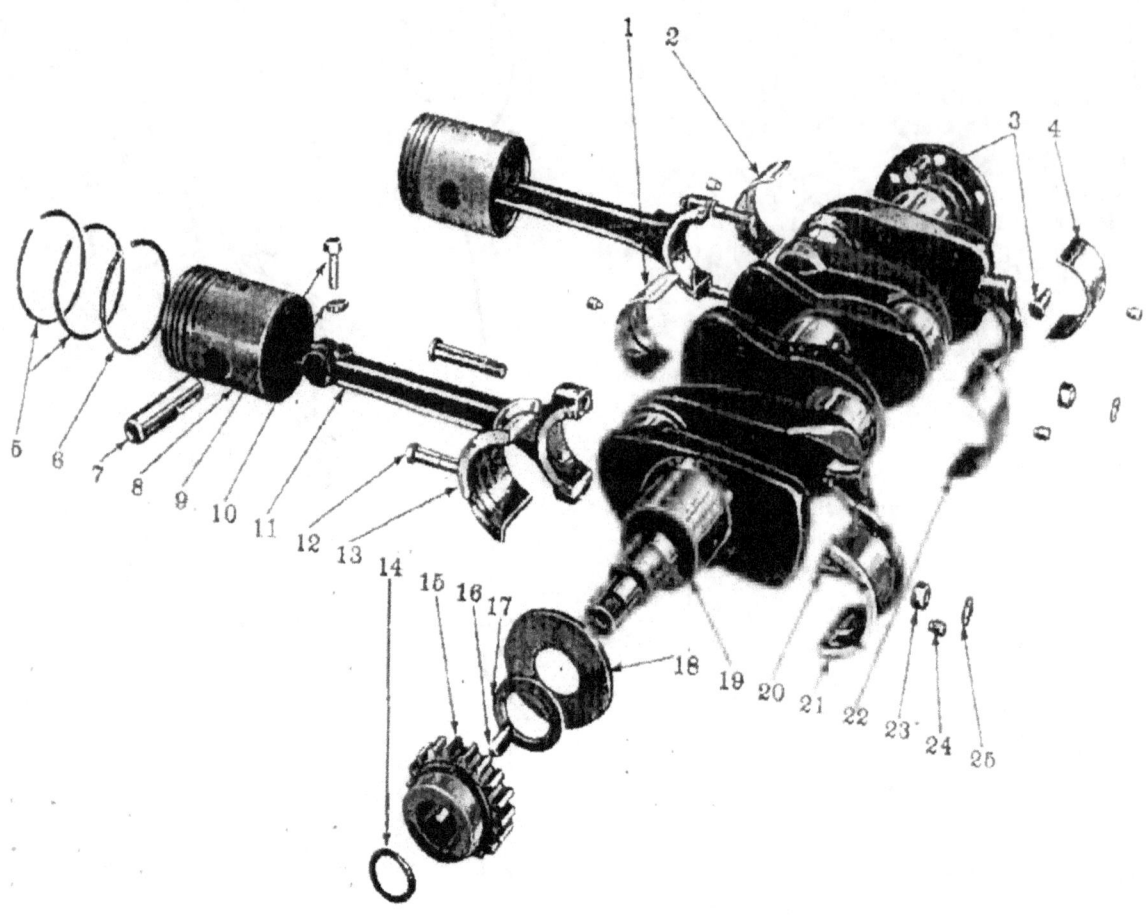

FIG. 16—CRANKSHAFT AND PISTONS

No.	Willys Part No.	Ford Part No.	Name
1	638730	GPW-6339-A	Crankshaft Bearing Center—Upper
2	638732	GPW-6331-A	Crankshaft Bearing Rear—Upper
3	632156	GPW-6387	Flywheel Crankshaft Dowel
4	638733	GPW-6337-A	Crankshaft Bearing Rear—Lower
5	116562	GPW-6155-A	Piston Ring—Compression (Lower)
	639864	GPW-6150	Piston Ring—Compression (Upper)
6	116566	GPW-6156-A	Piston Ring—Oil Regulating
7	636961	GPW-6135-A	Piston Pin
8	636957	GPW-6110-A	Piston—Grade "D"
9	632157	355407-S	Piston Pin Lock Screw
10	5010	34807-S	Piston Pin Lock Screw Lockwasher
11	639858	GPW-6200	Connecting Rod Assembly (No. 1 and No. 3 Cylinders)
12	637020		Connecting Rod Cap Bolt
13	637007	GPW-6333-A	Crankshaft Bearing Front—Upper
14	334103	GPW-6353	Crankshaft Oil Slinger Gasket
15	638459	GPW-6306	Crankshaft Sprocket
16	50917	74182-S	Crankshaft Sprocket Key
17	630727	GPW-6342-A	Crankshaft Sprocket Spacer
18	634798	GPW-6308	Crankshaft Thrust Washer
19	638121	GPW-6303-A	Crankshaft
20	639862	GPW-6211	Connecting Rod Bearing (Upper and Lower)
21	637008	GPW-6338-A	Crankshaft Bearing Front—Lower
22	638731	GPW-6341-A	Crankshaft Bearing Center—Lower
23	636962	356021-S	Connecting Rod Cap Bolt Nut
24	635377	GPW-6369	Crankshaft Bearing Dowel
25	52825	356023-S	Connecting Rod Cap Bolt Nut Lock

FIG. 17—CRANKSHAFT END PLAY

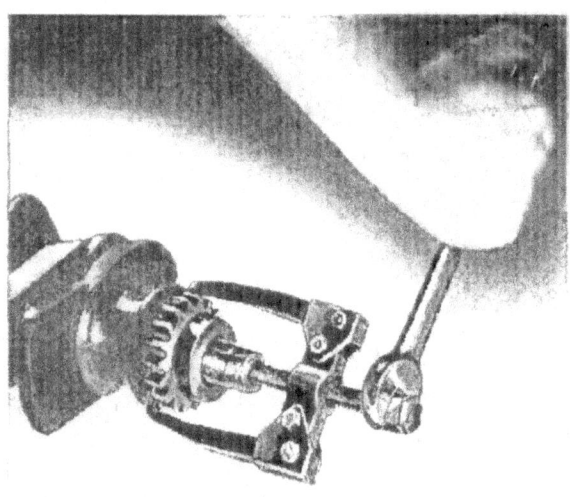

FIG. 18—CRANKSHAFT SPROCKET

CRANKSHAFT REAR BEARING SEAL

The rear main bearing is sealed by a wick type packing, installed in the groove machined in the crankcase, and rear main bearing cap, Fig. 19.

To install a new seal at the rear main bearing cap, insert the packing in the groove with the fingers. Then using a round piece of wood or steel, roll the packing into the groove. When rolling the packing, start at one end and roll the packing to the center of the groove. Then starting from the other end again roll towards the center. By following the above procedure you are sure that the wick is firmly pressed into the bottom of the groove.

The small portion of the packing which protrudes from the groove at each end should be cut flush with the surface of the bearing cap. To prevent the possibility of pulling the packing out of the groove while cutting off the ends it is recommended that a round block of wood, the same diameter as the crankshaft be used to hold the packing firmly in position while the ends are being cut off.

Should it be necessary to install a new seal in the crankcase, it will require the removal of the engine from the frame and the removal of the crankshaft.

The same procedure should be followed when installing a crankcase seal as when installing a seal in the bearing cap.

When installing rear main bearing cap to case a little sealer should be put on the faces of cap where it fits against the case. The rubber seal packing that goes between the main bearing cap and the case is cut to a given length and will protrude down from the case approximately ¼". When the oil pan is installed it will force this seal tightly into the holes and prevent any oil from leaking from the engine into the clutch housing. See Fig. 20.

If new crankshaft bearings are installed care should be taken to see that the drilled passages line up with drilled passages in the crankcase, and that the bearings set snugly over the dowel pins.

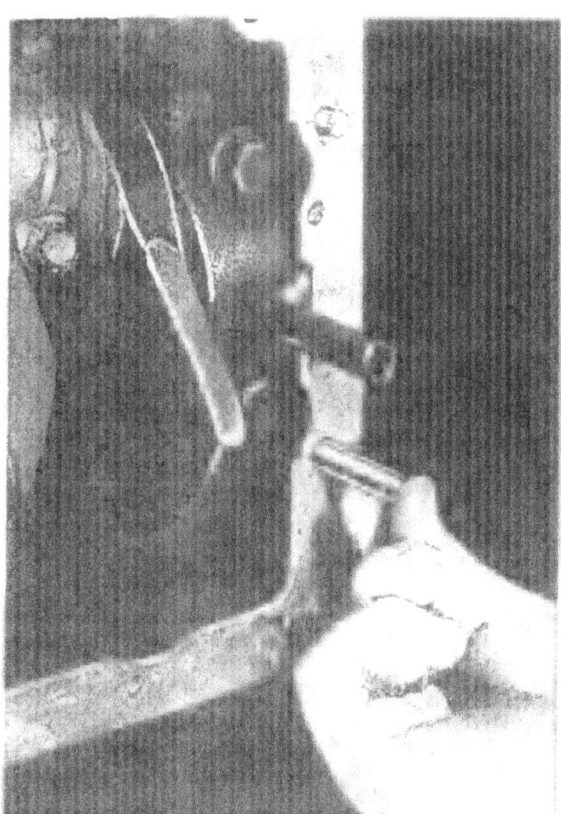

FIG. 20—REAR BEARING CAP SEAL

CONNECTING ROD

The connecting rods are drop forged and are of unusual length, measuring $9\frac{3}{16}$" from center to center. The babbitt bearings are of the replaceable type, steel-backed, babbitt lined, precision cut to size and no fitting is required.

When installing the upper half of rod bearing be sure that the oil spray hole lines up with the spray hole in the connecting rod. Undersize rod bearings are available in .010", .020" and .030".

The connecting rod and piston assembly is removed and installed from the top of the engine.

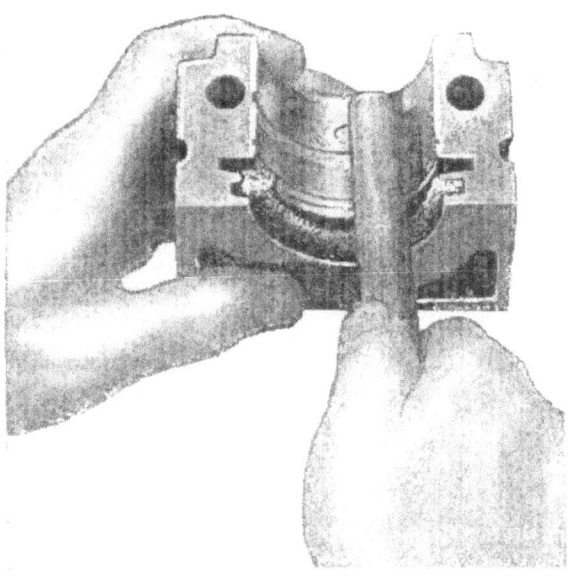

FIG. 19—REAR MAIN BEARING

When the rod is installed in the engine, the offset is away from the nearest main bearing. The oil spray hole in bearing end of rod should be on the follow side or away from camshaft, toward right side of vehicle. Torque wrench reading, 50-55 ft. lbs.

Clearance on crankshaft .0008" to .0023". Total side clearance .005" to .009". See Fig. 21.

Every time a connecting rod is removed from an engine or a new rod is to be installed, it should be checked for alignment on a connecting rod aligning fixture as shown in Fig. 22.

There are different types of connecting rod aligners. Follow the instructions issued by the manufacturer, when checking the connecting rod for twist or bend.

When straightening the rod, twist or bend in the opposite direction more than the original twist or bend then return the rod to true alignment. The rod will then retain correct alignment.

PISTON

The piston is aluminum alloy, "T" slotted, cam ground, tin plated, double ribbed at piston pin

FIG. 22—CONNECTING ROD ALIGNER

FIG. 21—CONNECTING ROD SIDE PLAY

bosses with a heat insulation groove above top ring. The clearance of the piston to the cylinder bore is .003". Check clearance with .003" feeler gauge ¾" wide; feeler gauge should have from 5 to 10 lbs. pull when being removed, see Fig. 23. The gauge should extend the entire length of the piston on the thrust side which is the opposite side from the "T" slot in the skirt.

Pistons are available in the following over-sizes: .010"; .020" and .030".

If it is ever found necessary to install an oversize piston, the cylinder bore must be honed with a regular cylinder honing tool and the manufacturers instructions should be carefully followed to get a true straight cylinder. Do not try to lap

in a new piston using compound, as in so doing it will ruin the tin plating on the piston and cause a scoring or wiping condition of both the piston and cylinder walls. See "Checking Cylinder Bores" and "Cylinder Boring", Pages 28 and 29.

PISTON RINGS

Width of compression rings ³⁄₃₂". Width of oil control ring ³⁄₁₆". The upper compression ring is installed with the inside beveled edge up. The face of the lower compression ring is tapered .005". The letters "T-O-P" on the upper edge of the ring indicates how the ring is installed, Fig. 24.

When fitting the rings to the cylinder bores, the end gap is .008"—.013". Fig. 25.

Fitting rings to piston grooves, Fig. 26 and 27. Compression rings .0005"—.001". Oil rings .001"—.0015".

Oversize rings are available in the following sizes: .010"; .020"; .030". Use standard rings up to .010" oversize cylinder bores.

PISTON PIN

The piston pin is anchored in the rod with a lock screw and fitted with a clearance of .0001" to .0009" in piston which is equivalent to a light thumb push fit at room temperature of 70°; pin diameter ¹³⁄₁₆" (.8117"). See Fig. 28.

Piston pins are available in oversizes .001"; .002"; .003".

ASSEMBLING CONNECTING ROD TO PISTON

Clamp connecting rod in vise using vise jaw protector shields of a soft metal or two pieces of hardwood on each side of connecting rod three inches from piston pin end.

Start piston pin in piston with groove facing down. Assemble piston to connecting rod with the slot in piston, No. 2, Fig. 29 opposite oil spray hole in bearing end of connecting rod, No. 1. Install piston pin clamp screw.

Center piston on pin and place assembly on connecting rod aligning fixture. Tilt piston to left with piston resting against surface plate. With feeler gauge measure clearance between piston skirt and surface plate. See Fig. 30. Tilt piston to right and check clearance. See Fig. 32. If clearance is within .003" on both left and right positions connecting rod is in alignment. A difference greater than .003" indicates connecting rod is twisted.

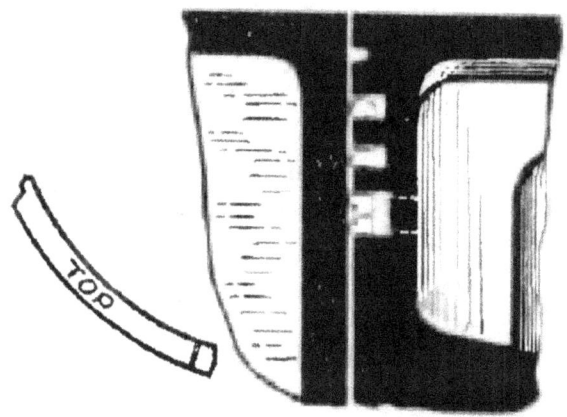

FIG. 24—PISTON RINGS

CHECKING CYLINDER BORES

The best method to be used in determining the condition of the cylinder bores preparatory to reconditioning is the use of a dial gauge such as shown in Fig. 31.

The dial gauge hand will instantly and automatically indicate the slightest variation of the cylinder bores.

To use the dial gauge simply insert in the cylinder bores and move up and down its full length. It is then turned spirally or completely rotated at different points, taking readings at each point. In this manner all variations in the cylinder bores from top to bottom may be determined.

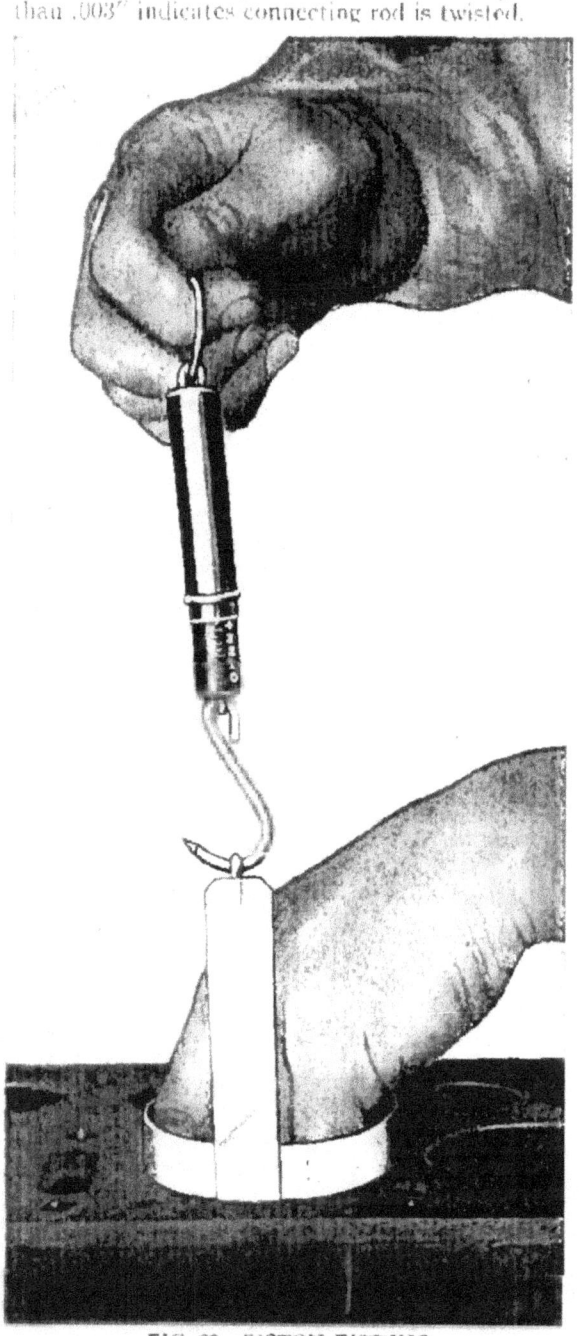

FIG. 23—PISTON FITTING

FIG. 25—PISTON RING GAP

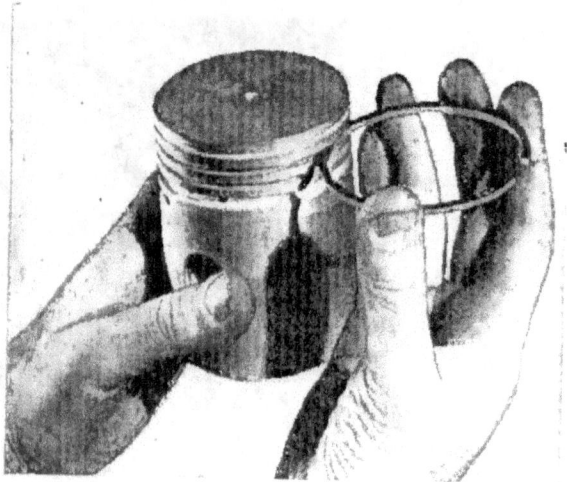

FIG. 26—COMPRESSION RING FITTING

CYLINDER BORING

When cylinders are more than .005" out of true it is best to rebore the cylinders. The instructions furnished by the manufacturer of the equipment should be carefully followed.

After the cylinder has been rebored within .002" of the size desired it should be finished or polished with a cylinder hone. Do not use a piston as a hone. In operating, the hone is placed in the cylinder bore and run up and down the full length of the cylinder wall. This procedure should be followed until the piston can be forced through the bores with a .003" feeler gauge ¾" wide on the thrust side and show a pull on the feeler gauge of five to ten pounds. See Fig. 23.

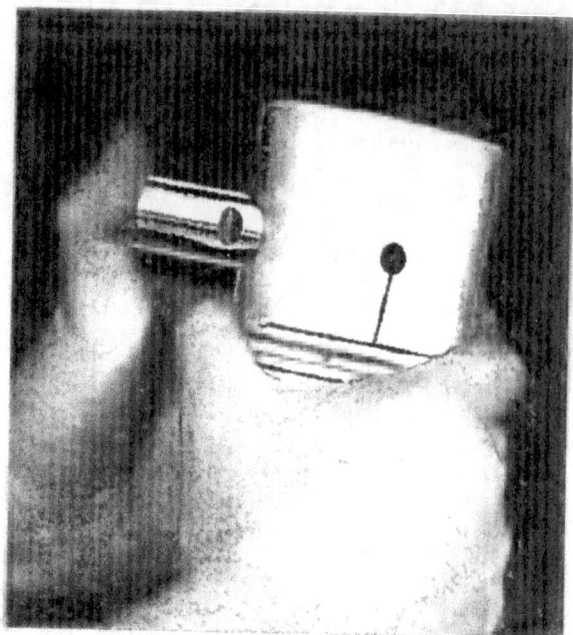

FIG. 28—PISTON PIN FITTING

OIL PUMP ASSEMBLY

The oil pump is the planetary gear type. It consists of two spur gears enclosed in a one piece housing. It is provided with a relief valve to control maximum oil pressure at all speeds. In operation the oil is drawn from the crankcase through the floating oil intake, Fig. 33. The oil then passes through a drilled passage in crankcase to the oil pump from which it passes to the oil distribution system or drilled passages in crankcase to crankshaft and camshaft bearings.

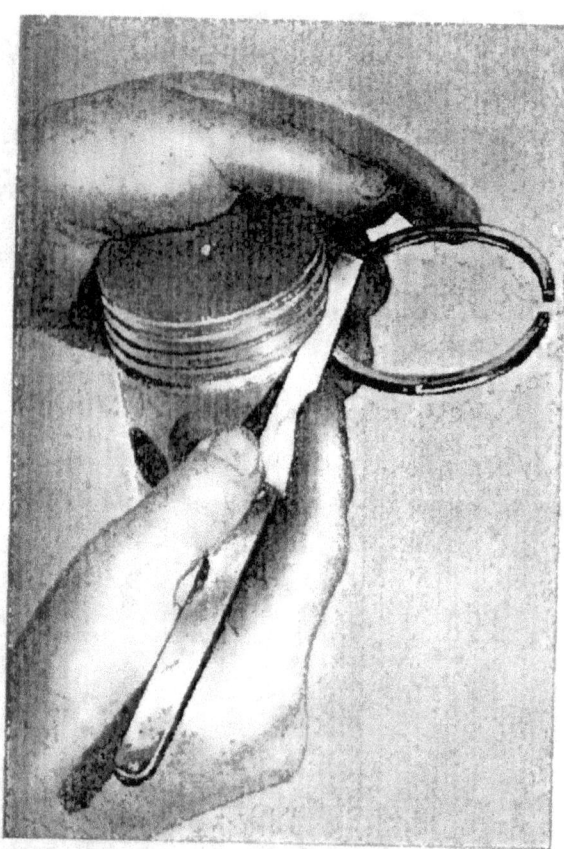

FIG. 27—OIL RING FITTING

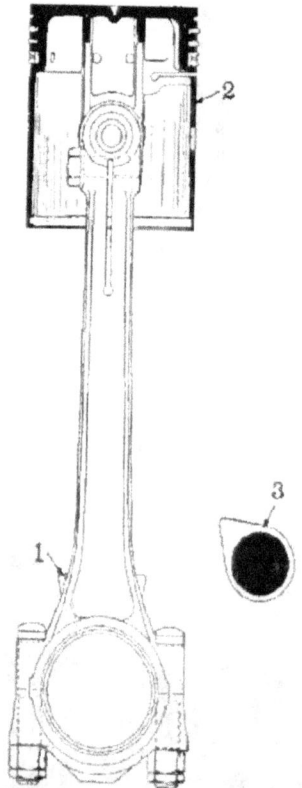

FIG. 29—CONNECTING ROD AND PISTON

The oil pump is driven from a spiral gear on the camshaft which is located at the center of the engine block on left hand side. See No. 22, Fig. 1.

To remove oil pump from engine for dismantling; remove the three nuts on studs holding oil pump to crankcase. Slide oil pump from studs. Remove screw No. 6, Fig. 34, in oil pump cover plate which will allow cover to be removed from housing.

To remove driven gear No. 16, file off one end of straight pin No. 17, with a small drift drive pin through the shaft. The oil pump shaft and rotor No. 12 can be removed from the body in an assembly.

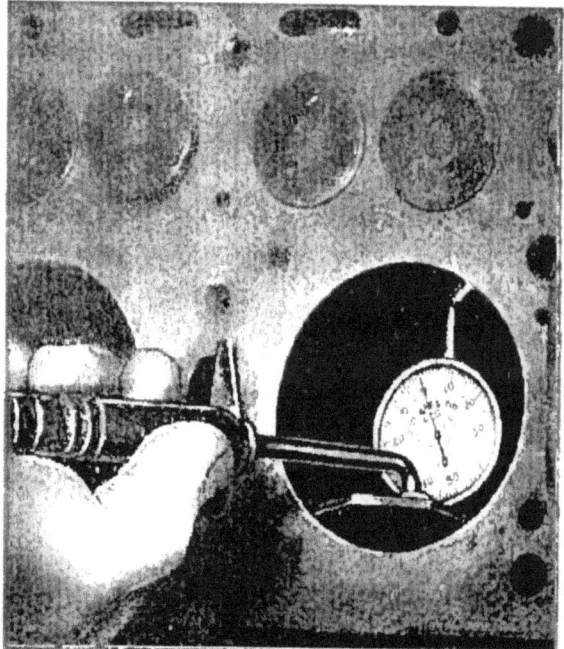

FIG. 31—CYLINDER BORE GAUGE

Set distributor rotor at No. 1 terminal tower in distributor cap and with the points just breaking.

Hold the oil pump in one hand with the oil relief valve retainer in the same position as it

FIG. 30—CONNECTING ROD TWIST-RIGHT

When removing spring retainer No. 1, care must be taken not to lose the small shims No. 3 which govern the spring tension of the relief valve No. 5. Adding shims increases the oil pressure, removing of shims decreases the pressure. The pressure at which the relief valve opens is 40 lbs. actual, however, on instrument panel gauge it will register between 75 and 80. The idle reading should not be less than 10.

When replacing the oil pump on engine the following procedure should be followed in order to have correct timing for the ignition.

Set No. 1 piston coming up on the compression stroke, then turn flywheel so that the timing mark "IGN" appears on the flywheel in the center of the hole in the flywheel housing on the right hand side, Fig. 15.

FIG. 32—CONNECTING ROD TWIST-LEFT

WILLYS MODEL "MB" ¼-TON 4 x 4 GOVERNMENT TRUCK

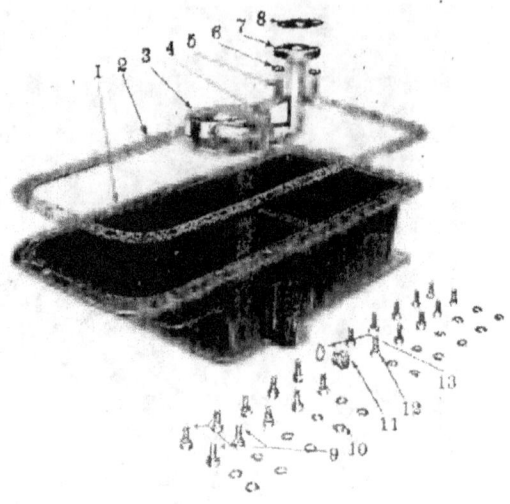

would be when installed in the engine; turn shaft so that the narrow side of slot in driven gear end is toward you, then line up the pin holding driven gear to shaft so that it will fall in line with the right hand side of the slot in pump body. Slide the assembly on studs in the crankcase, feed gear slowly into cam shaft gear, noting when fully set, if the rotor on distributor has moved from its original setting. If so remove oil pump and turn one tooth to obtain the correct setting.

FLOATING OIL INTAKE

The floating oil intake, Fig. 33 is attached to the crankcase with two cap screws. The construction of the float and screen cause it to float on top of the oil, raising and lowering in relation to the amount of oil in the crankcase.

This construction does not permit water or dirt to circulate, which may have accumulated in the bottom of the oil pan, because the oil is drawn horizontally from the top surface.

Whenever removed the float, screen and tube should be cleaned thoroughly in a suitable cleaning fluid to remove any accumulation of dirt.

FIG. 33—FLOATING OIL INTAKE AND PAN

No.	Willys Part No.	Ford Part No.	Name
1	A-1167	GPW-6675	Oil Pan Assembly
2	639980	GPW-6710	Oil Pan Gasket
3	630396	GPW-6615	Oil Float Assembly
4	5108	72053-S	Oil Float to Support Cotter Pin
5	636796	355806-S	Oil Float Support to Crankcase Screw
6	51833	34806-S	Oil Float Support to Crankcase—Screw Lockwasher
7	630397	GPW-6617	Oil Float Support
8	630398	GPW-6627	Oil Float Support Gasket
9	51485	20326-S	Oil Pan to Front Engine—Cover Screw
10	51833	34806-S	Oil Pan Screw Lockwasher
11	639979	GPW-6727	Oil Pan Drain Plug
12	51485	20326-S	Oil Pan to Cylinder Block Screw
13	314338	GPW-6734	Oil Pan Drain Plug Gasket

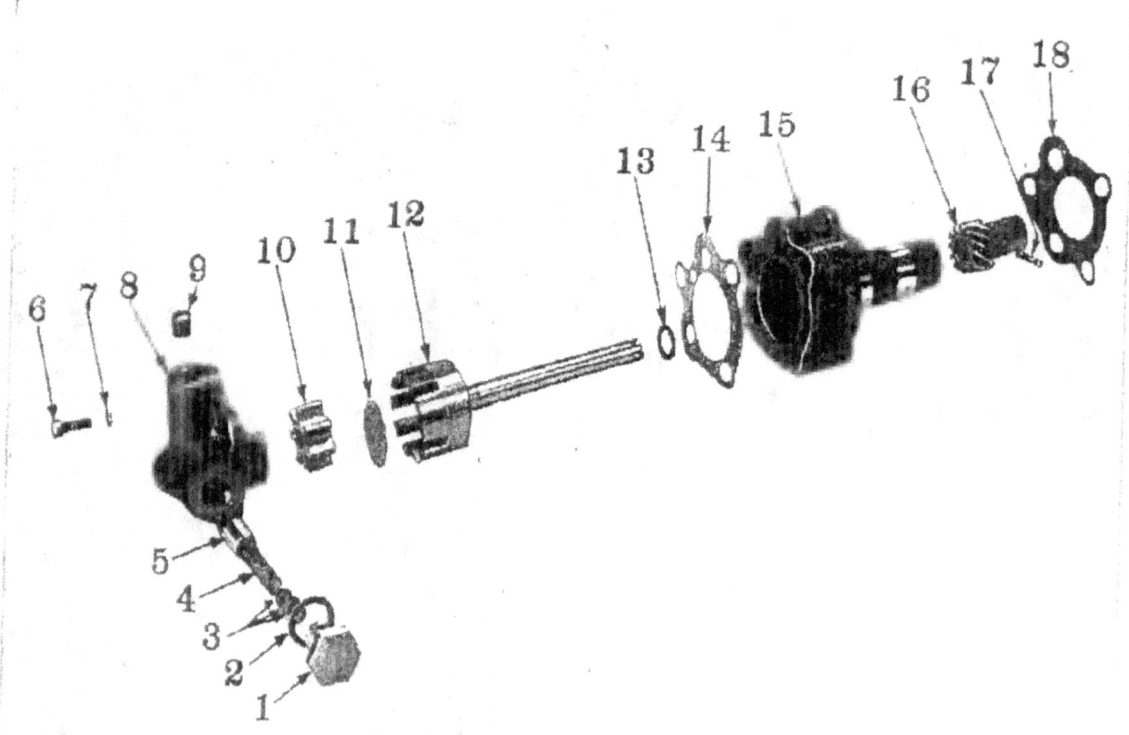

FIG. 34—OIL PUMP

No.	Willys Part No.	Ford Part No.	Name
1	630390	GPW-6644	Oil Pump Oil Relief Spring Retainer
2	634813	GPW-6649	Oil Pump Oil Relief Spring Retainer Gasket
3	630389	GPW-6658	Oil Pump Relief Spring Shim
4	356155	GPW-6651	Oil Pump Relief Plunger Spring
5	630518	GPW-6663	Oil Pump Oil Relief Plunger
6	51819	31079-S	Oil Pump Cover to Body Screw
7	350197		Oil Pump Cover to Body Screw Lockwasher
8	630387	GPW-6664	Oil Pump Cover Assembly
9	52525	353052-S	Oil Pump Cover Plug
10	343368	GPW-6614	Oil Pump Pinion
11	636600	GPW-6673	Oil Pump Rotor Disc
12	636599	GPW-6608	Oil Pump Shaft Assembly
13	375927	GPW-6625	Oil Pump Shaft Gasket
14	630870	GPW-6659	Oil Pump Cover Gasket
15	630384	GPW-6604	Oil Pump Body
16	637125	GPW-6610	Oil Pump Driven Gear
17	330064	GPW-6684	Oil Pump Driven Gear Pin
18	630394	GPW-6630	Oil Pump to Cylinder Block Gasket

FLYWHEEL

The flywheel is made of cast steel, machined all over and balanced to insure smooth engine performance. A steel ring gear is shrunk on the outer edge to mesh with the starter Bendix gear when starting engine.

The flywheel is attached to the crankshaft flange by two dowel bolts and four special head cap screws. When assembling the flywheel to crankshaft, be sure it is properly installed in relation to No. 1 crank throw and that it fits properly to crankshaft flange, to avoid runout or looseness. To check runout use dial indicator attached to the rear engine plate. The runout should not exceed .008" on the rear face near the rim. Torque wrench reading 36-40 ft. lbs.

When installing a new crankshaft or flywheel in service, it is the general practice to replace the tapered dowel bolts with straight snug fitting bolts. The crankshaft and flywheel should be assembled in proper relation, then install the straight bolts previously used and tighten securely. Next use a $\frac{35}{64}$" drill to enlarge the tapered bolt holes and then ream the holes with a $\frac{9}{16}$" (.5625") straight reamer and install two flywheel to crankshaft bolts No. 116295 with nut No. 52804 and lockwasher No. 52330, instead of the two dowel bolts formerly used. This procedure overcomes the difficulty in correctly tapering holes in the field.

OIL FILTER

The oil filter, No. 35 is so designed that it will effectively control contamination of engine oil. The filter element removes particles of dust, carbon and other foreign material from the oil which cause discoloration and sludge.

The inlet line to the filter is connected to the oil distribution line at the front plug on left hand side of engine. The outlet or oil return line to engine connects to the timing chain cover.

When the oil on the level indicator in the engine filler tube becomes dark, remove the oil filter cover; remove the drain plug and drain out the sludge after which replace the drain plug. Next, remove the element and install a new element. Install new cover gasket; reinstall cover, start engine and check for leaks; then check oil level; add to oil supply if necessary.

OIL PRESSURE GAUGE

The oil pressure gauge is of the Bourdon or hydraulic type and measures the pressure of the oil applied to the engine bearings. It does not indicate the amount of oil in the engine crankcase or the need for changing of the engine oil.

A pressure tube connects the gauge unit to the engine. It requires no special attention other than to see that the connection to the engine is tight.

If the unit becomes inoperative, it should be replaced as its construction does not permit repair or adjustment.

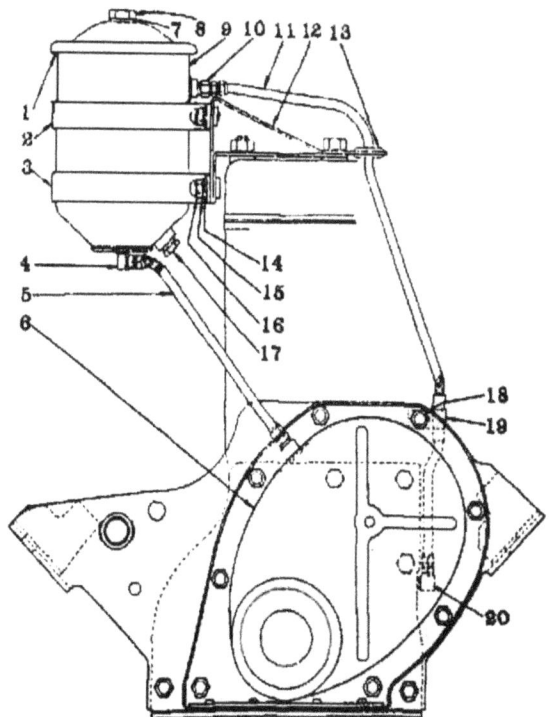

FIG. 35—OIL FILTER

No.	Willys Part No.	Ford Part No.	Name
1	A-1235	GPW-18883	Cover Gasket
2	A-1251	GPW-18658	Clamp Assembly
3	A-1251	GPW-18658	Clamp Assembly
4	384569	GPW-9628	Inverted Flared Tube Ell
5	A-1198	GPW-18666	Oil Filter Outlet Tube
6	A-1190		Chain Cover Assembly
7	A-1233	GPW-18675	Cover Bolt Gasket
8	A-1232	GPW-18691	Cover Bolt
9	A-1230	GPW-18680	Oil Filter Assembly
10	387801	9N-18579	Inverted Flared Tube Connector
11	A-1197	GPW-18667	Oil Filter Inlet Tube
12	A-1247	GPW-18663	Oil Filter Bracket Assembly
13	345961	GPW-13434-A2	Rubber Grommet
14	52274	34746-S2	Plain Washer
15	51833	34809-S	Lockwasher
16	51396	24347-S2	Screw (Filter to Bracket)
17	A-1237	358040-S	Drain Plug
18	5919	24389-S2	Screw (Timing Chain Cover)
19	A-1289	GPW-14585	Inlet Tube Clip
20	384569	GPW-9628	Inverted Flared Tube Ell
—	A-1236	GPW-18632	Filter Element Assembly

ENGINE SUPPORT PLATE AND MOUNTING

The front engine support plate is bolted to the front face of the cylinder block and forms the back panel for the attachment of the timing chain cover.

The rubber engine mountings, which are attached to the frame side rail brackets and to the support plate, prevent fore-and-aft motion of engine yet allow free side wise and vertical oscillation which has the effect of neutralizing vibration at its source, No. 9, Fig. 36.

The rear engine plate is attached to the rear of the cylinder block and provides a means for attaching the flywheel bell housing.

The engine is attached to the center cross member of the frame on a mounting which attaches to the bottom of the transmission. Torque wrench reading, 38-42 ft. lbs.

WILLYS MODEL "MB" ¼-TON 4 x 4 GOVERNMENT TRUCK

0100-33

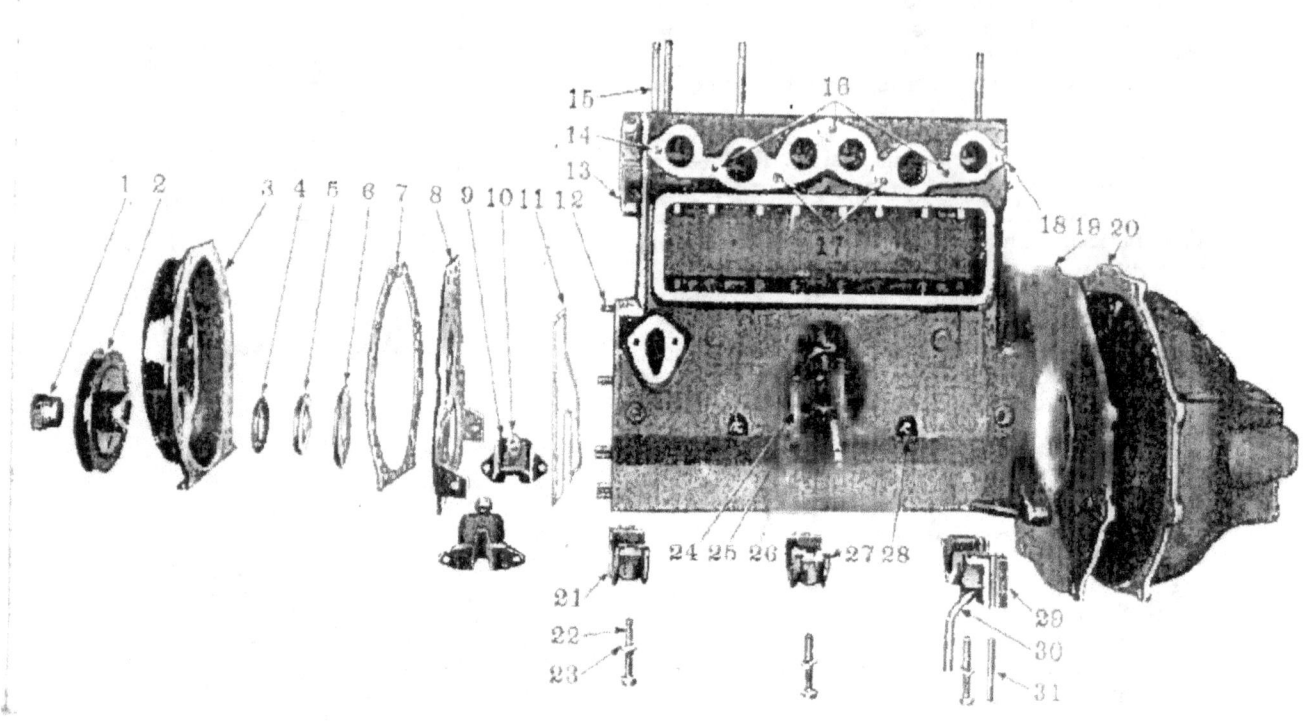

FIG. 36—TIMING CHAIN COVER, CYLINDER BLOCK AND BELL HOUSING

No.	Willys Part No.	Ford Part No.	Name
1	387633	GPW-6319	Starting Crank Nut Assembly
2	638113	GPW-6312	Fan Drive Pulley
3	A-1190		Chain Cover Assembly
4	637098	GPW-6700	Crankshaft Packing—Front End
5	375920	GPW-6287	Crankshaft Packing Retainer
6	375877	GPW-6310	Crankshaft Oil Slinger
7	630365	GPW-6288	Chain Cover Gasket
8	A-1463	GPW-6031	Engine Plate Front Assembly
9	A-542	GPW-6038	Engine Support Insulator—Front
10	5916	33846-S	Engine Support Insulator Nut—to Engine Plate
11	630359	GPW-6020	Cylinder Block Gasket—Front
12	384958	88032-S	Chain Cover and Front Plate Stud
13	A-1272		Cylinder Block and Bearings Assembly
14	349712	88082-S	Inlet and Exhaust Manifold Stud—1⅜"
15	349368	GPW-6066	Cylinder Head Stud
16	300143	88042-S	Inlet and Exhaust Manifold Stud—1½"
17	632159	88057-S7	Inlet and Exhaust Manifold Stud—1⅞"
18	349712	88082-S	Inlet and Exhaust Manifold Stud—1⅜"
19	A-5121	GPW-7007	Engine Plate—Rear
20	A-439	GPW-6392	Transmission Bell Housing Assembly
21	630285	GPW-6329	Crankshaft Bearing Cap—Front
22	581519	GPW-6345	Bearing Caps to Crankcase Screw
23	5609	34809-S	Bearing Cap to Crankcase Screw Lockwasher
24	375981	88141-S	Oil Pump to Cylinder Block Stud
25	51833	34806-S	Oil Pump Stud Nut Lockwasher
26	5910	33798-S	Oil Pump Stud Nut
27	630288	GPW-6330	Crankcase Bearing Cap—Center
28	5053		Oil Passage Plug
29	637236	GPW-6325	Crankshaft Bearing Cap—Rear
30	630204	GPW-6326	Crankshaft Rear Bearing Drain Pipe
31	637700	GPW-6701	Crankshaft Rear Bearing Cap Packing

ENGINE TROUBLES AND CAUSES

Poor Fuel Economy

Ignition Timing Slow or Spark Advance Stuck
Carburetor Float High
Accelerator Pump Not Properly Adjusted
Gasoline Leakage
Leaky Fuel Pump Diaphragm
Loose Engine Mounting causing high gasoline level in Carburetor.
Low Compression
Valves Sticking
Spark Plugs Bad
Weak Coil or Condenser
Improper Valve Tappet Clearance
Carburetor Air Cleaner Dirty
Clogged Muffler or bent Exhaust Pipe

Lack of Power

Low Compression
Ignition System (Timing Late)
Improper Functioning Carburetor or Fuel Pump
Gasoline Lines Clogged
Air Cleaner Restricted
Engine Temperature High
Improper Tappet Clearance
Sticking Valves—Valve Timing
Leaky Gaskets
Muffler Clogged
Bent Exhaust Pipe
Old Gasoline

Low Compression

Leaky Valves
Poor Piston Ring Seal
Sticking Valves
Valve Spring Weak or Broken
Cylinder Scored or Worn
Tappet Clearance Incorrect
Piston Clearance too Large
Leaky Cylinder Head Gasket

Burned Valves and Seats

Sticking Valves or too loose in Guides
Improper Timing
Excessive Carbon around Valve Head and Seat
Overheating
Valve Spring Weak or Broken
Valve Tappet Sticking or not set to .014"
Clogged Exhaust System

Valves Sticking

Warped Valve
Improper Tappet Clearance
Carbonize or scored Valve Stems
Insufficient Clearance Valve Stem to Guide
Weak or Broken Valve Spring
Valve Spring Cocked
Contaminated Oil

Overheating

Defective Cooling
Thermostat Inoperative
Improper Ignition Timing
Improper Valve Timing
Excessive Carbon Accumulation
Fan Belt too Loose
Clogged Muffler or Bent Exhaust Pipe
Oiling System Failure
Scored or Leaky Piston Rings

Popping-Spitting-Detonation

Improper Ignition
Improper Carburetion
Excessive Carbon Deposit in Combustion Chamber
Poor Valve Seating
Sticking Valves—Broken Valve Spring
Tappets Adjusted too Close
Spark Plug Electrodes Burnt
Water or Dirt in Fuel—Clogged Lines
Improper Valve Timing

Excessive Oil Consumption

Piston Rings Stuck in Groove, Worn or Broken
Piston Rings Improperly fitted or Weak
Piston Ring Oil Return Holes Clogged
Excessive Clearance Main and Connecting Rod Bearings
Oil Leaks at Gasket or Oil Seals
Excessive Clearance Valve Stem to Valve Guide (Intake)
Cylinder Bores Scored-Out-of-Round or Tapered
Too Much Clearance Piston to Cylinder Bore
Misaligned Connecting Rods
High Road Speeds or Temperature

Bearing Failure

Crankshaft Bearing Journal Out of Round
Crankshaft Bearing Journals Rough
Lack of Oil—Oil Leakage
Dirty Oil
Low Oil Pressure or Oil Pump Failure
Drilled Passages in Crankcase or Crankshaft Restricted
Oil Screen Dirty
Connecting Rod Bent

WILLYS MODEL "MB" ¼-TON 4 x 4 GOVERNMENT TRUCK

ENGINE SPECIFICATIONS

Type..L. Head
    Number of Cylinders..4
    Bore...3⅛"
    Stroke..4⅜"
    Piston Displacement.......................................134.2 cu. in.
    Compression Ratio...6.48 to 1
    Horse Power Max. Brake....................................60 @ 4000
    Compression...111 lbs. @ 185 R.P.M.
    S.A.E. Horse Power..15.63
    Maximum Torque......................................105 Lbs.-Ft. @ 2000 R.P.M.
    Firing Order...1-3-4-2

Cylinder Block:
    Bore Size..3.125"—3.127"

Cylinder Head:
    Torque Wrench Pull
    Cylinder Head Screw................................65-75 Lbs.-Ft.
    Cylinder Head Stud Nut..............................60-65 Lbs.-Ft.

Crankshaft:
    Counter Weights..4

Crankshaft Main Bearings:
    Bearing Journals...3
        Front..2.3340"x1.920"
        Center..2.3340"x1 13/16"
        Rear..2.3340"x1¾"

    Thrust..Front

    End-play...004"-.006"
    Bearing Clearance..001"
    Type...Steel back, babbitt lined
    Non-Adjustable...........................Replaceable Without Reaming
    Torque Wrench Pull...............................65-70 Lbs.-Ft.

Connecting Rod:
    Center to Center Length..9 3/16"
    Upper End............................Piston Pin Locked in Rod
    Lower Bearing Type........Steel back, babbitt lined, replaceable
    Lower Bearing Diameter and Length.................1 15/16"x1 5/16"
    Clearance on Crankshaft.............................0008"-.0023"
    Side Clearance...005"-.009"
    Torque Wrench Pull.............................50-55 Lbs.-Ft.
    Installation..From Top
        Offset away from nearest main bearing
        Oil spray hole away from camshaft

Piston:
    Lo-Ex Lynite T-Slot—Oval Ground—Tin Plated—Heat dam
    Length..3¾"
    Clearance Top Land................................0205"-.0225"
    Clearance Skirt..003"
    Oversize pistons Available.................010", .020", .030"
    Number Rings...3
        Compression Ring.............................2-Width 5/32"
        Oil Ring....................................1-Width 3/16"
        Ring Gap...008"-.013"
        Ring to Groove Clearance.................0005"-.0015"
    Piston Pin Hole
        Diamond Bored.................13/16" (.8007"-.8119")

Piston Pin:
    Length..2 25/32"
    Diameter...13/16"
    Type..Locked in Rod
    Clearance in Piston...............................0001"-.0009"
    Oversize Pins Available.......................001", .002", .003"

Engine Specifications—Continued

Camshaft:
    Number of Bearings..4
    Bearing Journal Diameter:
        Front..2 5/16"
        Front Intermediate............................2 1/4"
        Rear Intermediate............................2 3/16"
        Rear..1 3/4"
    Thrust Taken..Front
    End Play Control............................Plunger and Spring

Camshaft Bearings:
    Front............................Steel Back Babbitt Lined
    Clearance..002"-.0035"

Intake Valve:
    Tappet Clearance Cold..014"
    Seat Angle..45°
    Diameter Head..1 17/32"
    Length Over-all..5 3/4"
    Stem Diameter..373"
    Stem to Guide Clearance............................0015" to .00325"
    Intake Opens............9° BTC Flywheel (.039" Piston travel)
    Intake Closes............50° ABC Flywheel (3.772" Piston travel)
    Lift..23/64"

Exhaust Valve:
    Tappet Clearance Cold..014"
    Seat Angle..45°
    Diameter Head..1 15/32"
    Length Over-all..5 3/4"
    Stem Diameter..3725"
    Stem to Guide Clearance............................002"-.00375"
    Exhaust Opens............47° BBC Flywheel (3.799" Piston travel)
    Exhaust Closes............12° ATC Flywheel (.054" Piston travel)
    Lift..23/64"

Valve Spring:
    Free Length..2 1/2"
    Spring Pressure Valve Closed............50 lbs. length 2 7/64"
    Spring Pressure Valve Open............116 lbs. length 1 3/4"
    Closed Coil End of Spring............Installed Up against Block

Valve Tappet:
    Overall Length..2 7/8"
    Stem Diameter..6240"-.6245"
    Clearance To Guide..0005"-.002"
    Adjusting Screw............................3/8"-24 thd. x 1 1/32"

Timing Chain:
    Link-Belt
    Number Links..47
    Width..1"
    Pitch..1/2"
    Type..Non-adjustable

Fan Belt:
    Type.."V"
    Angle of Vee..42°
    Length outside..44 1/8"
    Width..11/16"

Oil Pump:
    Type..Planetary Gear
    Driven from Camshaft..Gear

Oil Pressure Relief:
    Pressure 40 lbs. actual—75 gauge at 30 miles per hour.
    Adjustable............................Shims in Spring Retainer.

Oil Filter..Purolator No. 27078

Spark Plugs..Champion QM-2

CLUTCH

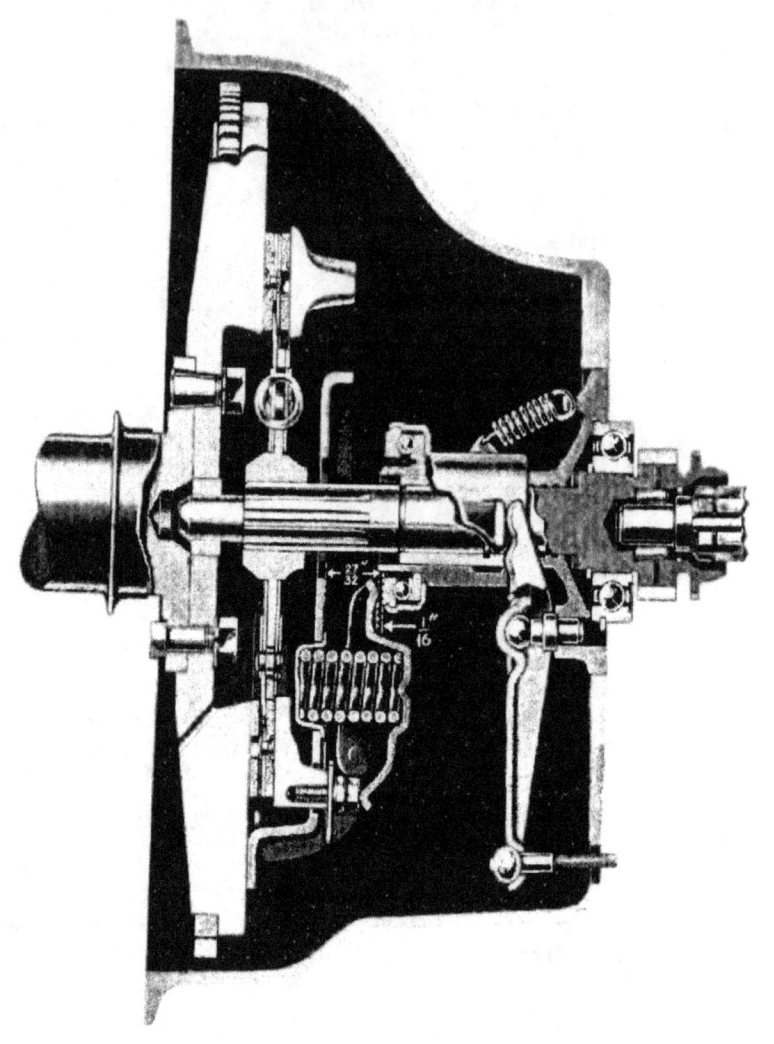

FIG. 1—CLUTCH SECTIONAL VIEW

The clutch is a single plate 8" dry disc type. The driven plate has a spring center vibration neutralizer, two facings $\frac{1}{8}$" thick, outside diameter $7\frac{7}{8}$", inside diameter $5\frac{1}{8}$".

There are two adjustments, the free pedal play which should be maintained at $\frac{3}{4}$" and the other, the pressure plate finger adjustment. See Fig. 1.

As the clutch facings wear it diminishes the free pedal travel. When clutch pedal rests tightly against the toe board it is necessary to adjust the clutch cable.

Clutch Pedal Adjustment

Lengthening or shortening clutch control lever cable No. 11, Fig. 2 governs the clearance of the clutch release bearing to clutch fingers, Fig. 1, which should be maintained at $\frac{1}{16}$". This represents $\frac{3}{4}$" free pedal travel. This also disengages the clutch release bearing and prevents unnecessary wear while the engine is running.

Loosen Clutch control lever cable adjusting yoke locknut, No. 24, Fig. 2. With a wrench unscrew cable No. 11, then tighten locknut.

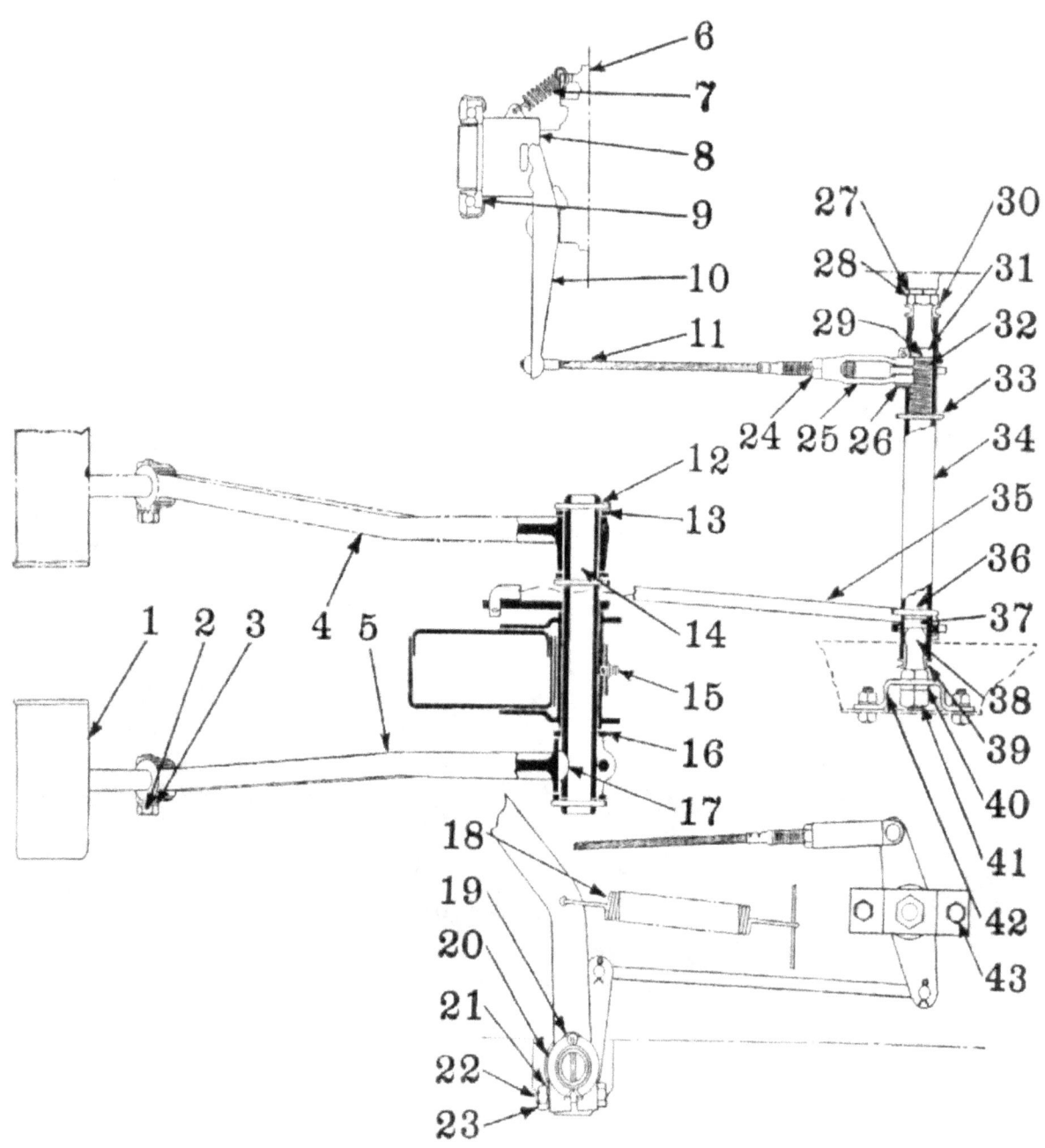

FIG. 2—CLUTCH CONTROL

No.	Willys Part No.	Ford Part No.	Name	No.	Willys Part No.	Ford Part No.	Name
1	A-1360	GPW-7525	Clutch Pedal Pad Assembly	23	5016	33798-S	Clutch Pedal Clamp Bolt Nut
2	6157	24626-S	Clutch Pedal Clamp Bolt	24	5016	33798-S	Clutch Control Lever Cable Yoke End Locknut
3	51833	34806-S	Clutch Pedal Clamp Bolt Lockwasher	25	632177	GPW-7532	Clutch Control Lever Cable Yoke End
4	A-1386	GPW-2452	Brake Pedal	26	339044	73880-S	Clutch Control Lever Cable Clevis Pin
5	A-405	GPW-7520	Clutch Pedal	27	5059	34808-S	Clutch Control Ball Stud Lockwasher
6	640017	GPW-7650	Transmission Main Drive Gear Bearing Retainer	28	A-181	GPW-7511	Clutch Control Ball Stud
7	630117	GPW-7562	Clutch Release Bearing Carrier Spring	29	A-177	GPW-7517	Clutch Control Tube Washer
8	639654	GPW-7564	Clutch Release Bearing Carrier	30	A-887	GPW-7512	Clutch Control Tube Dust Washer
9	635529	GPW-7580	Clutch Release Bearing	31	A-176	GPW-7519	Clutch Control Tube Felt Washer
10	630112	GPW-7515	Clutch Control Lever	32	A-178	GPW-7515	Clutch Control Spring
11	A-5102	GPW-7538	Clutch Control Lever Cable	33	5108	72063-S	Clutch Control Tube Spring Cotter Pin
12	52941	72063-S	Pedal Shaft Cotter Pin	34	A-1355	GPW-7503	Clutch Control Lever & Tube Assembly
13	A-1354	GPW-2138	Master Cylinder Tie Bar	35	A-499	GPW-7521	Clutch Control Pedal Rod
14	A-495	GPW-2473	Pedal Shaft Assembly	36	5108	72063-S	Clutch Control Tube Cotter Pin
15	638752	353043-S7	Pedal Shaft Hydraulic Grease Fitting	37	A-176	GPW-7519	Clutch Control Tube Felt Washer
16	A-498	356561-S	Pedal Shaft Washer	38	A-181	GPW-7511	Clutch Control Ball Stud
17	5036	74178-S	Clutch Pedal to Shaft Woodruff Key	39	A-887	GPW-7512	Clutch Control Tube Dust Washer
18	630593	GPW-7523	Clutch Pedal Retracting Spring	40	5059	34808-S	Clutch Control Ball Stud Lockwasher
19	52941	72063-S	Pedal Shaft Cotter Pin	41	5336	33801-S	Clutch Control Ball Stud Nut
20	A-498	356561-S	Pedal Shaft Washer	42	A-486	GPW-7509	Clutch Control Frame Bracket
21	51833	34806-S	Clutch Pedal Clamp Bolt Lockwasher	43	52132	24327-S	Clutch Control Frame Bracket Screw
22	50992	24505-S	Clutch Pedal Clamp Bolt				

Reconditioning

When it is necessary to recondition the clutch, follow the procedure outlined under "Transmission" for the removal of transmission and transfer case from Engine. Remove the bell housing, then the following procedure is suggested in disassembling the clutch.

Prick punch both the pressure plate and the flywheel so that the assembly can be installed in the same position after the repairs are performed.

The cap screws holding the clutch pressure plate to the flywheel should be loosened in sequence a little at a time so as to prevent distortion of the clutch bracket No. 10, Fig. 3. It is advisable where the clutch facings have worn that a new driven plate assembly be used in preference to relining the old plate as in many cases it is found that the torque springs in the hub of the driven plate have somewhat weakened or the center plate No. 13 has become distorted and if relined would not give satisfactory performance.

The clutch pressure plate fingers are properly adjusted before vehicle leaves the factory and should not require adjustment excepting where it is necessary to install new springs, fingers or pressure plate. When new parts are installed, it will be necessary to readjust the fingers, which adjustment can be made after the assembly is attached to the flywheel.

To assemble the clutch to the flywheel first put a small amount of light cup grease in the clutch shaft bushing; install driven plate, with short end of hub towards the flywheel then place pressure plate assembly in position. With clutch pilot arbor or a clutch shaft, align driven plate leaving arbor in place while tightening the pressure plate cap screws. Remove clutch pilot arbor and then adjust clutch fingers.

Adjustment of Clutch Fingers

Adjustment of the clutch fingers is accomplished by loosening the lock nut on adjusting screws, then turning screws clockwise or counter-clockwise until the measurement from the face of the fingers (release bearing contacts) to the face of the clutch bracket measures $27/64''$. Fig. 5. All fingers must be adjusted so that clutch release bearing will contact the three fingers simultaneously.

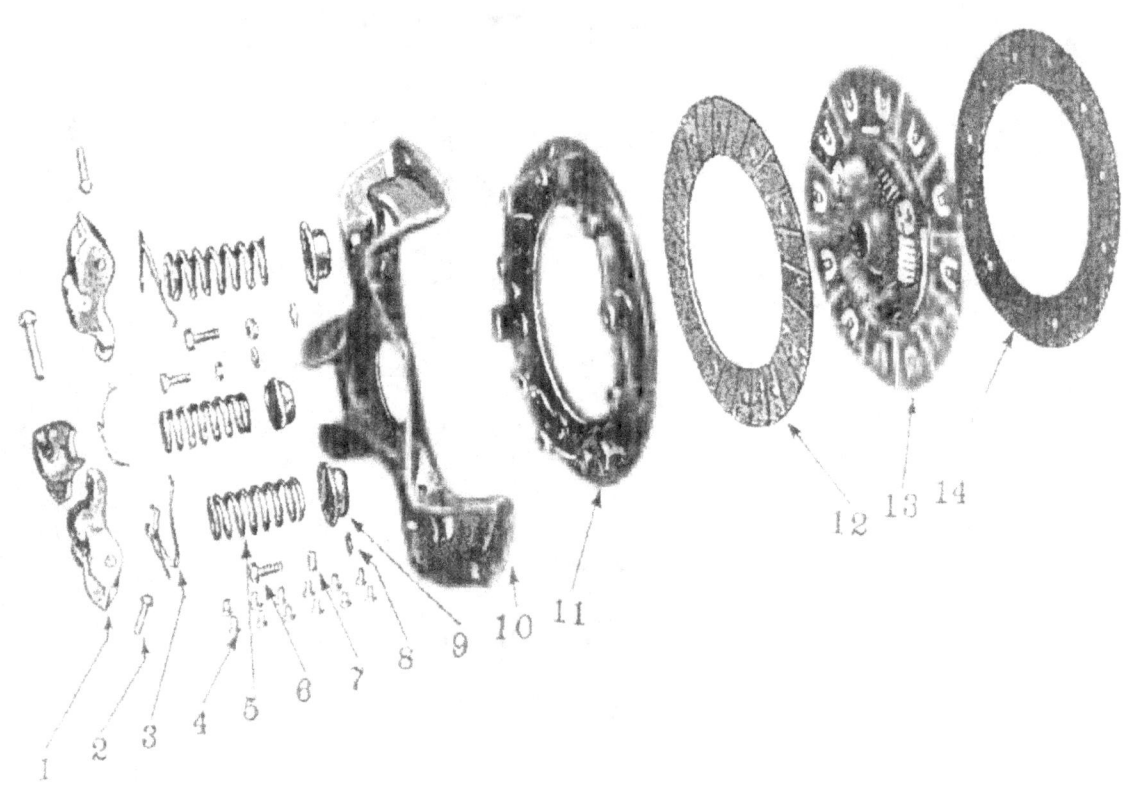

FIG. 3—CLUTCH

No.	Willys Part No.	Ford Part No.	Name
1	638158	GPW-7591	Clutch Lever
2	638159	GPW-7565	Clutch Lever Pivot Pin
3	638153	GPW-7590	Clutch Pressure Plate Return Spring
4	374681	351928-S	Clutch Facing Rivet—Tubular
5	638993	GPW-7572	Clutch Pressure Spring
6	638154	24325-S	Clutch Adjusting Screw
7	638155	33921-S	Clutch Adjusting Screw Lock Nut
8	638305	34745-S	Clutch Adjusting Screw Washer
9	638157	GPW-7567	Clutch Pressure Spring Cup
10	638151	GPW-7570	Clutch Bracket
11	638152	GPW-7568	Clutch Pressure Plate
12	636779		Clutch Facing—Rear
13	636755	GPW-7550	Clutch Driven Plate and Hub (with Facings)
14	371567	GPW-7549	Clutch Facing—Front

Next assemble the bell housing to the engine.

The clutch release bearing No. 9, Fig. 2 is pre-lubricated and the lubricant lasts the life of the bearing. If the bearing is rough, a new bearing should be installed.

Make sure that the clutch release bearing carrier return spring No. 7, Fig. 2 is hooked into place. For the balance of the assembly, reverse the operations that were used in the disassembly, referring to instructions given under "Transmission." Finally adjust the clutch release cable so there is ¾" free pedal travel before release bearing contacts the fingers.

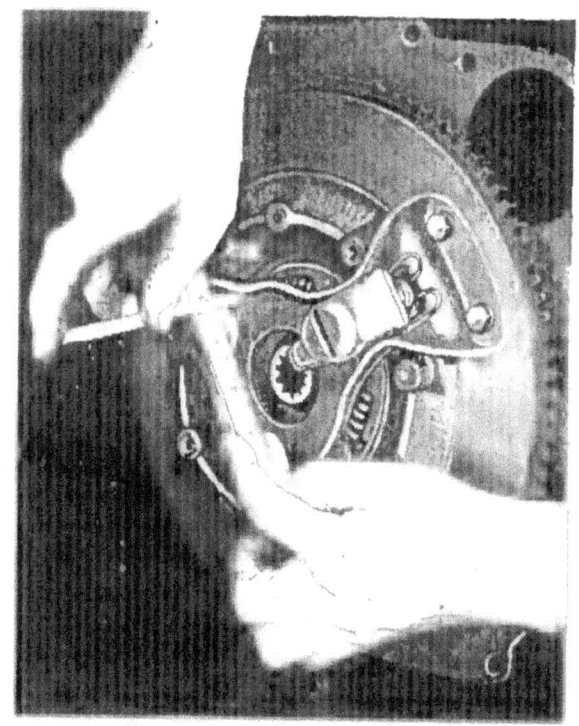

FIG. 5—CLUTCH FINGER ADJUSTMENT

FIG. 4—CLUTCH PRESSURE PLATE REMOVAL

Clutch Pressure Plate

When it is found necessary to install a new clutch pressure plate or clutch spring dismantle the assembly as follows: See Fig. 4.

Select a board or a surface plate of sufficient length and width to support the pressure plate bracket at all points. Place board or surface plate on arbor press table. Place a piece of wood 2½" square on top of clutch fingers, then depress springs, holding down while the adjusting screws are removed from pressure plate.

After screws No. 6, Fig. 3 are removed, the clutch pressure plate return springs, No. 3, may be removed.

Release press arbor slowly to prevent clutch pressure springs, No. 5, flying out from under the clutch fingers No. 1.

In checking assembly make sure that the clutch pressure spring cups, No. 9 are assembled in bracket No. 10 with indentation in bottom towards the center.

The assembly of the pressure plate is the reverse of operations used in dismantling.

CLUTCH TROUBLES AND REMEDIES

SYMPTOMS	PROBABLE REMEDY
Slipping:	
Improper Adjustment	Adjust Pedal Free Travel ¾"
Weak Pressure Springs	Replace
Lining Oil Soaked	Install New Driven Plate
Worn Linings or Torn Loose from Plate	Install New Driven Plate
Burned Clutch	Replace
Grabbing:	
Gummy or Worn Linings	Install New Driven Plate
Loose Engine Mountings	Tighten
Scored Pressure Plate	Install New Plate
Improper Clutch Lever (Finger) Adjustment	Readjust
Clutch Plate Crimp or Cushion Flattened Out	Replace Driven Plate
Dragging:	
Too much Pedal Play	Adjust
Improper Finger Adjustment	Readjust
Pressure Plate Binds in Bracket	Adjust
Warped Pressure or Driven Plate	Replace
Torn Clutch Facing	Replace
Rattling:	
Broken or Weak Return Springs in Driven Plate	Replace
Worn Throw Out Bearing	Replace
Fingers Improperly Adjusted	Readjust
Worn Driven Plate Hub or Clutch Spline Shaft	Replace
Pilot Bushing in Flywheel Worn	Replace

CLUTCH SPECIFICATIONS

Type......................Single, Dry Plate

Driven Plate:

    Make......................Borg & Beck
    Size......................7⅞"
    Facings.......1 Woven and 1 Molded Asbestos
    Diameter..........Inside 5⅛"-Outside 7⅞"
    Thickness....................⅛"-(.125")
    Torque Capacity...............132 Lbs.-Ft.

Pressure Plate:

    Make......................Atwood
    Number Springs.....................3
    Spring Pressure at 1 9/16"..........220-230 lbs.

Clutch Release Bearing:

    Type.......Sealed Ball Bearing—Prelubricated

Clutch Shaft Bushing:

    Location....................In Crankshaft
    Material. Bronze Bushing (Impregnated with graphite)
    Size........................(Inside) .628"

Clutch Pedal Adjustment:

    Pedal Adjustment..¾" Free Pedal Travel Before Release Bearing Contacts Clutch Fingers.

FUEL SYSTEM

FIG. 1—CARBURETOR

No.	Willys Part No.	Ford Part No.	Name
1	116537	GPW-9529	Pump Operating Lever Assembly
2	116181	GPW-9528	Pump Arm and Collar Assembly
3	116199	GPW-9527	Pump Connecting Link
4	116187	GPW-9570	Pump Arm Spring
5	116195	GPW-9631	Pump Plunger and Rod Assembly
6	116188	GPW-9596	Pump Plunger Spring
7	116204	GPW-9594	Discharge Disc Check Assembly
8	116205	GPW-9578	Intake Ball Check Assembly
9	116175	GPW-9575	Pump Check Strainer
10	116163	GPW-9696	Pump Check Strainer Nut
11	116180	GPW-9940	Pump Jet
12	116154	GPW-9585	Throttle Valve
13	116162	GPW-9579	Idle Port Rivet Plug
14	116183	GPW-9573	Idle Adjustment Screw Spring
15	116176	GPW-9541	Idle Adjustment Screw
16	116164	GPW-9028	Pump Jet also Nozzle Passage Plug and Gasket Assembly
17	116540	GPW-9906	Metering Rod
18	116541	GPW-9914	Metering Rod Jet and Gasket Assembly
19	116179	GPW-9544	Idle Well Jet
20	116539	GPW-9533	Low Speed Jet Assembly
21	116172	GPW-9550	Float and Lever Assembly
22	116174	GPW-9567	Needle, Pin, Spring and Seat Assembly
23	116157	GPW-9549	Choke Valve Assembly
24	116545	GPW-9546	Choke Shaft and Lever Assembly
25	116538	GPW-9907	Metering Rod Spring
26	116166	GPW-9922	Nozzle
27	116161	GPW-9532	Nozzle Retainer Plug
28	116206	GPW-9905	Metering Rod Disc

The Fuel System, Fig. 2, consists of the Fuel Tank, Fuel Lines, Fuel Filter, Fuel Pump, Carburetor and Air Cleaner.

The most important attention necessary to the fuel system is to keep it clean and free of water.

It should be periodically inspected for leaks.

The fuel tank capacity is given on Page 3. The tank sets in a sump in the floor pan and two drain holes are incorporated in this sump to allow for flushing. When the vehicle leaves the factory a cap is placed over the front drain hole to keep out stones and dirt and another is placed in the glove compartment. Should maneuvers in water be necessary, install the second cap over the rear drain hole from the left side of the vehicle. After passing through the water remove cap and return it to the glove compartment.

CAUTION—Whenever the vehicle is to be stored for an extended period, the fuel system should be completely drained. The engine started and allowed to run until carburetor is emptied. This will avoid oxidation of the gasoline, resulting in the formation of gum in the units of the Fuel System.

Information pertaining to the operation and servicing of the units contained in fuel system are covered in the succeeding paragraphs.

Carburetor

The Carter Carburetor, Model WO-539S, Fig. 1, is the plain tube type with a throttle operated accelerator pump and economizing device.

Since carburetion is dependent in several ways upon both compression and ignition, it should always be checked last in an engine tune-up.

The carburetor delivers the proper fuel and air ratios for all speeds of the engine. By proper cleaning and replacing all worn parts, the carburetor will function correctly.

The carburetor can be divided into five circuits which are:

1. Float Circuit
2. Low Speed Circuit
3. High Speed Circuit
4. Pump Circuit
5. Choke Circuit

By treating each circuit separately, the study and repair of the carburetor is made much easier.

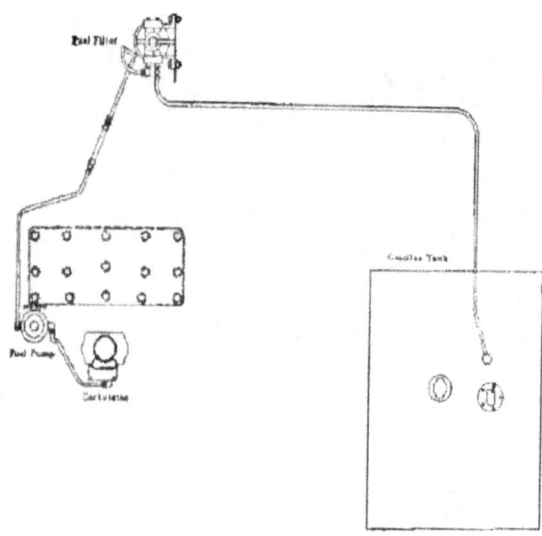

FIG. 2—FUEL SYSTEM

Float Circuit or Fuel Level

The float circuit Fig. 3, is important because it controls the height of the fuel level in the bowl and in the nozzle. If the fuel level is too high, it will cause trouble in the low and the high speed circuits.

The float bowl No. 3, acts as a reservoir to hold a supply of fuel. The level of the fuel in the bowl is controlled by a combination of parts which are: float and lever assembly No. 2, float bowl cover No. 4, needle valve and seat assembly No. 1.

Low Speed Circuit

The idle or low speed circuit, Fig. 4, controls the supply of fuel to the engine during idle and light load speeds up to approximately 20 miles per hour, and it feeds a small amount of fuel during the entire operation of the high speed circuit (gradually decreasing as speed is increased, above 20 m.p.h.).

During idling and low speed operation of the engine, fuel flows from the float bowl through the idle jet No. 8, to the point where it combines with a stream of air coming in through the by-pass, No. 9. The combining of the air with the fuel atomizes or breaks up the fuel into a vapor.

This mixture of air and fuel continues on through the economizer No. 10 until it begins to pass the point where it is further combined with a stream of air coming in through the lower bleed No. 11. This mixture of fuel and air then flows downward to the idle port chamber and thence into the engine at the port No. 12 and through the idle adjusting screw seat just below. This mixture is richer than the engine requires but when mixed with the air coming past the throttle valve it forms a combustible mixture of the right proportion for idle speeds.

The idle port is slotted so that as the throttle valve is opened it will not only allow more air to come in past it, but will also uncover more of the idle port allowing a greater quantity of fuel and air mixture to enter the intake manifold.

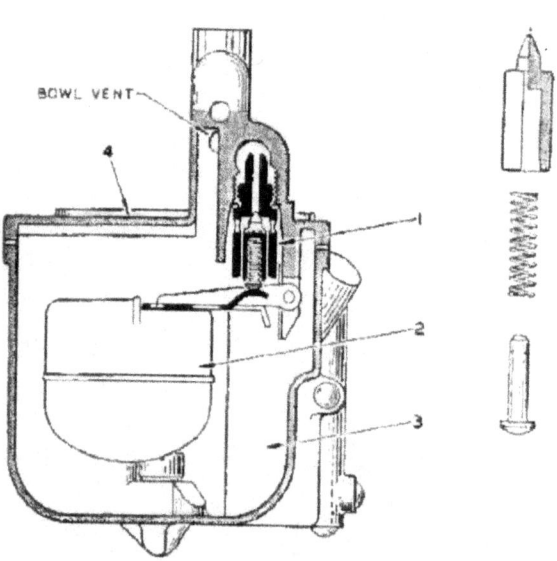

FIG. 3—FLOAT CIRCUIT

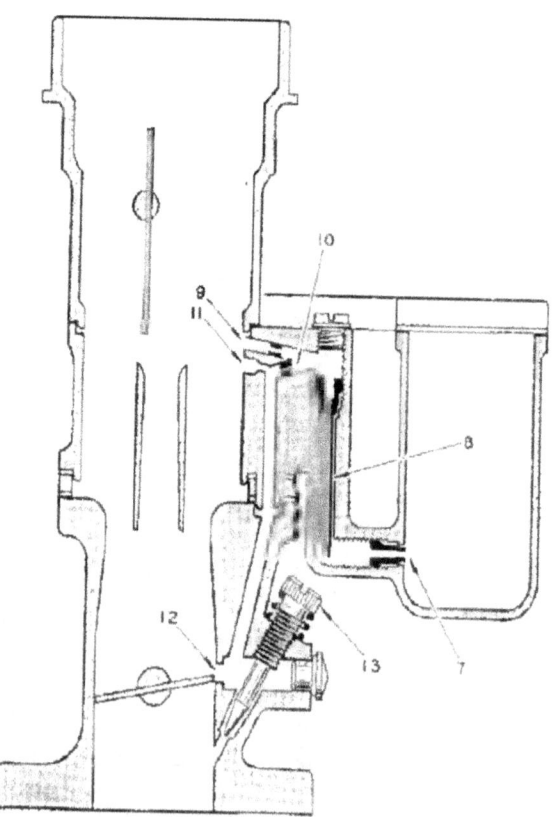

FIG. 4—LOW SPEED CIRCUIT

When the idle speed position of the throttle is fixed at an idle speed of 8 miles per hour, it leaves enough of the slotted port as reserve to cover the range in speed between idle and the time when the high speed system begins to cut in.

The idle adjusting screw No. 13 varies the quantity of the idle mixture.

High Speed Circuit

The high speed circuit, Fig. 5 cuts in as the throttle is opened wide enough for a speed of a little more than 20 miles per hour. The velocity of the air flowing down through the carburetor throat creates a pressure slightly less than atmospheric pressure at the tip of the main nozzle, No. 20.

Since the fuel in the float bowl is acted upon by atmospheric pressure, the difference in pressure between the two points causes fuel to flow from the bowl through the metering jet and out the main nozzle into the throat of the carburetor.

At higher speeds the area of the opening between the jet No. 17 and the metering rod No. 16 governs the amount of fuel going into the engine. At top speeds, the smallest section of the rod is in the jet.

Accelerating Pump Circuit

As the accelerator pedal is depressed, the pump plunger and lever are forced downward. This causes the fuel to leave the cylinder; closes the intake check valve No. 29, Fig. 6, opens the discharge check valve No. 30, and forces fuel into the throat of the carburetor at No. 33

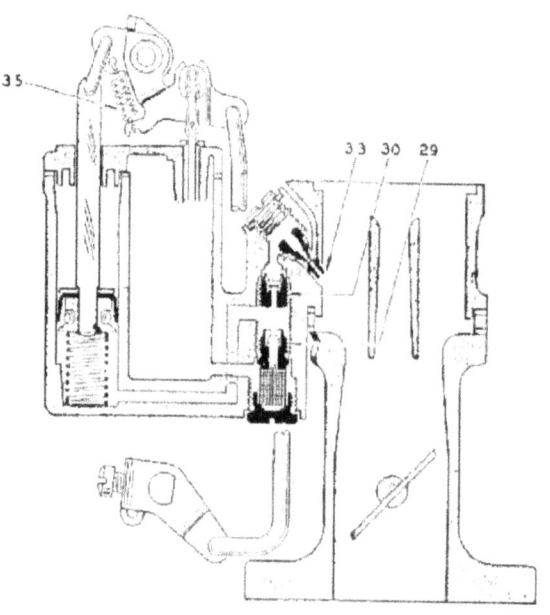

FIG. 6—PUMP CIRCUIT

The action is prolonged by the pump arm spring, No. 35, Fig. 6, since the hole in the top of the pump jet No. 33 restricts the flow of fuel so long as it is being forced out by the pump cylinder. The prolonging of the pump discharge gives the fuel in the high speed circuit sufficient time to flow fast enough to satisfy the demands of the engine.

As the accelerator pedal is allowed to return to its original position, the pump plunger is lifted upward. This creates a reduced pressure in the pump cylinder which opens the intake check valve No. 29 and closes the discharge check valve No. 30, thereby drawing in a new charge of fuel from the bowl.

Choke Circuit

This circuit, Fig. 7 is used only in starting and the warming up of the engine, by reducing the amount of air allowed to enter the carburetor and, thereby producing a richer mixture. It consists of the choke shaft and lever assembly No. 39, choke operating lever and spring No. 40, choke valve No. 37, and screws No. 38.

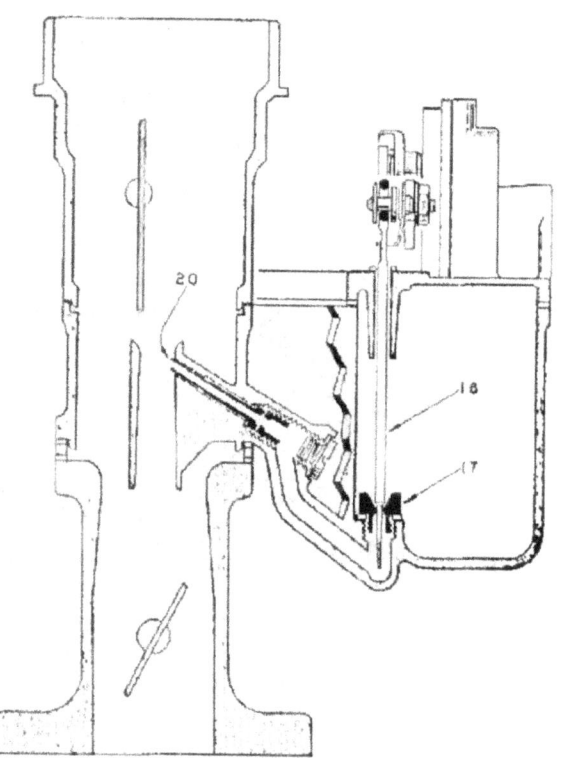

FIG. 5—HIGH SPEED CIRCUIT

Low Speed Circuit

In the low speed circuit Fig. No. 4 it will be found that the fuel for the low speed circuit does not come through the main metering jet, but through the well jet No. 7, and the low speed jet No. 8, the openings of which are carefully calibrated, so if they are damaged or worn they should be replaced. The jets should always be tightly seated.

The by-pass and air bleed holes, No. 9 and 11, may be restricted. Carbon deposit which forms in the throat of the carburetor may restrict the air bleed holes to the extent that insufficient air will be supplied to mix the fuel before it reaches the idle port, No. 12.

This condition will generally be indicated if it is necessary to screw the idle mixture adjusting screw, No. 13, in closer than the minimum limit of ½ turn. If the condition is bad, a rolling idle may continue even after the idle mixture adjusting screw is screwed entirely in against the seat. These air bleed holes may be cleaned with a soft copper wire.

The idle port must be clean and unrestricted. If it is damaged, the engine will not perform properly at low speeds and a new casting will be necessary.

A letter "C" enclosed within a circle is stamped on the face of the throttle valve. When installed in the carburetor, this side should be toward the idle port, and facing the intake manifold as viewed from the bottom.

To properly center the valve in the throat of the carburetor, the screws should be started in the shaft, and then with the valve tightly closed, (throttle lever adjusting screw backed out), it should be lightly tapped. This will centralize the valve in the bore. Pressure should then be maintained with the fingers until the screws are tightened.

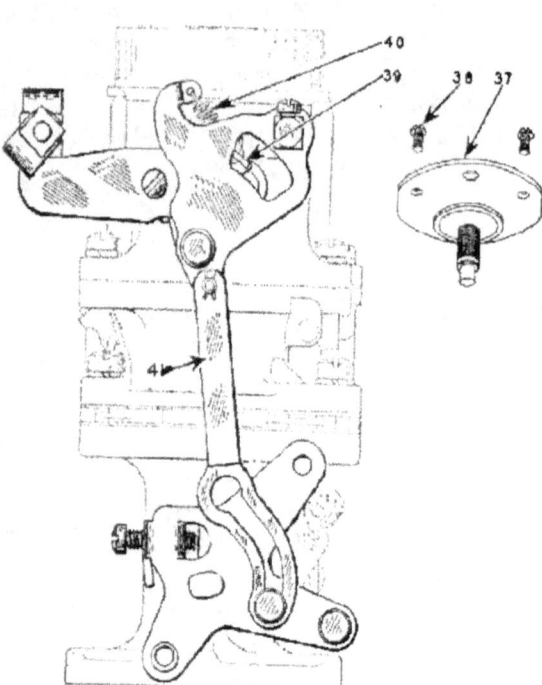

FIG. 7—CHOKE CIRCUIT

SERVICING AND ADJUSTMENT

Float Circuit

The Float Circuit is illustrated in Fig. 3.

If float is loaded with fuel or damaged, or if the holes for the pin are worn, the carburetor will flood. Poor action of float needle results if the lip of the float bracket is worn. In this event, it should be smoothed with emery cloth.

The needle and seat may leak because of wear, damage or sticking and will cause the carburetor to flood. Needles and seats are available only in matched sets. Never replace the needle without replacing the seat.

In determining the float level, Fig. 8, first turn the bowl cover gasket around and with the bowl cover in the position as shown, the float by its own weight, should rest at (⅜") as indicated by the gauge.

To make a change in the float level, it is best to press down with a screw driver on the brass lip of the float, holding up on the float while assembled to the cover of the carburetor. Bending the lip in this way allows it to retain its curvature which is necessary for the correct operation of the float valve.

Be sure the spring and pin in valve are in position and that the spring has not been stretched.

FIG. 8—FLOAT LEVEL SETTING

If the carburetor bore is restricted with carbon deposit it will be necessary to open the throttle wider than the specified opening to obtain the proper idle speed. Opening the throttle more than the specified amount in order to obtain the proper idle will then uncover more of the slotted idle port than was intended. This will result in leaving an insufficient amount of the idle port as a reserve to cover the period between idle and 20 miles per hour, where the high speed system begins to cut in. A flat spot on acceleration will result. Clean by scraping or with emery cloth.

High Speed Circuit

It is rarely necessary to remove the main nozzle No. 20, Fig. 5. It can usually be cleaned by removing plug and blowing out with compressed air. If it is damaged and requires replacing make sure, upon installation, only one gasket is between nozzle and its seat in the casting.

If the carburetor has been in service for a long time or has been tampered with, it may be found the metering rod is improperly adjusted or worn. A worn metering rod will have the effect of a rich mixture above 20 miles per hour. If the metering rod is worn, the metering rod jet will also be worn and both should be replaced.

To adjust metering rod, back out throttle lever adjusting screw "C" Fig. 9, and close throttle tight. Using gauge T-109-26, loosen nut "B" Fig. 9, and move pin until it seats in notch of gauge. Tighten nut securely. Remove gauge and install metering rod, disc, and connect spring through hole in metering rod.

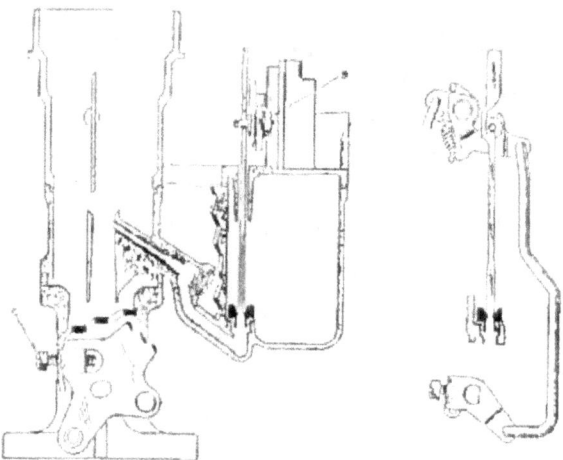

FIG. 9—METERING ROD GAUGING

Accelerating Pump Circuit

If the pump plunger is worn, sticks, or the spring under the leather has lost its tension, replace the plunger assembly No. 5, Fig. 1.

If the Accelerator Pump Intake Valve (ball check) No. 29, Fig. 6, leaks, part of the pump discharge will be forced back through the valve into the float bowl, thereby causing an insufficient amount of fuel to be discharged from the Jet No. 33. If the valve cannot be cleaned with compressed air, it must be replaced.

If the Accelerator Pump Discharge Valve (Disc check) No. 30 leaks, air will be drawn into the pump cylinder on the upstroke of the plunger. This gives an insufficient charge of fuel into the throat of the carburetor upon acceleration causing a flat spot. If the valve cannot be cleaned with compressed air so that it works properly, it must be replaced.

If the Accelerating Pump Arm Spring No. 35 is weak or damaged, it will cause poor acceleration.

If the hole in the Accelerating Pump Jet No. 33 is too large, the accelerating charge will be allowed to pass too fast and will make the mixture too rich. An enlarged jet must be replaced. A jet loose on its seat gives the same effect. A clogged jet will result in a stumble on acceleration.

To adjust the pump stroke, the pump gauge T-109-117S should be used. First back out the throttle adjusting screw "C", Fig. 9, so that it does not touch the casting. In gauging the pump stroke, place the gauge on top of the bowl cover, open the throttle wide then measure to the top of the pump rod. Close throttle tight and measure again. The difference should be $17/64"$. To adjust the stroke, bend the throttle connector rod at "A" Fig. 10. ALWAYS SET THE PUMP BEFORE SETTING THE METERING ROD. If set afterwards the metering rod will be thrown out of adjustment.

Throttle Connector Rod and Throttle Shaft Arm Assembly may be worn, and allow the throttle valve to be opened by the accelerator pedal before the pump jet begins to discharge gasoline, resulting in a flat spot. Replace all worn parts, because the operation of the metering rod is also affected.

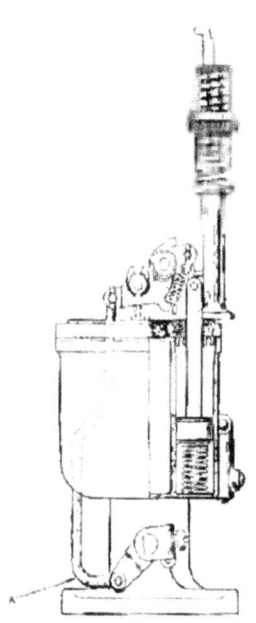

FIG. 10—PUMP TRAVEL GAUGING

Choke Circuit

The choke connector link No. 41, Fig. 7 connects the choke lever and the throttle lever and causes the throttle to be opened slightly when the choke valve is closed, thus insuring quick starting and freedom from stalling during the warm up period.

Carburetor information reprinted by permission of Carter Carburetor Corp., St. Louis, Mo., U. S. A. Holders of the copyright.

Accelerator and Linkage

The accelerator linkage, Fig. 11 is properly adjusted when vehicle leaves the factory, however, in time component parts will become worn and require adjusting to maintain a smooth even control of engine speed.

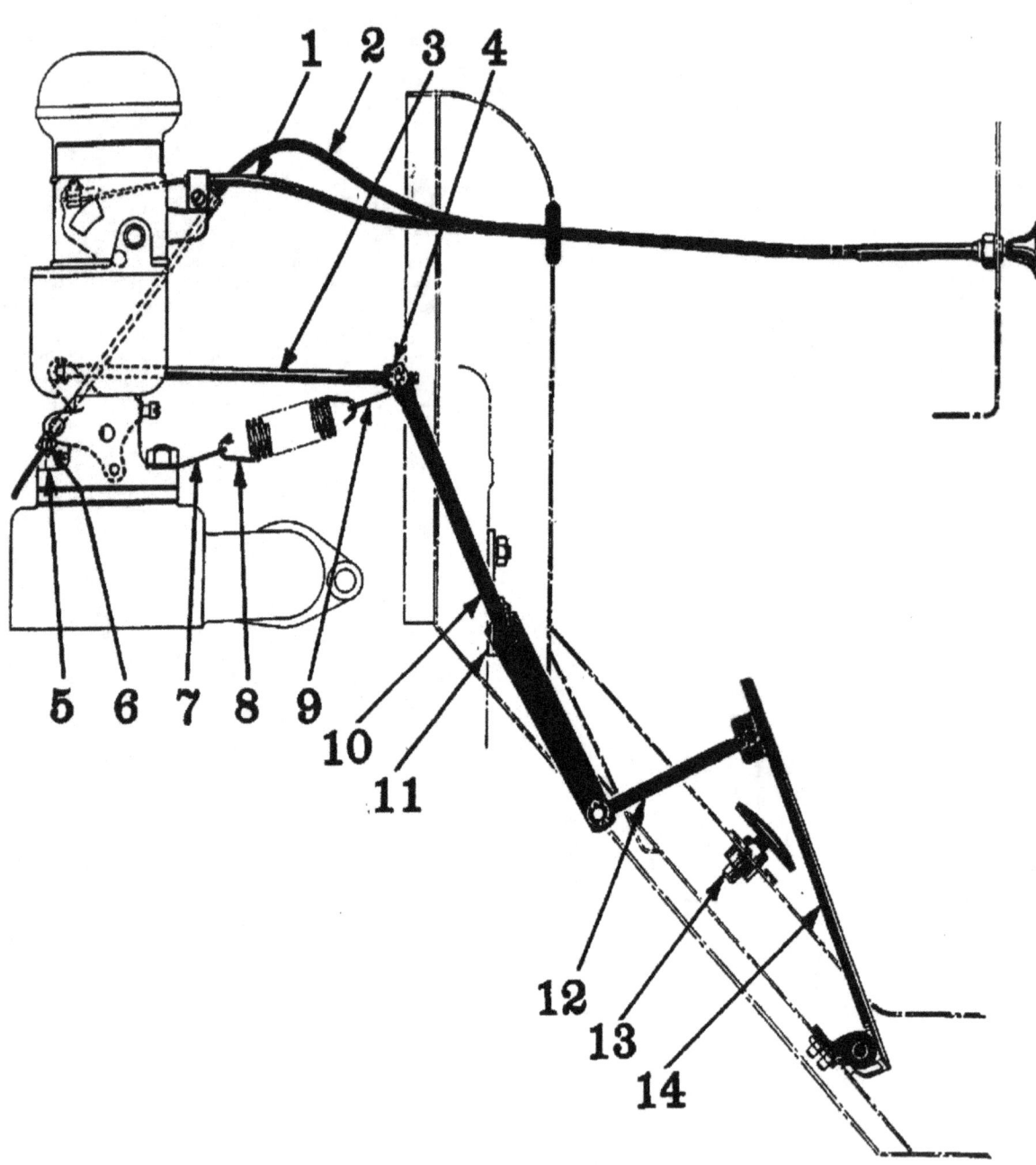

FIG. 11—ACCELERATOR, THROTTLE and CHOKE CONTROL

No.	Willys Part No.	Ford Part No.	Name
1	A-1301	GPW-9700	Choke Control Assembly
2	A-5106	GPW-9775	Throttle Control Assembly
3	A-1175	GPW-9742	Throttle Rod
4	633013	GPW-9752	Throttle Rod Adjusting Block
5	50922	33800-S2	Carburetor to Inlet Manifold Stud Nut
6	372438	GPW-11474	Throttle Wire Stop
7	A-1173	GPW-9751	Accelerator Retracting Spring Clip
8	633011	GPW-9799	Accelerator Spring
9	639610	GPW-9745	Accelerator Retracting Spring Clip
10	A-1243	GPW-9730	Accelerator Cross Shaft and Lever
11	639607	GPW-9728	Accelerator Cross Shaft Bracket
12	A-1174	GPW-9727	Accelerator Connector Link
13	A-1225	GPW-9716	Accelerator Foot Rest Assembly
14	A-1083	GPW-9735	Accelerator Treadle Assembly

Adjust the length of throttle rod No. 3 so when carburetor throttle valve is wide open, the accelerator treadle No. 14 will not strike the toe board. Tighten lock nut on adjusting block No. 4.

Fuel Pump

The Fuel Pump, Fig. 12 delivers a pressure of 1½ to 2½ lbs. maximum pressure at 1800 r.p.m. 16" above pump outlet.

The rotation of camshaft eccentric actuates

forcing fuel from chamber "F" through outlet valve "K" and out through "L" to the carburetor.

When the carburetor bowl is full, the float in the carburetor will shut off the needle valve, thus creating a pressure in pump chamber "F". This pressure will hold diaphragm assembly "D" downward against spring pressure "E" where it will remain inoperative until the carburetor requires further fuel and the needle valve opens. Spring "M" is merely for the

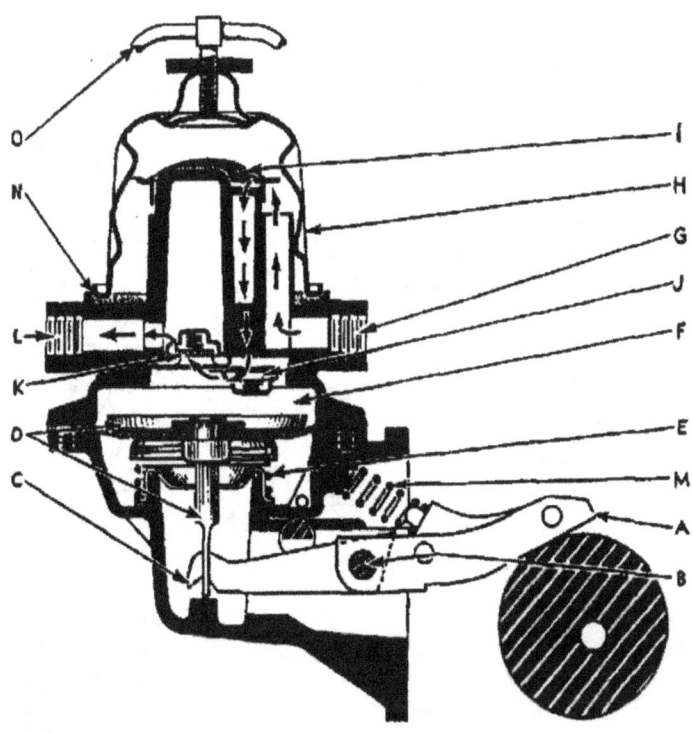

FIG. 12—FUEL PUMP

No.	Willys Part No.	Ford Part No.	Name	No.	Willys Part No.	Ford Part No.	Name
A	115641	GPW-9399	Fuel Pump Rocker Arm Assembly	H	A-1494	GPW-9355	Fuel Pump Bowl
B	A-1046	GPW-9378	Fuel Pump Rocker Arm Pin	I	115654	GPW-9365	Fuel Pump Filtering Screen Assembly
C	115880	GPW-9381	Fuel Pump Rocker Arm Link	J	115651	GPW-9352	Fuel Pump Inlet Valve Assembly
D	115644	GPW-9398	Fuel Pump Diaphragm and Pull Rod Assembly	K	115651	GPW-9352	Fuel Pump Outlet Valve Assembly
E	115868	GPW-9396	Fuel Pump Diaphragm Spring	L			Fuel Outlet
F			Pump Chamber	M	115643	GPW-9380	Fuel Pump Rocker Arm Spring
G			Fuel Inlet	N	115656	GPW-9364	Fuel Pump Bowl Gasket
				O	115657	GPW-9387	Strainer Ball Assembly

rocker arm "A" about ¼", pivoted at "B" which pulls link "C" and diaphragm assembly "D" downward against spring pressure "E" which creates a vacuum in pump chamber "F".

On the suction stroke of the pump, fuel from the tank enters inlet "G" into sediment bowl "H" and passes through strainer "I" and inlet valve "J" into pump chamber "F". On the return stroke spring pressure "E" pushes diaphragm "D" upward, purpose of keeping the rocker arm in constant contact with eccentric.

A lever and spring located on rear side of the fuel pump body is used for priming of the carburetor.

Moving the lever up and down operates the fuel pump diaphragm manually and pumps the fuel from the tank, filling the filter and carburetor bowl.

This provides a means of filling the fuel lines and units without using the starting motor, which would create unnecessary drain on battery.

The lever should be placed in the downward position to make pump function normally.

Diaphragm "D" is composed of several layers of specially treated cloth, which is impervious to fuel.

The fuel pump has a large reservoir and surge chamber. The filter bowl is clamped to the cover assembly, making it a simple matter to clean any sediment from the fuel pump. The inlet and outlet valve assemblies are interchangeable, and each assembly is a self-contained unit made up of a valve cage, a fibre valve, and a valve spring. Both valve assemblies are held in place by a valve retainer, permitting easy and speedy removal of the assemblies.

To disassemble fuel pump release thumb nut holding clamp of filter bowl, "H" and remove bowl. Remove strainer "I" from center tower, remove cork gasket, remove the six screws holding the cover flange to the pump body. Scratch a line across the two castings so that on assembly these two parts can be correctly assembled. Lift off top cover which brings into view the diaphragm assembly "D". Remove spring "M" that holds the rocker arm "A" against the camshaft eccentric.

To unhook the diaphragm pull rod "D" from the rocker arm link "C" press down and away from the rocker arm side. Remove oil seal and washer.

Remove the two screws holding inlet and outlet valve retainer.

Wash all parts thoroughly in cleaning solution and make the assembly as follows:

Install oil seal, (rubber cup) on body, then steel washer and spring that fits on the diaphragm assembly, holding the rocker arm "A" down, press the diaphragm assembly "D" down, tilt the diaphragm away from rocker arm and hook into place. Install inlet valve assembly "J" with new gasket. The inlet valve is installed in the body with the spring facing down. Install the outlet valve assembly "K" with the outlet valve spring up. Install valve retaining plate and two screws. Assemble upper and lower castings so that scratch marks line up. Install the six retaining screws and tighten them evenly. Install cam lever spring "M". Install new bowl gasket, filter screen "I" and bowl "H", tightening into place with thumb nut.

Fuel Filter

The fuel filter, Fig. 13 is of the multiple disc type bolted to the right side of the cowl and located in the fuel line between the fuel tank and the fuel pump. This is an added precaution against water or dirt reaching the carburetor. Drain the filter every few days to remove accumulated dirt and water. Be sure to tighten drain plug securely after draining.

To clean filter, remove the cover cap screw, No. 2, and remove the bowl, No. 10. Remove the filter unit, No. 8, and wash in any suitable cleaning solution. Blow out lightly with an air hose and clean out filter bowl. When replacing the unit be sure spring, No. 9, is placed over center post at the bottom of the bowl. Be sure that the gasket, No. 7, at top of filter unit and the bowl gasket, No. 6, are in good condition and in place. If the bowl gasket leaks, air will enter the fuel supply. Check for any fuel leaks with the engine stopped. Clean the filter unit every 500 operating hours or more often under severe conditions.

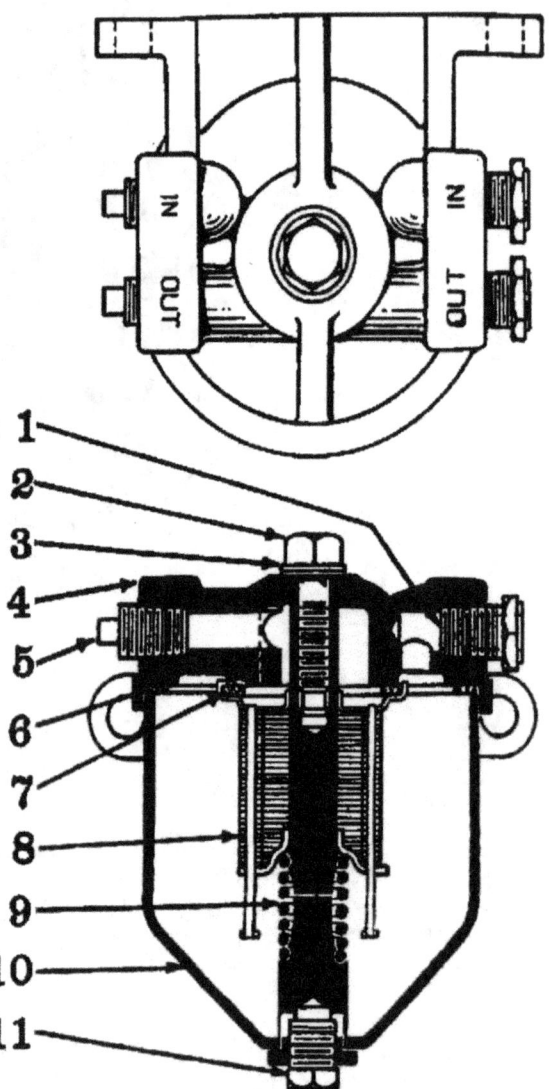

FIG. 13—FUEL FILTER

No.	Willys Part No.	Ford Part No.	Name
1	A-1265	GPW-9154	Reducing Pipe Bushing—¼ x ⅛ Pipe
2	A-1256	GPW-9183	Strainer Cover Cap Screw
3	A-1257	GPW-9184	Strainer Cover Cap Screw Gasket
4	A-1258	GPW-9149	Strainer Cover
5	5138	353055-S	Pipe Plug—⅛"
6	A-1259	GPW-9150	Strainer Bowl Gasket
7	A-1260	GPW-9186	Strainer Unit Gasket
8	A-1261	GPW-9140	Strainer Unit Assembly
9	A-1262	GPW-9182	Strainer Unit Spring
10	A-1263	GPW-9182	Strainer Bowl and Center Stud
11	A-1264	GPW-9185	Strainer Drain Plug

Diaphragm "D" is composed of several layers of specially treated cloth, which is impervious to fuel.

The fuel pump has a large reservoir and surge chamber. The filter bowl is clamped to the cover assembly, making it a simple matter to clean any sediment from the fuel pump. The inlet and outlet valve assemblies are interchangeable, and each assembly is a self-contained unit made up of a valve cage, a fibre valve, and a valve spring. Both valve assemblies are held in place by a valve retainer, permitting easy and speedy removal of the assemblies.

To disassemble fuel pump release thumb nut holding clamp of filter bowl, "H" and remove bowl. Remove strainer "I" from center tower, remove cork gasket, remove the six screws holding the cover flange to the pump body. Scratch a line across the two castings so that on assembly these two parts can be correctly assembled. Lift off top cover which brings into view the diaphragm assembly "D". Remove spring "M" that holds the rocker arm "A" against the camshaft eccentric.

To unhook the diaphragm pull rod "D" from the rocker arm link "C" press down and away from the rocker arm side. Remove oil seal and washer.

Remove the two screws holding inlet and outlet valve retainer.

Wash all parts thoroughly in cleaning solution and make the assembly as follows:

Install oil seal, (rubber cup) on body, then steel washer and spring that fits on the diaphragm assembly, holding the rocker arm "A" down, press the diaphragm assembly "D" down, tilt the diaphragm away from rocker arm and hook into place. Install inlet valve assembly "J" with new gasket. The inlet valve is installed in the body with the spring facing down. Install the outlet valve assembly "K" with the outlet valve spring up. Install valve retaining plate and two screws. Assemble upper and lower castings so that scratch marks line up. Install the six retaining screws and tighten them evenly. Install cam lever spring "M". Install new bowl gasket, filter screen "I" and bowl "H", tightening into place with thumb nut.

Fuel Filter

The fuel filter, Fig. 13 is of the multiple disc type bolted to the right side of the cowl and located in the fuel line between the fuel tank and the fuel pump. This is an added precaution against water or dirt reaching the carburetor. Drain the filter every few days to remove accumulated dirt and water. Be sure to tighten drain plug securely after draining.

To clean filter, remove the cover cap screw, No. 2, and remove the bowl, No. 10. Remove the filter unit, No. 8, and wash in any suitable cleaning solution. Blow out lightly with an air hose and clean out filter bowl. When replacing the unit be sure spring, No. 9, is placed over center post at the bottom of the bowl. Be sure that the gasket, No. 7, at top of filter unit and the bowl gasket, No. 6, are in good condition and in place. If the bowl gasket leaks, air will enter the fuel supply. Check for any fuel leaks with the engine stopped. Clean the filter unit every 500 operating hours or more often under severe conditions.

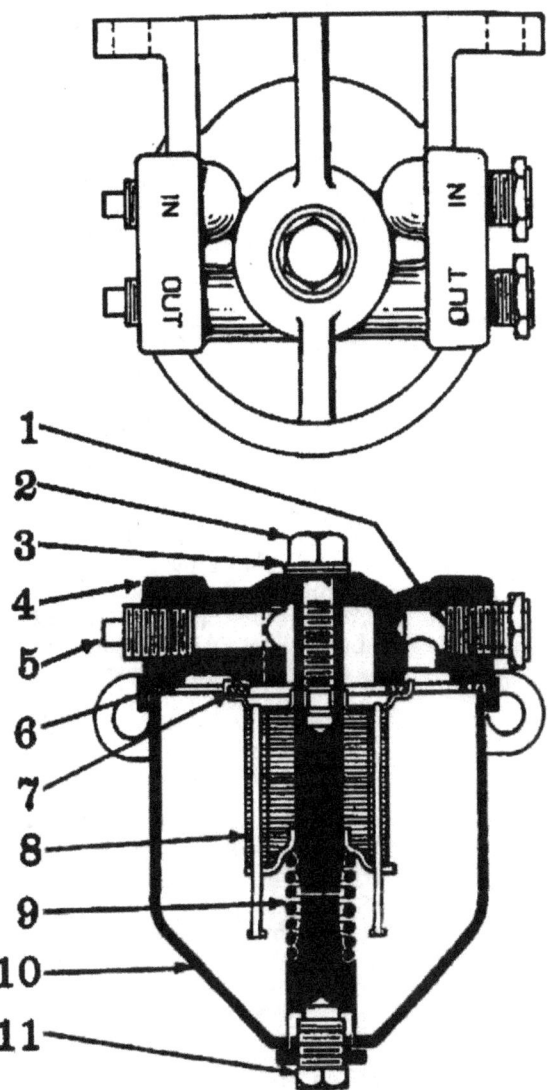

FIG. 13—FUEL FILTER

No.	Willys Part No.	Ford Part No.	Name
1	A-1265	GPW-9154	Reducing Pipe Bushing—¼ x ½ Pipe
2	A-1256	GPW-9183	Strainer Cover Cap Screw
3	A-1257	GPW-9184	Strainer Cover Cap Screw Gasket
4	A-1258	GPW-9149	Strainer Cover
5	5133	353055-S	Pipe Plug—¼"
6	A-1259	GPW-9150	Strainer Bowl Gasket
7	A-1260	GPW-9186	Strainer Unit Gasket
8	A-1261	GPW-9140	Strainer Unit Assembly
9	A-1262	GPW-9182	Strainer Unit Spring
10	A-1263	GPW-9152	Strainer Bowl and Center Stud
11	A-1264	GPW-9185	Strainer Drain Plug

Air Cleaner

The air which is taken into the carburetor, to mix with the fuel, is thoroughly cleaned when passed through the oil bath air cleaner Fig. 14. The cleaner is mounted on the right hand side of the dash and can be readily removed.

To clean the filter, loosen the hose clamp and two wing nuts at the center of the dash; remove the two wing nuts on the right side and remove air cleaner assembly from vehicle. Next, unfasten both clamps holding the oil cup. Unscrew element wing nut and pull out filter unit.

Wash the filter in cleaning solution by slushing it back and forth, then with an air hose dry off the unit. Do not reoil. Fill the oil cup to the indicated level. (See Capacity Chart, Page 3 and Lubrication Chart, Page 12). To assemble, reverse the dismantling procedure.

Fuel Tank Cap

The fuel tank cap is of the pressure type which keeps up to 1½ to 2 lbs. vapor pressure on the fuel. This reduces leakage and fire hazard, also eliminates depreciation of fuel qualities by evaporation.

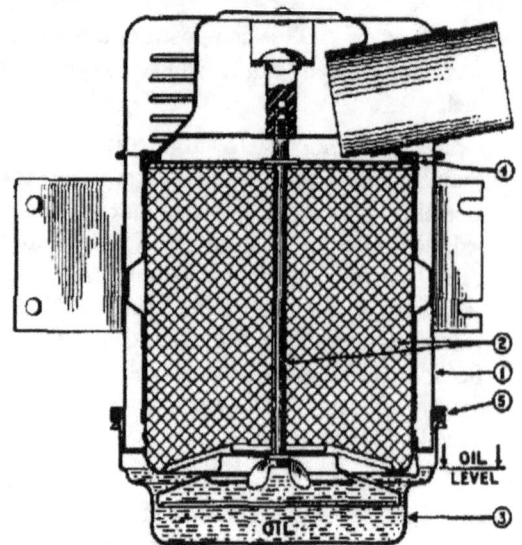

FIG. 14—AIR CLEANER (HEAVY DUTY—OIL TYPE)

No.	Willys Part No.	Ford Part No.	Name
1	A-3629	GPW-9609	Air Cleaner Body
2	A-3630	GPW-9617	Cleaner Element & Wing Bolt
3	A-3631	GPW-9658	Cleaner Cup (Oil)
4	A-3632	GPW-9621	Body Gasket (Upper)
5	A-3633	GPW-9623	Oil Cup Gasket (Lower)

FUEL SYSTEM TROUBLES AND REMEDIES

SYMPTOM	PROBABLE REMEDY
Excessive Fuel Consumption:	
Tires Improperly Inflated	Inflate to 30 lbs.
Brakes Drag	Adjust
Engine Operates too Cold	Check Thermostat
Heat Control Valve Inoperative	Check Thermostatic Spring
Leak in Fuel Line	Check All Connections
Carburetor Float Level High	See Carburetor Section
Accelerator Pump Not Properly Adjusted	Adjust
Leaky Fuel Pump Diaphragm	Replace
Loose Engine Mountings (High carburetor fuel level)	Tighten
Ignition Timing Slow or Spark Advance Stuck	See Distributor Section
Low Compression	Check Valve Tappet Clearance
Air Cleaner Dirty	Remove and Clean
Engine Hesitates on Acceleration:	
Accelerator Pump does not Function Properly	Replace Piston and Rod or Adjust
Carburetor Float Level	Adjust
Spark Plugs	Replace or Clean and Adjust
Low Compression	Check Valves
Distributor Points—Dirty or Pitted	Replace
Weak Condenser or Coil	Replace
Carburetor Jets Restricted	Remove and Clean
Excessive Engine Heat	See Engine Section
Engine Stalls—Won't Idle:	
Improper Condition of Carburetor	See Carburetor Section
Low Speed Jet Restricted	Remove and Clean
Dirty Fuel Sediment Bowl Screen	Remove and Clean
Air Cleaner Dirty	Remove and Clean
Leaky Manifold or Gasket	Replace
Fuel Pump Diaphragm Porous	Replace
Loose Carburetor	Tighten Flange Nuts
Water in Fuel	Drain System
Improper Ignition	See Distributor Section
Spark Plugs	Clean and Adjust
Valves Sticking	Grind Valves

FUEL SYSTEM SPECIFICATIONS

Carburetor:
- Make..Carter
- Model...........................W.O.-539 S
- Flange...................................1"
- Primary Venturi...................11/32" I.D.
- Main Venturi.........................1" I.D.
- Float Setting.............................3/8"
- Fuel Intake..... Square Vertical Spring Loaded Needle No. 53 Drill Size in Needle seat.
- Fuel Line Connection....1/8" pipe thread—3/16" inverted flared tube elbow.
- Low Speed Jet Tube
  - Jet Size.......................No. 71 Drill
  - Idle Well Jet..................No. 61 Drill
- Idle Screw Seat...................No. 46 Drill
- Main Nozzle Discharge Jet......096" Diameter
- Metering Rod.....................No. 75-547
  - Jet Size.....................070" Diameter
  - Setting (Use gauge No. T-109-26).....2.718"
- Accelerating Pump
  - Discharge Jet..................No. 73 Drill
  - Intake Ball Check..............No. 40 Drill
  - Discharge Disc Check..........No. 40 Drill
  - Relief Passage to Outside........No. 42 Drill
  - Adjustment (Use gauge T-109-117 S)....17/64"

Air Cleaner:
- Make..................................Oakes
- Model...........................613300
- Type...........................Oil Bath
- Oil Capacity.......See Capacity Chart, Page 3

Fuel Pump:
- Make..................................AC
- Model...............................AF
- Type............................Camshaft
- Pressure..1½ to 2½ lbs. at 16" above outlet @ 1800 R.P.M.

Fuel Tank:
- Make...............................Own
- Capacity.........See Capacity Chart, Page 3
- Location...............Under Driver's Seat
- Filler Cap...................AC No. 850018

Fuel Filter:
- Make..................................AC
- Model...............................T-2
- Type...............................Disc
- Mounting................Mounted on Dash

EXHAUST SYSTEM

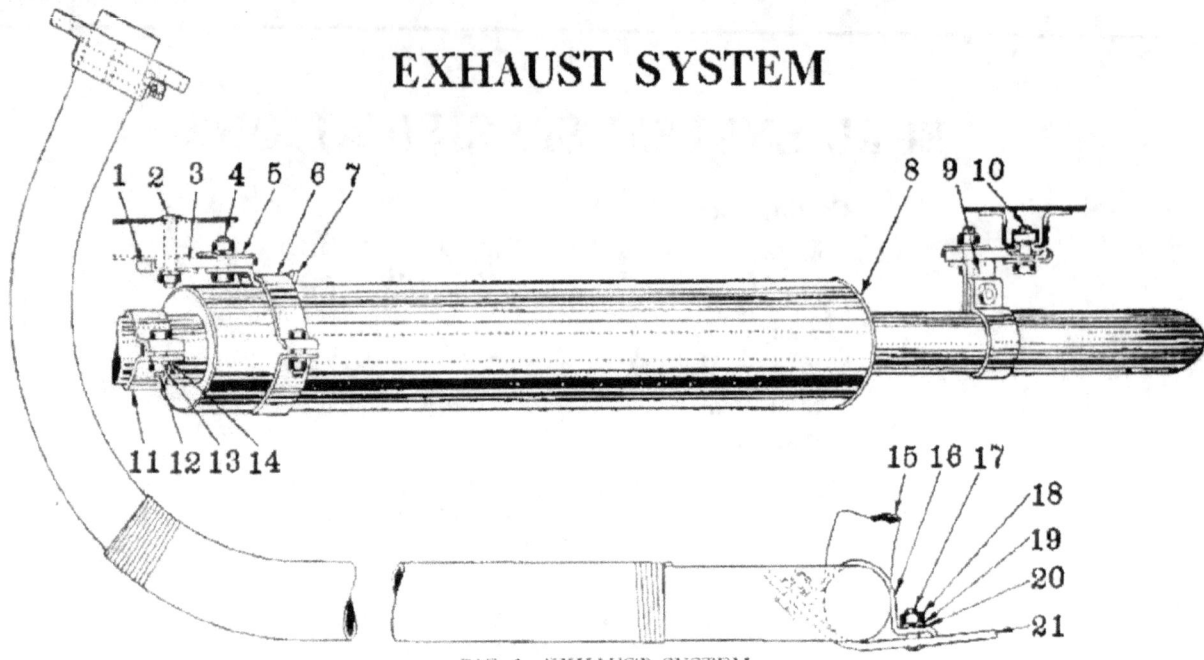

FIG. 1—EXHAUST SYSTEM

No.	Willys Part No.	Ford Part No.	Name
1	638058	GPW-5274	Muffler Support Insulator Plate
2	52372	23498-S	Body Sill to Muffler Support Bolt
3	A-558	GPW-5283	Muffler Support Insulator
4	50929	20367-S7	Muffler Support Screw
5	638058	GPW-5274	Muffler Support Insulator Plate
6	A-5753		Muffler Support Clamp
7	A-657	GPW-5262	Muffler Support Strap
8	A-6118		Muffler Assembly
9	A-6119		Muffler Tail Pipe Clamp
10	50929	24367-S	Muffler Support Screw
11	636004	GPW-5270	Exhaust Pipe to Muffler Clamp
12	5922	24407-S	Exhaust Pipe to Muffler Clamp Screw
13	5010	33798-S	Exhaust Pipe to Muffler Clamp Screw Nut
14	51833	34806-S	Exhaust Pipe to Muffler Clamp Screw Nut Lockwasher
15	A-1296	GPW-5246	Exhaust Pipe Assembly
16	A-1300	GPW-5251	Exhaust Pipe Extension Clamp (to Skid Plate)
17	52983	23393-S	Exhaust Pipe Extension to Skid Plate Bolt
18	6107	33797-S	Exhaust Pipe Extension to Skid Plate Bolt Nut
19	51833	34806-S	Exhaust Pipe Extension to Skid Plate Bolt Lockwasher
20	52274	34746-S2	Exhaust Pipe Extension to Skid Plate Bolt Plain Washer
21	A-1253	GPW-5291-B	Underframe Skid Plate

Seamless exhaust pipe and flexible metal tubing connect the Manifold to the Muffler.

The Muffler, No. 8, Fig. 1 is designed for straight through exhaust to minimize back pressure. The Muffler is attached to the under side of the vehicle on the right hand side by means of support straps and flexible Insulators.

The exhaust pipe slides into the nipple on the front end of the Muffler and is held in place by means of a clamp, No. 11.

Exhaust and Intake Manifold

The exhaust and intake Manifolds, make a unit in which the hot exhaust gases are thermostatically controlled, and directed around the intake Manifold to assist in vaporizing the fuel when engine is cold, thereby aiding in warming up the engine and reducing oil dilution. It also minimizes the use of the carburetor choke control and results in proper temperature of the incoming gases under all operating conditions.

When the engine is cold, the counterweight lever No. 6, Fig. 2, closes the valve and directs the hot exhaust gases against the intake Manifold. As the engine warms up, the thermostatic (or Bimetal) spring No. 7 expands and opens the valve directing the exhaust gases into the exhaust pipe.

When assembling the Manifolds to the cylinder block, new gaskets should be installed, and the nuts drawn up evenly until they are all tight to avoid gas leakage. Torque wrench reading, 31-35 ft. lbs.

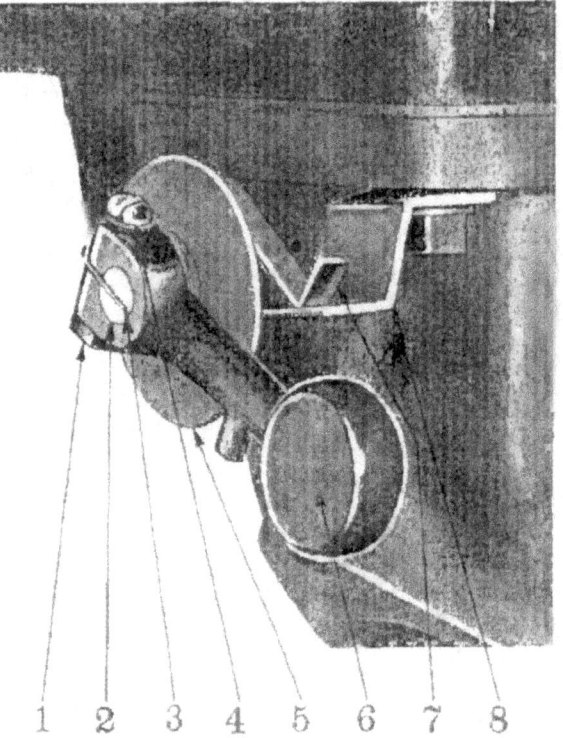

FIG. 2—HEAT CONTROL VALVE

No.	Willys Part No.	Ford Part No.	Name
1	6362	355836-S7	Heat Control Valve Lever Clamp Screw Nut
2	637206	GPW-9456	Heat Control Valve Shaft
3	637211	GPW-9465	Heat Control Valve Lever Key
4	5272	355160-S	Heat Control Valve Lever Clamp Screw
5	637209	GPW-9484	Heat Control Valve Bi-Metal Spring Washer
6	637210	GPW-9458	Heat Control Valve Counterweight Lever
7	637208	GPW-9467	Heat Control Valve Bi-Metal Spring
8	639743	GPW-9463	Heat Control Valve Bi-Metal Spring Stop

COOLING SYSTEM

The satisfactory performance of an engine is controlled to a great extent by the proper operation of the cooling system. The engine block is full length water jacketed which avoids distortion of the cylinder walls. Directed cooling and large water holes, properly placed in cylinder head gasket, cause more water to flow past the valve seats (which is the hottest part of the block) and carry the heat away from the valves, giving positive cooling of valves and seats.

To quickly warm up the engine and hold the cooling fluid to the maximum efficient temperature, there is a thermostat installed in the water outlet on the cylinder head.

Radiator

The radiator is designed to cool the water under all operating conditions, however, the radiator core must be kept free from corrosion and scale in addition to the maintenance of other cooling units to obtain satisfactory service.

At least every 20,000 miles remove the radiator and clean it inside and out in a cleaning solution. At the same time examine core for leaks or damaged cells.

After radiator and cooling system have been cleaned and flushed out, it is advisable to use a corrosion preventative. Rust and scale may eventually clog up water passages in both the radiator and water jacket of the engine unless a rust inhibitor is used. This condition is aggravated in some localities by the water available.

Emergency repairs in case of puncture by bullet or shrapnel; if a tube is not completely severed, cut it or break it off with a pair of pliers. With pliers strip fins from tube above and below break for ½" or necessary distance to enable bending of the tube around itself and flatten, both above and below the break thereby stopping the flow of water.

Radiator Filler Cap

The cap is of the pressure type which prevents evaporation and loss of cooling solution. A pressure up to 3¼ to 4¼ pounds makes the engine more efficient by running at a slightly higher temperature. Vacuum in the radiator is relieved by a vacuum valve opening at ½ to 1 pound vacuum.

Draining Cooling System

To drain the cooling system open the drain cock located at the lower left hand corner of the radiator, just under the water outlet, also the drain cock at right front lower corner of cylinder block.

Remove radiator cap to break any vacuum and thoroughly drain system.

Filling the Cooling System

Close drain cocks in the cylinder block and radiator. Fill the radiator with clean water or during cold weather with an anti-freeze solution. Do not overfill the radiator while anti-freeze solution is being used, because the solution expands when heated and an appreciable amount of liquid would be lost through the overflow. The solution should be 1" from the bottom of the filler neck.

Should water be lost from the cooling system and the engine overheats, do not add water immediately but allow the engine to cool, then add water slowly while the engine is running.

If cold water is poured into the radiator while the engine is overheated, there is danger of cracking the cylinder block and head.

Thermostat

The cooling system is designed to provide adequate cooling under the most adverse conditions; however, it is necessary to employ some device to prevent overcooling during normal operations. This is accomplished by use of a thermostat, (No. 39, Fig. 1, "Engine" Section) which is located in the water outlet on top of the cylinder

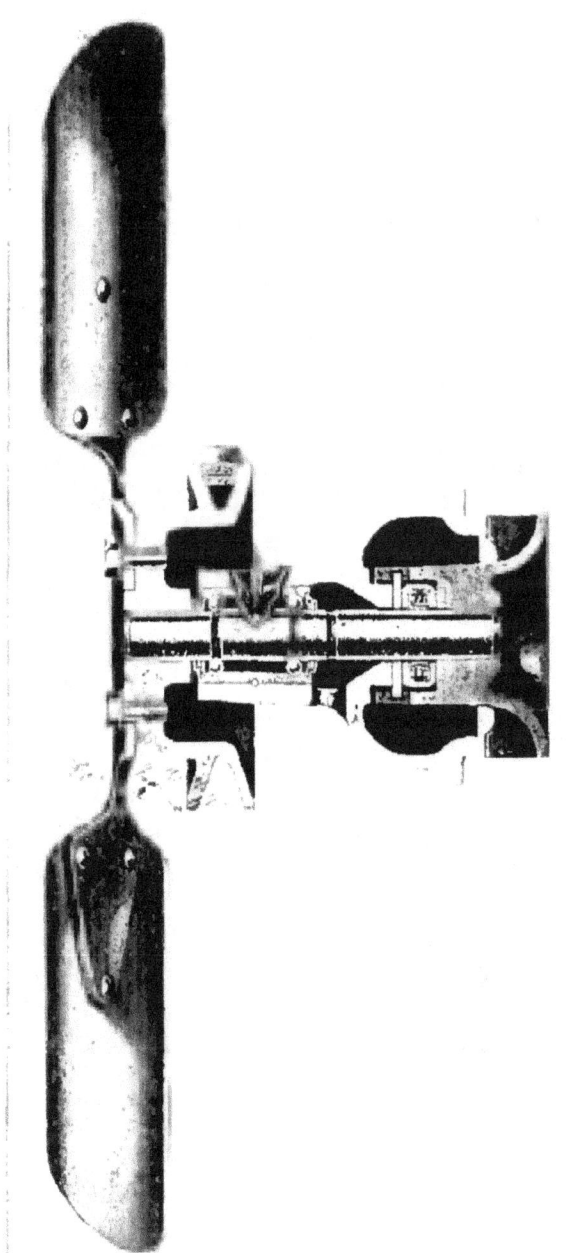

FIG. 1—WATER PUMP AND FAN ASSEMBLY

head. The thermostat opening is set by the manufacturer and cannot be altered. The thermostat opens at a temperature of 145° to 155° Fahr. To test thermostat, heat sufficient water to 170° Fahr. and submerge thermostat. The valve should open to the limit at this temperature. If valve fails to open, a new thermostat will be required.

Heat Indicator

The heat indicator is of the Bourdon type and is connected to a bulb in the engine block by means of a capillary tube.

If the unit becomes inoperative, it should be replaced as it is not practical to either repair or adjust this unit.

Water Pump

The water pump, Fig. 1 is a centrifugal impeller type of large capacity to circulate the water in the entire cooling system.

The double row ball bearing is integral with the shaft, No. 2, Fig. 2 and is packed with a special high melting point grease at the time of manufacture so it requires no lubrication. The ends of the bearings are sealed to retain the lubricant and prevent dust and dirt from entering.

The bearing is retained in the housing by a retaining wire No. 4, which snaps between the bearing and the water pump body. The seal washer No. 5 has four lugs which fit into the slots in the end of the impeller No. 8. One side of the seal washer bears against the ground surface of the pump housing and the other against the seal No. 7. The rubber seal bears against the machined surface on the inside of the impeller. The seal maintains a constant pressure against the seal washer and impeller assuring positive seal. The drain hole in the bottom of housing prevents any water seepage past the seal washer No. 5 entering the bearing.

The impeller and fan pulley are pressed on to the straight shaft under 2500 pounds pressure.

Dismantling of Water Pump

Remove fan belt and fan assembly and then the water pump from the engine.

Remove bearing retaining wire No. 4, Fig. 2.

Place water pump body on arbor press face plate and press water pump shaft through impeller No. 8, and pump body No. 3.

Remove the seal washer No. 5 and seal No. 7.

Place pump shaft No. 2 and fan pulley No. 1 on press so that the bearing will clear in the opening and press shaft from fan pulley.

To reassemble the water pump, install the long end of the shaft No. 2, in the pump body, No. 3 from the front end, until the outer end of bearing is flush with the front end of the pump body.

Dip seal No. 7 and seal washer No. 5 in brake fluid and install in the impeller No. 8. Place the impeller on an arbor press and press the long end of shaft into the impeller, until the end of the shaft is flush with the impeller.

Support assembly on impeller end of shaft and press the fan pulley on to shaft so the end of shaft is flush with the face of fan pulley—move the shaft in the pump body so grooves in the bearing and pump body line up and install bearing retaining wire No. 4.

Anti-Freeze Solution

Where air temperatures require, it is necessary to protect the cooling system with some type of anti-freeze solution so as to prevent damage resulting from freezing.

When alcohol is used as an anti-freeze solution care must be taken not to spill any of the solution on the finished portions of the vehicle; if so, it should be washed off immediately with a good supply of cold water, without wiping or rubbing.

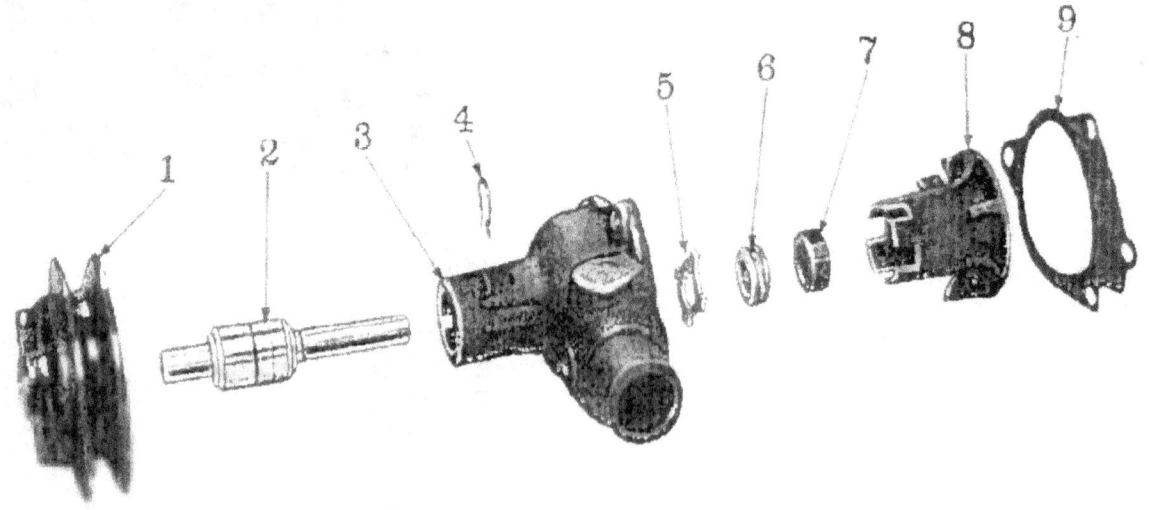

FIG. 2—WATER PUMP ASSEMBLY

No.	Willys Part No.	Ford Part No.	Name	No.	Willys Part No.	Ford Part No.	Name
1	636299	GPW-8509-A	Fan and Water Pump Pulley	5	639994	GPW-8557	Water Pump Seal Washer
2	636297	GPW-8530	Water Pump Bearing and Shaft Assembly	6-7	639663	GPW-8524	Water Pump Seal Assembly
3	637052	GPW-8505	Water Pump Body	8	639993	GPW-8512	Water Pump Impeller
4	636298	GPW-8576	Water Pump Bearing Retaining Wire	9	637053	GPW-8543	Water Pump to Cylinder Block Gasket

The distillation or evaporating point of water and alcohol is approximately 170° Fahrenheit, therefore, when the engine is operated in warm weather with alcohol solution, the solution must be checked regularly as there will be considerable loss of the alcohol through vaporization, and the freezing point raised in the solution, this might cause freezing of the solution at a sudden drop in temperature.

Ethylene glycol anti-freeze solutions have the distinct advantage of possessing a higher point of distillation than alcohol and consequently may be operated at higher temperatures without loss of the solution through evaporation.

Ethylene glycol has the further advantage that in a tight system only water is required to replace evaporation losses, however, any solution lost mechanically through leakage or foaming must be replaced by additional new solution. Under ordinary conditions Ethylene glycol solutions are not injurious to the car finish.

Rust and scale forms in every cooling system, therefore, we recommend that the cooling system be flushed out twice a year preferably before and after using anti-freeze. There are a number of flushing solutions and the instructions of the manufacturer should be closely followed when they are used.

Remove the thermostat when flushing the engine block, so the water and air pressure can get by and avoid possible damage to the thermostat.

When the cooling system is being conditioned it is good policy to tighten the cylinder head bolts to prevent the possibility of water leaking into the cylinders and lubrication oil. The radiator hoses should be inspected regularly for any indication of leakage which might be caused by loose clamps or deteriorated hose.

Fan Belt

The fan is driven by a "V" Belt. Angle of "V" -42°. Length outside $44\frac{1}{8}''$. Width maximum $\frac{11}{16}''$.

To install Fan Belt loosen clamp bolt on slotted bracket at generator and move generator towards engine. Slide belt over crankshaft pulley, up through fan blade assembly and over fan pulley, then over generator pulley. Adjust the fan belt by bringing the generator away from the engine to a point where the fan belt can be depressed 1" midway between fan pulley and generator pulley. The drive of the fan and generator is on the sides of the "V" belt, therefore it is not necessary to have the fan belt tight which might cause excessive wear on generator and water pump bearings.

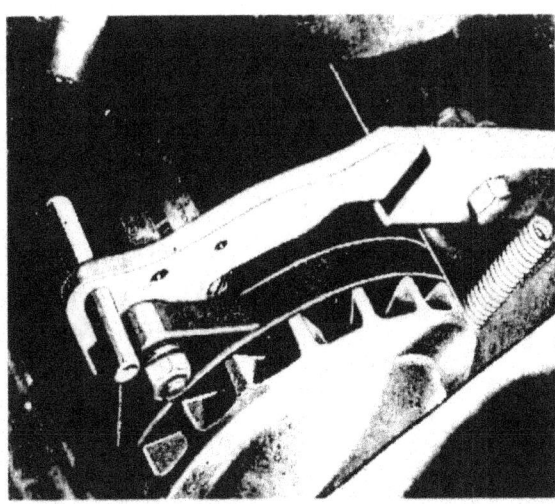

FIG. 2—GENERATOR BRACE

When there is a possibility of water being thrown over the engine by fan action in crossing streams, pull up on the handle of the generator brace, then remove the fan belt. As soon as possible the belt should be replaced, then pull out on the generator. The generator will lock in place by spring action of the brace.

COOLING TROUBLES AND REMEDIES

SYMPTOMS	PROBABLE REMEDY
Overheating	
Lack of Water	Refill Radiator
Thermostat Inoperative	Replace
Water Pump Inoperative	Overhaul or Replace
Incorrect Ignition or Valve Timing	Set Timing
Excessive Piston Blowby	Check Pistons, Rings and Cylinder Walls
Fan Belt Broken	Replace
Radiator Clogged	Reverse Flush
Air Passages in Core Clogged	Clean With Water and Air Pressure
Excessive Carbon Formation	Remove
Muffler Clogged or Bent Exhaust Pipe	Replace
Loss of Cooling Liquid	
Loose Hose Connections	Tighten
Damaged Hose	Replace
Leaking Water Pump	Replace
Leak in Radiator	Remove and Repair
Leaky Cylinder Head Gasket	Replace
Crack in Cylinder Block	Small Crack Can Be Closed with
Crack in Cylinder Head	Radiator Anti-Leak

COOLING SYSTEM SPECIFICATIONS

Cooling Capacity....See Capacity Chart, Page 3

Radiator..........................Jamestown
   Radiator Filler Cap....................A C

Fan—4 Blade—15" Dia................Hayes

Fan Belt
   Type................................"V"
   Length..............................44 1/8"
   Width...............................11/16"
   Angle of Vee.........................42°

Water Pump
   Type..........................Centrifugal
   Location.............Front of Cylinder Block
   Drive...................................Belt
   Bearing...Permanently Sealed-Lubricated Ball

Thermostat
   Location......Water outlet Top Cylinder Head
   Starts To Open At..................145°-155°
   Fully Open..........................170°

Anti-Freeze

Temp. Fahr.	Alcohol Qts	Ethylene Glycol Qts	Temp. Cent.
30°	1	1	— 1.1°
20°	2 1/8	2	— 6.6°
10°	3 1/4	3	—12.2°
0°	4 1/4	3 3/4	—17.7°
—10°	5	4 1/2	—23°
—20°	5 1/2	4 3/4	—29°
—30°	6 3/4	5 1/2	—34°
—40°	7 1/4	6	—40°

To convert quantities into Imperial quarts multiply by .833.

To convert quantities into Metric liters multiply by .946.

ELECTRICAL SYSTEM

The wiring diagrams, Fig. 1 and 3 show the general arrangement of all chassis electrical circuits, together with units in correct relation to position in which they will be found on the chassis.

Regular inspection of all electrical connections avoids failures in the electrical system. When tracing any one particular circuit refer to Wiring Harness, Page 59 for color of wire and tracer.

Radio Interference Suppression

The vehicle is equipped with filters in several of the electrical circuits and resistor type suppressors have been placed in the ignition high tension system, which together with bonding and shielding prevents interference with radio communication. (See Page 72.)

Filterettes, consisting of a coil in series with the line and two condensers across the line to ground, have been placed in the line from the primary of the high tension coil to the ignition switch; from ammeter to the "B" terminal of the voltage regulator and, from ammeter to the battery. Condenser type filters have been placed from the generator armature terminal to ground, and, from the regulator field terminal to ground. See detail wiring diagram in Fig. 1, Page 58.

Failure of the ignition or charging system because of open circuit filters can be checked by shorting across the terminals of the suspected filterette unit. If operation is not restored filterette may be considered to be in proper condition. These filterettes are combined in a container on the drivers side of the dash. There are two types of containers. Terminals are exposed on one type by opening cover which is held by a latch on the top, and on the other type by removing the cover. In an emergency, operation of the vehicle can be resumed by placing a shorting wire across the open circuit of the filterette.

To check filterette for short to ground, remove case of unit from dash. If short is NOT removed filterette may be considered satisfactory. If short IS removed filterette is defective. Temporary operation can be obtained by leaving case of the filterette ungrounded.

The regulator field filter (condenser type) which is mounted on the regulator itself can be checked by the same procedure as in the test for a shorted filterette.

Broken suppressors (resistor type) will affect the operation of the vehicle. Temporary operation can be obtained by removing the broken suppressor and making direct connection.

On vehicles equipped with radio suppression, the letter "S" appears on the cowl between the rear end of the hood and the windshield, 1¼" above bottom edge of hood.

Battery

The battery is a 6 Volt, 15 Plate, 116 Ampere Hour battery. It is located under the hood on a bracket attached to the right hand side rail of the frame and held solidly on the base with a battery hold down assembly over the top of the battery by two studs and wing nuts.

The battery should be checked once a week with a hydrometer and at the same time check the electrolyte level in each cell; add distilled water if necessary. Do not fail to replace filler caps and tighten securely. Battery fumes or acid coming in contact with any metal parts causes corrosion and eating away of these parts. To safeguard against this difficulty avoid overfilling.

The negative terminal of the battery (smallest post) is grounded by a cable bolted to the frame.

The engine ground cable located on the right hand side of engine, which connects the front engine support plate with the frame, is required due to the engine being mounted on rubber insulators.

If the terminal connections of this cable are loose or dirty, it will cause hard starting of the engine. Attention should be given to this ground cable during each inspection, and also at the time the engine requires a tuneup.

Fuel Gauge

The fuel gauge circuit Fig. 2, is composed of two units. The indicating unit or dash unit which is mounted in the instrument panel, and the tank unit, which is mounted in the fuel tank. These units are connected by a single wire. The circuit for this instrument passes through the ignition switch, therefore, the fuel gauge operates only when the ignition switch is on.

The dash unit is of the balanced coil type and is designed so that its operation is not affected by variations in the voltage of the electrical system.

The tank unit consists of a resistance wire wound on an insulator and a contact arm which is moved by the float arm. As the depth of the fuel in the tank varies, the contact arm is moved across the resistance wire and so varys the resistance. As this resistance is varied, there results a proportionate variation of current in the coils of the dash unit which is calibrated to accurately indicate the fuel level in the tank.

If the gauge does not register properly, first check all wire connections to be sure that they are clean and tight. Then make sure that the dash unit is grounded to the dash and that the tank unit is grounded to the fuel tank.

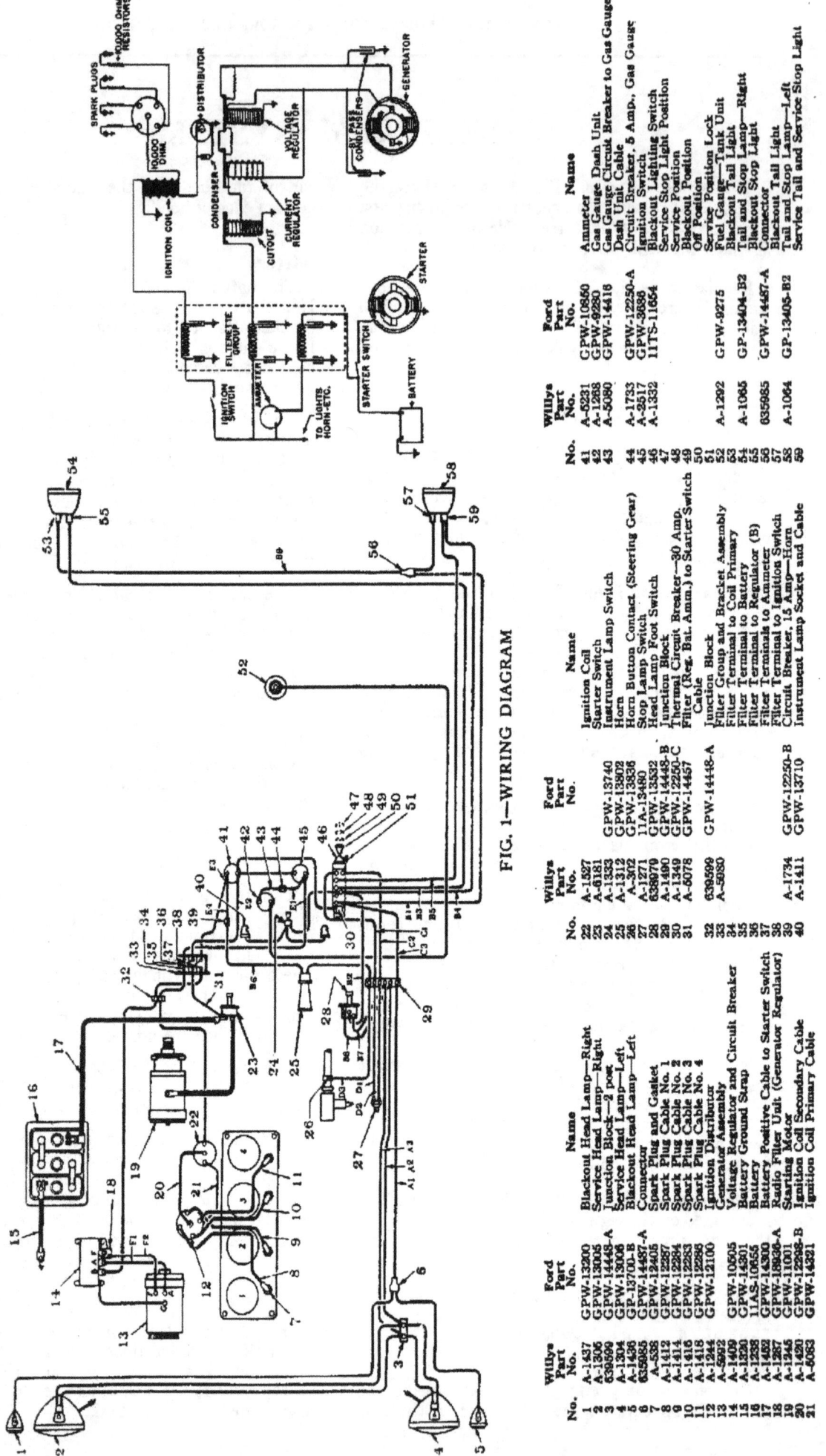

FIG. 1—WIRING DIAGRAM

WIRING HARNESS

Willys Part No.	Ford Part No.	Name
A-1665	GPW-14425	Head lamp wiring harness

The following (3) cables are included in this harness but not supplied individually:
- A-1 Blackout head lamp junction block cable (Yellow with two black tracings)
- A-2 Head lamp junction block to junction block cable (Upper beam) (Red, three white tracings)
- A-3 Head lamp junction block to junction block cable (Lower beam) (Black, white tracings)

A-5048	GPW-14401-B	Body wiring harness—Long

The following (9) cables are included in this harness but not supplied individually:
- B-1 Light switch (Terminal marked B.H.T.) to blackout tail light Conn. Cable (Yellow, two black tracings)
- B-2 Light switch (Terminal marked H.T.) to foot dimmer switch center terminal cable (Blue, three white tracings)
- B-3 Light switch (Terminal marked B.S.) to blackout stop light cable (White, two black tracings)
- B-4 Light switch (Terminal marked H.T.) to service tail light and inst. lamp switch cable (Blue, two white tracings)
- B-5 Light switch (Terminal marked S.) to service stop light cable (Black, two white tracings)
- B-6 Horn circuit breaker to horn cable (Black, two red tracings)
- B-7 Junction block to foot dimmer switch (Lower beam) Cable (Black, two white tracings)
- B-8 Junction block to foot dimmer switch (Upper beam) Cable (Red, three white tracings)
- B-9 Connector to blackout tail light cable (Yellow, two black tracings)

A-1551	GPW-14402	Body wiring harness—Left side—Short

The following (3) cables are included in above harness but not supplied separately:
- C-1 Junction block to light switch (Stop switch) (Red, two white tracings)
- C-2 Junction block to light switch cable (Stop switch) (Green, two black tracings)
- C-3 Junction block to light switch cable (Blackout head) (Yellow, two black tracings)

A-5061		Chassis wiring harness—Left side

The following (3) cables are included in above harness but not supplied separately:
- D-1 Stop light switch to junction block cable (Red, two white tracings)
- D-2 Stop light switch to junction block cable (Green, two black tracings)
- D-3 Steering gear horn terminal to junction block cable (Three black, two white tracings)

A-5981		Filter wiring harness

The following (4) cables are included in this harness but not supplied individually:
- E-1 Filter (Coil ign. switch) to ignition switch to gas gauge circuit breaker (Black, two white tracings)
- E-2 Filter (Battery, ammeter) to ammeter cable (Red, three white tracings)
- E-3 Filter (Reg. battery ammeter) to ammeter cable (Black, three white tracings)
- E-4 Ammeter to horn circuit breaker cable (Black, two red tracings)

A-5074	GPW-14305	Generator to voltage regulator and filter harness

- F-1 Generator to regulator armature cable (Red, three white tracings)
- F-2 Generator to regulator field cable (Green, two black tracings)

Willys Part No.	Ford Part No.	Name
A-719	GPW-13410	Blackout tail lamp to connector cable, left
A-5072	GPW-14458	Ignition switch to ammeter to blackout switch cable
A-5080	GPW-14416	Fuel gauge to circuit breaker cable
A-1731	GPW-14436	Headlamp ground cable
A-5070	GPW-14406	Fuel gauge (Inst. board) to fuel gauge (Tank unit) cable
A-5081	GPW-14409	Horn to junction block cable
A-5073	GPW-14459	Filter (Reg. bat.) to junction block cable
A-5078	GPW-14457	Filter (Bat.) to starter switch cable
A-5079	GPW-14456	Filter (Coil pri.) to junction block cable
A-5041	GPW-18846	Voltage regulator to generator ("G") cable (Bond No. 8)
A-5082	GPW-14465	Voltage regulator to junction block cable
A-1733	GPW-12250-A	Circuit breaker (Between ignition switch and fuel gauge on inst. board)
A-1734	GPW-12250-B	Circuit breaker (Between ammeter and horn)
635985	GPW-14487-A	Connector (3 wire)

If, after checking all grounds and wire connections, gauge does not indicate properly, remove wire from tank gauge unit and ground it to frame while ignition switch is on. Gauge should then read FULL. Remove wire from frame (with ignition switch on) and gauge should read EMPTY. If this is not the case, dash gauge should be replaced with a new one. If the gauge indicates as described, the trouble is probably in the tank unit, which should be replaced. Do not attempt to repair either gauge or tank unit, replacement is the only practical procedure.

Lighting System

The wiring of the lighting system is shown in Fig. 3. The lights are controlled by switches within easy reach of the driver. The lighting circuit is protected by an overload circuit breaker, which clicks off and on in the event of a short circuit in the wiring. The circuit breaker is located at the rear end of the main light switch and no replaceable fuse is required.

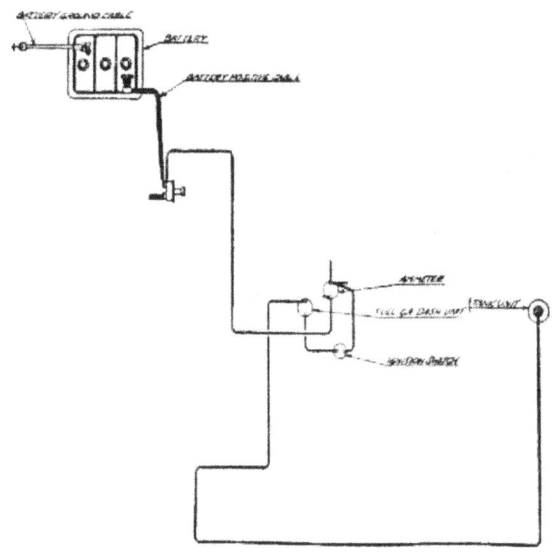

FIG. 2—FUEL GAUGE CIRCUIT

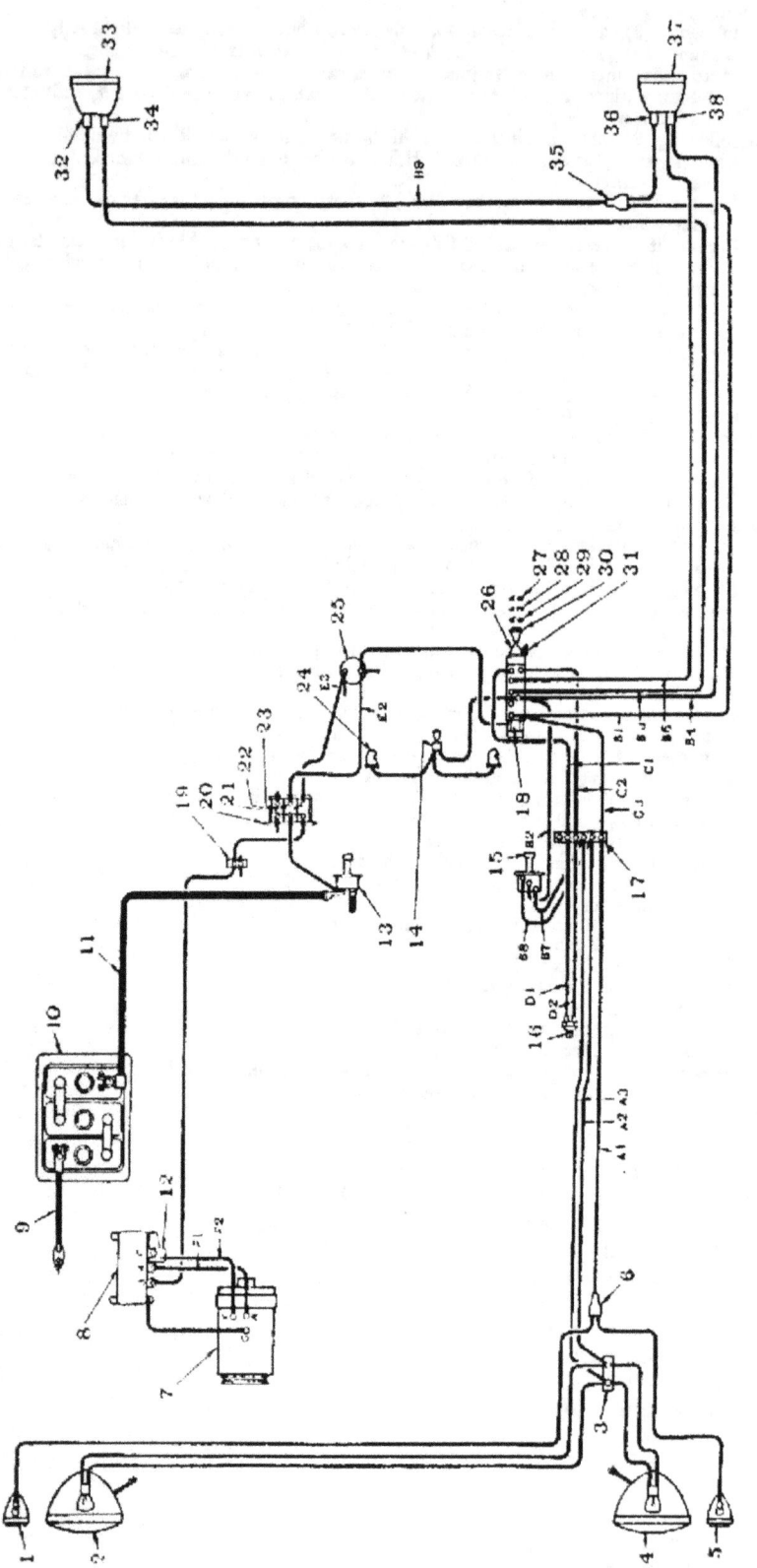

FIG. 3—LIGHTING SYSTEM

Main Light Switch

The main light switch Fig. 4 has four positions. When the switch button is all the way in, all lights are turned off. Pulling the switch out to the first position turns on the blackout lamps, the blackout tail lamp and also connects the circuit with a blackout stop lamp on the right side which is operated through the stop light switch when the brakes are applied.

To turn on the service headlamps, it is necessary to push down on the lock-out control button and while holding it down, pull the switch button out to the next position.

During the day to cause the service stop light only, to operate upon brake application, pull the knob out to the last position. This should be done whenever the vehicle is used under ordinary driving conditions.

When installing a new light switch refer to Fig. 4 and wiring diagram, Fig. 3 which will be helpful in determining the proper wires to install on terminals as marked.

The upper and lower headlight beams are controlled by a foot switch located on the toe board at the left side.

The instrument panel lights can only be turned on when main light switch is in the service position.

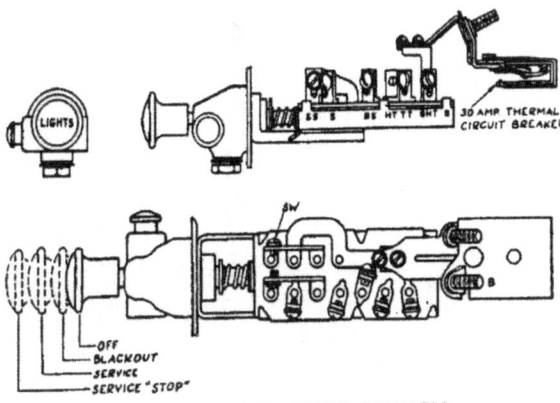

FIG. 4—MAIN LIGHT SWITCH

Stop Light Switch

Stop light switch is located in front end of brake master cylinder. The switch is a diaphragm type and closes the circuit when pressure is applied to the brakes forcing the fluid against the diaphragm which closes the circuit. When switch becomes inoperative it is necessary to install a new switch.

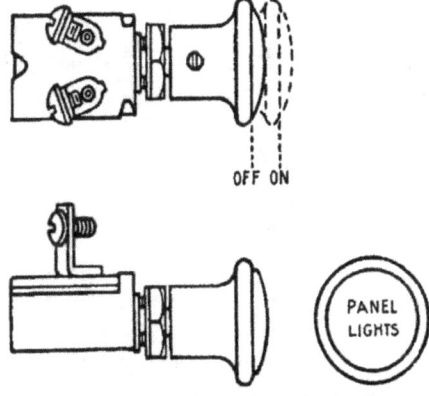

FIG. 5—PANEL LIGHT SWITCH

Panel Light Switch

The panel light switch, Fig. 5, controls the panel lights only when the main light switch is in "Service Position." Pull out on the switch knob to light the panel lights.

Head Lamps

The head lamps, Fig. No. 6, are the sealed beam type, in which the reflector, bulb and lens form a sealed unit and can only be replaced as a unit.

The lower beam filament is positioned slightly to one side of the focal point in the reflector, this results in deflecting the lower beam to the right side to illuminate the side of the road when meeting other vehicles on the highway.

To replace a burned out sealed beam unit remove door clamp screw and remove the door, No. 2, remove sealed beam assembly No. 1 and remove from connector at the rear of the unit. Install a new unit by reversing the above operations.

When a sealed beam unit has been replaced, check the aim of the head lamps.

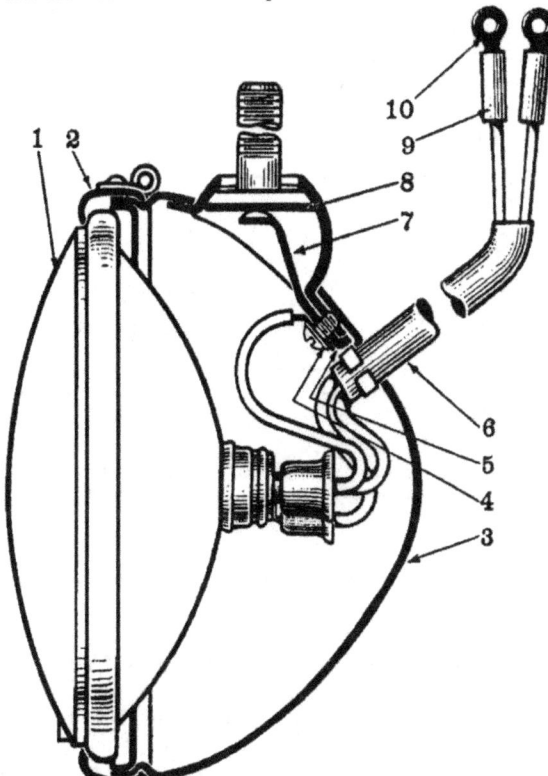

FIG. 6—HEADLAMP

No.	Willys Part No.	Ford Part No.	Name
1	A-1033	GPW-13007	Headlamp Seelite Unit Assembly
2	A-1036	GPW-13043	Headlamp Door Assembly
3	A-5586	GPW-13012	Headlamp Housing Sub—Assembly
4	A-1032	27693-S	Mounting Bolt Retainer Screw
5	52221	34803-S2	Mounting Bolt Retainer Lockwasher
6	A-1362	GPW-13076	Wire Assembly—Left (A-1363—Right)
7	A-1031	GPW-13015	Mounting Bolt Retainer
8	A-1361	GPW-13022	Mounting Bolt
9	306688	B-14455	Insulating Sleeve
10	307556	B-14463	Terminal

Head Lamp Aiming

Headlights may be aimed by use of an aiming screen or wall, Figure 7, providing a clear space of 25 feet from the front of the headlights to the screen or wall is available.

The screen should be made of a light colored material and should have a black center line for use in centering the screen with the vehicle. The screen should also have two vertical black lines, one on each side of the center line at a distance equal to lamp centers.

Place the vehicle on a level floor with the tires inflated to recommended specification. Set the vehicle 25 feet from the front of the screen or wall so that the center line of the truck is in line with the center line on the screen. To determine the center line of the vehicle, stand at the rear and sight through the windshield down across cowl and hood.

Measure from the floor to the center of the head lamp and mark a horizontal line on the screen 7" less.

Turn on the headlight upper beam, cover one lamp and check the location of the upper beam on the screen. The center of the hot spot should be centered on the intersection of the vertical and horizontal lines on the screen as shown in Fig. 7.

If aim is incorrect, loosen the nut on the mounting bolt and move the head lamp body on its ball and socket joint until the beam is aimed as described, then tighten.

Cover the head lamp just aimed and adjust the other in the same manner.

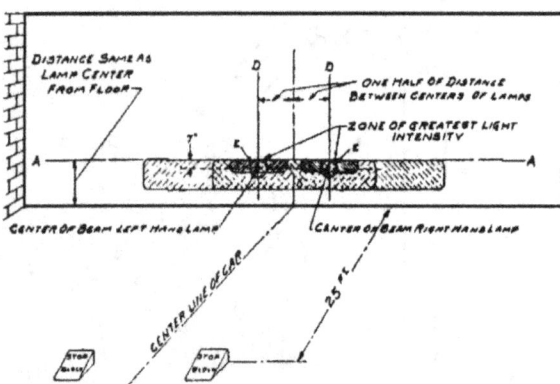

FIG. 7—HEADLIGHT AIMING CHART

Blackout Lamps

The blackout light, Fig. 8, is based on the principle of polarized light. The lens is so designed that only horizontal light beams are allowed to penetrate or pass thru the lens. This means the vertical light beams are blocked by the lens, therefore light rays cannot be seen from a point above the horizontal.

To replace lamp bulb remove door Screw No. 2 in lower side of rim—remove door No. 3 by slipping off bottom and tilt outward and up from lamp body. The door and lens are one unit. Replace Bulb (Mazda No. 63) and inspect gasket; if damaged replace and install door.

Tail and Stop Lamps

The tail and stop lamps, Fig. 9, consist of two separately sealed units placed in the Lamp Body.

The upper stop light or service unit consists of lens, gasket, reflector and (21-3 C.P. Bulb, L.H., No. 8, Fig. 9—R.H. 3 C.P., No. 4, Fig. 9) sealed as a unit. When Bulb fails entire Service unit must be replaced.

The Lower Tail lamp unit, No. 3 and 7, consists of lens, gasket, reflector and 3 C.P. Bulb sealed as a unit. When Bulb fails entire unit must be replaced.

To replace a unit remove the two screws in lamp door. Remove door then each unit can be pulled out of socket in lamp Body.

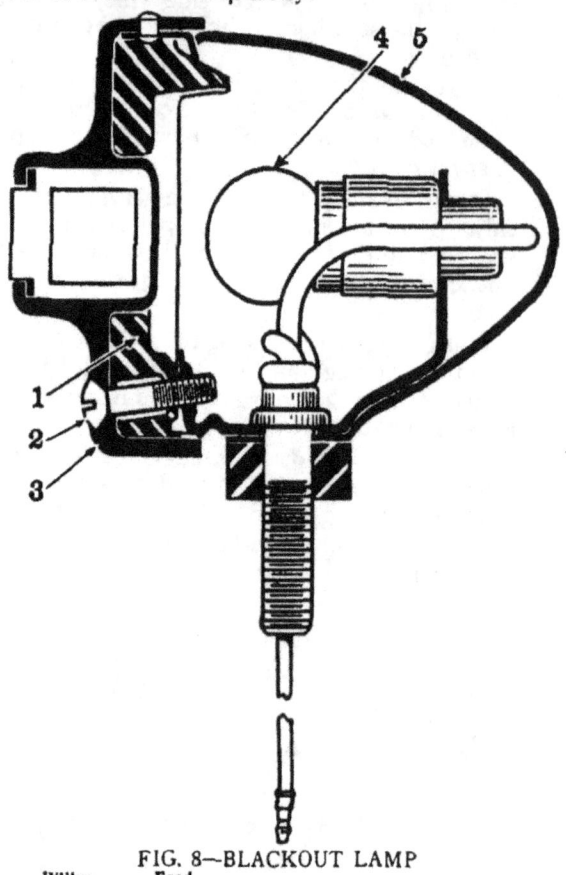

FIG. 8—BLACKOUT LAMP

No.	Willys Part No.	Ford Part No.	Name
1	A-1071	GP-13209-B2	Door Gasket
2	A-1072	28378-S2	Door Screw
3	A-1070	GP-13210-B	Door Assembly
4	51804	B-13466	Bulb
5	A-1439	GPW-13217	Housing Assembly—Left (Includes Wire Assembly) (Willys A-1440—Ford GPW-13216-B Right)

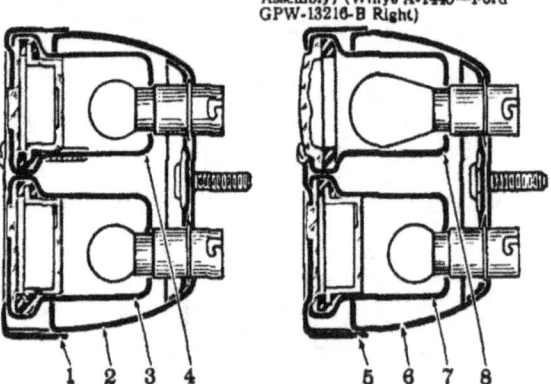

FIG. 9—TAIL LAMPS

No.	Willys Part No.	Ford Part No.	Name
1	A-1079	GP-13449-A	Door—Tail and Stop Lamp Assembly (Right)
2	A-1073	GPW-13408-B2	Housing Sub-assembly
3	A-1075	GP-13491-A2	Lower Tail Lamp Unit Assembly
4	A-1078	GP-13485-A2	Upper Stop Lamp Unit Assembly—Tail and Stop Lamp Assembly (Right)
5	A-1076	GP-13448-B2	Door—Tail and Stop Lamp Assembly (Left)
6	A-1073	GPW-13408-B	Housing Sub-assembly
7	A-1075	GP-13491-A	Lower Tail Lamp Unit Assembly
8	A-1074	GPW-13494-A2	Upper Service Assembly—Tail and Stop Lamp Assembly (Left)

IGNITION SYSTEM

The power in an internal combustion engine is derived from burning a fuel and air mixture in the engine cylinders under compression. In order to ignite these gases a spark is made to jump a small gap in the spark plug within each combustion chamber. The ignition system furnishes this spark. The spark must occur in each cylinder at exactly the proper time and the spark in the various cylinders must follow each other in sequence of firing order. To accomplish this the following parts are used:

The battery, which supplies the electrical energy;

The ignition coil, which transforms the battery current to high-tension current which can jump the spark plug gap in the cylinders under compression;

The distributor, which delivers the spark to the proper cylinders and incorporates the mechanical breaker, which opens and closes the primary circuit at the proper time;

The spark plugs, which provide the gap in the engine cylinders;

The wiring, Fig. 10, which connects the various units;

The ignition switch to control the battery current when it is desired to start or stop the engine.

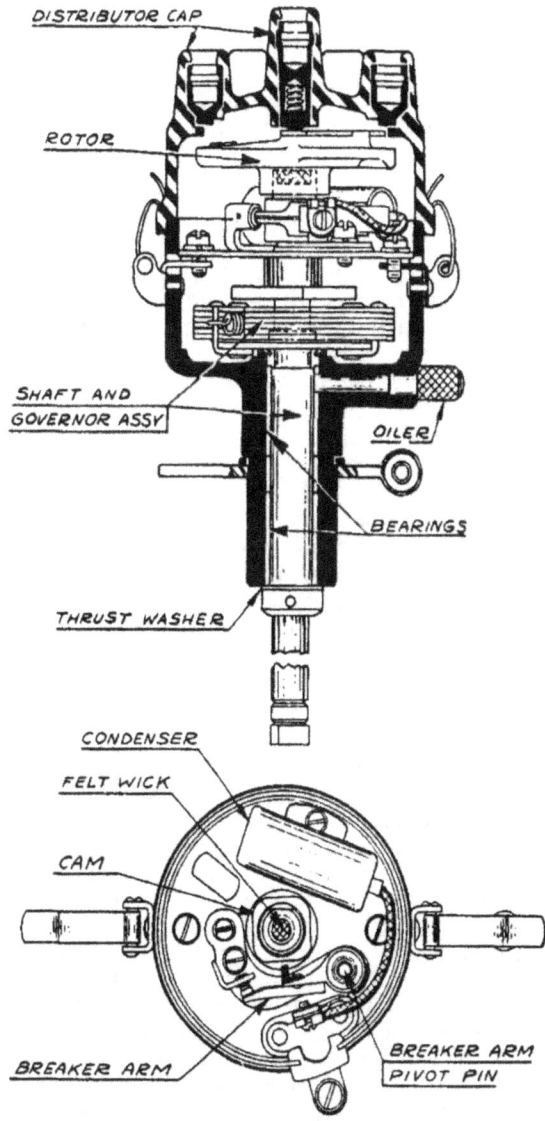

FIG. 11—DISTRIBUTOR

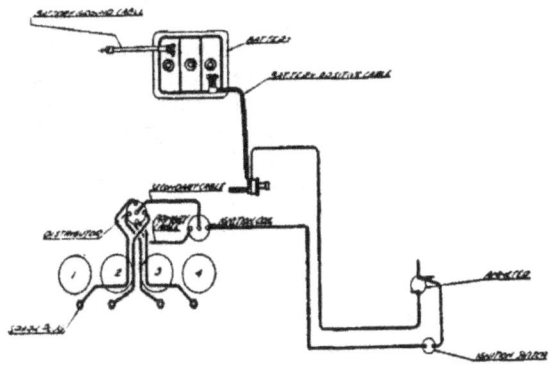

FIG. 10—IGNITION WIRING

Distributor

The distributor, Fig. 11 is mounted on the right hand side of the engine and is operated by a coupling on the oil pump shaft, driven by a spiral gear on the camshaft. The spark control is fully automatic, being operated by two counterweights pivoted on a plate which advances the timing automatically as the engine speed increases.

Distributor Overhaul

To remove distributor from engine the following procedure should be followed:

1. Remove high-tension wires from the distributor cap terminal towers, noting the order in which they are assembled to assure proper installation on reassembling. No. 1 spark plug terminal tower in distributor cap is the lower right hand tower at distributor cap spring clip. Starting with this tower the wires should be installed in a counter-clockwise direction 1-3-4-2.

2. Remove the primary lead from the terminal post at the side of the distributor.

3. Snap off the two distributor cap springs and lift the distributor cap off of the distributor housing.

4. Note the position of the rotor in relation to the base. This should be remembered to facilitate reinstalling and timing.

5. Remove the screw holding the distributor to the crankcase and lift the distributor from the engine.

6. Wash all parts thoroughly in a suitable cleaning fluid.

Distributor Cap

The distributor cap should be visually inspected for cracks, carbon runners, evidence of arcing, and corroded high-tension terminals. If any of these conditions exist, the cap should be replaced.

Rotor

Inspect the rotor for cracks or evidence of excessive burning at the end of the metal strip.

After a distributor rotor has had normal use, the end of the contact will become burned. If burning is found on top of the strip, it indicates the rotor is too short and needs replacing. Usually when this condition is found, the distributor cap insert will be burned on the horizontal face and the cap will also need replacing.

Distributor Points

The contacts should be clean and not burned or pitted. The contact gap should be set at .020" and should be checked with a wire gauge and readjusted if necessary by loosening the lock screw then turn the eccentric head screw. After adjusting, tighten the lock screw and then recheck the gap. If new contacts are installed they should be aligned so as to make contact near the center of the contact surfaces. Bend the stationary contact bracket to be sure of proper alignment and then recheck the gap.

The contact point spring pressure is very important and should be between 17 to 20 ounces. Check with spring scale hooked in the breaker arm at the contact and pull in a line perpendicular to the breaker arm. Make the reading just as the points separate. This pressure should be within the limits, too low a pressure will cause missing at high speeds, too high a pressure will cause excessive wear on the cam, block and points. Adjust the point pressure by loosening the screw holding the end of the contact arm spring and slide the end of the spring in or out as necessary. Retighten the screw and recheck the pressure.

Check the condenser, it should show a capacity of .18 to .26 microfarads. Check the condenser lead for broken wires or frayed insulation, clean and tighten the connections of the terminal posts. Be sure the condenser is firmly mounted to the distributor plate.

Governor Mechanism

The governor should be checked for free operation holding the distributor shaft and turn the cam to the left as far as it will go and release. The cam should immediately return to its original position with no drag or restrictions. Inspect the distributor shaft bearing in housing, also the shaft friction spring on end of shaft inserted into the coupling on the oil pump shaft; if damaged replace.

Setting Ignition Timing

Remove all spark plugs from engine, reinstall No. 1 spark plug finger tight. Loosen screw holding timing hole cover to flywheel housing which is located just under the starting motor on the right hand side of the engine, slide cover to side. Rotate engine crankshaft until No. 1 piston is coming up on the compression stroke, remove spark plug and rotate crankshaft slowly until the marking on flywheel "IGN" appears in the center of the timing hole in flywheel housing. Fig. 15 in Engine Section.

Place distributor rotor at No. 1 tower in distributor cap so that the points are just breaking.

Place the distributor in place on engine. When end of shaft enters driving collar on oil pump, rotate distributor shaft back and forth until driving lug on end of shaft enters the slot in coupling, then push distributor assembly down. Install holddown screw. Connect primary wire from coil to distributor. Rotate distributor body until points are just breaking, then lock in place by clamp screw. Install spark plugs and wires to distributor cap terminal towers starting with No. 1, installing in counter-clockwise direction, in the following order: 1-3-4-2. Start engine and run until it is fully warmed up, then recheck timing with Neon Timing Light. Accelerate engine and note automatic advance action.

Note: For 68 octane fuel (gasoline) set timing at top center (TC).

Generator

The generator, Fig. 12 is an air-cooled, 40 ampere, two brush type, and cannot be adjusted to increase or decrease output, as this is accomplished by use of a three-unit voltage regulator, consisting of a cutout relay, current limiting regulator and voltage regulator.

A periodic inspection should be made of the charging circuit. Under normal conditions an inspection of the generator should be made each 6,000 miles, however, the interval between these checks will vary depending upon the type of service. Dirt, dust and high speed operation are factors which contribute to increased wear of the bearings, brushes and commutator. Before assuming that any difficulty lies in the generator, a visual inspection should be made of all wiring, Fig. 13 to be sure that there are no broken wires and that all connections are clean and tight. Due attention should also be given to the Voltage Regulator, as covered under heading "Regulator" in this section. Bracket bolt torque wrench reading, 31-35 ft. lbs.

MAINTENANCE PROCEDURE

1. Commutator

If the commutator is dirty or discolored, it can be cleaned by holding a piece of No. 00 sandpaper against it while running the generator slowly. Blow the sand out of the generator after cleaning the commutator. If the commutator is rough or worn, the generator should be removed, the armature taken out, and the commutator turned down in a lathe. After turning the commutator, the mica should be undercut to a depth of $\frac{1}{32}$".

To test the armature for ground connect one prod of test set to the core or shaft, (not on bearing surfaces) and touch a commutator segment with the other. If the lamp lights, the armature winding is grounded and the armature should be replaced.

To test for short in armature coils a growler is necessary. Place the armature on the growler and hold a thin steel strip on the armature core. The armature is then rotated slowly by hand, and if a shorted coil is present, the steel strip will vibrate.

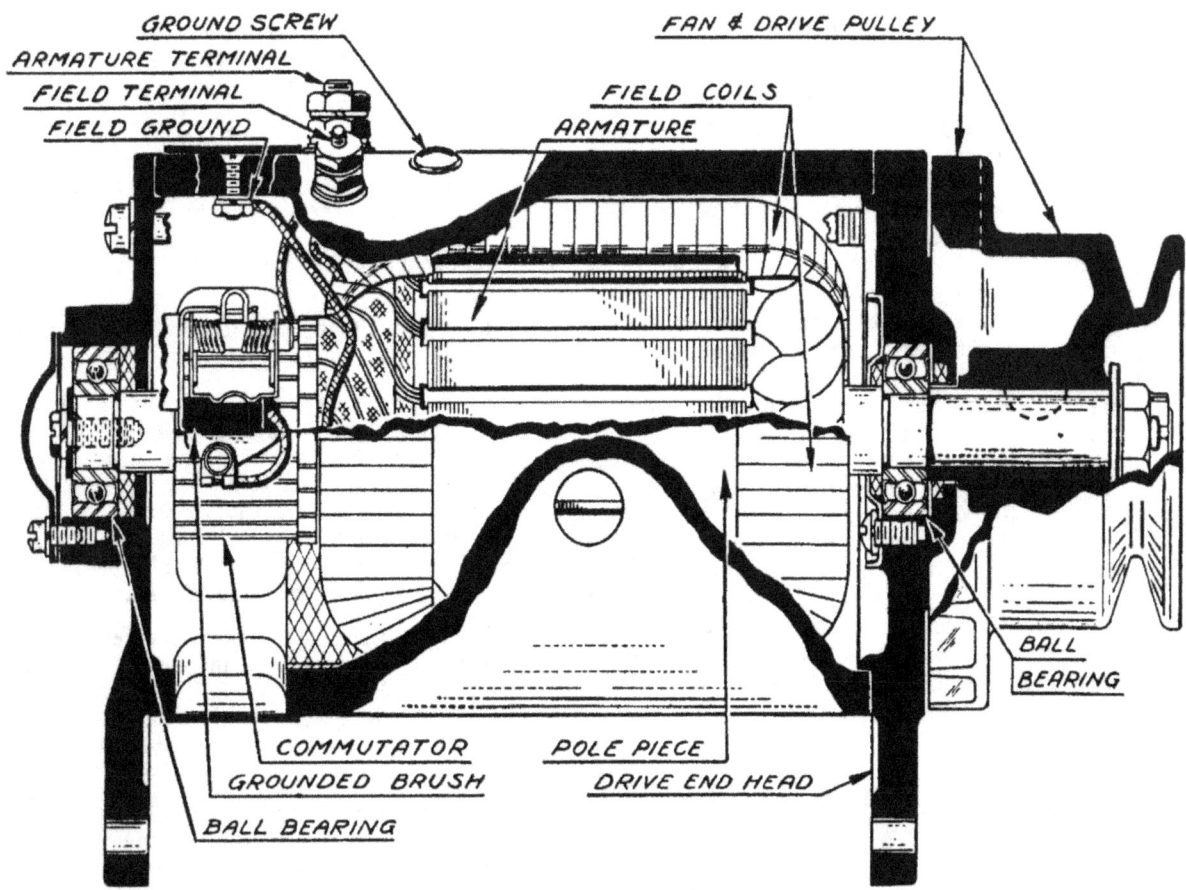

FIG. 12—GENERATOR

2. Brushes

The brushes should slide freely in their holders. If the brushes are oil soaked or if they are worn to less than one half of their original length, they should be replaced.

When replacing brushes, it is necessary to seat them so that they have 100% surface contact on the commutator. The brushes should be sanded to obtain this fit. This can be done by drawing a piece of No. 00 sandpaper around the commutator with the sanded side against the brush. After sanding the brushes, blow the sand and carbon dust out of the generator.

3. Brush Spring Tension

The brush spring tension should be checked. If the tension is excessive, the brushes and commutator will wear rapidly, while if the tension is low, arcing between the brushes and commutator will burn the commutator and reduce output. The brush spring tension is 64-68 ounces with new brushes.

4. Field Coils

Using test prods check the field coils for both open and ground. To test for open coil, connect the prods to the two leads of each coil. If the lamp fails to light, the coil is open and should be replaced.

To test for grounds, disconnect field coil ground terminal, place one prod on ground and the other on the field coil terminal. If a ground is present the lamp will light and the coil should be replaced.

5. Brush Holders

With test prods, check the insulated brush holder to be sure it is not grounded.

Touch the insulated brush holder with one prod and a convenient ground on the end plate, with the other prod. If the lamp lights, it indicates a grounded brush holder.

Inspect the brush holders for distortion and improper alignment. The brushes should swing or slide freely and should be perfectly in line with the commutator segments.

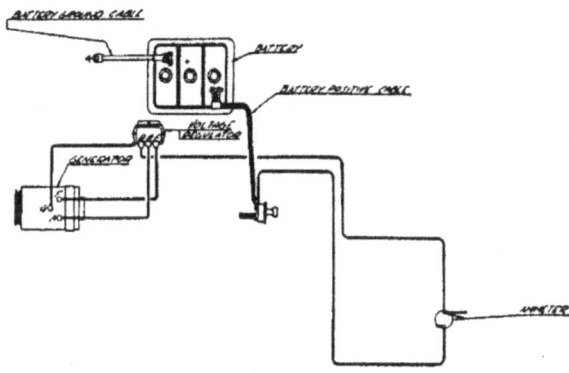

FIG. 13—GENERATOR WIRING CIRCUIT

REGULATORS

Regulators

The generator on this vehicle is controlled by a regulator unit, Fig. 14-15 which contains a voltage regulator, current limiting regulator, and circuit breaker.

The voltage regulator controls the generator voltage and does not allow it to rise above a value determined by the voltage regulator setting. This prevents overcharging of the battery.

The current regulator controls the maximum generator output of 40 amperes and does not allow the output to exceed the value determined by the current regulator setting. This prevents damage to the generator due to an overload.

The circuit breaker automatically closes the circuit between the generator and battery when the generator voltage rises above that of the battery, and automatically opens the circuit when the generator voltage falls below that of the battery.

The terminals of the regulator unit are marked and care should be used in making connections, otherwise serious damage may result.

Quick Check

The following checks may be made to determine whether or not the units are operating normally. If not, the checks will indicate whether the generator or regulator is at fault, so that proper correction can be made:

A Fully Charged Battery and a Low Charging Rate

A fully charged battery and a low charging rate indicate normal current regulator operation. To check the current regulator, remove the battery wire from the battery terminal of the regulator. Connect the positive lead of an Ammeter to the battery terminal of the regulator and the negative lead to the battery wire with the ignition switch in the off position. Push in on the starting switch and crank the engine about 30 seconds. Then start engine and with it running at a medium speed turn on the lights and other electrical accessories and note quickly the generator output, which should be the value for which the current regulator is set.

Turn off the lights and other electrical accessories and allow the engine to continue to run. As soon as the generator has replaced in the battery the current used in cranking, the voltage regulator, if operating properly, will taper the output down to a few amperes.

A Fully Charged Battery and a High Charging Rate

Disconnect the field wire from the field terminal of the regulator. This opens the generator field circuit and the output should immediately drop off. If it does not, the generator and field wires are shorted together in the wiring harness. If the output drops off to zero with the field lead disconnected, the trouble has been isolated in the regulator. Reconnect the field lead on the field terminal of the regulator.

Remove the regulator cover and depress the voltage regulator armature manually to open the points. If the output now drops off, the voltage regulator unit has been failing to reduce the output as the battery came up to charge, a voltage regulator adjustment is indicated.

If separating the voltage regulator contact points does not cause the output to drop off, the field circuit within the regulator is shorted and the regulator should be replaced.

With a Low Battery and Low or no Charging Rate

Check the entire generator wiring circuit for loose connections, corroded battery terminals, loose or corroded ground straps. High resistance resulting from these conditions will prevent normal charge from reaching the battery and possibly cause burned out lamp bulbs when in use. If the entire charging circuit is in good condition, then either the regulator or generator is at fault.

With a jumper wire connect the field and armature terminals together, increase the generator speed and check the output. If the output increases, the regulator requires attention. If the output does not increase, a further check is necessary.

If the generator output remains at a few amperes with the field and armature terminals connected together, the generator is at fault and should be checked.

If the generator does not show any output at all, either with or without the field and armature terminals together, flash the armature terminal on the generator to ground with a screw driver or a pair of pliers, with the generator operating at medium speeds. If a spark does not occur the trouble has been isolated in the generator and it should be removed and repaired. If a spark does occur, likely the generator can build up, but the circuit breaker is not operating to let the current flow to the battery, due to burnt points, points not closing, open voltage windings, or too high a voltage setting of the cutout.

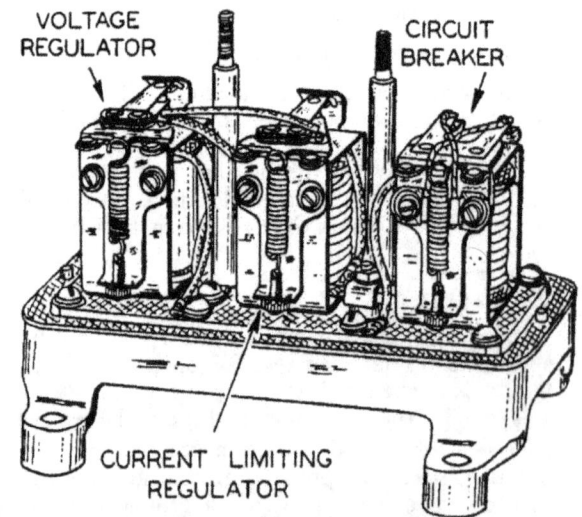

FIG. 14—VOLTAGE REGULATOR

Adjustment

To accurately adjust the regulator, it is essential to use a precision ammeter, voltmeter, thermometer and a fully charged battery. If battery in vehicle is not fully charged, it should be substituted with a fully charged battery in order to obtain a satisfactory setting of the unit.

The engine should be started and allowed to run for a period of fifteen minutes at approximately 25 to 30 miles per hour with the hood up before taking any meter readings.

The thermometer should be placed so that the bulb is approximately two inches from the side of the regulator so that the temperature can be taken while checking units. Set engine speed so that the generator charges 20 amperes, the voltmeter should show a reading according to the following specifications:

Temperature Fahr.	Volts
50°	7.41
60°	7.38
70°	7.35
80°	7.32
90°	7.29
100°	7.26
110°	7.23
120°	7.20

Tolerance—plus or minus .15 volts.

Circuit Breaker

Disconnect the battery wire from the battery terminal of the regulator. Connect the positive lead of the ammeter to the battery terminal of the regulator and the negative lead to the battery wire. Connect the positive lead of the voltmeter to the armature terminal of the regulator and the negative lead to the ground or regulator base. Gradually increase the engine speed, noting the voltage at which the points close. This should be 6.4 to 6.6 volts. Slowly decrease the engine speed, noting the discharge of current necessary to open the cutout points. This should be 0.5 to 6.0 amperes.

Check the armature air gap with the points open. Use a flat gauge .0595" to .0625". Insert between magnet core and the armature on contact side of brass pin in core. To adjust, bend armature stop.

Check the gap of the contact points when open. This should be .015" minimum, but will possibly be more than this in actual adjustment. Adjust by bending the supporting arms of the stationary points, be sure that the points are perfectly aligned.

The closing voltage of the circuit breaker may be adjusted by adjusting the screw holding the lower end of the spring. The point opening amperage can be adjusted by raising or lowering the stationary points by bending the supporting arms of the points. Be sure that there is a minimum point gap of .015".

Voltage Regulator

Connect the positive lead of voltmeter to the battery terminal and the negative lead to a ground on regulator base.

Run the engine at a speed equivalent to approximately 30 miles per hour and check the voltmeter reading, which should be in accordance with the specifications of temperature and volts (under

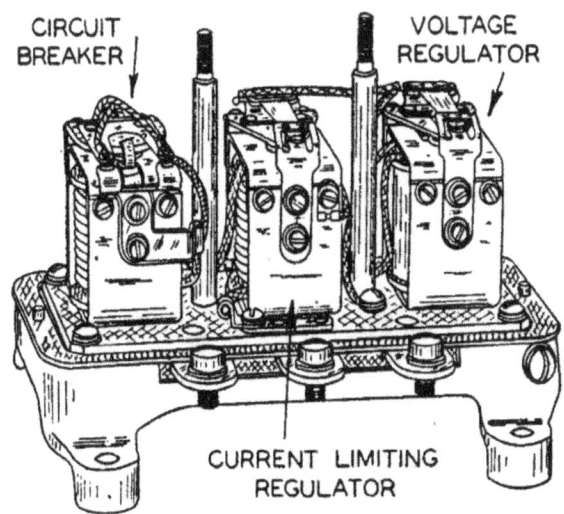

FIG. 15—VOLTAGE REGULATOR

heading "Adjustments") with a generator charging rate of 20 amperes.

Check armature air gap using .040" to .042" pin gauge. Test with pin gauge between magnet core and armature. This measurement should be taken on the contact side and next to the brass armature stop pin. To test connect a 3 candlepower test light in series with the armature and field terminals and a battery. With the low limit pin gauge in place depress the armature and the light should go out. With the high limit pin gauge in place depress the armature and the light should stay lit. To adjust slightly loosen the screw holding the upper point bracket, raise or lower bracket until correct gap is obtained. Keep points in perfect alignment when adjusting.

Check and see that the spring upon which the movable contact is mounted is straight and that it is approximately parallel with the armature.

The gap between the contact spring (upper) and armature stop is .010" to .016" with armature depressed.

Check the point gap with armature against stop pin. Hold the armature down with two fingers being careful not to apply pressure to the spring supporting the upper point. A .010" (minimum) feeler gauge should be used between points. Too much variation indicates wrong length of the brass armature stop pin and a new unit will be required.

To adjust the regulator increase or decrease the armature spring tension by adjusting the screw which holds the lower end of the spring.

Current Limiting Regulator

To adjust current limiting regulator, remove the battery wire from the battery terminal of the regulator and connect the positive lead of ammeter to the battery terminal of the regulator and the negative lead to the battery wire. Turn on the lights and other electrical accessories, then increase the engine speed until output remains constant. The ammeter reading with the unit at operating temperatures should be 40 amperes.

To check the armature air gap and point gap—refer to instructions under heading "Voltage Regulator." The armature air gap is .047" to .049".

The contact point gap is .010" minimum. The

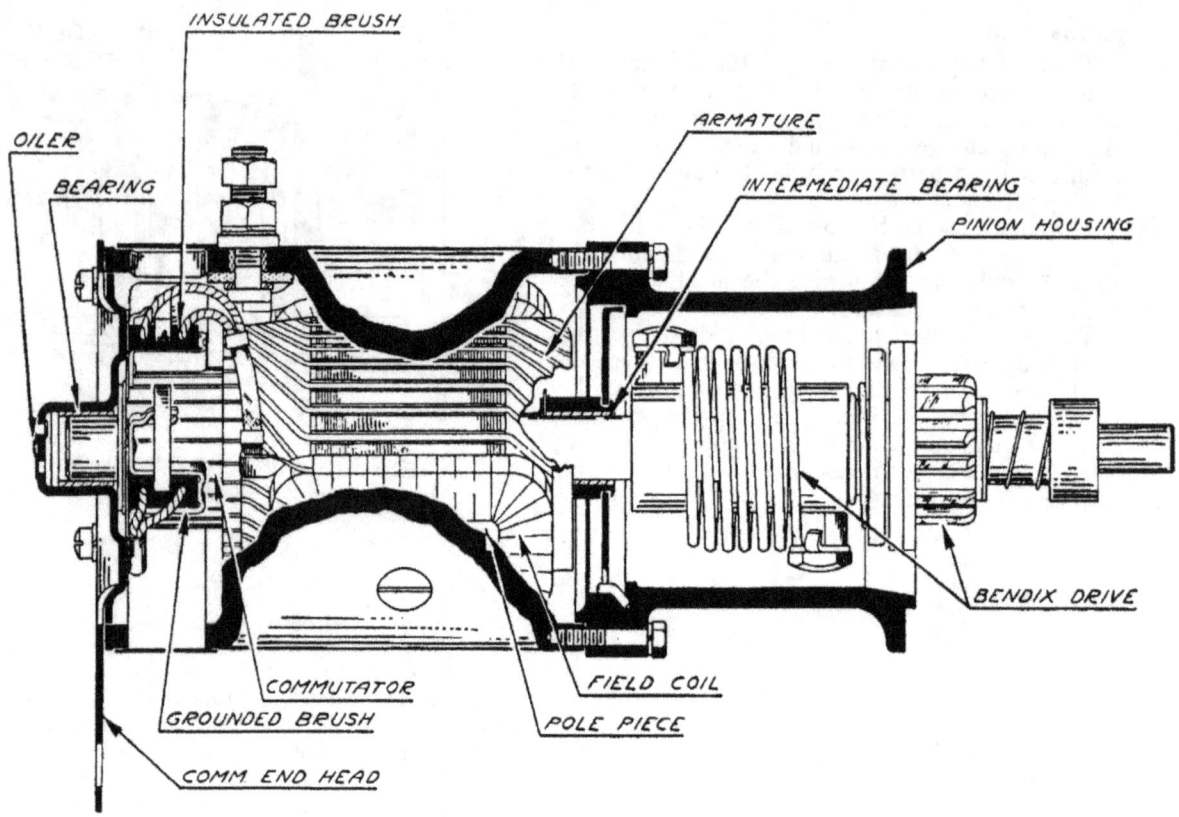

FIG. 16—STARTING MOTOR

gap between the contact spring and armature stop is .010" to .016" with armature depressed.

To adjust current setting vary the armature spring tension by adjusting the screw which holds the lower end of the armature spring.

STARTING MOTOR

The starting motor Fig. 16 is similar in construction and in appearance to the generator, but the design of the parts are different. Both motor and generator require a frame, field coils, armature, brushes.

A starting motor of this type requires very little attention except regular lubrication and periodic inspection of the brushes and commutator.

A visual inspection should be made of all wires and see that all connections in the circuit are clean and tight, Fig. 17. Mounting screws, torque wrench reading, 31-35 ft. lbs.

1. **Commutator**

   Check the commutator for wear or discoloration. If found to be dirty or discolored, it can be cleaned with No. 00 sandpaper. Blow the sand out of the motor after cleaning the commutator. If the commutator is rough or worn, the armature should be removed and the commutator turned down in a lathe.

2. **Brushes**

   The brushes should slide or swing freely in their holders and make full contact on the commutator. Worn brushes should be replaced.

3. **Brush Spring Tension**

   This tension should be 42 to 53 ounces with new brushes. Measure the tension with a spring scale hooked under the brush spring at end, and pull on a line parallel to the face of the brush taking the reading just as the spring leaves the brush.

4. **Armature**

   The armature should be visually inspected for mechanical defects before being checked for shorted or grounded coils.

   For testing armature circuits, it is advisable to use a set of test prods.

   To test the armature for grounds, touch one prod to a commutator segment and touch the core or shaft with the other prod. Do not touch the points or prods to the bearing or brush surface, as the arc formed will burn the smooth finish. If the lamp lights, the coil connection to the commutator segment is grounded.

   To test for shorted armature coils, a growler is necessary. Place the armature on growler with a steel strip held on the armature core, rotate the armature slowly by hand. If a shorted coil is present, the steel strip will become magnetized and vibrate.

   If an armature is shorted or grounded, it will be necessary to install a new armature.

5. **Field Coils**

   Using same test prods, check the field coils for both open circuit and ground. To test for grounds, place one prod on the motor frame or pole piece and touch the other to the field coil terminal. If a ground is present, the lamp will light.

   To test for open circuit, place the prods on the field coil terminal and across each coil separately. If the lamp does not light, the coil circuit is open.

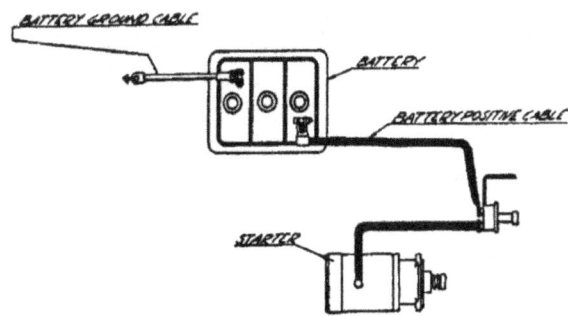

FIG. 17—STARTING MOTOR WIRING CIRCUIT

6. Brush Holder

Using test prods, touch the insulated brush holder with one prod and a convenient ground on the plate with the other. If the lamp lights, it indicates a grounded brush holder and a new brush holder will have to be installed.

Bendix Assembly

The Bendix Drive Fig. 18 is designed so that when the starting motor is energized, centrifugal force sends the counter-weighted drive pinion gear into engagement with the teeth on the flywheel. When the engine starts and the speed of the engine exceeds the comparable speed of the starting motor, the Bendix Drive pinion is forced out of engagement with the flywheel.

There are two types of Bendix Drives and springs, right hand and left hand. The type used on this starting motor is of the right hand type.

To determine right or left hand Bendix Drive, turn drive pinion so that the threads on shaft will show, note the spiral of the thread; right hand spiral, right hand drive; left hand spiral, left hand drive.

To determine a right or left hand spring, note the spiral of the coil; if to the right, it is a right hand spring; if to the left, it is a left hand spring.

If upon inspection of the Bendix Drive, the spring shows signs of being distorted, a new spring should be installed.

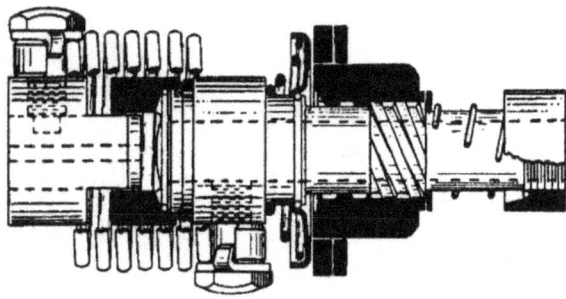

FIG. 18—BENDIX ASSEMBLY

Starting Switch

The starting switch is mounted on the toe board to the right of the accelerator pedal; pressing the starter switch closes the starter circuit and operates the starter. If the starting motor does not rotate, then the difficulty is probably a loose wire, poor ground, low battery or poor brush contact.

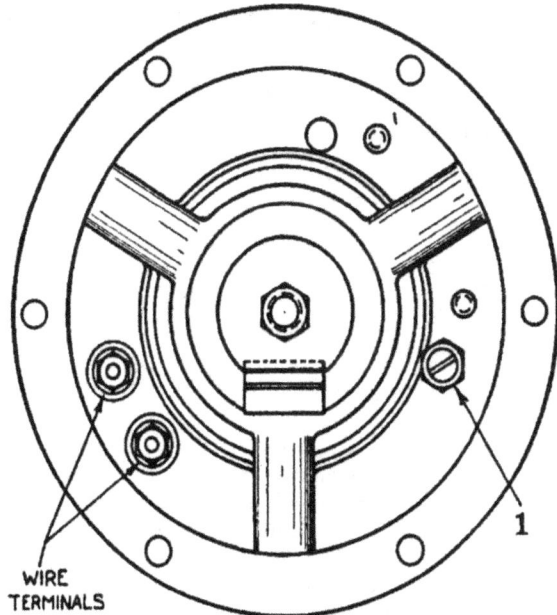

FIG. 19—HORN (Rear)

Horn and Horn Wire

The horn, Fig. 19 is the micro-vibrating type mounted on the dash under the hood. No. 1 indicates the horn adjusting screw. To adjust tone of horn, loosen the lock nut and turn the screw until the proper tone is obtained. It is best to have the engine running so the generator is charging when making this adjustment because the generator delivers 8 volts as compared to the battery 6 volts. This affects the horn tone.

The horn wire through the steering post connects to an insulator sleeve with brush contact where horn wire attaches to jacket tubing.

Whenever it is necessary to replace the horn wire, it will be necessary to remove the steering post jacket tubing. The wire may be removed by unsoldering it from the contact sleeve on the steering tube. When replacing the wire be sure to use a non-corrosive soldering flux when soldering the wire to the contact sleeve.

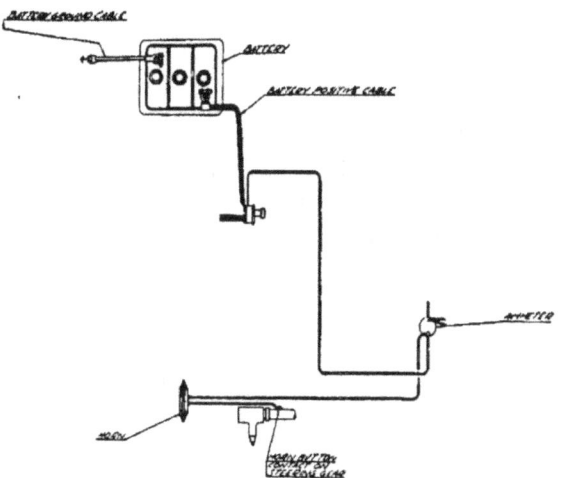

FIG. 20—HORN WIRING CIRCUIT

ELECTRICAL TROUBLES AND REMEDIES

| SYMPTOMS | PROBABLE REMEDY |

Battery Discharged:

Short in Battery Cell	Replace Battery
Short in Wiring	Check Wiring Circuit
Generator Not Charging	Inspect Generator and Fan Belt
Loose or Dirty Terminals	Clean and Tighten
Excessive Use of Starter	Tune Engine
Excessive Use of Lights	Check Battery

Generator:

Low Charging Rate—

Dirty Commutator	Clean Commutator
Poor Brush Contact	Install new Brushes
Voltage Regulator Improperly Adjusted	Adjust
High Resistance in Charging Circuit	Clean and Tighten Terminals
Ground Strap Engine to Frame Broken	Replace
Loose or Dirty Terminals	Clean and Tighten

Too High Charging Rate

| Current Regulator Improperly Adjusted | Adjust |
| Short in Armature | Replace |

Starting Motor:

Slow Starter Speed

Discharged Battery or Shorted Cell	Recharge
Ground Strap Engine to Frame	Clean Terminals and Tighten
Loose or Dirty Terminals	Clean and Tighten
Dirty Commutator	Clean With No. 00 Sand Paper
Poor Brush Contact	Install New Brushes
Worn Bearings	Replace
Burnt Starter Switch Contacts	Replace Switch

Distributor:

Hard Starting:

Distributor Points Burnt or Pitted	Clean Points or Replace
Breaker Arm Stuck on Pivot Pin	Clean and Lubricate
Breaker Arm Spring Weak	Replace
Points Improperly Adjusted	Adjust .020"
Spark Plug Points Improperly Set	Adjust .030"
Spark Plug Wire Terminals in Distributor Cap Corroded	Clean
Loose Terminals	Check Circuit
Loose or Dirty Terminals Ground Strap Engine to Frame	Clean and Tighten
Condenser Defective	Replace Condenser
Improper Ignition Timing	Set Timing

Lights:

Burn Dim

Loose or Dirty Terminals	Clean and Tighten
Leak in Wires	Check Entire Circuit for Broken Insulation
Poor Switch Contact	Install New Switch
Poor Ground Connection	Clean and Tighten
Aim Headlamp Beams	Use Chart

Horn Fails to Blow

1. Broken or loose electrical connection............ 1. Check wiring and connections at horn button and battery, making sure all are clean and tight.

2. Battery low or dead............................. 2. Check battery with hydrometer, should read at least 1200.

3. Contact points in horn not adjusted............. 3. Loosen locknut and turn adjusting screw to right or left until a clear steady tone is obtained, then tighten locknut, holding screw in proper position.

4. Contact points burnt or broken off............. 4. Replace parts necessary and adjust horn.

Horn Blows Unsatisfactory Tone

1. Poor electrical connection.....................
2. Battery low.............................
3. Loose cover and bracket screws................
4. Voltage at horn too high or too low ....
5. Contact points are not properly adjusted........

1. Check wiring and connections at horn, horn button and battery.
2. Check with hydrometer, should read 1200.
3. Draw cover screws and center nut tight, tighten bracket screws solidly both at horn and dash.
4. Check with voltmeter, should measure 5.5-6.5 volts at horn with horn sounding and engine running so generator is charging battery.
5. Loosen locknut and turn contact adjusting screw to right or left until a clear steady tone is obtained, then tighten locknut.

Excessive Radio Interference

1. Due to Ignition ....
   1. Check distributor, spark plugs and suppressors.
   2. Tighten braided bonding straps.
   3. Tighten radiator and fender supporting bolts.

2. Due to Generator ....
   1. Tighten regulator to generator bond.
   2. Defective commutator, brushes or holders.
   3. Discharge battery causing high discharging rate.

3. Due to Erratic Noises ....
   1. Failure of high tension insulation.
   2. Loose wiring connections or corroded distributor cap towers.
   3. Defective switches or gauges.

ELECTRICAL SYSTEM SPECIFICATIONS

Battery:
- Make......................Auto-Lite or Willard
- Model..................TS-2-15 or SW-2-119
- Plates per Cell...........................15
- Capacity........................116 Amp. Hr.
- Volts...6
- Length..............................Approx. 10"
- Width................................Approx. 7"
- Height.............................Approx. 8 5/16"
- Specific Gravity:
  - Fully Charged....................1225-1300
  - Recharge at........................1175
- Ground Terminal....................Negative
- Location............Under Hood Right Side

Starting Motor:
- Make..........................Auto-Lite
- Model..........................MZ-4113
- Drive...........Right hand outboard Bendix
- No Load Draw.............70 amps. max.; 5.5 volts—4300 R.P.M. Min.
- Stall torque....420 amps.; 3.0 volts—7.8 ft. lbs.
- Volts...6
- Armature End Play..................1/16" Max.
- Brushes.......................................4
- Brush Spring Tension..............42-53 Oz.
- Normal Engine Cranking Speed....185 R.P.M.
- Bearings....................3 absorbent bronze

STARTER SWITCH:
- Make.....Auto-Lite   Model.....SW-4001

Generator:
- Make..........................Auto-Lite
- Model..........................GEG-5002D
- Volts.......................................6-8
- Ground Polarity....................Negative
- Controlled Output................40 Amps.
- Rotation (Drive End)..............Clockwise
- Control. Vibrating type current-voltage regulator
- Air Cooled.................................Yes
- Armature End Play.................010" Max.
- Brushes.......................................2
- Brush Spring Tension..............64-68 Oz.
- Bearings....................................Ball
- Field Coil Draw..1.60 to 1.78 Amps.—6.00 V.
- Motorizing Draw..4.7 to 5.2 Amps.—6.0 volts (Have field and armature terminals connected).
- Output.8.0 amps.; 7.6 volts; 955 Max. R.P.M.
  - 40.0 amps.; 7.6 volts; 1460 Max. R.P.M.
  - 40.0 amps.; 8.0 volts; 1465 Max. R.P.M.

Current—Voltage Regulator
- Make.....Auto-Lite   Model..VRY-4203A
- Volts...6
- Amperes.....................................40
- Ground Polarity....................Negative

Voltage Regulator:
- Voltage Setting Open Circuit.........7.20-7.41
- Air Gap...........................040"-.042"
- Point Gap.........................010"-.012"

Circuit Breaker:
- Points Close (Hot)................6.4-6.6 Volts
- Points Open—Reverse Current...0.5-6.0 Amps.
- Air Gap.........................0595"-.0625"
- Points Gap...........................015"

Current Limiting Regulator:
- Air Gap...........................047"-.049"
- Point Gap.........................030"-.033"

ELECTRICAL SYSTEM SPECIFICATIONS
(Continued)

Distributor:

- Make.....................................Auto-Lite
- Model......................................IGC-4705
- Type Advance...................Centrifugal
- Firing Order.........................1-3-4-2
- Breaker Point Gap..................020"
- Breaker Arm Spring Tension.........17-20 Oz.
- Cam Angle (Time points are closed).......47°
- Max. Automatic Advance 1500 R.P.M. (dist.)..11°
- Condenser Capacity...............18-.26 Mfd.
- Timing—72 octane fuel (gasoline)
  5° BTC Flywheel (.0103" Piston travel)
- Timing—68 octane fuel (gasoline)
  TC Flywheel (Zero Piston travel)
- Timing Mark.........................Flywheel
- Location..Right Side Bell Housing under Starter
- Ignition Switch (Lock).......Douglas No. 5941

Coil:

- Make...Auto-Lite
- Model......................................IG-4070-L
- Draw Engine Stopped.....5 Amps. @ 6.4 Volts
- Draw Engine Idling................2.5 Amps.

Gauges:

- Fuel Gauge...............................Auto-Lite
- Oil Pressure.............................Auto-Lite
- Temperature............................Auto-Lite
- Ammeter..................................Auto-Lite

Spark Plugs:

- Make....................................Champion QM-2
- Size..14MM
- Gap..030"

Radio Filters:

- Generator Filter Unit.......Tobe Deutschmann
- Regulator Filter............Tobe Deutschmann
- Filter Group..............Tobe Deutschmann

Lamps:

- Light Switch Make....................Douglas
- Foot Beam Switch Make.......Clum No 9654
- Head Lamps.....Corcoran-Brown Sealed Beam
- Black Out Lamps...............Corcoran-Brown
- Tail and Stop Lamp..........Corcoran-Brown
- Head Lamp Bulbs (Seelite unit). 6-8V-45 C.P. DC.
  Mazda No. 2400
- Blackout Bulbs... 6-8V-3C.P. SC Mazda No. 63
- Tail and Stop Lamp Bulbs. 6-8V 3-21CP Mazda No. 1154
- Instrument Lamp Bulbs... 6-8V 3CP SC Mazda No. 63

Horn:

- Type..................................Micro-Vibrator
- Make.............................Sparks-Withington
- Model..B-9427

BONDED POINTS

Bond No.	Name
1.	Hood to Dash—Right Hand
2.	Hood to Dash—Left Hand
3.	Cylinder Head Stud to Dash
4.	Brake Cable, Speedometer Cable, Heat Indicator Cable to Dash
5.	Gas Line to Dash
6.	Choke Control, Throttle Control and Oil Gauge Line to Dash Stud
7.	Generator Mounting Bolt to Starting Motor Bracket
8.	Generator Voltage Regulator Filter & Ground
9.	Coil to Cylinder Block
10.	Right Hand Front Motor Bracket to Frame
11.	Left Hand Front Motor Bracket to Frame
12.	Exhaust Pipe to Frame
13.	Radiator Right Hand to Frame
14.	Radiator Left Hand to Frame
15.	Rear Engine Support to Frame Cross Member Stud
16.	Transfer Case to Body Floor Stud
17.	Right Hand Body Bracket Ground to Frame
18.	Left Hand Body Bracket Ground to Frame
19.	Right Hand Fender Ground to Frame
20.	Left Hand Fender Ground to Frame
21.	Left Hand Hood Ground to Grill
22.	Right Hand Hood Ground to Grill
23.	Headlamp Wiring Harness to Left Fender
24.	Cylinder Head Stud—Front
25.	Left Hand Fender to Cowl—Lower
26.	Right Hand Fender to Cowl—Lower

IMPORTANT: Where parts are grounded, particular attention must be given to any special position of lockwashers on bolts and screws. Tinned spots should be clean but not painted, to assure satisfactory bond.

TRANSMISSION

The transmission, Fig. 1 is of the three speed, synchromesh type with synchronized 2nd and high speed gears. See shifting diagram, Fig. 4 in Driver's Instructions Section.

The transmission is bolted to the rear face of the flywheel bell housing with four cap screws and is supported on a rubber insulator at the center frame cross member which is the rear engine support or mounting.

Removal of Transmission and Transfer Case from Engine

1. Remove front and rear propeller shafts at universal joint in accordance with instructions under the "Propeller Shafts and Universal Joints".

2. Disconnect speedometer cable at transfer case.

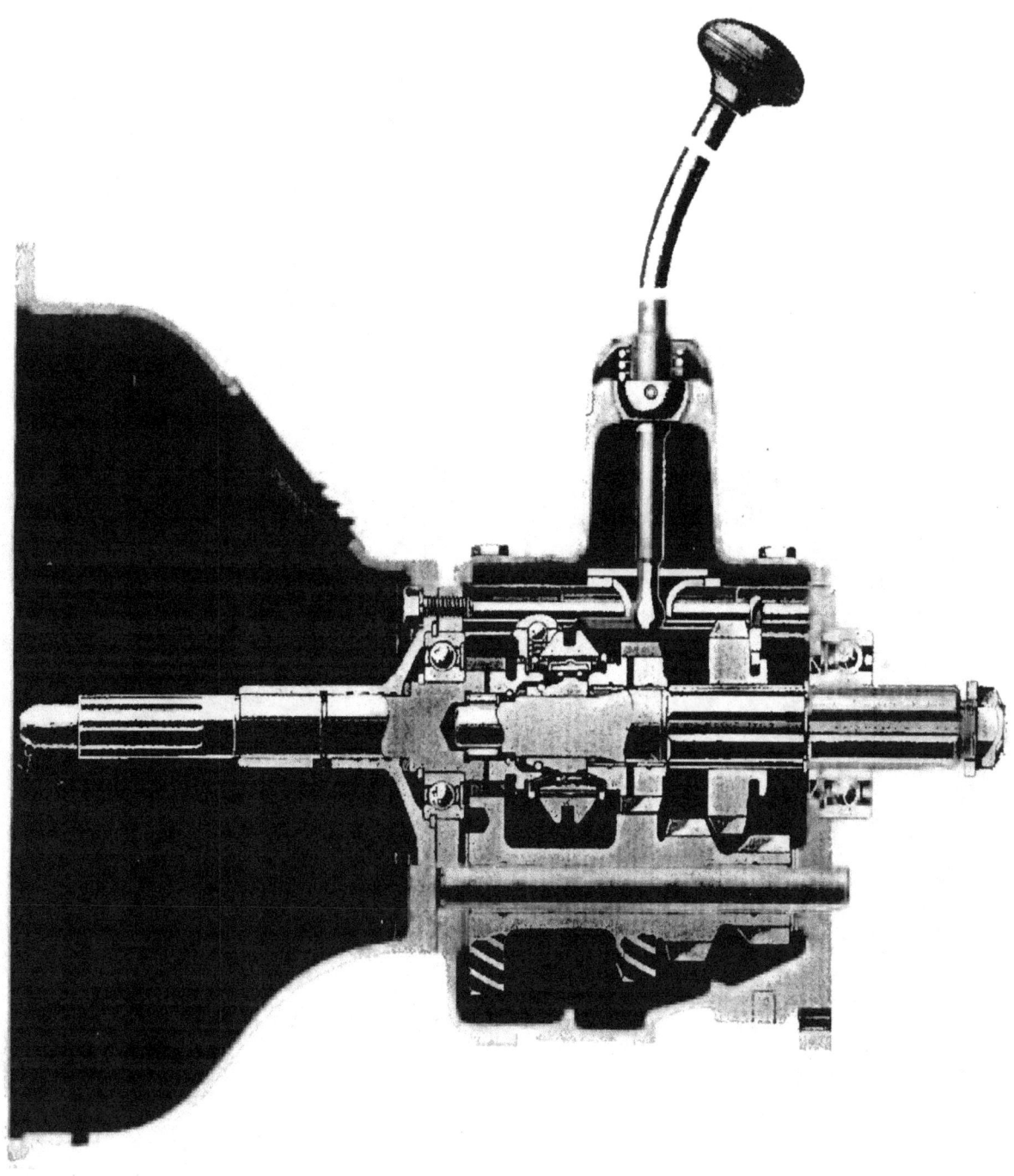

FIG. 1—TRANSMISSION

3. Disconnect brake and engine snubbing cables.
4. Remove nuts holding rear mounting to frame cross member.
5. Remove transfer case snubbing rubber bolt nut at cross member.
6. Remove transmission shift lever by unscrewing retainer collar at top of shift housing.
7. Disconnect the clutch release cable at bell crank and remove; also, remove clutch release lever No. 10, Fig. 2, "Clutch Section" through the inspection hole in the flywheel bell housing.
8. Place jacks under engine and transmission.
9. Remove floor board inspection plate, drain radiator and remove upper hose.
10. Remove transfer shift lever pivot pin screw and lubricator.
11. Remove shift lever pin and remove levers.
12. Remove bolts holding center cross member to frame side rail and remove cross member.
13. Remove bolts holding transmission to flywheel bell housing.
14. Force transmission to right and disconnect clutch control lever tube ball joint.
15. Lower jacks under engine and transmission; slide transmission assembly towards rear of vehicle until clutch shaft clears bell housing.
16. Lower jack under transmission and remove assembly from under chassis.

FIG. 2—REMOVING BEARING SNAP RING

Disassembly of Transmission

Drain lubricants from both the transmission and transfer case through drain plug holes in bottom of each case. It is advisable to clean the outside of the cases thoroughly with water or other suitable cleaning fluid before attempting to disassemble the units.

To disassemble the unit the following procedure is recommended:

1. Remove cap screws and lock washers holding rear cover. No. 37, Fig. 3 in Transfer Case Section.
2. Remove cotter pin, nut and washer permitting removal of main shaft gear No. 57.

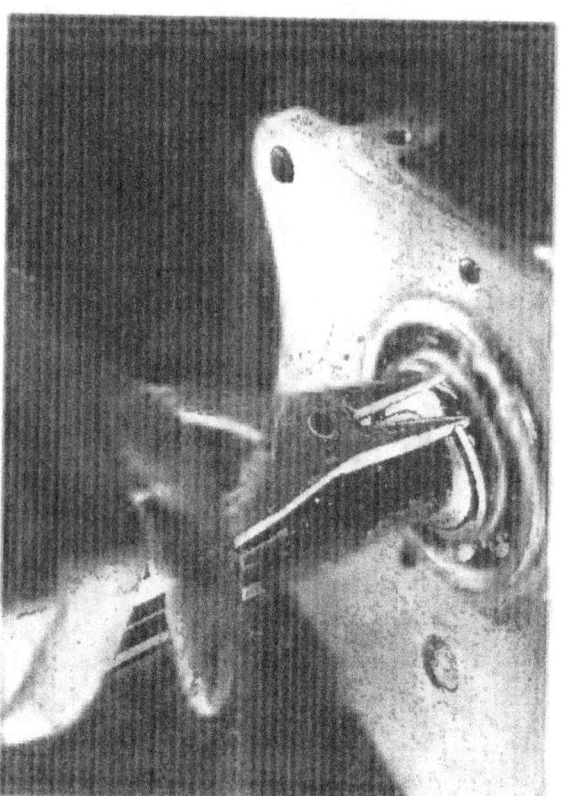

FIG. 3—REMOVING SNAP RING

3. Remove the four cap screws holding control housing to top of transmission and remove housing.
4. Remove shifter plate spring and take off shifter plate, No. 11, Fig. 4.
5. Loop a piece of wire around main shaft, just rear of main shaft second speed gear, twist wire and attach one end to the right hand front cover screw and the other end to the left hand cover screw, drawing the wire tightly to prevent the main shaft from sliding out of the case when transfer case is removed.
6. Remove the five cap screws holding the transfer case to the rear face of the transmission.

7. Support transfer case and with a rawhide mallet or brass rod and hammer, tap lightly on end of shaft and at the same time draw the transfer case away from the transmission. Be careful not to lose transmission gear shift interlock plunger. The transmission main shaft rear bearing No. 34, Fig. 4 should slide out of transfer case and remain in transmission.

8. Remove three screws holding main drive gear bearing retainer, No. 1 and remove retainer and gasket.

9. Remove shift fork guide pin, No. 20 through front of transmission.

10. Remove shift fork set screws with special wrench furnished in tool kit and remove shift shafts and forks. Be careful not to lose poppet springs and balls.

11. Remove lock plate at rear of transmission holding countershaft and reverse idler gear shaft.

12. With a drift, drive out the countershaft.

13. Remove main drive gear bearing, shaft and synchronizer blocking ring.

14. Remove snap rings from main drive gear shaft and bearing, Fig. 2 and 3 and remove bearing from shaft.

15. Remove main shaft assembly.

16. Remove countershaft gear set and three thrust washers, two bushings and a spacer.

17. Remove reverse idler gear shaft and gear.

To remove the gears on main shaft, first remove snap ring No. 27, Fig. 4, on end of shaft holding transmission high and intermediate clutch hub, No. 28. After the removal of the snap ring, the gears will slide off the shaft without difficulty. To disassemble synchronizer unit, push apart.

Wash all parts in suitable cleaning fluid and inspect for wear and damaged parts, replacing any parts which show excessive wear or damage.

Assembly of Transmission

The assembly of the parts in the transmission should be performed in the reverse manner in which it was dismantled making reference to exploded views of parts as shown in Fig. 4, for sequence of assembly.

When assembling synchronizer unit assembly place the right end of a synchronizer spring No. 14 in one shifting plate. Turn the unit around and make exactly the same installation with the other spring in the same shifting plate. This will actually place the spring action opposed to each other.

The bushings in countershaft gear set are of the floating type, being free to turn within the gear as well as on the shaft. When making assembly of countershaft to transmission case, dip these bushings in lubricant of S.A.E. 90 grade and be sure the spacer is installed between the two bushings. The steel thrust washer, No. 43 at the rear of countershaft gear is pinned in the case and the bronze washer, No. 42 is installed between steel washer and gear. Only one bronze washer, No. 37 is used at the front. The main shaft ball bearing, No. 34 is assembled to shaft so that the sealed side is in transmission, open side to transfer case.

TRANSMISSION TROUBLE AND REMEDIES

SYMPTOMS	PROBABLE REMEDY
Slips Out of High Gear	
Transmission misaligned with Bell Housing	Align Transmission Case to Bell Housing and Bell Housing to Engine
End play in Main Drive Gear	Tighten Front Retainer
Damaged Pilot Bearing or Front Bearing	Replace
Bent Shifting Fork	Replace
Slips Out of Second	
Bent Shifting Fork	Replace
Worn Gear	Replace
Weak Poppet Spring	Replace
Noise in Low Gear	
Rear Ball Bearing Broken	Replace
Gear Teeth Pitted or Worn	Replace gears
Shifting Fork Bent	Replace
Lack of Lubrication	Drain and Refill
Grease Leak into Bell Housing	
Gasket Broken Front Bearing Retainer	Replace
Transmission Case Overfilled with Lubricant	Drain off to proper level

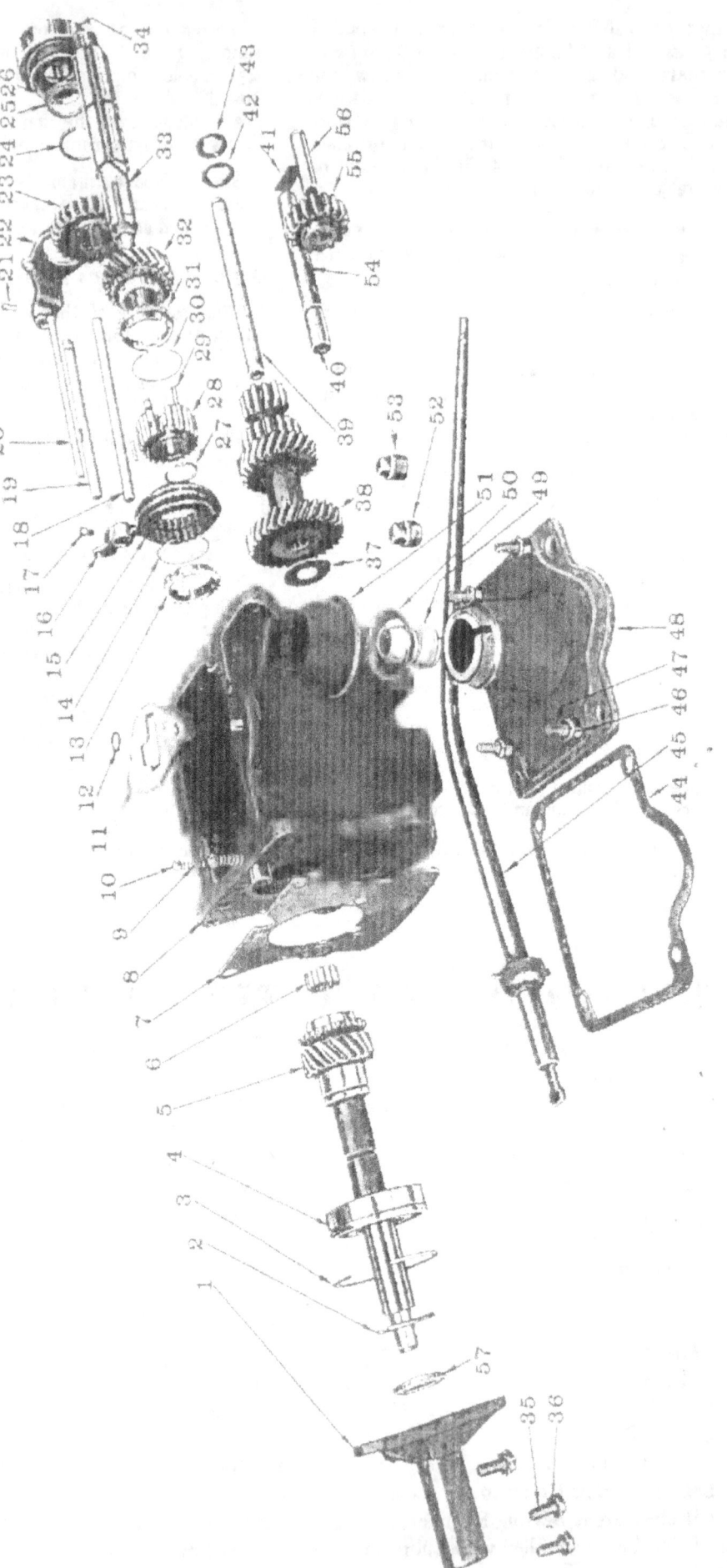

FIG. 4—TRANSMISSION

FIG. 4—TRANSMISSION (EXPLODED)

No.	Willys Part No.	Ford Part No.	Name
1	640017	GPW-7050	Transmission Main Drive Gear Bearing Retainer
2	635844	GPW-7064	Transmission Main Drive Gear Snap Ring
3	635846	B-7070	Transmission Main Drive Gear Bearing Snap Ring
4	636885	GPW-7025	Transmission Main Drive Gear Bearing
5	A-5554	GPW-7017	Transmission Main Drive Gear
6	639422	GPW-7120	Transmission Main Shaft Pilot Roller Bearing
7	637495	GPW-7051-B	Transmission Main Drive Gear Bearing Retainer Gasket
8	A-1148	GPW-7005	Transmission Case
9	635837	GPW-7234	Transmission Poppet Spring
10	635838	353081-S7	Transmission Shift Rail Poppet Ball
11	635841	GPW-7216	Transmission Shift Plate
12	635839	GPW-7208	Transmission Shift Plate Spring
13	637834	GPW-7107	Transmission Synchronizer Blocking Ring
14	637831	GPW-7109	Transmission Synchronizer Spring
15	637833	GPW-7106	Transmission Second and Direct Speed Clutch Sleeve
16	636196	GPW-7230	Transmission Shift Fork—High and Intermediate
17	636200	GPW-7245	Transmission Shift Fork Lock Screw
18	A-1155	GPW-7241	Transmission Shift Rail—High and Intermediate
19	A-1156	GPW-7240	Transmission Shift Rail—Low and Reverse
20	635836	GPW-7206	Transmission Shift Fork Guide Pin
21	636200	GPW-7245	Transmission Shift Fork Lock Screw
22	636197	GPW-7231	Transmission Shift Fork Low and Reverse
23	636879	GPW-7100	Transmission Sliding Gear—Low and Reverse
24	635844	GPW-7064	Transmission Main Shaft Snap Ring
25	A-738	GPW-7062	Transmission Main Shaft Bearing Spacer
26	A-410	GPW-7080	Transmission Oil Retaining Washer
27	637835	GPW-7059	Transmission High and Intermediate Clutch Hub Snap Ring
28	637830	GPW-7105	Transmission High and Intermediate Clutch Hub
29	637832	GPW-7116	Transmission Synchronizer Shifting Plate
30	637831	GPW-7109	Transmission Synchronizer Spring
31	637834	GPW-7107	Transmission Synchronizer Blocking Ring
32	638798	GPW-7102	Transmission Main Shaft Second Speed Gear Assembly
33	A-519	GPW-7061	Transmission Main Shaft
34	A-916	GP-7065	Transmission Main Shaft Bearing
35	635868	20366-S	Hex. Head Screw (Bearing Retainer)
36	52510	34941-S	Lockwasher
37	635812	GPW-7119	Transmission Countershaft Thrust Washer—Front
38	A-739	GPW-7113	Transmission Countershaft Gears
39	638948	GPW-7111	Transmission Countershaft
40	A-878	GPW-7121	Transmission Countershaft Gear Bushing
41	638949	GPW-7135	Transmission Countershaft and Idler Lock Plate
42	635811	GPW-7129	Transmission Countershaft Thrust Washer—Rear (Bronze)
43	A-879	GPW-7126	Transmission Countershaft Thrust Washer—Rear (Steel)
44	635861	GPW-7223	Transmission Control Housing Gasket
45	A-1380	GPW-7210	Transmission Control Lever Assembly (Gear Shift Lever)
46	635868	20366-S	Hex. Head Screw (Control Housing)
47	52045	34806-S	Lockwasher
48	635857	GPW-7204	Transmission Control Housing Assembly
49	392328	GPW-7227	Transmission Control Lever Support Spring
50	635863	BB-7228	Transmission Control Housing Cap Washer
51	A-1379	BB-7220	Transmission Control Housing Cap
52	5140	353064-S	Transmission Drain Plug
53	5140	353064-S	Transmission Filler Plug
54	A-880	GPW-7115	Transmission Countershaft Bearing Spacer
55	636882	GPW-7142	Transmission Reverse Idler Gear Assembly
56	638952	GPW-7140	Transmission Reverse Idler Gear Shaft
57	640018	GPW-7052	Front Bearing Retainer Oil Seal

TRANSMISSION SPECIFICATIONS

Transmission

Make	Warner
Model	T 84 J
Type	Synchronous Mesh
Mounting	Unit Power Plant
Shift Lever Location	On Transmission
Speeds	3 Forward—1 Reverse
Ratio	
Low	2.665
Second	1.564
High	1:1
Reverse	3.554

Bearings

Clutch Shaft (Flywheel)	Bushing
Clutch Release	Ball
Clutch Shaft Rear (Main Drive Gear)	Ball
Main Shaft Front	13 Rollers
Main Shaft Rear	Ball
Counter Shaft Gear	Bushings (2)
Reverse Idler Gear	Bushing

Transmission Oil

Capacity—(Pts.)	See Capacity Chart, Page 3
S.A.E. Viscosity	See Lubrication Chart, Page 12

TRANSFER CASE

The transfer case, Fig. 1 is an auxiliary unit located at the rear of the transmission. The transfer case is essentially a two speed transmission, which provides a low gear ratio and a means of connecting the transmission to the front axle.

The shifting mechanism, Fig. 2 is located on the transfer case for engaging and disengaging the drive to the front axle, also for shifting into the low gear ratio.

On hard surface and flat roads, disengage front axle drive by placing center shift lever in forward position. The right hand lever controls the gear ratio; low and high. The low gear ratio can only be used when left hand lever is in the engaged (rear) position for front drive. Proper position for disengaging axles to use power take-off unit is shown as "N" in Fig. 4, Page 6.

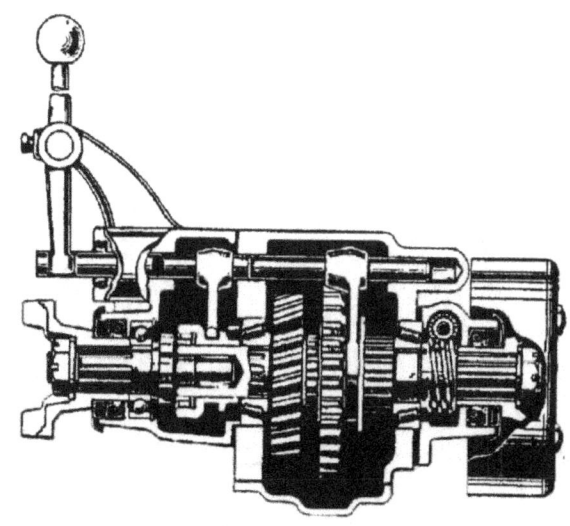

FIG. 2—TRANSFER CASE SHIFTING

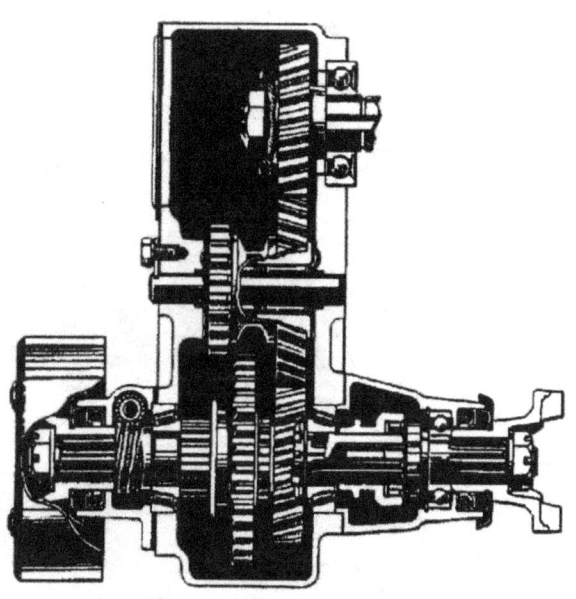

FIG. 1—TRANSFER CASE

Disassembling Transfer Case

To remove the gears and bearings from the transfer case on the bench, the following procedure is recommended.

1. Remove brake band assembly and linkage.

2. Remove cap screws and lockwashers holding lower cover, No. 55, Fig. 3.

3. Remove lock plate screw, lockwasher and lockplate, No. 50.

4. With a punch, drive out intermediate shaft No. 62, to rear of case.

5. Intermediate gear No. 64 with Thrust washers No. 63 and No. 65, and roller bearings No. 61 and 66 can be removed through the bottom of case.

6. Remove poppet plugs No. 18, springs, No. 17, and balls No. 16 on both sides of output bearing cap, No. 14. Shift front wheel drive to engaged position (shaft forward).

7. Remove cap screws holding front output bearing cap No. 14 and remove cap as an assembly, with universal joint end yoke No. 26, clutch shaft No. 12, bearing No. 10, clutch gear No. 3, fork No. 2, and shift rod No. 8. Taking care not to lose inter-lock No. 7.

8. Remove output shaft snap ring No. 70 and thrust washer No. 69.

9. Remove cap screws holding rear output bearing cap No. 43, and remove cap as an assembly with universal joint flange No. 33, speedometer gears No. 40 and 45, bearing No. 51, and output shaft No. 67. This will allow sliding gear No. 56 and output shaft gear No. 68 to slide off the output shaft No. 67 and come out through the bottom of case.

10. Remove set screw No. 48 in sliding gear shift fork No. 49. This will allow shift rod No. 9 to slide through the fork and hole in case. The fork can then be removed through bottom of case.

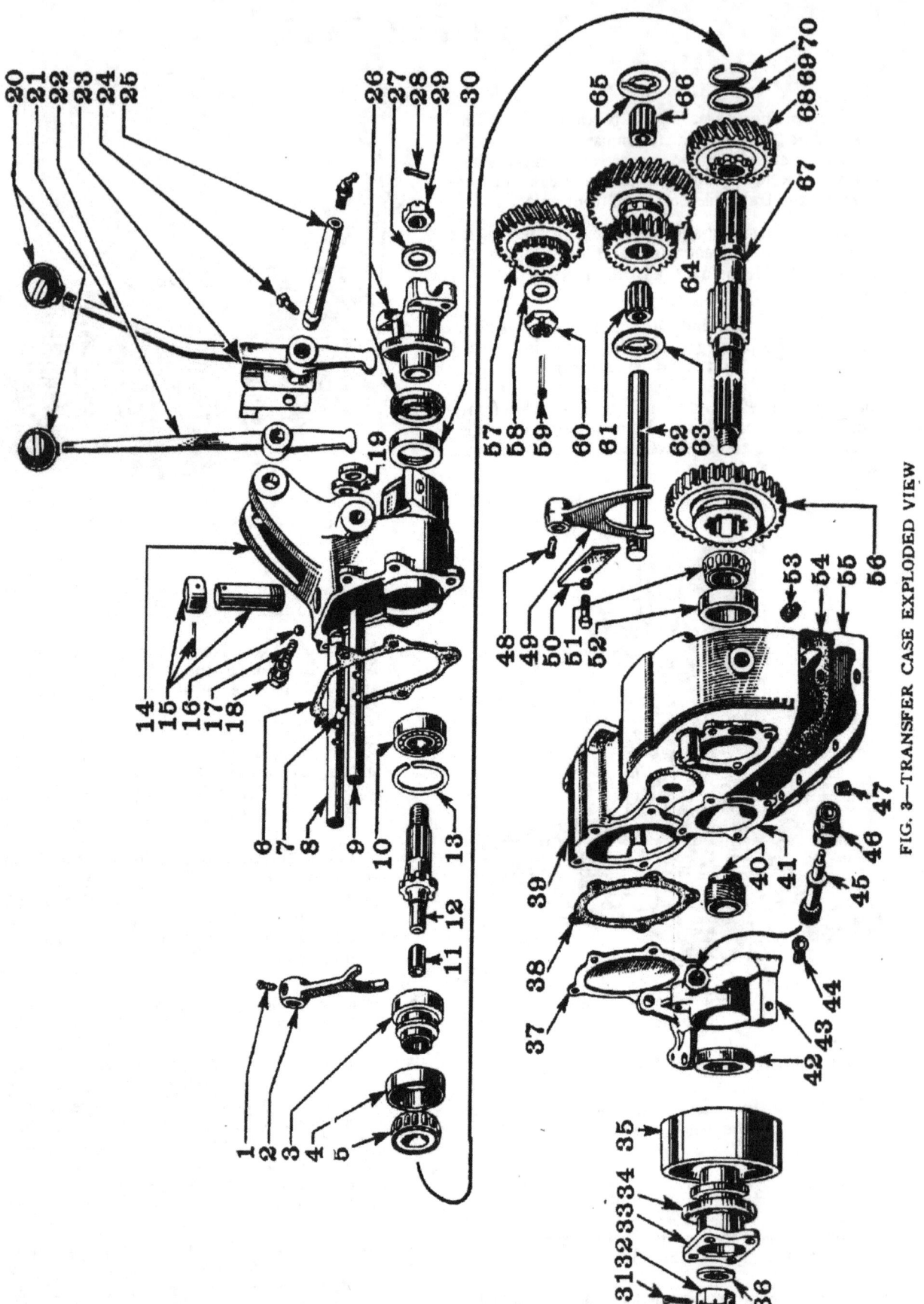

FIG. 3—TRANSFER CASE EXPLODED VIEW

FIG. 3—TRANSFER CASE EXPLODED VIEW

No.	Willys Part No.	Ford Part No.	Name
1	A-963	355550-S	Shift Fork to Rod Set Screw
2	A-960	GP-7711	Front Wheel Drive Shift Fork
3	A-992	GP-7702	Output Shaft Clutch Gear
4	52883	O1Y-1202	Output Shaft Bearing Cup
5	51575	GP-7723	Output Shaft Bearing Cone and Rollers
6	A-957	GPW-7773	Output Shaft Bearing Cap Gasket—Front
7	A-965	GP-7789	Shift Rod Interlock
8	A-962	GP-7787	Front Wheel Drive Shift Rod
9	A-1504	GPW-7786	Under Drive Shift Rod
10	A-1007	GP-7719	Output Clutch Shaft Bearing
11	A-987	GP-7777	Output Clutch Shaft Pilot Bushing
12	A-976	GP-7751	Output Clutch Shaft
13	A-976	GP-7783	Output Clutch Shaft Bearing Snap Ring
14	A-968	GPW-7774	Output Shaft Bearing Cap—Front
15	A-934	GP-7754	Transfer Case Breather Assembly
16	5599	353075-S	Shift Rod Poppet Ball
17	A-966	GP-7788	Shift Rod Poppet Spring
18	A-967		Shift Rod Poppet Plug
19	A-974	GP-7708	Shift Rod Oil Seal
20	A-971	GP-7213	Shift Lever Handle
21	A-1505	GPW-7793	Under Drive Shift Lever
22	A-1506	GPW-7710	Front Wheel Drive Shift Lever
23	A-970	GP-7799	Shift Lever Spring
24	A-973	355378-S	Shift Lever Pivot Pin Set Screw
25	A-972	GP-7796	Shift Lever Pivot Pin
26	A-1106	GP-7729	Universal Joint End Yoke—Front
27	A-1028	356504-S	Companion Flange Nut Washer
28	5108	72033-S	Companion Flange Nut Cotter Pin
29	A-980	356125-S	Companion Flange Nut
30	A-958	GP-7770-A	Output Shaft Oil Seal
31	5108	72053-S	Companion Flange Cotter Pin
32	A-980	356125-S	Companion Flange Nut
33	A-1105	GP-4603	Companion Flange—Rear
34	A-1111	GP-7776	Dust Shield
35	A-1002	GP-2814	Brake Drum
36	A-1028	356504-S	Companion Flange Washer
37	A-1508	GPW-7706	Rear Cover
38	A-1509	GPW-7707	Rear Cover Gasket
39	A-1503	GPW-7705	Transfer Case
40	A-1511	GP-17285	Speedometer Drive Gear
41	A-982	GP-7782-A	Output Shaft Bearing Shim
42	A-958	GP-7770-A	Output Shaft Oil Seal
43	A-1507	GPW-7789	Output Shaft Bearing Cap—Rear
44	A-985	GP-17277	Speedometer Drive Pinion Bushing
45	A-1512	GPW-17271	Speedometer Driven Gear
46	636396	GP-17333	Speedometer Driven Gear Sleeve
47	A-1104	358059-S	Transfer Case Drain Plug
48	A-963	355550-S	Shifting Fork to Rod Set Screw
49	A-959	GP-7712	Under Drive Shift Fork
50	A-1001	GP-7767	Intermediate Shaft Lock Plate
51	51575	GP-7723	Output Shaft Bearing Cone and Rollers
52	52883	O1Y-1202	Output Shaft Bearing Cup
53	5140	353064-S	Filler Plug
54	A-954	GP-7709	Transfer Case Cover Gasket—Bottom
55	A-953	GP-7708	Transfer Case Cover—Bottom
56	A-988	GP-7765	Output Shaft Sliding Gear
57	A-1510	GP-7722	Main Shaft Gear
58	A-1410	356580-S	Main Shaft Washer
59	5397	72071-S	Main Shaft Nut Cotter Pin
60	A-520	356134-S18	Main Shaft Nut
61	A-924	GP-7718	Intermediate Gear Bearing
62	A-999	GP-7743	Intermediate Shaft
63	A-1000	GP-7744	Intermediate Gear Thrust Washer
64	A-998	GP-7742	Intermediate Gear
65	A-1000	GP-7744	Intermediate Gear Thrust Washer
66	A-924	GP-7718	Intermediate Gear Bearing
67	A-1764		Output Shaft
68	A-989	GP-7766	Output Shaft Gear
69	A-990	GP-7771	Output Shaft Gear Thrust Washer
70	A-991	GP-7784	Output Shaft Gear Snap Ring

Disassembly of Front Cap Assembly

1. Remove cotter pin, No. 28, nut, No. 29 and washer No. 27.
2. Remove universal joint yoke No. 26.
3. Remove oil seal No. 30.
4. Remove set screw No. 1 and shifting rod No. 8.
5. Clutch gear No. 3 and fork No. 2 can now be removed together.
6. Remove output clutch shaft No. 12 carefully pressing through the bearing No. 10.
7. Remove snap ring No. 13.
8. Remove bearing No. 10.

Disassembly of Rear Cap Assembly

1. Remove cotter pin No. 31, nut, No. 32, and washer No. 36.
2. Remove companion flange No. 33.
3. Remove oil seal No. 42.
4. Remove speedometer driven gear, No. 45.
5. Output shaft No. 67 can now be removed from cap No. 43 after which bearing cone No. 51 and the speedometer driving gear No. 40 can be pressed off the shaft.

Shims No. 41 provided between the rear cap No. 43 and case No. 34 are for adjustment of the roller bearings No. 51 and 5. Bearings should be adjusted so that there is not more than .003" end movement of the shaft No. 67.

Reassembling is merely a reversal of the foregoing. When assembling the transfer case to the transmission be sure that the countershaft lock plate No. 41, Fig. 4, Transmission Section is properly located between the two shafts and fits into transfer case. Make certain that all parts are carefully washed and free from all dirt and foreign matter.

Assembly of Transmission and Transfer Case to Vehicle

The installation of the assembly to the engine is the reversal in the operations for disassembly as covered under heading "Removal of Transmission and Transfer Case". For illustration of snubbing rubber and rear engine mounting see Fig. 4.

After assembling to the engine be sure that the clutch pedal has 3/4" free pedal travel; refer to Page 37 for "Clutch Pedal Adjustment." Fill both the transmission and the transfer cases with the proper lubricant. See Lubrication Chart, Page 12.

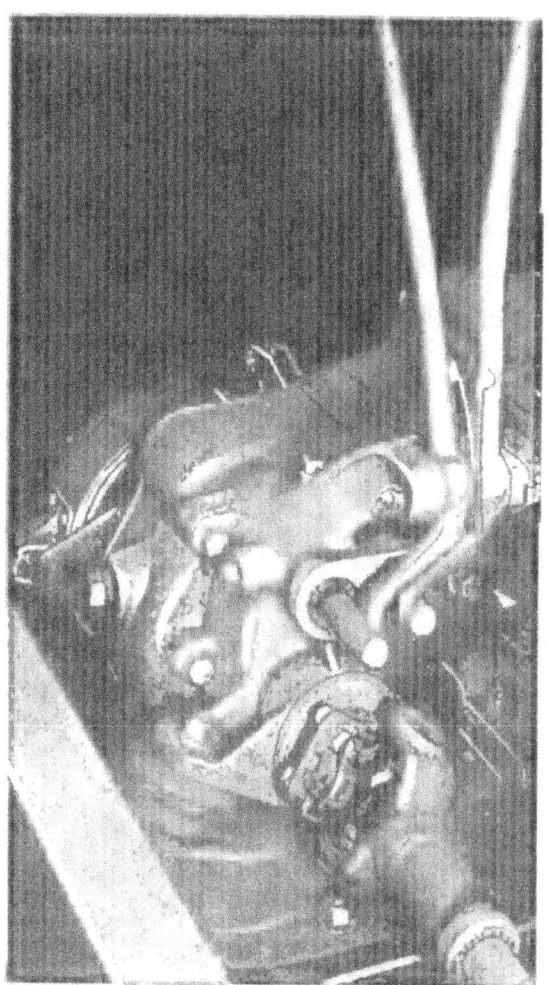

FIG. 4—TRANSFER CASE SNUBBING RUBBER

TRANSFER CASE TROUBLES AND REMEDIES

SYMPTOMS PROBABLE REMEDY

Slips Out of Gear (High-Low)

Shifting Lock Spring Weak	Replace spring
Bearing Broken or Worn	Replace
Shifting Fork Bent	Replace

Slips Out Front Wheel Drive

Shifting Lock Spring Weak	Replace
Bearing Worn or Broken	Replace
End Play in Shaft	Adjust (See Instructions)
Shifting Fork Bent	Replace

Hard Shifting

Lack Lubrication	Drain and Refill—3 pints
Shift Lever Stuck on Shaft	Remove, clean-lubricate
Shifting Lock Ball Scored	Replace Ball
Shifting Fork Bent	Replace Fork
Low Tire Pressure	Inflate all tires—30 lbs.

Grease Leak at Front or Rear Drive

Grease leak at covers	Install new Gaskets
Grease leak between Trans. and Transfer cases	Install new Gaskets
Grease leak at Output Shafts	Install new oil Seal

TRANSFER CASE SPECIFICATIONS

Transfer Case

Make	Spicer
Model	18
Mounting	Unit with Transmission
Shift Lever	Floor
Ratio	High 1:1 / Low 1.97:1

Transfer Case Bearings

Transmission Mainshaft	Ball
Idler Gear	2 Roller
Out Put Shaft	Taper Roller
Front Axle Clutch Shaft	
Front Bearing	Ball
Rear Pilot in Output Shaft	Bronze Bushing I.D. .627"

Transfer Case Oil

Capacity Pts.	See Capacity Chart, Page 3
S.A.E. Viscosity	See Lubrication Chart, Page 12

Speedometer Drive

Drive Gear Teeth	4
Driven Gear Teeth	14

PROPELLER SHAFTS AND UNIVERSAL JOINTS

The drive from the transmission to the front and rear axles is accomplished through a propeller shaft and two universal joints. Fig. 1.

The splined slip joint at one end of each shaft allows for variations in distance between the transfer case and the front and rear axle units due to spring action.

The slip joint is marked with arrows at the spline and the sleeve yoke. Note markings to facilitate proper assembly so the yokes of the universal joints at front and rear of shaft are in the same plane when assembled, Fig. 2.

The propeller shaft connecting the transfer case with the front axle has the "U" bolt type universal joint at both ends.

The rear propeller shaft is equipped with the "U" bolt type joint at the rear where it attaches to the rear axle. The front universal joint is the snap ring type.

These universal joints are the Needle Bearing type and are so designed that correct assembly is a very simple matter. No hand fitting or special tools are required.

The journal trunnion and needle bearing assemblies are the only parts subject to wear, and when it becomes necessary to replace these parts, the propeller shaft should be removed from the vehicle.

FIG. 1—PROPELLER SHAFT ASSEMBLY

Disassembling of Snap Ring Universal Joints

To remove snap rings pinch ends together with a pair of pliers. If the ring does not readily snap out of the groove tap the end of the bearing lightly, this will relieve pressure against the ring. See Fig. 3.

Drive on the end of one bearing until the opposite bearing is pushed out of the yoke. Turn the joint over and drive the first bearing back out of its lug by driving on the exposed end of the journal shaft. Use a soft round drift with a flat face about $\frac{1}{32}$" smaller in diameter than the hole in the yoke, otherwise there is danger of damaging the bearing.

Repeat this operation for the other two bearings, then lift out journal assembly, sliding to one side and tilting over the top of the yoke lug.

Wash all parts in cleaning solution and if parts are not worn, lubricate with a good grade of semi-fluid lubricant, see Lubrication Chart, Page 12. Make sure the reservoir in each journal trunnion is filled. Put the rollers in the race and fill the race about one-third full. It is advisable to install new gaskets, No. 2, Fig. 4 on the journal assembly.

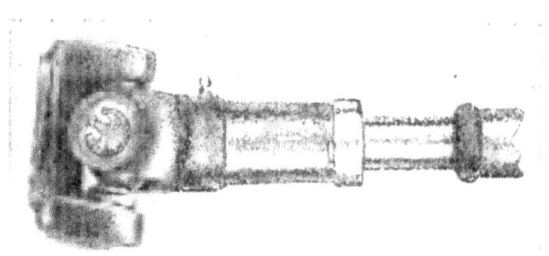

FIG. 2—ARROW MARKING

Reassembling of Snap Ring Universal Joints

Reassembling is merely a reversal of the dismantling operation. Hold the bearing in a vertical position to prevent needles from dropping out of bearing race when installing in joint.

When assembled, if joint appears to bind tap the lugs lightly with a hammer which will relieve any pressure on the bearings at the end of the journal.

When inserting the spline of the propeller shaft into the universal joint be sure that the arrows on the propeller shaft and yoke sleeve are in line. See Fig. 2.

Disassembly of "U" Bolt Type Universal Joint

Removal of the "U" Bolts at axle and transmission end yoke allows the complete propeller shaft assembly to be removed.

After removing "U" bolt slide sleeve yoke (slip joint) towards the shaft which will allow the bearing race to come out from behind the shoulders on end yoke. Care should be taken to hold bearing races in place to avoid losing the rollers.

Now remove snap lock ring, No. 1, Fig. 4 in the sleeve yoke at front and stud ball yoke at rear end of shaft by pinching ends together with a pair of pliers. If a ring does not snap readily out of the groove, tap the end of the bearing lightly, which will relieve the pressure against the ring.

Drive on the end of one bearing until the opposite bearing is pushed out of the yoke. Turn the universal joint over and drive the first bearing out by driving on the exposed end of the journal assembly. Use a soft round drift with a flat face about $\frac{1}{32}$" smaller in diameter than the hole in the yoke, otherwise there is danger of damaging the bearing.

Now lift out journal assembly by sliding to one side and tilting over the top of the yoke lug. Clean all parts and if parts are not worn, repack with a good grade of semi-fluid lubricant, see Lubrication Chart, Page 12. Make sure the reservoir in the end of each trunnion is filled. With the rollers in the race, fill the race about one-third full. It is advisable to install new gaskets on journal assembly.

Reassembling of "U" Bolt Type Universal Joint

Reassembling is merely a reversal of the dismantling operation.

Be sure to hold the bearing in a vertical position to prevent the needles from dropping out of the bearing race.

When assembled, if joints appear to bind tap the lugs lightly with a hammer which will relieve any pressure on the bearings at the end of the journal.

When assembling the bearings into the end yoke the use of a "C" Clamp over the extreme ends of the bearing races to draw the bearings into correct position will greatly facilitate seating them inside of the bearing shoulders on the end yokes. "U" bolt, torque wrench reading, 15-18 ft. lbs.

When inserting the propeller shaft spline into the universal joint be sure that the arrows on the propeller shaft and yoke sleeve are in line. See Fig. 2.

Lubrication

Do not use grease in the needle bearings.

At each 1,000 mile lubrication job, lubricate the Universal Joints, using a hand gun. See Lubrication Chart for oil specifications.

The sliding spline shaft should be lubricated with a good grade of grease or oiled every 1,000 miles, or every time the chassis is lubricated. A hydraulic pressure fitting is provided for this purpose on the side of the sleeve yoke.

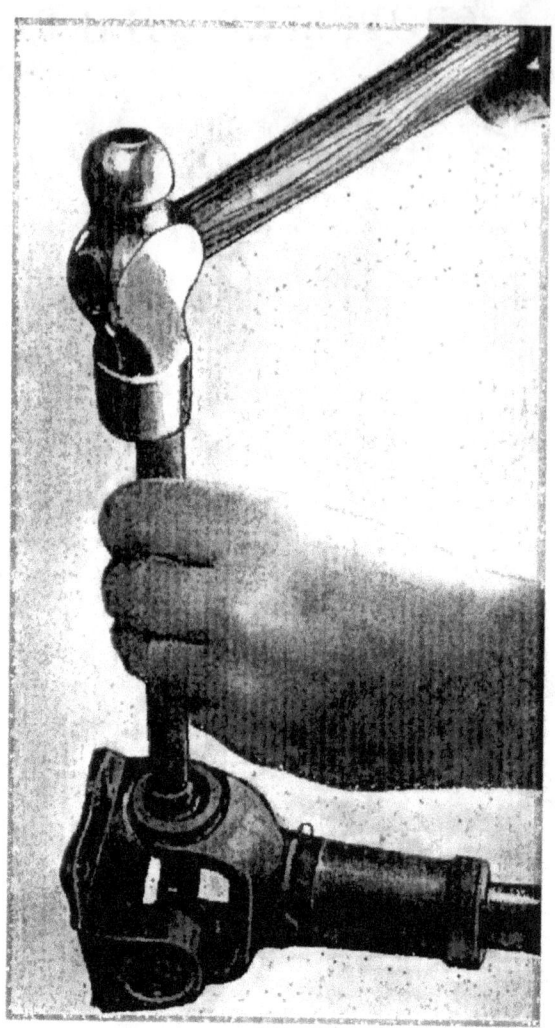

FIG. 3—REMOVING UNIVERSAL JOINT BEARING

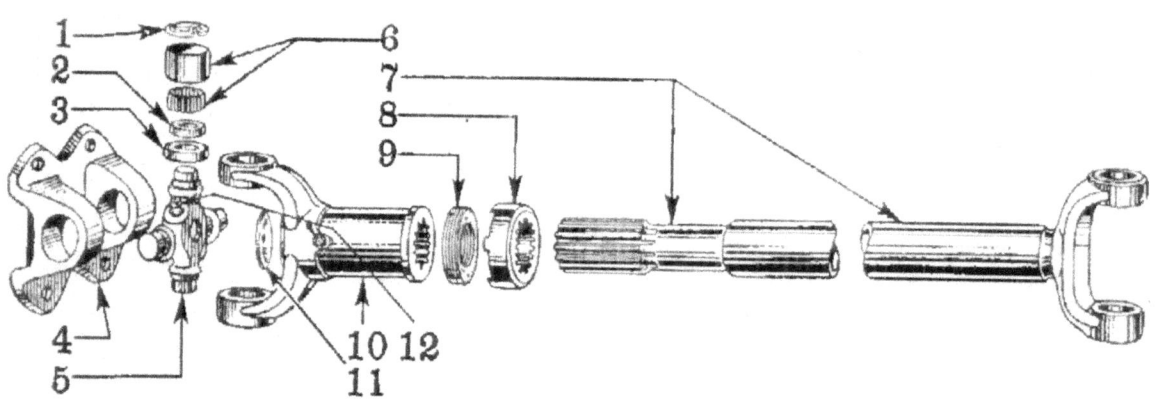

FIG. 4—PROPELLER SHAFT—REAR

No.	Willys Part No.	Ford Part No.	Name	No.	Willys Part No.	Ford Part No.	Name
1	A-945	O1Y-7096	Universal Joint Bearing Snap Ring	7	A-1429	GPW-4605	Propeller Shaft Tube Assembly—Rear
2	A-941	O1T-7078-A	Trunnion Gasket	8	A-942	GP-7077	Dust Cap
3	A-940	O1Y-7083	Trunnion Gasket Retainer	9	A-943	GP-7097	Cork Washer
4	A-950	GP-4866	Universal Joint Flange Yoke	10	A-935	GP-7093	Universal Joint Sleeve Yoke Assembly
5	A-1426	GPW-7084	Universal Joint Journal Assembly	11	A-937		Sleeve Yoke Plug
6	A-1425	GPW-7099	Universal Joint Bearing Race	12	638792	353043-S7	Hydraulic Fitting

PROPELLER SHAFT AND UNIVERSAL JOINT SPECIFICATIONS

Propeller Shaft

Make	Spicer
Shaft Diameter	1¼"
Length (Front) (Joint center to center)	21¹¹⁄₁₆"
Length (Rear) (Joint center to center)	20½₂"

Universal Joint Front Drive .... Front

Make	Spicer
Type	U Bolt and Snap Ring
Model	1268
Bearings	Needle Roller Spicer 98-851

Universal Joint Front Drive .... Rear

Make	Spicer
Type	Snap Ring and U Bolt
Model	1261
Bearings	Needle Roller Spicer 98-851

Universal Joint Rear Drive .... Front

Make	Spicer
Type	Snap Ring Slip Joint
Model	1261
Bearings	Needle Roller Spicer 98-851

Universal Joint Rear Drive .... Rear

Make	Spicer
Type	U Bolt and Snap Rings
Model	1268
Bearing	Needle Roller Spicer 98-851

Lubricant .... See Lubrication Chart, Page 12

FRONT AXLE

The front axle assembly is a front wheel driving unit with specially designed steering knuckles, Fig. 1 and a conventional type differential with hypoid drive gears.

The front wheels are driven by axle shafts equipped with constant velocity universal joints which are enclosed in the steering knuckle housing.

The differential is mounted in the housing similar to that used in rear axle, except that the drive pinion shaft is toward the rear instead of the front and to the right of the center of the axle. This design allows placing front propeller shaft along right side of engine oil pan without reducing road clearance under engine.

The differential parts are interchangeable with those of the rear axle.

The axle is the full-floating type and can be removed without disassembling the steering knuckle.

Axle Shaft and Universal Joint Assembly

To remove axle shaft and universal joint assembly the following operations should be performed. See Fig. 2.

1. Remove wheel assembly.
2. Remove hub cap by inserting two screw drivers from opposite sides behind inner flange on cap and pry off.
3. Remove axle shaft cotter pin, nut and washer.
4. Remove axle shaft drive flange bolts and lock washers.
5. Apply the foot brakes and remove the axle shaft flange with puller furnished in tool kit, see Fig. 3.
6. Remove wheel bearing nuts and washers, No. 22, Fig. 2. First bend the lip on lock washer, No. 21 away from the nut with a chisel, remove the outer nut, lock washer, adjusting nut and bearing lock washer, Fig. 4. Wrench is furnished in tool kit.

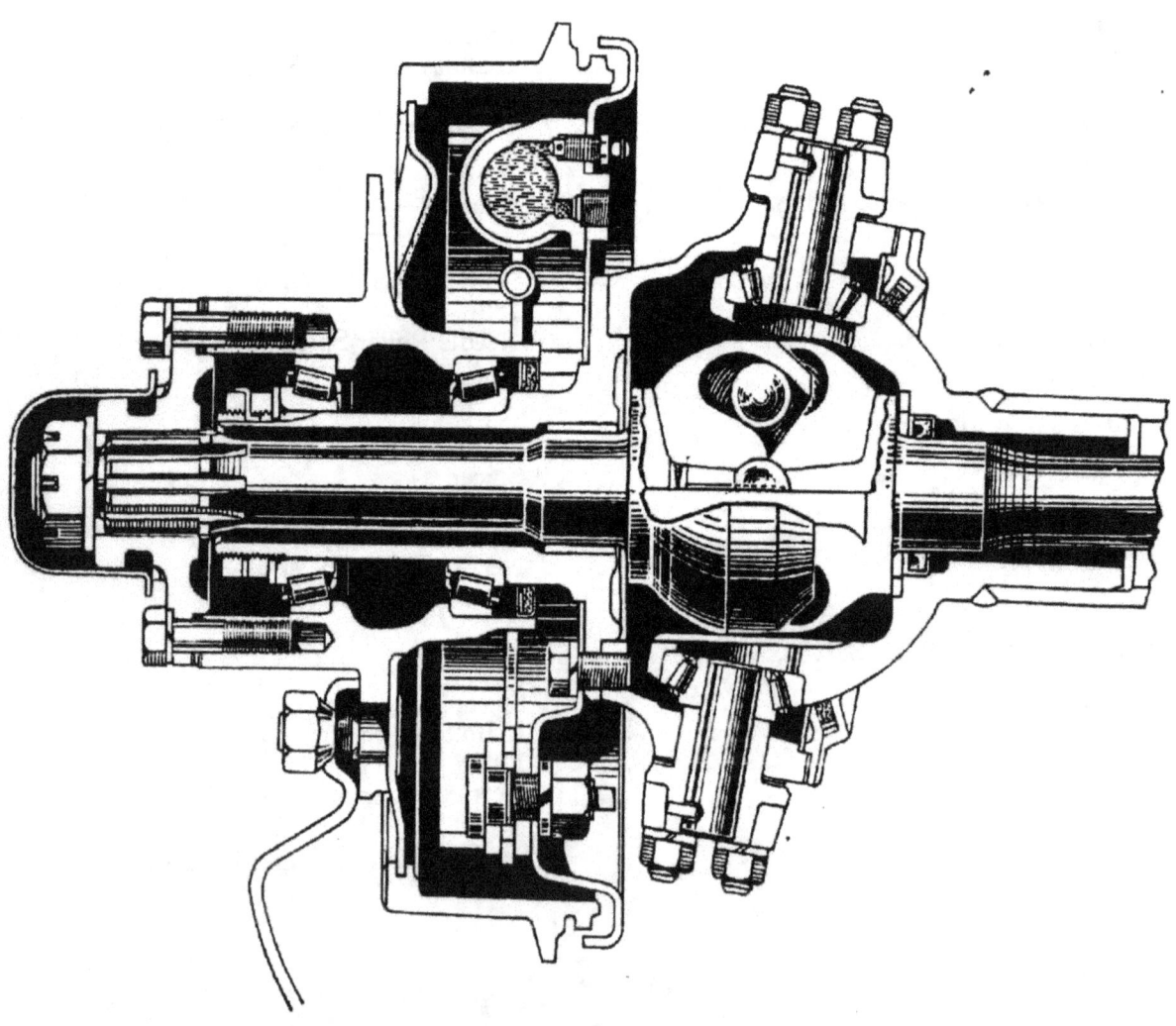

FIG. 1—FRONT WHEEL

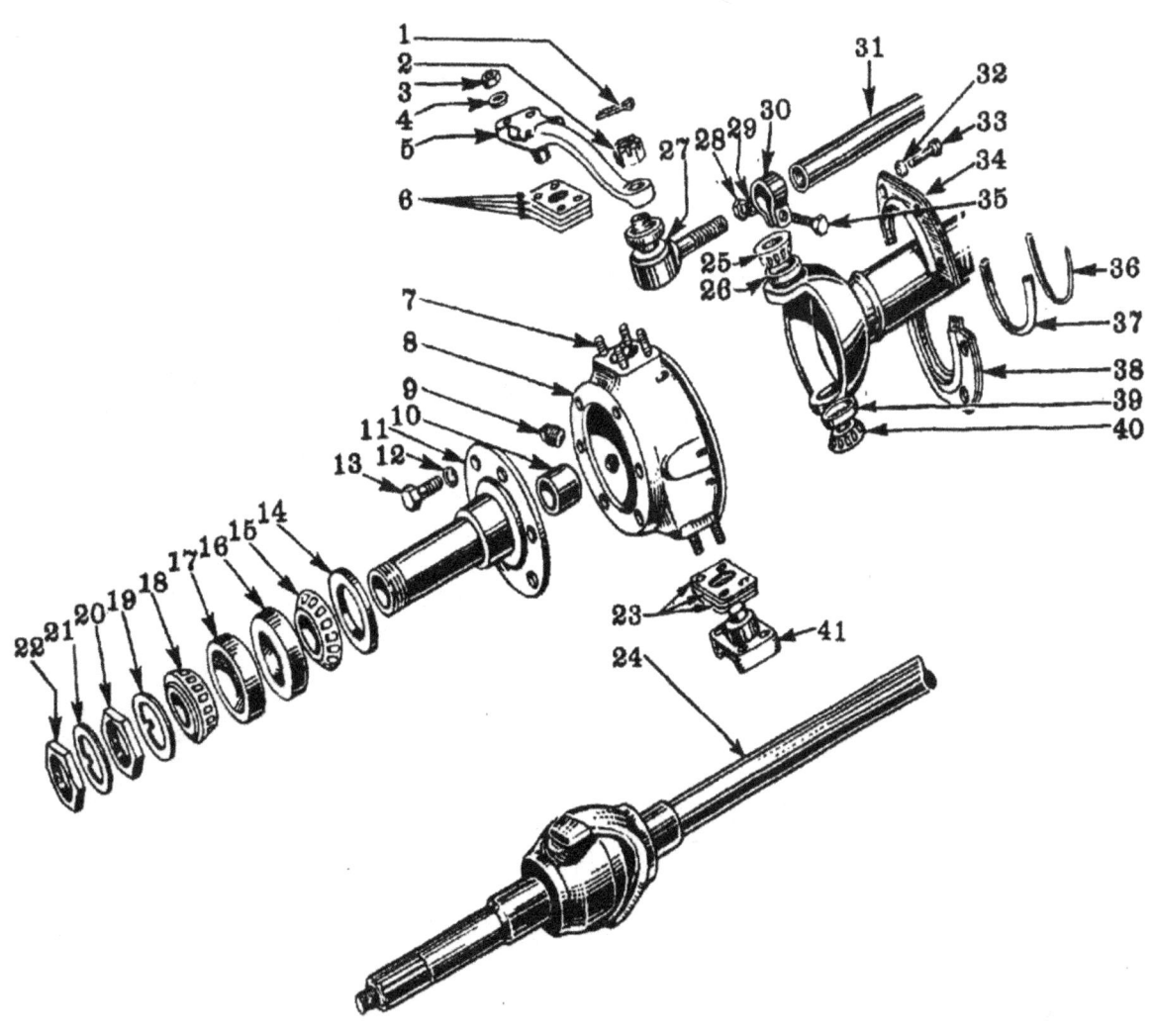

FIG. 2—FRONT AXLE, STEERING KNUCKLE AND WHEEL BEARINGS

(Bendix Universal Joint)

No.	Willys Part No.	Ford Part No.	Name	No.	Willys Part No.	Ford Part No.	Name
1	5152	72025-S	Tie Rod Stud Nut Cotter Pin	24	A-809	GPW-3206-A	Axle Shaft and Universal Joint Assembly (Bendix type) — Right Hand (Ford GPW-3207-A; Willys A-810 Left Hand)
2	10558	351059-S7	Tie Rod Stud Nut				
3	630698		Steering Arm Nut	25	52940	GP-3161	King Pin Bearing Cone and Rollers
4	5010	34807-S	Steering Arm Nut Lockwasher	26	52941	GP-3162	King Pin Bearing Cup
5	A-1712	GPW-3113	Upper Steering Arm—Left Hand (Ford GPW-3112; Willys A-1710 Right Hand)	27	A-847	GP-3290	Tie Rod Socket Assembly Left Hand (Ford GP-3289; Willys A-838 Right Hand)
6	A-830	GP-3117-A	King Pin Adjusting Shims	28	636575	34083-S2	Tie Rod Socket Clamp Nut
7	A-1714	357703-S	Steering Arm Stud—Upper (A-5504 Dowel Stud—Upper Outside Front and Inside Rear)	29	5010	34807-S	Tie Rod Socket Clamp Nut Lockwasher
				30	A-1706	51-3287	Tie Rod Socket Clamp
8	A-811	GP-3148-A2	Steering Knuckle Right Hand (Ford GP-3149-A2; Willys A-812 Left Hand)	31	A-1705	GPW-3281	Tie Rod Tube Right Hand (Ford GPW-3282; Willys A-1709 Left Hand)
9	5140	353064-S	Steering Knuckle Filler Plug	32	52510	34941-S	Knuckle Oil Seal Screw Lockwasher
10	A-853	GP-3205	Wheel Bearing Spindle Bushing	33	A-872	355483-S	Knuckle Oil Seal Screw
11	A-851	GP-3105	Wheel Bearing Spindle Assembly	34	A-813		Steering Knuckle Oil Seal Assembly—Half
12	5010	34807-S	Brake Disc Screw Lockwasher	35	A-1707	24916-S2	Tie Rod Socket Clamp Screw
13	A-877	355552-S	Brake Disc Screw	36	A-818	GP-3139	Steering Knuckle Oil Seal Felt Pressure Strip
14	A-864	GP-1177	Hub Oil Seal Assembly				
15	52942	GP-1201	Wheel Bearing Cone and Rollers	37	A-819	GP-3135	Steering Knuckle Oil Seal Felt—Half
16	52943	GP-1202	Wheel Bearing Cup	38	A-813	GPW-1088	Steering Knuckle Oil Seal Assembly—Half
17	52943	GP-1202	Wheel Bearing Cup				
18	52942	GP-1201	Wheel Bearing Cone and Rollers	39	52941	GP-3162	King Pin Bearing Cup
19	A-865	GP-1218	Wheel Bearing Lockwasher	40	52940	GP-3161	King Pin Bearing Cone and Rollers
20	A-866	GP-4252	Wheel Bearing Nut	41	A-828	GP-3140	Lower King Pin Bearing Cap
21	A-867	GP-1124	Wheel Bearing Nut Lockwasher				
22	A-866	GP-4252	Wheel Bearing Nut				
23	A-830	GP-3117-A	King Pin Adjusting Shims				

7. Remove wheel hub and drum assembly with bearings taking care not to damage oil seal.
8. Remove brake tube and brake backing plate screws, No. 13, Fig. 2.
9. Remove spindle No. 11.
10. The complete axle shaft and universal joint assembly No. 24 can now be pulled out of the axle housing. Care should be taken not to injure the outer oil seal assembly in axle housing.

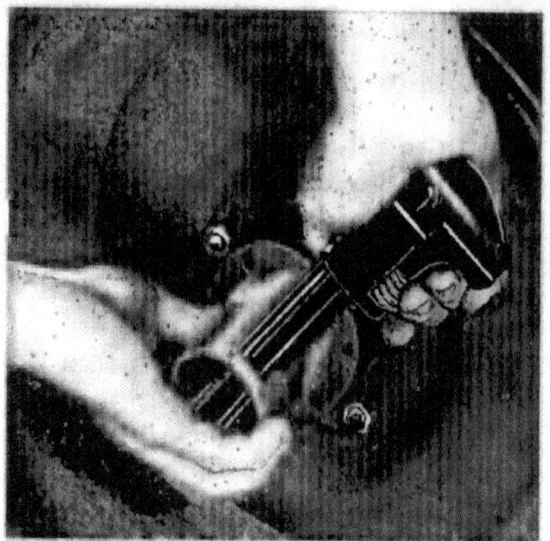

FIG. 4—REPLACING HUB NUT

Inspect the ball raceways for excessive wear. Fig. 6. If a raceway is badly worn the complete axle and universal joint assembly should be replaced. If the center ball pin is worn, it should be replaced. Inspect the center ball and the four driving balls for scratches, grooves or flat spots and replace if necessary. The driving balls (.875" diameter) are available from .003" undersize to .003" oversize in steps of .001" to permit selective fitting. If any or all of the driving balls are to be replaced the old ball or balls should be measured with a micrometer and the same size new balls used. Selective assembly is not required when installing a new center ball.

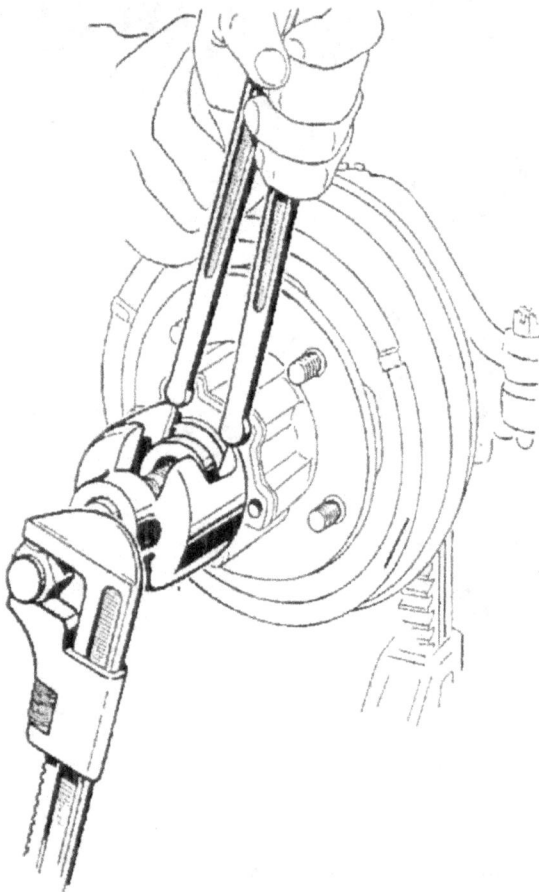

FIG. 3—PULLING DRIVING FLANGE

Disassembly (Bendix Joint)

After the axle shaft assembly has been removed, the universal joint may be disassembled as follows:

1. Wash the axle shaft and universal joint thoroughly in cleaning fluid.
2. Using a drift and hammer, drive out the retainer pin which locks the center ball pin in wheel end of shaft. See Fig. 5.
3. Bounce the wheel end of the shaft on a block of wood to cause the center ball pin to move into the drilled passage in the wheel end of the shaft.
4. Pull the two halves of the joint apart and then bend sharply at the universal. Rotate the center ball until grooved side lines up with ball raceway. This permits the adjacent ball to be moved past the center ball and removed from the joint. The remaining three driving balls and center ball will then drop out.

FIG. 5—REMOVING RETAINER PIN
(Bendix Joint)

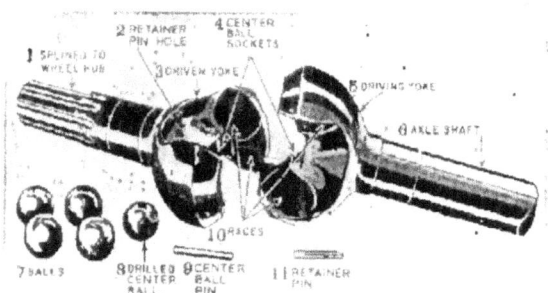

FIG. 6—AXLE SHAFT UNIVERSAL JOINT

Reassembly—(Bendix Joint)

1. Place the differential half of the axle shaft in a bench vise, with the ground portion of the shaft above the vise jaws.
2. Install the center ball (one with hole drilled in it) in its socket in the shaft, hole and groove facing you.
3. Drop the center ball pin into the drilled passage in the wheel half of the shaft.
4. Place the wheel half of the shaft on the center ball. Then slip three balls into the raceways.
5. Turn the center ball until the groove in it lines up with the raceway for the remaining ball as shown in Fig. 7. Slip the ball into the raceway and straighten up the wheel end of the shaft.
6. Turn the center ball until the center ball pin drops into the hole drilled in the ball.
7. Install the retainer pin (lock pin) and prick punch both ends to securely lock in place.

Disassembly (Rzeppa Joint)

After the shaft has been removed, the universal joint may be disassembled as follows, Fig. 4, Pg. 115:

1. Remove the three screws holding the front axle shaft to the joint and pull the shaft out of the splined inner race. To remove the axle shaft retainer, remove the retainer ring on the axle shaft.
2. Clean the universal joint in a suitable cleaning solution and lift out the axle centering pin.
3. Push down on various points of the inner race and cage until the balls can be taken out with the help of a small screw driver. Be careful not to damage parts.
4. After all the balls have been removed the inner race and cage can be turned over so the pilot cup is up, then remove the pilot cup.
5. There are two large elongated holes in the cage as well as four small holes. Turn the cage so

FIG. 7—ASSEMBLING UNIVERSAL JOINT BALLS

two bosses in the spindle shaft will drop into the elongated holes and lift out cage.
6. To remove the inner race turn it so one of the bosses will drop into an elongated hole in the cage, shift the race to one side, and lift out opposite side.

Reassembly (Rzeppa Joint)

1. Reassembly of the joint is the reverse of dismantling. Care should be exercised not to damage parts and see that they are clean of all dirt and grit.

To Reassemble Axle Shaft and Universal Joint Assembly to Housing

1. Clean all parts so that they are free from dust and foreign matter.
2. Enter universal joint and axle shaft assembly in the housing, taking care not to injure the outer and inner oil seals. Enter spline end of axle into the differential and push in until the shoulder on the universal joint stops against the axle.
3. Install wheel bearing spindle.

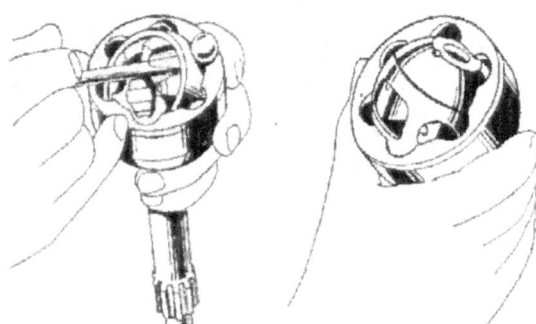

FIG. 8—DISMANTLING RZEPPA JOINT

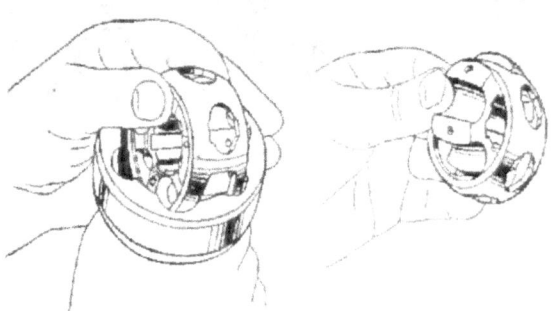

FIG. 9—REMOVING CAGE—RZEPPA JOINT

4. Install brake tube and bolt backing plate in position.
5. Grease wheel bearings and assemble bearings, wheel hub and drum on the wheel bearing spindle. Install wheel bearing washer, No. 19, Fig. 2, and adjusting nut, No. 20. Tighten nut until there is a slight drag on the bearings, when the wheel is turned, then back off approximately one-quarter turn. Install lock washer No. 21 and nut No. 22, tightening nut into place and then bending the lock washer over on the lock nut.
6. With Bendix joint install axle drive flange on axle splines, without shims.

Measure the space with a Feeler Gauge between the outer end of the wheel hub and the inner face of the drive flange, Fig. 10. This will determine the amount of shims to be installed. In order to have proper clearance in the universal joint, it is necessary to add a .040" shim to those required as measured by the Feeler Gauge.

Remove driving hub and install the correct amount of shims replacing driving hub on spline shaft and install six cap screws.

With Rzeppa joint be sure to install all the shims as removed when dismantling the axle drive flange. (.060" shims in each side).

7. Assemble axle shaft washer, nut and cotter pin.
8. Install the Hub Cap.
9. Assemble Wheel.
10. Check front wheel alignment, which is covered under "Steering".
11. Bleed Brake.

Make certain the steering knuckle universal joint is lubricated through the filler plug in the knuckle housing. See Lubrication Chart, Page 12.

FIG. 10—CHECKING FLANGE END PLAY

Replacing Steering Knuckle Bearing

Replacement of the bearings or bearing cups on the king pins necessitates removal of the hub and brake drum assembly, wheel bearings, axle shaft, wheel bearing spindle and the steering knuckle. The steering knuckle should be disassembled as follows:

1. Remove the eight screws No. 33, Fig. 2 which hold the oil seal retainers in place No. 34 and 38.
2. Remove the four nuts holding the lower king pin bearing cap, No. 41.
3. Remove the four nuts, No. 3 holding the upper steering arm in place, and remove brake hose shield also arm No. 5. The steering knuckle No. 8 can now be removed from the axle.
4. Wash all parts in cleaning solution and inspect bearings and races for scores, cracks or chips. All damaged parts should, of course, be replaced.

In the event the bearing cups are damaged, they can be removed by the use of a driver or a suitable drift.

Reassembling Steering Knuckle

Reverse the procedure outlined above to reassemble the unit. When reinstalling the steering knuckle, sufficient shims must be installed under the arm and lower bearing cap so the proper tension will be maintained on the bearing. The shims are available in thicknesses of .003", .005", .010" and .030".

Install one each of the .003", .005", .010" and .030" shims over studs on the steering knuckle, top and bottom. Install the arm, and lower bearing cap, lock washers, and nuts, and tighten securely. Check the tension of the bearings by hooking checking scale in the hole in the arm at tie rod and socket, either remove or add shims until the load is approximately 25 to 35 inch pounds, without oil seal assembly in position. Make sure there are the same thickness of shims between arm and knuckle as between lower cap and knuckle.

Steering Knuckle Oil Seal

Replacement of the oil seal No. 34 and 38 can be made very easily be merely removing the eight screws which hold the oil seal in place. Before reinstalling the oil seal examine the spherical surface of the axle for scores or scratches which might damage the seal. Roughness of any kind should be smoothed down with emery cloth.

Reinstall both upper and lower halves of the oil seal, making sure that the felt fits snugly at the points where the upper and lower halves come together, Fig. 11.

After driving in wet, freezing weather swing the

FIG. 11—STEERING KNUCKLE OIL SEAL

front wheels from right to left to remove moisture adhering to the oil seal and the spherical surface of the universal joint housing. This will prevent freezing with resulting damage to oil seal felts. Should the car be stored for any period of time, coat these surfaces with light grease to prevent rusting.

Axle Shaft Outer Oil Seal

In the event it should be necessary to replace the axle shaft outer oil seal, remove the axle shaft and universal joint as described under the "Axle Shaft and Universal Joint".

The oil seal is a light press fit in the housing and will require a tool or puller for removal. Insert the ends of the puller behind the oil seal and tap the end of the puller lightly with a hammer. See Fig. 12.

Before installing a new seal make sure it has been soaked thoroughly in oil. This will not only make the leather more pliable but will avoid it being burned by friction with the axle shaft when the vehicle is driven.

After placing the oil seal in position in the housing, it can easily be driven in place by using a driver or a block of hard wood and a hammer.

When installing axle shaft and universal joint assembly exceptional care should be used to prevent damage to the oil seal.

Removing and Overhauling Differential

Inasmuch as the front axle differential assembly is identical with that of the rear axle assembly, refer to the section under "Rear Axle" for the proper procedure to follow in dismantling and assembling differential.

Steering Tie Rod and Bell Crank

These parts being part of the steering mechanism, they are covered in the section under "Steering."

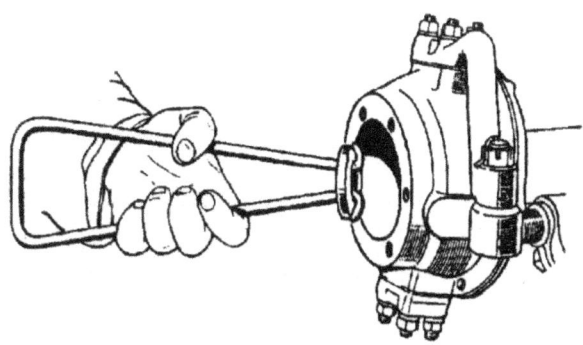

FIG. 12—REMOVING OIL SEAL

FRONT AXLE TROUBLES AND REMEDIES

SYMPTOMS	PROBABLE REMEDY
Hard Steering	
Lack of lubrication	Lubricate
Tires soft	Inflate to 30 lbs.
Tight steering	Adjust. See "Steering" Section
Low Speed Shimmy or Wheel Fight	
Spring Clips and Shackles loose	Readjust or replace
Front axle shifted	Broken spring center bolt
Insufficient toe-in	Adjust
Improper caster	Reset
Steering System loose or worn	Adjust or overhaul steering gear, front axle or steering parts
Twisted Axle	Straighten or adjust
High Speed Shimmy or Wheel Fight	
Check conditions under "Low Speed Shimmy"	
Tire pressures low or not equal	Inflate to 30 lbs.
Wheels out of balance	Balance—Check for patch
Wheel runout	Straighten
Radial runout of tires	Mount properly
Wheel camber	Same on both wheels
Front springs settled or broken	Repair or replace
Bent steering knuckle arm	Straighten or replace
Shock absorbers not effective	Replace
Steering gear loose on frame	Tighten
Front springs too flexible	Over lubricated
Tramp	
Wheels unbalanced	Check and balance
Wandering	
Improper toe-in	Adjust—Check for bent steering knuckle arm
Broken front spring main leaf	Replace
Axle shifted	Spring center bolt broken
Loose spring shackles or clips	Adjust or replace
Improper caster	Reset
Tire pressure uneven	Inflate to 30 lbs.
Tightness in steering system	Adjust
Loose wheel bearings	Adjust
Front spring settled or broken	Repair or replace

FRONT AXLE TROUBLES AND REMEDIES—Continued

SYMPTOMS	PROBABLE REMEDY

Axle Noisy on Pull
- Pinion and Ring gear adjusted too tight............Readjust
- Pinion bearings rough...............................Replace

Axle Noisy on Coast
- Excessive back lash at ring and pinion gear.........Readjust
- End play in pinion shaft............................Readjust
- Rough bearing.......................................Replace

Axle Noisy on Coast and Pull
- Ring and pinion adjusted too tight..................Readjust
- Pinion set too deep in ring gear....................Readjust
- Pinion bearing loose or worn........................Readjust or replace

Back Lash
- Axle shaft universal joint worn.....................Replace
- Axle shaft improperly adjusted......................Readjust
- Worn differential pinion washers....................Replace
- Worn propeller shaft universal joints...............Replace

Emergency
Where difficulty is experienced with front axle differential making the vehicle inoperative, remove axle driving flanges. This will allow bringing vehicle in under its own power. Be sure front wheel drive shift lever is in the forward (disengaged) position.

FRONT AXLE SPECIFICATIONS

Front Axle
- Make...Spicer
- Drive..Through springs
- Type...Full Floating
- Road Clearance.......................................8 7/16"

Differential
- Drive..Hypoid
- Gear Ratio...4.88:1
- Bearings...Timken Roller 2
- Adjustment...Shims
- Gears (Pinion).......................................2

Oil Capacity (Pts.)..................................See Lubrication Chart, Page 12

Steering Knuckle Thrust Up and Down
Adjusted by shims, should have 25 to 35 inch-pounds pull without oil seal assembly in position.

Steering Knuckle
Bearings Upper and Lower......Timken Roller

Turning Arc...26°

Tie Rods
- Number...2
- Right hand length center to center........24 1/4"
- Left hand length center to center.........17 11/32"
- Tie rod ends................Serviced as a unit

Steering Geometry
- King Pin inclination.................................7 1/2°
- Wheel camber...1 1/2°
- Wheel caster...3°
- Wheel toe-in...3/64"-1/32"

Bearings
- Cone and Roller......................................24780
- Differential side....................................Timken
- Cup..24721
- Shims................003", .005", .010", .030"
- Pinion Shaft...Timken
  - Cone and roller......Front 31593, Rear 02872
  - Cup..................Front 31520, Rear 02820
  - Shims................003", .005", .010", .030"
- Wheel Hub..Timken
  - Cone—Roller......Inner 18590, Outer 18590
  - Cup..............Inner 18520, Outer 18520
- Steering Knuckle.....................................Timken
  - Cone and roller...Upper 11590, Lower 11590
  - Cup..............Upper 11520, Lower 11520
- Steering Bell Crank
  - Bearing..........Needle, Torrington B1210

REAR AXLE

The rear axle Fig. 1 is the full-floating type designed so that the axle shafts can be removed without disturbing the wheels. The differential drive is of the hypoid type, having a ratio of 4.88 to 1; 8 tooth drive pinion, 39 tooth drive gear.

The axle shafts are splined at the inner end to fit the splines in the differential side gears. The outer ends of the axle shafts are equipped with integral driving flange which bolts to the rear wheel hubs. The wheels are each supported on two taper roller bearings on the axle housing. The bearing races are pressed into the wheel hub and the adjustment of the bearings made by adjusting nuts on end of housing.

A steel cover is used on the rear of the axle housing to permit inspection and flushing of the differential assembly.

It is necessary to use a hypoid gear lubricant. See Lubrication Chart, Page 12. Various types of hypoid lubricants must not be mixed. If the brand is changed it is best to drain and flush the rear axle housing before installing the new lubricant. The rear axle lubricant level should be checked every 1,000 miles. The lubricant should be drained and axle refilled to the bottom level of the filler hole every 6,000 miles.

Removing Rear Axle from Vehicle

To remove the rear axle first raise the rear end of vehicle with a hoist and support frame ahead of rear springs, then remove wheels and disconnect propeller shaft at rear universal joint by removing U-bolts. Disconnect brake line from hose at frame and remove lock clip. Remove spring clips, then remove spring front bolts, the rear axle can then be removed.

Axle Shaft

To remove the axle shaft, No. 29, Fig. 1, the following procedure should be followed:

1. Remove the six cap screws, No. 35 holding driving flange to wheel hub.

2. Remove axle shaft, two flange screws can be used in the threaded holes in the axle flange.

If rear axle shaft is broken, use a piece of stiff wire and make a loop on one end, slide the wire into axle housing and over broken end of shaft for a sufficient distance that when the wire is pulled out, the loop will bind on shaft and remove it from the differential side gear.

To replace axle shaft, reverse of the above operations are necessary, however, care should be taken when installing shaft that the inner oil seal, No. 28 at differential is not damaged.

Removing and Overhauling Differential

Before disassembling differential, it is advisable to determine through inspection the cause of difficulty or failure of the parts.

Drain lubricant from gear carrier housing and then remove gear carrier cover, No. 6, Fig. 1, flushing out unit thoroughly so that the parts can be carefully inspected.

After the inspection if it is determined that the differential should be completely overhauled, the rear axle assembly should be removed from the vehicle and the following procedure followed:

1. Remove axle shaft as covered in the foregoing paragraph.

2. Remove the four bolts No. 38 which hold the two differential side bearing caps No. 36 in position.

3. Using two pry bars, one on each side of the ring gear parallel with the tube of the axle housing, pry out the differential assembly as shown in Fig. 2.

4. Remove the cap screws No. 3, Fig. 1, holding the bevel drive gear No. 22 on the differential case No. 19.

5. Remove the drive gear from the differential case by tapping it lightly with a lead hammer.

6. The differential shaft No. 13 is held in place by a lock pin No. 25, using a small punch, drive out the lock pin to allow the differential shaft to be removed, Fig. 3.

7. Remove differential pinion gears, No. 15 and 24, Fig. 1, care being taken not to lose the pinion thrust washers, No. 16 and 23.

8. Remove axle shaft gears No. 12 and No. 26 and thrust washers No. 11 and 27.

9. Remove universal joint end yoke assembly No. 59 with puller shown in Fig. 4.

10. With a lead hammer drive on end of pinion shaft which will force the pinion into the gear carrier housing.

11. With bearing race removing tool drive out the front pinion shaft bearing cup No. 65 and oil seal No. 61.

12. To remove the drive pinion rear bearing cone use bearing removing tool or press off in an arbor press, Fig. 5.

    When replacing the cone, select a sleeve the diameter of the cone, so the rollers or cage will not be damaged.

Wash all parts in suitable cleaning fluid, taking care not to lose any of the shims No. 39, Fig. 1 which adjust the pinion shaft bearing running tolerance.

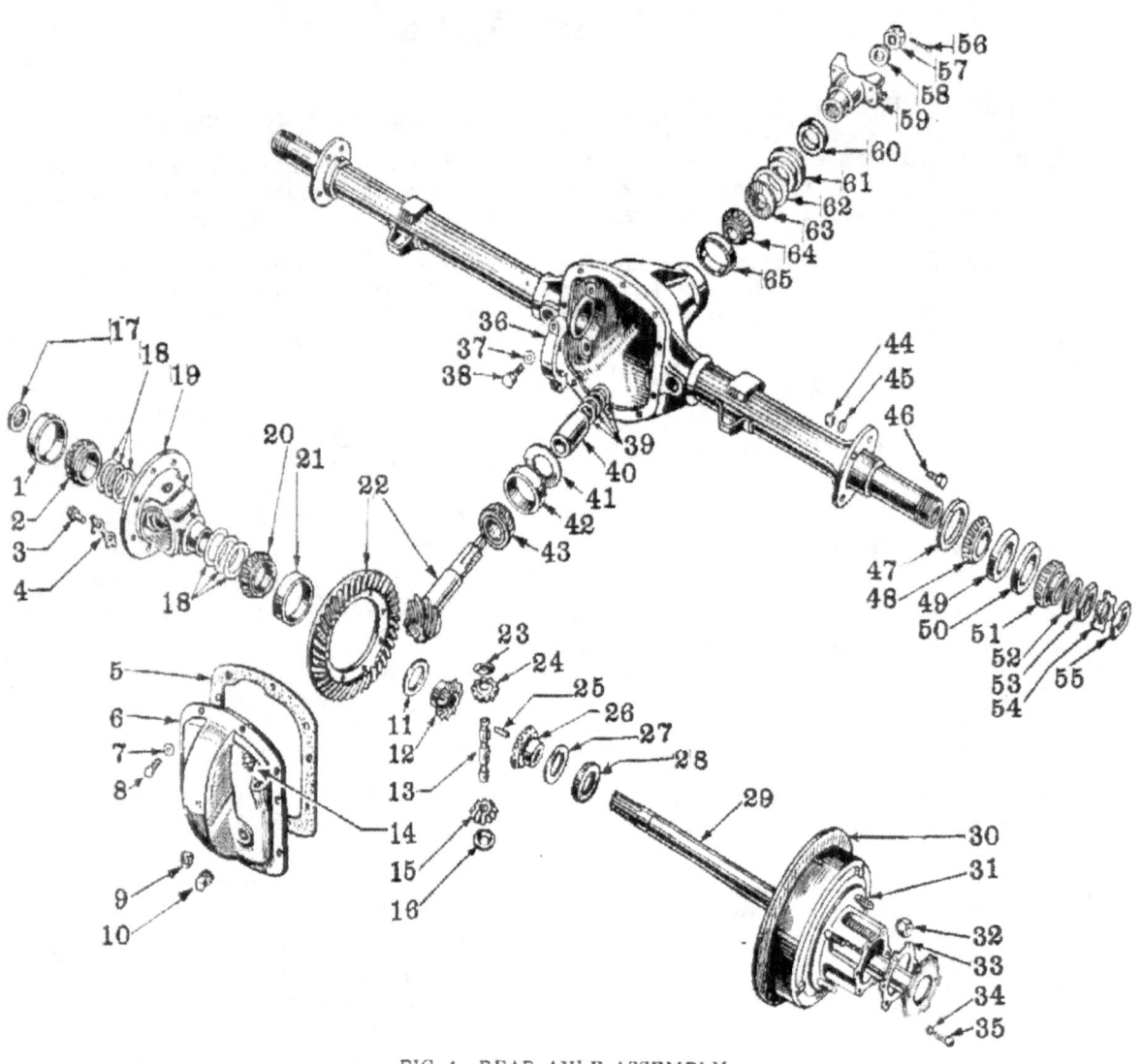

FIG. 1—REAR AXLE ASSEMBLY

No.	Willys Part No.	Ford Part No.	Name
1	52881	GP-4222	Differential Bearing Cup
2	52880	GP-4221	Differential Bearing Cone and Rollers
3	A-871	355511-S	Hypoid Bevel Drive Gear Screw
4	A-792	GP-4231	Drive Gear Screw Lock Strap
5	A-782	GP-4035	Gear Carrier Cover Gasket
6	A-781	GP-4016	Gear Carrier Cover
7	52510	34941-S	Gear Cover Screw Lockwasher
8	51623	20346-S2	Gear Cover Screw
9	636577	358048-S	Axle Housing Drain Plug
10	636538	353051-S	Gear Cover Filler Plug
11	A-795	GPW-4228	Differential Bevel Side Gear Thrust Washer
12	A-794	GP-4236	Differential Bevel Side Gear
13	A-798	GP-4211	Differential Bevel Pinion Mate Shaft
14	A-870	GP-4022	Differential Vent Plug
15	A-796	GPW-4215	Differential Bevel Pinion Mate
16	A-797	GP-4230	Differential Bevel Pinion Mate Thrust Washer
17	A-779	GP-3034	Oil Seal—Carrier End
18	A-784	GP-4229-A	Differential Adjusting Shims
19	A-793	GP-4206	Differential Case
20	52880	GP-4221	Differential Bearing Cone and Rollers
21	52881	GP-4222	Differential Bearing Cup
22	A-780	GPW-4209	Hypoid Bevel Drive Gear and Pinion Set
23	A-797	GP-4230	Differential Bevel Pinion Mate Thrust Washer
24	A-796	GP-4215	Differential Bevel Pinion Mate
25	636360	GP-4241	Differential Bevel Pinion Mate Shaft Lock Pin
26	A-794	GP-4236	Differential Bevel Side Gear
27	A-795	GP-4228	Differential Bevel Side Gear Thrust Washer
28	A-779	GP-3034	Oil Seal Carrier End
29	A-901	GPW-4234	Rear Axle Shaft—Right (Ford GP-4235; Willys A-902—Left)
30	A-472	GP-1111	Brake Drum
31	A-474	GP-1107	Wheel Hub Bolt—R.H. Thread (Ford GP-1108; Willys A-473 L.H. Thread)
32	A-476	GP-1012	Wheel Hub Bolt Nut—R.H. Thread (Ford GP-1013; Willys A-475 L.H. Thread)
33	A-904	GP-4032	Axle Shaft Gasket
34	5010	34807-S	Rear Axle Drive Shaft Screw Lockwasher
35	A-760	GP-1110	Rear Axle Drive Shaft Screw
36	A-764	GP-4224	Differential Bearing Cap
37	636528	34922-S	Differential Bearing Cap Screw Lockwasher
38	636527	355699-S	Differential Bearing Cap Screw
39	A-803	GP-4659-A	Pinion Bearing Adjusting Shim (Front)
40	A-799	GP-4668	Drive Pinion Bearing Spacer
41	A-800	GP-4660-A	Pinion Bearing Adjusting Shim (Rear)
42	52877	86H-4616	Drive Pinion Bearing Cup—(Rear)
43	52876	86H-4621	Drive Pinion Bearing Cone and Rollers—(Rear)
44	636575	34083-S2	Brake Disc Screw Nut
45	5010	34807-S	Brake Disc Screw Lockwasher
46	A-903	355578-S	Brake Disc Screw
47	A-864	GP-1177	Hub Oil Seal Assembly
48	52942	GP-1201	Hub Bearing Cone and Rollers
49	52943	GP-1202	Hub Bearing Cup
50	52943	GP-1202	Hub Bearing Cup
51	52942	GP-1201	Hub Bearing Cone and Rollers
52	A-865	GP-1218	Outer Wheel Bearing Washer
53	A-866	GP-4253	Outer Wheel Bearing Nut
54	A-867	GP-1124	Outer Wheel Bearing Lockwasher
55	A-866	GP-4252	Outer Wheel Bearing Nut
56	636571	357202-S	Drive Pinion Nut Cotter Pin
57	636569	356126-S	Drive Pinion Nut
58	636570	356504-S	Drive Pinion Nut Washer
59	A-1445	GP-4342	Universal Joint End Yoke Assembly
60	636568	GP-4606	Universal Joint End Yoke Dust Shield
61	636565	GP-4676	Pinion Leather Oil Seal
62	636565	GP-4661	Pinion Leather Oil Seal Gasket
63	636566	GP-4619	Drive Pinion Oil Slinger
64	52878	GP-4620	Drive Pinion Bearing Cone and Rollers (Front)
65	52879	GP-4628	Drive Pinion Bearing Cup (Front)

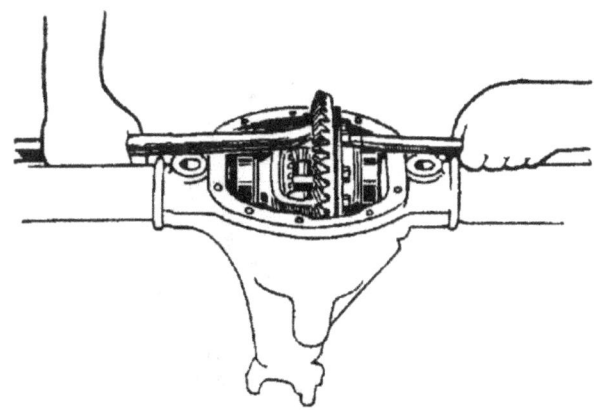

FIG. 2—REMOVING DIFFERENTIAL

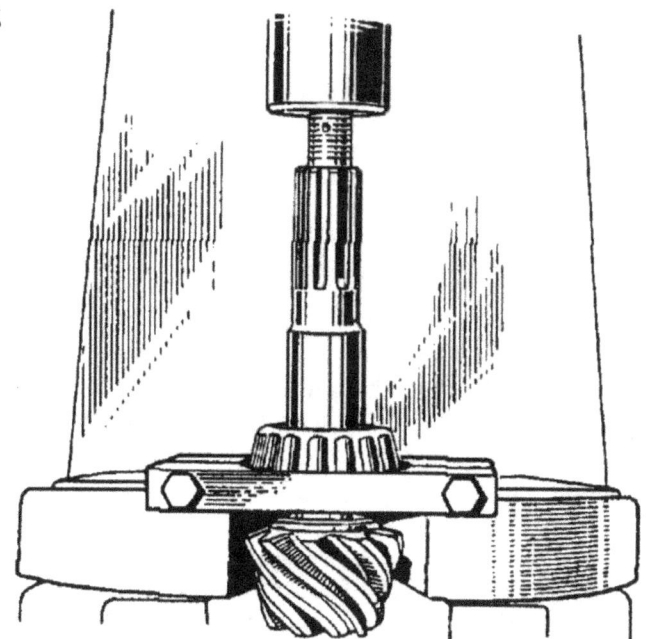

FIG. 5—REMOVING PINION BEARING CONE

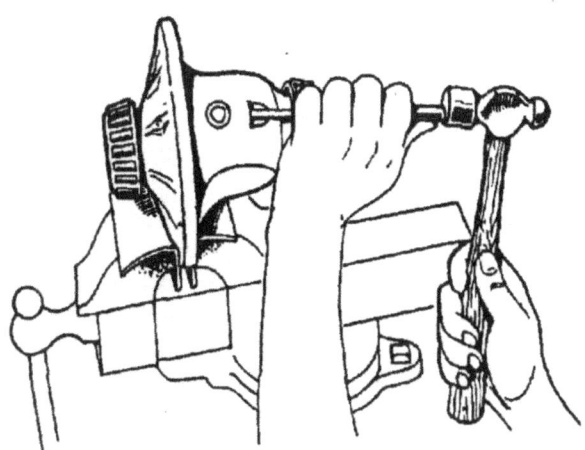

FIG. 3—REMOVING LOCK PIN

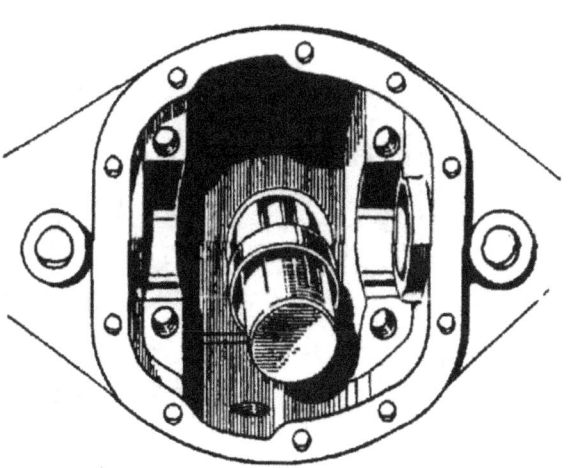

FIG. 6—REMOVING PINION BEARING CUP

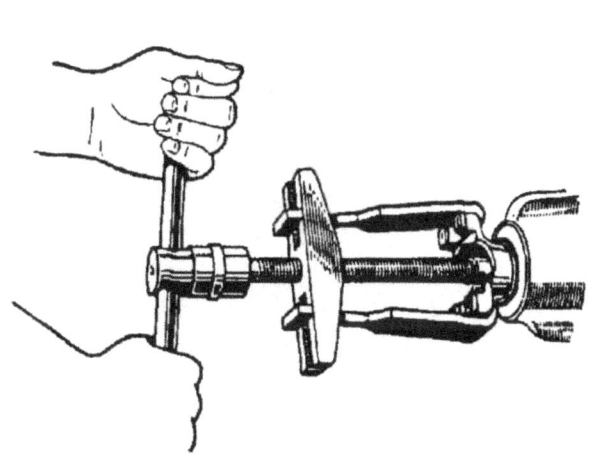

FIG. 4—REMOVING U. J. END YOKE

Adjusting the Drive Pinion

Before attempting to adjust the ring gear or differential parts the drive pinion should be carefully checked and adjusted. The setting of the pinion is accomplished by the use of shims No. 41, Fig. 1 between the rear bearing cup No. 42 and the housing. These shims are available in thickness of .003", .005" and .010".

If the rear bearing cup is to be replaced or if the pinion setting is to be changed, a suitable tool for removing and installing the drive pinion bearing cup in the differential housing should be used, Fig. 6 and 7.

Adjusting Pinion Bearings

The correct pinion bearing adjustment is obtained by shims between the pinion bearing spacer and the front bearing cone, Fig. 8, until a slight drag is obtained when pinion flange is turned by hand.

Install the pinion and the rear bearing in the housing, place the front bearing in position and then install the propeller shaft flange. This operation can be performed very easily by using a block of wood to support the pinion, Fig. 9. Do not install the pinion oil seal until the pinion setting has been checked with the pinion setting gauge.

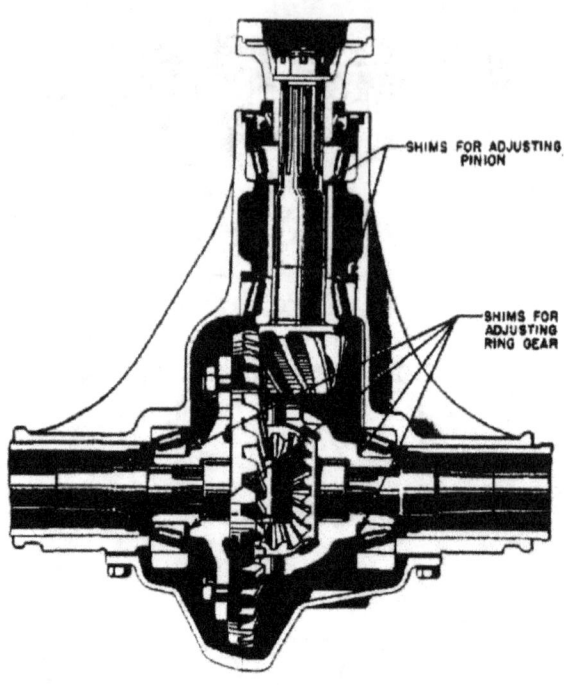

FIG. 8—ADJUSTING SHIMS

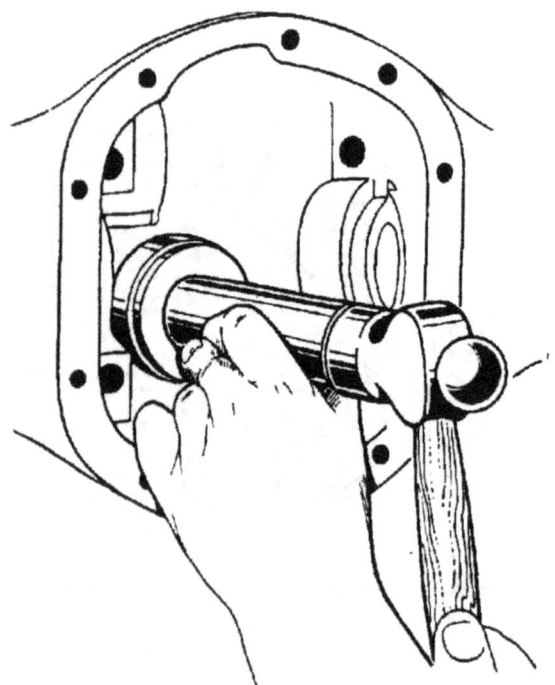

FIG. 7—INSTALLING PINION BEARING CUP

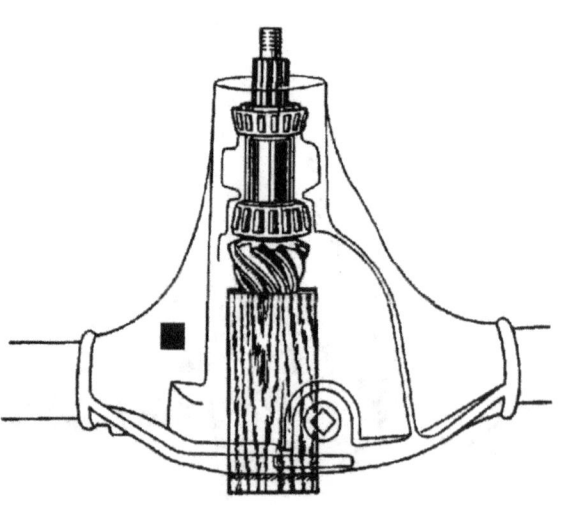

FIG. 9—INSTALLING PINION

Adjusting the Drive Pinion Setting

Proper adjustment of the drive pinion is facilitated by the use of a pinion setting gauge. See Fig. No. 10 and 11. This gauge is fitted with a micrometer for measuring the thickness of the shims required to properly locate the pinion in the differential housing so it will have correct tooth contact with the bevel drive gear.

All axle drive pinions are marked with an electric needle on the back face to show the correct setting. A pinion marked zero will show a reading .719" on the micrometer when properly adjusted. The dimension .719" represents the standard setting from the back face of the pinion to the center line of the differential case bearing. Therefore, a pinion marked +2 is .002" longer than the standard and will show a micrometer reading of .717" when properly adjusted. Likewise a pinion marked —4 is .004" shorter than the standard and will show a micrometer reading of .723" when properly adjusted.

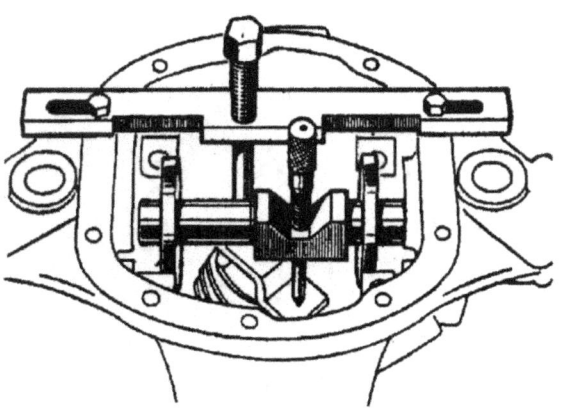

FIG. 10—PINION SETTING GAUGE

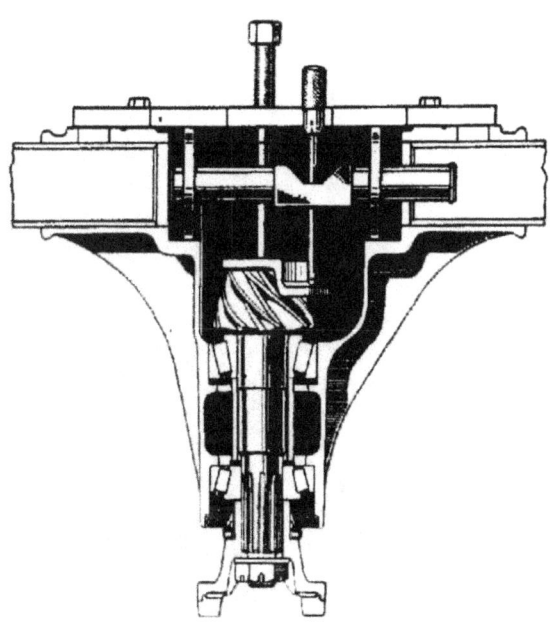

FIG. 11—PINION SETTING GAUGE

Assembling Differential Unit

Carefully examine the surfaces of the differential case and bevel gear to make sure there are no foreign particles or burrs on the two contacting surfaces. Line up the cap screw holes in the bevel gear with those on the differential case and then put it into position on the case by tapping it lightly with a lead hammer. Install the cap screws which hold the bevel gear to the differential case. After the cap screws have been tightened securely, make certain that the cap screw locks are bent around the cap screw heads so there is no possibility of the screws working loose.

The relative assembling position of the internal parts of the differential are shown in Fig. 8. Reassemble the differential pinions, sidegears, thrust washers and shaft in place and install differential shaft lock pin. In order to prevent the lock pin from working out, use a punch to peen over some of the metal of the differential case.

The adjustment of Differential bearings is maintained by the use of Shims between differential case and bearing cones with an .008" pinch fit when assembled in the axle housing.

Remove bearing cones and shims as shown in Fig. 13, reinstall bearing cones without shims, place assembly in Housing with bearing cups and force assembly to one side and check the clearance between bearing cup and case with a feeler gauge as shown in Fig. 12.

After clearance has been determined add .008" this will give thickness of shims required for proper bearing adjustment.

Remove differential bearings and install equal thickness of shims on each side and replace bearings.

Install the differential assembly in the housing. This operation can be facilitated by cocking the bearing cups slightly when the differential is placed in the housing and then tapping them lightly with a lead hammer, see Fig. 14.

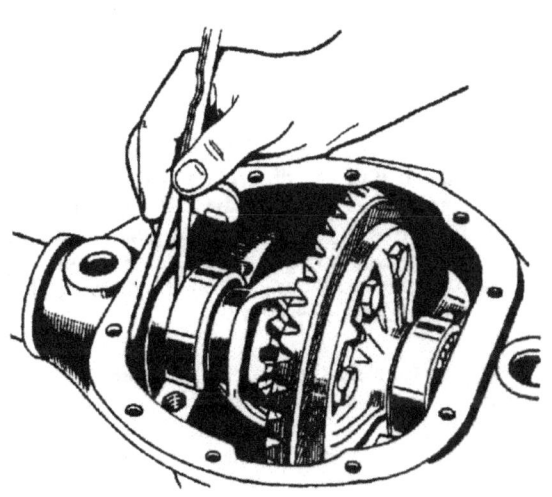

FIG. 12—CHECKING DIFFERENTIAL BEARINGS

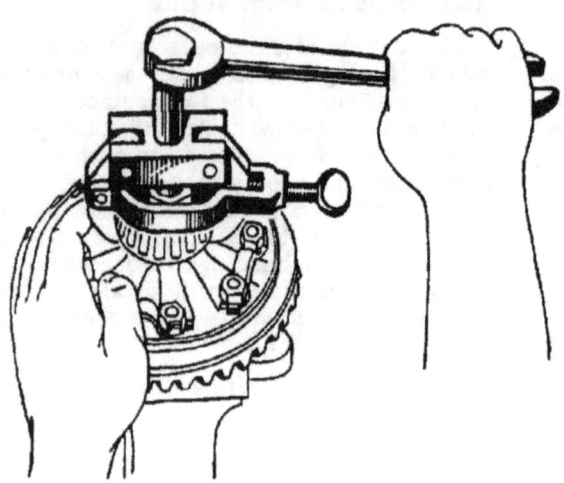

FIG. 13—REMOVING DIFFERENTIAL BEARING CONE

Total backlash between the bevel gear and pinion should be within .005" to .007". This can be checked by mounting a dial indicator on the rear axle housing with the button of the indicator against one of the gear teeth, Fig. 16. Moving the ring gear by hand will indicate the amount of backlash.

In the event the backlash is not within the limits mentioned above, it will be necessary to change the shims back of the differential case bearings. Changing the position of a .005" shim from one side to the other will change the amount of backlash approximately .0035"

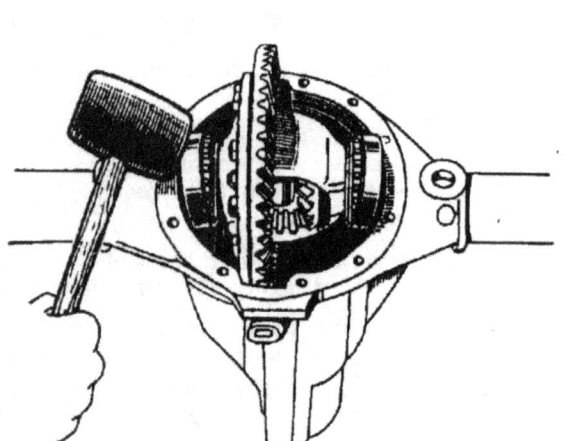

FIG. 14—INSTALLING DIFFERENTIAL

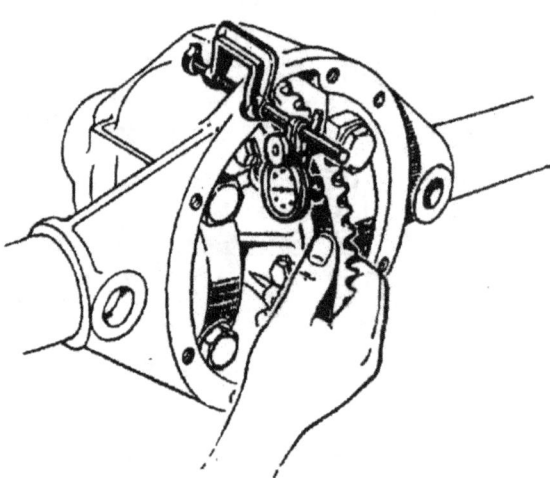

FIG. 15—CHECKING RUNOUT

After the bearing cups are firmly in place in the housing, install the differential bearing caps. It is important that the caps be installed in the same position in which they were originally assembled. Each cap should be installed so numeral corresponds with the numeral on the housing. Torque wrench reading, 38-42 ft. lbs.

After securely tightening the differential bearing caps, check the back face of the ring gear for runout, Fig. 15. Total indicator reading in an excess of .003" indicates a sprung differential case or an improperly installed bevel gear. In either case the assembly must be taken apart and rechecked thoroughly.

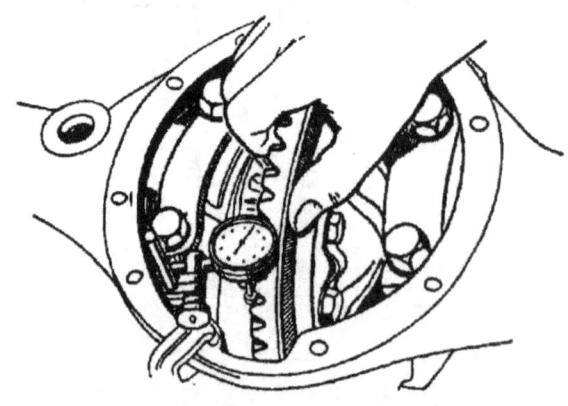

FIG. 16—CHECKING BACK-LASH

In order to assist in determining whether the gears are properly adjusted, paint the bevel gear with red lead or similar substance and turn the bevel gear so the pinion will make an impression on the teeth. Correct procedure to follow in the event of an unsatisfactory tooth contact is shown in Fig. 17.

After the differential has been assembled and adjusted, the pinion shaft oil seal should be installed. Remove universal joint flange and with oil seal replacing tool, Fig. 18 install oil seal. Fig. 19 gives dimensions of oil seal replacing tool. Install universal flange and tighten nut solidly in place, then install cotter pin.

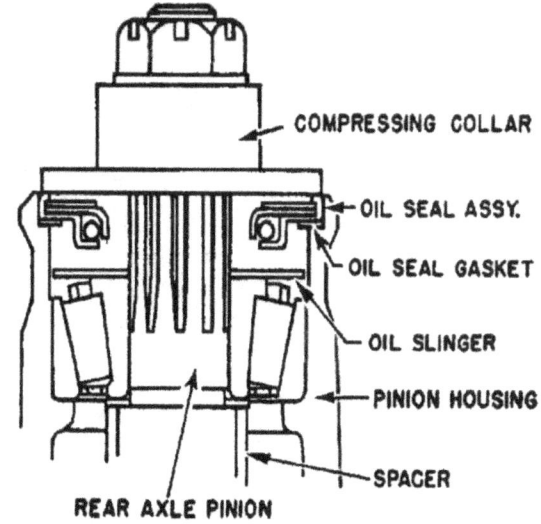

FIG. 18—INSTALLING PINION OIL SEAL

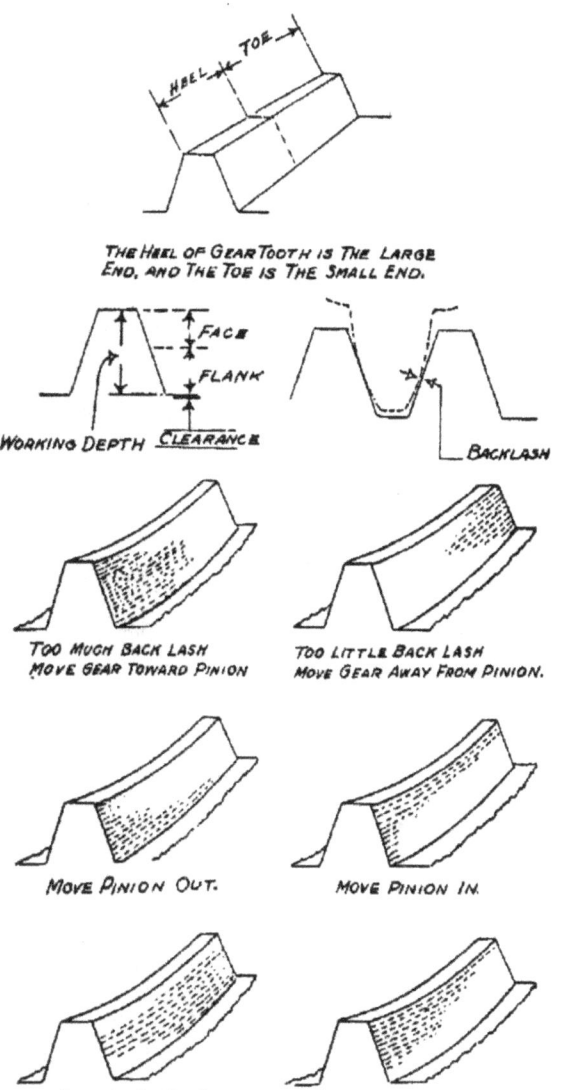

Install axle shafts as instructed under "Axle Shaft" and replace housing cover with new gasket. Fill differential housing with proper amount of hypoid lubricant. See Lubrication Chart, Page 12.

Install axle under vehicle in reverse order of removal, after which bleed the rear brake cylinders to remove any air from the lines, first making certain that there is an ample supply of fluid in the brake master cylinder reservoir. See Section "Brakes" for further instructions.

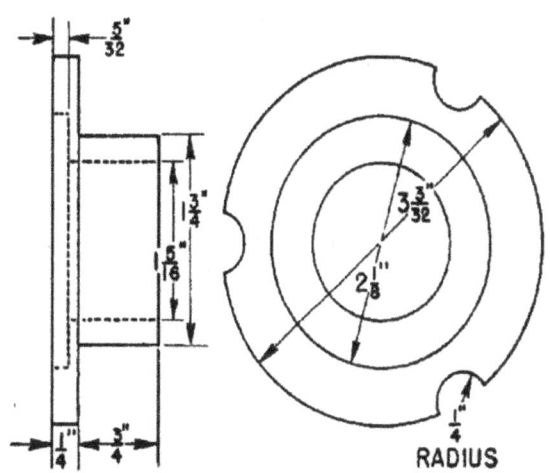

FIG. 17—TOOTH CONTACT

FIG. 19—OIL SEAL COMPRESSING COLLAR

REAR AXLE TROUBLES AND REMEDIES

                SYMPTOMS   PROBABLE REMEDY

Axle Noisy on Pull and Coast
- Excessive back lash bevel gear and pinion......Adjust
- End play pinion shaft......................Adjust
- Worn pinion shaft bearing.................Replace
- Pinion set too deep in ring gear.............Adjust
- Pinion and bevel gear too tight.............Adjust

Axle Noisy on Pull
- Pinion and bevel gear improperly adjusted.....Adjust
- Pinion bearings rough....................Replace
- Pinion bearings loose.....................Adjust

Axle Noisy on Coast
- Excessive lash in bevel gear and pinion........Adjust
- End play in pinion shaft....................Adjust
- Improper tooth contact.....................Adjust
- Rough bearings...........................Replace

Backlash
- Worn differential pinion gear washers.........Replace
- Excessive lash in bevel gear and pinion........Adjust
- Worn universal joints.......................Replace

Emergency

Should difficulty be experienced with differential or propeller shaft the vehicle may be driven in by removing the rear axle shafts and propeller shaft.

Place front wheel drive lever in rear (engaged) position. This will allow front wheel drive to propel the vehicle.

REAR AXLE SPECIFICATIONS

Rear Axle
- Type.............................Full floating
- Make.............................Spicer
- Drive.............................Thru springs
- Road Clearance...................8 3/16"

Differential
- Type.............................Hypoid
- Ratio............................4.88:1
- Bearings.........................Timken Roller
- Differential Pinion Gears........2
- Oil capacity.....................See Lubrication Chart, Page 12
- Adjustment......................Shims .003", .005", .010", .030"

Pinion Shaft
- Bearings........................Two Timken Roller
- Adjustment.....................Shims .003", .005", .010"

Bevel and Pinion Gear
- Back Lash.......................005"—.007"
- Adjustment.....................Shims .003", .005", .010", .030"

Bearings
- Make—Differential Side.......... Timken
- Cone and roller.................. 24780
- Cup.............................. 24721
- Make—Pinion Shaft.............. Timken
- Cone and roller..................Front 02872 Rear 31593
- Cup.............................Front 02820 Rear 31520
- Shims...........................003", .005", .010", .030"
- Make—Wheel Hub................ Timken
- Cone and Roller.................Inner 18590 Outer 18590
- Cup.............................Inner 18520 Outer 18520

BRAKES

The brake system is comparatively simple, Fig. 1. The foot or service brakes are of the internal expanding type hydraulically actuated in all 4 wheels. The hand brake is mechanically operated through a cable and conduit to an external type brake mounted at the rear of the transfer case on the propeller shaft. The foot brakes are of the Bendix, two shoe, double anchor type and have nickle-chromium alloy iron drums.

In order to thoroughly understand the operation of the hydraulic brake system, it is necessary to have a good knowledge of the various parts and their functions, and to know what takes place throughout the system during the application and the release of the brakes.

The piston in the master cylinder, Fig. 4 receives mechanical pressure from the brake pedal and exerts pressure on the fluid in the lines, building up the hydraulic pressure which moves the wheel cylinder pistons. The primary cup is held against the piston by the piston return spring which also holds the check valve against the seat. The spring maintains a slight fluid pressure in the line and in the wheel cylinders to prevent the possible entrance of air into the system. The secondary cup which is secured to the opposite end of the piston, prevents the leakage of fluid into the rubber boot. The holes in the piston head are for the purpose of allowing the fluid to flow from the angular space around the piston into the space between the primary cup and the check valve, keeping sufficient fluid in the line at all times. The holes in the check valve case allow the fluid to flow through the case, around the lips of the rubber valve cup and out into the line during the brake application. When the brakes are released the valve is forced off the seat permitting the fluid to return to the master cylinder. The piston assembly is held in the opposite end of the housing by means of a snap ring. The rubber boot that fits around the push rod and over the end of the housing prevents dirt or any foreign matter from entering the master cylinder.

The wheel cylinder is a double piston cylinder, the purpose of the two pistons being to distribute the pressure evenly to each of the two brake shoes. Rubber piston cups maintain pressure on the pistons to prevent the leakage of fluid. The rubber boots over the end of the cylinder prevent dust and dirt or foreign material from entering the cylinder.

When pressure is applied to the brake pedal, the master cylinder forces fluid through the lines and into the wheel cylinders. The pressure forces the pistons in the wheel cylinder outward, expanding the brake shoes against the drum. As the pedal is further depressed higher pressure is built up within the hydraulic system, causing the brake shoes to exert a greater force against the brake drums.

As the brake pedal is released, the hydraulic pressure is released and the brake shoe retracting spring draws the shoes together, pushing the wheel cylinder pistons inward and forcing the fluid out of the cylinder back into the line towards the master cylinder. The piston return spring in the master cylinder returns the piston to the piston stop faster than the brake fluid is forced back into the line, which creates a slight vacuum in that part of the cylinder ahead of the piston. The vacuum causes a small amount of fluid to flow through the holes in the piston head, past the lip of the primary cup and into the forward part of the cylinders. This action keeps the cylinder filled with fluid at all times, ready for the next brake application. As fluid is drawn from the space behind the piston head it is replenished from the reservoir through the intake port. When the piston is in the fully released position the primary cup clears the bypass port, allowing the excess fluid to flow from the cylinder into the reservoir as the brake shoe retracting springs force the fluid back into the master cylinder.

Brake Pedal Adjustment

There should always be at least ½" free pedal travel before the push rod engages the piston.

This adjustment is accomplished by the shortening or lengthening of the brake master cylinder eye bolt, No. 59, Fig. 1. This is done so the primary cup will clear the port No. 15, Fig. 4, when the piston is in the off position, otherwise the compensating action of the master cylinder for expansion and contraction of the fluid in the system, due to temperature changes, will be destroyed and cause the brakes to drag.

Brake Shoe Adjustment—Minor

When the brake lining becomes worn, as indicated by foot pedal going almost to the floor board, necessary adjustment can readily be made as described in the following paragraph; first making certain that there is ½" free brake pedal travel.

Jack up the wheels to clear the floor. Adjustment is made by rotating the eccentric No. 5, Fig. 1. With a wrench loosen lock nut No. 6 for forward brake shoe, hold lock nut and with another wrench turn eccentric towards the front of the car until brake shoe strikes drum, then turning wheel with one hand release eccentric until wheel turns free, holding eccentric tighten lock nut. To adjust reverse shoe, repeat this operation only turn the eccentric towards the back of the car. Do this on all brakes. Check fluid in master cylinder.

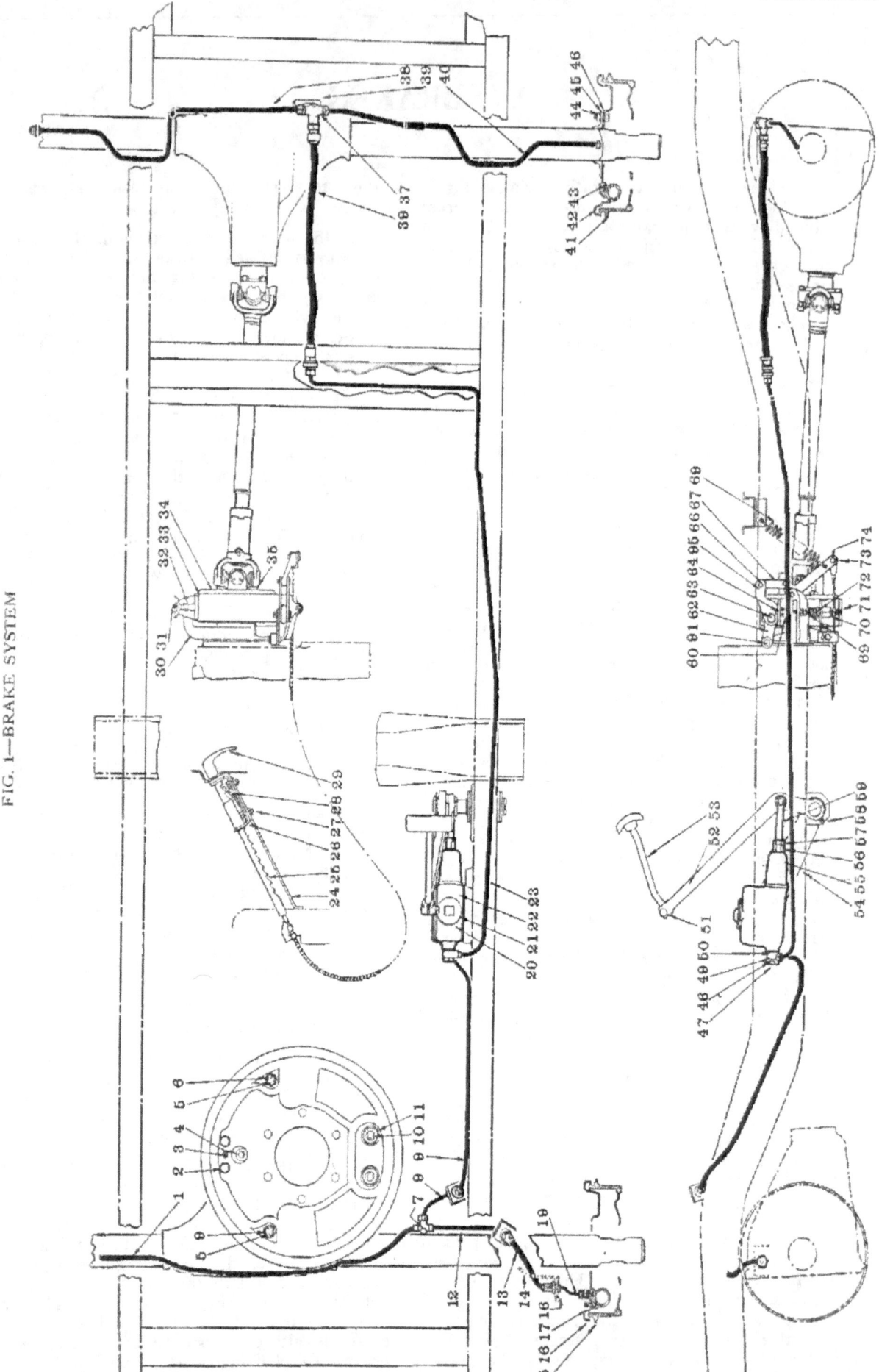

FIG. 1—BRAKE SYSTEM

FIG. 1—BRAKE SYSTEM

No.	Willys Part No.	Ford Part No.	Name
1	A-1376	GPW-2266	Brake Tube Assembly (Tee to Front Brake Hose—Right)
2	51738	20300-S7	Hex. Head Screw (Wheel Cylinder to Backing Plate)
3	637540	GP-2208	Wheel Brake Cylinder Bleeder Screw
4	A-1502		Front Wheel Brake Cylinder
5	A-754	GP-2038	Brake Shoe Eccentric
6	A-755	33800-S7-8	Hex. Nut (Eccentric)
7	637432	GP-2074	Axle Tee
8	A-1373	GPW-2078	Brake Hose—Front (Axle to Frame)
9	A-1377	GPW-2264	Brake Tube Assembly (Master Cylinder to Front Hose)
10	637899	91A-2027	Brake Shoe Anchor Pin
11	637924	33846-S7-8	Hex. Nut (Anchor Pin)
12	A-1501	GPW-2203	Brake Tube Assembly (Tee to Front Brake Hose—Left)
13	A-1460	GPW-2079	Brake Hose (Front Axle)
14	A-1457	GPW-3096	Front Wheel Brake Hose Guard
15	A-472	GPW-1125	Front Brake Drum
16	A-450	GP-2013	Front Brake Backing Plate
17	637540	GP-2208	Wheel Brake Cylinder Bleeder Screw
18	637427	78-2814-A	Spring Lock Clip (Brake Hose to Bracket)
19	A-1488	GPW-2298	Brake Tube Assembly (Wheel Cylinder to Axle Hose—Left)
20	637612	GP-2167	Master Cylinder Filler Cap Gasket
21	637608	GP-2162	Master Cylinder Filler Cap Assembly
22	637582	GP-2155	Master Cylinder and Supply Tank
23	A-5224		Brake Tube Assembly (Master Cylinder to Rear Hose)
24	A-2692	GPW-2852	Hand Brake Ratchet Tube Bracket Support
25	A-1242	GPW-2780	Hand Brake Handle Tube and Cable Assembly
26	639010	GPW-2348	Hand Brake Ratchet Tube Bracket Assembly
27	51396	24347-S	Hex. Head Screw (Bracket to Support)
28	635681	GPW-2793	Hand Brake Ratchet Tube Spring
29	639244	GPW-2782	Hand Brake Handle
30	A-1507	GPW-7769	Transfer Case Output Shaft Bearing Cap—Rear
31	A-1020	01T-2616	Hex. Head Screw (Anchor Clip)
32	A-1021	01T-2640	Transmission Brake Band Anchor Clip Screw Spring
33	A-1009	GP-2648	Transmission Brake Band and Lining Assembly
34	A-1002	GP-2614	Brake Drum (Transmission)
35	636575	33786-S	Hex. Nut (Brake Drum to Flange and Propeller Shaft)
36	637424	GP-2078	Brake Hose—Rear (Axle to Frame)
37	637432	GP-2074	Axle Tee
38	A-5226		Brake Tube Assembly (Tee to Rear Brake—Right)
39	A-5227		Rear Axle Brake Tube Tee Bracket
40	A-5225		Brake Tube Assembly (Tee to Rear Brake—Left)
41	A-472	GP-1111	Rear Brake Drum
42	A-450	GP-2013	Rear Brake Backing Plate Assembly
43	A-6111		Rear Wheel Brake Cylinder
44	A-903	355578-S	Brake Backing Plate Screw
45	636575	33786-S2	Brake Backing Plate Screw Nut
46	5010	34807-S7-8	Brake Backing Plate Screw Lockwasher
47	637505	GP-2077	Brake Master Cylinder Outlet Fitting Bolt
48	637606	91A-2151	Outlet Fitting Gasket—Large
49	A-557	GP-2076	Outlet Fitting
50	637604	91A-2152	Outlet Fitting Gasket—Small
51	6157	24426-S2	Clamp Screw (Pedal Shank to Pedal)
52	A-1388	GPW-2452	Brake Pedal Assembly
53	A-1369	GPW-2454	Brake Pedal Pad Assembly
54	A-1354	GPW-2138	Master Cylinder Tie Bar
55	637602	GP-2180	Brake Master Cylinder Boot
56	637599	GP-2143-A1-2	Brake Master Cylinder Push Rod Assembly
57	5939	33802-S	Brake Master Cylinder Eye Bolt Lock Nut
58	392909	353027-S7-8	Brake Pedal Hydraulic Fitting
59	A-163	GPW-2462	Brake Master Cylinder Eye Bolt
60	A-1017	01T-2634	Transmission Brake Releasing Spring (Hand)
61	A-1006	73889-S7	Transmission Brake Support Quadrant Pin (Hand)
62	A-1005	GPW-2630	Transmission Brake Support Quadrant (Hand)
63	A-1004	73928-S7-8	Transmission Brake Cam Pin (Hand)
64	A-1019	31218-S7	Transmission Brake Band Bracket Cap Screw (Hand)
65	A-1003	GPW-2632	Transmission Brake Cam (Hand)
66	311003	73904-S7	Clevis Pin (Link to Cam)
67	A-1228	GPW-2659	Hand Brake Relay Crank Link
68	A-5335	GPW-2035	Hand Brake Retracting Spring
69	5790	33795-S	Brake Band Cap Screw Nut
70	A-1018	01T-2805	Transmission Brake Band Adjusting Nut (Hand)
71	A-1016	01T-2842	Transmission Brake Adjusting Bolt (Hand)
72	A-1017	01T-2634	Transmission Brake Releasing Spring (Hand)
73	392468	357553-S18	Clevis Pin (Cable to Relay Crank)
74	A-1226	GPW-2656	Hand Brake Relay Crank Assembly

Brake Shoe Adjustment—Major

In the event the minor adjustment does not give adequate brakes or when it is necessary to reline the brakes it will be necessary to reset anchor pins, No. 10. The brake adjustments should be made as follows:

With the shoe and lining assemblies installed and the adjusting fixture or brake drum in place, loosen the anchor pin lock nuts No. 11 on the rear of the backing plate. Adjustment is made by turning the eccentric anchor pins towards each other and down until the shoes are set to the proper clearance, as determined by feeler gauges. The recommended shoe setting is .005" clearance at the heel (lower end), and .008" at the toe (upper end) of brake shoe lining. A slot is provided in the brake drum for checking these clearances.

Relining Brake Shoes

When necessary to reline the brakes, the car should be raised so that all four wheels are free from the floor.

Remove the wheels and then the hubs and drums which will then give access to the brake shoes.

Install wheel cylinder clamps or keepers to retain the wheel cylinder pistons in place and prevent leakage of brake fluid while replacing the shoes. Turn all eccentrics to the lowest side of the cam, and then remove the brake shoe contracting spring, No. 1, Fig. 2.

Remove anchor pin nuts, lock washers, and anchor pins from backing plate.

Remove rivets holding lining to shoes and install new linings through the use of a brake lining clamp.

Inspect the oil seals in the wheel hubs and if found that grease has been leaking, it is advisable to install new oil seals.

Install brake shoes to the brake backing plate, the shoe with the longest lining is the forward shoe on all four wheels. Install anchor pin No. 12, pin plate No. 2 and pin cam No. 13; then install anchor pins so the punch mark on the ends are facing each other. Install lock washer and nut. Install brake shoe return spring. Remove brake cylinder clamp or keeper.

Install the hubs and drums, then make the major adjustment of the brakes.

If it is found while the wheels are removed that there is brake fluid leakage at any of the wheel cylinders, it will be necessary to recondition that wheel cylinder, and bleed the brake lines. This subject is covered under the heading "Reconditioning Wheel Cylinders" and "Master Cylinder".

NOTE: Whenever the brake lining is renewed in one front or rear wheel be sure to perform the same operations in the opposite front or rear wheel, using the same brake lining as to color and part number, otherwise unequal brake action will result.

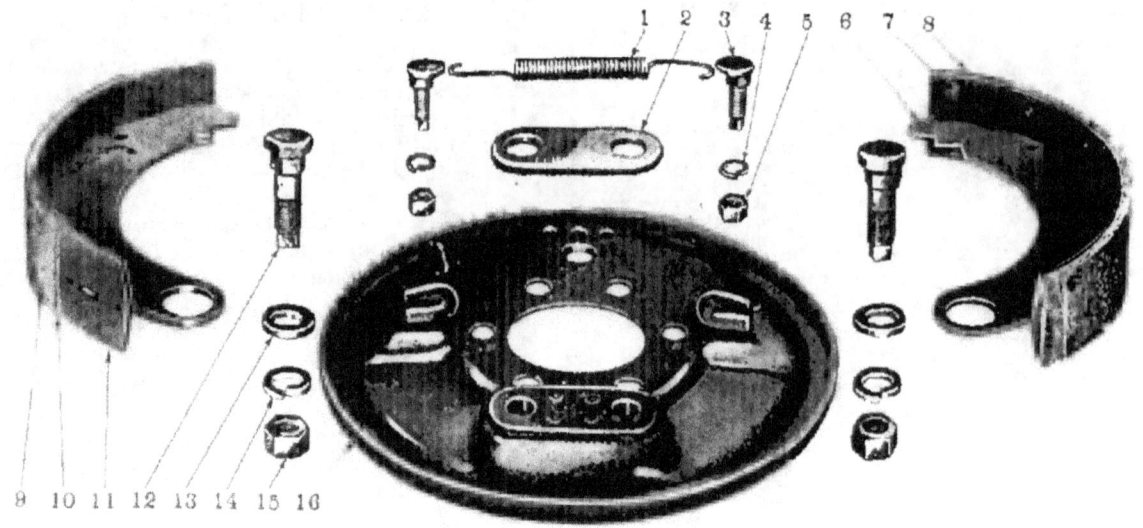

FIG. 2—BRAKE

No.	Willys Part No.	Ford Part No.	Name
1	637905	GP-2035	Brake Shoe Return Spring
2	637901	91A-2030	Brake Shoe Anchor Pin Plate
3	A-754	GP-2038	Brake Shoe Eccentric
4	5010	34807-S7-8	Brake Shoe Eccentric Lockwasher
5	A-755	33800-S7	Brake Shoe Eccentric Nut
6	116549	GP-2018	Brake Shoe and Lining Assembly—Forward
7	374586	351915-S	Brake Lining Tubular Brass Rivet
8	116551	GP-2021	Brake Shoe Lining—Forward
9	374586	351915-S	Brake Lining Tubular Brass Rivet
10	116552	GP-2022	Brake Shoe Lining—Reverse
11	116550	GP-2019	Brake Shoe and Lining Assembly—Reverse
12	637899	91A-2027	Brake Shoe Anchor Pin
13	637900	GP-2023	Brake Shoe Anchor Pin Cam
14	637923	351406-S24	Brake Shoe Anchor Pin Lockwasher
15	637924	33846-S7-8	Brake Shoe Anchor Pin Nut
16	A-450	GP-2013	Brake Backing Plate Assembly

Bleeding Brakes

The hydraulic brake system must be bled whenever a fluid line has been disconnected or air gets into the system. A leak in the system may sometimes be evidenced through the presence of a spongy brake pedal. Air trapped in the system is compressible and does not permit pressure applied to the brake pedal to be transmitted solidly through to the brakes. The system must be absolutely free from air at all times. When bleeding the brakes it is advisable that the longest fluid line from the master cylinder be bled first. The proper sequence of bleeding is right rear; right front; left rear; left front. During the bleeding operation the master cylinder must be kept at least ¾ full of hydraulic brake fluid.

To bleed the brakes first carefully clean all dirt from around the master cylinder filler plug. Remove filler plug and fill master cylinder to the lower edge of filler neck. Clean off all bleeder connections at all four wheel cylinders. Attach bleeder hose and fixture to right rear wheel cylinder bleeder screw and place end of tube in a glass jar, end submerged in fluid. Open the bleeder valve ½ to ¾ of a turn. See Fig. 3.

Depress the foot pedal by hand, allowing it to return very slowly. Continue this pumping action to force the fluid through the line and out the bleeder hose which carries with it any air in the system.

When bubbles cease to appear at the end of the bleeder hose, tighten the bleeder valve and remove the hose.

After the bleeding operation has been completed at all four wheels, fill the master cylinder reservoir and replace the filler plug.

It is not advisable to re-use the fluid which has been removed from the lines through the bleeding process.

Master Cylinder

When necessary to remove the master cylinder No. 22, Fig. 1 for reconditioning, the following procedure should be followed:

1. Raise front end of car with jack because the removal operation must all be performed from the under side of the vehicle.
2. Remove stop light switch wires.
3. Remove fitting bolt and switch No. 47.
4. Remove front bolt holding master cylinder to frame which is installed from the outside of the frame and screws into master cylinder body.
5. Remove master cylinder tie bar cap screw which is the front inside screw on master cylinder.
6. Remove rear master cylinder bolt which has nut on inside of frame bracket.
7. Remove master cylinder boot No. 55.
8. Remove master cylinder.

The installation of master cylinder to frame is the reverse of the above operations.

After the master cylinder has been removed it should be dismantled and thoroughly washed in alcohol. (Never wash any part of the hydraulic braking system with gasoline (petrol) or kerosene.)

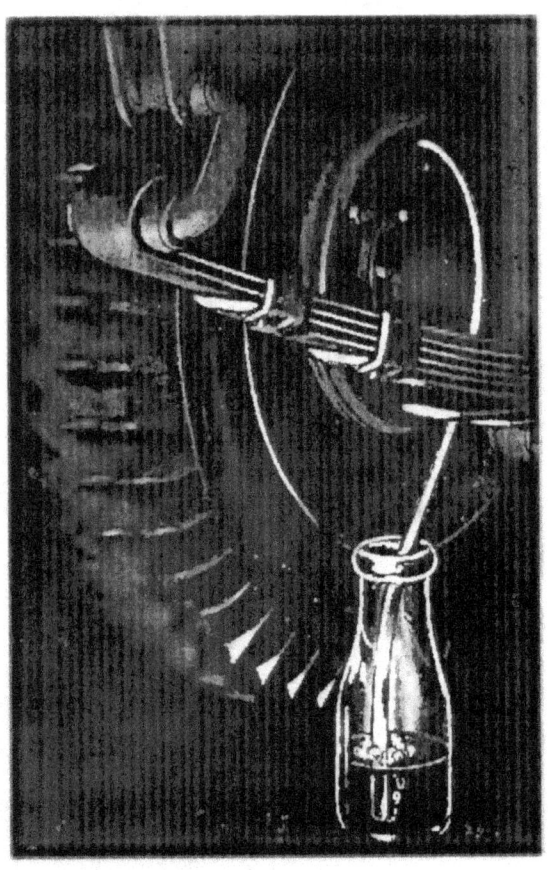

FIG. 3—BLEEDING BRAKE

Bleed lines as instructed under the heading, "Bleeding Brakes". See Fig. 3.

Recheck entire system to make sure that there are no leaks and if necessary make brake adjustments in order to have adequate brakes.

Filling Master Cylinder

The Master Cylinder reservoir should be checked each 1000 miles when vehicle is lubricated, be sure that there is a sufficient supply of brake fluid. The master cylinder can be reached by removing the five screws in the inspection cover on the toeboard below the steering post. After removing the cover thoroughly clean any dirt away from the filter cap on the master cylinder to prevent it from entering the brake system where it might cause a scored cylinder or possible failure of the brakes.

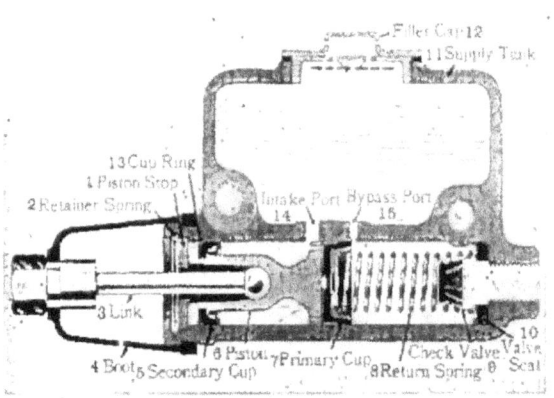

FIG. 4—MASTER CYLINDER

After the parts have all been thoroughly cleaned with alcohol, make careful inspection, replacing those parts which show signs of being deteriorated. Inspect cylinder bore and if found to be rough it should be honed out. The clearance between the piston and the cylinder bore should be .001" to .005". During the honing operation use hydraulic brake fluid on the hone, in order to obtain a polished surface in the cylinder bore. Wash out cylinder with alcohol and with a wire passed through the ports No. 14 and No. 15 that open from the supply reservoir into the cylinder bore, make sure that these passages are free and clear of any foreign matter. It is our recommendation that a new piston, primary cup, valve and valve seat be installed when rebuilding the master cylinder. See Fig. 4.

Install valve seat No. 10 in end of cylinder with flat surface toward valve. Install valve assembly No. 9. Install piston return spring No. 8. Install primary cup No. 7. Face of cup goes towards piston. Install piston No. 6, stop plate No. 1 and lock wire No. 2. Install fitting connection with new copper gaskets. Fill reservoir half full of brake fluid and operate the piston with piston rod until fluid is ejected at fitting. Install master cylinder to frame and make necessary connections and adjust pedal clearance to ½" free play.

Wheel Cylinders

To remove a hydraulic brake wheel cylinder Fig. 5, jack up the vehicle and remove the hub and drum. Disconnect brake line at fitting on brake backing plate. Remove brake shoe retracting spring which allows the brake shoes at the toe to fall clear of the brake cylinder. Remove two cap screws holding wheel cylinder to backing plate.

Remove rubber dust covers on end of cylinders, then pistons, piston cups and spring.

Wash the parts in clean alcohol. Examine the cylinder bore for roughness or scoring. Check fit of pistons to cylinder bore by using .002" feeler gauge.

In reassembling cylinder dip spring, pistons and piston cups in brake fluid. Install spring in center of wheel cylinder. Install piston cups with the cupped surface towards the spring so that the flat surface will be against piston. Install pistons and dust covers. Install wheel cylinder to backing plate, connect up brake line and install brake shoe retracting spring. Replace wheel hub and drum, then bleed the lines as instructed under "Bleeding Brakes."

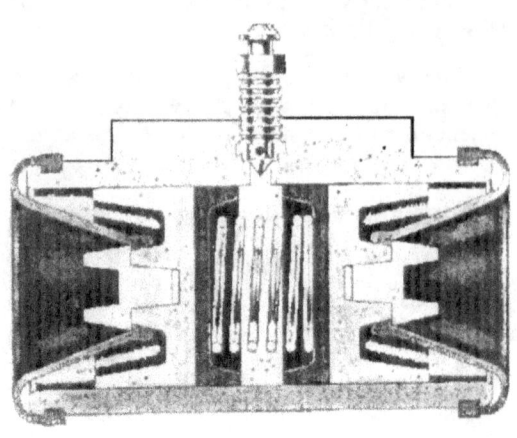

FIG. 5—WHEEL CYLINDER

Brake Hose—Front

To remove brake hose at the wheels the following procedure should be followed to prevent damage to hose and fitting. See Fig. 6.

1. Remove brake line connections at each end.
2. Slip brake hose spring lock clip off ends of hose fitting and remove brake hose from brackets.

To remove front brake hose, frame to axle the following procedure should be followed:

1. Remove brake line connection on frame bracket, top connection.
2. Remove brake hose spring lock clip from brake hose fitting at bracket.
3. Remove brake hose from bracket.
4. With open end wrench unscrew brake hose from tee connection on axle and remove.

FIG. 6—BRAKE HOSE

Brake Hose—Rear

To remove the rear brake hose, the following procedure should be followed:

1. Remove brake line from hose connection at frame.
2. Slip brake hose spring lock clip off of brake hose fitting.
3. Remove brake hose from frame.
4. With open end wrench unscrew brake hose from fitting on axle housing.

Whenever a brake line has been disconnected, it will be necessary to bleed the brakes. The bleeding of the brakes should be done in accordance with instructions given under "Bleeding Brakes".

Transmission Hand Brake

The hand brake is applied to the rear propeller shaft at transfer case, see Fig. 1. The operation of the brake is positive through a cable connection.

To adjust the hand brake Fig. 7, the following operations should be performed.

Have hand brake lever on dash in the released position. Adjust anchor screw No. 1, so that there will be .005"—.010" clearance between the band and the drum. Tighten nut No. 2 until band is brought tight against the drum. Adjust bolt No. 3 so that the head just rests on upper half of the band. Back off two turns on adjusting nut No. 2.

The length of the cable from the hand grip to the brake levers is of a predetermined length and cannot be changed. At the regular lubrication periods of 1,000 miles it is advisable to put a few drops of oil in the upper end of conduit tube at cable to keep it free to slide within the conduit.

This brake is designed for holding the car while parked.

To reline brake band, remove from bracket and adjusting linkage. Cut off rivets and remove lining, care being taken not to distort the band. Hold end of new lining flush with end of band, make lining hug band inside and cut off about $5/16"-1/2"$ long. Bring lining ends even with ends of band and install end rivets only. Then remove center bulge in lining with hammer, so lining will hug band tightly, then install balance of rivets and form band to drum, making regular adjustments.

FIG. 7—TRANSMISSION BRAKE

BRAKE TROUBLES AND REMEDIES

SYMPTOMS	PROBABLE REMEDY
Brakes Drag	
Brake shoes improperly adjusted...............	Readjust
Piston cups—Enlarged.........................	⎫ Flush all lines with alcohol—Install new cups in
Mineral oil or improper brake fluid in system...	⎭ Wheel and Master cylinders
Improper pedal adjustment.....................	Adjust master cylinder rod
Clogged master cylinder compensating port......	Clean master cylinder
One Brake Drags	
Brake shoe adjustment incorrect...............	Adjust
Brake hose clogged............................	Replace
Retracting spring broken.......................	Replace
Wheel cylinder piston or cups defective.........	Replace
Loose or damaged wheel bearings...............	Adjust or replace
Brake Grabs—Car Pulls to One Side	
Brake anchor pin adjustment incorrect..........	Adjust
Oil or brake fluid on lining....................	Replace lining
Dirt between lining and drum..................	Clean with wire brush
Drum scored or rough.........................	Turn drum and replace lining
Loose wheel bearings..........................	Adjust
Axle spring clips loose.........................	Tighten
Brake backing plate loose......................	Tighten
Brake lining..................................	Different kinds on opposite wheels
Brake shoes reversed..........................	Primary and secondary shoes reversed in one wheel
Tires under-inflated...........................	Inflate to 30 lbs. pressure
Tires worn unequally..........................	Replace or change around to opposite wheels
Excessive Pedal Travel	
Normal lining wear............................	Adjust
Lining worn out...............................	Replace
Leak in brake line.............................	Locate and repair
Scored brake drums...........................	Replace or regrind
Incorrect brake lining.........................	Replace
Air in hydraulic system........................	Fill master cylinder and Bleed lines
Spongy Brake Pedal	
Air in lines...................................	Bleed lines
Brake shoe adjustment incorrect...............	Adjust
Excessive Pedal Pressure	
Oil or brake fluid on lining....................	Replace lining
Shoes improperly adjusted.....................	Major adjustment
Warped brake shoes...........................	Replace
Distorted brake drums.........................	Replace or regrind
Squeaky Brakes	
Brake shoes warped or drums distorted.........	Replace
Lining loose..................................	Replace
Dirt imbedded in lining........................	Clean with wire brush or replace
Improper adjustment..........................	Adjust

BRAKE SPECIFICATIONS

Service Brakes:
- Type.................. 4 Wheel Hydraulic
- Size.................. 9" x 1¾"
- Fluid Capacity Pts. See Lubrication Chart, Pg. 12

Master Cylinder:
- Size.................. 1"
- Mounted.............. L.H. Frame Side Rail

Wheel Cylinder:
- Size.................. Front 1" Rear ¾"

Brake Shoes.................. Bendix
- Size.................. 9" x 1¾"
- Lining area.................. 117.8 Sq. in.
- Length Lining-Forward shoe............ 10 7/32"
- Length Lining-Reverse shoe............. 6 39/64"
- Width.................. 1¾"
- Thickness.................. 3/16"

Hand Brake
- Type.................. Mechanical
- Size.................. 6"
- Lining.................. Woven
- Length.................. 18 9/16"
- Width.................. 2"

Brake Return Springs:
- Brake Pedal
  - Free Length.................. 5 7/8"
  - Load when extended to 7 9/16"...... 23 lbs.
- Brake Shoe Return Spring
  - Free Length.................. 5 13/16"
  - Load when extended to 6 9/16"...... 40 lbs.
- Wheel Cylinder Spring
  - Length.................. 1 7/16"
  - Load when compressed......... 1 to 1¼ lbs.

WHEELS—WHEEL BEARINGS

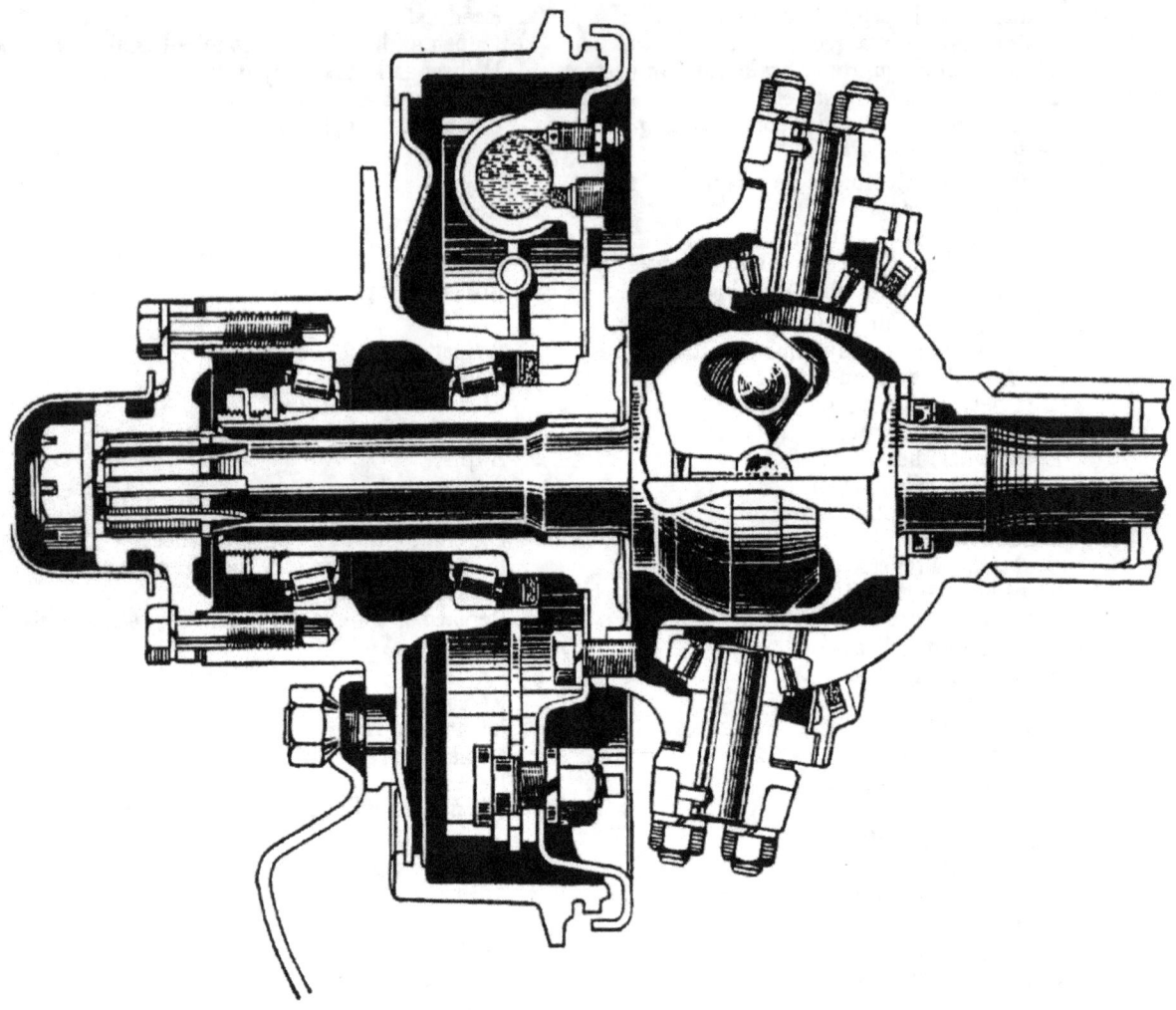

FIG. 1—FRONT WHEEL

The front and rear wheels are carried on two opposed tapered roller bearings. Bearings are adjustable for wear and their satisfactory operation and long life depends upon periodic attention and correct lubrication.

Wheel bearings cannot be checked for adjustment properly unless brakes are free from dragging on brake drums and are in fully released position.

Front Wheel Bearings

1. Raise front end of vehicle with jack so that tires clear the floor.
2. With hands test sidewise shake of the wheel. If bearings are correctly adjusted, shake of wheel will be just perceptible and wheel will turn freely with no drag. If bearing adjustment is too tight, the rollers may break or become overheated. Loose bearings may cause pounding.

If this test indicates adjustment is necessary, proceed as follows:

Adjustment

1. With wheels still on jack remove hub cap, axle shaft nut, washer and driving flange. Wheel bearing adjustment will then be accessible.
2. Bend lip of nut lock so that adjustment locknut and lock can be removed.
3. Tighten adjusting nut until wheel binds, at the same time rotating wheel to make sure all surfaces are in proper contact.
4. Then back off nut about $\frac{1}{8}$ turn or more if necessary making sure wheel rotates freely.
5. Replace lock and do not fail to bend over locknut.
6. Check adjustment and reassemble driving flange. When front hub is completely assembled, test wheel shake before removing jack.

Rear Wheel Bearings

Raise wheel on which adjustment is to be made by placing jack under axle housing. Test wheel for loose bearing. If adjustment is necessary proceed as follows:

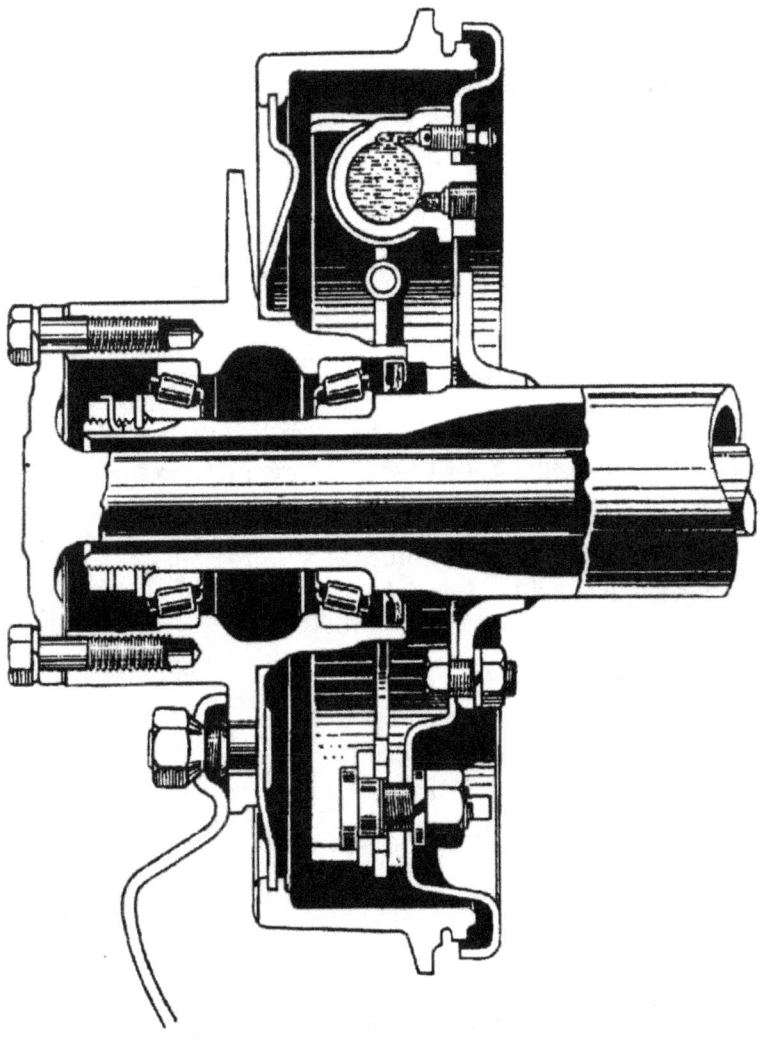

FIG. 1—REAR WHEEL

Adjustment

1. Remove axle shaft flange cap screws and axle shaft.
2. Bend lip of nut lock so that locknut can be removed.
3. Tighten inner adjusting nut until wheel binds, at the same time rotate wheel to make sure all surfaces are seating properly.
4. Back off nut $1/6$ turn or more if necessary until wheel turns freely.
5. Replace nut lock and locknut and be sure to bend over lock.
6. Replace axle shaft with gasket and install cap screws.

Lubrication Wheel Hub Bearings

Under normal operating conditions the hub bearings require lubrication only every 6,000 miles when hubs and bearings should be removed and thoroughly washed in suitable cleaning fluid.

Bearings should be given more than a casual cleaning. Use a clean stiff brush and remove all particles of old lubricant from bearings and hubs. After bearings are thoroughly cleaned inspect for pitted races and rollers, also check the hub oil seal.

Repack bearing cones and rollers with grease and reassemble hub in reverse order as that of dismantling, testing bearing adjustment as covered under "Adjustment"

When reinstalling hubs and drums the hubs with the right hand threaded studs are placed on the right hand side of vehicle. The left hand threaded studs are on the left hand side, viewing vehicle from the rear.

Brake Drum

The brake drums are attached to the wheel hubs by five serrated bolts. These bolts are also used for mounting the wheels to the hubs.

To remove a brake drum, drive out the serrated bolts and remove the drum from hub.

When placing drum on hub, make sure that the contacting surfaces are clean and flat. Line up holes in drum with those in hub and force drum over shoulder on hub. Insert five new serrated bolts through drum and hub and drive the bolts into place solidly. Place a round piece of stock in vise approximately the diameter of the head of the bolt and place hub and drum assembly over it so that it rests against head of the bolt then swedge bolt into countersunk section of hub with punch.

The runout of the face of the drum should be within .003". If runout is found to be greater than .003" it will be necessary to reset bolts to correct the condition.

Left hand hub bolts are identified with an "L" stamped on threaded end of bolt.

The left hand threaded nuts can be identified by a groove around the hexagon faces.

Hubs containing the left hand threaded hub bolts are installed on the left hand side of vehicle.

Tires

The most important factor of safe vehicle operation is systematic and correct tire maintenance. Tires must sustain the weight of a loaded vehicle, withstand more than ordinary rough service, provide maximum safety over all types of territory, and furnish the medium on which the vehicle can be maneuvered with ease.

Although there are other elements of tire service, inflation maintenance is the most important and in many instances the most neglected. Tire pressures should be consistently maintained for safe operation.

An under inflated tire is dangerous and too much flexing causes breakage of the fabric resulting in a failure. Over-inflation in time may cause a blowout.

To remove the tire from a drop center rim, first deflate completely and then force the tire away from the rim throughout its entire circumference until the bead falls into the center of the wheel rim, then with a heavy screw driver or tire removing tool, placed across the wheel from the valve, remove one side of the tire at a time and remove inner tube. (See "Combat Wheels").

Installation of tire is made in the same manner by first dropping one side of the tire into the center of the rim and with tire tool spring bead over the wheel rim using care not to damage the inner tube.

When tightening the wheel stud nuts, alternately tighten opposite nuts to prevent wheel runout. After nuts have been tightened with the wheel jacked up, lower jack so wheel rests on the floor and retighten the nuts.

Combat Wheels

Combat wheels are identified by eight bolts holding together the two halves of the tire rim. When removing a tire, first remove the wheel and be sure to deflate the tire before removing the rim nuts. After removing the rim nuts, remove the outer rim then remove the tire after which remove the bead locking ring and tube from the tire. Mounting the tire is the reverse procedure. Do not put too much air in the tube when mounting. Combat wheel rim bolt and hub bolt torque reading 60-70 ft. lbs.

WHEEL SPECIFICATIONS

Wheels:

Make........................Kelsey-Hayes
Rim 16x4.00 Drop Center-16x4.50 Combat Wheels
Tires................................16 x 6.00
Type.....Mud and Snow non-directional Tread
Tire Pressure........................30 lbs.

Bearings—F and R	Inner	Outer
Make...................	Timken	Timken
Cone and roller.........	18590	18590
Cup....................	18520	18520

STEERING

The stability and proper functioning of the steering system depends in a large measure upon correct alignment and a definite procedure for inspection of the steering system is recommended. In so doing, nothing is overlooked and the trouble is ascertained in the shortest possible time. It is suggested that the following sequence be used:

1. Equalize tire pressures and level car.
2. Inspect king pin and wheel bearing looseness.
3. Check wheel runout or wobble.
4. Test wheel balance.
5. Check for spring sag.
6. Inspect brakes and shock absorbers.
7. Check steering assembly and connecting rod.
8. Check caster.
9. Check toe-in.
10. Check toe-out on turns.
11. Check camber.
12. Check king pin inclination.
13. Check tracking of front and rear axle.
14. Check frame alignment.

The steering gear Fig. 1, is the cam and twin lever variable ratio type. The steering gear cam lever shaft is serrated for attachment to the steering pitman arm. The gear case is attached to the inside of the left frame side member by three bolts.

The cam thrust is taken at top and bottom by ball bearings which are adjustable through shims No. 13 at the upper housing cover No. 12.

When making adjustments free the steering gear of all load by disconnecting the steering connecting rod from the steering arm, loosen instrument panel bracket and steering gear frame bolts to allow the steering post to align itself.

Do not tighten the steering gear to dampen out steering troubles, adjust the steering only to remove play within the steering gear.

Adjustment of Ball Thrust Bearings on Cam

Adjust to a barely perceptible drag but allow the steering wheel to turn freely, with thumb and forefinger lightly gripping the rim.

Before making this adjustment, loosen the housing side cover adjusting screw No. 19 to free the pins in the cam groove.

To adjust, remove cap screws and move up the housing cover No. 12 to permit the removal of shims No. 13. Shims are of .002", .003", and .010" thickness.

Clip and remove a thin shim or more if required. Install cap screws and tighten. Test adjustment and if necessary remove or replace shims until adjustment is correct.

Adjustment of Tapered Pins in Cam Groove

Adjust so that a very slight drag is felt through the mid position when turning the steering wheel slowly from one extreme position to the other.

Backlash of the pins in the groove shows up as end play of lever shaft, also as backlash of steering at ball on steering arm.

Note that the groove is purposely cut shallow in the straight ahead driving position for each pin. Fig. 2. This feature permits a close adjustment for normal straight ahead driving thereby avoiding swaying in the road and also permits take-up of back-lash at this point after wear occurs without causing a bind elsewhere.

Adjust within the high range through the mid position of pin travel. Do not adjust the positions off "straight ahead". Backlash in turn positions is not objectionable.

Removal of Steering Gear from Chassis

To remove steering gear assembly from chassis, the following procedure should be followed:

1. Remove left front fender.
2. Remove horn button and steering wheel.
3. Remove steering post bracket at instrument board.
4. Remove steering post cover plates on toe board.
5. Remove horn wire contact brush, No. 37, Fig. 1.
6. Remove connecting rod at Pitman arm ball.
7. Remove three bolts holding steering gear housing to frame side rail.
8. Remove steering post by bringing it down through floor boards and over outside of the frame.

The installation of the steering gear assembly would be the reverse of the above operations. Frame bolts, torque wrench reading, 38-40 ft. lbs.

Disassembly of Steering Gear

First remove pitman arm No. 21, Fig. 1 with puller. Loosen lock nut No. 20 and unscrew adjusting screw No. 19 a few turns. Remove side cover screws and washers and remove side cover No. 18 with gasket. This will permit removal of lever shaft No. 16.

Remove upper cover plate screws and remove from housing the cam, wheel tube and bearing assembly.

When upper cup or upper cover plate requires replacement, the contact ring on wheel tube must be removed. To do this unsolder horn cable from ring and pull cable from tube, mark on wheel tube the location of ring and then press ring off of tube.

Inspect cam threads for wear, chipping and scoring, also the ball races on the cam ends and the separate ball cups. Existence of any of these conditions indicate necessity for replacement.

Inspect taper studs of lever shaft for flat spots and chipping. In the case of either, replacement is usually advisable. Inspect lever shaft for wear and test fit of shaft in bushing. Inspect condition of oil seal at outer end of lever shaft and the bearing in top end of jacket tube.

Assembling Steering Gear

Reassemble all parts to wheel tube in reverse order of disassembly and flatten cable and solder to ring. Assemble cam, wheel tube and bearing assembly in housing, seating well the lower bearing ball cup in the housing.

With adjusting shims in place, assemble upper cover plate with pin on top side of housing and adjust cam bearings.

Assemble lever shaft in housing and with gasket in place assemble side cover and make adjustment for a minimum backlash of studs in cam groove.

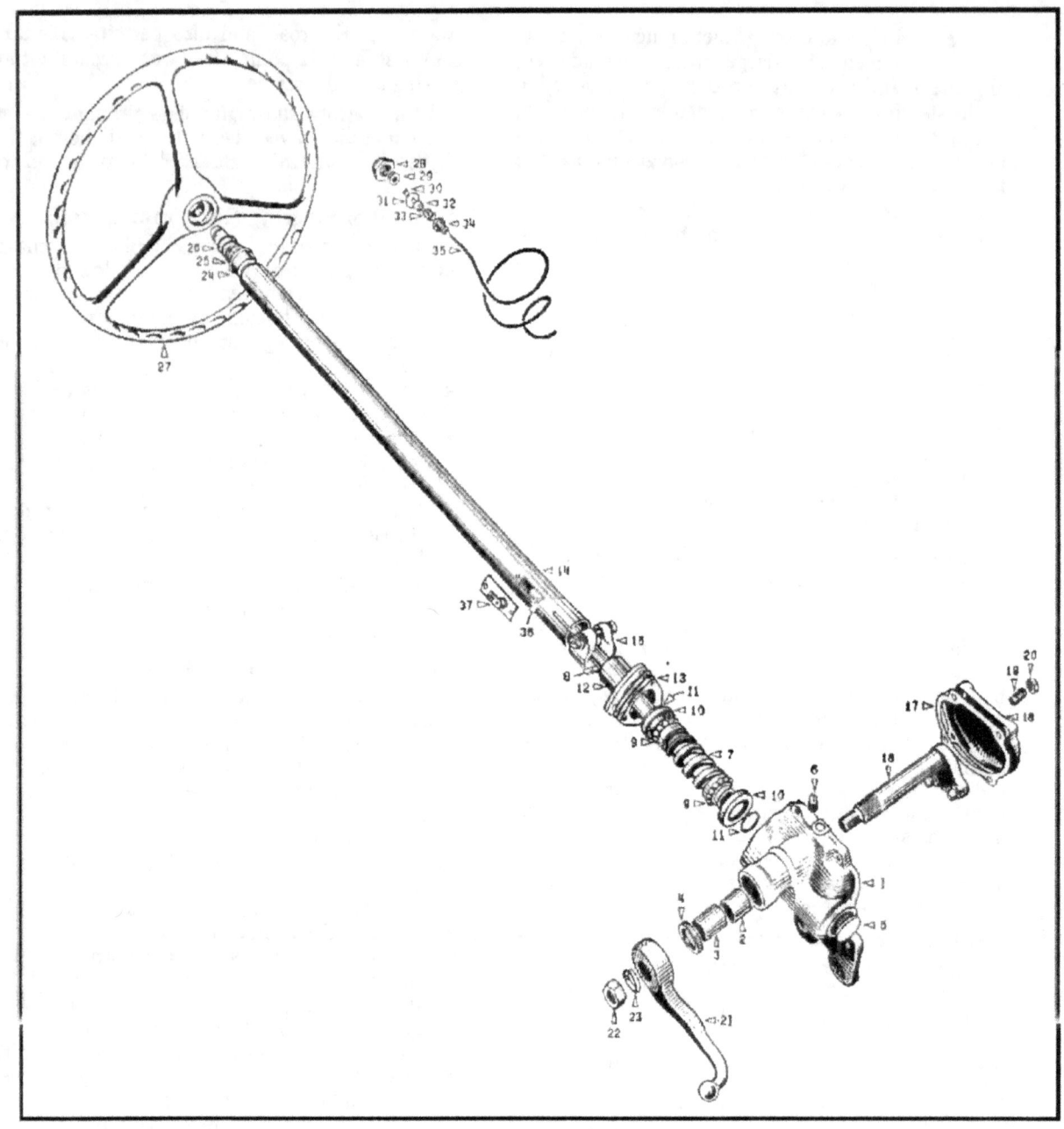

FIG. 1—STEERING GEAR

No.	Willys Part No.	Ford Part No.	Name	No.	Willys Part No.	Ford Part No.	Name
1	A-740	GPW-3548	Housing Assembly	20	52925	33927-S	Side Adjusting Screw Lock Nut
2	639090	GPW-3587	Housing Bushing—Inner	21	A-1116	GPW-3590	Steering Gear Arm
3	639091	GPW-3576	Housing Bushing—Outer	22	639115	350077-S8	Steering Gear Arm Nut
4	639095	GPW-3591	Housing Oil Seal	23	5038		Steering Gear Arm Nut Lockwasher
5	51091	74121-8	Housing Lower End Plug	24	639190	GPW-3517	Steering Column Bearing Assembly
6	5085	358064-S	Housing Oil Filler Plug	25	639192	GPW-3518	Steering Column Bearing Spring Seat
7-8	A-742	GPW-3524	Cam and Wheel Tube Assembly	26	639191	GPW-3520	Steering Column Bearing Spring
9	639104	GPW-3571	Cam Bearing Balls	27	A-635	GPW-3600	Steering Wheel
10	639102	GPW-3552	Ball Cup—Upper and Lower	28	A-653	GPW-3655	Steering Wheel and Horn Button Nut
11	639103	GPW-3589	Ball Cup Retaining Ring—Upper and Lower	29	A-634	GPW-3627	Horn Button
				30	638886	GPW-3630	Horn Cable Upper Terminal
12	A-1760	GPW-3568	Upper Housing Cover	31	A-750	GPW-3631	Contact Washer
13	639108	GPW-3593	Adjusting Shims	32	A-751	GPW-3635	Insulating Ferrule
14	A-1199	GPW-3509	Steering Column and Bearing Assembly	33	638884	GPW-3626	Horn Button Spring
15	A-635	GPW-3596	Steering Column Clamp Assembly	34	638885	GPW-3646	Horn Button Spring Cup
16	A-745		Lever Shaft Assembly	35	A-752	GPW-14171	Horn Cable Assembly
17	639119	GPW-3581	Side Cover Gasket	36	A-747	GPW-3652	Horn Wire Contact Ring Assembly
18	639117	GPW-3580	Side Cover	37	A-302	GPW-13836	Horn Wire Contact Brush Assembly
19	639118	GPW-3577	Side Adjusting Screw				

When assembling upper bearing spring No. 26, and spring seat No. 25 in jacket tube make sure that spring seat is positioned correctly. It must be placed with the lengthwise flange down against bearing and not up inside of spring coil.

Install pitman arm No. 21 to lever shaft No. 16 so that the line across the face of arm and end of shaft correspond, with the ball end down. Install lockwasher No. 23 and nut No. 22.

Install steering gear assembly in chassis in the reverse order in which it was removed.

When installing the steering wheel the steering gear should be in its mid position when the front wheels are in the straight ahead position. To check, turn the steering wheel as far to the right as possible then rotate the wheel in the opposite direction as far as possible and note the total number of turns. Turn the wheel back just one half of this total movement thus placing the gear in mid position at which point the front wheels should be straight ahead. The steering wheel spoke with moulded trade mark on underside will point down toward drivers seat and in line with the steering post. If not it will be necessary to remove the steering wheel and shift it on the serrations of the shaft.

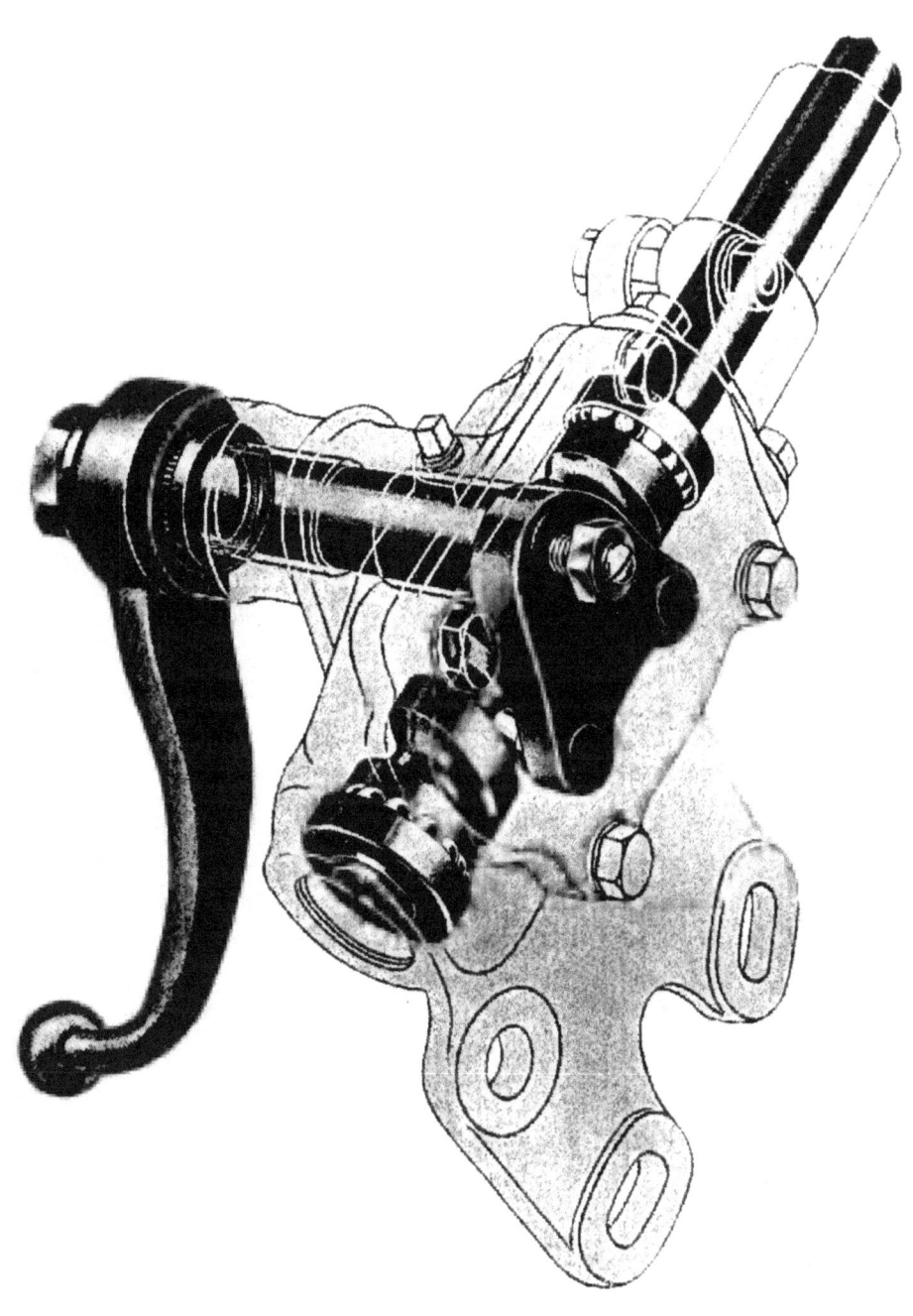

FIG. 2—SECTIONAL VIEW OF STEERING

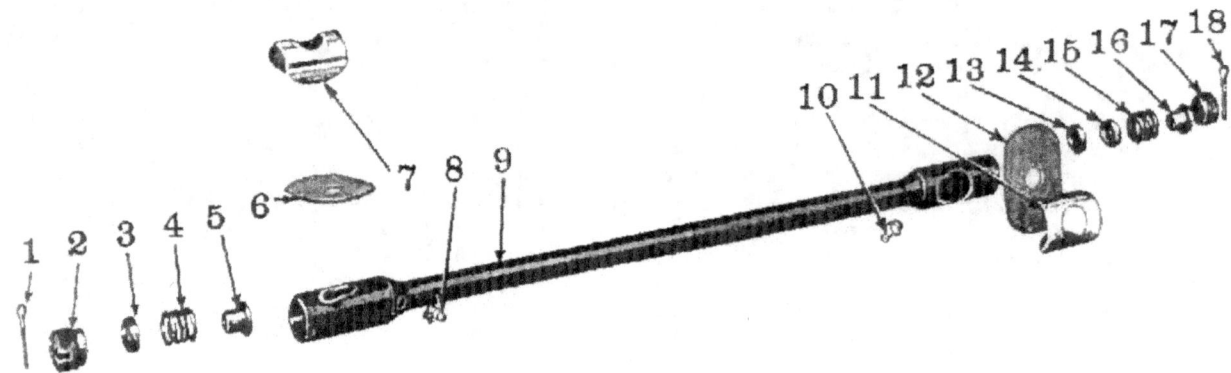

FIG. 3—STEERING CONNECTING ROD

No.	Willys Part No.	Ford Part No.	Name
1	6134	72089-S	Steering Connecting Rod Cotter Pin—Front
2	630756	GPW-3323	Steering Connecting Rod Adjusting Plug—Large
3	630755	GPW-3320	Steering Connecting Rod Ball Seat
4	630754	GPW-3327	Steering Connecting Rod Spring
5	630753	GPW-3328	Steering Connecting Rod Safety Plug
6	A-622	GPW-3332-A2	Steering Connecting Rod Dust Cover
7	A-623	GPW-3336	Steering Connecting Rod Dust Cover Shield
8	392909	353027-S7-8	Hydraulic Straight Grease Fitting
9	A-619	GPW-3304	Steering Connecting Rod Assembly
10	392909	353047-S7-8	Hydraulic Straight Grease Fitting
11	A-623	GPW-3336	Steering Connecting Rod Dust Cover Shield
12	A-622	GPW-3332	Steering Connecting Rod Dust Cover
13	630755	GPW-3320	Steering Connecting Rod Ball Seat
14	630754	GPW-3327	Steering Connecting Rod Spring
15	630753	GPW-3328	Steering Connecting Rod Safety Plug
16	630753	GPW-3328	Steering Connecting Rod Safety Plug
17	630757	GPW-3328	Steering Connecting Rod Adjusting Plug—Small
18	6134	72089-S	Steering Connecting Rod Cotter Pin—Rear

Steering Connecting Rod

The steering connecting rod is the ball and socket type. At front or axle end, the spring and spacer are assembled between rod (bottom of socket) and ball seat while at the steering gear end, spring and spacer are between ball seat and end plug. See Fig. 3.

When removing springs and seats for any reason make sure they are reassembled as above because this method of assembly relieves road shock from the steering gear in both directions. To adjust ball joint at axle, screw in plug firmly against the ball, then back off one half turn and lock with new cotter pin inserted through hole in tube and slot in adjusting plug.

To adjust ball joint at steering Pitman Arm, screw in end plug firmly against the ball, then back off one full turn and lock with new cotter pin inserted through hole in tube and slot in adjusting plug.

This will give the proper spring tension and avoid any tightness when swinging the wheels from maximum left to right turn.

Ball joints must be tight enough to prevent end play and yet loose enough to allow free movement.

Tie Rod

The tie rods, No. 11 and 14, Fig. 4 are of three piece construction consisting of rod and two ball and socket end assemblies. Ball and socket end assemblies are threaded into rod and locked with clamps around each end of tie rod. Right and left hand threads on tie rod end assemblies provide for toe-in adjustment without removing the tie rod ends from steering arms.

The length of the left hand tie rod No. 14 center to center of ball joint is $17\tfrac{11}{32}''$, the right hand tie rod No. 11 is $24\tfrac{1}{4}''$ center to center.

When wear takes place on tie rod end ball and socket, it will be necessary to replace the ball and socket assembly and rubber seal.

Front Wheel Alignment

Proper alignment of front wheels must be maintained in order to insure ease of steering and satisfactory tire life. Most important factors of front wheel alignment are wheel camber, axle caster and wheel toe-in.

These points should be checked at regular intervals particularly where the front axle has been subjected to heavy impact. When checking wheel alignment, it is important that wheel bearings and knuckle bearings be in proper adjustment. Loose bearings will affect reading of instruments when checking camber, knuckle pin inclination and toe-in.

Wheel toe-in is the distance the wheels are closer together at the front than at the rear.

Wheel camber is the amount wheels incline outward at the top from a vertical position.

Front axle caster is the amount in degrees that the steering knuckle pins are tilted toward the front or rear of the vehicle. Positive caster is inclination of top of knuckle pin toward rear of vehicle. Zero caster is vertical position of pin. Negative or reverse caster is the inclination of top of pin towards the front of the vehicle.

Front Wheel Toe-in

Toe-in Fig. 5, is the amount which wheels point inward at front and is necessary to offset the effect of camber.

Toe-in is usually measured at edge of rim, flange or at tire centers with wheels in straight ahead position, however in view of the tread being the same, front and rear, a straight edge or rope can be used.

It is highly important that the toe-in be checked regularly and if found to be excessively out of adjustment, correction should be made immediately.

To Adjust Toe-in

1. Set tie rod end of steering bell crank at right angles with front axle.
2. Place a straight edge or rope against the left rear wheel and left front wheel to determine if wheel is in straight ahead position. If the rear of tire on front wheel does not touch straight edge, it will be necessary to adjust the tie rod by loosening clamps on each end and turning the rod in a clockwise direction until the tire touches the straight edge both front and rear. If the front of the tire does not strike the straight edge, it will be necessary to lengthen the tie rod by turning the rod in a counter-clockwise direction.
3. Check the right hand side in the same manner adjusting the tie rod if necessary, making sure that the bell crank remains at right angles to the axle.
4. Set the toe-in to $3/64''$-$3/32''$ by shortening each tie rod approximately one half turn.

Front Wheel Camber

The purpose of camber Fig. 6 is to more nearly place the weight of the vehicle over the center of the steering knuckle pins and facilitate easy steering.

The result of excessive camber is irregular wear of tires on outside shoulders and is usually caused by bent axle parts.

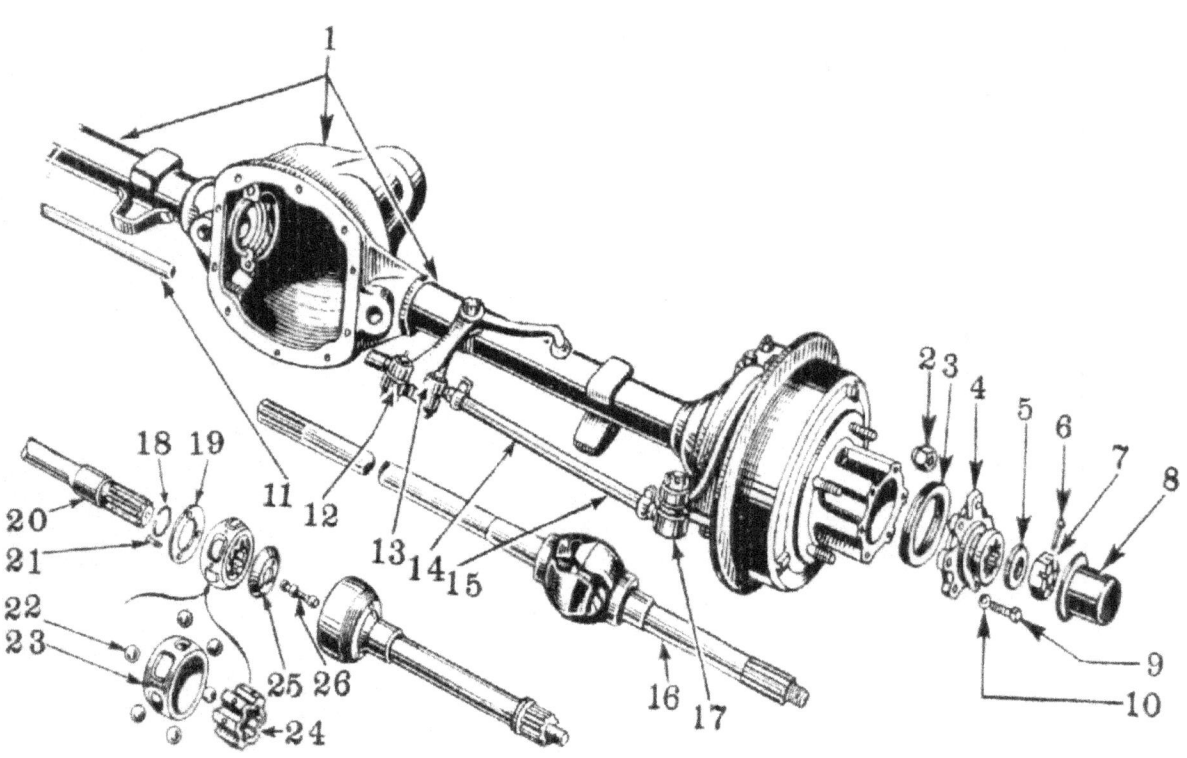

FIG. 4—FRONT AXLE ASSEMBLY

No.	Willys Part No.	Ford Part No.	Name
1	A-1703	GPW-3074	Gear Carrier Housing and Tube Assembly
2	A-476	GP-1012	Wheel Hub Bolt Nut R.H. Thread (Ford GP-1013; Willys A-475 L.H. Thd.)
3	A-862	GP-3208-A	Universal Joint Adjusting Shims
4	A-868	GP-3204	Axle Shaft Drive Flange
5	636570	356504-S	Axle Shaft Nut Washer
6	5397	72071-S	Axle Shaft Nut Cotter Pin
7	636569	356126-S	Axle Shaft Nut
8	A-869	GP-1139	Hub Cap
9	A-760	GP-1110	Axle Shaft Drive Flange Cap Screw
10	5010	34807-S	Axle Shaft Drive Flange Screw Lockwasher
11	A-1705	GPW-3281	Tie Rod Tube—Right
12	A-1211	GPW-3131	Drag Link Bell Crank
13	A-838	GP-3289	Tie Rod Socket Assembly—Right
14	A-1709	GPW-3282	Tie Rod Tube—Left
15	A-1708	GPW-3279	Tie Rod Assembly—Left
16	A-809	GPW-3206-A2	Axle Shaft and Universal Joint Assembly R.H. (Bendix) (Ford GPW-3207-A; Willys A-810 Left Hand)
17	A-847	GP-3290	Tie Rod Socket Assembly—Left
18	A-1726	GP-3216	Axle Shaft Retainer Snap Ring
19	A-1724	GP-3217	Axle Shaft Retainer
20	A-1729	GPW-3017-A	Axle Inner Shaft—Left (Ford GPW-3016-A; Willys A-1727- Right Inner)
21	A-1725	24622-S	Axle Shaft Retainer Screw
22	A-1721	358074-S	Axle Shaft Universal Joint Ball } Rzeppa Universal Joint
23	A-1719	GP-3215	Axle Shaft Universal Joint Cage
24	A-1720	GP-3221-A	Axle Shaft Universal Joint Inner Race
25	A-1722	GP-3219	Axle Shaft Universal Joint Pilot
26	A-1723	GP-3218	Axle Shaft Universal Joint Pilot Pin

The result of negative or reverse camber, if excessive, will be hard steering and possibly a wandering condition. Tires will also wear on inside shoulders. Negative camber is usually caused by excessive wear or looseness of front wheel bearings, axle parts or the result of a sagging axle.

Result of unequal camber may be any or a combination of the following conditions—unstable steering, wandering, kick-back or road shock, shimmy or excessive tire wear. The cause of unequal camber is usually a bent steering knuckle or axle center.

Correct wheel camber is 1½° and is set in the axle at the time of manufacture and cannot be altered by any adjustment. It is important that the camber be the same in both front wheels. Excessive heating of these parts to facilitate straightening destroys the heat treatment given them at the factory. Cold straightening of bad bends may cause a fracture of the steel and is unsafe. Replacement with new parts is recommended rather than any straightening of damaged parts.

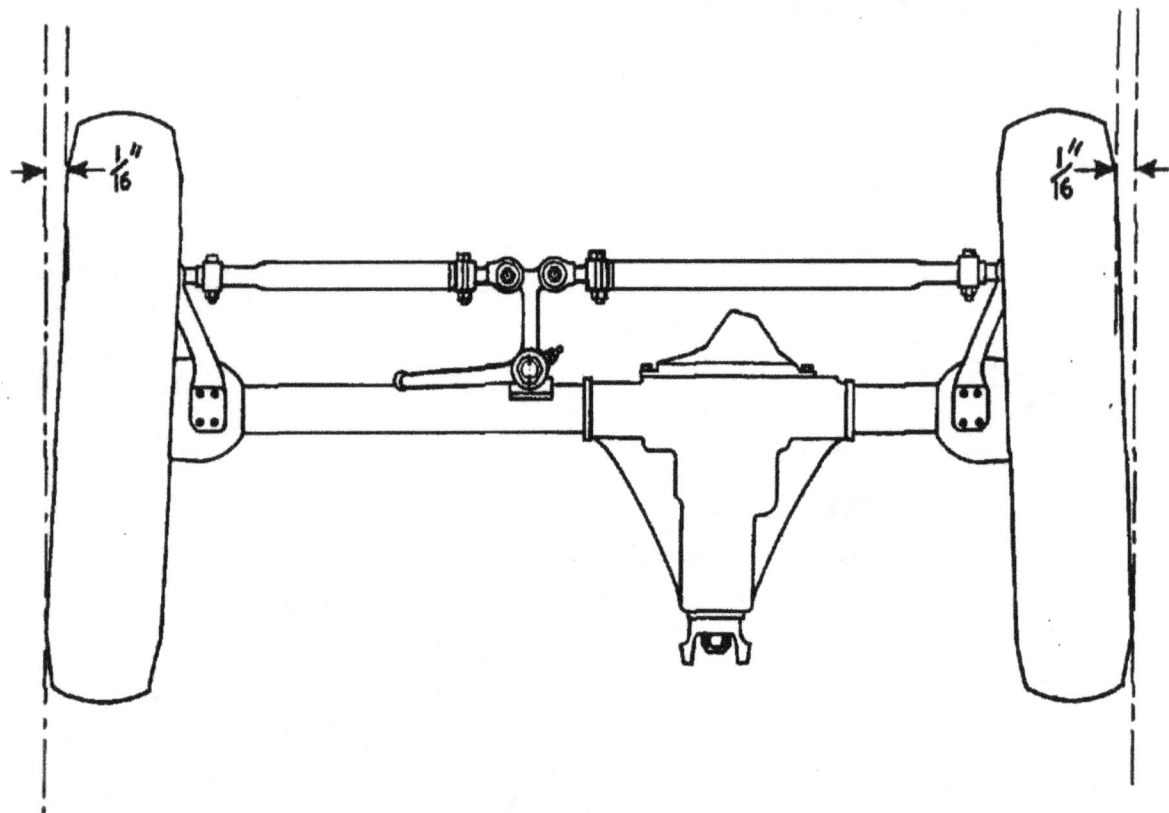

FIG. 5—FRONT WHEEL TOE-IN

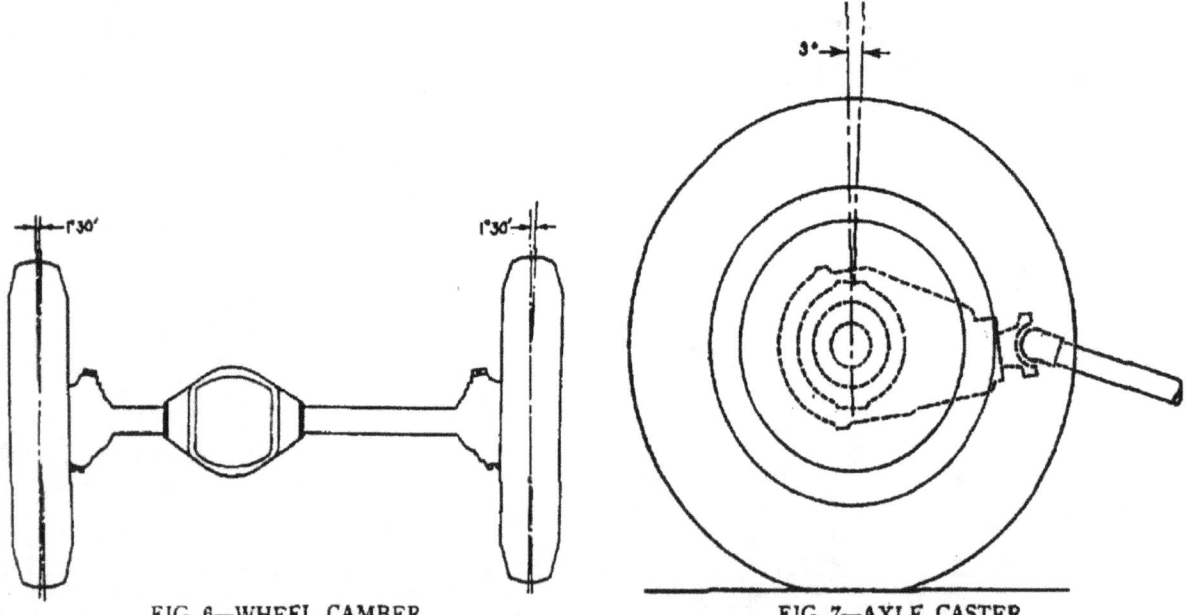

FIG. 6—WHEEL CAMBER

FIG. 7—AXLE CASTER

Axle Caster

The purpose of caster Fig. 7 is to provide steering stability which will keep front wheels in a straight ahead position and to assist in bringing wheels out of a turn on a curve.

The result of no caster is wandering or the vehicle will not come out of a turn normally.

No adjustment is provided for this angle but if checked with a suitable gauge and found to be incorrect, it should be investigated and if not excessive, correction made by the use of axle wedges or bending the axle cold.

If the camber and toe-in are correct and it is known that the axle is not twisted, a satisfactory check can be made by testing vehicle on the road. Before road testing make sure all tires are properly inflated, being particularly careful that both front tires are inflated to exactly the same pressure.

If vehicle turns easily to both sides but is hard to straighten out, this indicates insufficient caster for easy handling of vehicle.

Front Wheel Turning Angle

When the front wheels are turned the inside wheel on the turn travels in a smaller arc than the outside wheel; therefore it is necessary for the wheels to toe out. This change in wheel alignment is obtained through the length and angularity of the steering knuckle arms in relation to the front axle. When the wheels are turned so the inside wheel is on an angle of 20° as shown by "I" in Fig. 8, the outer wheel angle "O" should be 19°45'. The left steering knuckle arm controls the relationship of the front wheels on a left turn and the right arm controls the relation on a right turn. If a steering arm should be accidently bent it can be straightened cold if the bend is not excessive, otherwise the arm should be replaced.

Steering Bell Crank

The bell crank is a drop forging, heat treated for strength with removable ball ends, steering connecting rod and tie rods.

The bell crank is mounted on the front axle and swivels on two needle bearings.

If bell crank becomes damaged or bent do not attempt to heat and straighten, straighten it cold or install new parts.

The bell crank shaft is removable from axle by driving out tapered lock pin, driving pin out toward left front wheel.

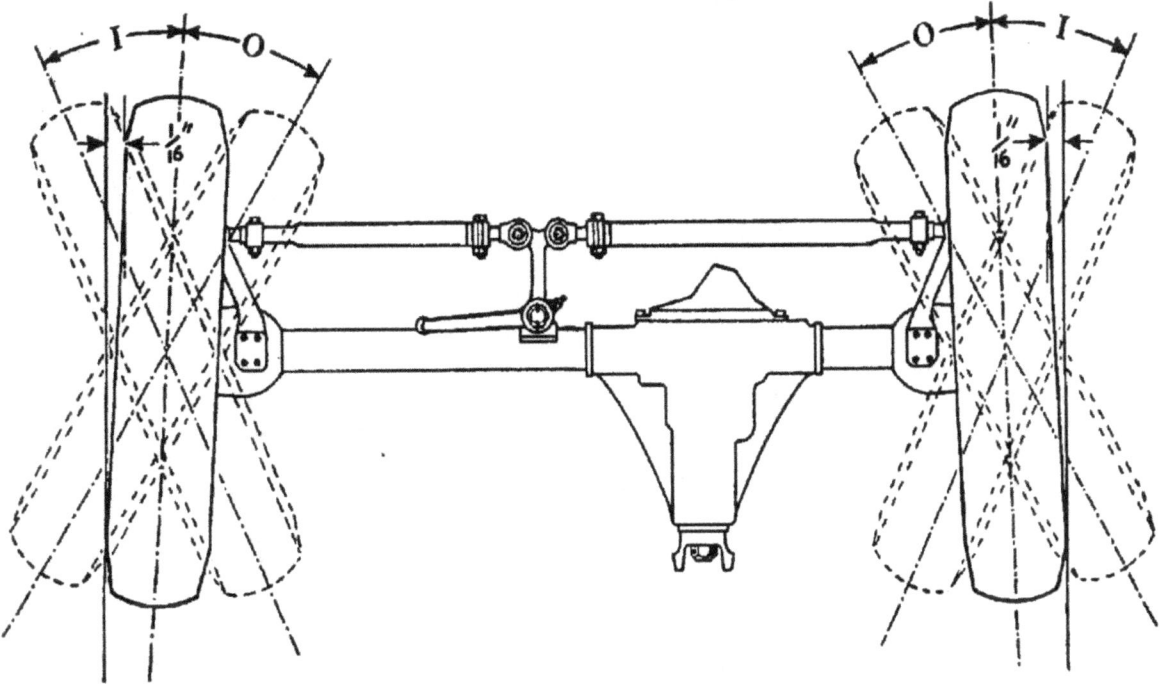

FIG. 8—FRONT WHEEL TURNING ANGLE

STEERING TROUBLES AND REMEDIES

SYMPTOMS	PROBABLE REMEDY

Hard Steering

Lack of Lubrication	Lubricate All Connections
Tie Rod Ends Worn	Replace
Connecting Rod Ball Joints Tight	Adjust
Cross Shaft Improperly Adjusted	Adjust
Steering Gear Parts Worn	Replace
Front Axle Trouble	See "Front Axle" Section

Steering Loose

Tie Rod Ends Worn	Replace
Connecting Rod Ball Sockets Worn	Replace
Steering Gear Parts Worn	Replace
Steering Gear Improperly Adjusted	Adjust

Road Shock .... Steering Connecting Rod too tight; Axle Spring Clips loose; Wheel Bearings loose; poor Shock Absorber control.

Turning Radius

Short one side .... Center Bolt in spring sheered off, Axle shifted, Steering Arm bent, Steering Arm not properly located on Steering Gear.

STEERING SPECIFICATIONS

Steering Gear:

Make	Ross
Type	Cam and Twin Pin Lever
Model	T-12
Ratio	Variable Ratio, 14-12-14 to 1
Wheel	3 spoke—17¼", Safety type

Bearings:

Cam Upper	Ball
Cam Lower	Ball
Levershaft	Bushing
Steering Column Upper	Ball

Lever Shaft:

Clearance to Bushing	.0005"-.0025"
End Play	.000"
Lash at Cam (Straight ahead)	Slight drag over high point

Steering Connecting Rod:

Make	Columbus Auto Parts
Type	Spring loaded
Adjustment	Threaded Plug

Steering Geometry:

Toe-in	3/64"-3/32"
Camber	1½°
Caster	3°

Toe out

Inside wheel	20°
Outside wheel	19°45'

FRAME

The frame is the structural center of any vehicle, for in addition to carrying the load, it provides and maintains correct relationship between other units to assure their normal functioning.

The frame is of rugged design and constructed of heavy channel steel side rails and cross members. Braces and brackets are used to maintain the proper longitudinal position of the side rails relative to each other, and at the same time provide additional resistance to torsional strains. Due to this rugged design the frame requires very little attention to maintain its dependability.

Vehicles which have been in an accident of any nature which may result in a swayed or sprung frame, should always be carefully checked for proper frame alignment in addition to steering geometry and axle alignment.

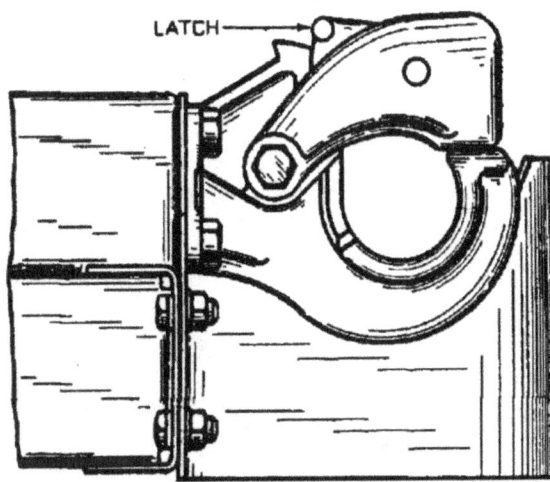

FIG. 1—PINTLE HOOK

A pintle hook for towing is provided on the rear frame crossmember. To open the hook lift up on the safety latch.

Checking Frame Alignment

When checking a frame for alignment Fig. 2, the most efficient method is "X" checking from given points on each side rail.

The most convenient way to check frame alignment, particularly when body is on chassis is by marking on floor all points from which measurements should be taken.

Select a space on the floor which is comparatively level. If cement floor, clean so that chalk marks will appear underneath the points of frame to be checked. If a wooden floor, it is advisable to lay a sheet of paper underneath the vehicle and tack in place, dropping a plumb-bob from each point indicated in Fig. 2, mark flooring directly underneath plumb-bob. Satisfactory checking depends upon the accuracy of marks with relation to the frame.

To reach points shown that have been marked, have vehicle carefully moved away from layout on floor, and proceed as directed in the following paragraphs:

1. Check frame width at front and rear ends using corresponding marks on floor. If widths correspond to specifications given draw center line full length of vehicle half way between marks indicating front and rear widths. If frame width is not correct lay out center line as follows:

    If center line cannot be laid out from checking points of ends of frame, it can be drawn through intersection of any two pairs of equal diagonals. If the extreme front end of frame is damaged, center of front end of frame can be located from point exactly midway between radiator support bolts.

2. With the center line properly laid out, measure distance from it to opposite point marked over entire length of chassis. If frame is in proper alignment, measurement should not vary more then 1/8" at any station.

3. To locate point at which frame is sprung measure diagonals marked "AB" "BC" "CD". If the diagonals in each pair are within 1/8", that part of the frame included between points of measurements may be considered in satisfactory alignment. These diagonals should intersect within 1/8" of center line. Any variations of more than 1/8" indicates misalignment. If the measurements do not agree within the above limits, it means that correction will have to be made between those measured points that are not equal.

Straightening Frame

In the case where the bending or twisting of the frame is not excessive, the frame may straighten. This should be done cold, excessive heat applied to the frame might change the structure of the metal and weaken the frame. For this reason it is recommended that badly damaged frame parts be replaced.

Front Axle Alignment

After it has been determined that the frame is properly aligned, the front axle alignment with frame can be checked as directed below.

The front axle is square with the frame if the distance between the front and rear axle is the same on both sides, and the distance from center of the upper spring bushing to front axle on both sides are equal.

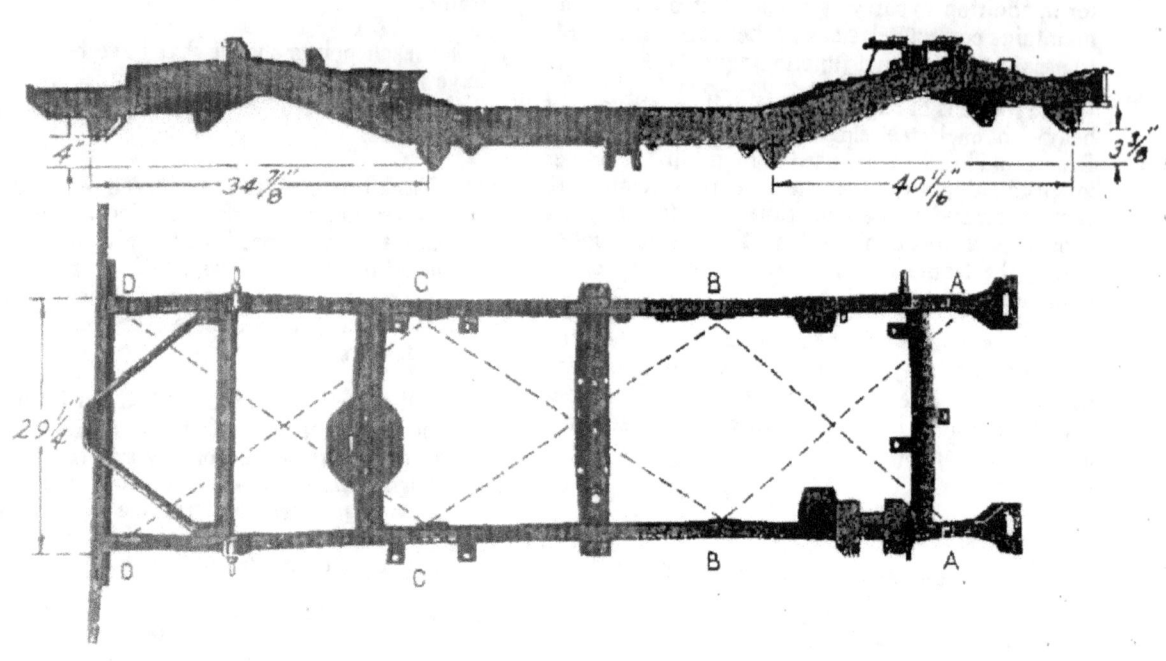

FIG. 2—FRAME ASSEMBLY

FRAME SPECIFICATIONS

Frame..SAE 1025
   Depth Maximum...4.186"
   Thickness Maximum....................................083"-.093"
   Flange Width...1¾"
   Length..122¾"

Width
   Front...29¼"
   Rear..29¼"

Number Cross Members.....................5-"K" member at Rear
Weight...140 lbs.

Wheel Base...80"

Tread
   Front.............................48¼"-with combat wheels 49"
   Rear..............................48¼"-with combat wheels 49"

SPRINGS

The specially designed springs used on this vehicle are constructed of chromium alloy to stand the severe service to which they may be subjected.

Front Springs

The front springs Fig. 1 are the semi-elliptic type, 36¼" long and 1¾" wide. There are eight leaves in each spring, seven leaves being of the parabolic shape with No. 2 leaf military wrapped over the eye of No. 1 leaf. The ends of the leaves are turned down to eliminate squeaks. Each spring is equipped with four rebound clips 1¼" wide.

The front springs appear to be identical in construction, nevertheless, they are different in load carrying ability.

The left spring requires a load of 525 lbs. for a $\frac{5}{16}$" camber. The right front spring requires a load of 390 lbs. for a $\frac{5}{16}$" camber. This difference is required due to the extra weight on left side of vehicle. The left spring can be identified by letter "L" painted on lower side at front on second leaf.

The front end of the front springs are shackled, using the "U" type shackle with threaded core bushing. The rear end of the spring is bronze bushed and is pivoted by a pivot bolt in the bracket on the frame. A torque reaction spring stabilizes the torque of the front axle.

The spring saddles on axle are welded in place to the underside of axle housing and springs are held in that position through "U" bolts, using the center spring bolt inserted in spring saddle to prevent the shifting of the axle.

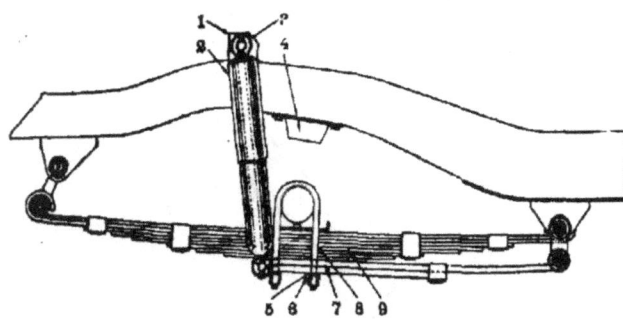

FIG. 1—LEFT FRONT SPRING AND SHOCK ABSORBER

No.	Willys Part No.	Ford Part No.	Name
1	A-1204		Front Shock Absorber Bracket and Shaft Assembly Left (A-1205 Right)
2	A-169	GPW-18045	Shock Absorber Assembly
3	637936	GPW-18060	Shock Absorber Mounting Pin Bushing (Rubber)
4	A-481	GPW-5783	Axle Bumper
5	5938	34848-S	Spring Clip Nut Lockwasher
6	339372	GPW-5456	Front Spring Clip Nut
7	A-6066	GPW-5588	Torque Reaction Spring Assembly
8	A-575	GPW-5705	Front Axle to Spring Clip, Left
9	A-612	GPW-5311	Front Spring—Left (Ford GPW-5310; Willys A-613 Right)

Rear Springs

The rear springs Fig. 2 are semi-elliptic, 42" long, 1¾" wide, 9 leaves with 4 rebound clips 1¼" wide. The spring leaves are the parabolic type with No. 2 leaf military wrapped around eye ends of No. 1 leaf. The ends of each leaf being turned down to eliminate squeaking.

The front end of the rear spring is bronze bushed and is pivoted by a pivot bolt at frame bracket, flexible "U" shackles are used at the rear.

The spring saddles are welded to the underside of rear axle housing and the center spring bolt is used to prevent shifting of the axle. The spring is held in position by two "U" bolts over the axle.

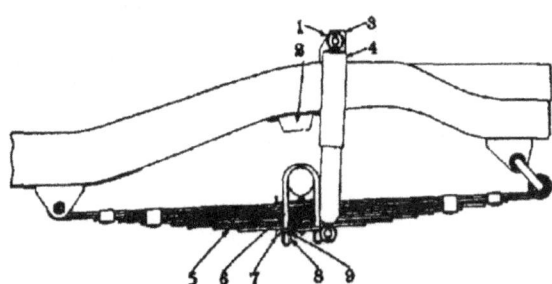

FIG. 2—REAR SPRING AND SHOCK ABSORBER

No.	Willys Part No.	Ford Part No.	Name
1	637936	GPW-18060	Shock Absorber Mounting Pin Bushing (Rubber)
2	A-481	GPW-5783	Axle Bumper
3	A-484		Rear Shock Absorber Bracket and Shaft Assembly (A-485—Right)
4	A-170	GPW-18080	Shock Absorber Assembly—Rear
5	A-614	GPW-5560	Rear Spring
6	A-575	GPW-5705	Rear Axle to Spring Clip
7	A-571	GPW-5460	Rear Spring Clip Plate and Shaft Assembly (Ford GPW-5459; Willys A-572 Right)
8	339372	GPW-5456	Rear Spring Clip Nut
9	5938	34848-S	Rear Spring Clip Nut Lockwasher

Spring Shackles and Pivot Bolts

The spring shackles are of the "U" type, Fig. 3 with threaded core bushings using right and left hand threads, depending at which position they are to be used in the chassis.

The bushings are anchored solidly in frame bracket and spring eyes and the oscillation taken between the threads of the "U" shackle and the inner threads of the bushing. The lubrication of the shackle bushings is very important, and should not be neglected, or excessive wear of the bushings and "U" shackles will occur.

There are six bushings used with right hand threads and two with left hand threads. The right hand threaded type bushings have plain hexagon head. The left hand threaded bushings have a groove around the head, Fig. 4.

The two left hand threaded "U" shackles can be identified by a forged boss on the lower shank of the shackle identifying the left hand thread Fig. 4. These two left hand threaded "U" shackles are used at the left front spring and the right rear spring, with the left hand threaded end at the spring eyes.

The "U" shackles are installed so that the bushing hexagon heads are to the outside of the frame. When making installation of a new "U" shackle or shackle bushing the following procedure should be followed.

Install shackle grease seal and retainer over threaded end of shackle up to the shoulder. Insert new shackle through frame bracket and eye of spring. Holding "U" shackle tightly against frame, start upper bushing on shackle, care being taken when it enters the thread in the frame that it is not cross threaded. Screw bushing on shackle about halfway, and then start lower bushing holding shackle tightly against spring eye and thread bushing in approximately halfway, then alternating from top bushing to lower bushing turn them in until the head of the bushing is snugly against the frame bracket, and the bushing in spring eye is $\frac{1}{32}$" away from spring measured from inside of hexagon head to spring.

Lubricate the bushings with high pressure lubricant and then try the flex of the shackle, which should be free. If shackle is tight it will be detrimental to the bushings as well as to the spring and it will be necessary to rethread the bushings on shackle.

Remove and Replace Spring

To remove a spring raise the vehicle, then place a stand jack under frame side rail, adjusted to a distance so that the load is relieved on the spring and yet the wheels still rest on the floor, remove the four "U" bolt nuts and lock washers. Remove spring plate or torque spring. Lower jack at side rail so that the spring is free from axle.

Remove pivot bolt nut and drive out pivot bolt from spring bracket and bushing Fig. 5.

Remove bushing from "U" shackle.

To install spring, replace pivot bolt first and then the "U" shackle bushing. Raise jack and place center bolt in spring saddle and install "U" bolts and nuts. "U" bolt nut, torque wrench reading, 50-55 ft. lbs., when torque reaction spring is used, 60-65 ft. lbs. Spring pivot bolt nut, 27-30 ft. lbs.

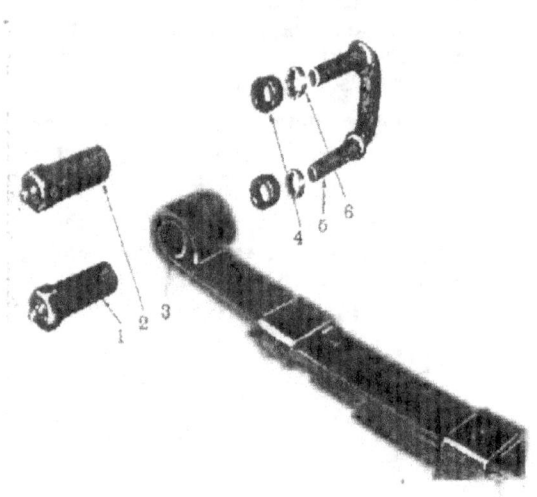

FIG. 3—SPRING SHACKLE—Left Front Spring

No.	Willys Part No.	Ford Part No.	Name
1	636532	GPW-5463	Spring Shackle Bushing Assembly (Left hand Thread)
2	634432	GPW-5464	Spring Shackle Bushing Assembly (Right hand Thread)
3	A-612	GPW-5311	Front Spring Assembly—Left
4	A-515	GPW-5481	Spring Shackle Grease Seal
5	A-513	GPW-5778	Spring Shackle U-Bolt (Left hand Thread)
6	A-1252	GPW-5482	Spring Shackle Grease Seal Retainer

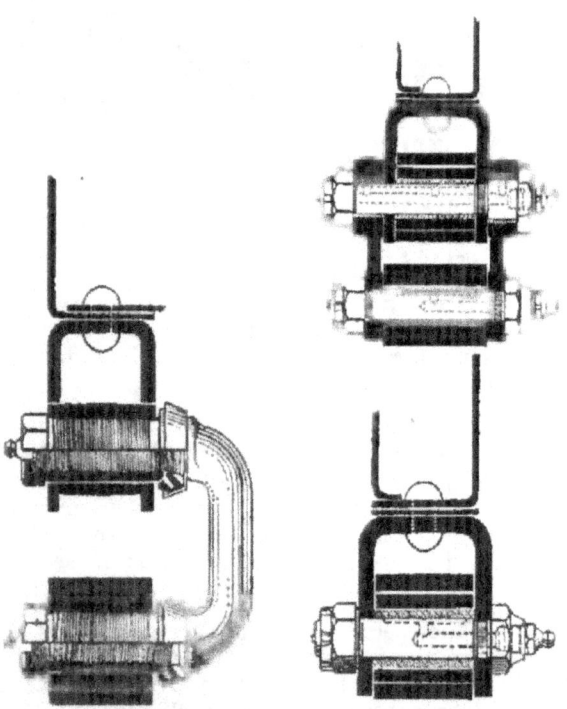

FIG. 4—SHACKLE & BOLT

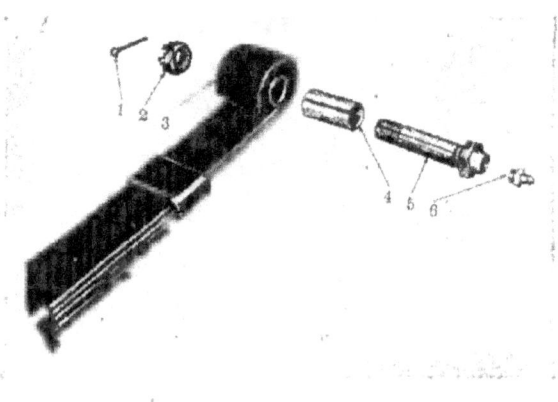

FIG. 5—SPRING BOLT—Right Rear Spring

No.	Willys Part No.	Ford Part No.	Name
1	5021	72034-S	Cotter Pin
2	6436	34033-S	Spring Bolt Nut
3	A-614	GPW-5560	Rear Spring Assembly—Right
4	359039	GPW-5781	Rear Spring Bolt Bushing
5	384228	GPW-5468	Spring Bolt
6	392909	353927-A1-S7-8	Grease Fittings

Shock Absorbers

The shock absorbers, Fig. 6 provide a much smoother ride by dampening the spring action as the vehicle passes over irregularities in the road.

The shock absorbers are the direct action type, two-way control and adjustable. The range of adjustment is four turns. To adjust the shock absorber, remove the lower end from the spring plate, push the unit together to engage the adjusting key and turn in a clockwise direction until the limit of the adjustment is reached. Holding adjusting key in slot, turn lower end anti-clockwise two turns. This is the average adjustment. Turning the adjustment to the right, or clockwise, gives firmer control for rough roads, while turning in the opposite direction gives a softer control, allowing faster spring action. Should squeaks occur in the rubber mounting bushings, do not use mineral oil or rubber lubricant, but add a flat washer on the mounting pins to place the bushing under pressure and prevent movement between the rubber and the metal part.

To install shock absorbers, install inner mounting rubber bushing on both upper and lower bracket pins, install shock absorbers, install outer bushings, flat washer and then compress, inserting cotter key and spreading it to hold washer in proper position.

The shock absorber is sealed at the factory with the proper amount of fluid and is non-refillable.

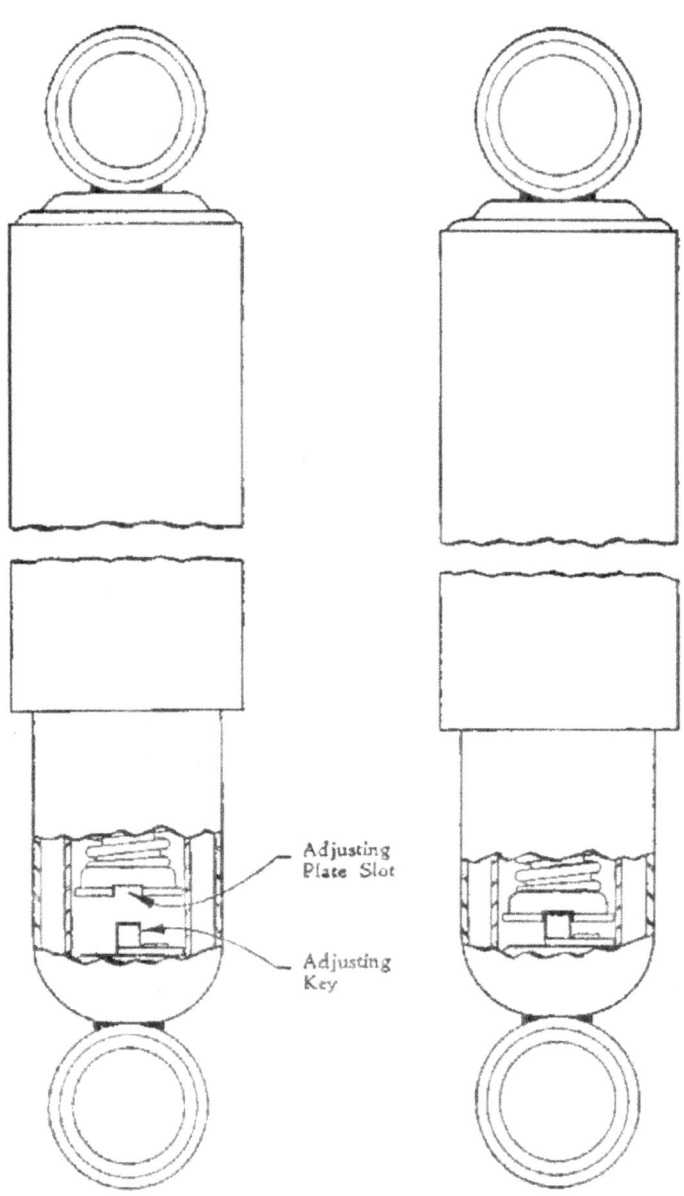

Sketch showing shock absorber before engaging adjusting slot and key.

Sketch showing shock absorber completely collapsed with adjusting key engaged in adjusting plate slot.

FIG. 6—SHOCK ABSORBER

SPRING TROUBLES AND CAUSES

SYMPTOMS	PROBABLE CAUSES
Spring Breakage—At center Bolt	Loose Spring to Axle Clips
Main Leaf Breakage on Ends	Tight Shackle or Pivot Bolt Shock Absorber Control Weak Poor Lubrication Spring Rebound too Great
Excessive Wear on Shackle Bushings	Inside Spring Eye Opened Up Bushing Improperly Installed Lack of Lubrication Worn Bushings
Shock Absorber Noise	Lack of fluid Damaged Cylinder Loose Mounting Brackets Mounting rubber bushings worn out
Shock Absorber Control	Adjust Lack of Fluid—replace shocks

SPRING SPECIFICATIONS

Front Spring:
    Make........................Mather
    Type Leaf....................Parabolic
    Length Center to Center of Eye........36¼"
    Width........................1¾"
    Number of Leaves.............8
    Front Eye Center to Center Bolt........18⅛"
    Rear Eye Center to Center Bolt.........18⅛"
    Left Camber under 525 lbs..............5/16"
    Right Camber under 390 lbs.............5/16"
    Rear Eye Bushed
      Bushing Size........1¾" long I.D., .5655"
      Rebound Clips.............4

Rear Spring:
    Make........................Mather
    Type Leaf....................Parabolic
    Length........................42"
    Width........................1¾"
    Number of Leaves.............9
    Rebound Clips.............4
    Camber under 800 lbs..............¼"
    Eye to Center Bolt.............21"
    Front Eye Bushed......1¾" long I.D., .5655"

Shock Absorber Specifications
    Make........Front-Monroe.....Rear-Monroe
    Type..........Hydraulic.....Hydraulic
    Action...........Double........Double
    Length Compressed. 10 9/16"........11 9/16"
    Length Extended ..16⅛"........18⅛"
    Adjustable........Yes..........Yes
    Mountings........Rubber.......Rubber

BODY

The body is of all-steel construction with mountings that provide a secure attachment to the frame.

All major panels are of No. 18 gauge steel. All open edges of panels are turned under, reinforced and flanged to give inherent strength. The panels are completely reinforced with "U" sections and welded. All component panels are seamed and welded together.

Body is insulated from frame with live rubber and fabric insulator shims placed between body and frame and held in place with body bolts.

The instruments and controls mounted on instrument panel are within clear view and easy reach.

Brass plugs have been placed in the left and right front corners to drain the floor.

Axe and Shovel

The axe or shovel can be removed or installed individually. The removal is apparent for they are held in place by fabric straps.

When installing the axe, turn the bit or blade up and place the handle in the front clamp, then insert the blade in the sheath after which pull up the clamp under the axe head and strap in place.

To install the shovel, turn the face against the cowl and place it in the strap on the cowl side. Next wrap the fabric strap through the handle and over the grip then between the grip and side of body; through the footman loop; over the outside of the grip and then buckle. This will hold the shovel forward securely in the strap on cowl side.

Windshield

The windshield and frame assembly provides for lowering entire assembly down on top of hood and also for opening and closing windshield for ventilation, when assembly is in upright position.

To lower windshield down on top of hood, loosen thumb screws on cowl so windshield can pivot and release two clamps holding tubular frame to instrument panel. When assembly is lowered onto hood fasten in place with hooks that are attached to the sides of hood.

To adjust windshield for ventilation, loosen the two wing nuts on upper brackets on each side of windshield, then entire frame assembly can be swung outward, anchor in position by tightening the wing nuts.

To Replace Windshield Glass

To replace windshield glass the following procedure should be followed:

1. Remove screws in each side of windshield adjusting bracket at top.
2. Bend down lip on left hand outer end of hinge at top.
3. Open windshield sufficiently to clear windshield frame and slide assembly off of hinge to left.
4. Remove the three nuts and bolts which hold the upper glass channel to frame.
5. Remove upper glass channel.
6. Withdraw glass from frame.

The replacement of windshield glass is the reverse of the above operations excepting that there should be new glazing tape used around the glass.

Windshield Seal

Windshield frame is sealed, when in closed position, by tension of special rubber seal against the tubular frame. The cowl seal, which is attached to the tubular frame is sealed when the clamps attaching frame to instrument panel are in locked position.

Top

To install the canvas top, it is necessary to loosen the two thumb screws at the pivot bracket, then slide tubing back out of front bracket, place in rear bracket and tighten thumb screws, allowing front bow to drop down over seats.

Attach canvas cover to top of windshield by the fasteners, then stretch canvas over bow and down to body back panel, placing the straps in the metal loops attached to body panel, stretch top and buckle straps. Next, raise front bow and assemble in the three bow flaps. The canvas top is carried under right front seat and held in place by straps.

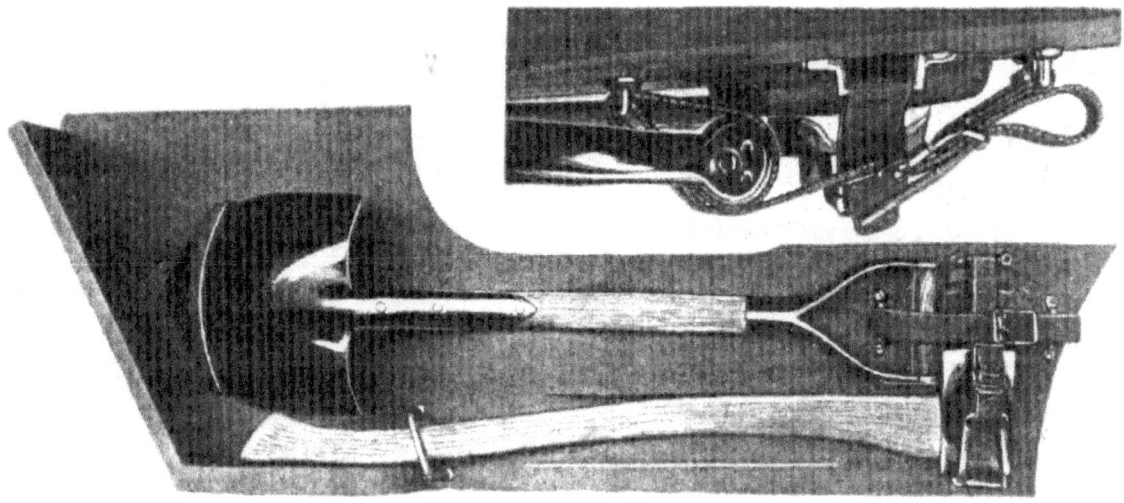

FIG. 1—SHOVEL AND AXE MOUNTING

TOOLS

Click here for Numerical Index of Tools

The manufacturers of the mechanical units used in this vehicle recommend the use of special precision tools, assembly jigs, gauges and close inspection of each part for assurance of proper operation and maximum service from each unit.

When necessary to perform a major operation on any mechanical unit special tools facilitate disassembling, checking and reassembling of the unit.

To aid the mechanic in performing satisfactory repairs, we suggest that tools as listed in this section or their equivalent be available when making major repairs.

OPERATING INSTRUCTIONS FOR SERVICE TOOLS

Supplied by Kent-Moore Organization
Detroit, Michigan

KMO-104—UNIVERSAL JOINT SNAP RING PLIERS. A special tool with jaws shaped to facilitate the removal and replacement of universal joint retainer rings.

J-270-1—DRIVER HANDLE. A heavy duty driving handle with a threaded end, on which can be mounted various adapters for removing and replacing bearing cups, oil seals, etc.

KMO-355—FEELER GAUGE SET. This feeler set consists of a number of blades, mounted in a suitable holder. The blades furnished total .040". The combination of blades provided herewith, will be found extremely useful in checking and reconditioning axle assemblies.

KMO-358—DRIVE PINION NUT WRENCH. This wrench is made to fit the retaining nuts that hold the several drive flanges to their splined shafts. The nuts are assembled very tightly and a heavy shank is necessary on the wrench in order that the nut can be loosened from its fit by a number of smart blows by a lead or copper hammer. This wrench consists of a 1¼" double broached hexagon socket, and a 15" hinged handle.

J-589-S—DRIVE PINION SETTING GAUGE SET—consists of:

  1 J-589-1—Spindle and Micrometer Assembly
  2 J-589-10-1—Locating Discs
  1 J-589-H-1—Clamping Plate for Hypoid Attachment
  1 J-589-H-5—Clamp Screw
  1 J-589-10-2—Offset Plate for checking Hypoids
  2 J-589-H-3—Hex Head Cap Screws
  2 5/16" Std. Plain Washers

  1 J-589-SX—Master Gauge for Micrometer checking
  1 J-589-B-1—Carrying Case

The rear axle pinion must be adjusted properly before any attempt is made to adjust the ring gear or differential. The ground face of the pinion is etched with its correct setting. The marking may be zero (0) minus (—) or plus (+). To determine the pinion setting, remove the differential and ring gear assembly. Bolt the clamping plate H-1 and screw H-5 across the open end of the axle housing in such manner that the hypoid (offset) plate, detail J-589-10-2 is clamped firmly to the end of the pinion. Next, place the locating discs, detail J-589-10-1 on each end of the micrometer spindle body and lower into position in the differential side bearing bores. Run the micrometer spindle down until the end contacts the free end of the hypoid offset plate, detail J-589-10-2. Rock the tool gently, adjusting the micrometer until it just drags slightly when rocked through a small arc, and note the reading on the micrometer spindle. A pinion marked "0" when properly adjusted, should show a micrometer reading of .719". A pinion marked "plus 2" when properly adjusted should show a micrometer reading of .717". A pinion marked (-4) should show a reading of .723" when properly adjusted. If pinion setting is not correct it will be necessary to remove the drive shaft flange, the pinion forward bearing cone, and the pinion, and remove or add shims as required between the rear bearing cup and the housing.

J-789—DRIVE PINION AXLE, AND TRANSFER UNIT FLANGE HOLDING TOOL. Used while removing the drive pinion flange to keep the pinion shaft from turning, and assists materially when removing or replacing the pinion shaft nut.

HM-872-S—DIFFERENTIAL SIDE BEARING AND DRIVE PINION FLANGE REMOVER SET. This tool consists of a Puller Body, with an adapter plug HM-872-4 for use when removing side bearings, and a screw-end adapter sleeve HM-872-S-3 for use in removing Drive Pinion Flange. The fingers of the tool can be adjusted to the part being removed by means of the hinged yoke and thumb screw. Keep screw threads thoroughly lubricated during operation of the tool, and tap head of screw with a lead hammer in order to assist in removal. This same tool, when used with HM-872-S-4 Adapter, will remove the dust shields from Transfer Case Brake Drum Flange and Drive Flange.

J-881-A—UNIVERSAL JOINT ASSEMBLY AND DISASSEMBLY TOOL consists of Tool Assembly, composed of C-clamp, Screw and Swivel, and detail J-881-6 Cup for receiving the roller bearing assemblies of the universal joint trunnion, while removing. To use the tool as a replacer, place bearing and retainer assemblies on each end of pin and press together until the assembly can be placed in the joint. Leave tool in place until the U-clamps at each end of the trunnion are pulled up just tight enough to keep bearings from slipping out of place. Then remove tool and finish tightening U-bolts. When disassembling a joint place J-881-6 cup on lower plug of tool. Remove snap rings from their seats. Assemble tool to joint and turn down on screw until bearing retainer and needle rollers come free into the cup. Then reverse the tool, and press on end of pin until opposite bearing retainer and rollers drop into the cup. By using the tool in this manner the danger of losing rollers or dropping them is eliminated.

HM-914—REMOVER PLATES FOR REMOVING BEARING CONE NEXT TO DRIVE PINION HEAD. The split halves of this tool are assembled to the drive pinion bearing, and bolted in place. By means of an arbor press, the pinion shaft can be pressed through the bearing. If an arbor press is not available, a suitable hand press can be devised by using Tool No. J-1759 with a pair of $3/8'' \times 8''$ standard thread bolts, that are furnished with the tool.

J-943—FRONT AXLE OUTER OIL SEAL REMOVER. The legs of this tool are made of tempered spring steel, and when compressed by hand, can be readily inserted behind the oil seal. A few taps by a hammer on the tie bar of the tool readily removes the seal.

SE-1066—RING GEAR BACK LASH CHECKING ATTACHMENTS—consists of:

1 SE-1066-2—Clamp

1 SE-1066-3—Connection for dial indicator with back mounting lugs

1 SE-1066-1—Sleeve

These attachments allow a dial indicator to be set up so that the contact button of a dial indicator will come in contact with one of the gear teeth. As the ring gear is rocked back and forth by hand, the dial indicator will show the amount of backlash between the ring gear and pinion.

SE-1094-5—DIAL INDICATOR. Not furnished with SE-1066 Attachment Set, but can be ordered extra if desired.

J-1375-S—FRONT PINION SHAFT FLANGE AND AXLE SHAFT FLANGE REPLACER. Under no circumstances should this flange be driven into place because of the possibility of damage to other parts of the unit. To operate place the flange in position on its shaft. Next, screw the socket adapter on end of pinion shaft, first placing the spacer washer between tool and end of shaft. Operating outer sleeve of the tool pushes the flange squarely and safely into place. Be sure the threads of tool are lubricated before each installation. Tool when used with adapters, S-1 and S-2 will also replace Transfer Case Brake Drum Flange and Main Transmission Drive Gear.

J-1436—WHEEL BEARING CUP AND OIL SEAL AND UTILITY PULLER. This item is a general utility tool with a wide range of uses such as removing oil seals, bearing cups, etc. Fingers are expanded or retracted by merely turning the handle right or left. A heavy sliding knocker that guides on the tool shaft and strikes against a lug welded to end of shaft, provides powerful leverage in removing parts pressed in place in various assemblies.

J-1735—FRONT AXLE SHAFT FLANGE PULLER. The puller body is so designed that its legs straddle the shaft and fit the under side of the flange. By turning down on the screw the flange is readily removed. In case of an exceptionally tight fit a few light blows with a lead hammer while the screw is under tension, will aid materially in the amount of effort required to free the flange.

J-1742—DRIVE PINION OIL SEAL REMOVER. This tool was designed to remove the oil seal without removing the pinion or differential. Consists of a body with a center puller screw, and four floating hooked fingers. To operate, turn the fingers so they will slide through the opening between the shaft and the oil seal, and when in position, turn the finger ends into the locking slots in the tool body. When fingers are in position the striking sleeve is propelled against the head of the center shaft until the oil seal is free of its seat in the housing. Illustration shows a screw to push against end of pinion shaft but this design has been changed to a headed shaft with a sliding knocker.

J-1743—DRIVE PINION OIL SEAL AND FRONT HUB OIL SEAL REPLACER. Designed to replace the oil seal without damage, and with the pinion in place.

J-1744—FRONT AXLE WHEEL BEARING ADJUSTING NUT WRENCH AND HANDLE. This hollow wrench is designed with a pilot guide ring on the inside of the body to prevent the wrench from slipping off the thin adjusting nuts. This construction permits tremendous pressure being applied without danger of the wrench slipping off and injuring the operator.

J-1751—KING PIN BEARING CUP REMOVER AND REPLACER AND FRONT AXLE SHAFT OUTER OIL SEAL REPLACER. Consists of a driver head and handle for removing the bearing cup, and an adapter ring, J-1751-3, for replacing the cups without damage. This tool is also used in replacing the front axle shaft outer oil seal.

J-1752—FRONT AND REAR AXLE SHAFT INNER OIL SEAL REMOVER. Consists of a screw with hinged ear, a cross bar and forcing nut. This oil seal is readily removed by slipping the hinged ear through the seal, and pulling it forward into position. A cross-bar assembled over the puller screw serves as a plate over the differential bore. As the forcing nut is turned down against the cross bar, the seal is removed from its position.

J-1753—FRONT AND REAR AXLE SHAFT INNER OIL SEAL REPLACER. This is a special driver designed to replace the oil seal without damage. The tool has a short shank to enable it to be operated in the confined area of the axle housing. A short mallet must be used because of the confined area in which to operate. See also J-1753-3 Adapter.

J-1753-3—FEED SCREW ADAPTER FOR USE WITH J-1753. In the event that a short mallet is not available, the Axle Shaft Inner Oil Seals can be installed in the following manner:— Place an oil seal on J-1753 pilot. Slip the socket end of feed screw over the shank of J-1753. Assemble disc J-270-14 to the threaded end of feedscrew. By turning the hexagon end of feedscrew with an open end wrench, the oil seal is forced into position, the thrust being taken by disc J-270-14 mounted in the opposite side of the differential opening.

J-1761—DRIVE PINION BEARING CONE REPLACER FOR CONE MOUNTED NEXT TO PINION HEAD. Use of this tool prevents the drive pinion bearing cone from being damaged while being installed on the pinion shaft, and eliminates the danger of scoring or shearing the shaft or of chipping the cone.

J-1763—DIFFERENTIAL SIDE BEARING CONE REPLACER HEAD. Designed to operate with J-270-1 Handle. This tool pilots in the hole in the differential case and is so designed that the pressure is taken directly on the cone. The cone can either be tapped or pressed into place, and when properly used, the tool will eliminate all danger of distorting the bearing roller cage.

J-1764—PAIR OF HOOKS FOR REMOVING FRONT SPINDLE LOCK WASHERS. The lock washer which is placed between the bearing adjusting nut and the lock nut has a tongued ear that rides in the spindle keyway. Removal is sometimes difficult because of housing interferences, and these hooks will materially assist in withdrawing the washer from the spindle.

J-1765—BRAKE ECCENTRIC ADJUSTING TOOL. This tool has two rectangular slots to fit the eccentric adjusting lugs on brake shoe anchor pins. The tool is designed to operate with box type wrenches such as are supplied with mechanics hand tool sets.

KMO-410—TRANSFER CASE MAIN SHAFT SNAP RING REMOVING PLIERS. These pliers with knurled lugs allow the snap ring to be expanded out of its groove in the shaft and readily removed.

HM-872-S-4—ADAPTER—TRANSFER CASE BRAKE DRUM FLANGE AND DRIVE FLANGE DUST SHIELD REMOVER. This adapter fits on the end of the main screw of HM-872-S puller and allows dust shields to be quickly removed.

J-1375-S-1—ADAPTER—TRANSFER CASE MAIN DRIVE GEAR REPLACER. The main drive gear should never be pounded into place because of danger of damage to the internal mechanism of the transmission. Use this adapter in connection with the regular shaft of tool J-1375-S to force the gear into place.

J-1375-S-2—ADAPTER—TRANSFER CASE BRAKE DRUM ASSEMBLY REPLACER. The brake drum assembly should never be pounded into place because of almost certain damage to connecting parts. Using this adapter in connection with the regular shaft of tool J-1375-S, the drum assembly is forced into place by screw feed safely and quickly.

J-1375-6—ADAPTER SLEEVE. Required as a spacer sleeve when using either J-1375-S-2 or J-1375-S-1 in replacing brake drum and main drive gear.

J-1748—TRANSFER CASE MAIN SHAFT FRONT BEARING REPLACER. The nose of this tool is specially designed to allow the bearing to be installed without damage. A copper, lead, or rawhide mallet should be used when replacing.

J-1749—TRANSFER CASE MAIN SHAFT FRONT CONE REMOVER. In order to disassemble the main shaft to remove from the case, use this wedge shaped tool. Insert the tool between the front bearing and the gear, and tap the tool with a lead, copper, or rawhide mallet until the bearing is wedged off the shaft.

J-1754—TRANSFER CASE REPLACER FOR MAIN SHAFT BEARING CUPS, FRONT AND REAR. This tool is designed so that the cups are seated in exactly the proper distance in the bore. Strike shank end of tool with lead, copper, or rawhide mallet until the stop shoulder bottoms, when replacing front bearing cup. Rear bearing cup protrudes slightly on outside of case when properly seated.

J-1755—TRANSFER CASE FRONT BALL BEARING REPLACER. This tool was designed to drive the ball bearing into place in the front bearing cap. The driver end of the tool contacts the outer race only and will not injure the bearing while installing. Use a lead, copper, or rawhide mallet with the tool. This tool can also be used for wheel hub inner bearing cone and oil seal.

J-1756—TRANSFER CASE OIL SEAL REPLACER. A specially designed tool with a pilot head to hold the oil seal while starting and to drive the seal in place without damage. Used for both front and rear bearing caps. A lead, copper, or rawhide mallet should be used with this tool.

J-1757 — TRANSFER CASE SHIFTER SHAFT OIL SEAL ASSEMBLY TOOLS. Consists of J-1757-1 Driver and J-1757-2 Pilot. The tapered pilot is placed on the end of the shifter shaft and allows the oil seal to be expanded gradually and uniformly as it is driven into position. Without the use of this tool, serious damage could result to the seal. When assembling shifter shaft, the pilot is also useful.

J-1758—SPEEDOMETER GEAR BUSHING REPLACER. The speedometer gear shaft rides in a small hardened bushing that is difficult to replace unless this special pilot driver is used.

J-1759—TRANSFER CASE BRAKE DRUM AND MAIN DRIVE GEAR PULLER SET. Tool can also be used as a hand press in connection with HM-914 for removing axle drive pinion bearing cone.

Tool consists of J-1759-1 Body; J-1759-2 Main Puller Screw; J-1759-3 Pair of Socket studs and check nuts for use in removing brake drum; two 3/8" x 8" bolts for use with HM-914 in removing drive pinion bearing; two 3/8" x 1¾" screws, nuts, and washers for removing Bantam type brake drum; and one pair of J-1759-5 Fingers and Nuts for removing transmission main drive gear. The body of this tool is slotted at each end, and various adapters can be added to make it a utility puller with a wide range of application.

J-1766—TRANSFER CASE FRONT AND REAR BEARING CAP OIL SEAL REMOVER. These oil seals are mounted tight against a shoulder and this specially designed drift provides a quick means of removal.

J-1767—TRANSFER CASE MAIN SHAFT SNAP RING INSTALLING SET. Consists of J-1767-1 pusher tube and J-1767-2 tapered thimble. The thimble pilots on the end of the shaft. The snap ring is started on the small end of the tapered part of the pilot. The forcing sleeve is used to force the ring on the shaft and into its groove.

J-1768—TRANSFER CASE IDLER GEAR THRUST WASHER LOCATOR. There is a thrust washer on each side of the case and without this tool it is difficult to assemble the idler gear cluster and at the same time keep the thrust washers from slipping out of position. To operate, place a film of grease on each thrust washer to help hold washer to case. Next, start the cluster gear shaft into the case, and it will help hold the washer in position on that side of the case. Place tool J-1768 through hole on opposite side with small end of tool centered into thrust washer. Install the gear cluster, push shaft through gear cluster, and it will pick up the opposite thrust washer and at the same time push the locator tool out of the way.

J-1769—TRANSFER CASE IDLER GEAR SHAFT DRIFT AND SHIFT LEVER PIN INSTALLING PILOT. This tool has a dual use. It is useful for forcing the idler gear shaft out of position. The same tool assists materially to locate the shift levers and springs as a pilot, which is pushed out of the way as the shift lever pin is pushed into position.

J-1770—TRANSFER CASE FRONT AND REAR DRIVE FLANGE DUST SHIELD REPLACER. This tool is a pilot driver correctly machined to allow the dust shields to be replaced without damage. Use a lead, copper, or rawhide mallet with this tool. This tool can also be used to replace wheel hub oil seals.

J-1771—TRANSFER CASE FRONT BEARING CAP SNAP RING REMOVER SET. Tool consists of J-1771-SA-1 tool to force snap ring out of groove, and J-1771-SA-2 hook to assist detail 1. This snap ring fits in an internal groove and is exceedingly difficult to remove with ordinary tools. To operate, pry one end of the snap ring out of its groove far enough to enter hook. When one end of the snap ring is free, it can be grasped by a pair of pliers and pulled out of the cap.

J-1772—TRANSFER CASE FRONT BEARING CAP SNAP RING REPLACER SET. Tool consists of the following parts:—J-1772-SA-1 subassembly into which ring is loaded and expelled into its groove; J-1772-SA-2 loading driver; J-1772-5 tapered loading sleeve.

To operate, place J-1772-SA-1 bottom side up in a vise, or with the handle protruding through a hole in the work bench. Next, place J-1772-5 tapered loading sleeve with large opening uppermost over the recessed collar of J-1772-SA-1. Then drop the snap ring into loading sleeve and use J-1772-SA-2 loading driver to put the ring through the tapered hole and into the recessed counterbore in J-1772-SA-1. With the snap ring installed in J-1772-SA-1, place the tool in position in the transfer case housing. Hold the outer case of the tool tight against the housing and strike the floating plunger smartly with a lead, copper, or rawhide mallet to expel the snap ring into its seat.

J-1773—TRANSFER CASE FRONT BEARING CUP REMOVER. After the main shaft cone has been wedged forward on the shaft (see tool J-1749 instructions) it is necessary to push the bearing cup out of the case in order to remove the shaft. Slide gear toward inside of case and insert tool J-1773 between the gear and the bearing cup. By driving on end of shaft with rawhide mallet, the cup is forced out of the case.

J-1774—TRANSFER CASE SPEEDOMETER GEAR SHAFT BUSHING REMOVER. (Not illustrated) No special tool has been provided for this purpose. Should service be required, any standard "Easy-Out", or tapered type stud extractor with a capacity of ¼" diameter, can be used to extract the bushing.

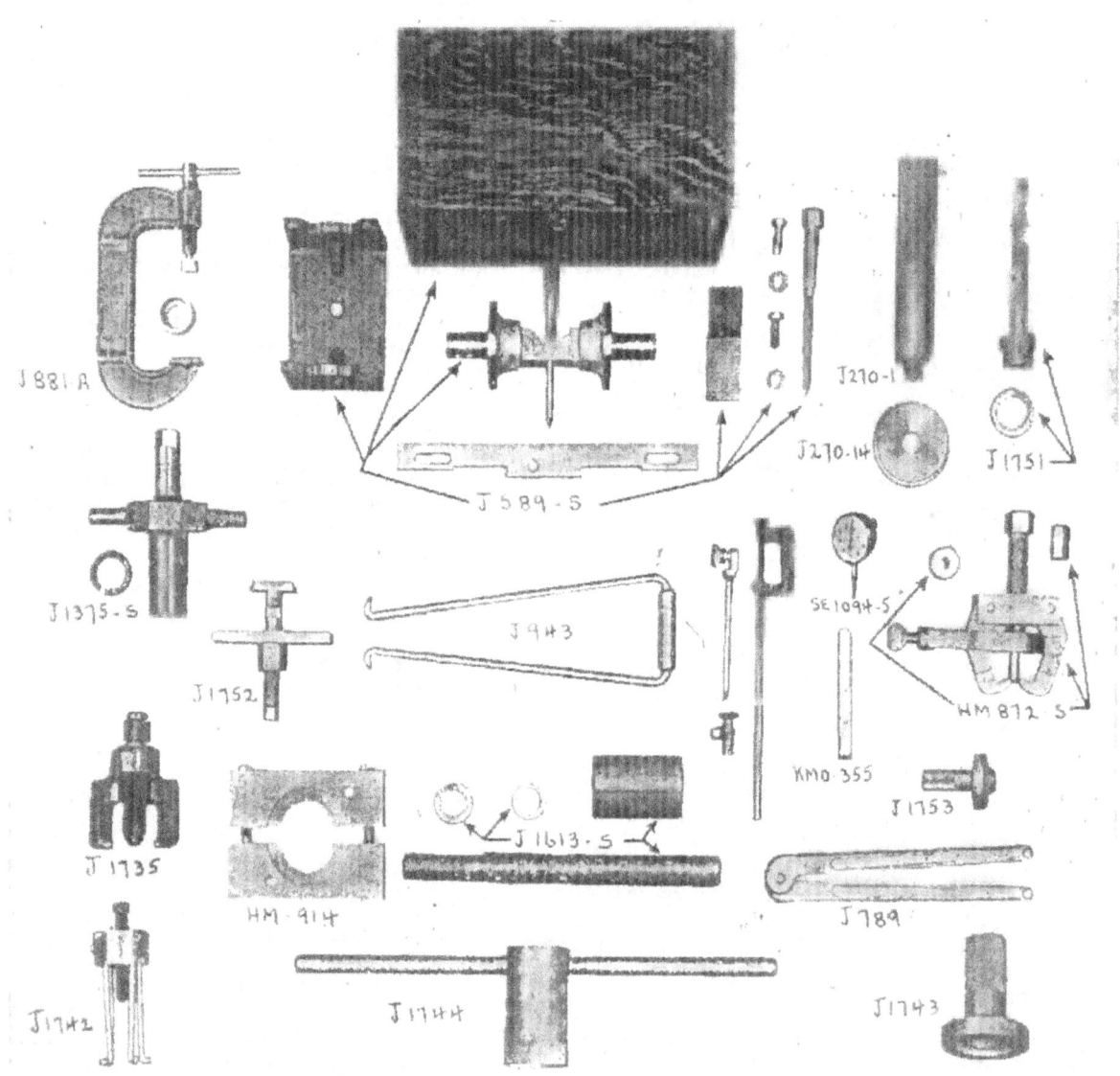

FIG. 1—SERVICE TOOLS

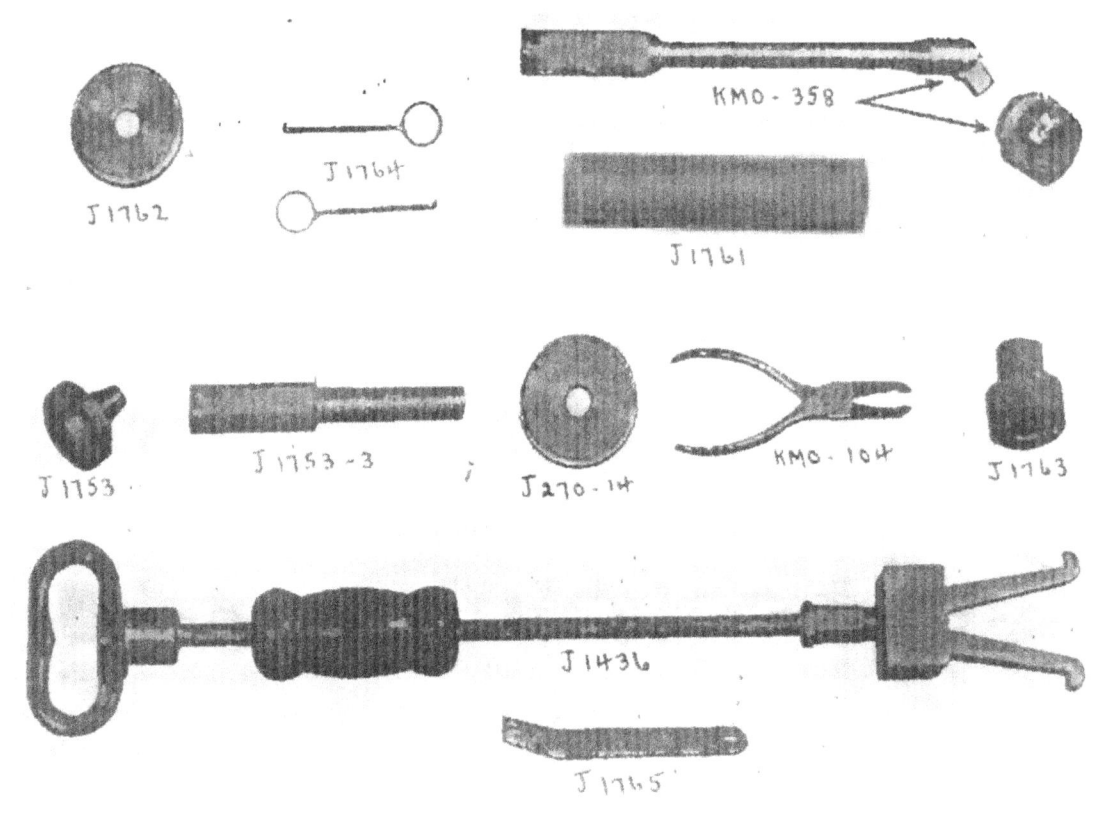

FIG. 2—SERVICE TOOLS

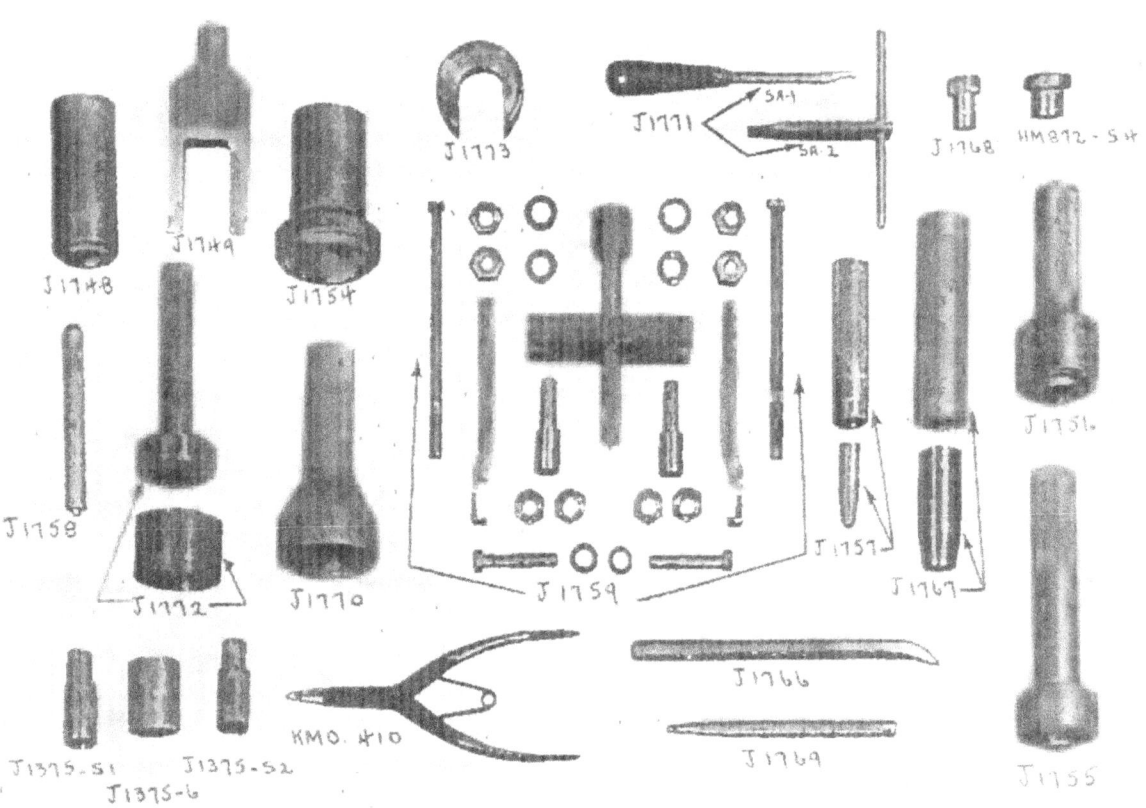

FIG. 3—SERVICE TOOLS

NUMERICAL INDEX

SERVICE TOOLS REQUIRED FOR TRANSFER CASE

Supplied by Kent-Moore Organization, Detroit, Mich.

Willys No.	K-M No.	
A-6218	KMO-410	Transfer Case Main Shaft Snap Ring Removing Pliers..............
A-6200	HM-872-S-4	Adapter—Transfer Case Brake Drum Flange and Drive Flange Dust Shield Remover—works with HM-872-S ....
A-6201	J-1375-S-1	Adapter—Transfer Case Main Drive Gear Replacer. Operates with tool J-1375-S..............................
A-6202	J-1375-S-2	Adapter—Transfer Case Brake Drum Assembly Replacer. Operates with tool J-1375-S............................
A-6203	J-1375-6	Sleeve required with J-1375-S-1 and J-1375-S-2 when replacing Transfer Case Main Drive Gear and Transfer Case Brake Drum............
A-6204	J-1748	Transfer Case Main Shaft Front Bearing Replacer.................
A-6205	J-1749	Transfer Case Main Shaft Front Cone Remover...................
A-6206	J-1754	Transfer Case Replacer for Main Shaft Bearing Cups, Front and Rear...
A-6207	J-1755	Transfer Case Front Ball Bearing Cup Replacer..................
A-6208	J-1756	Transfer Case Bearing Cap Oil Seal Replacer....................
A-6209	J-1757	Transfer Case Oil Seal Assembly Tool Set for Shifter Shaft.........
A-6210	J-1758	Speedometer Gear Bushing Replacer...........................
A-6211	J-1759	Transfer Case Brake Drum and Main Drive Gear Puller Set. Tool can also be used as a hand press in connection with HM-914 for removing Drive Pinion Bearing Cone................................
A-6212	J-1766	Transfer Case Front and Rear Bearing Cap Oil Seal Remover........
A-6213	J-1767	Transfer Case Main Shaft Snap Ring Installing Set...............
A-6214	J-1768	Transfer Case Idler Gear Thrust Washer Locator.................
A-6215	J-1769	Transfer Case Idler Gear Shaft Drift and Shift Lever Pin Installing Pilot
A-6216	J-1770	Transfer Case Front and Rear Drive Flange Dust Shield Replacer.....
A-6217	J-1772	Transfer Case Front Bearing Cap Snap Ring Replacer Set...........

SERVICE TOOLS REQUIRED FOR FRONT AND REAR AXLES

Supplied by Kent-Moore Organization, Detroit, Mich.

Willys No.	K-M No.	
A-6243	KMO-104	Universal Joint Snap Ring Removing Pliers.....................
A-6221	J-270-1	Drive Handle...
A-6244	KMO-355	Feeler Gauge Set..
A-6245	KMO-358	Drive Pinion Nut Wrench Socket and Hinge Handle................
A-6222	J-589-S	Drive Pinion Setting Gauge Set..............................
A-6223	J-789	Drive Pinion Axle and Transfer Unit Flange Holding Tool..........
A-6219	HM-872-S	Differential Side Bearing and Drive Pinion Flange Remover Set......
A-6224	J-881-A	Universal Joint Assembly and Disassembly Tool..................
A-6220	HM-914	Remover Plates for Bearing Cone next to Drive Pinion Head........
A-6225	J-943	Front Axle Outer Oil Retainer Remover........................
A-6246	SE-1066	Ring Gear Back Lash Checking Attachments less Dial Indicator or Feeler Gauge...
A-6247	SE-1094-5	Dial Indicator...
A-6226	J-1375-S	Front Pinion Shaft Flange and Axle Shaft Replacer...............
A-6227	J-1436	Wheel Bearing Cup, Oil Seal and Utility Puller..................
A-6228	J-1735	Front Axle Shaft Flange Puller...............................
A-6229	J-1742	Drive Pinion Oil Seal Remover...............................
A-6230	J-1743	Drive Pinion Oil Seal and Front Hub Oil Seal Replacer............
A-6231	J-1744	Front Axle Wheel Bearing Adjusting Nut Wrench and Handle.......
A-6232	J-1751	King Pin Bearing Cup Remover and Replacer, Front Axle Shaft Outer Oil Seal Replacer, Transfer Case Output Shaft Bearing Remover....
A-6233	J-1752	Front and Rear Axle Shaft Inner Oil Seal Remover................
A-6234	J-1753	Front and Rear Axle Shaft Inner Oil Seal Replacer...............
A-6235	J-1753-3	Feed Screw for use with J-1753 Rear Axle Shaft Inner Oil Seal Replacer.
A-6236	J-1761	Drive Pinion Bearing Cone Replacer for Cone mounted next to Pinion Head..
A-6237	J-1763	Differential Side Bearing Cone Replacer Head (works with J-270-1 Handle)..
A-6238	J-1764	Pair of Hooks for Removing Front Spindle Lock Washer............
A-6239	J-1765	Brake Eccentric Adjusting Tool...............................
A-6240	J-1783	Differential Case Assembly Studs (Set of 4)....................
A-6241	J-1784	Drive Pinion Bearing Cup Replacer Set.:......................
A-6242	J-1785	Drive Pinion Bearing Cup Removing Set.......................

WILLYS MODEL "MB" ¼-TON 4 x 4 GOVERNMENT TRUCK

NUMERICAL INDEX

SERVICE TOOLS REQUIRED FOR ENGINE

Supplied by KENT-MOORE ORGANIZATION, Detroit, Michigan

Willys No.	K-M No.	
A-6248	KMO. 375 EX	Valve Guide Expansion Reamer—Size .375″
A-6249	KMO-213	Cylinder Compression Indicator
A-6250	J-1950	Valve Guide Removing and Installing Tool
A-6251	HM-593-0	Piston Fitting Gauge and Scale .003″ x ¾″ x 12″
A-6252	HM-593-10	Piston Fitting Gauge (Less Scale)
A-6253	KMO-402-B	Special ½″ Tappet Wrench (Double End)
A-6254	KMO-402-BA	Special ¹⁷⁄₃₂″ Tappet Wrench (Double End)
A-6255	KMO. 812 Ex.	Piston Pin Reamer (Spec. Floating Pilot Expansion Type)—Size .8125″
A-6256	KMO. 913	Cylinder Bore Test Indicator
A-6257	J-1876-0	Connecting Rod and Piston Aligning Fixture
A-6258	J-1951	Piston Ring Installing and Removing Tool
A-6259	J-1952	Universal Clutch Shaft Pilot Arbor
A-6260	KMO-357	Universal Piston Ring Compressor
A-6262	J-1955	Valve Lifter
A-6261	KMO-144	Fuel Pump Checking Gauge & Vacuum Meter
A-6263	J-1953	Valve Lock Installing Tool
A-6264	J-1313	Tension Indicator Wrench Double Acting Beam Type
A-6265	C 537	Voltmeter
A-6266	J-1954	Stud Remover and Installing Tool

INDEX

Accelerating Pump, Carburetor.............. 46
Adjustment, Brake.......................... 101
Adjustment, Carburetor Idling.............. 45
Adjustment, Clutch Pedal................... 37
Adjustment, Current Regulator.............. 67
Adjustment, Cutout Relay................... 67
Adjustment, Distributor Breaker Points..... 64
Adjustment, Hand Brake..................... 106
Adjustment, Front Axle Differential........ 91
Adjustment, Front Wheel Bearing............ 108
Adjustment, Generator Regulator............ 66
Adjustment, Rear Axle Differential......... 96
Adjustment, Rear Wheel Bearings............ 108
Adjustment, Steering Gear.................. 111
Adjustment, Valve.......................... 21
Adjustment, Voltage Regulator.............. 67
Air Cleaner................................ 13-50
Alignment, Frame........................... 119
Alignment, Front Wheels.................... 114
Ammeter.................................... 6
Anti-Freeze Solutions...................... 54
Armature Tests, Generator.................. 64
Axle, Caster Front......................... 117
Axle, Front................................ 86
Axle Shaft, Front.......................... 86
Axle Shaft Oil Seal........................ 91
Axle Shaft, Rear........................... 93
Axle Shaft, Universal Joint................ 86

B

Battery.................................... 57
Bearing, Camshaft.......................... 23
Bearing Seal, Crankshaft Rear.............. 26
Bendix, Starter............................ 69
Body....................................... 125
Brake Adjustment........................... 101
Brake Drums................................ 109
Brake, Hand................................ 106
Brake, Foot................................ 101
Brake Fluid................................ 105
Brake Hose................................. 106
Brake, Master Cylinder..................... 104
Brake Pedal, Adjustment.................... 101
Brake Relining............................. 103
Brakes, Bleeding........................... 104
Brake Shoe Replacement..................... 103
Brake Specifications....................... 107
Brake Troubles and Remedies................ 107
Brake Wheel Cylinder....................... 105

C

Camshaft................................... 23
Capacity, Air Cleaner...................... 3
Capacity, Brake System..................... 3
Capacity, Cooling System................... 3
Capacity, Crankcase........................ 3
Capacity, Front Axle....................... 3
Capacity, Fuel Tank........................ 3
Capacity, Rear Axle........................ 3
Capacity, Transfer Case.................... 3
Capacity, Transmission..................... 3
Carburetor................................. 42
Carburetor Accelerating Pump............... 46

Carburetor Choke........................... 6-44
Carburetor Float Level..................... 45
Carburetor Idling Adjustment............... 45
Carburetor Metering Rod.................... 46
Chassis Lubrication........................ 10-11
Clutch..................................... 37
Clutch, Finger Adjustment.................. 39
Clutch Pedal Adjustment.................... 37
Clutch Specifications...................... 41
Clutch Troubles and Remedies............... 41
Connecting Rod............................. 26
Connecting Rod and Piston Alignment........ 27
Connecting Rod, Steering................... 114
Cooling System............................. 53
Cooling System Specifications.............. 56
Cooling System Troubles and Remedies....... 55
Crankcase Oil, When to Change.............. 9
Crankshaft................................. 24
Crankshaft Rear Bearing Seal............... 26
Cutout Relay............................... 67
Cylinder Boring............................ 29
Cylinder Checking.......................... 28
Cylinder Head.............................. 21
Cylinder Hone.............................. 29

D

Differential............................... 93
Distributor, Ignition...................... 63
Draining Cooling System.................... 53
Driver Instructions........................ 5

E

Electrical System.......................... 57
Electrical System Specifications........... 71
Electrical System Troubles and Remedies.... 70
Engine..................................... 17
Engine Compression......................... 3
Engine Lubrication......................... 9
Engine Mountings........................... 32
Engine Special Tools....................... 133
Engine, Starting........................... 7
Engine Troubles and Causes................. 34
Engine Tune-Up............................. 18
Exhaust System............................. 52

F

Fan Belt................................... 55
Fender Lamp................................ 62
Field Test, Generator...................... 65
Filter, Fuel............................... 49
Filter, Oil................................ 32
Float Level, Carburetor.................... 45
Floating Oil Intake........................ 31
Frame...................................... 119
Frame Alignment............................ 119
Freight Car Unloading...................... 4
Front Axle Specifications.................. 92
Front Axle Troubles and Remedies........... 91
Front Springs.............................. 121
Front Wheel Alignment...................... 114
Fuel Gauge................................. 57
Fuel Pump.................................. 48
Fuel System................................ 42
Fuel System Specifications................. 51
Fuel System Troubles and Remedies.......... 50

G

Gasoline Gauge	57
Gauge, Oil Pressure	32
Gauge, Oil Level	7
Gauge, Temperature	54
Gearshift Lever, Transmission	7
Generator	64
Generator Regulator Adjustments	66
Glass, Windshield	125
Grinding Valves	18
Guides, Valves	21

H

Headlamps	61
Headlamp Aiming	61
Heat Indicator	54
Heat Valve, Manifold	52
Horn	69

I

Ignition Distributor	63
Ignition System Wiring	64
Ignition Timing	64
Inflation Pressures, Tires	110
Inspections	14
Instructions, Driver	5
Instructions, Shifting Transfer Case	6-8
Instructions, Shifting Transmission	7-8
Instructions, Starting Engine	7
Instructions, Unloading Freight Car	4

L

Lever, Hand Brake	106
Lever, Transfer Case Shifting	6
Light Switch	5-61
Lubricant Specifications	12
Lubrication, Chassis	10
Lubrication, Engine	9
Lubrication, Front Axle	13
Lubrication, Rear Axle	13
Lubrication, Steering Gear	13
Lubrication, Transfer Case	13
Lubrication, Transmission	13

M

Main Bearings	24
Main Bearing Replacement	24
Master Cylinder, Brake	104
Manifold Heat Control	52
Manifold, Intake and Exhaust	52
Metering Rod, Carburetor	46
Mountings, Engine	32
Muffler	52

O

Oil Filter	32
Oil Gauge, Pressure	32
Oil, Floating Intake	31
Oil Level Gauge	7
Oil Pump	29
Oil Seal, Axle Shaft	91
Oil Seal, Crankshaft	26
Operating Instructions	5

P

Pintle Hook	119
Piston, Engine	27
Piston Pins	27
Piston Rings	27
Propeller Shaft, Front	83
Propeller Shaft, Rear	83
Pre-operation Instructions	4

R

Radiator	53
Rear Axle	93
Rear Axle Shaft	93
Rear Axle Specifications	100
Rear Axle Troubles and Remedies	100
Rear Springs	121
Relining Brakes	103
Repair Operations, Brakes	101
Repair Operations, Carburetor	45
Repair Operations, Clutch	39
Repair Operations, Engine	18
Repair Operations, Front Axle	86
Repair Operations, Fuel Pump	48
Repair Operations, Generator	64
Repair Operations, Ignition Distributor	63
Repair Operations, Propeller Shafts	83
Repair Operations, Rear Axle	93
Repair Operations, Steering Gear	111
Repair Operations, Starting Motor	67
Repair Operations, Transfer Case	79
Repair Operations, Transmission	73
Repair Operations, Universal Joints	83

S

Serial Number Plate	3
Shackles, Rear Spring	121
Shifting Transfer Case Gears	6-8
Shifting Transmission Gears	7-8
Shock Absorbers	123
Spark Plugs	18-21
Special Tools, Engine	133
Special Tools, Front Axle	126
Special Tools, Rear Axle	126
Special Tools, Transfer Case	126
Special Tools, Transmission	126
Specifications, Brakes	107
Specifications, Clutch	41
Specifications, Cooling System	56
Specifications, Electrical System	71
Specifications, Engine	35
Specifications, Frame	120
Specifications, Front Axle	92
Specifications, Fuel System	51
Specifications, Lubricant	12
Specifications, Propeller Shaft and Universal Joints	85
Specifications, Rear Axle	100
Specifications, Shock Absorbers	124
Specifications, Steering Gear	118
Specifications, Transfer Case	82
Specifications, Transmission	78
Specifications, Wheel and Tire	110
Speedometer	6
Springs, Front	121
Spring Shackles, Front	121
Springs, Rear	121
Springs, Valve	21
Starting Engine	7
Starting Motor	67
Starting Vehicle	7
Starting Switch	5-69

Steering Connecting Rod	114
Steering Gear	111
Steering Gear, Troubles and Remedies	118
Steering Geometry	114
Steering Knuckle	90
Switch, Ignition	5
Switch, Light	5-59
Switch, Starting Motor	5-69

T

Tail and Stop Light	62
Temperature Indicator	6
Thermostat	53
Throttle, Hand	5
Tie Rod	114
Timing Chain	23
Timing Ignition	64
Timing, Valve	24
Tires	110
Torque Reaction Spring	121
Towing Vehicle	8
Transfer Case	79
Transfer Case Shifting Lever	6-8
Transfer Case Special Tools	126
Transfer Case Specifications	82
Transfer Case Troubles and Remedies	82
Transmission	73
Transmission Gearshift Lever	7
Transmission Specifications	78
Transmission Troubles and Remedies	75
Troubles and Remedies, Brake	107
Troubles and Remedies, Clutch	41
Troubles and Remedies, Cooling System	55
Troubles and Remedies, Electrical	70
Troubles and Remedies, Engine	34
Troubles and Remedies, Front Axle	91
Troubles and Remedies, Fuel System	60
Troubles and Remedies, Rear Axle	100
Troubles and Remedies, Steering Gear	118
Troubles and Remedies, Springs	124
Troubles and Remedies, Transfer Case	82
Troubles and Remedies, Transmission	75

U

Universal Joint, Axle Shaft, Front	86
Universal Joints, Propeller Shaft	83
Unloading Instructions	4

V

Valves	18
Valves, Grinding	18
Valves, Refacing	20
Valve Springs	21
Valve Tappet	23
Valve Timing	24
Voltage Regulator	67

W

Water Pump	54
Wheel	108
Wheel Bearings	108
Wheel, Camber Front	115
Wheel Cylinder, Brake	105
Wheel, Toe-in Front	114
Windshield Glass	125
Windshield Weatherstrip	125
Wiring Diagram, Complete	58
Wiring Diagram, Gas Gauge	59
Wiring Diagram, Generator	65
Wiring Diagram, Horn	69
Wiring Diagram, Ignition	63
Wiring Diagram, Lights	60
Wiring Diagram, Starter	69

LUBRICATION SPECIFICATIONS
CANADIAN & BRITISH

See Lubrication Chart Page 11

ITEM TO BE LUBRICATED See Page 10 & 11	HOW APPLIED	Capac. Imper.	B.W.D. Specified	B.W.D. Emg'y	D.N.D. Summer	D.N.D. Winter	Nearest Commercial Equivalent Summer	Nearest Commercial Equivalent Winter	Miles
Engine Crankcase (39)	Filler pipe R. side check level daily	3½ qts.	M-120	M-160	51	45	Auto Eng. Oil— S.A.E. 20W	Auto Eng. Oil— S.A.E. 10W	2000
*Transmission Case (28)	Filler plug R. side—add Oil to level of plug	1¾ pts.	C-600	M-600	390	360	Truck Hypoid— S.A.E. 90	Truck Hypoid— S.A.E. 80	5000
Transfer Case (29)	Filler plug—add oil to level of plug	2½ pts.	M-600	C-600	140	100	Aviation Eng. Oil— No. 140	Aviation Eng. Oil— No. 100	5000
Differential F. & R. (30)	Filler plug in cover—add Hypoid oil to level of plug	2 pts.	Hypoid 90		390	360	Truck Hypoid— S.A.E. 90	Truck Hypoid— S.A.E. 80	5000
Propeller Shaft Universal Joints F & R (6)	Fitting	Fill	C-600	M-600	390	360	Truck Hypoid— S.A.E. 90	Truck Hypoid— S.A.E. 80	1000
Air Cleaner	Remove Cover	1 pt.	M-120	M-160	51	45	Auto Eng. Oil— S.A.E. 20W	Auto Eng. Oil— S.A.E. 10W	1000
Front Axle Shaft Universal Joint & Steering Knuckle Bearings (27)	Filler plug outer casing	½ lb.	Grease G.S.		632	632	Wheel Bearing Lub.	Wheel Bearing Lub.	1000
F & R Wheel Bearings (19)	Remove and Repack	Fill	Grease G.S.		632	632	Wheel Bearing Lub.	Wheel Bearing Lub.	5000
Steering Gear Housing (22)	Remove Plug	6½ oz.	C-600	M-600	390	360	Truck Hypoid— S.A.E. 90	Truck Hypoid— S.A.E. 80	1000
Steering Bell Crank (10) Steering Tie Rods (3) Steering Connecting Rod (4) Spring Shackles F & R (1) Spring Pivot Bolts F & R (2) Clutch & Brake Pedal Shaft (10) Propeller Shaft Slip Joints (7)	Fitting Fitting each end Fitting each end Fittings 8 Fittings 5 Fittings 2 Fitting 1 each		Grease G.S.		632	632	Chassis Grease	Chassis Grease	1000
Starter Front (36)	Oil Hole	5 drops	M-120	M-160	51	45	Auto Eng. Oil— S.A.E. 20W	Auto Eng. Oil— S.A.E. 10W	1000
Distributor (34)	Oil Cup on side	5 drops	M-120	M-160	51	45	Auto Eng. Oil— S.A.E. 20W	Auto Eng. Oil— S.A.E. 10W	1000
Distributor Shaft Wick (34)	Oil Can	1 drop	M-120	M-160	51	45	Auto Eng. Oil— S.A.E. 20W	Auto Eng. Oil— S.A.E. 10W	2000
Distributor Arm Pivot (34)	Oil Can	1 drop	M-120	M-160	51	45	Auto Eng. Oil— S.A.E. 20W	Auto Eng. Oil— S.A.E. 10W	2000
Distributor Cam (34)	Wipe with grease		Grease G.S.		665	665	Mineral Jelly	Mineral Jelly	2000
Clevis Pins and Yokes (21)	Oil Can	5 drops	M-160	M-120	390	360	Truck Hypoid— S.A.E. 90	Truck Hypoid— S.A.E. 80	1000
Pintle Hook (40)	Oil Can	5 drops	M-160	M-120	390	360	Truck Hypoid— S.A.E. 90	Truck Hypoid— S.A.E. 90	1000

Hydraulic Brake Fluid No. 2   510   510   Lockheed 21, Delco Super 9 or 150

*Remove Skid Plate to drain Transmission
Shock Absorbers Non-Refillable
☐ Emergency

WAR DEPARTMENT TECHNICAL MANUAL

TM 9-1803A

ORDNANCE MAINTENANCE

Engine and Engine Accessories For ¼-Ton 4x4 Truck

(Willys-Overland Model MB and Ford Model GPW)

WAR DEPARTMENT   24 FEBRUARY 1944

CONTENTS

	Paragraphs	Pages
CHAPTER 1. INTRODUCTION	1-2	4-7
CHAPTER 2. ENGINE	3-24	8-64
SECTION I. Description and data	3-4	8
II. Engine removal from vehicle	5	8-14
III. Disassembly of engine into sub-assemblies	6-7	14-21
IV. Disassembly, cleaning, inspection, repair, and assembly of sub-assemblies	8-17	21-43
V. Assembly of engine	18-19	43-57
VI. Installation of engine	20-21	57-61
VII. Fits and tolerances	22-24	62-64
CHAPTER 3. CLUTCH ASSEMBLY	25-28	65-71
REFERENCES		72-74
INDEX		75–77

TM 9-1803A
1

ORDNANCE MAINTENANCE — ENGINE AND ENGINE ACCESSORIES FOR ¼-TON 4x4 TRUCK (WILLYS-OVERLAND MODEL MB AND FORD MODEL GPW)

CHAPTER 1

INTRODUCTION

	Paragraph
Scope	1
MWO and major unit assembly replacement record	2

1. SCOPE.

a. The instructions contained in this manual are for the information and guidance of personnel charged with the maintenance and repair of the 4-cylinder engine used in the Willys MB and Ford GPW ¼-ton 4 x 4 Trucks. These instructions are supplementary to field and technical manuals prepared for the using arms. This manual does not contain information which is intended primarily for the using arms, since such information is available to ordnance maintenance personnel in 100-series TM's or FM's.

b. This manual contains a description of, and procedure for inspection, removal, disassembly, repair, and rebuilding of the engine.

c. TM 9-803 contains information and guidance for the using arms and first and second echelons.

d. TM 9-1803B contains information for removal, inspection, repair, rebuild, assembly, and installation of the power train and chassis.

e. TM 9-1825B contains information for the maintenance of the Auto-Lite electrical equipment used on this vehicle.

f. TM 9-1826A contains information for the maintenance of the Carter carburetor used on this vehicle.

g. TM 9-1827C contains information for the maintenance of the Wagner hydraulic brake system used on this vehicle.

h. TM 9-1828A contains information for the maintenance of the A. C. fuel pump used on this vehicle.

i. TM 9-1829A contains information for the maintenance of the speedometer used on this vehicle.

j. This manual includes engine ordnance maintenance instructions from the following Quartermaster Corps 10-series technical manuals. Together with TM 9-803 and TM 9-1803B, this manual supersedes them:

(1) TM 10-1103, 20 August 1941.
(2) TM 10-1207, 20 August 1941.
(3) TM 10-1349, 3 January 1942.
(4) TM 10-1513, Change 1, 15 January 1943.

INTRODUCTION

Figure 1 — Front View of Engine

2. MWO AND MAJOR UNIT ASSEMBLY REPLACEMENT RECORD.

a. **Description.** Every vehicle is supplied with a copy of AGO Form No. 478 which provides a means of keeping a record of each MWO (FSMWO) completed or major unit assembly replaced. This form includes spaces for the vehicle name and U.S.A. registration number, instructions for use, and information pertinent to the work accomplished. It is very important that the form be used as directed, and that it remain with the vehicle until the vehicle is removed from service.

TM 9-1803A
2

ORDNANCE MAINTENANCE — ENGINE AND ENGINE ACCESSORIES FOR ¼-TON 4x4 TRUCK (WILLYS-OVERLAND MODEL MB AND FORD MODEL GPW)

RA PD 28665

Figure 2 — Left Side View of Engine

RA PD 28664

Figure 3 — Right Side View of Engine

INTRODUCTION

description of the work completed, and must initial the form in the columns provided. When each modification is completed, record the date, hours and/or mileage, and MWO number. When major unit assemblies, such as engines, transmissions, transfer cases, are replaced, record the date, hours and/or mileage, and nomenclature of the unit assembly. Minor repairs and minor parts and accessory replacements need not be recorded.

c. **Early Modifications.** Upon receipt by a third or fourth echelon repair facility of a vehicle for modification or repair, maintenance personnel will record the MWO numbers of modifications applied prior to the date of AGO Form No. 478.

TM 9-1803A
3-5

ORDNANCE MAINTENANCE — ENGINE AND ENGINE ACCESSORIES FOR ¼-TON 4x4 TRUCK (WILLYS-OVERLAND MODEL MB AND FORD MODEL GPW)

CHAPTER 2
ENGINE

Section I
DESCRIPTION AND DATA

	Paragraph
Description	3
Data	4

3. DESCRIPTION.

a. The engine used in the ¼-ton 4 x 4 Truck is the 4-cylinder, L-head, gasoline-type (figs. 1, 2, and 3), equipped with a counter-balanced crankshaft. The camshaft is operated off the crankshaft through a timing chain (fig. 40). The oil pump and distributor operate off the camshaft.

4. DATA.

Type	L-head
Numbers of cylinders	4
Bore and stroke	3.125 x 4.375 in.
Piston displacement	134.2 cu in.
Compression ratio	6.48 to 1
Max. brake horsepower	54 at 4,000
Compression (lb per sq in. at 185 rpm)	111
SAE horsepower	15.63
Maximum torque	105 ft-lb at 2,000 rpm
Firing order	1-3-4-2

Section II
ENGINE REMOVAL FROM VEHICLE

	Paragraph
Removal from vehicle	5

5. REMOVAL FROM VEHICLE.

a. **General.** Unhook the two hood clamps, raise the hood, and lay it against the windshield. Drain the coolant from the radiator and the engine by opening the radiator drain cock and the drain cock

TM 9-1803A
5

ENGINE

Figure 4 — Underside View of Engine Installed in Vehicle

RA PD 28748

Labels (top): RADIATOR, BOND STRAP, RADIATOR BOLT, RADIATOR DRAIN COCK, OIL PAN, OIL PAN DRAIN PLUG, EXHAUST PIPE, STAY CABLE

Labels (bottom): BOND STRAP, RADIATOR BOLT

9

TM 9-1803A
5

ORDNANCE MAINTENANCE — ENGINE AND ENGINE ACCESSORIES FOR ¼-TON 4x4 TRUCK (WILLYS-OVERLAND MODEL MB AND FORD MODEL GPW)

Figure 5 — Right Side View of Engine Installed in Vehicle

TM 9-1803A
5

ENGINE

Figure 6 — Side View of Engine Compartment

TM 9-1803A
5

ORDNANCE MAINTENANCE — ENGINE AND ENGINE ACCESSORIES FOR ¼-TON 4x4 TRUCK (WILLYS-OVERLAND MODEL MB AND FORD MODEL GPW)

located on the right-hand side of the engine. Remove the oil pan drain plug and drain the engine oil. Some variation exists in the location of the various bond straps used to eliminate radio interference on these vehicles. Disregard references to bond straps in the following instructions if they are not present on the particular vehicle being worked on. If bond straps are found in locations other than those mentioned in the following procedure, they should be disconnected, if they prevent removal of the engine.

b. Remove Battery. Loosen the two battery cable bolts, and disconnect both cables. Loosen the battery brace wing nut on the fender. Remove the two battery hold-down frame wing nuts (fig. 5). Move the battery brace to one side, and remove the battery hold-down frame. Lift the battery from the vehicle.

c. Remove Radiator. Remove the nut and lock washer from the front and rear of the radiator brace, and remove the brace. Loosen the two front outlet radiator hose clamps, and slide the hose back on the metal tubing. Loosen the rear radiator outlet hose clamp, and remove the hose. Loosen the radiator hose clamps on the inlet hose at the water pump, also the one on the radiator and remove the radiator inlet hose. Working from underneath the vehicle, remove the two nuts, flat washers, and bond straps from the radiator bolts (fig. 4). Remove the two nuts and flat washers from the radiator bolts. Lift the radiator from the vehicle and remove the two radiator pads.

d. Disconnect Oil and Water Temperature Gages. Disconnect the oil gage line at the flexible oil line, located at the left-hand side of the engine. Disconnect the water temperature gage (engine unit) at the right-hand side of the cylinder head (fig. 5).

e. Remove Air Cleaner Hose (fig. 6). Loosen the hose clamps on the carburetor air cleaner and oil filler pipe, and remove the air cleaner hose.

f. Disconnect Electrical Wires and Bond Straps. Disconnect the field, armature, and ground wires at the generator. Disconnect the primary wire running from the dash to the coil, at the coil. Disconnect the bond strap at the rear of the cylinder head. Disconnect the ground strap at each front engine support. Disconnect the cranking motor cable at the cranking motor.

g. Remove Cranking Motor. Remove the cap screw that holds the cranking motor bracket to the cylinder block. Remove the two cap screws that hold the cranking motor to the clutch housing, and slip the cranking motor from the engine.

ENGINE

Figure 7 — Lifting Engine from Vehicle

h. Disconnect Choke and Throttle Controls. Remove the nut and bolt on the choke and throttle hold-down bracket. Loosen the set screw on the carburetor choke lever, and remove the choke control cable. Loosen the set screw on the throttle control cable and remove the throttle control cable. Disconnect the throttle control at the accelerator pedal in the driver's compartment.

i. Disconnect Exhaust Pipe. Remove the nut, bolt, and cap screw that hold the exhaust pipe to the exhaust manifold. Pry the exhaust pipe from the exhaust manifold.

j. Disconnect Front Engine Supports. Remove the two nuts and bolts from each front engine support.

TM 9-1803A
5—6

ORDNANCE MAINTENANCE — ENGINE AND ENGINE ACCESSORIES FOR ¼-TON 4x4 TRUCK (WILLYS-OVERLAND MODEL MB AND FORD MODEL GPW)

k. Remove Stay Cable and Clutch Housing Bolts. Remove the two engine stay cable nuts at the front crossmember, and remove the stay cable (fig. 4). Remove the 10 cap screws and bolts from the clutch housing.

l. Remove Engine From Vehicle. Install a suitable lifting sling or rope on the engine (fig. 7). Raise the engine high enough to release the weight on the front engine supports. Pull the engine forward until it is free from the clutch housing, and lift the engine from the vehicle (fig. 7).

Section III

DISASSEMBLY OF ENGINE INTO SUBASSEMBLIES

	Paragraph
Preliminary operations	6
Disassembly of stripped engine	7

6. PRELIMINARY OPERATIONS.

a. General. If the clutch housing was removed with the engine, start the procedure beginning with subparagraph b below. If the clutch housing was not removed with the engine, remove the cranking motor (par. 5 g), remove the rest of the clutch housing bolts or cap screws, and remove the clutch housing from the engine.

b. Remove Carburetor (fig. 6). Remove the fuel line connecting the carburetor and fuel pump. Remove the accelerator return spring from the careburetor and accelerator lever. Remove the two carburetor hold-down nuts, lock washers, and accelerator return spring clip.

c. Remove Fuel Pump (fig. 6). Disconnect the other fuel line at the fuel pump. Remove the two cap screws and lock washers that hold the fuel pump to the cylinder block, and remove the fuel pump.

d. Remove Distributor (fig. 5). Pull the spark wires off the spark plugs, and slide the wires out of the air filter tube bracket. Remove the primary and secondary wires from the ignition coil. Remove the distributor hold-down screw, and lift the distributor and wires from the cylinder block.

e. Remove Oil Filter (fig. 5). Disconnect the inlet oil line on the left-hand side of the cylinder block, and the outlet oil line on the engine front cover. Remove the cap screw that holds the oil filler pipe to the oil filter bracket. Remove the three cylinder head nuts

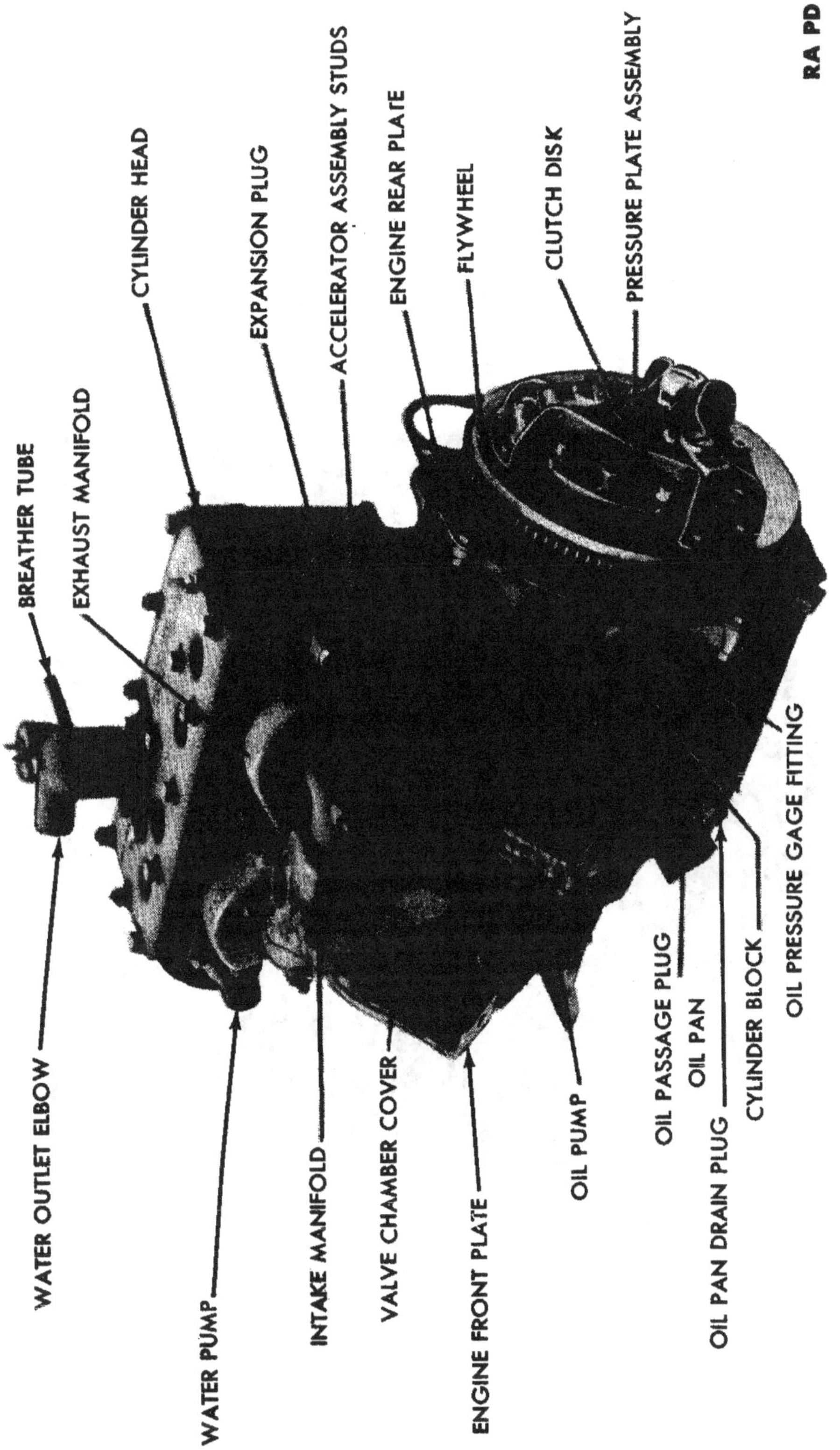

TM 9-1803A
6

ENGINE

Figure 8 — Three-quarter Left Rear View of Stripped Engine

TM 9-1803A
6

ORDNANCE MAINTENANCE — ENGINE AND ENGINE ACCESSORIES FOR ¼-TON 4x4 TRUCK (WILLYS-OVERLAND MODEL MB AND FORD MODEL GPW)

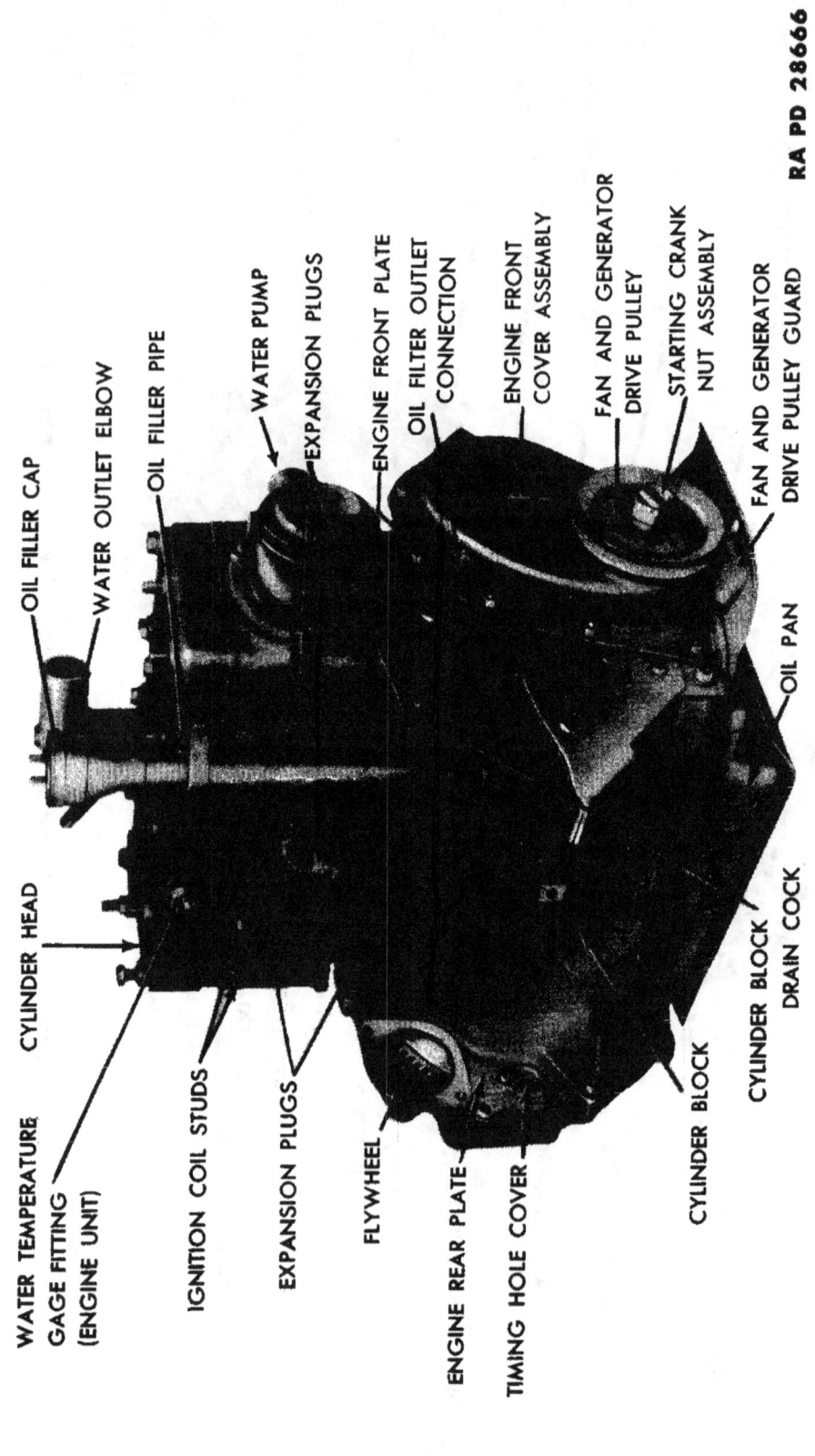

Figure 9 — Three-quarter Right Front View of Stripped Engine

TM 9-1803A
6–7

ENGINE

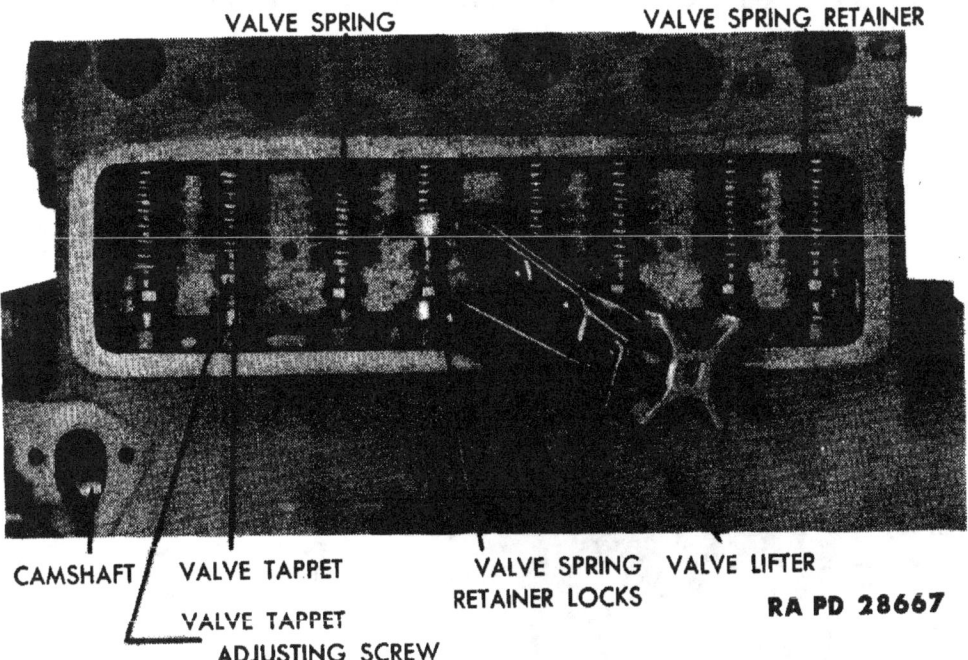

Figure 10 — Removing Valve Spring Retainer Locks, Using Valve Lifter (41-L-1410)

that hold the oil filter bracket to the cylinder head, and remove the oil filter.

f. Remove Generator and Generator Support Bracket. Pull up on the generator adjusting bracket, raise the generator to release the tension on the fan belt, and remove the belt. Remove the two bolts that hold the generator to the support bracket, and remove the generator. Remove the two cap screws that hold the generator support bracket to the cylinder block, and remove the generator support bracket.

g. Remove Ignition Coil. Remove the two nuts and lock washers that hold the ignition coil to the cylinder block, and remove the ignition coil and bond strap.

h. Remove Fan. Remove the four cap screws and lock washers that hold the fan to the water pump, and remove the fan.

7. DISASSEMBLY OF STRIPPED ENGINE.

a. Remove Water Pump (fig. 9). Remove the four cap screws and lock washers that hold the water pump to the cylinder block, and remove the water pump.

b. Remove Intake and Exhaust Manifold (fig. 8). Remove the ventilating tube that connects the intake manifold and valve chamber cover. Remove the seven nuts, and lift the intake and exhaust manifold off the engine.

TM 9-1803A
7

ORDNANCE MAINTENANCE — ENGINE AND ENGINE ACCESSORIES FOR ¼-TON 4x4 TRUCK (WILLYS-OVERLAND MODEL MB AND FORD MODEL GPW)

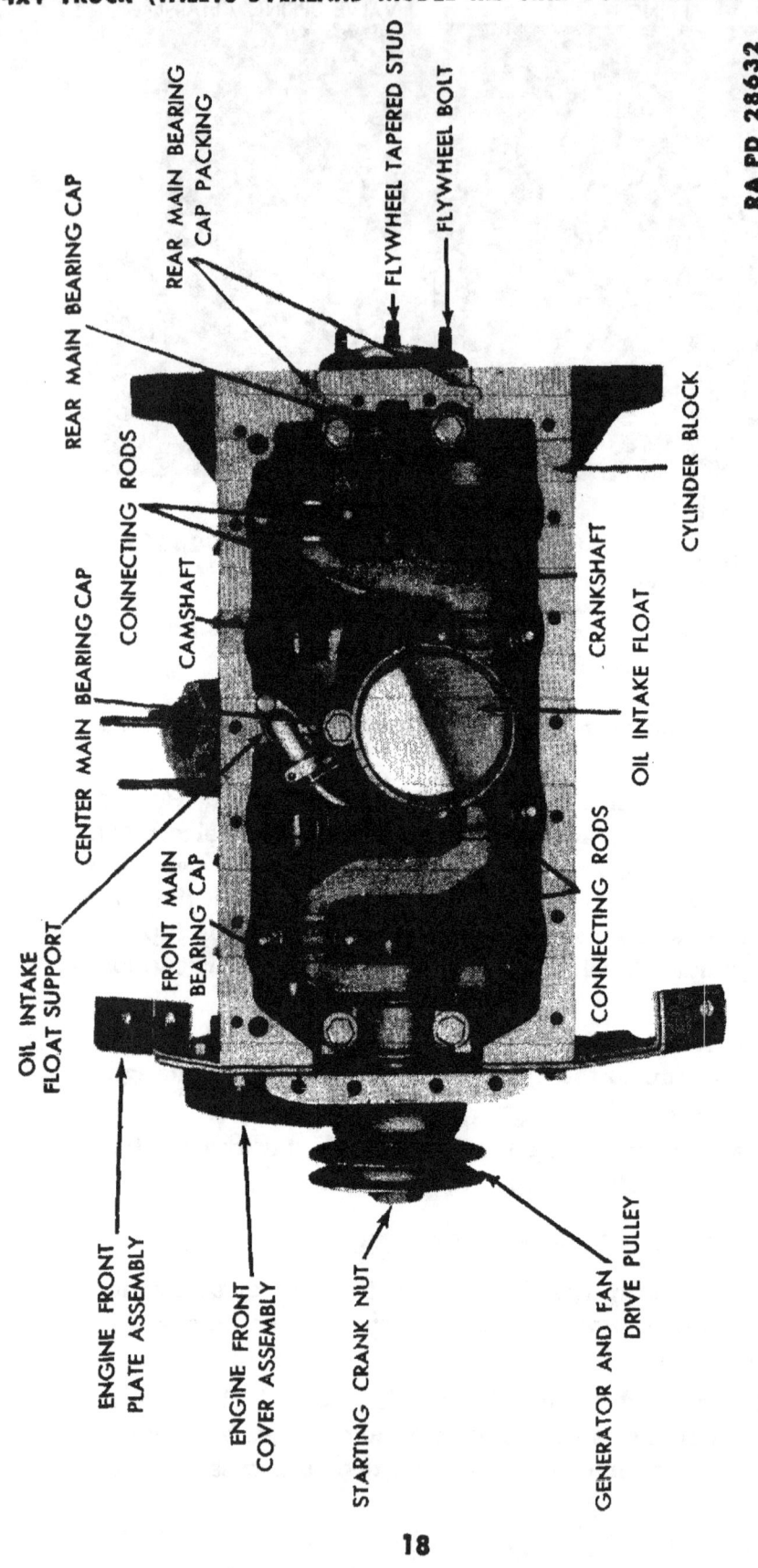

Figure 11 — Underside View of Engine with Oil Pan Removed

ENGINE

c. Remove Water Outlet Elbow (fig. 8). Remove the three nuts that hold the water outlet elbow to the cylinder head, and remove the water outlet elbow and thermostat. Remove the thermostat retainer and thermostat.

d. Remove Clutch Disk (fig. 8). Loosen the six pressure plate bracket cap screws in sequence, a little at a time, to prevent distortion of the pressure plate bracket. Remove the six cap screws, pressure plate, and clutch disk.

e. Remove Flywheel (fig. 8). Remove the six nuts and lock washers that hold the flywheel to the crankshaft. Tap the flywheel off the crankshaft with a brass hammer. Lift the rear engine plate from the engine.

f. Remove Cylinder Head (fig. 9). Remove the remaining cap screws that secure the head to the cylinder block, and remove the cylinder head.

g. Remove Valves and Springs (fig. 10). Remove the two cap screws and crankcase ventilator assembly from the valve chamber cover, and remove the cover. With a valve lifter (41-L-1410) inserted between the valve tappet and valve spring retainer, raise the valve springs that are in closed position, and remove the valve spring retainer locks (fig. 10). Turn the crankshaft until those valves which are open become closed, and remove the rest of the valve spring retainer locks. Remove the valves and place them in a valve carrying board, so that they can be identified as to cylinders from which they were removed. Compress the valve spring with the valve lifter on each valve tappet that is in the closed position, and pull the spring off the valve guide. Turn the crankshaft until the tappets are in a closed position, and remove the rest of the valve springs.

h. Remove Oil Pan and Oil Intake Float. Turn the engine on its side, and remove the cap screws that secure the oil pan and fan pulley guard to the cylinder block. Remove the fan pulley guard and oil pan. Remove the two cap screws from the oil intake float (fig. 11), and remove the oil intake float.

i. Remove Camshaft Sprocket and Camshaft. Remove the eight nuts and bolts that secure the engine front cover and engine front plate to the cylinder block and remove the cover. Remove the camshaft thrust plunger and spring. Straighten the tabs on the four camshaft sprocket cap screw lock washers (fig. 25), and remove the four cap screws and lock washers. Lift the camshaft sprocket and the camshaft drive link chain off the camshaft. Remove the camshaft thrust washer. Lay the cylinder block on its side. Pull all the valve tappets toward the top of the cylinder block. Pull the cam-

TM 9-1803A
7

ORDNANCE MAINTENANCE — ENGINE AND ENGINE ACCESSORIES FOR ¼-TON 4x4 TRUCK (WILLYS-OVERLAND MODEL MB AND FORD MODEL GPW)

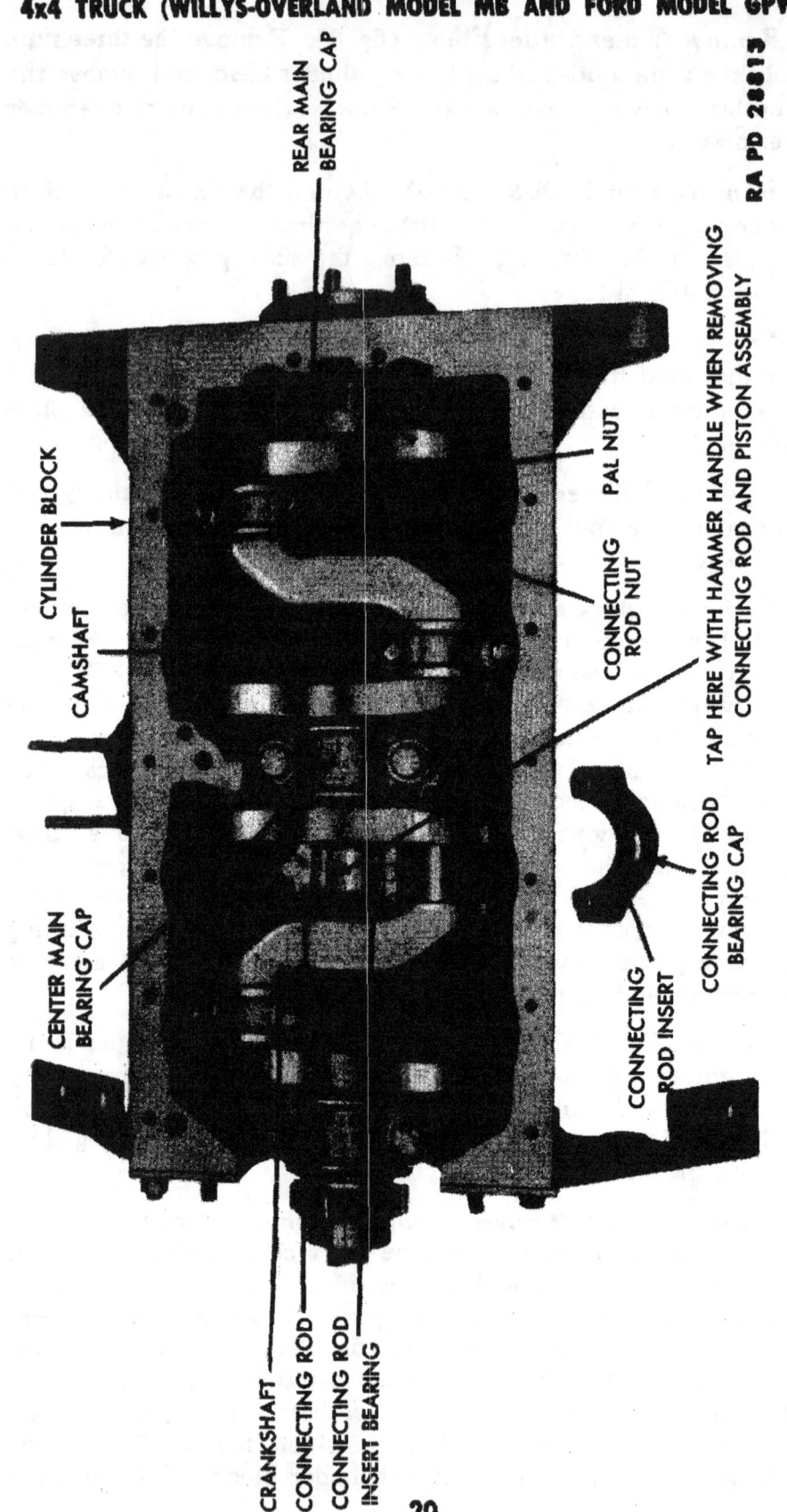

Figure 12 — Connecting Rod and Piston Assembly Removal

ENGINE

shaft out of the cylinder block, and remove the valve tappets. Remove the three cap screws that hold the engine front plate to the cylinder block, and remove the plate.

j. **Remove Piston and Connecting Rod Assemblies** (fig. 12). Remove the two pal nuts, connecting rod nuts, and connecting rod bearing cap from each connecting rod. Remove all carbon from the top of the cylinder walls. Tap the connecting rod and piston assembly out of the cylinder block with the handle end of a hammer (fig. 12). Install the connecting rod bearing caps on the rods in same position as originally installed, to prevent later improper mating of parts.

k. **Remove Crankshaft.** Remove the two cap screws from each main bearing cap (fig. 11), and remove the three main bearing caps. Lift the crankshaft from the cylinder block.

Section IV

DISASSEMBLY, CLEANING, INSPECTION, REPAIR, AND ASSEMBLY OF SUBASSEMBLIES

	Paragraph
Cylinder block, head, and oil pan	8
Water pump	9
Connecting rod and piston assembly	10
Camshaft assembly	11
Valve and valve springs	12
Valve tappets	13
Oil pump and oil intake float	14
Crankshaft assembly	15
Flywheel assembly	16
Intake and exhaust manifolds	17

8. CYLINDER BLOCK, HEAD, AND OIL PAN.

a. **Cleaning.** Strip off all old gaskets and sealing compound from all machined surfaces. Remove plugs, and clean all oil passages in the cylinder block with steam or compressed air. Scrape the carbon from the cylinder block and head. Clean the cylinder block, head, and oil pan thoroughly with dry-cleaning solvent.

b. **Inspection and Repair.**

(1) OIL PAN (fig. 13). An oil pan with stripped threads in the drain plug opening, or an oil pan that is badly dented or deformed, must be replaced.

TM 9-1803A
8

ORDNANCE MAINTENANCE — ENGINE AND ENGINE ACCESSORIES FOR ¼-TON 4x4 TRUCK (WILLYS-OVERLAND MODEL MB AND FORD MODEL GPW)

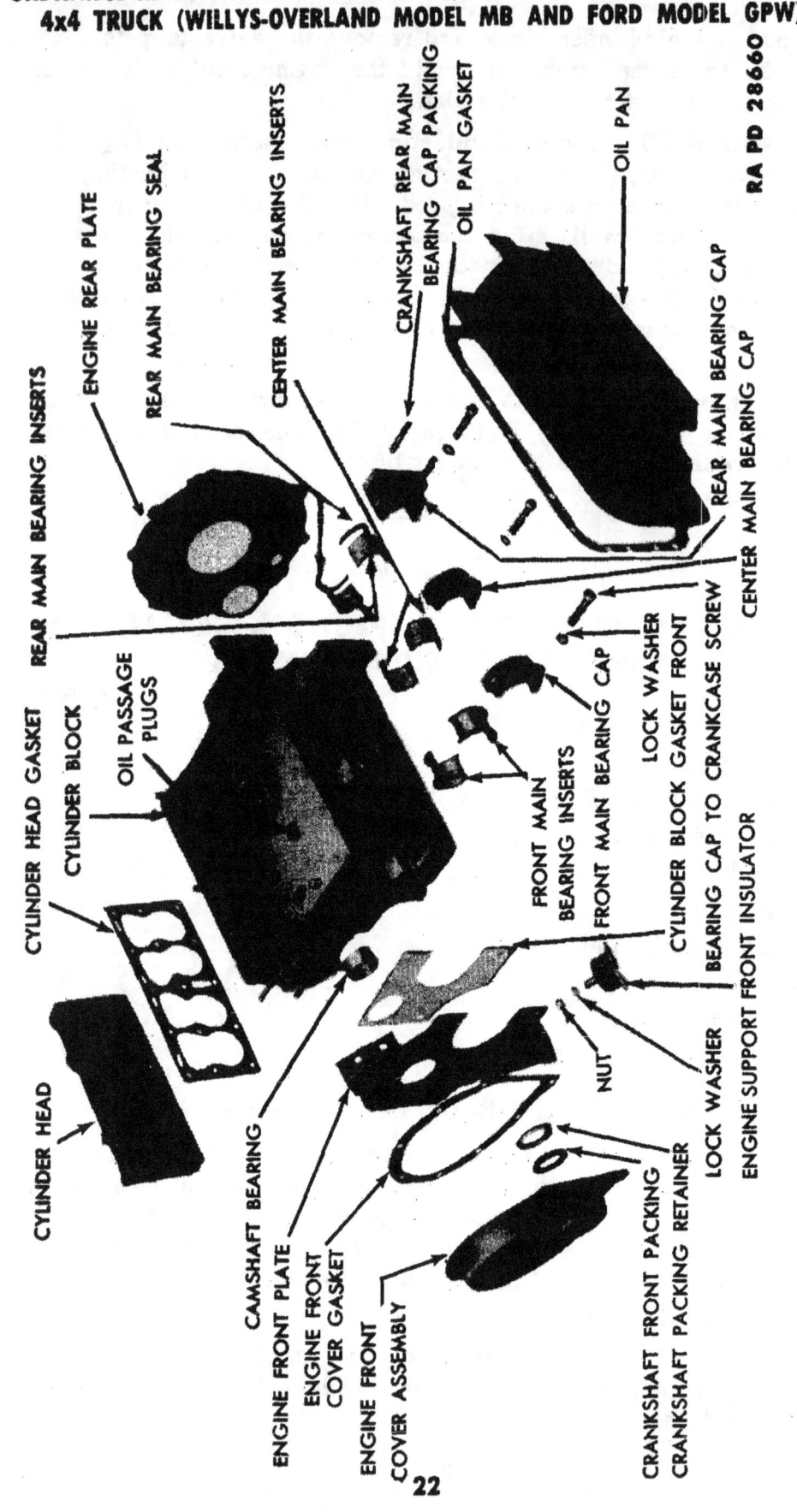

Figure 13 — Cylinder Block, Head, Oil Pan, and Bearings, Disassembled

ENGINE

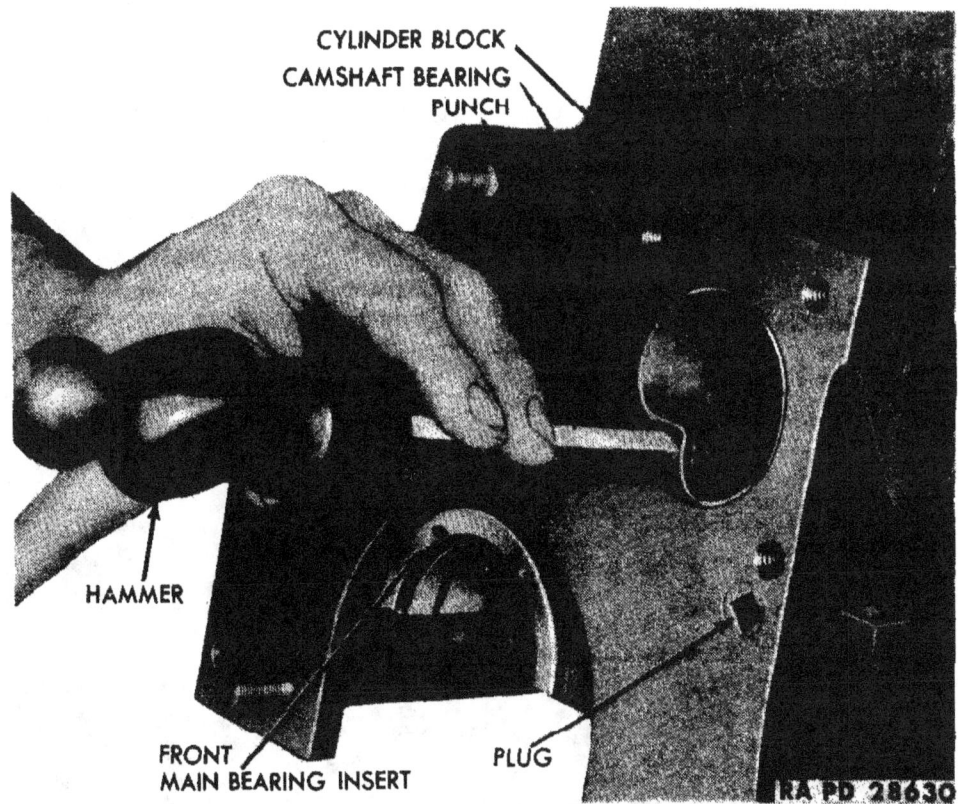

Figure 14 — Driving Camshaft Bearing from Cylinder Block

(2) CYLINDER HEAD (fig. 13). A cracked or warped cylinder head, or a cylinder head with stripped threads in the spark plug holes, must be replaced.

(3) CYLINDER BLOCK (fig. 13). A cracked or damaged cylinder block must be replaced. All loose expansion plugs (fig. 9) or damaged studs must be replaced (step (4) below). A scored, ridged, discolored, or excessively worn, front camshaft bearing (fig. 13) (worn to more than 2.190 in. inside diameter) must be replaced (step (5) below). Measure the other three camshaft bearings with a micrometer caliper. If the bearings are larger than 2.128 inches for the front intermediate, 2.1395 inches for the rear intermediate, or 1.628 inches for the rear bearing, the cylinder block must be replaced. Measure the cylinder bores with a micrometer caliper and telescope gage. If any of the cylinders has a taper of more than 0.010 inch, or an out-of-round condition of more than 0.005 inch, the cylinders must be rebored to 0.020 or 0.030 inch oversize. If cylinder walls will not clean up at 0.030 inch, the cylinder block must be replaced. Pitted, burned, or nicked valve seats must be reseated. Check the clearances of the valve guides with new valves. If the clearance ex-

TM 9-1803A
8

ORDNANCE MAINTENANCE — ENGINE AND ENGINE ACCESSORIES FOR ¼-TON 4x4 TRUCK (WILLYS-OVERLAND MODEL MB AND FORD MODEL GPW)

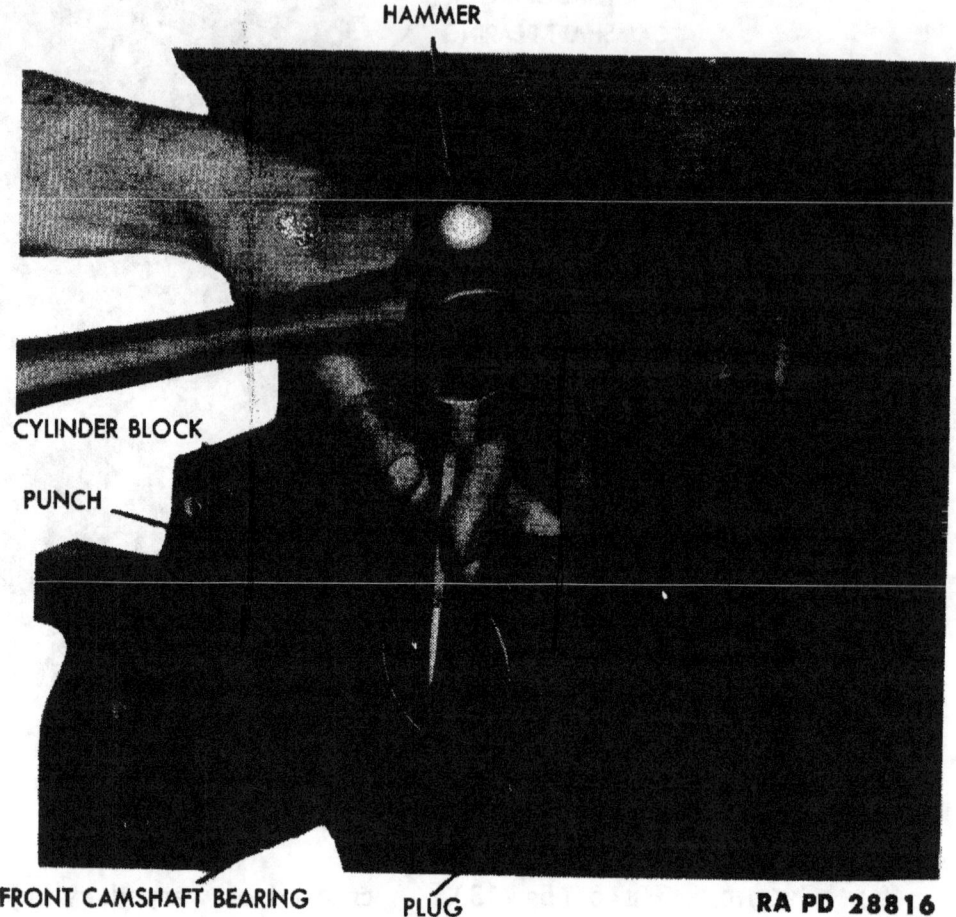

Figure 15 — Staking Camshaft Bearing in Place

ceeds 0.0045 inch in an intake valve guide (using a new intake valve as a gage), or 0.005 inch in an exhaust valve guide (using an exhaust valve as a gage), the valve guides must be replaced (step (6) below). If the clearance exceeds 0.003 inch between valve tappet and valve tappet bore, the valve tappet bores must be reamed to 0.004 inch oversize, and 0.004-inch oversize valve tappets must be installed when assembling engine. If valve tappet bore will not clean up at 0.004 inch oversize, the cylinder block must be replaced.

(4) REPLACE STUDS. Remove all damaged studs with a standard stud puller. To remove a broken stud, indent the end of the broken stud exactly in the center with a center punch. Drill approximately two-thirds through the broken stud with a small drill, then follow up with a larger drill. However, the drill selected must leave a wall thicker than the depth of the threads. Select an extractor (EZ-Out) of the proper size, insert it into the drilled hole, and screw out the

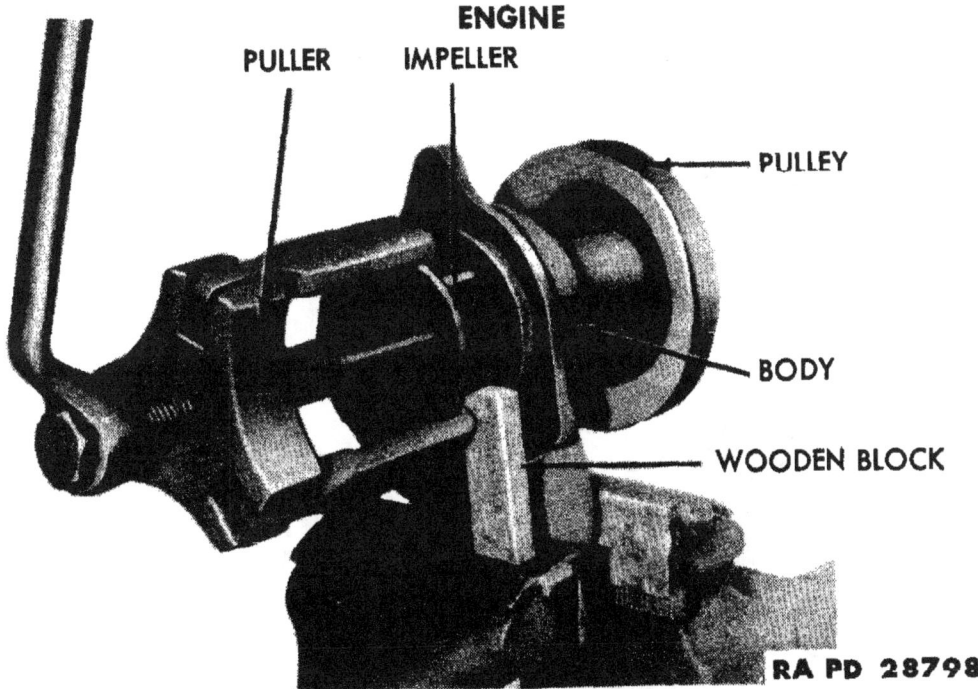

Figure 16 — Removing Water Pump Impeller, Using Puller (41-P-2912)

remaining part of the broken stud. Install the studs with a standard stud driver. Drive all studs until no threads show at the bottom of the studs.

(5) REPLACE CAMSHAFT BEARING. Drive a punch between the camshaft bearing and cylinder block (fig. 14), and tap the camshaft bearing from the cylinder block. To install the camshaft bearing, drive it in place with a fiber block, making sure the oil hole in the bearing is in line with the oil passage in the cylinder block. Stake the camshaft bearing in place with a punch (fig. 15). Line-ream the camshaft bearing to 2.3145 inches.

(6) REPLACE VALVE GUIDES. Remove the guides with a suitable valve guide remover. When installing valve guides, drive all intake and exhaust valve guides into the block with a valve guide replacer, leaving a distance of 1 inch from the top of the guide to the top of the cylinder block for exhaust valve guides, and a distance of $1\frac{5}{16}$ inches for the intake guides.

9. WATER PUMP.

a. **Disassembly.** Pull the water pump bearing retaining wire (fig. 17) from the water pump. Remove the water pump impeller with a puller (41-P-2912) as in figure 16, or press it off in an arbor press. Remove the water pump seal assembly, and water pump seal washer. Press the water pump bearing and shaft assembly, and water pump

TM 9-1803A
9

ORDNANCE MAINTENANCE — ENGINE AND ENGINE ACCESSORIES FOR ¼-TON 4x4 TRUCK (WILLYS-OVERLAND MODEL MB AND FORD MODEL GPW)

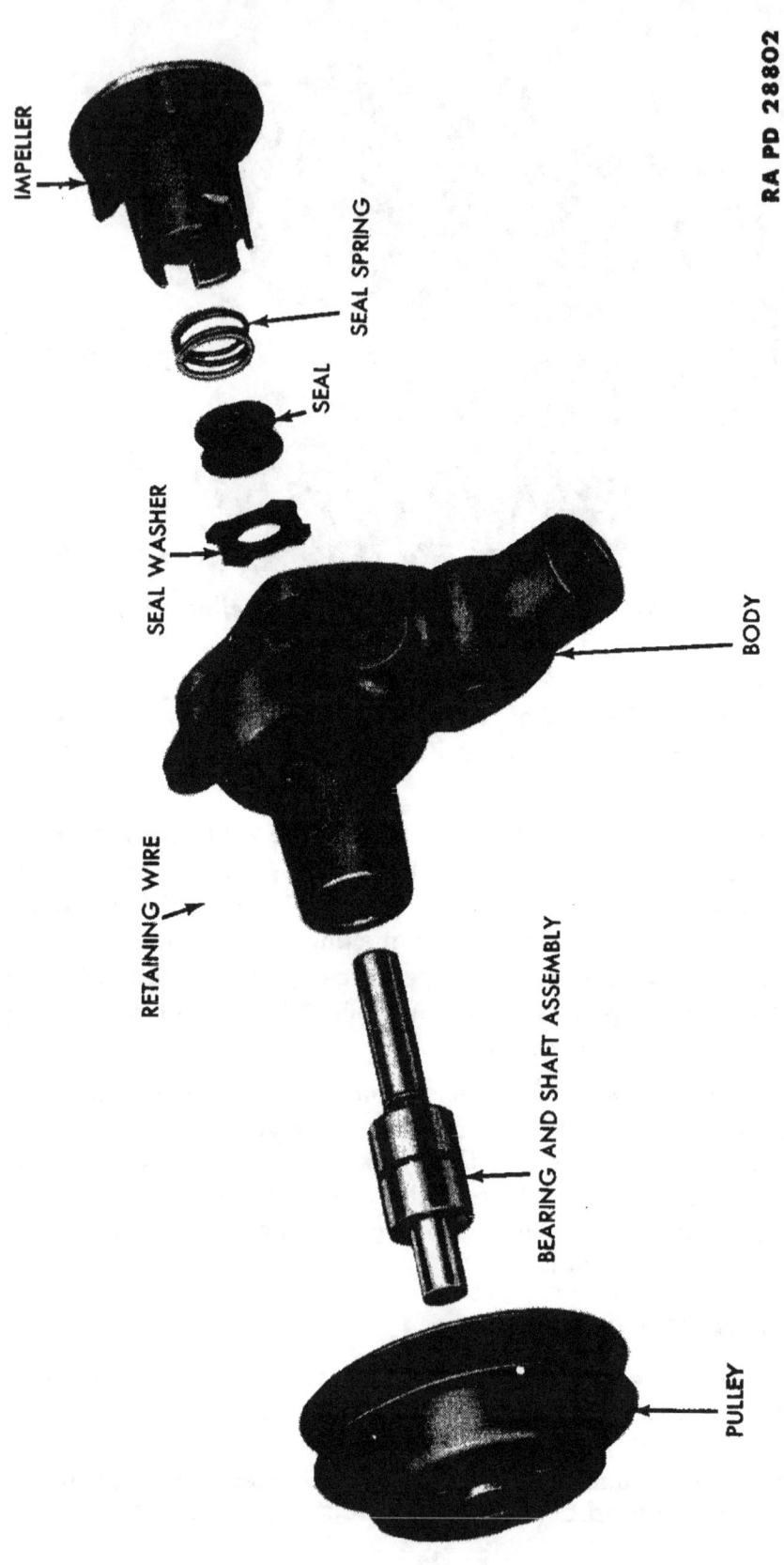

Figure 17 — Water Pump Disassembled

ENGINE

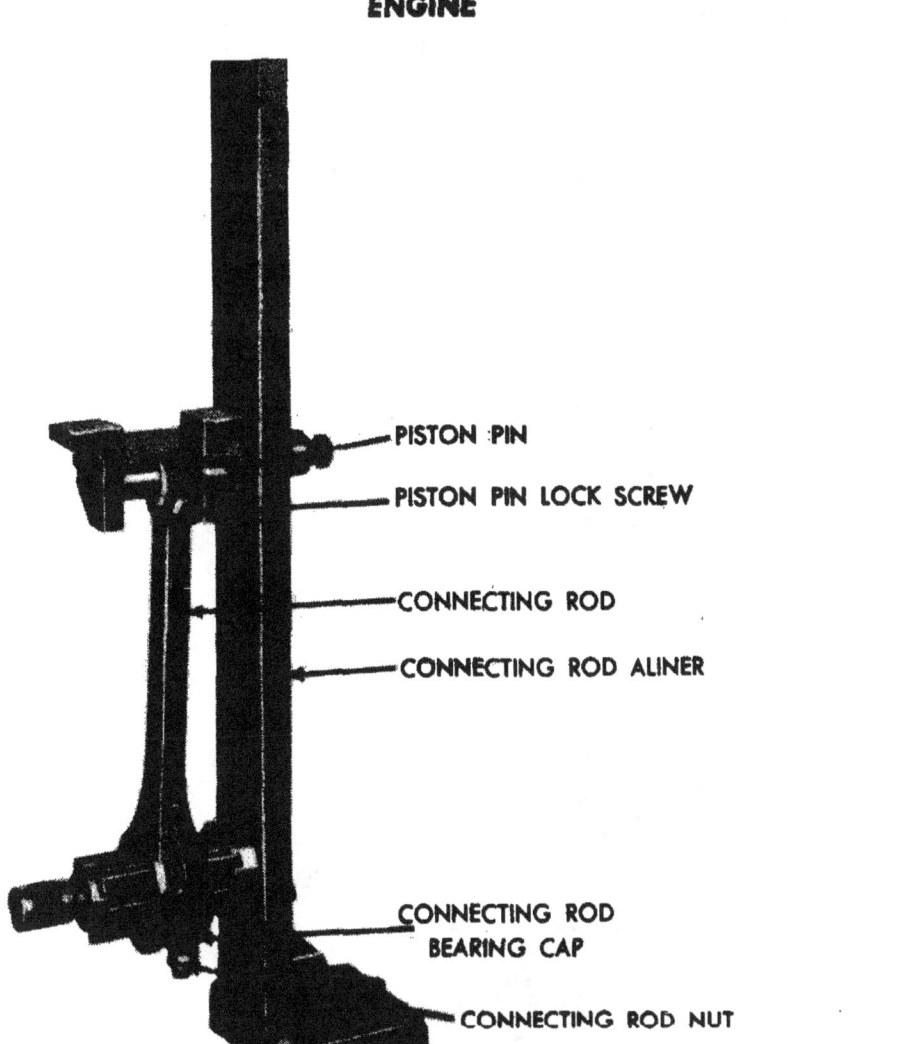

Figure 18 — Checking Connecting Rod Alinement for Twist, using Aliner (41-A-135)

pulley from the water pump body. Press the water pump pulley off the water pump bearing and shaft assembly.

b. **Cleaning.** Clean all parts thoroughly in dry-cleaning solvent.

c. **Inspection and Repair.**

(1) WATER PUMP BODY (fig. 17). A cracked or damaged water pump body must be replaced.

(2) WATER PUMP IMPELLER (fig. 17). A water pump impeller that is cracked or that has a broken fin must be replaced.

(3) WATER PUMP PULLEY (fig. 17). A distorted or damaged water pump pulley must be replaced.

TM 9-1803A
9

ORDNANCE MAINTENANCE — ENGINE AND ENGINE ACCESSORIES FOR ¼-TON 4x4 TRUCK (WILLYS-OVERLAND MODEL MB AND FORD MODEL GPW)

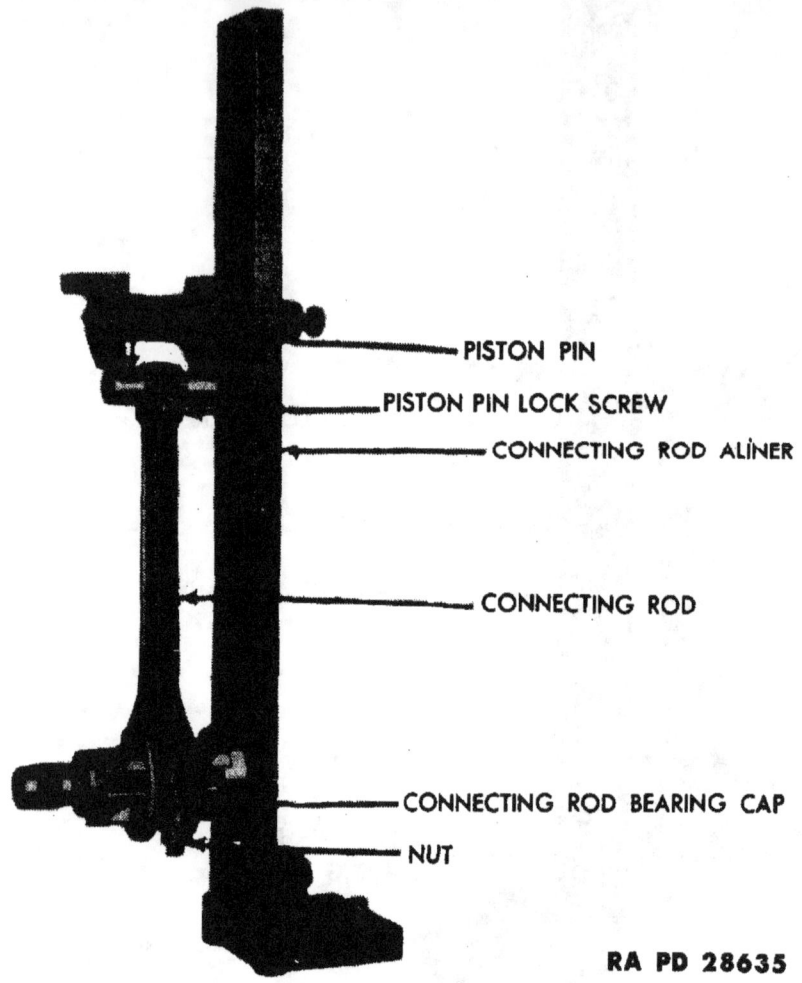

Figure 19 — Checking Connecting Rod Alinement for Bend, Using Aliner (41-A-135)

(4) WATER PUMP BEARING AND SHAFT ASSEMBLY (fig. 17). Rotate the water pump bearing; if the bearing binds or has a tendency to stick, it must be replaced. Bearings that have side or end play must be replaced.

d. Assembly. Press the front (short) end of the water pump bearing and shaft assembly into the water pump pulley. Press the water pump pulley and water pump bearing and shaft assembly into the front end of the water pump body until the groove on the bearing is in line with the small slot in the water pump body. Dip a new water pump seal assembly and water pump seal washer in hydraulic brake fluid, and install them in the water pump impeller. Place the impeller in a press, and press the shaft into the impeller until the end of the shaft is flush with the end of the water pump impeller. Install the water pump bearing retaining wire in place.

TM 9-1803A
10

ENGINE

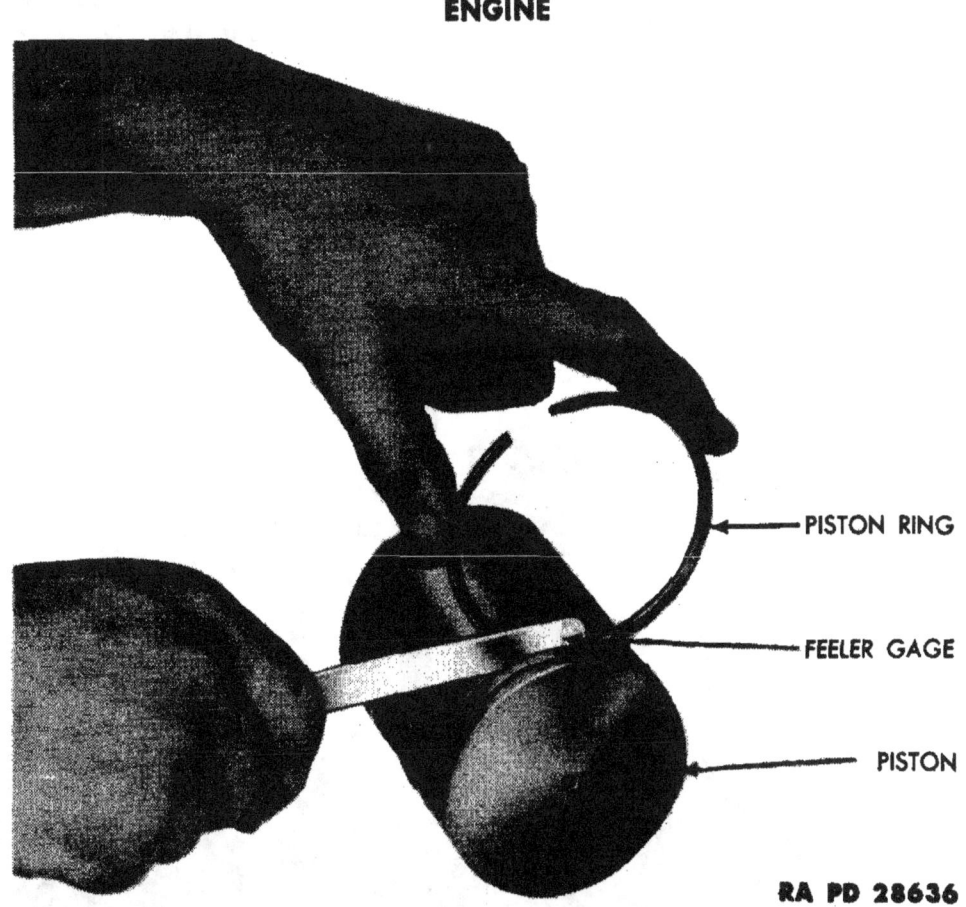

Figure 20 — Checking Clearance of Ring Groove with Feeler Gage

10. CONNECTING ROD AND PISTON ASSEMBLY.

a. Disassembly. Remove the piston rings with a standard ring remover. Remove the piston pin lock screw, and push the piston pin out of the piston.

b. Cleaning. Scrape the carbon from the ring grooves in the piston, and from the dome. Remove all foreign matter from the oil holes in the oil ring (lower) groove. Clean the complete assembly in dry-cleaning solvent.

c. Inspection and Repair. Pistons with cracks, scores, or damage of any kind must be replaced. Determine the wear on the skirt of each piston at the bottom at right angles to the piston pin. If the wear is 0.010 inch less than the original size, or if the piston is out-of-round more than 0.005 inch, the piston must be replaced. Check the width of the ring grooves with new rings and a feeler gage (fig. 20). If the piston ring groove wear exceeds 0.003-inch clearance between the piston ring and ring groove, the piston must be replaced. Measure the piston pin hole. If the inside diameter of the piston pin hole is more than 0.813 inch, the piston must be replaced. Piston pins worn to less than 0.8115-inch

TM 9-1803A
10

ORDNANCE MAINTENANCE — ENGINE AND ENGINE ACCESSORIES FOR ¼-TON 4x4 TRUCK (WILLYS-OVERLAND MODEL MB AND FORD MODEL GPW)

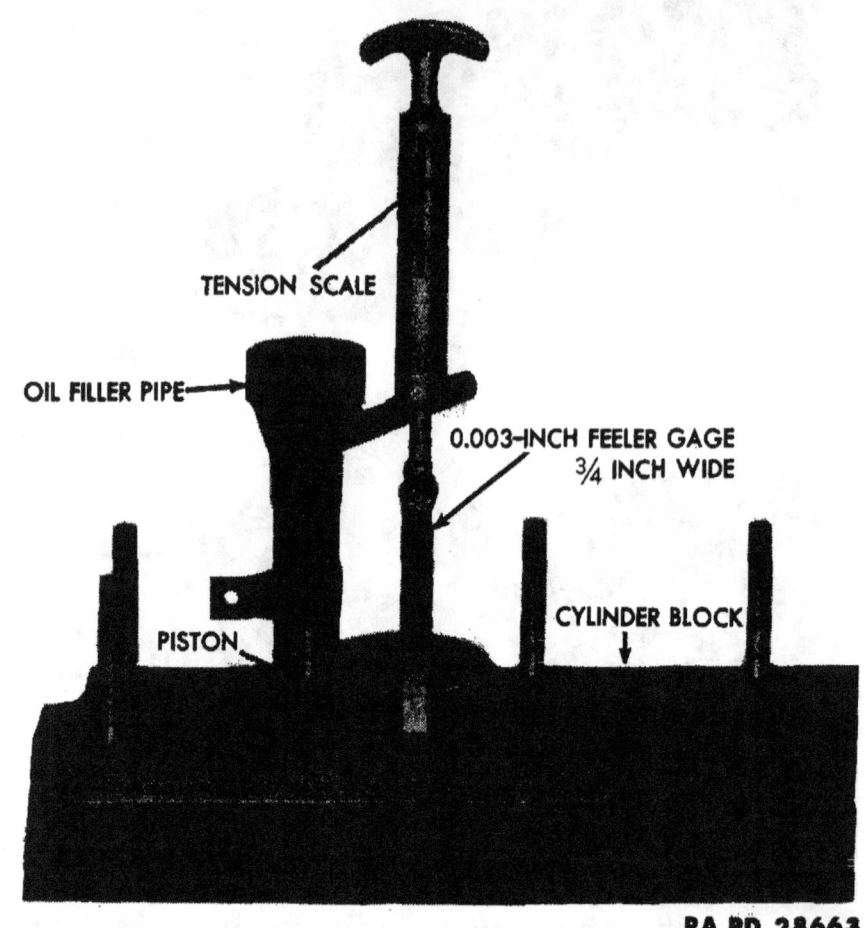

Figure 21 — Fitting Piston in Cylinder Bore, Using Scale w/feelers (41-S-498)

diameter must be replaced. Check the connecting rods for alinement, using aliner (41-A-135) (figs. 18 and 19). Bent or twisted connecting rods must be correctly alined. Damaged connecting rod bolts must be replaced. If connecting rods are fitted with studs, and studs are damaged, the complete connecting rod must be replaced. Excessively worn, scored, discolored, or pitted connecting rod insert bearings must be replaced.

d. Fit Piston. The normal clearance of the piston to the cylinder bore is 0.003 inch. Place a piston fitting scale with feelers (41-S-498) into the cylinder bore, making sure the feeler gage is long enough to cover the entire length of a piston. Push a piston into the cylinder bore with the T-slot in the piston opposite the feeler gage (fig. 21). Lift up on the tension scale; if more than 10 pounds is required to pull the feeler gage from the cylinder bore, the piston is too tight. Select a

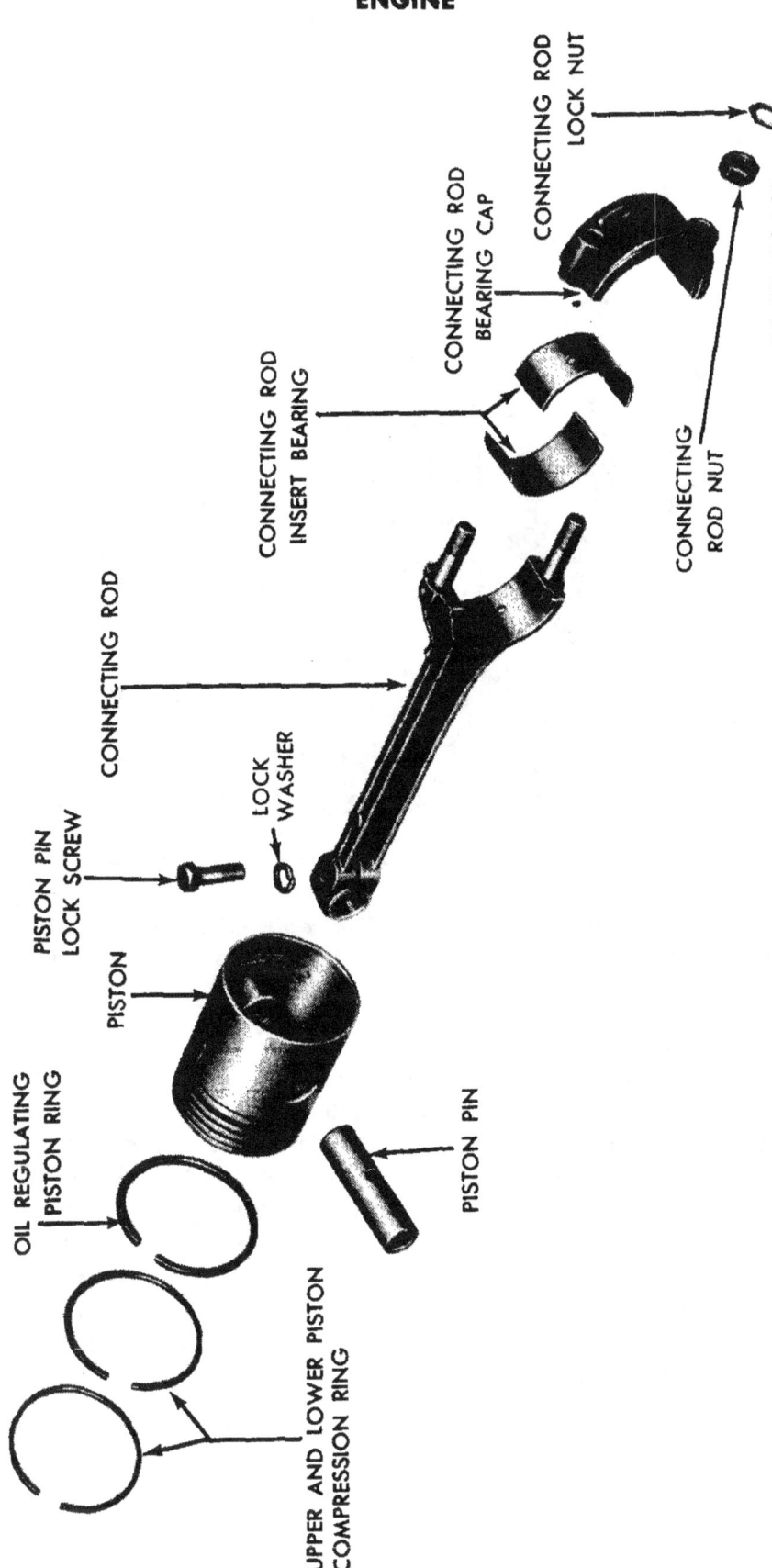

Figure 22 — Connecting Rod and Piston Assembly, Disassembled

TM 9-1803A
10

ORDNANCE MAINTENANCE — ENGINE AND ENGINE ACCESSORIES FOR ¼-TON 4x4 TRUCK (WILLYS-OVERLAND MODEL MB AND FORD MODEL GPW)

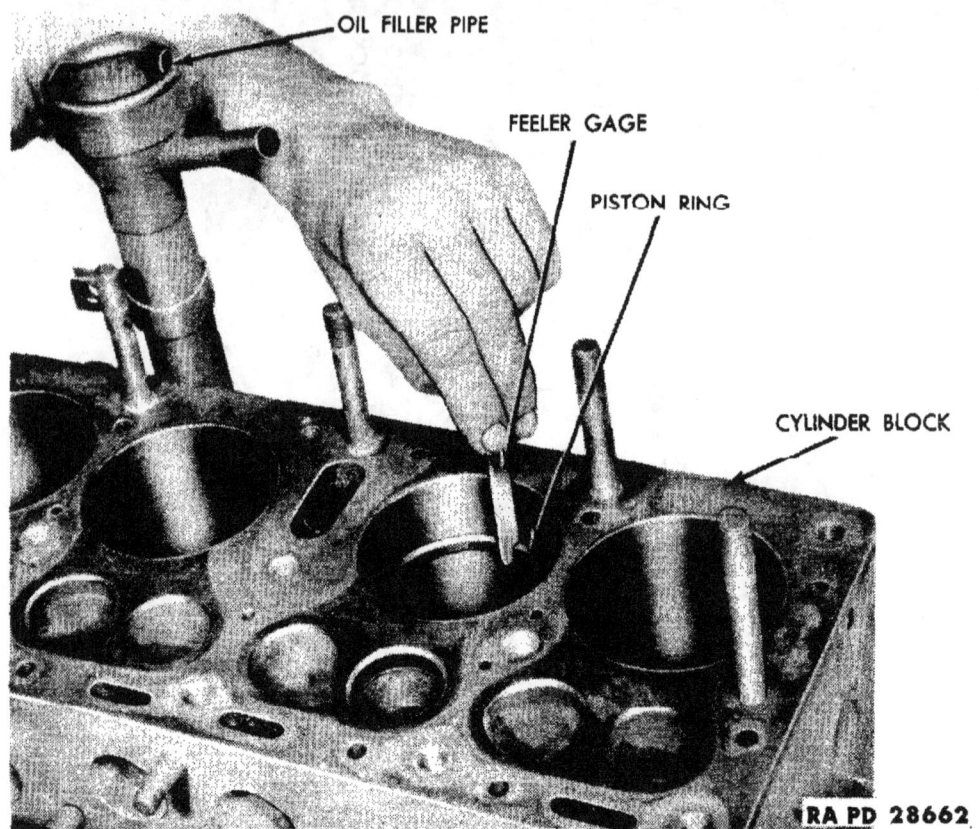

Figure 23 — Measuring Piston Ring End Gap with Feeler Gage

smaller piston. If less than 5 pounds pull is required to remove the gage, the piston is too loose. Select a larger piston. Mark the cylinder number on each piston after fitting.

 e. **Assemble Piston, Piston Pin, and Connecting Rod.** When installing connecting rods on pistons, make sure the oil squirt hole in the connecting rod is opposite the T-slot in the piston (fig. 36). If assembled in this manner, the off-set on the connecting rods will be in the correct position when installed in the cylinder block (par. 18 f). Select a piston pin which can be inserted in the piston with a light "push" fit (piston temperature at 70° F), and push it part way into the piston pin hole, with the groove in the piston pin facing downward. Hold the connecting rod in line with the piston pin hole, and push the piston pin the rest of the way into the piston. Install and tighten the piston pin lock screw in the connecting rod.

 f. **Fit and Install Piston Rings.** Place a new piston ring in the cylinder bore, and press it about halfway down into the cylinder bore with the bottom of a piston, so that the ring will be square with the cylinder wall. Measure the piston ring end gap with a feeler gage (fig.

ENGINE

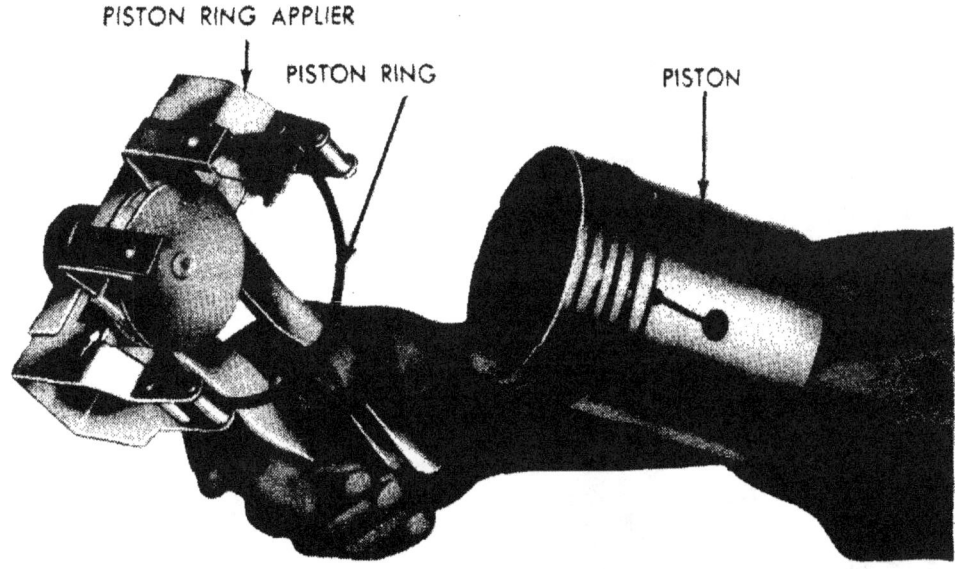

Figure 24 — Installing Piston Ring on Piston, Using Applier (41-A-329-500)

23). If the gap is less than 0.008 inch, remove the ring, and file with a fine-cut file until the correct gap (0.008 to 0.013 inch) is obtained. If end gap exceeds 0.013 inch, an oversize ring must be used. Repeat the same procedure for all piston rings. Roll the new piston ring around its particular groove in the piston. The ring should roll freely, and not have a clearance of more than 0.003 inch (fig. 20). Repeat the same procedure on each piston ring. Install the piston rings on the piston with a piston ring applier (41-A-329-500) (fig. 24), making sure that the beveled edge of both compression rings are towards the top.

11. CAMSHAFT ASSEMBLY.

a. **Cleaning.** Clean the camshaft, camshaft sprocket, camshaft thrust washer, and camshaft thrust spring and plunger, in dry-cleaning solvent.

b. **Inspection and Repair.** A camshaft with excessively scored or damaged cams, or with worn, corroded, scored, or discolored journals, must be replaced. Inspect the camshaft oil pump drive gear. If the teeth are worn, broken, or chipped, the camshaft must be replaced. Measure the four camshaft journals (fig. 25), and record the readings. If reading is less than 2.185 inches for the front journal, 2.122 inches for the front intermediate journal, 2.0595 inches for the rear intermediate journal, and 1.622 inches for the rear journal, the camshaft must be replaced. A camshaft gear with worn, broken, or chipped teeth must be replaced. Small nicks can be honed, and then polished

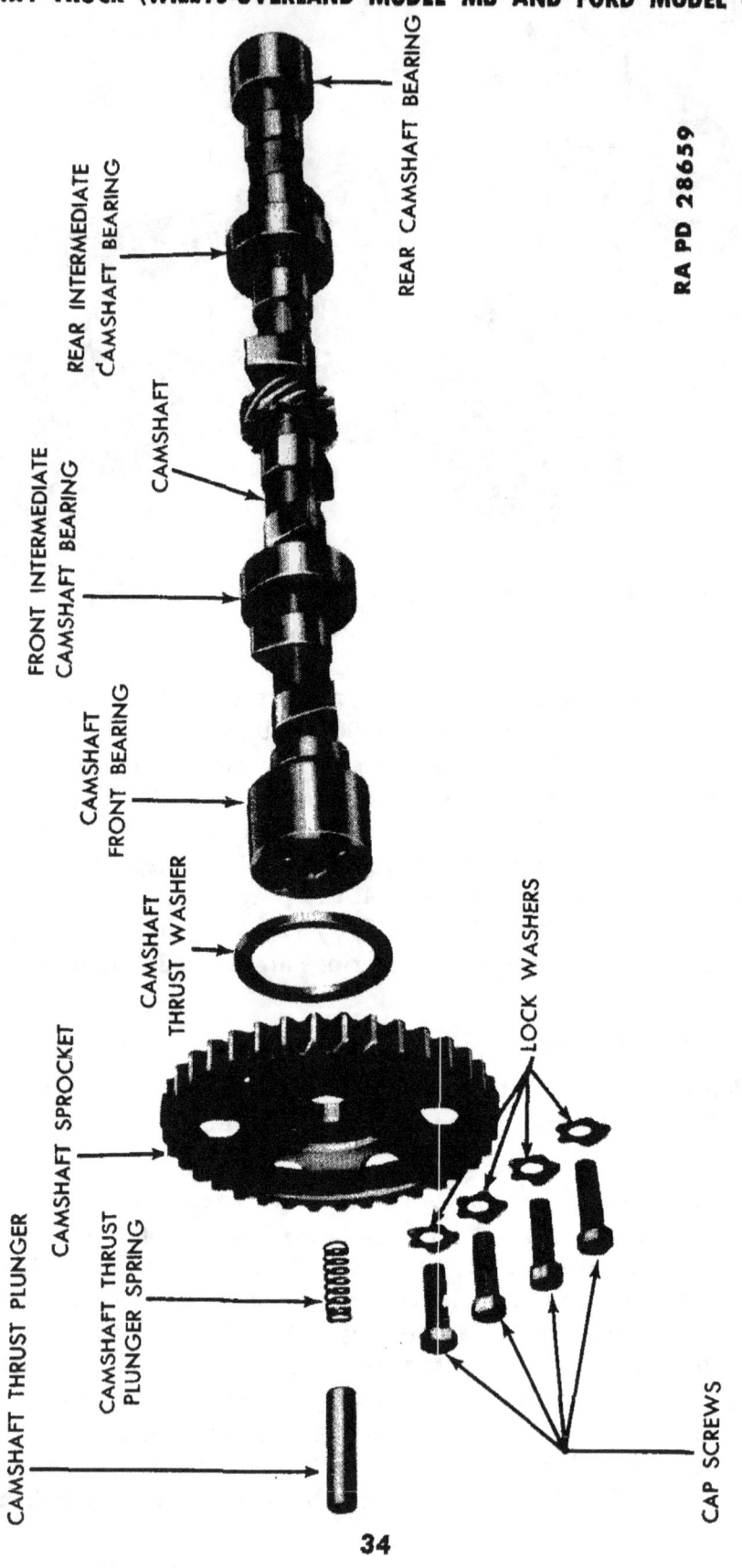

Figure 25 — Camshaft Assembly Disassembled

TM 9-1803A
11

ENGINE

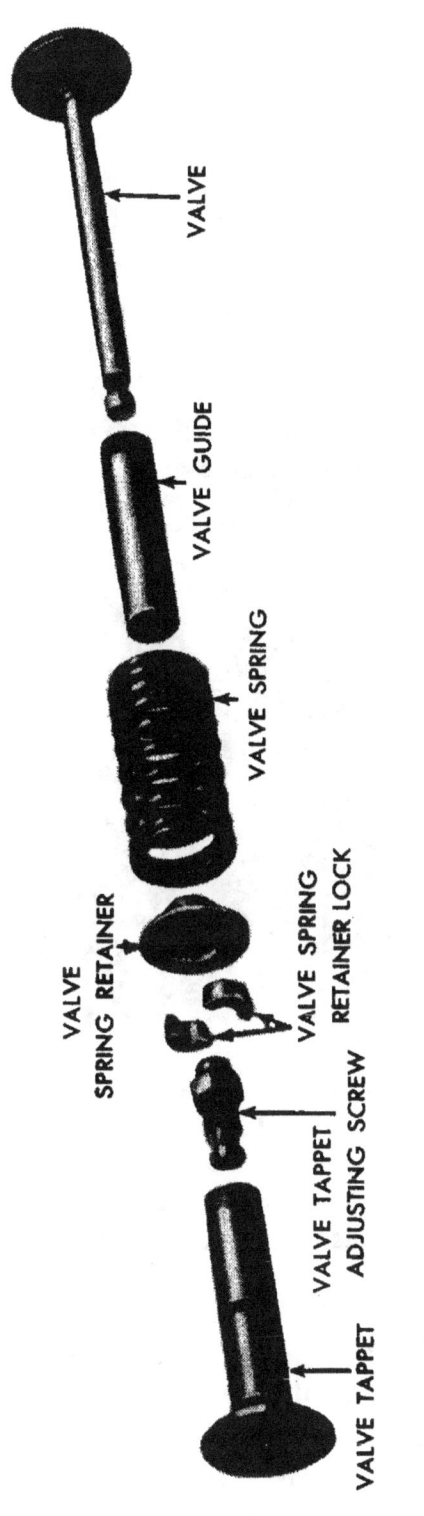

Figure 26 — Valve Assembly Disassembled

TM 9-1803A
11–12

ORDNANCE MAINTENANCE — ENGINE AND ENGINE ACCESSORIES FOR ¼-TON 4x4 TRUCK (WILLYS-OVERLAND MODEL MB AND FORD MODEL GPW)

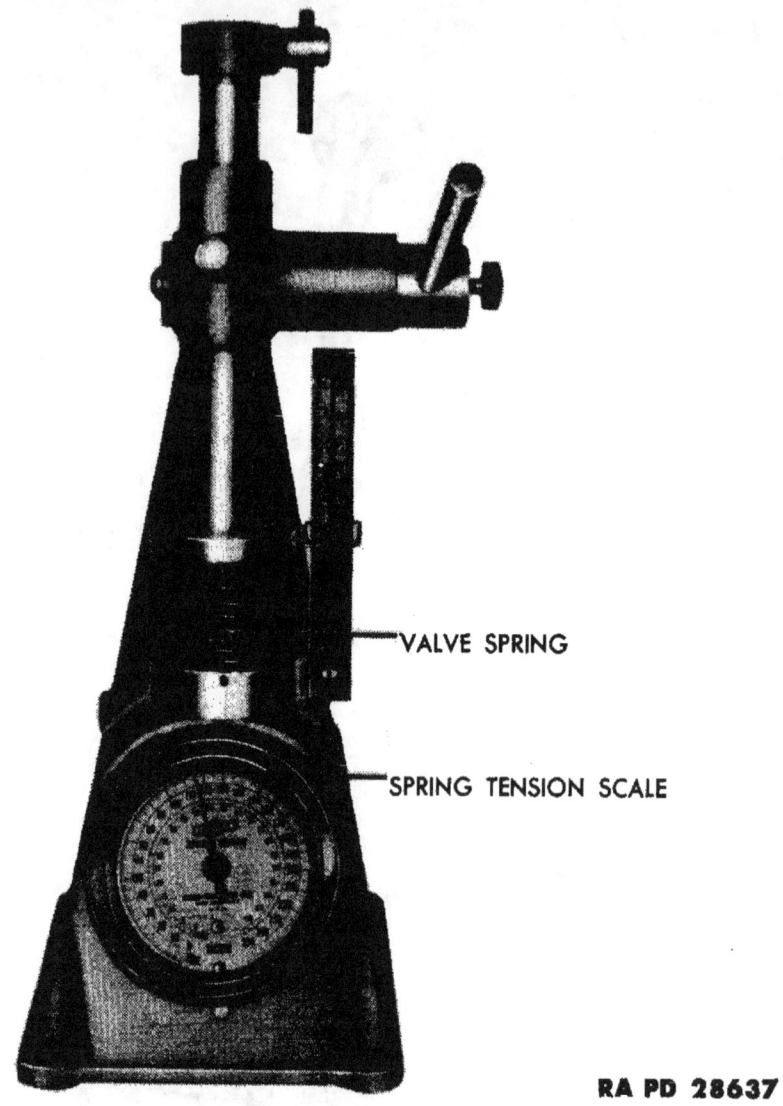

RA PD 28637

Figure 27 — Checking Tension of Valve Spring, Using Tester (41-T-1600)

with a fine stone. A weak (less than 15 pounds compressed to $^{29}\!/_{32}$ inch) or broken camshaft thrust plunger spring must be replaced.

12. VALVE AND VALVE SPRINGS.

a. Cleaning. Scrape the carbon off the valve heads and stems. Clean the valves and valve springs thoroughly in dry-cleaning solvent.

b. Inspection and Repair. Valves with bent or scored stems must be replaced. Measure the outside diameter of each valve stem (fig. 26). If measurement is less than 0.3685 inch for the exhaust valve, or 0.368 inch for the intake valve, the valves must be replaced. Pitted,

ENGINE

corroded, or burned valves must be refaced. Valves that are burned, warped, or pitted, and will not clean up with a light cut of the grinding wheel, must be replaced. Measure the free length of each valve spring; if less than 2½ inches in length, the spring must be replaced. Check the tension of each valve spring (fig. 27), using tester (41-T-1600). If the valve spring registers less than 50 pounds when compressed to 2 1/16 inches, or 116 pounds when compressed to 1¾ inches in length, it must be replaced.

13. VALVE TAPPETS.

a. **Cleaning.** Clean the valve tappets thoroughly in dry-cleaning solvent.

b. **Inspection and Repair.** Cracked, scored, or excessively worn valve tappets (fig. 26) must be replaced. Valve tappets, or valve tappet adjusting screws (fig. 26) with worn or damaged threads, must be replaced.

c. **Disassembly.** Unscrew the valve tappet adjusting screw from the tappet.

d. **Assembly.** Screw the valve tappet adjusting screw approximately three-quarters of the way into the valve tappet.

14. OIL PUMP AND OIL INTAKE FLOAT.

a. **Disassembly.** Remove the screw that holds the oil pump cover assembly to the oil pump, and remove the cover. Remove the oil pump relief spring retainer, gasket, shims, spring, and plunger from the oil pump cover (fig. 28). File either side of the oil pump driven gear pin (fig. 28), until the pin is flush with the driven gear sleeve. Drive the pin out of the sleeve and shaft with a small punch. Pull the oil pump shaft assembly out of the housing. Remove the cotter pin that holds the intake oil float to the oil float support, and remove the float (fig. 11). Straighten the four tabs on the oil intake float sump, and remove the sump. Lift the screen from the oil intake float.

b. **Cleaning.** Clean all parts and drilled passages thoroughly with dry-cleaning solvent, and blow out the oil intake float screen and all oil passages in the oil pump and oil intake float.

c. **Inspection and Repair.** A cracked or damaged oil pump housing or cover must be replaced. Measure the small pinion shaft on the oil pump cover. If less than 0.372 inch, the cover must be replaced. Measure the inside diameter of the oil pump housing (shaft end) (fig. 28). If larger than 0.505 inch, the oil pump housing must be replaced. An oil pump shaft assembly with broken teeth, or with a shaft measuring under 0.495 inch, must be replaced. An oil pump shaft assembly with a distributor slot worn more than three-sixteenths inch, must be

TM 9-1803A
14
ORDNANCE MAINTENANCE — ENGINE AND ENGINE ACCESSORIES FOR ¼-TON 4x4 TRUCK (WILLYS-OVERLAND MODEL MB AND FORD MODEL GPW)

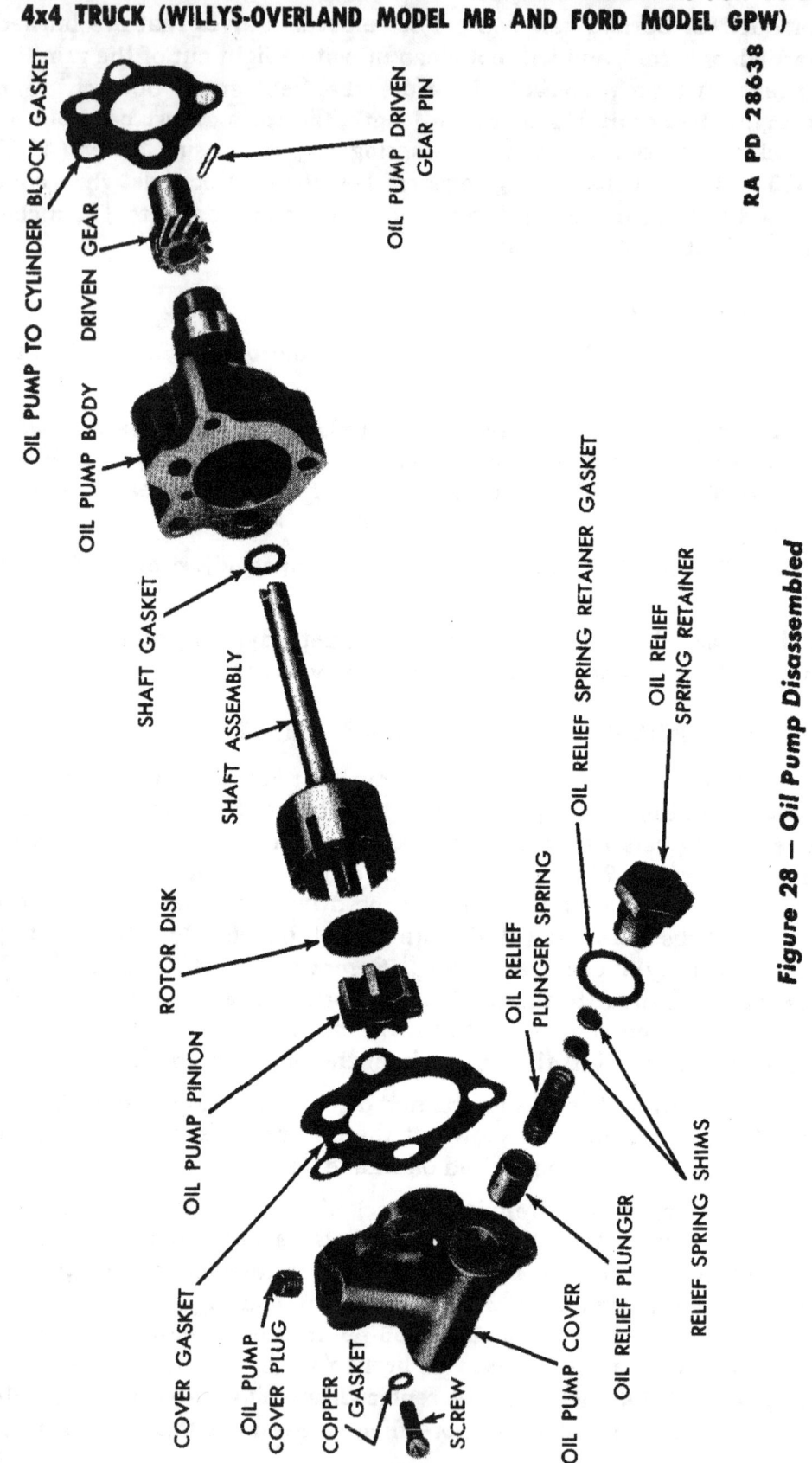

Figure 28 — Oil Pump Disassembled

ENGINE

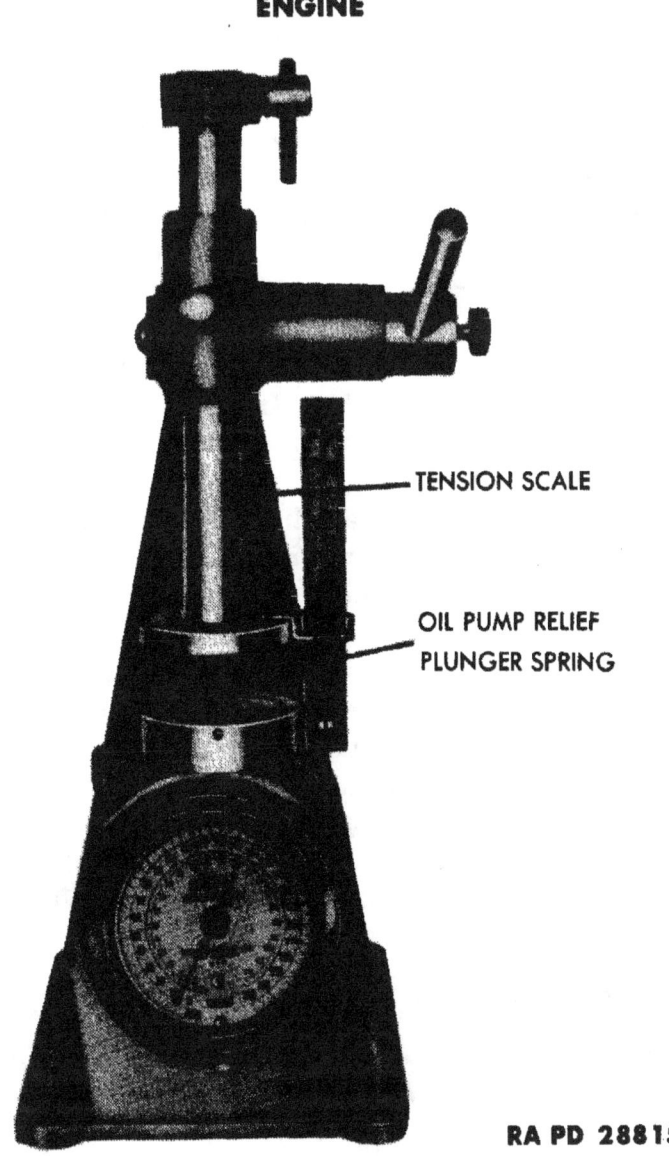

Figure 29 — Checking Oil Pump Relief Valve Spring Tension, Using Tester (41-T-1600)

replaced. An oil pump pinion gear with broken or worn teeth, or with an inside diameter of more than 0.378 inch, must be replaced. Measure the rotor disk (fig. 28); if less than 0.069 inch thick, it must be replaced. An oil pump driven gear with broken or chipped teeth must be replaced. Compress the oil pump relief valve spring to $1\frac{1}{16}$ inches (fig. 29), using tester (41-T-1600). If the tension is less than $5\frac{1}{2}$ pounds, the spring must be replaced. Replace a broken or cracked oil intake float support; also a distorted or leaking intake float support.

d. Assembly. Place the screen in the oil intake float. Place the sump on the oil intake float, and bend the four tabs to lock the sump

ORDNANCE MAINTENANCE — ENGINE AND ENGINE ACCESSORIES FOR ¼-TON 4x4 TRUCK (WILLYS-OVERLAND MODEL MB AND FORD MODEL GPW)

to the float. Slide the oil intake float support onto the float, making sure the tongue on the support is in the recess. Install a cotter pin in the support. Slide a new oil pump shaft gasket on the shaft assembly. Slide the shaft assembly into the oil pump housing. Tap the driven gear onto the shaft with the gear toward the oil pump, until there is 0.0312-inch clearance between the gear and oil pump body. If installing a new shaft, drill a hole for the pin, and install a new driven gear pin through the gear and shaft. Peen both ends of the driven gear pin. Install the rotor disk in the shaft assembly. Install the pinion gear on the oil pump cover. The pinion gear must have from 0.001- to 0.003-inch end play, measured from the end of the pinion shaft. Place a new gasket on the oil pump cover, and install the cover onto the housing. Install the copper gasket and hold-down screw in the cover. Tap the oil pump shaft into the housing, and check the clearance between the driven gear and housing. Insert a screwdriver between the gear and housing, and pry on the shaft. Remove screwdriver and again measure clearance. The difference represents the end play, and must be 0.002- to 0.004 inch. If sufficient, remove cover and add sufficient gaskets. Drop the oil relief plunger and spring (fig. 28) into the opening in the oil pump cover. Place two oil relief spring shims into the oil relief spring retainer (fig. 28). Place a new gasket on the oil relief spring retainer, and install and tighten the retainer to the cover. Install and tighten the oil pump cover plug.

15. CRANKSHAFT ASSEMBLY.

a. Cleaning. Clean out the drilled holes on the crankshaft journals with a piece of wire. Clean the crankshaft thoroughly with dry-cleaning solvent.

b. Inspection and Repair. Inspect all crankshaft journals. If worn or scored, the crankshaft must be replaced or reworked. Measure the outside diameter of each crankshaft journal. If the diameter is less than 1.9365 inches on the crankpin journals (fig. 30), or 2.3325 inches on the main bearing journals (fig. 30), or if any of the journals are out-of-round more than 0.0005 inch, the crankshaft must be reworked to 0.010-, 0.020-, or 0.030-inch undersize, whichever the case may be. Light scores and scratches can be honed, and then polished with crocus cloth. Crankshafts that will not clean up at 0.030-inch undersize must be replaced. If a new crankshaft or flywheel is being used, it must be fitted as outlined in paragraph 16 e.

c. Remove Crankshaft Sprocket (fig. 30). Install a standard puller on the crankshaft sprocket and remove the sprocket. Remove the Woodruff key, spacer, thrust washer, and shims.

ENGINE

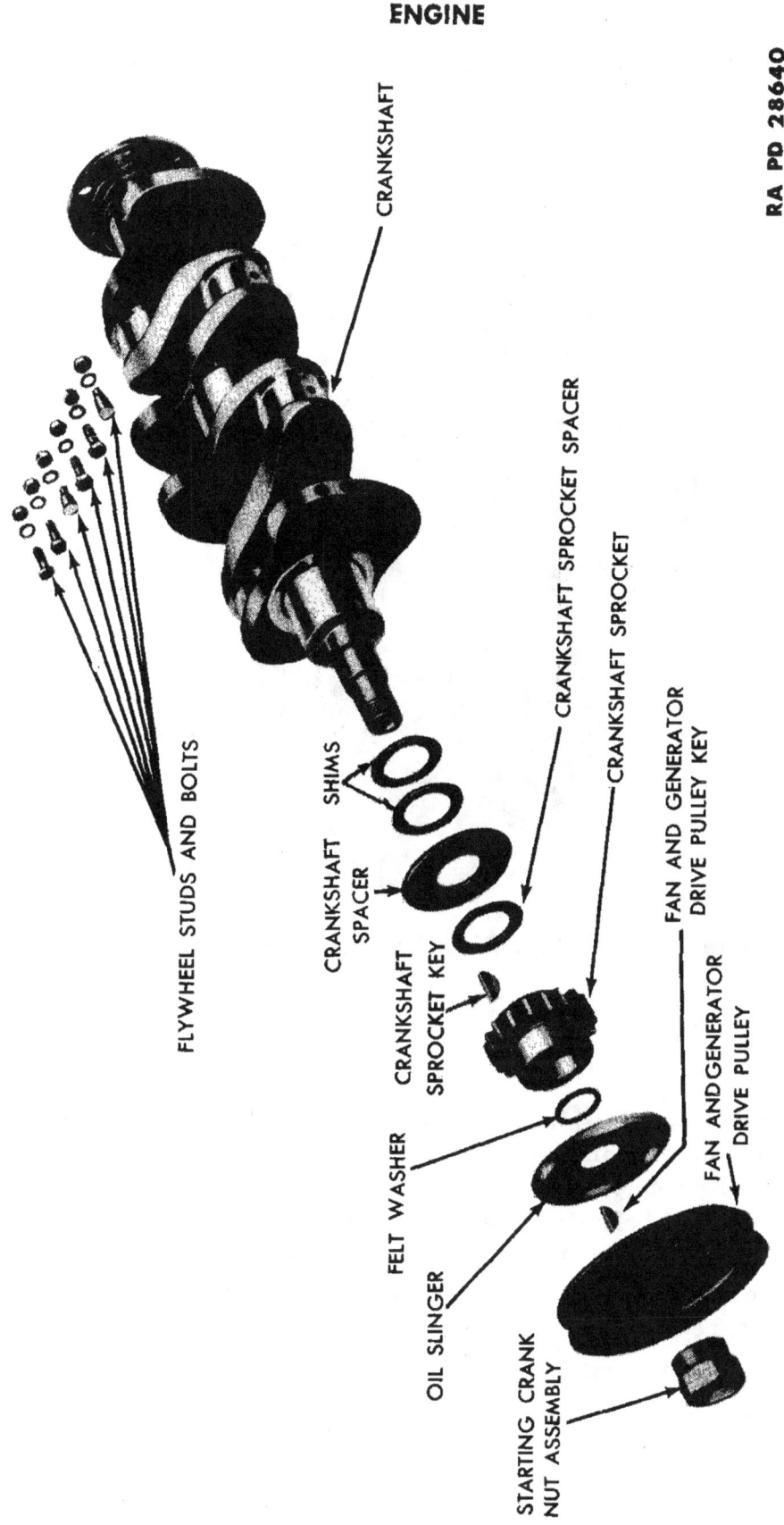

Figure 30 — Crankshaft Assembly Disassembled

TM 9-1803A
16

ORDNANCE MAINTENANCE — ENGINE AND ENGINE ACCESSORIES FOR ¼-TON 4x4 TRUCK (WILLYS-OVERLAND MODEL MB AND FORD MODEL GPW)

RA PD 28694

Figure 31 — Flywheel Ring Gear and Pilot Bushing

16. FLYWHEEL ASSEMBLY.

a. Cleaning. Wash the flywheel thoroughly in dry-cleaning solvent.

b. Inspection and Repair. A flywheel (fig. 31) with an excessively scored or worn friction face must be replaced. A flywheel ring gear with broken, chipped, or excessively worn teeth must be replaced (subpars. c and d below). Measure the inside diameter of the main drive gear pilot bushing. If more than 0.632 inch, it must be replaced (subpars. c and d below). If a new crankshaft or flywheel is being used, it must be fitted as outlined in subparagraph e below.

c. Disassembly. Drive the main drive gear pilot bushing out of the flywheel. Heat the flywheel ring gear until it can be driven off the flywheel.

d. Assembly. Clean the flywheel ring gear recess on the flywheel. Apply heat evenly to the ring gear. When the ring gear is thoroughly

42

ENGINE

heated, place it on the cold flywheel, making sure it is firmly seated in its recess. Drive a main drive gear pilot bushing in place with a fiber block.

e. **Fit Crankshaft to Flywheel When Either Part Is New.** Install the flywheel onto the crankshaft with the four crankshaft bolts, lock washers and nuts, making sure the index mark on the crankshaft is in line with the index mark on the flywheel. Drill the two tapered (stud) holes with a $35/64$-inch drill, and ream the two holes with a $9/16$-inch (0.5625-inch) reamer. Install the two bolts that are supplied with each crankshaft and/or flywheel.

17. INTAKE AND EXHAUST MANIFOLDS.

a. **Disassembly.** Remove the four cap screws that secure the intake manifold to the exhaust manifold, and separate the two manifolds. Remove the nut and bolt that hold the heat control valve shaft. Pull the counterweight lever, heat control lever, washer, and spring off the shaft.

b. **Cleaning.** Scrape all the old gaskets and carbon from the manifolds. Wash the manifold and parts in dry-cleaning solvent.

c. **Inspection and Repair.** Cracked or broken manifolds must be replaced. Damaged or broken studs must be replaced. An exhaust manifold with a damaged exhaust valve control or shaft must be replaced (par. 7 b).

d. **Assembly.** Slide the heat control valve spring onto the shaft, making sure the end of the heat control valve spring is resting on top of the stop. Slide the washer, counterweight, and control lever onto the shaft; install the nut and bolt through the counterweight. Place a new gasket between the two manifolds, and install the four cap screws and heat control spring stop.

Section V

ASSEMBLY OF ENGINE

	Paragraph
Assembly	18
Installation of accessories	19

18. ASSEMBLY.

a. **Install Valves.** Place a valve tappet in each valve tappet bore. Slide the camshaft into the cylinder block. Install a valve spring and valve spring retainer on each tappet, making sure the closed coils of the valve springs are against the cylinder block. Install the valves

TM 9-1803A
18

ORDNANCE MAINTENANCE — ENGINE AND ENGINE ACCESSORIES FOR ¼-TON 4x4 TRUCK (WILLYS-OVERLAND MODEL MB AND FORD MODEL GPW)

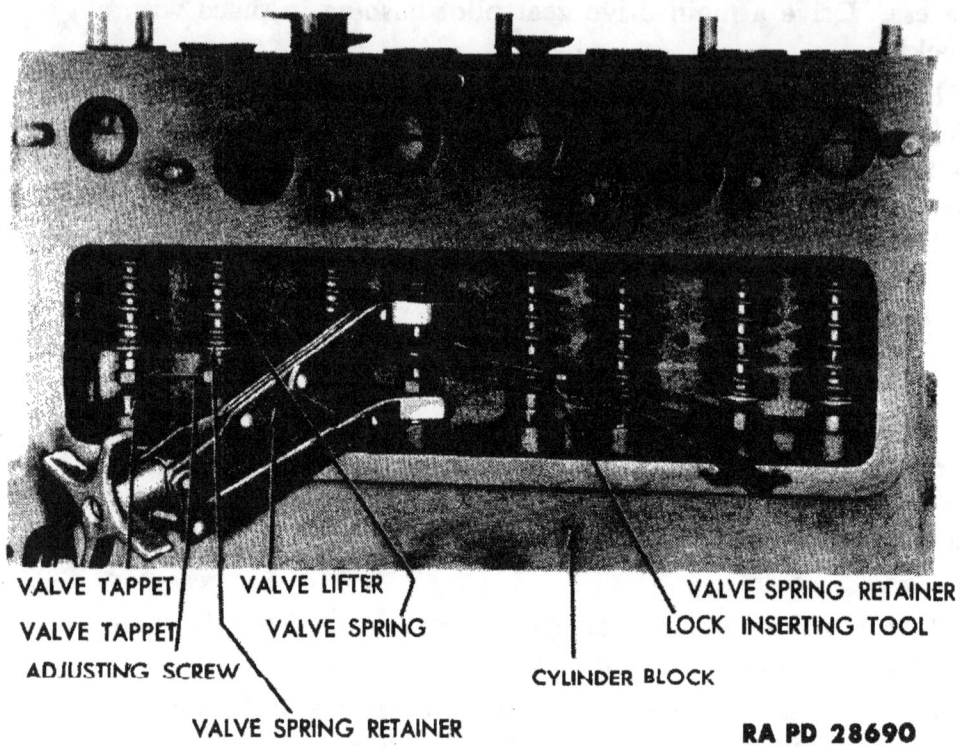

Figure 32 — Installing Valve Spring Retainer Locks, Using Lifter (41-L-1410) and Replacer (41-R-2398)

in their respective valve guides. Compress the valve springs on all valves that are in closed position using valve lifter (41-L-1410), and install the lower valve spring retainer locks (fig. 32), using replacer (41-R-2398). Turn the camshaft to close the other valves, and install the lower valve spring retainer locks on the rest of the valves.

b. Adjust Valve Tappets. Turn the camshaft until No. 1 valve is in a closed position, and the tappet is on the heel of the cam. Hold the valve tappet with one wrench, and turn the valve tappet adjusting screw with another wrench (fig. 33) clockwise or counterclockwise until 0.014-inch clearance is established between the valve and the valve tappet adjusting screw. Repeat the same procedure on each valve.

c. Install Crankshaft. If a new crankshaft or flywheel is being used, refer to paragraph 16 e. Install the three upper halves of the main bearing inserts in the cylinder block (fig. 13). Press the rear main bearing crankshaft packing into the recess provided at the rear main bearing (fig. 13), and in the rear main bearing cap. Cut the ends of the crankshaft packing flush with the crankcase and with the bearing cap. Install the four bolts and the two tapered studs in the

ENGINE

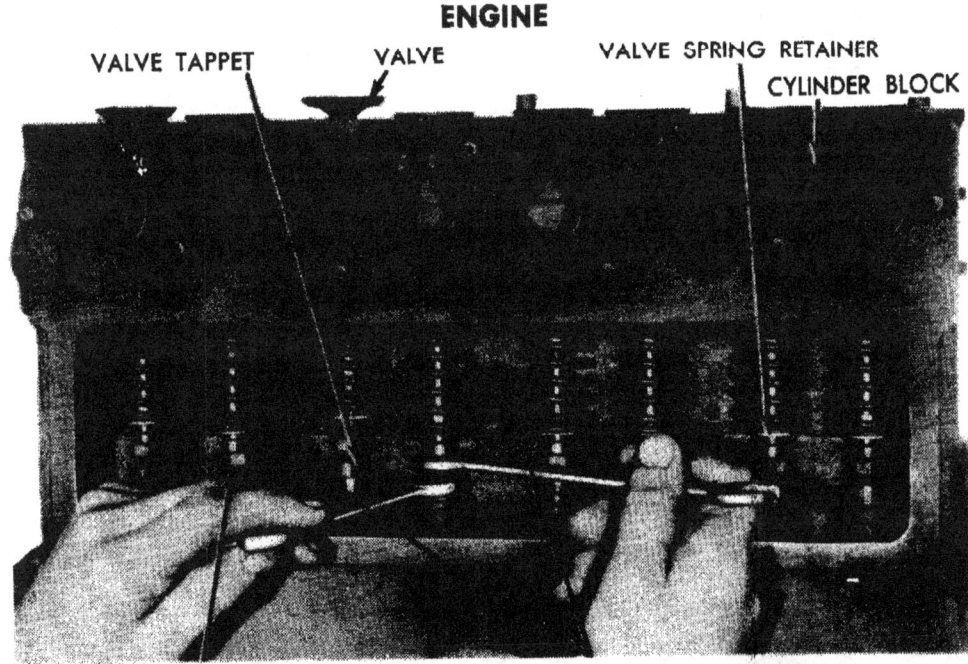

Figure 33 — Adjusting Valve Tappets, Using Wrenches (41-W-3575)

flywheel flange on the crankshaft. Install the three lower halves of the main bearing inserts in the three main bearing caps. Oil the main bearing inserts with a light oil. Place the crankshaft in place in the cylinder block. Install the front and center bearing caps, and tighten the bolts until they are just snug. Coat the rear bearing cap with joint and thread compound on both sides and top. Install the rear bearing cap in the cylinder block. Tighten the six main bearing bolts with a torque wrench to from 65 to 70 foot-pounds. Slip the rear bearing cap packing into the hole on each side of the rear main bearing cap, leaving ¼ inch of the packing to protrude from the crankcase.

d. Fit Crankshaft. Place a 0.006-inch feeler gage between the front main bearing cap and the crankshaft, and pull the crankshaft toward the front of the engine as far as possible. Place a straightedge across the front main bearing, and measure the distance between the straightedge and crankshaft to determine the amount of shims to be used (fig. 34).

e. Check Crankshaft End Play. Install the necessary amount of shims on the crankshaft to take up the space between the straightedge and crankshaft (fig. 35). Install the crankshaft thrust washer and spacer washer (fig. 30). Tap the large Woodruff key in the crankshaft, and slide the crankshaft sprocket, felt, and crankshaft oil slinger (fig. 30) on the shaft. Tap the small Woodruff key in the crankshaft, install the generator and fan belt drive pulley and cranking nut,

TM 9-1803A
18

ORDNANCE MAINTENANCE — ENGINE AND ENGINE ACCESSORIES FOR ¼-TON 4x4 TRUCK (WILLYS-OVERLAND MODEL MB AND FORD MODEL GPW)

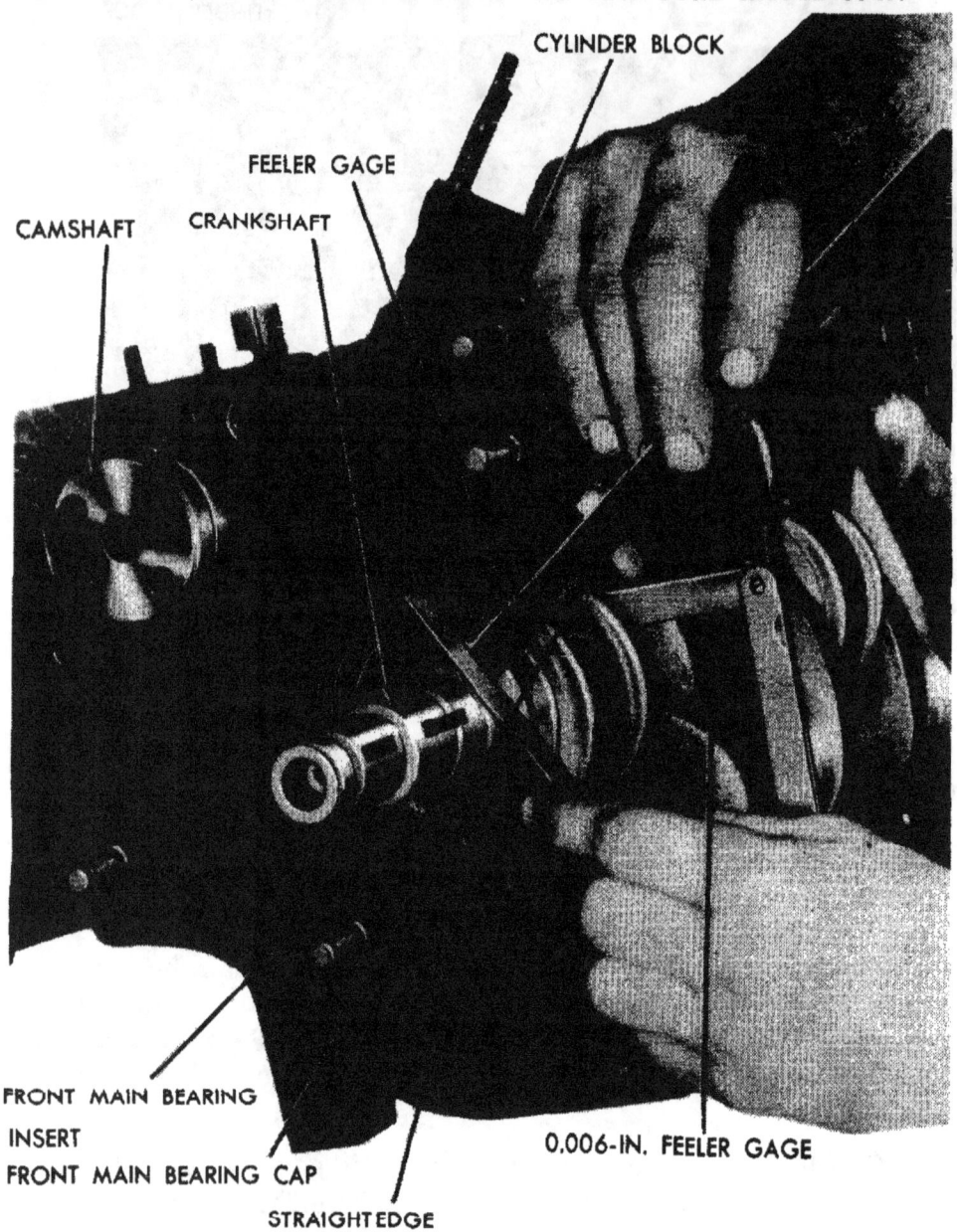

RA PD 28695

Figure 34 — Measuring Crankshaft End Play

and tighten the cranking nut. Place a feeler gage between the front main bearing cap and crankshaft. If more than 0.006-inch end play exists, shims must be removed. If less than 0.004 inch, shims must be added (fig. 35).

    f. Install Connecting Rod and Piston Assemblies. (Piston assemblies will have previously been selected for each cylinder as outlined in paragraph 10 d). Oil the piston rings and install a ring compres-

46

ENGINE

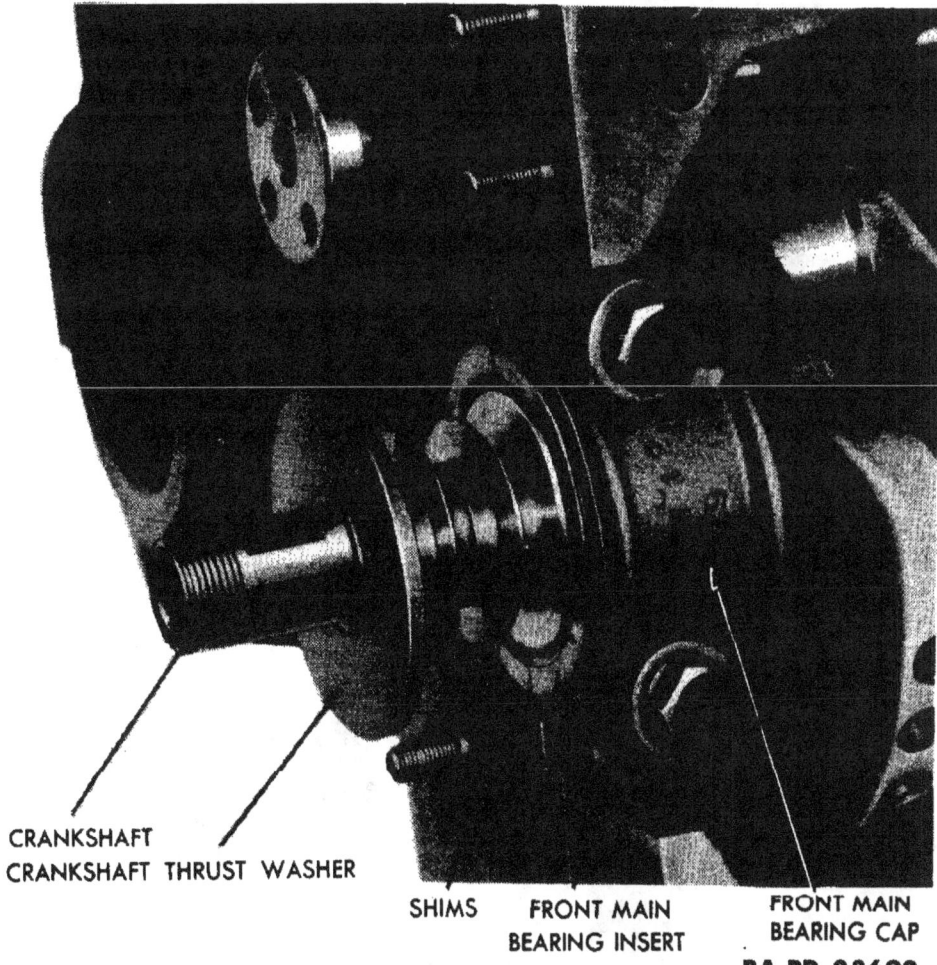

Figure 35 — Shims in Place on Crankshaft

sor (41-C-2550) on the piston rings. Place the No. 1 connecting rod and piston assembly in the No. 1 cylinder with the offset on the connecting rod away from the nearest main bearing (fig. 36). With the T-slot of the piston to the left, and the oil squirt hole in the connecting rod facing toward the right-hand side of the engine, tap the piston down into the cylinder with the handle end of a hammer (fig. 37). Place one-half of a connecting rod insert bearing in the connecting rod, and the other half in the connecting rod bearing cap. Coat the connecting rod insert bearings with a light film of oil. Connect the rod to the crankshaft and install, but do not tighten, the two connecting rod nuts. Repeat the same procedure when installing the other rods, making sure the offset on each connecting rod is away from the nearest main bearing, and the oil squirt hole facing toward the left-hand side of the engine. Tighten all the connecting rod nuts to from 50 to 55 foot-pounds pull with a torque wrench. Install a

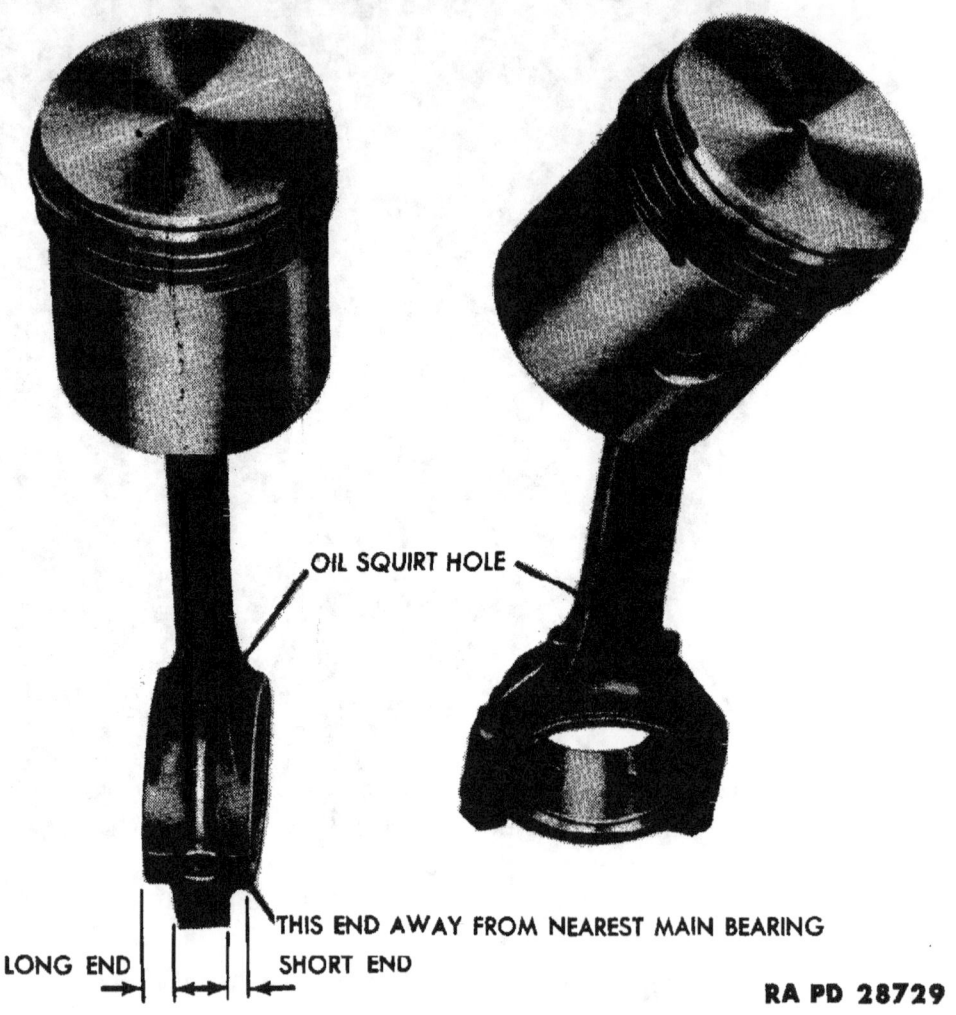

Figure 36 — Position of Connecting Rod Off-set and Oil Squirt Hole When Installed in Engine

pal nut on each connecting rod stud or bolt. Turn the pal nuts down on the stud or bolt until sealed, then turn one complete turn.

g. Install Flywheel. If installing a new flywheel or crankshaft, fit the crankshaft to the flywheel as outlined in paragraph 16 e. Fasten the engine rear plate temporarily to the engine with two bolts. Turn the crankshaft until the No. 1 and No. 4 pistons are at top center. Place the flywheel on the crankshaft flange so that the letters "TC" on the flywheel are lined up with the index mark at the center of the timing hole (fig. 38) in the engine rear plate, and the index mark on the crankshaft flange and on the flywheel are in line with each other. Install and tighten the six lock washers and nuts on the flywheel from 36 to 40 foot-pounds with a torque wrench. Check run-

TM 9-1803A
18

ENGINE

RA PD 28696

Figure 37 — Installing Piston and Connecting Rod Assembly in Cylinder Block, Using Ring Compressor (41-C-2550)

out on the flywheel with a dial gage. If the run-out exceeds 0.008 inch at the outer edge, the flywheel or crankshaft flange must be refaced.

h. Install Clutch Disk and Pressure Plate. Hold the clutch disk on the flywheel, and install a clutch pilot tool in the flywheel and the disk. Hold the pressure plate on the flywheel and install, but do not tighten, six lock washers and cap screws (fig. 39). Tighten the six cap screws evenly to prevent bending the pressure plate frame. Remove the clutch pilot.

i. Install Camshaft Sprocket. Place a gasket and the engine front plate on the engine, and install the three cap screws. Turn the crankshaft until No. 1 piston is at top center (fig. 38). Install the camshaft

49

TM 9-1803A
18

ORDNANCE MAINTENANCE — ENGINE AND ENGINE ACCESSORIES FOR ¼-TON 4x4 TRUCK (WILLYS-OVERLAND MODEL MB AND FORD MODEL GPW)

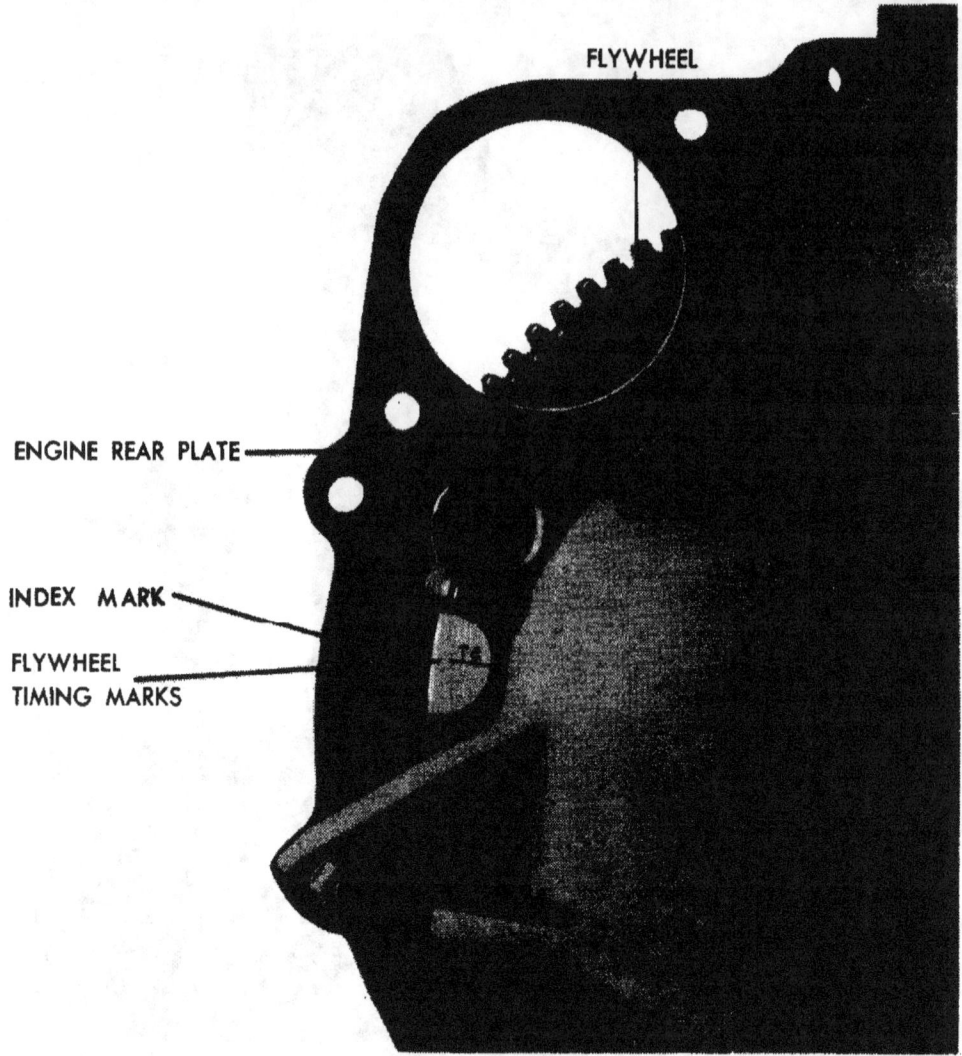

Figure 38 — Flywheel Timing Marks, T.C. (Top Center)

sprocket on the camshaft temporarily with two cap screws. Turn the camshaft sprocket until the punch mark on the camshaft sprocket is opposite the punch mark on the crankshaft sprocket (fig. 40). Remove the camshaft sprocket from the camshaft, being careful not to move the camshaft. Place the camshaft thrust washer on the camshaft. Place the camshaft drive chain on the crankshaft sprocket and camshaft sprocket, and install the camshaft sprocket on the camshaft with four lock washers and cap screws. Tighten the four cap screws, and bend the lock washer tabs down on the cap screws.

j. Install Front Engine Cover. Place the camshaft thrust plunger spring and plunger in the camshaft (fig. 25). Place a gasket on the engine front cover, and also install an oil seal in the recess provided in the cover. Install the cover on the engine.

ENGINE

Figure 39 — Installing Clutch Disk and Pressure Plate on Flywheel

k. Install Oil Pan. Hold a gasket and the oil intake float in place (fig. 11), and install the two lock washers and cap screws. Coat the bottom (machined surface) of the crankcase with grease, and install the oil pan gasket. Hold the oil pan in place, and install all the lock washers and cap screws except the six front cap screws. Hold the generator and fan drive pulley guard in place, and install the remaining six lock washers and cap screws. Tighten all the oil pan cap screws.

l. Install Cylinder Head. Install a cylinder head gasket on the cylinder block. Making sure there is no foreign material in the cylinders, place the cylinder head on the cylinder block, and install and tighten the cylinder head bolts to from 65 to 75 foot-pounds with a torque wrench. (Start with a centrally located bolt, and work alternately each way.)

m. Install Intake and Exhaust Manifold. Place an intake and exhaust manifold gasket in place on the cylinder block. Install the intake and exhaust manifold on the cylinder block. Install the seven

TM 9-1803A
18

ORDNANCE MAINTENANCE — ENGINE AND ENGINE ACCESSORIES FOR ¼-TON 4x4 TRUCK (WILLYS-OVERLAND MODEL MB AND FORD MODEL GPW)

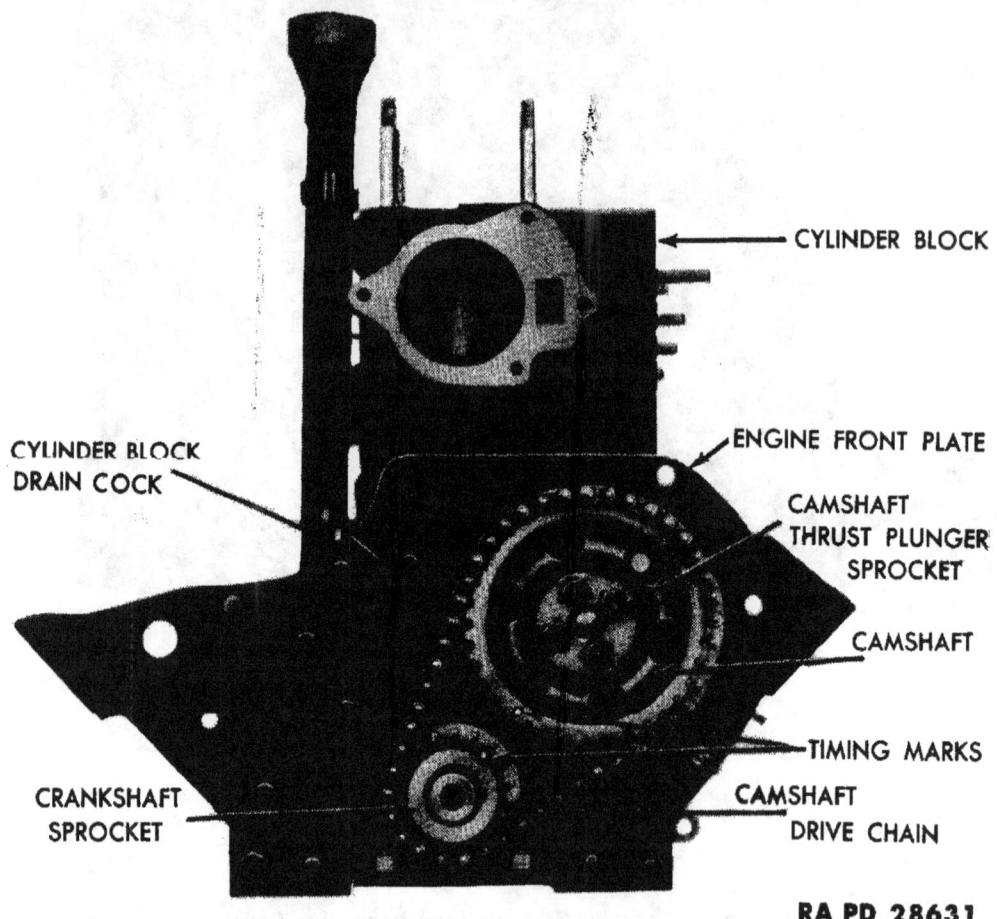

Figure 40 — Camshaft Timing Marks

flat washers and nuts. Connect the crankcase ventilation tube at the intake manifold, and at the crankcase ventilator assembly (fig. 41).

n. Install Oil Pump. Place a finger in No. 1 spark plug hole, and turn the crankshaft until No. 1 piston is coming up on compression stroke. Continue turning the crankshaft until the timing mark "IGN" appears on the flywheel and is in line with the index mark in the center of the timing hole on the engine rear plate (fig. 42). Install the distributor in the cylinder block (par. 19 e) temporarily. Set the rotor on No. 1 firing position (fig. 43) with the ignition points just breaking. Immerse the oil pump in a container of oil (same grade as used in engine), and turn the oil pump shaft assembly until the oil flows from the outlet hole in the oil pump body. Place a gasket on the oil pump and, with the wide side of pump shaft up, install the oil pump on the engine, making sure the slot in the oil pump shaft engages with the distributor shaft while the rotor is on No. 1 firing position with ignition

ENGINE

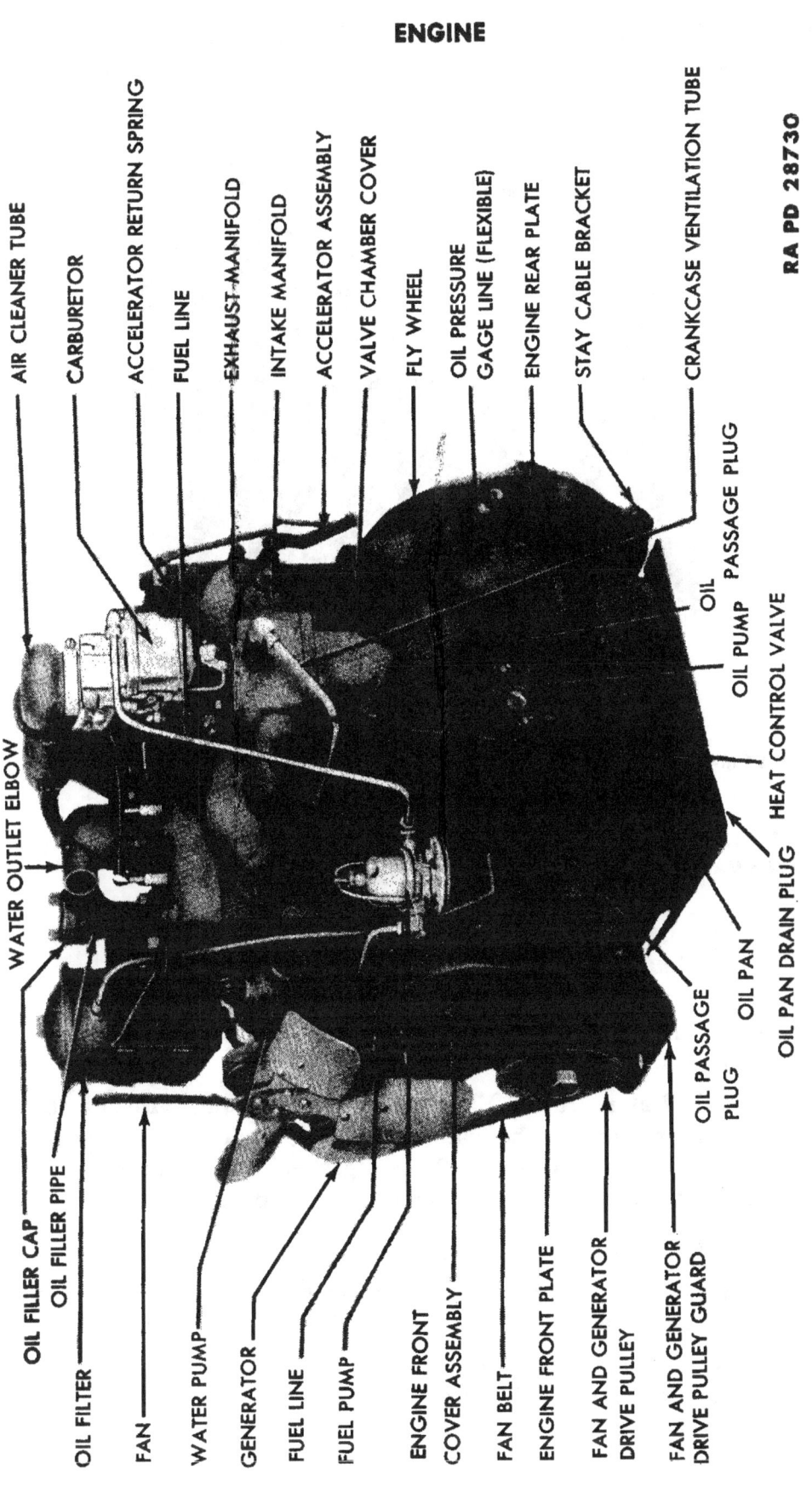

Figure 41 — Left Front View of Engine

Figure 42 — IGN Timing Marks on Flywheel

points just breaking. Install the lock washers and nuts on the oil pump. Remove the distributor.

o. Install Water Outlet Elbow and Thermostat. Install the thermostat and retainer in the water outlet elbow with the bellows of the thermostat facing downward. Place a gasket on the cylinder head and install the water outlet elbow, lock washers, and cap screws.

19. INSTALLATION OF ACCESSORIES.

a. Install Water Pump. Hold a gasket and the water pump in place on the engine, and install the three lock washers and cap screws.

b. Install Carburetor. Place a carburetor gasket and diffuser on the intake manifold and install the carburetor, accelerator return spring clip, lock washers, and nuts (fig. 41).

c. Install Oil Filter (fig. 43). Hold the oil filter in place, and in-

ENGINE

stall and tighten the three head nuts with a torque wrench to from 60 to 65 foot-pounds pull. Connect the oil filler pipe bracket to the oil filter bracket with a cap screw. Connect the outlet oil line to the engine front cover, and the inlet line to the elbow fitting located on the left-hand side of the engine in front of the fuel pump opening.

d. Install Fuel Pump. Place a gasket on the fuel pump. Hold the fuel pump in place on the engine, making sure the fuel pump rocker arm is on top of the camshaft. Install the two lock washers and cap screws in the fuel pump. Install the fuel line that connects the fuel pump and carburetor. Connect the generator brace and fuel line to the engine front plate. Connect the fuel line to the fuel pump.

e. Install Distributor. Place a thumb over No. 1 spark plug hole, and turn the crankshaft until No. 1 piston is coming up on compression stroke, and the timing mark "IGN" on the flywheel is in line with the index mark in the center of the timing hole on the engine rear plate (fig. 42). Install the distributor in the engine, and rotate the rotor until the distributor shaft engages in the oil pump shaft. Install the cap screw in the distributor hold-down clamp. Loosen the bolt in the distributor hold-down clamp, and turn the distributor until the points are just breaking. Tighten the bolt in the distributor hold-down clamp.

f. Install Ignition Coil. Install the ignition coil on the engine, making sure the bond strap is in place behind the ignition coil bracket (fig. 43). Connect the primary wire to the coil and distributor.

g. Install Fan. Hold the fan in place on the water pump pulley, and install the four lock washers and cap screws.

h. Generator. Install the generator bracket on the cylinder block with two lock washers and cap screws. Install the generator on the engine with the generator bolts, making sure there is a flat washer on each side of the rubber bushing in the generator bracket and engine front plate. Install a flat washer, lock washer, and nut on the generator front mount. Install a flat washer, bond strap, lock washer, and nut on the generator rear mount.

i. Install Spark Plugs, Wires, and Air Cleaner Tubing. Install air cleaner tube and bracket assembly, and tighten the head nuts from 60 to 65 foot-pounds. Connect the air cleaner tubing at the oil filler pipe (fig. 43) and carburetor. Install the distributor cap on the distributor. Pull the spark plug wires through the air cleaner tube bracket (fig. 43). Set the spark plug gap at 0.030 inch, and install the spark plugs and new spark plug gaskets in the cylinder head. Connect the spark plug wires.

TM 9-1803A
19

ORDNANCE MAINTENANCE — ENGINE AND ENGINE ACCESSORIES FOR ¼-TON 4x4 TRUCK (WILLYS-OVERLAND MODEL MB AND FORD MODEL GPW)

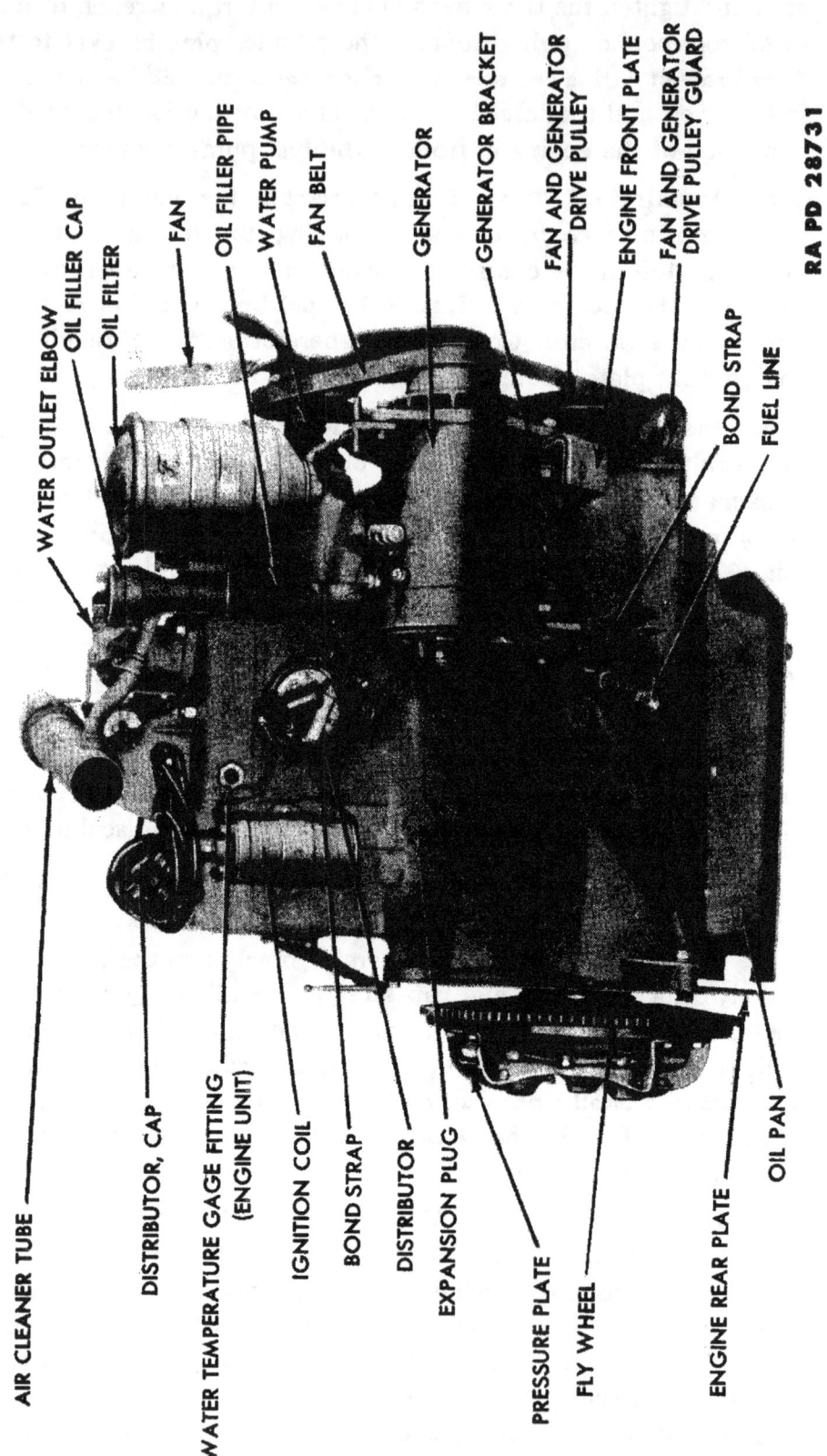

Figure 43 — Right Side View of Engine

TM 9-1803A
19—20

ENGINE

j. Install Accelerator Linkage. Install the accelerator (throttle linkage) on the engine with two lock washers and nuts. Connect the accelerator linkage to the carburetor throttle lever with a cotter pin. Connect the accelerator return spring (fig. 41) to the accelerator return spring clip.

k. Install Engine Front Supports. Install an engine front support on each side of the engine front plate.

Section VI

INSTALLATION OF ENGINE

	Paragraph
Installation	20
Adjustments and tests in vehicle	21

20. INSTALLATION.

a. General. If the clutch housing was not removed with the engine, start the installation procedure, beginning with subparagraph b below. If the clutch housing was removed with the engine, assemble the clutch housing onto the engine, and install the clutch housing bolts; then proceed with the installation as outlined, starting with subparagraph b below. Some variation exists in the location of the various bond straps used to eliminate radio interference on these vehicles. Disregard references to bond straps in the following instructions, if they are not present on the particular vehicle being worked on. If bond straps are found in locations other than those mentioned in the following instructions, they must be connected when installing the engine.

b. Place Engine in Vehicle. Install a suitable engine sling or a rope on the engine (fig. 7). Lift the engine into the vehicle with a hoist, and lower the engine until the clutch disk in the engine is in line with the main drive gear shaft in the transmission. Place the gearshift lever in low speed position, and roll the vehicle forward and backward, at the same time pushing in on the engine until the splines on the main drive gear shaft are engaged with the splines in the clutch disk. Push the engine back onto the clutch housing, and install the clutch housing bolts. Lower the engine until the two holddown bolts and bond strap can be installed loosely in the engine front support insulator. Lower the engine the rest of the way, and remove the engine sling or rope. Slide the stay cable through the bracket on the engine rear plate (fig. 4), and through the front crossmember. Install a nut on the stay cable, tighten it until all slack is removed, and install and tighten the stay cable lock nut. Tighten the two hold-down bolts in each engine front support insulator. Tighten the nut on each engine front support insulator.

c. Install Clutch Control Lever and Cable. Install the four lock washers and cap screws that hold the clutch housing to the transmission. Working through the inspection opening on top of the clutch housing, install the clutch control lever on the clutch release bearing carrier, and on the ball joint located on the main drive gear bearing retainer (fig. 44). Slide the clutch control lever cable through the hole in the clutch housing, and connect the clutch control lever cable yoke end to the clutch control lever and tube assembly with a clevis pin and cotter pin. Press the ball and socket joint end of the clutch control lever inward, and slide the clutch control lever cable in place on the clutch control lever (fig. 44).

d. Install Cranking Motor. Hold the cranking motor in place on the clutch housing, and install and tighten the two cranking motor cap screws on the clutch housing. Hold the cranking motor support bracket in place on the engine and install a lock washer, flat washer, generator bond strap, and cap screw; do not tighten the cap screw. Install and tighten a lock washer, flat washer, and cap screw in the cranking motor and cranking motor support bracket. Tighten the cap screw in the cranking motor support bracket. Connect the cranking motor cable to the cranking motor.

e. Connect Oil Pressure and Water Temperature Gages. Connect the water temperature gage (engine unit) at the right-hand side of the cylinder head (fig. 5). Connect the oil pressure gage line at the flexible oil line on the left-hand side of the engine.

f. Connect Electrical Wires and Bond Straps. Connect the field wire on the small post of the generator, and the armature wire and condenser on the large post. Connect the ground wire at the rear of the generator at the fillister-head cap screw. Connect the bond strap at the rear of the cylinder head. Connect the primary wire to the ignition coil.

g. Connect Choke and Throttle Controls. Slide the choke control cable and conduit through the choke lever carburetor bracket assembly on the carburetor, and the choke control cable through the collar on the choke lever. Push in the choke control button on the instrument panel. Pull the choke lever forward as far as possible, and tighten the set screw in the collar. Connect the throttle control cable and conduit to the choke lever carburetor bracket assembly with the carburetor air cleaner clamp, nut, and bolt. Run the throttle control cable and conduit to the left of the carburetor choker link between the link and the carburetor. Run the throttle control cable

TM 9-1803A
20

ENGINE

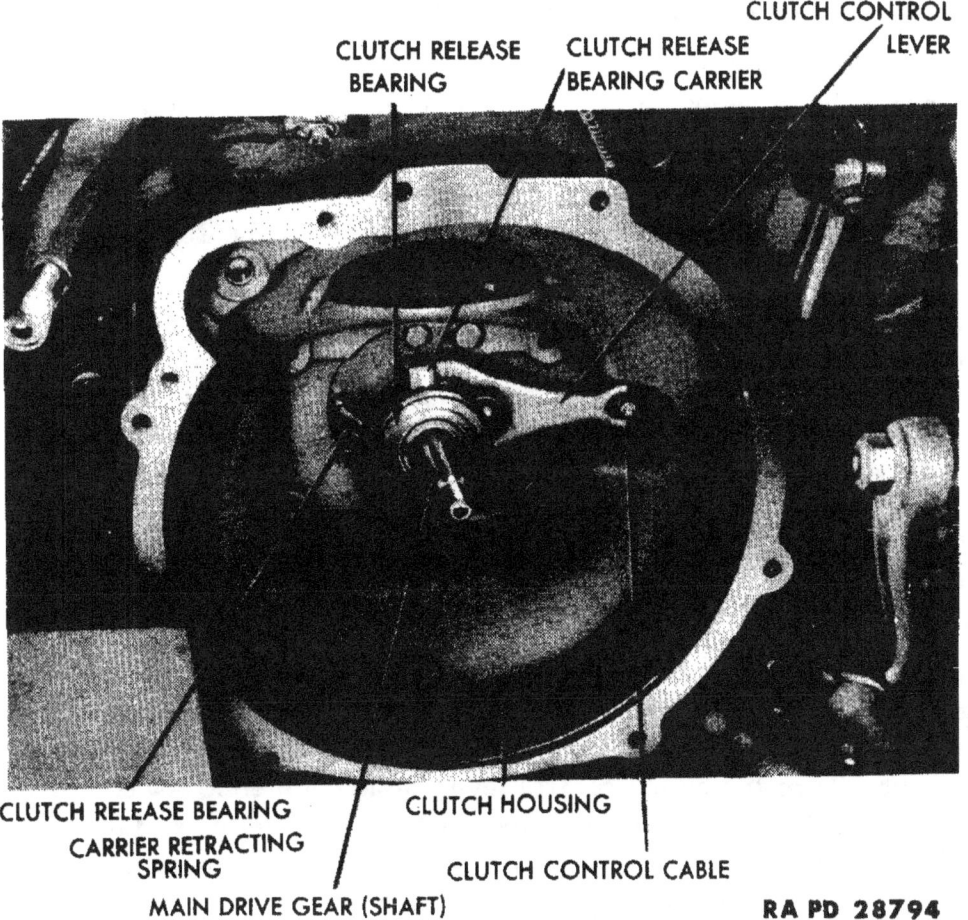

Figure 44 — Clutch Housing Installed in Vehicle

through the carburetor throttle shaft arm and screw assembly. Push the throttle control cable on the instrument panel all the way in. Tighten the screw on the carburetor throttle shaft arm.

h. Install Air Cleaner Hose. Slide the crankcase ventilation flexible hose on the oil filler pipe, and tighten the hose clamp. Slide the air cleaner flexible hose onto the metal tube, and tighten the clamp.

i. Connect Exhaust Pipe. Place a new exhaust gasket on the exhaust manifold, and attach the pipe to the manifold with a cap screw, lock washer, and bolt.

j. Install Radiator. Install the two carriage head bolts in the bottom of the radiator. Place a radiator pad on each of the two radiator mounting brackets. Lower the radiator into place in the vehicle. Install and tighten a flat washer and nut on each radiator bolt. Place a radiator bond strap on each radiator bolt, and install a flat washer and nut. Slide the radiator hose in place on the engine and radiator, and tighten the radiator hose clamps. Slide the straight end of the

59

TM 9-1803A
20–21

ORDNANCE MAINTENANCE — ENGINE AND ENGINE ACCESSORIES FOR ¼-TON 4x4 TRUCK (WILLYS-OVERLAND MODEL MB AND FORD MODEL GPW)

radiator brace through the radiator bracket mounted on top of the radiator; press the other end through the hole in the cowl (fig. 5), and install the lock washers and nuts.

k. Install Battery. Set the battery in the battery tray with the negative post toward the front of the vehicle. Install the battery hold-down frame, wing nuts, and battery cables.

l. Final Operations. Make sure the radiator and engine drain cocks are closed, and install the specified coolant. Tighten the oil pan drain plug, and install the specified amount and grade of oil. Start the engine. If the oil pressure does not register immediately on the oil pressure gage, stop the engine. Remove the oil pump relief valve retainer and prime the oil pump. Make adjustments and tests as outlined in paragraph 21.

21. ADJUSTMENTS AND TESTS IN VEHICLE.

a. Adjust Clutch. The free travel of the clutch pedal must be adjusted so that the pedal will have ¾-inch free travel before the clutch starts to disengage. Loosen the clutch control lever cable adjusting yoke lock nut. Turn the clutch control lever cable counterclockwise to decrease the pedal free travel, and clockwise to increase the pedal free travel. When a ¾-inch free pedal travel is obtained, tighten the clutch control lever cable adjusting yoke lock nut.

b. Set Timing With Neon Light.

(1) PRELIMINARY WORK. Loosen the screw in the timing hole cover, and move the cover to one side. Make certain the ignition switch is in the "OFF" position, then turn the engine with a crank until the timing mark "IGN" (fig. 42) on the flywheel is in line with the index mark on the engine rear plate. Mark the line under the timing mark "IGN" with chalk or white paint.

(2) CONNECT TIMING LIGHT (fig. 45). Attach the high tension lead of the timing light to the terminal of No. 1 spark plug. Attach the positive low tension lead to the positive terminal of the battery, and connect the negative low tension lead to the negative battery terminal.

(3) USE THE TIMING LIGHT. Start the engine, and allow it to warm up. Set the engine idle speed at 600 revolutions per minute. Point the timing light at the timing mark opening so that it can flash on the flywheel. If the timing mark "IGN" on the flywheel appears at the index mark on the opening in the engine rear plate, the timing is correct. If the timing mark "IGN" on the flywheel appears lower than the index mark, the timing is too far advanced.

ENGINE

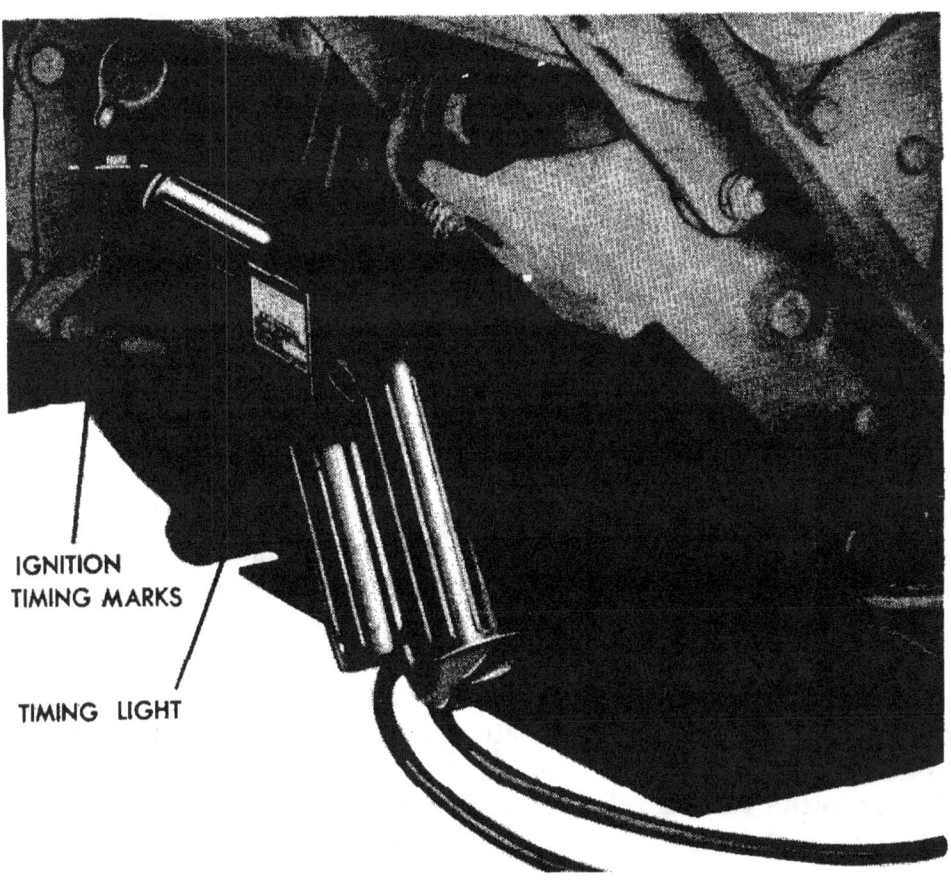

Figure 45 — Timing Engine, Using Timing Light (41-L-1440)

(4) ADJUST DISTRIBUTOR TO CORRECT TIMING. Loosen the bolt in the advance control arm, and turn the distributor clockwise to advance the ignition timing. Turn the distributor counterclockwise to retard the ignition timing. When the correct timing is obtained, tighten the bolt in the advance control arm. If the correct timing cannot be obtained by turning the distributor, set the timing as outlined in paragraph 19 e, but do not remove the distributor.

c. Adjust Carburetor. Start the engine, and allow it to run until it reaches normal operating temperature. Turn the idle fuel adjustment screw (fig. 1) clockwise, or counterclockwise, until all indications of vibration and roll are eliminated from the engine. Set the idle speed adjustment screw so that engine will idle at 400 revolutions per minute.

TM 9-1803A
22

ORDNANCE MAINTENANCE — ENGINE AND ENGINE ACCESSORIES FOR ¼-TON 4x4 TRUCK (WILLYS-OVERLAND MODEL MB AND FORD MODEL GPW)

Section VII

FITS AND TOLERANCES

	Paragraph
Definition of fits	22
Fits and tolerances	23
Torque wrench readings	24

22. DEFINITION OF FITS.

a. General. The table of fits and tolerances (par. 23) gives the original clearance established between various parts at the time of manufacture, as well as wear and limit clearances that indicate to what point the clearance may increase before the parts must be replaced. These clearances all are based on the assumption that the parts involved are both at a temperature of 70° F. The following definitions of the various types of fits are given to assist in arriving at the correct amount of clearance required between parts not included in paragraph 23, as well as to give a better appreciation of the necessity for adhering to the various tolerances. Generally speaking, all bores are made to a standard size (so that standard reamers, plug gages, etc. may be used) with a plus tolerance. The maximum size of male parts is usually a standard size, less the minimum clearance required for the type of fit desired. The minimum size for male parts is the maximum size, minus the tolerance.

b. Ring Fit. A ring fit is the type of fit obtained when the two parts are of identical size. This is the type of fit required between a bore and a plug gage when using the plug gage to determine the inside diameter of the bore. With a ring fit, it is necessary to turn or ring the plug gage or part to force it through the bore. This type of fit does not provide space for film of oil.

c. Slip Fit. A slip fit exists when the male part is slightly smaller than the female part, and involves less clearance than a running fit (subpar. d below). An example of the minimum allowable clearance for a slip fit would be a piston pin that from its own weight would pass slowly through the connecting rod bushing (bushing and pin both in a vertical position). In most cases (except where only a limited movement of the parts is involved) slip fits are specified, when, due to anticipated expansion (subpar. g below) of the female part, enough additional clearance will result to change this type of fit to a running fit (subpar. d below), and provide adequate clearance for a film of oil.

d. Running Fit. A running fit is a fit providing enough clearance

ENGINE

for a continuous film of oil between the two parts. A running fit usually requires 0.001 inch for the oil film plus a minimum of 0.001 inch clearance for each inch of diameter (subpar. g below).

e. Press Fit. A press fit is one that requires force to enter the male part into the bore. Accepted practice for press fits is to have the male part larger by 0.001 inch for each inch of diameter, than the bore into which it is to be pressed.

f. Shrink Fit. Generally speaking, a shrink fit is tighter than a press fit. The amount of the shrink ranges from 0.001 inch to 0.002 inch for each inch of diameter, and in some cases even more. There are two methods of shrinking two parts together, either one of which may be used (both may be used in some instances). One method involves expansion of the female member by heating. The second method involves contracting the male member by chilling with dry ice or liquid air.

g. Effect of Expansion on Fits. Allowances are made in establishing fits on parts that are exposed to high temperatures in order to provide for the anticipated expansion of the part during operation, and still provide adequate clearance for the type of fit required. Absolute minimum allowances for expansion of parts exposed to flame or exhaust gases (pistons, piston rings, and valves) is 0.001 inch for each inch of diameter or length. In anticipating the expansion of a valve stem or piston ring to make allowances for the additional gap required between the end of the valve and the push rod, or between the ends of the piston ring, 0.001 inch for each linear inch of the part is added.

23. FITS AND TOLERANCES.

CYLINDER BLOCK

Fit Location Name	Manufacturers Fit Tolerance	Fit Wear Limit	Type of Fit
Cylinder bore out-of-round	—	0.005 in.	—
Cylinder bore taper	—	0.010 in.	—
Clearance between camshaft and (front) bearing	0.002 in. to 0.0035 in.	0.006 in.	Running
Clearance between camshaft and intermediate and rear bearings	0.002 in. to 0.0035 in.	0.008 in.	Running
Valve guide and cylinder block	—	—	Press
Clearance between valve stem and valve guide (exhaust)	0.002 in. to 0.004 in.	0.005 in.	—
(intake)	0.0015 in. to 0.0035 in.	0.0045 in	—
Clearance between valve tappet and valve tappet bore	0.0005 in. to 0.002 in.	0.003 in.	—

TM 9-1803A
23-24

ORDNANCE MAINTENANCE — ENGINE AND ENGINE ACCESSORIES FOR ¼-TON 4x4 TRUCK (WILLYS-OVERLAND MODEL MB AND FORD MODEL GPW)

CONNECTING ROD AND PISTON ASSEMBLY

Fit Location Name	Manufacturers Fit Tolerance	Fit Wear Limit	Type of Fit
Connecting rod side clearance	0.005 in. to 0.009 in.	0.013 in.	—
Connecting rod clearance on crankshaft	0.0005 in. to 0.001 in.	0.001 in.	Running
Piston pin clearance in connecting rod.	Locked in connecting rod	—	—
Piston pin clearance in piston	0.0001 in. to 0.0005 in.	0.0015 in.	Slip
Piston and cylinder at skirt	5-pound to 10-pound pull with 0.003-in. feeler	5-pound to 10-pound pull with 0.010-in. feeler	—
Piston top land and cylinder	0.0205 in. to 0.0225 in.	—	—
Piston ring to groove clearance (all rings)	0.0005 in. to 0.0015 in.	0.003 in.	—
Piston ring gap (all rings)	0.008 in. to 0.013 in.	0.013 in.	—

VALVES AND VALVE SPRINGS

Intake valve stem diameter	0.373 in.	0.368 in.	—
Valve seat angle	45 degree		
Exhaust valve stem diameter	0.3725 in.	0.3685 in.	—
Valve seat angle	45 degree		
Spring tension at 2 7/64 inches (all springs)	50 pounds	—	—
Spring tension at 1¾ inches (all springs)	116 pounds	—	—

OIL PUMP

Clearance between pinion shaft and pinion gear	0.001 in. to 0.003 in.	0.003 in.	Running
Clearance between oil pump housing and oil pump shaft	0.002 in. to 0.004 in.	0.005 in.	Running
Oil pump relief plunger spring tension at 1 1/16 in.	5½ pounds	—	—

CRANKSHAFT

Crankshaft end play	0.004 in. to 0.006 in.	0.006 in.	—
Main bearing clearance (all main bearings)	0.0005 in. to 0.001 in.	0.001 in.	Running

24. TORQUE WRENCH READINGS.

Main bearing nuts	65 to 70 ft-lb
Connecting rod nuts	50 to 55 ft-lb
Flywheel to crankshaft cap screws	36 to 40 ft-lb
Cylinder head nuts	60 to 65 ft-lb
Cylinder head bolts	65 to 75 ft-lb

CHAPTER 3

CLUTCH ASSEMBLY

	Paragraph
Description and data	25
Pressure plate disassembly	26
Cleaning, inspection, and repair	27
Assembly of pressure plate	28

25. DESCRIPTION AND DATA.

a. Description. The clutch is of the single-plate automotive type, composed of two major units (fig. 46); the pressure plate assembly, and the driven plate or disk. The pressure plate is adjusted at the factory, and requires no other adjustments, except where it is necessary to install new clutch pressure springs, clutch fingers, or pressure plate.

b. Data.

Type	Single dry plate
Torque capacity	132 ft-lb
Clutch disk:	
Make	Borg and Beck
Facings	1 woven, 1 molded asbestos
Facing diameter	Inside 5 1/8 in.
	Outside 7 7/8 in.
Facing thickness	0.125 in.
Pressure plate:	
Make	Atwood
Number of springs	3
Spring pressure at 1 9/16 in.	220 to 230 lb

26. PRESSURE PLATE DISASSEMBLY.

a. Remove Clutch Adjusting Screw (fig. 47). Place the pressure plate in a press with a wood block 2 1/2 inches square on top of the clutch fingers (fig. 48). Depress the clutch fingers, and remove the three clutch adjusting screws, lock nuts, and lock washers. Release the pressure on the clutch fingers slowly to prevent the clutch pressure springs from flying out from under the clutch fingers.

TM 9-1803A
26

ORDNANCE MAINTENANCE — ENGINE AND ENGINE ACCESSORIES FOR ¼-TON 4x4 TRUCK (WILLYS-OVERLAND MODEL MB AND FORD MODEL GPW)

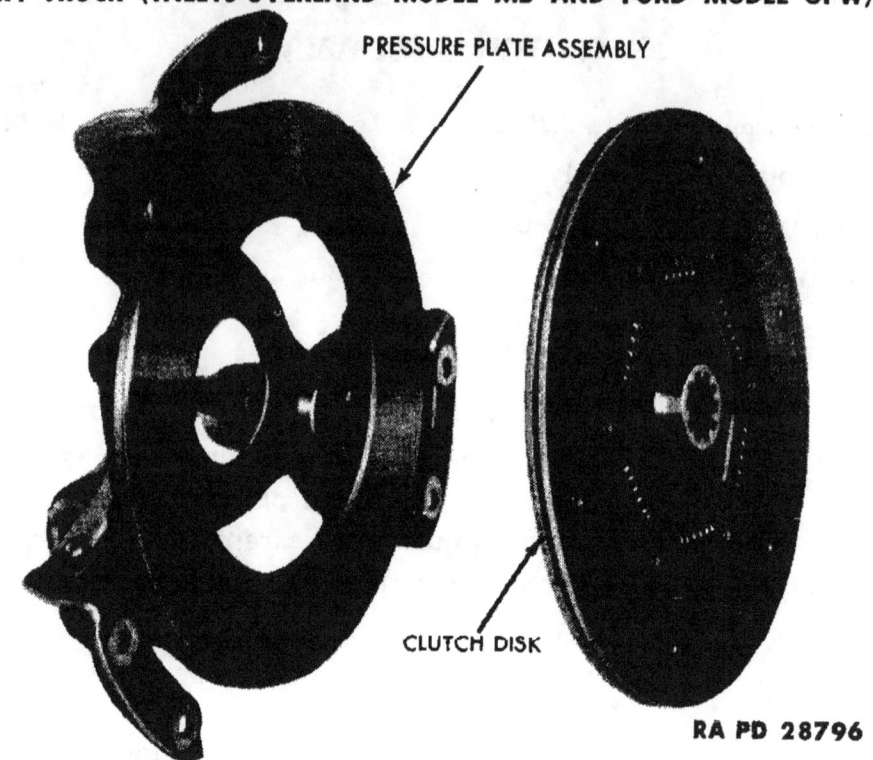

Figure 46 — Clutch Disk and Pressure Plate

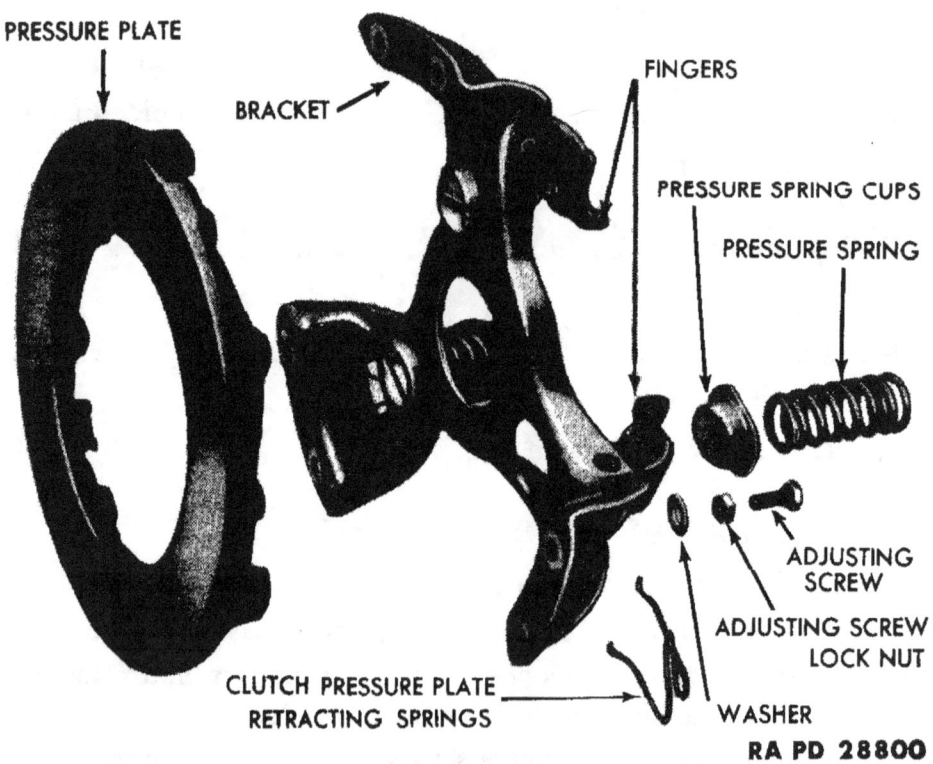

Figure 47 — Pressure Plate Disassembled

CLUTCH ASSEMBLY

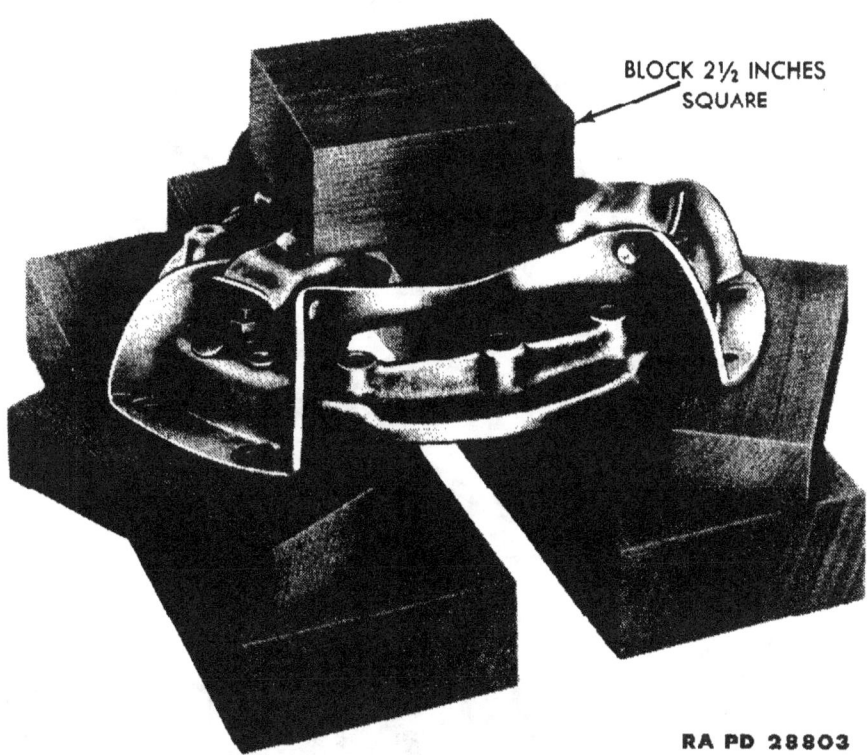

Figure 48 — Pressure Plate Blocked Up in Press for Disassembling or Assembling

b. Remove Clutch Pressure Spring Cups and Springs. Push the clutch pressure spring cups and clutch pressure springs from the pressure plate bracket with a screwdriver or punch (fig. 49). Remove the clutch pressure plate return springs from the clutch pressure plate bracket.

27. CLEANING, INSPECTION, AND REPAIR.

a. Cleaning. Clean all parts thoroughly in dry-cleaning solvent.

b. Inspection and Repair.

(1) CLUTCH PRESSURE PLATE (fig. 47). A ridged, scored, radial cracked, or burned pressure plate must be replaced.

(2) PRESSURE PLATE BRACKET (fig. 47). A distorted pressure plate bracket or a pressure plate bracket with worn clutch fingers must be replaced.

(3) CLUTCH PRESSURE SPRINGS. Place each clutch pressure spring in a tension scale, and depress it to $1\frac{9}{10}$ inches (fig. 50). If

TM 9-1803A
27–28

ORDNANCE MAINTENANCE — ENGINE AND ENGINE ACCESSORIES FOR ¼-TON 4x4 TRUCK (WILLYS-OVERLAND MODEL MB AND FORD MODEL GPW)

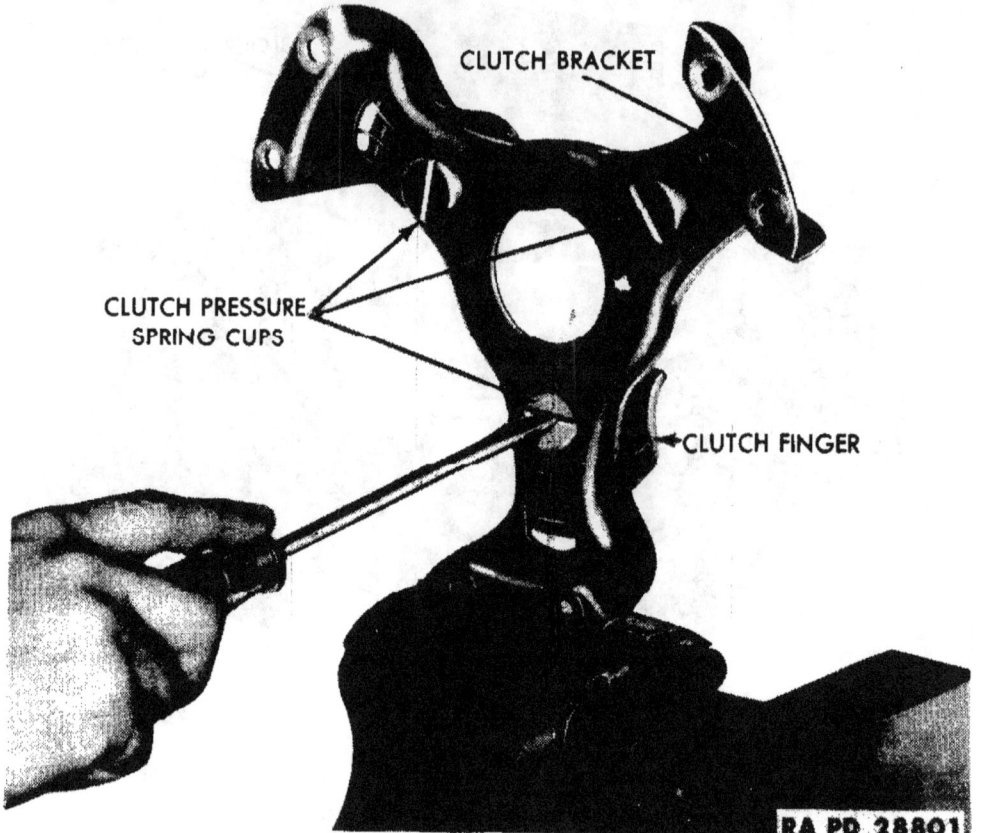

Figure 49 — Removing Clutch Pressure Spring Cups and Clutch Pressure Springs

the spring tension is less than 220 pounds on any clutch pressure spring, it must be replaced.

28. ASSEMBLY OF PRESSURE PLATE.

a. Install Clutch Pressure Spring Cups and Springs. Install the clutch pressure spring cups and clutch pressure springs in the pressure plate bracket (fig. 51), making sure the indentation of each clutch pressure spring cup is toward the center of the pressure plate bracket (fig. 51).

b. Install Pressure Plate on Pressure Plate Bracket. Slide a pressure plate return spring in place under each clutch finger on the pressure plate bracket (fig. 47). Place the pressure plate bracket in a press, blocking it up as shown in figure 48. Place a wood block 2½ inches square on top of the clutch fingers, and depress the fingers (fig. 48). Install the three clutch adjusting screws, lock nuts, and flat washers in the pressure plate.

CLUTCH ASSEMBLY

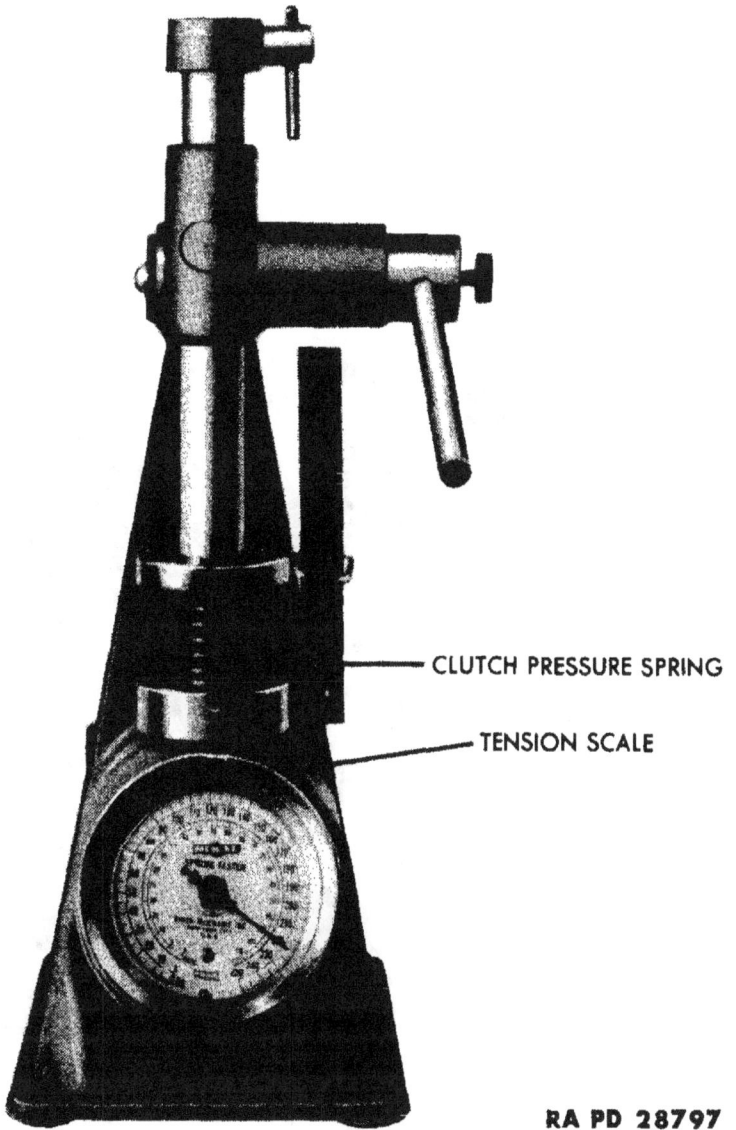

Figure 50 — Testing Spring Tension of Clutch Pressure Spring, Using Tester (41-T-1600)

c. Adjust Pressure Plate. Install the clutch disk and pressure plate onto the flywheel (par. 18 h). Hold a straightedge across the clutch fingers, and measure the distance from the straightedge to the face of the pressure plate bracket (fig. 52). Turn each clutch adjusting screw until a distance of $2\frac{7}{32}$ inches is established between the straightedge and the face of the pressure plate bracket. Hold the clutch adjusting screws with a wrench, and tighten each clutch adjusting screw lock nut. Recheck the distance between the straightedge and the face on the pressure plate bracket.

TM 9-1803A
28

ORDNANCE MAINTENANCE — ENGINE AND ENGINE ACCESSORIES FOR ¼-TON 4x4 TRUCK (WILLYS-OVERLAND MODEL MB AND FORD MODEL GPW)

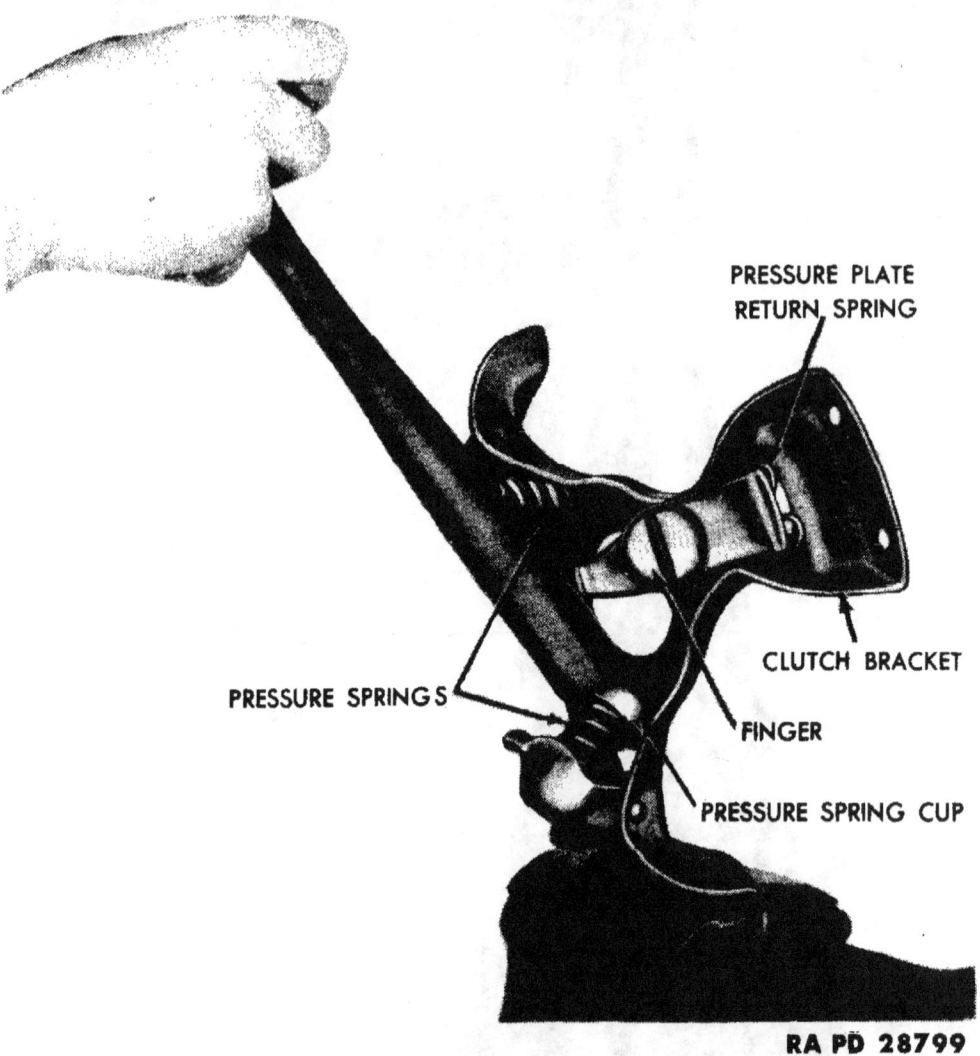

Figure 51 — Installing Clutch Pressure Spring Cups and Pressure Springs in Pressure Plate Bracket

TM 9-1803A
28

CLUTCH ASSEMBLY

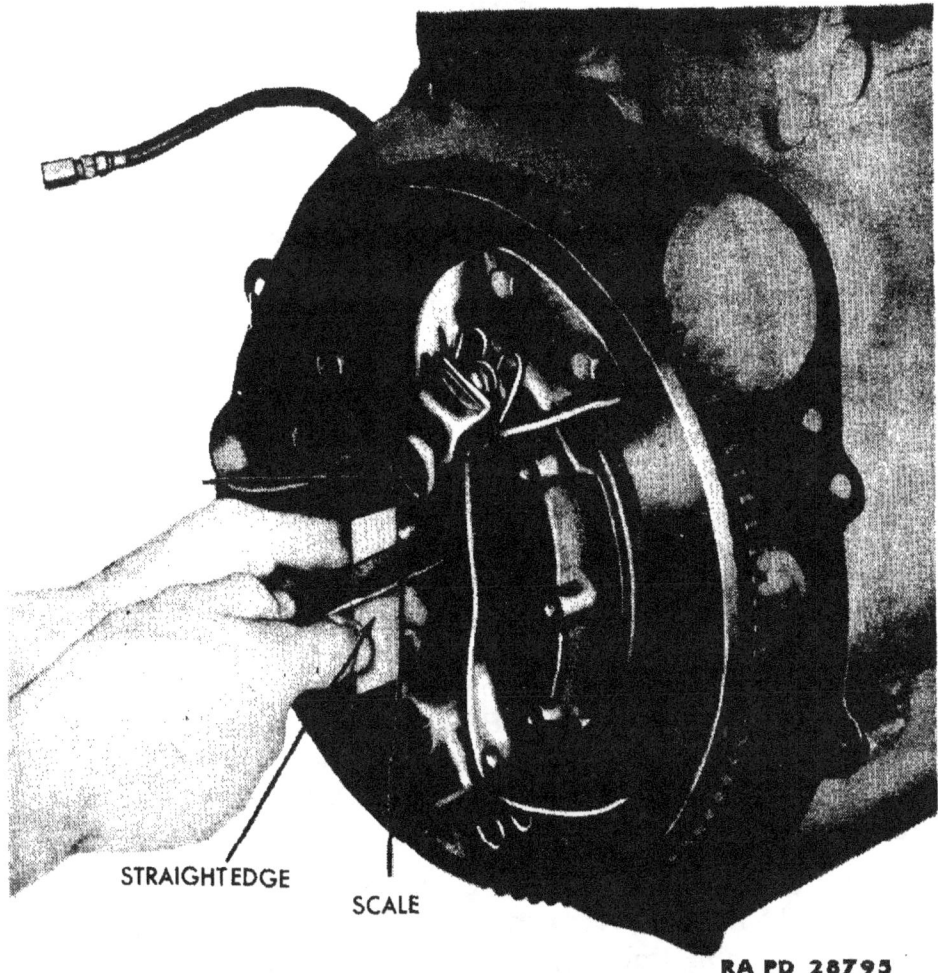

Figure 52 — Adjusting Pressure Plate

TM 9-1803A

ORDNANCE MAINTENANCE — ENGINE AND ENGINE ACCESSORIES FOR ¼-TON 4x4 TRUCK (WILLYS-OVERLAND MODEL MB AND FORD MODEL GPW)

REFERENCES

PUBLICATIONS INDEXES.

The following publications indexes should be consulted frequently for latest changes to or revisions of the publications given in this list of references and for new publications relating to materiel covered in this manual:

Introduction to Ordnance Catalog (explaining SNL system)	ASF Cat. ORD1 IOC
Ordnance Publications for Supply Index (index to SNL's)	ASF Cat. ORD2 OPSI
Index to ordnance publications (listing FM's, TM's, TC's, and TB's of interest to ordnance personnel, MWO's, OPSR, BSD, S of SR's, OSSC's and OFSB's and includes Alphabetical List of Major Items with Publications Pertaining Thereto)	OFSB 1-1
List of Publications for Training (listing MR's, MTP's, T/BA's, T/A's, and FM's, TM's, and TR's concerning training)	FM 21-6
List of Training Films, Film Strips, and Film Bulletins (listing TF's, FS's, and FB's by serial number and subject)	FM-21-7
Military Training Aids (listing Graphic Training Aids, Models, Devices, and Displays)	FM 21-8

STANDARD NOMENCLATURE LISTS.

Cleaning, preserving and lubrication materials, recoil fluids, special oils, and miscellaneous related items	SNL K-1
Soldering, brazing and welding materials, gases and related items	SNL K-2
Tools, maintenance for repair of automotive vehicles	SNL G-27 Volume 1.

TM 9-1803A

REFERENCES

Tool-sets, for ordnance service command automotive shops	SNL N-30
Tool-sets, motor transport	SNL N-19
Truck, ¼-ton, 4x4, command reconnaissance (Ford and Willys)	SNL G-503

EXPLANATORY PUBLICATIONS.

Fundamental Principles.

Automotive electricity	TM 10-580
Automotive lubrication	TM 10-540
Basic Maintenance Manual	TM 38-250
Driver's Manual	TM 10-460
Electrical fundamentals	TM 1-455
Military motor vehicles	AR 850-15
Motor vehicle inspections and preventive maintenance service	TM 9-2810
Precautions in handling gasoline	AR 850-20
Standard Military Motor Vehicles	TM 9-2800
The internal combustion engine	TM 10-570

Maintenance and Repair.

Cleaning, preserving, lubricating and welding materials and similar items issued by the Ordnance Department	TM 9-850
Cold weather lubrication and service of combat vehicles and automotive materiel	OFSB 6-11
Maintenance and care of pneumatic tires and rubber treads	TM 31-200
Ordnance Maintenance: Power train, chassis, and body for ¼-ton 4x4 truck (Ford and Willys)	TM 9-1803B
Ordnance Maintenance: Electrical equipment (Auto-Lite)	TM 9-1825B

73

TM 9-1803A

ORDNANCE MAINTENANCE — ENGINE AND ENGINE ACCESSORIES FOR ¼-TON 4x4 TRUCK (WILLYS-OVERLAND MODEL MB AND FORD MODEL GPW)

Ordnance Maintenance: Hydraulic Brake System (Wagner)	TM 9-1827C
Ordnance Maintenance: Carburetors (Carter)	TM 9-1826A
Ordnance Maintenance: Fuel Pumps	TM 9-1828A
Ordnance Maintenance: Speedometers and Tachometers (Stewart-Warner)	TM 9-1829A
Tune-up and adjustment	TM 10-530

Protection of Materiel.

Camouflage	FM 5-20
Chemical decontamination, materials and equipment	TM 3-220
Decontamination of armored force vehicles	FM 17-59
Defense against chemical attack	FM 21-40
Explosives and demolitions	FM 5-25

Storage and Shipment.

Ordnance storage and shipment chart, group G— Major items	OSSC-G
Registration of motor vehicles	AR 850-10
Rules governing the loading of mechanized and motorized army equipment, also major caliber guns, for the United States Army and Navy, on open top equipment published by Operations and Maintenance Department of Association of American Railroads.	
Storage of motor vehicle equipment	AR 850-18

74

TM 9-1803A

INDEX

A

	Pages
Accelerator, installation of linkage..	57
Accessories, installation	54

Adjustment
carburetor	61
clutch	60
clutch pressure plate	69
in vehicle	60
set timing with Neon light	60
valve tappets	44

AGO Form No. 478, description and instructions for use 5

Air cleaner
installation
hose	59
tubing	55
removal of hose	12

Assembly
clutch pressure plate	68
engine	43
flywheel assembly	42
intake and exhaust manifolds	43
oil pump and intake float	39
valve tappets	37
water pump	28

B

Battery
installation	60
removal	12

Bearings, camshaft, replacement..... 25

Bond straps
connect with wires	58
disconnect	12

C

Camshaft assembly
cleaning, inspection, and repair....	33
installation of sprocket	49
removal of camshaft and sprocket	19
replacement of bearings	25

Carburetor
adjustment	61
installation	54
removal	14

Choke
	Pages
connect with throttle controls	58
disconnect (controls)	13

Cleaning
camshaft assembly	33
clutch assembly	67
connecting rod and piston assembly	29
crankshaft assembly	40
cylinder block and head	21
flywheel assembly	42
intake and exhaust manifolds	43
oil pump and intake float	37
valve tappets	37
valves and valve springs	36
water pump	27

Clutch
adjustment	60
pressure plate	69
assembly of pressure plate	68
cleaning, inspection, and repair..	67
description, data, and disassembly of pressure plate	65
installation	
control lever and cable	58
disk and pressure plate	49
pressure plate on bracket	68
pressure spring cup and springs	68
removal	
disk	19
housing bolts	14
pressure spring cups and springs	67

Connecting rod
cleaning, disassembly, inspection, and repair	29
fits and tolerances	64
installation	46
removal	21

Crankshaft assembly
cleaning, inspection, repair, and removal of sprocket	40
fit and end play	45
fits and tolerances	64
installation	44
removal	21

TM 9-1803A

ORDNANCE MAINTENANCE — ENGINE AND ENGINE ACCESSORIES FOR ¼-TON 4x4 TRUCK (WILLYS-OVERLAND MODEL MB AND FORD MODEL GPW)

C — Cont'd

Cylinder
- cleaning (block and head) 21
- inspection and repair (block and head) 23
- installation (head) 51
- fits and tolerances (block) 63
- removal (head) 19

D

Description and data
- clutch assembly 65
- engine 8

Disassembly
- clutch and pressure plate 65
- connecting rod and piston assembly 29
- flywheel assembly 42
- intake and exhaust manifolds 43
- oil pump and intake float 37
- valve tappets 37
- water pump 25

Distributor
- adjust to correct timing 61
- installation 55
- removal 14

E

Engine
- assembly 43
- description, data, and removal from vehicle 8
- disassembly of stripped engine 17
- installation
  - front cover 50
  - front supports 57
- removal from vehicle 14

Exhaust pipe
- connect 59
- disconnect 13
- (See also Manifolds, intake and exhaust)

F

Fan
- installation 55
- removal 17

Filter, oil
- installation 54
- removal 14

Fits and tolerances 62, 63, 64

Flywheel assembly
- assembly, cleaning, disassembly, inspection, and repair 42
- installation 48
- removal 19

G

Gages
- connect oil pressure and water temperature gages 58
- disconnect oil and water temperature gages 12

Generator
- installation 55
- removal 17

I

Ignition coil
- installation 55
- removal 17

Inspection and repair
- camshaft assembly 33
- clutch assembly 67
- connecting rod and piston assembly 29
- crankshaft assembly 40
- cylinder block and head 23
- flywheel assembly 42
- intake and exhaust manifolds 43
- oil pan 21
- oil pump and intake float 37
- valve and valve springs 36
- valve tappets 37
- water pump 27

M

Major unit assembly replacement 5

Manifolds, intake and exhaust
- assembly, cleaning, disassembly, inspection, and repair 43
- installation 51
- removal 17
- (See also Exhaust pipe)

Motor, cranking
- installation 58
- removal 12

MWO (Modifications Work Order), description and instructions for use 5

INDEX

O
	Pages
Oil pan	
inspection and repair	21
installation	51
removal	19
Oil intake float, removal	19

P
Piston assembly
- assemble piston, piston rod, and connecting rod ... 32
- cleaning, disassembly, inspection, and repair ... 29
- fit and install piston rings ... 32
- fit piston ... 30
- fits and tolerances ... 64
- installation ... 46
- removal ... 21

Press fit, definition ... 63

Pump, fuel
- installation ... 55
- removal ... 14

Pump, water
- assembly ... 28
- cleaning, inspection, and repair ... 27
- disassembly ... 25
- installation ... 54
- removal ... 17

Pump (and intake float), oil
- assembly ... 39
- cleaning, disassembly, inspection, and repair ... 37
- fits and tolerances (oil pump) ... 64
- installation (oil pump) ... 52

R
Radiator
- installation ... 59
- removal ... 12

Repair (See Inspection and repair)

Reports and records, AGO Form No. 478) ... 5

Ring fit, definition ... 62
Running fit, definition ... 62

S
	Pages
Shrink fit, definition	63
Slip fit, definition	62
Spark plugs, installation	55
Springs, removal	19
Stay cable, removal	14
Studs, replacement	24
Subassemblies, disassembly of engine into	14
Supports, front engine, disconnect	13

T
Thermostat, installation ... 54

Throttle controls
- connect with choke ... 58
- disconnect ... 13

Timing light, set timing with Neon light ... 60

Torque wrench readings ... 64

V
Valves
- adjustment of tappets ... 44
- assembly and disassembly of tappets ... 37
- cleaning, inspection, and repair
  - valve tappets ... 37
  - valves and springs ... 36
- fits and tolerances ... 64
- installation ... 43
- removal ... 19
- replacement (valve guides) ... 25

W
Water outlet elbow
- installation ... 54
- removal ... 19

Wires, electrical
- connect wires and bond straps ... 58
- disconnect ... 12
- installation ... 55

WAR DEPARTMENT TECHNICAL MANUAL

ORDNANCE MAINTENANCE

Power Train, Body, and Frame for 1/4-Ton 4x4 Truck

(Willys-Overland Model MB and Ford Model GPW)

WAR DEPARTMENT

WAR DEPARTMENT
Washington 25, D. C., 8 April 1944

TM 9-1803B, Ordnance Maintenance: Power Train, Body, and Frame for ¼-ton 4 x 4 Truck (Willys-Overland Model MB and Ford Model GPW), is published for the information and guidance of all concerned.

$$\begin{bmatrix} \text{A.G. 300.7 (17 Nov 43)} \\ \text{O.O.M. 461/(TM-9) Rar. Ars. (4-15-44)} \end{bmatrix}$$

By order of the Secretary of War:

G. C. MARSHALL,
Chief of Staff.

Official:
    J. A. ULIO,
        Major General,
            The Adjutant General.

Distribution: R 9 (4); Bn 9 (2); C 9 (5).

(For explanation of symbols, see FM 21-6.)

*TM 9-1803B

CONTENTS

			Paragraphs	Pages
CHAPTER	1.	INTRODUCTION	1–2	4–6
CHAPTER	2.	POWER TRAIN	3–35	7–98
SECTION	I.	Power train description	3	7
	II.	Transmission	4–9	7–24
	III.	Transfer case	10–15	24–38
	IV.	Propeller (drive) shafts and universal joints	16–21	38–45
	V.	Front axle	22–28	45–86
	VI.	Rear axle	29–34	86–97
	VII.	Fits and tolerances	35	98
CHAPTER	3.	BODY AND FRAME	36–47	99–136
SECTION	I.	Springs and shock absorbers	36–38	99–110
	II.	Steering gear and drag link	39–40	111–122
	III.	Body	41–42	122–128
	IV.	Frame	43–46	128–136
	V.	Fits and tolerances	47	136
CHAPTER	4.	SPECIAL TOOLS	48–49	137–139
REFERENCES				140–142
INDEX				143–148

★This Technical Manual supersedes TB 1803-1, dated 8 December 1943. For supersession of Quartermaster Corps 10-series Technical Manuals, see paragraph 1 j.

TM 9-1803B
1

ORDNANCE MAINTENANCE — POWER TRAIN, BODY, AND FRAME FOR ¼-TON 4 x 4 TRUCK (WILLYS-OVERLAND MODEL MB AND FORD MODEL GPW)

CHAPTER 1

INTRODUCTION

1. SCOPE.

a. The instructions contained in this manual are for the information and guidance of personnel charged with the maintenance and repair of the power train, body, and frame of the ¼-ton 4 x 4 truck. These instructions are supplementary to field and technical manuals prepared for the using arms. This manual does not contain information which is intended primarily for the using arms, since such information is available to ordnance maintenance personnel in 100-series TM's or FM's.

b. This manual contains a description of, and procedure for, removal, disassembly, inspection, and repair of the transmission, transfer case, axles, body, and frame.

c. TM 9-803 contains operating instructions and information for the using arms.

d. TM 9-1803A contains instructions for the information and guidance of personnel charged with the maintenance and repair of the 4-cylinder engine used in these vehicles.

e. TM 9-1825B contains information for the maintenance of the Auto-Lite electrical equipment.

f. TM 9-1826A contains information for the maintenance of the Carter carburetor.

g. TM 9-1827C contains information for the maintenance of the Wagner hydraulic brake system.

h. TM 9-1828A contains information for the maintenance of the A. C. fuel pump.

i. TM 9-1829A contains information for the maintenance of the speedometer.

j. This manual includes pertinent ordnance maintenance instructions from the following Quartermaster Corps 10-series Technical Manuals. Together with TM 9-803 and TM 9-1803A, this manual supersedes them:

    (1) TM 10-1103, dated 20 August 1941.

    (2) TM 10-1207, dated 20 August 1941.

    (3) TM 10-1349, dated 3 January 1942.

    (4) TM 10-1513, Changes 1, dated 15 January 1943.

TM 9-1803B
1

INTRODUCTION

RA PD 28742

Figure 1 — ¼-ton Truck 4 x 4 — Three-quarter Front View

ORDNANCE MAINTENANCE — POWER TRAIN, BODY, AND FRAME FOR ¼-TON 4 x 4 TRUCK
(WILLYS-OVERLAND MODEL MB AND FORD MODEL GPW)

2. MWO AND MAJOR UNIT ASSEMBLY REPLACEMENT RECORD.

a. **Description.** Every vehicle is supplied with a copy of AGO Form No. 478 which provides a means of keeping a record of MWO's completed or major unit assemblies replaced. This form includes spaces for the vehicle name and U. S. A. Registration Number, instructions for use, and information pertinent to the work accomplished. It is very important that this form be used as directed and that it remain with the vehicle until the vehicle is removed from service.

b. **Instructions for Use.** Personnel performing modifications or major unit assembly replacements must record clearly on the form, a description of the work completed, and must initial the form in the columns provided. When each modification is completed, record the date, hours and/or mileage, and MWO number. When major unit assemblies, such as engine, transmission, transfer case, are replaced, record the date, hours and/or mileage and nomenclature of the unit assembly. Minor repairs and minor parts and accessory replacements need not be recorded.

c. **Early Modifications.** Upon receipt of a vehicle for modification or repair, by a third or fourth echelon repair facility, maintenance personnel will record the MWO numbers of modifications applied prior to the date of AGO Form No. 478.

CHAPTER 2
POWER TRAIN

Section I
POWER TRAIN DESCRIPTION

3. POWER TRAIN DESCRIPTION.

a. The power from the engine is transmitted to the driving wheels through a transmission and a transfer case, each of which provides a means of selecting the gear reduction. The power from the transfer case is transmitted to the front and rear axles through propeller shafts equipped with universal joints. The transmission is located at the rear of the engine and is secured to the clutch housing (fig. 2). The various gears in the transmission (par. 4) are controlled by a shift lever. The transfer case is mounted directly onto the rear of the transmission. The transmission output shaft extends from the rear of the transmission into splines of the main drive gear in the transfer case. The transfer case is provided with two levers, one to select the transfer case ratio, and the other to engage or disengage the front axle (fig. 5). A hand brake drum is mounted on the rear axle output shaft. Each axle is of the spiral bevel hypoid gear full-floating type, equipped with the conventional differential.

Section II
TRANSMISSION

4. DESCRIPTION AND DATA.

a. **Description.** The transmission (fig. 3) is of the 3-speed type with synchronized second and high speed gears. The transmission and transfer case are mounted on rubber on the frame center crossmember. The gearshift lever is incorporated in the gearshift housing.

b. **Data.**

Make ....Warner
Model T84J
Type .... Synchronous Mesh
Speeds:
    Forward 3
    Reverse 1
Ratios:
    Low .... 2.665 to 1
    Second .... 1.564 to 1

Figure 2 — Power Train

POWER TRAIN

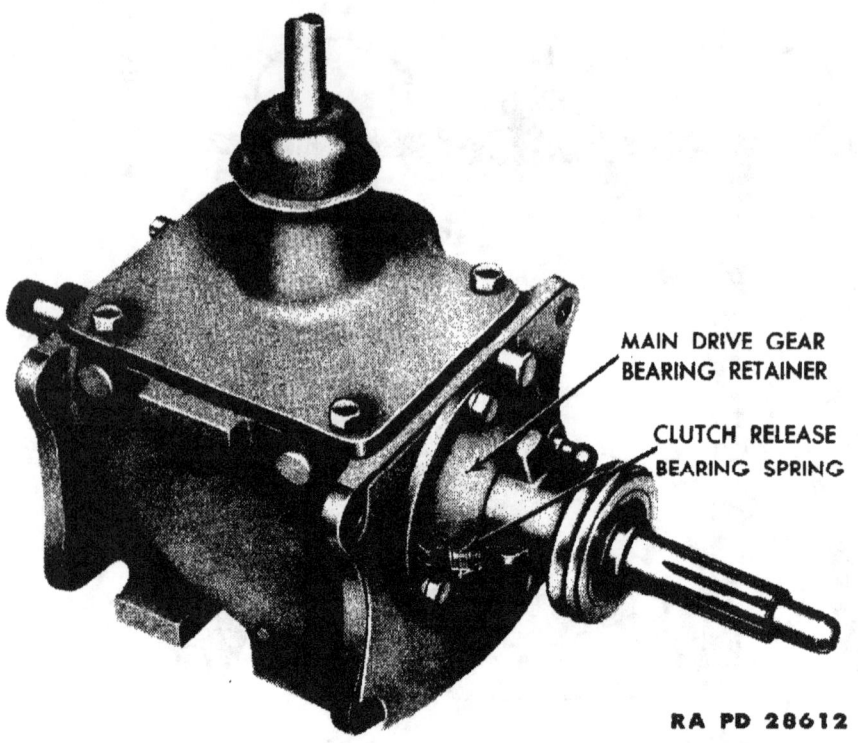

Figure 3 — Transmission — Three-quarter Front View

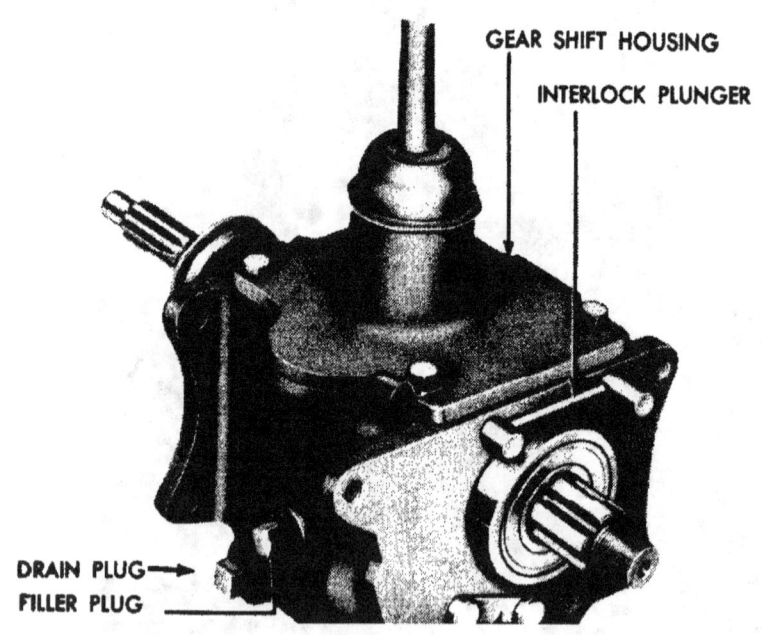

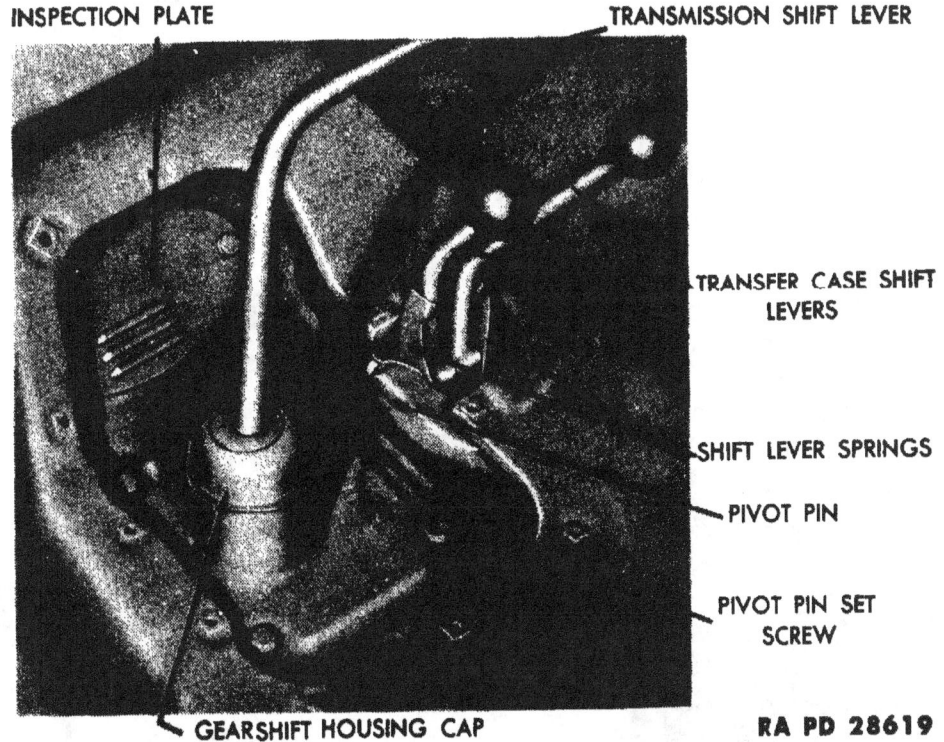

Figure 5 — Transmission and Transfer Case Shift Levers

High	1 to 1
Reverse	3.554 to 1

Bearings:
Clutch shaft (flywheel)	Bushing
Clutch release	Ball
Clutch shaft rear (main drive gear)	Ball
Mainshaft front	13 rollers
Mainshaft rear	Ball
Countershaft gear	Bushings (2)
Reverse idle gear	Bushing

5. REMOVAL.

a. **Remove Floor Plate and Shift Lever** (fig. 5). Remove the cap screws from the floor plate at the transmission, and remove the floor plate. Remove the gearshift housing cap and remove the shift lever from the transmission. Remove the set screw that secures the shift lever pivot pin on the transfer case and, with a suitable drift, remove the shift lever pivot pin. Remove the two shift levers and shift lever springs from the transfer case. Remove the two cap screws that secure the clutch housing inspection plate and remove the inspection plate.

TM 9-1803B
5

POWER TRAIN

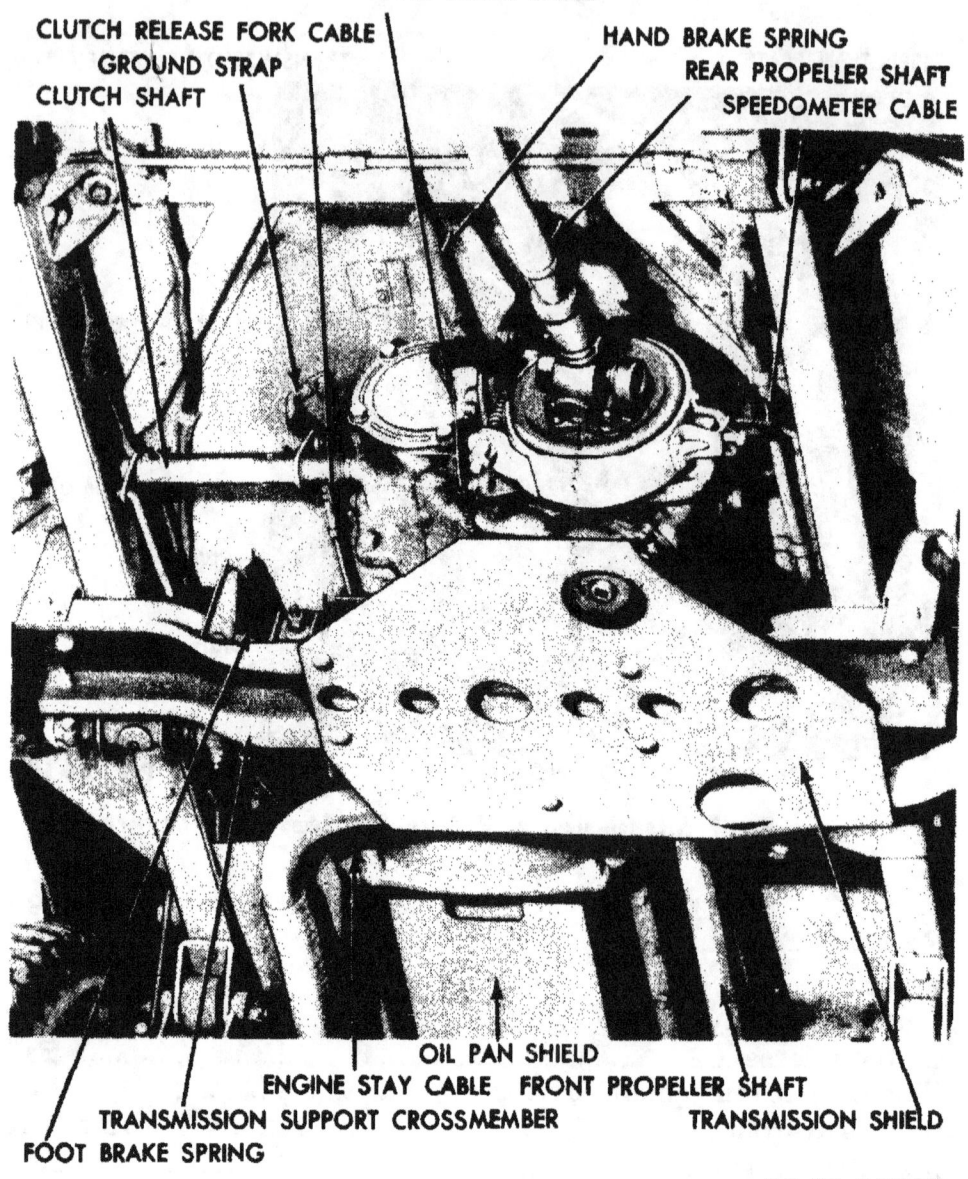

Figure 6 — Under Side of Chassis

b. **Remove Transmission Shield** (fig. 6). Remove the cap screws that secure the exhaust pipe clamp to the shield, and remove the clamp. Remove the five bolts that secure the transmission shield to the transmission support crossmember. Remove the transmission shield.

c. **Remove Brake Springs and Speedometer Cable** (fig. 6). Remove the hand brake spring. Remove the foot brake spring leading from the bottom of the brake pedal to the transmission support crossmember. Disconnect the speedometer cable at the transfer case.

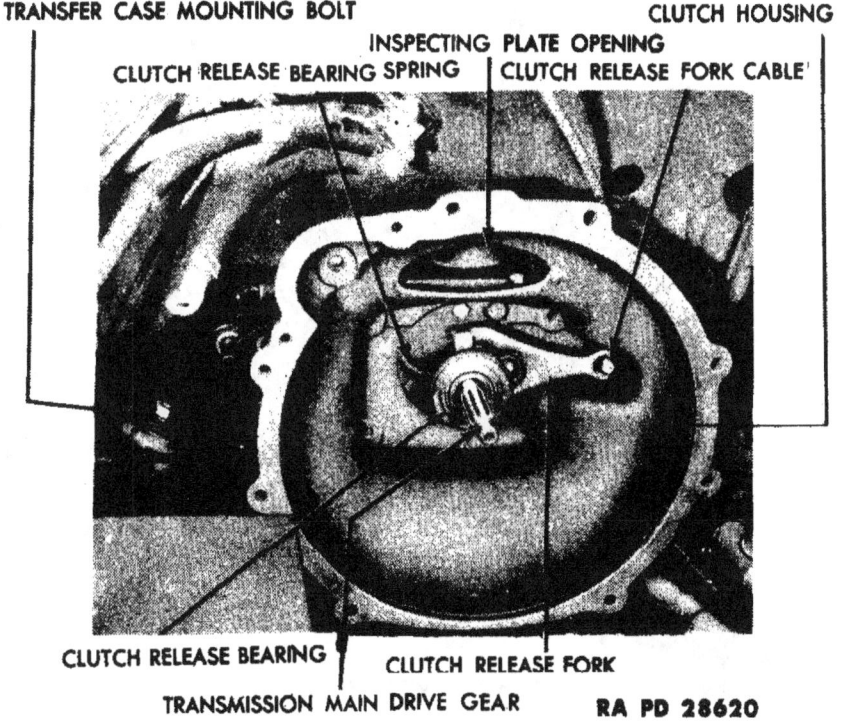

Figure 7 — Clutch Release Fork

d. Remove Hand Brake Cable, Clutch Cable, and Engine Stay Cable (fig. 6). Remove the clevis pin that secures the hand brake cable to the brake band. Remove the hand brake cable clamp at the transfer case. Disconnect the clutch cable at the clutch shaft. Remove the two nuts from the engine stay cable on the transmission support crossmember and remove the engine stay cable.

e. Remove Propeller Shafts (fig. 6). Disconnect the front propeller shaft at the transfer case (par. 17 a). Disconnect the rear propeller shaft at the transfer case (par. 17 b).

f. Remove Ground Strap (fig. 6). Remove the ground strap leading from the transfer case to the floor plate.

g. Remove Clutch Release Fork (fig. 7). Working through the inspection plate opening on the clutch housing, remove the clutch cable from the clutch release fork, and remove the clutch release fork from the clutch housing.

h. Disconnect Radiator Hose. Drain the coolant from the radiator. Loosen the radiator hose clamp at the radiator end, and remove the hose from the radiator.

i. Disconnect Transmission at Clutch Housing (fig. 6). Place a jack under the oil pan shield at the rear of the engine. Remove

POWER TRAIN

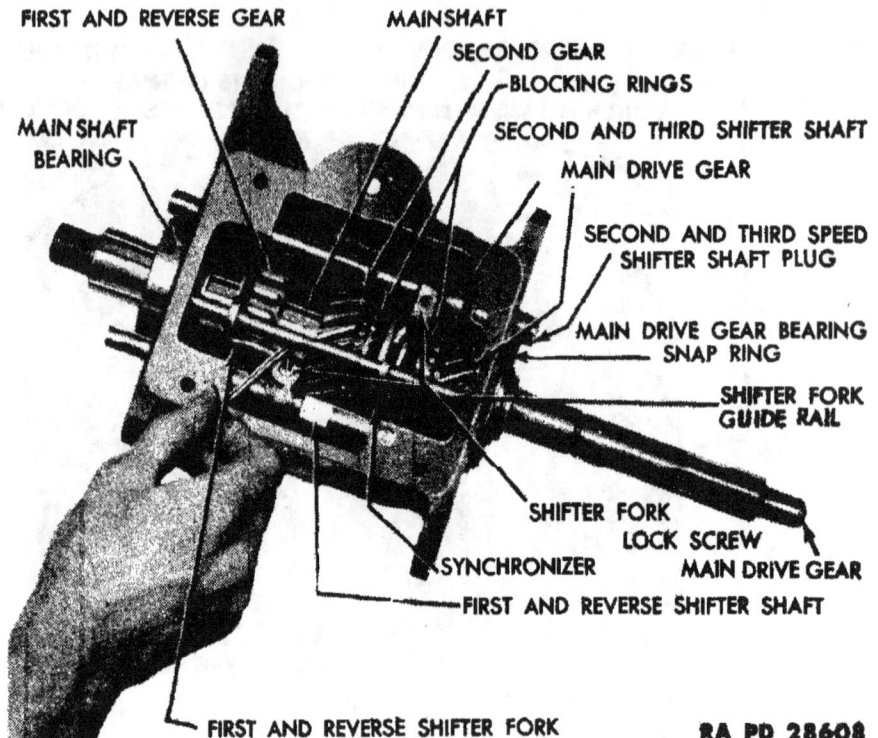

Figure 8 — Removing Shifter Fork Lock Screws

three cap screws from each side of the transmission support crossmember. Place another jack under the transmission. Remove the four bolts that secure the transmission to the clutch housing. Lower both jacks evenly until the transmission support crossmember is approximately 2 inches from the frame. Push the transmission and transfer case to the right so as to free the clutch shaft from the ball joint on the transfer case. Pull the transfer case with transmission straight back until the transmission main drive gear is out of the clutch housing and remove the transfer case and transmission.

j. Remove Transmission Support Crossmember (fig. 6). Remove the five mounting bolts that secure the transmission and transfer case to the transmission support crossmember. Remove the transmission support crossmember.

k. Remove Transmission From Transfer Case (fig. 27). Drain the oil from the transmission and transfer case. Remove the rear cover from the transfer case. Remove the castellated nut and flat washer that secure the drive gear on the transmission mainshaft and remove the drive gear and oil baffle from the transmission mainshaft, using a suitable puller, if necessary. NOTE: *Vehicles of early manufacture were not supplied with this oil baffle.*

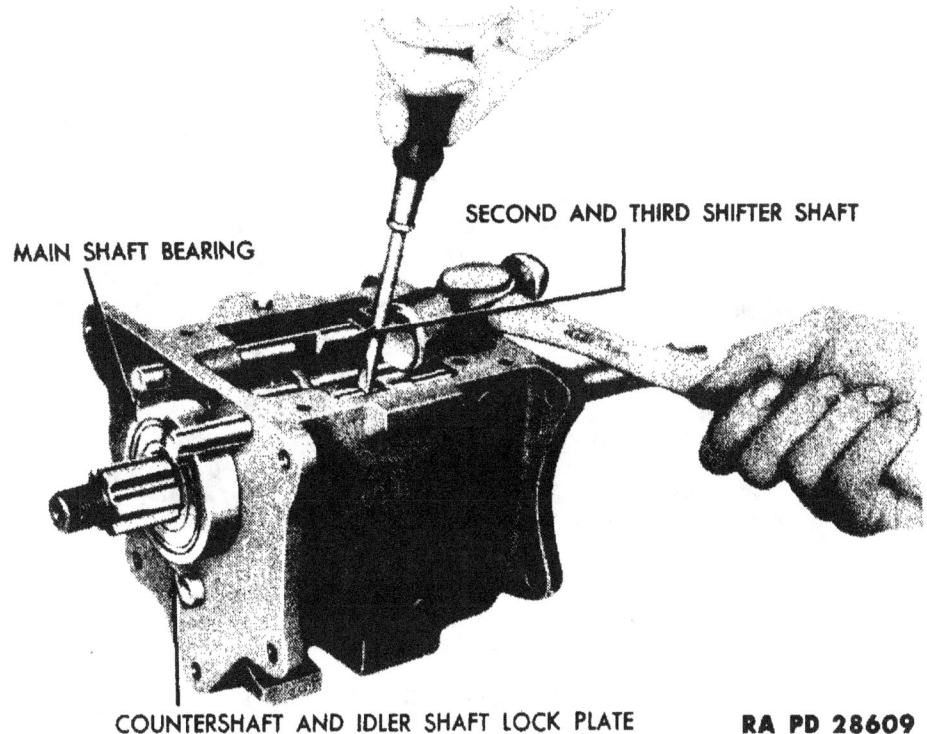

Figure 9 — Removing Shifter Shafts

6. DISASSEMBLY.

a. **Remove Gearshift Housing.** Remove the four cap screws that secure the gearshaft housing to the transmission (fig. 4). Lift the housing, shifter shaft plate, and spring washer from the transmission (fig. 17).

b. **Remove Main Drive Gear Bearing Retainer** (fig. 3). Unhook the clutch release bearing return spring and slide the bearing assembly off the bearing retainer. Remove the three cap screws from the bearing retainer. Slide the bearing retainer and cork gasket off the main drive gear.

c. **Remove Shifter Fork Guide Rail** (fig. 8). Push the shifter fork guide rail out of the transmission.

d. **Remove the Low and Reverse, and the Second and High Shifter Forks.** Remove the shifter fork lock screw from each fork (fig. 8). Tap the shifter shafts part way out of the transmission (fig. 9), being careful not to lose the interlocking ball in each shaft. Hold the shifter fork and pull the shafts from the transmission.

e. **Remove Main Drive Gear.** Tap the countershaft and idle reverse shaft lock plate out of the two shafts (fig. 4). With a long

POWER TRAIN

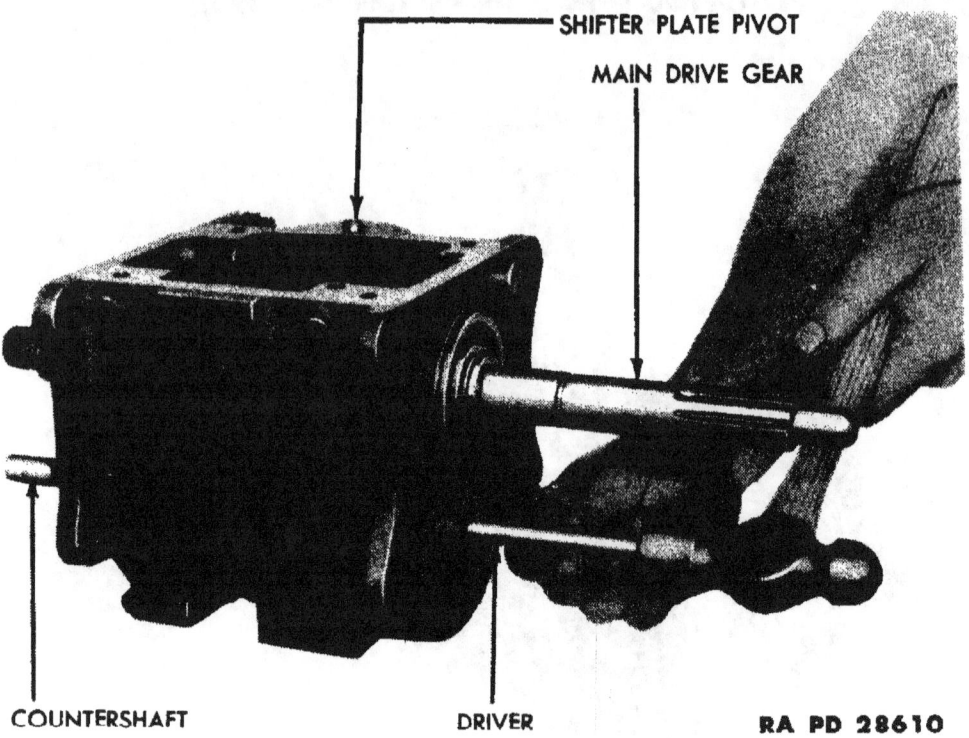

Figure 10 — Removing Countershaft

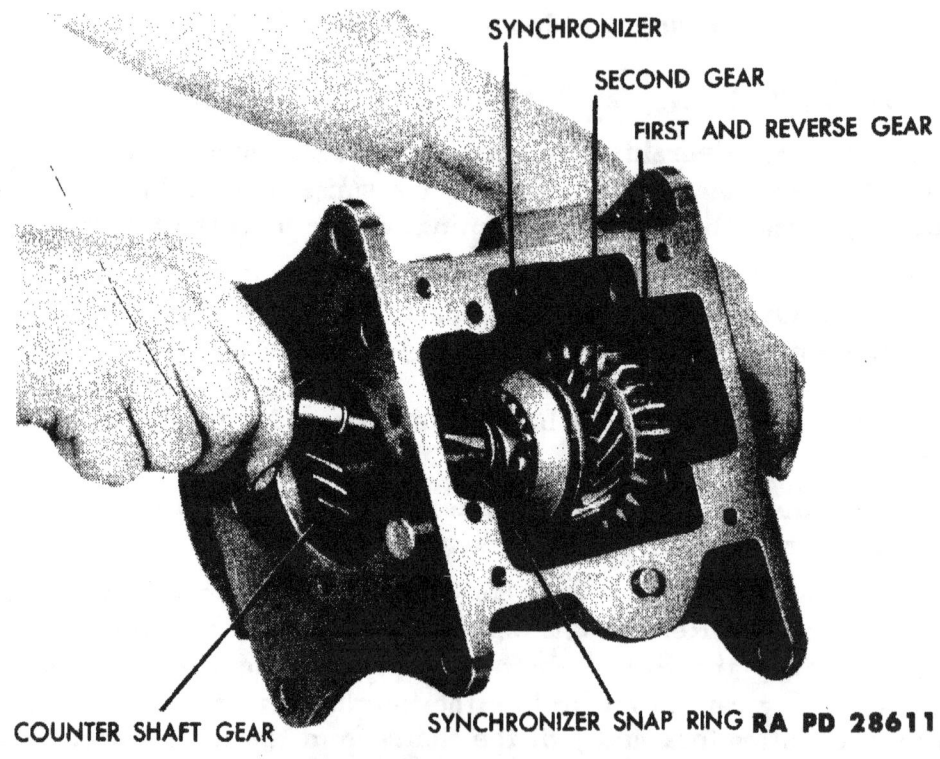

Figure 11 — Removing Synchronizer Hub Snap Ring

ORDNANCE MAINTENANCE — POWER TRAIN, BODY, AND FRAME FOR ¼-TON 4 x 4 TRUCK
(WILLYS-OVERLAND MODEL MB AND FORD MODEL GPW)

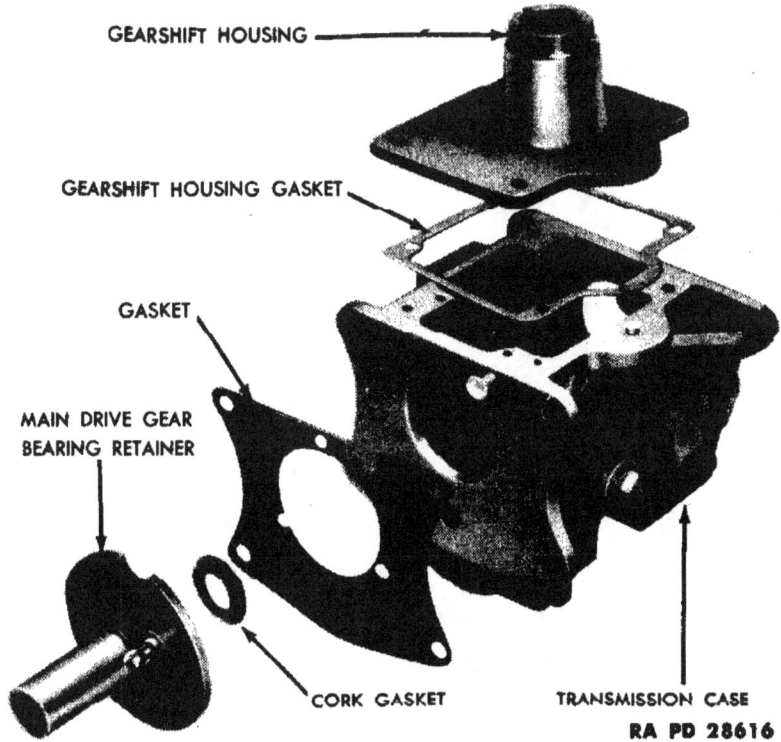

Figure 12 — Transmission Case and Gearshift Housing — Exploded View

drift, tap the countershaft out of the transmission (fig. 10). This will allow the countershaft gear to drop to the bottom of the case for clearance to remove the main drive gear. Pull the main drive gear assembly from the transmission.

    f. **Remove Mainshaft** (fig. 11). Remove the synchronizer hub snap ring. Slide the synchronizer assembly, second and first and reverse gear off the mainshaft. Remove the shaft.

    g. **Remove Idle Reverse Gear.** Tap the idle reverse gear shaft out of the transmission and remove the gear. Lift the countershaft gear and both thrust washers out of the transmission.

    h. **Disassemble Countershaft Gear** (fig. 14). Remove the two bushings and spacer from the countershaft gear.

    i. **Disassemble Main Drive Gear** (fig. 13). Remove the snap ring and the 13 rollers from the main drive gear.

    j. **Disassemble Synchronizer** (fig. 13). Slide the synchronizer sleeve off the synchronizer hub and remove the two lock rings.

7. CLEANING, INSPECTION, AND REPAIR.

    a. **Cleaning.** Wash all parts thoroughly in dry-cleaning solvent until all trace of old lubricant has been removed. Oil the bearings

TM 9-1803B
7

POWER TRAIN

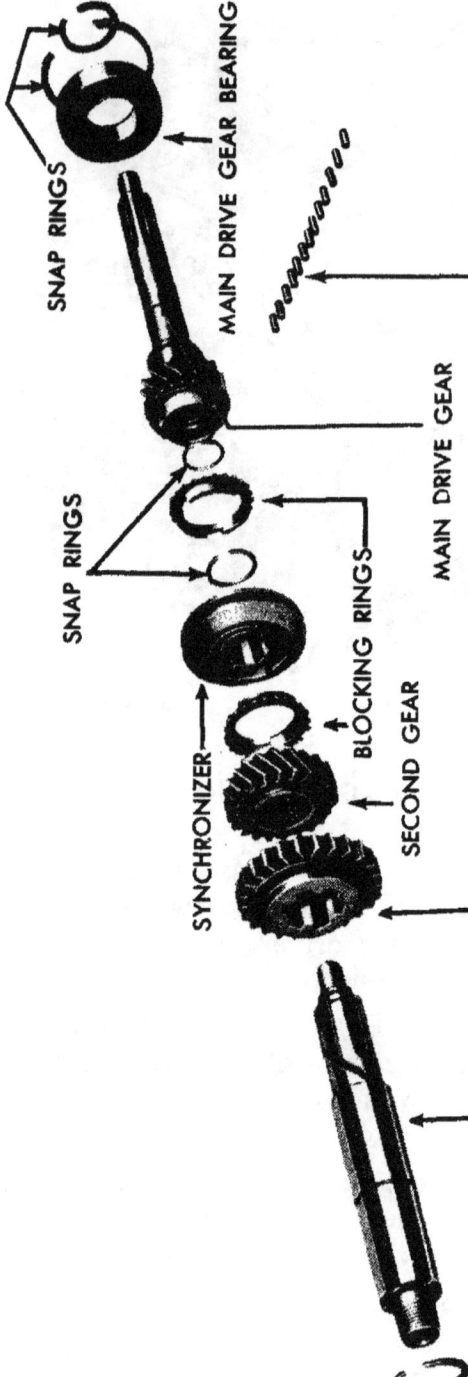

Figure 13 — Mainshaft Assembly — Exploded View

immediately after cleaning to prevent corrosion of the highly polished surfaces.

b. Inspection and Repair.

(1) TRANSMISSION CASE ASSEMBLY (fig. 12). Inspect the case and gearshaft housing for cracks or damage of any kind. Cracked or damaged units must be replaced.

(2) MAIN DRIVE GEAR ASSEMBLY (fig. 13). Replace the main drive gear (clutch shaft) if the following conditions are apparent: Broken teeth or excessive wear; pitted or twisted shaft; discolored bearing surfaces due to overheating. Small nicks can be honed and then polished with a fine stone. Measure the roller bearing recess in the gear end of the shaft. If more than 0.974 inch, replace the main drive gear. Measure the pilot end of the shaft. If it is less than 0.595 inch at the pilot end, replace the main drive gear.

(3) MAINSHAFT (fig. 13). A mainshaft excessively worn, or with pitted or discolored bearing surfaces due to overheating, must be replaced. Measure the diameter of the pilot end of the shaft and the diameter of the second speed gear bearing surface. If they are less than 0.595 inch at the pilot end, or less than 1.126 inches at the second speed gear bearing surface, replace the mainshaft.

(4) FIRST AND REVERSE GEAR (fig. 13). A first and reverse gear with excessively worn teeth or splines, or with broken or chipped teeth must be replaced. Slide the gear onto the mainshaft. If the backlash between the gear and the shaft exceeds 0.005 inch, either the gear or the shaft, or both, must be replaced. A gear with small nicks can be honed and then polished with a fine stone.

(5) SECOND GEAR (fig. 13).

(a) Inspection. A second gear with excessively worn, broken, or chipped teeth, or scored bearing surface must be replaced. Measure the inside diameter of the gear. If more than 1.129 inches the gear bushing must be replaced (step *(b)*, below). Small nicks can be honed and then polished with a fine stone.

(b) Second Gear Bushing Replacement. Place the second gear in an arbor press and, with a suitable driver, press the bushing out of the gear. Use a suitable driver to press a new bushing in the gear. Ream the bushing to from 1.1275 to 1.1280 inches.

(6) COUNTERSHAFT GEAR (fig. 14). Replace excessively worn gears, and gears with broken or chipped teeth, or with pitted or discolored bearing surface due to overheating. Measure the front and rear bearing surfaces of the countershaft gear. If more than 0.7625 inch on either end, replace.

TM 9-1803B
7

POWER TRAIN

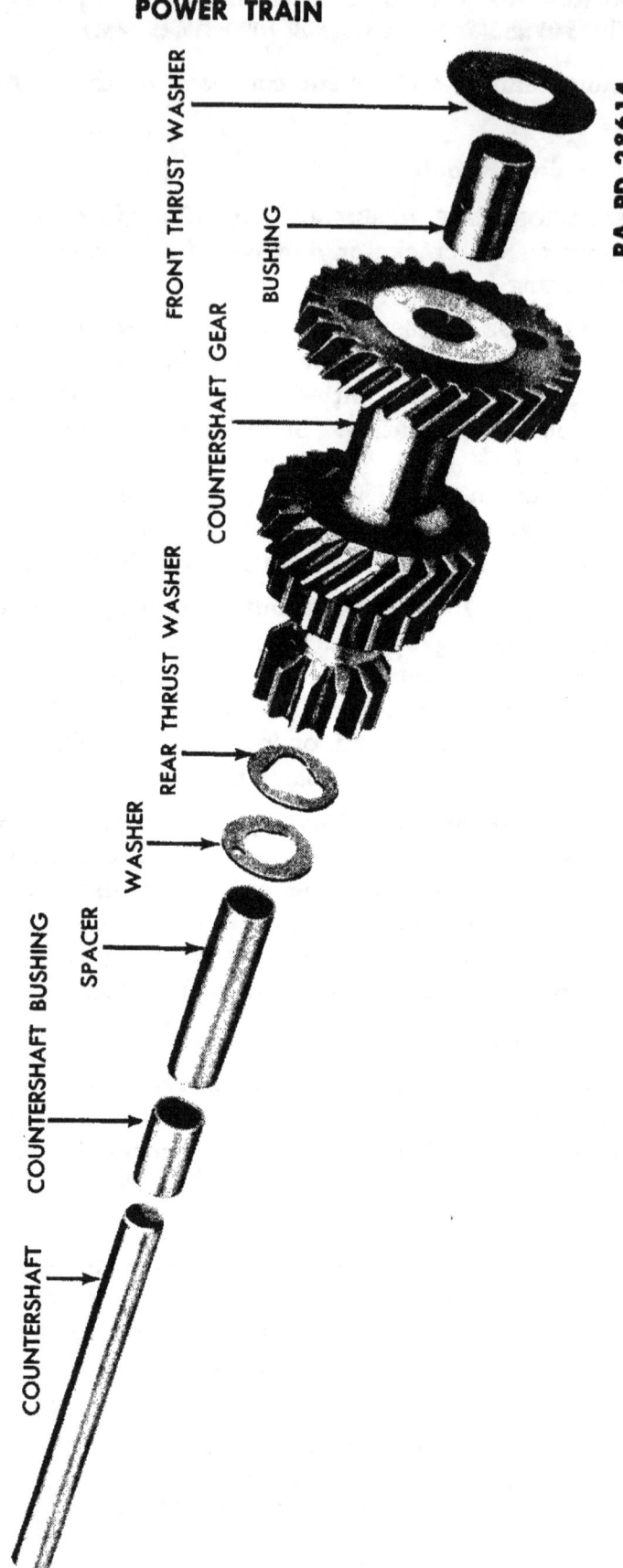

Figure 14 — Countershaft Gear Assembly — Exploded View

TM 9-1803B
7

ORDNANCE MAINTENANCE — POWER TRAIN, BODY, AND FRAME FOR ¼-TON 4 x 4 TRUCK (WILLYS-OVERLAND MODEL MB AND FORD MODEL GPW)

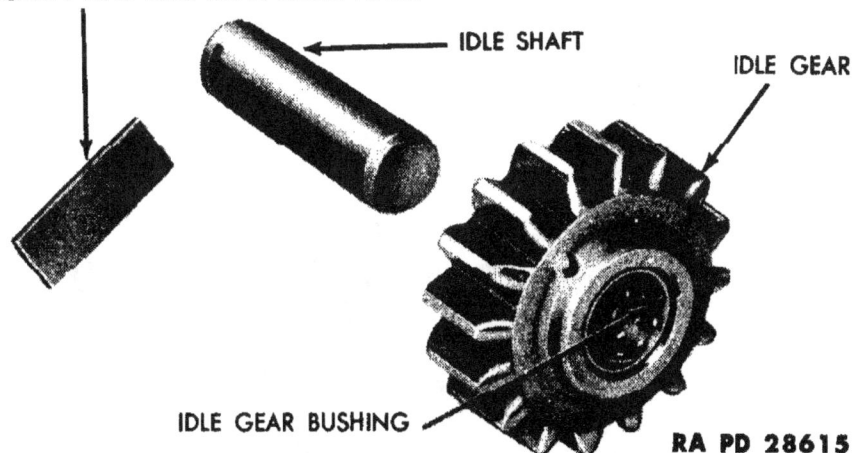

Figure 15 — Idle Gear Assembly — Exploded View

(7) IDLE GEAR (fig. 15).

(a) *Inspection.* A gear with excessively worn or broken teeth, or with a scored bearing surface must be replaced. Small nicks can be honed and then polished with a fine stone. Measure the inside diameter of the idle gear bushing. If more than 0.626 inch, the bushing must be replaced (step (b), below).

(b) *Idle Gear Bushing Replacement.* Place the idle gear in an arbor press and, with a suitable driver, press the bushing out of the gear. Use a suitable driver to press a new bushing in the idle gear. Ream the bushing to from 0.623 to 0.624 inch.

(8) IDLE GEAR SHAFT AND COUNTERSHAFT (figs. 14 and 15). Ridged, scored, or excessively worn, shafts must be replaced. An idle gear shaft measuring under 0.6185 inch or countershaft measuring under 0.7490 inch must be replaced.

(9) SYNCHRONIZER (fig. 13). Blocking rings with worn, broken, or nicked teeth, must be discarded. Hubs with excessively worn splines must be replaced. Sleeves with broken, nicked, or worn teeth, or excessively worn splines, must be replaced.

(10) MAIN DRIVE GEAR BEARING ROLLERS (fig. 13). Needle bearing rollers with flat spots, pitted, or discolored surfaces must be replaced. Measure the diameter of each roller. If less than 0.187 inch, the rollers must be replaced.

(11) BALL BEARINGS (fig. 13). Ball bearings with loose or discolored balls, or with pitted or cracked races must be replaced.

(12) COUNTERSHAFT THRUST WASHERS (fig. 14). Replace excessively worn or ridged thrust washers. Measure each thrust wash-

POWER TRAIN

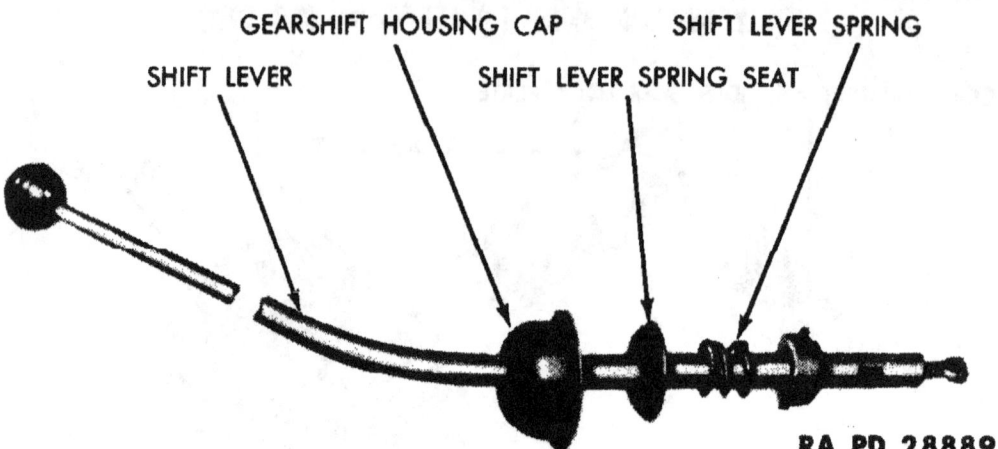

Figure 16 — Transmission Shift Lever

er. If the front washer is less than 0.029 inch, or if either of the rear washers are less than 0.060 inch, they must be replaced.

(13) COUNTERSHAFT BUSHINGS (fig. 14). Excessively worn, scored, or ridged countershaft bushings must be replaced. Measure the inside and outside diameter of the bushings. If the outside diameter is less than 0.759 inch, or if the inside diameter is more than 0.6225 inch, the bushings must be replaced.

(14) SHIFT LEVER (fig. 16). Replace the shift lever if it is excessively worn or bent. Check the gearshift housing cap for stripped threads. Replace the shift lever spring, if it is cracked.

8. ASSEMBLY.

a. Install Idle Gear. Hold the idle gear (fig. 15) in place in the case with the cone end of the hub toward the front, and push the idle gear shaft into the case.

b. Install Countershaft Gear (fig. 14). Dip the countershaft bearings into SAE 90 oil. Slide the spacer into the countershaft gear and install a bushing in each end of the countershaft gear. Coat the front thrust washer, rear thrust washer, and steel washer with a light film of grease to hold them in place while installing the gear. Lay the countershaft gear in the case with the large gear toward the front.

c. Install Mainshaft Assembly (fig. 13). Insert the mainshaft in the case through the opening in the rear of the case. Slide the first and reverse gear onto the shaft, with the shifter fork channel toward the rear. Slide the second gear onto the mainshaft with the tapered end of the gear toward the front. Install a blocking ring onto the second gear. Slide the synchronizer onto the mainshaft with the long end of the hub toward the front and install the snap ring.

d. Install Main Drive Gear Assembly (fig. 13). Place the other blocking ring in the synchronizer and install the main drive gear assembly in the case.

ORDNANCE MAINTENANCE — POWER TRAIN, BODY, AND FRAME FOR ¼-TON 4 x 4 TRUCK
(WILLYS-OVERLAND MODEL MB AND FORD MODEL GPW)

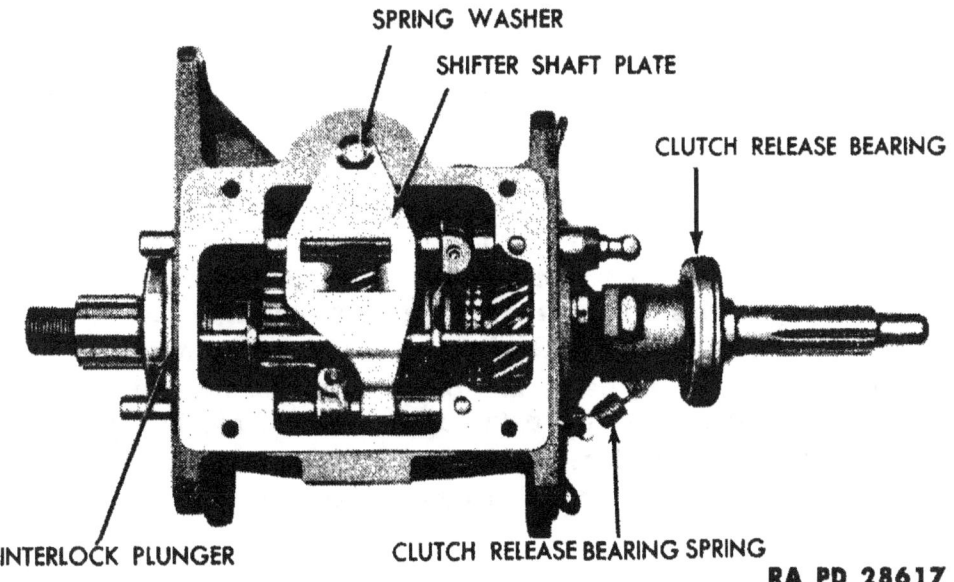

Figure 17 — Gears Installed in Transmission — Top View

e. Install Countershaft. Raise the countershaft gear into position. Making sure the three washers are in line, push the countershaft into the case and tap the lock plate between the countershaft and idle gear shaft (fig. 4).

f. Install First and Reverse Shifter Fork (fig. 8). Hold the first and reverse shifter fork in position on the first and reverse gear, and slide the low and reverse shifter shaft (short shaft) into the case about half way. Drop an interlock spring and ball in the pocket. Press down on the ball and push the shifter shaft all the way in the case. Line up the groove of the shaft with the shifter fork and install the lock screw.

g. Install Second and Third Shifter Fork (fig. 8). Repeat the same procedure as used in installing the low and reverse shifter fork, and then push the guide rail into the case and through both shifter forks.

h. Install Gearshift Housing on Case (fig. 17). Place the transmission in neutral position. Lay the shifter shaft plate on the pivot and on the shifter shafts. Lay the spring washer on the pivot. Place a new gearshift housing gasket on the case. Place the shift lever in neutral position. Lay the housing on the transmission and install the four lock washers and cap screws in the housing.

i. Install Clutch Release Bearing (fig. 3). Slide the clutch release bearing assembly onto the main drive gear bearing retainer and install the clutch release bearing return spring.

TM 9-1803B
9

POWER TRAIN

9. INSTALLATION.

a. Install Transmission to Transfer Case. Place the transmission in position on the transfer case. Be sure the interlock plunger (fig. 4) is in position between the two shifter shafts on the transmission. Install the bolts that secure the transmission to the transfer case. Slide the oil baffle and mainshaft gear on the transmission mainshaft through the rear cover opening on the transfer case. (The oil baffle was not supplied on vehicles of early manufacture. If grease is found to have been leaking from the transfer case into the transmission on vehicles without this baffle, reverse the rear mainshaft bearing (fig. 13) so that the open side of the bearing faces the front of the transmission. Leave the oil baffle in front of the bearing in its original position. Install another oil baffle at the rear of the bearing.) Install the flat washer and nut that secure the mainshaft gear to the transmission mainshaft. Install a new gasket and the rear cover on the transfer case (fig. 27).

b. Place Transmission in Position on Vehicle. Place a jack under the transmission and raise the transmission and transfer case up until the shaft of the main drive gear is lined up with the splines in the clutch disk.

c. Install Transmission Main Drive Gear to Clutch Housing. Insert the shaft of the main drive gear into the clutch splines carefully, do not use force. Slide the transmission in flush with the clutch housing. Install the four bolts that secure the transmission to the clutch housing.

d. Install Clutch Shaft to Transfer Case (fig. 6). Push the transfer case to the right until the clutch shaft has enough clearance to enter the ball joint on the transfer case.

e. Install Transmission Support Crossmember (fig. 6). Place the transmission support crossmember in position on the transmission. Install the four bolts that secure the crossmember to the transmission. Raise the transmission up with a jack until the crossmember is flush with the frame. With a long nosed drift, line up the holes on the crossmember with the holes in the frame. Install the three nuts and bolts on each end of the crossmember and remove the jack. Install the transfer case mounting bolt.

f. Install Clutch Release Fork (fig. 7). Working through the inspection plate opening on the clutch housing, insert the clutch release fork in the clutch housing. Place the release fork behind the clutch release bearing. Slide the clutch release fork cable in the slot on the opposite end of the clutch release fork. Install the clutch release fork cable to the clutch shaft at the transfer case.

TM 9-1803B
9-10

ORDNANCE MAINTENANCE — POWER TRAIN, BODY, AND FRAME FOR ¼-TON 4 x 4 TRUCK (WILLYS-OVERLAND MODEL MB AND FORD MODEL GPW)

g. **Install Hand Brake Cable (fig. 6).** Install the hand brake cable to the brake band at the transfer case. Install the hand brake spring leading from the brake band linkage to the body floor plate. Install the clamp that secures the hand brake cable to the transfer case.

h. **Install Engine Stay Cable and Ground Strap (fig. 6).** Install the engine stay cable leading from the engine rear plate to the transmission support crossmember. Install the ground strap leading from the transmission to the floor plate.

i. **Install Propeller Shafts and Speedometer Cable (fig. 6).** Install the rear propeller shaft to the transfer case (par. 21 a). Install the front propeller shaft to the transfer case (par. 21 b). Install the speedometer cable to the transfer case.

j. **Install Transmission Shield (fig. 6).** Install the five nuts and bolts that secure the shield to the transmission support crossmember. Install the clamp that secures the exhaust pipe to the shield.

k. **Lubricate and Adjust Clutch.** Fill both the transmission and transfer case to proper oil level with specified oil. Adjust the clutch pedal free travel (refer to TM 9-803).

Section III

TRANSFER CASE

10. DESCRIPTION AND DATA.

a. **Description.** The transfer case (figs. 28 and 29) is located at the rear of the transmission. The transfer case is essentially a 2-speed transmission, which provides two gear ratios and a means of distributing the power from the transmission to the two axles.

b. **Data.**

Make	Spicer
Model	18
Mounting	Unit with transmission
Shift lever	Floor
Ratio:	
High	1 to 1
Low	1.97 to 1

POWER TRAIN

Bearings:
- Transmission mainshaft .... Ball
- Idle gear .... 2 rollers
- Output shaft .... Taper rollers
- Front axle clutch shaft front bearing .... Ball
- Rear pilot in output shaft .... Bronze bushing

11. REMOVAL.

a. Remove Transmission Shield (fig. 6). Remove the two cap screws that secure the exhaust pipe clamp to the shield. Remove the exhaust pipe clamp. Remove the five bolts that secure the transmission shield to the transmission support crossmember and remove the shield.

b. Remove Hand Brake Cable and Clutch Cable (fig. 6). Remove the hand brake spring at the transfer case. Remove the clevis pin that secures the hand brake cable at the brake on the transfer case. Remove the hand brake cable clamp on the transmission. Remove the clevis pin from the clutch cable at the transmission support crossmember.

c. Remove Mounting Bolt and Rear Cover (figs. 7 and 27). Remove the mounting bolt that secures the transfer case to the transmission support crossmember at the right side of the transfer case. Remove the five cap screws that secure the rear cover to the transfer case.

d. Remove Rear Propeller Shaft (fig. 7). Disconnect the rear propeller shaft at the transfer case (par. 17 b).

e. Remove Mainshaft Gear (fig. 27). Through the opening at the rear of the transfer case, remove the castellated nut that secures the mainshaft gear to the transmission mainshaft. Remove the flat washer mainshaft gear and oil retainer.

f. Remove Transfer Case. Place a jack under the transfer case. Remove the five cap screws that secure the transfer case to the transmission. Move the transfer case straight back until the transmission mainshaft is out of the transfer case. Remove the transfer case.

12. DISASSEMBLY.

a. Remove Brake Band and Drum Assembly (fig. 28). Remove the two anchor screws from the brake band. Remove the brake band adjusting nut and adjusting screw. Remove the clevis pin from the hand brake linkage. Remove the brake band assembly. Remove the castellated nut that secures the universal joint flange to the output shaft. Install puller 41-P-2912 on the universal joint flange and remove the flange and brake drum (fig. 18). NOTE: *The puller illustrated in figure 18 is similar to puller 41-P-2912.*

TM 9-1803B
12

ORDNANCE MAINTENANCE — POWER TRAIN, BODY, AND FRAME FOR ¼-TON 4 x 4 TRUCK (WILLYS-OVERLAND MODEL MB AND FORD MODEL GPW)

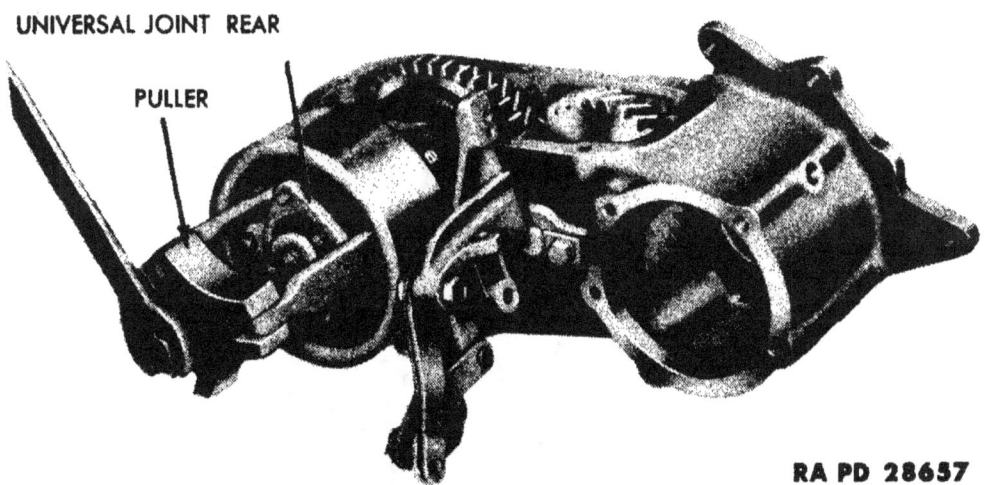

Figure 18 — Removing Rear Universal Joint Flange With Puller Similar to Puller 41-P-2912

b. Remove Rear Output Shaft Bearing Cap (fig. 26). Remove the four cap screws that secure the rear output shaft bearing cap to the transfer case housing. Remove the rear output shaft bearing cap. Remove the rear bearing cap shims. Remove the speedometer drive gear from the output shaft.

c. Remove Intermediate Gear and Bottom Cover (figs. 25 and 27). Remove the 10 cap screws that secure the bottom cover to the transfer case and remove the bottom cover. Remove the cap screw that secures the lock plate. Remove the lock plate. With a suitable driver, remove the intermediate gear shaft. Remove the intermediate gear, thrust washers, and roller bearings through the bottom of the transfer case.

d. Remove Shifter Shaft and Front Output Shaft Bearing (fig. 29). Shift front axle drive to the engaged position. Remove the poppet plug, spring, and ball on both sides of the output shaft bearing cap. Remove the five cap screws that secure the front output shaft bearing cap to the transfer case. Remove the front output shaft bearing cap as an assembly with the universal joint flange, clutch shaft, bearing, clutch gear, shifter fork, and shifter rod. Be careful not to lose the interlock in the front bearing cap.

e. Remove Output Shaft (fig. 19). Insert a screwdriver between the snap ring and output shaft bearing and pry the output shaft bearing away from the snap ring. Remove the snap ring from the groove in the output shaft. Pull the output shaft out from the rear of the housing. The output shaft bearing, snap ring thrust washer, output shaft sliding gear, and output shaft gear can now be removed through the bottom of the transfer case.

POWER TRAIN

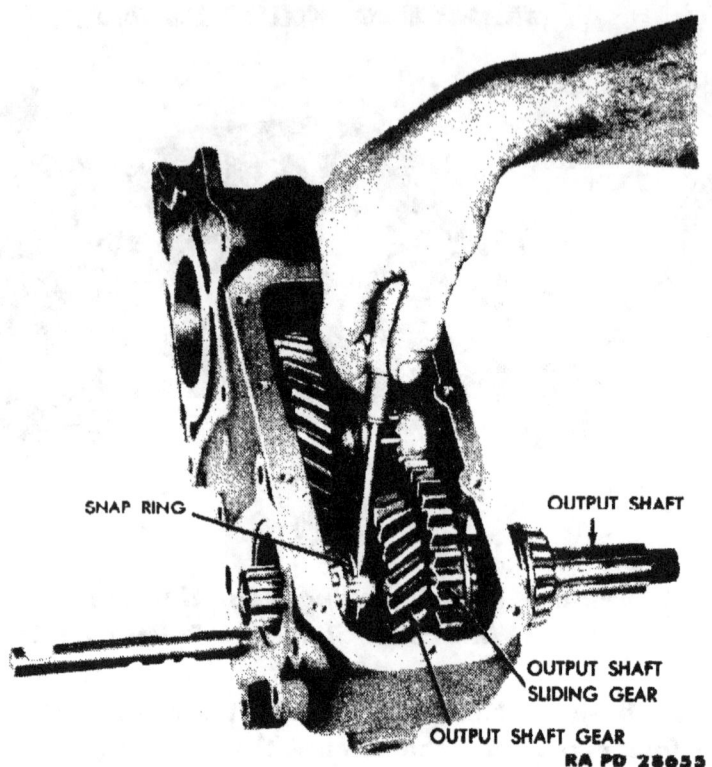

Figure 19 — Removing Snap Ring From Output Shaft

f. Disassemble Front Output Shaft Bearing Cap (fig. 21). Remove the set screw that secures the shifter fork to the front wheel drive shifter shaft. Slide the shifter shaft out of the shifter fork. Remove the shifter fork and clutch gear from the bearing cap. Remove the snap ring that secures the output shaft bearing and remove the output shaft bearing from the bearing cap.

13. CLEANING, INSPECTION, AND REPAIR.

a. Cleaning. Cleaning all parts thoroughly in dry-cleaning solvent. Clean the bearings by rotating them while immersed in dry-cleaning solvent until all trace of lubricant has been removed. Oil the bearings immediately to prevent corrosion of the highly polished surface.

b. Inspection.

(1) TRANSFER CASE ASSEMBLY (fig. 27). Inspect the transfer case housing for cracks or damage of any kind. Inspect the bottom and rear cover for bent or damaged condition. Replace the gaskets on the bottom and rear covers.

(2) FRONT OUTPUT SHAFT BEARING CAP ASSEMBLY (fig. 21).

(a) Front Output Shaft Bearing Cap Housing (fig. 20). Replace the front bearing cap, if it is cracked or damaged. Shifter shaft and output shaft oil seals must be replaced (subpar. c, below).

TM 9-1803B
13

ORDNANCE MAINTENANCE — POWER TRAIN, BODY, AND FRAME FOR ¼-TON 4 x 4 TRUCK (WILLYS-OVERLAND MODEL MB AND FORD MODEL GPW)

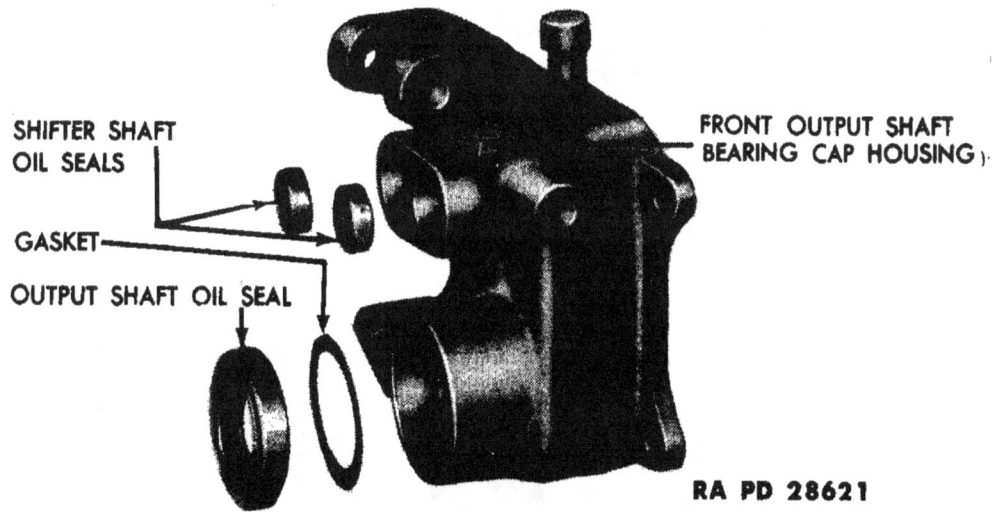

Figure 20 — Front Output Shaft Bearing Cap Housing and Oil Seals

(b) *Front Wheel Drive Shifter Shaft and Fork* (fig. 21). Replace the front wheel drive shifter shaft, if bent or damaged. Replace the fork if it has stripped set screw threads, if it is cracked or has bent forks.

(c) *Clutch Shaft and Gear* (fig. 21). Replace the clutch shaft if the splines or gear teeth are chipped or worn, if the gear has any teeth missing. Check the diameter of the pilot end of the clutch shaft. If the diameter is less than 0.625 inch, replace the clutch shaft. Replace the clutch gear, if it is worn or has any broken teeth.

(d) *Output Shaft Bearing* (fig. 21). Ball bearings with loose or discolored balls or with pitted or cracked races must be replaced.

(3) INTERMEDIATE GEAR ASSEMBLY (fig. 25). Replace the intermediate gear if excessively worn, or if any teeth are damaged. Check the thickness of the thrust washers. If the thrust washers are less than 0.093 inch in thickness, replace them. Check the diameter of the intermediate gear shaft. If the diameter is less than 0.750 inch, replace the intermediate gear shaft. Replace the roller bearing, if the rollers are scored or have flat spots.

(4) REAR OUTPUT SHAFT BEARING CAP ASSEMBLY (fig. 26). Replace the output shaft bearing cap if cracked or damaged. Replace the speedometer drive gear if it is worn or has damaged teeth. Replace the oil seal in the output shaft bearing cap housing (subpar. c, below). Replace the brake drum if it is worn or bent. Replace the universal joint rear flange, if the splines are worn. Replace the dust shield on the flange if bent.

(5) OUTPUT SHAFT ASSEMBLY (fig. 24). Replace the output shaft if the splines are worn. Small nicks can be removed by honing and then polishing with a fine stone. Measure the inside diameter of

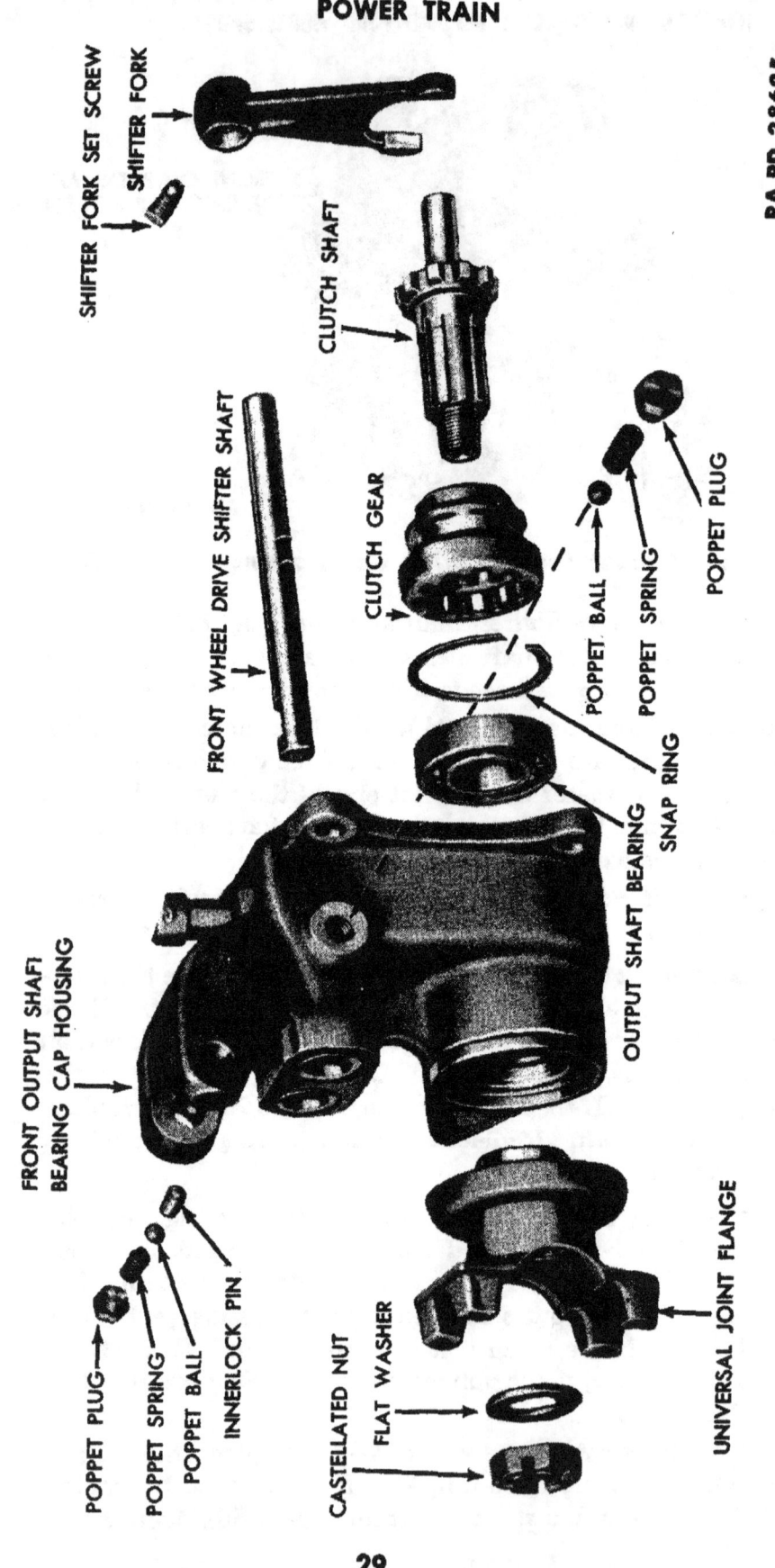

Figure 21 — Front Output Shaft Bearing Cap — Exploded View

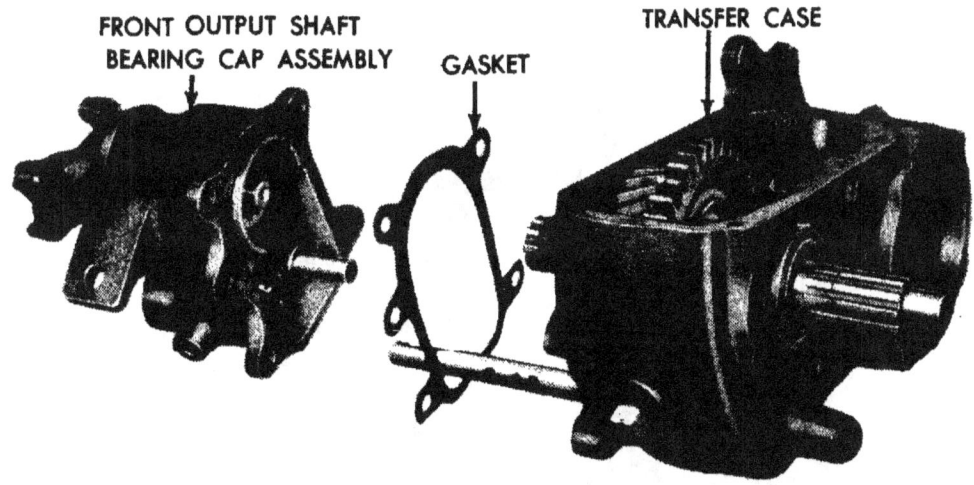

Figure 22 — Installing Front Output Shaft Bearing Cap to Transfer Case

the bushing in the output shaft. If it is greater than 0.627 inch, replace the output shaft. Replace the output shaft gear if it is worn or has any damaged teeth. Replace the sliding gear, if it is worn or has damaged teeth. Measure the thickness of the thrust washer. If the thrust washer thickness is less than 0.103 inch, replace it. Replace the roller bearings if they are scored or have flat spots, or if the races are nicked or cracked.

(6) UNDER DRIVE SHIFTER FORK ASSEMBLY (fig. 24). Check the fork for stripped set screw threads, cracked or bent forks. Replace if in any of these conditions. Replace the under drive shifter shaft if it is bent.

(7) SHIFT LEVER ASSEMBLY (fig. 29). Replace the shift levers if found bent or damaged. Replace the shift lever spring if bent or cracked. Measure the diameter of the shift lever pivot pin. If the diameter is less than 0.500 inch, replace the pivot pin.

c. **Output Shaft Bearing Cap Oil Seal Replacement** (fig. 20). Drive the old oil seal out of the output shaft bearing cap housing, using a suitable driver. Drive the oil seals out, working from the inside of the cap housing. To install a new oil seal, use a driver the size of the oil seal and drive the new seal in the output shaft bearing cap housing.

14. ASSEMBLY.

a. **Assemble the Front Output Shaft Bearing Cap** (fig. 21). Insert the bearing in the output shaft bearing cap. Install the snap ring that secures the bearing in the output shaft bearing cap. Insert the clutch shaft through the bearing from the inside of the output shaft bearing cap. Insert the front wheel drive shifter shaft in the output

POWER TRAIN

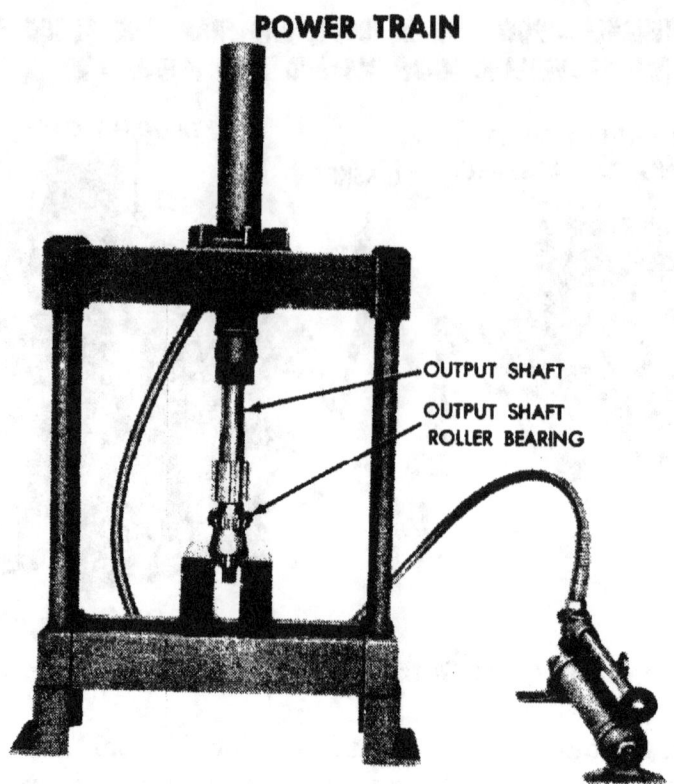

Figure 23 — Pressing Output Shaft Bearing on Output Shaft

shaft bearing cap through the outer side of the output shaft bearing cap. Place the front wheel drive shifter fork in position on the clutch gear. Slide the shifter fork on the shifter shaft and clutch gear on the clutch shaft together. Install the set screw in the shift fork and secure with a lock wire. Install the universal joint flange on the clutch shaft. Install the washer and castellated nut that secure the universal joint flange to the clutch shaft.

b. Install Under Drive Shifter Fork (fig. 20). Place the under drive shifter fork in the transfer case housing. Insert the under drive shifter shaft in the transfer case and shifter fork. Install the shifter fork set screw that secures the fork to the shifter shaft. Secure the set screw with lock wire.

c. Install Output Shaft in Transfer Case (figs. 23 and 24). Press the rear output shaft bearing on the output shaft (fig. 23). Set the output shaft sliding gear in the transfer case with the shifter fork in the channel of the sliding gear. Place the output shaft gear in the transfer case with the shoulder of the output shaft gear facing the sliding gear. Insert the output shaft in the transfer case and through the gears. Slide the thrust washer on the output shaft. Install the snap ring that secures the output shaft gear on the shaft. Slide the front output shaft roller bearing on the output shaft and, using a suitable driver, tap the roller bearing snug against the snap ring. Tap the front roller bearing cup

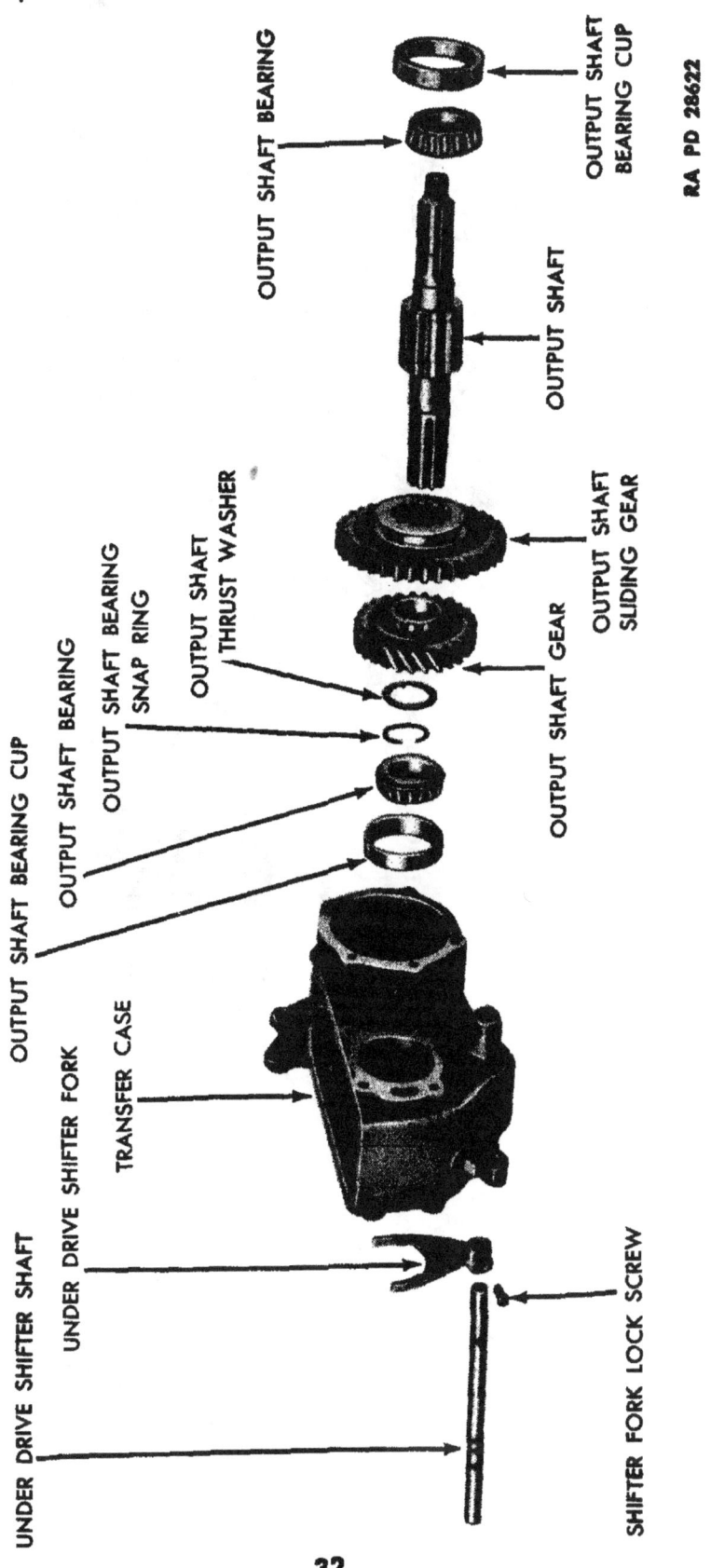

Figure 24 — Output Shaft — Exploded View

POWER TRAIN

TM 9-1803B
14

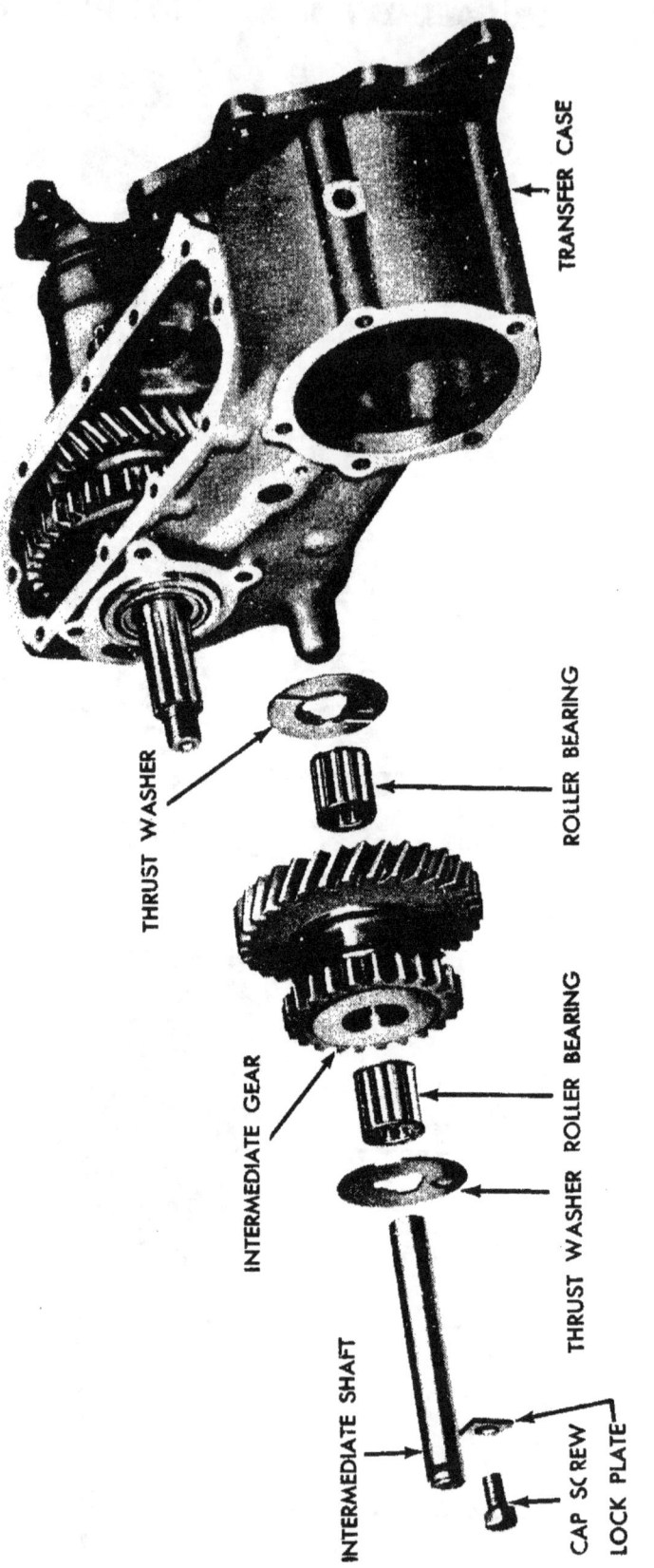

Figure 25 — Intermediate Gear Assembly — Exploded View

RA PD 28624

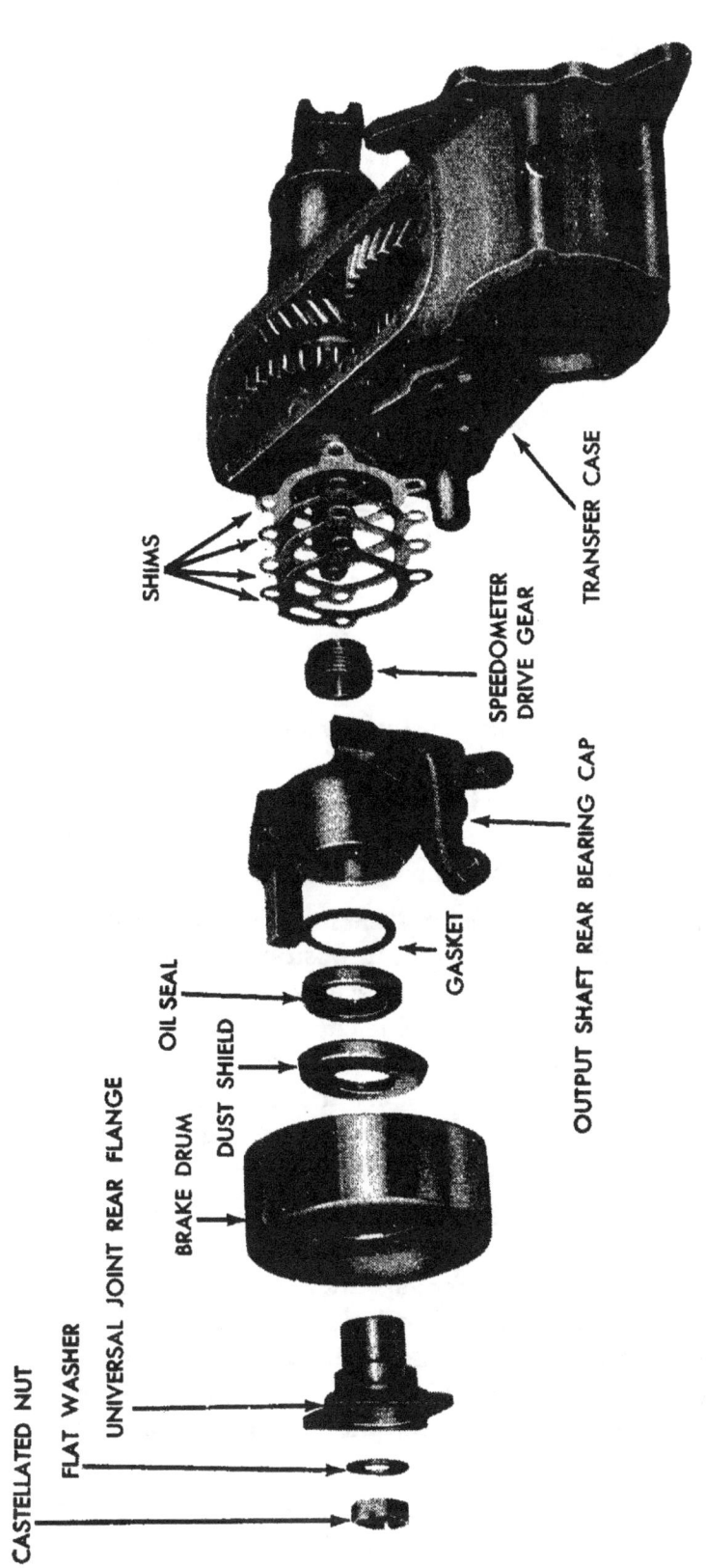

Figure 26 — Rear Output Shaft Cap — Exploded View

POWER TRAIN

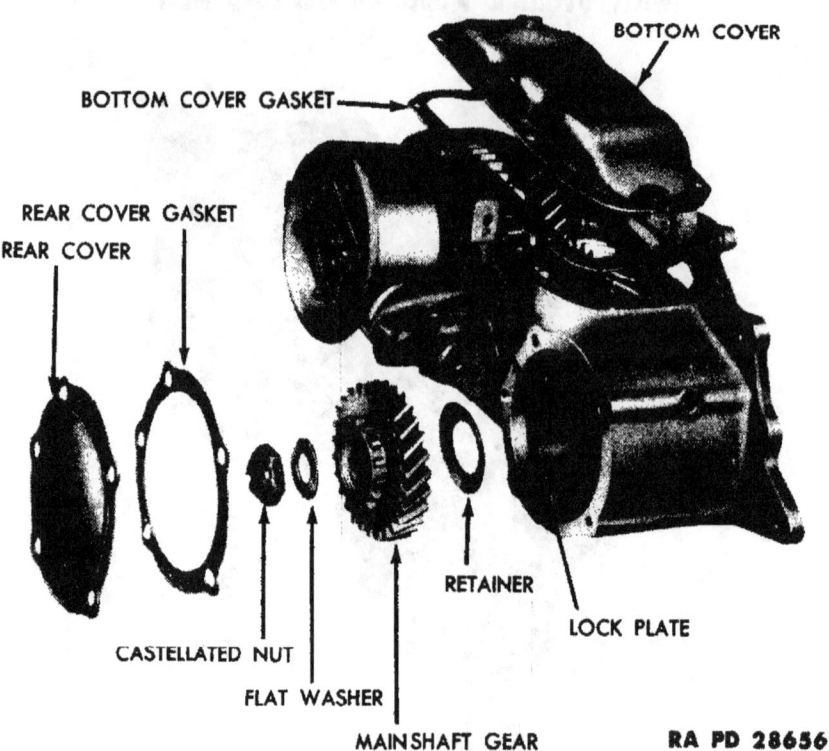

Figure 27 — Bottom Cover and Mainshaft Gear — Exploded View

in the transfer case until the cup is slightly below flush with the transfer case. Tap the rear bearing cup in the transfer case until the cup is approximately ⅛ inch from the transfer case surface.

d. Install Front Output Shaft Bearing Cap to Transfer Case (figs. 21 and 22). Place a new gasket in position on the transfer case. Install the interlock (fig. 21) in the interlock opening on the bearing cap. Slide the front output shaft bearing cap on the under drive shifter shaft, being careful not to damage the oil seal in the output shaft bearing cap. Install the five bolts that secure the front bearing cap to the transfer case. Install the poppet ball, poppet spring and poppet plug on both sides of the front bearing cap (fig. 21).

e. Install Intermediate Gear (fig. 25). Insert the roller bearings in the intermediate gear. Place the thrust washers in the transfer case, with the side having the bronze facing, toward the intermediate gear. Apply grease to the thrust washers to hold them in position, if necessary. Place the intermediate gear between the thrust washers in the transfer case. Install the intermediate gear shaft in the transfer case. Install the lock plate that secures the intermediate gear shaft to the transfer case.

f. Install Rear Output Shaft Cap to Transfer Case (fig. 26). Slide the speedometer drive gear on the output shaft. Install the oil seal in

TM 9-1803B
14

ORDNANCE MAINTENANCE — POWER TRAIN, BODY, AND FRAME FOR ¼-TON 4 x 4 TRUCK
(WILLYS-OVERLAND MODEL MB AND FORD MODEL GPW)

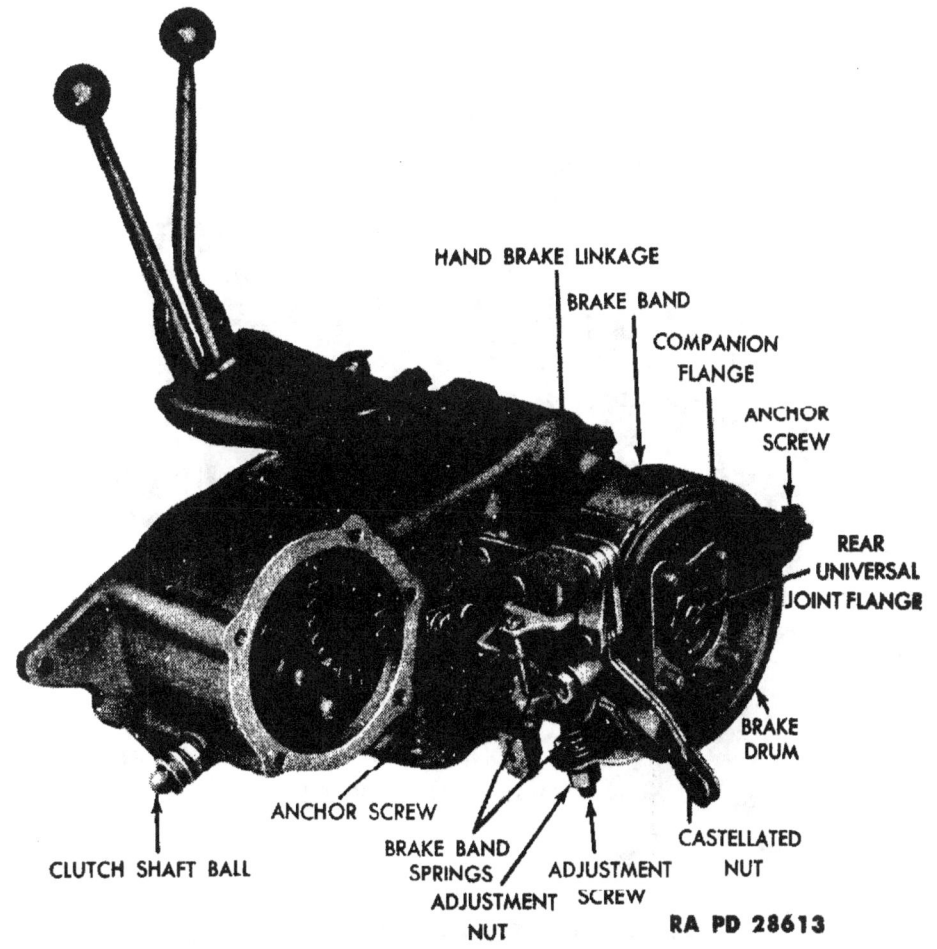

Figure 28 — Transfer Case

the rear output shaft cap (par. 13 c). Install the rear output shaft cap, shims and gasket on the transfer case. Tighten the four cap screws evenly to prevent cracking the output shaft cap. Shims are to be added or removed until the output shaft has no end play, but turns freely. When adjusting the bearings, each time shims are added, the shaft must be free before attempting to tighten the output shaft cap again. Insert the rear universal joint flange in the brake drum. Place the four cap screws in the brake drum and universal joint flange, using a suitable driver, drive the dust shield on the universal joint flange. Install the rear universal joint flange on the output shaft, and install the flat washer and nut.

g. Install Bottom Cover to Transfer Case (fig. 27). Install a new gasket in position on the transfer case. Place the bottom cover on the transfer case. Install the cap screws that secure the bottom cover to the transfer case.

POWER TRAIN

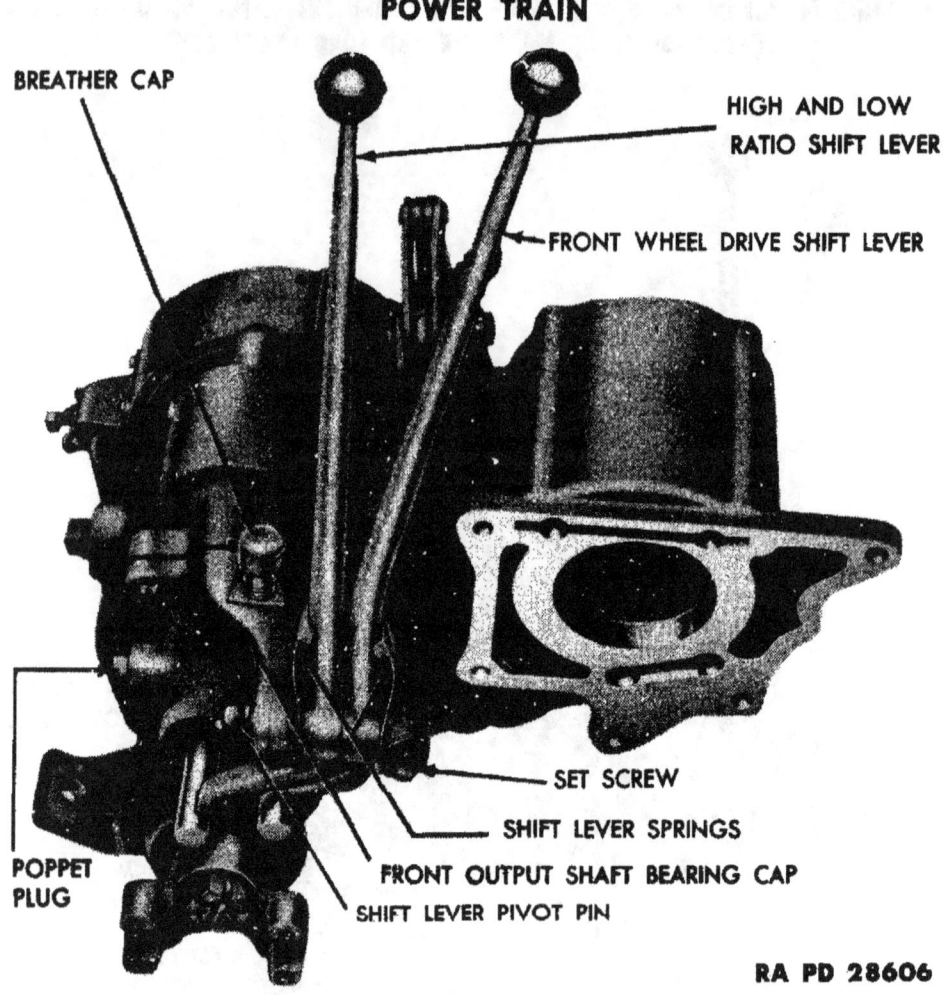

Figure 29 — Transfer Case Shift Levers

h. Install Brake Band to Transfer Case (fig. 28). Place the brake band on the brake drum. Place the brake band springs between the rear output shaft bearing cap and the ends of the brake band. Install the nut and bolt that secure the hand brake linkage to the rear output shaft bearing cap. Insert the adjusting screw through the brake band linkage, brake band springs, and install the adjusting nut. Install the two anchor screws on the brake band.

15. INSTALLATION.

a. Raise Transfer Case. Raise the transfer case and line up the clutch shaft ball joint in the transfer case. Line up the transfer case with the transmission. Be sure the interlock is in position on the rear of the transmission case before installing the transfer case to the transmission (fig. 4). Install the five cap screws that secure the transfer case to the transmission. Install the mounting bolt that secures the transfer case to the transmission support crossmember.

b. Install Mainshaft Gear (fig. 27). Insert the retainer and mainshaft gear on the transmission mainshaft. Install the flat washer and castellated nut that secure the mainshaft gear on the transmission mainshaft. Place a new gasket and the rear cover on the transfer case and install the cap screws that secure the cover to the case.

c. Install Clutch, Hand Brake and Speedometer Cables (fig. 6). Install the clevis that secures the clutch release fork cable to the clutch shaft. Install the clevis pin that secures the hand brake cable to the brake band. Install the cap screw that secures the hand brake clamp to the transfer case rear output shaft cap. Install the speedometer cable to the transfer case at the top of the rear output shaft cap.

d. Install Propeller Shaft and Transfer Case Shield (fig. 6). Connect the rear propeller shaft to the transfer case (par. 17 b). Place the transmission shield in position and install the five cap screws that secure the shield to the transmission support crossmember. Install the exhaust pipe clamp to the transmission shield. Fill the transfer case with specified oil to the proper level. Adjust the hand brake band (refer to TM 9-803).

Section IV

PROPELLER (DRIVE) SHAFTS AND UNIVERSAL JOINTS

16. DESCRIPTION AND TABULATED DATA.

a. Description (fig. 2). The power from the transfer case is carried through two propeller shafts. One propeller shaft runs from the front of the transfer case to the front axle, and a second propeller shaft runs from the rear of the transfer case to the rear axle. Each is equipped with two universal joints. The splined slip joint at one end of each shaft allows for variations in distance between the transfer case and the axle units due to spring action. Two types of universal joints are used; the U-bolt type and the solid yoke type.

b. Tabulated Data.

(1) PROPELLER SHAFTS.

Make	Spicer
Shaft diameter	1½ in.
Length (front)	$21\frac{11}{16}$ in.
Length (rear)	$20\frac{1}{32}$ in.

TM 9-1803B
16—17

POWER TRAIN

(2) Front Propeller Shaft Forward Universal Joint.

Make	Spicer
Type	U-bolt and solid yoke
Model	1268
Bearings	Needle roller

(3) Front Propeller Shaft Rear Universal Joint.

Make	Spicer
Type	U-bolt and solid yoke
Model	1261
Bearings	Needle roller

(4) Rear Propeller Shaft Forward Universal Joint.

Make	Spicer
Type	Solid yoke slip joint
Model	1261
Bearings	Needle roller

(5) Rear Propeller Shaft Rear Universal Joint.

Make	Spicer
Type	U-bolt and solid yoke
Model	1268
Bearings	Needle roller

17. REMOVAL.

a. **Front Propeller Shaft** (fig. 33). Bend the ears of the lock plates off the U-bolt nuts. Remove the two nuts from each of the two U-bolts at the front axle and at the transfer case. Remove the U-bolts from the propeller shaft. Take care to hold the bearing races in place on the universal joint to avoid losing the rollers.

b. **Rear Propeller Shaft** (fig. 34). The rear propeller shaft is similar to the front propeller shaft with the exception of the solid yoke type connection at the transfer case. Remove the nuts from the U-bolts at the rear axle end. Remove the U-bolts. Slide the universal joint out of the universal joint rear flange. Care must be taken to hold the bearing races on the universal joint to avoid losing the rollers. Remove the four nuts that secure the universal joint flange yoke to the rear flange at the transfer case. Remove the rear propeller shaft from the vehicle.

TM 9-1803B
18

ORDNANCE MAINTENANCE — POWER TRAIN, BODY, AND FRAME FOR ¼-TON 4 x 4 TRUCK (WILLYS-OVERLAND MODEL MB AND FORD MODEL GPW)

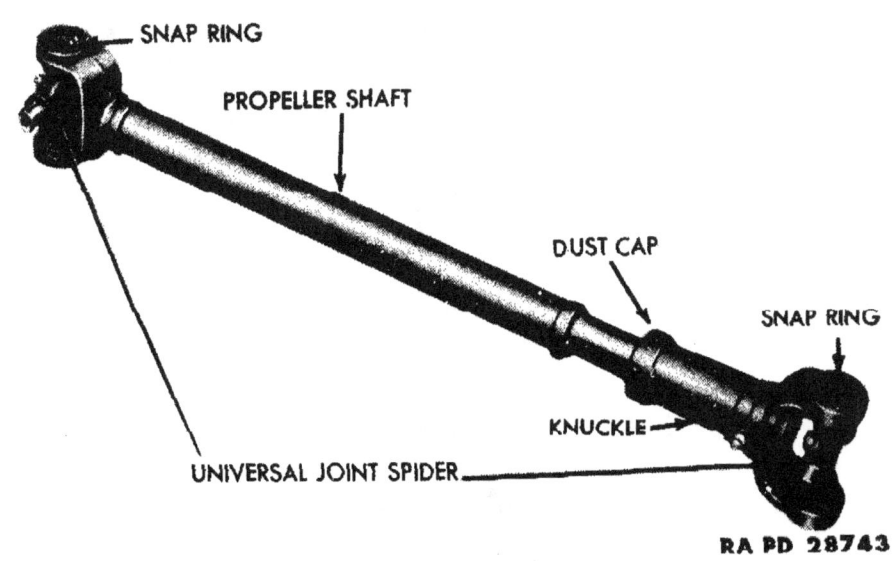

Figure 30 — Front Propeller Shaft

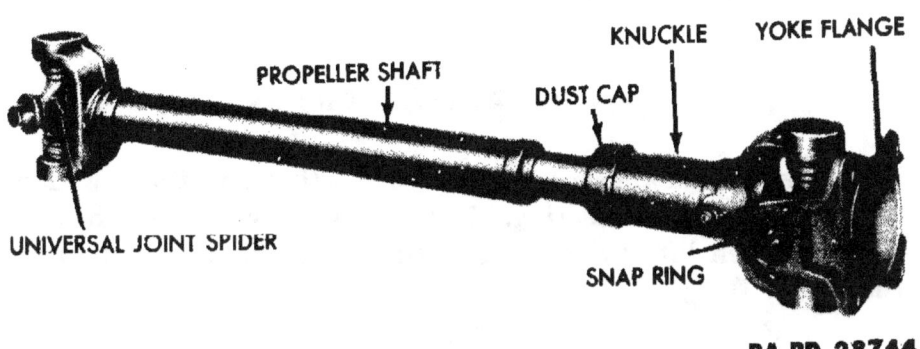

Figure 31 — Rear Propeller Shaft

18. DISASSEMBLY.

a. Front Propeller Shaft (fig. 30).

(1) REMOVE SNAP RINGS FROM YOKE (fig. 30). Place the propeller shaft in a vise. Remove the snap rings that secure the spider bearings in the yoke flange with a pair of pliers. If the snap ring does not snap out of the groove, tap the end of the bearing lightly. This will relieve the pressure against the snap ring.

(2) REMOVE SPIDER FROM YOKE (fig. 32). Drive lightly on the end of the spider bearing until the opposite bearing is pushed out of the yoke flange. Turn the assembly over in the vise and drive the first spider bearing back out of its lug by driving on the exposed end of the spider. Use a brass drift with a flat face about $\frac{1}{32}$ inch smaller

POWER TRAIN

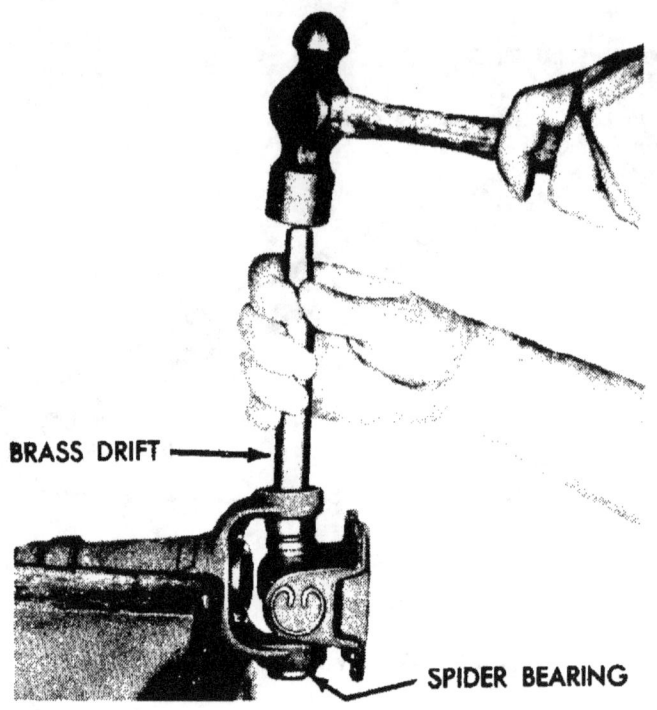

Figure 32 — Removing Spider Bearing

in diameter than the hole in the yoke, otherwise there is danger of damaging the spider bearing. Repeat this operation for the other two bearings, then lift out the spider, sliding to one side and tilting over the top of the yoke.

(3) REMOVE KNUCKLE FROM SHAFT (fig. 30). Bend the ears of the dust cap off the knuckle. Slide the knuckle off the drive shaft. Remove the split cork gasket from the bearing cap. Line up the slots in the dust cap with the splines on the drive shaft and remove the cap from the shaft.

b. **Rear Propeller Shaft.**

(1) REMOVE YOKE FLANGE (fig. 31). Place the propeller shaft in a vise. Remove the four snap rings that secure the spider bearings in the yoke flange and knuckle. Using a brass drift with a flat face about $\frac{1}{32}$ inch smaller than the hole in the yoke, drive lightly on the end of the bearing until the opposite bearing is out of the yoke flange. Turn the assembly over in the vise and drive the first bearing out of its lug by driving on the exposed end of the spider. Remove the yoke flange from the spider.

(2) REMOVE SPIDER AND KNUCKLE (fig. 32). Remove the spider from the knuckle (subpar. a (2), above). Remove the knuckle from the propeller shaft (subpar. a (3), above).

TM 9-1803B
18

ORDNANCE MAINTENANCE — POWER TRAIN, BODY, AND FRAME FOR ¼-TON 4 x 4 TRUCK (WILLYS-OVERLAND MODEL MB AND FORD MODEL GPW)

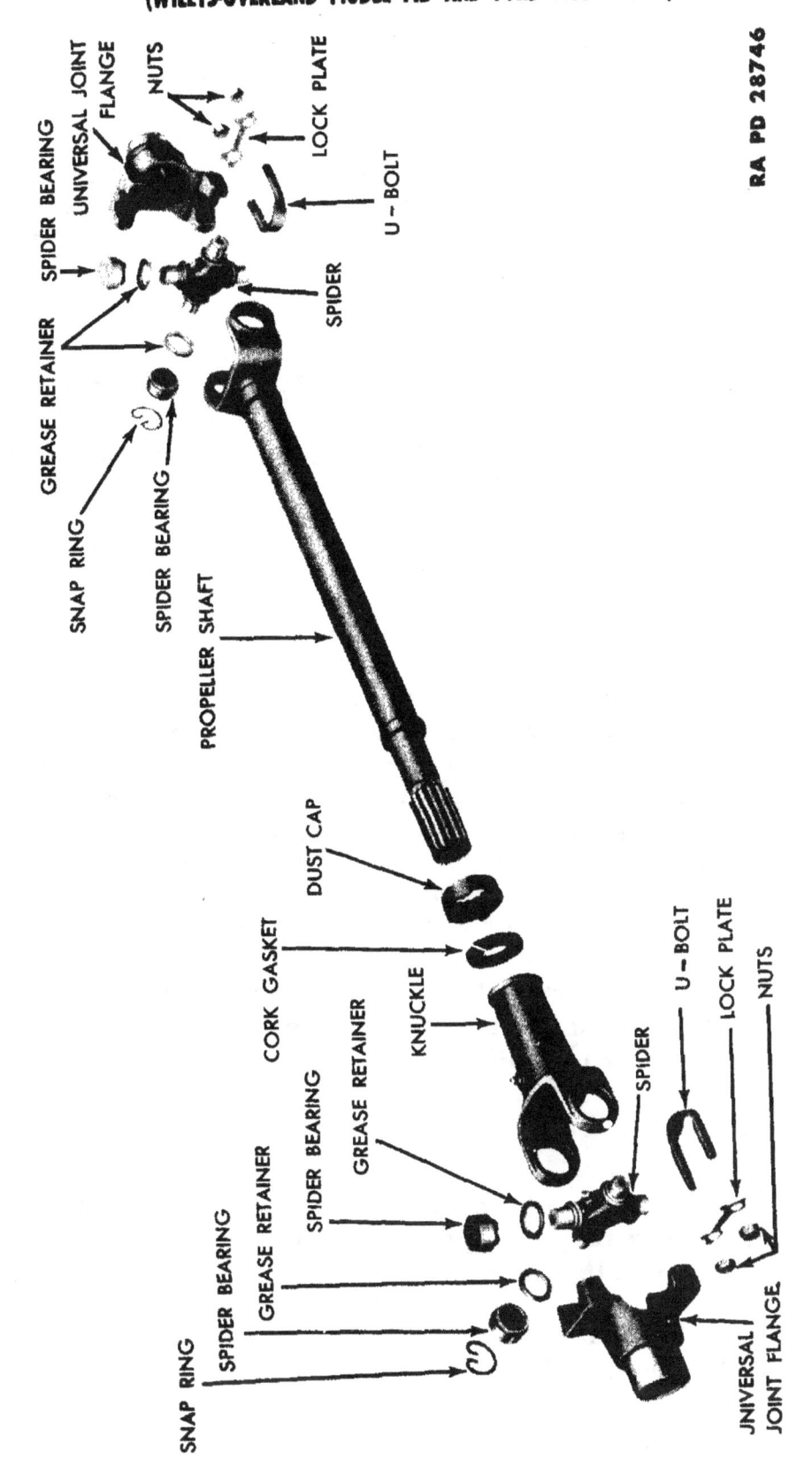

Figure 33 — Front Propeller Shaft — Exploded View

42

POWER TRAIN

19. CLEANING, INSPECTION, AND REPAIR.

a. Clean all parts thoroughly with dry-cleaning solvent. Inspect the drive shafts for cracks, broken welds, scored spider bearing surfaces, or bent shafts. Parts with any of these faults must be replaced. Inspect the knuckle for worn splines, worn bearing surfaces and bearings and plugged lubricant fittings. Check the diameter of the machined surface of the spiders. If the diameter is less than 0.595 inch, replace the spider. Replace all grease seals regardless of their condition.

20. ASSEMBLY.

a. **Front Propeller Shaft** (fig. 33). Place the propeller shaft in a vise. Slide the dust cap on the drive shaft. Place a new cork gasket in the cap. Slide the knuckle on the shaft splines, being sure that the knuckle on the shaft is in the same angle as the yoke at the opposite end of the propeller shaft. Slide the dust cap on the shoulder of the knuckle and bend the ears of the cap over the shoulder of the knuckle.

b. **Rear Propeller Shaft** (fig. 34).

(1) INSTALL SPIDER IN YOKE FLANGE (fig. 34). Insert the spider into the yoke flange. Tap the spider bearing approximately ¼ inch into the yoke flange, using a brass drift approximately 1/32 inch smaller than the hole in the yoke. Tap the other bearing into the opposite end of the yoke flange until the bearing is in line with the snap ring grooves. With a pair of pliers, install the snap rings on both ends of the yoke flange. Insert the flange assembly in the knuckle. Tap the bearing approximately ¼ inch into the yoke. Place the other bearing into the opposite end of the yoke, and tap this bearing into the yoke until the bearing is in line with the snap ring groove. Install the snap rings on both ends of the yoke.

(2) INSTALL KNUCKLE AND SPIDERS (fig. 34). Install the knuckle on the propeller shaft (subpar. a (1), above).

21. INSTALLATION.

a. **Rear Propeller Shaft.** Place the propeller shaft with the yoke flange end toward the transfer case (fig. 6). Install the four nuts that secure the yoke flange to the transfer case. Insert the two spider bearings on the spider at the rear axle end. Place the spider in the universal joint rear flange. Install the two U-bolts that secure the propeller shaft to the rear axle flange. Lubricate the propeller shaft with specified lubricant.

b. **Front Propeller Shaft.** Place the propeller shaft with the knuckle end at the transfer case. Insert the bearings on the spider

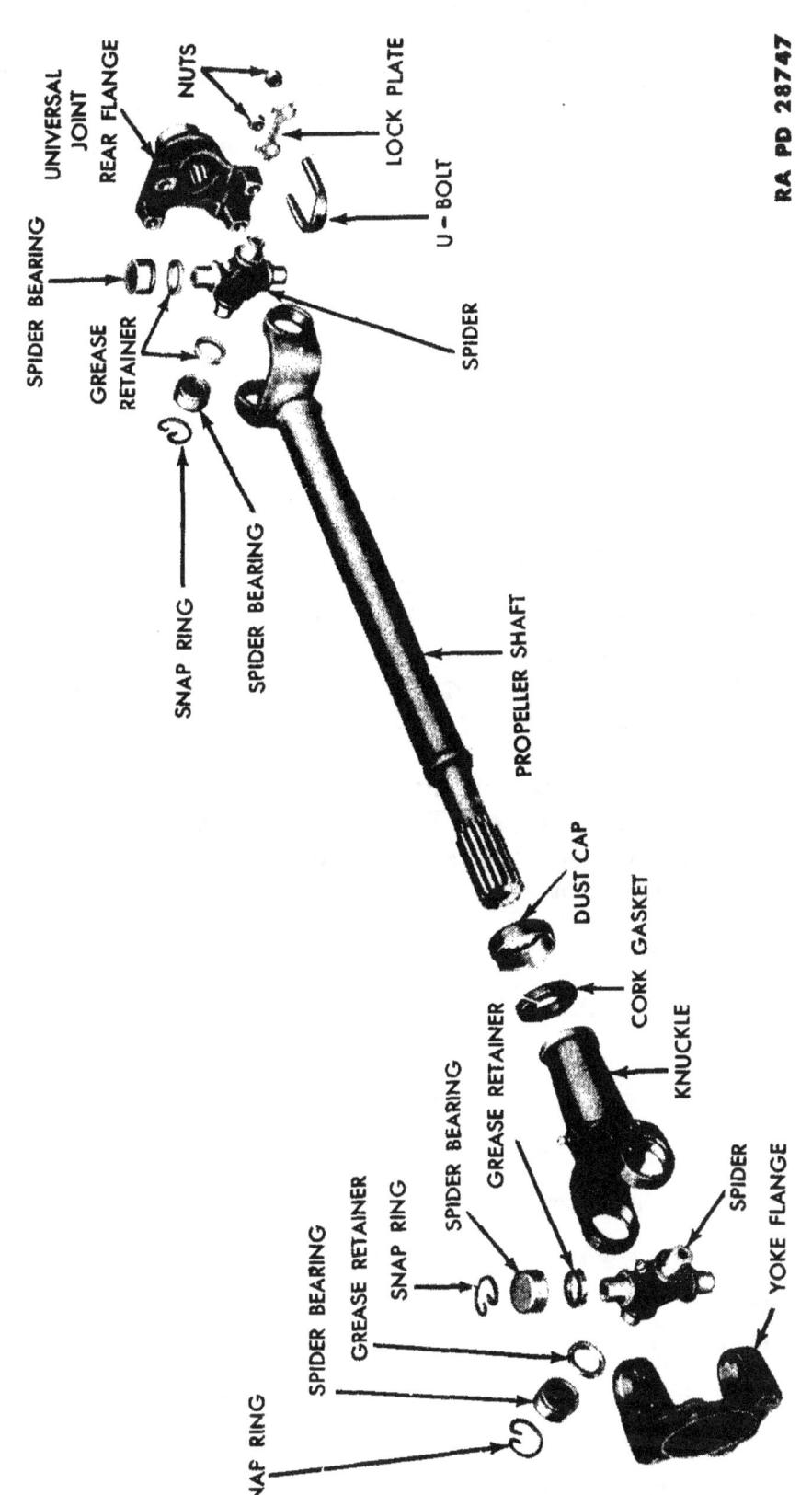

Figure 34 — Rear Propeller Shaft — Exploded View

POWER TRAIN

and place the propeller shaft in the universal joint flange on the transfer case. Install the two U-bolts that secure the propeller shaft to the transfer case. Insert the two spider bearings on the spider at the front axle end. Place the propeller shaft in the front axle flange. Install the two U-bolts that secure the propeller shaft to the universal joint flange. Lubricate the propeller shaft with specified lubricant.

Section V

FRONT AXLE

22. DESCRIPTION AND DATA.

a. Description (fig. 2). The front axle assembly is a front wheel driving unit, with specially designed spindle housings, and has a conventional type differential with hypoid drive gears. The differential parts are interchangeable with those of the rear axle. The axle shafts are of the full-floating type. The differential is mounted in the housing similar to the rear axle, except that the drive pinion shaft is toward the rear instead of the front and to the right of the center of the axle. Three types of axle shafts and universal joints have been used (Rzeppa, Bendix, and Tracta). The vehicles using the different types of shafts are identified by an identification tag attached to the spindle housing (fig. 35).

b. Data.

(1) FRONT AXLE.

Make	Spicer
Drive	Through springs
Type	Full-floating

(2) DIFFERENTIAL.

Drive	Hypoid
Gear ratio	4.88 to 1
Bearings	Timken roller 2
Adjustment	Shims
Gears (pinion)	2

(3) OIL CAPACITY .... 2½ pt

TM 9-1803B
22

**ORDNANCE MAINTENANCE — POWER TRAIN, BODY, AND FRAME FOR ¼-TON 4 x 4 TRUCK
(WILLYS-OVERLAND MODEL MB AND FORD MODEL GPW)**

Figure 35 — Front Axle Assembly in Vehicle

POWER TRAIN

TM 9-1803B
22

RA PD 329205-B

A—SPRING SHACKLE
B—SHOCK ABSORBER
C—TIE ROD
D—BREATHER CAP
E—TIE ROD ENDS
F—SPRING SHACKLE
G—TIE ROD
H—LEFT FRONT SPRING
J—SHOCK ABSORBER
K—TIE ROD CLAMP
L—TORQUE REACTION SPRING
M—DRAG LINK
N—DRAG LINK PLUG
O—PIVOT ARM
P—DRAIN PLUG
Q—FRONT PROPELLER SHAFT
R—SPRING SEAT PLATE
S—TIE ROD CLAMP
T—TIE ROD ENDS
U—RIGHT FRONT SPRING
V—TIE ROD CLAMPS
W—AXLE SHAFT IDENTIFICATION TAG

Legend for Figure 35 — Front Axle Assembly in Vehicle

TM 9-1803B
23

ORDNANCE MAINTENANCE — POWER TRAIN, BODY, AND FRAME FOR ¼-TON 4 x 4 TRUCK (WILLYS-OVERLAND MODEL MB AND FORD MODEL GPW)

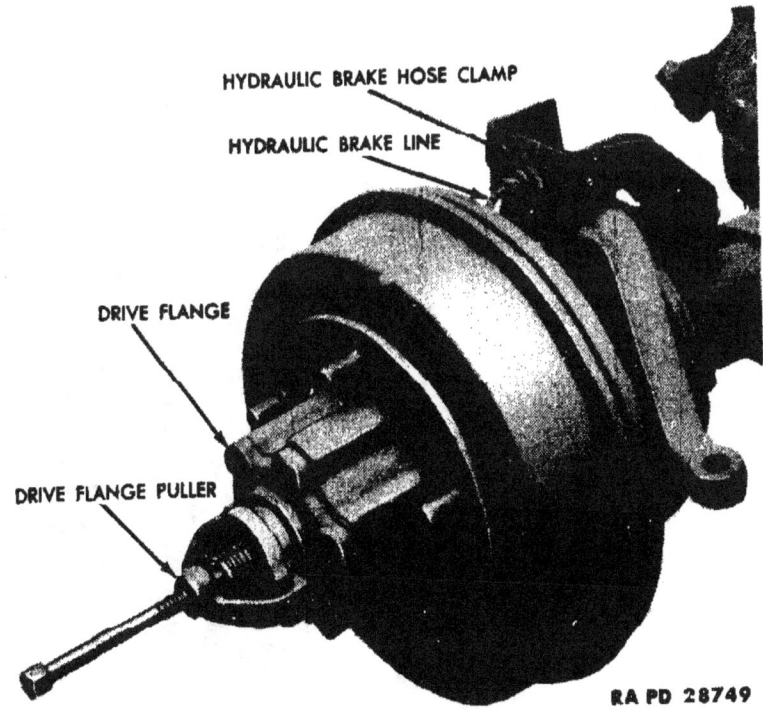

Figure 36 — Removing Drive Flange With Puller Similar to Puller 41-P-2912

23. REMOVAL.

a. Preliminary Work. Remove the drain plug at the differential housing and drain the oil. Raise the vehicle until the weight is off the front springs.

b. Disconnect Shock Absorbers and Drag Link (fig. 35). Remove the cotter pin and flat washer that secure the shock absorber to the spring seat plate at both front shock absorbers. Remove the drag link plug at the pivot arm. Remove the drag link from the pivot arm.

c. Disconnect Front Propeller Shaft and Spring U-bolts (fig. 35). Disconnect the front propeller shaft at the front anxle (par. 17 a). Remove the four nuts from the two U-bolts that secure the spring seat plate. Remove the spring seat plate and U-bolts. Remove the four nuts from the U-bolts at the torque reaction spring. Remove the two U-bolts.

d. Disconnect Spring Shackles (fig. 35). Remove the lower spring shackle bushing at the forward end of the front springs. Pull both springs out of the spring shackles and drop the forward end of the springs to the floor. Roll the front axle assembly from the vehicle.

POWER TRAIN

Figure 37 — Removing Bearing Lock Nut With Wrench 41-W-3825-200

24. DISASSEMBLY.

a. **Remove Wheels.** Place the front axle assembly on two blocks. Remove the five nuts that secure the wheels to the brake drum. Remove the wheels.

b. **Remove Axle Shaft Assembly.** Using a screwdriver, pry the hub cap off the drive flange. Remove the cotter pin and castellated nut from the axle shaft. Remove the six cap screws that secure the drive flange to the hub. Install the puller 41-P-2912 or similar on the drive flange and remove the drive flange (fig. 36). Bend the ear of the lock washer off the bearing lock nut. Remove the bearing lock nut, lock washer, and bearing adjustment nut, using the wheel bearing nut wrench 41-W-3825-200 furnished with the vehicle (fig. 37). Slide the brake drum and hub assembly, including the wheel bearings, off the spindle. Disconnect the hydraulic brake line at the brake hose guard (fig. 36). Remove the six cap screws that secure the brake plate to the spindle housing. Remove the brake plate from the spindle. Slide the spindle off the axle shaft. The axle shaft can now be removed from the housing. If equipped with a Tracta universal joint axle shaft, see subparagraph c, below. Use the same procedure to disassemble the other end of the front axle shaft.

c. **Axle Shaft Disassembly.** Three types of axle shaft universal joints, as shown in figures 38, 42, and 44, are used in the front axle.

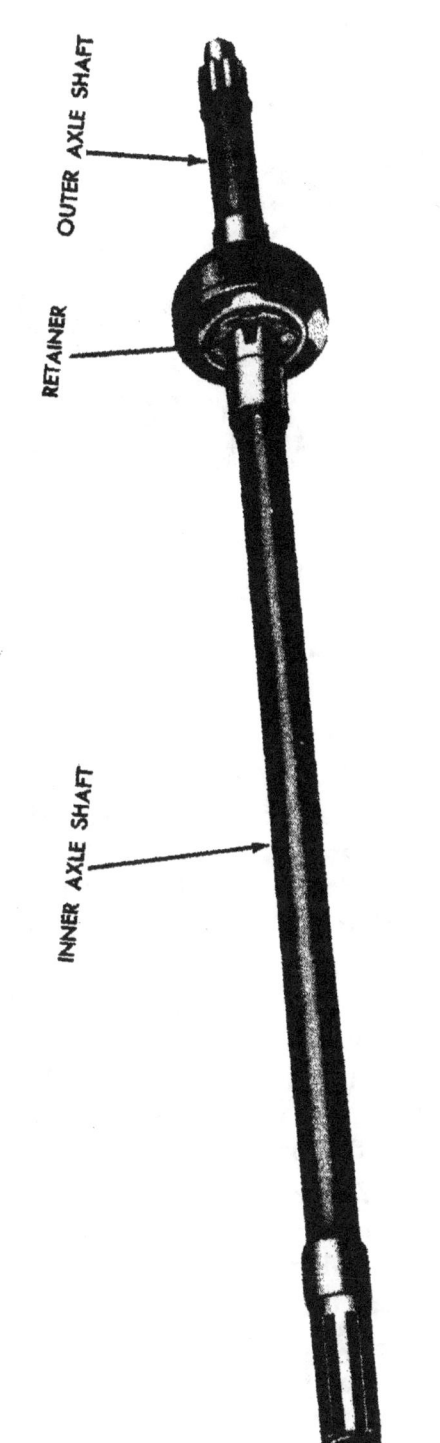

Figure 38 — Front Axle Shaft (Rzeppa Joint)

POWER TRAIN

RA PD 28752

Figure 39 — Removing Balls From Cage

Disassembly procedures for each are given in steps (1), (2), and (3), below.

(1) RZEPPA UNIVERSAL JOINT.

(a) Remove Inner Axle Shaft (fig. 59). Remove the three flat head screws that secure the retainer to the inner ball race. Slide the inner axle shaft out of the universal joint. Remove the pilot pin from the outer axle shaft. If the pilot pin does not drop out of the outer axle shaft, hold the shaft upside down and tap the shaft on a piece of wood.

(b) Remove Balls From Cage (fig. 39). Tilt the cage in the axle shaft cup until the opposite side of the cage is out of the housing. It may be necessary to use a brass drift and hammer to tilt the cage. Use a screwdriver to pry the steel ball out of the cage. Repeat this operation until all the balls are removed.

(c) Remove Cage and Inner Race From Axle Shaft (fig. 40). Turn the cage in the axle shaft cup in line with the shaft and with the two larger elongated holes between two bosses in the shaft. Lift the cage and inner race from the axle shaft cup.

(d) Remove Inner Race From Cage (fig. 41). Turn the inner race in the cage so that one of the bosses on the inner race can be dropped into one of the two elongated holes in the cage. Remove the inner race from the cage.

(2) BENDIX UNIVERSAL JOINT (figs. 42 and 43). Place the axle

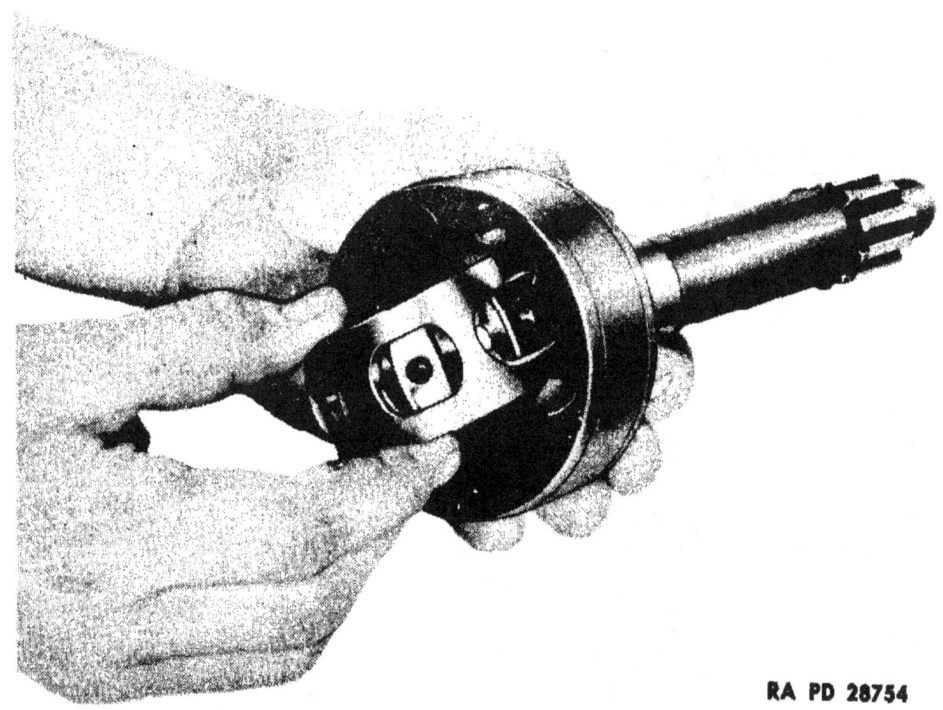

RA PD 28754

Figure 40 — Removing Cage and Inner Race From Axle Shaft (Rzeppa Joint)

RA PD 28753

Figure 41 — Removing Inner Race From Cage (Rzeppa Joint)

POWER TRAIN

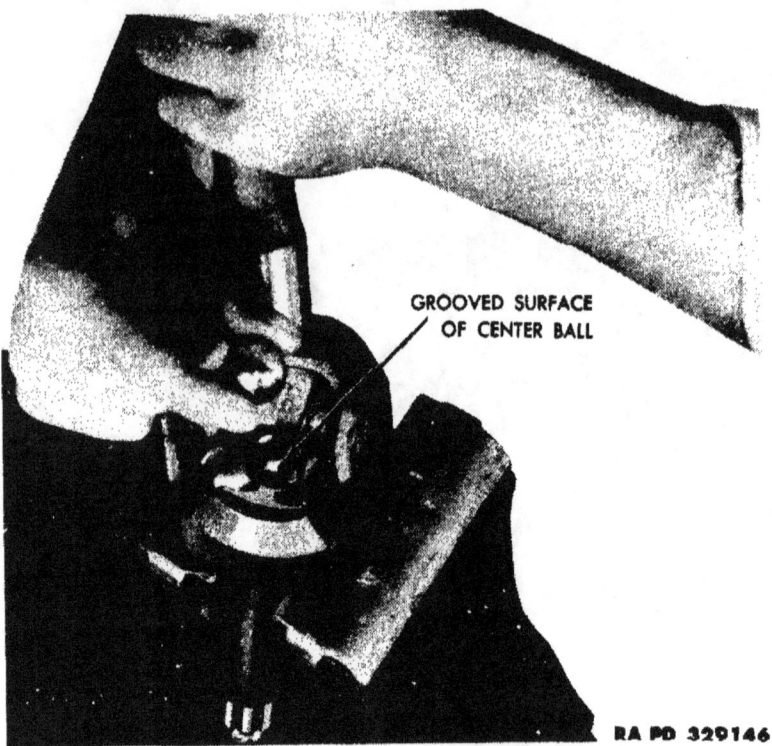

Figure 42 — Front Axle Shaft (Bendix Type)

shaft in a vise and with a long nosed drift remove the groove pin from the universal joint knuckle. Remove the axle shaft from the vise. Tap the knuckle end of the axle shaft on a wood block until the center ball pin drops in the groove pin hole. Place the axle shaft with the knuckle end (short end) in a vise. Bend the axle shaft so that the center ball can be rotated until the grooved surface of the center ball is facing the first ball that is to be removed. Holding the axle shaft in a bent position, raise the shaft until the first ball to be removed slides into the groove of the center ball, and remove the ball. Remove the axle shaft from the knuckle. The three remaining balls will drop out of the knuckle.

(3) TRACTA UNIVERSAL JOINT (fig. 45). Remove the outer portion of the axle shaft and the outer portion of the universal joint from the axle housing. Pull the inner portion of the axle shaft and the inner portion of the universal joint out of the housing.

d. **Remove Spindle Housing** (fig. 46). Remove the castellated nut that secures the tie rod ends to the two spindle arms. Remove the two castellated nuts that secure the two tie rod ends to the steering pivot arm and remove the two tie rods. Remove the hydraulic brake hose clamp from the hydraulic brake line at the brake hose guard. Remove the four nuts that secure the brake hose guard and spindle arm to the spindle housing. Remove the spindle arm and

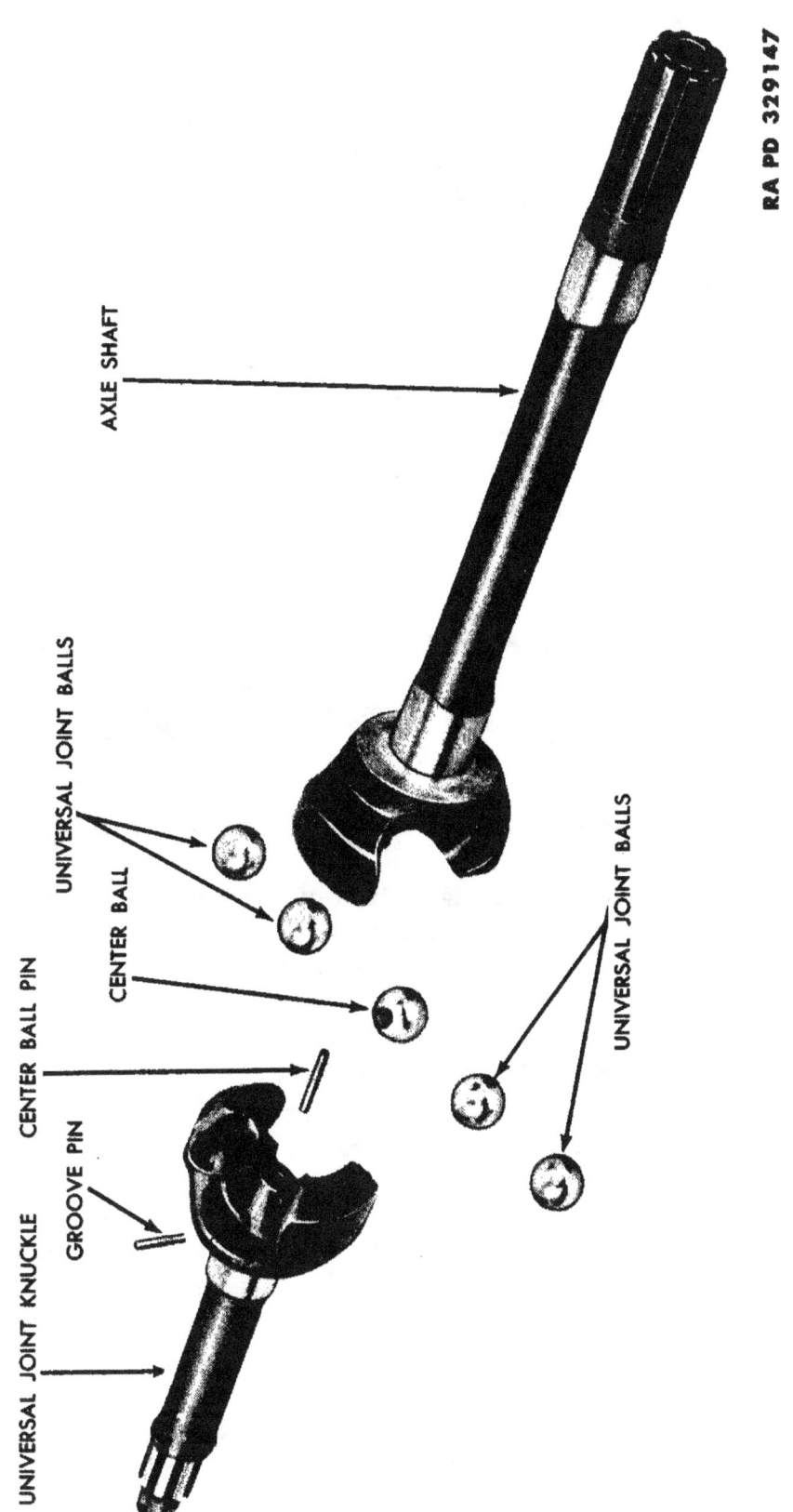

Figure 43 — Front Axle Shaft — Exploded View (Bendix Type)

POWER TRAIN

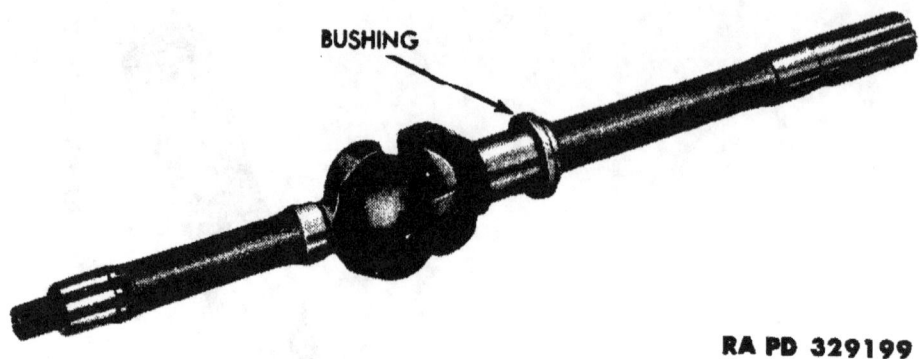

Figure 44 — Front Axle Shaft (Tracta Type)

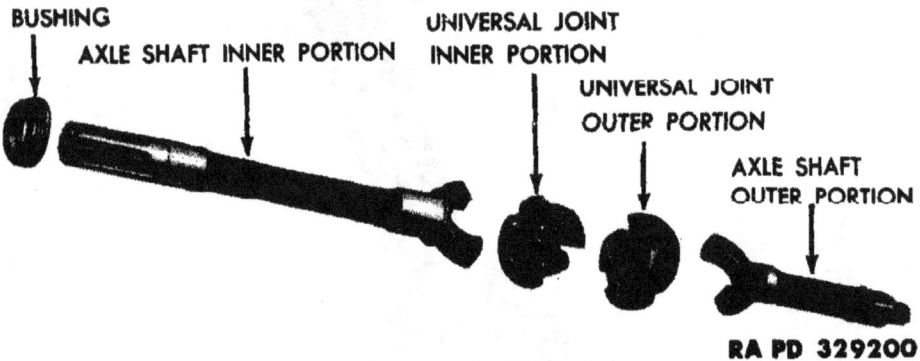

Figure 45 — Front Axle Shaft — Exploded View (Tracta Type)

shims from the spindle housing. Remove the four cap screws that secure the lower bearing cap to the spindle housing. Remove the bearing cap and shims. Remove the eight cap screws that secure the spindle housing oil seal retainer to the spindle housing. Remove the spindle housing from the axle housing. Use the same procedure for disassembling the other spindle housing.

e. **Remove Differential** (fig. 47). Remove the ten cap screws that secure the differential cover to the differential housing. Remove the differential cover and gasket. Remove the two cap screws from the bearing cap at each end of the differential gears and remove the caps. Remove the differential gear assembly from the housing, using a pry bar, if necessary. Reinstall the bearing caps in the housing, noting the markings (fig. 47) to assure their being installed in their correct location.

f. **Disassemble Differential.**

(1) REMOVE DIFFERENTIAL PINION GEARS AND AXLE SHAFT GEARS (fig. 48). Place the differential assembly in a vise equipped with brass jaws. With a long nosed drift, drive the differential

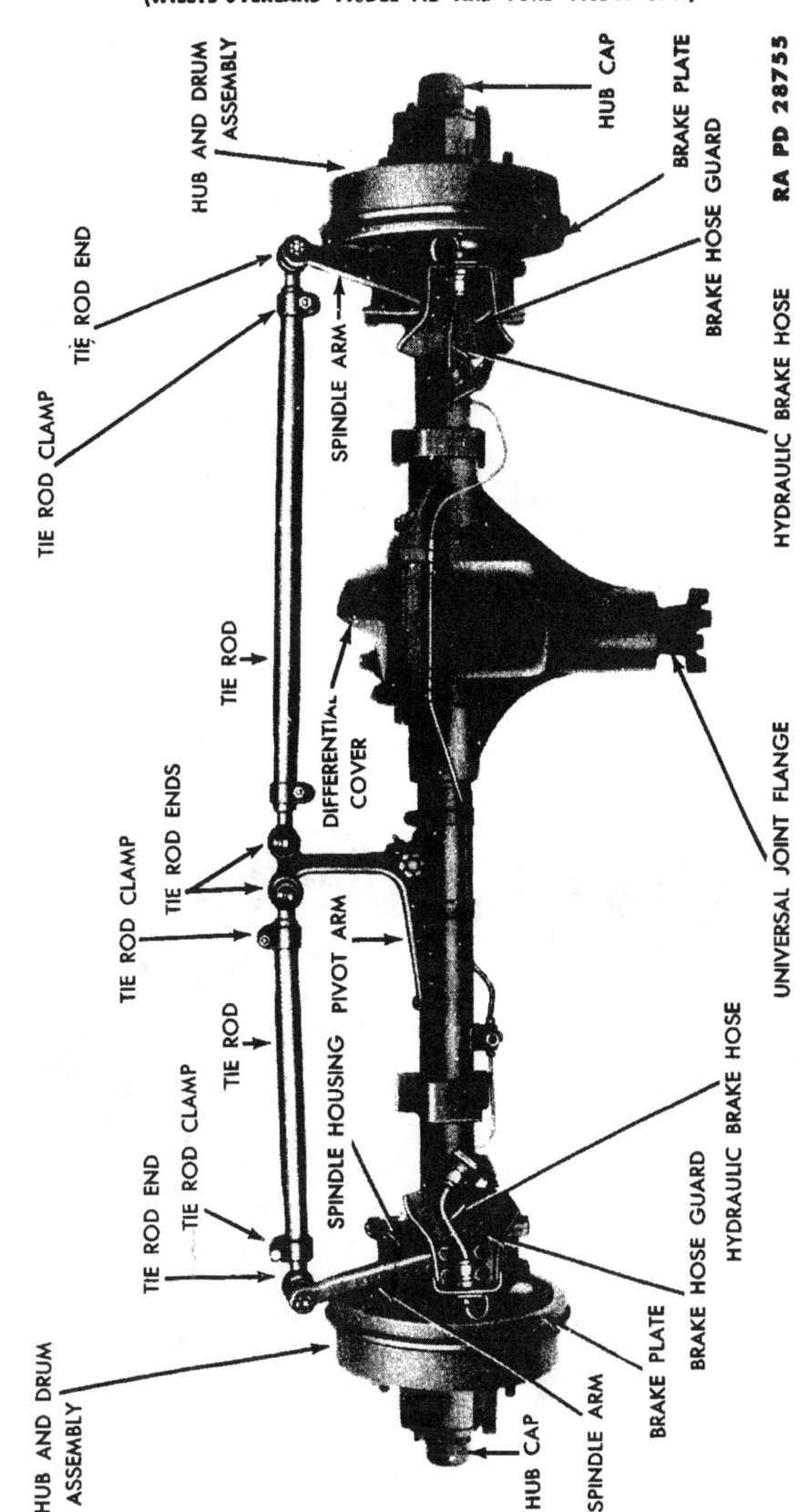

Figure 46 — Front Axle Assembly

POWER TRAIN

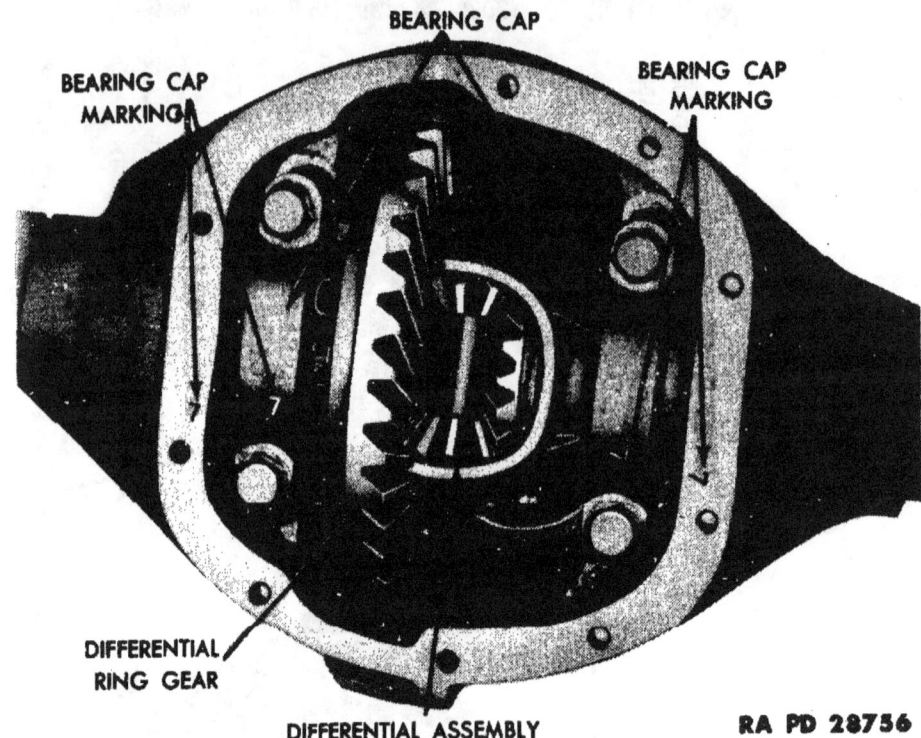

Figure 47 — Differential Assembly

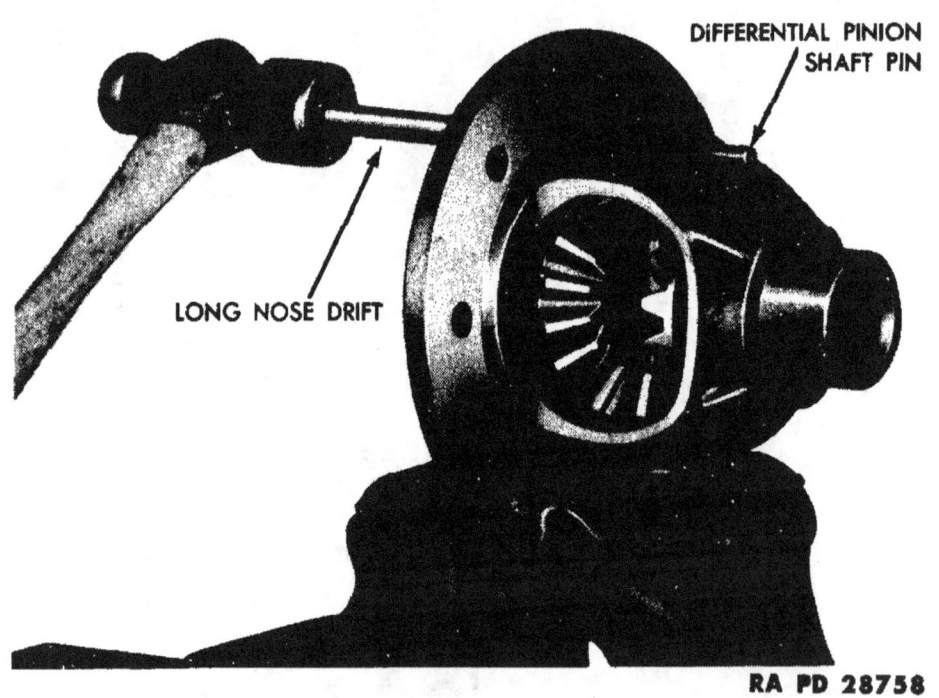

Figure 48 — Removing Pinion Shaft Lock Pin

TM 9-1803B
24

ORDNANCE MAINTENANCE — POWER TRAIN, BODY, AND FRAME FOR ¼-TON 4 x 4 TRUCK (WILLYS-OVERLAND MODEL MB AND FORD MODEL GPW)

RA PD 28757

Figure 49 — Removing Bearings From Differential Case With Special Tool 41-R-2378-30

pinion shaft tapered pin out of the differential gear case. Tap the differential pinion shaft from the case with a brass drift and hammer. Remove the two differential pinion gears and thrust washers and the two axle shaft gears and thrust washers from the case.

(2) REMOVE RING GEAR FROM CASE (fig. 47). Bend the ears of the lock plates off the cap screws. Remove the cap screws that secure the ring gear to the case, and remove the ring gear.

(3) REMOVE ROLLER BEARING FROM DIFFERENTIAL CASE (fig. 49). Place the differential case in a vise. Install the bearing remover 41-R-2378-30 to the roller bearing. Remove the roller bearing from each end of the differential case. Remove the shims, noting the thickness of the shims removed from each end.

g. Remove Drive Pinion. Remove the nut and flat washer that secure the universal joint flange to the drive pinion. Install the puller 41-P-2905-60 to the universal joint flange (fig. 50) and remove the flange. Using a brass drift and hammer, drive the drive pinion out of the axle housing (fig. 51). Remove the shims and spacer from the drive pinion, noting the thickness of the shims removed from the pinion.

POWER TRAIN

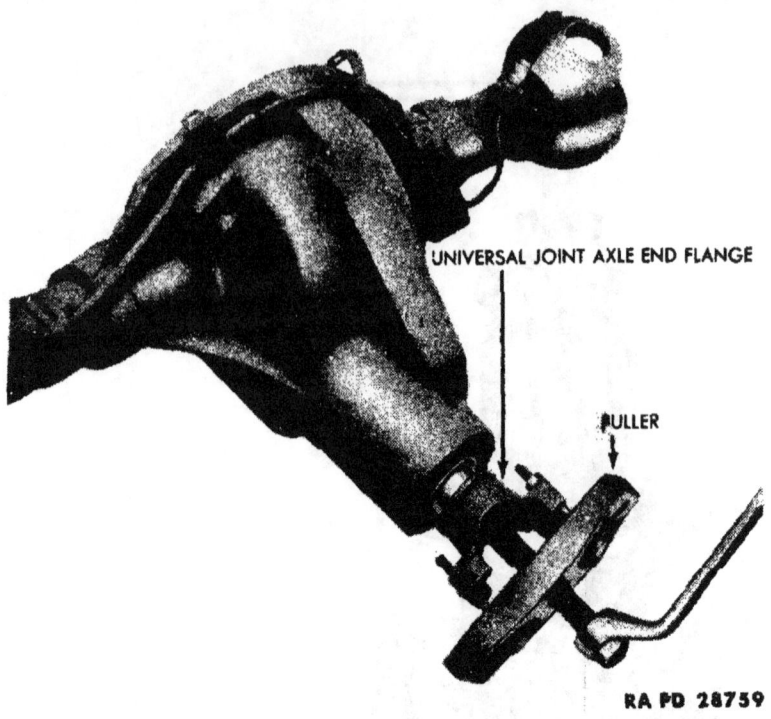

Figure 50 — Removing Universal Joint Axle End Flange With Puller 41-P-2905-60

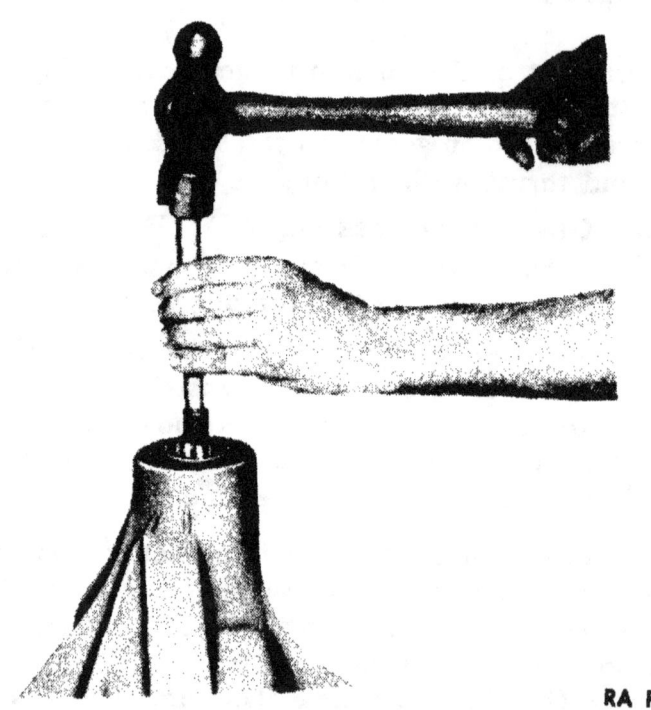

Figure 51 — Removing Drive Pinion

RA PD 28761

Figure 52 — Installing Pinion Outer Bearing Cup

25. CLEANING, INSPECTION, AND REPAIR.

a. Cleaning. Clean all parts in dry-cleaning solvent. Rotate the bearings while immersed in the dry-cleaning solvent until all trace of lubricant has been removed. Oil the bearings to prevent corrosion of the highly polished surface unless they are to be used immediately.

b. Inspection and Repair.

(1) AXLE HOUSING (fig. 53).

(a) Inspection. Replace the axle housing if it is bent or has any broken welds or cracks. Drive pinion bearing cups that are pitted, corroded or discolored due to overheating must be replaced (step *(c)*, below). Spindle housing bearing cups that are pitted or corroded must be replaced (step *(d)*, below). Replace the oil seals in the axle housing regardless of their condition (step *(e)*, below). Replace the differential cover, if cracked or if it has damaged threads in the filler plug hole. Check the cover for missing or damaged breather cap. Check the steering pivot arm shaft. If the diameter is less than 0.747 inch, replace the pivot shaft (step *(b)*, below). If the front axle is equipped with a Tracta type axle shaft, measure the

POWER TRAIN

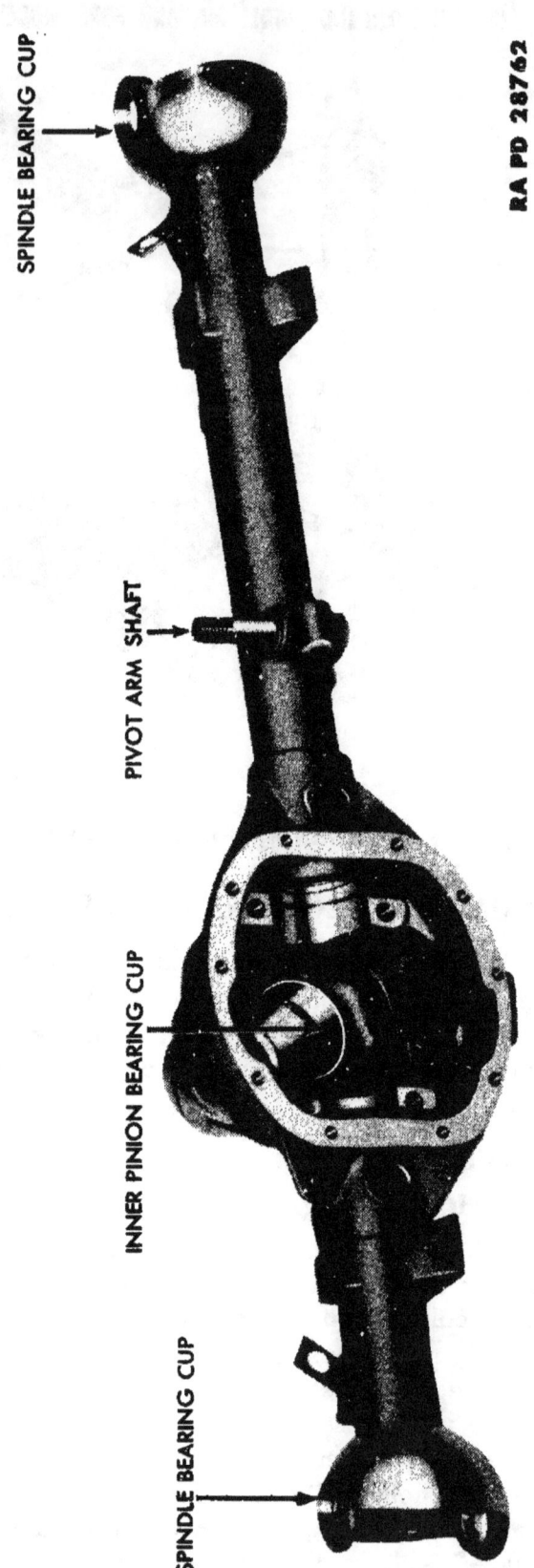

Figure 53 — Front Axle Housing

TM 9-1803B
25

ORDNANCE MAINTENANCE — POWER TRAIN, BODY, AND FRAME FOR ¼-TON 4 x 4 TRUCK (WILLYS-OVERLAND MODEL MB AND FORD MODEL GPW)

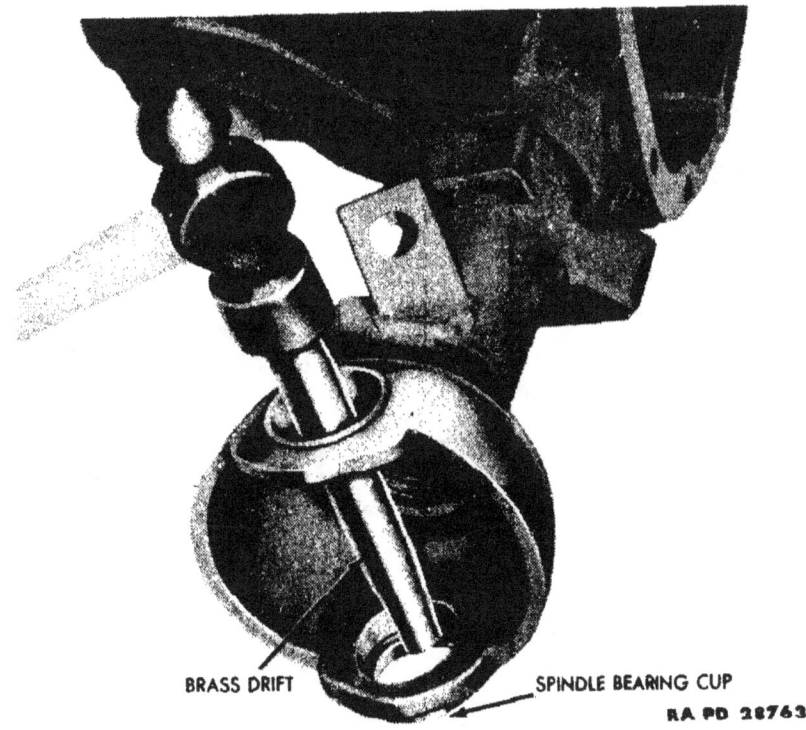

Figure 54 — Removing Spindle Bearing Cup From Axle Housing

inside diameter of the housing at each end of the axle housing. If the bushing is worn to more than 1.285 inch, replace the bushing (step *(f)*, below).

(b) Pivot Arm Shaft Replacement (fig. 53). With a long nosed drift, drive out the dowel that secures the pivot arm shaft to the axle housing. Tap the shaft out of the housing. To install a new pivot arm shaft, insert it in the bracket on the housing with the dowel slot in line with the dowel hole. Drive dowel in place.

(c) Drive Pinion Bearing Cup Replacement. Remove the inner and outer bearing cups, using a standard puller, noting the thickness of the shims when removing the inner bearing cup. To install new bearing cups, use a brass drift and hammer. Place the original thickness of shims behind the inner bearing cup and tap the bearing cups lightly around the entire circumference until flush with the shoulder in the axle housing (fig. 52).

(d) Spindle Housing Bearing Cup Replacement. Working through one of the bearing cups, tap the opposite bearing cup out of the axle housing, using a brass drift and hammer (fig. 54). To install new bearing cups, place the bearing cup in position and tap the cup lightly until it is flush with the shoulder in the axle housing.

POWER TRAIN

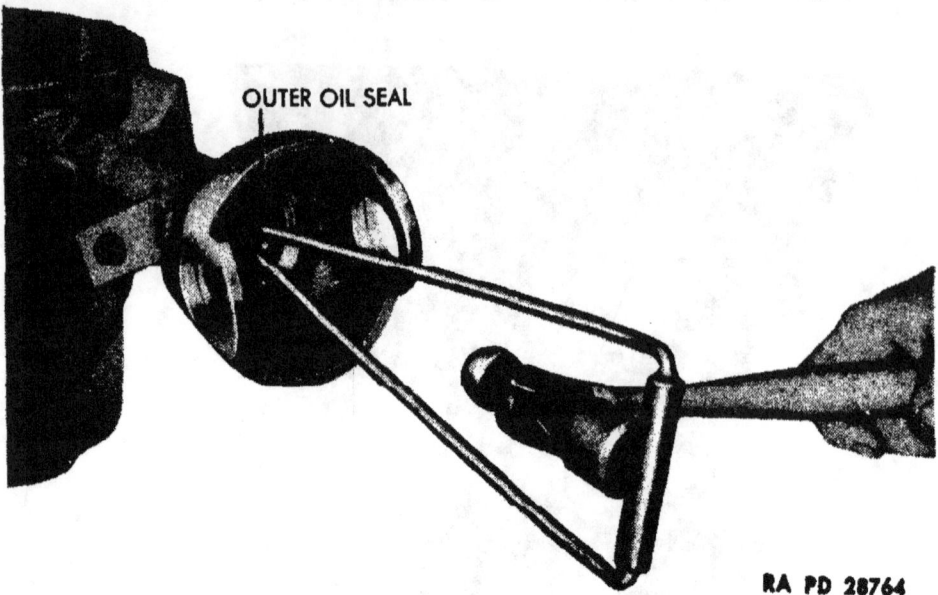

Figure 55 — Removing Oil Seal From Outer End of Axle Housing With Remover 41-R-2384-38

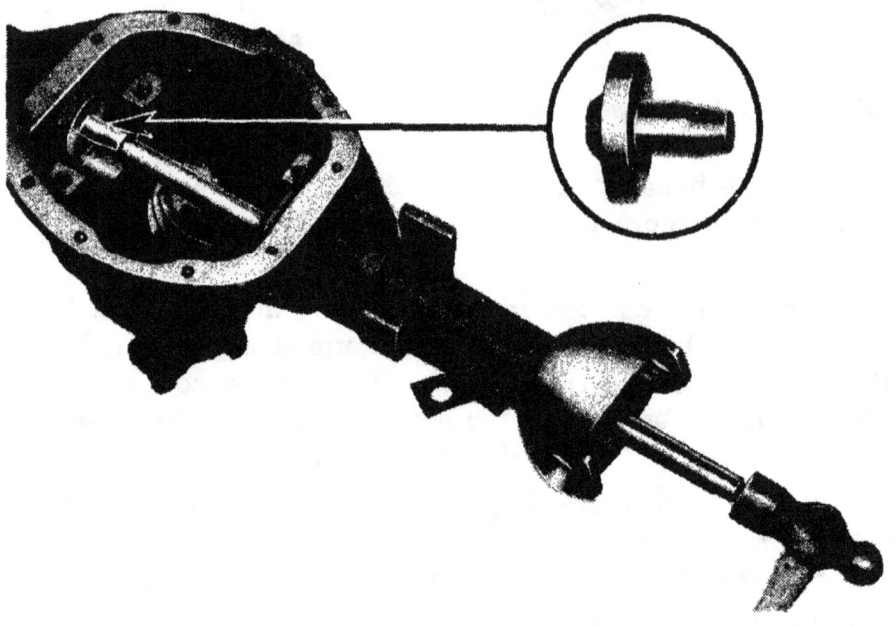

Figure 56 — Installing Oil Seal, With Replacer 41-R-2391-20

(e) *Oil Seal Replacement* (fig. 55). To remove the outer axle shaft oil seal, remove the oil seal retainer. Use a screwdriver to pry the retainer out of the housing. Use the oil seal remover 41-R-2384-38 to remove the inner and outer oil seals (figs. 55 and 80). To install the inner and outer oil seals, use the oil seal replacer

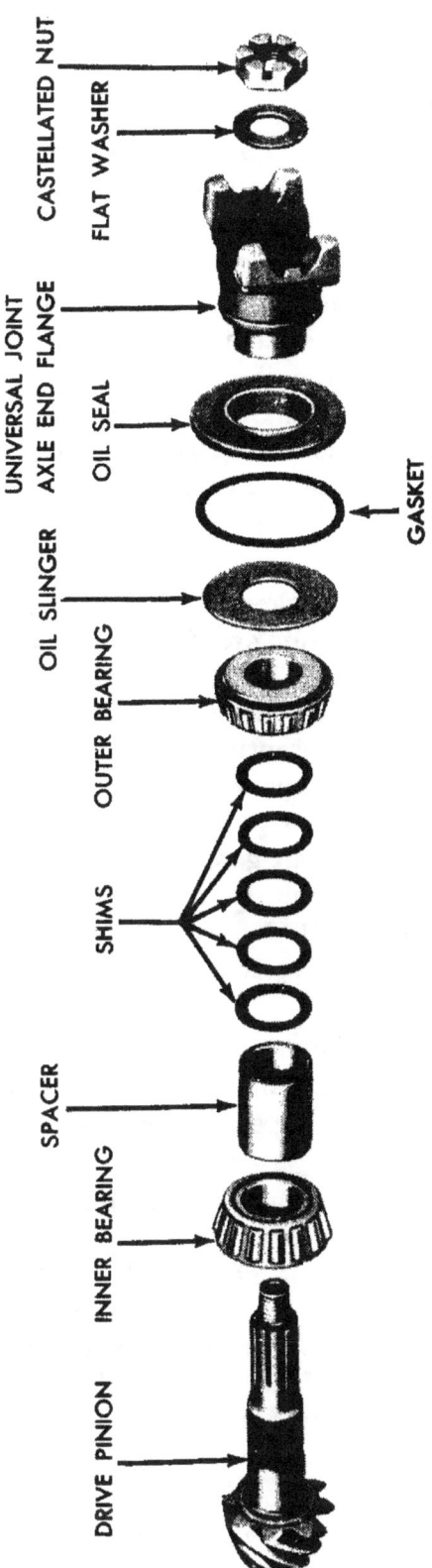

Figure 57 — Drive Pinion Assembly — Exploded View

POWER TRAIN

41-R-2391-20 and tap the oil seals in the inner and outer ends of the axle housing (fig. 56). Using a brass hammer, tap the oil seal retainer in the outer end of the axle shaft housing.

(f) Axle Housing Bushing Replacement (For Tracta Type Axle Shafts Only). Remove the bushing from the outer end of the axle housing, using a standard puller. To install a new bushing, place the bushing in position in the axle housing and using a suitable driver, drive the bushing in the housing until it is flush with the shoulder in the axle housing.

(2) DRIVE PINION ASSEMBLY (fig. 57). Roller bearings that are pitted, corroded or discolored due to overheating must be replaced. Replace the drive pinion if it has worn or broken teeth. The differential ring gear and the drive pinion assembly are furnished only in matched sets and if either is found damaged, both must be replaced. Small nicks can be removed from the pinion gear with a fine stone.

(3) DIFFERENTIAL ASSEMBLY (fig. 58). Replace any gear that is excessively worn or has any broken teeth. The differential ring gear and the drive pinion assembly are furnished only in matched sets and if either is found damaged, both must be replaced. Replace the differential pinion gears, if the inside diameter is worn to more than 0.627 inch. Replace the differential pinion shaft if the diameter is less than 0.623 inch. Replace the axle shaft gears if the outside diameter of the hub is worn to less than 1.498 inches. Replace the differential pinion gear and axle shaft gear thrust washers if the thickness is worn to less than 0.32 inch. Roller bearings and races that are pitted, corroded or discolored due to overheating must be replaced. All shims that were damaged during the disassembly must be replaced.

(4) AXLE SHAFTS. Three different types of axle shaft universal joints as shown in figures 38, 42, and 44 are used in front axles. Inspection of each of these three types is covered in steps *(a), (b),* and *(c),* below.

(a) Rzeppa Universal Joint (fig. 59). Replace the inner axle shaft if it is bent or has worn splines. Using a new axle shaft gear as a gage, slip it on the inner axle shaft and check the backlash. If the backlash is more than 0.005 inch, replace the axle shaft. Replace the outer axle shaft if it has worn splines or nicked ball bearing surfaces. Replace the inner race if it is found to be excessively worn. Small nicks or scratches can be removed with a fine stone. Replace steel balls that have flat spots. Replace the cage if it is cracked.

(b) Bendix Universal Joint (fig. 43). Replace the inner axle shaft if it is bent or has worn splines or worn universal joint ball surface. Replace the universal joint knuckle if it has worn splines or

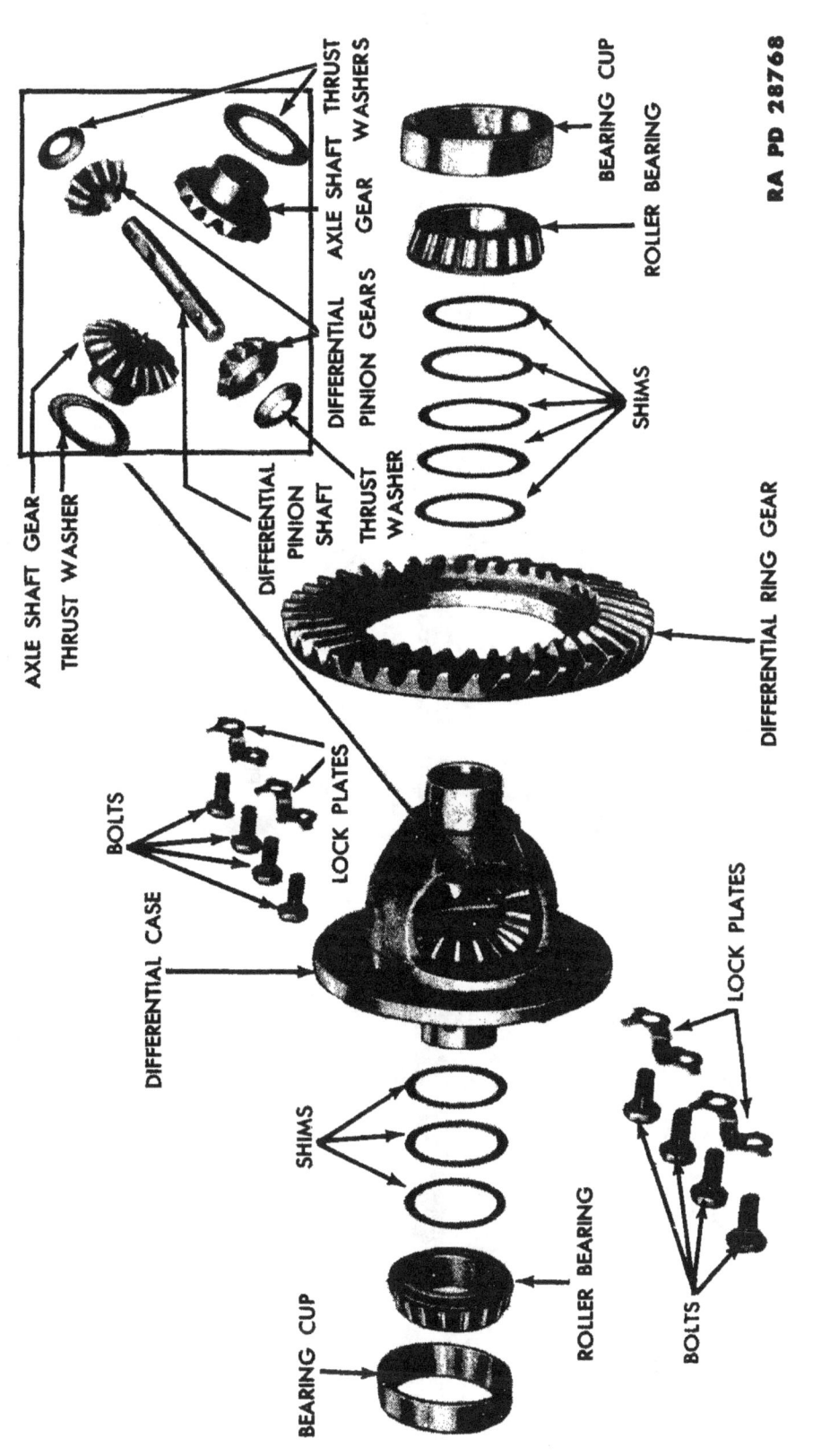

Figure 58 — Differential Assembly — Exploded View

POWER TRAIN

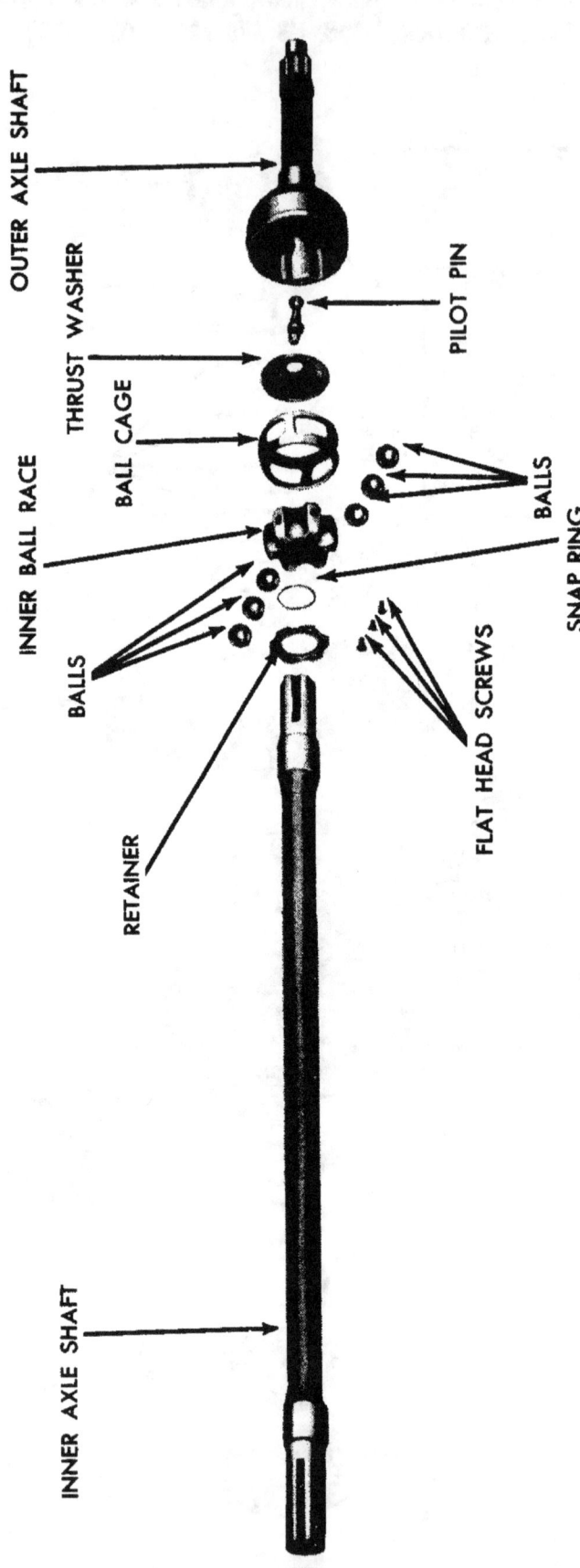

Figure 59 — Axle Shaft — Exploded View (Rzeppa Type)

TM 9-1803B
25

ORDNANCE MAINTENANCE — POWER TRAIN, BODY, AND FRAME FOR ¼-TON 4 x 4 TRUCK (WILLYS-OVERLAND MODEL MB AND FORD MODEL GPW)

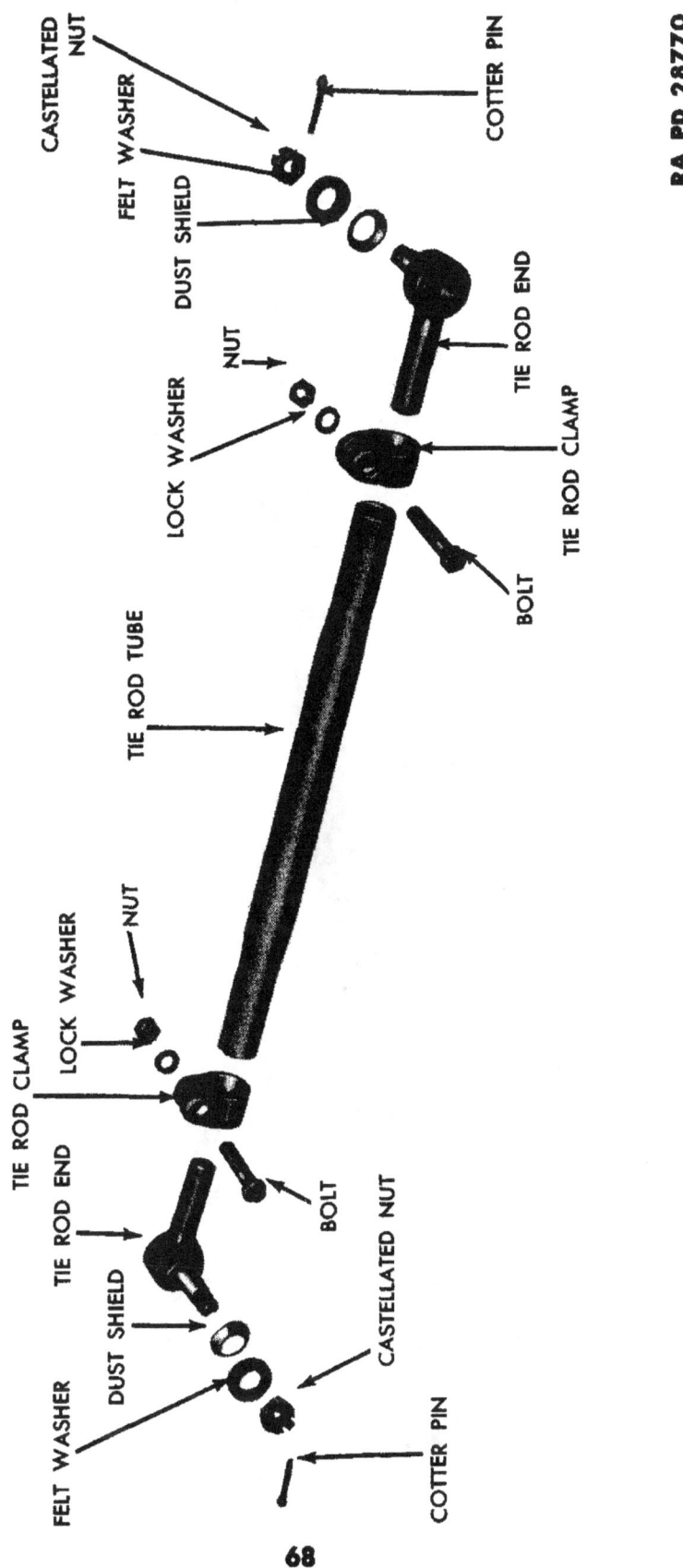

Figure 60 — Tie Rod, Right Side — Exploded View

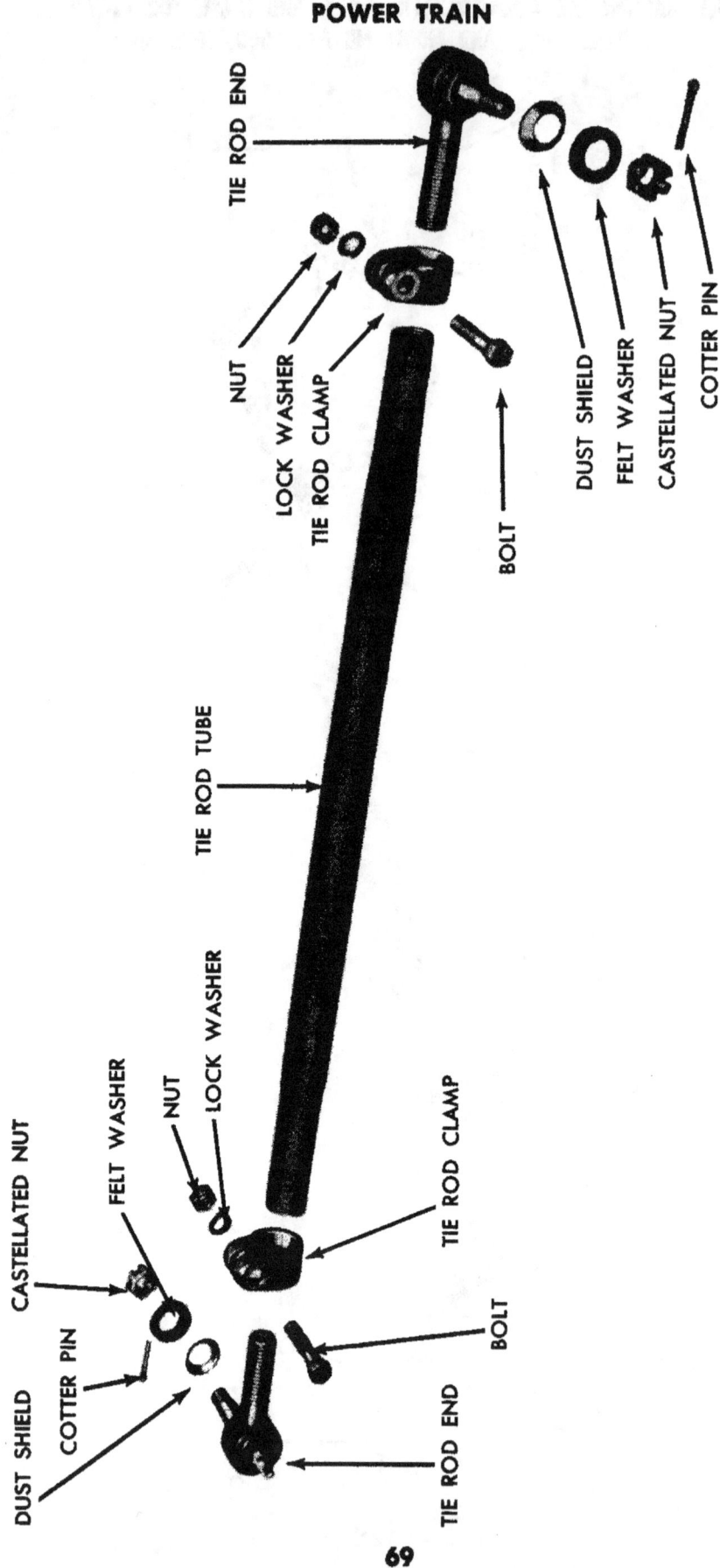

Figure 61 — Tie Rod, Left Side — Exploded View

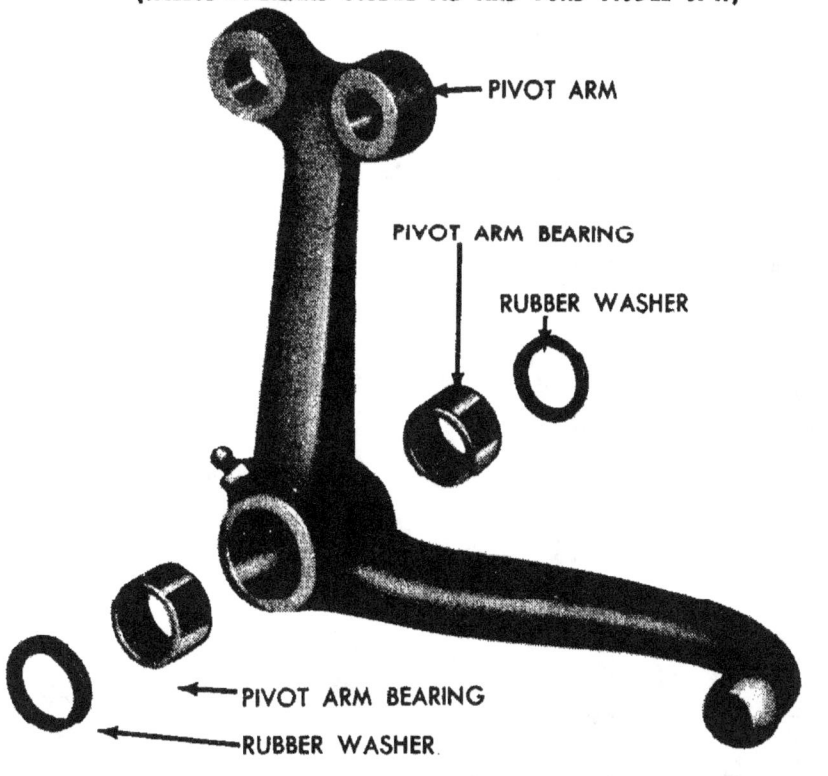

Figure 62 — Pivot Arm — Exploded View

worn ball surfaces. Small nicks or scratches can be removed with a fine stone. Replace universal joint balls, if they are excessively worn or have any flat spots.

(c) *Tracta Universal Joint* (fig. 45). Replace the inner portion or the outer portion of the axle shafts, if they are bent or have worn splines. Replace the inner portion or the outer portion of the universal joints, if they are cracked or excessively worn. Small nicks or scratches can be removed with a fine stone.

(5) TIE RODS AND PIVOT ARM (figs. 60, 61, and 62).

(a) *Inspection.* Replace the tie rods if bent or damaged. Replace the tie rod ends if the sockets are loose (step (b), below). Replace the pivot arm, if it is bent or has a worn ball joint. Replace the needle roller bearings in the pivot arm if they are loose or excessively worn (step (c), below).

(b) *Tie Rod End Replacement* (figs. 60 and 61). Loosen the tie rod clamps at both ends of the tie rod. Remove the tie rod ends from the tie rods. To install tie rod ends, place the tie rod clamps on the tie rod. Install the tie rod ends.

(c) *Pivot Arm Needle Bearing Replacement* (fig. 62). Place the pivot arm in a press and with a suitable driver, press out the two

POWER TRAIN

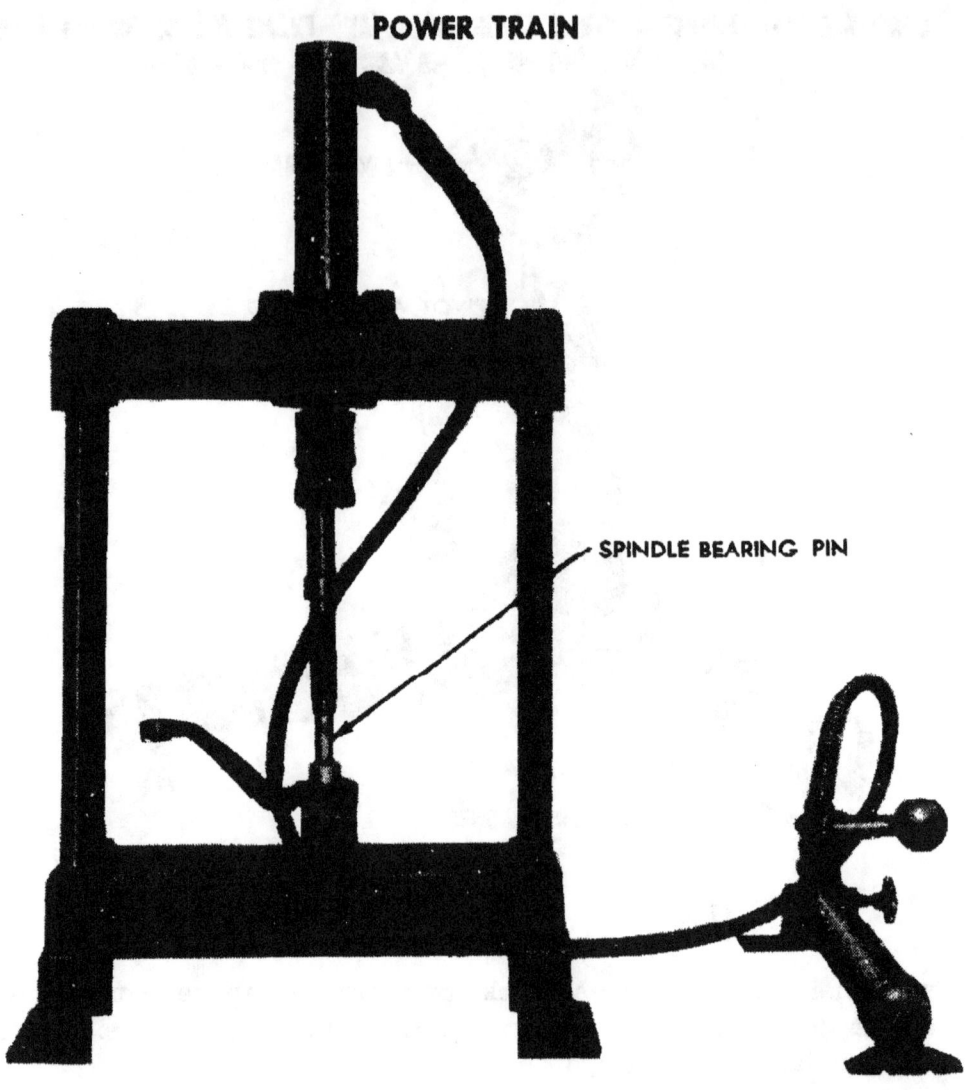

Figure 63 — Replacing Spindle Bearing Pin in Spindle Arm

needle bearings. To install the needle bearings, press one needle bearing in the pivot arm about $1/16$ inch below the shoulder of the pivot arm, then turn the pivot arm over and press the other bearing in the arm about $1/16$ inch below the shoulder of the pivot arm.

(6) SPINDLE ARM AND SPINDLE HOUSING BEARING CAP.

(a) Inspection. Replace the spindle arm if bent. Replace the spindle arm bearing pin if the diameter of the pin is worn to less than 0.623 inch. Replace the spindle housing bearing cap pin if it is worn to less than 0.625 inch (step *(b)*, below).

(b) Spindle Bearing Pin Replacement. Place the spindle housing bearing cap or the spindle arm (fig. 76) in a press and with a suitable driver, press out the bearing pin. To install a new pin, use a suitable driver and press the pin in until it is flush with the outer

TM 9-1803B
25

ORDNANCE MAINTENANCE — POWER TRAIN, BODY, AND FRAME FOR ¼-TON 4 x 4 TRUCK (WILLYS-OVERLAND MODEL MB AND FORD MODEL GPW)

RA PD 28774

Figure 64 — Removing Spindle Bushing

RA PD 28773

Figure 65 — Pressing New Bushing in Spindle

POWER TRAIN

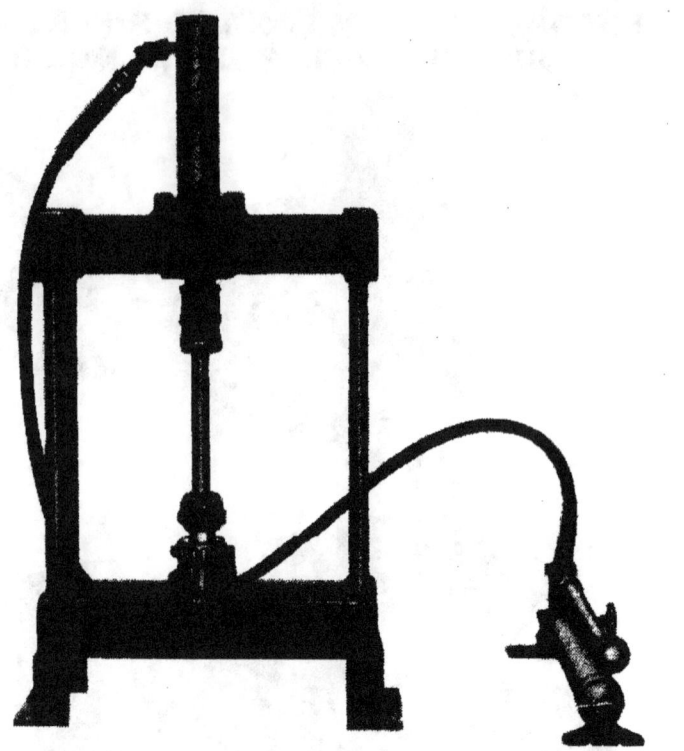

RA PD 28776

Figure 66 — Installing Inner Bearing on Pinion

shoulder. The same procedure applies to both the spindle housing bearing cap and the spindle arm (fig. 63).

(7) SPINDLE HOUSING AND SPINDLE (fig. 76).

(a) Inspection. Replace the spindle housing if it is cracked. If the studs on the spindle housing are bent, broken or damaged, replace them (step *(b)*, below). Replace the spindle, if it has damaged threads or grooved bearing surfaces. If the inner diameter of the spindle bushing is more than 1.225 inch, replace the bushing (subpar. *(c)*, below).

(b) Broken Stud Replacement. Indent the end of the broken stud exactly in the center with a center punch. Drill approximately two-thirds through the broken stud, using a small drill, then follow up with a larger drill (the size of the drill depending on the size of the stud to be removed). The drill selected, however, must leave a wall thicker than the depth of the threads. Select an extractor of the proper size. Insert it into the drilled hole and screw out the remaining part of the broken stud. To install a new stud, use a standard stud driver and drive all studs until no threads show at the bottom of the stud. If the stud is too tight or too loose in the stud hole, select another stud.

(c) Spindle Bushing Replacement. The spindle bushing can be removed with a center punch as shown in figure 64. To install a

Figure 67 — Checking Pinion in Differential Housing With Gage 41-G-176

new bushing, use a suitable driver and press the bushing in the spindle (fig. 65).

26. ASSEMBLY.

a. **Install Inner Bearing on Pinion** (fig. 66). Press the inner bearing on the pinion, using an arbor press. Make sure the bearing is seated against the shoulder of the pinion gear when installed.

b. **Adjust Pinion in Housing** (fig. 67). Place the pinion in the differential housing. Install the 41-G-176 gage to check the setting from the back face of the pinion to the center line of the differential case bearing. The standard setting is 0.719 inch. If the gage reading is more than 0.179 inch, shims will have to be added to the inner bearing cup (par. 25 b (1) (c)). If the gage reading is less than 0.719 inch, shims will have to be removed from the inner bearing cup (par. 25 b (1) (c)).

c. **Install Outer Bearing on Pinion** (fig. 57). After the correct pinion setting has been obtained, install the spacer and the original amount of shims on the pinion. If the thickness of the original shims is unknown, install the shims totaling approximately 0.060 inch thick

POWER TRAIN

Figure 68 — Installing Differential Bearing, Using Special Replacer 41-R-2391-65

before installing the outer bearing. Start the outer bearing on the pinion. Install the oil slinger on the pinion.

d. Adjust the Outer Bearing. Place the universal joint flange on the pinion. Install the nut on the universal joint flange and draw the flange down tight. Turn the universal joint flange, if there is a slight drag, the pinion bearing adjustment is correct. If the pinion turns with difficulty or can't be turned by hand, shims will have to be added behind the outer bearing. If the pinion turns loosely, shims will have to be removed. If the pinion bearing adjustment is not correct, remove the universal joint flange, and add or remove shims until the correct adjustment is obtained. After the correct adjustment is obtained, again remove the universal joint flange and install the oil seal on the pinion. Install the universal joint flange. Install the nut and cotter pin.

e. Install Gears in Differential Case (fig. 58). Place the axle shaft gear thrust washers on the two axle shaft gears. Place the axle shaft gears in the case. Place the two differential pinion thrust washers and gears in the case. Install the differential pinion gear shaft in the case. Install the pinion shaft lock pin in the case and stake the pinion shaft lock pin to prevent the pin from coming out.

TM 9-1803B
26

ORDNANCE MAINTENANCE — POWER TRAIN, BODY, AND FRAME FOR ¼-TON 4 x 4 TRUCK (WILLYS-OVERLAND MODEL MB AND FORD MODEL GPW)

Figure 69 — Checking Clearance Between Differential Case and Bearing

f. Install Differential Ring Gear (fig. 58). Place the differential ring gear in position on the case. Install the lock plates and cap screws that secure the ring gear to the case. Bend the ears of the lock plates on the cap screws.

g. Install Roller Bearings on Case (fig. 68). If all the original parts have been used in the differential assembly, add the same thickness of shims as originally used and press the roller bearings on the case, then proceed with subparagraph h, below. If the original parts are not being used, or if the original shim thickness is not known, install the roller bearings on the case without the shims, and proceed with subparagraph i, below.

h. Install Differential Assembly in Housing (fig. 47). Place the bearing cups on the roller bearings. Tilt the bearing cups in order to start the assembly in the housing. Tap the bearing cups lightly until the assembly is seated firmly in the housing. Install the two bearing caps so the numbers on the caps and the housing face the same way and match in every way as shown in figure 47. If the differential assembly being used is not the one originally in the axle, proceed with subparagraph i, below.

i. Adjust Differential Assembly (fig. 69). Place the bearing cups on the differential assembly and place the assembly in the housing.

TM 9-1803B
26
POWER TRAIN

RA PD 28780

Figure 70 — Checking Ring Gear Backlash With Dial Indicator 41-I-100

Slide the assembly to one side of the housing. Check the clearance between the bearing cup and differential housing with a feeler gage. After this clearance has been determined, add 0.008 inch. This will give the thickness of shims required for proper bearing adjustment. Remove the differential assembly from the housing. Remove the bearings from the differential case (par. 24 f (3)). Install the amount of shims determined above in equal amounts on each side of the case and install the bearings back on the case (subpar. g, above). Tilt the bearing cups and place the differential in the housing. Tap the bearing cups lightly until the assembly is seated firmly in the housing. Install the two bearing caps so the numbers on the caps and the housing face the same way, and match in every way.

j. Check Backlash (fig. 70). Install a dial indicator 41-I-100 on the differential housing so that the indicator contact is resting on the surface of a ring gear tooth as shown in figure 70. Rotate the ring gear back and forth to determine the backlash. If the backlash is less than 0.005 inch or more than 0.007 inch, remove the differential from the housing (par. 24 e) and remove the bearings from the differential case (par. 24 f (3)). If the backlash was more than 0.007 inch, the ring gear must be brought closer to the pinion. If the backlash was less than 0.005 inch, the ring gear must be moved away from the pinion. This is accomplished by moving shims equal to the

77

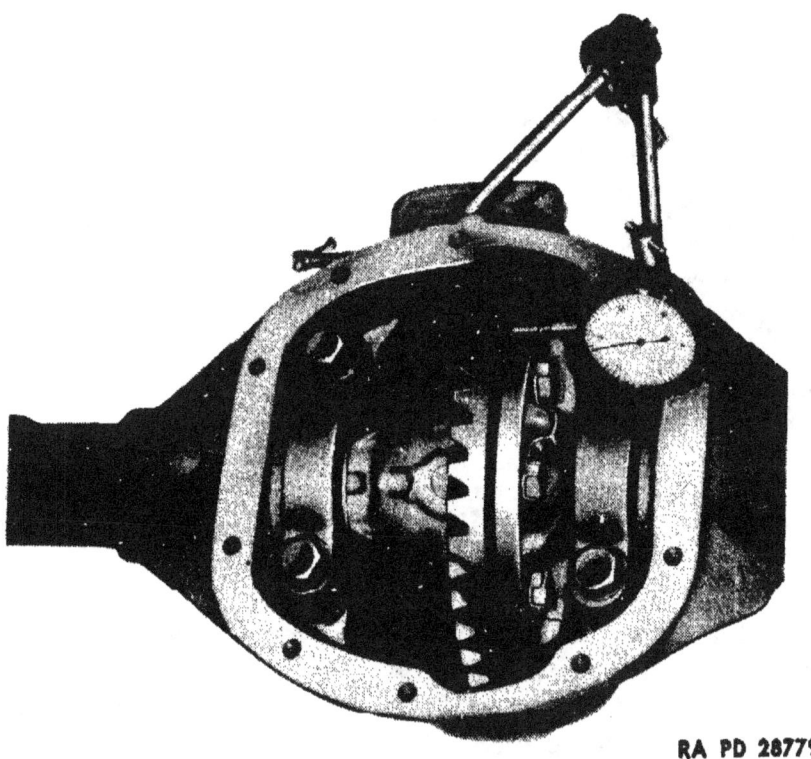

Figure 71 — Checking Ring Gear Run-out With Dial Indicator 41-I-100

error in backlash from one side of the case and adding them to the other side. Install the bearings on the case (subpar. g, above). Install differential in housing (subpar. h, above) and recheck the backlash.

k. Check Ring Gear Run-out (fig. 71). Install a dial indicator on the differential housing so that the indicator contact is resting on the flat side of the ring gear as shown in figure 71. Turn the pinion drive flange by hand to determine the run-out on the ring gear. The run-out should not exceed 0.003 inch. If the run-out is more than 0.003 inch, remove the differential assembly from the housing (par. 24 e), and remove the ring gear from the differential case. Check the surface of the differential case and the ring gear for chips or small nicks, which might have occurred during the assembly of the differential. If any small nicks are found, remove them with a fine stone, also check the flange on the differential case for being sprung. Reinstall the differential assembly in the housing (subpar. h, above) and recheck the ring gear run-out.

l. Install Differential Cover (fig. 46). Place a new gasket and the differential cover in place on the axle housing. Install the 10 cap screws that secure the cover to the housing.

POWER TRAIN

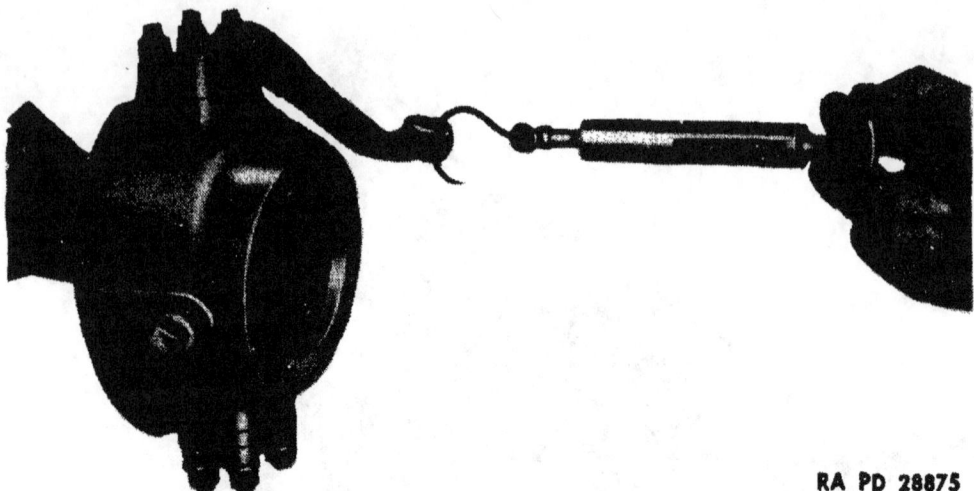

RA PD 28875

Figure 72 — Checking Tension of Spindle Housing

m. Install Pivot Arm (fig. 46). Insert the two rubber seals in the pivot arm. Place the flat washer on the pivot arm shaft. Place the pivot arm on the shaft with the ball joint of the arm facing downward. Place the flat washer and dust shield on the shaft. Install the castellated nut and cotter pin.

n. Install Spindle Housing (fig. 76). Dip the two spindle housing bearings in grease. Place the bearings in the bearing cups on the axle housing. Place the spindle housing on the axle housing with the grease plug to the rear of the vehicle. Install shims totaling 0.048 inch thick on the spindle bearing cap and the spindle arm. Shims are available in thicknesses of 0.003 inch, 0.005 inch, 0.010 inch and 0.030 inch. Install one of each size on the top and bottom of the spindle housing. Place the lower bearing cap on the spindle housing and install the four nuts that secure the cap to the spindle housing. Place the spindle arm on the spindle housing and install the four nuts that secure the spindle arm to the spindle housing.

o. Adjust Spindle Housing (fig. 72). Check the tension of the spindle housing by hooking a scale to the end of the spindle arm. The tension should not be more than 6 pounds or less than 4 pounds. If the tension is over 6 pounds, shims must be removed from the spindle housing. If the tension is less than 4 pounds, shims must be added. When removing or adding shims, be sure the same thickness is removed from, or added to, both ends of the spindle housing. Remove, or add, shims until the correct tension is obtained.

p. Install Spindle Housing Oil Seal (fig. 76). Place a new gasket on the spindle housing. Place the upper and lower halves of the oil seal on the spindle housing. Install the four cap screws that secure the lower half of the oil seal to the spindle housing. Place the axle shaft identification tag on the upper half of the oil seal. Install the four cap screws that secure the upper half of the oil seal to the spindle housing.

q. **Assemble Axle Shafts.** Three different types of front axle universal joints (figs. 38, 42, and 44) are used. Assembly procedures for the Rzeppa and Bendix types are covered in subparagraphs (1) and (2), below. The Tracta type front axle universal joint (fig. 43) requires no assembly before installation (subpar. r (2), below).

(1) RZEPPA JOINT (fig. 38). Hold the cage in a horizontal position and hold the inner race in a vertical position (fig. 41). Insert the inner race in the cage, dropping one of the inner race bosses into one of the larger elongated holes. When the race is entered in the cage, turn the race so that it is entirely in the cage. Line up the two larger elongated holes with the bosses on the axle shaft (fig. 40). Slide the cage in the axle shaft. Holding the cage in this position, insert the thrust washer (fig. 59) behind the cage. Tilt the cage so that it is flush with the shaft. Tilt the cage so that a steel ball can be inserted in the elongated hole (fig. 39). After the steel ball is in position, push the cage down until the opposite side of the cage is exposed. Insert another steel ball in the elongated hole on the cage and push the cage down. Repeat this operation until all the steel balls are in the cage. Insert the pilot pin (fig. 59) in position. Insert the retainer on the axle shaft and secure the retainer with the snap ring. Insert the inner shaft in the outer shaft. Install the three flat head screws that secure the retainer to the inner race.

(2) BENDIX JOINT (figs. 42 and 43). Place the universal joint knuckle in an upright position in a vise. Insert the center ball in the hole of the knuckle. Place the center ball in its race on the center ball pin hole. Arrange the center ball so that the grooved side of the center ball is away from the pin hole. Insert the three universal joint balls in their races. Arrange the center ball so that the grooved side is in line with the race of the last ball to be installed as shown in figure 42 and drop the ball in its race. Rotate the center ball in its race until the hole in the ball is in line with the center ball pin. Remove the assembly from the vise. Turn the assembly over so that the pin may drop in the hole of the center ball. Install the grooved pin in the knuckle and stake the pin to prevent it from coming out.

r. **Install Axle Shaft.**

(1) BENDIX AND RZEPPA JOINTS. Slide the axle shaft in the axle housing. It will be necessary to turn the axle shaft until the splines on the axle shaft are in line with the axle shaft gear in the differential.

TM 9-1803B
26

POWER TRAIN

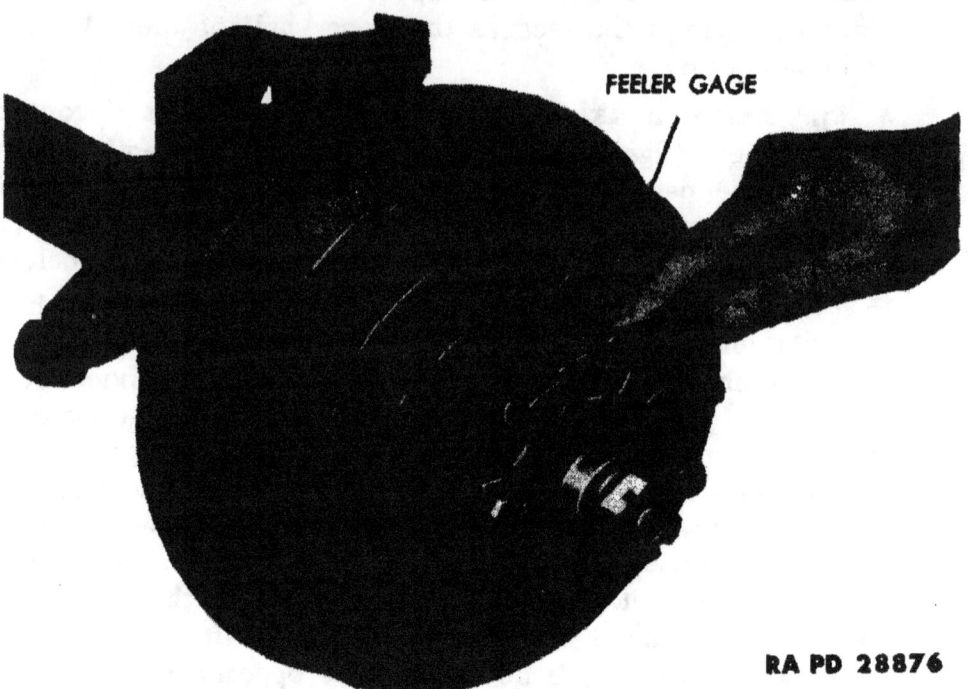

RA PD 28876

Figure 73 — Checking Clearance Between Drive Flange and Hub

(2) TRACTA JOINT (fig. 45). Slide the inner portion of the axle shaft and the inner portion of the universal joint into the axle housing. Turn the axle shaft so as to line up the splines of the axle shaft with the axle shaft gear in the differential. Slide the outer portion of the universal joint on the outer portion of the axle shaft. Line up the slots of the two universal joints and slide the outer axle shaft in place on the axle.

s. Install Brake Plate and Spindle. Place the spindle on the spindle housing. Place the brake plate on the spindle with the wheel cylinder toward the top of the brake plate. Line up the holes in the brake plate and spindle with the spindle housing. Install the six cap screws that secure them to the spindle housing.

t. Install Hydraulic Brake Hose (fig. 46). Install the brake hose to the brake line on the axle housing. Install the clamp to the brake hose at the bracket on the axle housing. Insert the brake hose through the guard and connect the hose to the brake line on the brake plate. Install the brake hose clamp at the guard.

u. Install Hub and Brake Drum (fig. 46). Pack the wheel bearings with the specified lubricant. Insert the hub and brake drum on the spindle with the inner wheel bearing and grease retainer in the hub. Insert the smaller thrust washer on the spindle and install the bearing adjusting nut. Tighten the adjusting nut until the brake

TM 9-1803B
26

ORDNANCE MAINTENANCE — POWER TRAIN, BODY, AND FRAME FOR ¼-TON 4 x 4 TRUCK (WILLYS-OVERLAND MODEL MB AND FORD MODEL GPW)

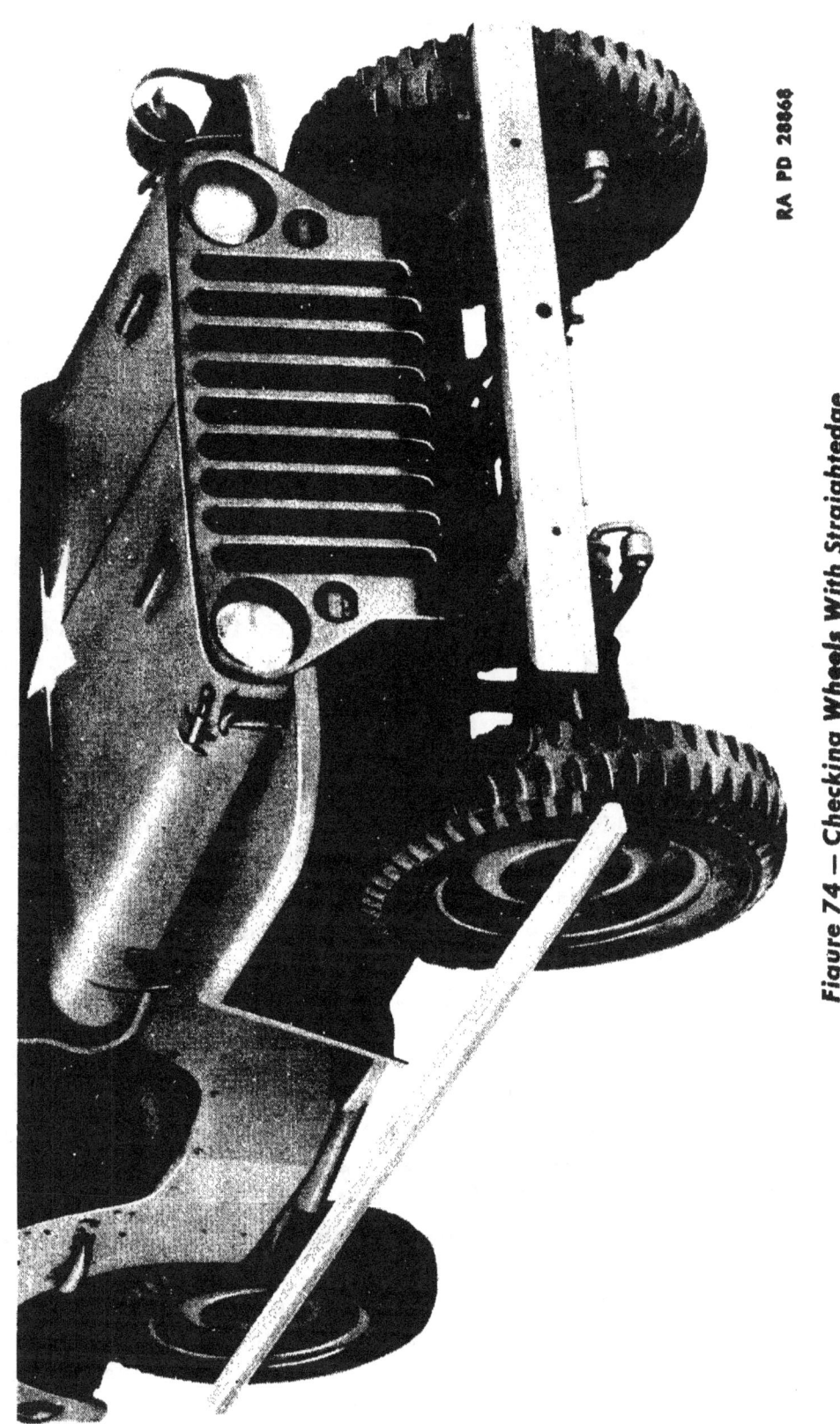

Figure 74 — Checking Wheels With Straightedge

POWER TRAIN

drum binds when turned; then back off the adjusting nut one-eighth turn. This will give the correct wheel bearing adjustment. Install the lock washer and lock nut on the spindle. Bend the ears of the lock washer over the lock nut.

v. **Install Drive Flange** (fig. 73).

(1) RZEPPA TYPE AXLE SHAFTS. Install a 0.060-inch thickness of shims between the drive flange and the hub. Place the drive flange on the axle shaft. Install the six cap screws that secure the drive flange to the hub. Install the castellated nut on the axle shaft. Install the hub cap on the drive flange.

(2) BENDIX OR TRACTA TYPE AXLE SHAFT (fig. 73). Place the drive flange on the axle shaft. Install the castellated nut on the axle shaft and draw it down tight. Turn the front wheels to the maximum left or right and measure the space between the drive flange and hub with a feeler gage (fig. 73) to determine the number of shims to be installed. Remove the nut from the axle shaft and remove the drive flange. Add the required thickness of shims between the drive flange and the hub. Install the six cap screws that secure the drive flange to the hub. Install the castellated nut on the axle shaft. Back off the nut on the axle shaft until a 0.50-inch feeler gage can pass between the nut and drive flange. Tap the nut on the axle shaft lightly. The axle shaft will move inward. Again check the space between the nut and drive flange. The space should not be less than 0.015 inch or more than 0.035 inch. If the space is less than 0.015 inch, add shims behind the drive flange and hub until this limit is obtained. If the space is more than 0.035 inch, remove shims from the drive flange until the above limit is obtained. Draw the nut on the axle shaft up tight. Install the hub cap.

w **Install Tie Rods** (fig. 46). Insert the ends of the tie rods in the spindle arms and pivot arm. Be sure the dust shield and felt washer are on the tie rod ends. Install the castellated nuts that secure the tie rod ends to the spindle arms and to the pivot arm.

27. INSTALLATION.

a. **Preliminary Work.** Place a hydraulic jack under the front axle assembly. Roll the assembly under the vehicle. Raise the assembly until the front springs can be raised and secure to the spring shackles. Lower the jack to allow the axle assembly to rest on the springs.

b. **Install Spring U-bolts** (fig. 35). Place the spring U-bolts in position on the axle housing. Install the spring seat plate on the U-bolts at the right side of the axle. Install the four nuts that secure the spring seat to the U-bolts. Raise the torque reaction spring in position on the U-bolts on the left side. Install the nuts that secure the torque reaction spring to the U-bolts.

TM 9-1803B
27

ORDNANCE MAINTENANCE — POWER TRAIN, BODY, AND FRAME FOR ¼-TON 4 x 4 TRUCK
(WILLYS-OVERLAND MODEL MB AND FORD MODEL GPW)

Figure 75 — Adjusting Toe-in, Using Wheel Alining Gage 41-G-510

POWER TRAIN

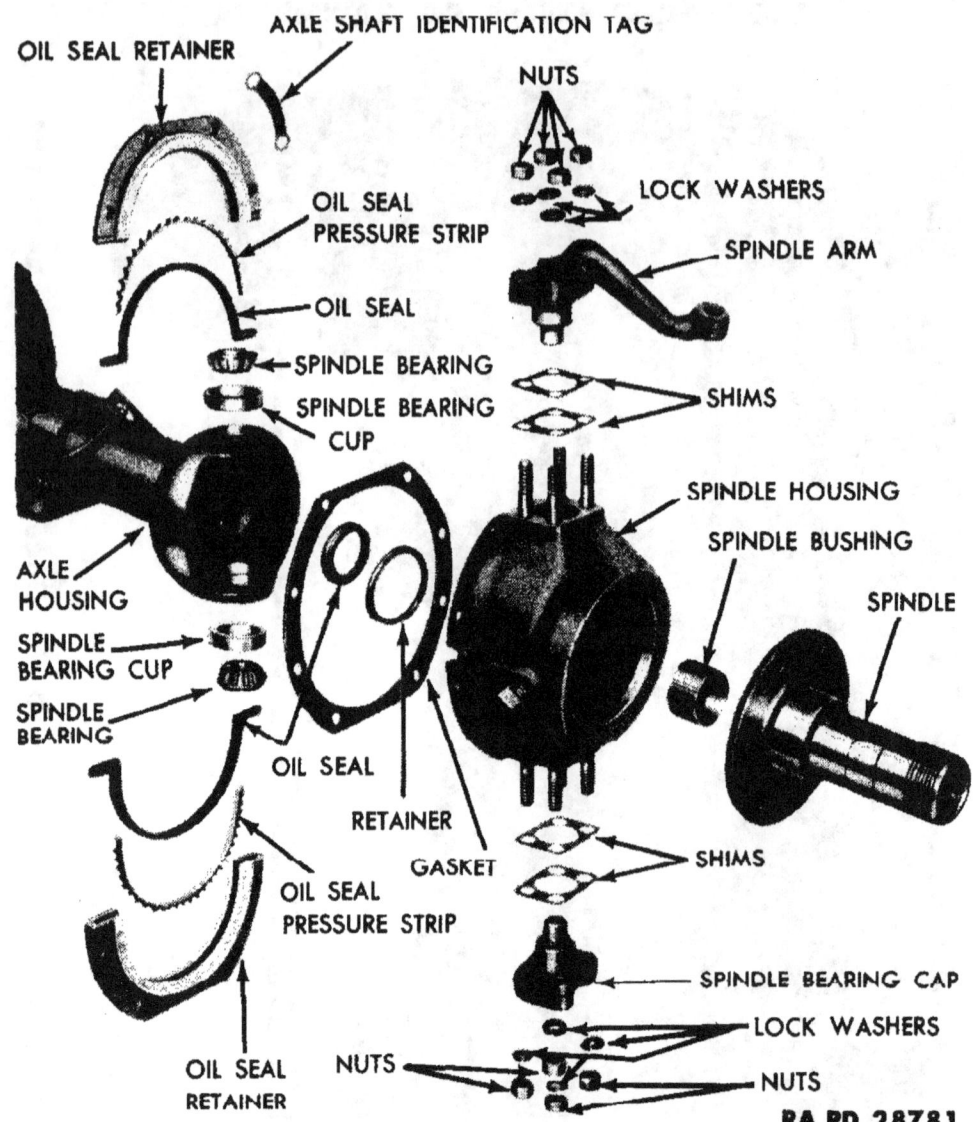

Figure 76 — Spindle Housing — Exploded View

c. **Install Shock Absorbers** (fig. 35). Insert a rubber mounting in each side of each shock absorber eye. Place the shock absorber on the mounting bracket at the spring seat plate. If new rubber mountings are being used, compress them with compressor 41-C-2554-400. Install the flat washer and cotter pin that secure the shock absorber to the spring seat plate. Place the left shock absorber on the mounting bracket at the torque reaction spring. Install the flat washer and cotter pin that secure the shock absorber to the torque reaction spring.

d. **Install Propeller Shaft, Drag Link, and Wheels** (fig. 35). Install the propeller shaft to the front axle (par. 21 b). Place the drag link in the ball joint on the pivot arm. Install the drag link plug on the drag link. Install the cotter pin in the drag link plug. Install the wheels.

e. **Lubricate.** Fill the differential to proper level with specified oil. Apply specified grease in each spindle housing and to all fittings. Bleed the hydraulic brake system. Refer to TM 9-803.

28. WHEEL ALINEMENT.

a. **Caster and Camber.** The caster and camber is established at the time of manufacture and cannot be altered by any adjustment.

b. **Toe-in.**

(1) EQUALIZE TIE RODS (fig. 74). Set the pivot arm parallel to the axle and place a straightedge against the left rear and left front wheel. If the rear or front sides of the front tire do not touch the straightedge, the tie rods must be adjusted. Loosen the tie rod clamps at both ends of the left tie rod. Turn the tie rod tube clockwise to bring the forward side of the front wheel inward, or counterclockwise to bring the rear side of the front wheel inward. Adjust until the straightedge touches the side of the front tire at both front and rear. Repeat this procedure on the right-hand side of the vehicle.

(2) ADJUST TOE-IN (fig. 75). After the tie rods have been equalized, pull the vehicle forward at least three feet to remove the backlash and place the telescoping wheel alining gage 41-G-510 between the wheels in front of the axle so that the chains on both ends of the gage are barely touching the floor. Set the scale so the pointer registers zero. Pull the vehicle forward until the gage is brought to a position back of the axle with both chains barely touching the floor. The reading at this point will be the amount of toe-in or toe-out. Adjust the right-hand tie rod until a toe-in of $\frac{1}{16}$ inch is obtained. Recheck the toe-in after making the adjustment. Tighten the tie rod clamps.

Section VI

REAR AXLE

29. DESCRIPTION AND DATA.

a. **Description** (fig. 78). The rear axle is the full-floating type, designed so that the axle shafts can be removed without disturbing the wheels. The differential drive is of the hypoid type. The differential parts are identical and are interchangeable with the front axle.

POWER TRAIN

b. Data.

Rear Axle:

Type	Full-floating
Make	Spicer
Drive	Through springs
Road clearance	8⅞ in.

Differential:

Type	Hypoid
Ratio	4.88 to 1
Bearings	Timken roller
Oil capacity (pt)	2.5

Pinion Shaft:

Bearings	Timken
Adjustment	Shims
Backlash	0.005 to 0.007 in.

30. REMOVAL.

a. **Preliminary Work.** Remove the differential drain plug and drain the oil. Raise the rear of the vehicle until the weight of the vehicle is off the rear springs.

b. **Disconnect Propeller Shaft** (fig. 77). Remove the four nuts and two U-bolts that secure the propeller shaft to the universal joint flange at the rear axle. Slide the propeller shaft off the universal joint flange. Wrap a piece of tape around the bearings on the propeller shaft to prevent losing the bearings.

c. **Disconnect Shock Absorbers and Hydraulic Brake Line** (fig. 77). Remove the cotter pin and flat washer that secure the two rear shock absorbers to the bracket on the spring seat plates. Pull the shock absorbers off the bracket. Disconnect the hydraulic brake line leading to the rear axle at the differential housing.

d. **Remove Spring U-bolts** (fig. 77). Remove the four nuts that secure the spring U-bolts at both rear springs. Remove the U-bolts and the spring seat plates from the axle.

e. **Disconnect Springs** (fig. 77). Remove the lower spring shackle bushings at the rear of both springs. Pull both springs off the spring shackles. Drop the springs to the floor and roll the rear axle assembly out from the vehicle.

Figure 77 — Rear Axle Assembly in Vehicle

TM 9-1803B
31

POWER TRAIN

31. DISASSEMBLY.

a. Remove Wheels. Remove the five nuts that secure each wheel to the hub. Remove the wheels.

b. Remove Axle Shafts (fig. 82). Remove the six cap screws that secure the drive flange to the hub. Install two of the cap screws that were removed from the drive flange in the two threaded holes on the drive flange. Draw the cap screws down until the drive flange is free from the hub. Remove the axle shafts from the axle housing.

c. Remove Hub and Drum Assembly (fig. 82). Bend the ears of the flat washer off the bearing lock nut. Remove the bearing lock nut and bearing adjusting nut off the housing, using the wrench furnished with the vehicle. Slide the hub and drum assembly with the wheel bearings off the housing.

d. Remove Brake Plate (fig. 82). Disconnect the hydraulic brake line at the brake plate. Remove the six cap screws that secure the brake plate to the axle housing. Remove the brake plate from the axle housing.

e. Remove Differential Assembly. Remove the 10 cap screws that secure the differential cover to the housing (fig. 78). Remove the differential cover. Remove the 4 cap screws from the 2 differential bearing caps (fig. 47), and remove the caps. Remove the differential assembly from the housing, using a pry bar if necessary. Reinstall the bearing caps in the housing, noting the markings (fig. 47) to assure their being installed in their correct location.

f. Remove Differential Pinion Gears and Axle Shaft Gears From Differential (fig. 48). Place the differential assembly in a vise equipped with brass jaws. With a long-nosed drift, drive the differential pinion shaft tapered pin out of the differential gear case (fig. 48). Tap the differential pinion shaft from the case with a brass drift and hammer. Remove the two differential pinion gears and thrust washers and the two axle shaft gears and thrust washers from the case.

g. Remove Ring Gear From Case (fig. 58). Bend the ears of the lock plates off the cap screws. Remove the cap screws that secure the ring gear to the case, and remove the ring gear.

h. Remove Roller Bearings From Differential Case (fig. 49). Place the differential case in a vise. Install the bearing remover 41-R-2378-30 to the roller bearing. Remove the roller bearings from both ends of the differential case. Remove the shims. Note the thickness of the shims removed from each side to assist in reassembly.

i. Remove Drive Pinion. Remove the nut and flat washer that secure the universal joint axle end flange to the drive pinion. Install

TM 9-1803B
31

ORDNANCE MAINTENANCE — POWER TRAIN, BODY, AND FRAME FOR ¼-TON 4 x 4 TRUCK (WILLYS-OVERLAND MODEL MB AND FORD MODEL GPW)

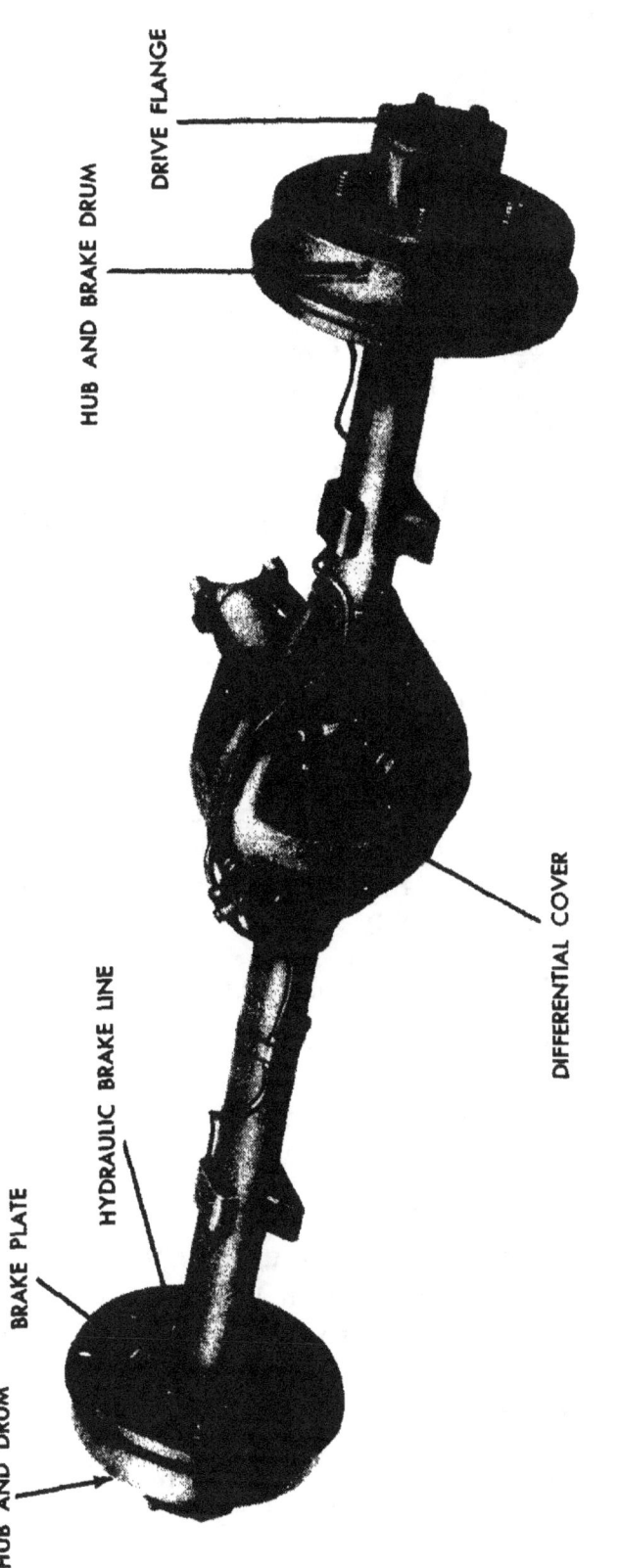

Figure 78 — Rear Axle Assembly

POWER TRAIN

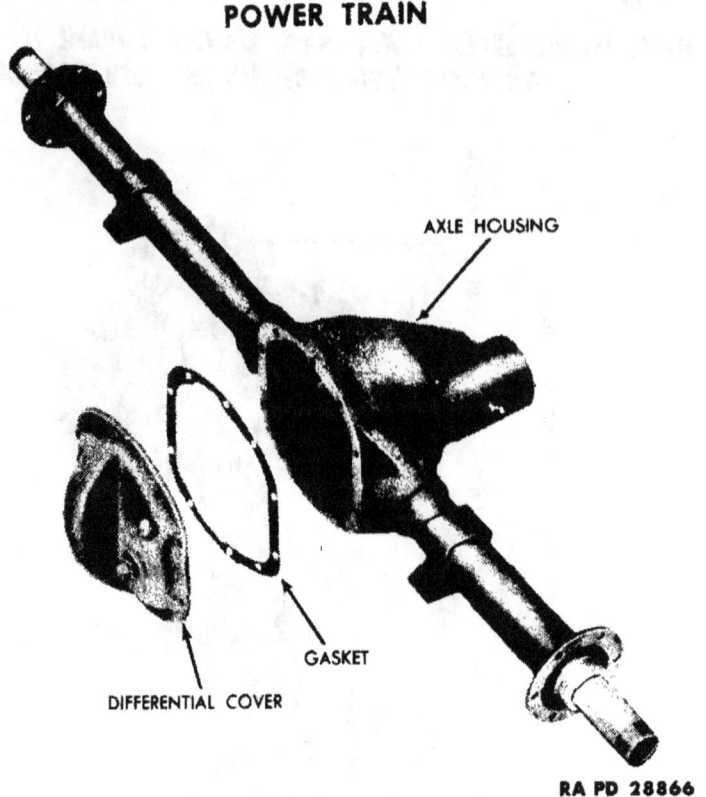

Figure 79 — Rear Axle Housing

the flange puller to the universal joint flange (fig. 50) and remove the flange. Using a brass drift and hammer, drive the drive pinion out of the axle housing (fig. 51). Remove the shims and spacer from the drive pinion. Note the thickness of shims removed from the pinion to assist in reassembly.

32. CLEANING, INSPECTION AND REPAIR.

a. Cleaning. Clean all parts in dry-cleaning solvent. Rotate the bearings in dry-cleaning solvent until all trace of lubricant has been removed. Oil the bearings immediately to prevent corrosion of the highly polished surface.

b. Inspection and Repair.

(1) AXLE HOUSING AND COVER (fig. 79).

(a) Inspection. Replace the axle housing if it is broken at any of the welds or if it is cracked or bent. Replace the drive pinion bearing cups if they are pitted, corroded, or discolored due to overheating (subpar. *(b)*, below). Replace the oil seals in the axle housing regardless of their condition (step *(c)*, below). Replace the differential cover if cracked or if it has damaged threads in the filler plug hole. Replace the breather cap on the cover, if it is missing or damaged.

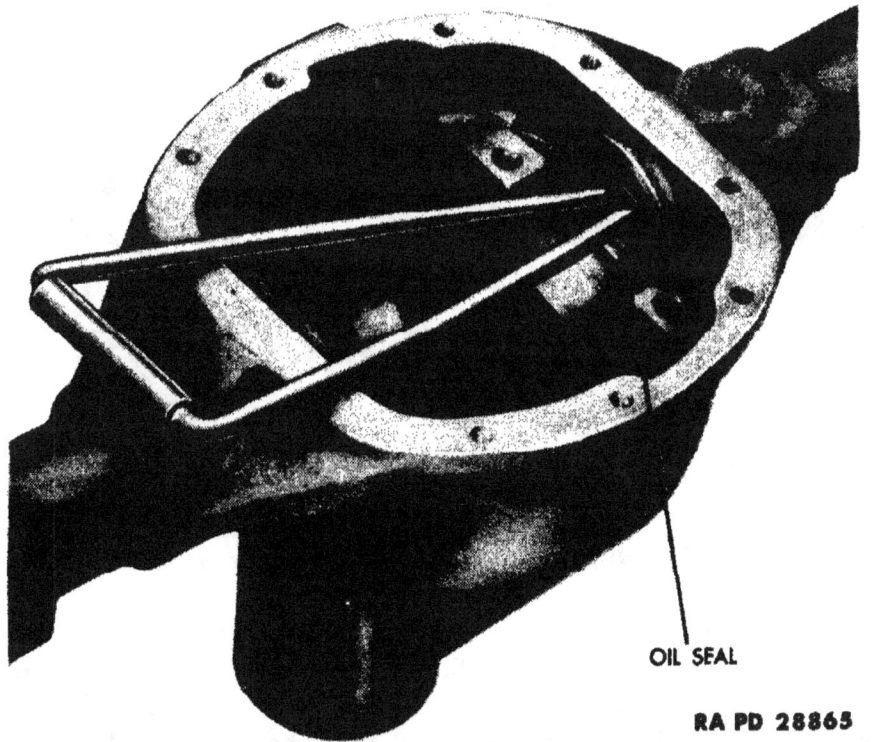

Figure 80 — Removing Oil Seal From Axle Housing, With Remover 41-R-2384-38

(b) *Drive Pinion Bearing Cap Replacement.* Remove the inner and outer bearing cups, using a standard puller. To assist in assembly, note the thickness of shims when removing the inner bearing cup. To install a new bearing cup, use a brass drift and hammer. Place the original thickness of shims behind the inner bearing cup and tap the bearing cup lightly around the entire circumference of the cup until it is flush with the shoulder in the axle housing (fig. 52).

(c) *Oil Seal Replacement* (fig. 80). Remove the inner oil seal with the remover 41-R-2384-38. To install a new oil seal, use special replacer 41-R-2391-20 and tap the oil seals in place (fig. 81).

(2) DRIVE PINION ASSEMBLY (fig. 57). Replace any roller bearings that are pitted, corroded, or discolored due to overheating. Replace the drive pinion gear if it has excessively worn, or broken teeth, or if the splines are worn or the threads damaged. The differential gear and the drive pinion are furnished in matched sets only, and if either is found damaged, both must be replaced. Small nicks can be removed from the pinion gear with a fine stone.

(3) DIFFERENTIAL ASSEMBLY (fig. 58). Replace any gears that are excessively worn or have any missing teeth. The differential ring gear and the drive pinion are furnished in matched sets only, and if

POWER TRAIN

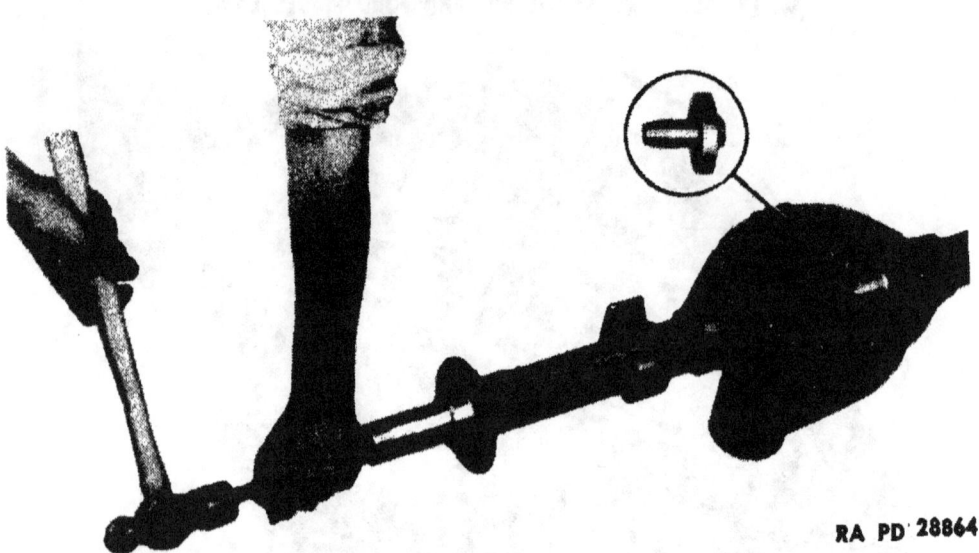

Figure 81 — Installing Oil Seal With Replacer 41-R-2391-20

either is found damaged, both must be replaced. Replace the differential pinion gear if its inside diameter is more than 0.625 inch. Replace the differential pinion shaft, if the inside diameter is worn to less than 0.625 inch. Replace the axle shaft gear if the hub is worn to less than 1.500 inches. Replace the differential pinion gear and the axle shaft gear thrust washer if the thickness is worn to less than 0.032 inch. Roller bearings and cups that are pitted, corroded, or discolored due to overheating must be replaced.

(4) AXLE SHAFT (fig. 82). Replace the axle shafts if they are bent or have any worn or broken splines.

33. ASSEMBLY.

a. Install Inner Bearing on Pinion (fig. 66). Press the inner bearing on the pinion, using an arbor press. Make sure the bearing is firmly seated on the shoulder of the pinion gear when installed.

b. Adjust Pinion in Housing (fig. 67). Place the pinion in the differential housing. Install the gage 41-G-176 to check the setting from the back face of the pinion to the center line of the differential case bearing. The standard setting is 0.719 inch. If the gage reading is more than 0.719 inch, shims will have to be added to the inner bearing cup (par. 32 b). If the reading is less than 0.719 inch, shims will have to be removed from the inner bearing cup (par. 32 b).

c. Install Outer Bearing on Pinion (fig. 57). After the correct pinion setting has been obtained, install the spacer and the original amount of shims on the pinion. If the thickness of the original shims is unknown, install shims totaling approximately 0.060 inch thick.

Start the outer bearing on the pinion. Install the oil slinger on the pinion.

d. Adjust the Outer Bearing. Place the universal joint flange on the pinion. Install the nut on the pinion and draw the universal joint flange down tight. Turn the universal joint flange. If there is a slight drag, the pinion bearing adjustment is correct. If the pinion turns with difficulty or cannot be turned by hand, shims should be added behind the outer bearing. If the pinion is too loose, shims should be removed. Remove the universal joint flange and add, or remove, shims, until the correct adjustment is obtained. After the correct adjustment is obtained, again remove the universal joint flange and install the oil seal on the pinion. Install the universal joint flange. Install the nut and cotter pin.

e. Install Gears in Differential Case (fig. 58). Place the axle shaft gear thrust washers on the two axle shaft gears. Place the axle shaft gears in the case. Place the two differential pinion thrust washers and gears in the case. Install the differential pinion gear shaft that secures the two differential pinion gears in the case. Install the pinion shaft lock pin in the case.

f. Install Differential Ring Gear (fig. 58). Place the differential ring gear in position on the case. Install the lock plates and cap screws that secure the ring gear to the case. Bend the ears of the lock plate on the cap screws.

g. Install Roller Bearings on Case (fig. 68). If all the original parts have been used in the differential assembly, add the same thickness of shims as originally used, and press the roller bearings on the case, then proceed with subparagraph h, below. If the original parts are not being used, or if the original shim thickness is not known, install the roller bearings on the case without the shims, and proceed with subparagraph i, below.

h. Install Differential Assembly in Housing (fig. 47). Place the bearing cups on the bearings. Tilt the bearing cups in order to start the assembly in the housing. Tap the bearing cups lightly until the assembly is seated firmly in the housing. Install the two bearing caps so that the numbers on the caps, and the housing face the same way, and match in every way as shown in figure 47. If the differential assembly being used is not the one originally in the axle, proceed with subparagraph i, below.

i. Differential Assembly Adjustment (fig. 69). Place the differential assembly in the housing with the bearing cups on the assembly. Slide the assembly to one side of the housing. Check the clearance between the bearing cup and differential housing with a

TM 9-1803B
33
POWER TRAIN

feeler gage. After this clearance has been determined, add 0.008 inch. This will give the thickness of shims required for proper bearing adjustment. Remove the differential assembly from the housing. Remove the bearings from the differential case (par. 24 f (3)). Install the number of shims, determined above, on each side of the case and install the bearings back on the case (par. 26 e (3)). Tilt the bearing cups in order to start the assembly in the housing. Tap the bearing cups lightly until the assembly is seated firmly in the housing. Install the two bearing caps so the numbers on the bearing caps and housing face the same way and match in every way.

j. **Check Backlash** (fig. 70). Install a dial indicator on the differential housing so that the indicator contact is resting on the surface of a ring gear tooth as shown in figure 70. Rotate the ring gear back and forth to determine the backlash. If the backlash is less than 0.005 inch or more than 0.007 inch, remove the differential from the housing (par. 24 e) and remove the bearings from the differential case (par. 24 f (3)). If the backlash was more than 0.007 inch, the ring gear must be brought closer to the pinion. If the backlash was less than 0.005 inch, the ring gear must be moved away from the pinion. This is accomplished by removing the shims, equal to the error in backlash, from one side of the case, and adding them to the other side of the case. Install the bearings on the case (subpar. g, above). Install the differential in the housing (subpar. h, above).

k. **Check Ring Gear Run-out** (fig. 71). Install a dial indicator on the differential housing so that the indicator contact is resting on the flat side of the ring gear as shown in figure 71. Turn the pinion drive flange by hand to determine the run-out of the ring gear. The run-out should not exceed 0.003 inch. If the run-out is more than 0.003 inch, remove the differential assembly from the housing (par. 24 e) and remove the ring gear from the differential case. Check the surface of the differential case and the ring gear for chips or small nicks which might have occurred during the assembly of the differential. If any small nicks are found, remove them with a fine stone; also check the flange on the differential case for being sprung. Reinstall the differential assembly in the housing (subpar. h, above) and recheck the ring gear run-out.

l. **Install Differential Cover** (fig. 78). Place a new gasket and the differential cover in place on the axle housing. Install the ten cap screws that secure the cover to the housing.

m. **Install Brake Plate** (fig. 82). Place the brake plate on the housing, with the brake cylinder on the brake plate toward the top. Line up the holes in the brake plate with the axle housing. Install the six cap screws that secure it to the axle housing. Install the hydraulic

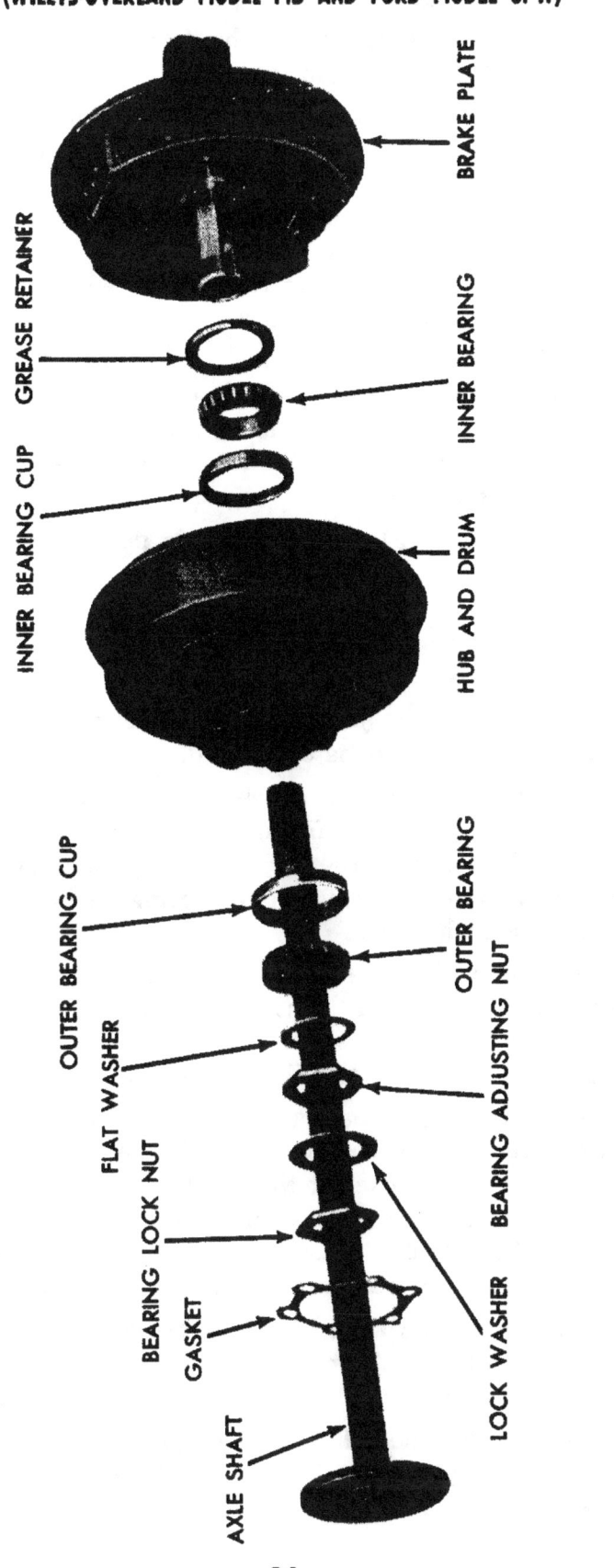

Figure 82 — Axle Shaft — Exploded View

POWER TRAIN

brake line at the connection on the brake plate. Install the flexible hydraulic brake line leading from the frame crossmember at the connection at the differential housing.

n. Install Hub and Brake Drum (fig. 82). Pack the wheel bearings with the specified lubricant. Install the inner bearing in place in the hub and install the hub and brake drum on the housing. Install the outer wheel bearing and thrust washers. Install and tighten the bearing adjusting nut until the brake drum binds, then back it off one-sixteenth turn. This will give the correct wheel bearing adjustment. Install the lock washer and lock nut. Bend the ears of the lock washer over the lock nut.

o. Install Axle Shafts (fig. 82). Insert the axle shaft in the axle housing. Turn the axle shaft to line up the splines on the axle shaft with the gear in the differential. Install the six cap screws that secure the drive flange to the hub.

p. Install Wheels. Place the wheel in position on the hub and secure it with five cap screws.

34. INSTALLATION.

a. Preliminary Work. Place the rear axle assembly under the vehicle. With a hydraulic jack, raise the rear axle high enough so that the spring shackles can be connected.

b. Install Springs (fig. 77). Raise the two rear springs and install them on the spring shackles. Install the spring shackle bushings in the spring shackles. Lower the jack until the axle assembly is resting on the springs, making sure that the spring tie bolt is in line with the hole on the axle housing.

c. Install Spring U-bolts (fig. 77). Place the spring U-bolts in position on the axle housing. Install the spring seat plate on the U-bolts and secure it to the spring with four nuts. The same procedure applies for installing the U-bolts on the other spring.

d. Install Shock Absorbers (fig. 77). Insert a rubber mounting in each side of each shock absorber eye. Place the lower end of the shock absorber on the bracket at the spring seat plate. If new shock absorber rubber mountings are being used, compress them with compressor 41-C-2554-400. Install the flat washer and cotter pin that secure the shock absorber to the bracket on the spring seat plate.

e. Install Hydraulic Brake Line and Propeller Shaft. Install the flexible hydraulic line to the connection at the differential housing (fig. 77). Connect the propeller shaft at the axle (par. 21 a).

f. Lubricate. Fill the differential to proper level with specified oil. Apply specified grease to all fittings. Bleed the hydraulic brake system. Refer to TM 9-803.

ORDNANCE MAINTENANCE — POWER TRAIN, BODY, AND FRAME FOR ¼-TON 4 x 4 TRUCK (WILLYS-OVERLAND MODEL MB AND FORD MODEL GPW)

Section VII

FITS AND TOLERANCES

35. FITS AND TOLERANCES.

Fits Location and Name	Manufacturers Fit Tolerance	Fit Wear Limit	Type of Fit
a. Transmission.			
Second speed gear bushing	—	—	Press
Second speed gear and mainshaft	0.001-0.002 in.	0.004 in.	Running
Idle gear bushing	—	—	Press
Idle gear and idle gear shaft	0.003-0.0045 in.	0.005 in.	Running
Countershaft end play	0.004-0.016 in.	0.016 in.	—
Countershaft gear bushings and countershaft gear	0.0015-0.003 in.	0.005 in.	Running
Countershaft gear bushings and countershaft	0.0015-0.0025 in.	0.005 in.	Running
b. Transfer Case.			
Intermediate gear end play	0.006-0.017 in.	0.017 in.	—
Output shaft bushing and clutch shaft	0.0015-0.003 in.	0.003 in.	Running
Shift lever pivot pin and shift levers	0.001-0.005 in.	0.010 in.	Slip
Output shaft and output shaft gear	0.0015-0.0025 in.	0.003 in.	Running
c. Front Axle.			
Differential pinion gears and differential pinion shaft	0.0019-0.0044 in.	0.005 in.	Running
Axle shaft gear and differential case	0.003-0.006 in.	0.006 in.	Running
Differential pinion adjustment	0.719 in.	0.719 in.	—
Differential ring gear backlash	0.005-0.007 in.	0.005-0.007 in.	—
Differential ring gear run-out	0.003 in.	0.003 in.	—
Spindle housing tension	4 to 6 lb	4 to 6 lb pull scale	—
Bendix or Tracta axle shaft backlash	0.015-0.035 in.	0.015-0.035 in.	—
d. Rear Axle.			
Differential pinion gears and differential pinion shaft	0.0019-0.004 in.	0.005 in.	Running
Axle shaft gear and differential case	0.003-0.006 in.	0.006 in.	Running
Differential ring gear backlash	0.005-0.007 in.	0.005-0.007 in.	—
Differential pinion adjustment	0.719 in.	0.719 in.	—
Differential ring gear run-out	0.003 in.	0.003 in.	—

TM 9-1803B

CHAPTER 3

BODY AND FRAME

Section I

SPRINGS AND SHOCK ABSORBERS

36. **SPRINGS.**

   a. **Description and Data.**
   (1) Description. The front and rear springs are the semi-elliptic type. The front end of the front springs and the rear end of the rear springs are shackled, using the U-bolt type shackle with a threaded core bushing. The rear ends of the front springs and the front ends of the rear springs each have a bronze bushing and are each pivoted by a pivot bolt mounted to a bracket on the frame. A torque reaction spring, mounted on the left front spring, stabilizes the torque of the front axle. The front springs appear to be identical in construction but are different in load carrying ability. The left spring can be identified by the letter "L" stamped on the No. 8 leaf.

   (2) DATA.

Front spring:

Make	Mather
Type leaf	Parabolic
Length (center to center of eye)	36¼ in.
Width	1¾ in.
Number of leaves	8
Front eye (center to center bolt)	$18^1/_8$ in.

Rear eye (center to center bolt) 18¹/a in.
Left camber under 525 lb •~ in.
Right camber under 390 lb 6 in. Rear eye Bushing

Rebound clips   *4*

Rear springs:

Make	Mather
Type leaf	Parabolic
Length	42 in.
Width	**1¾ in.**
Number of leaves	9
Rebound clips	4

Camber under **800** lb ¾ in.
Eye to center bolt 21 **in.** Front eye Bushing

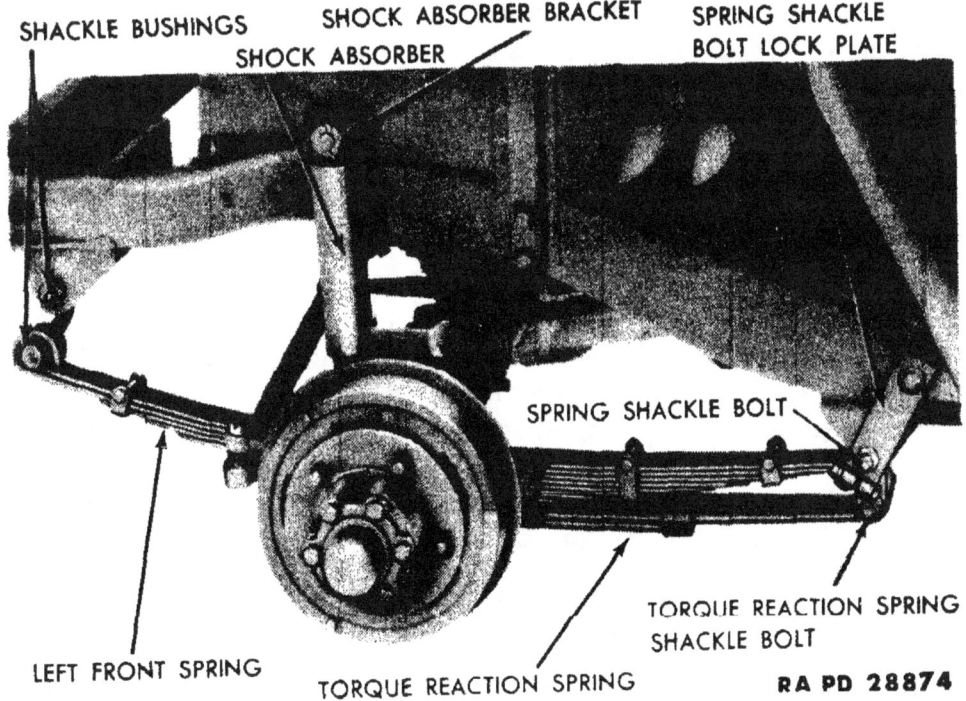

Figure 83 — Left Front Spring With Torque Reaction Spring

b. Removal.

(1) RIGHT FRONT SPRINGS (fig. 35). Raise the vehicle frame until the weight is off the springs but the wheels are still on the floor. Remove the cotter pin and flat washer that secure the shock absorber to the spring seat plates. Remove the shock absorbers from the spring seat plates. Remove the four nuts from the spring U-bolts and remove the U-bolts and spring seat plates. Remove the two front shackle bushings from the spring shackles at the forward end of the frame. Remove the cotter pin and nut from the shackle bolt at the rear of the spring. Remove the shackle bolt from the spring. Remove the spring from the vehicle.

(2) LEFT FRONT SPRING (fig. 83). Raise the vehicle frame until the weight is off the springs but the wheels are still on the floor. Remove the cap screw that secures the shackle bolt lock plate to the left side of the frame. Remove the nut and bolt from the clamping end of the lock plate and remove the lock plate from the shackle bolt. Remove the cotter pin and flat washer that secure the lower end of the shock absorber to the torque reaction spring. Pull the shock absorber off the reaction spring. Remove the cotter pin and nut from the reaction spring shackle bolt and remove the shackle bolt. Remove the cotter pin and nut from the spring shackle bolt and remove

BODY AND FRAME

the shackle bolt and shackles from the spring. Remove the four nuts from the U-bolts and remove the torque reaction spring. Remove the two spring shackle bushings from the spring shackle at the forward end of the spring. Remove the spring from the vehicle.

(3) REAR SPRINGS (fig. 77). Raise the rear of the vehicle frame until the weight is off the spring but the wheels still are on the floor. Remove the cotter pin and flat washer that secure each shock absorber to the spring seat plate. Remove the shock absorbers from the spring seat plates. Remove the four nuts from the spring U-bolts at both springs. Remove the U-bolts and spring seat plates. Remove the two shackle bushings from the spring shackle at the rear of the spring. Remove the spring shackles from the spring. Remove the cotter pin and castellated nut from the two shackle bolts at the front of the rear spring. Remove the two shackle bolts from the springs. Remove the rear springs from the vehicle.

c. **Cleaning, Inspection, and Repair.**

(1) CLEANING AND INSPECTION (figs. 85 and 86). Clean all parts in dry-cleaning solvent. Replace spring leaves or spring clips that are cracked or bent (step (2) (b), below). Replace spring shackles or shackle bolts that are bent or excessively worn. Replace the shackle bolt if the diameter is worn to less than 0.055 inch. Replace the spring bushing in the spring if the inside diameter is worn to more than 0.565 inch (step (2) (a), below). Replace the torque reaction leaves if they are cracked or bent. Replace the bushing in the torque reaction spring if worn to more than 0.566 inch (step (2) (a), below). Replace the inner shackle bushing if the inside diameter is worn to more than 0.570 inch. Replace the outer shackle bushing if the inside diameter is worn to more than 0.630 inch.

(2) REPAIR.

(a) *Front and Rear Spring and Torque Reaction Spring Bushing Replacement* (fig. 84). Place the spring in a press and, with a suitable driver, press out the bushing. Press a new bushing in the spring, using the same driver.

(b) *Spring Leaf Replacement* (fig. 86). Remove the nut and bolt from each of the four spring clips and remove the clips. Install a C-clamp next to the spring tie bolt to hold the tension of the spring leaves before removing the tie bolt. Remove the nut from the spring tie bolt and remove the spring tie bolt from the spring. Remove the C-clamp and separate the spring leaves. Replace the damaged or broken spring leaves. To reassemble the spring, place the spring leaves on the spring tie bolt, starting with the shortest leaf. Pull the leaves together in a vise or a suitable press and install the nut on the tie bolt. Install the four spring leaf clips on the spring.

Figure 84 — Pressing Bushing Out of Spring

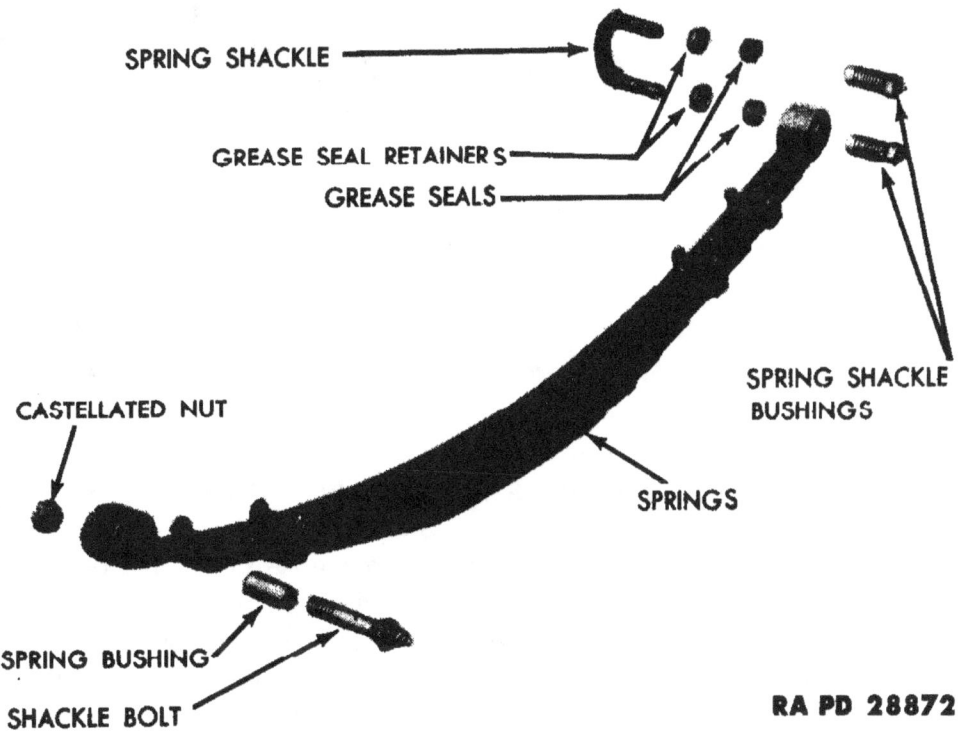

Figure 85 — Rear Spring and Shackles

TM 9-1803B
36

BODY AND FRAME

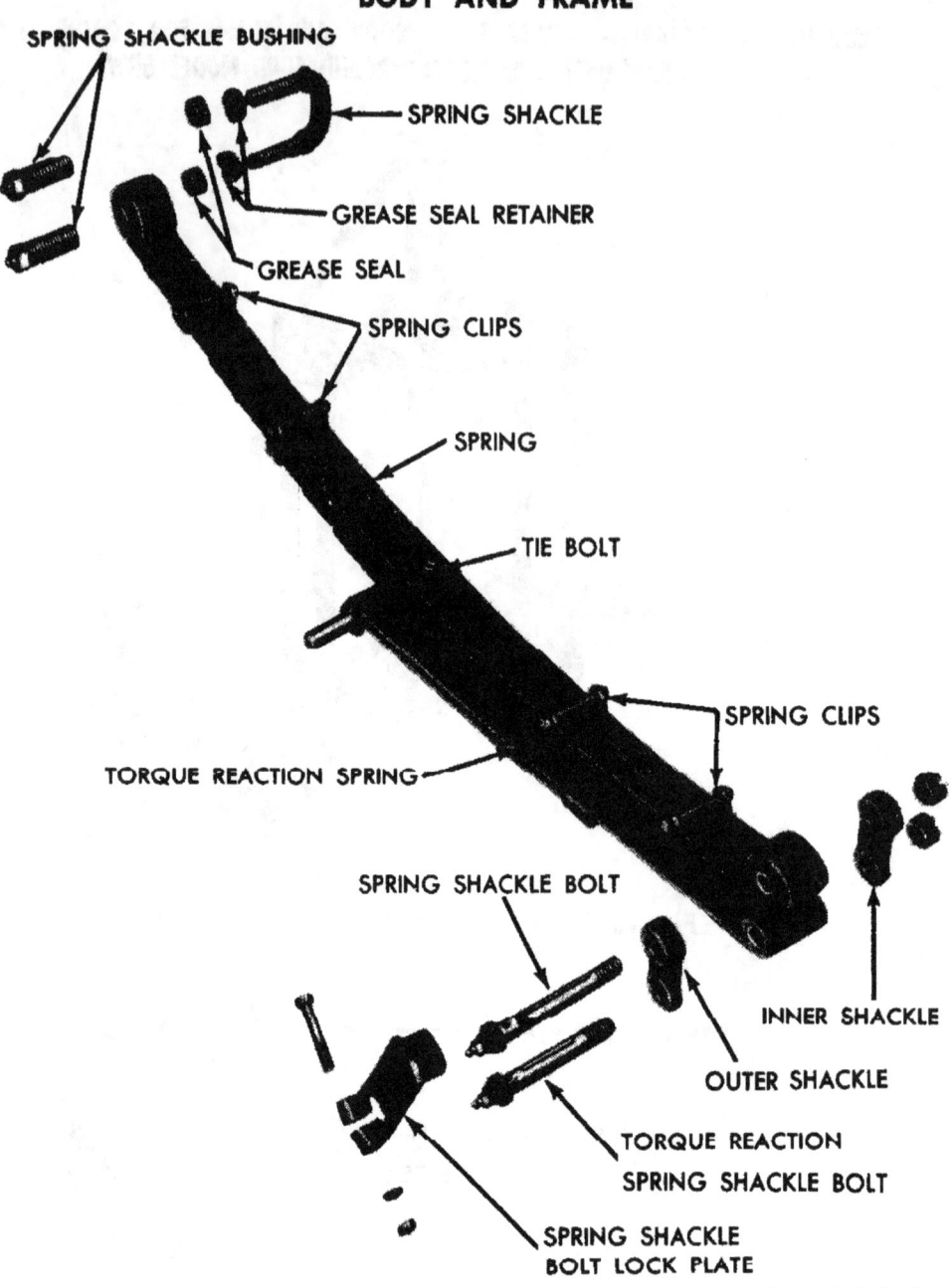

Figure 86 — Front Spring — Exploded View

d. Installation.

(1) RIGHT FRONT SPRING (fig. 35). Place the front spring with the bushing end in the spring bracket on the frame. Insert the spring shackle bolt in the spring with the grease fitting facing outward. Install the nut and cotter pin on the shackle bolt. Raise the forward end of the spring and insert the spring shackle in the bracket on the frame and in the spring. Install the spring shackle bushing with the grease fittings facing outward. Place the spring U-bolts in position

on the axle. Place the spring seat plate on the U-bolts. Install the the four nuts that secure the U-bolts to the axle housing. Install the lower end of the shock absorbers to the spring seat plate. Apply specified lubricant to all fittings.

(2) LEFT FRONT SPRING (fig. 83). Place the spring with the bushing end in the spring bracket on the frame. Insert the outer shackle on the shackle bolt. Insert the shackle bolt in the spring. Place the inner shackle on the shackle bolt and install the nut and cotter pin. Place the torque reaction spring between the inner and outer shackles. Insert the shackle bolt through the shackle and spring. Install the nut and cotter pin on the shackle bolt. Raise the forward end of the spring and insert the spring shackle in the spring. Install the spring shackle bushings with the grease fittings facing outward on the spring shackles. Place the spring U-bolts on the axle housing. Raise the torque reaction spring onto the U-bolts. Install the four nuts to the U-bolts. Install the lower end of the shock absorber to the torque reaction spring. Apply specified lubricant to all fittings.

(3) REAR SPRINGS (fig. 77). Place the rear spring with the bushing end in the spring bracket on the frame. Insert the spring shackle bolt in the spring with the grease fitting facing outward. Raise the rear end of the spring and insert the spring shackle in the spring and in the bracket on the frame. Install the two spring shackle bushings with the grease fitting facing outward. Place the spring U-bolts in position on the axle housing. Place the spring seat plate on the U-bolts. Install the four nuts that secure the spring seat plate to the U-bolts. Insert a rubber mounting in each side of each shock absorber eye. Place the lower end of the shock absorber on the bracket at the spring seat plate. If new shock absorber rubber mountings are being used, compress them with compressor 41-C-2554-400. Install the flat washer and cotter pin that secure the shock absorber to the spring seat plate.

37. GABRIEL SHOCK ABSORBER.

a. Description and Data.

(1) DESCRIPTION. The Gabriel shock absorbers used on some of the vehicles can be distinguished from the Monroe type (par. 38) in that the upper tube has no cutaway section (fig. 87). Four of these direct-acting shock absorbers are used, one at each side of each axle. These shock absorbers are sealed at the factory with the proper amount of fluid and are non-refillable. These shock absorbers are adjustable (subpar. e, below).

TM 9-1803B
37

BODY AND FRAME

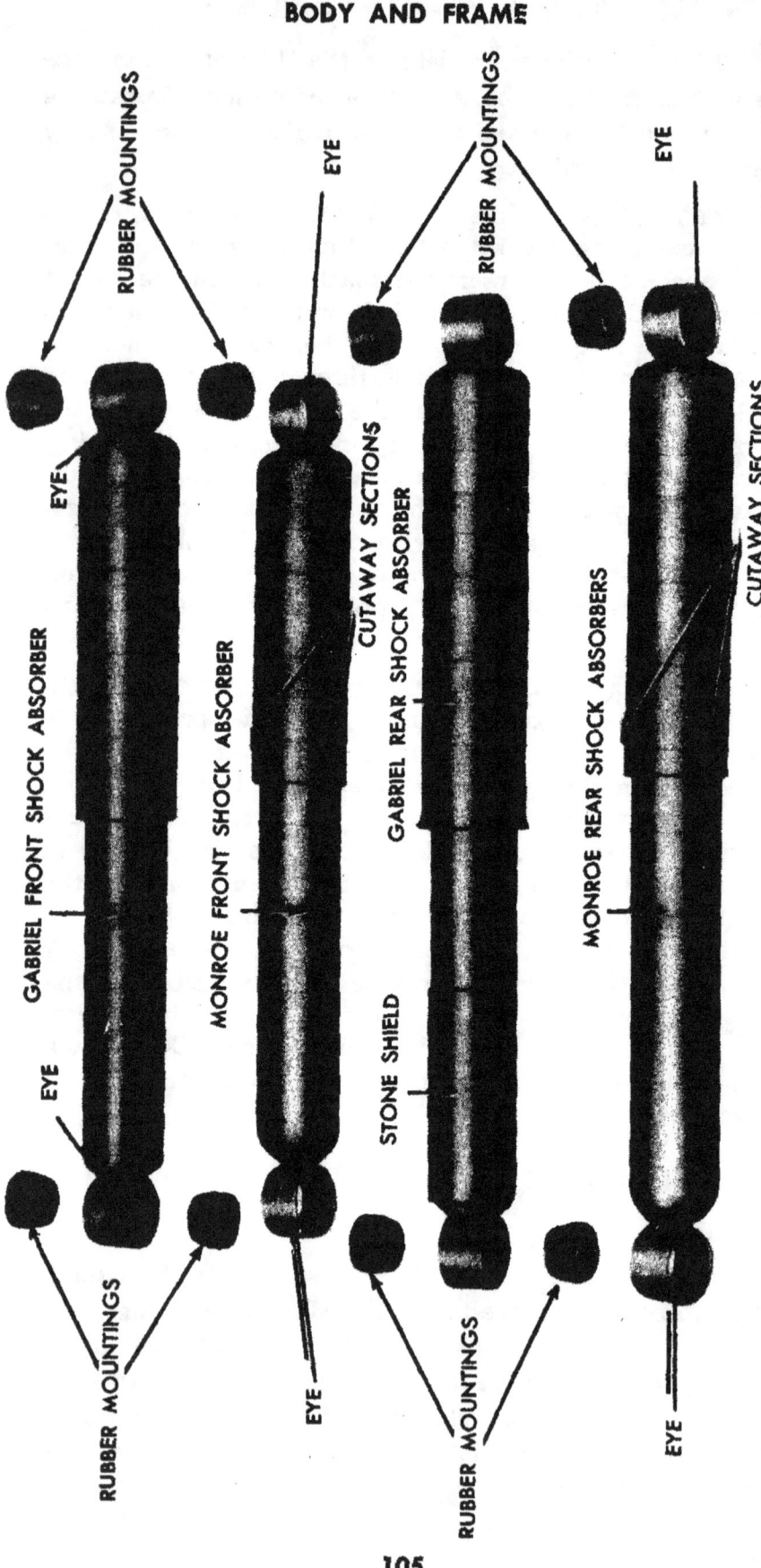

Figure 87 — Monroe and Gabriel Shock Absorbers

(2) DATA.

Make	Gabriel
Type	Hydraulic
Action	Double
Length compressed:	
Front	10 3/16 in.
Rear	11 5/16 in.
Length extended:	
Front	16 3/16 in.
Rear	18 5/16 in.
Mountings	Rubber

b. Removal (figs. 35 and 77). Remove the cotter pin and flat washer that secure the upper end of the shock absorber to the bracket on the frame. Remove the cotter pin and flat washer that secure the lower end of the shock absorber to the spring seat plate. Remove the shock absorber and rubber mountings from the vehicle.

c. Cleaning and Inspection. Wash the shock absorber with dry-cleaning solvent. If the shock absorber is cracked, excessively worn or is leaking fluid, replace the shock absorber. Replace the rubber mountings if they are excessively worn. Do not clean the shock absorber rubber mountings in dry-cleaning solvent.

d. Installation (figs. 35 and 77). Insert a rubber mounting in each side of the upper and lower eye of the shock absorber. Place the shock absorber onto the spring seat plate and onto the bracket on the frame. Install the flat washer and cotter pin that secure the upper end of the shock absorber to the frame. Install the flat washer and cotter pin that secure the lower end of the shock absorber to the spring seat plate. Install the rear shock absorbers so that the stone shield (fig. 87) on the shock absorber is facing forward on the vehicle.

e. Adjustment. Remove the cotter pin and flat washer from the lower end of the shock absorber and remove the lower end from the bracket. Push the unit together to engage the adjusting key, turn the lower half of the shock absorber clockwise until the limit of the adjustment is reached. Holding the unit together to keep the adjusting key still in the slot, turn the lower end of the shock absorber back (counterclockwise) one-half turn. This is the standard adjustment. Turning the adjustment to the right (clockwise) gives a firmer control for rough terrain, turning the adjustment counterclockwise establishes a softer control.

BODY AND FRAME

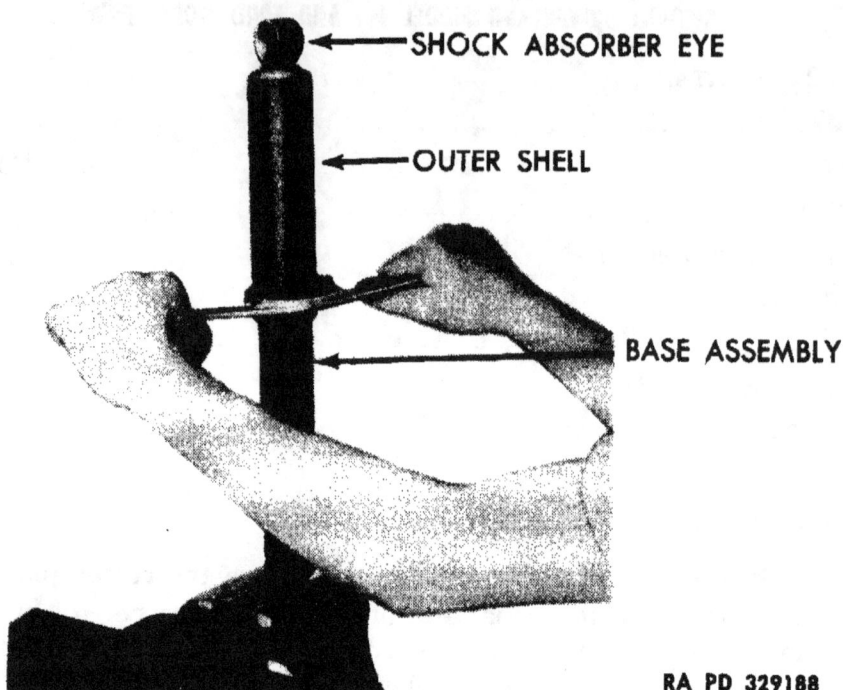

Figure 88 — Removing Seal Assembly With Special Spanner Wrench 41-W-3336-745

38. MONROE SHOCK ABSORBER.

a. Description and Data.

(1) DESCRIPTION. The Monroe type shock absorber used on some of the vehicles can be distinguished from the Gabriel shock absorber by the cutaway sections on the outer shell of the shock absorber (fig. 87). Four of these direct-acting shock absorbers are used, one on each spring. These shock absorbers are refillable (subpar. e (5), below) and can be disassembled for repairs. They are also adjustable (subpar. e (3), below).

(2) DATA.

Make	Monroe
Type	Hydraulic
Action	Double
Length, compressed:	
Front	10 9/16 in.
Rear	11 9/16 in.
Length, extended:	
Front	16 1/8 in.
Rear	18 1/8 in.
Mountings	Rubber

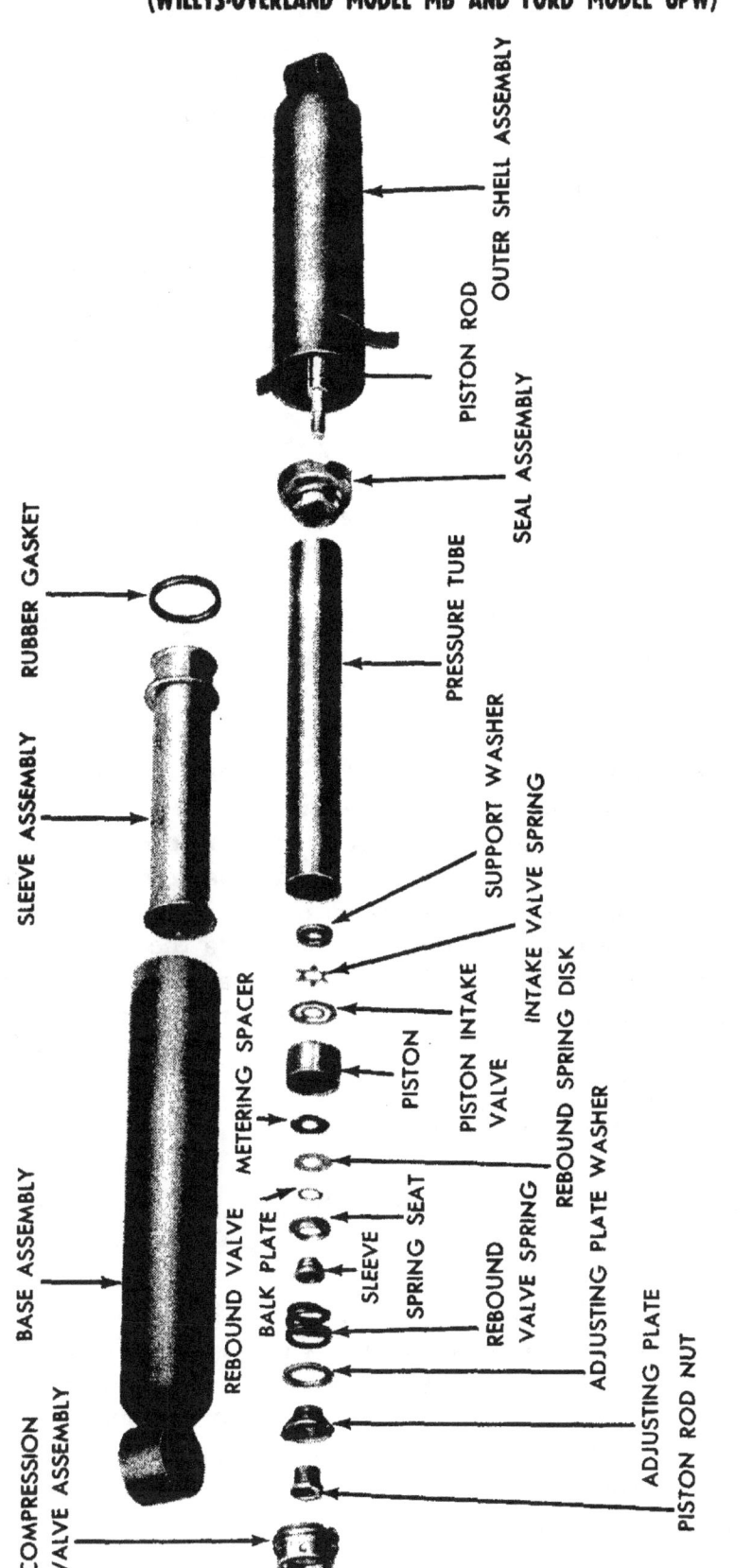

Figure 89 — Monroe Shock Absorber — Exploded View

BODY AND FRAME

b. Removal. Remove the cotter pin and flat washer that secure the upper end of the shock absorber to the bracket on the frame. Remove the cotter pin and flat washer that secure the lower end of the shock absorber to the spring seat plate. Remove the shock absorber and rubber mountings from the vehicle. The same procedure applies to all four shock absorbers.

c. Disassembly (figs. 88 and 89). Place the eye of the base assembly in a vise. Pry open the two metal punch-out openings at the lower end of the outer shell. Install the special spanner wrench 41-W-3336-745 in the slots of the seal assembly and unscrew the seal assembly from the base. Pull the outer shell and pressure tube out of the base. Remove the base assembly from the vise and install the eye of the outer shell in the vise. Pry the compression valve assembly off the pressure tube, using a pair of pliers. Remove the shock absorber from the vise. Turn the pressure tube upside down and remove the fluid. Place the eye of the outer shell back in the vise in its original position. Push the pressure tube down into the outer shell. Remove the piston rod nut. Pull the pressure tube off the piston rod. Remove all of the internal parts from the pressure tube. Place a long drift in the pressure tube and tap the seal assembly out of the pressure tube.

d. Cleaning and Inspection (fig. 89). Clean all parts in drycleaning solvent. Replace the rubber gasket and seal assembly regardless of its condition. Replace all parts that are cracked, bent or excessively worn. Replace the piston if the diameter is worn to less than 0.997 inch. Replace the pressure tube if the inside diameter is worn to more than 1.001 inches. Replace the outer shell if the piston rod is bent or excessively worn. Replace the compression valve assembly if the valve spring is broken or if the adjustment slots are excessively worn. Replace the base assembly if there are any bad dents in the casing or if the threads are damaged. Replace the sleeve assembly if it is bent or out of shape.

e. Assembly.

(1) INSTALL PRESSURE TUBE AND INTERNAL PARTS (fig. 89). Place the eye of the outer shell in a vise. Place the seal assembly at either end of the pressure tube and press the seal assembly in the tube. Install the special thimble 41-T-1657 on the threaded end of the piston rod. Push the pressure tube down into the outer shell and remove the pilot tool from the piston rod. Place the following parts on the piston rod in the order given; piston support washer with the flat surface facing down; intake valve spring with the bent ends facing up; piston intake valve; piston with the skirt of the piston facing up; metering spacer; rebound spring disk; rebound valve back plate; spring seat with the flat surface facing down; sleeve with the tapered

end facing down; rebound valve spring; and adjusting plate washer. Screw the piston rod nut all the way into the adjusting plate. Install the nut and adjusting plate on the piston rod. Stake the rod and nut to prevent the nut from loosening.

(2) FILL SHOCK ABSORBER WITH FLUID. Pull the pressure tube out of the outer shell to its fullest extent. If working on a front shock absorber, measure 5 ounces of shock absorber fluid and put it in a clean container or 5¾ ounces if working on a rear shock absorber. Fill the pressure tube with fluid from this container to within ⅜ inch from the top. Pour the remaining amount of the measured fluid into the base assembly. Hold the pressure tube firmly and place the compression valve assembly on the tube. Tap the valve lightly until it is seated firmly in the pressure tube. Remove the outer shell from the vise and install the loop end of the base in the vise. Insert the sleeve assembly into the base assembly. Insert a new rubber gasket into the base assembly. Place the outer shell on the base. Using the special spanner wrench 41-W-3336-745, tighten the seal assembly into the base assembly (fig. 88).

(3) ADJUST. Push the unit together to engage the adjusting key, turn the base assembly (lower half) of the shock absorber clockwise until the limit of the adjustment is reached. Holding the unit together to keep the adjusting key still in the slot, turn the lower end of the shock absorber back (counterclockwise) two turns. This establishes the standard adjustment. Turning the adjustment to the right (clockwise) gives a firmer control for rough terrain, turning the adjustment counterclockwise establishes a softer control.

(4) INSTALL. Insert the rubber mountings in the upper and lower ends of the shock absorber. Install the shock absorber to the spring seat plate and to the frame. Install the flat washer and cotter pin that secure the upper end of the shock absorber to the frame. Install the flat washer and cotter pin that secure the lower end of the shock absorber to the spring seat plate.

(5) REFILL SHOCK ABSORBER. Remove the shock absorber (subpar. h, above). Place the eye end of the shock absorber base assembly in a vise. Pry open the two metal punch-out openings at the lower end of the outer shell. Install the special spanner wrench 41-W-3336-745 in the slots of the seal assembly (fig. 88). Unscrew the seal assembly from the base assembly. Pull the outer shell with the pressure tube out of the base assembly. Remove the base assembly from the vise and install the eye end of the outer shell in the vise. Pry the compression valve assembly off the pressure tube, using a pair of pliers. Fill the shock absorber with fluid as outlined in step (2), above.

BODY AND FRAME

Section II

STEERING GEAR AND DRAG LINK

39. STEERING GEAR ASSEMBLY.

a. Description. The Ross Model T-12 steering gear (fig. 90) is of the cam and twin lever, variable ratio type, having a ratio of 14-12-14 to 1. The steering gear sector shaft is serrated for attachment of the Pitman arm, and the steering wheel is serrated for attachment to the worm and shaft assembly. The steering wheel is of the safety type, having three spokes and is 17¼ inches in diameter.

b. Removal.

(1) REMOVE LEFT FRONT FENDER. Remove the 12 bolts that secure the left front fender to the body, frame and radiator guard. Remove the bolt that secures the fender to the top of the frame in the engine compartment. Remove the wing nut that secures the headlight bracket to the fender. Disconnect the wires leading from the fender to the junction block on the cowl. Disconnect the wires leading from the junction block on the fender to the headlight and blackout light. Remove the fender from the vehicle.

(2) REMOVE STEERING WHEEL. Remove the steering wheel nut, horn button nut and horn button. Pull the steering wheel off the shaft with a steering wheel puller.

(3) REMOVE STEERING COLUMN TUBE AND BEARING ASSEMBLY. Remove the two nuts and bolts that secure the steering column support clamp at the instrument panel and remove the clamp. Remove the four metal screws that hold the steering column cover plate to the floor plate in the driver's compartment. Remove the two screws that hold the horn wire contact brush to the steering column and remove the brush. Loosen the bolt at the steering column clamp and slide the steering column tube and bearing assembly off the shaft.

(4) DISCONNECT DRAG LINK AT PITMAN ARM (fig. 104). Remove the cotter pin from the Pitman arm end of the drag link. Loosen the drag link socket plug and lift the drag link off the Pitman arm.

(5) REMOVE STEERING GEAR (fig. 104). Remove the three bolts that hold the steering gear to the frame. Slide the steering gear assembly down through the floorboard and out over the frame.

c. Disassembly.

(1) REMOVE PITMAN ARM (fig. 92). Remove the nut and lock washer that hold the Pitman arm on the sector shaft assembly. Pull the Pitman arm off the steering sector shaft assembly with a standard Pitman arm puller.

TM 9-1803B
39

ORDNANCE MAINTENANCE — POWER TRAIN, BODY, AND FRAME FOR ¼-TON 4 x 4 TRUCK (WILLYS-OVERLAND MODEL MB AND FORD MODEL GPW)

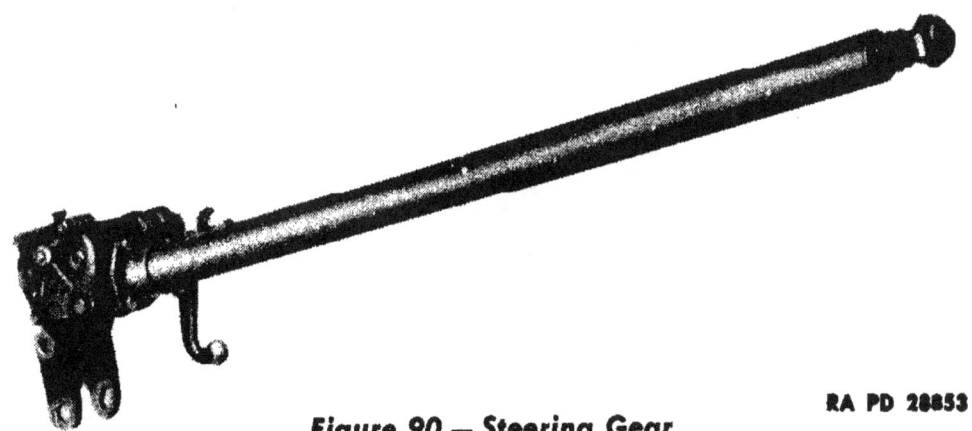

Figure 90 — Steering Gear

RA PD 28853

(2) REMOVE STEERING SECTOR SHAFT ASSEMBLY (fig. 92). Remove the four cap screws that hold the side cover to the housing and remove the side cover and gasket. Slide the sector shaft assembly from the housing.

(3) REMOVE STEERING GEAR WORM AND SHAFT ASSEMBLY FROM HOUSING (fig. 92). Remove the three cap screws that secure the housing end cover and shims to the housing. Slide the housing and shims off the worm and shaft assembly.

(4) REMOVE STEERING GEAR WORM BEARINGS (fig. 91). Remove the retainer ring that secures the steering gear worm lower bearing cup at the end of the shaft assembly. Remove the bearing cup and balls. Remove the retainer ring that secures the worm upper bearing cup to the shaft assembly. Slide the worm bearing cup up on the shaft and remove the balls.

d. Cleaning, Inspection, and Repair.

(1) CLEANING AND INSPECTION (figs. 91 and 97). Clean all parts thoroughly in dry-cleaning solvent. Replace a housing assembly or side cover that is cracked or damaged. Replace the expansion plug in the lower end of the housing if it is loose. Replace the inner and outer bushings in the housing (step (2) (c), below) if worn larger than 0.876 inch inside diameter. Replace a sector shaft assembly that has flat spots on the tapered studs or that has chipped studs. Replace the sector shaft if the shaft measures less than 0.870 inch at the bearing surfaces. Replace the worm and shaft assembly if the worm is excessively worn, ridged, scored, or chipped. Replace a worn, pitted, or cracked worm upper and lower bearing cup (step (2) (b), below). Replace a broken or damaged horn wire (step (2) (a), below). Replace a steering column tube that is bent or damaged. Replace the whole assembly if it is damaged. Replace

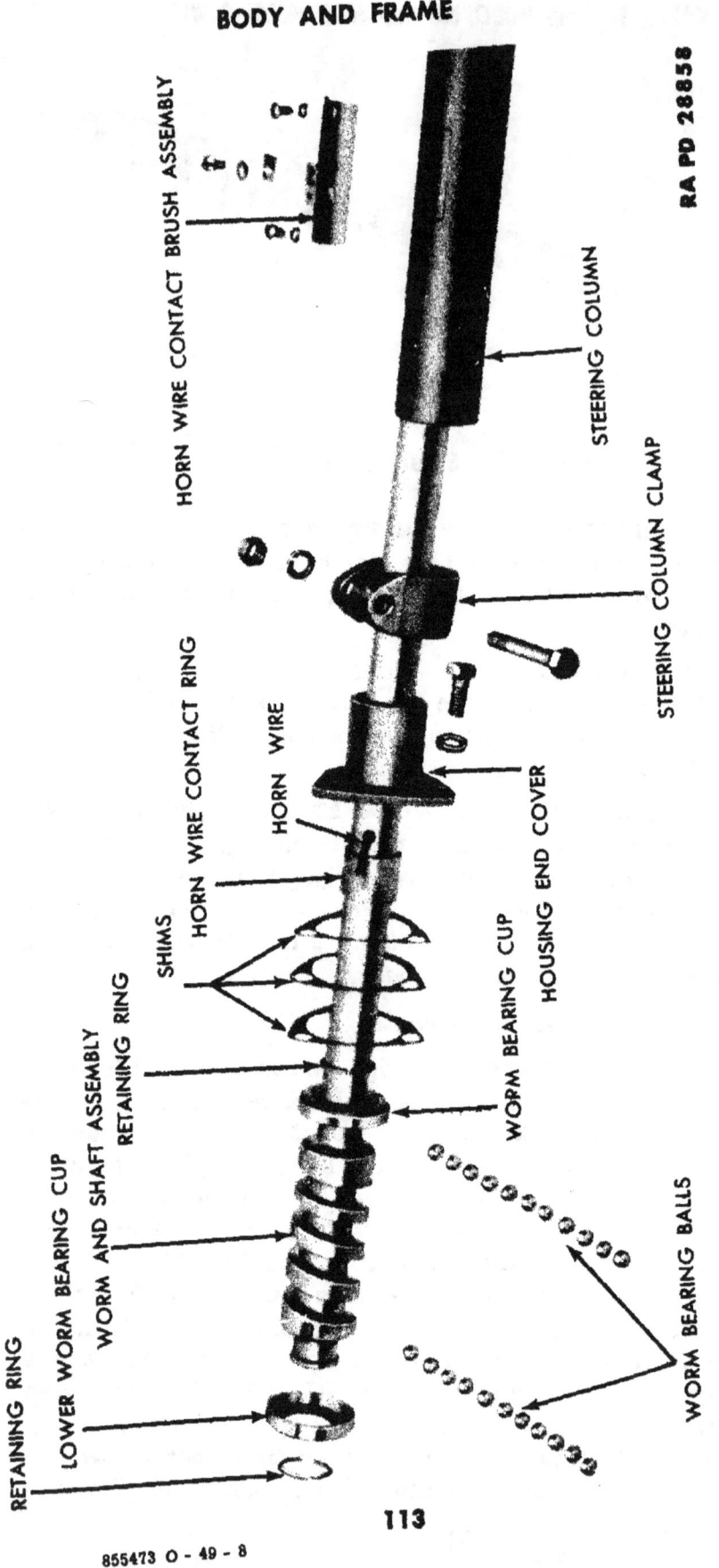

Figure 91 — Worm and Shaft Assembly — Exploded View

TM 9-1803B
39

ORDNANCE MAINTENANCE — POWER TRAIN, BODY, AND FRAME FOR ¼-TON 4 x 4 TRUCK (WILLYS-OVERLAND MODEL MB AND FORD MODEL GPW)

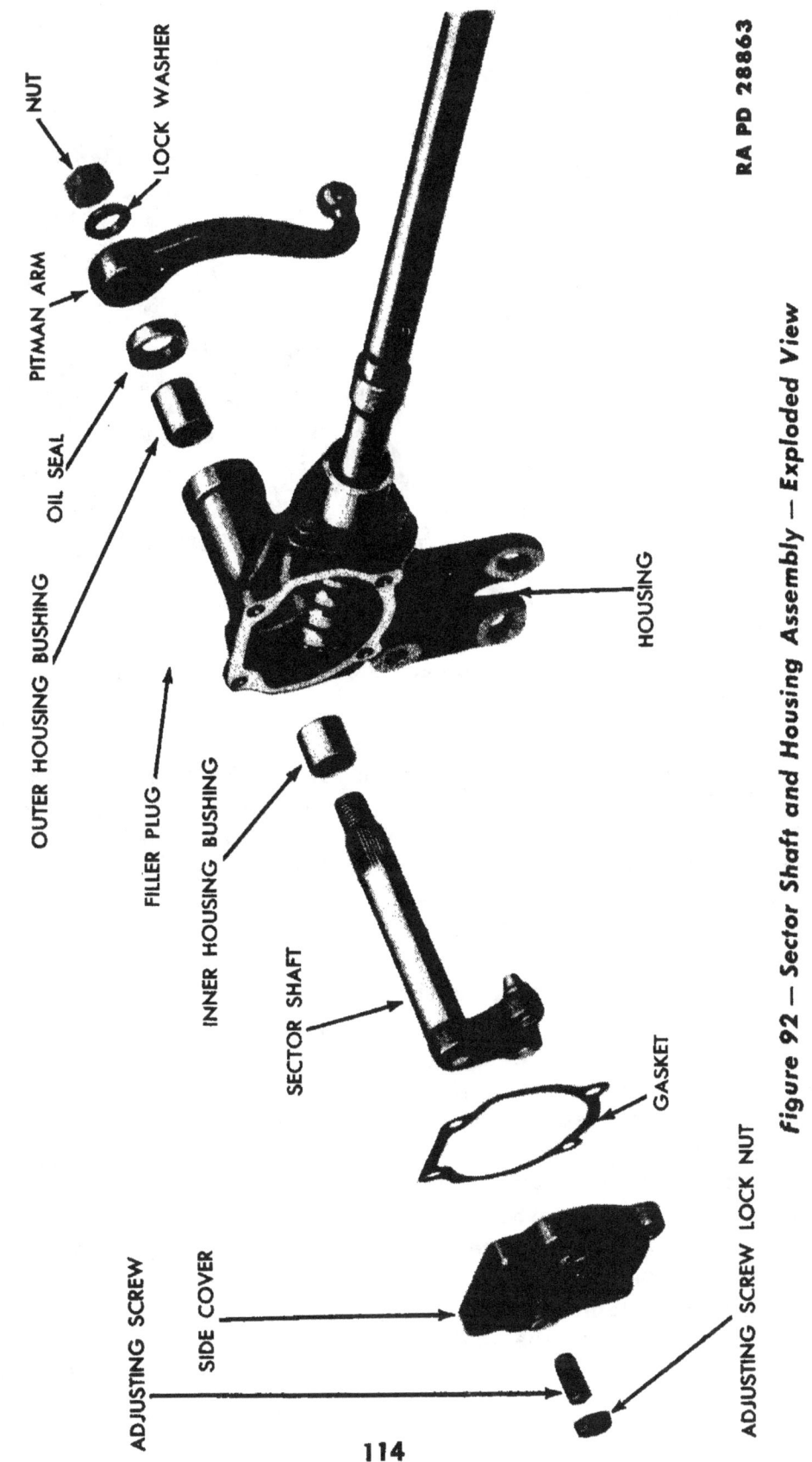

Figure 92 — Sector Shaft and Housing Assembly — Exploded View

BODY AND FRAME

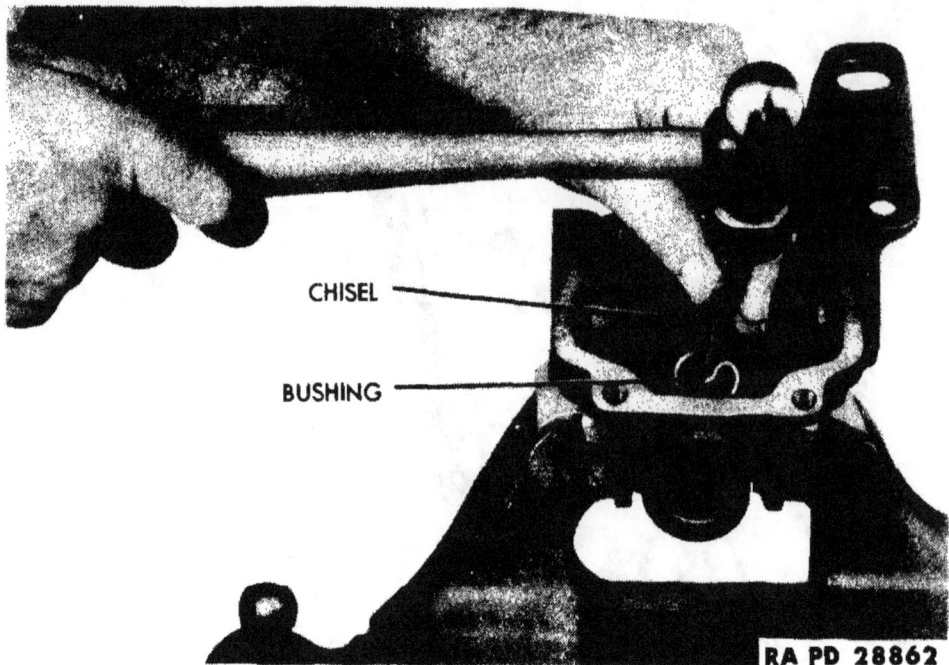

Figure 93 — Removing Housing Bushings

any balls with flat spots. Replace the steering column tube bearing if it is excessively worn.

(2) REPAIR.

(a) Horn Wire Replacement (fig. 96). Unsolder the horn wire at the horn wire contact ring and pull the wire assembly from the shaft. To install the horn wire, slide the contact washer, insulating ferrule, horn button spring and horn button spring cup onto the horn wire. Push the horn wire down through the shaft and out of the hole in the lower end of the shaft. Solder the horn wire onto the horn wire contact ring, making sure that none of the strands of wire or the solder are touching the shaft.

(b) Worm Upper Bearing Cup Replacement. Remove the horn wire (step *(a)*, above). Unsolder the horn wire from the contact ring (fig. 96) and slide the ring off the shaft. Slide the worm upper bearing cap and retaining ring off the shaft. To reinstall the worm bearing cup on the shaft, slide the cup onto the shaft with the concave side toward the worm. Slide the retaining ring onto the shaft. Select a heavy flat washer that will slide onto the shaft, but will not slide over the horn wire contact ring. Make a mark on the lower part of the shaft with a file ½ inch below the horn wire hole. Slide the horn wire contact ring onto the shaft. Place the flat washer on the shaft and place the shaft assembly loosely in a vise (fig. 98). Using a fiber block to protect the shaft, drive down on the shaft until the mark on the shaft is even with the upper part of the horn

TM 9-1803B
39
ORDNANCE MAINTENANCE — POWER TRAIN, BODY, AND FRAME FOR ¼-TON 4 x 4 TRUCK (WILLYS-OVERLAND MODEL MB AND FORD MODEL GPW)

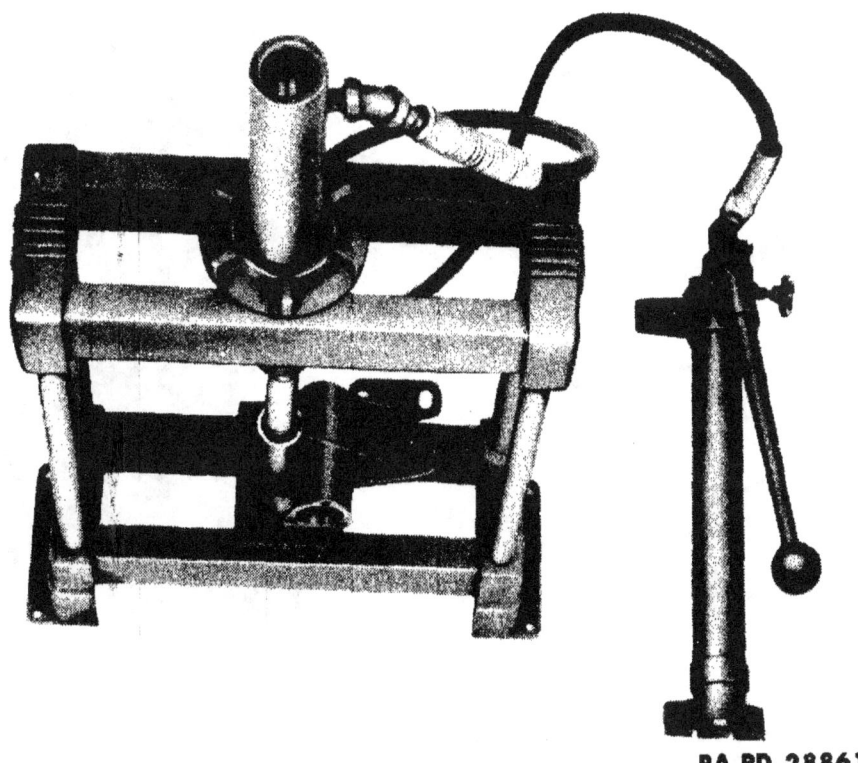

Figure 94 — Pressing Bushing in Steering Gear Housing

wire contact ring. Install the horn wire in the shaft (step *(a)*, above).

(c) Housing, Inner, and Outer, Bushing Replacement. Remove the housing oil seal with a small punch or chisel. Drive a small punch or chisel between the housing and the joint in the bushings (fig. 93), until the ends of the bushings overlap. Tap the bushings out of the housing. To install the bushings, press the outer bushing into the housing (fig. 94) until it is flush with the oil seal shoulder in the housing assembly. Press the inner bushing into the housing. Ream the bushings to 0.875 inch diameter (fig. 95) with reamer 41-R-1220.

e. **Assembly.**

(1) ASSEMBLE WORM AND SHAFT ASSEMBLY. Place the 11 worm bearing balls in the upper worm bearing cup. Slide the cup and retainer ring in place. Place the balls in the lower worm bearing cup and install the cup and retainer ring on the lower end of the shaft.

(2) INSTALL WORM AND SHAFT IN HOUSING. Slide the worm and shaft assembly in the housing. Install shims approximately 0.024 inch thick and the housing end cover on the shaft (fig. 91).

116

BODY AND FRAME

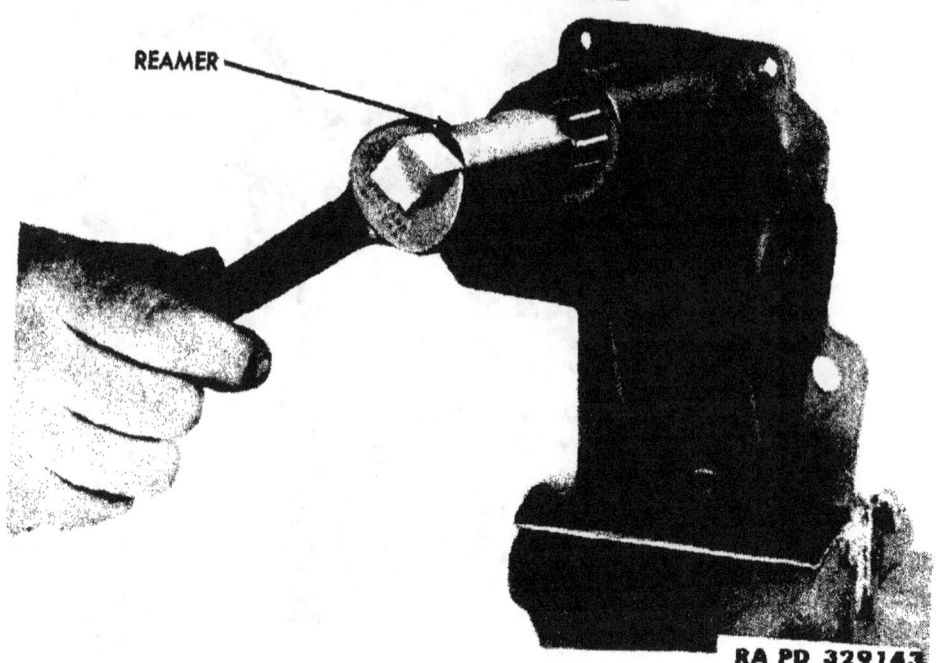

Figure 95 — Reaming Steering Gear Housing Bushing With Reamer 41-R-1220

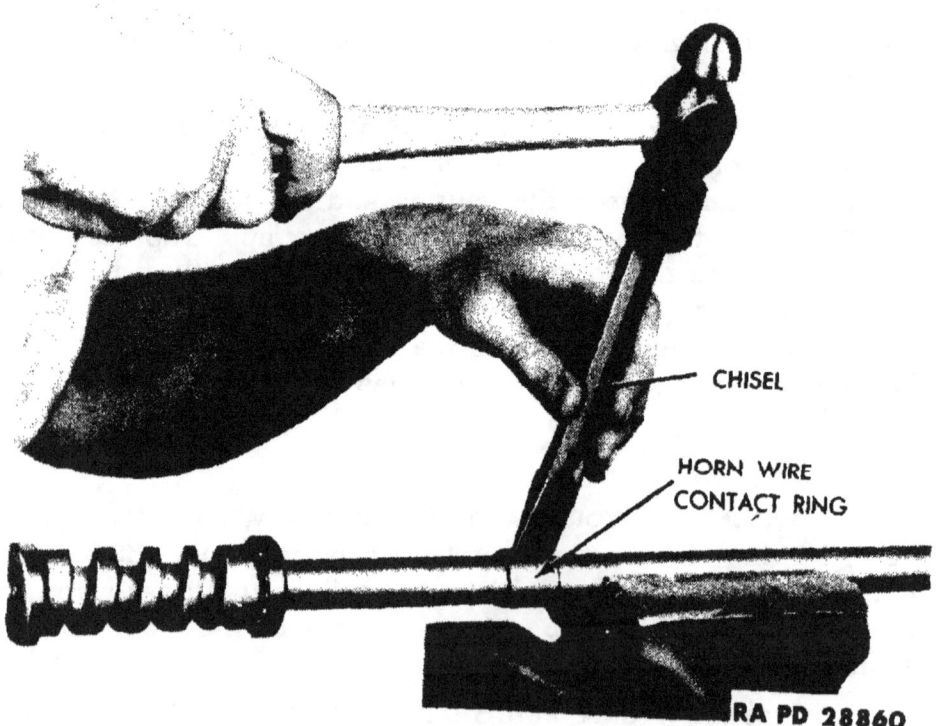

Figure 96 — Removing Horn Wire Contact Ring

TM 9-1803B
39

ORDNANCE MAINTENANCE — POWER TRAIN, BODY, AND FRAME FOR ¼-TON 4 x 4 TRUCK
(WILLYS-OVERLAND MODEL MB AND FORD MODEL GPW)

Install the three cap screws that secure the housing end cover. Turn the shaft by hand, if it is too tight, shims must be added, if too loose, shims must be removed. The correct adjustment is when shaft turns freely but has no end play. Shims are supplied in 0.0035, 0.0025 and 0.0065 inch thicknesses.

(3) INSTALL SECTOR SHAFT ASSEMBLY (fig. 92). Slide the sector shaft assembly into the housing, making sure the two tapered studs engage in the worm.

(4) INSTALL SIDE COVER ON HOUSING. Install a new side cover gasket and cover on the housing. Install the side adjusting screw and lock nut in the side cover (fig. 92). Turn the shaft counterclockwise until the shaft cannot be turned any farther. Turn the shaft clockwise until it cannot be turned any farther, counting the complete and part revolutions of the shaft and then turn the shaft counterclockwise half the number of turns. This will center the shaft. Turn the side adjusting screw in, until a slight drag on the shaft can be felt at this point. Tighten the adjusting screw lock nut and recheck the adjustment.

(5) INSTALL STEERING COLUMN TUBE ON SHAFT (fig. 91). Install the steering column clamp on the steering column with the bolt side of the clamp in line with the horn wire contact brush opening in the steering column. Slide the steering column tube down onto the housing end cover, making sure the horn wire contact brush opening in the steering column is in a vertical position when the housing base is in normal position. Tighten the clamp bolt.

(6) INSTALL PITMAN ARM (fig. 91). Place the Pitman arm on the sector shaft assembly, making sure the line on the Pitman arm is in line with the mark on the sector shaft assembly and that the ball joint on the Pitman arm is facing downward when the shaft is centered. Install the lock washer and nut that secure the Pitman arm to the sector shaft.

(7) INSTALL HORN WIRE CONTACT BRUSH ASSEMBLY. Hold the horn wire contact brush assembly in place on the steering column and install the two hold-down screws.

f. Installation.

(1) INSTALL STEERING ASSEMBLY IN VEHICLE. Slide the steering gear up through the floorboard. Install, but do not tighten, the three steering gear bolts.

(2) CONNECT STEERING GEAR TO BRACKET. Slide the steering column cover plate down the steering column and fasten it to the floorboard with four metal screws. Install the two bolts that secure the steering column clamp to the instrument panel in the driver's compartment.

BODY AND FRAME

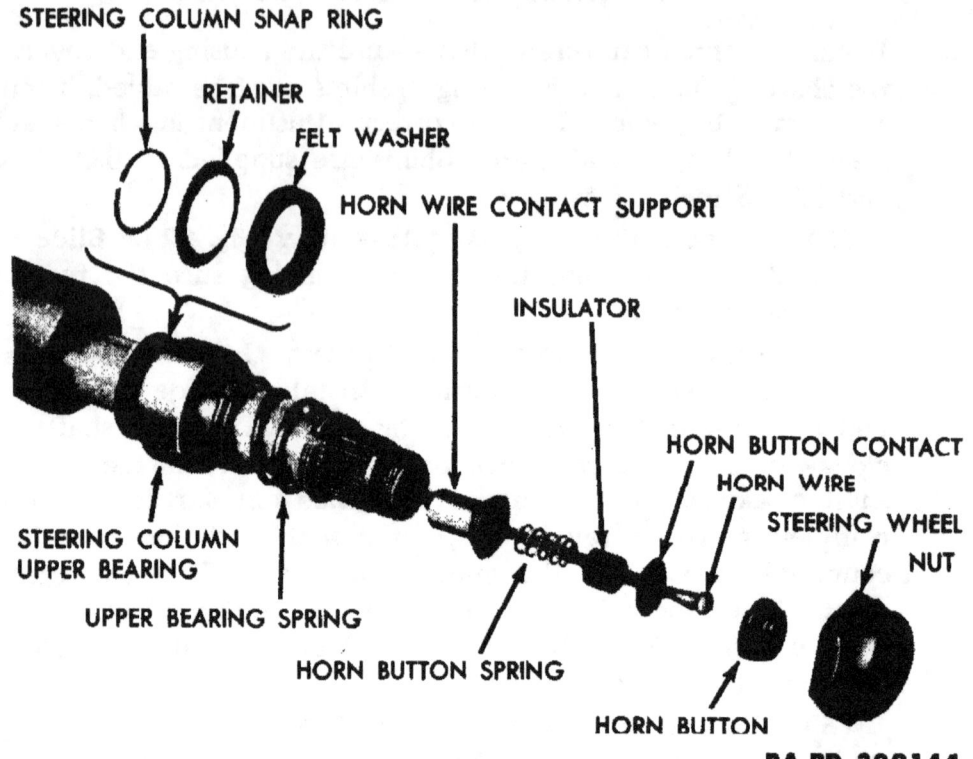

Figure 97 — Horn Button Assembly — Exploded View

(3) CONNECT DRAG LINK (fig. 104). With the front wheels in a straight ahead position, connect the drag link to the Pitman arm, making sure the ball socket on the Pitman arm is between the socket and ball seat (fig. 99). Tighten the socket plug and install the cotter pin. Tighten the three steering gear hold-down bolts at the frame.

(4) INSTALL HORN WIRE CONTACT BRUSH (fig. 91). Hold the horn wire contact brush in place on the steering column and install the two hold-down screws. Connect the horn (live) wire to the contact brush.

(5) INSTALL STEERING WHEEL (fig. 97). Slide the steering column bearing spring and spring seat on the shaft. With the front wheels in a straight ahead position, place the steering wheel on the shaft with the center spoke in a vertical position and facing downward. Install the horn button and horn nut.

(6) INSTALL FENDER AND REPLENISH LUBRICANT. Place the fender in position on the vehicle. Install the 12 bolts that secure the fender to the frame, body, and radiator guard. Install the bolt that secures the fender on the frame in the engine compartment. Install the wires leading from the fender to the junction block on the cowl. Install the wires leading from the junction block on the

TM 9-1803B
39–40

ORDNANCE MAINTENANCE — POWER TRAIN, BODY, AND FRAME FOR ¼-TON 4 x 4 TRUCK
(WILLYS-OVERLAND MODEL MB AND FORD MODEL GPW)

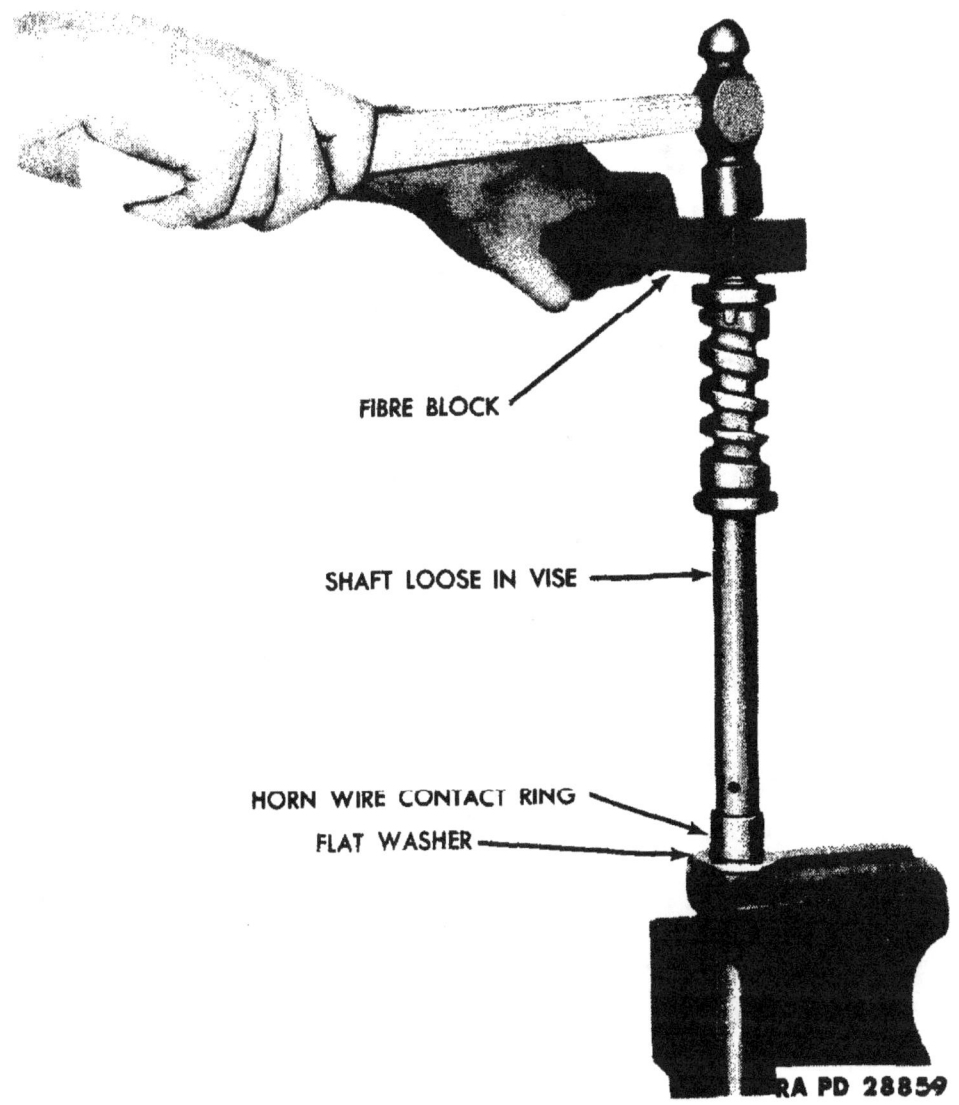

Figure 98 — Installing Horn Wire Contact Ring

fender to the headlight and blackout light. Replenish the lubricant, using the specified amount and grade (refer to TM 9-803).

40. DRAG LINK.

a. Removal. Remove the cotter pin from each end of the drag link. Loosen the adjusting plug at each end of the drag link. Lift the drag link from the vehicle.

b. Disassembly. Remove the dust seal shield and dust seal from each end of the drag link. Remove the adjusting plug, ball seat, spring, and spring seat from the front end of the drag link. Remove

TM 9-180
40

BODY AND FRAME

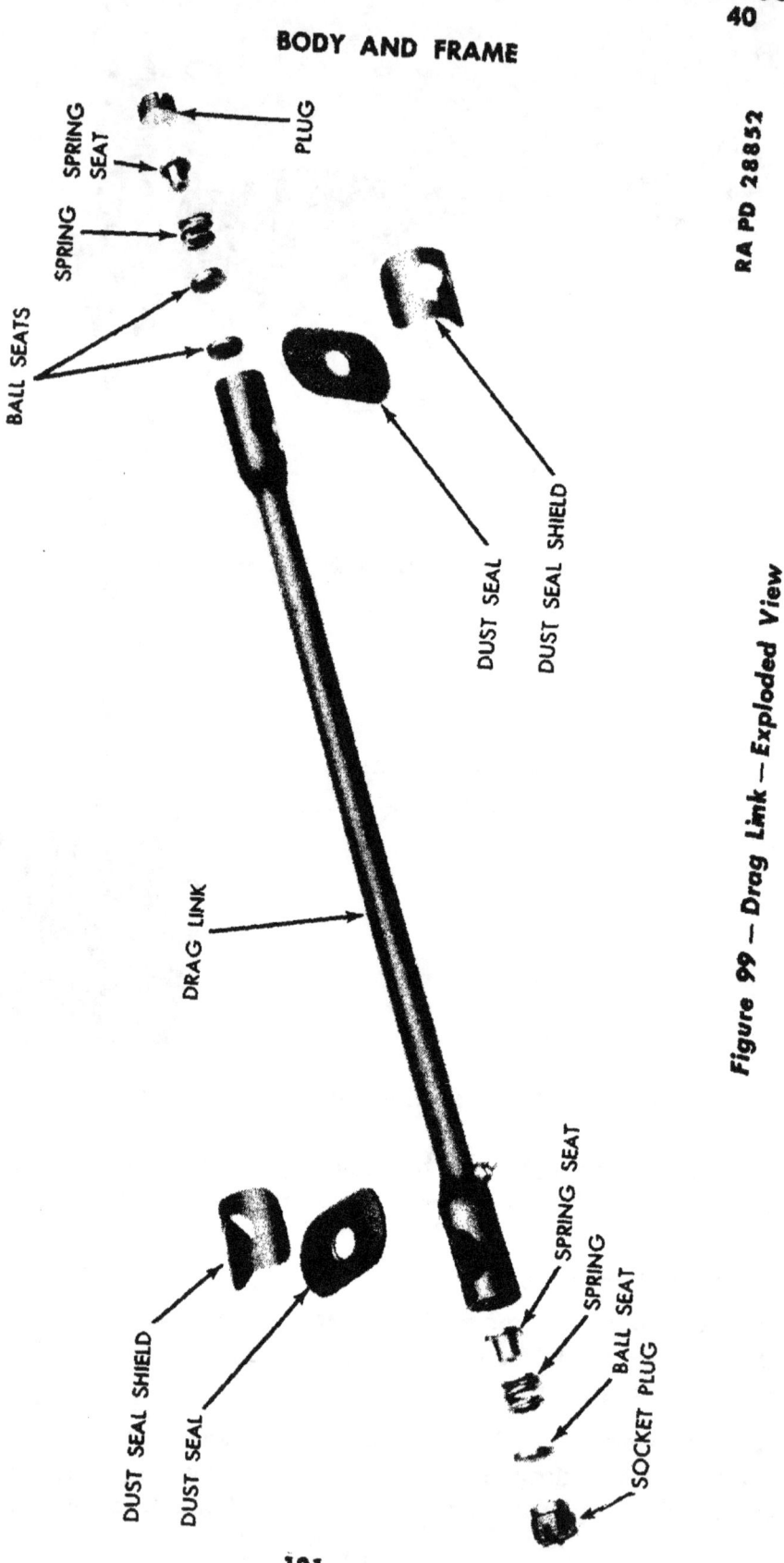

RA PD 28852

Figure 99 — Drag Link — Exploded View

121

TM 9-1803B
40-41

**ORDNANCE MAINTENANCE — POWER TRAIN, BODY, AND FRAME FOR ¼-TON 4 x 4 TRUCK
(WILLYS-OVERLAND MODEL MB AND FORD MODEL GPW)**

the adjusting plug, spring seat, spring, and the two ball seats from the rear end of the drag link.

c. Cleaning, Inspection, and Repair. Clean all parts thoroughly in dry-cleaning solvent. Replace or straighten the drag link if it is bent. Replace damaged grease fittings. With a small wire, clean all grease fittings that are clogged. Replace excessively worn adjusting socket plugs or broken springs. Replace ball seats that are excessively worn. Replace excessively worn spring seats. Replace damaged dust shields or seals.

d. Assembly (fig. 99). Place a spring seat, spring and ball seat in the front end of the drag link. Screw the socket plug approximately three or four turns in the front end of the drag link. Place two ball seats, spring and spring seat in the rear end of the drag link. Screw the socket plug approximately three or four turns in the drag link.

e. Installation (fig. 99). Hold a new dust seal and a dust seal shield in place on the rear end of the drag link. Place the drag link on the Pitman arm, making sure the ball joint on the Pitman arm is seated between the spring seat and socket plug. Screw the socket plug in firmly against the ball, then back the plug off one full turn and install a cotter pin. Hold a new dust seal and dust seal shield in place on the front end of the drag link. Place the drag link on the intermediate steering arm, making sure the ball joint is seated between the spring seat and socket plug (fig. 99). Screw the adjusting plug in firmly against the ball and back off the socket plug one-half turn and install a cotter pin.

Section III

BODY

41. REMOVAL.

a. Remove Hood and Windshield. Raise the hood and remove the five bolts that secure the hood to the cowl. Remove the hood from the vehicle. Remove the wing nuts that secure the windshield at each side of the cowl. Remove the windshield from the vehicle.

b. Remove Body Bolts From Frame (fig. 100). Remove the five bolts under each fender that secure the fender to the body. Remove the four bolts that secure the body to the rear crossmember of the frame. Remove the two bolts that secure the body to the pintle hook brace. Remove the two bolts at each side of the frame that secure the body to the rear body brackets on the frame side member. Remove the bolts that secure the body at each side of the

TM 9-1803B
41

BODY AND FRAME

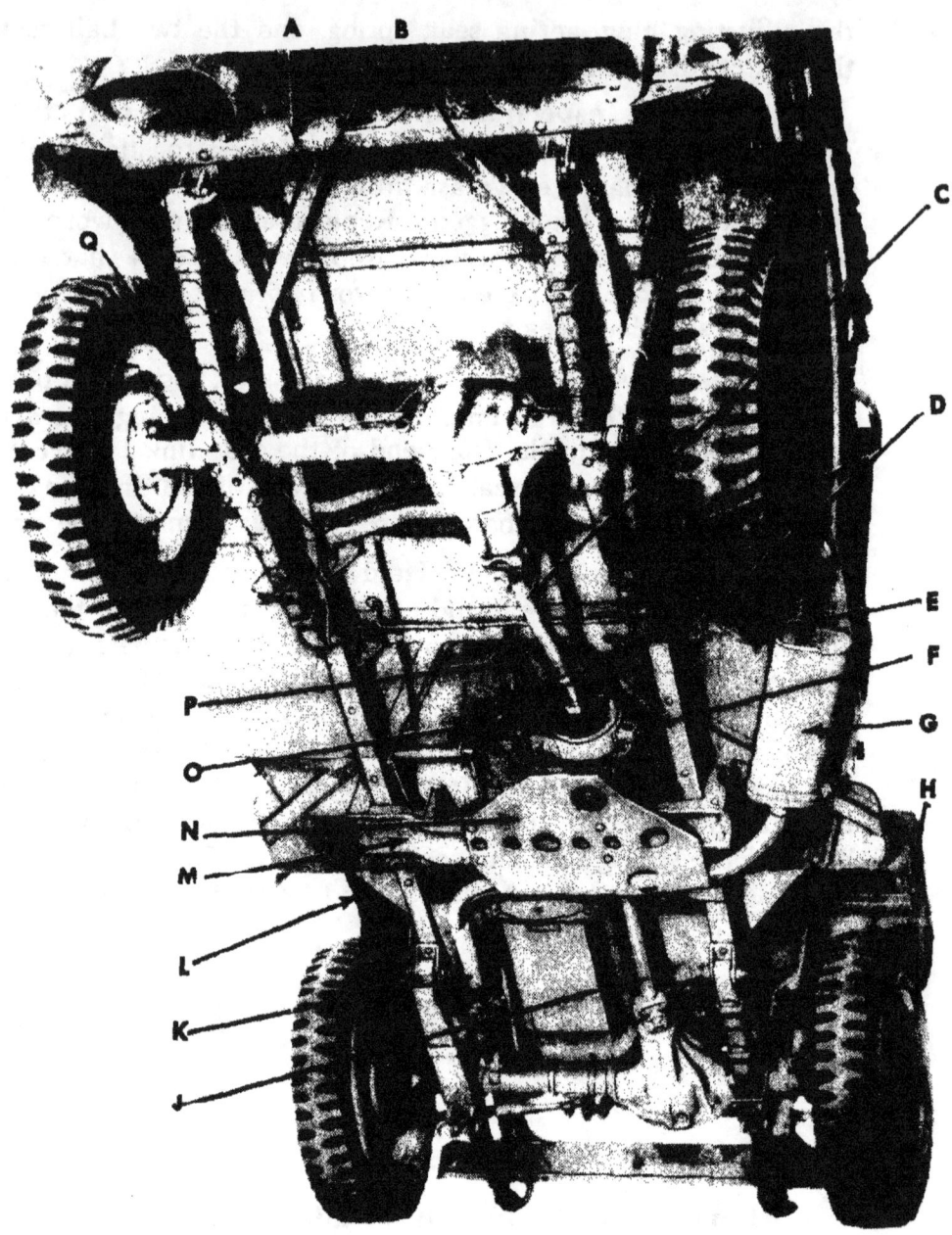

- **A** REAR CROSSMEMBER
- **B** PINTLE HOOK BRACE
- **C** REAR PROPELLER SHAFT
- **D** BODY BRACKETS
- **E** GROUND STRAP
- **F** SPEEDOMETER CABLE
- **G** MUFFLER
- **H** FRONT FENDER
- **J** FRONT PROPELLER SHAFT
- **K** EXHAUST PIPE
- **L** FRONT FENDER
- **M** TRANSMISSION SUPPORT CROSSMEMBER
- **N** TRANSMISSION SHIELD
- **O** HAND BRAKE CABLE
- **P** HAND BRAKE SPRING
- **Q** BODY BRACKETS

RA PD 329148

Figure 100 — Under Side of Body

TM 9-1803B
41

ORDNANCE MAINTENANCE — POWER TRAIN, BODY, AND FRAME FOR ¼-TON 4 x 4 TRUCK
(WILLYS-OVERLAND MODEL MB AND FORD MODEL GPW)

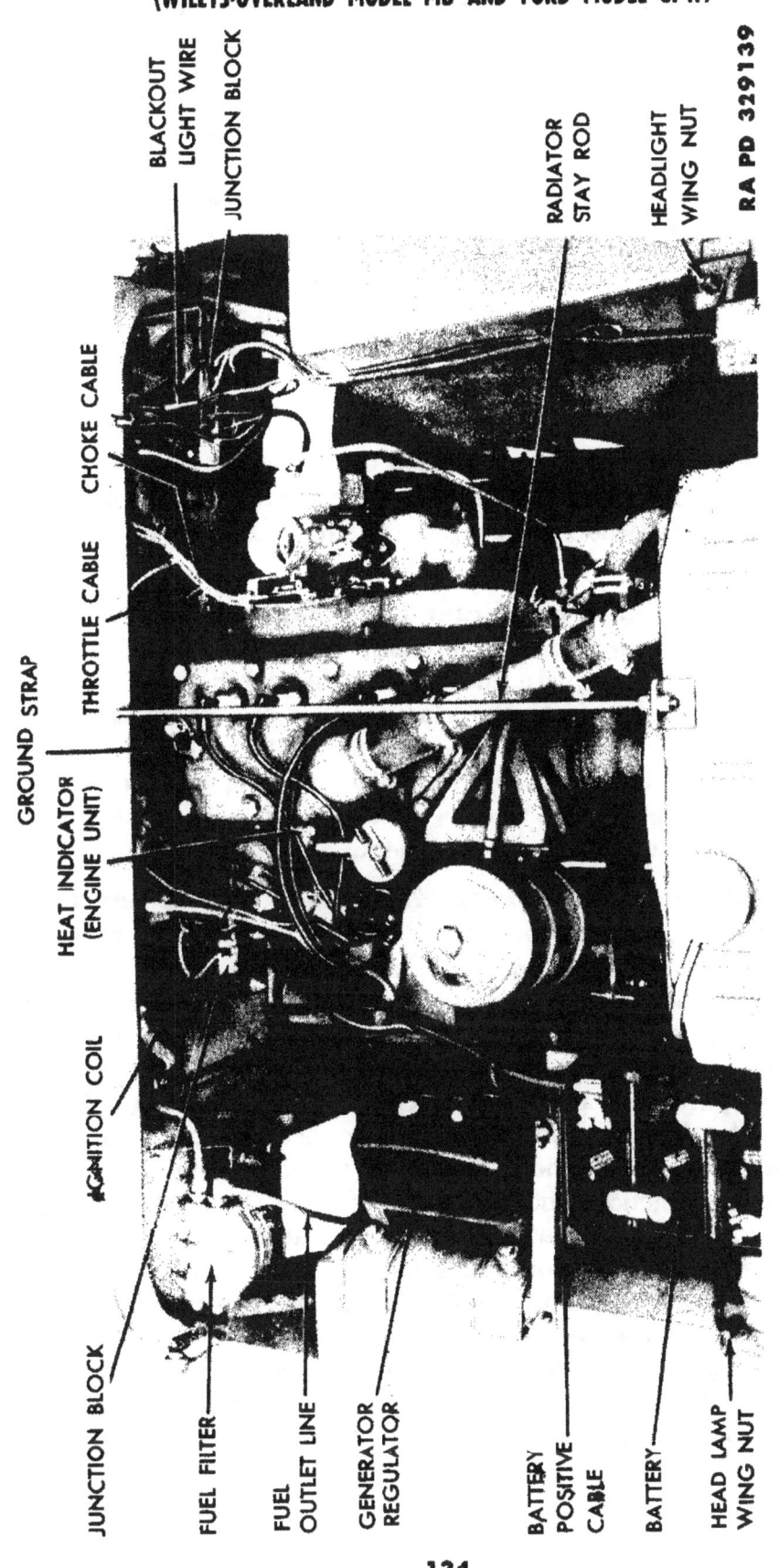

Figure 101 — View of Electrical Wires and Tubes on Cowl

TM 9-1803B
41

BODY AND FRAME

transmission support crossmember. Remove the bolts that secure the body at each side of the frame at the forward end of the body.

c. Disconnect Hand Brake Cable and Speedometer Cable (fig. 100). Remove the hand brake spring at the transfer case. Remove the clevis pin that secures the hand brake cable at the brake on the transfer case. Remove the hand brake cable clamp on the transmission. Disconnect the speedometer cable at the transfer case.

d. Disconnect Ground Straps and Muffler (fig. 100). Disconnect the ground strap at the left side of the transmission. Disconnect the ground strap leading from the body to the right side of the frame at the muffler. Remove the two nuts and bolts that secure the muffler to the body.

e. Remove Air Cleaner. Loosen the hose clamp that secures the air cleaner hose at the air cleaner. Remove the hose from the air cleaner. Remove the four wing nuts that secure the air cleaner to the brackets on the cowl.

f. Remove Clutch and Brake Pedal Pads and Steering Wheel. Remove the cap screw from the clamp at the bottom of the clutch pedal under the floor plate and remove the clutch pedal pads. Remove the cap screw from the clamp under the floor plate at the bottom of the brake pedal and remove the brake pedal. Remove the nut that secures the steering wheel to the steering shaft and remove the steering wheel, using a steering wheel puller. Disconnect the foot accelerator in the driver's compartment.

g. Disconnect Miscellaneous Wires and Tubes in Engine Compartment (fig. 101). Disconnect the positive cable at the battery. Disconnect the wire leading from the junction block at the left side of the cowl to the generator regulator. Disconnect the wire leading from the ignition coil to the junction block. Disconnect the battery cable at the starting switch. Disconnect the ground strap leading from the cylinder to the cowl. NOTE: *When removing the wires, tag them for later identification.* Disconnect the fuel outlet line at the fuel filter. Disconnect the three wires leading from the left fender to the junction block at the left side of the cowl. Disconnect the blackout light wire at the connection at the cowl. Disconnect the two wires at the hydraulic brake master cylinder. Disconnect the wire at the bottom of the steering tube. Disconnect the choke and throttle cable at the carburetor. Disconnect the oil line leading to the oil gage at the flexible connection at the left side of the engine. Disconnect the radiator stay rod at the radiator and cowl and remove the stay rod. Drain the coolant from the radiator and remove the heat indicator (engine unit) at the right-hand side of the cylinder head.

125

TM 9-1803B
41

ORDNANCE MAINTENANCE — POWER TRAIN, BODY, AND FRAME FOR ¼-TON 4 x 4 TRUCK (WILLYS-OVERLAND MODEL MB AND FORD MODEL GPW)

RA PD 28856

Figure 102 — Raising Body From Chassis

BODY AND FRAME

h. Remove Body From Frame (fig. 102). Install a rope on the two forward lifting handles on the body and raise the body slowly off the frame. While raising the body, move it toward the rear of the vehicle until the steering tube is clear of the body. Remove the body from the vehicle.

42. INSTALLATION.

a. Place Body in Position on Frame (fig. 102). Install a rope on the two lifting handles at the forward end of the body. Position the body over the chassis. Line up the steering post with the hole provided in the floor plate of the body. Lower the body slowly and at the same time roll the chassis under the body so as to follow the angle of the steering post until the body is resting on the frame.

b. Install Body Bolts (fig. 100). Install the two bolts that secure the body to each side of the frame at the forward end of the body. Install the four bolts that secure the body to the rear crossmember of the frame. Install the two bolts that secure the body to the pintle hook brace. Install the two bolts that secure the body to the two brackets on each frame side member. Install the bolt that secures the body to each side of the transmission support crossmember.

c. Connect Hand Brake Cable and Speedometer Cable (fig. 100). Connect the speedometer cable to the transfer case. Install the clevis pin that secures the hand brake cable at the brake linkage on the transfer case. Install the hand brake cable clamp to the transmission. Install the spring leading from the hand brake to the floor plate.

d. Connect Muffler and Ground Straps (fig. 100). Install the two nuts and bolts that secure the muffler to the right-hand side of the body. Connect the ground strap leading from the left side of the transmission to the floor plate. Connect the ground strap leading from the body to the right-hand side of the frame.

e. Install Miscellaneous Wires and Tubes in Engine Compartment (fig. 101). Install the heat indicator (engine unit) at the right-hand side of the cylinder head. Install the radiator stay rod to the radiator and at the cowl. Connect the gage oil line at the flexible connection on the left side of the cylinder block. Connect the choke and throttle cables at the carburetor. Connect the horn wire at the bottom of the steering post. Connect the two stop light wires at the hydraulic brake master cylinder. Connect the blackout light wire at the connection at the cowl. Connect the three headlight wires on the left fender to junction block at the left side of the cowl. Connect the ground strap leading from the rear of the cylinder head to the cowl. Connect the battery cable at the starting motor switch. Connect the wire leading from the coil to the junction block at the right-hand side of the cowl. Connect the wire leading from the junction block at the left side of the cowl to the generator regulator. Connect the positive cable at the battery. Connect the fuel line at the fuel filter.

f. Install Clutch and Brake Pedal Pads and Steering Wheel. Place the clutch and brake pedal pads in the clutch and brake pedals so that the raised ends of the pedal pads are toward the steering post. Install the two cap screws that secure the pedals to the levers. Install the steering wheel on the steering shaft. Connect the foot accelerator to the accelerator rod. Install the transmission gear shift lever on the transmission.

g. Install Air Cleaner. Install the air cleaner with four wing nuts securing it to the brackets on the cowl. Connect the air cleaner hose to the air cleaner.

h. Install Hood and Windshield. Place the hood in position on the vehicle and install the five cap screws that secure the hood to the cowl. Place the windshield in position on the cowl and install the wing nuts that secure the windshield at each side of the cowl.

Section IV

FRAME

43. INSPECTION BEFORE REMOVAL.

a. Position the vehicle on a clean level floor. Attach a plumb bob to the grease fittings at the forward ends of the front spring shackle brackets. Mark the floor at the point indicated by the plumb bob. Attach the plumb bob to the grease fittings at the rear of the rear spring shackle brackets and mark the floor at the point indicated by the plumb bob. Move the vehicle off the markings on the floor. Measuring from the markings on the floor, compare the distance between the front shackle and the rear shackle on the one side of the vehicle with the same measurement on the opposite side of the vehicle, and compare the diagonal distance between each of the front shackles and the rear shackles on the opposite side of the vehicle. Differences of more than ¼ inch in these measurements indicate misalinement that must be corrected. If these comparative measurements are the same within ¼ inch, the frame is not misalined.

44. REMOVAL.

a. **Remove Battery (fig. 101).** Disconnect the positive and negative cables from the battery. Remove the two wing nuts that secure the battery hold-down rack and remove the rack from the battery.

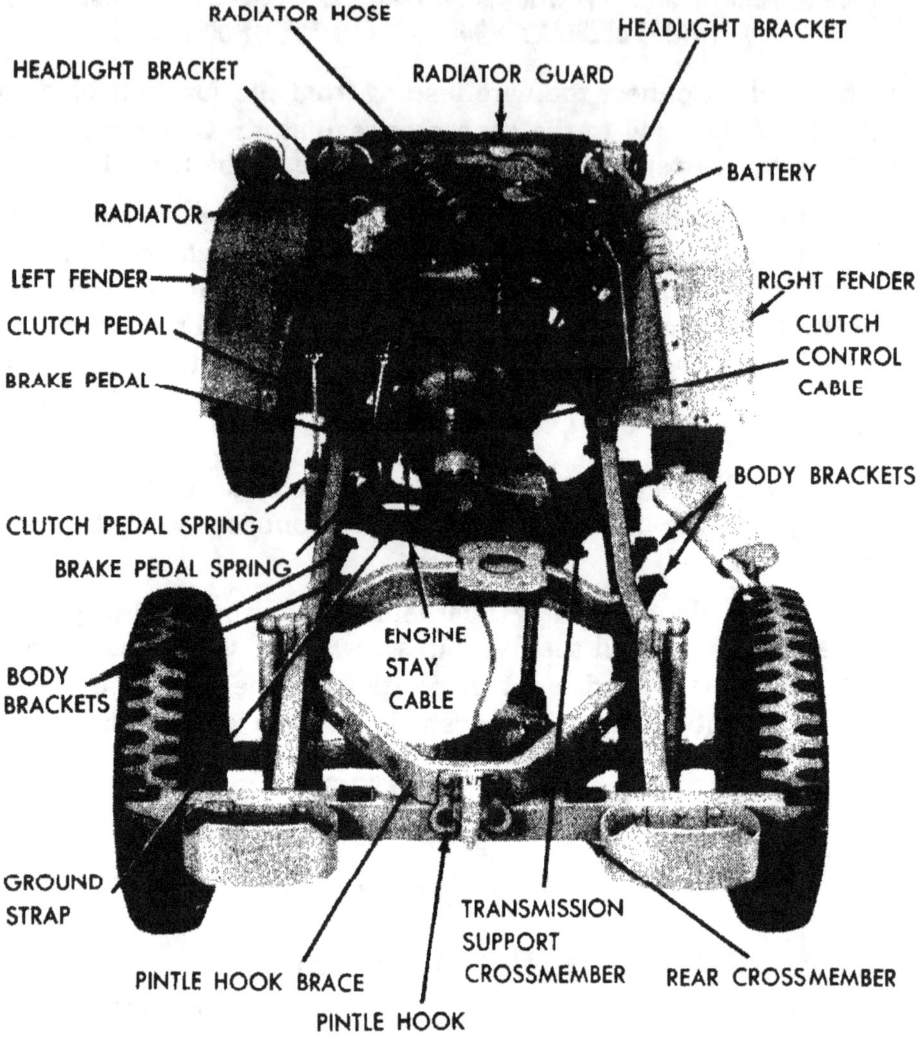

TM 9-1803B
44

BODY AND FRAME

Figure 103 — Chassis

b. Remove Body and Fenders. Remove the body (par. 40). Loosen the wing nuts that secure the headlight support bracket to the top of the fenders. Remove the seven bolts that secure the right-hand fender to the frame and to the radiator guard and remove the fender. Remove the eight bolts that secure the left fender to the frame and to the radiator guard, and remove the fender.

c. Remove Radiator and Radiator Guard (fig. 103). Drain the coolant from the radiator. Disconnect the hose connections at the top and bottom of the radiator. Remove the two bolts that secure the radiator to the frame front crossmember and remove the radiator. Remove the three nuts that secure the radiator guard to the frame front crossmember and remove the radiator guard.

TM 9-1803B
44

ORDNANCE MAINTENANCE — POWER TRAIN, BODY, AND FRAME FOR ¼-TON 4 x 4 TRUCK (WILLYS-OVERLAND MODEL MB AND FORD MODEL GPW)

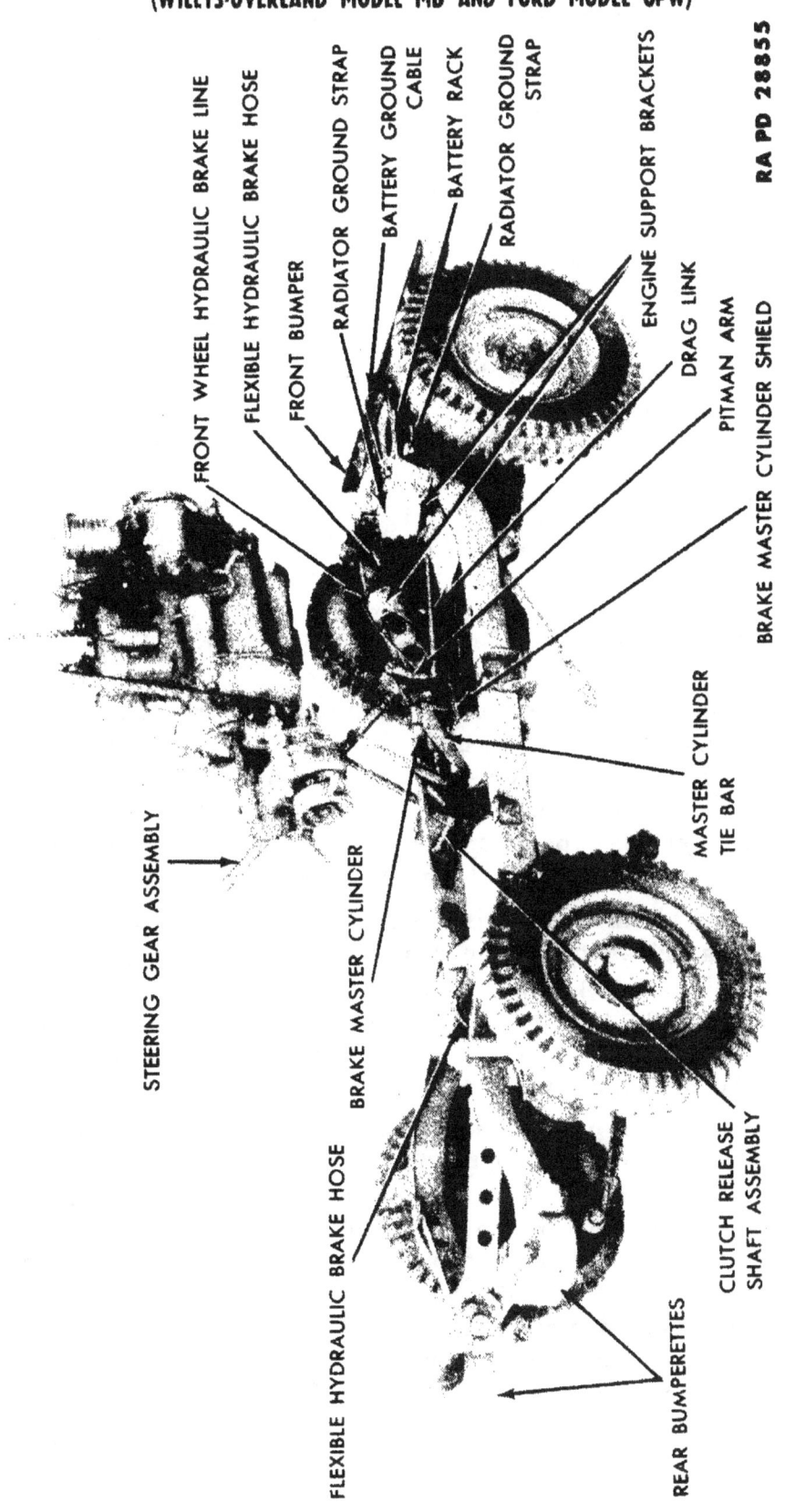

Figure 104 — Removing Engine From Chassis

TM 9-1803B
44

BODY AND FRAME

d. Remove Exhaust Pipe and Transmission Shield (fig. 100). Remove the two bolts that secure the exhaust pipe at the exhaust manifold. Remove the five bolts that secure the transmission shield to the transmission support crossmember. Remove the exhaust pipe and shield from the vehicle.

e. Disconnect the Propeller Shafts (fig. 100). Remove the two U-bolts that secure the front propeller shaft to the transfer case. Remove the propeller shaft from the universal joint flange on the transfer case. Remove the four nuts that secure the rear propeller shaft to the transfer case. Remove the propeller shaft from the transfer case.

f. Remove Front and Rear Bumpers (fig. 104). Remove the eight nuts and bolts that secure the two rear bumperettes to the rear crossmember of the frame and remove the bumperettes. Remove the four nuts and bolts that secure the front bumper bar to the frame and remove the bumper bar.

g. Remove Engine From Vehicle (figs. 103 and 104). Remove the two nuts that secure the transmission to the transmission support crossmember. Remove the ground strap leading from the transmission support crossmember to the transmission. Remove the two nuts that secure the engine stay cable to the transmission support crossmember. Remove the clevis pin that secures the clutch control cable to the clutch release shaft. Remove the two nuts and cap screws that secure the engine mounting to the engine support brackets on each side of the frame. Remove the two ground straps leading from both engine support brackets to the engine. Install a suitable lifting sling on the engine and remove the engine from the frame.

h. Remove Steering Gear Assembly (fig. 104). Disconnect the drag link at the Pitman arm. Remove the cap screw that secures the brake master cylinder shield to the master cylinder. Remove the three nuts and bolts that secure the steering gear assembly to the frame. Remove the steering gear assembly and master cylinder shield from the vehicle.

i. Remove Clutch and Brake Levers (fig. 103). Remove the cotter pin from the clutch and brake pedal shaft at the clutch end of the shaft. Remove the nut and bolt that secure the clutch pedal to the shaft. Remove the clutch pedal and Woodruff key from the shaft. Remove the clutch and brake pedal return springs leading from the pedals to the transmission support crossmember. Remove the two cap screws that secure the master cylinder tie bar to the brake master cylinder. Remove the clutch and brake pedal shaft with the brake pedal and the clutch release shaft from the frame as a unit.

j. Remove Brake Master Cylinder (fig. 104). Disconnect the hydraulic brake line leading to the flexible connection at the forward

TM 9-1803B
44

ORDNANCE MAINTENANCE — POWER TRAIN, BODY, AND FRAME FOR ¼-TON 4 x 4 TRUCK (WILLYS-OVERLAND MODEL MB AND FORD MODEL GPW)

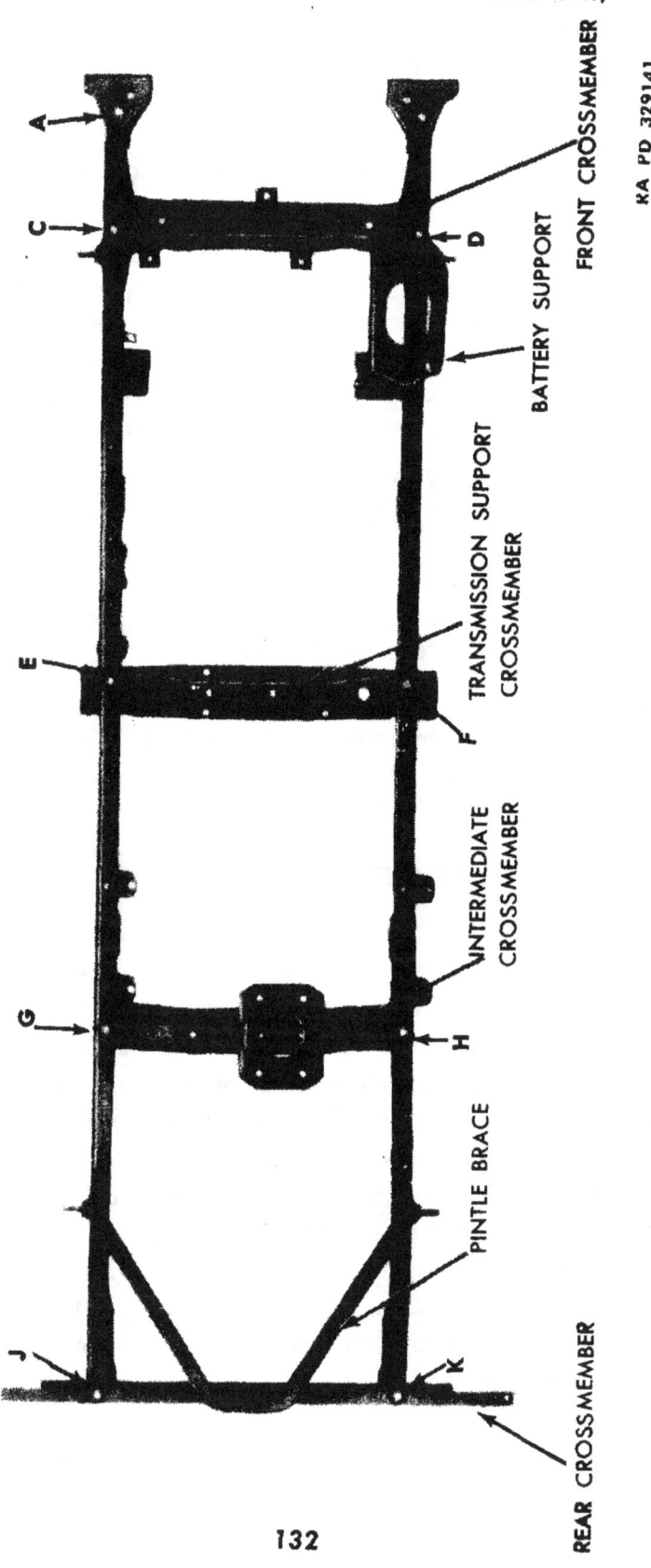

Figure 105 — Frame Assembly

BODY AND FRAME

end of the frame. Remove the lock plate that secures the flexible hose to the frame. Disconnect the hydraulic brake line leading to the rear wheels at the master cylinder. Remove the two bolts that secure the master cylinder to the frame and remove the master cylinder. Disconnect the hydraulic brake line at the flexible connection on the center crossmember. Remove the lock plate that secures the flexible hose to the center crossmember.

k. Remove Front Axle (fig. 35). Remove the cotter pin from the upper end of the two front shock absorbers. Pull the shock absorbers off the brackets on the frame. Remove the two spring shackle bushings from the forward end of each front spring and remove the spring shackles. Remove the shackle bolt that secures the rear ends of the front springs to the frame. Remove the front axle from the frame.

l. Remove Rear Axle (fig. 77). Remove the cotter pin from the upper ends of the two rear shock absorbers and pull the shock absorbers off the brackets on the frame. Remove the two spring shackle bushings at the rear end of the frame from each spring. Remove the shackle bolts that secure the forward end of the rear springs to the frame. Remove the rear axle from the frame.

m. Remove Ground Straps (fig. 104). Remove the two radiator ground straps at the front crossmember of the frame. Remove the ground strap at the transmission support crossmember. Remove the battery ground cable at the front crossmember.

45. INSPECTION AND REPAIR.

a. **Inspection.**

(1) CHECK FRAME ALINEMENT (fig. 105). The extent of misalinement of the frame can be determined by taking measurements at the various points as indicated in figure 105. Measure the distance from A to D and from B to C. The distance between these two points should not vary more than 1/8 inch. Measure the distance from C to F and D to E. The distance between these two points should not vary more than 1/8 inch. Measure the distance from E to H and F to G. The distance between these two points should not vary more than 1/8 inch. Measure the distance from H to J and G to K. The distance between these two points should not vary more than 1/8 inch. If the frame is found to be out of alinement, in most cases it can be corrected by straightening or replacing the damaged frame members or sections.

(2) INSPECT FRAME FOR LOOSE RIVETS. Replace any rivets that are loose or missing. Replace the battery support bracket if bent.

ORDNANCE MAINTENANCE — POWER TRAIN, BODY, AND FRAME FOR ¼-TON 4 x 4 TRUCK
(WILLYS-OVERLAND MODEL MB AND FORD MODEL GPW)

46. INSTALLATION.

a. Install Ground Straps (fig. 104). Install the two radiator ground straps at the front crossmember of the frame. Install the ground strap at the transmission support crossmember. Install the battery ground cable at the front crossmember.

b. Install Rear Axle (fig. 77). Place the rear axle in position under the frame. Install the rear spring forward shackle bolts with the grease fittings facing outward. Install the two castellated nuts and cotter pins that secure the shackle bolts. Install the rear spring rear shackles with the shackle ends facing outward. Install the two shackle bushings to each spring shackle. Install the shock absorbers to the brackets at each side of the frame.

c. Install Front Axle (fig. 35). Place the front axle in position under the frame. Install the two front spring rear shackle bolts with the grease fittings facing outward. Install the two castellated nuts and cotter pins that secure the shackle bolts. Install the spring shackle at the forward end of the frame with the shackle ends facing outward. Install the two spring shackle bushings to each spring shackle. Install the shock absorbers to the brackets at each side of the frame.

d. Install Brake Master Cylinder (fig. 104). Install the lock plate that secures the hydraulic brake flexible hose to the center crossmember of the frame. Connect the hydraulic brake line at the flexible connection on the center crossmember. Install the two bolts that secure the hydraulic brake master cylinder to the side of the frame. Install the brake line leading from the rear wheels to the brake master cylinder. Install the lock plate that secures the brake flexible hose at the bracket on the frame. Install the brake line leading from the flexible hose to the master cylinder.

e. Install Clutch and Brake Pedals (fig. 103). Install the clutch and brake shaft with brake pedal and clutch release shaft onto the frame. Install the two cap screws that secure the master cylinder tie bar to the master cylinder. Insert the brake pedal rod in the master cylinder. Install the clutch and brake return springs leading from the clutch and brake pedals to the transmission support crossmember. Install the Woodruff key on the clutch and brake pedal shaft. Slide the clutch pedal on the clutch shaft. Install the lock bolt that secures the clutch pedal to the shaft and install the cotter pin.

f. Install Steering Assembly and Bleed Brakes (fig. 104). Place the steering assembly in position on the frame. Install the three nuts and bolts that secure the steering assembly to the frame. Connect

BODY AND FRAME

the drag link at the Pitman arm. Bleed the hydraulic brake system. (Refer to TM 9-803.)

g. Install Engine in Vehicle (fig. 104). Install a suitable lifting sling on the engine and place the engine in position on the frame. Move the rear of the engine slightly to the right-hand side so as to allow the ball joint on the transfer case to enter in the clutch release shaft. Install the two nuts and bolts that secure the engine mounts to the engine support brackets on each side of the frame. Install the two ground straps leading from the engine support brackets to the engine. Install the two nuts that secure the transmission to the transmission support crossmember. Install the ground strap leading from the transmission to the transmission support crossmember. Install the clevis pin that secures the clutch control cable to the clutch release shaft. Install the engine stay cable to the transmission support crossmember.

h. Install Front and Rear Bumpers (fig. 104). Place the front bumper in position on the frame. Install the four nuts and bolts that secure the bumper to the ends of the frame. Place the two bumperettes in position on the rear crossmember of the frame and install the eight cap screws that secure the bumperettes to the rear crossmember of the frame.

i. Install Propeller Shafts (fig. 100). Place the front propeller shaft in the universal joint flange at the transfer case. Install the two U-bolts that secure the propeller shaft to the transfer case. Place the rear propeller shaft yoke flange in position on the transfer case. Install the four nuts that secure the propeller shaft to the transfer case.

j. Install Exhaust Pipe and Transmission Shield (fig. 100). Install the two nuts that secure the exhaust pipe to the exhaust manifold. Place the transmission shield in position under the vehicle. Install the five bolts that secure the transmission shield to the transmission. Install the clamp that secures the exhaust pipe to the shield.

k. Install Radiator and Radiator Guard (fig. 103). Place the insulators on the radiator brackets on the frame. Place the radiator in position in the brackets on the frame. Install the flat washers, ground straps, and flat washers on the two radiator studs. Install the nuts that secure the radiator to the frame. Install the upper and lower radiator hose connections. Install the radiator guard in the bracket in front of the radiator. Install the three nuts and bolts that secure the guard to the frame.

l. Install Fenders and Body (fig. 102). Place the fenders in position on the vehicle. Install the seven cap screws that secure each fender to the radiator guard and to the frame. Install the body on the chassis (par. 41).

m. Install Battery (fig. 101). Place the battery in the battery rack. Install the hold-down bracket on the battery. Install the two wing nuts that secure the hold-down bracket. Connect the negative and positive cables to the battery.

n. Lubricate. Lubricate the chassis of the vehicle with specified lubricants. Fill the radiator to proper level with coolant.

Section V

FITS AND TOLERANCES

47. FITS AND TOLERANCES.

Fit Location and Name	Manufacturer's Fit Tolerance	Fit Wear Limit	Type of Fit
a. Front Springs.			
Spring bushing and shackle bolt	—	0.010 in.	Running
Torque reaction spring and shackle bolt	—	0.010 in.	Running
Torque reaction spring shackle inner bushing and lower shackle bolt	—	0.010 in.	Running
Torque reaction spring shackle outer bushing and lower shackle bolt	—	0.010 in.	Running
b. Rear Springs.			
Rear spring bushing and shackle bolt	—	0.010 in.	Running
c. Steering Gear.			
Sector shaft bushing and sector shaft	—	0.001 in.	Running
d. Monroe Shock Absorber.			
Piston and pressure tube	0.002 in.	0.004 in.	Running

CHAPTER 4

SPECIAL TOOLS

48. PURPOSE.

a. The special tools required for maintenance and repair of the ¼-ton 4 x 4 Truck are listed in SNL G-27.

b. The following list, extracted from SNL G-27, contains those special tools required to perform the operations described in this manual. The list is supplied for identification purposes only; it is not to be used as a basis for requisition.

49. LIST OF SPECIAL TOOLS.

a. Special Tools for Power Train.

Name	Federal Stock Number	Mfr's Number	Figure Number
Gage, drive pinion setting (set)	41-G-176	KM-J-589-S	67 and 106
Locator, idle gear thrust washer	41-L-1570	KM-J-1758	106
Remover, drive pinion flange and side bearing	41-R-2378-30	KM-J-872-S	49 and 106
Remover, drive pinion oil seal	41-R-2378-40	KM-J-1742	106
Remover, front axle outer oil retainer	41-R-2384-38	KM-J-943	55-80
Remover, mainshaft front cone, transmission	41-R-2368-200	KM-J-1749	106
Replacer, axle shaft, front and rear oil seal	41-R-2391-20	KM-J-1753	56 and 106
Replacer, differential side bearing cone	41-R-2391-65	KM-J-1763	68 and 106
Tool, oil seal assembly shifter shaft (set)	41-T-3280	KM-J-1757	106
Tool, universal joint, assembly and disassembly	41-T-3379	KM-J-881-A	—
Wrench, wheel bearing nut	41-W-3825-200	GP-17033	—

b. Special Tools for Shock Absorbers (Monroe).

Name	Federal Stock Number	Mfr's Number	Figure Number
Compressor, shock absorber grommet	41-C-2554-400	MAS-1148	—
Filler cup, Monroe shock absorber	41-F-2985-200	—	106
Remover, base valve	41-R-2373-340	—	106
Remover, rod guide and seal assembly	41-R-2373-115	—	106
Replacer, pressure tube	41-R-2399-350	—	106
Spanner wrench, special piston rod guide and seal assembly	41-W-3336-745	—	88 and 106
Thimble, piston rod	41-T-1657	—	106

TM 9-1803B
49

ORDNANCE MAINTENANCE — POWER TRAIN, BODY, AND FRAME FOR ¼-TON 4 x 4 TRUCK
(WILLYS-OVERLAND MODEL MB AND FORD MODEL GPW)

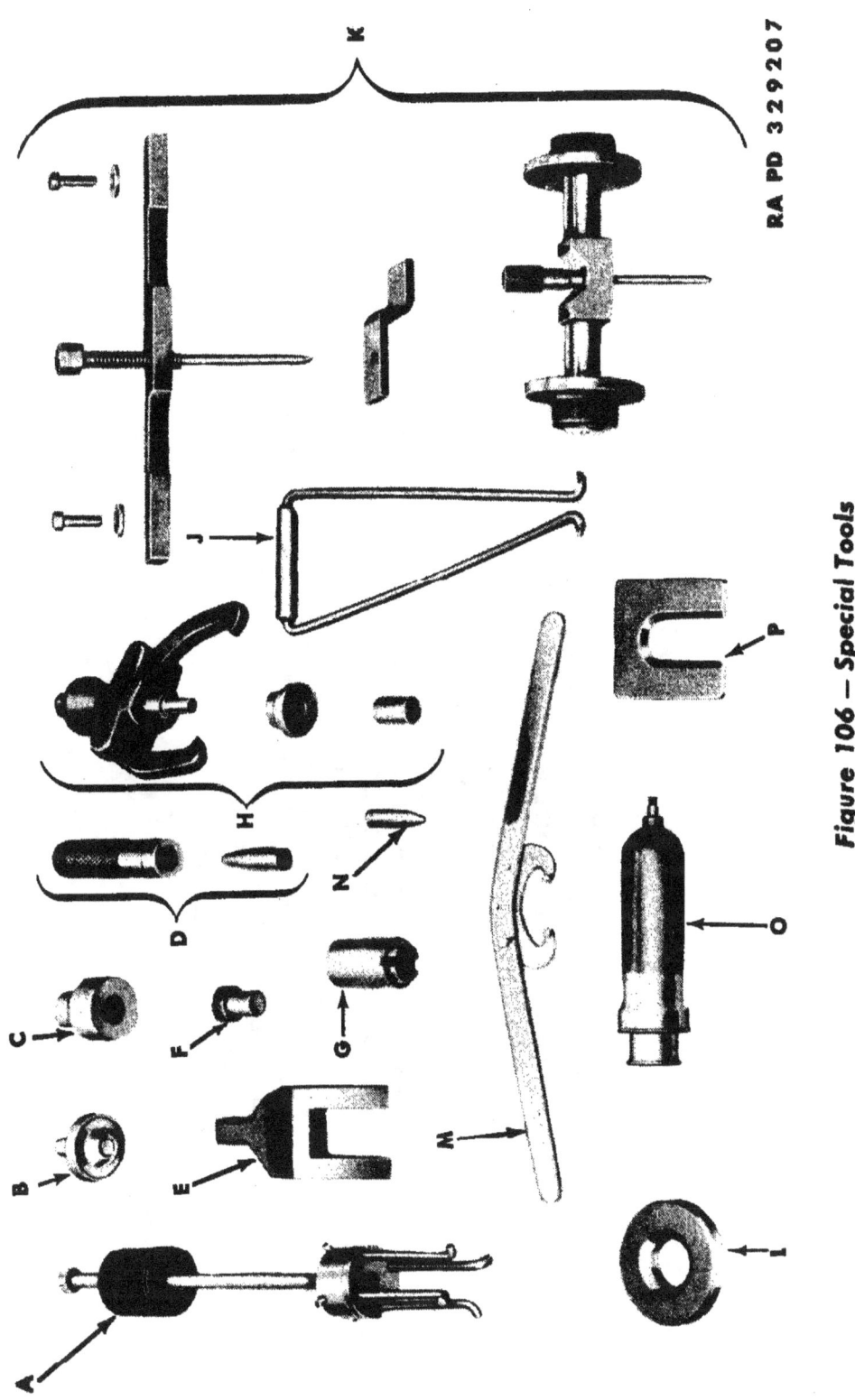

Figure 106 — Special Tools

RA PD 329207

SPECIFICATION

TM 9-1803B
49

SPECIAL TOOLS

	Federal Stock No.
A—REMOVER, DRIVE PINION OIL SEAL	41-R-2378-40
B—REPLACER, AXLE SHAFT FRONT AND REAR OIL SEAL	41-R-2391-20
C—REPLACER, DIFFERENTIAL SIDE BEARING CONE	41-R-2391-65
D—TOOL, OIL SEAL ASSEMBLY, TRANSFER CASE SHIFTER SHAFT SET	41-T-3280
E—REMOVER, MAINSHAFT FRONT CONE	41-R-2368-200
F—LOCATOR, IDLER GEAR THRUST WASHER	41-L-1570
G—REPLACER, PRESSURE TUBE	41-R-2399-350
H—REMOVER, DRIVE PINION FLANGE AND DIFFERENTIAL SIDE BEARING	41-R-2378-30
J—REMOVER, FRONT AXLE OUTER OIL RETAINER	41-R-2384-38
K—GAGE, DRIVE PINION SETTING (SET)	41-G-176
L—REMOVER, ROD GUIDE AND SEAL ASSEMBLY	41-W-2373-115
M—SPANNER WRENCH, SPECIAL PISTON ROD GUIDE AND SEAL ASSEMBLY	41-W-3336-745
N—THIMBLE, PISTON ROD	41-T-1657
O—FILLER CUP, SHOCK ABSORBER	41-F-2985-200
P—REMOVER, PRESSURE VALVE	41-R-2373-340

RA PD 329207-B

Legend for Figure 106 — Special Tools

TM 9-1803B

ORDNANCE MAINTENANCE — POWER TRAIN, BODY, AND FRAME FOR ¼-TON 4 x 4 TRUCK
(WILLYS-OVERLAND MODEL MB AND FORD MODEL GPW)

REFERENCES

PUBLICATIONS INDEXES.

The following publications indexes should be consulted frequently for latest changes or revisions of references given in this section and for new publications relating to materiel covered in this manual:

a. Introduction to Ordnance Catalog (explaining SNL system) .... ASF Cat. ORD 1 IOC

b. Ordnance Publications for Supply Index (index to SNL's) .... ASF Cat. ORD 2 OPSI

c. Index to Ordnance Publications (listing FM's, TM's, TC's, and TB's of interest to ordnance personnel, OPSR, MWO's, BSD, S of SR's, OSSC's, and OFSB's; and includes Alphabetical List of Major Items with Publications Pertaining Thereto) .... OFSB 1-1

d. List of Publications for Training (listing MR's, MTP's, T/BA's, T/A's, FM's, TM's, and TR's concerning training) .... FM 21-6

e. List of Training Films, Film Strips, and Film Bulletins (listing TF's, FS's, and FB's by serial number and subject) .... FM 21-7

f. Military Training Aids (listing Graphic Training Aids, Models, Devices, and Displays) .... FM 21-8

STANDARD NOMENCLATURE LISTS.

Cleaning, preserving and lubricating materials; recoil fluids; special oils, and miscellaneous related items	SNL K-1
Ordnance maintenance sets	SNL N-21
Soldering, brazing and welding materials, gases and related items	SNL K-2
Tools, maintenance for repair of automotive vehicles	SNL G-27 Volume 1
Tool-sets, for ordnance service command automotive shops	SNL N-30
Tool-sets, motor transport	SNL N-19
Truck, ¼-ton, 4 x 4, command reconnaissance (Ford and Willys)	SNL G-503

TM 9-1803B

REFERENCES

EXPLANATORY PUBLICATIONS.

Fundamental Principles.

Automotive brakes	TM 10-565
Automotive electricity	TM 10-580
Automotive power transmission units	TM 10-585
Basic maintenance manual	TM 38-250
Chassis, body, and trailer units	TM 10-560
Electrical fundamentals	TM 1-455
Military motor vehicles	AR 850-15
Motor vehicle inspections and preventive maintenance service	TM 9-2810
Precautions in handling gasoline	AR 850-20
Sheet metal work, body, fender, and radiator repairs	TM 10-450
Standard military motor vehicles	TM 9-2800
The body finisher, woodworker, upholsterer, painter, and glassworker	TM 10-455
The machinist	TM 10-445

Maintenance and Repair.

Cleaning, preserving, lubricating and welding materials and similar items issued by the Ordnance Department	TM 9-850
Cold weather lubrication and service of combat vehicles and automotive materiel	OFSB 6-11
Maintenance and care of pneumatic tires and rubber treads	TM 31-200
Ordnance Maintenance: Electric equipment (Auto-Lite)	TM 9-1825B
Ordnance Maintenance: Engine and engine accessories for ¼-ton 4 x 4 truck (Ford and Willys)	TM 9-1803A
Ordnance Maintenance: Hydraulic brake system (Wagner)	TM 9-1827C
Ordnance Maintenance: Speedometers and tachometers (Stewart-Warner)	TM 9-1829A

Operator's Manual.

¼-ton 4 x 4 truck (Willys-Overland model MB and Ford model GPW)	TM 9-803

141

ORDNANCE MAINTENANCE — POWER TRAIN, BODY, AND FRAME FOR ¼-TON 4 x 4 TRUCK
(WILLYS-OVERLAND MODEL MB AND FORD MODEL GPW)

Protection of Materiel.

Camouflage	FM 5-20
Chemical decontamination, materials and equipment	TM 3-220
Decontamination of armored force vehicles	FM 17-59
Defense against chemical attack	FM 21-40
Explosives and demolitions	FM 5-25

Storage and Shipment.

Ordnance storage and shipment chart, group G—Major items	OSSC-G
Registration of motor vehicles	AR 850-10
Rules governing the loading of mechanized and motorized army equipment, also major caliber guns, for the United States Army and Navy, on open top equipment published by Operations and Maintenance Department of Association of American Railroads.	
Storage of motor vehicle equipment	AR 850-18

INDEX

A

	Page
AGO Form No. 478	6
Adjustment	
clutch	24
differential assembly	
front axle	76
rear axle	94
Monroe shock absorber	110
outer bearing	
front axle	75
rear axle	95
pinion in housing	
front axle	74
rear axle	93
shock absorber	
Gabriel	106
Monroe	107
spindle housing	79
Air cleaner	
installation	128
removal	125
Assembly	
axles	
axle shafts	80
front axle	74
rear axle	93
drag link	122
output shaft bearing cap, front	30
propeller shafts	43
shock absorbers (Monroe)	109
steering gear assembly	116
transfer case	30
transmission	21
Axle, front	
assembly	74
cleaning, inspection, and repair	60
description and data	45
disassembly	49
installation	83, 134
removal	48, 133
Axle, rear	
assembly	93
description and data	86
installation	97, 134
removal	87, 133
Axle housing	
inspection and repair	60
replacement of bushing	65
Axle shafts	
assembly (front axle)	80

	Page
disassembly (front axle)	49
installation	
front axle	80
rear axle	97
removal	
front axle	49
rear axle	89

B

	Page
Backlash	
front axle	77
rear axle	95
Ball bearings, inspection and repair	20
Battery	
installation	136
removal	129
Bearings, outer and inner	
front axle	74
rear axle	93
Bearings, roller	
installation	
front axle	76
rear axle	94
removal	
front axle	58
rear axle	89
Bendix universal joint	
disassembly	51
replacement	65
Body	
installation	122, 135
place in position on frame	127
removal	122, 129
Bolts, body, remove from frame	122
Brake, band	
installation	37
removal	25
Brake, bleed, installation	134
Brake, hand, installation	38
Brake drum, installation	
front axle	81
rear axle	97
Brake lever removal	131
Brake pedal pads	
installation	128
removal	125

143

TM 9-1803B

B — Cont'd	Page
Brake plate	
installation	
front axle	81
rear axle	95
removal	89
Brake springs, removal	11
Bumpers	
installation	135
removal	131
Bushings, countershaft, inspection and repair	21

C

	Page
Cable, clutch	
transfer case	25
transmission	12
Cable, engine stay	
installation	24
removal	12
Cable, hand brake	
connect	127
disconnect	125
installation	24
removal	12, 25
Cable, speedometer	
connect	127
disconnect	125
installation	
transfer case	38
transmission	24
Chassis lubrication	137
Cleaning	
front axle	60
propeller shafts and universal joints	38
rear axle	91
springs	101
steering gear assembly	112
transmission parts	16
Clutch	
installation	38
removal	131
Clutch pedal, installation	134
Clutch pedal pads	
installation	128
removal	125
Clutch release bearing, assembly	22

	Page
Clutch release fork	
installation	23
removal	12
Clutch shaft, installation to transfer case	23
Countershaft	
assembly	22
inspection and repair	20
Crossmember, support, transmission	
installation	23
removal	13
Cylinder, master brake	
disconnect	131
installation	134

D

	Page
Data	
axles	
front	45
rear	87
Gabriel shock absorber	105
propeller shafts and universal joints	38
Monroe shock absorber	107
special tools	137
springs	99
transfer case	25
transmission	7
Description	
axles	
front	45
rear	86
power train	7
propeller (drive) shafts and universal joints	38
shock absorbers	
Gabriel	104
Monroe	107
springs	99
steering gear assembly	111
transfer case	25
transmission	7
Differential assembly	
installation and adjustment	
front axle	76
rear axle	94

INDEX

D -- Cont'd

Differential assembly — Cont'd
removal
- front axle 55
- rear axle 89

replacement
- front axle 65
- rear axle 92

Differential
- disassembly 55

Differential cover
installation
- front axle 78
- rear axle 95

Differential pinion gears
- removal from differential 89

Disassembly
- drag link 120
- front axle 49
- Monroe shock absorber 109
- rear axle 89
- steering gear assembly 111
- transmission 15

Drag link
- assembly, cleaning, inspection, installation, and repair 122
- connect 119
- disconnect 111
- removal and disassembly 120

Drive flange
- installation 83

Drive pinion
removal
- front axle 58
- rear axle 89

Drive pinion assembly
replacement
- front axle 65
- rear axle 92

Drive pinion bearing cap
- replacement 92

Drum assembly
- rear axle 89
- transfer case 25

E

Engine
- installation 135
- removal 131

Exhaust pipe
- installation 135
- removal 131

F

Fenders
- installation 119, 135
- removal 111, 129

Floor plate, transmission 11

Frame
- inspection and repair 128, 133
- installation 134
- removal 128

Frame alinement 133

G

Gear, countershaft
- assembly 21
- disassembly 16
- inspection and repair 18

Gear, differential ring 76

Gear, first and reverse 18

Gear, idle
- assembly 21
- inspection and repair 20

Gear, idle reverse 16

Gear, intermediate 35

Gear, main drive
- disassembly 16
- installation 23

Gear, mainshaft
- installation 38
- removal 25

Gear, steering (See Steering gear)

Gears, installation
- front axle 75
- rear axle 94

Gears, axle shaft 89

Gears, differential pinion (See Differential pinion gears)

Gearshift housing
- assembly 22
- removal 14

Ground straps
- connect 127
- disconnect 125
- installation 24, 134
- removal 12, 133

Guide rail, shifter fork (See Shifter fork guide rail)

TM 9-1803B

ORDNANCE MAINTENANCE — POWER TRAIN, BODY, AND FRAME FOR ¼-TON 4 x 4 TRUCK (WILLYS-OVERLAND MODEL MB AND FORD MODEL GPW)

H

	Page
Hood	
installation	128
removal	122
Horn wire	
installation	118
replacement	115
Housing, inner and outer, bushing	
replacement	116
Hub	
installation	
front axle	81
rear axle	97
removal	
rear axle	89
Hydraulic brake hose	
installation	81
Hydraulic brake line	
disconnect	87
installation	97

I

Idle reverse gear (See Gear, idle reverse)

	Page
Inspection and repair	
axles	
front	60
rear	91
propeller shafts and universal joints	43
steering gear assembly	112, 113
transmission	18

K

	Page
Knuckle, propeller shaft, removal	
front shaft	40
rear shaft	41

L

	Page
Lubrication	
chassis	137
clutch	24
front axle	86
rear axle	97

M

	Page
Main drive assembly	18
Main drive gear assembly	21
Main drive gear bearing retainer	14
Main drive gear bearing rollers	20
Mainshaft	
inspection and repair	18
removal	16
Mainshaft assembly	21
Mounting bolt, removal	25
Muffler	
connect	127
disconnect	125

O

	Page
Oil seal, replacement	
front axle	63
rear axle	92
Oil seal, spindle housing	79
Output shaft, installation	31
Output shaft bearing cap oil seal, replacement	30

P

	Page
Pedals, brake	134
Pitman arm	
disconnect	111
installation	118
Pivot arm	70, 79
Pivot arm shaft	62
Power train	7
Propeller shaft, front	
disconnect	48
removal	39
Propeller shaft, rear	
disconnect	25
removal	39
Propeller shafts	
disconnect	87
installation	24, 38, 85, 97, 135
removal	12, 131
Propeller shafts and universal joints	
assembly	43
cleaning, inspection and repair	43
data	39
description	39
disassembly	40
installation	43

R

	Page
Radiator	
installation	135
removal	129

INDEX

R — Cont'd

	Page
Radiator guard	
installation	135
removal	129
Radiator hose	12
Rear cover	25
Rear output shaft cap	35
Ring gear	
check run-out	78, 95
removal	
front axle	58
rear axle	89
Ring gear, differential	94
Rivets in frame, loose	133
Rzeppa universal joint	
disassembly	51
replacement	65

S

	Page
Shaft, idle gear	
inspection and repair	20
Shaft assembly	
removal from housing	112
Shaft assembly, steering sector	
removal	112
Shift lever	
inspection and repair	21
removal	11
Shift lever assembly	
replacements	30
Shifter fork guide rail	
removal	14
Shifter forks	
assembly	22
installation	31
removal	14
Shock absorber, Gabriel	
cleaning and inspection	106
description and data	104
installation and adjustment	106
removal	106
Shock absorber, Monroe	
adjustment	110
assembly and disassembly	109
cleaning, inspection, and removal	109
description and data	107
fill with fluid	110
installation	110

	Page
Shock absorbers	
disconnect	
front axle	48
rear axle	87
installation	
front axle	85
rear axle	97
Snap rings (front propeller shaft)	
removal from yoke	40
Spider (front propeller shaft)	
removal from yoke	
front shaft	40
rear shaft	41
Spider (rear propeller shaft)	
install in yoke flange	43
Spindle	
inspection and replacement	73
installation	81
Spindle arm	
inspection and replacement of spindle bearing pin	71
Spindle housing	
adjustment and installation	79
removal	53
Spindle housing bearing cap	
inspection and replacement	71
Spring shackles	
disconnect	48
Spring U-bolts	
disconnect	
front axle	48
rear axle	87
installation	
front axle	83
rear axle	87
Springs	
disconnect	87
installation	
front	103
rear	97, 104
removal	
front	100
rear	101
Steering assembly	
installation	134
Steering column, installation	118
Steering column tube, removal	111
Steering gear, removal	111

S — Cont'd

Steering gear assembly
- assembly ... 116
- cleaning and inspection ... 112
- description and disassembly ... 111
- removal ... 111, 131
- repair ... 115

Steering gear worm, removal ... 113

Steering sector shaft assembly, removal ... 112

Steering wheel
- installation ... 119, 128
- removal ... 111, 125

Synchronizer, disassembly ... 16

T

Tie rods
- inspection and replacement of end ... 70
- installation ... 83

Tools, special ... 137

Torque, reaction spring bushing ... 101

Tracta universal joint
- disassembly ... 53
- replacement ... 70

Transfer case
- assembly ... 30
- cleaning, inspection, and repair ... 27
- description and data ... 24
- installation ... 37
- raising of ... 37
- removal and disassembly ... 25

Transfer case shield ... 38

Transmission
- description and data ... 7
- disassembly ... 15
- disconnect at clutch housing ... 12
- inspection and repair ... 18
- installation ... 23
- removal ... 11, 13

Transmission shield
- installation ... 24, 135
- removal ... 11, 25

Tubes, engine compartment ... 125

U

Under drive shifter fork assembly ... 30

Universal joints (See Propeller shafts and universal joints, Bendix universal joint, and Tracta universal joint)

W

Washers, countershaft thrust ... 20

Wheel alinement ... 86

Wheels
- installation ... 85, 97
- removal
  - front axle ... 49
  - rear axle ... 89

Windshield
- installation ... 128
- removal ... 122

Wires, engine compartment ... 125

Worm upper bearing cup ... 115

Y

Yoke, front propeller shaft ... 40

Yoke flange, rear propeller shaft ... 41

SNL G-503

THE ORDNANCE CATALOG

STANDARD NOMENCLATURE LIST

FOR

TRUCK, ¼-TON, 4 x 4, COMMAND RECONNAISSANCE

WILLYS-OVERLAND MOTORS INC., MODEL MB

Parts formerly listed in TM-lo-1512 change No. 3, dated 1 Augurt 1943 and in TM-10-1348 change No. 1, dated 10 March 1943, are now included in this Standard Nomenclature List No. Q-503.

Supersedes Addendum, SNL G-503, dated 2 Feb. 1943, and Organizational Spare parts and Equipment List, SNL G-503, dated 8 June 1943.

SECTION 1

GENERAL INFORMATION

For full information relative to THE ORDNANCE PROVISION SYSTEM, of which the Standard Nomenclature List is a part, see the Introduction to the Ordnance Catalog (IOC), the Ordnance Provision System Regulations (OPSR), and the Index to Ordnance Publications (OFSB 1-1). (See OFSB 1-8 for distribution of publications.)

THE STANDARD NOMENCLATURE LIST of Spare Parts and Equipment is divided into eight sections. These are, besides this General Information Section:

Maintenance Parts Procurement List (Section 2)
Vehicular Spare Parts and Equipment List (Section 3)
Organizational Spare Parts and Equipment List (Section 4)
Ordnance Maintenance Unit Stockage List (Section 5)
Depot Stockage List (Section 6)
Geographical or Seasonal Maintenance Parts List (Section 7)
Indexes (Section 8)

MAINTENANCE PARTS PROCUREMENT LIST (Section 2)

This section lists all serviceable parts in the vehicle and also designates which of these parts were procured for maintenance of the vehicle. Column 6, Quantity Required Per Unit Assembly, shows the quantity of each part on the vehicle used in the location for which the listing is given. Column 7, 12 Mos. Field Maintenance, shows the quantity of each part that has been estimated to be required for the third and fourth echelons for the maintenance of one hundred (100) vehicles for twelve months. These figures are exclusive of the quantities estimated for Organization Maintenance Parts, which appear in Section 4. Column 8, Major Overhaul (5th echelon), shows the quantity of each part that has been estimated to be required for fifth echelon maintenance for one hundred (100) vehicles for twelve months. Column 9, Total First Year Procurement, shows the quantity of each part procured for the total or complete maintenance by all echelons of one hundred (100) vehicles for twelve months. Column 10, Estimated Requirements per 100 Rebuilds, shows the estimated quantity of each part required to rebuild one hundred (100) vehicles.

Using troops performing first and second echelon maintenance may requisition those parts marked by a % sign in Column 7 of Section 2. Those parts so marked which are not also listed in the Organizational Maintenance Parts and Equipment List (Section 4) are to be requisitioned for replacement purposes only when time and the tactical situation permit, and trained personnel and suitable tools are available. Normally only those parts shown in Section 4 are carried by using troops for first and second echelon maintenance.

VEHICULAR SPARE PARTS AND EQUIPMENT LIST (Section 3)

This section lists all maintenance parts, accessories, equipment, common and special tools, supplies, ammunition and sighting equipment that are carried in or on the vehicle. Also included are any other major items that are issued with, and as an integral part of the vehicle. For example, the armament (guns and gun mounts) is listed in this section, as well as the armament spare parts that are carried on the vehicle.

ORGANIZATIONAL SPARE PARTS AND EQUIPMENT LIST (Section 4)

This section lists all maintenance parts, accessories, common and special tools, equipment and supplies, fire control equipment, applicable cleaning and preserving materials, required by, and authorized for using troop organizations in connection with the use and maintenance of the vehicle. Column 6 through 13 in this section show the estimated quantities of parts required by each company or regiment or separate battalion to maintain the various quantities of the vehicle assigned to each respectively. The Organizational Accessories shows the tool sets allocated to these organizations, the figures in the quantity column indicating the quantities per tool set.

ORDNANCE MAINTENANCE UNIT STOCKAGE LIST (Section 5)

This section lists all maintenance parts, accessories, common and special tools, equipment and supplies required to be carried by Ordnance Maintenance Units. The quantities of the parts shown are those required for the maintenance of the various quantities of the vehicle on a basis of thirty (30) days field maintenance.

DEPOT STOCKAGE LIST (Section 6)

This section is an explanation of the method to be used for determining the quantities of parts to be stocked by Field, Base and Mobile Depots.

GEOGRAPHICAL OR SEASONAL MAINTENANCE PARTS LIST (Section 7)

This section consists of sub-divisions for each geographical location, season or particular terrain conditions. Each sub-division contains a list of maintenance parts, accessories, common and special tools, equipment and supplies, which, due to the peculiarities of the geographical location, season, or terrain condition, are required to be carried in stock by Ordnance Maintenance Units *in addition to* the stockage prescribed in Section 5. The quantities of parts shown in each sub-division are those of the *additional items* or *additional quantities* (of the items in Section 5) required for the maintenance of one hundred (100) vehicles for twelve months within the particular operating condition involved.

FEATURES OF THIS PUBLICATION

The following paragraphs explain various important details with which the user should be familiar for proper understanding of the catalog and practices peculiar to this individual book.

STANDARDIZED GOVERNMENT GROUPING: The parts listed in this catalog are divided into groups and sub-groups in accordance with the Standardized Government Grouping. This division into groups facilitates the location and identification of each part. The Standardized Government Grouping as adapted to this catalog is outlined on page 8.

ILLUSTRATIONS: Assembly views, exploded views and cross sectional drawings are used throughout this catalog to aid the user in locating and identifying parts. All illustrations are placed as closely as possible to the groups to which they apply. On each illustration or separate reference legend, the noun name, piece-mark and appropriate group number of each part illustrated are given. This promotes the quick identification of each part and accurate reference to the text for complete listing. Each illustration is identified by a figure number and title and in Section 8, Indexes, in the back of the book, will be found an index of all figures and the page on which they appear.

COMPONENTS OF KITS, SETS AND ASSEMBLIES: Whenever any doubt might exist as to what items are included as components of kits, sets or assemblies, an explanatory note or complete listing of such components is shown.

ATTACHING PARTS: Attaching parts such as bolts, nuts, rivets, screws and washers, etc., are indented immediately following the parts which they attach. This promotes quick selection of all parts to be requisitioned for use in installing a bracket, generator, etc. Whenever two similar parts, having the same attaching parts, such as right and left fenders are listed, they are placed one immediately following the other and the attaching parts for both are covered by a single listing under the second part. The quantities shown for the attaching parts are those necessary to attach both items.

STANDARD HARDWARE AND BULK MATERIAL: A complete resume of standard hardware and bulk materials will be found in Group 23, Section 2. For these items, which are also listed throughout Section 2 for each place of use, the quantity required per unit assembly (Column 6) is shown only for each listing in the various groups other than group 23 and the echelon breakdown (Columns 7, 8 and 9) is given only in group 23.

INDEXES: Particular attention is directed to the fact that all items are not included in all indexes. This is covered by explanatory notes at the beginning of each index. Section 8 contains these indexes and in this sequence: Illustrations, Alphabetical, Numerical (Ordnance Drawing Numbers, Manufacturers' and Vendors' Part Numbers, and Official Stockage Numbers).

INTERCHANGEABILITY OF COMPONENTS: The engine, transmission, steering gear, clutch and the differential of the rear axle are the same as or similar to the corresponding units installed in Tractor, Snow, M7 (reviewed in SNL G194).

ORDERING PARTS, ITEM STOCK NUMBERS: All piece marks, item stock number *and* complete nomenclature must be carefully included on the requisition when parts are ordered. This is extremely essential. The prefix letter and number of the item stock number signifies only that the item was originally coded under that major item and should not be interpreted as meaning that the part is used only with the one vehicle. This prefix continues unchanged when the item is listed again in catalogs for other vehicles in which it is used.

HOW TO FIND PARTS

There are several ways to find any part quickly, using the various indexes, illustrations and standardized government grouping. The following instructions and examples are given as a guide:

a. If only the noun name and description of the part is known—

1. Turn to alphabetical index and locate the item by the noun name and description.

2. Turn to the group shown opposite the name. Group numbers appear on the upper outside corner of each page.

3. Refer to the proper sub-heading under which the part is listed and locate the part in alphabetical order under the sub-heading.

b. If only the location or function of the part in the vehicle is known—

1. Turn to the standardized government grouping list, page 8, and determine the number of the group in which the part is most likely to be listed.

2. If found in the list, review the illustrations which apply to group or allied groups. When the proper illustration is found, using the code letter, refer to the legend to determine the ordnance or manufacturers' part number, description of the part, and group number.

3. Turn to the proper group and locate the part, either by part number or by description.

c. If only the item stock number of the part is known—

1. Refer to the item stock number index and locate the proper item stock number.

2. Turn to the group shown opposite the item stock number and locate the part.

d. If only the ordnance and or manufacturers' number of the part is known—

1. Refer to the numerical index of ordnance and or manufacturers' numbers and locate the proper part number.

2. Turn to the group shown opposite the ordnance and or manufacturers' part number and locate the part.

For example: To find information on a brake shoe one may refer to the Standardized Government Grouping List. Under the "Brake Group" major group 12, is listed sub-group 1200 "Brake Assembly." Turning to this group in the catalog section of the brake shoes are found in alphabetical sequence. If some doubt exists as to whether or not the correct shoe has been found, a check can be made against the illustration referred to in the plate number column.

The shoe also could be found by referring to the alphabetical index under Shoe, brake. The same sub-group reference is shown and by following the steps outlined in the paragraph above, the shoes are found under sub-group 1201.

Another way of finding the shoe is to determine the sub-group in which the parts may be found and refer to the illustrations which apply to that group. The part may be found on the illustration and part number, noun name, and group number will be found in the legend.

The quantities specified as first year procurement are based on actual procurement as of 15 January 1944, plus all changes thereto as included in Procurement List No. TRK 780R, dated 28 October 1943. There are a number of new items included in Procurement List No. TRK 780R which will not be available until approximately 1 May 1944.

MANUFACTURERS' IDENTIFICATION SYMBOLS

Listed below are the symbols assigned to the various original manufacturers and used as a prefix to their part numbers throughout this catalog.

SYMBOL	MANUFACTURERS' NAME	ADDRESS
AB	Aetna Ball Bearing Mfg. Co.	4600 Schubert Ave., Chicago, Illinois
AC	A. C. Spark Plug Division (General Motors Corp.)	Flint, Michigan
AD	Alemite Division (Stewart Warner Corp.)	Chicago, Illinois
AE	Appleton Electric Co.	1745 Wellington Ave., Chicago, Illinois
AL	Electric Auto-Lite Co.	Mulberry and Champlain St., Toledo, Ohio
AN	Ainsworth Mfg. Co.	2200 Franklin St., Detroit, Michigan
AVM	Atwood Vacuum Mfg. Co.	Rockford, Illinois
BA	Bassick Co. (The)	36 Austin St., Bridgeport, Conn.
B-B	Borg and Beck	Rockford, Illinois
BOW	Bower Roller Bearing Co.	3040 Hart Ave., Detroit, Michigan
BUE	Budd, Edw. G. Mfg. Co.	Philadelphia, Pennsylvania
BX	Bendix Products Corp.	401 Bendix Drive, South Bend, Indiana
CA	Columbus Auto Parts	Columbus, Ohio
CAR	Carter Carburetor Co.	2820-56 N. Spring Ave., St. Louis, Missouri
CCG	Chicago Screw Co.	1020 S. Homan Ave., Chicago, Illinois
CL	Clum Mfg. Co.	601 W. National Ave., Milwaukee, Wisconsin
CN	Copeland Gibson Products Corp.	7451 West 8 Mile Rd., Detroit, Michigan
CNC	Cinch Mfg. Co.	Chicago, Illinois
CP	Champion Spark Plug	Toledo, Ohio
DM	Douglas Mfg. Co. (H. A.)	Bronson, Michigan
DTC	Deutschmann, Tobe, Corp.	Washington Ave., Canton, Massachusetts
EAT	Eaton Mfg. Co. (The)	Cleveland, Ohio
ER	Erie Resistor Corp.	64 W. 12th St., Erie, Pennsylvania
FB	Federal Bearings Co.	Poughkeepsie, New York
FM	Ford Motor Co.	Dearborn, Michigan
FO	Flex-O-Tube Co.	752 14th St., Detroit, Michigan
FYR	Fyr-Fighter Co.	2221 Crane St., Dayton, Ohio
GM	General Motors Corp.	General Motors Bldg., Detroit, Michigan
HH	Oakes Products Division (Houdaile Hershey Corp.)	National Bank Bldg., Detroit, Michigan
HI	Hayes Industries	Jackson, Michigan
HLH	Holland Hitch Co.	Holland, Michigan
HO	Hoover Ball and Bearing Co.	Hoover Ave., Ann Arbor, Michigan
HP	Harris Products Co.	Detroit, Michigan
HR	Harrison Radiator Corp.	Washburn and Walnut Sts., Lockport, New York
HY	Hyatt Bearings Division (General Motors Corp.)	P.O. Box 71, Harrison, New Jersey
KHW	Kelsey Hayes Wheel Co.	Detroit, Michigan
KM	Kent-Moore Organization	485 W. Milwaukee, Detroit, Michigan
KS	King Seeley Corp.	305 1st St., Ann Arbor, Michigan

SYMBOL	MANUFACTURERS' NAME	ADDRESS
LK	Link Belt Co.	307 N. Michigan Ave., Chicago, Illinois
LO	Wagner Electric Co.	6420 Plymouth Ave., St. Louis, Missouri
MAE	Monroe Auto Equipment	1400 E. 1st St., Monroe, Michigan
MAS	Master Engineering Service	
MDR	McCord Radiator and Mfg. Co.	2588 E. Grand Blvd., Detroit, Michigan
MIL	Mallory and Co. (P. R.)	3029 E. Washington, Indianapolis, Indiana
MRC	Marlin-Rockwell Corp.	Jamestown, New York
MSP	Midland Steel Products Co.	Cleveland, Ohio
MZ	"Mazda"-General Electric Co.	Schenectady, New York
ND	New Departure Mfg. Co.	1940 Hughes St., Bristol, Connecticut
NP	New Process Gear, Inc.	500 Plum St., Syracuse, New York
PMA	Pressed Metals of America, Inc.	Port Huron, Michigan
PU	Purolator Products, Inc.	365 Frelinghuysen Ave., Newark, New Jersey
RG	Ross Gear and Tool Co.	Lafayette, Indiana
RMC	Rich Mfg. Co.	Battle Creek, Michigan
RZ	(Rzeppa) Gear Grinding Machine Co.	Detroit, Michigan
SA	Stant Mfg. Co.	Connersville, Indiana
SKF	SKF Industries, Inc.	Front St. and Erie Ave., Philadelphia, Pennsylvania
SOL	Solar Mfg. Co.	588 Avenue A., Bayonne, New Jersey
SP	Spicer Mfg. Corp.	4100 Bennett Rd., Toledo, Ohio
SPR	Sprague Specialties Co.	North Adams, Massachusetts
SPW	Sparks Withington Co.	Jackson, Michigan
SRM	Soreng Manegold Co.	1901 Clybourne Ave., Chicago, Illinois
SV	Schrader's Son, A.	470 Vanderbilt Ave., Brooklyn, New York
SW	Stewart-Warner Corp.	1828 Diversey Pkwy., Chicago, Illinois
SZE	Schwarze Electric Co.	1939 Berry St., Adrian, Michigan
SPS	Spun Steel	Canton, Ohio
TB	Thomas and Betts	36 Butler St., Elizabeth, New Jersey
TM	Timken Roller Bearing Co. (The)	1935 Kelley Ave., Canton, Ohio
TP	Thompson Products Inc.	2196 Clarkwood Rd., Cleveland, Ohio
TR	Torrington Co. (The)	Torrington, Connecticut
TRI	Trico Products Co.	817 Washington St., Buffalo, New York
TRU	American Chain and Cable Co. (TRU-Stop)	Wilkes-Barre, Pennsylvania
TSE	Taylor Sales Engineering Co.	125 E. Lexington Ave., Elkhart, Indiana
USL	USL Battery Corp.	Highland Ave., Niagara Falls, New York
VG	Victor Mfg. and Gasket Co.	Roosevelt Rd., Chicago, Illinois
WB	Willard Storage Battery Co.	E. 131 st. and St. Clair St., Cleveland, Ohio
WEB	Warner Electric Brake Co.	Beloit, Wisconsin
WG	Warner Gear Co.	Muncie, Indiana
WH	Weatherhead Co. (The)	30 E. 131st St., Cleveland, Ohio
WIL	Wilcox-Rich	9771 French Rd., Detroit, Michigan
WO	Willys-Overland Motors, Inc.	Toledo, Ohio
YA	Yale and Towne Mfg. Co.	Philadelphia, Pennsylvania

LIST OF ABBREVIATIONS

alloy-S	alloy steel
amp	ampere
b-hd	button-head
br	brass
brg	bearing
bz	bronze
c-pin	cotter pin
cd	cadmium
chr	chromium
C. I.	cast iron
ck-hd	countersunk head
contd	continued
cop	copper
cp	candlepower
dble-hd	double head
diam	diameter
dld	drilled
dld. f/c-pin	drilled for cotter pin
engrs	engineers
f/	for
fil	filament
fil-hd	fillister head
fin	finished
fl-hd	flat head
ft	foot or feet
hdls	headless
hex	hexagonal
hex-hd	hexagonal-head
I.D.	inner diameter
in	inch
incand	incandescent
lgh	length
mach	machine
mf. d	micro-farad
M. I.	malleable iron
mtg	mounting
N. C.	National coarse
N. F.	National fine
ni	nickel
No.	number
O.D.	outer diameter
ov	oval
ov-hd	oval head
Pkzd	parkerized
pltd	plated
rd	round
rd-hd	round head
S.	steel
s-fin	semi-finished
sq	square
sq-hd	square-head
thd	thread
thk	thick
tung	tungsten
w/	with
w/o	without
z.	zinc

GOVERNMENT GROUP INDEX

01 ENGINE GROUP

0100	Engine Assembly
0101	Crankcase and Cylinder Block
0101A	Cylinder Head
0102	Crankshaft
0102A	Crankshaft Bearing
0103	Pistons
0103A	Rings
0103B	Piston Pins
0104	Connecting Rods
0104A	Connecting Rod Bearings
0105	Valves, Seats, Guides and Spring
0105A	Tappets
0105C	Valve Covers and Gaskets
0106	Camshaft
0106A	Camshaft Bearings
0106C	Chains and Sprockets
0106D	Chain Case Cover
0107	Oil Pump
0107A	Oil Pan
0107A-1	Oil Float
0107E	Oil Filter
0107F	Oil Filter Attaching Parts
0107G	Oil Filler
0107H	Level and/or Float Gauges
0107J	Oil Distribution System
0107N	Crankcase Ventilator
0108	Manifolds as an Assembly
0108A	Intake Manifold
0108B	Exhaust Manifold
0108C	Heat Controls
0109	Flywheel
0109B	Ring Gear
0109C	Flywheel Housing
0110	Mountings

02 CLUTCH GROUP

0201A	Facings and Rivets
0201B	Clutch Driven Plate
0202	Cover, Pressure Plates, and Springs
0203	Release Yoke, Bearing Fork
0204	Pedal
0204A	Pedal Linkage

03 FUEL GROUP

0300	Fuel Tanks
0300A	Tank Gauge Unit
0301	Carburetor
0301A	Carburetor Attaching Parts
0301B	Choke
0301C	Air Cleaner
0302	Fuel Pump
0302A	Attaching Parts
0303	Accelerator and Linkage
0303A	Hand Throttle and Linkage
0304	Fuel Lines
0306B	Fuel Strainer

04 EXHAUST GROUP

0401	Muffler
0402	Exhaust Pipe

05 COOLING GROUP

0501	Radiator
0502	Thermostat
0503	Water Pump
0503A	Fan and Pulleys
0503B	Fan Belt

06 ELECTRICAL GROUP

0601	Generator
0601A	Generator Attaching Parts
0601B	Generator Regulator
0602	Cranking Motor
0603	Distributor
0604	Ignition Coil
0604A	Ignition Wiring
0604B	Spark Plugs
0604C	Ignition Lock
0605	Instruments
0605B	Instrument Wiring
0605C	Panel Light
0606	Switches
0606B	Chassis Wiring
0606D	Circuit Breakers
0606E	Junction Blocks
0606F	Trailer Electric Coupling
0607	Head Lights
0607A	Sealed Beam
0607B	Lamps
0607D	Blackout Lights
0608	Tail Light
0609	Horns
0609B	Horn Button and Wiring
0610	Battery
0610A	Battery Hangers

07 TRANSMISSION GROUP

0700	Transmission Assembly
0701	Case
0703	Main Drive Gear and Bearings
0704	Main Shaft and Bearings
0704B	Gears
0704C	Countershaft and Reverse Idler Shaft
0706A	Gear Shift Lever and Parts
0706B	Gearshifter Rails and Parts
0706C	Gearshifter Yokes or Forks

SNL G-503

GOVERNMENT GROUP INDEX—Cont'd

08 TRANSMISSION TRANSFER ASSEMBLY

0800	Transmission Transfer Assembly
0801	Case
0802	Drive Gear
0803	Driven Gears, Shafts, Bearings
0804	Idler Gear, Shaft, Bearings
0805	Shifter Rods
0805A	Yokes
0805B	Shift Lever and Parts
0809	Mountings
0810	Speedometer Drive Gears

09 PROPELLER SHAFT AND UNIVERSAL JOINT

0901	Propeller Shaft Assemblies
0902	Universal Joints

10 FRONT AXLE GROUP

1000	Front Axle Assembly
1001	Housing
1002	Differential Assembly
1003	Differential Gears, Pinion and Bearing
1006	Steering Knuckle, Flange and Arm
1007	Axle Shaft, Universal Joints

11 REAR AXLE GROUP

1100	Rear Axle Assembly
1101	Housing Assembly
1102	Axle Drive Shafts
1103	Differential and Carrier Assembly
1104	Differential Pinion Bearing

12 BRAKE GROUP

1200	Brake Assembly
1201	Hand Brake Emergency Brake Parts
1202	Shoes and Facings
1203	Brake Shoe Support
1203A	Guide Springs
1203B	Adjusting Pin and Anchor Plate
1204	Pedal and Spring
1205	Master Cylinder
1207	Wheel Cylinders
1209B	Hoses
1209C	Tubes and Clips

13 WHEELS, HUBS AND DRUMS GROUP

1301	Wheel Assembly
1301A	Bearings
1301B	Seals
1301C	Retainers
1302	Hubs and Drums
1302A	Studs and Bolts

14 STEERING GROUP

1401	Steering Connecting Rod
1402	Tie Rod
1403A	Gear Assembly
1404	Wheel Assembly
1405	Brackets

15 FRAME & BRACKETS GROUP

1500	Frame and Brackets
1502	Pintle Hook
1505	Spare Wheel

16 SPRINGS AND SHOCK ABSORBER GROUP

1601	Front Spring
1601A	Rear Springs
1601B	Spring Bumpers
1601C	Torque Reaction Spring
1602	Shackle and Spring Attaching Parts
1603	Shock Absorbers and Mountings

17 HOOD, FENDERS, APRONS GROUP

1701	Fenders
1702	Splash Apron
1704	Hood

18 BODY GROUP

1800	Body
1800A	Body Handles
1801	Body Mounting Parts
1802	Body Pillars
1802A	Body Panels
1803	Floor Mats and Seals
1804	Seat
1804A	Cushions
1809	Instrument Panel and Mounting Parts
1809A	Compartment and Parts
1811	Windshield Assembly
1811A	Windshield Glass
1811B	Windshield Frame, Moulding and Seals
1811D	Windshield Wiper Arm and Blades
1817	Tool Box
1817A	Spare Gas Carrier

21 BUMPERS AND GUARD GROUP

2101	Bumpers
2103	Radiator Guards
2103A	Lamp Guards

GOVERNMENT GROUP INDEX—Cont'd

22 MISCELLANEOUS BODY CHASSIS AND ACCESSORY GROUP

2201	Tarpaulins
2201A	Bows
2201F	Side Curtains
2202	Rear View Mirrors
2203	Identification and Caution Plates
2203A	Reflectors
2204	Speedometer and Parts

23 GENERAL USE STANDARDIZED PARTS GROUP

2301	Tools
2301A	Pioneering Equipment
2301B	Tire Chains
2302	Brake Fluid
2304B	Bolts
2304C	Chains
2304D	Clamps Hose
2304E	Elbows
2304F	Fittings
2304G	Nuts
2304H	Pins
2304Y	Plugs
2304J	Rivets
2304K	Screws
2304L	Washers
2304M	Wire Locking
2305	Miscellaneous Cotter Keys
2305A	Woodruff Keys
2310	Winterization Kits

24 FIRE EXTINGUISHER SYSTEM GROUP

2402	Portable System and Parts

26 RADIO GROUP

2601	Radio Installation Parts
2603	Radio Suppression System Parts

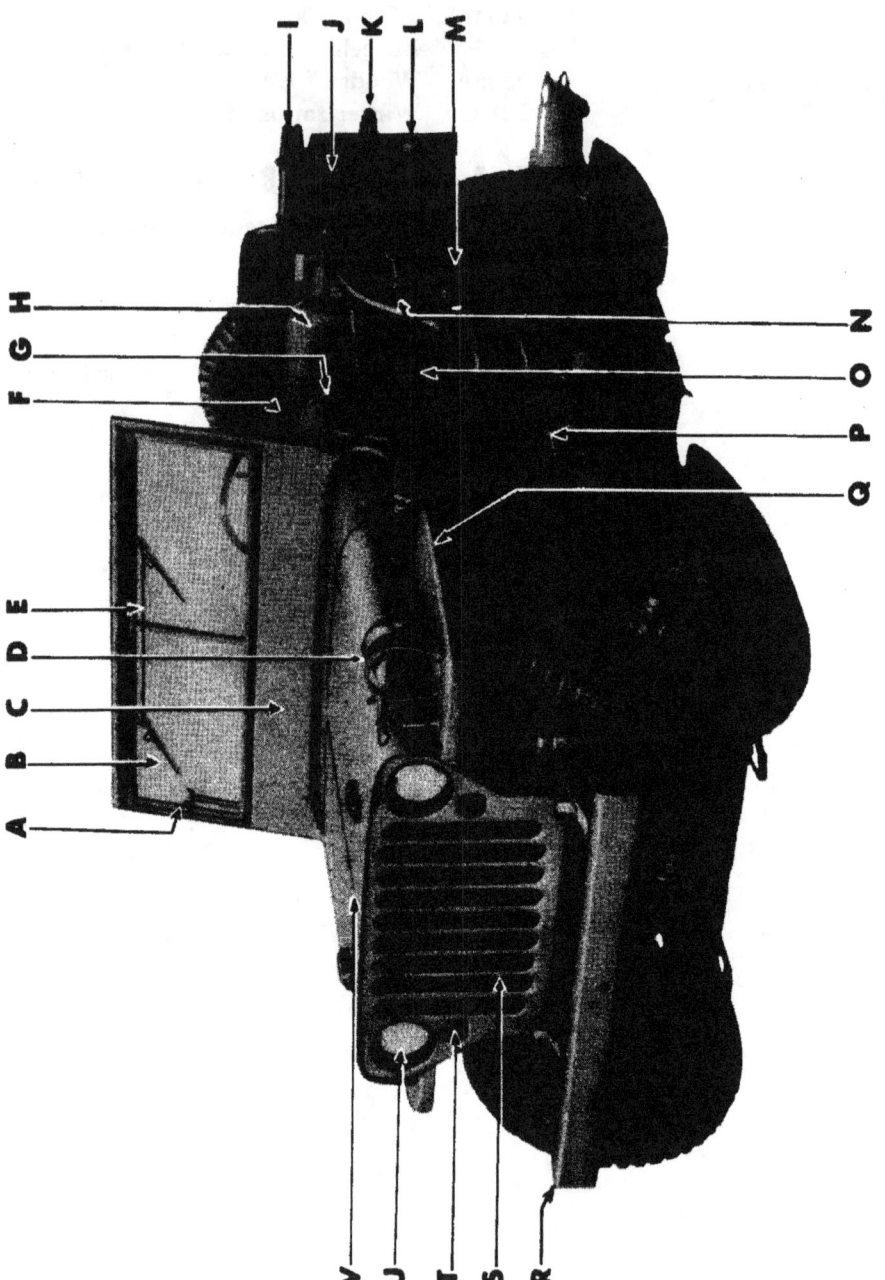

FIGURE A—¼ TON 4 x 4 RECONNAISSANCE TRUCK, ¾ FRONT VIEW

Key	Item	Willys Part No.	Ford Part No.	Gov't Group No.
A	WINDSHIELD assembly	WO-A-3210	FM-GPW-1103010	1811
B	GLASS	WO-A-2478	FM-GPW-1103100	1811A
C	PANEL	WO-A-3190	Willys only	1811
D	LAMP assembly	WO-A-6142	FM-GPW-43150	0607D
E	WIPER	WO-A-11433	FM-GPW-17500	1811D
F	WHEEL assembly	WO-A-6858	FM-GPW-3600-A-3	1404
G	MIRROR assembly	WO-A-2934	FM-21CS-17682-B	2202
H	SEAT	WO-A-3107	FM-GPW-1160016-B	1804A
I	BOW	WO-A-2897	FM-GPW-1151266	2201A
J	BRACKET	WO-A-2754	FM-GPW-1153030	2201A
K	HANDLE	WO-A-2389	FM-GPW-1129672	1800A
L	REFLECTOR	WO-A-1306	FM-GPW-13380-A	2203A
M	STRAP	WO-A-3139	FM-GPW-1128267	2301A
N	HANDLE	WO-A-2390	FM-GPW-1129670	1800A
O	CUSHION	WO-A-2986	FM-GPW-1162900-B	1804A
P	PANEL	WO-A-3008	FM-GPW-1102039	1802A
Q	FENDER	WO-A-2942	FM-GPW-16006	1701
R	BUMPER	WO-A-1117	FM-GPW-17750	2101
S	GUARD assembly	WO-A-3615	FM-GPW-8307	2103
T	LIGHT	WO-A-1437	FM-GPW-13200	0607D
U	LIGHT	WO-A-1305	FM-GPW-13005	0607
V	HOOD	WO-A-3225	FM-GPW-16610	1704

RA PD 305058-A

FIGURE B—¼ TON 4 x 4 COMMAND RECONNAISSANCE TRUCK, ¾ REAR VIEW

Key	Item	Willys Part No.	Ford Part No.	Gov't Group No.
A	BOW assembly	WO-A-2897	FM-GPW-1151266	2201A
B	SEAT assembly	WO-A-3107	FM-GPW-1160016-B	1804A
C	ARM	WO-A-2235	FM-GP-1103302	1811
D	CLAMP	WO-A-2227	FM-GP-1103482-A	1811
E	CATCH assembly	WO-A-2896	FM-GPW-16892	1704
F	SCREW	WO-A-2214	Willys only	1811
G	STRAP assembly	WO-A-2883	FM-GPW-1131414	1804
H	FENDER assembly	WO-A-2943	FM-GPW-16005	1701
I	HANDLE	WO-A-2390	FM-GPW-1129670	1800A
J	REFLECTOR	WO-A-1306	FM-GPW-13380-A	2203A
K	LIGHT assembly	WO-A-1065	FM-GPW-13404-B	0608
M	BUMPERETTE	WO-A-1157	FM-GPW-17775	2101
N	HOOK	WO-A-593	FM-GPW-5182	1502
O	SOCKET assembly	WO-A-6019	FM-11YS-18142-B	0606F
P	HANDLE	WO-A-2389	GPW-1129672	1800A

RA PD 305059-A

SNL G-503

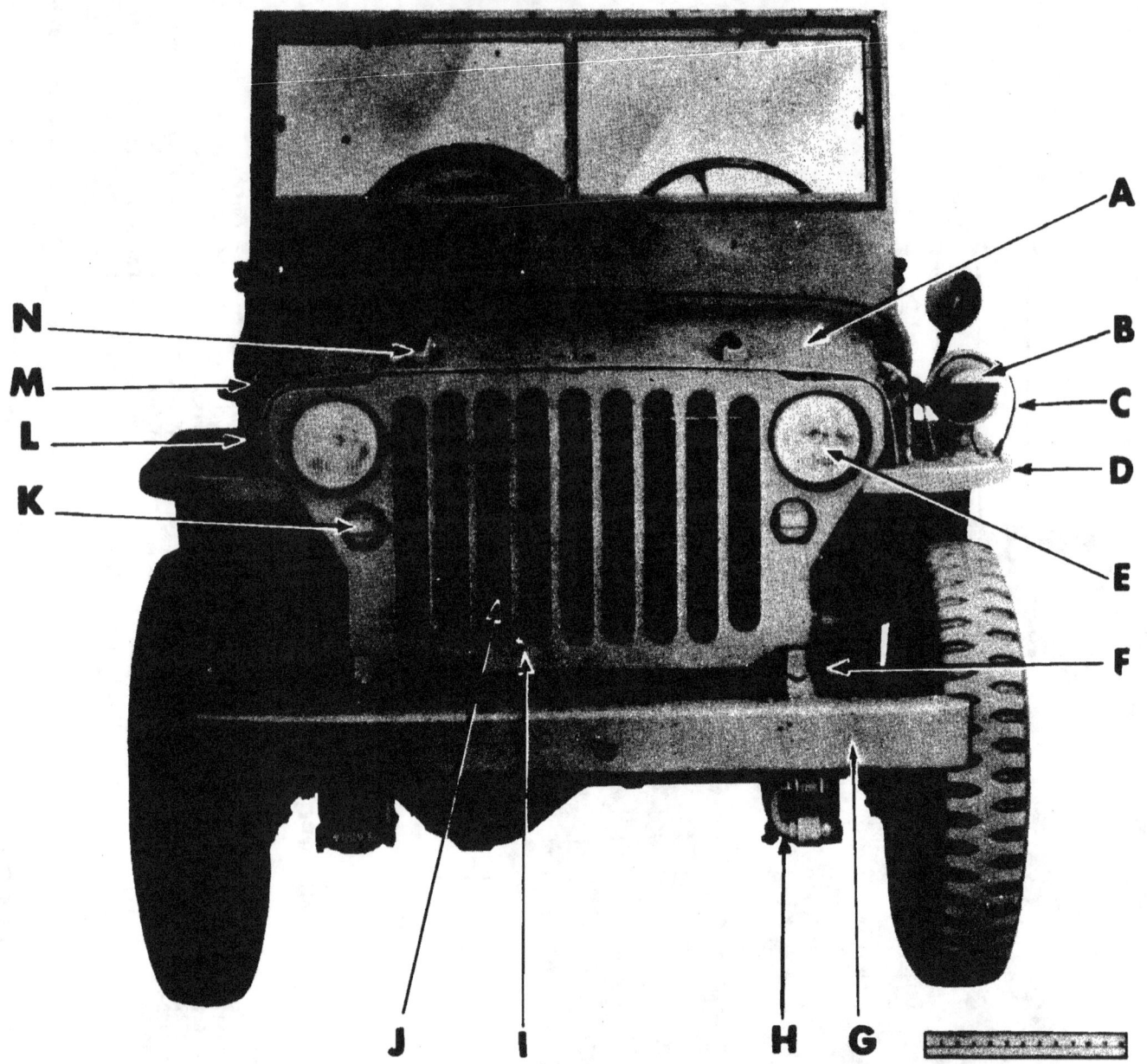

FIGURE C—¼ TON 4 x 4 COMMAND RECONNAISSANCE TRUCK, FRONT VIEW

Key	Item	Willys Part No.	Ford Part No.	Gov't Group No.
A	HOOD assembly	WO-A-3225	FM-GPW-16610	1704
B	LIGHT assembly	WO-A-6142	FM-GPW-13150	0607D
C	GUARD w/SUPPORT	WO-A-4118	FM-GPW-13176	2103A
D	FENDER w/SPLASHER	WO-A-2942	FM-GPW-16006	1701
E	LIGHT assembly	WO-A-1304	FM-GPW-13006	0607
F	SHOCK ABSORBER assembly	WO-A-6902	FM-GPW-18045	1603
G	BUMPER	WO-A-1117	FM-GPW-17750	2101
H	BOLT	WO-A-513	FM-GPW-5778	1602
I	GUARD assembly	WO-A-3615	FM-GPW-8307	2103
J	CORE w/TANK assembly	WO-A-1214	FM-GPW-8005	0501
K	LIGHT assembly	WO-A-1437	FM-GPW-13200	0607D
L	CATCH assembly	WO-A-2896	FM-GPW-16892	1701
M	CATCH assembly	WO-A-3197	FM-GPW-1103027	1704
N	BUMPER assembly	WO-A-4683	(Willys only)	1704

RA PD 305062-A

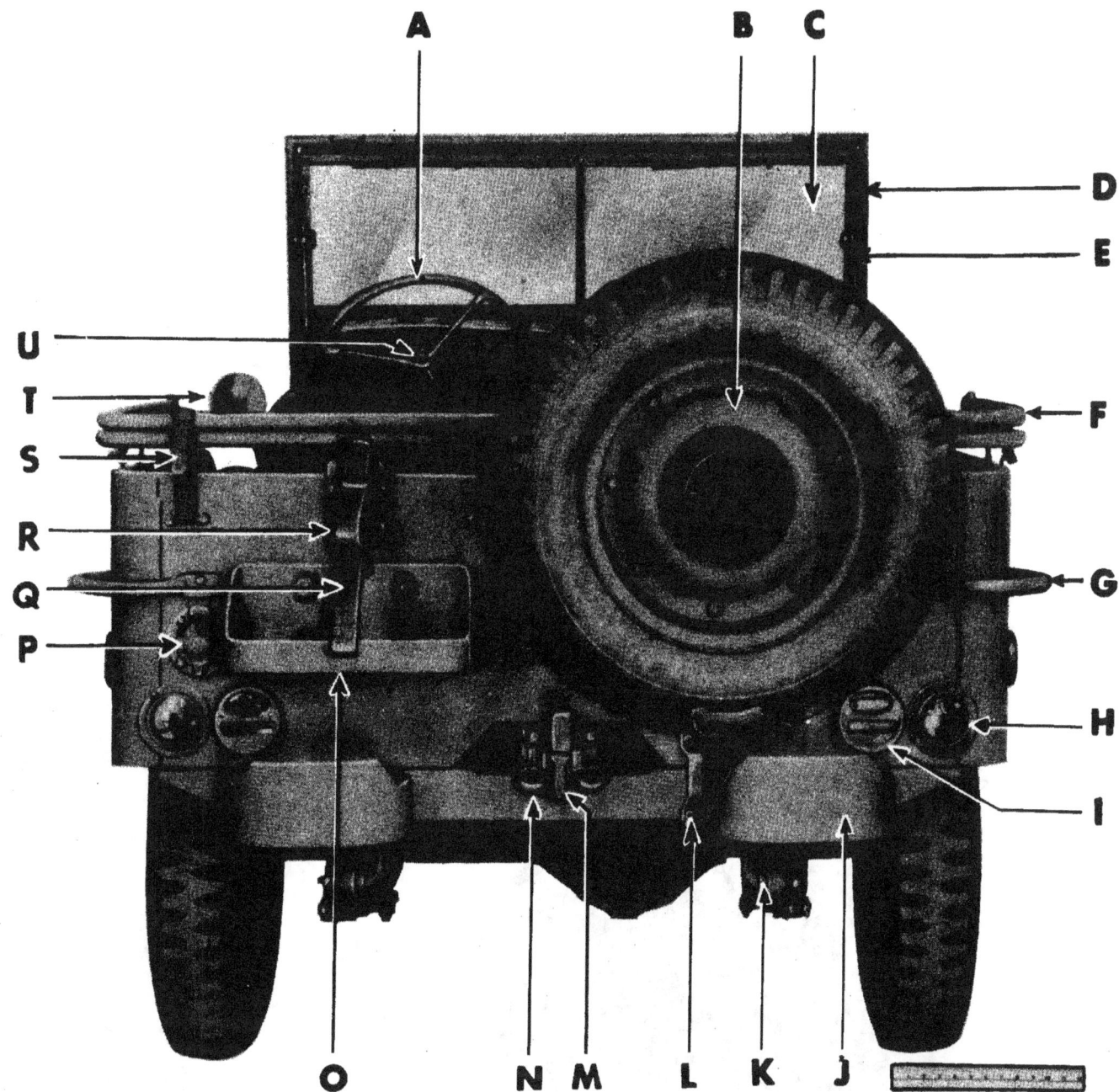

FIGURE D—¼ TON 4 x 4 COMMAND RECONNAISSANCE TRUCK, REAR VIEW

Key	Item	Willys Part No.	Ford Part No.	Gov't Group No.
A	WHEEL	WO-A-6858	FM-GPW-3600-A3	1404
B	WHEEL	WO-A-5467	FM-GPW-1025-C	1301
C	GLASS	WO-A-2478	FM-GP-1103100	1811A
D	WINDSHIELD assembly	WO-A-3210	FM-GPW-1103010	1811
E	ARM	WO-A-2235	FM-GP-1103302	1811
F	BOW assembly	WO-A-2897	FM-GPW-1151266	2201A
G	HANDLE	WO-A-2389	FM-GPW-1129672	1800A
H	REFLECTOR	WO-A-1306	FM-GPW-13380-A	2203A
I	LIGHT	WO-A-1065	FM-GPW-13404-B	0608
J	BUMPERETTE	WO-A-1157	FM-GPW-17775	2101
K	SHACKLE	WO-A-6069	FM-GPW-5605	1602
L	BRACKET	WO-A-2359	FM-GPW-1433	1505
M	HOOK	WO-A-593	FM-GPW-5182	1502
N	BOLT	WO-A-6393	FM-GPW-5186	1502
O	BRACKET	WO-A-4123	FM-GPW-1140330	1817A
P	SOCKET	WO-A-6586	FM-11YS-18151-B	0606F
Q	STRAP	WO-A-4127	(Willys only)	1817A
R	STRAP	WO-A-3110	FM-GPW-1152720	
S				
T	MIRROR	WO-A-2934	FM-21CS-17682-B	
U	BUTTON	WO-A-634	FM-GPW-3627	0609B

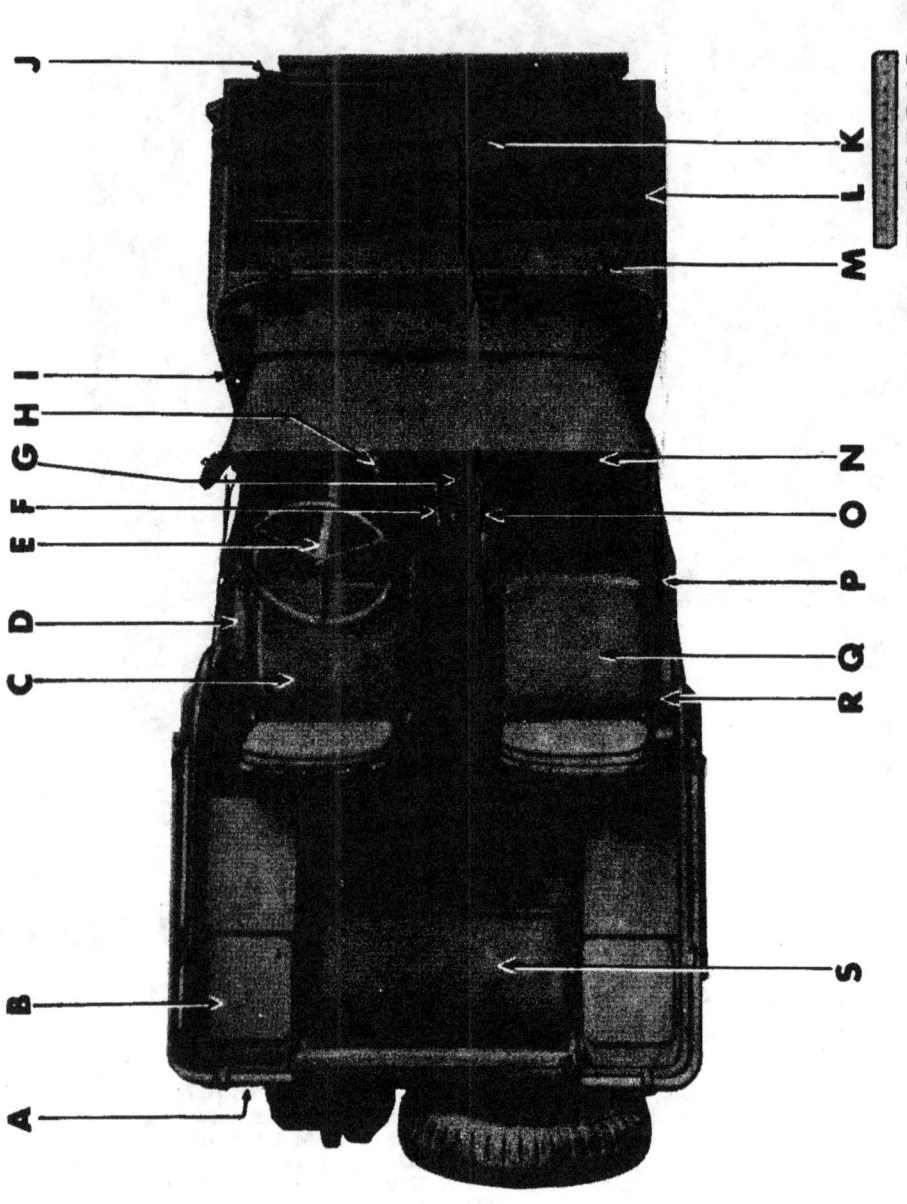

FIGURE E—¼ TON 4 x 4 COMMAND RECONNAISSANCE TRUCK, TOP VIEW

Key	Item	Willys Part No.	Ford Part No.	Gov't Group No.
A	BOW assembly	WO-A-2897	FM-GPW-1151266	2201A
B	LID w/HINGE	WO-A-3227	FM-GPW-1146100	1817
C	SEAT assembly	WO-A-3107	FM-GPW-1160016-B	1804A
D	TANK	WO-A-6618	FM-GPW-9002-B	0300
E	GEAR, steering	WO-A-1239	FM-GPW-3504	1403
F	LEVER assembly	WO-A-1380	FM-GPW-7210-A	0706A
G	HANDLE w/TUBE	WO-A-1242	FM-GPW-2780	1201
H	SOCKET assembly	WO-A-1411	FM-GPW-13710	0605C
I	SCREW	WO-A-2214	Willys only	1811
J	RETAINER	WO-A-3203	FM-GPW-1103028	1704
K	WIPER assembly	WO-A-11433	Willys only	1811D
L	WINDSHIELD assembly	WO-A-3210	FM-GPW-1103010	1811
M	CLAMP assembly	WO-A-2227	FM-GPW-1103482-A	1811
N	CATCH	WO-A-2791	FM-GPW-1104388	1811
O	LEVER	WO-A-1505	FM-GPW-7793	0805B
P	STRAP assembly	WO-A-2883	FM-GPW-1131414	1804
Q	SEAT assembly	WO-A-3106	FM-GPW-1160012-A	1804A
R	PAD assembly	WO-A-3115	FM-GPW-116400	1804A
S	SEAT assembly	WO-3108	FM-GPW-1160026	1804A

RA PD 305060-A

SNL G-503

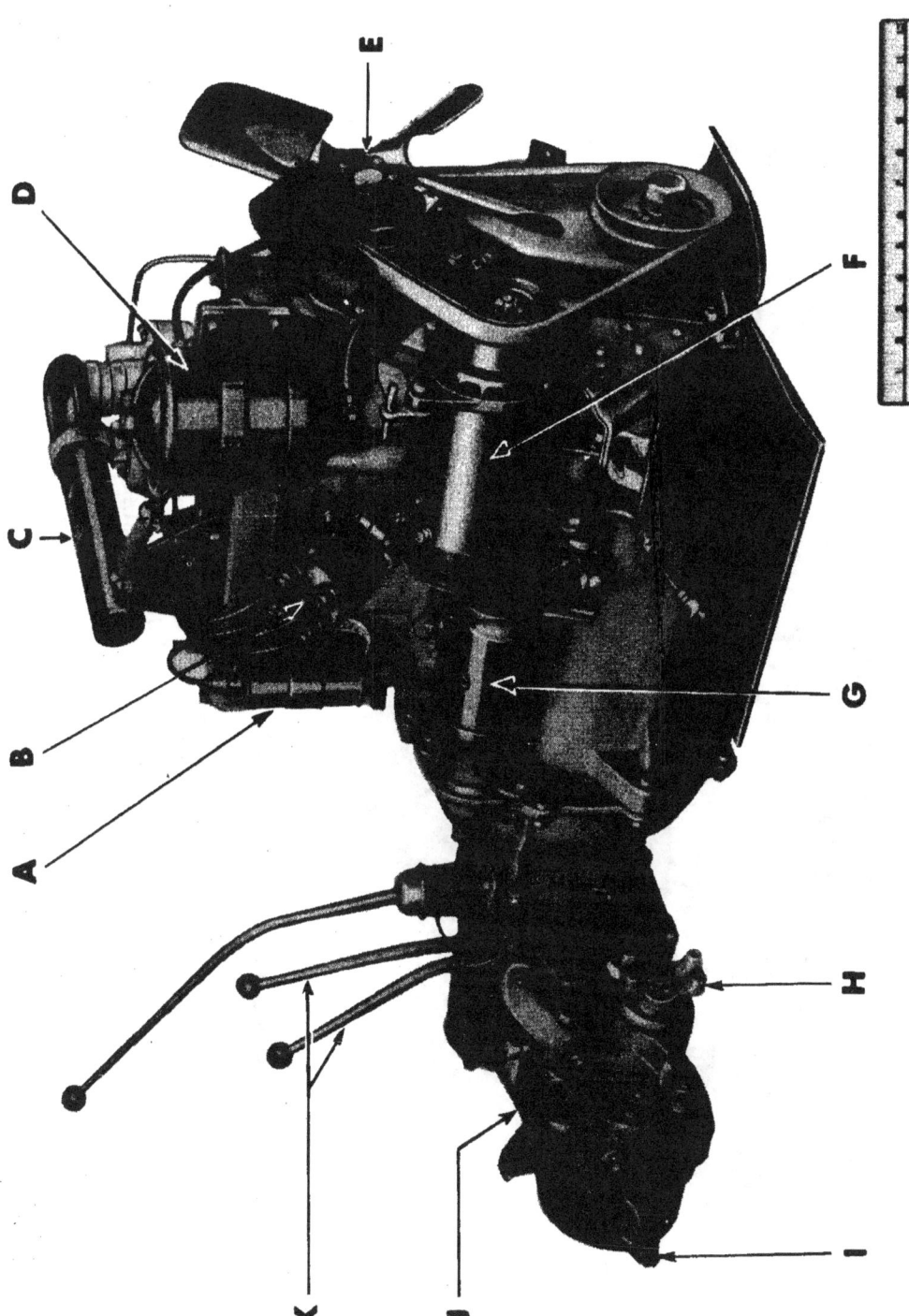

FIGURE F—POWER PLANT ASSEMBLY, RIGHT FRONT VIEW

Key	Item	Willys Part No.	Ford Part No.	Gov't Group No.
A	COIL assembly	WO-A-7792	FM-GPW-12000-B	0604
B	DISTRIBUTOR assembly	WO-A-1244	FM-GPW-1200	0603
C	TUBE w/BRACKET assembly	WO-A-6911	FM-GPW-9637-B	0301C
D	FILTER assembly	WO-A-1230	FM-GPW-18660-A	0107E
E	FAN assembly	WO-A-447	FM-GPW-8600	0503A
F	GENERATOR assembly	WO-A-5992	FM-GPW-10000-A	0601
G	CRANKING motor assembly	WO-A-1245	FM-GPW-11001-A	0602
H	YOKE	WO-A-1106	FM-GP-7709	0803
I	BRAKE assembly	WO-A-1008	FM-GPW-2529	1201
J	TRANSFER assembly	WO-A-1195	FM-GPW-7700	0800
K	LEVER	WO-A-1505	FM-GPW-7793	0805B

RA PD 305064-A

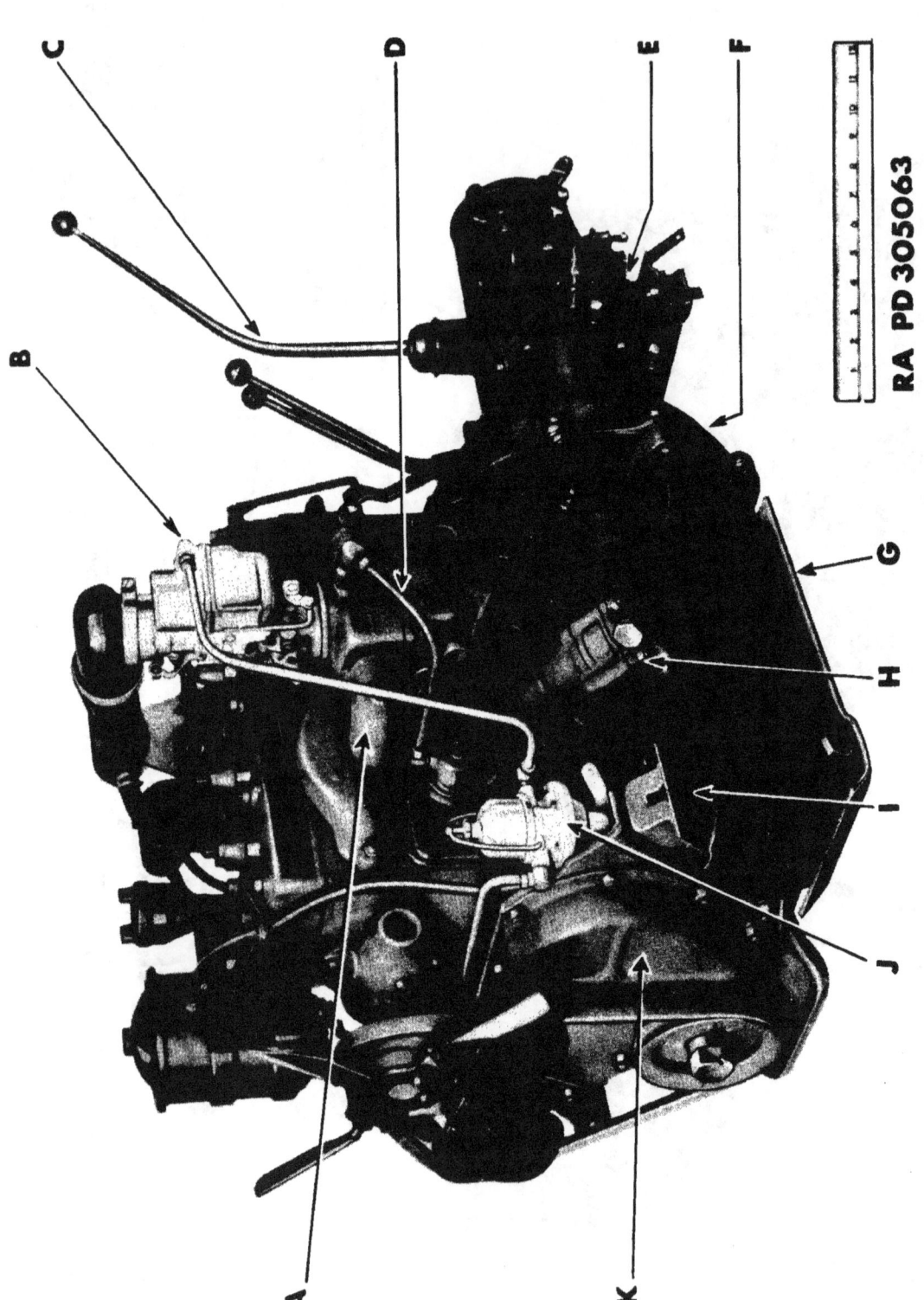

FIGURE G—POWER PLANT ASSEMBLY, LEFT FRONT VIEW

Key	Item	Willys Part No.	Ford Part No.	Gov't Group No.
A	MANIFOLD assembly	WO-A-1165	FM-GPW-9410-B	0108
B	CARBURETOR assembly	WO-A-1223	FM-GPW-9510	0301
C	LEVER assembly	WO-A-1380	FM-GPW-7210-A	0706A
D	TUBE assembly	WO-A-6922	FM-GPW-6756	0107N
E	TRANSMISSION assembly	WO-A-1145	FM-GPW-7000	0700
F	HOUSING assembly	WO-A-439	FM-GPW-6392	0109C
G	PAN assembly	WO-A-7238	FM-GPW-6675	0107A
H	PUMP assembly	WO-637636	FM-GPW-6600	0107
I	INSULATOR assembly	WO-A-7498	FM-GPW-6038-A	0110
J	PUMP assembly	WO-A-8323	FM-GPW-9350-B	0302
K	COVER assembly	WO-A-1190	FM-GPW-6016	0106C

RA PD 305063-A

17

SNL G-503

FIGURE H—ENGINE, CROSS SECTIONAL, END VIEW

Key	Item	Willys Part No.	Ford Part No.	Gov't Group No.
A	OILER	WO-107128	FM-B-10141	0603
B	DISTRIBUTOR	WO-A-1244	FM-GPW-12100	0603
C	COIL	WO-A-1424	FM-GPW-12000-A	0604
D	GUIDE	WO-375811	FM-GPW-6510-B	0105
E	MANIFOLD	WO-A-1166	FM-GPW-9424-B	0108A
F	COVER assembly	WO-630303	FM-GPW-6520	0105C
G	VALVE	WO-636439	FM-GPW-9460	0108C
H	BAFFLE	WO-630298	FM-GPW-6762	0107N
I	MANIFOLD	WO-A-912	FM-GPW-9428	0108E
J	BODY	WO-A-699	FM-FPW-6758-B	0107N
K	SPRING	WO-637615	FM-GPW-12083	0603
L	GEAR	WO-637425	FM-GPW-6610	0107
M	DISC	WO-636600	FM-GPW-6673	0107
N	PUMP assembly	WO-637636	FM-GPW-6600	0107
O	PINION	WO-343306	FM-GPW-6614	0107
P	RETAINER	WO-630390	FM-GPW-6644	0107
Q	SHIM	WO-630389	FM-GPW-6628	0107
R	SPRING	WO-356155	FM-GPW-6654	0107
S	PLUNGER	WO-630518	FM-GPW-6663	0107
T	SHAFT assembly	WO-636599	FM-GPW-6608	0107
U	OIL PAN assembly	WO-A-7238	FM-GPW-6675	0107A
V	PLUG	WO-639979	FM-GPW-6727	0107A
W	SUPPORT	WO-630397	FM-GPW-6617	0107A-1
X	DOWEL	WO-635377	FM-GPW-6369	0102
Y	BOLT	WO-381519	FM-GPW-6345	0102
Z	FLOAT assembly	WO-630396	FM-GPW-6615	0107A-1
AB	TUBE	WO-A-6915	FM-GPW-6763C	0107G
AC	INDICATOR	WO-A-5168	FM-GPW-6766-B	0107H

RA PD 305045-A

SNL G-503

SECTION 2

MAINTENANCE PARTS PROCUREMENT LIST

0100-0101

MAINTENANCE PARTS PROCUREMENT LIST

Truck, ¼-Ton, 4 x 4, Command Reconnaissance

MAJOR ITEMS AUTHORIZED FOR UNIT REPLACEMENT FOR 100 VEHICLES FOR 12 MONTHS - - - - - - - - 76

Figure Number	Official Stockage Number	Part Number — Willys	Part Number — Ford	ITEM	Quantity Reqd. per Unit Assy.	Per 100 Major Items — 12 Mos. Field Maintenance	Per 100 Major Items — Major Overhaul (5th Ech)	Total First Year Procurement	Estimated Reqmts. per 100 Rebuilds
Col. 1	Col. 2	Col. 3	Col. 4	Col. 5	Col. 6	Col. 7	Col. 8	Col. 9	Col. 10
				GROUP 01—ENGINE					
				0100—ENGINE ASSEMBLY					
		WO-A-5497	FM-GPW-6005	ENGINE assembly (includes CLUTCH, FLYWHEEL HOUSING, and all units ready to run)	1	18	—	18	—
		WO-A-1493		ENGINE assembly (Willys only) (less AIR CLEANER, FLYWHEEL HOUSING, CABLES, CARBURETOR, CLUTCH, COIL, DISTRIBUTOR, FAN, FAN BELT, FUEL PUMP, GENERATOR, OIL FILTER, SPARK PLUGS and CRANKING MOTOR)	1	—	—	—	—
F, G		WO-A-5338	FM-GPW-6002	PLANT, power (includes ENGINE, TRANSFER CASE and TRANSMISSION complete, ready to run)	1	—	—	—	—
				0101—CRANKCASE AND CYLINDER BLOCK					
01-2		WO-A-1272	FM-GPW-6010	BLOCK, cylinder, engine, w/BEARING, assembly	1	—	—	—	10
01-12		WO-A-6793	FM-GPW-6009	BLOCK, cylinder, engine, w/BEARING and PISTON, assembly	1	—	3	4	10
01-2		WO-A-1126	FM-9N-8115	COCK, drain, engine cylinder water jacket (¼ in., S., z-pltd., taper pipe thread) (WH-145-A)	1	% 12	4	18	13
		WO-A-1536	FM-GPW-18390	GASKET SET, engine overhaul	as req.	—	40	50	105
				(Includes:					
				1 WO-630365 FM-GPW-6288 GASKET, chain cover					
				1 WO-630359 FM-GPW-6020 GASKET, cylinder block					
				1 WO-A-8558 FM-GPW-6051-B GASKET, cylinder head					
				1 WO-634814 FM-GPW-9450 GASKET, exhaust pipe flange					
				1 WO-638737 FM-GPW-9417 GASKET, fuel pump					
				1 WO-A-6357 FM-GPW-9445 GASKET, insulator and diffuser					
				1 WO-638640 FM-GPW-9448 GASKET, intake and exhaust manifold					
				1 WO-634811 FM-GPW-9435 GASKET, intake to exhaust manifold					
				1 WO-630398 FM-GPW-6627 GASKET, oil float support					
				1 WO-639980 FM-GPW-6710 GASKET, oil pan					
				1 WO-639870 FM-GPW-6659 GASKET, oil pump cover					
				4 WO-630392 FM-GPW-6619 GASKET, oil pump cover					
				1 WO-314338 FM-GPW-6734 GASKET, oil drain plug					
				1 WO-375927 FM-GPW-6625 GASKET, oil pump shaft					

SNL G-503

FIGURE 01-1—ENGINE, SIDE VIEW

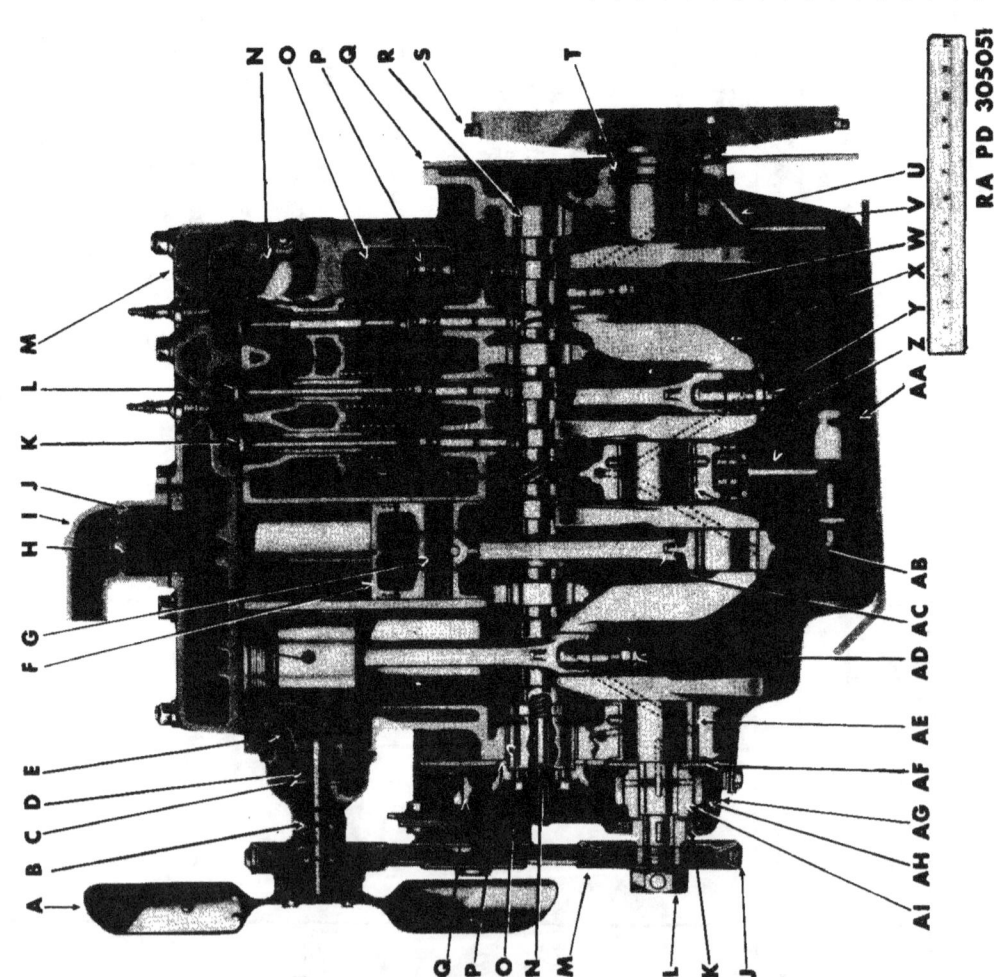

RA PD 305051-A

Key	Item	Willys Part No.	Ford Part No.	Gov't Group No.
A	FAN assembly	WO-A-447	FM-GPW-8600	0503A
B	SHAFT w/BEARING assembly	WO-636297	FM-GPW-8530	0503
C	WASHER	WO-644034	FM-GPW-8557-A	0503
D	SEAL assembly	WO-640031	FM-GAA-8524	0503
E	IMPELLER	WO-639993	FM-GPW-8512	0503
F	PISTON assembly	WO-637041	FM-GPW-6105-A	0103
G	PIN	WO-636961	FM-GPW-6135-A	0103B
H	THERMOSTAT	WO-637646	FM-GPW-8575	0502
I	ELBOW	WO-A-1192	FM-GPW-8250	0101A
J	RETAINER	WO-639651	FM-GPW-8578	0502
K	VALVE	WO-637183	FM-GPW-6505	0105
L	VALVE	WO-637182	FM-GPW-6507	0105
M	HEAD	WO-A-1534	FM-GPW-6050	0101A
N	MANIFOLD assembly	WO-A-912	FM-GPW-9428	0108B
O	SPRING	WO-638636	FM-GPW-6513	0105
P	SCREW	WO-640020	FM-GPW-6549-B	0105A
Q	PLATE assembly	WO-A-5121	FM-GPW-6044	0110
R	CAMSHAFT	WO-637065	FM-GPW-6250	0106
S	GEAR	WO-635394	FM-GPW-6384	0109B
T	PACKING	WO-637237	FM-GPW-6702	0102A
U	PIPE	WO-630294	FM-GPW-6326	0102A
V	BEARING	WO-638733	FM-GPW-6337-A	0102A
W	TAPPET	WO-637047	FM-GPW-6500-A	0105A
X	CRANKSHAFT	WO-A-7568	FM-GPW-18288	0102
Y	BOLT	WO-640070	(Willys only)	0104
Z	SUPPORT	WO-630397	FM-GPW-6167	0107A-1
AA	FLOAT assembly	WO-630396	FM-GPW-6615	0107A-1
AB	BEARING	WO-638731	FM-GPW-6341-A	0102A
AC	ROD	WO-640067	(Willys only)	0104
AD	NUT	WO-52825	FM-356028-S	0104
AE	BEARING	WO-637008	FM-GPW-6338-A	0102A
AF	WASHER	WO-634796	FM-GPW-6308	0102
AG	COVER assembly	WO-A-1190	FM-GPW-6016	0106D
AH	CHAIN	WO-638457	FM-GPW-6260	0106C
AI	SPROCKET	WO-638459	FM-GPW-6306	0106C
AJ	BELT	WO-A-1495	FM-GPW-8260	0503B
AK	PACKING	WO-637098	FM-GPW-6700	0102A
AL	NUT assembly	WO-387633	FM-GPW-6319	0102
AM	PULLEY	WO-638113	FM-GPW-6312	0102
AN	PLUNGER	WO-375907	FM-GPW-6243	0106A
AO	BUSHING	WO-639051	FM-GPW-6262-A	0106A
AP	WASHER	WO-375900	FM-GPW-6245	0106
AQ	SPROCKET	WO-638458	FM-GPW-6246	0106C

RA PD 305051-A

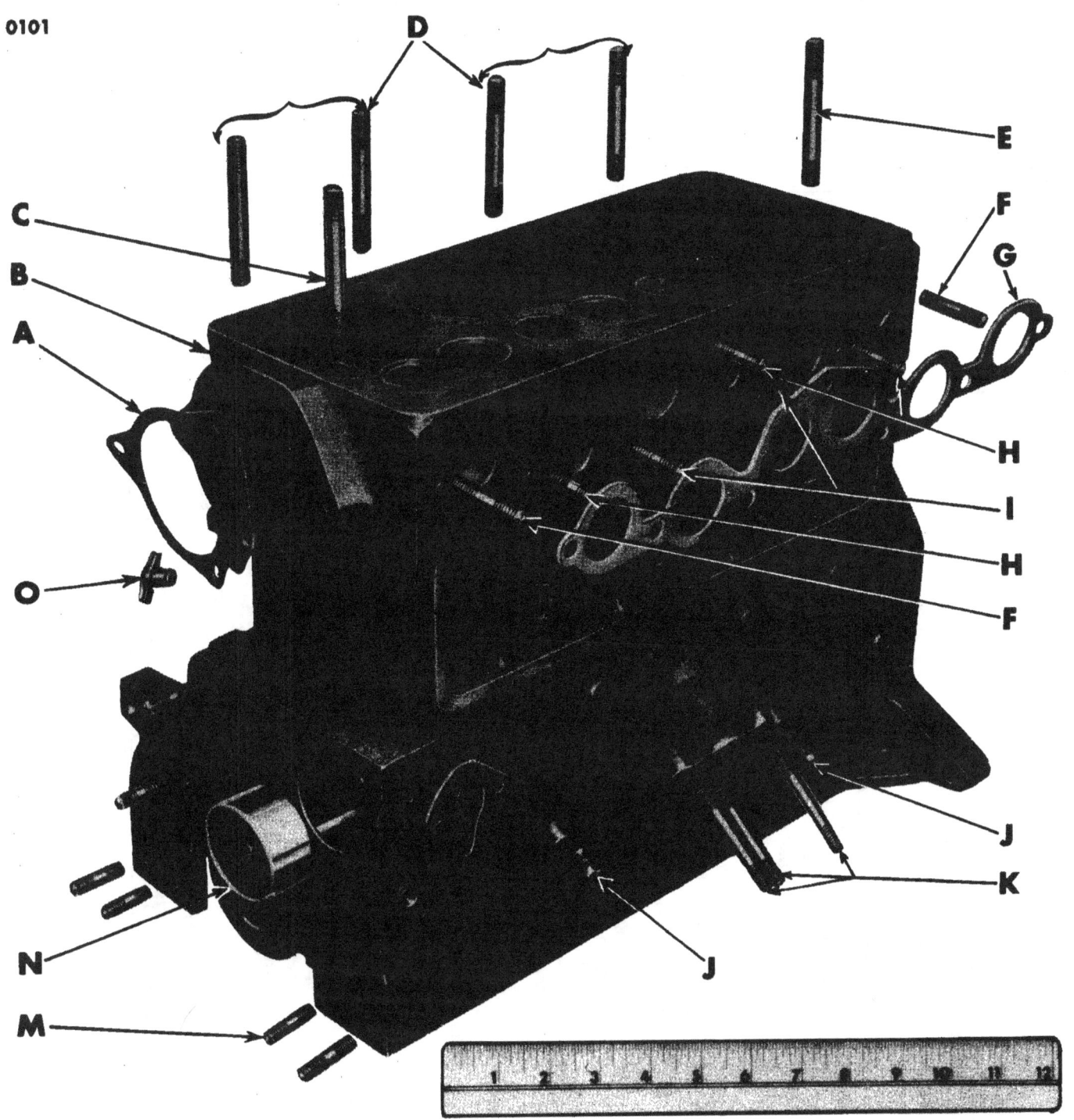

FIGURE 01-2—CYLINDER BLOCK ASSEMBLY

Key	Item	Willys Part No.	Ford Part No.	Gov't Group No.
A	GASKET	WO-637053	FM-GPW-8543	0503
B	BLOCK assembly	WO-A-1272	FM-GPW-6010	0101
C	STUD	WO-A-1549		0101A
D	STUD	WO-349368	FM-GPW-6066	0101A
E	STUD	WO-A-1548	FM-GPW-6067	0101A
F	STUD	WO-349712	FM-88082-S	0108
G	GASKET	WO-638640	FM-GPW-9448	0108
H	STUD	WO-300143	FM-88042	0108
I	STUD	WO-632159	FM-88057	0108
J	PLUG	WO-5085	FM-358064-S	0101
K	STUD	WO-375981	FM-88141-S	0107
M	STUD	WO-384958	FM-88022-S	0106D
N	BUSHING	WO-639051	FM-GPW-6262-A	0106A
O	COCK	WO-A-1126	FM-9N-8115	0101

RA PD 305047-A

GROUP 01—ENGINE (Cont'd)

Figure Number	Official Stockage Number	Part Number: Willys	Part Number: Ford	ITEM	Quantity Reqd. per Unit Assy.	Per 100 Major Items: 12 Mos. Field Maintenance	Per 100 Major Items: Major Overhaul (5th Ech)	Per 100 Major Items: Total First Year Procurement	Estimated Reqmts. per 100 Rebuilds
Col. 1	Col. 2	Col. 3	Col. 4	Col. 5	Col. 6	Col. 7	Col. 8	Col. 9	Col. 10
				0101—CRANKCASE AND CYLINDER BLOCK (Cont'd)					
				1 WO-630394 FM-GPW-6630 GASKET, oil pump to cylinder block					
				1 WO-634813 FM-GPW-6642 GASKET, oil relief spring retainer					
				1 WO-334103 FM-GPW-6353 GASKET, oil slinger					
				4 WO-637863 FM-OIA-12410 GASKET, spark plug					
				1 WO-630305 FM-GPW-6521 GASKET, valve spring cover					
				2 WO-51875 FM-GPW-6555 GASKET, valve cover screw					
				1 WO-630299 FM-GPW-6648 GASKET, ventilator to valve cover					
				1 WO-639650 FM-GPW-8255 GASKET, water outlet elbow					
				1 WO-637053 FM-GPW-8543 GASKET, water pump to cylinder block					
				1 WO-637098 FM-GPW-6700 PACKING, crankshaft					
				2 WO-637237 FM-GPW-6702 PACKING, rear bearing cap					
				2 WO-637790 FM-GPW-6701 PACKING, rear bearing cap					
		WO-A-1537	FM-GPW-18387	GASKET SET, engine, valve job	as req.	80	—	100	—
				(Includes:					
				1 WO-A-8558 FM-GPW-6051-B GASKET, cylinder head					
				1 WO-634814 FM-GPW-9450 GASKET, exhaust pipe flange					
				1 WO-A-6357 FM-GPW-9445 GASKET, insulator and diffuser assembly					
				1 WO-638640 FM-GPW-9448 GASKET, intake and exhaust manifold					
				1 WO-634811 FM-GPW-9435 GASKET, intake to exhaust manifold					
				4 WO-637863 FM-OIA-12410 GASKET, spark plug					
				2 WO-51875 FM-GPW-6555 GASKET, valve cover screw					
				1 WO-630305 FM-GPW-6521 GASKET, valve spring cover					
				1 WO-630299 FM-GPW-6648 GASKET, valve spring cover to ventilator					
				1 WO-639650 FM-GPW-8255 GASKET, water outlet elbow)					
	H6-02-82480	WO-51091	FM-74121-S	PLUG, expansion, S., 1¼ in. (BEDX1AT)	5	% 20	20	50	62
	H6-02-83805	WO-5138	FM-353055-S7	PLUG, pipe, sq-hd., black, ¼ in. (cylinder water jacket, drain) (CPMX1AB)	1	% —	—	—	—
	H6-02-83015	WO-376373	FM-358063-S	PLUG, pipe, alloy-S., ck., ⅜ in. (A164342C) (issue until stock exhausted)	2	% —	—	—	—
01-11	H6-02-83800	WO-5085	FM-358064-S	PLUG, pipe, sq-hd., black, ⅛ in. (CPMX1AA)	4	% —	—	—	—
				0101A—CYLINDER HEAD					
01-3		WO-A-1192	FM-GPW-8250	ELBOW, engine cylinder head water outlet (issue until stock exhausted)	1	% 4	—	—	—
01-3	H1-01-16026	WO-52911	FM-24408-S	BOLT, hex-hd., s-fin., alloy-S., ⅜-16NC-2 x 1⅛ (GM-106331) (elbow to cylinder head) (BANX1CC)	3	% —	—	—	37
01-3	H1-15-18113	WO-5010	FM-34807-S7	WASHER, lock, reg., S., ⅜ in. (13/32 I.D. x 21/32 O.D. x 3/32 thk.) (BECX1K)	3	% —	—	—	—

FIGURE 01-3 — CYLINDER HEAD

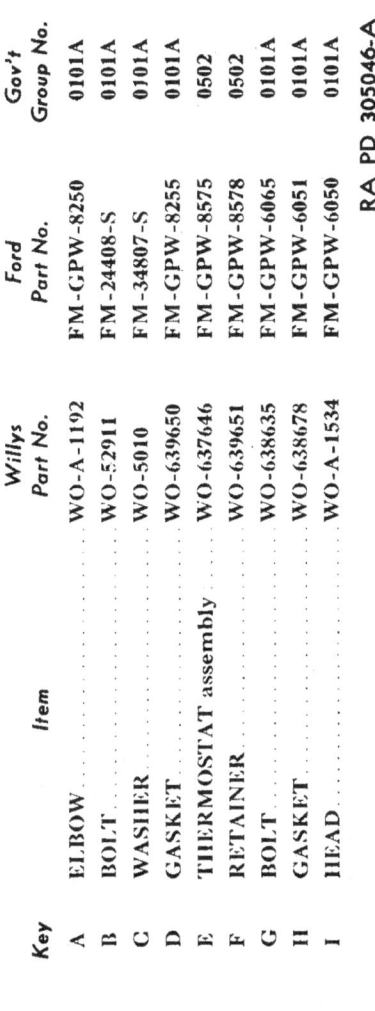

Key	Item	Willys Part No.	Ford Part No.	Gov't Group No.
A	ELBOW	WO-A-1192	FM-GPW-8250	0101A
B	BOLT	WO-52911	FM-24408-S	0101A
C	WASHER	WO-5010	FM-34807-S	0101A
D	GASKET	WO-639650	FM-GPW-8255	0101A
E	THERMOSTAT assembly	WO-637646	FM-GPW-8575	0502
F	RETAINER	WO-639651	FM-GPW-8578	0502
G	BOLT	WO-638635	FM-GPW-6065	0101A
H	GASKET	WO-638678	FM-GPW-6051	0101A
I	HEAD	WO-A-1534	FM-GPW-6050	0101A

RA PD 305046-A

RA PD 305046

GROUP 01—ENGINE (Cont'd)

Figure Number	Official Stockage Number	Part Number		ITEM	Quantity Reqd. per Unit Assy.	Per 100 Major Items			Estimated Reqmts. per 100 Rebuilds
		Willys	Ford			12 Mos. Field Maintenance	Major Overhaul (5th Ech)	Total First Year Procurement	
Col. 1	Col. 2	Col. 3	Col. 4	Col. 5	Col. 6	Col. 7	Col. 8	Col. 9	Col. 10
				0101A—CYLINDER HEAD (Cont'd)					
01-3		WO-A-8558	FM-GPW-6051-B	GASKET, engine cylinder head	1	%120	—	150	0
01-3		WO-639650	FM-GPW-8255	GASKET, engine cylinder head water outlet elbow	1	% 8	—	20	—
01-3		WO-A-1534	FM-GPW-6050	HEAD, cylinder engine (ratio 6.48)	1	% 3	3	8	10
01-3	H1-01-16045	WO-638635	FM-GPW-6065	BOLT, hex-hd., s-fin., alloy-S, 7/16-14NC-2 (rolled thd.) x 2¾ (BANX1DL)	9	%14	14	30	45
	H1-07-16265	WO-638539	FM-351025-S8	NUT, regular, hex., s-fin., alloy-S, 7/16-20NF-2 (cylinder head stud) before Willys serial 288835)	4	%16	10	90	30
				after Willys serial 288835)	5				
				all Ford trucks	5				
		WO-A-1550	FM-356025-S	NUT, regular, hex., S., cd-pltd., 7/16-20NF-2. cylinder head stud 10 and 12, before Willys serial 288835	2	% 8	5	50	15
				cylinder head stud 12 only, after Willys serial 288835	1				

01-2 12	WO-349368	FM-GPW-6066	STUD, S., $\frac{7}{16}$-14NC x 3$\frac{7}{16}$ x $\frac{7}{16}$-20NF		% 6	6	20	20
			cylinder head holes, No. 5, 7, 9, 15, before Willys serial 288835	4				
			cylinder head holes, No. 5, 7, 9, 10, 15, after Willys serial 28835, all Ford trucks	5				
01-2	WO-A-1549	FM-GPW-6060	STUD, S., cd-pltd., $\frac{7}{16}$-14NC x 3$\frac{7}{16}$ x $\frac{7}{16}$-20NF					
			cylinder head hole No. 10, before Willys serial 288835 only	1	% 2	2	8	5
01-2	WO-A-1548	FM-GPW-6067	STUD, S., cd-pltd., $\frac{7}{16}$-14NC x 3$\frac{7}{8}$ x $\frac{7}{16}$-20NF (Cylinder head hole No. 12)	1	% 2	2	8	5
			0102—CRANKSHAFT					
01-12	WO-381519	FM-GPW-6345	BOLT, hex-hd., S., $\frac{1}{2}$-13NC x 2$\frac{3}{8}$ (crankshaft bearing cap to crankcase)	6	—	10	12	30
01-4	WO-638121	FM-GPW-6303-A	CRANKSHAFT (issue until stock is exhausted, then use WO-A-7568)	1	4	12	—	—
01-4	WO-A-7568	FM-GPW-18288	CRANKSHAFT and DOWEL engine group assembly	1		3	4	10
			(Includes:					
			2 WO-116295 FM-GPW-6390 BOLT					
			1 WO-638121 FM-GPW-6303-A CRANKSHAFT					
			2 WO-52804 FM-33832-S2 NUT					
			2 WO-52330 FM-34909-S2 WASHER)					
01-4	WO-635377	FM-GPW-6369	DOWEL, engine crankshaft bearing	5	12	20	36	62
01-4	WO-334103	FM-GPW-6353	GASKET, engine crankshaft oil slinger	1	10	—	10	—
01-4	WO-5036	FM-74178-S	KEY, Woodruff, No. 9 (pulley to crankshaft)	1	—	8	10	25
01-1	WO-387633	FM-GPW-6319	NUT, engine crankshaft, w/PIN, assembly	1	2	2	6	8
	WO-28025	FM-357574	PIN, clutch, engine crankshaft nut (issue until stock exhausted)	1	4	8	—	—
01-1	WO-638113	FM-GPW-6312	PULLEY, engine crankshaft	1	10	5	18	14
01-12	WO-375920	FM-GPW-6287	RETAINER, engine crankshaft packing	1	8	32	48	100
	WO-630262	FM-GPW-6342-B	SHIM, engine crankshaft (.002 thk.)	as req.	—	32	36	100
01-12	WO-375877	FM-GPW-6310	SLINGER, oil, engine crankshaft	1	—	6	6	20
01-4	WO-634796	FM-GPW-6308	WASHER, thrust, engine crankshaft (S., 3$\frac{1}{8}$ O.D. x 1.251 I.D. x .156 thk.)	1	—	3	6	10
01-12	WO-5009	FM-34809-S	WASHER, lock, reg., S., $\frac{1}{2}$ ($\frac{7}{8}$ O.D. x $\frac{17}{32}$ I.D. x $\frac{1}{8}$ thk.) (bearing cap to crankcase bolt) (BECXIM)	6	% —	—	—	—
			0102A—CRANKSHAFT BEARING					
01-4	WO-637007	FM-GPW-6333-A	BEARING, engine crankshaft, front upper (std.)	1	—	—	—	—
01-4	WO-637008	FM-GPW-6338-A	BEARING, engine crankshaft, front lower (std.)	1	—	—	—	—
01-4	WO-638730	FM-GPW-6339-A	BEARING, engine crankshaft, center upper (std.)	1	—	—	—	—
01-4	WO-638731	FM-GPW-6341-A	BEARING, engine crankshaft, center lower (std.)	1	—	—	—	—
01-4	WO-638732	FM-GPW-6331-A	BEARING, engine crankshaft, rear upper (std.)	1	—	—	—	—
01-4	WO-638733	FM-GPW-6337-A	BEARING, engine crankshaft, rear lower (std.)	1	—	—	—	—
	WO-637724	FM-GPW-6333-B	BEARING, engine crankshaft, front upper (.010 U.S.)	as req.	—	—	—	—
	WO-637725	FM-GPW-6338-B	BEARING, engine crankshaft, front lower (.010 U.S.)	as req.	—	—	—	—
	WO-639237	FM-GPW-6339-B	BEARING, engine crankshaft, center upper (.010 U.S.)	as req.	—	—	—	—
	WO-639238	FM-GPW-6341-B	BEARING, engine crankshaft, center lower (.010 U.S.)	as req.	—	—	—	—
	WO-639239	FM-GPW-6331-B	BEARING, engine crankshaft, rear upper (.010 U.S.)	as req.	—	—	—	—
	WO-639240	FM-GPW-6337-B	BEARING, engine crankshaft, rear lower (.010 U.S.)	as req.	—	—	—	—
	WO-116522	FM-GPW-6333-C	BEARING, engine crankshaft, front upper (.020 U.S.)	as req.	—	—	—	—
	WO-116524	FM-GPW-6338-C	BEARING, engine crankshaft, front lower (.020 U.S.)	as req.	—	—	—	—
	WO-116526	FM-GPW-6339-C	BEARING, engine crankshaft, center upper (.020 U.S.)	as req.	—	—	—	—
	WO-116528	FM-GPW-6341-C	BEARING, engine crankshaft, center lower (.020 U.S.)	as req.	—	—	—	—
	WO-116530	FM-GPW-6331-C	BEARING, engine crankshaft, rear upper (.020 U.S.)	as req.	—	—	—	—

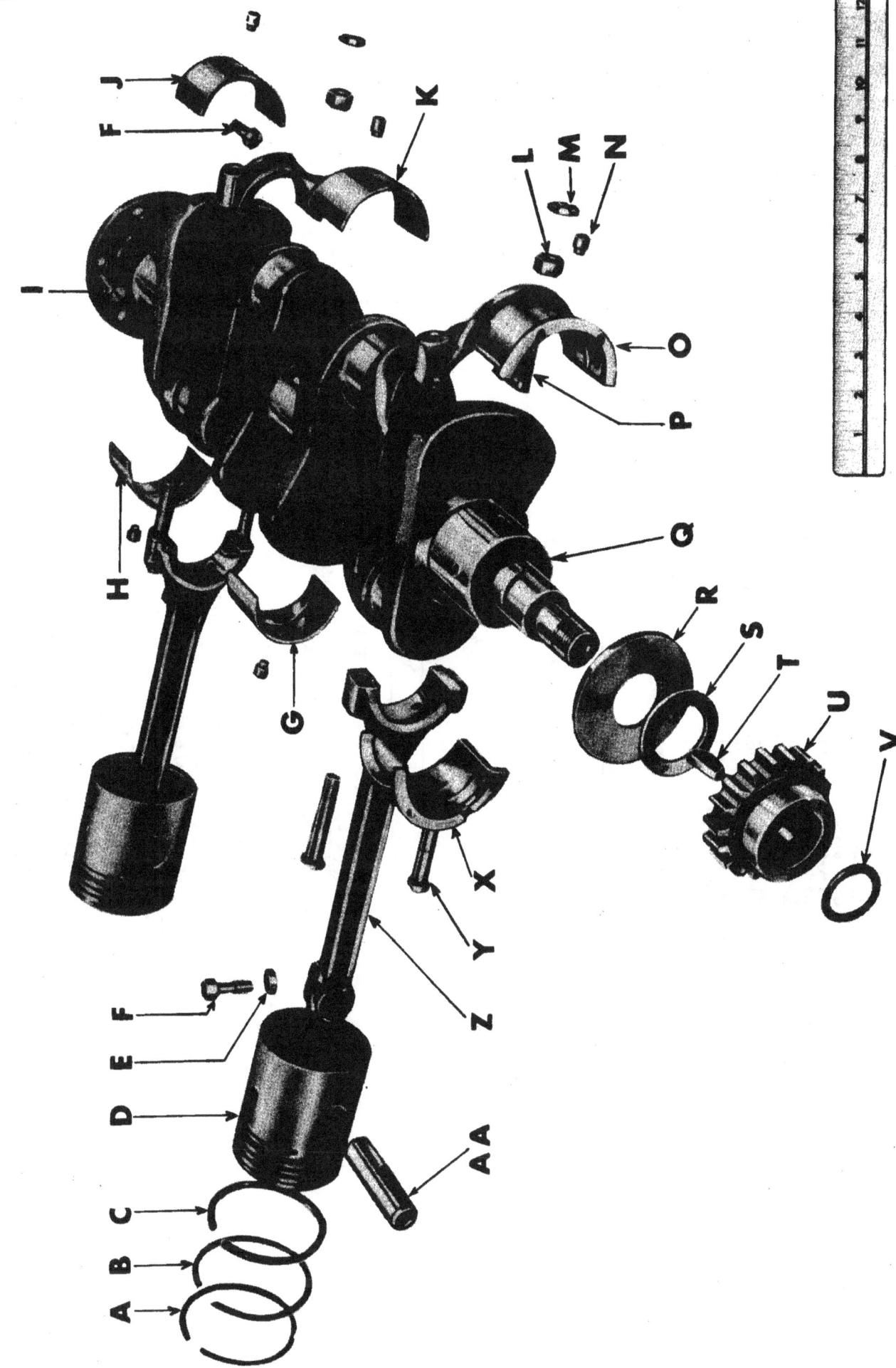

FIGURE 01-4—CRANKSHAFT AND PISTON ASSEMBLY

Key	Item	Willys Part No.	Ford Part No.	Gov't Group No.
A	RING	WO-639864	FM-GPW-6150-A	0103A
B	RING	WO-116562	FM-GPW-6152-A	0103A
C	RING	WO-116566	FM-GPW-6159-A	0103A
D	PISTON assembly	WO-637041	FM-GPW-6105-A	0103
E	WASHER	WO-5010	FM-34807-82	0103B
F	BOLT	WO-632157	FM-355497	0103B
G	BEARING	WO-638730	FM-GPW-6339-A	0102A
H	BEARING	WO-638732	FM-GPW-6331-A	0102A
I	DOWEL	WO-632156	FM-GPW-6387	0109
J	BOLT	WO-632157	FM-355497	0109
K	BEARING	WO-638733	FM-GPW-6337-A	0102A
L	BEARING	WO-638731	FM-GPW-6341-A	0102A
M	NUT	WO-636962	FM-356021-S	0104
N	NUT	WO-52825	FM-356028-S	0104
O	DOWEL	WO-635377	FM-GPW-6369	0102
P	BEARING	WO-637008	FM-GPW-6338-A	0102A
Q	BEARING	WO-639862	FM-GPW-6211-A	0104A
R	CRANKSHAFT	WO-A-7568	FM-GPW-18288	0102
S	WASHER	WO-634796	FM-GPW-6308	0106C
T	SPACER	WO-630727	FM-GPW-6342-A	0102
U	KEY	WO-50917	FM-74182-S	0106C
V	SPROCKET	WO-638459	FM-GPW-6306	0102
W	GASKET	WO-334103	FM-GPW-6353	0102A
X	BEARING	WO-637007	FM-GPW-6333-A	0102A
Y	BOLT	WO-381519	FM-GPW-6345	0102
Z	ROD assembly	WO-640066		0104
AA	PIN	WO-636961	FM-GPW-6135-A	0103B

RA PD 305052A

GROUP 01—ENGINE (Cont'd)

Figure Number	Official Stockage Number	Part Number Willys	Part Number Ford	ITEM	Quantity Reqd. per Unit Assy.	Per 100 Major Items 12 Mos. Field Maintenance	Per 100 Major Items Major Overhaul (5th Ech)	Per 100 Major Items Total First Year Procurement	Estimated Reqmts. per 100 Rebuilds
Col. 1	Col. 2	Col. 3	Col. 4	Col. 5	Col. 6	Col. 7	Col. 8	Col. 9	Col. 10
				0102A—CRANKSHAFT BEARING—Cont'd					
		WO-116532	FM-GPW-6337-C	BEARING, engine crankshaft, rear lower (.020 U.S.)	as req.	—	—	—	—
		WO-A-6798	FM-GPW-18347	BEARING SET, engine crankshaft (std.)	as req.	4	8	15	25
				(Includes:					
				1 WO-637008 FM-GPW-6338-A BEARING					
				1 WO-637007 FM-GPW-6333-A BEARING					
				1 WO-638730 FM-GPW-6339-A BEARING					
				1 WO-638731 FM-GPW-6341-A BEARING					
				1 WO-638732 FM-GPW-6331-A BEARING					
				1 WO-638733 FM-GPW-6337-A BEARING)					
		WO-A-6746	FM-GPW-18348	BEARING SET, engine crankshaft (.010 U.S.)	as req.	2	8	13	25
				(Includes:					
				1 WO-637724 FM-GPW-6333-B BEARING					
				1 WO-637725 FM-GPW-6338-B BEARING					
				1 WO-639237 FM-GPW-6339-B BEARING					
				1 WO-639238 FM-GPW-6341-B BEARING					
				1 WO-639239 FM-GPW-6331-B BEARING					
				1 WO-639240 FM-GPW-6337-B BEARING)					

SNL G-503

GROUP 01—ENGINE (Cont'd)

0102A-0103A

Figure Number	Official Stockage Number	Part Number - Willys	Part Number - Ford	ITEM	Quantity Reqd. per Unit Assy.	Per 100 Major Items - 12 Mos. Field Maintenance	Per 100 Major Items - Major Overhaul (5th Ech)	Total First Year Procurement	Estimated Reqmts. per 100 Rebuilds
Col. 1	Col. 2	Col. 3	Col. 4	Col. 5	Col. 6	Col. 7	Col. 8	Col. 9	Col. 10
				0102A—CRANKSHAFT BEARING (Cont'd)					
		WO-A-6747	FM-GPW-18349	BEARING SET, engine crankshaft (.020 U.S.) .... (Includes:	as req.	3	16	22	50
				1 WO-116522 FM-GPW-6333-C BEARING					
				1 WO-116524 FM-GPW-6338-C BEARING					
				1 WO-116526 FM-GPW-6339-C BEARING					
				1 WO-116528 FM-GPW-6341-C BEARING					
				1 WO-116530 FM-GPW-6331-C BEARING					
				1 WO-116532 FM-GPW-6337-C BEARING)					
01-12		WO-637098	FM-GPW-6700	PACKING, engine crankshaft, front end	1	8	—	10	—
01-1		WO-637237	FM-GPW-6702	PACKING, engine crankshaft, rear end (CN-type 310-0)	2	16	—	20	—
01-12		WO-637790	FM-GPW-6701	PACKING, engine crankshaft rear bearing cap	2	16	—	20	—
01-12		WO-337112	FM-357420-S	PIN, drain tube to engine crankshaft rear bearing cap (S., ⅛ x ⅞)	1	—	8	12	25
		WO-630294	FM-GPW-6326	PIPE, drain engine, crankshaft rear bearing cap	1	—	5	6	15
				0103—PISTONS					
		WO-637037		PISTON, engine, assembly (std.) (conventional type) (Willys only)	4	—	—	—	—
		WO-116017		PISTON, engine, assembly (.020 O.S.) (conventional type) (Willys only)	as req.	—	—	—	—
		WO-116018		PISTON, engine, assembly (.030 O.S.) (conventional type) (Willys only)	4	—	—	—	—
		WO-116558		PISTON, engine, assembly (std.) (re-ring type) (Willys only)	4	—	—	—	—
		WO-116560		PISTON, engine, assembly (.020 O.S.) (re-ring type) (Willys only)	as req.	—	—	—	—
		WO-116561		PISTON, engine, assembly (.030 O.S.) (re-ring type) (Willys only)	as req.	—	32	—	—
		WO-116612		PISTON, engine, semi-finished (use until stock is exhausted) (Willys only)	—	16	32	8	—
01-4		WO-637041	FM-GPW-6105-A	PISTON, engine, w/PIN, assembly (std. size re-ring)	4	6	—	8	250
		WO-116019	FM-GPW-6105-C	PISTON, engine, w/PIN, assembly (.020 O.S. re-ring)	as req.	3	80	92	250
		WO-116020	FM-GPW-6105-D	PISTON, engine, w/PIN, assembly (.030 O.S. re-ring)	as req.	3	48	60	150
				0103A—RINGS					
01-4		WO-639864	FM-GPW-6150-A	RING, engine, piston, conventional or re-ring (top groove) (std.)	4	—	—	—	—
		WO-116502	FM-GPW-6150-C	RING, engine, piston, conventional or re-ring (top groove) (.020 O.S.)	as req.	—	—	—	—
		WO-116503	FM-GPW-6150-D	RING, engine, piston, conventional or re-ring type (top groove) (.030 O.S.)	4	—	—	—	—
		WO-637042	FM-GPW-6155-A	RING, engine, piston, conventional type (second groove) (std.)	4	—	—	—	—
		WO-116023	FM-GPW-6155-G	RING, engine, piston, conventional type (second groove) (.020 O.S.)	as req.	—	—	—	—
		WO-116024	FM-GPW-6155-H	RING, engine, piston, conventional type (second groove) (.030 O.S.)	as req.	—	—	—	—
		WO-116616	FM-GPW-6156-F	RING, engine, piston, conventional type (bottom groove) (std.)	4	—	—	—	—
		WO-116116	FM-GPW-6156-H	RING, engine, piston, conventional type (bottom groove) (.020 O.S.)	as req.	—	—	—	—
		WO-116117	FM-GPW-6156-J	RING, engine, piston, conventional type (bottom groove) (.030 O.S.)	as req.	—	—	—	—
01-4		WO-116562	FM-GPW-6152-A	RING and SPRING, engine piston, 2nd groove (std. to .009 O.S.)	as req.	—	—	—	—
		WO-116564	FM-GPW-6152-C	RING and SPRING, engine piston, 2nd groove (.020 to .029 O.S.)	as req.	—	—	—	—

0103A-0105

01-4	WO-116565	FM-GPW-6152-D	RING and SPRING, engine piston, 2nd groove (.030 to .039 O.S.)	as req.	—	—	—	—
	WO-116566	FM-GPW-6159-A	RING and SPRING, engine piston, bottom groove (std. to .009 O.S.)	as req.	—	—	—	—
	WO-116568	FM-GPW-6159-C	RING and SPRING, engine piston, bottom groove (.020 to .029 O.S.)	as req.	—	—	—	—
	WO-116569	FM-GPW-6159-D	RING and SPRING, engine piston, bottom groove (.030 to .039 O.S.)	as req.	—	—	—	—
	WO-A-6794	FM-GPW-6149-E	RING SET, engine piston, conventional type (std.)	1	3	22	5	—
	WO-A-6796	FM-GPW-6149-G	RING SET, engine piston, conventional type (.020 O.S.)	as req.	2	14	28	70
	WO-A-6797	FM-GPW-6149-H	RING SET, engine piston, conventional type (.030 O.S.)	as req.	2	16	19	45
	WO-116110	FM-GPW-6149-A	RING SET, engine piston (std.) (re-ring type) (issue until stock exhausted)	as req.	8	8	—	—
	WO-116112	FM-GPW-6149-C	RING SET, engine piston (.020 O.S.) (re-ring type) (issue until stock exhausted)	as req.	4	8	—	—
	WO-116113	FM-GPW-6149-D	RING SET, engine piston (.030 O.S.) (re-ring type) (issue until stock exhausted)	as req.	4	4	—	—

0103B—PISTON PINS

01-4	WO-636961	FM-GPW-6135-A	PIN, engine piston (std. size)	4	—	6	8	20
01-4	WO-632157	FM-355497	SCREW, lock, engine piston pin	4	3	32	62	100
01-4	WO-5010	FM-34807-S2	WASHER, lock, reg., S., 3/8 in. (13/32 I.D. x 21/32 O.D. x 3/32 thk.) (piston pin lock screw) (BECX1K)	4	—	—	—	—

0104—CONNECTING RODS

01-4	WO-640070	FM-GPW-6200	BOLT, engine connecting rod bearing cap (S., 7/16—20NF x 2 3/8) (Willys only)	8	% 10	16	30	50
01-4	WO-636962	FM-GPW-6201	NUT, hex, S., 7/16-20NF-2 (connecting rod bolt)	8	% 32	32	100	100
01-4	WO-52825		NUT, hex, spring, S., 7/16-20NF-2, stamped (connecting rod bolt) (GM-107381)	8	% 96	256	400	800
01-4	WO-640071	FM-GPW-6211-A	ROD, connecting, engine, No. 1 and 3, assembly (less bearings)	2	2	2	6	5
	WO-640072	FM-GPW-6211-B	ROD, connecting, engine, No. 2 and 4, assembly (less bearings)	2	2	2	6	5
	WO-640066		ROD, connecting, engine, No. 1 and 3, w/BEARING, assembly (Willys only)	2	—	—	—	—
	WO-640067		ROD, connecting, engine, No. 2 and 4, w/BEARING, assembly (Willys only)	2	—	—	—	—

0104A—CONNECTING ROD BEARINGS

01-4	WO-639862	FM-GPW-6211-A	BEARING, engine connecting rod (std.)	8	—	—	—	—
	WO-116534	FM-GPW-6211-B	BEARING, engine connecting rod (.010 U.S.)	as req.	—	—	—	—
	WO-116535	FM-GPW-6211-C	BEARING, engine connecting rod (.020 U.S.)	as req.	—	—	—	—
	WO-A-7233	FM-GPW-18330-A	BEARING SET, engine connecting rod (std.) (Includes: 2 WO-639862 FM-GPW-6211-A BEARINGS)	4	48	32	91	100
	WO-A-7234	FM-GPW-18330-B	BEARING SET, engine connecting rod (.010 U.S.) (Includes: 2 WO-116534 FM-GPW-6211-B BEARINGS)	as req.	16	32	53	100
	WO-A-7235	FM-GPW-18330-C	BEARING SET, engine connecting rod (.020 U.S.) (Includes: 2 WO-116535 FM-GPW-6211-C BEARINGS)	as req.	26	64	99	200

0105—VALVES, SEATS, GUIDES AND SPRING

H	WO-375811	FM-GPW-6510-B	GUIDE, engine exhaust valve stem	4	—	19	24	60
01-5	WO-637045	FM-GPW-6511-B	GUIDE, engine intake valve stem	4	—	19	24	60
01-5	WO-375994	FM-GPW-6546	LOCK, engine valve spring retainer (lower)	16	144	48	300	150
01-5	WO-637044	FM-GPW-6514	RETAINER, engine valve spring (lower)	8	10	19	50	60
01-5	WO-638636	FM-GPW-6513	SPRING, engine valve	8	29	48	88	150
01-5	WO-637183	FM-GPW-6505	VALVE, exhaust, engine (RMC-A-366)	8	19	32	58	100
01-1	WO-637182	FM-GPW-6507	VALVE, intake, engine (RMC-A-365)	8	10	16	30	50

SNL G-503

FIGURE 01-5—CAMSHAFT, CHAIN, SPROCKETS, VALVES

Key	Item	Willys Part No.	Ford Part No.	Gov't Group No.
A	VALVE	WO-637183	FM-GPW-6505	0105
B	SPRING	WO-658636	FM-GPW-6513	0105
C	RETAINER	WO-637044	FM-GPW-6514	0105
D	LOCK	WO-375994	FM-GPW-6546	0105
E	SCREW	WO-640020	FM-GPW-6549-B	0105A
F	TAPPET	WO-637047	FM-GPW-6500-A	0105A
G	CAMSHAFT	WO-637065	FM-GPW-6250	0106
H	PLUNGER	WO-375907	FM-GPW-6243	0106
I	SPRING	WO-375908	FM-GPW-6244	0106
J	BOLT	WO-634850	FM-355499-S	0106C
K	WASHER	WO-315932	FM-GPW-6269	0106C
L	SPROCKET	WO-638459	FM-GPW-6342-A	0106C
M	SPROCKET	WO-638458	FM-GPW-6256	0106C
N	CHAIN	WO-638457	FM-GPW-6260	0106C
O	WASHER	WO-375900	FM-GPW-6245	0106
P	BUSHING	WO-639051	FM-GPW-6262-A	0106A

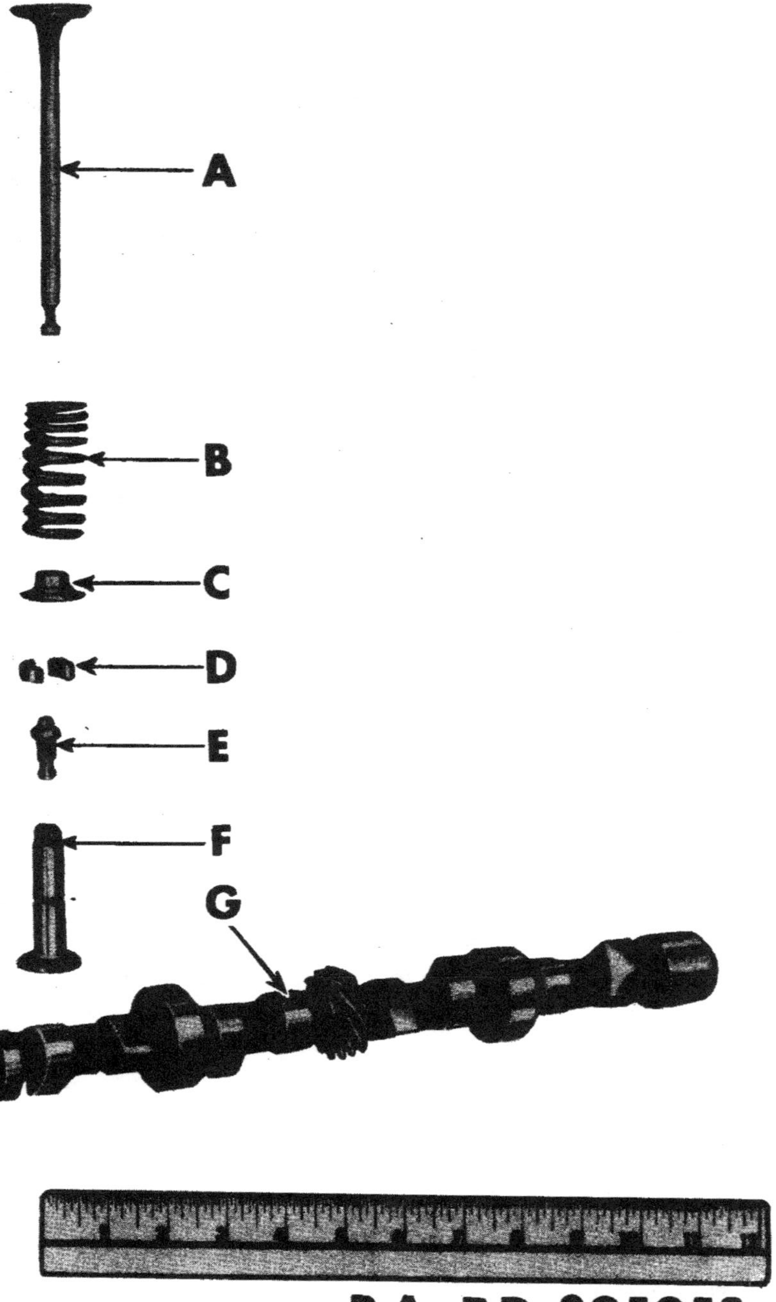

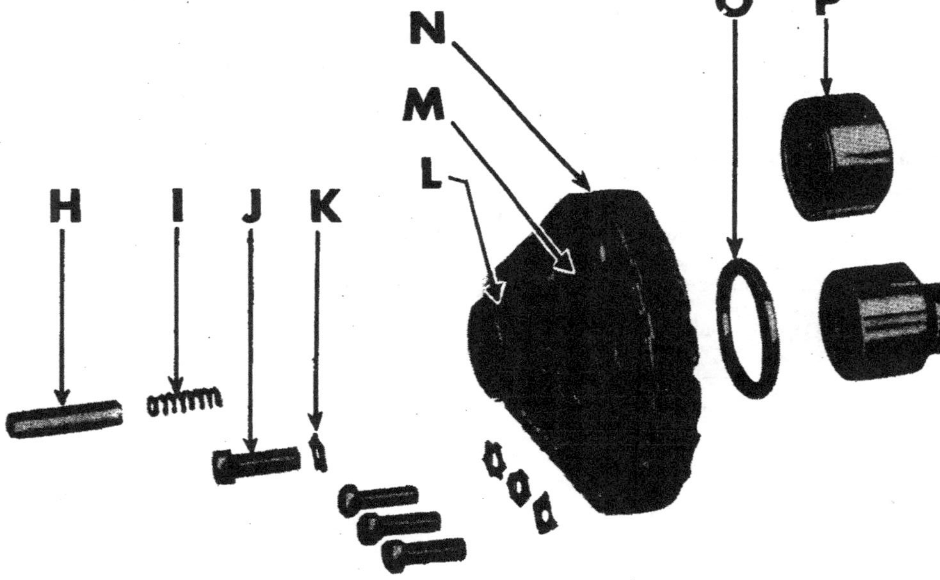

RA PD 305053-A

Figure Number	Official Stockage Number	Part Number: Willys	Part Number: Ford	ITEM	Quantity Reqd. per Unit Assy.	Per 100 Major Items: 12 Mos. Field Main-tenance	Per 100 Major Items: Major Overhaul (5th Ech)	Per 100 Major Items: Total FirstYear Procure-ment	Estimated Reqmts. per 100 Rebuilds
Col. 1	Col. 2	Col. 3	Col. 4	Col. 5	Col. 6	Col. 7	Col. 8	Col. 9	Col. 10
				0105A—TAPPETS					
01-5		WO-640020	FM-GPW-6549-B	SCREW, adjusting, engine valve tappet (self locking)	8	—	19	25	60
01-5		WO-637047	FM-GPW-6500-A	TAPPET, engine valve (WIL-ST-1265) (CCS-W-41-25)	8	—	13	14	40
		WO-115948	FM-GPW-6500-B	TAPPET, engine valve (.004 O.S.) (issue until stock exhausted)	as req.	28	52	—	—
				0105C—VALVE COVERS AND GASKETS					
01-6		WO-630303	FM-GPW-6520	COVER, engine valve spring	1	% 2	2	8	5
01-6		WO-632158	FM-355451-S	BOLT, hex-hd., S., 5/16-18NC-2 x 4 [illegible]/8 (front)	1	% 5	3	16	10
01-6		WO-639052	FM-355452-S	BOLT, hex-hd., S., 5/16-18NC-2 x 3 1/16 (rear)	1	% 5	3	16	10
01-6		WO-630305	FM-GPW-6521	GASKET, engine valve spring cover	1	% 32	11	60	—
01-6		WO-51875	FM-GPW-6555	GASKET, engine valve spring cover bolt	2	% 48	—	60	12
				0106—CAMSHAFT					
01-5		WO-637065	FM-GPW-6250	CAMSHAFT	1	—	2	3	5
01-5		WO-375907	FM-GPW-6243	PLUNGER, camshaft thrust	1	5	5	12	10
01-5		WO-375908	FM-GPW-6244	SPRING, camshaft thrust plunger	1	5	5	12	15
01-5		WO-375900	FM-GPW-6245	WASHER, camshaft thrust	1	—	3	5	10
				0106A—CAMSHAFT BEARINGS					
01-2, 5		WO-639051	FM-GPW-6262-A	BUSHING, camshaft, front	1	—	32	35	100
		WO-51460	FM-74127	PLUG, expansion, engine camshaft rear bearing, (1¾ in. O.D.)	1	—	3	5	10
				0106C—CHAINS AND SPROCKETS					
01-4		WO-50917	FM-74182-S	KEY, engine crankshaft sprocket	1	—	8	10	25
01-5		WO-638457	FM-GPW-6260	CHAIN, drive, engine camshaft (½ in. thk. x 1 in. wide) (LK-S-40936)	1	6	6	15	20
01-4		WO-630727	FM-GPW-6342-A	SPACER, engine crankshaft sprocket	1	—	3	3	10
01-4		WO-638459	FM-GPW-6306	SPROCKET, engine crankshaft, S., 18 teeth, 2.870 O.D. x 1.656 thk. (LK-S-35116-1)	1	3	3	9	10
01-5		WO-638458	FM-GPW-6256	SPROCKET, engine camshaft, S., 36 teeth, 5.780 O.D. x 1.013 thk. (LK-S-35117-1)	1	3	3	9	10
01-5	H1-01-16220	WO-634850	FM-355499-S	BOLT, hex-hd., s-fin., alloy-S., 3/8-24NF x 1¼ (to camshaft) (BAOXKD)	4	6	16	40	50
01-5		WO-315932	FM-GPW-6269	WASHER, lock, reg., S., 3/8 in. (0.390 I.D. x .640 O.D. x 3/32 thk.) (BECX1K)	4	16	112	150	350

0106C

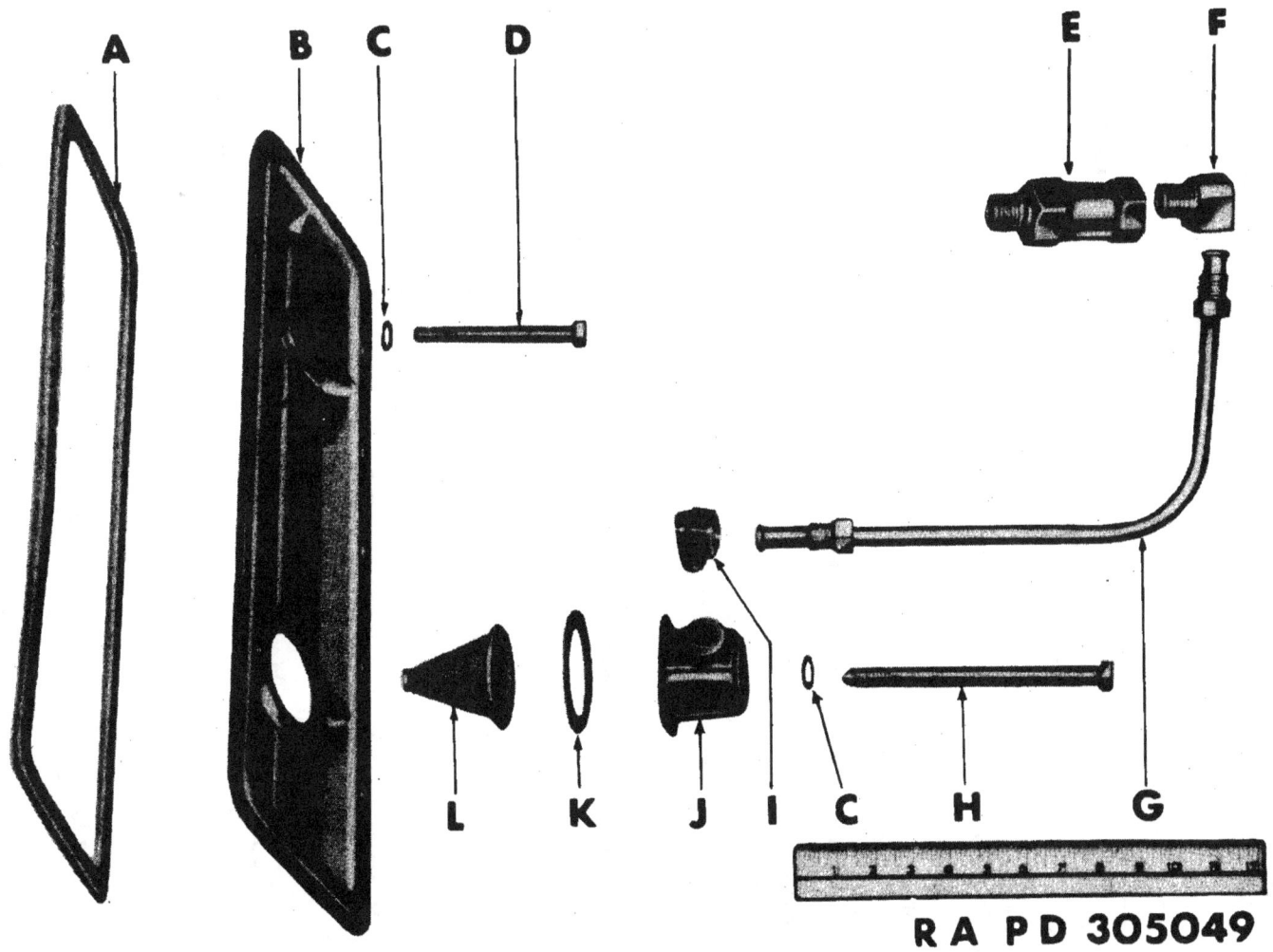

FIGURE 01-6—VALVE COVER PLATE WITH CRANKCASE VENTILATOR

Key	Item	Willys Part No.	Ford Part No.	Gov't Group No.
A	GASKET	WO-630305	FM-GPW-6521	0105C
B	COVER	WO-630303	FM-GPW-6519	0105C
C	GASKET	WO-51875	FM-GPW-6555	0105C
D	BOLT	WO-639052	FM-355452	0105C
E	VALVE assembly	WO-A-6895	FM-GPW-6769	0107N
F	ELBOW	WO-A-6885	FM-GPW-6722	0107N
G	TUBE assembly	WO-A-6922	FM-GPW-6756	0107N
H	BOLT	WO-632158	FM-355451-S	0105C
I	ELBOW	WO-384549	FM-GPW-9268	0107N
J	BODY assembly	WO-A-6919	FM-GPW-6758-B	0107N
K	GASKET	WO-630299	FM-GPW-6648	0107N
L	BAFFLE	WO-630298	FM-GPW-6762	0107N

RA PD 305049-A

SNL G-503

GROUP 01—ENGINE (Cont'd)

Figure Number	Official Stockage Number	Part Number Willys	Part Number Ford	ITEM	Quantity Reqd. per Unit Assy.	Per 100 Major Items 12 Mos. Field Maintenance	Per 100 Major Items Major Overhaul (5th Ech)	Total First Year Procurement	Estimated Reqmts. per 100 Rebuilds
Col. 1	Col. 2	Col. 3	Col. 4	Col. 5	Col. 6	Col. 7	Col. 8	Col. 9	Col. 10
				0106D—CHAIN CASE COVER					
G	H1-01-16217	WO-A-1190	FM-GPW-6016	COVER, engine timing chain, assembly	1	—	2	3	5
		WO-50163	FM-20349-S2	BOLT, hex-hd., s-fin., alloy-S., 3/8-24NF-3 x 3/4 (GM-100025) (cover to engine plate)	3	—	—	—	30
	H1-01-16219	WO-5919		BOLT, hex-hd., S., 3/8-24NF-3 x 1 (GM-100026) (cover to engine plate) (BAOX1CC) (Willys only)	1	—	—	—	—
			FM-350356-S2	BOLT, hex-hd., s-fin., alloy-S., 3/8-24NF-3 x 3/4 (Ford only)	1	—	—	—	—
		WO-5901	FM-33800-S2	NUT, reg., hex., s-fin., alloy-S., 3/8-24NF-2 (BBB1CA)	10	—	—	—	—
01-2, 12	H1-07-19003	WO-384958	FM-88022-S	STUD, S., 3/8-16NC x 1 5/16 x 3/8-24NF (GM-103026) (issue until stock exhausted) (4 used plate and chain cover to cylinder block 2 used plate to cylinder block)	6	92	188	—	—
		WO-5010	FM-34807-S2	WASHER, lock, S., 3/8 in. (21/32 O.D. x 13/32 I.D. x 3/32 thk.) (BECX1K) Willys	10	—	—	—	—
				Ford	6	8	—	10	—
01-12		WO-630365	FM-GPW-6288	GASKET, engine timing chain cover	1	8	—	10	—
01-12		WO-630359	FM-GPW-6020	GASKET, engine cylinder block, front (issue until stock exhausted)	1	40	80	—	—
		WO-375917	FM-GPW-6286	RING, retaining, engine timing chain cover packing (issue until stock exhausted)	1	4	4	—	—
		WO-630364	FM-GPW-6285	STUD, engine camshaft thrust plunger (issue until stock exhausted)	1	4	4	—	—
				0107—OIL PUMP					
01-7		WO-630384	FM-GPW-6604	BODY, engine oil pump	1	—	—	—	—
01-7		WO-630387	FM-GPW-6664	COVER, engine oil pump, assembly	1	—	4	—	—
01-7	H1-1026108	WO-51819	FM-31079-S	SCREW, machine, oval fil-hd., S., No. 12 (.216)-24NC-2 x 5/8 (cover to body) (BCFX2CG)	1	—	—	—	—
01-7		WO-636600	FM-GPW-6673	DISC, engine oil pump rotor	1	10	10	24	30
01-7		WO-630392	FM-GPW-6619	GASKET, engine oil pump cover	1	3	5	12	15
01-7		WO-639870	FM-GPW-6659	GASKET, engine oil pump cover (velumoid)	1	—	—	—	—
01-7		WO-380197	FM-355262	GASKET, engine oil pump cover to body (copper) (issue until stock exhausted)	1	32	64	—	—
01-7		WO-375927	FM-GPW-6625	GASKET, engine oil pump shaft	1	—	—	—	—
01-7		WO-630394	FM-GPW-6630	GASKET, engine oil pump to cylinder block	1	6	—	10	—
01-7		WO-634813	FM-GPW-6642	GASKET, engine oil pump relief spring retainer (paper)	1	—	—	—	—
01-7		WO-637425	FM-GPW-6610	GEAR, driven, engine oil pump (12 teeth, 1.1369 O.D. x 9/16 thk.)	1	—	—	—	—
01-7		WO-A-6750	FM-GPW-18380	GASKET SET, service, engine oil pump. (Includes: 1 WO-380197 FM-355262-S GASKET 1 WO-630398 FM-GPW-6627 GASKET, oil float support	as req.	10	37	50	115

33

SNL G-503

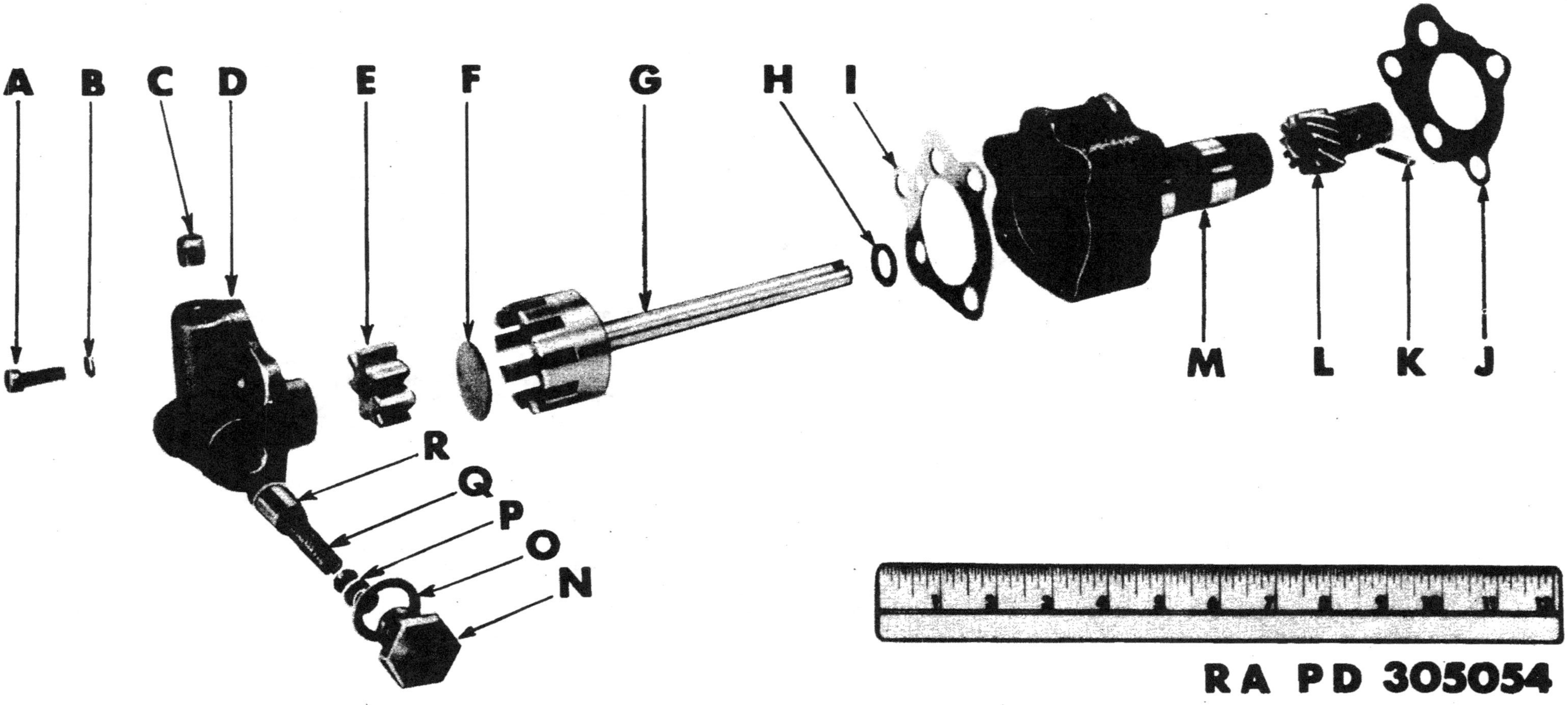

FIGURE 01-7—OIL PUMP ASSEMBLY

RA PD 305054-A

Key	Item	Willys Part No.	Ford Part No.	Gov't Group No.
A	SCREW	WO-15819	FM-31079-S	0107
B	GASKET	WO-380197	FM-355262	0107
C	PLUG	WO-52525	FM-353052	0107
D	COVER assembly	WO-630387	FM-GPW-6664	0107
E	PINION	WO-343306	FM-GPW-6614	0107
F	DISC	WO-636600	FM-GPW-6673	0107
G	SHAFT assembly	WO-636599	FM-GPW-6608	0107
H	GASKET	WO-375927	FM-GPW-6625	0107
I	GASKET	WO-639870	FM-GPW-6659	0107
J	GASKET	WO-630394	FM-GPW-6630	0107
K	PIN	WO-330964	FM-GPW-6684	0107
L	GEAR	WO-637425	FM-GPW-6610	0107
M	BODY	WO-630384	FM-GPW-6604	0107
N	RETAINER	WO-630390	FM-GPW-6644	0107
O	GASKET	WO-634813	FM-GPW-6642	0107
P	SHIM	WO-630389	FM-GPW-6628	0107
Q	SPRING	WO-356155	FM-GPW-6654	0107
R	PLUNGER	WO-630518	FM-GPW-6663	0107

GROUP 01—ENGINE (Cont'd)

Figure Number	Official Stockage Number	Part Number Willys	Part Number Ford	ITEM	Quantity Reqd. per Unit Assy.	Per 100 Major Items 12 Mos. Field Maintenance	Per 100 Major Items Major Overhaul (5th Ech)	Total First Year Procurement	Estimated Reqmts. per 100 Rebuilds
Col. 1	Col. 2	Col. 3	Col. 4	Col. 5	Col. 6	Col. 7	Col. 8	Col. 9	Col. 10

0107—OIL PUMP (Cont'd)

Figure Number	Official Stockage Number	Willys	Ford	ITEM	Qty	Col.7	Col.8	Col.9	Col.10
				1 WO-639870 FM-GPW-6659 GASKET, oil pump cover					
				4 WO-630392 FM-GPW-6619 GASKET, oil pump cover					
				1 WO-630394 FM-GPW-6630 GASKET, oil pump to cylinder block					
				1 WO-375927 FM-GPW-6625 GASKET, oil pump shaft					
				1 WO-634813 FM-GPW-6642 GASKET, oil relief spring retainer					
				1 WO-630299 FM-GPW-6648 GASKET, ventilator to valve spring cover					
				4 WO-630389 FM-GPW-6628 SHIM, oil relief spring)					
		WO-A-6749	FM-GPW-18379	KIT, repair, engine oil pump	as req.	—	5	5	15
				(Includes:					
01-7		WO-330964	FM-GPW-6684	1 PIN, engine oil pump driven gear, straight (5/32 x 31/32)	1	—	8	9	30
01-7		WO-343306	FM-GPW-6614	1 PINION, engine oil pump (7 teeth)	1	3	—	5	—
01-7		WO-52525	FM-353052-S	1 PLUG, pipe, slotted, 1/8 in. (oil pump cover plug) (issue until stock exhausted)	1	16	24	4	10
01-7		WO-630518	FM-GPW-6663	1 PLUNGER, engine oil pump relief	1	—	3	—	—
H		WO-637636	FM-GPW-6600	1 PUMP, oil, engine, assembly	1	3	2	8	5
01-12		WO-5910	FM-33798-S2	NUT, reg, hex, s-fin, alloy-S, 5/16-24NF-2 (pump assembly to block) (BBBX1BA)	3				
01-2, 12		WO-375981	FM-88141-S	STUD, S, 5/16-18NC-2 x 2¾ x 5/16-24NF-2 (issue until stock exhausted)	3	20	40	—	—
01-8, 12		WO-51833	FM-34806-S2	WASHER, lock, reg, S, 5/16 in. (19/32 O.D. x 11/32 I.D. x 1/16 thk.) (BECX1H)	3	—	—	—	—
01-7		WO-630390	FM-GPW-6644	RETAINER, engine oil pump oil relief spring	1	—	3	3	10
01-7		WO-636599	FM-GPW-6608	SHAFT, engine oil pump, assembly	1	10	—	—	—
01-7		WO-630389	FM-GPW-6628	SHIM, engine oil pump oil relief spring	as req.	—	—	12	—
01-7		WO-356155	FM-GPW-6654	SPRING, engine oil pump oil relief plunger (S., wire, 13 turns, .360 O.D. x 1⅛ free lgth.)	1	—	6	6	20

0107A—OIL PAN

Figure	Stockage	Willys	Ford	ITEM	Qty	Col.7	Col.8	Col.9	Col.10
01-8		WO-639980	FM-GPW-6710	GASKET, engine oil pan	1	—	—	—	—
01-8		WO-314338	FM-GPW-6734	GASKET, engine oil pan drain plug	1	198	—	260	—
	G503-01-94020	WO-A-1538	FM-GPW-18512	GASKET SET, engine oil pan	as req.	77	—	100	—
				(Includes:					
				1 WO-630398 FM-GPW-6627 GASKET, oil float support					
				1 WO-639980 FM-GPW-6710 GASKET, oil pan					
				1 WO-314338 FM-GPW-6734 GASKET, oil pan drain plug)					
		WO-A-1167		PAN, oil, engine, assembly (issue until stock is exhausted, then use WO-A-7238)	—	4	4	—	—

SNL G-503

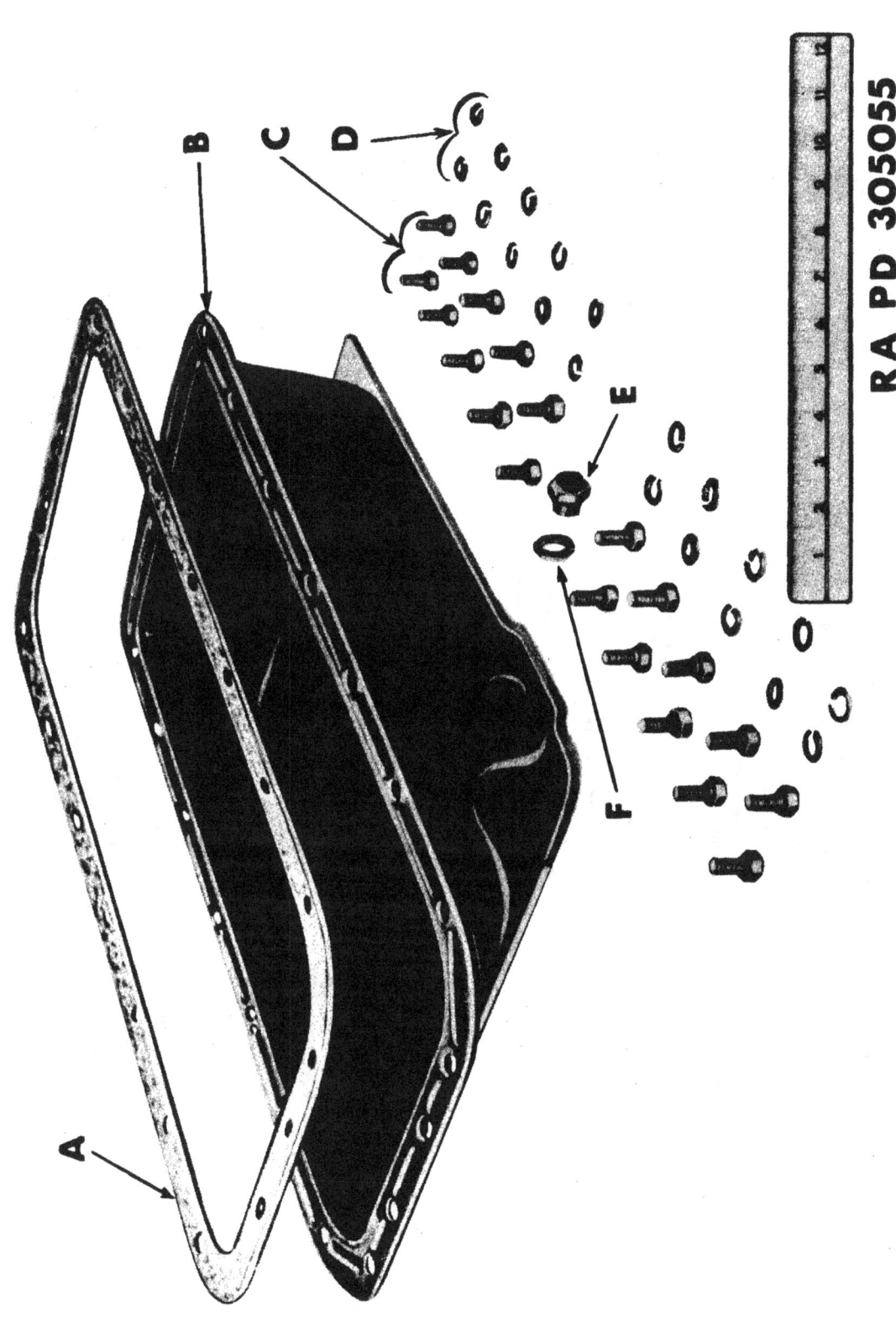

FIGURE 01-8—OIL PAN ASSEMBLY

Key	Item	Willys Part No.	Ford Part No.	Gov't Group No.
A	GASKET	WO-639980	FM-GPW-6710	0107A
B	PAN assembly	WO-A-7238	(Willys only)	0107A
C	BOLT	WO-51485	See text	0107A
D	WASHER	WO-51833	See text	0107A
E	PLUG	WO-639979	FM-GPW-6727	0107A
F	GASKET	WO-314338	FM-GPW-6734	0107A

RA PD 305055-A

GROUP 01—ENGINE (Cont'd)

Figure Number	Official Stockage Number	Part Number Willys	Part Number Ford	ITEM	Quantity Reqd. per Unit Assy.	12 Mos. Field Maintenance	Major Overhaul (5th Ech)	Total First Year Procurement	Estimated Reqmts. per 100 Rebuilds
Col. 1	Col. 2	Col. 3	Col. 4	Col. 5	Col. 6	Col. 7	Col. 8	Col. 9	Col. 10
				0107A—OIL PAN (Cont'd)					
01-8		WO-A-7238	FM-GPW-6675	PAN, oil, engine, assembly (after Willys serial 297089) (was WO-A-1167)........	1	2	2	7	5
01-8		WO-51485		BOLT, hex-hd., 5/16-18NC-2 x 5/8. (6 used oil pan and fan pulley shield to cylinder block) (Willys only) (14 used oil pan to cylinder block and front engine cover) (Willys only).......	20				
			FM-355403-S2	BOLT, hex-hd., S, 5/16-18NC-2 x 11/16, w/WASHER, assembly (Ford only) (BCYX51FG)..........	20				
01-8		WO-51833		WASHER, lock, reg., S, 5/16 in. (11/32 O.D. x 11/32 I.D. x 1/16 thk.) (BECX1H) (Willys only)........	20				
01-8		WO-639979	FM-GPW-6727	PLUG, drain, engine oil pan (7/8 in.)..........	1	19%	10	36	30
				0107A-1—OIL FLOAT					
H		WO-630396	FM-GPW-6615	FLOAT, strainer, oil, engine, assembly (TSE-215-B)..........	1	2		3	5
		WO-5108	FM-72053	PIN, cotter, S, 1/8 x 1 1/4 (float to support).......	1				
		WO-630398	FM-GPW-6627	GASKET, engine oil strainer float support........	1				
		WO-630397	FM-GPW-6617	SUPPORT, engine oil strainer float........	1	6	6	16	20
01-1		WO-636796	FM-355396-S	BOLT, support to crankcase (hex-hd., 5/16-18NC-2 x 3/4)........	2	8	8	24	25
		WO-51833	FM-34806-S2	WASHER, lock, reg, S, 5/16 in. (11/32 O.D. x 11/32 I.D. x 1/16 thk.) (BECX1H).........	2				
				0107E—OIL FILTER					
01-9		WO-A-6818		CASE, engine oil filter, assembly (Willys only).......	1				
01-9		WO-A-1231	FM-GPW-18687-A	COVER, engine oil filter, assembly (PU-25791).......	1	2%	2	8	5
01-9		WO-A-1232	FM-GPW-18691-A	BOLT, oil filter cover (hex-hd., S, 5/16-20NF-2) (PU-25755).......	1	5%	2	10	5
01-9		WO-A-1236	FM-GPW-18662-A	ELEMENT, engine oil filter, assembly (PU-26637)........	1	280%	40	400	125
F		WO-A-1230	FM-GPW-18660-A	FILTER, oil, engine assembly (PU-27078)........	1	5%	2	10	5
01-9	G503-01-94033	WO-A-1235	FM-GPW-18688-A	GASKET, engine oil filter cover (PU-25802).......	1	86%	32	120	100
01-9		WO-A-1233	FM-GPW-18675-A	GASKET, engine oil filter cover bolt (PU-25756)......	1	58%	32	120	100
01-9		WO-A-1237	FM-358040-S	PLUG, drain, engine oil filter (S, hex, 5/16, 1/4-18 thd.) (PU-25795).......	1	19%	10	36	30
01-9		WO-A-1234	FM-GPW-18685-A	SPRING, engine oil filter cover bolt (S-wire, tapered, I.D. 13/16 to 17/32) (PU-25757)........	1		3	3	10
				0107F—OIL FILTER ATTACHING PARTS					
		WO-51396	FM-24347-S2	BOLT, hex-hd., s-fin., alloy-S., 5/16-24NF-2 x 3/4 (oil filter to bracket) (BAOX1BA).......	3				
		WO-A-1247	FM-GPW-18663	BRACKET, engine oil filter, assembly........	1		4		
01-9		WO-A-1251	FM-GPW-18644-A	CLAMP, engine oil filter, assembly (PU-27081) (Includes: BOLT, NUT).......	2				

0107A-0107F

37

G-503

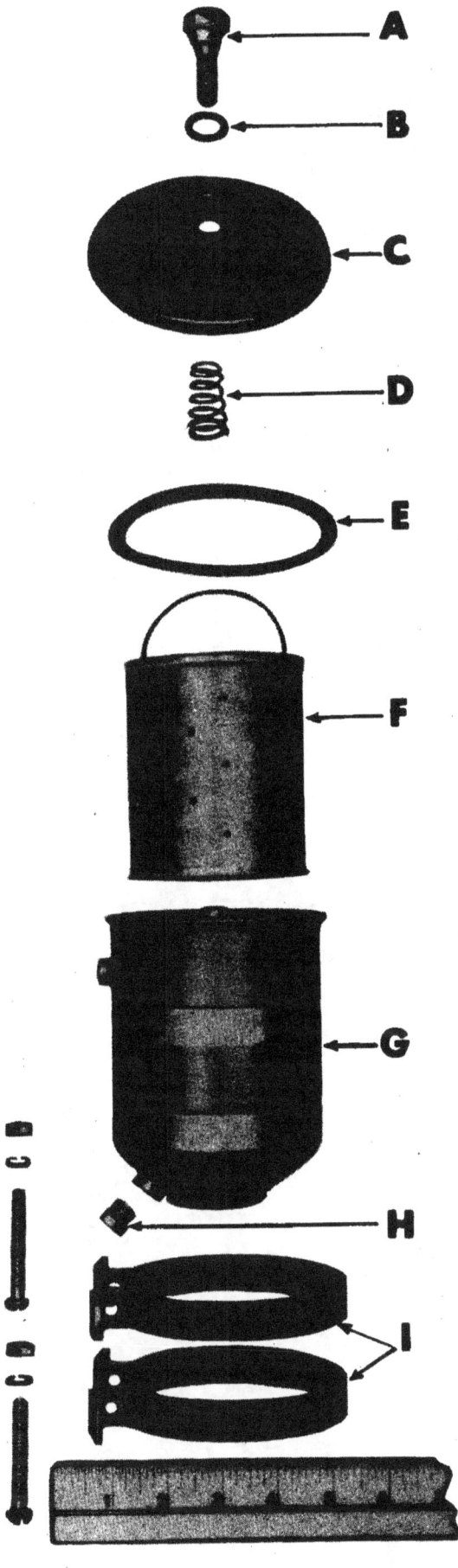

FIGURE 01-9—OIL FILTER ASSEMBLY

Key	Item	Willys Part No.	Ford Part No.	Gov't Group No.
A	BOLT	WO-A-1232	FM-GPW-18691-A	0107E
B	GASKET	WO-A-1233	FM-GPW-18675-A	0107E
C	COVER	WO-A-1231	FM-GPW-18687-A	0107E
D	SPRING	WO-A-1234	FM-GPW-18685-A	0107E
E	GASKET	WO-A-1235	FM-GPW-18688-A	0107E
F	ELEMENT assembly	WO-A-1236	FM-GPW-18662-A	0107E
G	CASE assembly	WO-A-6818	(Willys only)	0107E
H	PLUG	WO-A-1237	FM-358040-S	0107E
I	CLAMP, w/ BOLT and NUT	WO-A-1251	FM-GPW-18644-A	0107F

RA PD 305056-A

GROUP 01—ENGINE (Cont'd)

Figure Number	Official Stockage Number	Part Number		ITEM	Quantity Reqd. per Unit Assy.	Per 100 Major Items			Estimated Reqmts. per 100 Rebuilds
		Willys	Ford			12 Mos. Field Maintenance	Major Overhaul (5th Ech)	Total First Year Procurement	
Col. 1	Col. 2	Col. 3	Col. 4	Col. 5	Col. 6	Col. 7	Col. 8	Col. 9	Col. 10
				0107F—OIL FILTER ATTACHING PARTS (Cont'd)					
		WO-52274	FM-34746-S2	WASHER, plain, S., S.A.E. std., 5/16 in. (filter to bracket nut) (BEBX1H).......	4	--	--	--	--
		WO-51833	FM-34806-S2	WASHER, lock, reg, S., 5/16 in. (11/32 O.D. x 11/52 I.D. x 1/16 thk.) (BECX1H).......	4	--	--	--	--
				0107G—OIL FILLER					
		WO-A-5105	FM-GPW-6770	BRACKET, support, engine oil filler tube............	1	--	--	--	--
		WO-639655	FM-GPW-6763-A	TUBE, engine oil filler (before Willys engine 114550 only) (For straight type).	1	--	--	--	--

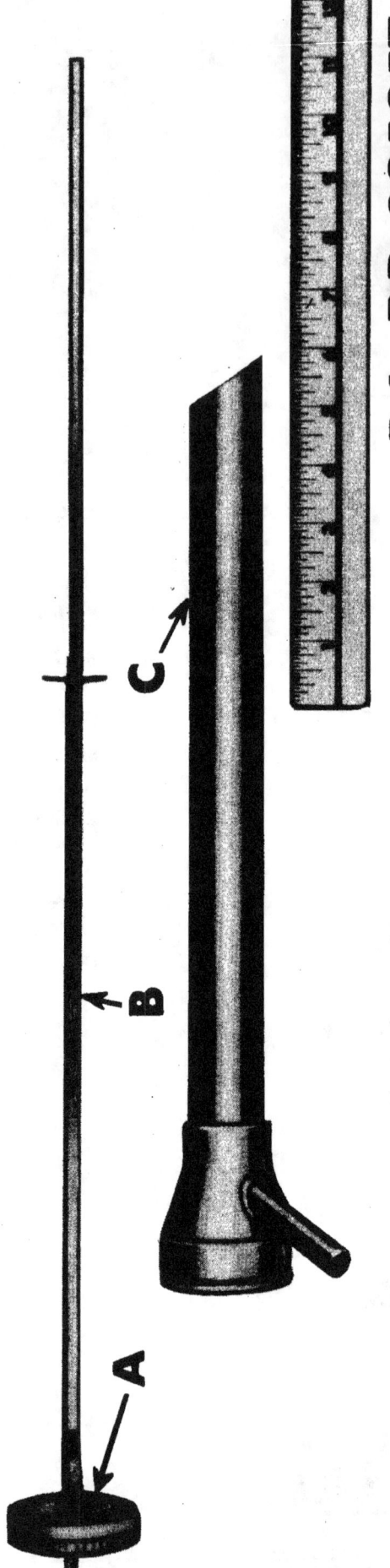

FIGURE 01-10—OIL FILLER TUBE AND LEVEL INDICATOR

Key	Item	Willys Part No.	Ford Part No.	Gov't Group No.
A	GASKET............	WO-A-7280	FM-GPW-6789	0107H
B	INDICATOR assembly......	WO-A-6525	FM-GPW-6766-C	0107H
C	TUBE assembly........	WO-A-6915	FM-GPW-6763-C	0107G

RA PD 305057-A

0107G–0107N

GROUP 01—ENGINE (Cont'd)

Figure Number	Official Stockage Number	Part Number Willys	Part Number Ford	ITEM	Quantity Reqd. per Unit Assy.	Per 100 Major Items 12 Mos. Field Maintenance	Per 100 Major Items Major Overhaul (5th Ech)	Total First Year Procurement	Estimated Reqmts. per 100 Rebuilds
Col. 1	Col. 2	Col. 3	Col. 4	Col. 5	Col. 6	Col. 7	Col. 8	Col. 9	Col. 10
				0107G—OIL FILLER (Cont'd)					
01-10		WO-A-5165	FM-GPW-6763-B	TUBE, engine oil filler, assembly (after Willys engine 114550 only) (For funnel type)					
		WO-A-6915	FM-GPW-6763-C	TUBE, engine oil filler, assembly (after Willys serial 208437 only)	1		2	3	5
				0107H—LEVEL AND/OR FLOAT GAUGES					
01-10		WO-A-7280	FM-GPW-6789	GASKET, engine oil filler cap	1	%8	8	20	25
		WO-639556	FM-GPW-6766-A	INDICATOR and breather CAP, engine oil filler, assembly (before Willys engine 114550 only)	1				
H	G503-01-32455	WO-A-5168	FM-GPW-6766-B	INDICATOR and breather CAP, engine oil filler, assembly (after Willys engine 114550 only)	1				
01-10		WO-A-6525	FM-GPW-6766-C	INDICATOR and breather CAP, engine oil filler, assembly (after Willys serial 208437 only)	1	%6	2	15	5
				0107J—OIL DISTRIBUTION SYSTEM					
		WO-A-1289	FM-GPW-14585	CLIP, tube to front engine plate (7/16 in. S, closed)	1				
		WO-A-5449	FM-GPW-2281	CLIP, oil gauge tube to dash (S., 5/16 in. 7/32 bolt hole)	1				
		WO-5182	FM-355158-S2	SCREW, rd-hd., S., No. 10 (.190)-24NC-2 x 3/8	1				
		WO-52221	FM-34803-S7	WASHER, lock, S., No. 10 (.190 in.)	1				
		WO-387891	FM-9N-18679	CONNECTOR, flared tube, br., 1/4 in. (tube to oil filter)	1	%6	3	12	10
		WO-384569	FM-9N-18686	ELBOW, (inverted flared tube, br., 1/4 in.) (1 used flexible connection to engine; 1 used tube to cylinder block)	2	%13	6	25	20
		WO-345961	FM-GPW-13434-A	GROMMET, rubber, 13/32 in. (issue until stock exhausted)	1	%16	24		
G503-02-17815		WO-A-1197	FM-GPW-18667	TUBE, engine oil filter inlet, assembly	1	%66	14	108	45
G503-02-17816		WO-A-1198	FM-GPW-18666	TUBE, flexible (1 used oil filter outlet; 1 used oil gauge tube)	2	%85	14	130	45
		WO-A-1450	FM-GPW-9316	TUBE, 1/4 in., oil gauge, assembly	1	%3		6	
		WO-A-1456	FM-GPW-9323	UNION, inverted flared tube, br., 1/4 in. (flexible connection to tube assembly)	1	%6	3	12	10
				0107N—CRANKCASE VENTILATOR					
01-6		WO-630298	FM-GPW-6762	BAFFLE, engine crankcase ventilator	1	3	3	9	10
01-6		WO-A-6919	FM-GPW-6758-B	BODY, engine crankcase ventilator, assembly (after Willys engine 204040)	1	2	2	7	5
01-6		WO-384549	FM-GPW-9268	ELBOW, pipe, 1/8 in. (engine crankcase ventilator to tube) (after Willys engine 204040)					
01-6		WO-A-6885	FM-GPW-6722	ELBOW, pipe, 1/4 in. (engine crankcase ventilator tube to valve) (after Willys engine 204040)	1	%6	3	10	10
					1	%6	3	10	10

SNL G-503

01-6		WO-630299	FM-GPW-6648	GASKET, engine crankcase ventilator to valve spring cover	1	% 38		45	
01-6		WO-A-6922	FM-GPW-6756	TUBE, engine crankcase ventilator valve, assembly (after Willys engine 204040)	1	% 3	3	9	10
01-6		WO-A-6895	FM-GPW-6769	VALVE, engine crankcase ventilator, assembly (after Willys engine 204040)	1	% 3	3	9	10
		WO-A-1061	FM-GPW-6758	VENTILATOR, engine crankcase, assembly (before Willys engine 204040)	1	—	—	—	—

0108—MANIFOLDS AS AN ASSEMBLY

01-2	G503-01-94021	WO-638640	FM-GPW-9448	GASKET, engine intake and exhaust manifold (issue until stock exhausted)	1	% 36	68	—	—
01-11		WO-634811	FM-GPW-9435	GASKET, engine intake to exhaust manifold	1				
		WO-A-7835	FM-GPW-18323	GASKET SET, engine manifold	as req.	% 38	—	46	—
				(Includes:					
				1 WO-638640 FM-GPW-9448 GASKET, intake and exhaust manifold					
				1 WO-634811 FM-GPW-9435 GASKET, intake to exhaust manifold					
				1 WO-634814 FM-GPW-9450 GASKET, exhaust pipe flange					
		WO-A-1165	FM-GPW-9410-B	MANIFOLD, engine intake and exhaust, assembly	1	%	—	—	—
G	H1-07-19003	WO-53288	FM-33800-S2	NUT, regular, hex., s-fin., alloy-S., Seez Pruf $\frac{3}{8}$-24NF-2 (intake and exhaust manifold)	7	—	—	—	—
01-2, 12		WO-349712	FM-88082-S	STUD, S., $\frac{3}{8}$-16NC x $1\frac{15}{16}$ x $\frac{3}{8}$-24NF-2 (issue until stock exhausted)	2	% 20	40	—	—
01-11		WO-632159	FM-88057	STUD, S., $\frac{3}{8}$-16NC x $1\frac{11}{16}$ x $\frac{3}{8}$-24NF-2	2	—	—	—	—
01-2, 12		WO-300143	FM-88042	STUD, S., $\frac{3}{8}$-16NC x $1\frac{1}{2}$ x $\frac{3}{8}$-24NF-2 (issue until stock exhausted)	3	% 48	92	—	—
		WO-344732	FM-GPW-9443	WASHER, manifold clamp, S., $\frac{13}{16}$ O.D. x $\frac{13}{32}$ I.D. x $\frac{1}{8}$ to $\frac{3}{16}$ thk. at I.D.	2	% 32	48	90	150

0108A—INTAKE MANIFOLDS

01-11		WO-A-1166	FM-GPW-9424-B	MANIFOLD, intake, engine assembly	1	% 1	1	4	21
01-11		WO-51486	FM-24386-S2	BOLT, hex-hd., s-fin., alloy-S., $\frac{5}{16}$-18NC-2 x 1 (intake to exhaust manifold) (BANX1BC)	4	—	—	—	—
01-11		WO-52428	FM-34906-S	WASHER, lock $\frac{5}{16}$ in.	4	—	—	—	—
01-11	H6-02-83805	WO-5138	FM-353055-S	PLUG, pipe, sq-hd., black, $\frac{1}{4}$ in. (engine intake manifold) (CPMX1AB)	3	—	—	—	—

0108B—EXHAUST MANIFOLDS

01-11		WO-A-912	FM-GPW-9428	MANIFOLD, exhaust, engine assembly	1	% 2	2	8	5

0108C—HEAT CONTROLS

		WO-636438	FM-GPW-9462	BEARING, engine heat control valve shaft	1	3	10	15	30
01-11		WO-637211	FM-GPW-9465	KEY, engine heat control lever	1	13	13	30	40
01-11		WO-637210	FM-GPW-9458	LEVER, engine heat control valve counterweight	1	2	5	10	15
01-11	H1-07-16540	WO-6352	FM-355836-S7	NUT, machine screw, hex., S., No. 10 (.190)-24NC-1 (clamp lever to shaft) (BBKX2C)	1	—	—	—	—
01-11	H1-10-30102	WO-5272	FM-355160-S2	SCREW, machine, rd-hd., S., No. 10 (.190)-24NC-2 x $\frac{3}{4}$ (BCNX2AG)	1	—	—	—	—
01-11		WO-637206	FM-GPW-9456	SHAFT, engine heat control valve	1	2	5	10	15
01-11		WO-637208	FM-GPW-9467-A	SPRING, engine heat control valve	1	13	13	30	40

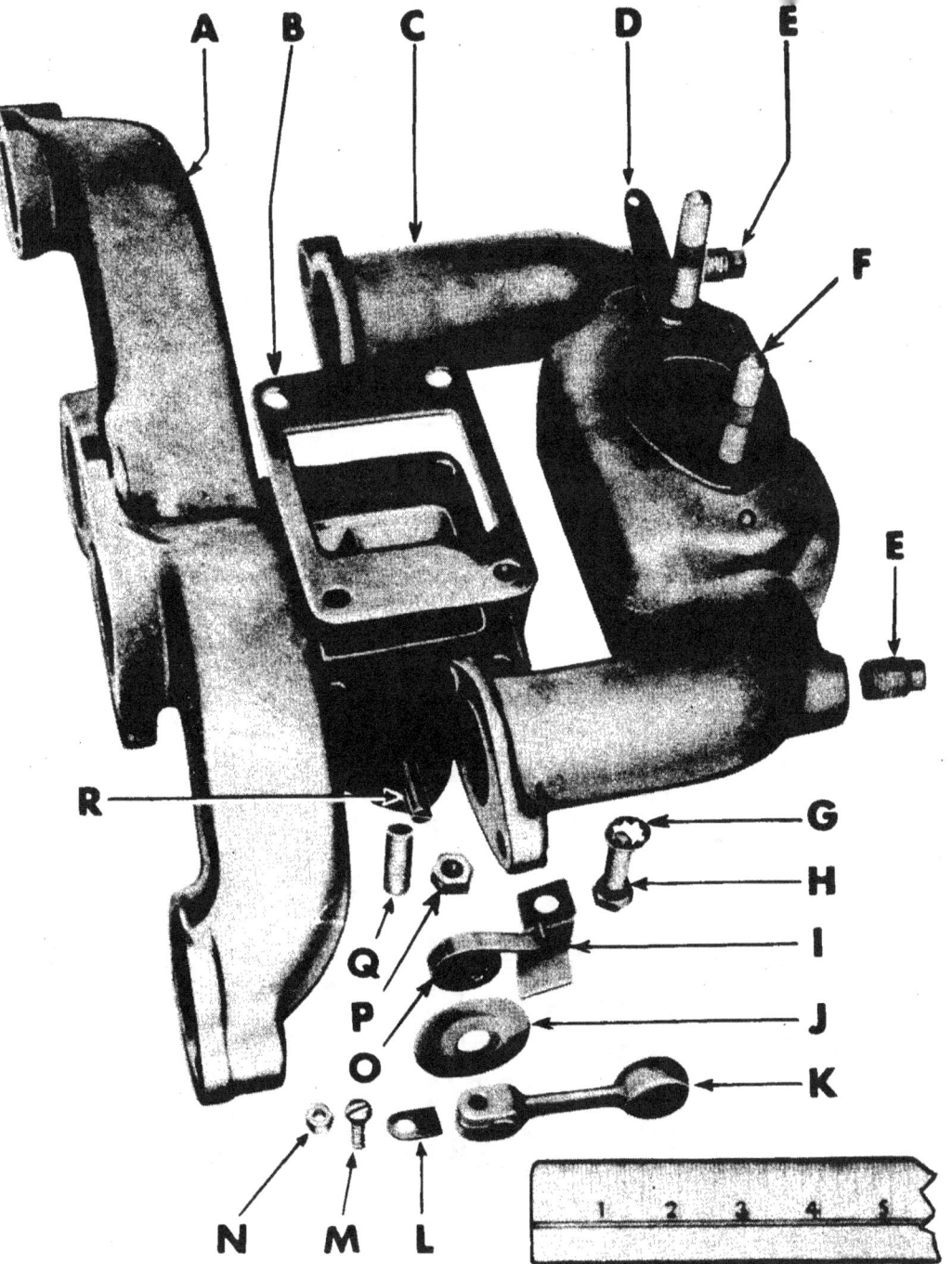

FIGURE 01-11—MANIFOLD ASSEMBLY

Key	Item	Willys Part No.	Ford Part No.	Gov't Group No.
A	MANIFOLD	WO-A-912	FM-GPW-9428	0108B
B	GASKET	WO-634811	FM-GPW-9435	0108
C	MANIFOLD	WO-A-1166	FM-GPW-9424-B	0108A
D	CLIP	WO-A-1173	FM-GPW-9751	0303A
E	PLUG	WO-5138	FM-353055-S	0108A
F	STUD	WO-632159	FM-88057	0301A
G	WASHER	WO-52428	FM-34906-S	0108A
H	BOLT	WO-51486	FM-24386-S2	0108A
I	STOP	WO-639743	FM-GPW-9463	0108C
J	WASHER	WO-637209	FM-GPW-9484	0108C
K	LEVER	WO-637210	FM-GPW-9458	0108C
L	KEY	WO-637211	FM-GPW-9465	0108C
M	SCREW	WO-5272	FM-355160-82	0108C
N	NUT	WO-6352	FM-355836-S7	0108C
O	SPRING	WO-637208	FM-GPW-9467-A	0108C
P	NUT	WO-6352	FM-355836-S7	0108C
Q	STUD	WO-332515	FM-88032-S	0402
R	SHAFT	WO-637206	FM-GPW-9456	0108C

GROUP 01—ENGINE (Cont'd)

Figure Number	Official Stockage Number	Part Number: Willys	Part Number: Ford	ITEM	Quantity Reqd. per Unit Assy.	Per 100 Major Items: 12 Mos. Field Maintenance	Per 100 Major Items: Major Overhaul (5th Ech)	Per 100 Major Items: Total First Year Procurement	Estimated Reqmts. per 100 Rebuilds
Col. 1	Col. 2	Col. 3	Col. 4	Col. 5	Col. 6	Col. 7	Col. 8	Col. 9	Col. 10
				0108C—HEAT CONTROLS (Cont'd)					
01-11		WO-639743	FM-GPW-9463	STOP, engine heat control valve spring	1	13	13	30	40
H		WO-636439	FM-GPW-9460	VALVE, engine heat control	1	2	5	10	15
01-11		WO-637209	FM-GPW-9484	WASHER, engine heat control valve spring (S., special, 1½ O.D. x .310 I.D. x ⅛ thk. at opening)	1	6	6	15	20
				0109—FLYWHEEL					
01-4		WO-632157	FM-355497-S	BOLT, hex-hd., S., ⅜-24NF-2 x 1$\frac{5}{32}$ (flywheel to crankshaft)	4	10	29	50	90
		WO-116295	FM-GPW-6390	BOLT, special hd., S., ½-20NF-2 x 1$\frac{3}{16}$ (flywheel to crankshaft dowel)	2	—	6	12	20
		WO-639578	FM-GPW-7600	BUSHING, flywheel to crankshaft	1	3	3	12	10
01-4		WO-632156	FM-GPW-6387	DOWEL, flywheel to crankshaft	2	—	14	15	45
		WO-A-1443	FM-GPW-6375	FLYWHEEL, engine (issue until stock is exhausted, then use WO-A-7503)	—	4	4	—	—
		WO-A-7503	FM-GPW-18289	FLYWHEEL and DOWEL, engine, group assembly (was WO-A-1443) (Includes: 2 WO-116295 FM-GPW-6390 BOLT, dowel; 1 WO-A-1443 FM-GPW-6375 FLYWHEEL; 2 WO-52804 FM-33832-S2 NUT, dowel bolt; 2 WO-52330 FM-34909-S2 WASHER, lock)	1	3	3	8	10
				NOTE: Parts numbered WO-116295—FM-GPW-6390, WO-52804—FM-33832-S2, WO-52330—FM-34909-S2, are required for straight doweling of flywheel to crankshaft. Eliminates need of special tapered reamer.					
	H1-07-19003	WO-5901	FM-33800-S2	NUT, reg., hex., s-fin., alloy-S., ⅜-24NF-2 (flywheel dowel and bolt) (GM-103026) (BBBX1CA)	6	—	—	—	—
	H1-07-16025	WO-52804	FM-33832-S2	NUT, hex., S., ½-20NF-2 (flywheel dowel bolt) (BBDX1E)	2	—	—	—	—
		WO-52332	FM-34907-S2	WASHER, lock, S., internal teeth, ⅜ in. ($\frac{25}{64}$ I.D. x $\frac{11}{16}$ O.D. x .035 thk.) (BEAX1L)	6	—	—	—	—
		WO-52330	FM-34909-S2	WASHER, lock, S., external teeth, ½ in. (.520 I.D. x ⅞ O.D. x .045 thk.) (BEAX4L)	2	—	—	—	—
				0109B—RING GEAR					
		WO-635394	FM-GPW-6384	GEAR, ring, engine flywheel (S., 97 teeth, 10.980 I.D. x 12.185 O.D. x ⅜ wide)	1	2	3	8	10
				0109C—FLYWHEEL HOUSING					
		WO-630101	FM-355597	BOLT, dowel (transmission to engine plate to cylinder block)	2	10	10	25	30
		WO-630103	FM-GPW-7518	COVER, transmission case inspection	1	—	3	5	10
		WO-51763	FM-24308-S	BOLT, hex-hd., s-fin., alloy-S., ¼-20NC-2 x ½ (cover to case) (BANX4AA)	2	—	—	—	—
		WO-52706	FM-34805-S	WASHER, lock, reg., S., ¼ in. (cover to case bolt) (BECX1G)	2	—	—	—	—

0109C

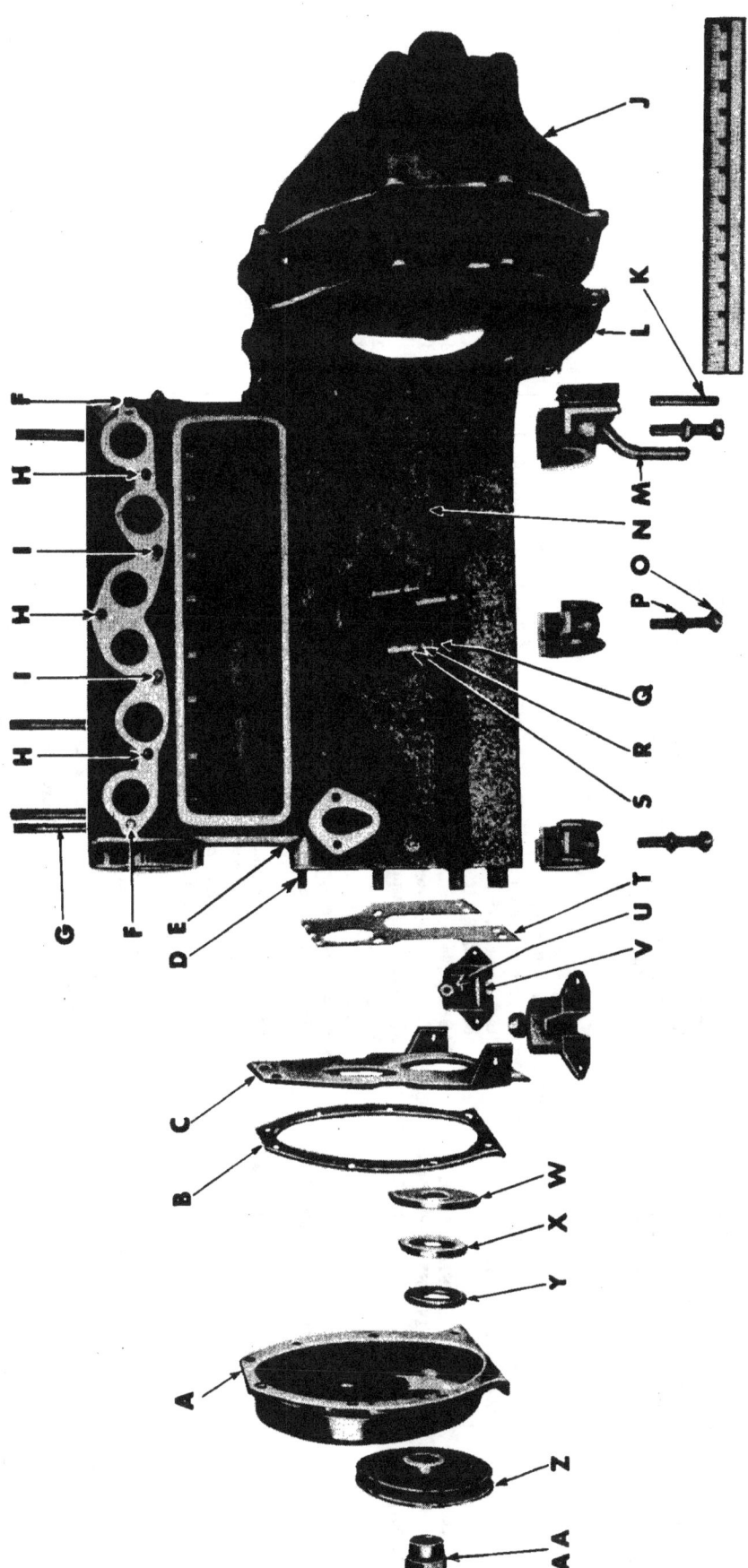

FIGURE 01-12—TIMING CHAIN COVER, CYLINDER BLOCK AND FLYWHEEL HOUSING

Key	Item	Willys Part No.	Ford Part No.	Gov't Group No
A	COVER assembly	WO-A-1190	FM-GPW-6016	0106D
B	GASKET	WO-630365	FM-GPW-6288	0106D
C	PLATE assembly	WO-A-1463	FM-GPW-6031-A3	0110
D	STUD	WO-384958	FM-88022-S	0106D
E	BLOCK assembly	WO-A-1272	FM-GPW-6010	0101
F	STUD	WO-349712	FM-88082-S	0108
G	STUD	WO-349368	FM-GPW-6066	0101A
H	STUD	WO-300143	FM-88042	0108
I	STUD	WO-632159	FM-88057	0109C
J	HOUSING assembly	WO-A-439	FM-GPW-6392	0102A
K	PACKING	WO-637790	FM-GPW-6701	0110
L	PLATE	WO-A-5121	FM-GPW-6044	0102A
M	PIPE	WO-630294	FM-GPW-6326	0102A
N	PLUG	WO-5085	FM-358064-S	0101

Key	Item	Willys Part No.	Ford Part No.	Gov't Group No.
O	BOLT	WO-381519	FM-GPW-6345	0102
P	WASHER	WO-5009	FM-34809-S	0102
Q	NUT	WO-5910	FM-33798-S2	0107
R	WASHER	WO-51833	FM-34806-S2	0107
S	STUD	WO-375981	FM-88141-S2	0106D
T	GASKET	WO-630359	FM-GPW-6020	0110
U	NUT	WO-5916	FM-33846-S	0110
V	INSULATOR	WO-A-7498	FM-GPW-6038-A	0102
W	SLINGER, oil	WO-375877	FM-GPW-6310	0102
X	RETAINER	WO-375920	FM-GPW-6287	0102A
Y	PACKING	WO-637098	FM-GPW-6700	0503A
Z	PULLEY	WO-638113	FM-GPW-6312	2215
AA	NUT assembly	WO-387633	FM-GPW-6319	

RA PD 305050-A

SNL G-503

GROUP 01—ENGINE (Cont'd)

Figure Number	Official Stockage Number	Part Number Willys	Part Number Ford	ITEM	Quantity Reqd. per Unit Assy.	Per 100 Major Items — 12 Mos. Field Maintenance	Per 100 Major Items — Major Overhaul (5th Ech)	Total FirstYear Procurement	Estimated Reqmts. per 100 Rebuilds
Col. 1	Col. 2	Col. 3	Col. 4	Col. 5	Col. 6	Col. 7	Col. 8	Col. 9	Col. 10
				0109C—FLYWHEEL HOUSING (Cont'd)					
		WO-375217	FM-GPW-7023	COVER, engine flywheel housing timing hole	1	---	3	4	10
		WO-52036	FM-26147-S7	SCREW, machine, rd-hd., S., 1/4-20NC-2 x 3/8 (BCNX2CC)	1	---	---	---	---
		WO-52706	FM-34805-S2	WASHER, lock, reg., S., 1/4 in. (15/32 O.D. x .260 I.D. x 1/16 thk.) (BECX1G)	1	---	---	---	---
01-12		WO-A-439	FM-GPW-6392	HOUSING, engine flywheel	1	---	3	4	10
		WO-52379	FM-24489-S	BOLT, hex-hd., s-fin., alloy-S., 3/8-24NF-2 x 1 5/8 (housing to bracket at hand brake cable clip) (BAOX4CH)	1	---	---	---	---
		WO-6606	FM-24409-S7	BOLT, hex-hd., s-fin., alloy-S., 3/8-24NF-2 x 1 1/8 (housing to brkt.)	3	---	---	---	---
H1-01-16039		WO-6184	FM-24430-S7	BOLT, hex-hd., s-fin., alloy-S., 7/16-14NC-2 x 1 1/4 (housing to transmission) (BANX1DD)	4	---	---	---	---
H1-07-19003		WO-5901	FM-33800-S	NUT, reg, hex., s-fin., alloy-S., 3/8-24NF-2 (BBBX1CA)	8	---	---	---	---
H1-15-18113		WO-5010	FM-34807-S	WASHER, lock, reg., S., 3/8 in. (BECX1K)	8	---	---	---	---
H6-02-82440		WO-51921	FM-74113-S	PLUG, expansion, S., 3/4 in. diam. (BEDX1AK)	1	---	3	10	10
				0110—MOUNTINGS					
		WO-A-146	FM-GPW-6043	BRACKET, mounting, engine, rear (issue until stock exhausted)	1	8	12	---	---
		WO-52189	FM-24328-S2	BOLT, hex-hd., s-fin., alloy-S., 3/8-16NC-2 x 5/8 (bracket to transmission)	3	---	---	---	5
		WO-6412	FM-24348-S	BOLT, hex-hd., S., 3/8-16NC-2 x 3/4 (bracket to transmission) (GM-100133) (BANX1CA)	1	---	---	---	---
H1-15-18113		WO-5010	FM-34807-S	WASHER, lock, reg., S., 3/8 in. (BECX1K)	4	5/8 2	2	8	5
H1-07-19003		WO-A-5125	FM-GPW-6044	CABLE, engine stay, assembly	1	---	---	---	---
		WO-5901	FM-33800-S2	NUT, reg., hex., s-fin., alloy-S., 3/8-24NF-2 (GM-103026) (BBBX1CA) Willys	2	---	---	---	---
		WO-52909		NUT, hex., S., stamped 3/8-24NF-2 (type "B") (GM-107322) (Willys only). Ford	1	---	---	---	---
		WO-52101	FM-34747-S2	WASHER, plain, S., S.A.E. std., 3/8 in. (13/16 O.D. x 13/32 I.D. x 1/32 thk.) (BEBX1K)	2	---	---	---	1
		WO-5010	FM-34807-S2	WASHER, lock, reg., S., 3/8 in. (BECX1K)	2	5/8 10	2	20	1
H1-01-16192		WO-A-7498	FM-GPW-6038-A	INSULATOR, engine support, front, assembly	1	---	---	---	---
		WO-5934		BOLT, hex-hd., s-fin., alloy-S., 5/16-24NF-2 x 1 (insulator to engine front support bracket) (GM-100014) (BAOX1BC) (Willys only)	4	---	---	---	---
H1-01-16193			FM-24427-S7	BOLT, hex-hd., s-fin., alloy-S., 5/16-24NF-2 x 1 1/4 (insulator to engine front support bracket) (Ford only) (BAOX1BD)	4	---	---	---	---
H1-07-19002		WO-5910	FM-33798-S2	NUT, reg., hex., s-fin., alloy-S., 5/16-24NF-2 (GM-103025) (BBBX1BA)	4	---	---	---	---
H1-07-19005		WO-5916	FM-33846-S	NUT, hex., S., 1/2-20NF-2 (insulator to front engine plate) (GM-103028) (BBBX1BD)	2	---	---	---	---
		WO-52274	FM-34746-S2	WASHER, plain, S., S.A.E. std., 5/16 in. (BEBX1H)	4	---	---	---	---
01-12		WO-51833	FM-34806-S2	WASHER, lock, reg., S., 5/16 in. (19/32 O.D. x 11/32 I.D. x 1/16 thk.) (BECX1H)	4	---	---	---	---

SNL G-503

0110

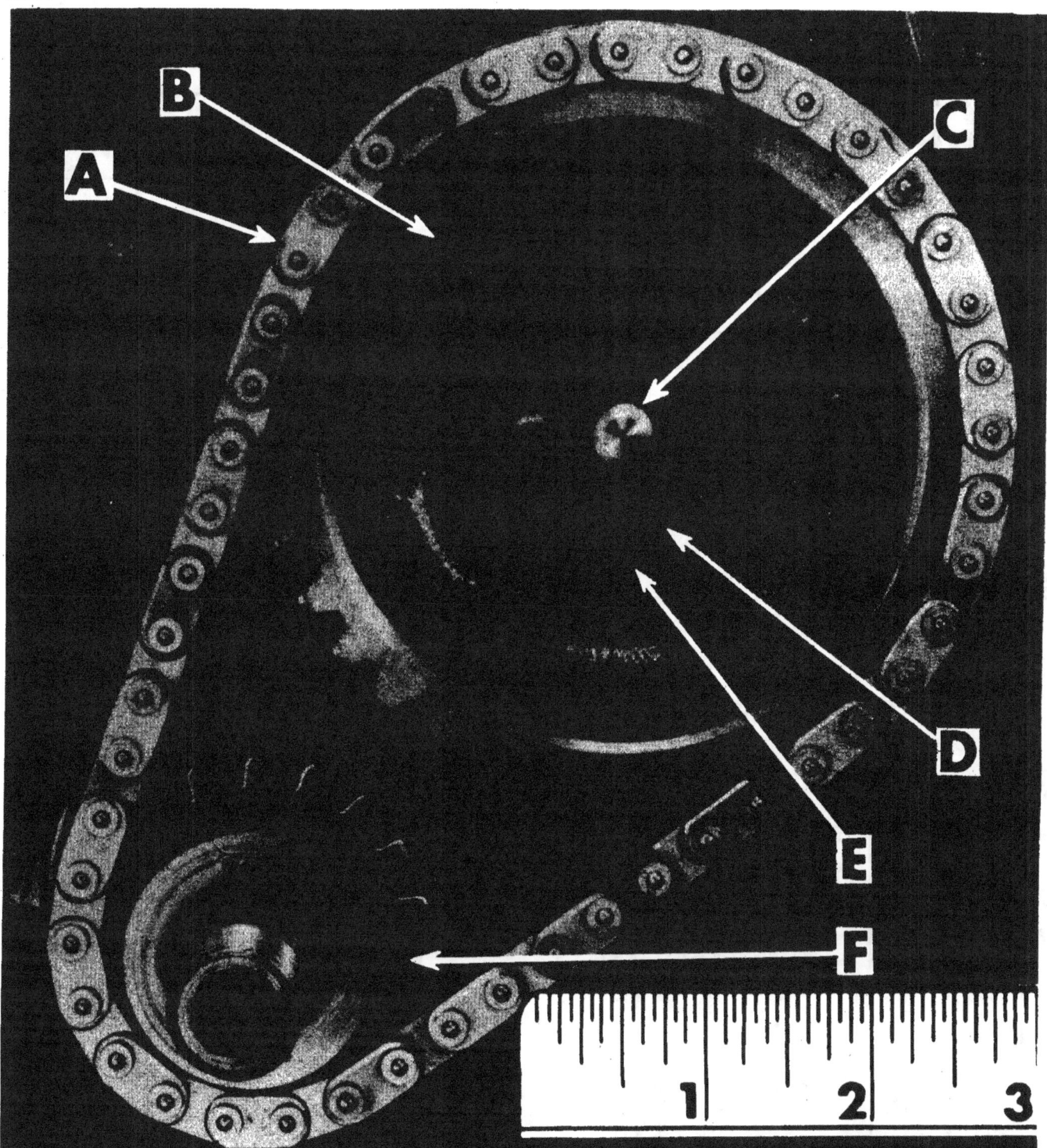

FIGURE 01-13—CRANKSHAFT CHAIN AND SPROCKETS

Key	Item	Willys Part No.	Ford Part No.	Gov't Group No.
A	CHAIN	WO-638457	FM-GPW-6260	0106C
B	SPROCKET	WO-638458	FM-GPW-6256	0106C
C	PLUNGER	WO-375907	FM-GPW-6243	0106
D	BOLT	WO-634850	FM-355499-S	0106C
E	WASHER	WO-315932	FM-GPW-6269	0106C
F	SPROCKET	WO-638459	FM-GPW-6306	0106C

RA PD 305156-A

SNL G-503

GROUP 01—ENGINE (Cont'd)

Figure Number	Official Stockage Number	Part Number: Willys	Part Number: Ford	ITEM	Quantity Reqd. per Unit Assy.	Per 100 Major Items: 12 Mos. Field Maintenance	Per 100 Major Items: Major Overhaul (5th Ech)	Per 100 Major Items: Total FirstYear Procurement	Estimated Reqmts. per 100 Rebuilds
Col. 1	Col. 2	Col. 3	Col. 4	Col. 5	Col. 6	Col. 7	Col. 8	Col. 9	Col. 10
				0110—MOUNTINGS (Cont'd)					
		WO-5009	FM-34809-S2	WASHER, lock, reg., S., 1/2 in. (7/8 O.D. x 17/32 I.D. x 1/8 thk.) (BECX1M)	2	—	—	—	—
		WO-A-6156	FM-GPW-6040-B	INSULATOR, engine support, rear	1	% 2	—	10	—
		WO-5901	FM-33800-S2	NUT, reg., hex., s-fin., alloy-S., 3/8-24NF-2 (insulator to bracket and cross member) (BBBX1CA) (GM-103026)	4	—	—	—	—
		WO-52101	FM-34747-S2	WASHER, plain, S., S.A.E. std., 3/8 in. (13/16 O.D. x 13/32 I.D. x 1/16 thk.) (BEBX1K)	2	—	—	—	—
		WO-5010	FM-34807-S2	WASHER, lock, reg., S., 3/8 in. (13/32 I.D. x 21/32 O.D. x 3/32 thk.) (BECX1K)	4	—	—	—	—
01-12		WO-A-1463	FM-GPW-6031-A3	PLATE, engine, front, assembly (mounts on cylinder block studs)	1	3	3	8	10
01-12		WO-A-5121	FM-GPW-7007	PLATE, engine, rear, assembly	1	—	2	4	5
		WO-A-1051	FM-GPW-6098	SHIM, engine support insulator (issue until stock exhausted)	as req.	4	4	—	—
				NOTE: For engine front support bracket see Group 1500.					
				GROUP 02—CLUTCH					
				0201A—FACINGS AND RIVETS					
02-2		WO-636778	FM-GPW-7577	FACING, engine clutch, rear (B-B-4324)	1	—	—	—	—
02-2		WO-371567	FM-GPW-7549	FACING, engine clutch, front (B-B-2940)	1	—	—	—	—
02-2		WO-374586	FM-351915-S	RIVET, tubular, S., fl-ck-hd., .143 O.D. x 1/4 lgth. (BMGX1)	12	—	—	—	—
		WO-A-6751	FM-GPW-18358	FACING SET, engine clutch disc (Includes: 1 WO-636778 FM-GPW-7577 FACING; 1 WO-371567 FM-GPW-7549 FACING; 15 WO-374586 FM-351926 RIVET)	as req.	6	13	22	40
				0201B—CLUTCH DRIVEN PLATE					
		WO-638992	FM-GPW-7563	PLATE, pressure, engine clutch, assembly (AVM-TP-28-7-1) Includes: LEVERS, PLATES, SPRINGS)	1	13	3	20	10
		WO-630129	FM-355476-S	BOLT, clutch to flywheel (special, hex-hd., S., 5/16-18NC-2 x 1 1/16)	6	24	24	60	75
		WO-5051	FM-34836-S7	WASHER, lock, hv., S., 5/16 in. (.328 I.D. x .578 O.D. x 3/32 thk.) (BECX3H)	6				
02-2		WO-636755	FM-GPW-7550	PLATE, driven, engine clutch, w/HUB, assembly (B-B-11123)	1	19	6	30	20
				0202—COVER, PRESSURE PLATES, AND SPRINGS					
02-2		WO-638151	FM-GPW-7580	COVER, engine clutch (AVM-TP-2811)	1	—	—	—	—
02-2		WO-638157	FM-GPW-7567	CUP, engine clutch pressure spring (AVM-TP-283)	3	—	—	—	—

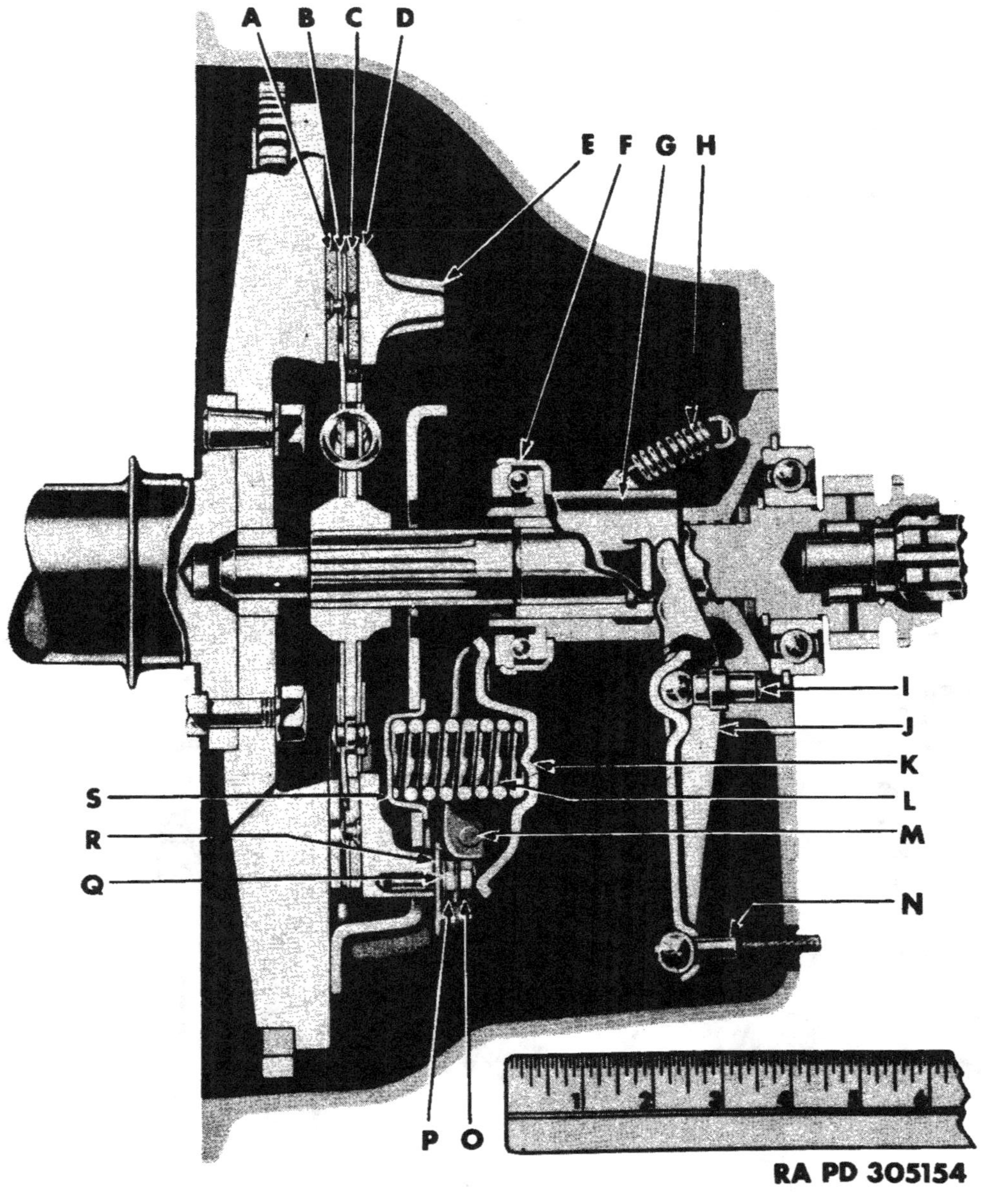

FIGURE 02-1—CLUTCH ASSEMBLY—SECTIONAL VIEW

Key	Item	Willys Part No.	Ford Part No.	Gov't Group No.
A	FACING	WO-371567	FM-GPW-7549	0201A
B	PLATE	WO-636755	FM-GPW-7550	0201B
C	FACING	WO-636778	FM-GPW-7577	0201A
D	RIVET	WO-374586	FM-351915-S	0201A
E	PLATE	WO-638152	FM-GPW-7566	0202
F	BEARING	WO-635529	FM-GPW-7580-B	0203
G	CARRIER	WO-639654	FM-GPW-7561	0203
H	SPRING	WO-630117	FM-GPW-7562	0203
I	FULCRUM	WO-630068	FM-GPW-7516	0204A
J	LEVER	WO-630112	FM-GPW-7515	0203
K	LEVER	WO-638158	FM-GPW-7591	0202
L	SPRING	WO-638993	FM-GPW-7572	0202
M	PIN	WO-638159	FM-GPW-7564	0202
N	CABLE	WO-A-5102	FM-GPW-7530	0203
O	SCREW	WO-638154	FM-24325-S	0202
P	NUT	WO-638155	FM-33921-S7	0202
Q	WASHER	WO-638305	FM-34745-S2	0202
R	SPRING	WO-638153	FM-GPW-7590	0202
S	CUP	WO-638157	FM-GPW-7567	0202

RA PD 305154-A

FIGURE 02-2—CLUTCH ASSEMBLY

Key	Item	Willys Part No.	Ford Part No.	Gov't Group No.
A	PIN	WO-638159	FM-GPW-7564	0202
B	LEVER	WO-638158	FM-GPW-7591	0202
C	SPRING	WO-638153	FM-GPW-7590	0202
D	SPRING	WO-638993	FM-GPW-7572	0202
E	CUP	WO-68157	FM-GPW-7567	0202
F	COVER	WO-638151	FM-GPW-7570	0202
G	PLATE	WO-638152	FM-GPW-7566	0202
H	FACING	WO-371567	FM-GPW-7549	0201A
I	PLATE w/HUB and SPRINGS	WO-636775	FM-GPW-7550	0201B
J	FACING	WO-636778	FM-GPW-7577	0201A
K	WASHER	WO-638305	FM-34745-S2	0202
L	NUT	WO-638155	FM-33921-S7	0202
M	SCREW	WO-638154	FM-24325-S	0202
N	RIVET	WO-374586	FM-351926	0201A

RA PD 305065-A

GROUP 02—CLUTCH (Cont'd)

Figure Number	Official Stockage Number	Part Number: Willys	Part Number: Ford	ITEM	Quantity Reqd. per Unit Assy.	Per 100 Major Items: 12 Mos. Field Maintenance	Per 100 Major Items: Major Overhaul (5th Ech)	Per 100 Major Items: Total First Year Procurement	Estimated Reqmts. per 100 Rebuilds
Col. 1	Col. 2	Col. 3	Col. 4	Col. 5	Col. 6	Col. 7	Col. 8	Col. 9	Col. 10
				0202—COVER, PRESSURE PLATES, AND SPRINGS (Cont'd)					
		WO-A-6752	FM-GPW-18359	KIT, repair, engine clutch cover (issue until stock is exhausted, then use WO-A-7833)	as req.	6	13	22	
		WO-A-7833	FM-GPW-18359-B	KIT, repair, engine clutch cover (was WO-A-6752) (Includes: 3 WO-638157 FM-GPW-7567 CUP 3 WO-638158 FM-GPW-7591 LEVER 3 WO-638155 FM-33921-S7 NUT 3 WO-638159 FM-GPW-7565 PIN 3 WO-638154 FM-24325-S SCREW 3 WO-638993 FM-GPW-7572 SPRING 3 WO-638153 FM-GPW-7590 SPRING 3 WO-638305 FM-34745-S2 WASHER)	as req.	3	6	12	20
02-2		WO-638158	FM-GPW-7591	LEVER, release, engine clutch (AVM-TP-2820)	3	—	—	—	—
02-1 02-2		WO-638155	FM-33921-S7	NUT, lock, engine clutch adjusting screw (hex., s-fin., S., thin $\frac{1}{4}$-28NF-2) (BBBX1B)	3	—	—	—	—
02-2		WO-638152	FM-GPW-7566	PLATE, pressure, engine clutch (AVM-TP-2851)	1	3	6	12	20
02-2		WO-638159	FM-GPW-7565	PIN, engine clutch release lever pivot (AVM-TP-287)	3	—	—	—	—
02-1 02-2		WO-638154	FM-24325-S	SCREW, adjusting, engine clutch (hex-hd., s-fin., alloy-S., $\frac{1}{4}$-28NF-2 x $\frac{5}{8}$) (AVM-TP-2818) (BAOX4AB)	3	—	—	—	—
02-2		WO-638993	FM-GPW-7572	SPRING, pressure, engine clutch (spring-steel, 1 in. O.D. x 1$\frac{9}{16}$ in. lgth.) (AVM-TP-2831)	3	—	—	—	—
02-2		WO-638153	FM-GPW-7590	SPRING, engine clutch pressure plate, return (AVM-TP-2817)	3	—	—	—	—
02-1 02-2		WO-638305	FM-34745-S2	WASHER, engine clutch adjusting screw, (S., flat, $\frac{5}{8}$ O.D. x $\frac{9}{32}$ I.D. x .050 thk.) (AVM-TP-2827)	3	—	—	—	—
				0203—RELEASE YOKE, BEARING FORK					
02-1, 3		WO-635529	FM-GPW-7580-B	BEARING, engine clutch release (AB-A-935-1)	1	10	10	23	30
02-1, 3		WO-A-5102	FM-GPW-7530	CABLE, engine clutch control lever	1	6	—	8	—
02-3		WO-639654	FM-GPW-7561	CARRIER, engine clutch release bearing	1	2	2	9	5
02-1 02-1		WO-630068	FM-GPW-7516	FULCRUM, engine clutch release lever (TRU-SA-2844-1)	1	3	—	4	—
02-3		WO-630112	FM-GPW-7515	LEVER, clutch control	1	% 5	2	9	—
		WO-5910	FM-33798-S	NUT, lock, engine clutch control lever yoke end (S., hex., $\frac{5}{16}$-24NF-2) (GM-103025) (BBBX1BA)	1	—	—	—	—
02-3		WO-339043	FM-73880-S	PIN, clevis, engine clutch yoke end to control cable	1	% 8	—	10	—
02-1		WO-630117	FM-GPW-7562	SPRING, engine clutch release bearing carrier	1	% 19	—	24	—
02-3		WO-632177	FM-GPW-7532	YOKE, adjusting, engine (clutch control lever cable)	1	% 6	—	12	—

				0204—PEDAL					
02-3		WO-638792	FM-353043-A-S7	FITTING, grease, straight, 1/4-28NF tapered thd. (AD-1641)	1	—	—	—	—
			FM-GPW-2476-A	GROMMET, engine clutch pedal shaft (Ford only)	2	—	—	—	—
02-3		WO-5036	FM-74178-S	KEY, Woodruff, No. 9 (pedal to shaft)	1	—	—	—	—
02-3		WO-A-1360	FM-GPW-7525	PAD, engine clutch pedal, assembly (issue until stock exhausted)	1	—	4	—	—
		WO-A-6359		PAD, engine clutch pedal shank (Willys only) (issue until stock exhausted)	2	% 4	8	—	—
02-3		WO-A-405	FM-GPW-7250	PEDAL, engine clutch (issue until stock exhausted)	1	—	4	—	—
02-3		WO-50992	FM-24505-S7	BOLT, hex-hd., s-fin., alloy-S., 5/16-24NF-2 x 1 3/4 (pedal clamp) (BAOX1BF)	1	—	—	—	—
02-3		WO-6157	FM-24426	BOLT, hex-hd., s-fin., alloy-S., 5/16-18NC-2 x 1 1/4 (shank to pedal) (BANX1BD)	2	—	—	—	—
02-3		WO-5910	FM-33798-S2	NUT, reg., hex., s-fin., S., 5/16-24NF-2 (GM-103025) (BBBX1BA)	1	—	—	—	—
02-3		WO-51833	FM-34806-S2	WASHER, lock, reg., S., 5/16 in. (19/32 O.D. x 11/32 I.D. x 1/16 thk.) (BECX1H)	1	—	—	—	—
02-3		WO-52944	FM-72063-S	PIN, cotter, S., 5/32 x 1 3/8 (shaft to pedals) (issue until stock exhausted)	3	% 20	40	—	—
02-3		WO-A-495	FM-GPW-2473	SHAFT, engine clutch pedal, assembly (issue until stock exhausted)	1	—	4	—	—
		WO-650684		SPRING, engine clutch pedal shank draft pad (escutcheon, black enamel, free lgth. 3 1/2 coils) (Willys only) (issue until stock exhausted)	2	% 4	8	—	—
02-3		WO-630593	FM-GPW-7523	SPRING, retracting, engine clutch pedal (free length 5 7/8 in., 40 coils)	2	% 19	30	—	—
02-3		WO-A-498	FM-356561-S	WASHER, engine clutch pedal shaft (special, S., 1 1/2 O.D. x 1 1/32 I.D. x 1/8 thk.)	3	% 10	40	—	—
		WO-A-6360		WASHER, engine clutch pedal shank pad (S., 1 3/4 O.D. x 17/32 I.D.) (Willys only) (issue until stock exhausted)	2	% 4	8	—	—
				0204A—PEDAL LINKAGE					
02-3		WO-A-180	FM-GPW-7508	BRACKET, frame, engine clutch control (included w/ball STUD assembly)	1	—	—	—	—
02-3		WO-52132	FM-24327-S2	BOLT, hex-hd., s-fin., alloy-S., 5/16-24NF-2 x 5/8 (bracket to frame) (GM-106279)	2	—	—	—	—
		WO-5910	FM-33798-S2	NUT, S., hex., 5/16-24NF-2 (GM-103025) (BBBX1BA)	2	—	—	—	—
		WO-51833	FM-34806-S2	WASHER, lock, S., 5/16 in. (19/32 O.D. x 11/32 I.D. x 1/16 thk.) (BECX1H)	2	—	—	—	—
02-3		WO-A-1355	FM-GPW-7503	LEVER, engine clutch control, w/TUBE, assembly	1	% 2	—	4	—
		WO-5354	FM-72004-S	PIN, cotter, S., 3/8 x 1/2 (engine clutch pedal rod)	2	—	—	—	—
02-3		WO-5108	FM-72053	PIN, cotter, S., 1/8 x 1 1/4 (engine clutch control tube) (BFAX1DH)	1	—	—	—	—
02-3		WO-A-176	FM-GPW-7539	PAD, engine clutch control tube (felt, 5/8 O.D. x 1/4 thk.) (issue until stock exhausted)	2	% 52	108	—	—
02-3		WO-A-177	FM-GPW-7517	RETAINER, engine clutch control tube spring (S., 19/32 O.D. x 3/32 thk.) (issue until stock exhausted)	2	% 8	16	—	—
02-3		WO-A-499	FM-GPW-7521	ROD, engine clutch release pedal	1	% 3	4	—	—
02-3		WO-A-887	FM-GPW-7512	SEAL, engine clutch control tube dust (1 in. O.D.)	2	% 6	—	12	—
02-3		WO-A-178	FM-GPW-7545	SPRING, engine clutch control tube (S., 9/16 O.D. x 1 3/4 free lgth.)	1	% 6	—	12	—
02-3		WO-A-181	FM-GPW-7514	STUD, engine clutch control, ball (S., 7/16-14NC)	1	% 3	—	6	—
		WO-A-179	FM-GPW-7507	STUD, engine clutch control, ball, w/BRACKET, assembly	1	—	4	—	—
02-3		WO-5336	FM-33801-S2	NUT, reg., hex., s-fin., S., 7/16-14NC-2 (stud to bracket) (BBAX1D)	1	% 3	—	50	—
02-3		WO-5059	FM-34808-S2	WASHER, lock, S., 7/16 in. (7/8 O.D. x 17/32 I.D. x 1/8 thk.) (BECX1L)	1	—	—	—	—
				GROUP 03—FUEL					
				0300—FUEL TANKS					
		WO-A-1738	FM-GPW-9069	ANTI-SQUEAK, fuel tank (1 x 6 in. long)	2	—	—	—	—
		WO-A-1739	FM-GPW-9079	ANTI-SQUEAK, fuel tank (1 x 7 1/4 in. long)	2	—	—	—	—

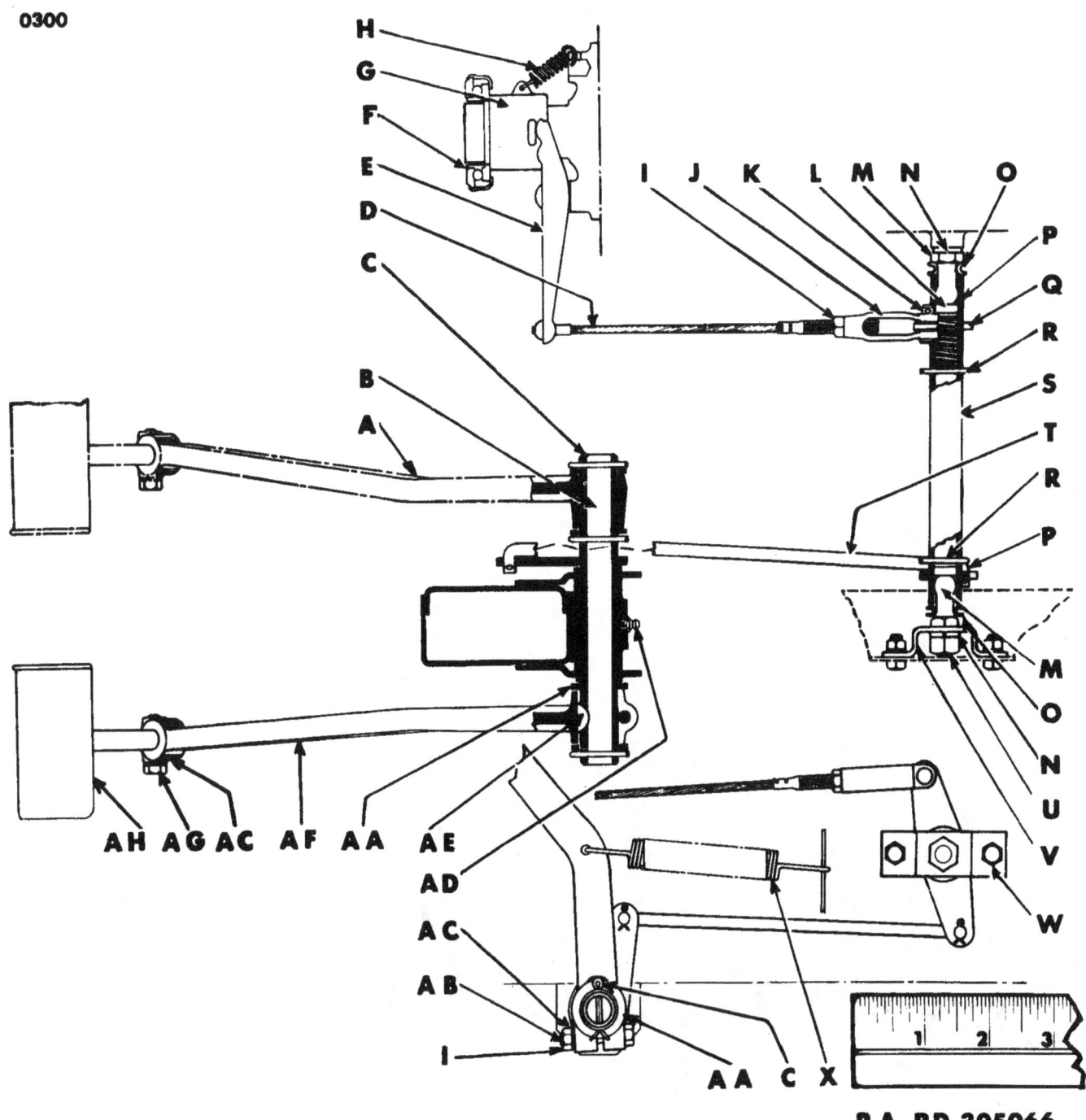

FIGURE 02-3—CLUTCH CONTROLS

Key	Item	Willys Part No.	Ford Part No.	Gov't Group No.		Key	Item	Willys Part No.	Ford Part No.	Gov't Group No.
A	PEDAL	WO-A-8253	FM-GPW-2452-B	1204		S	LEVER			
B	SHAFT	WO-A-495	FM-GPW-2473	0204			assembly	WO-A-1355	FM-GPW-7503	0204A
C	PIN	WO-52944	FM-72063-S	0204		T	ROD	WO-A-499	FM-GPW-7521	0204A
D	CABLE	WO-A-5102	FM-GPW-7530	0203		U	NUT	WO-5336	FM-33801-S2	0204A
E	LEVER	WO-630112	FM-GPW-7512	0203		V	BRACKET	WO-A-180	FM-GPW-7508	0204A
F	BEARING	WO-635529	FM-GPW-7580-B	0203					(See WO-A-179)	
G	CARRIER	WO-639654	FM-GPW-7561	0203		W	SCREW	WO-52132	FM-24327-S2	0204A
H	SPRING	WO-630117	FM-GPW-7562	0203		X	SPRING	WO-630593	FM-GPW-7523	0204
I	NUT	WO-5910	FM-33798-S2	0204		Y	WASHER	WO-A-498	FM-356561	0204
J	YOKE	WO-632177	FM-GPW-7532	0203		Z	BOLT	WO-50992	FM-24505-S7	0204
K	PIN	WO-339043	FM-73880-S	0203		AA	WASHER	WO-51833	FM-34806-S2	0204
L	RETAINER	WO-A-177	FM-GPW-7517	0204A		AB	GREASE			
M	STUD	WO-A-181	FM-GPW-7514	0204A			FITTING	WO-638792	FM-353043-A-S7	0204
N	WASHER	WO-5059	FM-34808-S2	0204A		AC	KEY	WO-5036	FM-74178-S	0204
O	SEAL	WO-A-887	FM-GPW-7512	0204A		AD	PEDAL	WO-A-405	FM-GPW-7250	0204
P	PAD	WO-A-176	FM-GPW-7539	0204A		AE	BOLT	WO-6157	FM-24426	0204
Q	SPRING	WO-A-178	FM-GPW-7545	0204A		AF	PAD	WO-A-1360	FM-GPW-7525	0204
R	PIN	WO-5108	FM-72053	0204A						

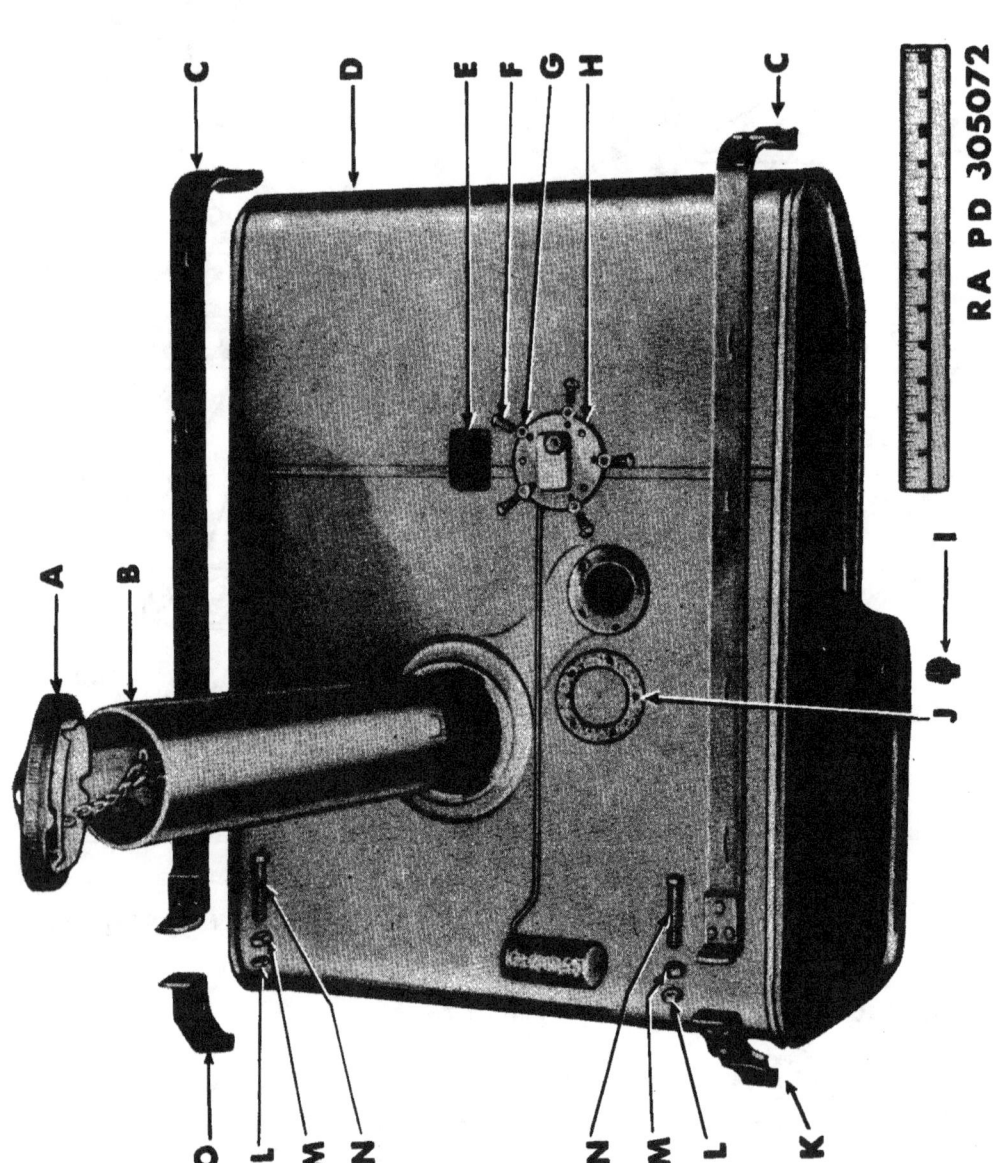

FIGURE 03-1—FUEL TANK ASSEMBLY

Key	Item	Willys Part No.	Ford Part No.	Gov't Group No.
A	CAP assembly	WO-A-6333	FM-GPW-9030-B	0300
B	EXTENSION	WO-A-6424	FM-GPW-9034-A	0300
C	STRAP assembly	WO-A-1472	FM-GPW-9095	0300
D	TANK assembly	WO-A-6618	FM-GPW-9002-B	0300
E	PROTECTOR	WO-A-1763	FM-355095-S7 (Willys only)	0300A
F	SCREW	WO-5556		0300A
G	WASHER	WO-52705		0300A
H	GAUGE assembly	WO-A-1292	FM-GPW-9275	0300A
I	STRAP assembly	WO-A-1472	FM-GPW-9095	0300
J	PLUG	WO-A-5120	FM-355055-S7	0300
K	STRAP assembly	WO-A-1477	FM-GPW-9057	0300
L	NUT	WO-52891	(Willys only)	0300
M	NUT	WO-5910	FM-33798-S2	0300
N	BOLT	WO-A-1483	FM-24505-S7	0300
O	GASKET	WO-A-1293	FM-GPW-9276	0300A
P	CLAMP assembly	WO-A-1481	FM-GPW-9066	0300

SNL G-503

GROUP 03—FUEL (Cont'd)

0300—FUEL TANKS (Cont'd)

Figure Number	Official Stockage Number	Part Number Willys	Part Number Ford	ITEM	Quantity Reqd. per Unit Assy.	12 Mos. Field Maintenance	Major Overhaul (5th Ech)	Total First Year Procurement	Estimated Reqmts. per 100 Rebuilds
Col. 1	Col. 2	Col. 3	Col. 4	Col. 5	Col. 6	Col. 7	Col. 8	Col. 9	Col. 10
		WO-A-1740	FM-GPW-9075	ANTI-SQUEAK, fuel tank (1 x 9½ in. long)	2	—	—	—	—
		WO-A-1741	FM-GPW-9071	ANTI-SQUEAK, fuel tank (1 x 18 in. long)	2	—	—	—	—
		WO-A-1480	FM-GPW-9078	ANTI-SQUEAK, fuel tank hold down strap	2	—	—	—	—
		WO-A-1476	FM-GPW-9074	ANTI-SQUEAK, fuel tank hold down strap, inner	2	—	—	—	—
		WO-A-2954	FM-GPW-9063	BRACKET, fuel tank hold down	2	—	—	—	—
		WO-51977	FM-62218-S	RIVET, rd-hd, S., ¼ x ⁷⁄₁₆ (BMCX1)	4	—	—	—	—
		WO-A-2953	FM-GPW-9065	BRACKET, fuel tank hold down, front inner	1	—	—	—	—
		WO-52168	FM-20344-S2	BOLT, hex-hd., s-fin., alloy-S., pkzd., ¼-20NC-2 x ⅝ (to front floor pan) (GM-121668) (BANX4AE-15)	2	—	—	—	—
		WO-52217	FM-33795-S2	NUT, reg, hex, S., pkzd., ¼-20NC-2 (GM-118613) (BBAX1A-15)	2	—	—	—	—
		WO-52031	FM-34805-S2	WASHER, reg, lock, S., pkzd, ¼ in. (BECX1G-15)	2	—	—	—	—
		WO-A-2952	FM-GPW-9062	BRACKET, fuel tank hold down, rear inner	1	—	—	—	—
		WO-A-2970	FM-GPW-9051	BRACKET, fuel tank hold down, rear inside, w/SUPPORT, assembly	1	—	—	—	—
		WO-A-3055	FM-GPW-1111322	CAP, fuel tank well drain (issue until stock exhausted)	2	% 4	8	—	—
		WO-A-1254	FM-GPW-9030-A	CAP, fuel tank, w/safety CHAIN and BAR (2⅞ in. O.D.) (before Willys serial 174739) (AC-850042)	1	—	—	—	—
03-1		WO-A-6333	FM-GPW-9030-B	CAP, fuel tank, w/safety CHAIN (4⅝ in. O.D.) (after Willys serial 174739)	1	% 10	—	20	—
03-1		WO-A-1481	FM-GPW-9066	CLAMP, fuel tank hold down, assembly	1	—	—	—	—
03-1		WO-A-1483	FM-24505-S7	BOLT, hex-hd., S., ⁵⁄₁₆-24NF-2 x 1¾ (BCBX1BF)	2	—	—	—	—
03-1		WO-5910	FM-33798-S2	NUT, reg, hex, S., ⁵⁄₁₆-24NF-2 (GM-103025) (BBBX1BA)	2	%	—	—	—
03-1		WO-52891		NUT, hex, S., stamped, ⁵⁄₁₆-24NF-2 (after serial 104433) (Willys only) (GM-147510)	2				
		WO-A-6424	FM-GPW-9034-A	EXTENSION, fuel tank filler tube, assembly (STN-6935-A)	1	—	—	—	—
		WO-A-1275	FM-GPW-9035-B	GASKET, fuel tank cap (before Willys serial 174739) (AC-850024)	1	—	—	—	—
		WO-A-3056	FM-1111323	NECK, fuel tank well drain hole	2	—	—	—	—
		WO-A-5120	FM-353055-S7	PLUG, pipe ¼ (fuel tank drain plug)	1	% 7	—	9	—
03-1		WO-A-2930	FM-GPW-1140474	SHIELD, front seat to rear floor fuel tank (driver)	1	—	—	—	—
		WO-52170	FM-24308-S2	BOLT, hex-hd., s-fin., alloy-S., pkzd., ¼-20NC-2 x ½ (shield to floor) (GM-119860) (BANX4AA-15)	3	—	—	—	—
		WO-52165	FM-34129-S2	NUT, hex, S., pkzd., ¼-20NC-2 (GM-118643)	3	—	—	—	—
		WO-52702	FM-34745-S	WASHER, plain, S., pkzd., ¼ in.	3	—	—	—	—
		WO-52031	FM-34805-S2	WASHER, reg, lock, S., pkzd., ¼ in. (BECX1G-15)	2	% 4	8	—	—
03-1		WO-A-1472	FM-GPW-9095	STRAP, hold down, fuel tank, inner, assembly (issue until stock exhausted)	2				
03-1		WO-A-1477	FM-GPW-9057	STRAP, hold down, fuel tank outer, assembly	1	—	—	—	—
		WO-A-1221	FM-GPW-9002-A	TANK, fuel, assembly (before Willys serial 174739) (use WO-A-6897 FM-GPW-9001)	1	—	—	—	—
03-1		WO-A-6618	FM-GPW-9002-B	TANK, fuel, assembly (after Willys serial 174739)	1	—	—	—	—

SNL G-503

		WO-A-6897	FM-GPW-9001	TANK, fuel, w/EXTENSION and CAP, assembly (as required to replace tanks on trucks built prior to Willys serial 174739) (issue until stock exhausted)	1	% 4	4	—	—
		WO-A-3497	FM-GPW-1111324	WELL, fuel tank, assembly	1	—	—	—	—

0300A—TANK GAUGE UNIT

03-1		WO-A-1293	FM-GPW-9276	GASKET, fuel tank gauge (cork) ($2\frac{5}{8}$ O.D. x $1\frac{5}{8}$ I.D. x $\frac{1}{8}$ thk.)	1	% 19	30	—	—
03-1		WO-A-1292	FM-GPW-9275	GAUGE, fuel, tank unit, assembly (AL-9979-A)	1	% 10	14	—	—
03-1		WO-5556	FM-355095-S7	SCREW, machine, rd-hd., S., No. 8 (.164 in.)-32NC-2 x $\frac{3}{8}$ (gauge to tank) (BCNX1FE)	5	—	—	—	—
03-1		WO-52705		WASHER, lock, S., No. 8 (.164 in.) (.2627 O.D. x .169 I.D. x $\frac{3}{64}$ thk.) (BECX1D) (Willys only)	5	—	—	—	—
03-1		WO-A-1763	FM-GPW-9211	PROTECTOR, fuel gauge, tank unit	1	% 3	4	—	—
		WO-5253		SCREW, machine, rd-hd., S., No. 8 (.164)-32NC-2 x $\frac{1}{4}$ (fuel gauge tank unit terminal) (BCNX1FC) (Willys only)	1	—	—	—	—
		WO-52705		WASHER, lock, S., No. 8 (.164 in.) (.2627 O.D. x .169 I.D. x $\frac{3}{64}$ thk.) (fuel gauge tank unit terminal screw) (BECX1D) (Willys only)	1	—	—	—	—

0301—CARBURETOR

03-2		WO-116181	FM-GPW-9528	ARM, carburetor pump, w/COLLAR, assembly (CAR-53A-168S)	1	% 3	3	8	10
03-2		WO-116197	FM-GPW-9583	ARM, carburetor throttle shaft, w/SCREW (CAR-114-21S)	1	% 2	2	8	5
		WO-116543	FM-GPW-6333-D	BRACKET, carburetor choke lever, assembly (before 4000 Willys trucks) (CAR-62-108S)	1	—	—	—	—
		WO-301232	FM-31037-S7	SCREW, choke lever bracket (CAR-101-6)	1	2	2	8	8
G		WO-A-1223	FM-GPW-9510	CARBURETOR assembly (CAR-539S)	1	% 6	3	13	10
03-2		WO-116204	FM-GPW-9594	CHECK, carburetor discharge disc, assembly (CAR-122-47S)	1	—	—	—	—
03-2		WO-116205	FM-GPW-9576	CHECK, carburetor intake ball, assembly (CAR-122-64S)	1	—	—	—	—
		WO-116548		CLAMP, carburetor tube, assembly (before Willys serial 103468) (CAR-62-131S) (Willys only)	1	—	—	—	—
03-2		WO-116587	FM-GPW-9595	CLAMP, carburetor tube, assembly (after Willys serial 103468) (CAR-62-134) (issue until stock exhausted)	1	—	4	—	—
03-2		WO-116208	FM-GPW-9515	COVER, carburetor bowl, w/PIN, assembly (CAR-146-95S)	1	2	2	8	5
03-2		WO-116213	FM-31061-S8	SCREW, carburetor bowl cover attaching (CAR-101-82)	4	—	—	—	—
03-2		WO-6330	FM-34802-S2	WASHER, lock, S., .294 O.D. x .169 I.D. x $\frac{3}{64}$ thk (carburetor bowl cover screw) (CAR-86-10)	4	—	—	—	—
03-2		WO-116206	FM-GPW-9905	DISC, carburetor metering rod (CAR-129-15)	1	10	10	24	30
03-2		WO-116584	FM-GPW-9518	FLANGE, carburetor body, assembly (CAR-1-407)	1	—	—	—	—
03-2		WO-116215	FM-31662-S	SCREW, carburetor body flange attaching (CAR-101-122)	1	—	—	—	—
03-2		WO-5045	FM-34805-S2	WASHER, lock, S., $\frac{15}{32}$ O.D. x $\frac{9}{32}$ I.D. x $\frac{1}{16}$ thk., (carburetor body flange screw) (CAR-86-11)	1	—	—	—	—
03-2		WO-116172	FM-GPW-9550	FLOAT, carburetor, w/LEVER, assembly (CAR-21-74S)	2	2	2	6	7
03-2		WO-116202	FM-GPW-9516	GASKET, carburetor body flange (CAR-121-56)	1	—	—	—	—
03-2		WO-116203	FM-GPW-9519	GASKET, carburetor bowl cover (CAR-121-73)	1	—	—	—	—
03-2		WO-116169	FM-GPW-9608	GASKET, carburetor metering rod jet and plug (CAR-20-26)	3	—	—	—	—
03-2		WO-116168	FM-GPW-9569	GASKET, carburetor needle seat and plug (CAR-20-22)	3	—	—	—	—
03-2		WO-116171	FM-GPW-9926	GASKET, carburetor nozzle (CAR-20-72)	1	—	—	—	—
03-2		WO-116170	FM-GPW-9574	GASKET, carburetor strainer plug (CAR-20-61)	1	—	—	—	—
		WO-A-6837	FM-GPW-18352	GASKET SET, carburetor	as req.	29	—	45	—

0301

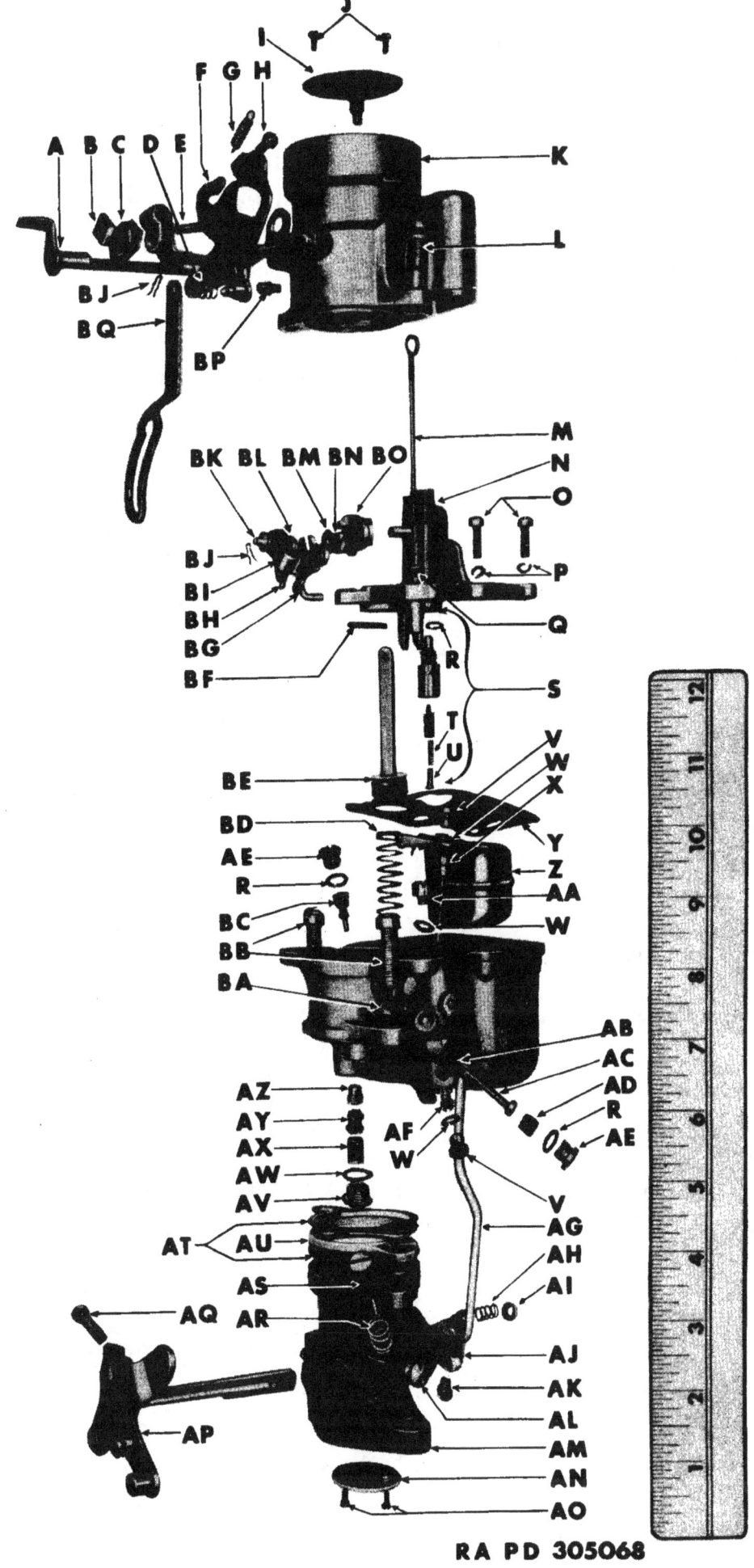

RA PD 305068

SNL G-503

56

FIGURE 03-2—CARBURETOR

Key	Item	Willys Part No.	Ford Part No.	Gov't Group No.
A	SHAFT assembly	WO-116545	FM-GPW-9546	0301
B	NUT	WO-116219	FM-355858-S	0301
C	CLAMP	WO-116587	FM-GPW-9595	0301
D	SPRING	WO-116189	FM-GPW-9587	0301
E	SCREW	WO-116588	FM-355152-S	0301
F	LEVER assembly	WO-116586	FM-GPW-9526	0301
G	SPRING	WO-116184	FM-GPW-9624	0301
H	SCREW	WO-116211	FM-31032-S7	0301
I	VALVE assembly	WO-116157	FM-GPW-9549	0301
J	SCREW	WO-116216	FM-GPW-9586	0301
K	HORN assembly	WO-116544	FM-GPW-9520	0301
L	SCREW	WO-116385	FM-355200-S7	0301
M	ROD	WO-116540	FM-GPW-9906	0301
N	COVER	WO-116208	FM-GPW-9515	0301
O	SCREW	WO-116213	FM-31061-S8	0301
P	WASHER	WO-6630	FM-34802-S2	0301
Q	DISC	WO-116206	FM-GPW-9905	0301
R	GASKET	WO-116168	FM-GPW-9569	0301
S	GASKET	WO-116174	FM-GPW-9567	0301
T	SPRING	WO-116191	FM-GPW-9935	0301
U	PIN	WO-116177	FM-GPW-9566	0301
V	PASSAGE	WO-116165	FM-GPW-9543	0301
W	GASKET	WO-116169	FM-GPW-9608	0301
X	JET	WO-116539	FM-GPW-9533	0301
Y	GASKET	WO-116203	FM-GPW-9519	0301
Z	FLOAT assembly	WO-116172	FM-GPW-9550	0301
AA	JET w/GASKET	WO-116541	FM-GPW-9914	0301
AB	GASKET	WO-116171	FM-GPW-9926	0301
AC	NOZZLE	WO-116166	FM-GPW-9922	0301
AD	PLUG	WO-116161	FM-GPW-9562	0301
AE	PLUG w/GASKET	WO-116164	FM-GPW-9928	0301
AF	JET	WO-116179	FM-GPW-9544	0301
AG	ROD	WO-116198	FM-GPW-9531	0301
AH	SPRING	WO-116185	FM-GPW-9615	0301
AI	RETAINER	WO-116194	FM-GPW-9614	0301
AJ	ARM assembly	WO-116197	FM-GPW-9583	0301
AK	SCREW	WO-52290	FM-31588-S7	0301
AL	PLUG	WO-116162	FM-GPW-9579	0301
AM	FLANGE assembly	WO-116584	FM-GPW-9518	0301
AN	VALVE	WO-116154	FM-GPW-9585	0301
AO	SCREW	WO-116217	FM-GPW-9588	0301
AP	SHAFT assembly	WO-116585	FM-GPW-9581	0301
AQ	SCREW	WO-116651	FM-GPW-9610	0301
AR	SPRING	WO-116183	FM-GPW-9578	0301
AS	SCREW	WO-116176	FM-GPW-9541	0301
AT	GASKET	WO-116202	FM-GPW-9516	0301
AU	INSULATOR	WO-116210	FM-GPW-9554	0301
AV	PLUG, w/GASKET	WO-116163	FM-GPW-9696	0301
AW	GASKET	WO-116170	FM-GPW-9574	0301
AX	STRAINER	WO-116175	FM-GPW-9575	0301
AY	CHECK assembly	WO-116205	FM-GPW-9576	0301
AZ	CHECK assembly	WO-116204	FM-GPW-9594	0301
BA	WASHER	WO-5045	FM-34805-S2	0301
BB	SCREW	WO-116215	FM-31662-S	0301
BC	JET	WO-116180	FM-GPW-9940	0301
BD	SPRING	WO-116188	FM-GPW-9636	0301
BE	PLUNGER assembly	WO-116195	FM-GPW-9631	0301
BF	PIN	WO-116173	FM-GPW-9558	0301
BG	LINK	WO-116199	FM-GPW-9527	0301
BH	SPRING	WO-116187	FM-GPW-9570	0301
BI	LEVER assembly	WO-116537	FM-GPW-9529	0301
BJ	SPRING	WO-116178	FM-GPW-9599	0301
BK	PIN	WO-116209	FM-GPW-9930	0301
BL	SPRING	WO-116538	FM-GPW-9907	0301
BM	WASHER	WO-116207	FM-34711-S	0301
BN	NUT	WO-52615	FM-34051-S7	0301
BO	ARM assembly	WO-116181	FM-GPW-9528	0301
BP	SCREW	WO-116384	FM-355067-S7	0301
BQ	LINK	WO-116542	FM-GPW-9598	0301

RA PD 305068-A
SNL G-503

GROUP 03—FUEL (Cont'd)

0301

Figure Number	Official Stockage Number	Part Number Willys	Part Number Ford	ITEM	Quantity Reqd. per Unit Assy.	Per 100 Major Items 12 Mos. Field Maintenance	Per 100 Major Items Major Overhaul (5th Ech)	Total First Year Procurement	Estimated Reqmts. per 100 Rebuilds
Col. 1	Col. 2	Col. 3	Col. 4	Col. 5	Col. 6	Col. 7	Col. 8	Col. 9	Col. 10
				0301—CARBURETOR (Cont'd)					
				(Includes:					
				2 WO-116202 FM-GPW-9516 GASKET, body flange					
				1 WO-116203 FM-GPW-9519 GASKET, bowl cover					
				1 WO-A-6357 FM-GPW-9445 GASKET, diffuser					
				1 WO-116171 FM-GPW-9926 GASKET, nozzle					
				3 WO-116169 FM-GPW-9608 GASKET, jet					
				3 WO-116168 FM-GPW-9569 GASKET, seat and plug					
				1 WO-116170 FM-GPW-9574 GASKET, strainer)					
03-2		WO-116544	FM-GPW-9520	HORN, air, carburetor, assembly (CAR-6-312S)	1	2	2	8	5
03-2		WO-116385	FM-355200-S7	SCREW, carburetor airhorn attaching, w/WASHER (CAR-101-150S)	2	10	10	24	30
03-2		WO-116210	FM-GPW-9554	INSULATOR, carburetor to choke valve (CAR-183-19)	1	2	2	8	5
03-2		WO-116179	FM-GPW-9544	JET, carburetor idle well (CAR-43-67)	1	—	—	—	—
03-2		WO-116539	FM-GPW-9533	JET, carburetor, low speed assembly (CAR-11-180S)	1	—	—	—	—
03-2		WO-116180	FM-GPW-9940	JET, carburetor pump (CAR-48-84)	1	—	—	—	—
03-2		WO-116541	FM-GPW-9914	JET, carburetor metering rod, w/GASKET, assembly (CAR-120-151S)	1	—	—	—	—
		WO-A-6840	FM-GPW-18357-B	KIT, repair, carburetor	as req.	—	—	—	—
				(Includes:					
				1 WO-116197 FM-GPW-9583 ARM					
				1 WO-116204 FM-GPW-9594 CHECK, discharge disc					
				1 WO-116205 FM-GPW-9576 CHECK, intake ball					
				1 WO-116206 FM-GPW-9905 DISC, metering rod					
				2 WO-116202 FM-GPW-9516 GASKET, body flange					
				1 WO-116203 FM-GPW-9519 GASKET, bowl cover					
				1 WO-A-6357 FM-GPW-9445 GASKET, diffuser					
				1 WO-116171 FM-GPW-9926 GASKET, nozzle					
				1 WO-116179 FM-GPW-9544 JET, idle well					
				1 WO-115539 FM-GPW-9533 JET, low speed assembly					
				1 WO-116541 FM-GPW-9914 JET, metering rod					
				1 WO-116180 FM-GPW-9940 JET, pump					
				1 WO-116199 FM-GPW-9527 LINK, pump connector					
				1 WO-116166 FM-GPW-9922 NOZZLE					
				1 WO-52615 FM-34051-S7 NUT, hex					
				1 WO-116209 FM-GPW-9930 PIN					
				4 WO-116178 FM-GPW-9599 PIN					
				1 WO-116161 FM-GPW-9562 PLUG					
				1 WO-116162 FM-GPW-9579 PLUG					
				5 WO-116160 FM-GPW-9523 PLUG					
				1 WO-116159 FM-GPW-9522 PLUG					

03-2			1	WO-116163	FM-GPW-9696 PLUG and GASKET				
03-2			2	WO-116164	FM-GPW-9928 PLUG and GASKET				
03-2			2	WO-116165	FM-GPW-9543 PLUG and GASKET				
03-2			1	WO-116195	FM-GPW-9631 PLUNGER and ROD				
03-2			1	WO-116194	FM-GPW-9614 RETAINER				
03-2			1	WO-116198	FM-GPW-9531 ROD				
03-2			1	WO-116540	FM-GPW-9906 ROD, metering				
03-2			2	WO-116216	FM-GPW-9586 SCREW				
03-2			2	WO-116217	FM-GPW-9588 SCREW				
03-2			2	WO-116213	FM-31061-S8 SCREW				
03-2			1	WO-116215	FM-31662-S SCREW				
03-2			1	WO-116385	FM-355200-S7 SCREW				
03-2			1	WO-116184	FM-GPW-9624 SPRING				
03-2			1	WO-116185	FM-GPW-9615 SPRING				
03-2			1	WO-116187	FM-GPW-9570 SPRING				
03-2			1	WO-116188	FM-GPW-9636 SPRING				
03-2			1	WO-116189	FM-GPW-9587 SPRING				
03-2			1	WO-116538	FM-GPW-9907 SPRING				
03-2			1	WO-116174	FM-GPW-9567 SPRING and SEAT				
03-2			1	WO-116175	FM-GPW-9575 STRAINER				
03-2			2	WO-6330	FM-34802-S2 WASHER				
03-2			1	WO-5045	FM-34805-S7 WASHER				
03-2			1	WO-116207	FM-34711-S WASHER)				
	WO-A-5501				KIT, repair, carburetor throttle shaft and lever (Willys only) (CAR-3-4664)..	as req.			
					(Includes:				
			1	WO-116587	FM-GPW-9595 CLAMP, tube				
			1	WO-116219	FM-355858-S7 NUT, clamp				
			1	WO-116540	FM-GPW-9906 ROD, metering				
			1	WO-116588	FM-355132-S SCREW, clamp				
			2	WO-116217	FM-GPW-9588 SCREW, attaching, valve				
			1	WO-116585	FM-GPW-9581 SHAFT and LEVER, throttle)				
03-2	WO-116537	FM-GPW-9529			LEVER, operating, carburetor pump, assembly (CAR-53A-251S)	1	% 3	3	8	10
03-2	WO-116586	FM-GPW-9526			LEVER, carburetor choke, w/BRACKET assembly (after 4000 trucks) (CAR-62-135-S)	1	% 2	2	8	8
03-2	WO-116545	FM-GPW-9546			LEVER, carburetor choke control, w/SHAFT, assembly (CAR-14-246S) (issue until stock exhausted)	1	% 4	4	—	—
03-2	WO-116542	FM-GPW-9598			LINK, carburetor choke (CAR-117-106)	1	—	—	—	—
03-2	WO-116199	FM-GPW-9527			LINK, connector, carburetor pump (CAR-117-58)	1	4	4	—	—
03-2	WO-116166	FM-GPW-9922			NOZZLE, carburetor (CAR-12-255) (issue until stock exhausted)	1	—	—	—	—
03-2	WO-50922	FM-33800-S2			NUT, carburetor flange stud (CAR-105A-13)	1	—	—	—	—
03-2	WO-52615	FM-34051-S7			NUT, hex. S., (carburetor metering rod pin) (CAR-105A-19)	1	4	4	—	—
03-2	WO-116219	FM-355858-S			NUT, carburetor tube clamp (CAR-105A-8) (issue until stock exhausted)	1	4	8	—	—
03-2	WO-116165	FM-GPW-9543			PASSAGE, carburetor low speed jet and idle well, w/GASKET, assembly (CAR-11B-129S)	1				
03-2	WO-116173	FM-GPW-9558			PIN, carburetor float lever (CAR-24-23)	1	10	5	12	15
03-2	WO-116177	FM-GPW-9566			PIN, carburetor intake needle (CAR-150-98)	1	5	5	12	—
03-2	WO-116209	FM-GPW-9930			PIN, carburetor metering rod (CAR-150-97)	1	—	—	—	—
03-2	WO-116204	FM-GPW-119594			PLUG, carburetor discharge disc check, assembly (CAR-122-47S)	1	—	—	—	—
03-2	WO-116162	FM-GPW-9579			PLUG, carburetor idle port rivet (CAR-11B-108)	1	—	—	—	—
03-2	WO-116161	FM-GPW-9562			PLUG, carburetor nozzle retainer (CAR-11B-105)	1	—	—	—	—

0301

GROUP 03—FUEL (Cont'd)

Figure Number	Official Stockage Number	Part Number Willys	Part Number Ford	ITEM	Quantity Reqd. per Unit Assy.	Per 100 Major Items 12 Mos. Field Maintenance	Per 100 Major Items Major Overhaul (5th Ech)	Total First Year Procurement	Estimated Reqmts. per 100 Rebuilds
Col. 1	Col. 2	Col. 3	Col. 4	Col. 5	Col. 6	Col. 7	Col. 8	Col. 9	Col. 10
				0301—CARBURETOR (Cont'd)					
03-2		WO-116160	FM-GPW-9523	PLUG, carburetor rivet (CAR-11B-79)	1	—	—	—	—
03-2		WO-116159	FM-GPW-9522	PLUG, carburetor rivet (CAR-11B-35)	2	—	—	—	—
03-2		WO-116164	FM-GPW-9928	PLUG, carburetor pump jet and nozzle passage, w/GASKET (CAR-11B-127-S)	2	10	—	12	—
03-2		WO-116163	FM-GPW-9696	PLUG, carburetor strainer passage, w/GASKET, assembly (CAR-11B-125S)	1	5	—	8	—
03-2		WO-116195	FM-GPW-9631	PLUNGER, carburetor pump, w/ROD, assembly (CAR-64-62S)	1	—	—	—	—
03-2		WO-116194	FM-GPW-9614	RETAINER, carburetor spring (CAR-63-35)	1	19	—	24	—
03-2		WO-116540	FM-GPW-9906	ROD, metering carburetor, (CARK75-547)	1	10	—	12	—
03-2		WO-116198	FM-GPW-9531	ROD, connector carburetor throttle (CAR-115-59)	1	—	—	—	—
03-2		WO-116176	FM-GPW-9541	SCREW, adjustment idler carburetor (CAR-30A-39)	1	6	3	12	10
03-2		WO-116651	FM-GPW-9610	SCREW, adjustment carburetor throttle lever (CAR-101-121) (issue until stock exhausted)	1	¾	8	—	—
		WO-52290		SCREW, carburetor throttle shaft arm clamp (CAR-101-28)	1	4	—	—	—
03-2		WO-116218	FM-31588-S7	SCREW, carburetor tube clamp (before Willys serial 103468) (CAR-105-11) (Willys only)	1	—	—	—	—
03-2		WO-116588	FM-355132-S	SCREW, carburetor tube clamp (after Willys serial 103468) (CAR-105-13) (issue until stock exhausted)	1	40	80	20	10
03-2		WO-116211	FM-31032-S7	SCREW, carburetor wire clamp (CAR-101-10)	1	6	3	—	—
03-2		WO-116384	FM-355067-S7	SCREW, fill-hd. No. 8 (.164)-32NC, w/WASHER (carburetor choke valve) (issue until stock exhausted)	1	4	8	—	—
03-2		WO-116545	FM-GPW-9546	SHAFT, carburetor choke control, w/LEVER, assembly	1	—	—	—	—
03-2		WO-116585	FM-GPW-9581	SHAFT, carburetor throttle, w/LEVER, assembly (after Willys serial 103468) (CAR-3-465-S)	1	¾	4	12	12
03-2		WO-116189	FM-GPW-9587	SPRING, carburetor connector link (CAR-61-190)	1	10	—	12	—
03-2		WO-116185	FM-GPW-9615	SPRING, carburetor connector rod (CAR-61-128)	1	10	—	12	—
03-2		WO-116184	FM-GPW-9624	SPRING, carburetor choke pull back (CAR-61-119)	1	3	3	9	10
03-2		WO-116183	FM-GPW-9578	SPRING, carburetor idle adjustment screw (CAR-61-57)	1	—	—	—	—
03-2		WO-116191	FM-GPW-9935	SPRING, carburetor intake needle (CAR-61-207)	1	—	—	—	—
03-2		WO-116538	FM-GPW-9907	SPRING, carburetor metering rod (CAR-61-272)	1	19	—	24	—
03-2		WO-116178	FM-GPW-9599	SPRING, carburetor pin (CAR-150-A-10)	1	—	—	—	—
03-2		WO-116186	FM-GPW-9650	SPRING, carburetor plunger (CAR-61-143)	1	—	—	—	—
03-2		WO-116188	FM-GPW-9636	SPRING, carburetor pump arm (CAR-61-171)	1	6	—	8	—
03-2		WO-116187	FM-GPW-9570	SPRING, carburetor pump (CAR-61-169)	1	8	—	12	—
03-2		WO-116174	FM-GPW-9567	SPRING, carburetor needle, w/SEAT, assembly (CAR-25-94S)	1	5	—	8	—
03-2		WO-116175	FM-GPW-9575	STRAINER, carburetor pump check (CAR-30-20)	1	—	—	—	—
03-2		WO-116157	FM-GPW-9549	VALVE, choker, carburetor, assembly (CAR-7-116S)	1	—	—	—	—
03-2		WO-116216	FM-GPW-9586	SCREW, attaching, carburetor choker valve (CAR-39-10)	2	—	—	—	—
03-2		WO-116154	FM-GPW-9585	VALVE, throttle, carburetor (CAR-2-89)	1	—	—	—	—
03-2		WO-116217	FM-GPW-9588	SCREW, attaching carburetor throttle valve (CAR-39-11)	2	—	—	—	—

SNL G-503

0301-0301C

03-2	WO-116207	FM-34711-S	WASHER, carburetor metering rod pin (CAR-136-39) (issue until stock exhausted)	1	—	—	—	1	—
	WO-5010	FM-34807-S2	WASHER, S, lock, carburetor flange stud ($^{21}/_{32}$ O.D. x $1^3/_{32}$ I.D. x $^3/_{32}$ thk.) (CAR-86-15)	1	8	12	13	—	—
			0301A—CARBURETOR ATTACHING PARTS						
03-5	WO-A-6357	FM-GPW-9445	GASKET, insulator, carburetor and intake manifold diffuser, assembly	1	—	—	—	—	—
	WO-50922	FM-33800-S2	NUT, hex, reg, hex, S, $^3/_8$-24NF-2 (carburetor to manifold stud) (GM-103026) (BBBX1CA)	1	$^{5}/_8$ 10	—	—	—	—
01-11	WO-632159	FM-88057-S7	STUD, S, $^3/_8$-16NC x $1^{11}/_{16}$ x $^3/_8$-24NF (carburetor to manifold) (issue until stock exhausted)	2	$^{3}/_8$ 20	40	—	—	—
			0301B—CHOKE						
	WO-345961	FM-GPW-13434-A2	BUSHING, rubber, carburetor choke control through dash (issue until stock exhausted)	2					
03-5	WO-A-7517	FM-GPW-9775-B	CONTROL, carburetor choke or throttle, assembly	1	$^{5}/_8$ 16	24	—	—	—
	WO-5901	FM-33925-S7	NUT, reg, hex, S, $^3/_8$-24NF-2 (choke control to dash) (GM-103026) (BBBX1CA)	2	$^{5}/_8$ 6	8	—	—	—
	WO-52332	FM-34907-S2	WASHER, lock, S, internal teeth, $^3/_8$ in. ($^{11}/_{16}$ O.D. x $^{23}/_{64}$ I.D. x .035 thk.) (BEAX3C)	2	$^{5}/_8$ —	—	—	—	—
18-1	WO-A-1307	FM-GPW-97303	KNOB, carburetor choke throttle w/PLUNGER and WIRE, assembly (issue until stock is exhausted, then use WO-A-7518)	1	$^{5}/_8$ 10	—	12	—	—
	WO-A-7518	FM-GPW-9778	KNOB, carburetor choke or throttle w/PLUNGER and WIRE, assembly (was WO-A-1307)	2	—	—	—	—	—
			0301C—AIR CLEANER						
03-3	WO-A-1315	FM-GPW-9609	BASE, air cleaner, assembly (before Willys serial 124309) (AC-1542510) (Willys only)	1	—	—	—	—	—
	WO-A-5629		BODY, air cleaner, assembly (after Willys serial 124309) (HH-613455) (issue until stock exhausted)	1	—	4	—	—	—
	WO-A-1313		BOLT, air cleaner cover (S, wing) (before Willys serial 124309) (AC-1542508) (Willys only)	1	—	—	—	—	—
	WO-A-1279	FM-GPW-9657	BRACKET, support, air cleaner, left, assembly	1	—	—	—	—	—
	WO-A-1278	FM-GPW-9656	BRACKET, support, air cleaner, right, assembly	1	—	—	—	—	—
	WO-52132	FM-20027-S2	BOLT, hex-hd, s-fin, alloy-S, $^5/_{16}$-24NF-3 x $^5/_8$ (GM-106279)	6	—	—	—	—	—
	WO-51833	FM-34806-S2	WASHER, lock, S, $^5/_{16}$ in. ($^{19}/_{32}$ O.D. x $^{11}/_{32}$ I.D. x $^1/_{16}$) (BECX1H)	6	—	—	—	—	—
	WO-53025	FM-356309-S7	WASHER, lock, S, cd-pltd, internal, external tooth, $^5/_{16}$ in. (GM-178532)	3	—	—	—	—	—
03-3	WO-A-1451	FM-GPW-9686-A	BUSHING, air cleaner tube to carburetor (3 ply rad. hose, $2^1/_4$ x $1^1/_8$)	1	$^{5}/_8$ 10	—	12	—	—
	WO-A-281	FM-GPW-9628	CLAMP, carburetor air horn, w/BOLT and NUT (clamp, S, $2^{13}/_{32}$ I.D. x $^1/_2$) (bolt and nut cd-pltd.) (before serial 104310) (issue until stock exhausted)	1	—	4	—	—	—
03-3	WO-53108	FM-GPW-6772	CLAMP, hose, S, $^3/_4$ in. (after Willys serial 208437)	2	$^{5}/_8$ 6	20	—	—	—
03-3	WO-635097	FM-GPW-9653	CLAMP, hose S, $2^{29}/_{32}$ in., I.D. (USL-461389)	4	$^{5}/_8$ 6	—	—	—	—
03-3	WO-A-1515	FM-GPW-2250	CLAMP, hose S, $2^1/_2$ in., I.D. (after Willys serial 104310) (1 used for WO-A-1451 BUSHING: 1 used carburetor to air cleaner tube)						
	WO-6273	FM-34114-S2	NUT, hex, S, No. 10 (.190)-32NF-2	2	$^{5}/_8$ 10	—	10	—	—
	WO-6383	FM-27145-S2	SCREW, machine, rd-hd, S, No. 10 (.190)-32NF-2 x 1 (GM-100768) (BCOX1.1AL)	2	—	—	—	—	—
03-3	WO-A-5621	FM-GPW-18205-B	CLEANER, air, assembly (after Willys serial 124309) (HH-613300)	2	$^{5}/_8$ 3	5	—	—	—
	WO-115905	FM-355253-S	SCREW, wing, $^1/_4$-20NC-2 x $^3/_4$ (cleaner to brkt.)	4	$^{5}/_8$ 13	—	40	—	—

61

SNL G-503

0301C

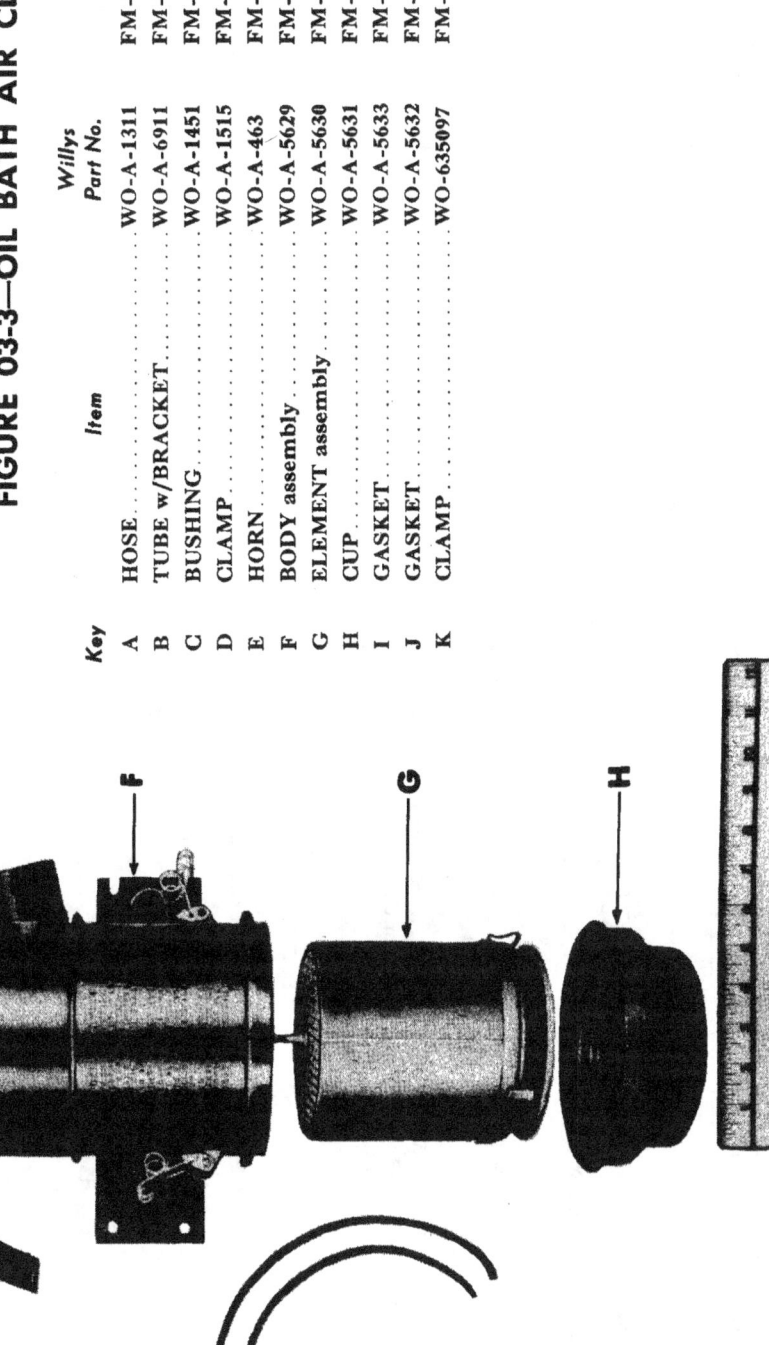

FIGURE 03-3—OIL BATH AIR CLEANER

Key	Item	Willys Part No.	Ford Part No.	Gov't Group No.
A	HOSE	WO-A-1311	FM-GPW-9652	0301C
B	TUBE w/BRACKET	WO-A-6911	FM-GPW-9637-B	0301C
C	BUSHING	WO-A-1451	FM-GPW-9686-A	0301C
D	CLAMP	WO-A-1515	FM-GPW-2250	0301C
E	HORN	WO-A-463	FM-GPW-9632	0301C
F	BODY assembly	WO-A-5629	FM-GPW-9609	0301C
G	ELEMENT assembly	WO-A-5630	FM-GPW-9617	0301C
H	CUP	WO-A-5631	FM-GPW-9658	0301C
I	GASKET	WO-A-5633	FM-GPW-9623	0301C
J	GASKET	WO-A-5632	FM-GPW-9621	0301C
K	CLAMP	WO-635097	FM-GPW-9653	0301C

RA PD 305070-A

Figure Number	Official Stockage Number	Part Number: Willys	Part Number: Ford	ITEM	Quantity Reqd. per Unit Assy.	Per 100 Major Items: 12 Mos. Field Maintenance	Per 100 Major Items: Major Overhaul (5th Ech)	Per 100 Major Items: Total First Year Procurement	Estimated Reqmts. per 100 Rebuilds
Col. 1	Col. 2	Col. 3	Col. 4	Col. 5	Col. 6	Col. 7	Col. 8	Col. 9	Col. 10
				0301C—AIR CLEANER (Cont'd)					
		WO-52702	FM-34745-S	WASHER, S., 1/4 in. (5/8 O.D. x 9/32 I.D. x 1/16 thk.)	4	—	—	—	—
		WO-53024		WASHER, lock, S., cd-pltd., internal, external, 1/4 in. (GM-174916)	2	—	—	—	—
03-3		WO-A-5631	FM-GPW-9658	CUP, oil, air cleaner (after Willys serial 124309) (HH-613306)	1	% 3	—	4	—
		WO-A-1314		ELEMENT, filter, air cleaner, assembly (after Willys serial 124309) (AC-1542509) (Willys only)	1	—	—	—	—
03-3		WO-A-5630	FM-GPW-9617	ELEMENT, air cleaner, w/wing BOLT (after Willys serial 124309) (HH-613387)	1	% 6	—	8	—
03-3		WO-A-5632	FM-GPW-9621	GASKET, air cleaner, body upper (after Willys serial 124309) (HH-613313)	1	% 8	—	15	—
03-3		WO-A-5633	FM-GPW-9623	GASKET, air cleaner, oil cup, lower (after Willys serial 124309) (HH-613314)	1	% 8	—	15	—
03-3		WO-A-463	FM-GPW-9632	HORN, air cleaner	1	5	—	8	—
		WO-A-6918	FM-GPW-6771	HOSE, air cleaner tube to oil filler tube	1	% 10	—	12	—
03-3		WO-A-1311	FM-GPW-9652	HOSE, flexible, air cleaner tube to cleaner	1	% 19	—	24	—
		WO-A-1224		OUTLET, air cleaner (first 24309 Willys trucks) (Willys only)	1	—	—	—	—
		WO-51523		SCREW, hex-hd., cap, S., 5/16-18NC-2 x 3/4 (outlet to brkt.) (GM-100121) (BCAX1BA) (Willys only)	2	—	—	—	—
		WO-A-642	FM-GPW-9647	SHIELD, air cleaner, assembly	1	—	—	—	—
		WO-51732	FM-20309-S7	BOLT, hex-hd., s-fin., alloy-S., 1/4-28NF-2 x 1/2 (shield to frame) (BAOX4AA)	3	—	—	—	—
		WO-5914	FM-33796-S2	NUT, hex., S., 1/4-28NF-2 (GM-103024) (BBBX1AA)	3	—	—	—	—
		WO-5121	FM-34706-S2	WASHER, S., 5/16 in. (3/4 O.D. x 5/16 I.D. x 1/16 thk.)	3	—	—	—	—
		WO-52706	FM-34805-S2	WASHER, lock, S., 1/4 in. -BECX1G)	3	—	—	—	—
		WO-A-7191	FM-GPW-9612	SPRING, air cleaner toggle (after Willys serial 124309) (HH-613380)	2	% 3	—	6	—
		WO-A-1290	FM-GPW-9637-A	TUBE, air cleaner, w/BRACKET, assembly (before Willys serial 208437)	1	—	—	—	—
03-2		WO-A-6911	FM-GPW-9637-B	TUBE, air cleaner, w/BRACKET, assembly (after Willys serial 208437)	1	% 3	—	4	—
				0302—FUEL PUMP					
03-4		WO-115641	FM-GPW-9399	ARM, rocker, fuel pump, assembly (AC-1521960)	1	3	3	10	10
03-4		WO-115657	FM-GPW-9387	BAIL, fuel pump strainer, assembly (AC-1523231)	1	% 3	—	5	—
03-4		WO-A-1045	FM-GPW-9386	BODY, fuel pump priming, w/LEVER, assembly (AC-1537812)	1	—	—	—	—
03-4		WO-A-1494	FM-GPW-9355	BOWL, fuel pump (metal) (AC-1537065)	1	% 3	—	4	—
03-4		WO-115653	FM-11A-9361	CLAMP, fuel pump valve (AC-1521956)	1	% 5	5	12	15
03-4		WO-51546	FM-26466-S7	SCREW, fuel pump valve clamp (AC-132629)	2	—	—	—	—
03-4		WO-115650	FM-GPW-9354	COVER, fuel pump, top (AC-132696)	1	—	—	—	—
03-4		WO-113439	FM-31628-S7	SCREW, fill-hd., No. 10 (.190)—32NF-2 x 1/2 (AC-855493)	6	—	—	—	—
03-4		WO-113440	FM-34803-S7	WASHER, lock, S., No. 10 (.190 in.) (AC-855064)	6	—	—	—	—
03-4		WO-116695	FM-GPW-9398	DIAPHRAGM, fuel pump, w/PULL ROD, assembly (AC-1538205)	1	—	—	—	—
03-4		WO-115656	FM-GPW-9364	GASKET, fuel pump bowl (AC-1523096)	1	% 48	—	70	—

FIGURE 03-4—FUEL PUMP ASSEMBLY

Key	Item	Willys Part No.	Ford Part No.	Gov't Group No.
A	BAIL assembly	WO-115657	FM-GPW-9387	0302
B	SEAT	WO-113460	FM-GPW-9388	0302
C	BOWL	WO-A-1494	FM-GPW-9355	0302
D	SCREEN assembly	WO-115654	FM-GPW-9365	0302
E	GASKET	WO-115656	FM-GPW-9364	0302
F	COVER	WO-115650	FM-GPW-9354	0302
G	SCREW	WO-113439	FM-31628-S7	0302
H	WASHER	WO-113440	FM-34803-S7	0302
I	DIAPHRAGM w/PULL ROD	WO-116695	Willys only	0302
J	CLAMP	WO-115653	FM-11A-9361	0302
K	VALVE assembly	WO-115651	FM-11A-9352	0302
L	SCREW	WO-51546	FM-26466-S7	0302
M	GASKET	WO-115652	FM-GPW-9362	0302
N	SPRING	WO-116694	Willys only	0302
O	SPRING	WO-115643	FM-GPW-9380	0302
P	ARM assembly	WO-115641	FM-GPW-9399	0302
Q	LINK	WO-115880	FM-INC-9381	0302
R	PIN	WO-A-1046	FM-GPW-9378	0302
S	BODY w/LEVER	WO-A-1045	FM-GPW-9386	0302
T	SEAL	WO-115870	FM-GPW-19469	0302
U	WASHER	WO-115869	FM-GPW-9468	0302

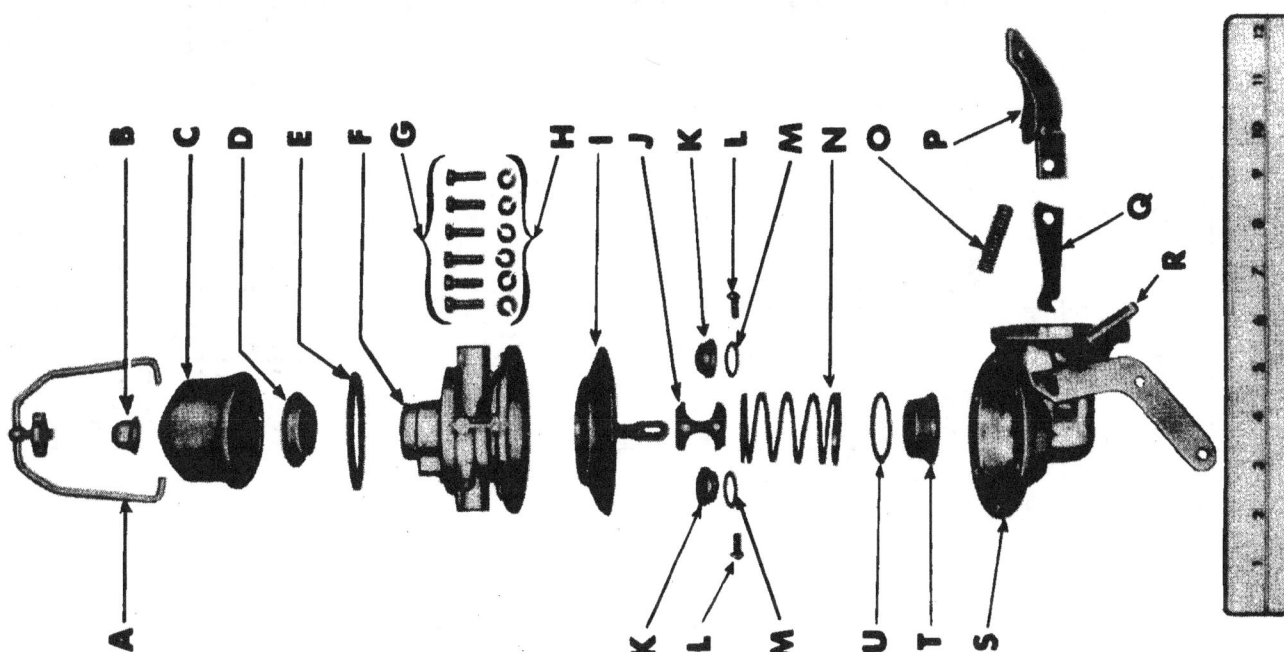

RA PD 305069-A

GROUP 03—FUEL (Cont'd)

0302

Figure Number	Official Stockage Number	Part Number Willys	Part Number Ford	ITEM	Quantity Reqd. per Unit Assy.	Per 100 Major Items 12 Mos. Field Maintenance	Per 100 Major Items Major Overhaul (5th Ech)	Total First Year Procurement	Estimated Reqmts. per 100 Rebuilds
Col. 1	Col. 2	Col. 3	Col. 4	Col. 5	Col. 6	Col. 7	Col. 8	Col. 9	Col. 10
				0302—FUEL PUMP (Cont'd)					
03-4		WO-115652	FM-GPW-9363	GASKET, fuel pump valve (AC-1521953)	2	—	—	—	—
03-4		WO-A-7834	FM-GPW-18373-D	KIT, repair, fuel pump	as req.	26	32	65	100
				(Includes:					
				1 WO-116695 FM-GPW-9398 DIAPHRAGM, w/ROD					
				1 WO-115656 FM-GPW-9364 GASKET, bowl					
				1 WO-638737 FM-GPW-9417 GASKET, fuel pump					
				2 WO-115652 FM-GPW-9363 GASKET, valve					
				1 WO-115880 FM-INC-9381 LINK, rocker arm					
				1 WO-A-1046 FM-GPW-9378 PIN, rocker arm					
				1 WO-115870 FM-GPW-9469 SEAL					
				1 WO-115654 FM-GPW-9365 SCREW, filtering, assembly					
				3 WO-51546 FM-26466-S7 SCREW, clamp					
				3 WO-113439 FM-31628-S7 SCREW, cover					
				1 WO-116694 FM-GPW-9396 SPRING, diaphragm					
				1 WO-115643 FM-GPW-9380 SPRING, rocker arm					
				2 WO-115651 FM-11A-9352 VALVE assembly					
				3 WO-113440 FM-34803-S7 WASHER, lock					
				1 WO-A-1047 FM-GPW-9377 WASHER, pin)					
	WO-A-6754			KIT, repair, fuel pump (issue until stock is exhausted, then use WO-A-7956)	as req.	4	8	—	—
	WO-A-7956			KIT, repair, fuel pump (was WO-A-6754)	as req.	—	—	—	—
				(Includes:					
				1 WO-115657 FM-GPW-9387 BAIL assembly					
				1 WO-115656 FM-GPW-9364 GASKET					
				1 WO-113461 FM-GPW-9373 NUT					
				1 WO-115654 FM-GPW-9365 SCREW, filtering					
				1 WO-113460 FM-GPW-9388 SEAT, bowl)					
03-4	WO-115880		FM-INC-9381	LINK, fuel pump rocker arm (AC-1521708)	1	—	—	—	—
03-4	WO-113461		FM-GPW-9373	NUT, thumb, fuel pump strainer (AC-855763)	1	—	—	—	—
Fig. 6	WO-A-1046		FM-GPW-9378	PIN, fuel pump rocker arm (AC-1521578)	1	—	—	—	—
03-4	WO-A-8323		FM-GPW-9350	PUMP, fuel, assembly (AC-1538312)	1	10	—	13	—
03-4	WO-115870		FM-GPW-9469	SEAL, fuel pump pull rod (rubber) (AC-1521880)	1	26	32	60	100
03-4	WO-113460		FM-GPW-9388	SEAT, fuel pump strainer bowl (AC-854005)	1	3	3	4	—
03-4	WO-115654		FM-GPW-9365	SCREEN, filtering, fuel pump, assembly (AC-1523099)	1	5 %	—	10	—
03-4	WO-116694		FM-GPW-9396	SPRING, fuel pump diaphragm (AC-1523068)	1	—	—	—	—
03-4	WO-115643		FM-GPW-9380	SPRING, fuel pump rocker arm (AC-1522046)	1	—	—	—	—
03-4	WO-115651		FM-11A-9352	VALVE, fuel pump, assembly (AC-1523106)	1	—	—	—	—
03-4	WO-115869		FM-GPW-9468	WASHER, fuel pump diaphragm spring seat (AC-1521985)	1	26	32	60	100
03-4	WO-A-1047		FM-GPW-9377	WASHER, fuel pump rocker arm pin (AC-1521288)	1	—	—	—	—

SNL G-503

0302A-0303

GROUP 03—FUEL (Cont'd)

Figure Number	Official Stockage Number	Part Number Willys	Part Number Ford	ITEM	Quantity Reqd. per Unit Assy.	Per 100 Major Items 12 Mos. Field Maintenance	Per 100 Major Items Major Overhaul (5th Ech)	Total First Year Procurement	Estimated Reqmts. per 100 Rebuilds
Col. 1	Col. 2	Col. 3	Col. 4	Col. 5	Col. 6	Col. 7	Col. 8	Col. 9	Col. 10

0302A—FUEL PUMP ATTACHING PARTS

Figure Number	Official Stockage Number	Willys	Ford	ITEM	Qty	12 Mos Field	Major Overhaul	Total First Year	Est. Reqmts
		WO-638737	FM-GPW-9417	GASKET, fuel pump to block (AC-38263)	1	—	—	—	—
		WO-6428	FM-20366-S2	SCREW, hex-hd., cap, S., 5/16-18NC-2 x 7/8 (fuel pump to cylinder block) (GM-106325) (BCAX1BB)	2	38%	—	60	—
		WO-51833	FM-34806-S2	WASHER, lock, S., 5/8 in. (fuel pump to cylinder block bolt) (BECX1H)	2	—	—	—	—

0303—ACCELERATOR AND LINKAGE

Figure Number	Official Stockage Number	Willys	Ford	ITEM	Qty	12 Mos Field	Major Overhaul	Total First Year	Est. Reqmts
03-5		WO-633013	FM-GPW-9752	BLOCK, adjusting, accelerator throttle rod	1	10	—	12	—
03-5		WO-639607	FM-GPW-9728	BRACKET, accelerator cross shaft	1	—	—	—	—
		WO-5914	FM-33796-S2	NUT, hex, S., 1/4-20NC-2 (GM-103024) (BBBX1AA) (issue until stock exhausted)	2	—	—	—	—
		WO-337304	FM-88350-S	STUD, S., 1/4-20NC-2 x 7/8 (cross shaft bracket)	2	8%	16	—	—
		WO-52768	FM-356305-S2	WASHER, S., plain, 21/64 (at each end of cross shaft)	2	—	—	—	—
		WO-52706	FM-34805-S2	WASHER, lock, S., 1/4 in. (BECX1G)	2	—	—	—	—
03-5		WO-A-1173	FM-GPW-9751	CLIP, throttle rod retracting spring (on manifold) (issue until stock exhausted) (GM-110633)	1	4%	4	—	—
		WO-6352	FM-355836-S8	NUT, hex, S., No. 10 (.190)-24NC-2 (to manifold) (BBKX2BC)	1	—	—	—	—
		WO-5067	FM-72003-S	PIN, cotter, 1/16 x 1/2 (BFAX1BC)	1	—	—	—	—
		WO-51662	FM-356201-S	WASHER, S., 13/64	1	10%	—	25	—
03-5		WO-639610	FM-GPW-9745	CLIP, throttle rod retracting spring (on throttle rod)	1	—	—	—	—
		WO-5067	FM-72003-S	PIN, cotter, 1/16 x 1/2 (BFAX1BC)	1	—	—	—	—
		WO-51662	FM-356201-S	WASHER, S., 13/64	1	—	—	—	—
		WO-651298	FM-GPW-9731-B	CONNECTION, accelerator pedal (female)	1	—	—	—	—
		WO-650482	FM-GPW-9711	HINGE, accelerator pedal (female)	1	—	—	—	—
03-5		WO-A-1174	FM-GPW-9727	LINK, connecting, accelerator pedal (before Willys serial 225209)	1	5%	—	6	—
		WO-A-6710	FM-GPW-9719	LINK, connecting, accelerator pedal (after Willys serial 225209)	1	—	—	—	—
		WO-5067	FM-72003-S	PIN, cotter, S., 1/16 x 1/2 (BFAX1BC)	1	—	—	—	—
		WO-52702	FM-34745-S2	WASHER, plain, S., S.A.E. std., 1/4 in. (BEBX1G)	1	3%	—	4	—
03-5		WO-A-1910	FM-GPW-9726	LEVER, accelerator cross shaft, w/BRACKET, assembly	1	8%	12	—	—
		WO-650484	FM-74019-S	PIN, accelerator pedal hinge (issue until stock exhausted)	2	—	—	—	—
		WO-5067	FM-72003-S	PIN, cotter, 1/16 x 1/2 (BFAX1BC)	1	—	—	—	—
03-5		WO-A-1225	FM-GPW-9716	REST, foot, accelerator (issue until stock exhausted)	1	—	4	—	—
		WO-52954	FM-33909-S	NUT, jam, hex, s-fin., alloy-S., 5/16-24NF-2 (foot rest) (GM-114493) (BBDX1BA)	1	—	—	—	—
03-5		WO-A-1084	FM-GPW-9732-A	PEDAL, accelerator, assembly (before Willys 116,790 trucks)	1	—	—	—	—
03-5		WO-A-6851	FM-GPW-9735-B	PEDAL, accelerator, w/HINGE and LINK, assembly	1	5%	—	5	—
03-5		WO-A-1175	FM-GPW-9742	ROD, throttle, accelerator	1	3%	—	6	—
03-5		WO-A-1243	FM-GPW-9739	SHAFT, cross, accelerator, w/LEVER	1	—	4	—	—

SNL G-503

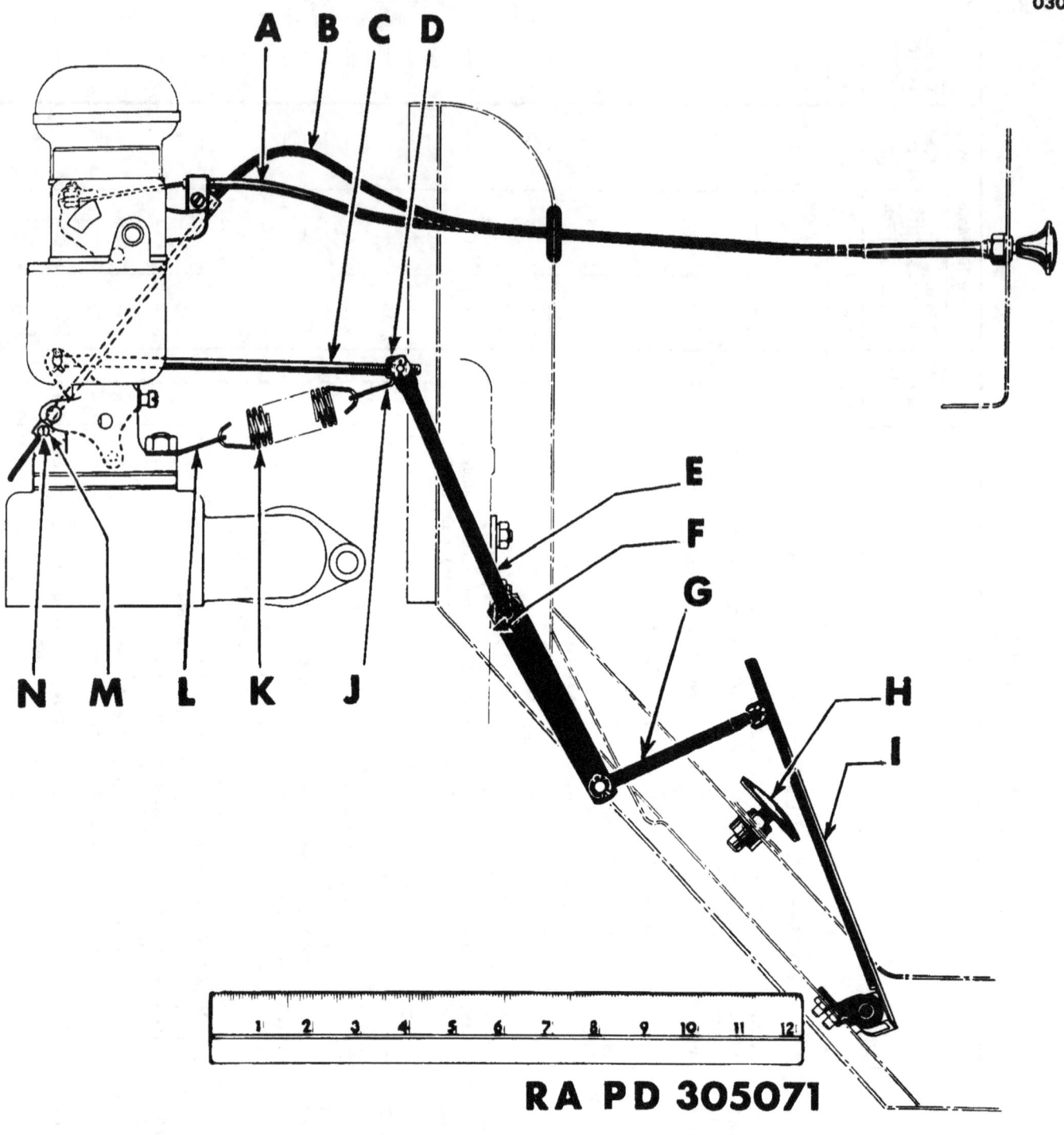

FIGURE 03-5—ACCELERATOR THROTTLE AND CHOKE CONTROLS

Key	Item	Willys Part No.	Ford Part No.	Gov't Group No.
A	CHOKE	WO-A-7517	FM-GPW-9775-B	0301B
B	THROTTLE assembly	(See WO-A-7517)	FM-GPW-9775-B	0301B
C	ROD	WO-A-1175	FM-GPW-9742	0303
D	BLOCK	WO-633013	FM-GPW-9752	0303
E	SHAFT w/LEVER	WO-A-1243	FM-GPW-9739	0303
F	BRACKET	WO-639607	FM-GPW-9728	0303
G	LINK	WO-A-6710	FM-GPW-9719	0303
H	REST assembly	WO-A-1225	FM-GPW-9716	0303
I	PEDAL w/LINK	WO-A-6851	FM-GPW-9735-B	0303
J	CLIP	WO-639610	FM-GPW-9745	0303
K	SPRING	WO-633011	FM-GPW-9799	0303
L	CLIP	WO-A-1173	FM-GPW-9751	0303
M	STOP	WO-372438	FM-GPW-11474	0303A
N	SCREW	WO-51040	FM-26457-S	0303A

GROUP 03—FUEL (Cont'd)

Figure Number	Official Stockage Number	Part Number Willys	Part Number Ford	ITEM	Quantity Reqd. per Unit Assy.	12 Mos. Field Maintenance	Major Overhaul (5th Ech)	Total First Year Procurement	Estimated Reqmts. per 100 Rebuilds
Col. 1	Col. 2	Col. 3	Col. 4	Col. 5	Col. 6	Col. 7	Col. 8	Col. 9	Col. 10
				0303—ACCELERATOR AND LINKAGE (Cont'd)					
03-5		WO-633011	FM-GPW-9799	SPRING, retracting, accelerator	1	% 29	—	33	—
		WO-632174	FM-GPW-9737	SPRING, accelerator cross shaft (issue until stock exhausted)	1	% 4	6	—	—
		WO-650483	FM-GPW-9795	SPRING, accelerator pedal hinge	1	% 29	—	35	—
		WO-6352	FM-355836-S8	NUT, hex, S., No. 10 (.190)-24NC-2 (accelerator pedal assembly to toe board) (GM-128854) (BBKX2BC)	2	—	—	—	—
		WO-51537	FM-355162-S2	SCREW, fl-hd., S., cd-pltd., No. 10 (.190)-24NC-2 x ½ (GM-123854) (BCKX2AE-15)	2	% 10	—	100	—
		WO-52221	FM-34803-S7	WASHER, lock, S., No. 10 (.190 in.)	2	—	—	—	—
				0303A—HAND THROTTLE AND LINKAGE					
				NOTE: For parts not shown here, see Group 0301B					
03-5		WO-A-1302	FM-GPW-9775-B	CONTROL, throttle, assembly (before Willys serial 103468)	1	—	—	—	—
18-1		WO-A-7517		KNOB, throttle, assembly (was WO-A-5106)	1	—	—	—	—
		WO-A-1308	FM-GPW-9778	KNOB, carburetor choke or throttle, w/PLUNGER and WIRE, assembly (Use until stock is exhausted, then use WO-A-7518)	—	% 10	—	12	—
03-5		WO-372438	FM-GPW-11474	STOP, throttle wire (issue until stock is exhausted, then issue WO-A-8834)	1	% 4	8	—	—
03-5		WO-A-8834	FM-GPW-11495	STOP, throttle wire, w/SCREW, assembly (was WO-372438)	1	% 19	—	24	—
		WO-51040	FM-26457-S	SCREW, rd-hd., S., No. 8 (.164)-32NC-2 x ¼ (throttle wire stop screw) (issue until stock is exhausted, then issue WO-A-8834)	1	% 68	132	—	—
				0304—FUEL LINES					
03-6		WO-A-1265	FM-GPW-9154	BUSHING, reducing, ¼ pipe thd. to ⅛ tapered pipe thd.	2	% 10	—	12	—
		WO-A-1325	FM-GPW-9288	CONNECTION, flexible (in fuel line) (FO-HA-8031)	1	% 29	—	40	—
		WO-384549	FM-GPW-9268	CONNECTOR, inverted flare tube, 5/16 in.	2	% 10	—	10	—
		WO-387249	FM-GPW-9267-A	CONNECTOR, inverted flared tube, br., 5/16 in.	3	% 20	—	22	—
		WO-384549	FM-GPW-9268	ELBOW, inverted flared tube (br.)	3	% 16	—	20	—
		WO-A-1368	FM-GPW-9289	TUBE, flexible connection to fuel pump, assembly	1	% 3	—	6	—
		WO-A-1367	FM-GPW-9282	TUBE, fuel filter to flexible connection, assembly	1	% 3	—	4	—
		WO-A-1289	FM-GPW-14585	CLIP, closed, 7/16, bolt hole 13/32 (tube to front engine plate, under generator brace bolt)	1	—	—	—	—
23-3		WO-A-1366	FM-GPW-9237	TUBE, fuel tank to filter, assembly (73⅝ long x 5/16 O.D.)	1	% 3	—	4	—
		WO-A-5450	FM-GPW-9295	CLIP, 7/16 in., open, S., 7/32 dia. hole	5	% 8	12	—	—
				(1 used tube to floor pan front cross sill					
				2 used tube to floor pan rear cross sill					
				1 used tube to fender splasher					
				1 used tube to dash)					

0304

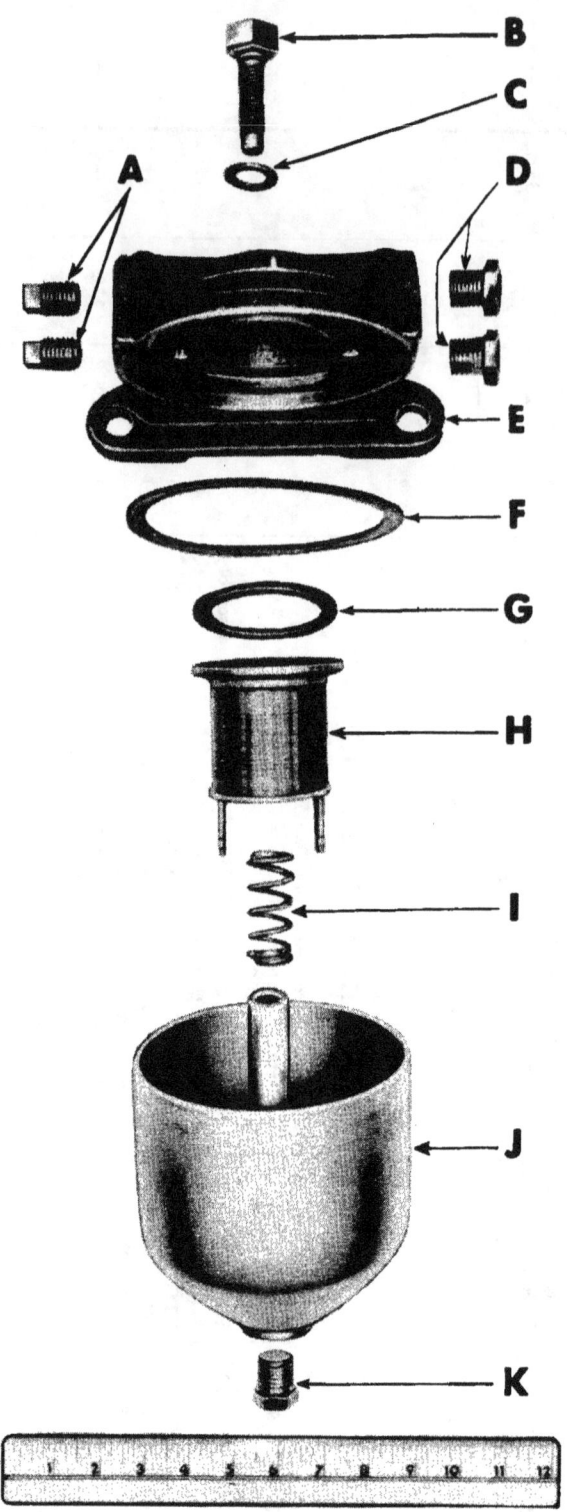

RA PD 305067

FIGURE 03-6—FUEL FILTER ASSEMBLY

Key	Item	Willys Part No.	Ford Part No.	Gov't Group No.
A	PLUG	WO-5138	FM-353055	0306B
B	SCREW	WO-A-1256	FM-GPW-9183	0306B
C	GASKET	WO-A-1257	FM-GPW-9184	0306B
D	BUSHING	WO-A-1265	FM-GPW-9154	0304
E	COVER	WO-A-1258	FM-GPW-9149	0306B
F	GASKET	WO-A-1259	FM-GPW-9160	0306B
G	GASKET	WO-A-1260	FM-GPW-9186	0306B
H	STRAINER-UNIT	WO-A-1261	FM-GPW-9140	0306B
I	SPRING	WO-A-1262	FM-GPW-9182	0306B
J	BOWL w/STUD assembly	WO-A-1263	FM-GPW-9162	0306B
K	PLUG	WO-A-1264	FM-GPW-9185	0306B

RA PD 305067-A

SNL G-503

GROUP 03—FUEL (Cont'd)

Figure Number (Col. 1)	Official Stockage Number (Col. 2)	Willys (Col. 3)	Ford (Col. 4)	ITEM (Col. 5)	Quantity Reqd. per Unit Assy. (Col. 6)	12 Mos. Field Maintenance (Col. 7)	Major Overhaul (5th Ech) (Col. 8)	Total First Year Procurement (Col. 9)	Estimated Reqmts. per 100 Rebuilds (Col. 10)
				0304—FUEL LINES (Cont'd)					
		WO-6352	FM-355836-S8	NUT, hex., S., No. 10 (.190)-24NC-2 (clip attaching) (GM-110633) (BBKX2BC)	2	—	—	—	—
		WO-52889	FM-32924	SCREW, binding-hd., S., type A, No. 10 (.190) x ⅝ (clip to cross sill) (GM-140363) (Willys only)	3	% 19	19	100	—
		WO-5113	FM-355130-S7	SCREW, rd-hd., mach., S., No. 10 (.190)-24NC-2 x ½ (clip to dash, clip to fender splasher) (GM-110500) (BCNX2AE)	2	—	—	—	—
		WO-52221	FM-34803-S	WASHER, lock, S., No. 10 (.190 in.) (BECX1E)	2	—	—	—	—
		WO-A-1369	FM-GPW-9369	TUBE, fuel pump to carburetor, assembly	1	% 6	—	8	—
				0306B—FUEL STRAINER					
03-6		WO-A-1263	FM-GPW-9162	BOWL, fuel strainer, w/STUD, center, assembly (AC-1504117)	1	—	—	—	—
03-6		WO-A-1258	FM-GPW-9149	COVER, mounting, fuel strainer (AC-1504212) (issue until stock is exhausted)	1	—	4	—	—
03-6		WO-A-1256	FM-GPW-9183	SCREW, attaching, fuel strainer cover (fill-hd., S.)	1	5	—	6	—
03-6		WO-A-1257	FM-GPW-9184	GASKET, fuel strainer cover cap screw (AC-853562)	1	—	—	—	—
03-6		WO-A-1259	FM-GPW-9160	GASKET, fuel strainer bowl (AC-853558)	1	—	—	—	—
03-6		WO-A-1260	FM-GPW-9186	GASKET, fuel strainer unit (AC-853572)	1	—	—	—	—
		WO-A-6883	FM-GPW-18337	GASKET SET, fuel strainer (Includes: 1 WO-A-1257 FM-GPW-9184 GASKET, cover cap screw; 1 WO-A-1259 FM-GPW-9160 GASKET, strainer bowl; 1 WO-A-1260 FM-GPW-9186 GASKET, strainer unit)	as req	% 77	—	120	—
03-6		WO-A-1264	FM-GPW-9185	PLUG, drain, fuel strainer (AC-127951)	1	% 5	—	6	—
03-6		WO-5138	FM-353055-S	PLUG, pipe, I., 1¼ in. std.	2	% 5	—	—	—
03-6		WO-A-1262	FM-GPW-9182	SPRING, fuel strainer unit (AC-1504118)	1	% 5	—	6	—
		WO-A-1255	FM-GPW-9155	STRAINER, fuel strainer unit (issue until stock is exhausted, then use WO-A-7850)	—	% 2	—	4	—
		WO-A-7850	FM-GPW-9155-A2	STRAINER, fuel, assembly (AC-1595235) (was WO-A-1255)	1	—	—	—	—
		WO-5919	FM-24389-S2	BOLT, hex-hd., S., ⅜-24NF-3 x ½ (fuel strainer to dash) (GM-100026)	2	—	—	—	—
		WO-5010	FM-34807-S2	WASHER, lock, S., ⅜ in. (²¹⁄₃₂ O.D. x ¹³⁄₃₂ I.D. x ³⁄₃₂ thk.) (BECX1K)	2	% 5	—	9	—
03-6		WO-A-1261	FM-GPW-9140	STRAINER, fuel, assembly (AC-1595823)	1	—	—	—	—
				GROUP 04—EXHAUST					
				0401—MUFFLER					
04-1		WO-A-655	FM-GPW-5264-A	CLAMP, support, muffler (before Willys serial 143507) (rd-type)	2	—	—	—	—
		WO-A-5753	FM-GPW-5264-B	CLAMP, support, muffler (after Willys serial 143507) (oval type) (issue until stock exhausted)	1	% 4	14	—	—

SNL G-503

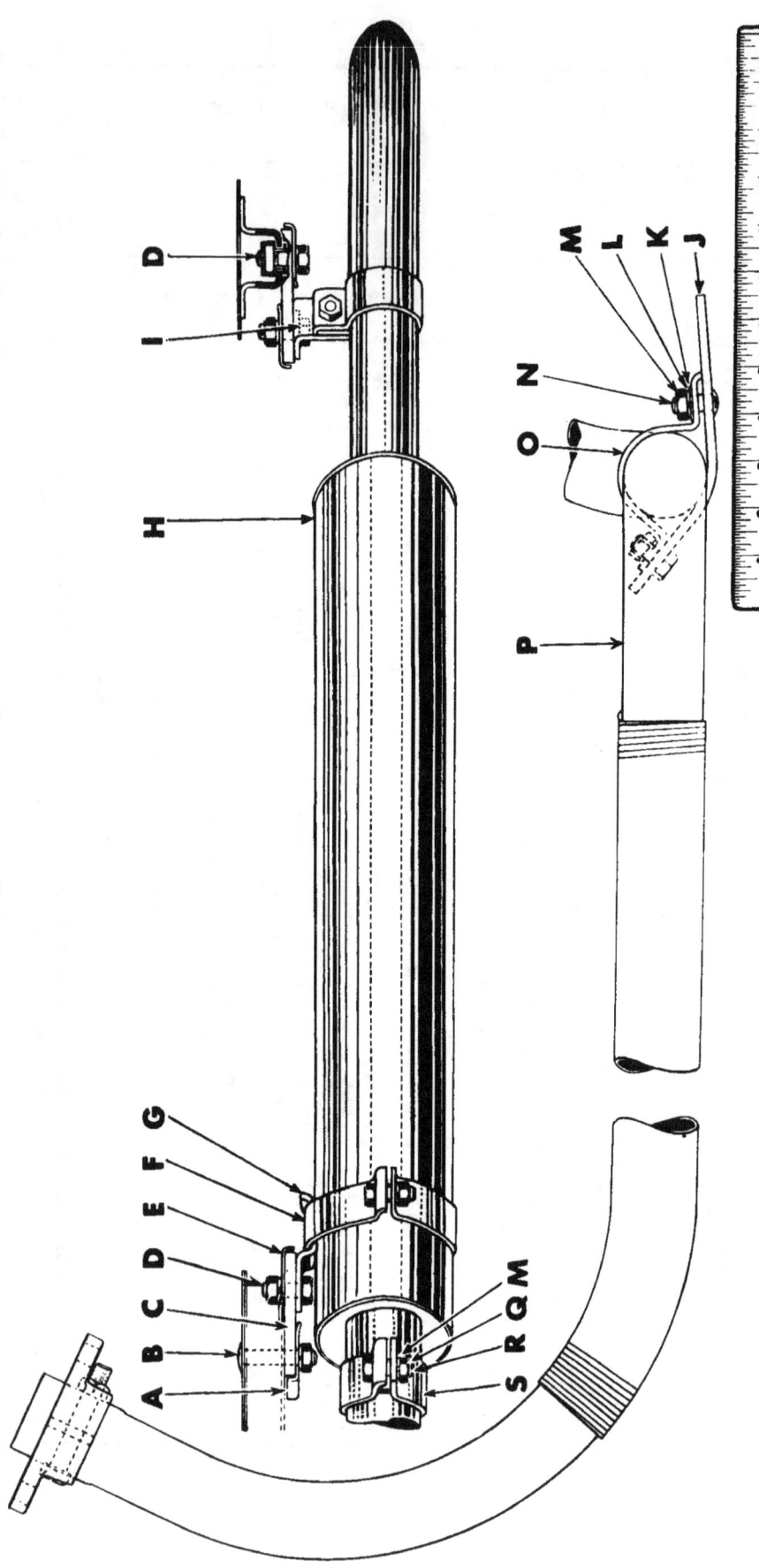

FIGURE 04-1—EXHAUST SYSTEM

Key	Item	Willys Part No.	Ford Part No.	Gov't Group No.
A	PLATE	WO-638058	FM-GPW-5274	0401
B	BOLT	WO-52372	FM-23498-S2	0401
C	INSULATOR	WO-A-658	FM-GPW-5283	0401
D	BOLT	WO-50929	FM-24367-S2	0401
E	PLATE	WO-638058	FM-GPW-5274	0401
F	CLAMP	WO-A-5753	FM-GPW-5264-B	0401
G	STRAP	WO-A-657	FM-GPW-5262	0401
H	MUFFLER assembly	WO-A-6118	FM-GPW-5230-B	0401
I	CLAMP	WO-A-6119	FM-GPW-5298	0401
J	PLATE	WO-A-1253	FM-GPW-9251-B	0402
K	WASHER	WO-52274	FM-34746-S2	0401
L	WASHER	WO-51833	FM-34806-S2	0401
M	NUT	WO-6167	FM-33797-S2	0402
N	BOLT	WO-52983	FM-23393-S2	0401
O	CLAMP	WO-A-1300	FM-GPW-5251	0402
P	PIPE assembly	WO-A-1296	FM-GPW-5246	0402
Q	NUT	WO-53285	See FM-33798-S2	0401
R	BOLT	WO-5922	FM-24407-S	0402
S	CLAMP	WO-636004	FM-GPW-5270	0402

RA PD 305105-A

SNL G-503

GROUP 04—EXHAUST (Cont'd)

0401-0402

Figure Number	Official Stockage Number	Part Number Willys	Part Number Ford	ITEM	Quantity Reqd. per Unit Assy.	Per 100 Major Items 12 Mos. Field Maintenance	Per 100 Major Items Major Overhaul (5th Ech)	Total First Year Procurement	Estimated Reqmts. per 100 Rebuilds
Col. 1	Col. 2	Col. 3	Col. 4	Col. 5	Col. 6	Col. 7	Col. 8	Col. 9	Col. 10
				0401—MUFFLER (Cont'd)					
04-1		WO-5934	FM-20387-S2	BOLT, hex-hd., s-fin., alloy-S., 5/16-24NF-2 x 1 (GM-100014) (BAOX1BC)	2	—	—	—	—
04-1		WO-53285		NUT, hex., reg., S., 5/16-24NF-2 (Seez-Pruf) (Willys only)	2	—	—	—	—
04-1		WO-51833	FM-33798-S2	NUT, hex., reg., S., 5/16-24NF-2 (BBBX1BA) (Ford only)	2	—	—	—	—
04-1		WO-A-6119	FM-34806-S2	WASHER, lock, S., 5/16 in. (19/32 O.D. x 11/32 I.D. x 1/16 thk.) (BECX1H)	2	—	—	—	—
04-1		WO-5922	FM-GPW-5298	CLAMP, muffler tail pipe (after Willys serial 143507, oval type muffler) (issue until stock exhausted)	1	% 4	12	—	—
04-1		WO-5437	FM-24407-S	BOLT, hex-hd., s-fin., alloy-S., 5/16-24NF-2 x 1 1/8 (GM-106281) (BAOX4BC)	1	—	—	—	—
04-1		WO-A-658	FM-34706-S2	WASHER, plain, S., 5/16 in. (GM-106262)	1	—	—	—	—
04-1		WO-52372	FM-GPW-5283	INSULATOR, muffler support	2	% 6	—	9	—
04-1		WO-50929	FM-23498-S2	BOLT, carriage, sq-nk., S., 5/16-18NC-2 x 1 3/4 (insulator to support strap and body still) (BADX1CF)	1	% 80	—	100	—
04-1		WO-53285		NUT, hex., reg., S., 5/16-24NF-2 x 7/8 (GM-106280) (BAOX1BB)	3	—	—	—	—
04-1		WO-51833	FM-33798-S2	NUT, hex., reg., S., 5/16-24NF-2 (Seez-Pruf) (Willys only)	2	—	—	—	—
04-1		WO-A-1146	FM-34806-S2	WASHER, lock, S., 5/16 in. (2 2/32 O.D. x 11/32 I.D. x 1/16 thk.) (BECX1H)	4	—	—	—	—
04-1		WO-A-6118	FM-GPW-5230-A	MUFFLER assembly (rd-type) (before Willys serial 143507) (use WO-A-6118 and attaching parts for replacement)	1	% 80	—	94	—
04-1		WO-638058	FM-GPW-5230-B	MUFFLER assembly (oval type) (after Willys serial 143507)	1	% 4	12	—	—
04-1		WO-A-657	FM-GPW-5274	PLATE, muffler support insulator (issue until stock exhausted)	4	% 12	16	—	—
			FM-GPW-5262	STRAP, muffler support round type muffler (issue until stock exhausted) (oval type muffler)	2	—	—	—	—
					1	—	—	—	—
				0402—EXHAUST PIPE					
04-1		WO-A-1300	FM-GPW-5251	CLAMP, exhaust pipe extension (issue until stock exhausted)	1	% 4	12	—	—
04-1		WO-52983	FM-23393-S2	BOLT, carriage, sq-nk., S., 5/16-18NC-2 x 7/8 (pipe extension to skid plate) (GM-119045)	1	% 40	—	50	—
04-1		WO-51523	FM-20346-S2	BOLT, hex-hd., s-fin., alloy-S., 5/16-18NC-2 x 3/4 (GM-100121) (BCAX1BA)	1	—	—	—	—
04-1		WO-6167	FM-33797-S2	NUT, hex, S., 5/16-18NC (GM-102634) (BBAX1B)	2	—	—	—	—
04-1		WO-52274	FM-34746-S2	WASHER, lock, S., 5/16 in. (11/32 I.D. x 11/16 O.D. x 1/16 thk.) (BEBX1H)	2	—	—	—	—
04-1		WO-51833	FM-34806-S2	WASHER, lock, S., 5/16 in.	2	—	—	—	—
04-1		WO-636004	FM-GPW-5270	CLAMP, exhaust pipe to muffler (issue until stock exhausted)	1	% 4	12	—	—
04-1		WO-5922	FM-24407-S	BOLT, hex-hd., s-fin., alloy-S., 5/16-24NF-2 x 1 1/8 (GM-106281) (BAOX4BC)	1	—	—	—	—
		WO-53285		NUT, hex, S., 5/16-24NF-2 (Seez-Pruf) (Willys only)	1	—	—	—	—
		WO-51833	FM-33798-S2	NUT, hex, S., 5/16-24NF-2 (BBBX1BA) (Ford only)	1	—	—	—	—
		WO-630526	FM-34806-S2	WASHER, lock, S., 5/16 in. (19/32 O.D. x 11/32 I.D. x 1/16 thk.) (BECX1H)	1	—	—	—	—
				FLANGE, exhaust pipe (Willys only)	1	% 4	12	—	—

SNL G-503

	WO-6486		BOLT, hex-hd., s-fin., alloy-S., 3/8-16NC-2 x 2 1/4 (clamp bolt) (GM-106333) (BCAX1BB) (Willys only)	1	—	—	—	—
	WO-50878	FM-24429-S7	BOLT, hex-hd., s-fin., alloy-S., 3/8-24NF-2 x 1 1/4 (flange to manifold) (GM-100027) (BCBX1CD)	1	—	—	—	—
04-1	WO-53289		NUT, hex., S., 3/8-16NC-2 (Seez-Pruf) (Willys only)	1	—	—	—	—
04-1		FM-33799	NUT, hex., reg., S., 3/8-16NC-2 (Ford only)	1	—	—	—	—
	WO-53287		NUT, hex., S., 3/8-24NF-2 (Seez-Pruf) (Willys only)	2	—	—	—	—
		FM-33800-S2	NUT, hex., S., 3/8-24NF-2 (Ford only)	1	—	—	—	—
	WO-5010	FM-34807-S7	WASHER, lock, reg., S., 3/8 in. (21/32 I.D. x 13/32 O.D. x 3/32 thk.) (BECX1K)	1	—	—	—	—
	WO-634814	FM-GPW-9450	GASKET, exhaust pipe flange	1	% 80	100	—	—
04-1	WO-A-1253	FM-GPW-5291-B	PLATE, skid, muffler guard, under frame	7	—	—	—	—
	WO-52945		BOLT, carriage, sq-nk., S., 3/8-16NC-2 x 7/8 (plate to cross member) (Willys only)	4	—	—	—	—
	WO-5544	FM-33799-S7	NUT, hex., S., 3/8-16NC-2 (GM-102635) (BBAX1C)	4	—	—	—	—
	WO-52101	FM-34747-S2	WASHER, plain, S., S.A.E. std., 3/8 in. (13/16 O.D. x 13/32 I.D. x 1/16 thk.) (BEBX1K)	4	—	—	—	—
	WO-A-1296	FM-GPW-5646	PIPE, exhaust, w/bond STRAP, assembly (issue until stock exhausted)	1	% 40	80	—	—
	WO-A-10198	FM-GPW-5246-B	PIPE, exhaust, w/o bond strap, assembly (for new radio suppression)	1	—	—	—	—
	WO-332515	FM-88032-S	STUD, S., 3/8-16NC-2 x 1 3/8 x 3/8-24NF-2 (exhaust pipe to manifold)	2	—	—	—	—

GROUP 05—COOLING

0501—RADIATOR

05-1	WO-A-1215	FM-GPW-8100-A	CAP, filler, radiator, assembly (AC-846709) (STN-6455A)	1	% 16	—	20	—
05-1	WO-A-1126	FM-9N-8115	COCK, drain, radiator (1/4 in.) (WH-145-A) (EAT-IE-1212)	1	% 6	—	6	—
05-1	WO-A-1214	FM-GPW-8005	CORE, radiator, w/TANK and SHROUD, assembly	1	% 10	—	12	—
05-1	WO-A-1546	FM-34084-S	NUT, radiator to support bracket (hex., S., cd-pltd., 7/16-20NF-2) (BBBX1DA-15) (issue until stock exhausted)	4	% 28	52	—	—
05-1	WO-A-1547	FM-34708-S	WASHER, plain, cd-pltd., S., 7/16 in. (1/2 I.D. x 1 1/4 O.D. x 5/64 thk.) (issue until stock exhausted)	2	% 16	24	—	—
	WO-53027	FM-356314-S7	WASHER, lock, S., internal, external tooth, 7/16 in.	2	—	—	—	—
05-1	WO-A-1216	FM-K-7129-B	GASKET, radiator filler cap (AC-846732)	1	% 38	—	60	—
	WO-A-1217	FM-GPW-8133	ROD, brace, radiator (issue until stock exhausted)	1	% 4	4	—	—
	WO-5910	FM-33798-S2	NUT, hex., S., 5/16-24NF-2 (GM-103025) (BBBX1BA)	3	—	—	—	—
	WO-51833	FM-34806-S2	WASHER, lock, S., 5/16 in. (19/32 O.D. x 11/32 I.D. x 1/16 thk.) (BECX1H)	2	—	—	—	—
05-1	WO-A-4413	FM-GPW-8125-A	SHIM, radiator to support bracket (issue until stock exhausted)	as req.	% 20	40	—	—

0502—THERMOSTAT

01-3	WO-639651	FM-GPW-8578	RETAINER, thermostat	1	% 3	—	6	—
01-3	WO-637646	FM-GPW-8575	THERMOSTAT assembly (HR-3108628)	1	% 10	8	25	25

0503—WATER PUMP

05-2	WO-636297	FM-GPW-8530	BEARING, water pump, w/SHAFT, assembly (double row, annular) (ND-885141) (MRC-D2-13567) (HO-88541)	1	—	—	—	—
	WO-640032	FM-GPW-8548	BELLOWS, water pump seal	1	—	—	—	—
05-2	WO-637052	FM-GPW-8505	BODY, water pump seal	1	—	—	—	—

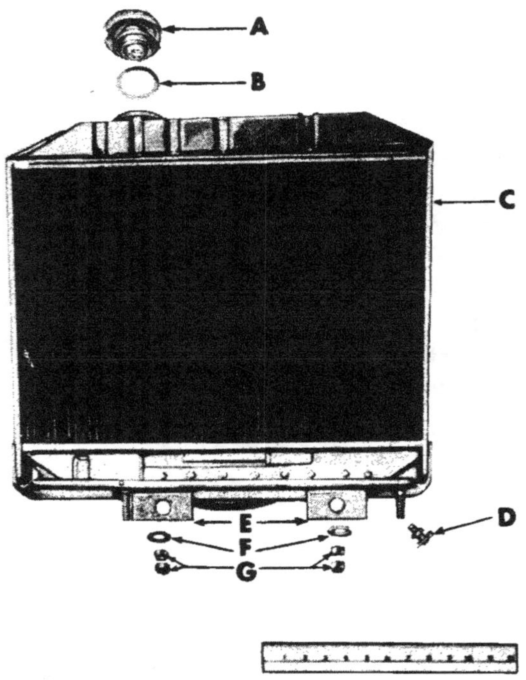

FIGURE 05-1—RADIATOR

Key	Item	Willys Part No.	Ford Part No.	Gov't Group No.
A	CAP	WO-A-1215	FM-GPW-8100-A	0501
B	GASKET	WO-A-1216	FM-GPW-8578	0501
C	CORE assembly	WO-A-1214	FM-GPW-8005	0501
D	COCK	WO-A-1126	FM-9N-8115	0501
E	SHIM	WO-A-4413	FM-GPW-8125-A	0501
F	WASHER	WO-A-1547	FM-34708-S	0501
G	NUT	WO-A-1546	FM-34084-S	0501

RA PD 305174-A

RA PD 305174

GROUP 05—COOLING (Cont'd)

Figure Number	Official Stockage Number	Part Number Willys	Part Number Ford	ITEM	Quantity Reqd. per Unit Assy.	Per 100 Major Items — 12 Mos. Field Maintenance	Per 100 Major Items — Major Overhaul (5th Ech)	Per 100 Major Items — Total First Year Procurement	Estimated Reqmts. per 100 Rebuilds
Col. 1	Col. 2	Col. 3	Col. 4	Col. 5	Col. 6	Col. 7	Col. 8	Col. 9	Col. 10
				0503—WATER PUMP (Cont'd)					
05-2		WO-637053	FM-GPW-8543	GASKET, water pump to cylinder block	1	% 38	—	45	—
05-2		WO-639993	FM-GPW-8512	IMPELLER, water pump	1	6	13	21	40
		WO-A-6839	FM-GPW-18515-B	KIT, repair, water pump	as req.	10	32	48	100
				(Includes:					
				1 WO-636297 FM-GPW-8530 BEARING, w/SHAFT					
				1 WO-637053 FM-GPW-8543 GASKET, pump to block					
				1 WO-640031 FM-GPW-8524-A2 SEAL assembly					
				1 WO-640034 FM-GPW-8557-A WASHER					
				1 WO-636298 FM-GPW-8576 SPRING, retaining)					
05-2		WO-639992	FM-GPW-8501	PUMP, water, assembly	1	% 13	16	—	—

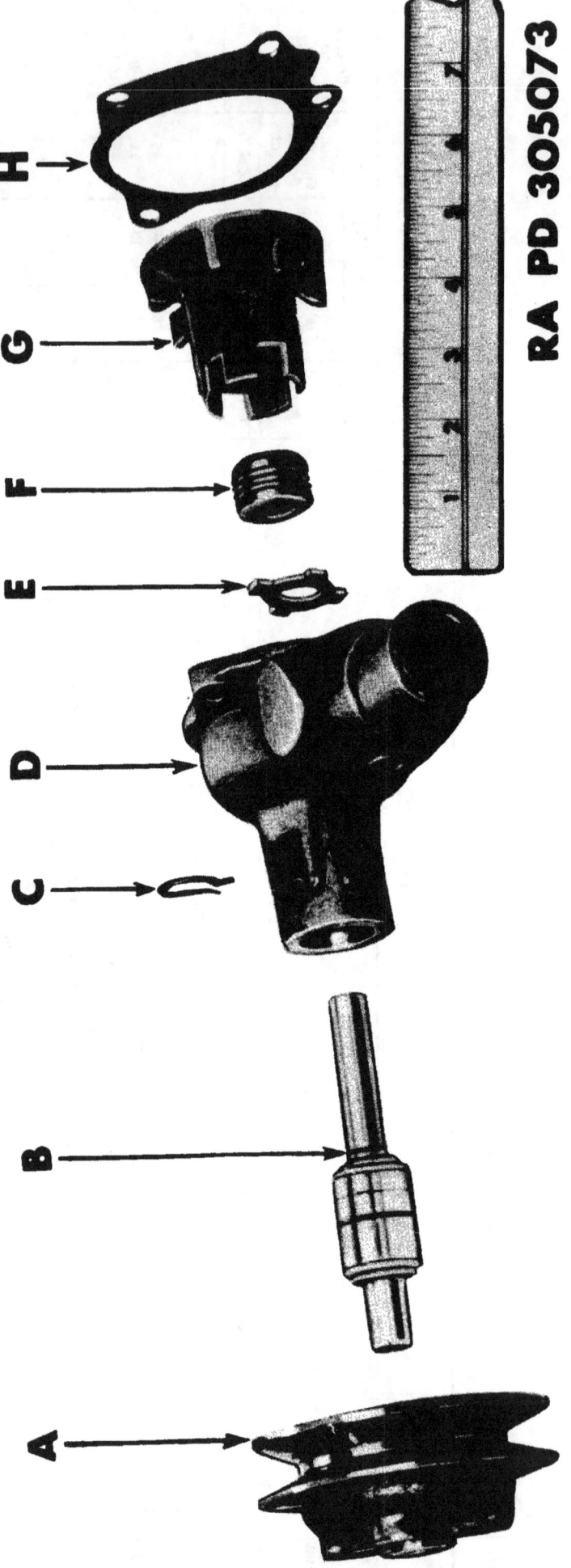

FIGURE 05-2—WATER PUMP ASSEMBLY

Key	Item	Willys Part No.	Ford Part No.	Gov't Group No.
A	PULLEY	WO-636299	FM-GPW-8509-A	0503A
B	SHAFT, w/BEARING	WO-636297	FM-GPW-8530	0503
C	SPRING	WO-636298	FM-GPW-8576	0503
D	BODY	WO-637052	FM-GPW-8505	0503
E	WASHER	WO-640034	FM-GPW-8557-A	0503
F	SEAL	WO-640031	FM-GAA-8524	0503
G	IMPELLER	WO-639993	FM-GPW-8512	0503
H	GASKET	WO-637053	FM-GPW-8543	0503

RA PD 305073-A

SNL G-503

0503-0505

GROUP 05—COOLING (Cont'd)

Figure Number	Official Stockage Number	Part Number - Willys	Part Number - Ford	ITEM	Quantity Reqd. per Unit Assy.	Per 100-Major Items - 12 Mos. Field Maintenance	Per 100-Major Items - Major Overhaul (5th Ech)	Total First Year Procurement	Estimated Reqmts. per 100 Rebuilds
Col. 1	Col. 2	Col. 3	Col. 4	Col. 5	Col. 6	Col. 7	Col. 8	Col. 9	Col. 10
				0503—WATER PUMP (Cont'd)					
		WO-6428	FM-20366-S	BOLT, hex-hd., s-fin., alloy-S., $\frac{5}{16}$-18NC-2 x $\frac{7}{8}$ (pump to block) (GM-106325) (BANX1BB)	3	—	—	—	—
		WO-51858	FM-355442-S	BOLT, hex-hd., s-fin., alloy-S., $\frac{5}{16}$-18NC x $2\frac{1}{2}$ (pump to block) (GM-100127) (BANX1BK)	1	—	—	—	—
		WO-51833	FM-34806-S2	WASHER, lock, reg., S., $\frac{5}{16}$ in. ($1\frac{9}{32}$ O.D. x $1\frac{1}{32}$ I.D. x $\frac{1}{16}$) (BECX1H)	4	—	—	—	—
05-2		WO-640031	FM-GPW-8524-A2	SEAL, water pump, assembly (issue until stock exhausted)	1	50	100	—	—
05-2		WO-636298	FM-GPW-8576	SPRING, retaining, water pump bearing	1	—	—	—	—
05-2		WO-640033	FM-GPW-8572-B	SPRING, water pump seal	1	—	—	—	—
		WO-640034	FM-GPW-8557-A	WASHER, water pump seal	1	—	—	—	—
				0503A—FAN AND PULLEYS					
Fig. 6		WO-A-447	FM-GPW-8600	FAN assembly (HI-88632)	1	3	2	8	5
		WO-51514	FM-20324-S2	BOLT, hex-hd., s-fin., alloy-S., $\frac{1}{4}$-20NC-2 x $\frac{5}{8}$ (fan to pulley) (GM-106319) (BANX4AE)	4	—	—	—	—
		WO-52706	FM-34805-S2	WASHER, lock, S., $\frac{1}{4}$ in. (BECX1G)	4	—	—	—	—
05-2		WO-636299	FM-GPW-8509-A	PULLEY, fan and water pump	1	3	3	10	10
		WO-A-1124	FM-GPW-8240	SHIELD, fan pulley	1	2	2	6	5
		WO-A-1125	FM-GPW-8240	WASHER, fan pulley shield (S., $\frac{5}{8}$ O.D. x $\frac{11}{32}$ I.D.)	6	10	10	40	30
		WO-6701	FM-35676-S	WASHER, fan pulley shield retaining bolt (S., $\frac{9}{16}$ O.D. x $\frac{1}{4}$ I.D. x .005 thk.)	6	10	10	25	30
				0503B—FAN BELT					
		WO-A-1495	FM-GPW-8620-A1	BELT, drive, fan and generator (issue until stock exhausted then use WO-A-9490)	1	%48	32	120	100
		WO-A-9490	FM-GPW-8620-A2	BELT, drive, fan and generator (Goodrich Neopreme) (was WO-A-1495)	1	—	—	—	—
				0505—ENGINE WATER FITTINGS AND HOSE					
01-1		WO-52226	FM-60-8287	CLAMP, hose, $1\frac{13}{16}$ in. dia. (used on all radiator hose)	8	—	—	—	—
		WO-A-6373	FM-GPW-8285	HOSE, radiator, upper (issue until stock exhausted)	2	%136	280	—	—
		WO-A-592	FM-GPW-8284	HOSE, radiator water outlet, lower	1	%200	—	260	—
		WO-630512	FM-IGT-8260	HOSE, radiator water outlet, upper	1	%64	132	—	—
		WO-A-6374	FM-GPW-8290	TUBE, connecting, water inlet	1	%6	—	10	—
		WO-636109	FM-GPW-8269	TUBE, connecting, water outlet	1	%3	—	5	—

SNL G-503

0601

GROUP 06—ELECTRICAL

0601—GENERATOR

06-1	WO-A-1629	FM-GPW-10105	ARM, generator brush (AL-CCE-54)	2			10
06-1	WO-A-1637	FM-GPW-10005	ARMATURE, generator, assembly (AL-GEG-2120-F)	1	3	3	6	10
06-1	WO-A-1649	FM-GPW-10142	BAND, generator, head (AL-GCE-24)	1	3	3	8	10
06-1	WO-51248	FM-GPW-10094	BEARING, ball, generator (before Willys engine 158007) (AL-X-295) (ND-1203)	2				
06-1	WO-A-6299	FM-B-10094	BEARING, ball, generator commutator end (after Willys engine 158007) (AL-X-1655) (ND-77503)	2	6	6	15	20
06-1	WO-A-1623	FM-355260-S7	BOLT, retaining, generator commutator end bearing (hex-hd, S., 1/4-20NF x 5/8) (AL-DA-60)	1	8	16	—	—
06-1	WO-A-1630	FM-GPW-10069	BRUSH, main, generator (AL-GCE-1012) (issue until stock exhausted)	2	132	268	—	30
06-1	WO-A-1651	FM-GPW-18274	BRUSH SET, generator (AL-GCE-2012S)	as req.	19	10	34	—
			(Includes:					
			2 WO-A-1630 FM-GPW-10069 BRUSH)					
06-1	WO-A-1599	FM-GPW-10206-A	BUSHING, insulating, generator field (.203 x .305 x 5/16) (AL-GCY-25) .... Willys	1				
06-1	WO-A-1598	FM-GPW-10104	BUSHING, insulating, generator field (.250 x .312 x 5/16) (AL-GCT-25) .... Ford	1				
				3				
06-1	WO-A-1604	FM-GPW-10175	COIL, field, generator left, right, assembly (AL-GEB-1005A) (issue until stock exhausted)	1	4	4		
06-1	WO-A-1606	FM-GPW-10192	COIL, field, generator left, assembly (AL-GEB-1007A)	1				
06-1	WO-A-1607	FM-GPW-10191	COIL, field, generator right, assembly (AL-GEB-1008B)	1				
06-1	WO-A-1626	FM-GPW-10118	COVER, generator commutator end cap (AL-GBJ-32A) (issue until stock exhausted)	1		4		
06-1	WO-A-1625	FM-GPW-10119	GASKET, generator commutator end cap cover (AL-GBJ-25) (issue until stock exhausted)	1	40	80		
F	WO-A-5992	FM-GPW-10000-A	GENERATOR assembly (AL-GED-5002D) (For attaching parts see Group 0601A)	1	% 10		13	5
06-1	WO-A-1622	FM-GPW-10124	GUARD, oil, generator (flat, S., .986 x 1.570 x .020) (AL-DA-39)	1				
06-1	WO-A-6298	FM-GPW-10050	HEAD, generator, commutator end, assembly (after Willys engine 158007) (AL-GCE-2118A)	1	2	2	6	5
06-1	WO-A-6301	FM-GPW-10139	HEAD, generator, drive end (after Willys engine 158007) (AL-GCE-125A)	1	2	2	6	5
06-1	WO-A-6300	FM-GPW-10138	HEAD, generator, drive end, assembly (after engine 158007) (AL-GCE-1125A)	1				
06-1	WO-A-1592	FM-01A-10193	INSULATOR, generator field connection (AL-GAL-44)	1				
06-1	WO-A-1595	FM-GPW-10208-A	INSULATOR, generator field terminal post, inner (AL-GBW-67)	1				
06-1	WO-A-1594	FM-GPW-10202-C	INSULATOR, generator armature terminal post, inner (AL-GBW-66)	1				
06-1	WO-A-1591	FM-GPW-10202-A	INSULATOR, generator armature terminal post top (17/64 x 3/4 x .062) (AL-GAA-32)	1				
06-1	WO-A-1641	FM-74144-S	KEY, Woodruff No. 5, S. (generator armature assembly) (AL-X-260)	1	5	5	20	15
06-1	WO-A-7840	FM-GPW-18342	KIT, repair, generator field coil	as req.	2	2	5	5
			(Includes:					
			1 WO-A-1599 FM-GPW-10206-A BUSHING					
			1 WO-A-1598 FM-GPW-10104 BUSHING					
			1 WO-A-1604 FM-GPW-10175 COIL assembly					
			1 WO-A-1592 FM-01A-10193 INSULATOR					
			1 WO-A-1595 FM-GPW-10208-B INSULATOR					

77

SNL G-503

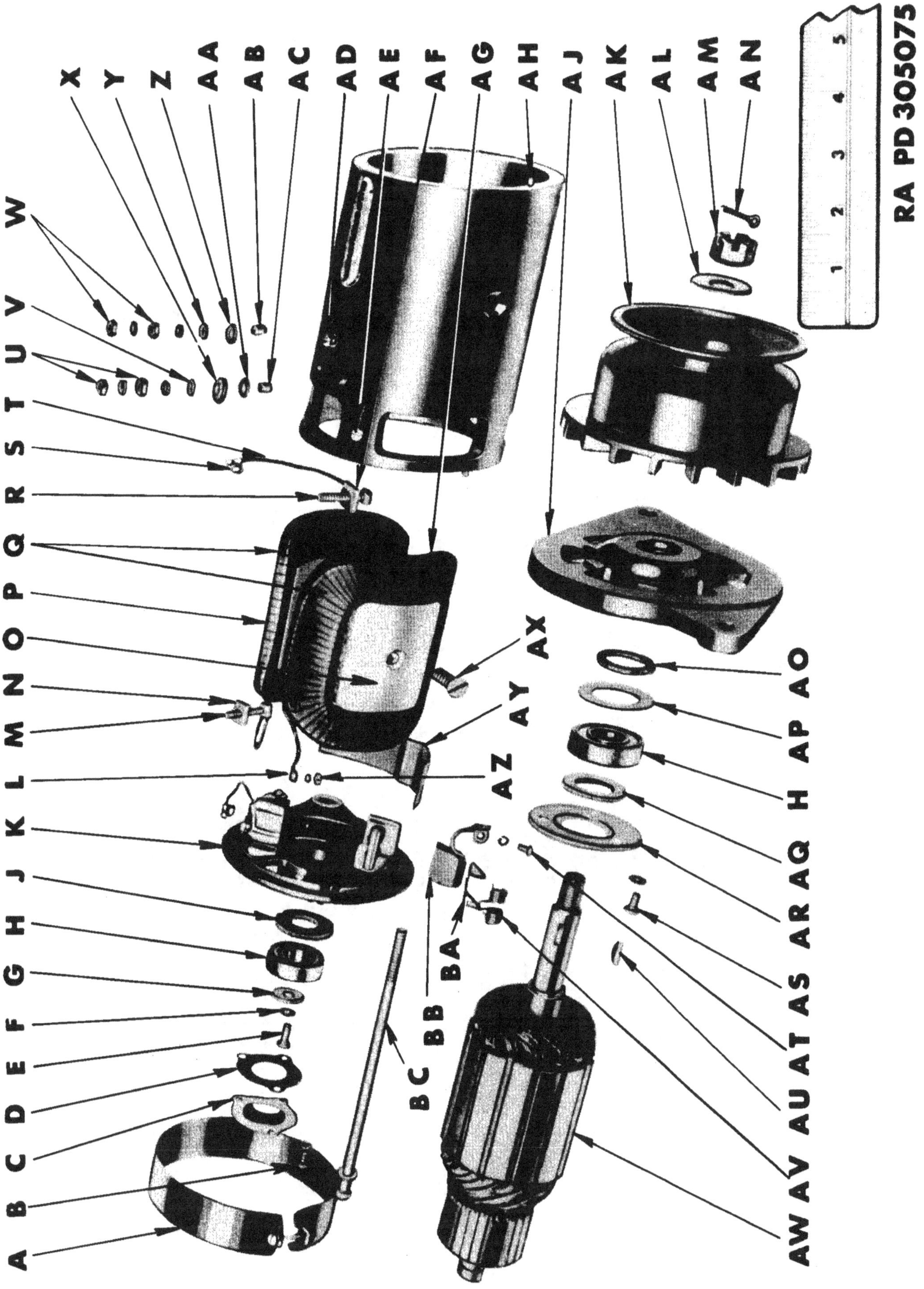

FIGURE 06-1—GENERATOR ASSEMBLY

Key	Item	Willys Part No.	Ford Part No.	Gov't Group No.	Key	Item	Willys Part No.	Ford Part No.	Gov't Group No.
A	BAND	WO-A-1649	FM-GPW-10142	0601	AC	BUSHING	WO-A-1598	FM-GPW-10104	0601
B	SCREW	WO-A-1636	FM-31588-S	0601	AD	SCREW	WO-A-6297	FM-37789-S7	0601
C	COVER	WO-A-1626	FM-GPW-10118	0601	AE	SCREW	WO-A-1618	FM-36009-S	0601
D	GASKET	WO-A-1625	FM-GPW-10119	0601	AF	INSULATION	WO-A-1594	FM-GPW-10106	0601
E	BOLT	WO-A-1623	FM-31588-S	0601	AG	COIL	WO-A-1607	FM-GPW-10191	0601
F	WASHER	WO-5045	FM-34805-S2	0601	AH	PIN	WO-A-1609	FM-GPW-1008-B	0601
G	WASHER	WO-A-1621	FM-GPW-10099	0601	AJ	HEAD assembly	WO-A-6300	(Willys only)	0601
H	BEARING	WO-A-6299	FM-B-10094	0601	AK	PULLEY	WO-A-1639	FM-GPW-10130	0601
J	WASHER	WO-A-1624	FM-GPW-10116	0601	AL	WASHER	WO-A-1638	FM-GPW-10134	0601
K	HEAD assembly	WO-A-6298	FM-GPW-10050	0601	AM	NUT	WO-A-1640	FM-34032-S7	0601
L	TERMINAL	WO-A-1608	FM-GPW-10218	0601	AN	PIN	WO-A-1642	FM-72043-S	0601
M	POST	WO-A-1605	FM-GPW-10210	0601	AO	WASHER	WO-A-1646	FM-GPW-10122	0601
N	INSULATION	WO-A-1595	FM-GPW-10208-A	0601	AP	GUARD	WO-A-1622	FM-GPW-10124	0601
O	POLE PIECE	WO-A-1600	FM-GPW-10041	0601	AQ	WASHER	WO-A-1644	FM-GPW-10212	0601
P	COIL	WO-A-1606	FM-GPW-10192	0601	AR	RETAINER	WO-106313	FM-GPW-10098	0601
Q	COIL assembly	WO-A-1604	FM-GPW-10175	0601	AS	SCREW	WO-A-1647	FM-36800-S	0601
R	POST	WO-A-1602	FM-GPW-10211-A	0601	AT	SCREW	WO-A-1632	FM-26457-S7	0601
S	TERMINAL	WO-A-1603	FM-GPW-10216	0601	AU	KEY	WO-A-1641	FM-74144-S	0601
T	LEAD assembly	WO-A-1601	FM-GPW-10100	0601	AV	SPRING	WO-A-1628	FM-GPW-10057	0601
U	NUT	WO-A-1611	FM-355883-S	0601	AW	ARMATURE	WO-A-1637	FM-GPW-10005	0601
V	WASHER	WO-A-1616	FM-34705-S2	0601	AX	SCREW	WO-A-1596	FM-355486-S7	0601
W	NUT	WO-A-1617	FM-350853-S7	0601	AY	INSULATION	WO-A-1592	FM-0IA-10193	0601
X	INSULATOR	WO-A-1591	FM-GPW-10202-A	0601	AZ	NUT	WO-A-1610	FM-34051-S7	0601
Y	WASHER	WO-A-1615	FM-34703-S7	0601	BA	ARM	WO-A-1629	FM-GPW-10105	0601
Z	WASHER	WO-A-1593	FM-GPW-10206-A	0601	BB	BRUSH	WO-A-1630	(Willys only)	0601
AA	WASHER	WO-A-1597	(Willys only)	0601	BC	SCREW	WO-A-1590	FM-GPW-10120	0601
AB	BUSHING	WO-A-1599	(Willys only)	0601					

RA PD 305075-A

SNL G-503

GROUP 06—ELECTRICAL (Cont'd)

Figure Number	Official Stockage Number	Part Number Willys	Part Number Ford	ITEM	Quantity Reqd. per Unit Assy.	Per 100 Major Items 12 Mos. Field Maintenance	Per 100 Major Items Major Overhaul (5th Ech)	Total First Year Procurement	Estimated Reqmts. per 100 Rebuilds
Col. 1	Col. 2	Col. 3	Col. 4	Col. 5	Col. 6	Col. 7	Col. 8	Col. 9	Col. 10
				0601—GENERATOR (Cont'd)					
				1 WO-A-1594 FM-GPW-10202-C INSULATOR (bottom)					
				1 WO-A-1591 FM-GPW-10202-B INSULATOR (top)					
				1 WO-A-1601 FM-GPW-10100 LEAD assembly					
				1 WO-A-1610 FM-34051-S7 NUT (grd. screw)					
				1 WO-A-1617 FM-350853-S7 NUT (terminal)					
				2 WO-A-1611 FM-355883-S NUT					
				2 WO-A-1618 FM-36009-S7 SCREW (grd.)					
				1 WO-A-1596 FM-355486-S7 SCREW (pole)					
				2 WO-A-1615 FM-34703-S7 WASHER					
				1 WO-A-1597 FM-GPW-10208-A WASHER (insulating)					
				1 WO-A-1593 FM-GPW-10206-A WASHER (insulating)					
				1 WO-A-1614 FM-34803-S7 WASHER, lock					
				1 WO-A-1616 FM-34705-S2 WASHER, lock					
				1 WO-A-1612 FM-356263-S7 WASHER, lock					
				1 WO-A-1613 FM-34801-S7 WASHER, lock					
	WO-A-7895	FM-GPW-18363-B		KIT, repair, generator (issue until stock is exhausted, then use WO-A-9055)	as req.	16	24	—	—
	WO-A-9055	FM-GPW-18363-C		KIT, repair, generator	as req.	6	6	15	20
				(Includes:					
				2 WO-A-1630 FM-GPW-10069 BRUSH					
				1 WO-A-1625 FM-GPW-10119 GASKET, cover					
				1 WO-A-1589 FM-34141-S7 NUT					
				3 WO-A-1636 FM-31588-S7 SCREW, cover					
				1 WO-A-1650 FM-27161-S7 SCREW					
				2 WO-A-1632 FM-26457-S7 SCREW, lead					
				3 WO-A-1647 FM-36800-S SCREW, retainer					
				1 WO-A-1623 FM-355260-S7 SCREW, retaining					
				2 WO-A-1628 FM-GPW-10057 SPRING, brush					
				1 WO-A-1624 FM-GPW-10116 WASHER, felt					
				1 WO-A-1644 FM-78-10212-A WASHER, felt					
				1 WO-A-1646 FM-GPW-10212 WASHER, felt					
				3 WO-A-1635 FM-34803-S7 WASHER, lock					
				1 WO-5045 FM-34805-S7-8 WASHER, lock					
				1 WO-A-5288 FM-34806-S7 WASHER, lock					
				2 WO-51532 FM-34802-S2 WASHER, lock					
				3 WO-52960 FM-34803-S2 WASHER, lock					
				1 WO-A-1621 FM-356208-S7 WASHER, retaining)					
06-1	WO-A-1601		FM-GPW-10100	LEAD, generator armature terminal, assembly (AL-GEB-44) (issue until stock exhausted)	1	12	16	—	—

SNL G-503

06-1	WO-A-1610	FM-34051-S7	NUT, generator field ground screw (hex., S., cd-pltd., No. 6 (.138)-32NC-V (AL-8X-140) (BBKX2A-15) (issue until stock exhausted)	1	8	16	—	—
06-1	WO-A-1617	FM-350853-S7	NUT, generator field terminal post (hex., S., cd-pltd., No. 10 (.190)-32NF-1) (BBMX1C-15) (issue until stock exhausted)	2	16	32	—	—
06-1	WO-A-1611	FM-355883-S	NUT, generator armature terminal (hex., S., cd-pltd., No. 14 (.242)-24NC) (AL-8X-177)	2	—	—	—	—
06-1	WO-A-1589	FM-34141-S7	NUT, generator head band screw (sq-hd., S., cd-pltd., No. 10 (.190)-32NF) (AL-X-794)	1	—	—	—	—
06-1	WO-302347	FM-B-10141	OILER, generator (press in type) (before Willys engine 158007) (AL-X-489)	2	—	—	—	—
06-1	WO-A-1600	FM-GPW-10041	PIECE, pole, generator (AL-GEB-29)	2	—	—	—	—
06-1	WO-A-1609	FM-GPW-10218	PIN, dowel, generator commutator end plate, assembly (S., 1/8 x 7/16) (AL-MN-21)	2	—	—	—	—
06-1	WO-A-6301	FM-GPW-10139	PLATE, generator, commutator end, assembly (after Willys engine 158007) (AL-GCE-125A)	1	—	—	—	—
06-1	WO-A-1645	FM-GPW-10138	PLATE, generator, drive end, assembly (before Willys engine 158007) (AL-GCE-1125)	1	—	—	—	—
06-1	WO-A-1605	FM-GPW-10211-B	POST, generator field terminal (No. 10 (.190)-32NF) (AL-GEB-58)	1	—	—	—	—
06-1	WO-A-1602	FM-GPW-10211-A	POST, generator armature terminal (No. 14 (.242)-24NC) (AL-GEB-27)	1	—	—	—	—
06-1	WO-A-1639	FM-GPW-10130	PULLEY, drive, generator, w/SPACER, assembly (Willys only)	1	2	2	6	5
06-1	WO-A-9492	FM-34032-S7	PULLEY, drive, generator (AL-SP-484-A)	—	—	—	—	—
06-1	WO-A-1640		NUT, generator pulley to shaft (hex., heavy, alloy-S., slotted, 1/2-20NF-2) (AL-X-156) (BBHX1AA) (issue until stock exhausted)	1	4	8	—	—
06-1	WO-A-1642	FM-72034-S	PIN, generator drive pulley nut (cotter, S., 3/32 x 1) (AL-X-404) (BFAX1CG)	1	—	4	—	—
06-1	WO-106313	FM-GPW-10098	RETAINER, generator drive end bearing (AL-CG-6) (issue until stock exhausted)	1	4	4	—	—
06-1	WO-A-1618	FM-36009-S	SCREW, ground, generator (machine fl-hd., S., cd-pltd., No. 6 (.138)-32NC-2 x 9/16) (AL-8X-1420) (issue until stock exhausted)	1	8	16	—	—
06-1	WO-A-1636	FM-31588-S	SCREW, generator commutator end cover (fill-hd., S., cd-pltd., No. 10 (.190)-32NF-2 x 5/16) (AL-8X-870) (issue until stock exhausted)	3	40	80	—	—
06-1	WO-A-1632	FM-26457-S7	SCREW, generator brush lead (rd-hd., S., No. 8 (.164)-32NC-2 x 1/4) (AL-8X-305) (issue until stock exhausted)	2	16	32	—	—
06-1	WO-A-1647	FM-36800-S	SCREW, generator drive end bearing retainer (rd-hd., S., cd-pltd., No. 10 (.190)-32NF-2 x 3/8) (AL-8X-311) (issue until stock exhausted)	3	20	40	—	—
06-1	WO-A-6297	FM-37789-S7	SCREW, generator ground (rd-hd., S., cd-pltd., No. 10 (.190)-32NF-2 x 7/16) (after Willys engine 158007) (AL-8X-1368)	1	6	6	20	—
06-1	WO-A-1650	FM-27161-S7	SCREW, generator head band screw (rd-hd., S., cd-pltd., No. 10 (.190)-32NF-2 x 1 1/4) (AL-8X-715) (issue until stock exhausted)	1	8	16	—	—
06-1	WO-A-1590	FM-GPW-10120	SCREW, generator frame (AL-DK-23)	2	—	3	3	10
06-1	WO-A-1596	FM-355486-S7	SCREW, generator pole piece (AL-GBY-38A) (issue until stock exhausted)	2	12	20	—	—
06-1	WO-A-8842		SPACER, generator shaft (AL-GEG-31) (Willys only)	1	—	—	—	—
06-1	WO-A-1628	FM-GPW-10057	SPRING, generator brush (AL-GCE-53) (issue until stock exhausted)	2	32	68	—	—
06-1	WO-A-1608		TERMINAL, generator field (br., tinned, .175 hole) (AL-X-959) (Willys only) (issue until stock exhausted)	1	8	16	—	—
06-1	WO-A-1603		TERMINAL, generator armature lead (br., tinned, .175 hole) (AL-X-847) (Willys only) (issue until stock exhausted)	1	8	16	—	—
06-1	WO-A-1615	FM-34703-S7	WASHER, S., plain, No. 10 (.190) (generator field terminal) (AL-8X-349) (issue until stock exhausted)	1	20	40	—	—
06-1	WO-A-1616	FM-34705-S2	WASHER, S., cd-pltd., 1/4 in. (generator armature terminal) (AL-8X-361)	1	—	—	—	—
06-1	WO-A-1638	FM-GPW-10134	WASHER, S., cd-pltd., 17/32 x 1 1/4 x .095 (generator armature shaft) (AL-GEW-31) (issue until stock exhausted)	1	5	5	20	15

0601-01601A

GROUP 06—ELECTRICAL (Cont'd)

Figure Number	Official Stockage Number	Part Number Willys	Part Number Ford	ITEM	Quantity Reqd. per Unit Assy.	Per 100 Major Items 12 Mos. Field Maintenance	Per 100 Major Items Major Overhaul (5th Ech)	Total First Year Procurement	Estimated Reqmts. per Unit 100 Rebuilds
Col. 1	Col. 2	Col. 3	Col. 4	Col. 5	Col. 6	Col. 7	Col. 8	Col. 9	Col. 10
				0601—GENERATOR (Cont'd)					
06-1		WO-A-1624	FM-GPW-10116	WASHER, generator commutator end, inner (felt, $\frac{3}{4}$ x $1\frac{9}{16}$ x $\frac{1}{8}$) (AL-DH-7)	1	—	—	—	—
06-1		WO-A-1644	FM-78-10212-A	WASHER, generator drive end head, inner (felt, $\frac{13}{16}$ x $1\frac{5}{16}$ x $\frac{1}{8}$) (AL-GAU-31)	1	—	—	—	—
06-1		WO-A-1646	FM-GPW-10212	WASHER, generator drive end head, outer (felt, $\frac{7}{8}$ x $1\frac{1}{4}$ x $\frac{1}{8}$) (AL-GT-78)	1	—	—	—	—
06-1		WO-A-1597	FM-GPW-10208-A	WASHER, insulating, generator armature terminal, outer ($\frac{17}{64}$ x $\frac{9}{16}$ x $\frac{1}{16}$) (AL-GC-26)	1	—	—	—	—
		WO-A-1593	FM-GPW-10206-A	WASHER, insulating, generator field terminal, outer ($\frac{13}{64}$ x $\frac{9}{16}$ x $\frac{1}{16}$) (AL-GBW-34)	1	—	—	—	—
		WO-A-1614	FM-34803-S7	WASHER, generator terminal and ground screw (lock, S, No. 10) (.190) (AL-12X-196)	7	—	—	—	—
		WO-A-1613	FM-34801-S7	WASHER, lock, generator field ground (S, tinned, No. 6 (.138)) (AL-12X-194) (issue until stock exhausted)	1	20	40	50	60
		WO-51532	FM-34802-S2	WASHER, lock, generator brush lead (S., No. 8 (.164)) (AL-X-195)	2	—	—	—	—
		WO-A-1635	FM-34803-S7	WASHER, lock, generator commutator end cover screw (S., tinned, No. 10 (.190)) (AL-12X-544)	3	—	—	—	—
		WO-5168	FM-34803-S7	WASHER, lock, generator bearing retaining bolt (S., No.10 (.190)) (AL-X-196)	3	—	19	50	—
		WO-A-1612	FM-356263-S7	WASHER, lock, generator armature terminal (S., tinned, No. 14 (.242)) (AL-12X-193)	1	—	—	—	—
06-1		WO-5045	FM-34805-S8	WASHER, lock, S, $\frac{1}{4}$ in. (generator bearing retaining screw) (AL-X-199)	1	—	—	—	—
		WO-A-1619	FM-34806-S2	WASHER, lock, $\frac{5}{16}$ in. (before Willys engine 158007) (AL-12X-203)	2	—	—	—	—
		WO-A-5288	FM-34806-S7	WASHER, lock, generator frame screw (S., tinned, $\frac{5}{16}$) (after Willys engine 158007) (AL-12X-1014)	2	—	—	—	—
06-1		WO-A-1621	FM-356208-S7	WASHER, retaining, generator commutator end bearing (S., plain, .254 x $\frac{13}{16}$ x .065) (AL-DA-22) (issue until stock exhausted)	1	4	8	—	—
				0601A—GENERATOR ATTACHING PARTS					
		WO-A-1399	FM-GPW-10143	BRACE, generator, short (issue until stock exhausted)	1	—	—	—	—
		WO-6157	FM-24426-S2	BOLT, hex-hd., s-fin., alloy-S., $\frac{5}{16}$-18NC-2 x $1\frac{1}{4}$ (brace to generator) (BANX1BD)	1	—	4	—	—
		WO-6167	FM-33797	NUT, reg., hex., S., $\frac{5}{16}$-18NC-2 (GM.102634) (BBAX1B)	1	—	—	—	—
		WO-53025	FM-356309-S7	WASHER, lock, S., cd-pltd, internal, external teeth, $\frac{5}{16}$ in. (GM-178378)	1	—	—	—	—
		WO-A-1491	FM-GPW-10153-A	BRACE, generator, w/HANDLE, assembly	1	% 3	—	4	—
		WO-A-1470	FM-GPW-10177	BUSHING, generator brace locking	1	—	—	—	—
		WO-A-1397	FM-355455-S	BOLT, generator support insulator (hex-hd., S., $\frac{5}{16}$-24NF-3$\frac{29}{32}$)	2	—	—	—	—
		WO-5910	FM-33798-S2	NUT, hex, S., $\frac{5}{16}$-24NF-2 (GM-103025) (BBBX1BA)	2	% 10	5	16	15
		WO-53025	FM-356309-S7	WASHER, lock, S., cd-pltd, internal, external tooth, $\frac{5}{16}$ in. (Bond No. 7) (GM-178378)	2	—	—	—	—

SNL G-503

82

		WO-A-1468	FM-GPW-10176	BOLT, pivot, generator brace (hex-hd., S., $\frac{3}{8}$-24NF-2 x $\frac{7}{8}$) (B19450CB) (issue until stock exhausted)	1	% 4	8	—	—
		WO-5901	FM-33800-S2	NUT, hex., S., $\frac{3}{8}$-24NF-2 (GM-103026) (BBBX1CA)	1	—	—	—	—
		WO-5010	FM-34807-S2	WASHER, lock, S., $\frac{3}{8}$ in. ($\frac{29}{32}$ O.D. x $\frac{13}{32}$ I.D. x $\frac{3}{32}$ thk.) (BECX1K)	1	—	—	—	—
		WO-A-1400	FM-GPW-10162	GUIDE, adjusting, generator brace (issue until stock exhausted)	1	—	4	—	—
		WO-6606	FM-24409-S7	BOLT, hex-hd., S., $\frac{3}{8}$-24NF-2 x $1\frac{1}{8}$ (guide to brace) (GM-106286) (BAQX4AA)	1	—	—	—	—
		WO-5901	FM-33800-S2	NUT, hex., S., $\frac{3}{8}$-24NF-2 (GM-103026) (BBBX1CA)	1	—	—	—	—
		WO-5455	FM-34707-S2	WASHER, plain, S., $\frac{3}{8}$ in.	1	—	—	—	—
		WO-5010	FM-34807-S2	WASHER, lock, S., $\frac{3}{8}$ in. ($\frac{29}{32}$ O.D. x $\frac{13}{32}$ I.D. x $\frac{3}{32}$ thk.) (BECX1K)	1	—	—	—	—
		WO-A-1395	FM-GPW-10178-A	INSULATOR, generator support (rubber, $1\frac{1}{16}$ O.D. x $\frac{1}{2}$ I.D.)	2	6	—	8	—
		WO-A-1469	FM-GPW-10155	SPRING, generator brace (issue until stock exhausted)	1	% 8	16	—	—
		WO-A-1392	FM-GPW-10166	SUPPORT, generator, assembly (issue until stock exhausted)	1	—	4	—	—
		WO-633949	FM-355496-S	BOLT, hex-hd., S., $\frac{3}{8}$-16NC-2 x $\frac{7}{8}$ (support to block)	2	% 19	—	24	—
		WO-5010	FM-34807-S2	WASHER, lock, S., $\frac{3}{8}$ in. ($\frac{29}{32}$ O.D. x $\frac{13}{32}$ I.D. x $\frac{3}{32}$ thk.) (BECX1K)	2	—	—	—	—
		WO-A-1401	FM-356371-S7	WASHER, generator support ($\frac{21}{64}$ hole)	2	—	—	—	—
		WO-A-1396	FM-356436-S	WASHER, generator support ($\frac{33}{64}$)	2	% 6	—	8	—
				0601B—GENERATOR REGULATOR					
06-2		WO-A-7800		ARMATURE, cutout relay coil, assembly (AL-VRY-1043) (Willys only)	1	—	—	—	—
06-2		WO-A-7808		ARMATURE, current regulator coil, assembly (AL-VRY-1061B) (Willys only)	1	—	—	—	—
06-2		WO-A-7809		ARMATURE, voltage regulator coil, assembly (AL-VRY-1080B) (Willys only)	1	—	—	—	—
06-2		WO-A-7807		BRACKET, current regulator coil, adjusting, assembly (AL-VRA-1060) (Willys only)	2	—	—	—	—
06-2		WO-A-7802		SCREW, mounting bridge (fil-hd., S., cd-pltd., No. 8 (.164)-32NC x $\frac{7}{16}$) (AL-8X-878) (Willys only)	4	—	—	—	—
06-2		WO-A-7803		SCREW, mounting bridge and bracket (fil-hd., S., cd-pltd., No. 8 (.164)-32NC x $\frac{5}{16}$ (AL-8X-888) (Willys only)	8	—	—	—	—
06-2		WO-A-1666		WASHER, plain, S., cd-pltd., No. 8 (.164 in.) (AL-8X-350) (Willys only)	8	—	—	—	—
		WO-51532		WASHER, lock, S., No. 8 (.164 in.) (AL-X-195) (Willys only)	12	—	—	—	—
		WO-A-8322		BRACKET, mounting, resistor (AL-VRA-67) (Willys only)	1	—	—	—	—
		WO-52781		SCREW, rd-hd., S., cd-pltd., No. 8 (.164)-32NC x $\frac{3}{8}$ (AL-8X-55) (Willys only)	6	—	—	—	—
06-2		WO-A-1666		WASHER, plain, S., cd-pltd., No. 8 (.164 in.) (AL-8X-350) (Willys only)	6	—	—	—	—
		WO-51532		WASHER, lock, S., No. 8 (.164 in.) (AL-X-195) (Willys only)	6	—	—	—	—
06-2		WO-A-7801		BRACKET, stationary, cutout relay, w/CONTACT (AL-VRH-1073) (Willys only)	1	—	—	—	—
06-2		WO-A-7802		SCREW, mounting, bridge (fil-hd., S., cd-pltd., No. 8 (.164)-32NC x $\frac{7}{16}$ (AL-8X-878) (Willys only)	2	—	—	—	—
06-2		WO-A-7803		SCREW, mounting, bridge (fil-hd., S., cd-pltd., No. 8 (.164)-32NC x $\frac{5}{16}$ (AL-8X-888) (Willys only)	4	—	—	—	—
06-2		WO-A-1666		WASHER, plain, S., cd-pltd., No. 8 (.164 in.) (AL-8X-350) (Willys only)	4	—	—	—	—
		WO-51532		WASHER, lock, S., No. 8 (.164 in.) (AL-X-195) (Willys only)	6	—	—	—	—
06-2		WO-A-9044		COIL, cutout relay, assembly (AL-VRY-3035) (Willys only)	1	—	—	—	—
06-2		WO-A-9051		COIL, current regulator, assembly (AL-VRY-3070D) (Willys only)	1	—	—	—	—
06-2		WO-A-9049		COIL, voltage regulator, assembly (AL-VRY-3071-B) (Willys only)	1	—	—	—	—
06-2		WO-A-9053		NUT, mounting, coil, hex., S., cd-pltd., $\frac{1}{4}$-20NC (AL-8X-1055) (Willys only)	3	—	—	—	—
		WO-A-8914		WASHER, lock, S., $\frac{1}{4}$ in. (AL-X-535) (Willys only)	3	—	—	—	—

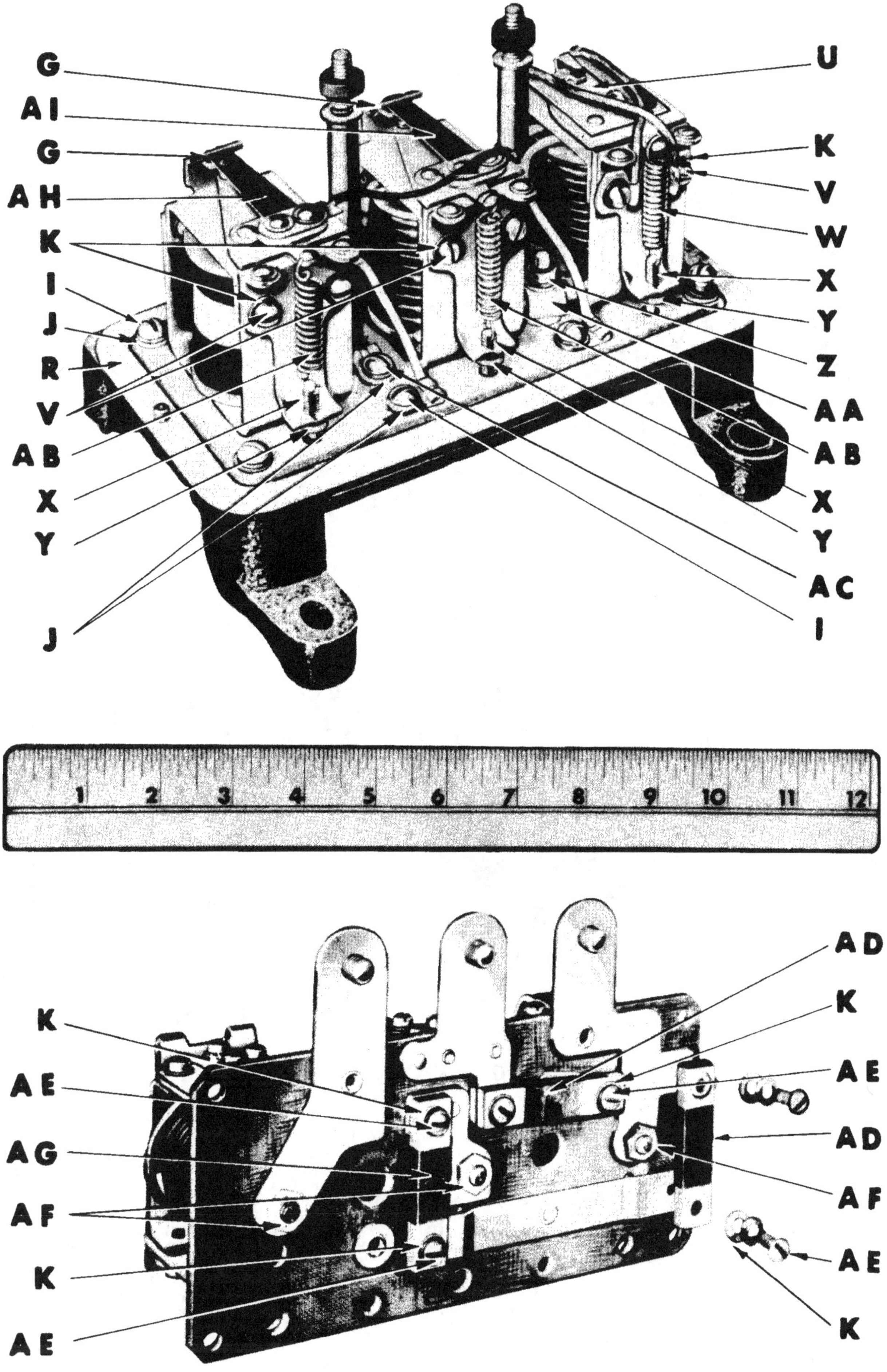

RA PD 305084

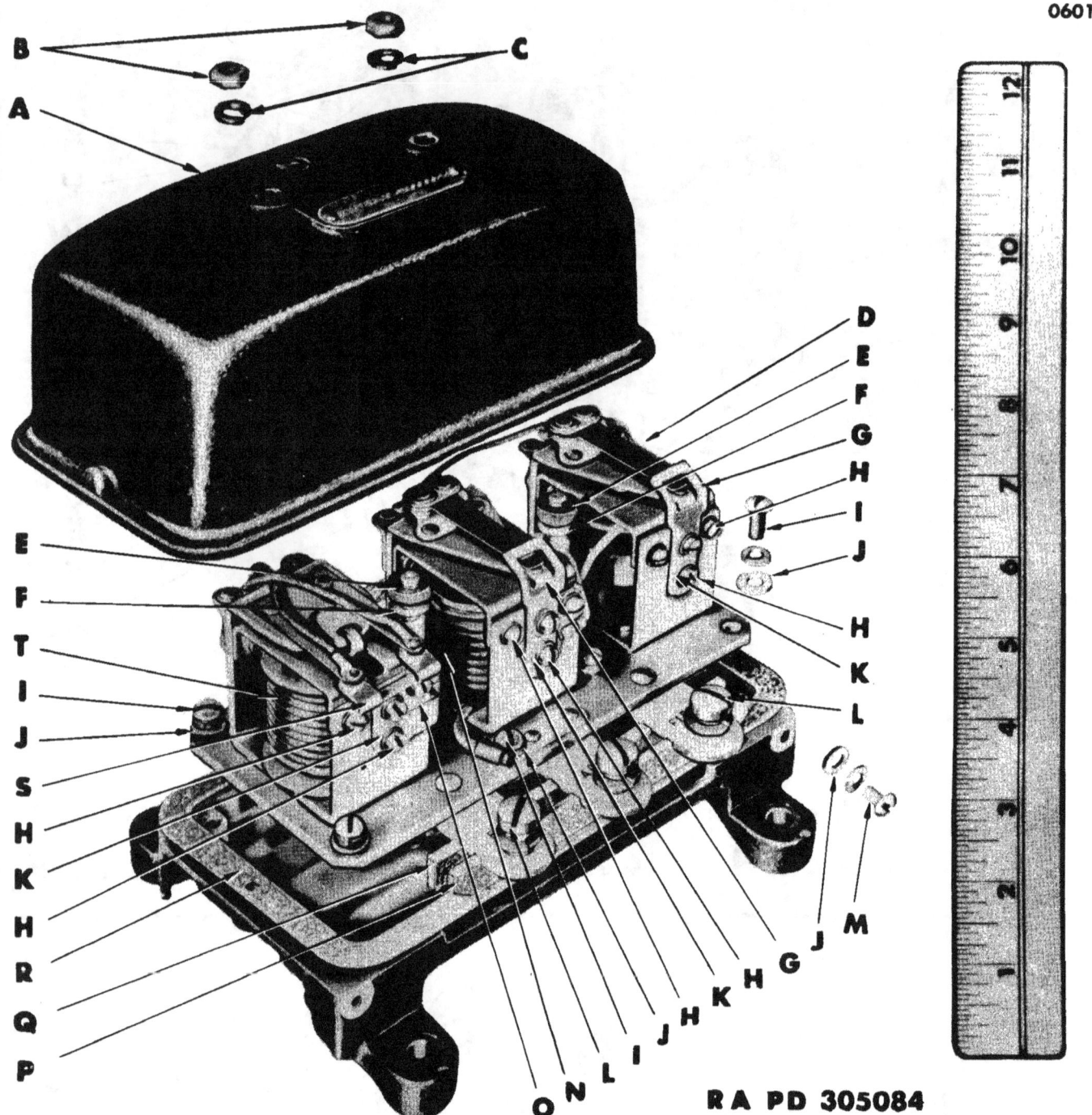

FIGURE 06-2—VOLTAGE REGULATOR

Key	Item	Willys Part No.	Ford Part No.	Gov't Group No.
A	COVER	WO-A-9042	(Willys only)	0601B
B	NUT	WO-A-9040	(Willys only)	0601B
C	WASHER	WO-A-9041	(Willys only)	0601B
D	COIL assembly	WO-A-9049	(Willys only)	0601B
E	WASHER	WO-A-9043	(Willys only)	0601B
F	WASHER	WO-A-1616	(Willys only)	0601B
G	BRACKET	WO-A-7807	(Willys only)	0601B
H	SCREW	WO-A-7803	(Willys only)	0601B
I	SCREW	WO-A-8993	(Willys only)	0601B
J	WASHER	WO-A-5262	(Willys only)	0601B
K	WASHER	WO-A-1666	(Willys only)	0601B
L	SCREW	WO-A-9050	(Willys only)	0601B
M	SCREW	WO-A-1620	(Willys only)	0601B
N	COIL assembly	WO-A-9051	(Willys only)	0601B
O	INSULATION	WO-A-7799	(Willys only)	0601B
P	INSULATION	WO-A-9048	(Willys only)	0601B
Q	GASKET	WO-A-9047	(Willys only)	0601B
R	GASKET	WO-A-9046	(Willys only)	0601B
S	BRACKET	WO-A-7801	(Willys only)	0601B
T	COIL assembly	WO-A-9044	(Willys only)	0601B
U	ARMATURE	WO-A-7800	(Willys only)	0601B
V	SCREW	WO-A-7802	(Willys only)	0601B
W	SPRING	WO-A-7798	(Willys only)	0601B
X	SCREW	WO-A-7797	(Willys only)	0601B
Y	NUT	WO-A-7796	(Willys only)	0601B
Z	NUT	WO-A-5260	(Willys only)	0601B
AA	CONNECTOR	WO-A-9052	(Willys only)	0601B
AB	SPRING	WO-A-7806	(Willys only)	0601B
AC	SCREW	WO-A-1620	(Willys only)	0601B
AD	RESISTANCE	WO-A-9054	(Willys only)	0601B
AE	SCREW	WO-52781	(Willys only)	0601B
AF	NUT	WO-A-9053	(Willys only)	0601B
AG	RESISTANCE	WO-A-5256	(Willys only)	0601B
AH	ARMATURE	WO-A-7809	(Willys only)	0601B
AI	ARMATURE	WO-A-7808	(Willys only)	0601B

RA PD 305084-A
SNL G-503

GROUP 06—ELECTRICAL (Cont'd)

Figure Number	Official Stockage Number	Part Number (Willys)	Part Number (Ford)	ITEM	Quantity Reqd. per Unit Assy.	Per 100 Major Items — 12 Mos. Field Maintenance	Per 100 Major Items — Major Overhaul (5th Ech)	Total First Year Procurement	Estimated Reqmts. per 100 Rebuilds
Col. 1	Col. 2	Col. 3	Col. 4	Col. 5	Col. 6	Col. 7	Col. 8	Col. 9	Col. 10
				0601B—GENERATOR REGULATOR (Cont'd)					
06-2		WO-A-9052		CONNECTOR, series coil (AL-VRA-46) (Willys only)	1	—	—	—	—
06-2		WO-A-9042		COVER, voltage regulator, assembly (AL-VRA-1002) (Willys only)	1	—	—	—	—
06-2		WO-A-9040		NUT, mounting, cover, hex, S, cd-pltd, ¼-20NC (AL-8X-163) (Willys only)	2	—	—	—	—
06-2		WO-A-1616		WASHER, plain, S., cd-pltd., ¼ in. (AL-8X-361) (Willys only)	4	—	—	—	—
06-2		WO-A-9041		WASHER, lock, S., tinned, ¼ in. (AL-12X-199) (Willys only)	2	—	—	—	—
06-2		WO-A-9043		WASHER, rubber (VRA-109) (Willys only)	2	—	—	—	—
06-2		WO-A-9046		GASKET, cover (AL-VRA-50) (Willys only)	1	10	10	24	30
06-2		WO-A-9047		GASKET, terminal (AL-VRA-51) (Willys only)	1	5	5	12	15
06-2		WO-A-7799		INSULATION, cutout relay stationary contact bracket (AL-VRA-76) (Willys only)	1	—	—	—	—
06-2		WO-A-9048		INSULATION, terminal (AL-VRA-52) (Willys only)	1	—	—	—	—
		WO-A-7805		KIT, repair, current, regulator	as req.	3	3	9	10
				(Composed of:					
				1 WO-A-7808 ARMATURE					
				1 WO-A-7796 NUT, adjusting					
				1 WO-A-7797 SCREW, adjusting					
				2 WO-A-7802 SCREW					
				1 WO-A-7806 SPRING					
				3 WO-A-1666 WASHER					
				3 WO-51532 WASHER, lock					
		WO-A-7794		KIT, repair, generator cutout relay	as req.	3	6	12	20
				(Composed of:					
				1 WO-A-7800 ARMATURE					
				1 WO-A-7801 BRACKET and CONTACT					
				1 WO-A-7799 INSULATION					
				1 WO-A-7796 NUT, adjusting					
				1 WO-A-7797 SCREW, adjusting					
				2 WO-A-7802 SCREW					
				2 WO-A-7803 SCREW					
				1 WO-A-7798 SPRING, armature					
				3 WO-A-1666 WASHER					
				3 WO-A-7795 WASHER					
				2 WO-A-7804 WASHER					
				3 WO-51532 WASHER)					
		WO-A-7810		KIT, repair, voltage regulator	as req.	3	6	12	20
				(Composed of:					
				1 WO-A-7809 ARMATURE					

0601B

				—	—	—	—	—	—	—	—	—
				—	—	—	—	—	—	—	—	—
				—	—	10	—	—	—	—	—	—
				—	—	—	—	—	—	—	—	—
				—	4	—	—	—	—	—	—	—
				—	8	20	—	—	—	—	—	—
				—	13	—	—	—	—	—	—	—

06-2	WO-A-8997		1 WO-A-7796 NUT, adjusting	1			
06-2	WO-A-5260		1 WO-A-7797 SCREW, adjusting	1			
			2 WO-A-7802 SCREW				
			1 WO-A-7806 SPRING				
			3 WO-A-1666 WASHER				
			3 WO-51532 WASHER)				
			LEAD, voltage regulator armature to base (AL-VRH-26) (Willys only)........	1			
06-2	WO-A-7796		NUT, series connection (hex, S, cd-pltd., No. 10 (.216)-32NC) (AL-8X-173) (Willys only)....................	1			
			NUT, adjusting screw (knurled, br., No. 10 (.216)-48NS) (AL-VRA-15) (Willys only)....................	3			
06-2	WO-A-5256		RESISTANCE, carbon (marked 7) (AL-TC-51U) (Willys only)........	1		3	4
06-2	WO-A-9054		RESISTANCE, carbon (marked 80) (AL-TC-51N) (Willys only)........	2		6	8
06-2	WO-A-1409	FM-GPW-10505	REGULATOR, generator, 6V., 40 amp, assembly (AL-VRY-4203A)........	1	% 10	—	13
			(Includes:				
			REGULATOR, current limiting				
			REGULATOR, voltage				
			RELAY, cutout)				
		FM-24427-S7	BOLT, hex-hd, s-fin, alloy-S., 5/16-24NF-2 x 1¼ (regulator to bracket)	2			
	WO-6609	FM-34706-S2	WASHER, plain, S., 5/16 in. (7/8 O.D. x 3/8 I.D. x 1/16) (GM-106262)....	2			
	WO-5437	FM-34806-S2	WASHER, lock, S., 5/16 in. (19/32 O.D. x 11/32 I.D. x 1/16 thk.) (BECX1H)	2			
	WO-51833		SEAL, lead and wire (AL-GAG-138) (Willys only)....................	1			
	WO-A-8498		SCREW, adjusting (br., No. 10 (.216)-48NS-2 x 19/32) (AL-VRA-16) (Willys only)....................	3			
06-2	WO-A-7797		SCREW, rd-hd., S., cd-pltd., No. 10 (.216)-32NF x 5/16 (AL-8X-321) (Willys only)....	3			
			(2 used lead connection				
			1 used ground)				
06-2	WO-A-1620		SCREW, mounting, resistor (rd-hd., S., cd-pltd., No. 8 (.164)-32NC x 3/8) (AL-8X-55) (Willys only)....................	6			
06-2	WO-52781		SCREW, rd-hd., S., cd-pltd., No. 10 (.216)-32NF x ½ (AL-8X-309) (Willys only)....................	7			
			(6 used base mounting				
			1 used lead connection)				
06-2	WO-A-8993		SCREW, terminal, fil-hd, S, cd-pltd., 5/16-24NF x ½ (AL-8X-137) (Willys only)	3	10	10	20
06-2	WO-A-9050		SPRING, cut out relay (S., wire, 16 turns) (AL-VRA-17) (Willys only)....	1			
06-2	WO-A-7798		SPRING, voltage regulator and current regulator coil (S., wire, 14½ turns) (AL-VRA-84) (Willys only)....	2			
06-2	WO-A-7806		WASHER, S., cd-pltd., No. 10 (.190 in.) (AL-8X-133A) (Willys only)....	11			
	WO-A-5262		(6 used base mounting				
			1 used ground				
			4 used lead connection)				
	WO-5168		WASHER, lock, S., No. 10 (.216 in.) (AL-X-196) (Willys only)........	11			30
			(6 used base mounting				
			1 used ground screw				
			4 used lead connection)				
	WO-A-5288		WASHER, lock, terminal screw (S., tinned 5/16 in.) (AL-12X-1014) (Willys only)	3			
	WO-A-7795		WASHER, insulating cutout relay stationary contact bracket (.187 x .250 x 1/32) (AL-GAA-35) (Willys only)....	2			
	WO-A-7804		WASHER, insulating cut out relay stationary contact bracket (21/64 x 3/8 x 1/32) (AL-X-1465) (Willys only)....	2			

SNL G-503

GROUP 06—ELECTRICAL (Cont'd)

0602—CRANKING MOTOR

Figure Number	Official Stockage Number	Part Number Willys	Part Number Ford	ITEM	Quantity Reqd. per Unit Assy.	Per 100 Major Items 12 Mos. Field Maintenance	Per 100 Major Items Major Overhaul (5th Ech)	Total First Year Procurement	Estimated Reqmts. per 100 Rebuilds
Col. 1	Col. 2	Col. 3	Col. 4	Col. 5	Col. 6	Col. 7	Col. 8	Col. 9	Col. 10
06-3		WO-A-1568	FM-GPW-11005	ARMATURE, cranking motor, assembly (AL-MZ-2089)	1	3	3	9	10
06-3		WO-109452	FM-GPW-11077	BAND, head, cranking motor, assembly (AL-MZ-1024G)	1	3	3	8	10
		WO-A-1589	FM-34141-S2	NUT, sq., S., No. 10 (.190)-32NF-2 (AL-8X-794)	1	%			
		WO-A-1588	FM-36954-S7	SCREW, machine, rd-hd., S., No. 10 (.190)-32NF-2 x 1½ (AL-X-714)	1	%			
		WO-A-1583	FM-GPW-11135	BEARING, cranking motor, .626 x .7535 x .735, absorbent bz. (AL-MG-77A) (issue until stock exhausted)	1		6	6	20
06-3		WO-A-1582	FM-GPW-11130	BEARING, cranking motor, intermediate, assembly (AL-MAB-2040A)	1	36	74	6	5
06-3		WO-109431	FM-GPW-11055	BRUSH, insulated cranking motor (AL-MZ-12)	2	2	2	6	20
06-3		WO-109446	FM-GPW-11056	BRUSH, grounded, cranking motor, assembly (AL-MZ-1034)	1	6	6	20	
06-3		WO-A-1552	FM-GPW-18535	BRUSH SET, cranking motor (AL-MZ-2012-S) (issue until stock exhausted) (Includes: 2 WO-109431 BRUSH; 2 WO-109446 BRUSH)	1	72	148		
06-3		WO-639734	FM-GPW-11134	BUSHING, cranking motor (in flywheel housing)	1		3	6	10
06-3		WO-109436	FM-GPW-11107	BUSHING, insulating, cranking motor terminal post (AL-MU-31) (issue until stock exhausted)	1				
06-3		WO-109427	FM-GPW-11082	COIL, cranking motor, lower left (AL-MZ-1009) (issue until stock exhausted)	1	8	12		
06-3		WO-109428	FM-GPW-11084	COIL, cranking motor, lower right (AL-MZ-1008) (issue until stock exhausted)	1	8	12		
06-3		WO-A-1560	FM-GPW-11083	COIL, cranking motor, upper left (AL-MZ-1007)	1				
06-3		WO-A-1563	FM-GPW-11085	COIL, cranking motor, upper right (AL-MZ-1010)	1				
06-3		WO-A-1558	FM-GPW-11090	CONNECTOR, cranking motor field coil (AL-MZ-32) (issue until stock exhausted)	2	4	8	8	10
06-3		WO-A-1573	FM-GPW-11350	DRIVE, Bendix, assembly (AL-EBA-46)	1	3	3	9	
06-3		WO-A-1576	FM-B-11381	HEAD, driving, cranking motor Bendix (AL-EB-8503) (issue until stock exhausted)	1		4		
06-3		WO-109442	FM-GPW-11061	HOLDER, cranking motor brush (AL-MZ-16)	2				
06-3		WO-A-1585	FM-GPW-11131	HOUSING, cranking motor Bendix drive pinion, assembly (AL-PS-1079A)	1	2	2	8	5
06-3		WO-A-1584	FM-355164-S	BOLT, hex-hd., S., No. 10 (.190)-32NF x 31/32 (AL-MZ-52)	4	13	13	100	40
		WO-5168	FM-34803-S	WASHER, lock, S., No. 10 (.190 in.) (AL-X-196)	4				
06-3		WO-A-1557	FM-GPW-11089	INSULATION, cranking motor field connection (AL-MZ-30A) (issue until stock exhausted)	1	4	8		
06-3		WO-5017	FM-74175-S7	KEY, Woodruff, No. 6 (AL-X-261)	1	5	5	20	15
		WO-A-6756	FM-GPW-18376	KIT, repair, cranking motor (issue until stock is exhausted, then use WO-A-7836)	as req.	12	20		
		WO-A-7836	FM-GPW-18376-B	KIT, repair, cranking motor (was WO-A-6756) (issue until stock exhausted) (Includes: 1 WO-A-1583 FM-GPW-11135 BEARING; 1 WO-A-1588 FM-36954-S7 SCREW, head band; 1 WO-A-1589 FM-34141-S2 NUT, head band	as req.	12			

SNL G-503

0602

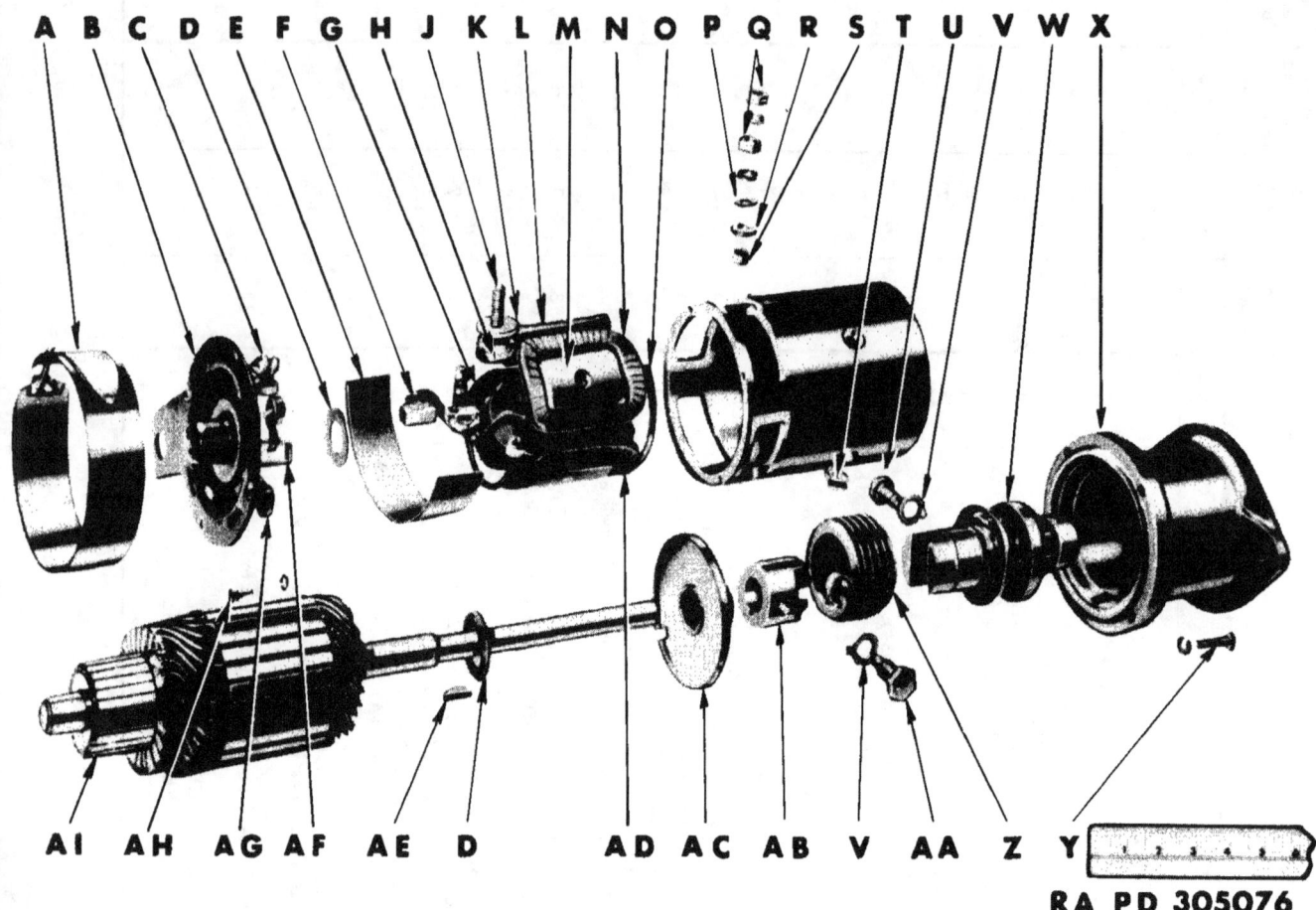

FIGURE 06-3—CRANKING MOTOR ASSEMBLY

Key	Item	Willys Part No.	Ford Part No.	Gov't Group No.
A	BAND	WO-109452	FM-GPW-11077	0602
B	HEAD assembly	WO-A-1566	FM-GPW-11049	0602
C	BRUSH	WO-109446	FM-GPW-11056	0602
D	WASHER	WO-109455	FM-GPW-11036-B	0602
E	INSULATION	WO-A-1557	FM-GPW-11089	0602
F	BRUSH	WO-109431	FM-GPW-11055	0602
G	COIL	WO-109427	FM-GPW-11082	0602
H	TERMINAL	WO-A-1554	FM-GPW-11102	0602
J	TERMINAL	WO-109433	FM-GPW-11103	0602
K	WASHER	WO-109437	FM-GPW-11095	0602
L	COIL	WO-A-1560	FM-GPW-11083	0602
M	PIECE	WO-A-1556	FM-GPW-11120	0602
N	COIL	WO-A-1563	FM-GPW-11085	0602
O	CONNECTOR	WO-A-1558	FM-GPW-11090	0602
P	WASHER	WO-A-1555	FM-34706-S2	0602
Q	NUT	WO-A-1565	FM-355944-S5	0602
R	WASHER	WO-A-1553	FM-GPW-11094	0602
S	BUSHING	WO-109436	FM-GPW-11107	0602
T	SCREW	WO-A-1559	FM-355485-S7	0602
U	SCREW	WO-A-1579	FM-GPW-11382	0602
V	WASHER	WO-A-1574	FM-B-11379	0602
W	SHAFT	WO-A-1581	FM-GPW-11354	0602
X	HOUSING assembly	WO-A-1585	FM-GPW-11131	0602
Y	BOLT	WO-A-1584	FM-355164-S	0602
Z	SPRING	WO-A-1577	FM-GPW-11375	0602
AA	SCREW	WO-A-1578	FM-GPW-11377	0602
AB	HEAD	WO-A-1576	FM-B-11381	0602
AC	BEARING	WO-A-1582	FM-GPW-11130	0602
AD	COIL	WO-109428	FM-GPW-11084	0602
AE	KEY	WO-5017	FM-74175-S7	0602
AF	HOLDER	WO-109442	FM-GPW-11060	0602
AG	SPRING	WO-109445	FM-B-11059	0602
AH	SCREW	WO-A-1572	FM-37364-S7	0602
AI	ARMATURE assembly	WO-A-1568	FM-GPW-11005	0602

RA PD 305076-A

SNL G-503

GROUP 06—ELECTRICAL (Cont'd)

0602—CRANKING MOTOR (Cont'd)

Figure Number	Official Stockage Number	Part Number Willys	Part Number Ford	ITEM	Quantity Reqd. per Unit Assy.	Per 100 Major Items			Estimated Reqmts. per 100 Rebuilds
						12 Mos. Field Maintenance	Major Overhaul (5th Ech)	Total First Year Procurement	
Col. 1	Col. 2	Col. 3	Col. 4	Col. 5	Col. 6	Col. 7	Col. 8	Col. 9	Col. 10
				2 WO-A-1572 FM-31596-S SCREW, head commutator					
				2 WO-A-1584 FM-35164-S SCREW, head					
				1 WO-A-1552 FM-GPW-11535 SET, brush					
				4 WO-109445 FM-B-11059 SPRING, brush					
				2 WO-A-1571 FM-34803-S7 WASHER, lock					
				2 WO-52960 FM-34803-S7 WASHER, lock, w/KIT only					
				2 WO-109455 FM-GPW-11036-B WASHER, thrust					
		WO-A-7842		KIT, repair, cranking motor Bendix drive	as req.	—	3	3	10
				(Includes:					
				1 WO-A-1576 FM-B-11381 HEAD					
				1 WO-5017 FM-74175-S7 KEY, Woodruff					
				1 WO-A-1578 FM-GPW-11377 SCREW, head spring					
				1 WO-A-1579 FM-GPW-11382 SCREW, shaft spring					
				1 WO-A-1575 FM-B-11357-A SLEEVE					
				1 WO-A-1577 FM-GPW-11375 SPRING, drive					
				2 WO-A-1574 FM-B-11379 WASHER, lock)					
		WO-A-7841	FM-GPW-18319	KIT, repair cranking motor field coil	as req.	2	2	8	5
				(Includes:					
				1 WO-109436 FM-GPW-11107 BUSHING, insulating					
				1 WO-109427 FM-GPW-11082 COIL					
				1 WO-109428 FM-GPW-11084 COIL					
				1 WO-A-1560 FM-GPW-11083 COIL					
				1 WO-A-1563 FM-GPW-11085 COIL					
				2 WO-A-1558 FM-GPW-11090 CONNECTOR					
				1 WO-A-1557 FM-GPW-11089 INSULATOR					
				2 WO-A-1565 FM-355944-S5 NUT					
				1 WO-109433 FM-GPW-11103 POST					
				4 WO-A-1559 FM-355485-S7 SCREW					
				1 WO-A-1554 FM-GPW-11102 TERMINAL					
				2 WO-5051 FM-34806-S2 WASHER					
				1 WO-A-1555 FM-34706-S2 WASHER					
				1 WO-A-1553 FM-GPW-11094 WASHER, insulating					
				1 WO-109437 FM-GPW-11095 WASHER, lock)					
F		WO-A-1245	FM-GPW-11001-A	MOTOR, cranking, assembly (AL-MZ-4113)	1	10	—	13	
		WO-51406	FM-24428-S	BOLT, hex-hd, s-fin., alloy-S., 3/8-16NC-2 x 1 1/4 (to bell housing) GM-100135) (BANX1CD)	2	—	—	—	—
		WO-5010	FM-34807-S2	WASHER, lock, S., 3/8 (29/32 O.D. x 13/32 I.D. x 3/32 thk.) (BECX1K)	2	—	—	—	—
06-3		WO-A-1556	FM-GPW-11120	PIECE, pole, cranking motor (AL-MZ-29)	4	—	—	—	—

		WO-A-1586	FM-72798-S7-8	PIN, locating, dowel, cranking motor intermediate bearing (AL-MAB-88)	1	—	—	—	—
06-3		WO-A-1566	FM-GPW-11049	PLATE, cranking motor, commutator end, assembly (AL-MZ-2156)	1	3	3	9	10
06-3		WO-A-1572	FM-37364-S7	SCREW, fill-hd., No. 10 (.190)-32NF x $\frac{3}{8}$ (plate screw) (AL-X-902)	4	13	13	40	40
06-3		WO-109433	FM-GPW-11103	POST, terminal, cranking motor ($\frac{5}{16}$-24NF) (AL-MU-28)	1	—	—	—	—
		WO-A-1565	FM-355944-S5	NUT, hex., $\frac{5}{16}$-24NF (AL-5X-1376) (issue until stock exhausted)	2	8	16	—	—
		WO-5051	FM-34806-S2	WASHER, lock, $\frac{5}{16}$ in. (AL-X-1014) (BECX3H)	2	—	—	—	—
		WO-A-1580	FM-B-11371	RING, takeup, cranking motor, Bendix drive (AL-EB-8734) (issue until stock exhausted)	1	4	4	—	—
		WO-A-1567	FM-GPW-11069	RIVET, cranking motor, commutator end plate (tubular, ov-hd., $\frac{1}{8}$ x $\frac{1}{4}$) (AL-X-532)	—	—	—	—	—
				(Willys)	4	—	—	—	—
				(Ford)	8	—	—	—	—
06-3		WO-A-1578	FM-GPW-11377	SCREW, cranking motor Bendix head spring (AL-EB-8506)	1	19	19	45	60
06-3		WO-A-1559	FM-355485-S7	SCREW, cranking motor pole piece (AL-MZ-38A)	4	—	—	—	—
06-3		WO-A-1579	FM-GPW-11382	SCREW, cranking motor Bendix shaft spring (AL-EB-8507)	1	19	19	45	60
06-3		WO-A-1581	FM-GPW-11354	SHAFT, cranking motor Bendix drive w/PINION, assembly (AL-EBA-4611) (issue until stock exhausted)	1	12	16	—	—
		WO-A-1575	FM-B-11357-A	SLEEVE, compression, cranking motor Bendix drive (AL-EB-7819S) (issue until stock exhausted)	1	4	4	—	—
		WO-A-1569	FM-GPW-11053	SPACER, thrust, cranking motor bearing (AL-MZ-51)	1	—	—	—	—
06-3		WO-109445	FM-B-11059	SPRING, cranking motor brush (AL-MZ-19) (issue until stock exhausted)	4	16	24	—	—
06-3		WO-A-1577	FM-GPW-11375	SPRING, Bendix drive, cranking motor (AL-EB-8505)	1	3	—	6	—
		WO-638646	FM-GPW-11140	SUPPORT, cranking motor (issue until stock exhausted)	1	—	4	—	—
		WO-A-1746		BOLT, hex-hd., S., $\frac{3}{8}$-16NC x $\frac{7}{8}$ (issue until stock exhausted, then use WO-51405)	—	4	4	—	—
		WO-51405		BOLT, hex-hd., s-fin., alloy-S., $\frac{3}{8}$-16NC-2 x 1 (to crankcase) (was WO-A-1746) (BCAX1CB) (Willys only)	1	—	—	—	—
		WO-52132	FM-24327-S7	BOLT, hex-hd., S., $\frac{5}{16}$-24NF-2 x $\frac{5}{8}$ (to cranking motor) (GM-106279)	1	—	—	—	—
		WO-5437	FM-34706-S2	WASHER, S., $\frac{5}{16}$ in. (GM-106262)	1	—	—	—	—
		WO-53026	FM-356200-S7	WASHER, lock, S., cd-pltd., internal, external, tooth, $\frac{3}{8}$ in. (GM-178551)	2	—	—	—	—
		WO-53025	FM-356309-S7	WASHER, lock, S., cd-pltd., internal tooth, $\frac{5}{16}$ in. (GM-178378)	1	—	—	—	—
		WO-51833	FM-34806-S2	WASHER, lock, reg., S., $\frac{5}{16}$ in. (BECX1H)	1	—	—	—	—
06-3		WO-A-1554	FM-GPW-11102	TERMINAL, cranking motor field coil (AL-MU-14)	1	—	—	—	—
		WO-A-1555	FM-34706-S2	WASHER, cranking motor terminal post, plain, S., $\frac{5}{16}$ in. (AL-MU-37)	1	—	—	—	—
06-3		WO-A-1553	FM-GPW-11094	WASHER, insulating, cranking motor terminal post, outer (AL-MAB-31)	1	—	—	—	—
06-3		WO-109437	FM-GPW-11095	WASHER, insulating, cranking motor terminal post, inner (AL-MU-39) (issue until stock exhausted)	1	8	16	—	—
		WO-A-1571	FM-34803-S7	WASHER, cranking motor, commutator end (lock, No. 10 (.190 in.) (AL-X-544)	1	—	—	—	—
06-3		WO-A-1574	FM-B-11379	WASHER, lock, cranking motor housing screw (AL-EB-108)	1	38	38	84	120
06-3		WO-109455	FM-GPW-11036-B	WASHER, thrust, cranking motor armature, commutator end (fibre, .645 x $1\frac{1}{4}$ x $\frac{1}{32}$) (AL-MU-54)	—	—	—	—	—
				(Willys)	2	10	10	20	30
				(Ford)	1	—	—	—	—

0603—DISTRIBUTOR

06-4		WO-A-1674	FM-GPW-12155	ARM, advance control, distributor, assembly (AL-IGS-1080) (issue until stock exhausted)	1	—	4	—	—
06-4		WO-A-1570	FM-GPW-12162	ARM, breaker, distributor, assembly (AL-IGP-3028)	1	—	—	—	—

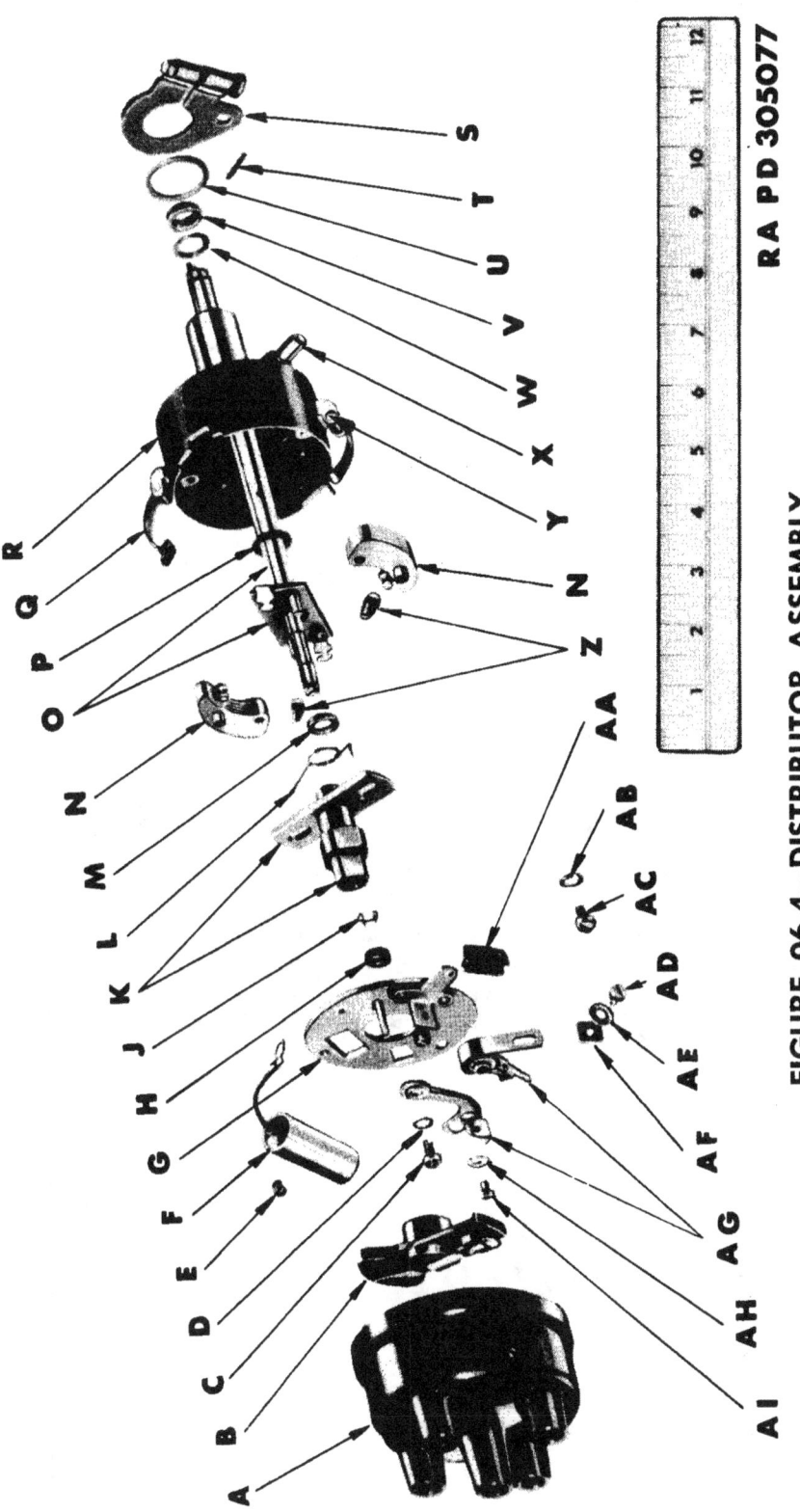

FIGURE 06-4—DISTRIBUTOR ASSEMBLY

Key	Item	Willys Part No.	Ford Part No.	Gov't Group No.
A	CAP assembly	WO-A-1655	FM-GPW-12106	0603
B	ROTOR	WO-A-1658	FM-GPW-12200	0603
C	SCREW	WO-A-1636	FM-31588-S7	0603
D	WASHER	WO-109453	FM-34803-S7	0603
E	SCREW	WO-A-1670	FM-31026-S7	0603
F	CONDENSER	WO-A-1631	FM-GPW-12300	0603
G	PLATE	WO-A-1664	FM-GPW-12010	0603
H	WICK	WO-A-1671	FM-GPW-12133	0603
I	WASHER	WO-A-1669	FM-34801-S7	0603
J	SNAP RING	WO-A-1653	FM-GPW-12177	0603
K	PLATE	WO-A-1661	FM-GPW-12176	0603
L	SPRING	WO-A-1684	FM-GPW-12191	0603
M	SPACER	WO-A-1672	FM-GPW-12120	0603
N	WEIGHT	WO-A-1676	FM-GPW-12188	0603
O	SHAFT	WO-A-1678	FM-GPW-12178	0603
P	WASHER	WO-A-1673	FM-GPW-12182	0603
Q	SPRING	WO-A-1682	FM-GPW-12144	0603
R	BASE assembly	WO-A-1679	FM-GPW-12139	0603
S	ARM	WO-A-1674	FM-GPW-12155	0603
T	RIVET	WO-A-1685	FM-72867-S7	0603
U	WASHER	WO-A-1654	FM-GPW-12267	0603
V	COLLAR	WO-A-1659	FM-GPW-12195	0603
W	WASHER	WO-106740	FM-GPW-12193	0603
X	OILER	WO-107128	FM-B-10141	0603
Y	PIN	WO-A-1683	FM-GPW-12145	0603
Z	SPRING	WO-A-1677	FM-GOW-42084	0603
AA	COVER	WO-A-1660	FM-GPW-12174	0603
AB	WASHER	WO-5168	FM-34803-S7	0603
AC	SCREW	WO-A-1686	FM-31583-S7	0603
AD	SCREW	WO-A-1633	FM-36787-S7	0603
AE	WASHER	WO-A-1667	FM-34701-S7	0603
AF	CLIP	WO-A-1663	FM-GPW-12217	0603
AG	ARM	WO-A-1570	FM-GPW-12162	0603
AH	WASHER	WO-A-1666	FM-34702-S7	0603
AI	SCREW	WO-A-1668	FM-31027-S8	0603

RA PD 305077-A

GROUP 06—ELECTRICAL (Cont'd)

Figure Number	Official Stockage Number	Part Number Willys	Part Number Ford	ITEM	Quantity Reqd. per Unit Assy.	Per 100 Major Items 12 Mos. Field Maintenance	Per 100 Major Items Major Overhaul (5th Ech)	Total First Year Procurement	Estimated Reqmts. per 100 Rebuilds
Col. 1	Col. 2	Col. 3	Col. 4	Col. 5	Col. 6	Col. 7	Col. 8	Col. 9	Col. 10
				0603—DISTRIBUTOR (Cont'd)					
06-4		WO-A-1679	FM-GPW-12139	BASE, distributor, assembly (AL-IGS-2135)	1	—	—	—	—
06-4		WO-A-1681	FM-GPW-12082	BEARING, distributor, absorbent bronze (AL-IG-579-A) (issue until stock exhausted)	2	—	—	—	—
06-4		WO-A-1655	FM-GPW-12106	CAP, distributor, assembly (AL-IG-1324)	1	16	32	30	15
06-4		WO-A-1663	FM-GPW-12217	CLIP, distributor breaker arm spring (AL-IG-676)	1	19	5	50	15
06-4		WO-A-1659	FM-GPW-12195	COLLAR, distributor drive shaft (AL-IGB-199)	1	19	5	50	10
06-4		WO-A-1631	FM-GPW-12300	CONDENSER, distributor, assembly (AL-1GW-2139)	1	% 32	3	45	20
60-4		WO-A-1670	FM-31026-S7	SCREW, fill-hd., No. 6 (.138)-32NC x 5/32 (AL-8X-1546) (issue until stock exhausted)	1	% 32	6	—	—
06-4		WO-A-1669	FM-34801-S7	WASHER, lock, S., No. 6 (.138) (AL-X-1012)	1	% 32	68	—	—
06-4		WO-A-1687	FM-GPW-18354	CONTACT SET distributor (AL-1GP-3028FS) (point set)	1	% 96	24	154	75
				(Includes:					
				1 WO-A-1570 FM-GPW-12162 ARM assembly					
				1 WO-A-1564 FM-GPW-12172 CONTACT, breaker)					
06-4		WO-A-1660	FM-GPW-12174	COVER, distributor, terminal slot (AL-1GC-117)	1	5	5	12	15
F		WO-A-1244	FM-GPW-12100	DISTRIBUTOR assembly (AL-1GC-4705)	1	% 5	5	14	15
		WO-51514	FM-20308-S2	BOLT, hex-hd., S., 1/4-20NC-2 x 5/8 (to cylinder block)	1	—	—	—	—
		WO-52702	FM-34745-S2	WASHER, S., 1/4 in. (BECX1G)	1	—	—	—	—
		WO-52706	FM-34805-S2	WASHER, lock, S., 1/4 in. (BECX1G)	1	—	—	—	—
		WO-A-7843	FM-GPW-17343	KIT, repair, distributor	as req.	6	6	16	20
				(Includes:					
				1 WO-A-1681 FM-GPW-12082 BEARING					
				1 WO-A-1663 FM-GPW-12217 CLIP, spring					
				1 WO-A-1660 FM-GPW-12174 COVER					
				1 WO-107128 FM-B-10141 OILER					
				2 WP-A-1683 FM-GPW-12145 PIN, hinge					
				1 PLUG (AL-1GS-32)					
				1 WO-A-1653 FM-GPW-12177 RING, lock					
				1 WO-A-1685 FM-72867-S7 RIVET					
				1 WO-A-1670 FM-31026-S7 SCREW, condenser					
				1 WO-A-1668 FM-31027-S8 SCREW					
				2 WO-A-1636 FM-31588-S7 SCREW					
				1 WO-A-1633 FM-36787-S7 SCREW, condenser					
				2 WO-A-1682 FM-GPW-12144 SPRING, cap					
				2 WO-A-1669 FM-34801-S7 WASHER, lock					
				3 WO-5168 FM-34803-S7 WASHER, lock					
				1 WO-A-1672 FM-GPW-12120 WASHER					
				1 WO-A-1667 FM-34701-S7 WASHER, No. 6					

0603

GROUP 06—ELECTRICAL (Cont'd)

Figure Number	Official Stockage Number	Part Number Willys	Part Number Ford	ITEM	Quantity Reqd. per Unit Assy.	12 Mos. Field Maintenance	Major Overhaul (5th Ech)	Total First Year Procurement	Estimated Reqmts. per 100 Rebuilds
Col. 1	Col. 2	Col. 3	Col. 4	Col. 5	Col. 6	Col. 7	Col. 8	Col. 9	Col. 10

0603—DISTRIBUTOR (Cont'd)

1 WO-A-1666   FM-34702-S2   WASHER, No. 8
1 WO-A-1673   FM-GPW-12182 WASHER, thrust
1 WO-106740   FM-GPW-12193 WASHER, thrust
1 WO-A-1671   FM-GPW-12133 WICK

Figure Number	Official Stockage Number	Willys	Ford	ITEM	Qty	Col. 7	Col. 8	Col. 9	Col. 10
06-4		WO-107128	FM-B-10141	OILER, distributor shaft (press in sleeve type) (AL-X-490)	1				
06-4		WO-A-1662	FM-GPW-12151	PLATE, breaker, distributor, assembly (AL-IGC-2148C)	1	5	—	5	5
06-4		WO-A-1636	FM-31588-S7	SCREW, fill-hd. S., No. 10 (.190)-32NF x 5/16 (AL-SX-870)	2	2	24	6	5
06-4		WO-109453	FM-34803-S7	WASHER, lock, S., No. 10 (.190 in.) (AL-X-1270)	2	—	—	—	—
06-4		WO-A-1664	FM-GPW-12010	PLATE, breaker part, distributor, assembly (AL-IGC-1148)	1	2	2	6	5
06-4		WO-A-1661	FM-GPW-12176	PLATE, distributor cam and stop, 4 cyl., left hand (AL-IGC-1132LB)	1	—	12	—	—
06-4		WO-A-1683	FM-GPW-12145	PIN, distributor cap hinge (AL-X-1448) (issue until stock exhausted)	2	8	32	—	—
06-4		WO-A-1656	FM-GPW-12011	PLUNGER, distributor contact spring (AL-IG-514) (issue until stock exhausted)	1	% 16	—	—	—
06-4		WO-A-1564	FM-GPW-12218	POINT, breaker distributor (AL-IGC-1149)	1	16	32	45	30
06-4		WO-A-1685	FM-72867-S7	RIVET, distributor, drive shaft collar (AL-SW-213) (issue until stock exhausted)	1	29	10	12	15
06-4		WO-A-1658	FM-GPW-12200	ROTOR, distributor (AL-IG-1657R)	1	% 5	5	—	—
06-4		WO-A-1686	FM-31583-S7	SCREW, condenser terminal (fill-hd., No. 10 (.190)-32NF x 1/4) (AL-8X-872)	1				
06-4		WO-A-1668	FM-31027-S8	SCREW, locking, distributor, stationary contact (fill-hd., S., No. 8 (.164)-32NC x 3/16) (AL-8X-884) (issue until stock exhausted)	1	% 4	8	—	—
06-4		WO-A-1633	FM-36787-S7	SCREW, distributor, breaker arm spring clip (hex-hd. S., slotted, No. 6 (.138)-32NC x 5/16) (AL-IGC-175)	1	—	—	—	—
06-4		WO-A-1678	FM-GPW-12178	SHAFT, drive, distributor assembly (AL-IGS-1134L)	1	—	—	—	—
06-4		WO-A-1675	FM-GPW-12175	SHAFT, drive, distributor, w/GOVERNOR, assembly (AL-IGS-2134L) (issue until stock exhausted)	1	—	4	—	—
06-4		WO-A-1653	FM-GPW-12177	SNAP RING, retaining, distributor cam (AL-IG-680) (issue until stock exhausted)	1	40	80	—	—
06-4		WO-A-1672	FM-GPW-12120	SPACER, distributor cam (AL-IGS-99) (issue until stock exhausted)	1	4	8	—	—
06-4		WO-A-1684	FM-GPW-12191	SPRING, distributor anti-rattle (AL-IGT-69) (issue until stock exhausted)	1	4	4	—	—
06-4		WO-A-1587	FM-GPW-12169	SPRING, distributor breaker arm (AL-IGP-30)	1	—	—	—	—
06-4		WO-A-1657	FM-GPW-12012	SPRING, distributor contact (AL-IG-515) (issue until stock exhausted)	1	16	32	—	—
06-4		WO-A-1682	FM-GPW-12144	SPRING, distributor cap (AL-IG-694) (issue until stock exhausted)	2	8	12	—	—
H		WO-637615	FM-GPW-12083	SPRING, friction distributor shaft	1	5	5	15	20
06-4		WO-A-1677	FM-GPW-12084	SPRING SET, distributor governor weight (AL-IGB-202S) (issue until stock exhausted)	1	20	40	—	—
06-4		WO-5168	FM-34803-S7	WASHER, lock, No. 10 (.190 in.) (condenser terminal screw) (AL-X-196)	4	—	—	—	—
06-4		WO-A-1666	FM-34702-S2	WASHER, distributor stationary contact (lock, No. 8 (.164)) (AL-8X-350) (issue until stock exhausted)	1	—	—	—	—
06-4		WO-A-1667	FM-34701-S7	WASHER, distributor breaker arm spring screw (No. 6 (.138)) (AL-8X-353)	1	20	40	—	—
06-4		WO-A-1654	FM-GPW-12267	WASHER, thrust, distributor advance arm (AL-IG-816C)	1	2	2	6	5

SNL G-503

0603-0604B

06-4	WO-A-1673	FM-GPW-12182	WASHER, thrust, distributor shaft, upper (AL-IGS-104) (issue until stock exhausted)	1				
06-4	WO-106740	FM-GPW-12193	WASHER, thrust, distributor drive shaft, lower (AL-IG-90) (issue until stock exhausted)	1		4	9	10
06-4	WO-A-1676	FM-GPW-12188	WEIGHT, distributor governor assembly (AL-IG-2456)	2	4	8		
06-4	WO-A-1671	FM-GPW-12133	WICK, oiling, distributor (felt) (AL-IGH-28) (issue until stock exhausted)	1	3	3		
					4	8		

0604—IGNITION COIL

	WO-A-1526	FM-GPW-12030	BRACKET, ignition coil (AL-IG-17980D)	1				
	WO-5914	FM-33796-S2	NUT, hex, S., 1/4-28NF-2 (bracket to stud) (GM-103024) (BBBX1AA)	2				10
	WO-635886	FM-357689-S	STUD, S., 1/4-20NC x 1 x 1/4-28NF-2	2	% 3	3	10	
	WO-631105	FM-GPW-12064	WASHER, S., black enamel, 9/32 in. (issue until stock exhausted)	2	% 8	16		10
	WO-53024	FM-351274-S7	WASHER, lock, S., cd-pltd., internal, external teeth, 1/4 in. (GM-174916)	4				
	WO-A-1424	FM-GPW-12000-A	COIL, ignition, w/BRACKET and WASHER, assembly (before Willys serial 288835)	1				10
	WO-A-7792	FM-GPW-12000-B	COIL, ignition, w/BRACKET and WASHER, assembly (after Willys serial 288835) (coil does not have bottom ground connection)	1	% 6	3	13	

0604A—IGNITION WIRING

	WO-301435	FM-GPW-12091-A	BUSHING, spark plug cable support (rubber, 11/16 in. hole) (issue until stock exhausted)	1	% 4	4		
06-5	WO-A-5083	FM-GPW-14321	CABLE, coil primary	1	%			
06-5	WO-A-1420	FM-GPW-12298-B	CABLE, coil secondary	1	%			
06-5	WO-A-1412	FM-GPW-12287	CABLE, spark plug, No. 1	1				
06-5	WO-A-1414	FM-GPW-12284	CABLE, spark plug, No. 2	1				
06-5	WO-A-1416	FM-GPW-12283	CABLE, spark plug, No. 3	1				
	WO-A-1418	FM-GPW-12286	CABLE, spark plug, No. 4	1				
	WO-A-6321		INSULATOR, ignition cable (Willys only)	4				
	WO-A-6757	FM-GPW-18363	KIT, repair, ignition wiring (issue until stock is exhausted. Then use WO-A-7844 for maintenance)	as req.	% 4	8		20
	WO-A-7844	FM-GPW-18363-B	KIT, repair, ignition wiring (was WO-A-6757) (Includes: 1 WO-A-1412 FM-GPW-12287 CABLE, No. 1 1 WO-A-1414 FM-GPW-12284 CABLE, No. 2 1 WO-A-1416 FM-GPW-12283 CABLE, No. 3 1 WO-A-1418 FM-GPW-12286 CABLE, No. 4 1 WO-A-5083 FM-GPW-14321 CABLE, coil primary 1 WO-A-1420 FM-GPW-12298-B CABLE, coil secondary)	as req.	% 5	6	24	
	WO-327257	FM-9N-12113-A	SEAL, weather, spark plug cables, distributor end (rubber)	5	% 10	10	40	30
	WO-A-1652	FM-GPW-12006	TERMINAL, high tension spark plug cables (AL-16-94)	5	% 80	80	200	250
	WO-314369	FM-B-14466	TERMINAL, spark plug cables and coil secondary cable	5	% 19	19	50	60
	WO-307556	FM-B-14463	TIP, coil primary cable	2				

0604B—SPARK PLUGS

	WO-A-1096	FM-11A-12425	CAP, spark plug insulator	4				
	WO-637863	FM-01A-12410	GASKET, spark plug (VC-2066C-C1) (MDR-AM-504K)	4	% 40	64	50	
	WO-538	FM-GPW-12405	PLUG, spark, w/GASKET, assembly (14MM) (CP type-AN-7)	4	% 416		610	200

95

SNL G-503

0604C-0605B

GROUP 06—ELECTRICAL (Cont'd)

Figure Number	Official Stockage Number	Part Number Willys	Part Number Ford	ITEM	Quantity Reqd. per Unit Assy.	Per 100 Major Items 12 Mos. Field Maintenance	Per 100 Major Items Major Overhaul (5th Ech)	Total First Year Procurement	Estimated Reqmts. per 100 Rebuilds
Col. 1	Col. 2	Col. 3	Col. 4	Col. 5	Col. 6	Col. 7	Col. 8	Col. 9	Col. 10
				0604C—IGNITION LOCK					
		WO-A-6814	FM-GPW-3685-B	HANDLE, ignition switch (after 104402 Willys trucks)	1	—	—	—	—
		WO-A-2518	FM-GPW-3685	KEY, ignition switch (before Willys serial 202023)	1	—	—	—	—
		WO-A-6813	FM-26457-S7	NUT, mounting (DM-53170A) (after Willys serial 202023) (Willys only)	1	—	—	—	—
		WO-52131		SCREW, switch terminal (machine rd-hd, S, No. 8 (.164)-32NC-2) (GM-122159)	2	—	—	—	—
18-1		WO-A-6811	FM-GPW-3686-B	SWITCH, keyless ignition, assembly (after Willys serial 202023)	1	% 3	—	5	—
		WO-A-1350	FM-356229-S	WASHER, lock, S, kzd, internal teeth .3350DK-16910D x .018 thk. (before Willys serial 202023)	2	% 19	10	30	—
				0605—INSTRUMENTS					
18-1		WO-A-8186	FM-GPW-10850	AMMETER, 50 amp. (AL-10311A)	1	% 6	—	8	—
		WO-5848		NUT, hex, S, No. 10 (.190)-32NF (to bracket) (AL-8170) (Willys only)	4	—	—	—	—
		WO-A-8129		WASHER, lock, S, No. 10 (.190) (AL-2229) (Willys only)	2	—	—	—	—
		WO-A-8130		BRACKET, mounting, ammeter (AL-9288-A) (Willys only)	1	—	—	—	—
		WO-A-8132		BRACKET, mounting, fuel gauge (AL-10063) (Willys only)	1	—	—	—	—
		WO-A-8131		BRACKET, mounting, oil pressure gauge and heat indicator (AL-21608) (Willys only)	2	—	—	—	—
18-1		WO-A-8184	FM-GPW-9280	GAUGE, fuel (AL-10313-A)	1	% 8	4	1	—
18-1		WO-A-8190	FM-GPW-9273	GAUGE, oil pressure, assembly (AL-10310-A)	1	% 8	4	10	—
		WO-6536		NUT, hex, S, No. 10 (.190)-32NF (to bracket) (AL-2235) (Willys only)	4	—	—	—	—
		WO-A-8129		WASHER, lock, S, No. 10 (.190) (AL-2229) (Willys only)	2	—	—	—	—
		WO-662420	FM-GPW-9319-A	GROMMET, heat indicator through dash (3/16 in., rubber) (issue until stock exhausted)	1	% 4	8	10	—
18-1		WO-A-8188	FM-GPW-10883	INDICATOR, heat, assembly (Moto Meter) (AL-10312-A)	1	% 8	—	—	—
		WO-6536		NUT, hex, S, No. 10 (.190)-32NF (AL-2235) (gauges to bracket) (Willys only)	6	—	—	—	—
		WO-A-8129		WASHER, lock, S, No. 10 (.190) (AL-2229) (Willys only)	4	—	—	—	—
				0605B—INSTRUMENT WIRING					
				NOTE: For cables not appearing here, see Group 0606B.					
		WO-A-5080	FM-GPW-14416	CABLE, fuel gauge to circuit breaker (issue until stock exhausted)	1	% 4	4	—	—
		WO-A-5070	FM-GPW-14406	CABLE, fuel gauge (instrument board) to fuel gauge tank unit (issue until stock exhausted)	1	% 4	4	—	—
06-5		WO-A-5072	FM-GPW-14458	CABLE, ignition switch to ammeter to blackout switch	1	% 4	—	—	—

SNL G-503

0605B-0606

WO-A-5981	FM-GPW-14432	HARNESS, filter wiring (includes the following (4) cables, w/CABLE, ammeter to horn circuit breaker (black—2 red tracings) CABLE, filter (battery ammeter) to ammeter (red—3 black tracings) CABLE, filter (coil ignition switch) to ignition switch to gasoline gauge circuit breaker (black—2 white tracings), CABLE, filter (regulator battery ammeter) to ammeter (black—2 red tracings))	1	2	7	7

0605C—PANEL LIGHT

WO-A-1748	FM-GPW-13713	ADAPTER, instrument light (DM-29392) (issue until stock exhausted)	2	%	4	—
WO-A-1334	FM-GPW-13704	SHIELD, instrument, lamp assembly	2	% 5	—	6
WO-A-1411	FM-GPW-13710	SOCKET, instrument lamp, w/CABLE, assembly (includes two SOCKETS)	1	% 2	2	8

0606—SWITCHES

WO-A-1347		CONTROL, lockout, blackout lighting switch assembly (DM-5943) (Willys only)	1	—	—	—
WO-A-6152		KNOB, w/set SCREW, blackout driving lamp switch assembly (Willys only)	1	—	—	—
WO-A-1348		KNOB, w/set SCREW, blackout lighting switch assembly (DM-5944) (Willys only)	1	—	—	—
WO-A-1352		KNOB, w/set SCREW, instrument panel light switch assembly (DM-5999) (Willys only)	1	—	—	—
WO-111062	FM-33798-S7	NUT, starting switch terminal (hex., S., thin, 5/16-24NF-2)	2	% 10	10	40
WO-52131	FM-26457-S7	SCREW, rd-hd., S., No. 8 (.164)-32NC x 1/4 (instrument light switch, blackout driving light switch terminal) (DM-50033)	3	—	—	—
WO-A-5197	FM-31037-S7	SCREW, fill-hd., S., No. 8 (.164)-32NC x 5/16 (lighting switch terminal)	7	—	—	—
WO-A-6149	FM-GTB-13739	SWITCH, blackout driving light, assembly (DM-D-398) (Includes: KNOB, w/set SCREW ATTACHING PARTS)	1	% 4	2	8
WO-A-1351		SWITCH, blackout driving light, assembly (after serial 163750) (DM-6000) (Willys only)	1	—	—	—
		(Less:				
		KNOB, w/set SCREW)				
WO-A-1353		NUT, mounting, hex., S., 3/8-24NF-2 (DM-53414) (Willys only)	1	—	—	—
WO-52332		WASHER, lock, shakeproof, 3/8 (DM-52947) (Willys only)	1	10	5	24
WO-638979	FM-GPW-13532	SWITCH, dimmer, headlight foot assembly (CL-9654) (use until exhausted then use No. A-12056)	1	%	—	—
A-12056		SWITCH, dimmer headlight, foot, assembly (was WO-638979)	1	%	—	—
WO-51492	FM-26483-S	SCREW, rd-hd., S., 1/4-20NC-2 x 1/2 (GM-113955)	3	—	—	—
WO-52706	FM-34805-S	WASHER, lock, S., 1/4 in. (BECX1G)	3	—	—	—
WO-A-1345		SWITCH, lighting (DM-5969) (Willys only)	1	—	—	—
WO-A-1332	FM-GPW-11649	SWITCH, lighting, assembly (DM-5970)	1	% 6	3	15
		(Includes:				
		BREAKER, circuit				
		CONTROL, lockout				
		KNOB, w/set SCREW				
		ATTACHING PARTS)				
WO-A-7225	FM-9N-11450-A	SWITCH, starting (AL-SW-4015)	1	% 29	—	36
WO-51732	FM-20309-S7	BOLT, hex-hd., S., 1/4-28NF-2 x 1/2 (switch to floor pan) (BAOK4AA)	2	—	—	—
WO-5914	FM-33796-S2	NUT, hex., S., 1/4-28NF-2 (GM-103024) (BBBX1AA)	2	—	—	—

97

SNL G-503

GROUP 06—ELECTRICAL (Cont'd)

Figure Number	Official Stockage Number	Part Number: Willys	Part Number: Ford	ITEM	Quantity Reqd. per Unit Assy.	Per 100 Major Items: 12 Mos. Field Maintenance	Per 100 Major Items: Major Overhaul (5th Ech)	Per 100 Major Items: Total First Year Procurement	Estimated Reqmts. per 100 Rebuilds
Col. 1	Col. 2	Col. 3	Col. 4	Col. 5	Col. 6	Col. 7	Col. 8	Col. 9	Col. 10
				0606—SWITCHES (Cont'd)					
		WO-52031	FM-34805-S2	WASHER, lock, S., 1/4 in. (BECX1G)	2	—	—	—	—
		WO-A-1346		NUT, mounting, special, hex., S., 3/8-24NF-2 (DM-53175) (Willys only)	1	—	—	—	—
18-1		WO-A-1333	FM-GPW-13740	SWITCH, panel light (DM-5995)	1	% 2	2	6	—
		WO-A-1353		NUT, mounting, hex., S., 3/8-24NF-2 (DM-53414) (Willys only)	1	—	—	—	—
		WO-52332		WASHER, lock, S., 3/8 in. (DM-52947) (Willys only)	1	—	—	—	—
12-3		WO-A-1271	FM-11A-13480	SWITCH, stop light	1	% 4	2	12	—
		WO-A-1350	FM-356229-S	WASHER, lock, keyed, shakeproof, S., pk, zd., 17/16 in. (2 used blackout driving lamp switch terminal screw) (7 used blackout lighting switch) (DM-51104)	10	—	—	—	—
		WO-52424	FM-34806-S7	WASHER, lock, shakeproof, 5/16 in. (starting switch terminal)	2	% 10	10	100	—
				0606B—CHASSIS WIRING					
06-5		WO-A-6153	FM-GPW-13181	CABLE, blackout driving light connector to switch (black—2 white tracings) (after serial Willys 163750)	1	—	—	—	—
		WO-A-5078	FM-GPW-14457	CABLE, filter (batt.) to starting switch (before Willys serial 288835)	1	—	—	—	—
		WO-A-5079	FM-GPW-14456	CABLE, filter (coil pri.) to junction block (before Willys serial 288835)	1	—	—	—	—
		WO-A-5073	FM-GPW-14459	CABLE, filter (reg. batt.) to junction block (before Willys serial 288835)	1	—	—	—	—
		WO-A-5041	FM-GPW-18846	CABLE, voltage regulator to generator ("G" cable bond No. 6)	1	% 3	—	4	—
		WO-A-5082	FM-GPW-14465	CABLE, voltage regulator to junction block (before Willys serial 000000)	1	% 3	—	5	—
		WO-A-5598	FM-GPW-14561	CLIP, S., open 1/4 in., bolt hole 7/32 in.	8	—	—	—	—
		WO-A-5449	FM-GPW-2281	CLIP, S., open 5/16 in., bolt hole 7/32 in. (on body floor)	13	—	—	—	—
		WO-6352	FM-355836-S7	NUT, reg., hex., S., No. 10 (.190)-24NC-2 (GM-110633) (BBKX2BC)	11	—	—	—	—
		WO-5064	FM-355131-S	SCREW, machine, rd-hd., S., No. 10 (.190)-24NC-2 x 5/8 (BCNX2AC)	11	—	—	—	—
		WO-52889	FM-32866-S2	SCREW, sheet metal, pkzd., No. 10 (.190) x 5/8 clip to body floor	1	% 10	10	100	—
		WO-5580	FM-31866-S	SCREW, rd-hd., wood, No. 10 (.190) x 5/8 clip to floor	1	—	—	—	—
		WO-52221	FM-34803-S2	WASHER, lock, S., No. 10 (.190) (BECX1E)	11	—	—	—	—
		WO-A-5450	FM-GPW-9295	CLIP, S., open, 7/16 in. bolt hole 7/32 in.	6	—	—	—	—
		WO-6352	FM-355836-S7	NUT, hex., S., No. 10 (.190)-24NC-2 (GM-110633) (BBKX2BC)	6	—	—	—	—
		WO-5064	FM-355131-S	SCREW, rd-hd., S., No. 10 (.190)-24NC-2 x 5/8 (BCNX2BC)	6	—	—	—	—
		WO-52221	FM-34803-S2	WASHER, lock, S., No. 10 (.190) (BECX1E)	6	—	—	—	—
		WO-A-1289	FM-GPW-14589	CLIP, S., closed 7/16, bolt hole 13/32 (anchored at foot rest to floor pan bolt)	1	—	—	—	—
		WO-52768	FM-356305-S	WASHER, S., plain, 5/16 in.	1	—	—	—	—
		WO-78932	FM-GPW-14566	CLIP, S., z-pltd., 5/8, bolt hole 1/4 (1 req'd before 24,308 Willys trucks)	2	—	—	—	—
		WO-6352	FM-355836-S7	NUT, hex., S., No. 10 (.190)-24NC-2	2	—	—	—	—
		WO-5064	FM-355131-S	SCREW, machine, rd-hd., S., No. 10 (.190)-24NC x 5/8	2	—	—	—	—
		WO-52221	FM-34803-S2	WASHER, lock, S., No. 10 (.190) (BECX1E)	2	—	—	—	—
06-5		WO-635985	FM-GPW-14487-A	CONNECTOR (3 wire)	2	% 5	5	12	—
06-5		WO-635981	FM-GPW-14487-B	CONNECTOR (2 wire)	1	% 5	5	12	—

	WO-662276	FM-GPW-13437-A	GROMMET, ⅜ in. (wiring harness through dash) (issue until stock exhausted)	1	% 4	8	—	—
	WO-345961	FM-GPW-13434-A	GROMMET, 13/32 in. (wiring harness through dash) (trailer coupling socket cable through floor pan) (issue until stock exhausted)	4	% 16	24	—	—
06-5	WO-A-7824	FM-GPW-14305	HARNESS, wiring generator to voltage regulator and filter	1	% 3	5	—	—
	WO-A-5048	FM-GPW-14401-D	HARNESS, wiring body, long (issue until stock exhausted) Then use WO-A-9220 for maintenance before Willys serial 288835	1	% —	3	4	—
	WO-A-9220		HARNESS, wiring, body, long (after Willys serial 288835)	1	—	—	—	—
06-5	WO-A-6154	FM-GPW-14402-B	HARNESS, wiring, body, left side, short (was WO-A-5048)	1	% 2	2	5	—
	WO-A-7823	FM-GPW-14432-B	HARNESS, wiring, body, right side (after Willys serial 288835)	1	—	—	—	—
06-5	WO-A-5061	FM-GPW-14446	HARNESS, wiring, chassis, left side (after Willys serial 163750)	1	% 2	2	5	—
	WO-A-7824	FM-GPW-14305-C	HARNESS, wiring, generator to voltage regulator (after Willys serial 238835 (Willys only)	1	—	—	—	—
06-5	WO-A-1665	FM-GPW-14425	HARNESS, wiring, head light (before Willys serial 238835 (issue until stock exhausted)	1	% 4	8	—	—
	WO-A-7845	FM-GPW-18361-B	KIT, repair, wiring harness	as req.	—	—	—	—
			(Includes:					
			2 WO-A-5078 FM-GPW-14457 CABLE (filter to starter switch)					
			1 WO-A-5079 FM-GPW-14456 CABLE (filter to junction block)					
			1 WO-A-5073 FM-GPW-14459 CABLE (filter to junction block)					
			1 WO-A-5070 FM-GPW-14406 CABLE (gas gauge to tank)					
			1 WO-A-5080 FM-GPW-14416 CABLE (gas gauge to circuit breaker)					
			1 WO-A-5081 FM-GPW-14409 CABLE (horn to junction block)					
			1 WO-A-5072 FM-GPW-14458 CABLE (ignition switch to ammeter)					
			1 WO-A-5082 FM-GPW-14465 CABLE (voltage regulator to junction block)					
			1 WO-A-5074 FM-GPW-14305 HARNESS, wiring (filter)					
			1 WO-A-5048 FM-GPW-14401-C HARNESS, wiring (body, long)					
			1 WO-A-5061 FM-GPW-14446 HARNESS, wiring (chassis)					
			1 WO-A-1665 FM-GPW-14425 HARNESS, wiring (head light)					

0606D—CIRCUIT BREAKERS

	WO-A-1733	FM-GPW-12250A	BREAKER, circuit, 5 amp. (between ignition switch and gas gauge on instrument board)	1	% 4	2	10	—
	WO-A-1734	FM-GPW-12250-B	BREAKER, circuit, 15 amp. (between ammeter and horn)	1	% 4	2	10	—
	WO-A-1349	FM-GPW-12250-C	BREAKER, thermal, circuit, 30 amp. (included w/lighting SWITCH, assembly (DM-53097-A)	1	% 4	2	9	—
	WO-52131	FM-26496-S2	SCREW, S., rd-hd., mach., cd. or z. pltd., No. 8 (.164)-32NC-2 x ¼ (breaker terminal) (GM-122159) (BCNX1FC-15)	4	—	—	—	—
	WO-52652	FM-34802-S2	WASHER, lock, S., cd-pltd., No. 8 (.164) (breaker terminal screw) (BCEXID-15)	4	—	—	—	—

0606E—JUNCTION BLOCKS

	WO-639599	FM-GPW-14448-C	BLOCK, junction, 2 post (headlight wires) (before Willys serial)	2	% 2	2	6	—
	WO-A-1490	FM-GPW-14448-D	BLOCK, junction, 6 post (body wiring harness)	1	% 2	2	6	—
	WO-6352	FM-355836-S7	NUT, hex., S., No. 10 (.190)-24NC-2 (GM-110633) (BBKX2BC)	6	—	—	—	—
	WO-5064	FM-355131-S-	SCREW, rd-hd., S., No. 10 (.190)-24NC-2 x ⅝ (BCNX2BC) (block to fender splasher)	6	—	—	—	—

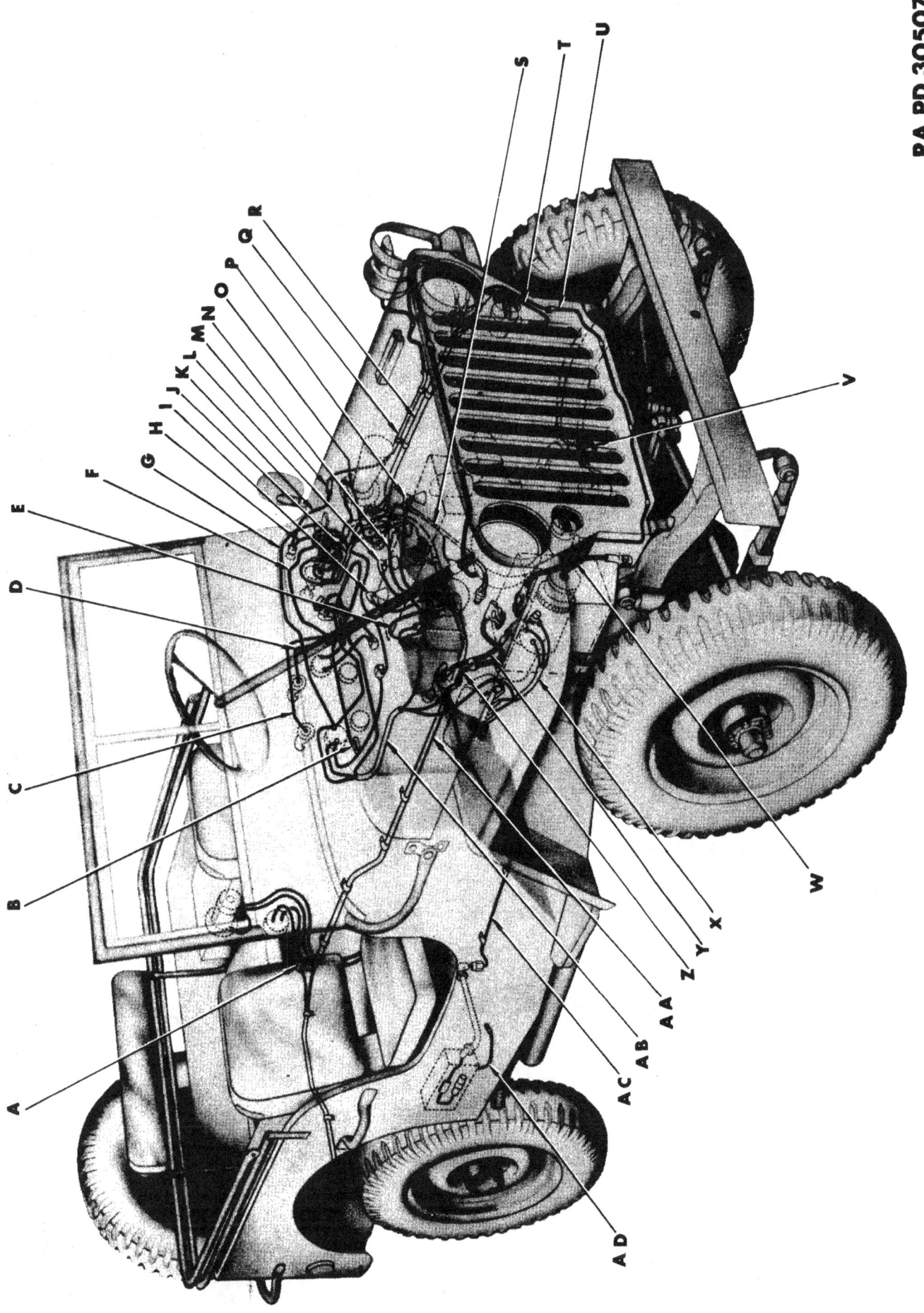

FIGURE 06-5—WIRING INSTALLATION

Key	Item	Willys Part No.	Ford Part No.	Gov't Group No.
A	CONNECTOR	WO-635985	FM-GPW-14487-A	0606B
B	GAUGE	WO-A-1292	FM-GPW-9276	0300A
C	SOCKET w/CABLE	WO-A-1411	FM-GPW-13710	0605C
D				
E	CABLE	WO-A-1420	FM-GPW-12298-B	0604A
F	CABLE	WO-A-5054	(included w/WO-A-5048)	
G	CABLE	WO-A-5072	FM-GPW-14458	0606B
H	CABLE	See WO-A-6154	FM-GPW-14402-B	0605B
I	CABLE	WO-A-1418	FM-GPW-12286	0606B
J	HARNESS	WO-A-6154	FM-GPW-14402B	0604A
K	CABLE	WO-A-6153	FM-GPW-13181	0606B
L	CABLE	WO-A-1416	FM-GPW-12283	0604A
M	CONNECTOR	WO-635981	FM-GPW-14487-B	0606B
N	CABLE	WO-A-1414	FM-GPW-12284	0604A
O	CABLE	WO-A-5081	FM-GPW-14409	0609B
P	CABLE	WO-A-1412	FM-GPW-12287	0604A
Q	CABLE	WO-A-6146	FM-GPW-13175	0607D
R	HARNESS	WO-A-1665	FM-GPW-14425	0606B
S	HARNESS	WO-A-5061	FM-GPW-14446	0606B
T	CABLE	See WO-A-1436	FM-GPW-13201	0604D
U	CABLE	See WO-A-1437	FM-GPW-13200	0607D
V	STRAP	WO-A-1098	FM-GPW-14303	2602
W	STRAP	WO-635883	FM-GPW-14301	0610A
X	HARNESS	WO-A-5074	FM-GPW-14305	0606B
Y	CABLE	WO-A-1454	FM-GPW-14431	0610A
Z	CABLE	WO-A-1452	FM-GPW-14300	0610A
AA	HARNESS	WO-A-5048	FM-GPW-14401-D	0606B
AB	HARNESS	WO-A-5981	FM-GPW-14432	0605B
AC	CABLE	WO-A-8113	FM-GPW-14480-B	2601
AD	CABLE	WO-A-7640	FM-GPW-14513	2601

RA PD 305078-A

FIGURE 06-6—TRAILER ELECTRIC COUPLING SOCKET

Key	Item	Willys Part No.	Ford Part No.	Gov't Group No.
A	NUT	WO-53061	FM-34079	0606F
B	WASHER	WO-53024	FM-351279-S7	0606F
C	WASHER	WO-53313	FM-34805-S8	0606F
D	BOLT	WO-53069	(Willys only)	0606F
E	BOLT	WO-52921	(Willys only)	0606F
F	COVER assembly	WO-A-6587	FM-11YS-18149-B	0606F
G	RETAINER	WO-A-6588	FM-11YS-18198-B	0606F
H	SHIELD	WO-A-6589	FM-11YS-18193-B	0606F
I	SOCKET assembly	WO-A-6586	FM-11YS-18151-B	0606F

RA PD 305079-A

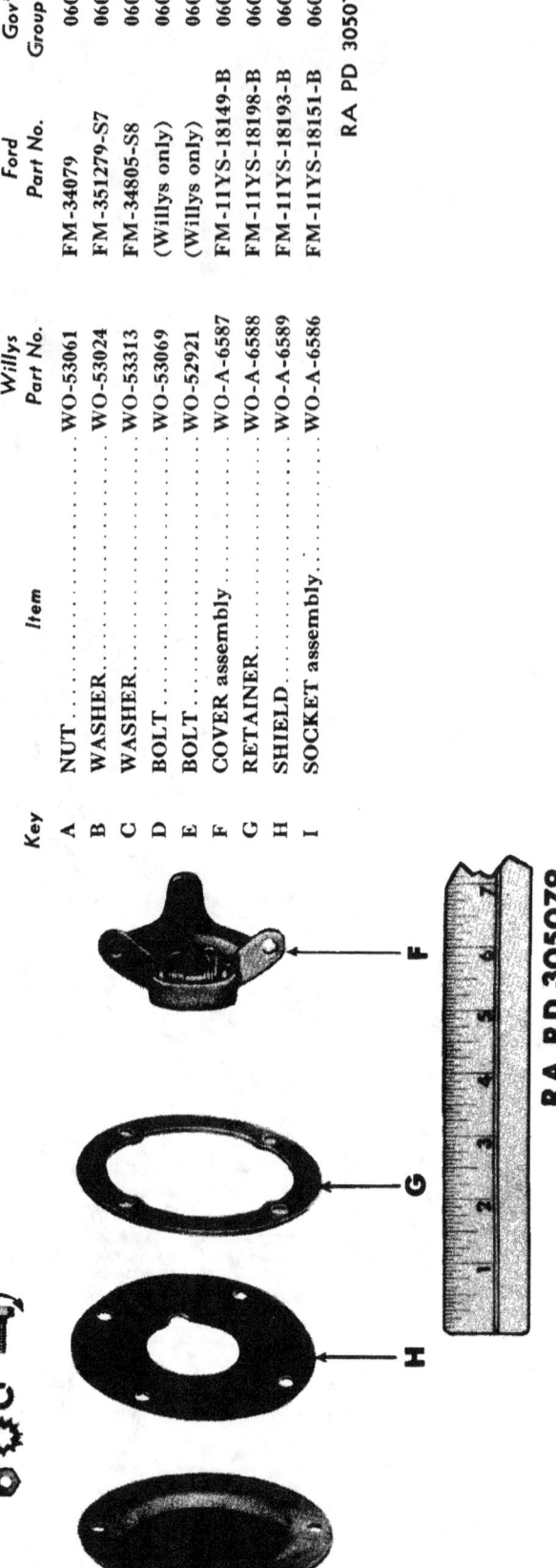

GROUP 06—ELECTRICAL (Cont'd)

Figure Number	Official Stockage Number	Part Number: Willys	Part Number: Ford	ITEM	Quantity Reqd. per Unit Assy.	Per 100 Major Items: 12 Mos. Field Maintenance	Per 100 Major Items: Major Overhaul (5th Ech)	Per 100 Major Items: Total First Year Procurement	Estimated Reqmts. per 100 Rebuilds
Col. 1	Col. 2	Col. 3	Col. 4	Col. 5	Col. 6	Col. 7	Col. 8	Col. 9	Col. 10
				0606E—JUNCTION BLOCKS (Cont'd)					
		WO-5272	FM-355132-S	SCREW, rd-hd., S., No. 10 (.190)-24NC-2 x ¾ (block to dash) (GM-110502) (BCNX2AG) (before utility serial 28835)	2	—	—	—	—
		WO-52994	FM-27698-S	SCREW, machine, rd-hd., S., No. 10 (.190)-32NF-2 x 5/16 (GM-117627) (BCCX1.1AC) (terminal to block)	10	—	—	—	—
		WO-A-1089	FM-356299-S	WASHER (terminal to block screw, special, S., ¼)	10	% 10	10	100	—
		WO-52221	FM-34803-S2	WASHER, lock, S., No. 10 (.190) (BECX1E)	16	—	—	—	—
				0606F—TRAILER ELECTRIC COUPLING					
		WO-A-6356	FM-09B-14362	CABLE, trailer coupling socket to body ground	1	—	—	—	—
06-6		WO-A-6587	FM-11YS-18149-B	COVER assembly (WEB-11935-B)	1	—	—	—	—
		WO-A-8088		COVER, terminal (WEB-110242) (Willys only)	3	—	—	—	—
06-6		WO-53061	FM-34079-S7	NUT, hex., S., cd-pltd., No. 10 (.190)-32NF-2 (socket terminal) (WEB-110477) (GM-120614)	8	—	—	—	—
06-6		WO-A-6588	FM-11YS-18198-B	RETAINER, dust shield (WEB-20099)	—	—	—	—	—
06-6		WO-A-6589	FM-11YS-18193-B	SHIELD, dust (WEB-20098)	1	—	—	—	—
06-6		WO-A-6586	FM-11YS-18151-B	SOCKET, trailer electric coupling, assembly (WEB-3529) (less COVER)	1	—	—	—	—
		WO-A-6019	FM-11YS-18142-B	SOCKET, trailer electric coupling, assembly (to accommodate trailer) (WEB-3604) (Includes: COVER, INTERNAL PARTS, ATTACHING PARTS)	1	—	—	—	—
06-6		WO-53069		BOLT, hex-hd., S., cd-pltd., ¼-28NF-2 x 1 (socket to body) (120364) (Willys only)	2	—	—	—	—
		WO-52921		BOLT, hex-hd., S., cd-pltd., ¼-28NF-2 x ¾ (socket to body) (GM-123450) (Willys only)	2	—	—	—	—
			FM-20364-S7	BOLT, hex-hd., S., cd-pltd., ¼-28NF-2 x ⅞ (Ford only)	4	—	—	—	—
		WO-52847	FM-33795-S7	NUT, hex., S., cd-pltd., ¼-28NF-2 (GM-120367)	4	—	—	—	—
		WO-53313	FM-34805-S8	WASHER, lock, S., z-pltd., ¼ in. S.A.E. (15/32 O.D. x .254 I.D. x 1/16 thk.)	4	—	—	—	—
		WO-53024	FM-351279-S7	WASHER, lock, S., cd-pltd., ¼ in. (internal, external tooth) (GM-174916)	4	—	—	—	—
		WO-A-8087	FM-34753-S	WASHER, terminal screw br., No. 10 (190) (WEB-110110)	4	—	—	—	—
		WO-53194	FM-34903-S2	WASHER, lock, S., pkzd., shakeproof, No. 10 (.190) (GM-138534)	4	—	—	—	—

0607-0607D

06-7	WO-A-1361	FM-GPW-13022	**0607—HEAD LIGHTS** BOLT, mounting headlight (CB-CB-6012)	2				—	
06-7	WO-A-2873	FM-GPW-13020	BOLT, hold down, headlight, w/NUT and WASHER, assembly (S., 5/16-18NC)	2				—	
	WO-A-1036	FM-GPW-13043	DOOR, headlight assembly (CB-CB-8523)	2	%	3	4	6	
	WO-A-1731	FM-GPW-14436	CABLE, ground headlight, (issue until stock exhausted)	2	%	4	10	100	
	WO-52993	FM-32852-S	SCREW, rd-hd., sheet metal, No. 10 (.190) x 3/8 (cable to bracket)	2	%	10	2	8	
	WO-A-1304	FM-GPW-13006-A	LIGHT, service head, left, assembly (CB-215142)	1	%	4	2	8	
	WO-A-1305	FM-GPW-13005-A	LIGHT, service head, right, assembly (CB-215242)	1	%	4			
06-7	WO-5916	FM-33846-S	NUT, hex, S., 1/2-20NF-2 (to support bolt) (GM-103028) (BBBX1EA)	2				—	
06-7	WO-5009	FM-34809-S	WASHER, lock, S., 1/2 in. (BECX1M)	2				—	
	WO-A-2878	FM-GPW-13032	HINGE, head light support (issue until stock exhausted)	2	%	4	4	—	
	WO-51485	FM-20326-S2	BOLT, hex-hd., S., 5/16-18NC-2 x 5/8 (to radiator guard) (GM-106324)	4				—	
	WO-52350	FM-33797-S2	NUT, hex, S., 5/16-18NC-2	4				—	
	WO-51840	FM-34806-S2	WASHER, lock, S., 5/16 in.	4				—	
06-7	WO-A-5586	FM-GPW-13012	HOUSING, headlight, assembly (CB-CB-10436)	2	%	4	12	—	
	WO-A-2466	FM-33896-S2	NUT, wing, S., 5/16-18NC-2 (headlight mounting bolt) (included w/WO-A-2873)	2				—	
	WO-A-2875	FM-357419-S2	PIN, clevis, S., 25/64 lgth. x .248 diam., dld. f/c-pin (head light support bolt)	2				—	
06-7	WO-A-1031	FM-GPW-13015	RETAINER, headlight mounting bolt (CB-CB-8010) (issue until stock exhausted)	2	%	4			
06-7	WO-A-1032	FM-37206-S	SCREW, rd-hd., S. (CB-CB-4057)	2				—	
06-7	WO-52221	FM-34803-S	WASHER, lock, S. (CB-CB-300) (BCEX1E)	2				—	
06-7	WO-A-1037	FM-38095-S	SCREW, headlight door (CB-CB-5099)	2	%	10	4	12	
	WO-A-2870	FM-GPW-13071	SUPPORT, left, headlight, assembly (issue until stock exhausted)	1			4	—	
	WO-A-2871	FM-GPW-13070	SUPPORT, right, headlight, assembly (issue until stock exhausted)	1			4	—	
	WO-52768	FM-34706-S2	WASHER, S. plain 5/16 in. (BECX1G)	2				—	
06-7	WO-A-1362	FM-GPW-13076	WIRE, left, headlight assembly	1	%	2	2	6	
06-7	WO-A-1363	FM-GPW-13075	WIRE, right, headlight assembly	1	%	2	2	6	
			NOTE: For support Bracket, see Group 1701.						
06-8	WO-A-6145	FM-GPW-13152	**0607A—SEALED BEAM** LAMP-UNIT, blackout, driving headlight, 6-8V., sealed, assembly (CB-CB-11267)	1	%	10	3	17	
06-7	WO-A-1033	FM-GPW-13007	LAMP-UNIT, headlight, sealed 6V, assembly (CB-CB-8494) (Seelight unit)	2	%	29	10	46	
06-10	WO-A-1074	FM-GPW-13494-A	LAMP-UNIT, service tail and stop, sealed, 21-3C.P., 6V., assembly (CB-CB-9218)	1	%	21	—	30	
06-10	WO-A-1078	FM-GPW-13485-A	LAMP-UNIT, blackout stop, sealed 3 C.P., 6V, assembly (CB-CB-9234)	1	%	19	1	25	
06-10	WO-A-1075	FM-GPW-13491-A	LAMP-UNIT, blackout tail, sealed 4 opening, 6V, assembly (CB-CB-9225)	2	%	19	—	30	
06-9	WO-51804	FM-B-13466	**0607B—LAMPS** LAMP, elec. incand., 6-8V., 3 C.P. single tung. fil. (MZ-63)	2	%	32	8	60	
06-9	WO-52837	FM-48-15021	LAMP, elec. incand., 6-8V., single tung. fil. (MZ-51)	2	%	16	8	60	
06-8	WO-A-6146		**0607D—BLACKOUT LIGHTS** NOTE: For blackout tail light see Group 0608. CABLE, blackout driving light, assembly (before Willys serial 288835) (CB-CB-11503)	1	%	2	2	4	

SNL G-503

0607D

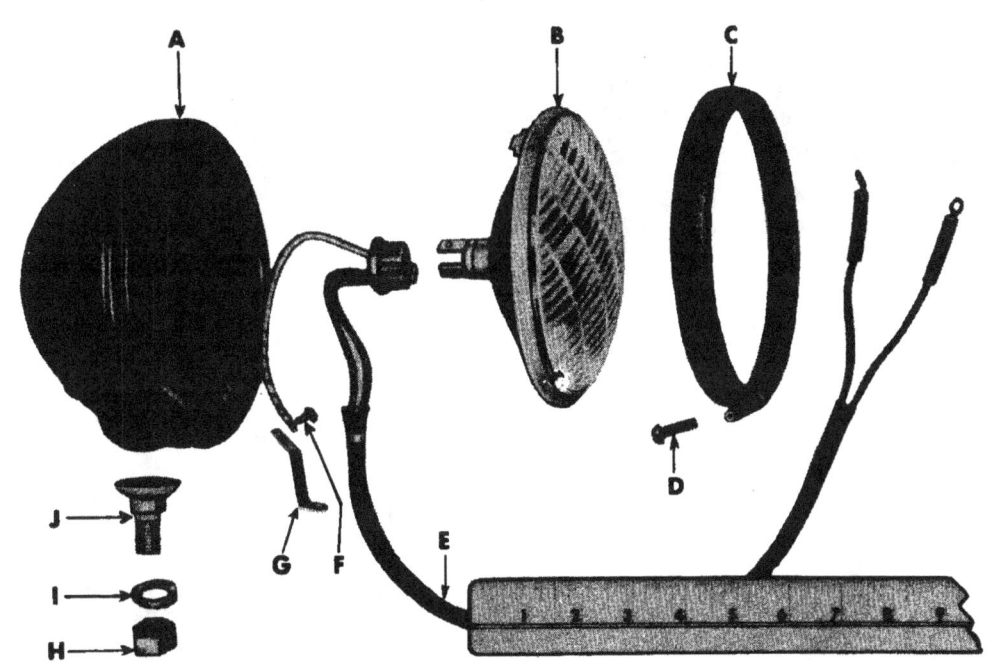

FIGURE 06-7—HEADLIGHT ASSEMBLY

Key	Item	Willys Part No.	Ford Part No.	Gov't Group No.
A	HOUSING SUB assembly	WO-A-5586	FM-GPW-13012	0607
B	SEELITE-UNIT	WO-A-1033	FM-GPW-13007	0607A
C	DOOR assembly	WO-A-1036	FM-GPW-13043	0607
D	SCREW	WO-A-1037	FM-38095-S	0607
E	WIRE assembly	WO-A-1362	FM-GPW-13076	0607
F	SCREW	WO-A-1032	FM-32706-S	0607
G	RETAINER	WO-A-1031	FM-GPW-13022	0607
H	NUT	WO-5916	FM-33846-S2	0607
I	WASHER	WO-5009	FM-34809-S	0607
J	BOLT	WO-A-1361	FM-GPW-13022	0607

RA PD 305082-A

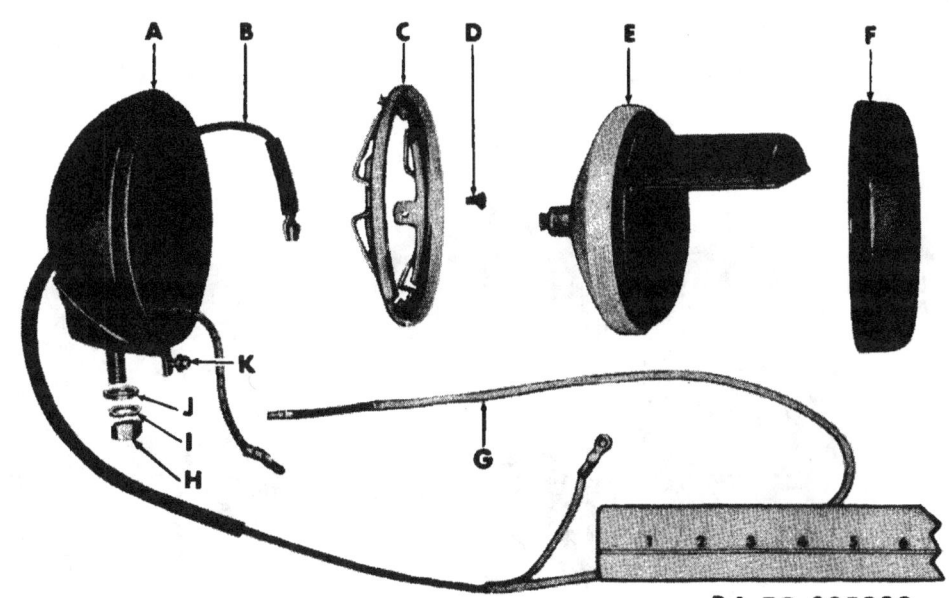

FIGURE 06-8—BLACKOUT DRIVING LIGHT ASSEMBLY

Key	Item	Willys Part No.	Ford Part No.	Gov't Group No.
A	HOUSING assembly	WO-A-6143	FM-GPW-13170	0607D
B	WIRE	WO-A-6147	FM-GPW-13174	0607D
C	RING assembly	WO-A-6783	FM-GPW-13166	0607D
D	SCREW	WO-53071	FM-40631-S7	0607D
E	SEALED-UNIT assembly	WO-A-6145	FM-GPW-13153	0607A
F	DOOR assembly	WO-A-6144	FM-GPW-13162	0607D
G	CABLE assembly	WO-A-6146	FM-GPW-13175	0607D
H	NUT	WO-52033	FM-33799-S2	0607D
I	WASHER	WO-52046		0607D
J	WASHER	WO-A-6313	FM-GPW-13180	0607D
K	SCREW w/WASHER	WO-A-6148	FM-131045-S2	0607D

RA PD 305080-A

SNL G-503

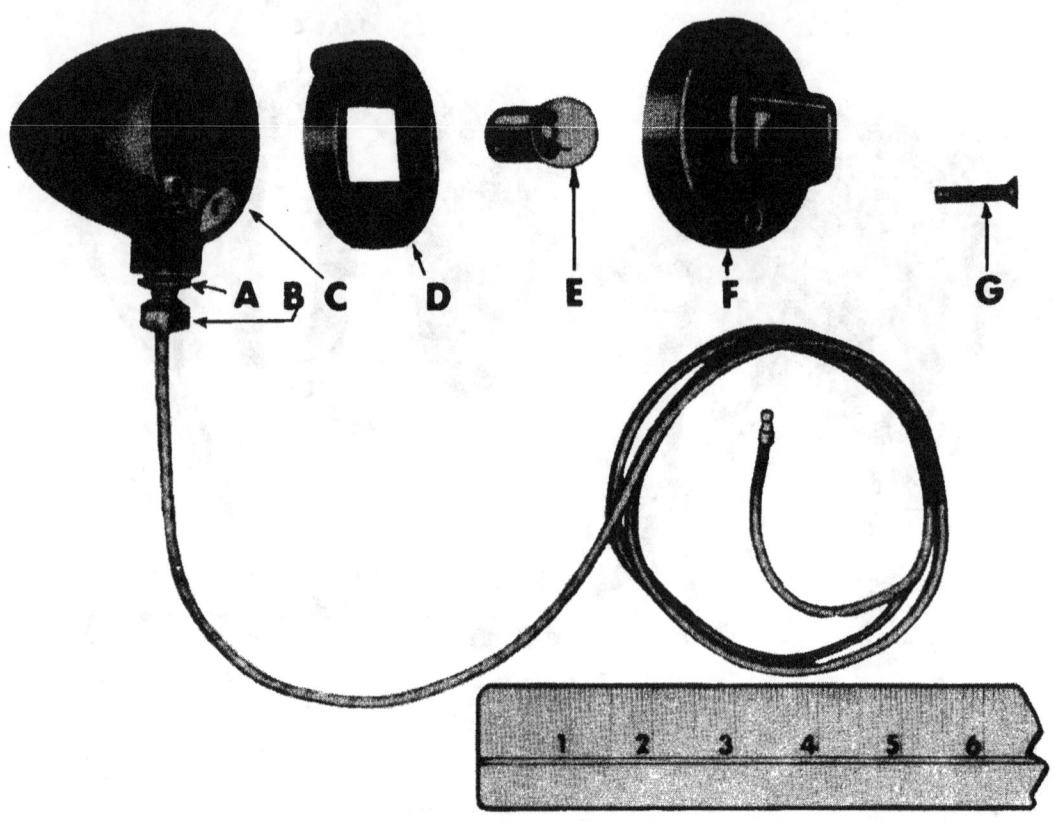

FIGURE 06-9—MARKER LIGHT

Key	Item	Willys Part No.	Ford Part No.	Gov't Group No.	Key	Item	Willys Part No.	Ford Part No.	Gov't Group No.
A	NUT	WO-5910	FM-33798-S	0607D	D	GASKET	WO-A-1071	FM-GPW-13209-B2	0607D
B	WASHER	WO-52510	FM-34941-S	0607D	E	LAMP	WO-51804	FM-B-13466	0607B
C	HOUSING assembly	WO-A-1439	FM-GPW-13217	0607D	F	DOOR	WO-A-1070	FM-GPW-13210-B	0607D
					G	SCREW	WO-A-1072	FM-28378-S2	0607D

RA PD 305081-A

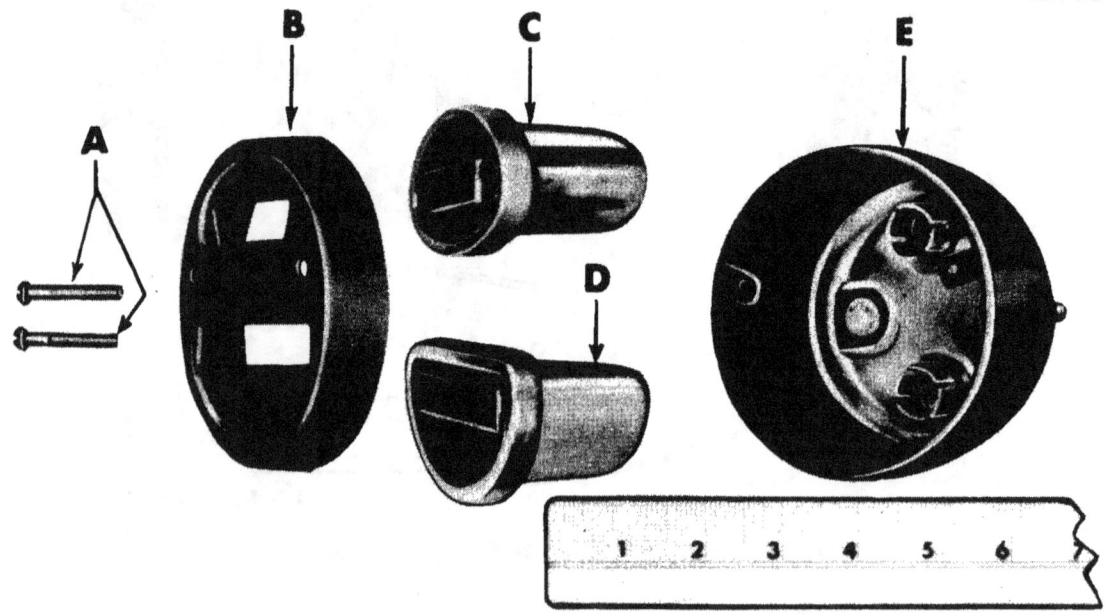

FIGURE 06-10—TAIL AND STOP LIGHT ASSEMBLY

Key	Item	Willys Part No.	Ford Part No.	Gov't Group No.	Key	Item	Willys Part No.	Ford Part No.	Gov't Group No.
A	SCREW	WO-A-1077	FM-36931-S	0608	D	LAMP-UNIT	WO-A-1075	FM-GPW-13491-A	0607A
B	DOOR	WO-A-1076	FM-GPW-13448-B	0608	E	HOUSING SUB assembly	WO-A-1073	FM-GPW-13408-B	0608
C	SERVICE UNIT	WO-A-1074	FM-GPW-13494	0607A					

RA PD 305083-A

SNL G-503

GROUP 06—ELECTRICAL (Cont'd)

Figure Number	Official Stockage Number	Part Number Willys	Part Number Ford	ITEM	Quantity Req'd per Unit Assy.	Per 100 Major Items 12 Mos. Field Maintenance	Per 100 Major Items Major Overhaul (5th Ech)	Total FirstYear Procurement	Estimated Reqmts. per 100 Rebuilds
Col. 1	Col. 2	Col. 3	Col. 4	Col. 5	Col. 6	Col. 7	Col. 8	Col. 9	Col. 10
				0607D—BLACKOUT LIGHTS (Cont'd)					
06-9		WO-A-11770	FM-GPW-13175-C	CABLE, blackout driving light, assembly (after Willys serial 288835)........	1	—	—	—	—
06-8		WO-A-1070	FM-GP-13210-B	DOOR, marker light, assembly (CB-9204) (issue until stock exhausted)......	2	—	4	—	—
06-8		WO-A-6144	FM-GPW-13162	DOOR, blackout driving light, assembly (CB-CB-11193)...................	1	⅔	2	4	—
06-9		WO-A-1071	FM-GPW-13209	GASKET, marker light door (CB-CB-9281) (issue until stock exhausted).....	2	⅔	4	—	—
06-9		WO-A-1439	FM-GPW-13217	HOUSING, marker light, w/WIRE, left, assembly (CB-CB-10461)...........	1	—	—	—	—
06-9		WO-A-1440	FM-GPW-13216	HOUSING, marker light, w/WIRE, right, assembly (CB-CB-10462)..........	1	—	—	—	—
06-8		WO-A-6143	FM-GPW-13170	HOUSING, blackout driving light, assembly (CB-CB-11500)................	1	⅔	2	4	—
		WO-A-1436	FM-GPW-13201	LIGHT, marker, left, assembly (CB-415342)............................	1	⅔	2	4	—
		WO-A-1437	FM-GPW-13200	LIGHT, marker, right, assembly (CB-415442)..........................	1	—	—	—	—
06-9		WO-5910	FM-33798-S	NUT, hex, S., 5/16-24NF-2 (to support) (GM-103025) (BBBX1BA).........	2	—	—	—	—
06-9		WO-52510	FM-34941-S	WASHER, lock, S., 5/16 in. (after 8,430 Willys trucks)...................	2	—	—	—	—
		WO-A-6142	FM-GPW-13150	LIGHT, driving, blackout assembly (CB-CB-11495) (after Willys serial 163750) (before Willys serial 288835)..	1	⅔	2	7	—
		WO-A-11768	FM-GPW-13150-C	LIGHT, driving, blackout assembly (after Willys serial 288835).............	1	—	—	—	—
		WO-52033	FM-33799-S7	NUT, hex, S., 3/8-16NC-2 (GM-123228)...............................	1	—	—	—	—
		WO-52046	FM-39807-S7	WASHER, lock, S., 3/8 in. (GM-115093)...............................	2	—	—	—	—
		WO-A-5806		NIPPLE, rubber marker (Willys only)..................................	2	—	—	—	—
06-8		WO-A-6783	FM-GPW-13166	PAD, mounting, marker (CB-B-9276) (Willys only).......................	1	⅔	2	8	—
06-9		WO-A-1072	FM-28378-S2	RING, mounting, blackout driving light unit mounting (CB-CB-11199)......	2	6	—	12	—
06-8		WO-53071	FM-40631-S7	SCREW, marker light door (CB-7861).................................	1	—	—	—	—
06-8		WO-A-6148	FM-131045-S2	SCREW, blackout driving light ground (CB-CB-7783) (GM-125760).........	1	⅔	3	12	—
06-8		WO-6313	FM-GPW-13180	SCREW, marker, light door, w/WASHER, assembly (CB-CB-11263)..........	1	—	—	—	—
		WO-53070		WASHER, adjusting, blackout driving light.............................	1	—	—	—	—
06-8		WO-A-6147	FM-GPW-13174	WASHER, cd-pltd., S., internal teeth #8, ground screw (GM-121752) (Willys only)..	1	—	—	—	—
				WIRE, ground, blackout driving light, assembly (CB-CB-11200)...........	1	—	—	—	—
				0608—TAIL LIGHT					
06-10		WO-A-719	FM-GPW-13410	CABLE, blackout tail lamp to connector.................................	1	⅔	2	4	—
		WO-A-1076	FM-GP-13448-B2	DOOR, blackout, tail and service tail and stop light, assembly (CB-CB-9231)..	1	⅔	4	—	—
		WO-A-1079	FM-GP-13449-A	DOOR, blackout tail and service tail and stop light (CB-CB-9232) (issue until stock exhausted)..	1	⅔	4	—	—
06-10		WO-A-1073	FM-GPW-13408-B	HOUSING, tail and stop light, assembly (CB-CB-9212)....................	2	—	—	—	—
		WO-A-1064	FM-GP-13405-B	LIGHT, blackout tail and service tail and stop, 6V, 7/8 std. lgh., assembly (CB-60142)...	1	⅔	2	8	—

SNL G-503

		WO-A-1065	FM-GP-13404-B	LIGHT, blackout tail and blackout stop, 6V., 7/8 std. lgh., assembly (CB-60242)	1	% 4	2	8	—
		WO-5790	FM-33795-S7	NUT, hex., S., 1/4-20NC-2 (to bracket) (GM-109084) (BBAX1A)	4	—	—	—	—
		WO-52706	FM-34805-S7	WASHER, lock, S., 1/4 in. (BECX1G)	4	—	—	—	—
		WO-352760	FM-34905-S	WASHER, lock, shakeproof, S., 1/4 in.	4	—	—	—	—
06-10		WO-A-1077	FM-36931-S2	SCREW, rd-hd., S. (door screw) (CB-CB-9233)	2	—	—	—	—

0609—HORNS

		WO-A-1389	FM-GPW-13831	BRACKET, horn (issue until stock exhausted)	1	% 4	8	—	—
		WO-51396	FM-24347-S2	BOLT, hex-hd., S., 5/16-24NF-2 x 3/4 (bracket to dash) (GM-100013)	2	—	—	—	—
		WO-52954	FM-33909-S7	NUT, hex., S., 5/16-24NF-2 (GM-114493) (BBDX1BA)	2	—	—	—	—
		WO-5437	FM-34706-S7	WASHER, S., 5/16 in. (plain) (GM-106262)	2	—	—	—	—
		WO-51833	FM-34806-S2	WASHER, lock, S., 5/16 in. (BECX1H)	2	—	—	—	—
		WO-A-1312	FM-GPW-13802	HORN assembly (SPW-B-9427) (SZE-61400)	1	% 4	4	—	—
		WO-5920	FM-20325-S7	BOLT, hex-hd., S., 1/4-28NF-2 x 5/8 (horn to bracket) (GM-106274)	2	—	—	—	—
		WO-52706	FM-34805-S2	WASHER, lock, S., 1/4 in. (BECX1G)	2	—	—	—	—
		WO-52182	FM-34052-S2	NUT, hex., S., No. 8 (.164)-32NC-2 (adjusting)	2	—	—	—	—
		WO-52705	FM-34802-S2	WASHER, lock, S., No. 8 (.164) (adjusting nut) (BECX1D)	2	—	—	—	—

0609B—HORN BUTTON AND WIRING

NOTE: For Wiring not appearing here see Group 0606B.

		WO-A-302	FM-GPW-13836	BRUSH, horn wire contact (RG-032087) (GM-263549)	1	% 2	2	4	—
		WO-51040	FM-26457-S7	SCREW, rd-hd., mach., S., No. 8 (.164)-32NC-2 x 1/4 (brush to column) (GM-100749)	2	—	—	—	—
		WO-52754	FM-34902-S	WASHER, lock, S., No. 8 (.164)	2	—	—	—	—
18-1		WO-A-634	FM-GPW-3627	BUTTON, horn (RG-450054)	1	—	—	—	—
		WO-A-5081	FM-GPW-14409	CABLE, horn to junction block (issue until stock exhausted)	1	% 4	8	—	—
		WO-A-752	FM-GPW-14171	CABLE, horn, assembly (RG-8287-32)	1	% 2	2	6	—
		WO-638885	FM-GPW-3646	CUP, horn button spring (RG-029046)	1	—	—	—	—
		WO-A-751	FM-GPW-3635-B	FERRULE, insulating (under horn button) (RG-051035)	1	—	—	—	—
		WO-A-6742	FM-GPW-18382	KIT, repair, horn button	as req.	% 2	2	7	—
				(Includes:					
				1 WO-A-634 FM-GPW-3627 BUTTON					
				1 WO-638885 FM-GPW-3646 CUP, spring					
				1 WO-A-751 FM-GPW-3653-B FERRULE					
				1 WO-638884 FM-GPW-3626 SPRING					
				1 WO-A-750 FM-GPW-3631 WASHER, contact)					
				NUT, horn button (also steering wheel nut. See Group 1404)	—	—	—	—	—
		WO-A-747	FM-GPW-3652	RING, horn contact, assembly (RG-063991)	1	% 2	2	6	—
		WO-638884	FM-GPW-3626	SPRING, horn button (RG-401107) (issue until stock exhausted)	1	% 4	8	—	—
		WO-A-750	FM-GPW-3631	WASHER, horn contact (RG-029049)	1	—	—	—	—

0610—BATTERY

		WO-A-1238		BATTERY, 6V., 15 plate, 116 amp. hr. (wet, charged) (U.S.L. type TS-2-15) (WB-SW-2-119) (Willys only)	1	—	—	—	—
		WO-A-1767	FM-11AS-10658	BATTERY, 6V., 15 plate, 116 amp. hr. (dry, charged) (AL-TSR-2-15) (WB-SR-2-119)	1	% 40	40	103	—

0610-0610A

GROUP 06—ELECTRICAL (Cont'd)

Figure Number	Official Stockage Number	Part Number (Willys)	Part Number (Ford)	ITEM	Quantity Reqd. per Unit Assy.	Per 100 Major Items — 12 Mos. Field Maintenance	Per 100 Major Items — Major Overhaul (5th Ech)	Total First Year Procurement	Estimated Reqmts. per 100 Rebuilds
Col. 1	Col. 2	Col. 3	Col. 4	Col. 5	Col. 6	Col. 7	Col. 8	Col. 9	Col. 10

0610—BATTERY (Cont'd)

		Willys	Ford	ITEM	Qty	12 Mos	5th Ech	1st Yr	per 100
		WO-A-5433	FM-11AS-10657	BATTERY, 6V., 15 plate, 116 amp. hr. (wet dump) (unfilled and uncharged) (AL-TS-2-15)	1	—	—	—	—
		WO-A-11760		ELECTROLYTE, battery, and CONTAINER, assembly (use with dry charged battery, WO-A-1767, FM-11AS-10658) (PL-26027)	as req.	% 48	51	111	—

0610A—BATTERY HANGERS

NOTE: For battery support see Group 1500.

		Willys	Ford	ITEM	Qty	12 Mos	5th Ech	1st Yr	per 100
		WO-A-1291	FM-GPW-5165	FRAME, hold down, battery	1	—	—	—	—
		WO-A-1164	FM-GPW-5175	BOLT, S., 5/16-18NC-2 x 9 (battery clamp) (issue until stock exhausted)	2	% 2	2	6	—
		WO-300329	FM-33797-S2	NUT, hex., S., 5/16-18NC-2 (Ford only)	2	% —	4	—	—
		WO-52768	FM-356305-S	NUT, wing, S., 5/16-18NC-2 (Willys only)	2	—	—	—	—
		WO-51833	FM-34806-S2	WASHER, S., 21/64 in. (plain)	2	—	—	—	—
		WO-A-1757	FM-GPW-5168	WASHER, lock, S., 5/16 in. (BECX1H)	2	—	—	—	—
		WO-A-3728	FM-33896-S2	STRAP, battery to front fender	1	—	—	—	—
		WO-52768	FM-356305-S	NUT, strap to fender bolt (wing, S., 5/16-18NC)	1	—	—	—	—
		WO-51833	FM-34806-S2	WASHER, S., 5/16 in. plain	1	—	—	—	—
		WO-323397	FM-23017-S16	WASHER, lock, S., 5/16 in. (BECX1H)	1	—	—	—	—
				BOLT, terminal clamp (hex-hd., S., ld-pltd., 5/16-18NC x 1¼) (issued until stock exhausted)	2	% 8	16	—	—
		WO-A-9093	FM-350343-S16	BOLT, terminal clamp, w/NUT, assembly	2	—	—	—	—
		WO-335912	FM-350343-S16	BOLT, terminal clamp w/NUT, assembly, (use until stock exhausted, then use WO-A-9093) (was WO-A-355912)	2	% 96	24	200	—
		WO-A-1452	FM-GPW-14300	CABLE, battery to starting switch, positive	1	% 3	3	9	—
		WO-A-1454	FM-GPW-14431	CABLE, starting switch to cranking motor, (issue until stock exhausted)	1	% 4	4	—	—
		WO-A-9094	FM-34056-S16	NUT, hex., S., ld-pltd., 5/16-18NC (terminal clamp bolt)	2	—	—	—	—
		WO-A-1320		STRAP, battery ground (before Willys serial 120700) (Willys only)	1	—	—	—	—
		WO-635883	FM-GPW-14301	STRAP, battery ground (after Willys serial 120700) (all Ford trucks)	1	% 3	3	8	—
		WO-50151	FM-24369-S	BOLT, hex-hd., S., 3/8-24NF-2 x 7/8 (strap to frame) (GM-106285)	1	—	—	—	—
		WO-5901	FM-33800-S	NUT, hex., S., 3/8-24NF-2 (GM-103026) (BBBX1CA)	1	—	—	—	—
		WO-A-1680	FM-34747-S7	WASHER, S., cd-pltd. (plain) 3/8 in.	1	—	—	—	—
		WO-5010	FM-34807-S2	WASHER, lock, S., 3/8 in. (BECX1K)	1	—	—	—	—
		WO-371400	FM-11A-14452	TERMINAL, battery cable (1½2 in.)	1	% 5	5	12	—
		WO-372668	FM-19B-14451	TERMINAL, battery, positive cable clamp	1	% 5	5	12	—

SNL G-503

GROUP 07—TRANSMISSION

0700—TRANSMISSION ASSEMBLY

	Part No.	Ref. No.	Description	Qty				
G	WO-A-7596		TRANSMISSION, w/gear shift LEVER, assembly (Willys only) (Includes: clutch release BEARING, clutch bearing CARRIER, clutch bearing carrier SPRING, clutch control LEVER, and clutch control lever CABLE)	1	6		8	
	WO-A-1145	FM-GPW-7000	TRANSMISSION, w/gearshift LEVER, assembly (WG-ASI-T-84-J) (issue until stock exhausted)	1	4	8		

0701—CASE

07-1	WO-A-1148	FM-GPW-7005	CASE, transmission, assembly (WG-T-84J-1A)	1			4	10
			(Includes:					
			CUP, oil retaining					
			PIN, countershaft washer)					
07-1	WO-637503	FM-GPW-7056	CUP, oil retaining, transmission (high and intermediate shift rail)	1	3	3	10	10
07-1	WO-637495	FM-GPW-7051-B	GASKET, transmission case to flywheel housing (Victorite, .015 thk.) (WG-T84D-145C)	1	38		40	
	WO-A-1542	FM-GPW-13356	GASKET, SET, transmission (issue until stock is exhausted, then use WO-A-7832)	as req.	50	100		100
	WO-A-7832	FM-GPW-18356-B	GASKET, SET transmission (was WO-A-1542)	as req.	18	26	48	
			(Includes:					
			1 WO-637495 FM-GPW-7051-B GASKET					
			1 WO-635861 FM-GPW-7223 GASKET					
			1 WO-640018 FM-GPW-7052 OIL SEAL					
			1 WO-635862 FM-GPW-7207 PACKING)					
	WO-5140	FM-353064-S	PLUG, pipe, I, sq-hd., ½ in. (drain and reqll)	2				

0703—MAIN DRIVE GEAR AND BEARINGS

07-1	WO-636885	FM-GPW-7025	BEARING, ball, annular, single row, single oil shield, snap ring groove (bore 2.834 O.D. x 1.378 I.D.) (transmission drive gear) (FB-1207-MGF) (WG-X-3204-ML)	1	3	10	15	30
07-1	WO-639422	FM-GPW-7120	BEARING, roller, S, spherical end type .527 lgth. x .1875 diam. (transmission drive gear pilot) (BN-C-1110-Q) (WG-T-84G-26)	13	21	229	300	705
07-1	WO-A-5554	FM-GPW-7017	GEAR, main drive, S, 17 teeth (2.048 O.D.) (WG-T-84J-16A)	1		6	7	20
07-1	WO-A-5553	FM-GPW-7015	GEAR, main drive, w/BEARINGS, assembly (WG-AT-84J-16A)	1				
			SNAP RINGS					
07-1	WO-640018	FM-GPW-7052	OIL SEAL, transmission front bearing retainer (WG-T-84J-54)	1	3	6	15	20
07-1	WO-640017	FM-GPW-7050	RETAINER, main drive gear bearing (WG-T-84J-6)	1		3	4	10
07-1	WO-639689	FM-20366-S	BOLT, hex-hd., S, 5/16-18NC-2 x 7/8 (WG-X-802)	3				
07-1	WO-52510	FM-34941-S2	WASHER, lock, S, 5/16 in.	3				
07-1	WO-635844	FM-GPW-7064	SNAP RING, main drive gear, 1-9/32 in. I.D. (WG-T-84-17)	1	3	22	30	70
07-1	WO-635846	FM-B-7070	SNAP RING, main drive gear bearing, 2-11/16 in. I.D. (WG-B-7070)	1	10	26	39	80
07-1	WO-639423	FM-GPW-7063	SNAP RING, main drive gear pilot roller bearing, 1.062 O.D. (WG-T-84G-25)	1	3	26	33	80

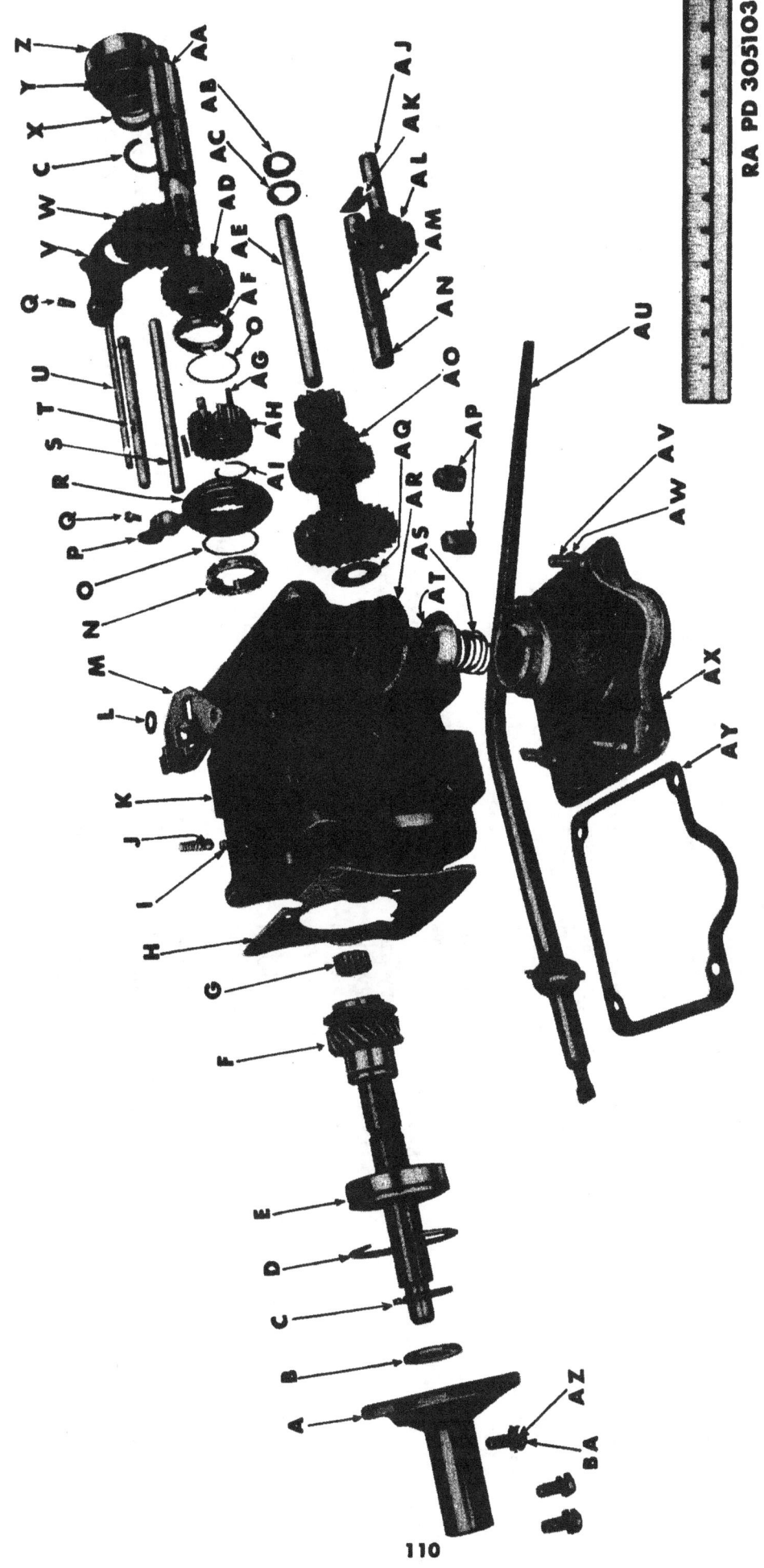

FIGURE 07-1—TRANSMISSION ASSEMBLY

Key	Item	Willys Part No.	Ford Part No.	Gov't Group No.	Key	Item	Willys Part No.	Ford Part No.	Gov't Group No.
A	RETAINER	WO-640017	FM-GPW-7050	0703	AB	WASHER	WO-A-879	FM-GPW-7126	0704C
B	OIL seal	WO-640018	FM-GPW-7052	0703	AC	WASHER	WO-635811	FM-GPW-7129-A	0704C
C	SNAP ring	WO-635844	FM-GPW-7064	0703	AD	GEAR assembly	WO-638798	FM-GPW-7102	0704B
D	SNAP ring	WO-635846	FM-B-7070	0703	AE	COUNTERSHAFT	WO-638948	FM-GPW-7111	0704C
E	BEARING	WO-636885	FM-GPW-7025	0703	AF	RING	WO-637834	FM-GPW-7107	0704B
F	GEAR	WO-A-5554	FM-GPW-7017	0703	AG	PLATE	WO-637832	FM-GPW-7116	0704B
G	BEARING	WO-639422	FM-GPW-7120	0703	AH	HUB	WO-A-6319	FM-GPW-7105	0704B
H	GASKET	WO-637495	FM-GPW-7051-B	0701	AI	SNAP ring	WO-637835	FM-GPW-7059	0704B
I	SPRING	WO-635837	FM-GPW-7234	0706B	AJ	SHAFT	WO-638952	FM-GPW-7140	0704C
J	BALL	WO-635838	FM-35081-S	0706B	AK	PLATE	WO-638949	FM-GPW-7155	0704B
K	CASE	WO-A-1148	FM-GPW-7005	0701	AL	GEAR assembly	WO-636882	FM-GPW-7141	0704C
L	SPRING	WO-635839	FM-GPW-7208	0706C	AM	SPACER	WO-A-880	FM-GPW-7115	0704C
M	PLATE assembly	WO-A-7260	FM-GPW-7211-B	0706C	AN	BUSHING	WO-A-878	FM-GPW-7121	0704C
N	RING	WO-637834	FM-GPW-7107	0704B	AO	GEAR assembly	WO-A-739	FM-GPW-7113	0704B
O	SPRING	WO-637831	FM-GPW-7109	0704B	AP	PLUG	WO-5140	FM-353064	0701
P	FORK	WO-636196	FM-GPW-7230	0706C	AQ	WASHER	WO-635812	FM-GPW-7119	0704C
Q	SCREW	WO-636200	FM-GPW-7245	0706B	AR	CAP	WO-A-1379	FM-BB-7220	0706A
R	SLEEVE	WO-637833	FM-GPW-7106	0704B	AS	SPRING	WO-392328	FM-GPW-7227	0706A
S	RAIL	WO-A-1155	FM-GPW-7241	0706B	AT	WASHER	WO-635863	FM-BB-7228	0706A
T	RAIL	WO-A-1156	FM-GPW-7240	0706B	AU	LEVER assembly	WO-A-1380	FM-GPW-7210-A	0706A
U	PIN	WO-635836	FM-GPW-7206	0706C	AV	BOLT	WO-639689	FM-20366-S	0706A
V	FORK	WO-636197	FM-GPW-7231	0706C	AW	WASHER	WO-52045	FM-34806-S2	0706A
W	GEAR	WO-636879	FM-GPW-7100	0704B	AX	HOUSING assembly	WO-635857	FM-GPW-7204	0706A
X	SPACER	WO-A-738	FM-GPW-7062	0704	AY	GASKET	WO-635861	FM-GPW-7223	0706A
Y	WASHER	WO-A-410	FM-GPW-7080	0704	AZ	WASHER	WO-52510	FM-34941-S2	0703
Z	BEARING	WO-A-916	FM-GP-7065	0704	BA	BOLT	WO-639689	FM-20366-S	0703
AA	SHAFT	WO-A-519	FM-GPW-7061	0704					

RA PD 305103-A

SNL G-503

GROUP 07—TRANSMISSION (Cont'd)

Figure Number	Official Stockage Number	Part Number Willys	Part Number Ford	ITEM	Quantity Reqd. per Unit Assy.	Per 100 Major Items 12 Mos. Field Maintenance	Per 100 Major Items Major Overhaul (5th Ech)	Total First Year Procurement	Estimated Reqmts. per 100 Rebuilds
Col. 1	Col. 2	Col. 3	Col. 4	Col. 5	Col. 6	Col. 7	Col. 8	Col. 9	Col. 10
				0704—MAIN SHAFT AND BEARINGS					
08-2		WO-A-916	FM-GP-7065	BEARING, ball annular, single row, bore 1.378 x .8268 (main shaft rear end) (MRC-307) (SKF-6307-Z) (ND-7607)	1	3	10	15	30
		WO-640006	FM-GPW-7104-A2	BUSHING, main shaft second speed gear (WG-T-84C-19)	1	—	6	6	20
07-1		WO-A-519	FM-GPW-7061	SHAFT, main (dld. f/c-pin) (WG-T-84H-2)	1	2	5	9	15
07-1		WO-A-6317	FM-GPW-7060	SHAFT, main, w/GEARS, assembly (WG-1AT-84H-2)	1	—	—	—	10
07-1		WO-A-738	FM-GPW-7062	SPACER, main shaft bearing (WG-T-84J-28)	1	—	3	3	10
07-1		WO-A-410	FM-GPW-7080	WASHER, retaining, main shaft, oil (WG-T-184H-137)	2	3	10	15	30
				0704B—GEARS					
07-1		WO-A-739	FM-GPW-7113	GEARS, cluster, countershaft, w/BUSHING and SPACER, assembly (WG-AT-84J-8)	1	—	6	7	20
07-1		WO-636879	FM-GPW-7100	GEAR, sliding, mainshaft, low and reverse (O.D. 3.265 x 1¼ thk., 25 teeth) (WG-T-84J-12A)	1	—	6	7	20
07-1		WO-638798	FM-GPW-7102	GEAR, main shaft, second speed, w/BUSHING, assembly (O.D. 2.090, 24 teeth) (WG-T-84F-11A)	1	—	6	7	20
07-1		WO-636882	FM-GPW-7141	GEAR, reverse idler, w/BUSHING, assembly (O.D. 2.167 x .996 thk., 16 teeth) (WG-AT-84F-10A)	1	—	6	7	20
07-1		WO-A-6319	FM-GPW-7105	HUB, intermediate and high clutch (WG-T-84J-2½) (issue until stock exhausted)	1	4	8	—	60
07-1		WO-637832	FM-GPW-7116	PLATE, synchronizer shifting (WG-T-84F-13)	3	—	19	21	60
07-1		WO-637834	FM-GPW-7107	RING, synchronizer blocking (WG-T-84F-14)	2	—	10	11	30
07-1		WO-637833	FM-GPW-7106	SLEEVE, second and direct speed clutch (WG-T-84F-15) (issue until stock exhausted)	1	4	8	—	70
07-1		WO-637835	FM-GPW-7059	SNAP RING, high and intermediate clutch hub, I.D. 1½ (WG-4686)	1	3	22	27	70
07-1		WO-637831	FM-GPW-7109	SPRING, synchronizer (WG-4682-K)	2	—	10	10	30
07-1		WO-A-6318	FM-GPW-7124	SYNCHRONIZER UNIT, w/blocking RING, assembly (WG-1AT-84J-2½)	1	—	14	15	45
				0704C—COUNTERSHAFT AND REVERSE IDLER SHAFT					
07-1		WO-A-878	FM-GPW-7121	BUSHING, countershaft gear (WG-T-84J-167)	2	—	19	24	60
07-1		WO-635804	FM-GPW-7143	BUSHING, reverse idler gear (WG-T-84-85A)	1	—	6	6	20
07-1		WO-638948	FM-GPW-7111	COUNTERSHAFT (WG-T-84G-3)	1	—	3	4	10
		WO-A-524	FM-357417-S	PIN, countershaft washer (WG-T-84J-31) (issue until stock exhausted)	1	8	16	—	—
07-1		WO-638949	FM-GPW-7155	PLATE, lock, countershaft and idler shaft (WG-T-84C-48)	1	—	3	3	10
07-1		WO-638952	FM-GPW-7140	SHAFT, reverse idler gear (WG-T-84G-35)	1	—	3	3	10
07-1		WO-A-880	FM-GPW-7115	SPACER, countershaft bushing (WG-T-84J-28) (issue until stock exhausted)	1	4	12	—	—
07-1		WO-635812	FM-GPW-7119	WASHER, thrust, countershaft, front (WG-T-84B-30A)	1	—	19	20	60

0704C

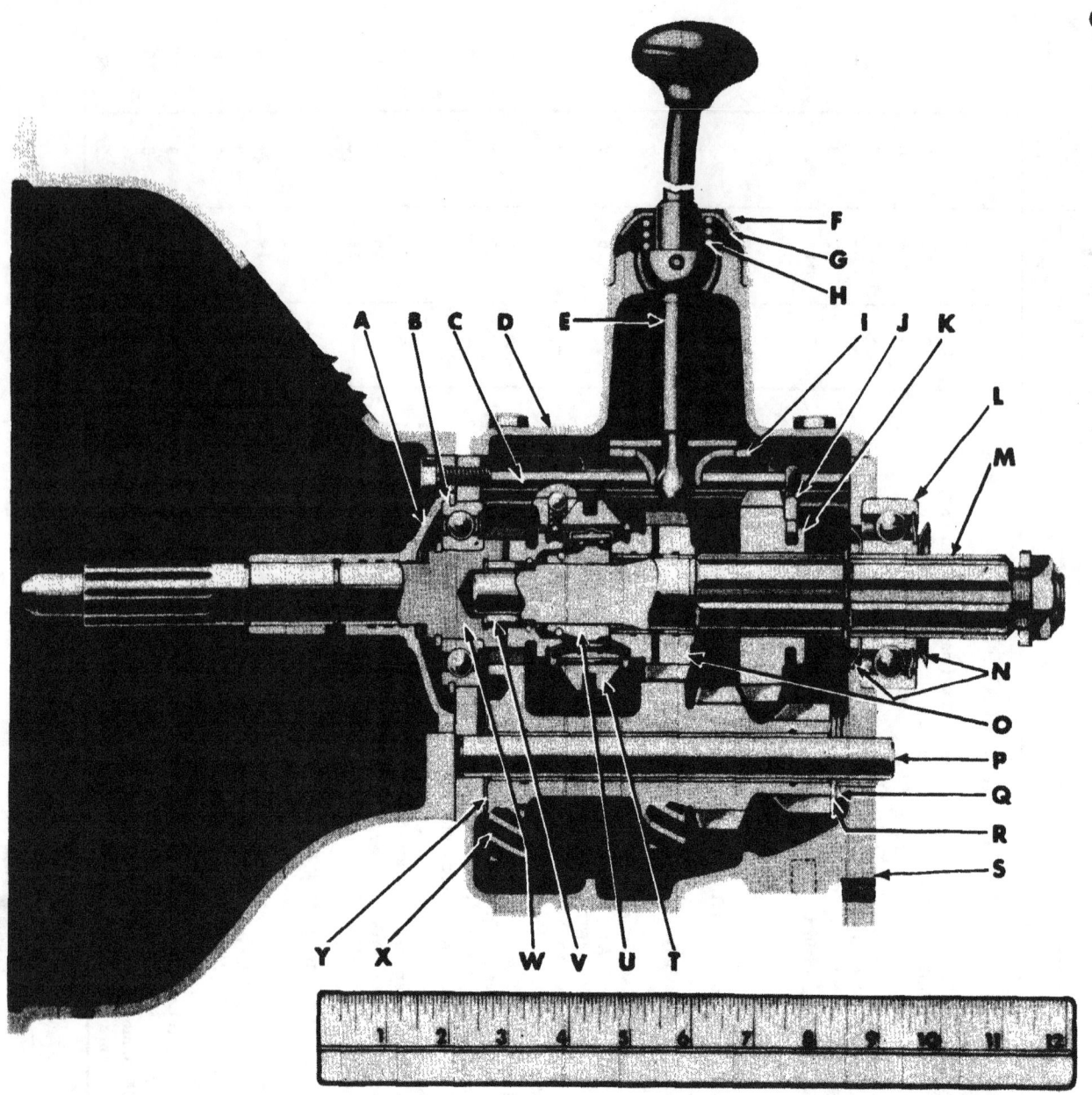

FIGURE 07-2—TRANSMISSION ASSEMBLY—CROSS-SECTIONAL VIEW

Key	Item	Willys Part No.	Ford Part No.	Gov't Group No.
A	RETAINER	WO-640017	FM-GPW-7050	0703
B	BEARING	WO-636885	FM-GPW-7025	0703
C	RAIL	WO-A-1156	FM-GPW-7240	0706B
D	HOUSING	WO-635857	FM-GPW-7204	0706A
E	LEVER	WO-A-1380	FM-GPW-7210-A	0706A
F	CAP	WO-A-1379	FM-BB-7220	0706A
G	WASHER	WO-635863	FM-BB-7228	0706A
H	SPRING	WO-392328	FM-GPW-7227	0706A
I	PLATE	WO-A-7260	FM-GPW-7214	0706A
J	FORK	WO-636197	FM-GPW-7231	0706C
K	GEAR	WO-636879	FM-GPW-7100	0704B
L	BEARING	WO-A-916	FM-GP-7065	0704
M	SHAFT	WO-A-519	FM-GPW-7061	0704
N	WASHER	WO-A-410	FM-GPW-7080	0704
O	GEAR	WO-638798	FM-GPW-7102	0704B
P	COUNTERSHAFT	WO-638948	FM-GPW-7111	0704C
Q	WASHER	WO-A-879	FM-GPW-7126	0704C
R	WASHER	WO-635811	FM-GPW-7129A	0704C
S	CASE	WO-A-1148	FM-GPW-7005	0701
T	SLEEVE	WO-637833	FM-GPW-7106	0704B
U	HUB	WO-A-6319	FM-GPW-7105	0704B
V	BEARING	WO-639422	FM-GPW-7120	0703
W	GEAR	WO-A-5554	FM-GPW-7017	0703
X	GEAR	WO-A-739	FM-GPW-7113	0704B
Y	WASHER	WO-635812	FM-GPW-7119	0704C

RA PD 305155-A

SNL G-503

GROUP 07—TRANSMISSION (Cont'd)

Figure Number	Official Stockage Number	Part Number Willys	Part Number Ford	ITEM	Quantity Reqd. per Unit Assy.	Per 100 Major Items 12 Mos. Field Maintenance	Per 100 Major Items Major Overhaul (5th Ech)	Total First Year Procurement	Estimated Reqmts. per 100 Rebuilds
Col. 1	Col. 2	Col. 3	Col. 4	Col. 5	Col. 6	Col. 7	Col. 8	Col. 9	Col. 10
				0704C—COUNTERSHAFT AND REVERSE IDLER SHAFT (Cont'd)					
07-1		WO-635811	FM-GPW-7129-A	WAHSER, thrust, countershaft, rear (bz.) (WG-T-84-29)	1	—	19	20	60
07-1		WO-A-879	FM-GPW-7126	WASHER, thrust, countershaft, rear (S) (WG-T-84J-29)	1	—	10	12	30
				0706A—GEAR SHIFT LEVER AND PARTS					
07-1		WO-A-971	FM-GPW-7213	BALL, gearshift lever (rubber) (issue until stock exhausted)	1	16	24	—	—
07-1		WO-A-1381	FM-GP-W-7217	BALL, gearshift lever fulcrum (WG-C-8-2½)	1	—	3	4	10
07-1		WO-A-1379	FM-BB-7220	CAP, gearshift lever housing (WG-4496-K)	1	2	2	6	5
		WO-A-3783	FM-GPW-1111286-A3	COVER, gearshift lever housing (rubber)	1	—	—	—	—
		WO-635868		BOLT, hex-hd., S., 5/16-18NC-2 x ¾ (issue until stock is exhausted, then use WO-639689)	4	32	68	—	—
07-1		WO-639689	FM-20366-S	BOLT, hex-hd., S., 5/16-18NC-2 x ¾ (was WO-635868)	4	—	—	—	—
07-1		WO-52045	FM-34806-S2	WASHER, lock, S., 5/16 in. (housing cover bolt)	4	—	—	—	—
07-1		WO-635861	FM-GPW-7223	GASKET, gearshift lever housing (WG-T-84-115)	1	19	10	30	—
07-1		WO-635857	FM-GPW-7204	HOUSING, gearshift lever, assembly (Includes: PLATE RIVET)	1	—	—	—	—
07-1		WO-A-1380	FM-GPW-7210-A	LEVER, gearshift, assembly (WG-AC-84J-2A)	1	—	3	3	10
07-1		WO-635862	FM-GPW-7207	PACKING, gearshift lever housing (issue until stock exhausted)	1	32	74	12	30
07-1		WO-A-1382	FM-GPW-7221	PIN, gearshift lever fulcrum (WG-C-84B-12A)	1	—	10	12	30
07-2		WO-A-7260	FM-GPW-7214	PLATE, guide, gearshift lever (WG-T-84B-32)	1	—	3	4	10
07-1		WO-635860	FM-62216-S	RIVET, S., fl-hd., 3/16 x 7/16 (WG-X-3089BH)	2	—	10	25	30
07-1		WO-392328	FM-GPW-7227	SPRING, support, gearshift lever (WG-4498-K)	1	—	3	10	10
07-1		WO-635863	FM-BB-7228	WASHER, gearshift lever housing cap (WG-T-84H-137)	1	—	3	3	10
				0706B—GEARSHIFTER RAILS AND PARTS					
07-1		WO-635838	FM-353081-S	BALL, poppet, shift rail (S., 5/16 O.D.) (WG-X-2136)	2	—	10	10	30
07-1		WO-A-1385	FM-GPW-7233	PLUNGER, gearshift interlock (WG-T-84J-86A)	1	19	19	42	60
07-1		WO-A-1155	FM-GPW-7241	RAIL, shift, high and intermediate (WG-T-84J-20A)	1	—	6	7	20
07-1		WO-A-1156	FM-GPW-7240	RAIL, shift, low and reverse (WG-T-84J-21)	1	—	6	7	60
07-1		WO-636200	FM-GPW-7245	SCREW, lock, gearshift fork (WG-4418-M)	2	10	19	30	60
07-1		WO-635837	FM-GPW-7234	SPRING, shift rail, poppet ball (WG-T-84-42)	2	—	10	10	30

SNL G-503

0706C-0801

0706C—GEARSHIFTER YOKES OR FORKS

WO-636196	FM-GPW-7230	FORK, shift, high and intermediate (WG-T-84C-23A)	1	3	3	10	10	07-1
WO-636197	FM-GPW-7231	FORK, shift, low and reverse (WG-T-84C-24A)	1	3	3	10	10	07-1
WO-A-7260	FM-GPW-7211-B	PLATE, shifter, assembly (WG-AT-84J-25)	1	—	6	5	20	07-1
WO-635836	FM-GPW-7206	PIN, shift fork guide (WG-T-84-22)	1	—	3	5	10	07-1
WO-635840	FM-357418-S	PIN, shifter plate fulcrum (WG-T-84-30)	1	3	3	10	10	07-1
WO-635839	FM-GPW-7208	SPRING, shifter plate (WG-T-84-31)	1	10	19	30	60	07-1

GROUP 08—TRANSMISSION TRANSFER

0800—TRANSMISSION TRANSFER ASSEMBLY

WO-A-1195	FM-GPW-7700	CASE, transfer, assembly	1	6	—	8	—	F
WO-A-1543	FM-GPW-18355	GASKET SET, transfer case overhaul (issue until stock is exhausted, then use WO-A-7443)	as req.	40	80	—	—	
WO-A-7443	FM-GPW-18355-B	GASKET SET, transfer case, overhaul (was WO-A-1543) (Includes:	as req.	—	26	28	100	
		1 WO-A-957 FM-GPW-7773 GASKET						
		1 WO-A-1509 FM-GPW-7707 GASKET, cover, rear						
		1 WO-A-954 FM-GP-7709 GASKET, cover						
		2 WO-A-1134 FM-GP-7746 GASKET, oil seal						
		1 WO-A-1435 FM-GPW-7756 GASKET, transfer case to transmission)						
WO-A-7445	FM-GPW-18317-B	OIL SEAL SET, transfer case, (Includes:	as req.	—	26	28	100	
		2 WO-A-958 FM-GP-7770-A OIL SEAL, output shaft						
		2 WO-A-974 FM-GP-7798-A OIL SEAL, shift rod)						

0801—CASE

WO-A-934	FM-GP-7754	BREATHER, transfer case, assembly (SP-5951X)	1	2	2	5	5	08-1
WO-A-956	FM-GPW-7774	CAP, output clutch shaft bearing, front (SP-18-19-8)	1	2	2	6	5	08-2
WO-A-1136	FM-355554-S	BOLT, hex-hd., S., head dld. f/c-pin, $\frac{3}{8}$-16NC-3 x 1 (SP-122-D) (5 used front cap to case, 1 used rear cap to case)	6	—	29	40	90	
WO-A-1137	FM-355533	BOLT, hex-hd., S. (head, dld. f/c-pin) $\frac{3}{8}$-16NC-3 x 2 (rear cap to case) (SP-634-D)	3	3	10	40	30	
WO-A-1135	FM-356205-S	WASHER, cap., $\frac{5}{8}$ O.D. x $\frac{25}{64}$ I.D. x $\frac{1}{32}$ thk. (SP-477-W)	9	29	230	280	720	
WO-A-1507	FM-GPW-7768	CAP, output shaft bearing, rear (SP-6452X)	1	2	2	8	5	08-1
WO-A-1503	FM-GPW-7705	CASE, transfer (SP-18-15-9)	1	—	3	4	10	08-2
WO-A-953	FM-GP-7708	COVER, transfer case, bottom (SP-18-16-3)	1	—	3	4	10	08-1
WO-51485	FM-20326-S7	BOLT, hex-hd., S., $\frac{5}{16}$-18NC-2 x $\frac{5}{8}$ (cover to case)	10	—	—	—	—	
WO-51833	FM-34806-S2	WASHER, lock, S., $\frac{5}{16}$ in.	10	—	—	—	—	
WO-A-1508	FM-GPW-7706	COVER, transfer case (rear) (SP-18-267-4)	1	—	3	4	10	08-1
WO-6412	FM-24348-S	BOLT, hex-hd., S., $\frac{3}{8}$-16NC-2 x $\frac{3}{4}$ (cover to case)	5	—	—	—	—	
WO-5010	FM-34807-S	WASHER, lock, S., $\frac{3}{8}$ in.	5	—	—	—	—	
WO-A-957	FM-GPW-7773	GASKET, output clutch shaft bearing cap (SP-18-223-2)	1	3	—	15	—	08-1
WO-A-1509	FM-GPW-7707	GASKET, rear cover (SP-18-324-2)	1	13	—	15	—	08-1
WO-A-954	FM-GP-7709	GASKET, transfer case, bottom cover (SP-18-155-2) (issue until stock exhausted)	1	50	100	—	—	

115

SNL G-503

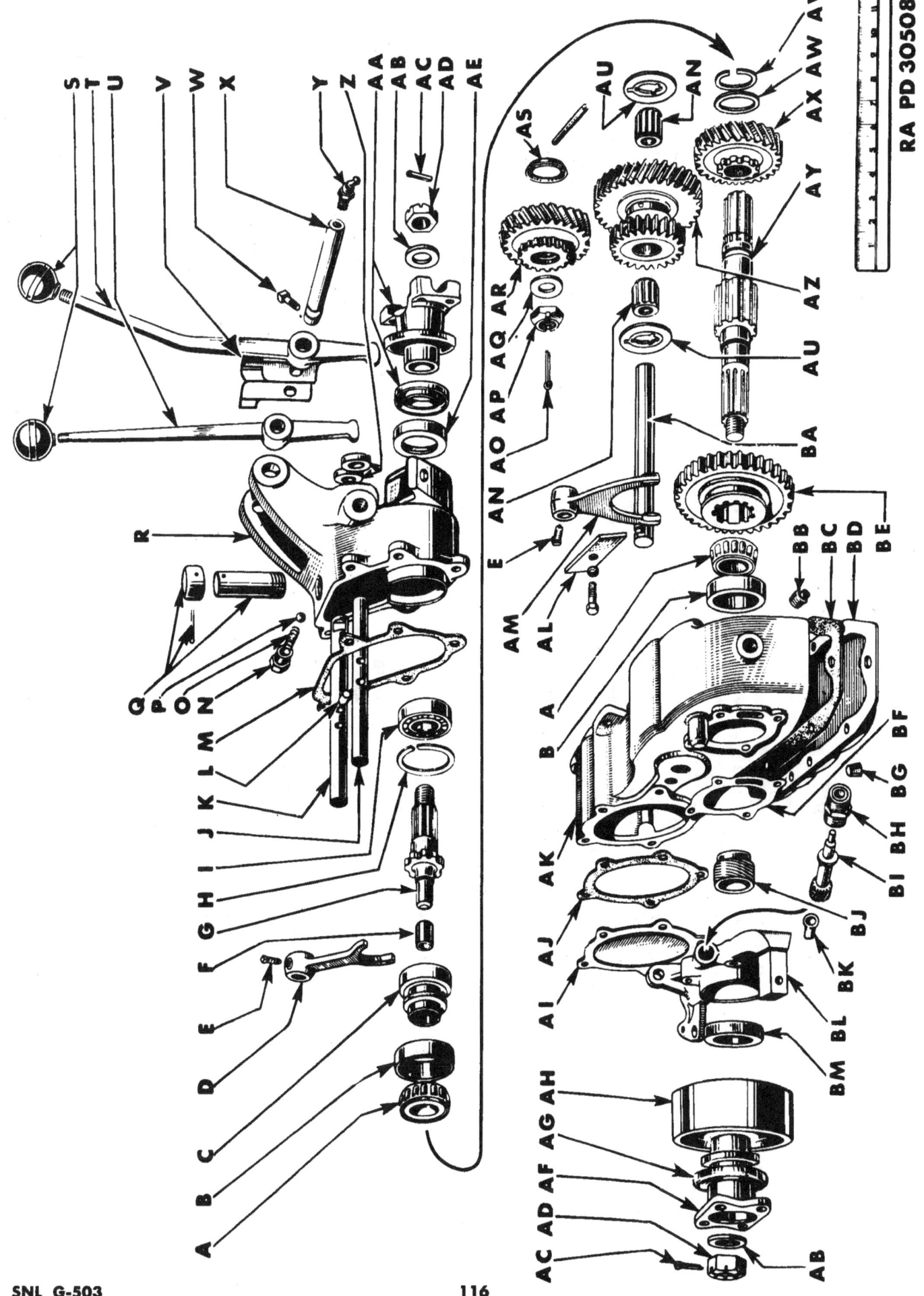

RA PD 305086

FIGURE 08-1—TRANSFER CASE ASSEMBLY

Key	Item	Willys Part No.	Ford Part No.	Gov't Group No.	Key	Item	Willys Part No.	Ford Part No.	Gov't Group No.
A	ROLLERS	WO-51575	FM-GP-7723	0803	AG	SHIELD	WO-A-1111	FM-GP-7776	0803
B	CUP	WO-52883	FM-OIY-1202	0803	AH	DRUM	WO-A-1002	FM-GP-2614	1201
C	GEAR	WO-A-992	FM-GP-7762	0803	AI	COVER	WO-A-1508	FM-GPW-7706	0801
D	FORK	WO-A-960	FM-GP-7711	0805A	AJ	GASKET	WO-A-1509	FM-GPW-7707	0801
E	SCREW	WO-A-963	FM-355550	0805A	AK	CASE	WO-A-1503	FM-GPW-7768	0801
F	BUSHING	WO-A-987	FM-GP-7777-A	0803	AL	PLATE	WO-A-1001	FM-GP-7767	0804
G	SHAFT	WO-A-975	FM-GP-7761	0803	AM	FORK	WO-A-959	FM-GP-7712	0805A
H	RING	WO-A-976	FM-GP-7783	0803	AN	BEARING	WO-A-924	FM-GP-7718-A	0804
I	BEARING	WO-A-1007	FM-GP-7719	0802	AO	PIN	WO-5397	FM-72071	0802
J	ROD	WO-A-1504	FM-GPW-7786	0805	AP	NUT	WO-A-520	FM-356134-S-18	0802
K	ROD	WO-A-962	FM-GP-7787	0805	AQ	WASHER	WO-A-1410	FM-356519-S	0802
L	INTERLOCK	WO-A-965	FM-GP-7789	0805	AR	GEAR	WO-A-1510	FM-GP-7722	0802
M	GASKET	WO-A-957	FM-GPW-7773	0801	AS	WASHER	WO-A-410	FM-356519-S	0802
N	PLUG	WO-A-967	FM-355698-S	0805	AU	WASHER	WO-A-1000	FM-GP-7744	0804
O	SPRING	WO-A-966	FM-GP-7788	0805	AV	RING	WO-A-991	FM-GP-7784	0803
P	BALL	WO-5599	FM-353075-S	0805	AW	WASHER	WO-A-990	FM-GP-7771	0803
Q	BREATHER assembly	WO-A-934	FM-GP-7754	0801	AX	GEAR	WO-A-989	FM-GP-7776	0803
R	CAP	WO-A-956	FM-GPW-7774	0801	AY	SHAFT	WO-A-1764	FM-GP-7763	0803
S	HANDLE	WO-A-971	FM-GP-7213	0805B	AZ	GEAR	WO-A-999	FM-GP-7742	0804
T	LEVER	WO-A-1505	FM-GPW-7795	0805B	BA	SHAFT	WO-A-998	FM-GP-7743	0804
U	LEVER	WO-A-1506	FM-GPW-7710	0805	BB	PLUG	WO-5140	FM-353064-S	0801
V	SPRING	WO-A-970	FM-GP-7796	0805B	BC	GASKET	WO-A-954	FM-GP-7709	0801
W	SCREW	WO-A-973	FM-355378-S	0805B	BD	COVER	WO-A-953	FM-GP-7708	0801
X	PIN	WO-A-972	FM-GP-7796	0805B	BE	GEAR	WO-A-988	FM-GP-7765	0803
Y	FITTING	WO-638224	FM-353035-A-S7	0805B	BF	SHIM	WO-A-982	FM-GP-7782-A	0801
Z	SEAL	WO-A-974	(Willys only)	0805B	BG	PLUG	WO-A-1104	FM-353053	0801
AA	YOKE	WO-A-1106	FM-GP-7729	0803	BH	SLEEVE	WO-636396	FM-GP-17333	0810
AB	WASHER	WO-A-1028	FM-356504-S	0803	BI	GEAR	WO-A-1512	FM-GPW-17271	0810
AC	PIN	WO-5108	FM-72053-S	0803	BJ	GEAR	WO-A-1511	FM-GP-17285	0810
AD	NUT	WO-A-980	FM-356125-S	0803	BK	BUSHING	WO-A-985	FM-GP-17277	0810
AE	OIL seal	WO-A-958	(Willys only)	0803	BL	CUP	WO-A-1507	FM-GPW-7768	0801
AF	FLANGE	WO-A-1105	FM-GP-4863	0803	BM	OIL seal	WO-A-958	(Willys only)	0803

SNL G-503

0801-0803

GROUP 08—TRANSMISSION TRANSFER (Cont'd)

Figure Number	Official Stockage Number	Part Number Willys	Part Number Ford	ITEM	Quantity Reqd. per Unit Assy.	Per 100 Major Items 12 Mos. Field Maintenance	Per 100 Major Items Major Overhaul (5th Ech)	Total First Year Procurement	Estimated Reqmts. per 100 Rebuilds
Col. 1	Col. 2	Col. 3	Col. 4	Col. 5	Col. 6	Col. 7	Col. 8	Col. 9	Col. 10
				0801—CASE (Cont'd)					
08-1		WO-A-1435	FM-GPW-7756	GASKET, transfer case to transmission (SP-18-352-5)	1	26	—	30	—
08-1		WO-A-1104	FM-353053	PLUG, pipe, I., ½ in. (drain) (SP-18-39-2) (issue until stock exhausted)	1	4	8	—	—
08-1		WO-5140	FM-353064-S	PLUG, pipe, I., ½ in. (filler)	1	—	—	—	—
		WO-A-982	FM-GP-7782-A	SHIM, output shaft, rear bearing cap (.003 thk.) (SP-18-228-7)	as req.	—	—	—	—
		WO-A-983	FM-GP-7782-B	SHIM, output shaft, rear bearing cap (.010 thk.) (SP-18-228-8)	as req.	—	—	—	—
		WO-A-984	FM-GP-7782-C	SHIM, output shaft, rear bearing cap (.031 thk.) (SP-18-228-9)	as req.	—	—	—	—
		WO-A-6753	FM-GPW-18360	SHIM SET, output shaft (Includes: 4 WO-A-982 FM-GP-7782A SHIM; 2 WO-A-983 FM-GP-7782B SHIM; 2 WO-A-984 FM-GP-7782C SHIM)	as req.	—	6	5	20
				0802—DRIVE GEAR					
08-1		WO-A-1510	FM-GP-7722	GEAR, drive (O.D. 4.104 in., 27 teeth) (SP-18-8-7)	1	—	3	4	10
08-1		WO-A-520	FM-356134-S18	NUT, mainshaft (S, hex, slotted, ⅞-16NC-2) (WG-T-9-50P)	1	6	13	20	40
08-1		WO-5397	FM-72071	PIN, cotter, S, ⅛ x ½ (main shaft nut)	1	—	—	—	—
08-1		WO-A-1410	FM-356519-S	WASHER, main shaft nut (S., plain, 1⁵⁄₁₆ O.D. x ²¹⁄₃₂ I.D. x ⁵⁄₃₂ thk.) (WG-T-84J-50½A)	1	6	3	10	10
				0803—DRIVEN GEARS, SHAFTS, BEARINGS					
08-1		WO-A-1007	FM-GP-7719	BEARING, ball, annular, single row, bore 1.1811 x .6299 (output clutch shaft) (MRC-206) (SKF-6206) (ND-3206)	1	—	6	7	30
08-1		WO-A-987	FM-GP-7777-A	BUSHING, output clutch shaft pilot (SP-18-24-1)	1	—	3	3	10
08-1		WO-51575	FM-GP-7723	CONE and ROLLERS, output shaft bearing (1.3125 diam. x .740 wide) (TIM-14131)	2	—	13	14	40
08-1		WO-52883	FM-O1Y-1202	CUP, output shaft bearing (TIM-14276) (BT-BT-14276)	2	—	13	14	40
08-1		WO-A-1105	FM-GP-4863	FLANGE, companion, rear (SP-K2-1-28)	1	3	3	8	10
08-1		WO-A-980	FM-356125-B	NUT, hex, S, slotted, ¾-20NF-3 (companion flange) (SP-124-J)	2	6	3	24	10
08-1		WO-5108	FM-72053-S	PIN, cotter, ⅛ x 1¼ (flange nut)	2	—	—	—	—
08-1		WO-A-1134	FM-GP-7746	GASKET, output clutch shaft bearing cap oil seal (SP-18-223-3)	2	13	—	22	—
08-1		WO-A-989	FM-GP-7766	GEAR, output shaft, O.D. 4.101 bore 1.250 x 1.713, 27 teeth (SP-18-8-1)	1	—	3	4	10
08-1		WO-A-988	FM-GP-7765	GEAR, sliding, output shaft O.D. 5.571 x 1.192 wide, 27 teeth (SP-18-8-2)	1	—	3	4	10
		WO-A-992	FM-GP-7762	GEAR, clutch, output shaft (10 internal teeth, bore 1.088 diam. x 1.755 wide, face 2.411 diam.) (SP-18-466-1)	1	—	—	—	—
08-1		WO-A-958		OIL SEAL, output shaft (front and rear bearing cap) (SP-62-463-2) (Willys only)	2	13	3	20	10

SNL G-503

118

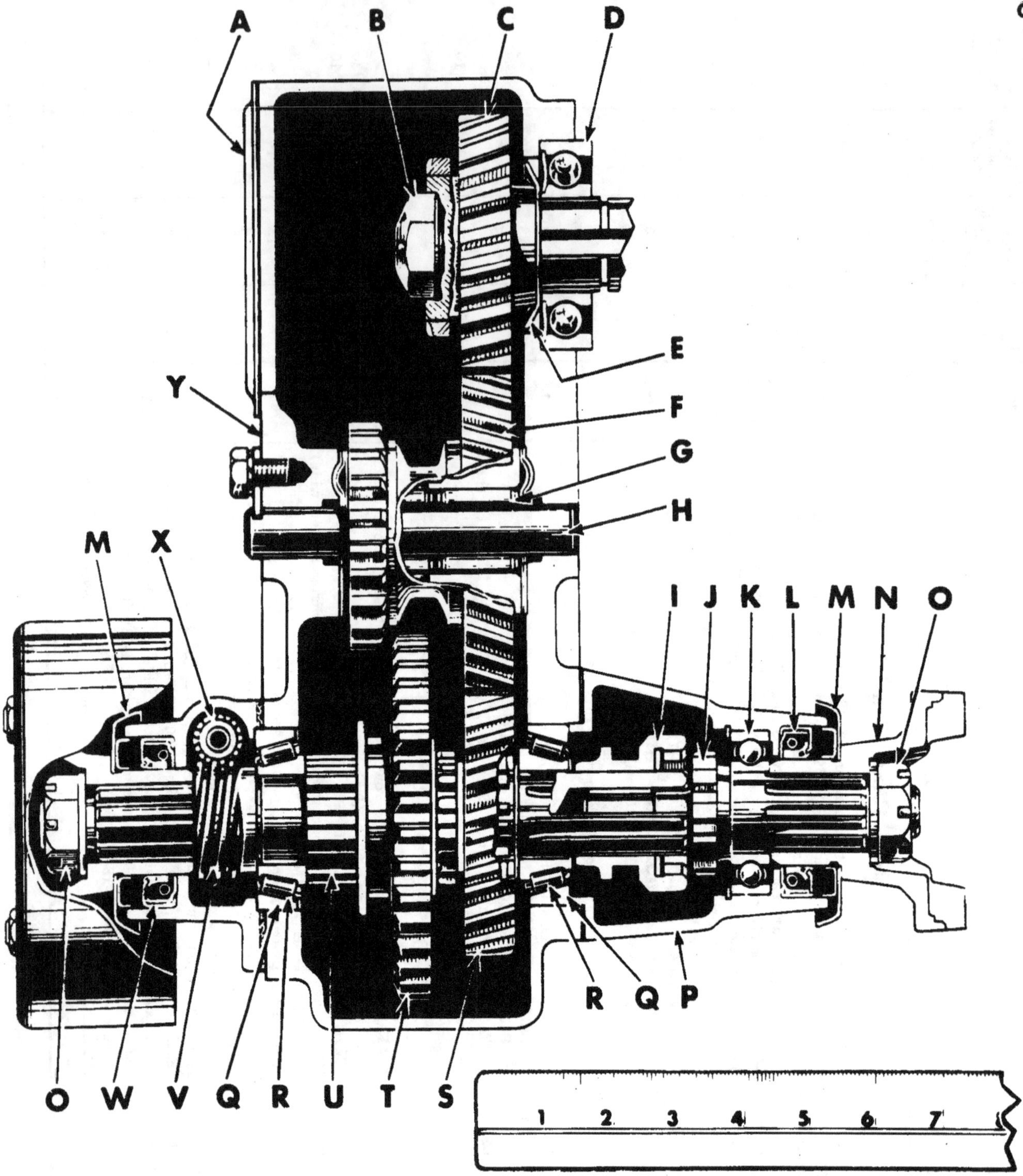

FIGURE 08-2—TRANSFER CASE—CROSS-SECTIONAL VIEW

Key	Item	Willys Part No.	Ford Part No.	Gov't Group No.	Key	Item	Willys Part No.	Ford Part No.	Gov't Group No.
A	COVER	WO-A-1508	FM-GPW-7706	0801	N	YOKE	WO-A-1106	FM-GP-7729	0803
B	NUT	WO-A-520	FM-356134S-18	0802	O	NUT	WO-A-980	FM-356125-S	0803
C	GEAR	WO-A-1510	FM-GP-7722	0802	P	CAP	WO-A-956	FM-GPW-7774	0901
D	BEARING	WO-A-916	FM-GP-7065	0704	Q	CUP	WO-52883	FM-01Y-1202	0803
E	SHIELD	WO-A-1111	FM-GP-7776	0803	R	CONE and ROLLERS	WO-51575	FM-GP-7723	0803
F	GEAR	WO-A-999	FM-GP-7742	0804	S	GEAR	WO-A-989	FM-GP-7766	0803
G	BEARING	WO-A-924	FM-GP-7718-A	0804	T	GEAR	WO-A-988	FM-GP-7765	0803
H	SHAFT	WO-A-998	FM-GP-7742	0804	U	SHAFT	WO-A-1764	FM-GP-7763	0803
I	GEAR	WO-A-992	FM-GP-7762	0803	V	GEAR	WO-A-1511	FM-GP-17285	0810
J	SHAFT	WO-A-975	FM-GP-7761	0803	W	OIL SEAL	WO-A-958	(Willys only)	0803
K	BEARING	WO-A-1007	FM-GP-7719	0803	X	GEAR	WO-A-1512	FM-GPW-17271	0810
L	SEAL	WO-A-958	(Willys only)	0803	Y	CASE	WO-A-1503	FM-GPW-7705	0801
M	SHIELD	WO-A-1111	FM-GP-7776	0803					

RA PD 305157-A

SNL G-503

0803-0805A

GROUP 08—TRANSMISSION TRANSFER (Cont'd)

Figure Number	Official Stockage Number	Part Number Willys	Part Number Ford	ITEM	Quantity Reqd. per Unit Assy.	Per 100 Major Items 12 Mos. Field Maintenance	Per 100 Major Items Major Overhaul (5th Ech)	Total First Year Procurement	Estimated Reqmts. per 100 Rebuilds
Col. 1	Col. 2	Col. 3	Col. 4	Col. 5	Col. 6	Col. 7	Col. 8	Col. 9	Col. 10
				0803—DRIVEN GEARS, SHAFTS, BEARINGS (Cont'd)					
08-1		WO-A-975	FM-GP-7761	SHAFT, output clutch (SP-18-362-3) (issue until stock exhausted. Then use WO-A-8835 for maintenance)	1				
		WO-A-8835		SHAFT, output clutch, w/NUT and WASHER (was WO-A-975)	1	4	8		10
08-1		WO-A-1764	FM-GP-7763	SHAFT, output, w/BUSHING, assembly (SP-18-362-4-5866-X) (issue until stock exhausted)	1		3	4	
08-1		WO-A-8841	FM-GPW-7736	SHAFT, output, w/NUT and WASHER, assembly	1	4	4	—	10
08-1		WO-A-1111	FM-GP-7776	SHIELD, dust (SP-5962-X) (1 used companion flange, 1 used universal joint end yoke)	2			3	4
08-1		WO-A-991	FM-GP-7784	SNAP RING, output shaft (SP-18-381-2)	1	4	—	9	5
08-1		WO-A-976	FM-GP-7783	SNAP RING, output clutch shaft bearing (SP-22-381-19)	1		26	27	80
08-1		WO-A-1028	FM-356504-S	WASHER, companion flange nut (SP-145-W)	2		26	28	80
08-1		WO-A-990	FM-GP-7771	WASHER, thrust, output shaft (SP-710-W)	1		10	20	30
08-1		WO-A-1106	FM-GP-7729	YOKE, end, universal joint, front (SP-K2-4-88X)	1	2	3	5	10
							2	5	5
				0804—IDLER GEAR, SHAFT, BEARINGS					
08-1		WO-A-924	FM-GP-7718-A	BEARING, intermediate gear (HY-94322)	2		13	14	40
08-1		WO-A-999	FM-GP-7742	GEAR, intermediate (O.D. 4.962, 33 teeth) (SP-18-5-1)	1		3	4	10
08-1		WO-A-1001	FM-GP-7767	PLATE, lock, intermediate (SP-31-246-2)	1		3	3	10
		WO-52189	FM-20328-S2	BOLT, hex-hd, S., 3/8-16NC-2 x 5/8	1			—	
		WO-5010	FM-34807-S2	WASHER, lock, S., 3/8 in.	1			—	
08-1		WO-A-998	FM-GP-7743	SHAFT, intermediate (SP-18-187-1)	1		3	4	10
08-1		WO-A-1000	FM-GP-7744	WASHER, thrust, intermediate gear (SP-690-W)	1		13	12	40
				0805—SHIFTER RODS					
08-1		WO-5599	FM-353075-S	BALL, poppet, shift rod (SP-50-80-1)	2		6	36	20
08-1		WO-A-974	FM-GP-7798-A	OIL SEAL, shift rod (SP-13-463-1) (Willys only)	2	13	—	20	—
08-1		WO-A-967	FM-355698-S	PLUG, shift rod poppet (SP-854-D)	2		6	24	60
08-1		WO-A-962	FM-GP-7787	ROD, shift, front wheel drive (SP-18-67-2)	1		3	4	10
08-1		WO-A-1504	FM-GPW-7786	ROD, shift, underdrive and direct (SP-18-67-5)	2		3	4	10
08-1		WO-A-966	FM-GP-7788	SPRING, shift rod poppet (SP-22-72-5)	2		6	15	20
08-1		WO-A-964		WIRE, lock, No. 16 ga. (shift fork) (SP-LW-1) (Willys only)	2		—	—	—
				0805A—YOKES					
08-1		WO-A-960	FM-GP-7711	FORK, front wheel drive shift (SP-18-66-7)	1		3	4	10
08-1		WO-A-959	FM-GP-7712	FORK, underdrive and direct shift (SP-18-66-1)	1		3	4	10
08-1		WO-A-965	FM-GP-7789	INTERLOCK, shift rod (SP-18-21-1)	1		3	5	10
08-1		WO-A-963	FM-355550	SCREW, set, shift fork (S., 3/8-24NF, head dld. f/c pin, 3/64 in. diam.) (SP-745D)	2		13	36	40

SNL G-503

0805B-0901

0805B—SHIFT LEVER AND PARTS

08-1	WO-A-971	FM-GP-7213	BALL, shift lever handle (SP-39-Q-20) (issue until stock exhausted)	2				
08-1	WO-638224	FM-353035-S7	FITTING, grease, 45° (shift lever pivot pin) (AD-1636) (issue until stock exhausted)	1	16	24		
		FM-GPW-1101735-A	GROMMET, transfer shift levers (leather)	1	36	74		
08-1	WO-A-3784	FM-GPW-7710	LEVER, shift, front wheel drive (SP-25-Q-1056)	1	6		6	10
08-1	WO-A-1506	FM-GPW-7793	LEVER, shift, underdrive and direct (SP-25-Q-1055)	1		3	4	10
08-1	WO-A-1505	FM-GP-7796	PIN, pivot, shift lever (SP-363-SP)	1		3	4	10
08-1	WO-A-972	FM-355378-S	SCREW, set, shift lever pivot pin (SP-433-D)	1		3	3	10
08-1	WO-A-973	FM-GP-7799	SPRING, shift lever (SP-97-Q-13)	1		13	20	40
08-1	WO-A-970			2		3	6	10

0809—MOUNTINGS

	WO-52911	FM-24408-S7	BOLT, hex-hd., S., 3/8-16NC-2 x 1 1/4 (transfer case to transmission)	5				
	WO-51405		BOLT, hex-hd., S., 3/8-16NC-2 x 1 (transfer case to transmission lower left) (Willys only)	1	6		24	
	WO-A-147	FM-355741-S	BOLT, support, transfer case (hex-hd., S., 1/2-20NF x 3)	1	3		5	
	WO-634758	FM-74-6038	INSULATOR, transfer case support	1				
	WO-5916	FM-33846-S	NUT, hex, S., 1/2-20NF-2 (support bolt)	1				
	WO-634759	FM-GPW-7781-A	SNUBBER, transfer case support insulator (issue until stock exhausted)	1	16		10	
	WO-634762	FM-356524-S	WASHER, S., 1 1/2 O.D. x 21/32 I.D. x 1/16 thk. (support bolt)	1	6	24		
	WO-5009	FM-34809-S2	WASHER, lock, S., 1/2 (7/8 O.D. x 17/32 I.D. x 1/8 thk.) (support bolt) (Willys only)	1				
		FM-34807-S7	WASHER, lock, S., 3/8 S.A.E. (21/32 O.D. x 13/32 I.D. x 3/32 thk.) (case to transmission)	4				

0810—SPEEDOMETER DRIVE GEARS

08-1	WO-A-985	FM-GP-17277	BUSHING, speedometer drive gear (SP-475-2) (issue until stock exhausted)	1	4	4		
08-1	WO-A-1511	FM-GP-17285	GEAR, drive, speedometer (SP-18-452-2) (issue until stock exhausted)	1	4	4		
08-1	WO-A-1512	FM-GPW-17271	GEAR, driven, speedometer (SP-18-453-3) (issue until stock exhausted)	1	4	4		
	WO-A-7837	FM-GPW-18314	KIT, repair, speedometer drive gear (Includes: 1 WO-A-985 FM-GP-17277 BUSHING; 1 WO-A-1511 FM-GP-17285 GEAR, drive; 1 WO-A-1512 FM-GPW-17271 GEAR, driven)	as req.				
08-1	WO-636396	FM-GP-17333	SLEEVE, speedometer driven gear (SP-18-454-4) (issue until stock exhausted)	1	4	4		

GROUP 09—PROPELLER SHAFT AND UNIVERSAL JOINT

0901—PROPELLER SHAFT ASSEMBLIES

	WO-A-1326	FM-GPW-3365	SHAFT, propeller, front, assembly (SP-8996-SF)	1	2	2	7	
	WO-A-1327	FM-GPW-4602	SHAFT, propeller, rear, assembly (SP-8997-SF)	1	2	2	7	

121

SNL G-503

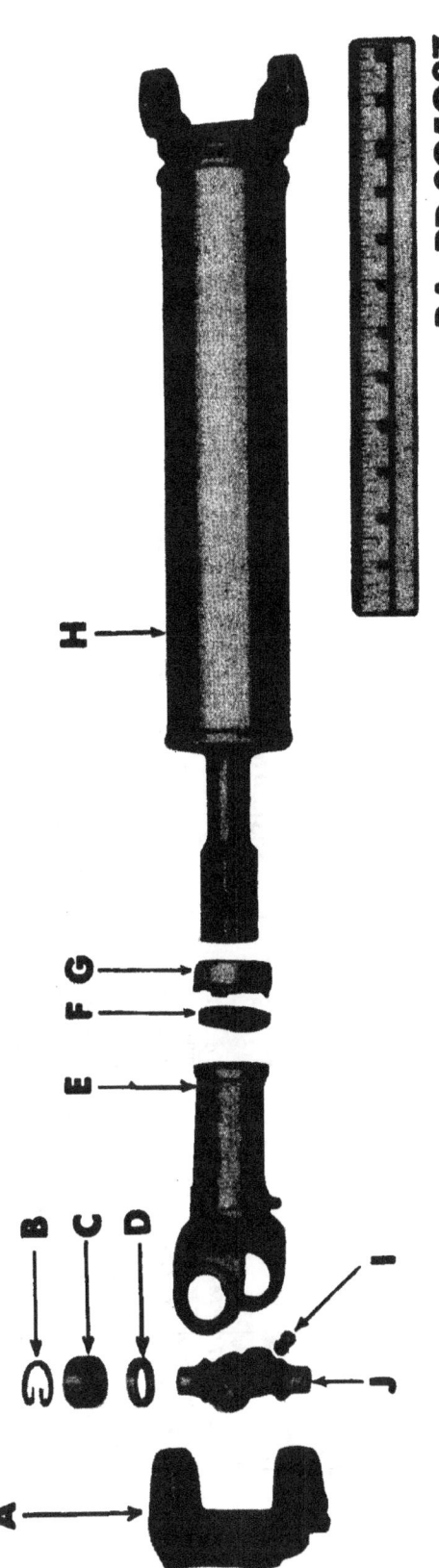

FIGURE 09-1—REAR PROPELLER SHAFT

Key	Item	Willys Part No.	Ford Part No.	Gov't Group No.	Key	Item	Willys Part No.	Ford Part No.	Gov't Group No.
A	YOKE	WO-A-950	FM-GP-4866	0902	E	YOKE	WO-A-935	FM-GP-4841	0902
B	SNAP ring	WO-A-945	FM-O1Y-7096	0902	F	WASHER	WO-A-943	FM-GP-7097	0902
C	BEARING	WO-A-1434	FM-GPW-7074	0902	G	CAP	WO-A-942	FM-GP-7077	0902
D	RETAINER	WO-A-940	FM-O1Y-7083	0902	H	TUBE	WO-A-1429	FM-GPW-4605	0902

RA PD 305087-A

GROUP 09—PROPELLER SHAFT AND UNIVERSAL JOINT (Cont'd)

Figure Number	Official Stockage Number	Part Number		ITEM	Quantity Reqd. per Unit Assy.	Per 100 Major Items			Estimated Reqmts. per 100 Rebuilds
		Willys	Ford			12 Mos. Field Maintenance	Major Overhaul (5th Ech)	Total First Year Procurement	
Col. 1	Col. 2	Col. 3	Col. 4	Col. 5	Col. 6	Col. 7	Col. 8	Col. 9	Col. 10
				0902—UNIVERSAL JOINTS					
		WO-A-1434	FM-GPW-7074	BEARING, universal joint, assembly	4	—	—	—	—
		WO-A-490	FM-O1Y-4529	BOLT, "U", S., 5/16-24NF-3 (universal joint bearing) (SP-K2-94-29)	6	13	13	20	—
		WO-A-491	FM-33784-S2	NUT, hex, S., 5/16-24NF-3 (SP-5-74-11) (issue until stock exhausted)	12	28	52	—	—
		WO-51833	FM-34806-S2	WASHER, lock, S., 5/16 (SP-5-75-19)	12	—	—	—	—
09-1		WO-A-942	FM-GP-7077	CAP, dust, universal joint yoke sleeve (SP-K2-14-69)	2	3	3	6	10
		WO-638792	FM-353043-S7	FITTING, grease 1/4-28NF (AD-1641) (GM-110347)	6	—	—	—	—
09-1		WO-A-941	FM-O1T-7078-A	GASKET, universal joint trunnion (SP-K3-86-89)	16	—	—	—	—

0902-1001

				4	120	40	187	125
WO-A-1433	FM-21C-18397-B	KIT, repair, universal joint journal (SP-K5-21-X) (Includes: 4 WO-A-1434 FM-GPW-7074 BEARING assembly, 1 WO-A-1426 FM-GPW-7084 JOURNAL assembly, 4 WO-A-945 FM-O1Y-7096 SNAP RING)						
WO-A-940	FM-O1Y-7083	RETAINER, trunnion gasket (SP-K2-76-17)		16				
WO-636568	FM-GP-4666	SHIELD, dust, front universal end yoke (SP-15009)	09-1	1	3	3	8	10
WO-A-945	FM-O1Y-7096	SNAP RING, trunnion bearing (SP-K2-7-29)	09-1	16	19	19	40	60
WO-A-1428	FM-GPW-3370	TUBE, front, propeller shaft, assembly (SP-K2-62-210-212-1907)	09-1	1				
WO-A-1429	FM-GPW-4605	TUBE, rear, propeller shaft, assembly (SP-K2-62-210-212-1718) (issue until stock exhausted)	09-1	1		3		
WO-A-943	FM-GP-7097	WASHER, universal joint yoke sleeve (cork, split, 1$\frac{37}{64}$ O.D. x 1 I.D. x $\frac{5}{16}$) (SP-K2-16-53)	09-1	2	6	6	20	20
WO-A-950	FM-GP-4866	YOKE, propeller shaft flange (SP-K2-2-329)	09-1	1	2	2	6	5
WO-A-935	FM-GP-4841	YOKE, universal joint sleeve, w/PLUG, assembly (SP-K2-3-198X)	09-1	2		3	8	10

GROUP 10—FRONT AXLE

1000—FRONT AXLE ASSEMBLY

WO-A-1212		AXLE, front, assembly (Bendix Weiss joints) (SP-2058-1) (Willys only)		1				
WO-A-1387		AXLE, front, assembly (Rzeppa joints) (SP-2058-2) (Willys only)		1				
		NOTE: Front axles to be tagged at joints for identification.						
WO-A-1765		AXLE, front, assembly (Spicer joints) (SP-SKA-34172) (Willys only)		1				
		NOTE: Identified by yellow paint on lower side of differential assembly.						
WO-A-6442		AXLE, front, assembly (Tracta joints) (Willys only)		1				
		NOTE: The above (4) axles are interchangeable.						
WO-A-6029	FM-GPW-3001	AXLE, front, w/BRAKES, DRUMS and HUBS, assembly (Willys only) (This part number to be used when either Tracta or Bendix universal joints are to be supplied with this assembly)		1	3		4	
		AXLE, front, w/BRAKES, DRUMS and HUBS, assembly (Ford only)		1				
WO-A-5498		AXLE, front, w/BRAKES, DRUMS and HUBS, assembly (Bendix joints) (Willys only)		1				
WO-A-5499		AXLE, front, w/BRAKES, DRUMS and HUBS, assembly (Rzeppa joints) (Willys only)		1				
WO-A-6470		AXLE, front, w/BRAKES, DRUMS and HUBS, assembly (Tracta joints) (Willys only)		1				
		NOTE: The above (3) Willys parts are interchangeable.						

1001—HOUSING

WO-636527	FM-355699-S2	BOLT, differential bearing cap (hex-hd., S., $\frac{1}{2}$-13NC x 2$\frac{1}{4}$)		4		6	8	20
WO-A-1486	FM-GPW-2084	BRACKET, front brake hose, left (issue until stock exhausted)		1		4		
WO-A-1487	FM-GPW-2082	BRACKET, front brake hose, right (issue until stock exhausted)		1		4		
WO-A-781	FM-GP-4016	COVER, axle housing (SP-16976)		1	1	1	4	2
WO-51523	FM-20046-S2	BOLT, hex-hd., S., $\frac{5}{16}$-18NC-2 x $\frac{7}{8}$ (cover to axle housing)	10-3	10				
WO-52510	FM-34941-S2	WASHER, lock, S., $\frac{5}{16}$ in.	10-3	10				
		HOUSING, axle, w/TUBE, assembly (SP-17310-X) (Willys only)	10-3	1		3	4	10
WO-A-1703	FM-GPW-3074	HOUSING, axle, w/TUBE, assembly (Ford only)	10-1	1				

123

SNL G-503

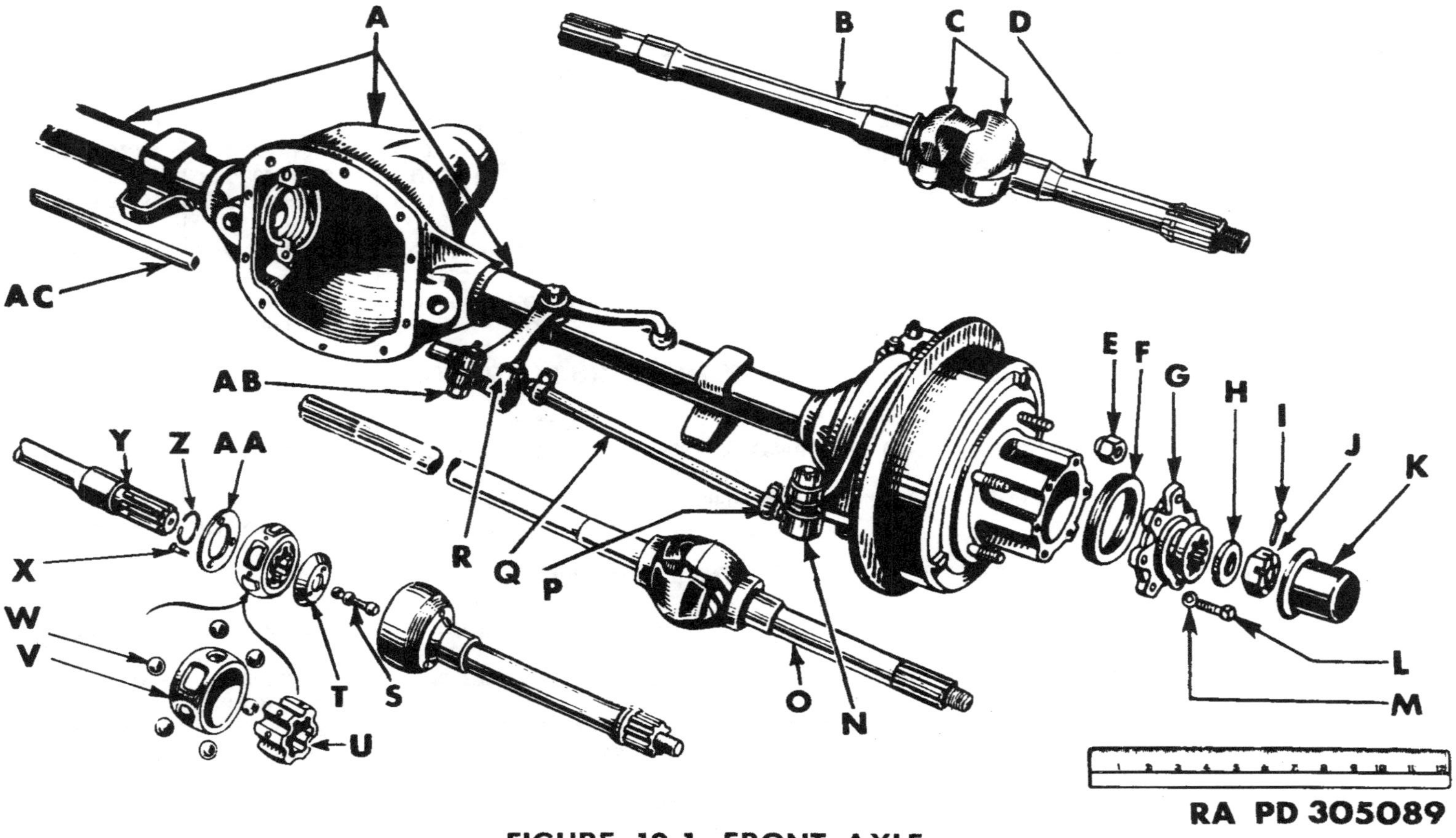

1001

FIGURE 10-1—FRONT AXLE

Key	Item	Willys Part No.	Ford Part No.	Gov't Group No.
A	HOUSING	WO-A-1703	(Willys only)	1001
B	SHAFT	WO-A-6383	(Willys only)	1007
C	UNIVERSAL joint	WO-A-6361	(Willys only)	1007
D	SHAFT	WO-A-6382	FM-GP-1013	1302A
E	NUT	WO-A-476	FM-GP-3208-A	1007
F	SHIMS	WO-A-862	FM-GP-3204	1007
G	FLANGE	WO-A-868	FM-356504-S	1003
H	WASHER	WO-636570	FM-72071	1301C
I	PIN	WO-5397	FM-356126-S	1301C
J	NUT	WO-636569	FM-GP-1139	1301C
K	CAP	WO-A-869	FM-GP-1110	1007
L	BOLT	WO-A-760	FM-34807-S	1402
M	WASHER	WO-5010	FM-GP-3292	1007
N	END assembly	WO-A-847	(Willys only)	
O	SHAFT assembly	WO-A-809		
P	CLAMP	WO-A-1706	FM-51-3287	1402
Q	TUBE	WO-A-1709	FM-GPW-3282	1402
R	CRANK	WO-A-8249	(Willys only)	1401
S	PIN	WO-A-1723	FM-GP-3218	1007
T	PILOT	WO-A-1722	FM-GP-3219	1007
U	RACE	WO-A-1720	FM-GP-3221-A	1007
V	CAGE	WO-A-1719	FM-3215-A	1007
W	BALL	WO-A-1721	FM-358074-S	1007
X	SCREW	WO-A-1725	FM-24622-S	1007
Y	SHAFT	WO-A-1727	FM-3106-A	1007
Z	SNAP ring	WO-A-1726	FM-GP-3216	1007
AA	RETAINER	WO-A-1724	FM-GP-3217	1007
AB	END assembly	WO-A-838	FM-GPW-3291	1402
AC	TUBE	WO-A-1705	FM-GPW-3281	1402

RA PD 305089-A

SNL G-503

GROUP 10—FRONT AXLE (Cont'd)

Figure Number	Official Stockage Number	Part Number Willys	Part Number Ford	ITEM	Quantity Reqd. per Unit Assy.	Per 100 Major Items 12 Mos. Field Maintenance	Per 100 Major Items Major Overhaul (5th Ech)	Total First Year Procurement	Estimated Reqmts. per 100 Rebuilds
Col. 1	Col. 2	Col. 3	Col. 4	Col. 5	Col. 6	Col. 7	Col. 8	Col. 9	Col. 10
				1001—HOUSING (Cont'd)					
10-3		WO-A-782	FM-GP-4035	GASKET, axle housing cover (SP-16409)	1	40	—	50	—
		WO-A-7830	FM-GPW-18365-B	GASKET SET, front axle	as req.	6	13	20	40
				(Includes:					
				2 WO-A-820 FM-GP-1092 GASKET, oil seal, assembly					
				4 WO-A-819 FM-GP-3135 SEAL, pivot					
				4 WO-A-818 FM-GP-3139 STRIP, pressure					
				2 WO-A-858 FM-GPW-3167 SEAL, steering					
				1 WO-A-782 FM-GP-4035 GASKET, housing					
				1 WO-636565 FM-GP-4061 GASKET, oil seal					
12-5		WO-A-1457	FM-GPW-2096	GUARD, front wheel brake hose, assembly (issue until stock exhausted)	2	—	4	—	—
		WO-636577	FM-358048-S	PLUG, drain, axle housing (SP-S-780) (issue until stock exhausted)	1	8	16	—	—
		WO-636538	FM-353051-S	PLUG, filler, axle housing (SP-50-39-4) (issue until stock exhausted)	1	8	16	—	—
		WO-A-870	FM-GP-4022	PLUG, vent, axle housing (SP-16979)	1	5	—	5	—
		WO-636528	FM-34922-S	WASHER, lock, internal, S., ½ in. (differential bearing cap bolt) (SP-11586) (issue until stock exhausted)	4	68	132	—	—
		WO-5010	FM-34807-S7	WASHER, lock, S., ⅜ in.	12	—	—	—	—
				1002—DIFFERENTIAL ASSEMBLY					
		WO-A-793	FM-GP-4206	CASE, differential (SP-16383)	1	—	3	4	10
		WO-A-788		DIFFERENTIAL, front assembly (SP-16968-X) (Willys only) (includes CASE, GEARS, and PINIONS)	1	—	3	4	10
			FM-GPW-4212	DIFFERENTIAL, front assembly (Ford only) (includes CASE, GEARS, PINIONS)	1	—	—	—	—
				1003—DIFFERENTIAL GEARS, PINION AND BEARINGS					
10-3		WO-52880	FM-GP-4221	CONE and ROLLERS, differential bearing (bore 1.625 x 1.000) (TIM-24780)	2	—	13	14	40
10-3		WO-52876	FM-86H-4621	CONE and ROLLERS, drive pinion bearing, inner (bore 1.375 x 1.125) (TIM-31593)	1	—	3	4	10
10-3		WO-52878	FM-GP-4630	CONE and ROLLERS, drive pinion bearing, outer (bore 1.125 x .875) (TIM-02872)	1	—	3	4	10
10-3		WO-52881	FM-GP-4222	CUP, differential bearing (3.000 O.D. x .8125 wide) (TIM-24721)	1	—	13	14	40
10-3		WO-52877	FM-86H-4616	CUP, drive pinion bearing, inner (3.00 O.D. x .9375 wide) (TIM-31520)	1	—	3	4	10
10-3		WO-52879	FM-GP-4628	CUP, drive pinion bearing, outer (2.875 O.D. x .6875 wide) (TIM-02820)	1	—	3	4	10
10-3		WO-636565	FM-GP-4661	GASKET, drive pinion oil seal (SP-S-171)	1	26	6	30	20
10-3		WO-A-789	FM-GPW-4209	GEAR, drive, differential, w/PINION (SP-16412-X) (Willys only)	1	—	—	6	—
				GEAR, drive, differential, w/PINION (Ford only)	1	—	—	—	—

125

SNL G-503

FIGURE 10-3 — DIFFERENTIAL ASSEMBLY — CROSS-SECTIONAL VIEW

Key	Item	Willys Part No.	Ford Part No.	Gov't Group No.
A	GEARS	WO-A-789	(Willys only)	1003
B	SEAL	WO-639265	FM-GP-4676-A	1003
C	YOKE	WO-A-1445	FM-GP-4842	1104
D	NUT	WO-639569	FM-356125-S	1003
E	CONE	WO-52878	FM-GP-4630	1003
F	CUP	WO-52879	FM-GP-4628	1003
G	SHIMS	WO-A-803	FM-GP-4659-A	1003
H	CONE	WO-52876	FM-86H-4621	1003
I	CUP	WO-52877	FM-86H-4616	1003
J	PIN	WO-636360	FM-GP-4241	1003
K	SHIMS	WO-A-784	FM-GP-4229-A	1003
L	CONE	WO-52880	FM-GP-4221	1003
M	CUP	WO-52881	FM-GP-4222	1003
N	SEAL	WO-A-779	FM-GP-3034	1006
O	SHAFT	WO-A-901	FM-GP-4234	1101
P	WASHER	WO-52510	FM-34941-S2	1001
Q	BOLT	WO-51523	FM-20046-S2	1001
R	GEAR	WO-A-794	FM-GPW-4236	1003
S	GEAR	WO-A-796	FM-GPW-4215	1003
T	SHAFT	WO-A-798	FM-GP-4211	1103
U	COVER	WO-A-781	FM-GP-4016	1001
V	CASE	WO-A-793	FM-GP-4206	1103
W	GASKET	WO-A-782	FM-GP-4085	1001
X	SHAFT	WO-A-902	FM-GP-4235	1102
Y	SCREW	WO-A-871	FM-355511-S	1103
Z	STRAP	WO-A-792	FM-GP-4281	1103

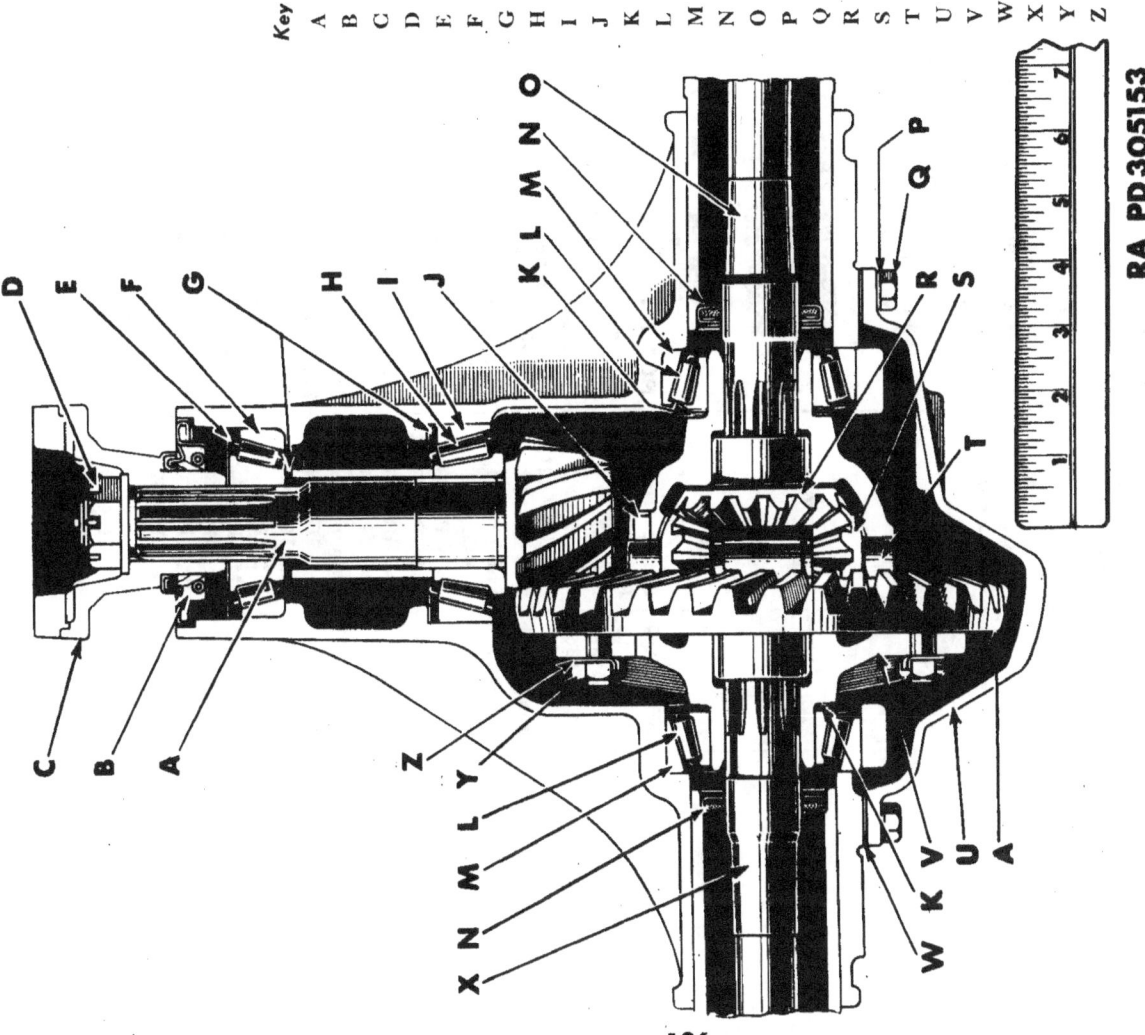

RA PD 305153-A

GROUP 10—FRONT AXLE (Cont'd)

1003—DIFFERENTIAL GEARS, PINION AND BEARINGS (Cont'd)

Figure Number	Official Stockage Number	Part Number Willys	Part Number Ford	ITEM	Quantity Reqd. per Unit Assy.	Per 100 Major Items 12 Mos. Field Maintenance	Per 100 Major Items Major Overhaul (5th Ech)	Total First Year Procurement	Estimated Reqmts. per 100 Rebuilds
Col. 1	Col. 2	Col. 3	Col. 4	Col. 5	Col. 6	Col. 7	Col. 8	Col. 9	Col. 10
10-3		WO-A-794	FM-GPW-4236	GEAR, side, differential, 2.758 O.D. x 1⅜ in. wide, 16 teeth (SP-16385)	2	—	—	—	—
		WO-A-6743	FM-GPW-18389	KIT, repair, front axle differential gear	as req.	—	3	4	10
				(Includes:					
				2 WO-A-794 FM-GPW-4236 GEAR, side					
				2 WO-A-796 FM-GPW-4215 PINION					
				1 WO-A-798 FM-GP-4211 SHAFT, pinion					
				2 WO-A-797 FM-GP-4230 WASHER, thrust					
				2 WO-A-795 FM-GP-4228 WASHER, thrust)					
		WO-A-6816	FM-GPW-18384	KIT, repair, drive gear and pinion	as req.	—	3	4	10
				(Includes:					
				1 WO-636569 FM-356126-S NUT, drive pinion					
				8 WO-A-871 FM-355511-S SCREW, drive gear					
				1 WO-A-789 FM-GPW-4209 SET, gear and pinion					
				4 WO-A-792 FM-GP-4281 STRAP, lock)					
13-1		WO-636569	FM-356126-S	NUT, hex, S, slotted, ¾-16NF-3 (drive pinion) (issue until stock exhausted)	1	20	40	20	—
10-3		WO-639265	FM-GP-4676-A	OIL SEAL, drive pinion (leather) (SP-14223)	1	13	—	—	—
10-3		WO-636571	FM-357202-S	PIN, cotter (S.) drive pinion nut (SP-7-72-39)	1	—	—	—	—
10-3		WO-636360	FM-GP-4241	PINION, differential pinion shaft (SP-13449)	1	—	—	—	—
10-3		WO-A-796	FM-GPW-4215	PINION, differential (1.140 O.D., 10 teeth) (SP-15926)	2	—	—	—	—
		WO-A-871	FM-355511-S	SCREW, differential, drive gear (SP-6-73-414)	8	—	13	20	40
		WO-A-798	FM-GP-4211	SHAFT, differential pinion (SP-16075)	1	—	—	—	—
		WO-636568	FM-GP-4666	SHIELD, dust, universal joint end yoke (SP-15099)	1	—	—	—	—
10-3		WO-A-784	FM-GP-4229-A	SHIM, adjusting, differential bearing (.003 thk.) (SP-S-58)	as req.	—	—	—	—
		WO-A-785	FM-GP-4229-B	SHIM, adjusting, differential bearing (.005 thk.) (SP-S-59)	as req.	—	—	—	—
		WO-A-786	FM-GP-4229-C	SHIM, adjusting, differential bearing (.010 thk.) (SP-S-74)	as req.	—	—	—	—
		WO-A-787	FM-GP-4229-E	SHIM, adjusting, differential bearing (.015 thk.) (Ford only)	as req.	—	—	—	—
		WO-A-800	FM-GP-4229-D	SHIM, adjusting, differential bearing (.030 thk.) (SP-S-75)	as req.	—	—	—	—
		WO-A-801	FM-GP-4660-A	SHIM, adjusting, drive pinion, large (.003 thk.) (SP-112)	as req.	—	—	—	—
10-3		WO-A-802	FM-GP-4660-B	SHIM, adjusting, drive pinion, large (.005 thk.) (SP-113)	as req.	—	—	—	—
		WO-A-803	FM-GP-4660-C	SHIM, adjusting, drive pinion, large (.010 thk.) (SP-S-114)	as req.	—	—	—	—
		WO-A-804	FM-GP-4659-A	SHIM, adjusting, drive pinion, bearing, small (.003 thk.) (SPS-638)	as req.	—	—	—	—
		WO-A-805	FM-GP-4659-B	SHIM, adjusting, drive pinion, bearing, small (.005 thk.) (SPS-638-1)	as req.	—	—	—	—
10-3		WO-A-806	FM-GP-4659-C	SHIM, adjusting, drive pinion, bearing, small (.010 thk.) (SPS-638-2)	as req.	—	—	—	—
		WO-636566	FM-GP-4659-D	SHIM, adjusting, drive pinion, bearing, small (.030 thk.) (SPS-638-3)	as req.	—	—	—	—
		WO-A-799	FM-GP-4619	SLINGER, oil, drive pinion bearing (SP-13575)	1	—	3	4	10
		WO-A-792	FM-GP-4668	SPACER, drive pinion bearing (SP-15367)	1	—	3	3	10
		WO-636570	FM-GP-4281	STRAP, lock, differential, drive gear screw (SP-16866)	4	—	12	—	—
10-1			FM-356504-S	WASHER, drive pinion nut (SP-S-1056) (issue until stock exhausted)	1	8	—	—	—

1003-1006

GROUP 10—FRONT AXLE (Cont'd)

Figure Number	Official Stockage Number	Part Number Willys	Part Number Ford	ITEM	Quantity Reqd. per Unit Assy.	Per 100 Major Items 12 Mos. Field Maintenance	Per 100 Major Items Major Overhaul (5th Ech)	Total First Year Procurement	Estimated Reqmts. per 100 Rebuilds
Col. 1	Col. 2	Col. 3	Col. 4	Col. 5	Col. 6	Col. 7	Col. 8	Col. 9	Col. 10
				1003—DIFFERENTIAL GEARS, PINION AND BEARINGS (Cont'd)					
		WO-A-795	FM-GP-4228	WASHER, thrust, differential side gear (SP-16323-2)	2	—	—	—	—
		WO-A-797	FM-GP-4230	WASHER, thrust, differential pinion (SP-16322-2)	2	—	—	—	—
		WO-A-1445	FM-GP-4842	YOKE, end, universal joint, assembly (SP-K2.4-108-X)	1	—	3	4	10
				1006—STEERING KNUCKLE, FLANGE AND ARM					
		WO-A-1712	FM-GPW-3113	ARM, knuckle steering, upper left (SP-17302-X) (Includes PIN, king; PIN, lock)	1	2	2	8	5
		WO-A-1710	FM-GPW-3112	ARM, knuckle steering, upper right (SP-17301-X) (Includes PIN, king and PIN, lock)	1	2	2	8	5
		WO-A-857	FM-GPW-3171	BEARING, bell crank (SP-17212) (TR-B-1210) (issue until stock exhausted)	2	22	148	—	—
		WO-A-8249	FM-GPW-3131-B	BELL CRANK, drag link front (SP-17307) (was WO-A-1211) (issue until stock exhausted) (FM-GPW-3131)	1	8	16	—	—
10-2		WO-A-872	FM-24327-S2	BOLT, hex-hd. S., 5/16-24NF-3 x 5/8 (oil seal) (SP-5-73-310) (issue until stock exhausted)	16	116	234	50	20
		WO-A-821	FM-355526-S	BOLT, hex-hd. S., 3/8-24NF-3 x 1 1/16 (stop screw) (SP-6-73-1117)	2	6	6	50	20
10-2		WO-A-853		BUSHING, wheel bearing spindle bz, 1.376 O.D. x 1.251 I.D. x 7/8 wide	2	6	6	6	20
		WO-A-828	FM-GP-3140	CAP, king pin bearing, lower assembly (SP-17048-X)	2	—	—	—	—
				(Includes:					
				PIN					
				PIN, king					
				PIN, lock)					
10-2		WO-52940	FM-GP-3161	CONE and ROLLERS, king pin bearing (bore .625 O.D. x 9/16 wide (TIM-11590)	4	6	6	15	20
		WO-A-814	FM-GP-1089	CONTAINER, steering knuckle, oil seal, assembly (half) (SP-17133-X)	4	6	—	—	20
10-2		WO-52941	FM-GP-3162	CUP, king pin bearing, 1.687 O.D. x 3/8 wide (TM-11520)	4	6	6	15	20
		WO-A-820	FM-GP-1092	GASKET, steering knuckle, oil seal, assembly (SP-17041)	2	80	—	100	—
		WO-A-6882	FM-GPW-18388	SHIM SET, king pin bearing (issue until stock exhausted)	as req.	4	12	—	—
				(Includes:					
				4 WO-A-830 FM-GP-3117-A SHIM, .003 thk.					
				4 WO-A-831 FM-GP-3117-B SHIM, .005 thk.					
				8 WO-A-832 FM-GP-3117-C SHIM, .010 thk.					
				6 WO-A-833 FM-GP-3117-D SHIM, .030 thk.)					
		WO-A-812	FM-GP-3149-A	KNUCKLE, steering, left (SP-17222)	1	2	2	7	5
		WO-A-811	FM-GP-3148-A	KNUCKLE, steering, right (SP-17221)	1	2	2	7	5
		WO-A-873	FM-33911-S2	NUT, hex, S., 3/8-24NF-3 (stop screw lock) (SP-6-74-101)	2	6	6	50	20
10-2		WO-630598	FM-33786-S2	NUT, hex, S., 3/8-24NF-2 (steering arm stud)	16	51	51	120	160
10-3		WO-A-779	FM-GP-3034	OIL SEAL, carrier end (SP-17036)	2	3	26	40	80

SNL G-503

128

10-2		WO-A-819	FM-GP-3135	OIL SEAL, felt, steering knuckle (half) (SP-17019)	4	%160	—	220	—
10-2		WO-A-813	FM-GP-1088	OIL SEAL, steering knuckle, assembly (half) (SP-17135-X)	4	% 6	6	20	20
13-1		WO-A-824	FM-GP-3115	PIN, king (SP-16991) (issue until stock exhausted)	4	28	52	—	—
13-1		WO-A-825	FM-GP-3122	PIN, lock, king pin (S., 1/8 O.D. x 3/8 lgth.) (SP-S-957) (issue until stock exhausted)	4	148	292	—	—
10-2		WO-5140	FM-353064-S	PLUG, filler, steering arm knuckle (SP-50-39-2)	2	—	—	—	—
10-2		WO-A-830	FM-GP-3117-A	SHIM, adjusting, king pin bearing (.003 thk.) (SP-16992-1)	as req.	—	—	—	—
		WO-A-831	FM-GP-3117-B	SHIM, adjusting, king pin bearing (.005 thk.) (SP-16992-2)	as req.	—	—	—	—
		WO-A-832	FM-GP-3117-C	SHIM, adjusting, king pin bearing (.010 thk.) (SP-16992-3)	as req.	—	—	—	—
		WO-A-833	FM-GP-3117-D	SHIM, adjusting, king pin bearing (.030 thk.) (SP-16994-4)	as req.	—	—	—	—
10-2		WO-A-851	FM-GP-3105	SPINDLE, wheel bearing, w/BUSHING, assembly (SP-17202-X)	2	3	3	9	10
10-2		WO-A-818	FM-GP-3139	STRIP, pressure, felt, steering knuckle, oil seal (SP-16983)	4	160	—	220	—
10-2		WO-A-1714	FM-357703-S	STUD, S., 3/8-24NF-3 x 2 (steering arm) (SP-S-962)	12	19	19	40	60
		WO-A-5504	FM-GPW-3325	STUD, S., 3/8-24NF-3 x 2 (steering arm, dowel)	4	6	6	35	20
10-2		WO-5010	FM-34807-S	WASHER, lock, S., 21/32 O.D. x 13/32 I.D. x 3/32 thk. (steering arm stud)	16	—	—	—	—
10-2		WO-52510	FM-34941-S2	WASHER, lock, oil seal bolt (SP-529-W)	16	—	—	—	—
				1007—AXLE SHAFTS, UNIVERSAL JOINTS					
10-2		WO-A-1721	FM-358074-S	BALL, S., 11/16, universal joint (SP-17232) (with Rzeppa joints WO-A-1715, FM-GPW-3206-A1 and WO-A-1728, FM-GPW-3207-A1)	12	—	—	—	—
		WO-A-1725	FM-24622-S2	BOLT, axle shaft retainer (hex-hd., S., No. 8) (.164-32NC-2) (included w/Rzeppa joints) (SP-17217) WO-A-1715, FM-GPW-3206-A1 and WO-A-1728, FM-GPW-3207-A1	6	14	14	50	45
10-2		WO-A-6362		BUSHING, axle shaft (SP-17361) (NP-38493) (required when replacing Bendix or Rzeppa joints with Tracta type) (Willys only)	2	2	2	8	5
10-1		WO-A-1719	FM-GP-3215-A	CAGE, universal joint (SP-17230) (w/Rzeppa joints WO-A-1715, FM-GPW-3206-A and WO-A-1728, FM-GPW-3207-A1)	2	—	—	—	—
10-1		WO-A-868	FM-GP-3204	FLANGE, drive, axle shaft (SP-17153)	2	3	3	9	10
10-1		WO-A-760	FM-GP-1110	BOLT, hex-hd., s-fin., alloy S., 3/8-16NC-2 x 1 1/2	12	19	19	50	60
10-1		WO-5010		WASHER, lock, S., 3/8 in.	12	—	—	—	—
10-2		WO-A-6361		JOINT, universal, axle shaft, assembly (SP-17356-X) (NP-38487) (Willys only)	2	—	—	—	—
		WO-A-6472		KIT, repair, axle shaft and universal joint, left (Tracta joint) (Willys only) (Includes: BUSHING; JOINT, universal)	as req.	—	—	—	—
10-1		WO-A-1722	FM-GP-3219	PILOT, universal joint (SP-17233) (included w/Rzeppa joints A-1715, and A-1728)	2	3	3	8	10
10-1		WO-A-1723	FM-GP-3218	PIN, universal joint pilot (SP-17224) (included w/Rzeppa joints A-1715 and A-1728)	2	3	3	8	10
10-1		WO-A-1720	FM-GP-3221-A	RACE, universal joint, inner (SP-17231) (included w/Rzeppa joints A-1715 and A-1728)	2	—	—	—	—
10-1		WO-A-1724	FM-GP-3217	RETAINER, axle shaft (SP-17216) (included w/Rzeppa joints A-1715, and A-1728)	2	—	—	—	—
10-1		WO-A-1725	FM-24622-S2	SCREW, fl-hd., S., No. 8 (.164)-32NC-2 x 11/32 (SP-17217) (included with Rzeppa joints A-1715 and A-1728)	6	—	—	—	—
10-1		WO-A-1729	FM-GP-3017-A	SHAFT, axle, left (SP-17122-4) (included w/Rzeppa joints A-1715, and A-1728)	1	2	2	7	5
		WO-A-1727	FM-GP-3016-A	SHAFT, axle, right (SP-17122-3) (included w/Rzeppa joints A-1715 and A-1728)	1	2	2	7	5
10-1, 2		WO-A-6383		SHAFT, axle, inner left (SP-17360-2) (NP-38491) (Willys only)	1	—	—	—	—

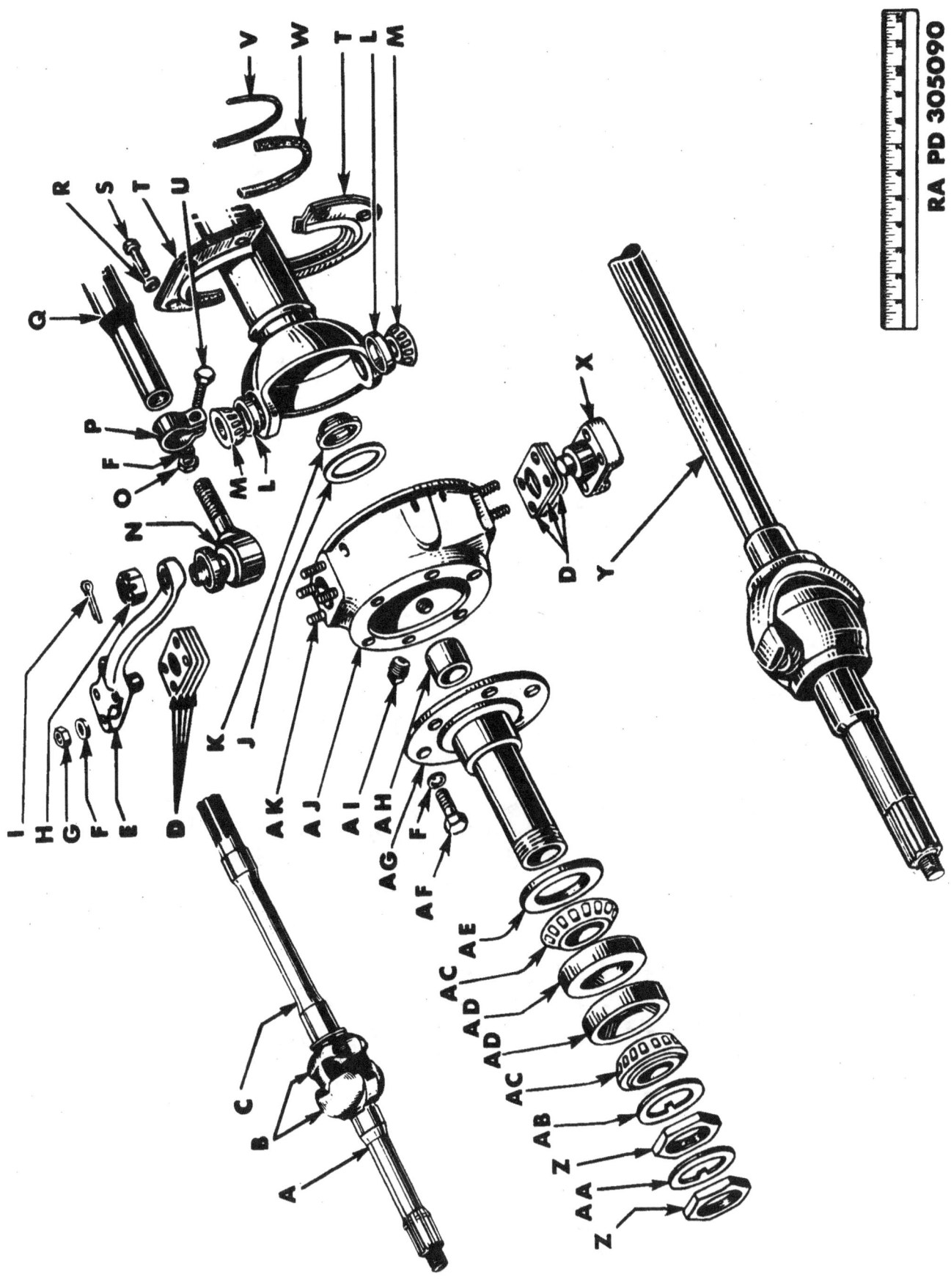

RA PD 305090

FIGURE 10-2—FRONT AXLE, STEERING KNUCKLE AND WHEEL BEARING

Key	Item	Willys Part No.	Ford Part No.	Gov't Group No.
A	SHAFT	WO-A-6382	(Willys only)	1007
B	UNIVERSAL joint assembly	WO-A-6361	(Willys only)	1007
C	SHAFT	WO-A-6383	(Willys only)	1007
D	SHIMS	WO-A-830	FM-GP-3117-A	1006
E	STEERING arm assembly	WO-A-1712	FM-GPW-3113	1006
F	WASHER	WO-5010	FM-34807-S	1006
G	NUT	WO-630598	FM-33786-S2	100
H	NUT	WO-10558	FM-351059-S7	1402
I	COTTER pin	WO-5152	FM-72025-S	1402
J	WASHER	WO-A-780	FM-CP-3374	1007
K	BUSHING	WO-A-6362	(Willys only)	1007
L	CUP	WO-52941	FM-GP-3162	1006
M	CONE and ROLLERS	WO-52940	FM-GP-3161	1006
N	END assembly	WO-A-847	FM-GP-3292	1402
O	NUT	WO-636575	FM-34083-S2	1402
P	CLAMP	WO-A-1706	FM-51-3287	1402
Q	TUBE—right	WO-A-1705	FM-GPW-3281	1402
R	WASHER	WO-52510	FM-34941-S2	1006
S	BOLT	WO-A-872	FM-24327-S2	1006
T	OIL seal assembly	WO-A-813	FM-GP-1088	1006
U	SCREW	WO-A-1707	FM-24916-S2	1402
V	FELT	WO-A-818	FM-GP-3139	1006
W	OIL seal	WO-A-819	FM-GP-3135	1006
X	CAP assembly	WO-A-828	FM-GP-3140	1006
Y	SHAFT w/JOINT assembly	WO-A-809	(Willys only)	1007
Z	NUT	WO-A-866	FM-GP-4252	1301C
AA	WASHER	WO-A-867	FM-GP-1124	1301C
AB	WASHER	WO-A-865	FM-GP-1218	1301C
AC	CONE and ROLLERS	WO-52942	FM-GP-1201	1301A
AD	CUP	WO-52943	FM-GP-1202	1301A
AE	OIL seal assembly	WO-A-864	FM-GP-1177	1301B
AF	SCREW	WO-A-877	FM-355552-S	1203
AG	SPINDLE and BUSHING assembly	WO-A-851	FM-GP-3105	1006
AH	BUSHING	WO-A-853		1006
AI	PLUG	WO-5140	FM-353064-S	1006
AJ	STEERING knuckle—right	WO-A-811	FM-GP-3148-A	1006
AK	STUD	WO-A-1714	FM-357703	1006

RA PD 305090-A

GROUP 10—FRONT AXLE (Cont'd)

1007—AXLE SHAFTS, UNIVERSAL JOINTS (Cont'd)

Figure Number	Official Stockage Number	Part Number Willys	Part Number Ford	ITEM	Quantity Reqd. per Unit Assy.	12 Mos. Field Maintenance	Major Overhaul (5th Ech)	Total First Year Procurement	Estimated Reqmts. per 100 Rebuilds
Col. 1	Col. 2	Col. 3	Col. 4	Col. 5	Col. 6	Col. 7	Col. 8	Col. 9	Col. 10
10-2		WO-A-6384		SHAFT, axle, inner right (SP-17360-1) (NP-38492) (Willys only)	1	—	—	—	—
		WO-A-6382		SHAFT, axle, outer (SP-17359) (NP-38373) (Willys only)	2	—	—	—	—
		WO-A-810		SHAFT, axle, left, universal JOINT, assembly (Bendix) (SP-17128-2X) (Willys only)	1	—	—	—	—
10-2		WO-A-809		SHAFT, right, axle, w/universal JOINT, assembly (Bendix) (SP-17128-3X) (Willys only)	1	—	—	—	—
		WO-A-6030		SHAFT, axle, left, w/universal JOINT, assembly (Bendix or Tracta) (Willys only) (issue until stock exhausted)	1	4	8	—	—
		WO-A-6031		SHAFT, axle, right, w/universal JOINT, assembly (Bendix or Tracta) (Willys only) (issue until stock exhausted)	1	4	8	7	5
		WO-A-1728	FM-GPW-3207-A	SHAFT, axle, left, w/universal JOINT, assembly (Rzeppa) (SP-17120-4X)	1	2	2	7	5
		WO-A-1715	FM-GPW-3206-A1	SHAFT, axle, right, w/universal JOINT, assembly (Rzeppa) (SP-17120-3X)	1	2	2	—	—
		WO-A-1716	FM-GP-3200-A	SHAFT, axle, outer, w/universal JOINT, assembly (included w/Rzeppa JOINTS A-1715, and A-1728) (SP-17121-X)	2	3	3	8	10
10-1		WO-A-862	FM-GP-3208-A	SHIM, adjusting, universal joint (.010 thk.) (SP-17155-1)	as req.	—	—	—	—
		WO-A-863	FM-GP-3208-B	SHIM, adjusting, universal joint (.030 thk.) (SP-17155-2)	as req.	—	—	—	—
		WO-A-6881	FM-GPW-18336	SHIM SET, universal joint adjusting	as req.	6	6	18	20
				(Includes: 6 WO-A-862 FM-GP-3208-A SHIM, .010 thk. 3 WO-A-863 FM-GP-3208-B SHIM, .030 thk.)					
10-2		WO-A-1726	FM-GP-3216	SNAP RING, axle shaft retainer (SP-17218)	2	5	5	12	15
		WO-A-780	FM-GP-3374	WASHER, thrust, axle shaft (SP-S-953)	2	3	3	12	10

GROUP 11—REAR AXLE

1100—REAR AXLE ASSEMBLY

Figure Number	Official Stockage Number	Willys	Ford	ITEM	Qty	Col.7	Col.8	Col.9	Col.10
		WO-A-445		AXLE, rear, assembly (4.88 ratio) (Willys only)	1	—	—	—	—
		WO-A-5500	FM-GPW-4001	AXLE, rear, w/BRAKES, DRUMS and HUBS, assembly (issue until stock exhausted)	1	3	8	4	—
		WO-A-575	FM-GPW-5705	CLIP, rear spring (axle to spring)	4	—	—	—	—
		WO-339372	FM-GPW-5456	NUT, hex, S., 7/16-20NF-2 (clip nut) (issue until stock exhausted)	8	200	400	—	—
		WO-5938	FM-34838-S	WASHER, lock, S., 7/16 in.	8	—	—	—	—
		WO-A-1545	FM-GPW-18366	GASKET SET, rear axle (Use until exhausted, then use WO-A-1545)	as req.	—	—	—	—
		WO-A-7831	FM-GPW-18366-B	GASKET SET, rear axle (was WO-A-1545)	as req.	6	13	30	40
				(Includes: 2 WO-A-904 FM-GP-4032 GASKET, axle shaft; 1 WO-A-782 FM-GP-4035 GASKET, housing cover; 1 WO-636565 FM-GP-4661 GASKET, oil seal)					

1101-1103

10-3	WO-636527	FM-355699-S2	BOLT, hex-hd., S., ½-13NC-2 x 2¼ (differential bearing cap)	4	—	6	8	20
10-3	WO-A-781	FM-GP-4016	COVER, axle housing (SP-16976)	1	1	1	4	2
10-3	WO-51523	FM-20046-S2	BOLT, hex-hd., S., ⁵⁄₁₆-18NC-2 x ¾ (cover to housing)	8	—	—	—	—
10-3	WO-52510	FM-34941-S2	WASHER, lock, S., ⁵⁄₁₆ in.	10	—	—	—	—
10-3	WO-A-782	FM-GP-4035	GASKET, axle housing cover (SP-16409)	1	40%	2	50	5
	WO-A-888	FM-GPW-4004	HOUSING, axle, w/TUBE, assembly (SP-17226-X)	1	8%	16	3	—
11-1	WO-636577	FM-358048-S	PLUG, drain, axle housing (SP-S-780) (issue until stock exhausted)	1	—	—	—	—
11-1	WO-636538	FM-353051-S	PLUG, filler, axle housing (SP-50-39-4)	1	5%	—	5	10
11-1	WO-A-870	FM-GP-4022	PLUG, vent, axle housing (SP-16979)	1	—	—	—	—
11-1	WO-636528	FM-34922-S	WASHER, lock, S., internal, ½ in. (differential bearing cap screw) (issue until stock exhausted)	4	68	132	—	—

1102—AXLE DRIVE SHAFTS

11-1	WO-A-760	FM-GP-1110	BOLT, hex-hd., S., ⅜-16NC-3 x 1½ (drive shaft bolt) (SP-6-73-124)	12	19	19	50	60
11-1	WO-A-904	FM-GP-4032	GASKET, axle shaft (SP-17146)	2	160%	—	200	35
	WO-A-6439	FM-GPW-4259	GUIDE, axle shaft (SP-17377) (issue until stock exhausted)	2	7	8	—	—
11-1	WO-A-779	FM-GP-3034	OIL SEAL, axle inboard (SP-17036)	2	3%	26	40	80
10-3	WO-A-902	FM-GP-4235	SHAFT, rear axle, left (SP-17144-4)	1	2%	2	9	5
11-1	WO-A-901	FM-GPW-4234	SHAFT, rear axle, right (SP-17144-3)	1	2%	2	9	5
11-1	WO-5010	FM-34807-S7	WASHER, lock, S., ⅜ in. (axle shaft to hub bolt)	12	—	—	—	—

1103—DIFFERENTIAL AND CARRIER ASSEMBLY

11-1	WO-A-793	FM-GP-4206	CASE, differential (SP-16383)	1	—	3	4	10
11-1	WO-52880	FM-GP-4221	CONE, and ROLLERS, differential bearing, bore 1.625 x 1.000 (TIM-24780)	2	—	13	14	40
11-1	WO-52881	FM-GP-4222	CUP, differential bearing, .8125 wide x 3.000 O.D. (TIM-24721)	2	—	13	14	40
	WO-A-788	FM-GPW-4212	DIFFERENTIAL, assembly (Ford only)	1	—	3	4	10
11-1	WO-A-794	FM-GP-4236	DIFFERENTIAL, assembly (SP-16968-X) (Willys only)	2	—	—	—	—
11-1	WO-A-6816	FM-GPW-18384	GEAR, side, O.D. 2.758 x 1⅜ wide, 16 teeth (SP-16385)	2	—	3	—	10
			KIT, repair, drive gear and pinion	as req.				
			(Includes:					
			1 WO-636569 FM-356126-S NUT, drive pinion					
			8 WO-A-871 FM-355511-S SCREW, drive gear					
			1 WO-A-789 FM-GPW-4209 SHAFT, pinion					
			4 WO-A-792 FM-GP-4281 STRAP, lock)					
	WO-A-6743	FM-GPW-18389	KIT, repair, rear axle differential gear	as req.	—	3	4	10
			(Includes:					
			2 WO-A-794 FM-GP-4236 GEAR, side					
			2 WO-A-796 FM-GPW-4215 MATE, pinion					
			1 WO-A-798 FM-GP-4211 SHAFT, pinion					
			2 WO-A-797 FM-GP-4230 WASHER, thrust, pinion					
			2 WO-A-795 FM-GP-4228 WASHER, thrust side gear)					
11-1	WO-636360	FM-GP-4241	PIN, lock, pinion shaft (SP-13449)	1	—	6	—	60
11-1	WO-A-796	FM-GPW-4215	PINION, differential (O.D. 1.140, 10 teeth) (SP-15926)	2	—	—	—	—
11-1	WO-A-871	FM-355511-S	SCREW, drive gear (SP-73-414)	8	—	13	20	40
	WO-A-789	FM-GPW-4209	SET, drive gear and pinion (matched gear w/PINION) (SP-16412-X)	1	—	—	—	—

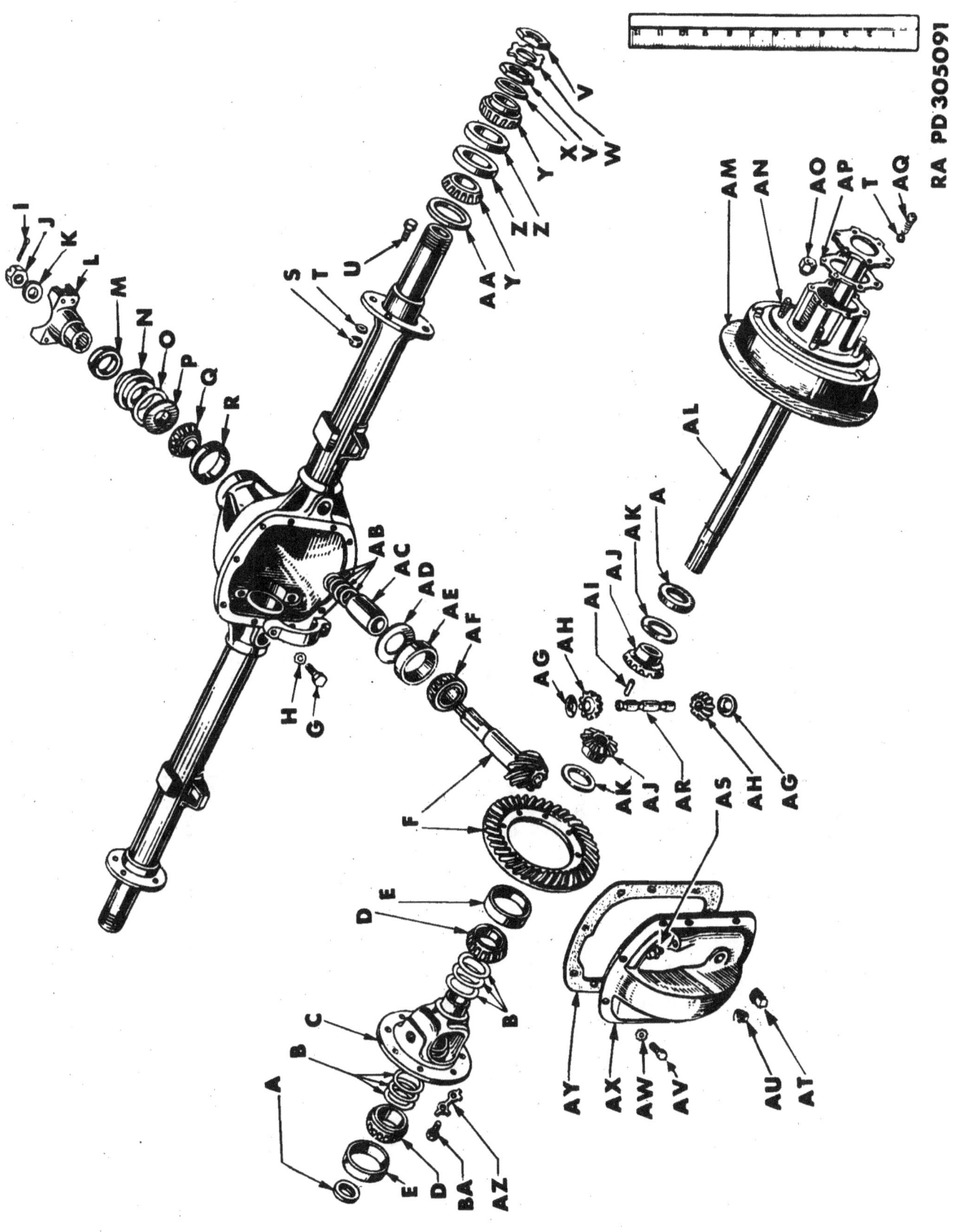

FIGURE 11-1—REAR AXLE ASSEMBLY

Key	Item	Willys Part No.	Ford Part No.	Gov't Group No.	Key	Item	Willys Part No.	Ford Part No.	Gov't Group No.
A	SEAL	WO-A-779	FM-GP-3034	1102	AB	SHIM	WO-A-803	FM-GP-4659-A	1104
B	SHIMS	WO-A-784	FM-GP-4229-A	1103	AC	SPACER	WO-A-799	FM-GP-4668	1104
C	CASE	WO-A-793	FM-GP-4206	1103	AD	SHIM	WO-A-800	FM-GP-4660-A	1104
D	CONE and ROLLERS	WO-52880	FM-GP-4221	1103	AE	CUP	WO-52877	FM-86H-4621	1104
E	CUP	WO-52881	FM-GP-4222	1103	AF	CONE and ROLLERS	WO-52876	FM-86H-4621	1104
F	GEAR and PINION	WO-A-789	FM-GPW-4209	1103	AG	WASHER	WO-A-797	FM-GP-4230	1103
G	BOLT	WO-636527	FM-355699-S2	1101	AH	PINION	WO-A-796	FM-GPW-4215	1103
H	WASHER	WO-636528	FM-34922-S	1101	AI	PIN	WO-636360	FM-GP-4241	1103
I	PIN	WO-636571	FM-357202-S	1104	AJ	GEAR	WO-A-794	FM-GPW-4236	1103
J	NUT	WO-636569	FM-356126	1104	AK	WASHER	WO-A-795	FM-GP-4228	1103
K	WASHER	WO-636570	FM-356504-S	1104	AL	SHAFT	WO-A-901	FM-GPW-4234	1102
L	YOKE assembly	WO-A-1445	FM-GP-4842	1104	AM	DRUM	WO-A-472	(Willys only)	1302
M	SHIELD	WO-636568	FM-GP-4666	1104	AN	BOLT	WO-A-474	FM-GP-1107	1302A
N	SEAL	WO-639265	FM-GP-4676-A	1104	AO	NUT	WO-A-476	FM-GP-1012	1302A
O	GASKET	WO-636565	FM-GP-4661	1104	AP	GASKET	WO-A-904	FM-GP-4032	1102
P	SLINGER	WO-636566	FM-GP-4619	1104	AQ	BOLT	WO-A-760	FM-GP-1110	1102
Q	CONE and ROLLERS	WO-52878	FM-GP-4630	1104	AR	SHAFT	WO-A-798	FM-GPW-4209	1103
R	CUP	WO-52879	FM-GP-4628	1104	AS	PLUG	WO-A-870	FM-GP-4022	1101
S	NUT	WO-636575	FM-33786-S2	1203	AT	PLUG	WO-636538	FM-353051-S	1101
T	WASHER	WO-5010	FM-34807-S7	1102	AU	PLUG	WO-636577	FM-358048-S	1101
U	SCREW	WO-A-903	FM-355578-S	1203	AV	SCREW	WO-51523	FM-20046-S2	1101
V	NUT	WO-A-866	FM-GP-4252	1301C	AW	WASHER	WO-52510	FM-34941-S2	1101
W	WASHER	WO-A-867	FM-GP-1124	1301C	AX	COVER	WO-A-781	FM-GP-4016	1101
X	WASHER	WO-A-865	FM-GP-1218	1301C	AY	GASKET	WO-A-782	FM-GP-403511	1101
Y	CONE and ROLLERS	WO-52942	FM-GP-1201	1301A	AZ	STRAP	WO-A-792	FM-GP-4281	1103
Z	CUP	WO-52943	FM-GP-1202	1301A	BA	SCREW	WO-A-871	FM-355511-S	1103
AA	RETAINER	WO-A-864	FM-GP-1177	1301B					

RA PD 305091-A

SNL G-503

GROUP 11—REAR AXLE (Cont'd)

Figure Number	Official Stockage Number	Part Number (Willys)	Part Number (Ford)	ITEM	Quantity Reqd. per Unit Assy.	12 Mos. Field Maintenance	Major Overhaul (5th Ech)	Total First Year Procurement	Estimated Reqmts. per 100 Rebuilds
Col. 1	Col. 2	Col. 3	Col. 4	Col. 5	Col. 6	Col. 7	Col. 8	Col. 9	Col. 10
				1103—DIFFERENTIAL AND CARRIER ASSEMBLY (Cont'd)					
11-1		WO-A-798	FM-GP-4211	SHAFT, pinion (SP-16075)	1	—	—	—	—
10-3		WO-A-784	FM-GP-4229-A	SHIM, adjusting (.003 thk.) (SP-S-58)	as req.	—	—	—	—
		WO-A-785	FM-GP-4229-B	SHIM, adjusting (.005 thk.) (SP-S-59)	as req.	—	—	—	—
		WO-A-786	FM-GP-4229-C	SHIM, adjusting (.010 thk.) (SP-S-74)	as req.	—	—	—	—
		WO-A-787	FM-GP-4229-E	SHIM, adjusting (.015 thk.) (Ford only)	as req.	—	—	—	—
		WO-A-6744	FM-GPW-18388	SHIM, adjusting (.030 thk.) (SP-S-75)	as req.	—	—	—	10
				SHIM SET, rear axle, differential bearing.	as req.	—	3	5	—
				(Includes:					
				2 WO-A-784 FM-GP-4229-A SHIM, .003					
				2 WO-A-785 FM-GP-4229-B SHIM, .005					
				2 WO-A-786 FM-GP-4229-C SHIM, .010					
				2 WO-A-787 FM-GP-4229-D SHIM, .030)					
11-1		WO-A-792	FM-GP-4281	STRAP, lock, drive gear screw (SP-16866)	4	—	—	—	—
11-1		WO-A-797	FM-GP-4230	WASHER, thrust, pinion (SP-16322-2)	2	—	—	—	—
11-1		WO-A-795	FM-GP-4228	WASHER, thrust, side gear (SP-16323-2)	2	—	—	—	—
				1104—DIFFERENTIAL PINION BEARING					
11-1		WO-52876	FM-86H-4621	CONE and ROLLERS, bearing (bore 1.375 x 1.125) drive pinion, inner (TM-31593)	1	1	3	4	10
11-1		WO-52878	FM-GP-4630	CONE and ROLLERS, bearing (bore 1.125 x .875) drive pinion, outer (TM-02872)	1	1	3	4	10
11-1		WO-52877	FM-86H-4616	CUP, drive pinion bearing, inner (3.00 O.D. x .9375 wide) (TM-31520)	1	1	3	4	10
11-1		WO-52879	FM-GP-4628	CUP, drive pinion bearing, outer (2.875 O.D. x .6875 wide) (TM-02820)	1	1	3	4	10
11-1		WO-636565	FM-GP-4661	GASKET, pinion oil seal, leather (SP-S-171)	1	26	—	30	—
11-1		WO-636569	FM-356126-S	NUT, hex, S, slotted, ¾-16NF-3, drive pinion (SP-S-1135) (issue until stock exhausted)	1	—	—	—	—
11-1		WO-639265	FM-GP-4676-A	OIL SEAL, pinion, leather (SP-14223)	1	20	40	20	—
11-1		WO-636571	FM-357202-S	PIN, cotter, S, 1.540 x .109 (SP-7-72-39)	1	13	—	—	—
11-1		WO-636568	FM-GP-4666	SHIELD, dust, universal joint end yoke (SP-15099)	1	—	—	—	—
11-1		WO-A-800	FM-GP-4660-A	SHIM, adjusting, bearing (large, .003) (SP-S-112)	as req.	—	—	—	—
11-1		WO-A-801	FM-GP-4660-B	SHIM, adjusting, bearing (large, .005) (SP-S-113)	as req.	—	—	—	—
11-1		WO-A-802	FM-GP-4660-C	SHIM, adjusting, bearing (large, .010) (SP-S-114)	as req.	—	—	—	—
11-1		WO-A-803	FM-GP-4659-A	SHIM, adjusting, bearing (small, .003) (SP-S-638)	as req.	—	—	—	—
		WO-A-804	FM-GP-4659-B	SHIM, adjusting, bearing (small, .005) (SP-S-638-1)	as req.	—	—	—	—
		WO-A-805	FM-GP-4659-C	SHIM, adjusting, bearing (small, .010) (SP-S-638-2)	as req.	—	—	—	—
		WO-A-806	FM-GP-4659-D	SHIM, adjusting, bearing (small, .030) (SP-S-638-3)	as req.	—	—	—	—

		WO-A-6745	FM-GPW-18386	SHIM SET, rear axle drive pinion	as req.	—	3	5	10
				(Includes:					
				4 WO-A-800 FM-GP-4660-A SHIM, large, .003					
				4 WO-A-801 FM-GP-4660-B SHIM, large, .005					
				4 WO-A-802 FM-GP-4660-C SHIM, large, .010					
				4 WO-A-803 FM-GP-4659-A SHIM, small, .003					
				4 WO-A-804 FM-GP-4659-B SHIM, small, .005					
				4 WO-A-805 FM-GP-4659-C SHIM, small, .010					
				4 WO-A-806 FM-GP-4659-D SHIM, small, .030)					
11-1		WO-636566	FM-GP-4619	SLINGER, oil drive pinion bearing (SP-13575)	1	—	3	4	10
11-1		WO-A-799	FM-GP-4668	SPACER, drive pinion bearing (SP-15367)	1	—	3	3	10
11-1		WO-636570	FM-356504-S	WASHER, S., 1½ O.D. x 49/64 I.D. x ⅛ thk., drive pinion nut (SP-S-1056) (issue until stock exhausted)	1	8	12	—	—
11-1		WO-A-1445	FM-GP-4842	YOKE, universal joint end, w/SHIELD assembly (SP-K2-4-108X)	1	—	3	4	10

GROUP 12—BRAKE

1200—BRAKE ASSEMBLY

12-2		WO-A-8894	FM-GPW-2011	BRAKE, front, left, assembly (BX-47047) (issue until stock exhausted)	1	% —	4	—	—
		WO-A-8895	FM-GPW-2010	BRAKE, front, right, assembly (BX-47048) (issue until stock exhausted)	1	% —	4	—	—
		WO-A-927	FM-GPW-2211	BRAKE, rear, left, assembly, ⅞ in. (before Willys serial 134356) (BX-452322)	1	—	—	—	—
		WO-A-8896	FM-GPW-2211-B	BRAKE, rear, left, assembly, ¾ in. (after Willys serial 134356) (BX-47050) (issue until stock exhausted)	1	% —	4	—	—
		WO-A-928	FM-GPW-2210	BRAKE, rear, right, assembly, ⅞ in. (before Willys serial 134356) (BX-45323)	1	—	—	—	—
		WO-A-8897	FM-GPW-2210-B	BRAKE, rear, right, assembly, ¾ in. (after Willys serial 134356) (BX-47051) (issue until stock exhausted)	1	%	4	—	—
				NOTE: WO-A-8894, FM-GPW-2011 and WO-A-8895, FM-GPW-2010 may be used for service in place of WO-A-927, FM-GPW-2211 and WO-A-928, FM-GPW-2210, providing both LEFT and RIGHT brakes are changed.					

1201—HAND BRAKE EMERGENCY BRAKE PARTS

12-1		WO-A-1009	FM-GP-2648	BAND, hand brake, assembly (SP-5900-X)	1	% 2	2	6	5
12-1		WO-A-1019	FM-355352-S7	BOLT, brake band brkt. (hex-hd., S., ¼-20NC-2 x 2¾) (SP-856D) (WG-X-2428-A)	1	% 5	5	40	—
12-1		WO-5790	FM-33795-S7	NUT, hex., S., ¼-20NC-2	1	—	—	—	—
12-1		WO-52706	FM-34805-S2	WASHER, lock, S., ¼ in.	1	—	—	—	—
12-5		WO-639010	FM-GPW-2848	BRACKET, hand brake ratchet tube, w/guide SPRING and SCREW, assembly (issue until stock exhausted)	1	—	4	—	—
12-5		WO-51396	FM-24347-S2	BOLT, hex-hd., S., 5/16-24NF-2 x ¾ (brkt. to support)	2	—	—	—	—
		WO-5910	FM-33798-S2	NUT, hex-hd., S., 5/16-24NF-2	2	—	—	—	—
		WO-5437	FM-34706-S2	WASHER, S. (plain) 5/16 in.	2	—	—	—	—
		WO-51833	FM-34806-S2	WASHER, lock, S., 5/16 in.	2	—	—	—	—
12-1		WO-A-1016	FM-O1T-2642	BOLT, adjusting, hand brake (SP-10-B-18)	1	% 2	2	8	5
12-1		WO-A-1020	FM-O1T-2616	BOLT, anchor clip (hex-hd., S., 5/16-18NC-2 x 1⅝) (SP-887-D)	1	% 10	10	25	30
F		WO-A-1008	FM-GPW-2598	BRAKE, hand, assembly (SP-5815-X) (issue until stock exhausted)	1	—	4	—	—
		WO-A-1241	FM-GPW-2853	CABLE, hand brake ratchet, w/TUBE, assembly (issue until stock exhausted)	1	% 4	8	—	—

1201

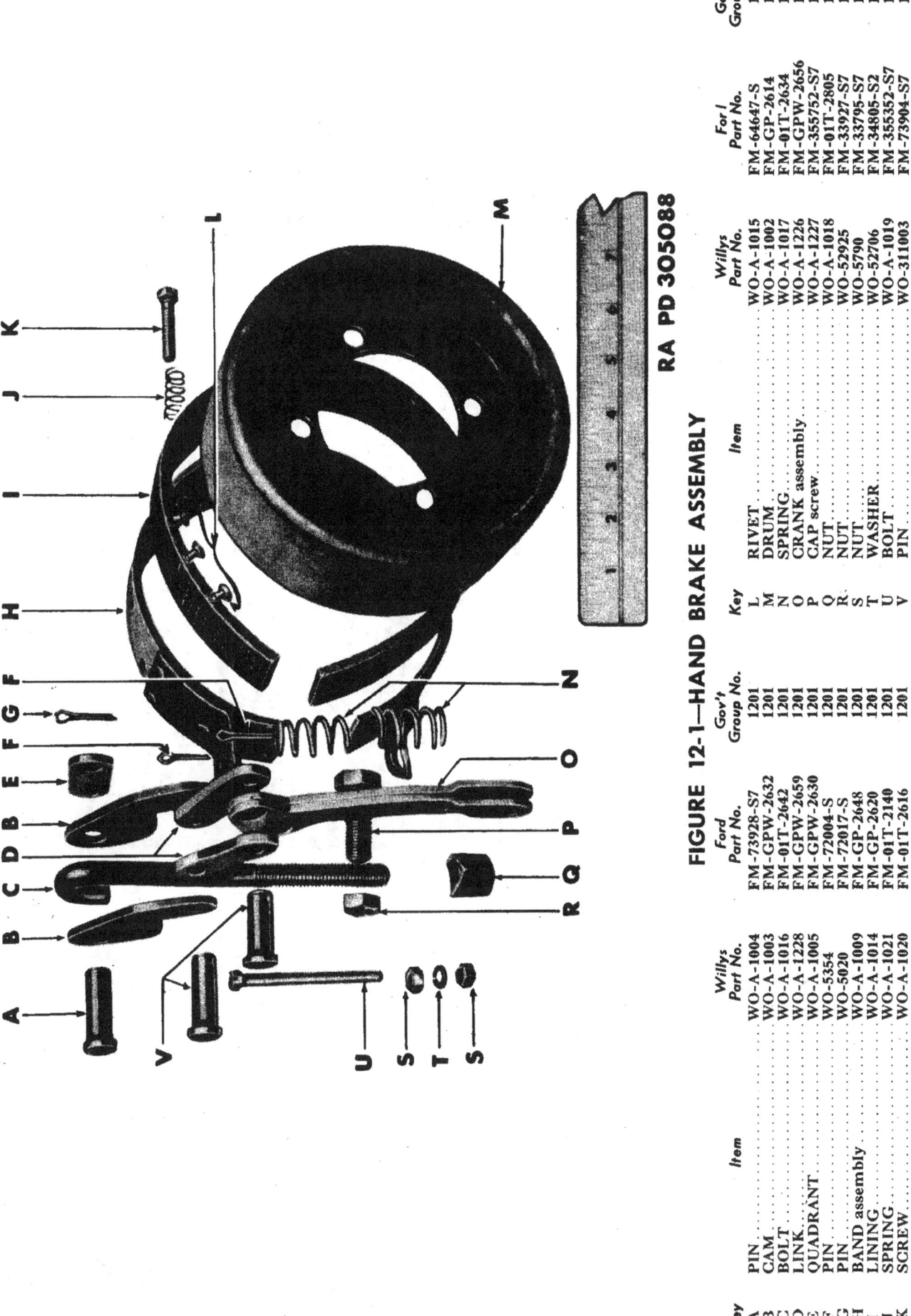

FIGURE 12-1—HAND BRAKE ASSEMBLY

Key	Item	Ford Part No.	Willys Part No.	Gov't Group No.
A	PIN	FM-73928-S7	WO-A-1004	1201
B	CAM	FM-GPW-2632	WO-A-1003	1201
C	BOLT	FM-01T-2642	WO-A-1016	1201
D	LINK	FM-GPW-2659	WO-A-1228	1201
E	QUADRANT	FM-GPW-2630	WO-A-1005	1201
F	PIN	FM-72004-S	WO-5354	1201
G	PIN	FM-72017-S	WO-5020	1201
H	BAND assembly	FM-GP-2648	WO-A-1009	1201
I	LINING	FM-GP-2620	WO-A-1014	1201
J	SPRING	FM-01T-2140	WO-A-1021	1201
K	SCREW	FM-01T-2616	WO-A-1020	1201
L	RIVET	FM-64647-S	WO-A-1015	1201
M	DRUM	FM-GP-2614	WO-A-1002	1201
N	SPRING	FM-01T-2634	WO-A-1017	1201
O	CRANK assembly	FM-GPW-2656	WO-A-1226	1201
P	CAP screw	FM-355752-S7	WO-A-1227	1201
Q	NUT	FM-01T-2805	WO-A-1018	1201
R	NUT	FM-33927-S7	WO-52925	1201
S	NUT	FM-33795-S7	WO-5790	1201
T	WASHER	FM-34805-S2	WO-52706	1201
U	BOLT	FM-355352-S7	WO-A-1019	1201
V	PIN	FM-73904-S7	WO-311003	1201

RA PD 305088-A

SNL G-503

1201

GROUP 12—BRAKE (Cont'd)

Figure Number	Official Stockage Number	Part Number Willys	Part Number Ford	ITEM	Quantity Reqd. per Unit Assy.	Per 100 Major Items 12 Mos. Field Maintenance	Per 100 Major Items Major Overhaul (5th Ech)	Total First Year Procurement	Estimated Reqmts. per 100 Rebuilds
Col. 1	Col. 2	Col. 3	Col. 4	Col. 5	Col. 6	Col. 7	Col. 8	Col. 9	Col. 10
				1201—HAND BRAKE EMERGENCY BRAKE PARTS (Cont'd)					
12-1		WO-A-1003	FM-GPW-2632	CAM, hand brake (SP-5-B-15) (issue until stock exhausted)	2		4		—
		WO-A-1735	FM-GPW-2272	CLAMP, hand brake cable tube (at air cleaner brkt.) (issue until stock exhausted)	1		4		—
		WO-A-1533	FM-34746-S7	WASHER, S, 5/16 in. (issue until stock exhausted)	1	% 16	24		1
		WO-638780	FM-GPW-2279	CLAMP, hand brake cable tube (at transfer case bearing cap) (issue until stock exhausted)	1		4		—
		WO-50929	FM-20367-S7	BOLT, hex-hd., S, 5/16-24NF-2 x 7/8	2		—		—
		WO-5910	FM-33798-S2	NUT, hex, S, 5/16NF-2	1		—		—
		WO-51833	FM-34806-S7	WASHER, lock, S, 5/16 in.	2		—		—
		WO-A-5393	FM-GPW-2270-B	CLIP, 5/16 in., S, closed type (9/32 bolt hole) (cable tube to bell housing) (after engine Willys 114550) (All Ford trucks) (issue until stock exhausted)	1	% 4	8		—
		WO-51763	FM-20308-S2	BOLT, hex-hd., S, 1/4-20NC-2 x 1/2 (to bell housing) (after Willys engine 114550)	1		—		—
		WO-52706	FM-34805-S2	WASHER, lock, S, 1/4 in. (after Willys engine 114550)	1		—		—
		WO-A-1795		CLIP, support, hand brake cable (cable to rear engine insulator) (before Willys engine 114550) (Willys only)	1		—		—
12-5		WO-A-1226	FM-GPW-2656	CRANK, hand brake cable, assembly (SP-5-B-16)	1	% 2	2	8	5
12-1		WO-52925	FM-33927-S7	NUT, hex, S, 7/16-20NF (crank to transfer bearing cap) (issue until stock exhausted)	1	% 4	8	10	—
		WO-A-1227	FM-355752-S	SCREW, hex-hd., S, 7/16-20NF x 1 1/4	1	% 3	3	7	—
12-1		WO-A-1002	FM-GP-2614	DRUM, hand brake (SP-20-B-62)	1	% 2	2	100	5
		WO-A-997	FM-355551-S	BOLT, hex-hd., S, 3/8-20NF-3 x 1 1/8 (SP-6-73-218)	4	29	29		90
12-5		WO-636575	FM-33786-S2	NUT, hex, S, 3/8-24NF-2	4		—		—
		WO-5010	FM-34807-S2	WASHER, lock, S, 3/8 in.	4		—		—
12-1		WO-A-1014	FM-GP-2620	LINING, hand brake (SP-4-B-26)	1		—		—
12-1		WO-A-1015	FM-64647-S	RIVET, tubular, ck-hd., br. (lining to band) (SP-238-R)	14		—		—
		WO-A-6759	FM-GPW-18377	LINING SET, hand brake, w/RIVETS (Includes: 1 WO-A-1014 FM-GP-2620 LINING; 15 WO-A-1015 FM-64647-S RIVET)	as req.	6	6	15	20
12-5		WO-639244	FM-GPW-2782	HANDLE, hand brake (issue until stock exhausted)	1		4		—
12-5		WO-51904	FM-92047-S	SCREW, rd-hd., S, br-plted., type "U", No. 6 (.138) x 5/16 (handle to tube)	2	% 10	10	50	30
		WO-A-1242	FM-GPW-2780	HANDLE, hand brake, w/TUBE and CABLE, assembly (issue until stock exhausted)	1		—		—
12-5		WO-392468	FM-357553-S18	PIN, clevis, S, 1/4 O.D. x 21/32 lgth. (dld. f/c-pin)	1	% 4	12	24	—
		WO-5067	FM-72003-S	PIN, cotter 1/16 x 1/2	1	% 10	10		—
12-1		WO-A-1228	FM-GPW-2659	LINK, hand brake cable crank (issue until stock exhausted)	2	% 4	4		—

SNL G-503

GROUP 12—BRAKE (Cont'd)

Figure Number	Official Stockage Number	Part Number		ITEM	Quantity Reqd. per Unit Assy.	Per 100 Major Items			Estimated Reqmts. per 100 Rebuilds
		Willys	Ford			12 Mos. Field Maintenance	Major Overhaul (5th Ech)	Total First Year Procurement	
Col. 1	Col. 2	Col. 3	Col. 4	Col. 5	Col. 6	Col. 7	Col. 8	Col. 9	Col. 10
				1201—HAND BRAKE EMERGENCY BRAKE PARTS (Cont'd)					
12-1		WO-311003	FM-73904-S7	PIN, clevis, S., .3115 diam. x 1 15/32 lgth. (crank to link and link to cam) dld. f/c-pin (SP-3-Q-76)	2	% 10	10	40	30
12-1		WO-5354	FM-72004-S	PIN, cotter, 3/32 x 1/2 (SP-42-G)	2	—	—	—	—
12-1		WO-A-1018	FM-01T-2805	NUT, adjusting, hand brake band (SP-373-J) (issue until stock exhausted)	1	8	16	—	—
12-1		WO-A-1004	FM-73928-S7	PIN, hand brake cam (dld. f/c-pin) (SP-232-SP) (issue until stock exhausted)	1	8	16	—	—
12-5		WO-A-1006	FM-73889-S7	PIN, hand brake support quadrant (dld. f/c-pin) (SP-153-SP)	1	3	3	24	10
12-1		WO-5020	FM-72017-S	PIN, cotter, 3/32 x 3/4 (cam pin)	1	—	—	—	—
12-1		WO-52967	FM-72016-S	PIN, cotter, 3/32 x 5/8 (quadrant pin)	1	—	—	—	—
12-1		WO-A-1005	FM-GPW-2630	QUADRANT, support, hand brake (SP-3-0-74) (issue until stock exhausted)	1	% 4	8	9	10
12-1		WO-A-1021	FM-01T-2640	SPRING, anchor clip bolt (SP-14-B-6)	1	% 3	3	6	—
12-1		WO-A-1017	FM-01T-2634	SPRING, releasing, hand brake (SP-12-B-5)	2	% 2	2	6	—
12-5		WO-635681	FM-GPW-7291	SPRING, hand brake ratchet tube	1	% 2	2	12	—
12-5		WO-A-5335	FM-GPW-2635	SPRING, retracting, hand brake (after Willys serial 102731)	1	% 5	5	12	—
12-5		WO-A-2892	FM-GPW-2852	SUPPORT, hand brake ratchet tube bracket (issue until stock exhausted)	1	—	4	—	—
		WO-52207	FM-60371-2	RIVET, rd-hd, 1/4 x 9/16	2	—	—	—	—
		WO-A-1241	FM-GPW-2853	TUBE, hand brake ratchet, w/CABLE and CONDUIT, assembly	1	% 5	5	12	—
		WO-5437	FM-34706-S2	WASHER, plain, S, SAE std., 5/16 in. (ratchet tube support)	2	—	—	—	—
				1202—SHOES AND FACINGS					
12-2		WO-116551	FM-GP-2021	LINING, brake shoe, forward, ground (BX-1067-S-3)	4	—	—	—	—
12-2		WO-116552	FM-GP-2022	LINING, brake shoe, reverse, ground (BX-1067-S-4)	4	—	—	—	25
12-2		WO-116600	FM-GPW-18367	LINING SET, brake shoe, w/RIVETS (BX-44517)	as req.	72	8	93	25
				(Includes: 2 WO-116551 FM-GP-2021 LINING, forward; 2 WO-116552 FM-GP-2022 LINING, reverse; 42 WO-374586 FM-351915-S RIVETS)					
				NOTE: This lining set for two (2) wheels.					
12-2		WO-374586	FM-351915-S	RIVET, tubular, br. (BX-179-S-6)	80	—	—	—	—
12-2		WO-116549	FM-GP-2018	SHOE, brake, forward, w/LINING, assembly (BX-1141-S-1)	4	% 13	—	16	4
12-2		WO-116550	FM-GP-2019	SHOE, brake, reverse, w/LINING, assembly (BX-1141-S-2)	4	% 13	—	16	4
				1203—BRAKE SHOE SUPPORT					
12-2		WO-A-450		PLATE, backing, front and rear, brake, assembly (issue until stock exhausted then use WO-A-8898 for maintenance)	4	—	3	4	10
12-2		WO-A-8898	FM-GP-2013	PLATE, backing, front and rear brake, assembly (BX-47054) (was WO-A-450)	4	—	—	—	—

SNL G-503

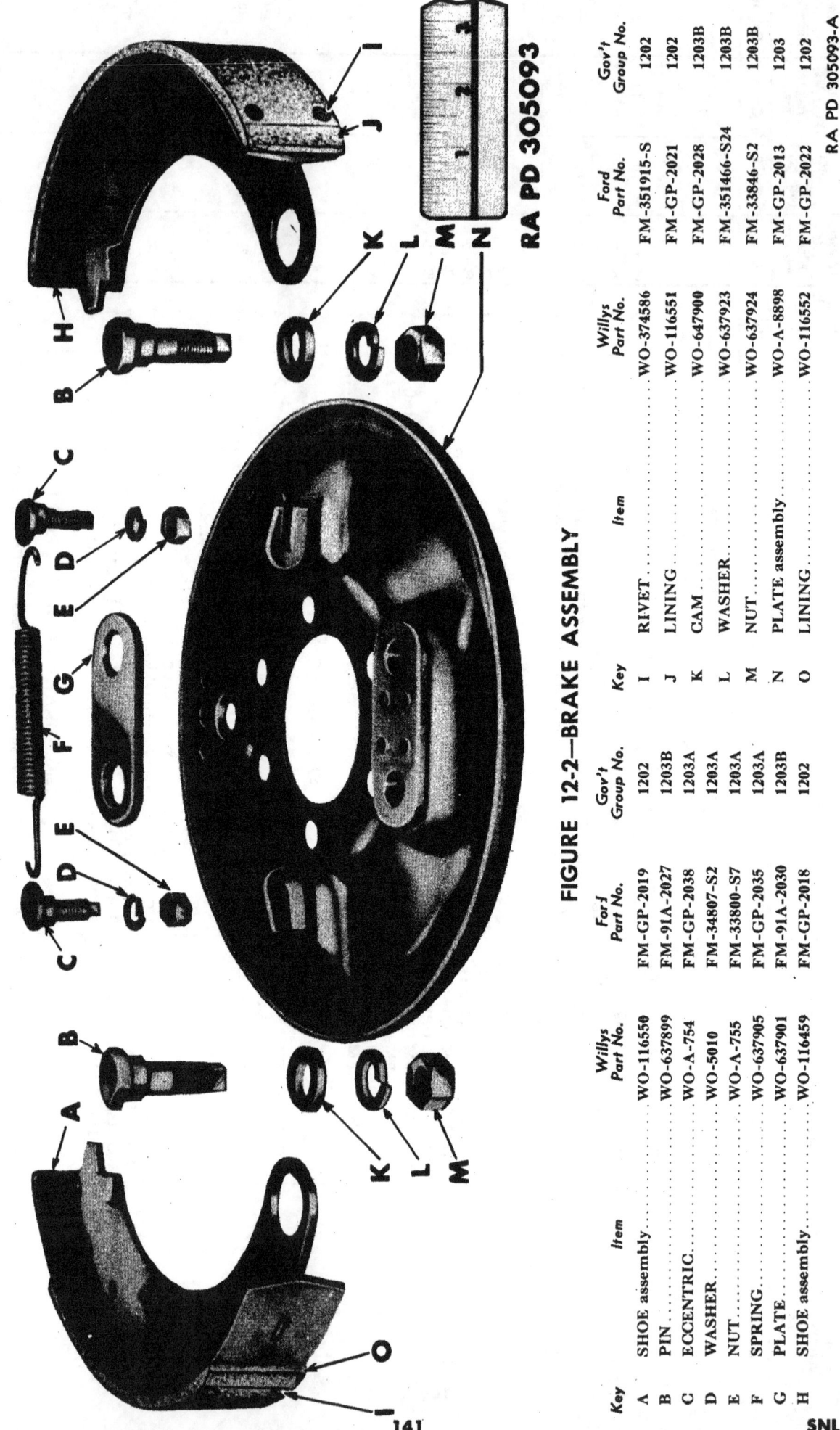

FIGURE 12-2—BRAKE ASSEMBLY

Key	Item	Willys Part No.	Ford Part No.	Gov't Group No.
A	SHOE assembly	WO-116550	FM-GP-2019	1202
B	PIN	WO-637899	FM-91A-2027	1203B
C	ECCENTRIC	WO-A-754	FM-GP-2038	1203A
D	WASHER	WO-5010	FM-34807-S2	1203A
E	NUT	WO-A-755	FM-33800-S7	1203A
F	SPRING	WO-637905	FM-GP-2035	1203A
G	PLATE	WO-637901	FM-91A-2030	1203B
H	SHOE assembly	WO-116459	FM-GP-2018	1202
I	RIVET	WO-374586	FM-351915-S	1202
J	LINING	WO-116551	FM-GP-2021	1202
K	CAM	WO-647900	FM-GP-2028	1203B
L	WASHER	WO-637923	FM-351466-S24	1203B
M	NUT	WO-637924	FM-33846-S2	1203B
N	PLATE assembly	WO-A-8898	FM-GP-2013	1203
O	LINING	WO-116552	FM-GP-2022	1202

SNL G-503

GROUP 12—BRAKE (Cont'd)

Figure Number	Official Stockage Number	Part Number (Willys)	Part Number (Ford)	ITEM	Quantity Reqd. per Unit Assy.	Per 100 Major Items — 12 Mos. Field Maintenance	Per 100 Major Items — Major Overhaul (5th Ech)	Total First Year Procurement	Estimated Reqmts. per 100 Rebuilds
Col. 1	Col. 2	Col. 3	Col. 4	Col. 5	Col. 6	Col. 7	Col. 8	Col. 9	Col. 10
				1203—BRAKE SHOE SUPPORT (Cont'd)					
12-5		WO-A-903	FM-355578-S	BOLT, hex-hd., S., 3/8-24NF x 1 11/16 (rear brake plate to housing)(SP-6-73-513)	12	19	19	66	60
10-2		WO-A-377	FM-353352-S	BOLT, hex-hd., 3/8-24NF-2 x 3/4 (front brake plate to steering knuckle) (SP-IS-1310)	12	19	19	70	60
11-1		WO-636575	FM-33786-S2	NUT, hex., 3/8-24NF-2 (rear brake plates only)	12	—	—	—	—
12-5		WO-5010	FM-34807-S7	WASHER, lock, S., 3/8 in. (front brake plates only)	12	—	—	—	—
				1203A—GUIDE SPRINGS					
12-2		WO-A-754	FM-GP-2038	ECCENTRIC, brake shoe (BX-45771)	8	6	6	20	20
12-2		WO-A-755	FM-33800-S7	NUT, hex., S., 3/8-24NF-2 eccentric (BX-46752)	8	—	—	—	—
12-2		WO-5010	FM-34807-S2	WASHER, lock, S., 3/8 in. (BX-40-S-33)	8	—	—	—	—
12-2		WO-637905	FM-GP-2035	SPRING, return, brake shoe (BX-41545)	4	% 4	5	45	15
				1203B—ADJUSTING PIN AND ANCHOR PLATE					
12-2		WO-637900	FM-GP-2028	CAM, anchor pin (BX-41876)	8	13	13	28	40
12-2		WO-637924	FM-33846-S2	NUT, anchor pin (hex., S., 1/2-20NF-2) (BX-41708)	8	—	—	—	—
12-2		WO-637899	FM-91A-2027	PIN, anchor (BX-39953)	8	5	5	14	15
12-2		WO-637901	FM-91A-2030	PLATE, anchor pin (BX-39956) (issue until stock exhausted)	8	20	40	80	30
12-2		WO-637923	FM-351466-S24	WASHER, anchor pin (lock, S., 1/2 in.)	8	10	10	—	—
				1204—PEDAL AND SPRING					
				NOTE: For parts not listed here see Groups 0204 and 0204A.					
12-5		WO-640038	FM-358006-S8	FITTING, grease, straight (AD-1980)	1	—	—	—	—
12-5		WO-A-1359	FM-GPW-2454	PAD, brake pedal, assembly (issue until stock exhausted)	1	—	4	—	—
		WO-A-1386	FM-GPW-2452-A	PEDAL, brake, assembly (issue until stock is exhausted, then use WO-A-8253 for maintenance)	1	—	4	—	—
12-5		WO-A-8253	FM-GPW-2452-B	PEDAL, brake, assembly (was WO-A-1386)	1	—	4	—	—
		WO-A-495	FM-GPW-2473	SHAFT, brake pedal, assembly (issue until stock exhausted)	2	—	4	—	—
				1205—MASTER CYLINDER					
		WO-A-1354		BAR, tie, master cylinder	1	—	—	—	—
		WO-51523	FM-20046-S	BOLT, hex-hd., S., 5/16-18NC-2 x 3/4 (tie bar and shield to master cylinder)	1	—	—	—	—
12-5		WO-52836	FM-355444-S	BOLT, hex-hd., S., 5/16-24NF-2 x 3 (tie bar and shield to cylinder)	1	% 10	10	24	30
		WO-5910	FM-33798-S	NUT, hex., S., 5/16-24NF-2	1	—	—	—	—
		WO-51833	FM-34806-S2	WASHER, lock, S., 5/16 in.	2	—	—	—	—

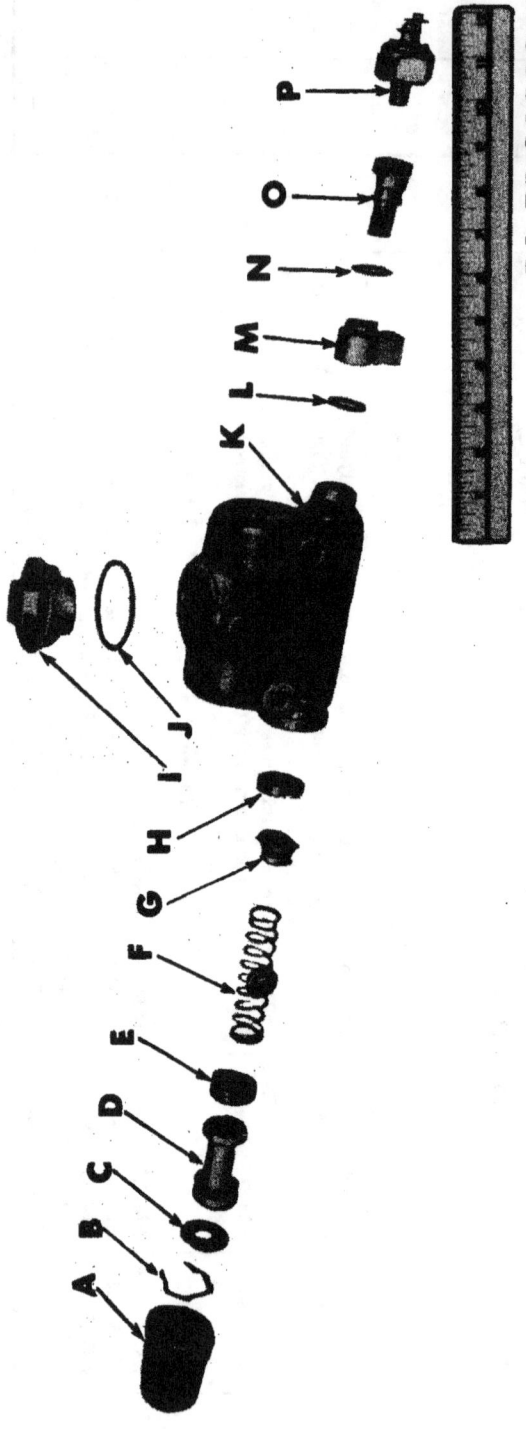

FIGURE 12-3—BRAKE MASTER CYLINDER

Key	Item	Willys Part No.	Ford Part No.	Gov't Group No.
A	BOOT	WO-637602	FM-GP-2180	1205
B	WIRE	WO-637598	FM-GP-2174	1205
C	PLATE	WO-637597	FM-GP-2188	1205
D	PISTON	WO-637591	FM-GP-2169	1205
E	CUP	WO-637590	FM-GP-2173	1205
F	SPRING	WO-637587	FM-GP-2145	1205
G	VALVE	WO-637584	FM-GP-2175	1205
H	SEAT	WO-637583	FM-GP-2160	1205
I	CAP	WO-637608	FM-GP-2162	1205
J	GASKET	WO-637612	FM-GP-2167	1205
K	TANK	WO-637582	FM-GP-2155	1205
L	GASKET	WO-637604	FM-91A-2152	1205
M	FITTING	WO-A-557	FM-GP-2076	1205
N	GASKET	WO-637606	FM-91A-2151	1205
O	BOLT	WO-637605	FM-GP-2077	1205
P	SWITCH	WO-A-1271	FM-11A-1348	0606

RA PD 305094-A

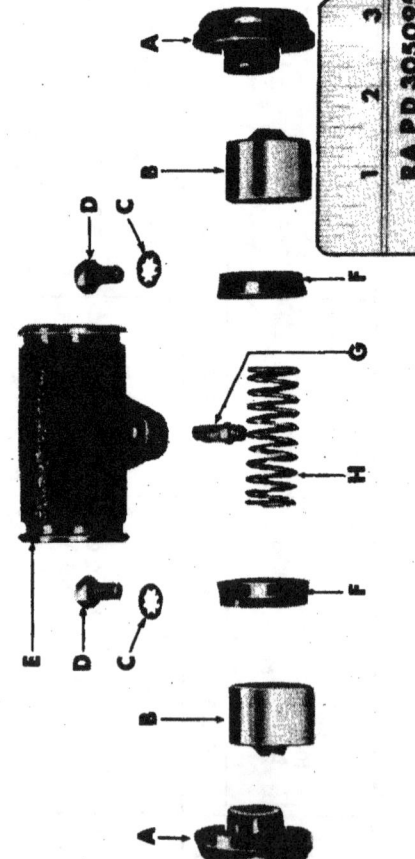

FIGURE 12-4—WHEEL BRAKE CYLINDER ASSEMBLY

Key	Item	Willys Part No.	Ford Part No.	Gov't Group No.
A	BOOT	WO-637546	FM-GP-2206-A	1207
B	GUIDE	WO-637577	FM-GP-2194	1207
C	WASHER	WO-52483	FM-34905-S2	1207
D	BOLT	WO-51738	FM-20300-S7	1207
E	CYLINDER	WO-A-1502	FM-GPW-2063	1207
F	CUP	WO-637579	FM-91A-2201	1207
G	SCREW	WO-637540	FM-GP-2208	1207
H	SPRING	WO-637580	FM-GP-2205	1207

RA PD 305095-A

GROUP 12—BRAKE (Cont'd)

1205—MASTER CYLINDER (Cont'd)

Figure Number	Official Stockage Number	Part Number (Willys)	Part Number (Ford)	ITEM	Quantity Reqd. per Unit Assy.	Per 100 Major Items — 12 Mos. Field Maintenance	Per 100 Major Items — Major Overhaul (5th Ech)	Total First Year Procurement	Estimated Reqmts. per 100 Rebuilds
Col. 1	Col. 2	Col. 3	Col. 4	Col. 5	Col. 6	Col. 7	Col. 8	Col. 9	Col. 10
12-5		WO-A-183	FM-GPW-2462	BOLT, master cylinder to pedal (eye, S., 7/16-20NF-2) (issue until stock exhausted)	1				
12-5		WO-5939	FM-33802-S2	NUT, hex, S., 7/16-20NF-2 (lock nut)	1				
		WO-5020	FM-72016-S	PIN, cotter, 3/32 x 3/4	1				
		WO-52835	FM-356394-S2	WASHER, S., (plain) 7/16 in.	1	% 4	4	8	
12-3		WO-637605	FM-GP-2077	BOLT, master cylinder outlet fitting	1			10	10
12-3		WO-637602	FM-GP-2180	BOOT, master cylinder (LO-S-FC-6011)	1	% 3	3	8	10
12-3		WO-637608	FM-GP-2162	CAP, filler, master cylinder, assembly (LO-S-FC-6018-E)	1	% 3	3	10	10
12-3		WO-637586	FM-GP-2183	CUP, master cylinder, check valve (rubber) 19/32 in. diam. (LO-S-FD-2108-F)	1				
		WO-637590	FM-GP-2173	CUP, master cylinder primary (rubber 1 in. bore) (LO-S-FD-2108-F)	1				
		WO-637595	FM-GP-2170	CUP, master cylinder, secondary (rubber 1 in. piston) (LO-FE-1444)	1	% 10	10	13	
		WO-A-556	FM-GP-2140	CYLINDER, master cylinder, assembly (LO-FE-1444)	1	% 10	10	24	
		WO-51798	FM-355398-S2	BOLT, hex-hd., S., 5/16-18NC-2 x 3 (cylinder to brkt.)	1				
		WO-52274	FM-34706-S2	WASHER, S. (plain) 5/16 in.	1	% 10	10	25	
		WO-51833	FM-34806-S2	WASHER, lock, S., 5/16 in.	1	% 5	5	12	15
12-3		WO-A-557	FM-GP-2076	FITTING, outlet, master cylinder (LO-S-FC-5727-A)	1	% 5	5	20	15
12-3		WO-637612	FM-GP-2167	GASKET, master cylinder filler cap (LO-S-FC-6019)	1	% 6	6	20	20
12-3		WO-637604	FM-91A-2151	GASKET, master cylinder outlet fitting (LO-S-FC-602)	1	% 6	6	20	20
12-3		WO-637606	FM-91A-2151	GASKET, master cylinder outlet fitting bolt (LO-S-FC-603)	1				
		WO-A-6836		KIT, repair, brake master cylinder (issue until stock is exhausted, then use WO-A-7838 for maintenance)	1	38	72		
		WO-A-7838	FM-GPW-18370-B	KIT, repair, brake master cylinder (was WO-A-6836) (Includes:	as req.	10	3	16	10
				1 WO-637602 FM-GP-2180 BOOT					
				1 WO-637590 FM-GP-2173 CUP, primary					
				1 WO-637604 FM-91A-2152 GASKET, outlet					
				1 WO-637606 FM-91A-2151 GASKET, outlet bolt					
				1 WO-637591 FM-GP-2169 PISTON					
				1 WO-637583 FM-GP-2160 SEAT					
				1 WO-637584 FM-GP-2175 VALVE					
				1 WO-637598 FM-GP-2174 WIRE, lock)					
12-3		WO-637591	FM-GP-2169	PISTON, master cylinder, assembly (LO-S-FC-6007)	1				
12-3		WO-637597	FM-GP-2188	PLATE, stop, master cylinder piston (LO-S-FC-2926)	1	3	3	8	10
		WO-637585	FM-GP-2176	RETAINER, master cylinder, check valve cup (one in. diam. cylinder) (LO-S-FC-2918-A) (issue until stock exhausted)	1			4	
12-5		WO-637599	FM-GP-2143-A	ROD, push, master cylinder, assembly (LO-S-FC-6014) (issue until stock exhausted)	1			4	
12-3		WO-637583	FM-GP-2160	SEAT, master cylinder valve (LO-S-FC-6010)	1				

1205-1207

1207—WHEEL CYLINDERS

Fig.	Part No.	Ref. No.	Description	Qty				
	WO-A-647	FM-GPW-5118	SHIELD, master cylinder, assembly (issue until stock exhausted)	1	—	—	4	—
12-3	WO-637587	FM-GP-2145	SPRING, return, master cylinder piston, w/RETAINER, assembly (LO-S-FC-6009)	1	—	—	8	10
12-3	WO-637582	FM-GP-2155	TANK, supply, master cylinder (LO-S-FD-4564)	1	3	3	—	—
12-3	WO-637584	FM-GP-2175	VALVE, check, master cylinder, assembly (LO-S-FC-2917)	1	—	—	—	—
12-3	WO-637598	FM-GP-2174	WIRE, lock, piston stop, master cylinder (LO-S-FC-2927)	1	3	3	10	10
12-4	WO-637546	FM-GP-2206-A2	BOOT, front and rear wheel cylinder, (⅞ in. cylinder) (up to serial 137915 rear wheel only) (LO-FC-5994)	8	—	—	—	—
12-3	WO-A-6117	FM-GPW-2206	BOOT, rear wheel cylinder, ¾ in., (after Willys serial 137915) (LO-FC-8779)	4	—	—	—	—
	WO-637579	FM-91A-2201	CUP, front wheel cylinder, (1 in. diam.) (LO-S-FC-1499)	4	—	—	—	—
	WO-637544	FM-GP-2201	CUP, rear wheel cylinder, (⅞ in. diam.) (up to Willys serial 137915) ,LO-S-FC-3023)	4	—	—	—	—
12-4	WO-A-6116	FM-GPW-2201	CUP, rear wheel cylinder, (¾ in. diam.)(after Willys serial 137915) (LO-FC-4158)	4	—	—	—	—
	WO-A-1502	FM-GPW-2063	CYLINDER, front wheel brake, (LO-FD-8547)	2	—	—	—	—
12-4	WO-637789	FM-GP-2192	CYLINDER, rear wheel brake, (⅞ in. diam.) (before Willys serial 137915) (LO-FD-4664)	2	—	—	—	—
12-5	WO-A-6111	FM-GPW-2135	CYLINDER, rear wheel brake, (¾ in. diam.) (after Willys serial 137915) (LO-FC-8782)	2	—	—	—	—
13-1	WO-A-1484	FM-GPW-2061	CYLINDER, front wheel brake, assembly (1 in. diam.) (LO-FD-7379) (BX-45908)	2	5%	5	14	15
	WO-637787	FM-GP-2261	CYLINDER, rear wheel brake, assembly (⅞ in. diam.) (before Willys serial 137915) (LO-FD-4665) (BX-41887)	2	—	—	—	—
	WO-A-6110	FM-GPW-2261	CYLINDER, rear wheel brake, assembly (¾ in. diam.) (after Willys serial 137915) (LO-FD-7568-A) (BX-46491)	2	5%	5	14	15
12-4	WO-51738	FM-20300-S7	BOLT, hex-hd., S., ¼-20NC-2 x ⅜ (to backing plate) (BX-62-S-176) (issue until stock exhausted)	8	—	—	—	—
12-4	WO-52483	FM-34905-S2	WASHER, lock, S., internal, flat, ¼ in. (BX-76-S-25) (issue until stock exhausted)	8	—	—	—	—
12-4	WO-637577	FM-GP-2194	GUIDE, front wheel cylinder, piston and shoe, assembly (LO-FC-5997)	4	6	6	16	20
	WO-637541	FM-GP-2196	GUIDE, rear wheel cylinder, piston and shoe, assembly (before Willys serial 137915) (LO-S-FC-5998)	4	—	—	—	—
	WO-A-6113	FM-GPW-2196	GUIDE, rear wheel cylinder, piston and shoe, assembly (after Willys serial 137915) (LO-FC-8778)	4	19	6	8	20
	WO-115962	FM-GPW-18371	KIT, repair, front wheel cylinder (LO-FC-5381) (Includes: 2 WO-637546 FM-GP-2206-A2 BOOT; 2 WO-637579 FM-91A-2201 CUP)	as req.	—	6	28	20
	WO-115963	FM-GPW-18372	KIT, repair, rear wheel cylinder (before Willys serial 137915) (Includes: 2 WO-637546 FM-GP-2206-A2 BOOT; 2 WO-637544 FM-GP-2201 CUP)	as req.	—	—	—	—
	WO-A-6133	FM-GPW-18368	KIT, repair, rear wheel cylinder (after Willys serial 137915) (Includes: 2 WO-A-6117 FM-GPW-2206 BOOT; 2 WO-A-6116 FM-GPW-2201 CUP)	as req.	19	6	28	20
12-4	WO-637540	FM-GP-2208	SCREW, bleeder, wheel cylinder (LO-S-FC-5993)	4	6%	5	20	15

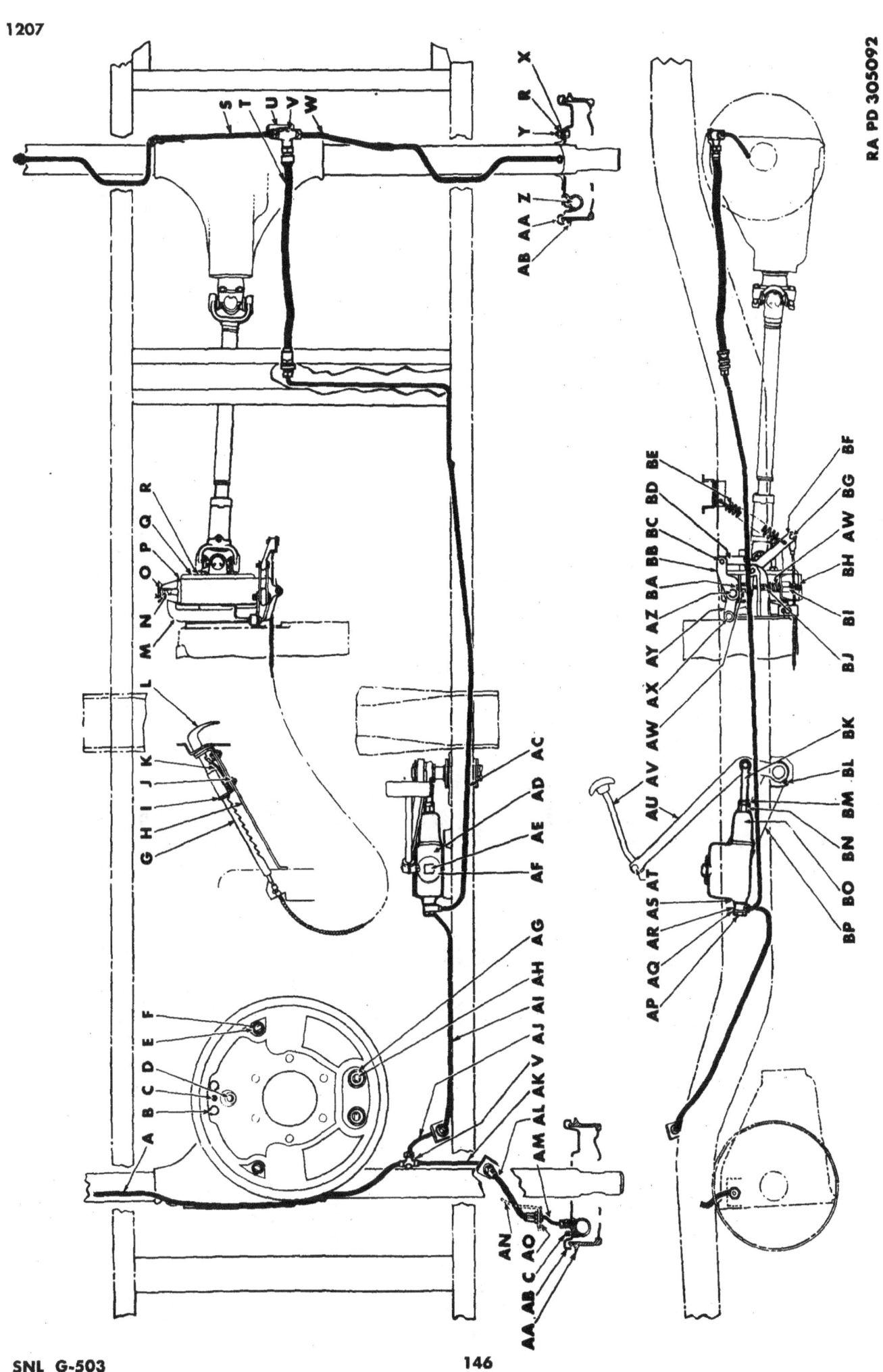

FIGURE 12-5—BRAKE SYSTEM

Key	Item	Willys Part No.	Ford Part No.	Gov't Group No.	Key	Item	Willys Part No.	Ford Part No.	Gov't Group No.
A	TUBE	WO-A-1376	FM-GPW-2266	1209C	AI	TUBE assembly	WO-A-1377	FM-GPW-2264	1209C
B	SCREW	WO-51738	FM-20300-S7	1207	AJ	HOSE	WO-A-1373	FM-GPW-2078	1209B
C	SCREW	WO-637540	FM-GP-2208	1207	AK	TUBE assembly	WO-A-1501	FM-GPW-2263	1209C
D	CYLINDER	WO-A-1502	FM-GPW-2063	1207	AL	HOSE	WO-A-1460	FM-GPW-2079	1209B
E	NUT	WO-A-755	FM-33800-S7	1203A	AM	TUBE assembly	WO-A-1488	FM-GPW-2298	1209C
F	ECCENTRIC	WO-A-754	FM-GP-2038	1203A	AN	GUARD	WO-A-1457	FM-GPW-2096	1001
G	HANDLE w/TUBE and CABLE	WO-A-1242	FM-GPW-2780	1201	AO	CLIP	WO-637427	FM-78-2814-A	1209C
H	SUPPORT	WO-A-2892	FM-GPW-2852	1201	AP	BOLT	WO-637605	FM-GP-2077	1205
I	BRACKET assembly	WO-639010	FM-GPW-2848	1201	AQ	GASKET	WO-637606	FM-91A-2151	1205
J	NUT	WO-51396	FM-24347-S2	1201	AR	FITTING	WO-A-557	FM-GP-2076	1205
K	SPRING	WO-635681	FM-GPW-7291	1201	AS	GASKET	WO-637604	FM-91A-2152	1205
L	HANDLE	WO-639244	FM-GPW-7282	1201	AT	SCREW	WO-6157	FM-24426	0204
M	CAP	WO-A-1507	FM-GPW-7768	0801	AU	PEDAL assembly	WO-A-8253	FM-GPW-2452-B	1204
N	SPRING	WO-A-1021	FM-01T-2634	1201	AV	PAD assembly	WO-A-1359	FM-GPW-2454	1204
O	SCREW	WO-A-1020	FM-01T-2616	1201	AW	SPRING	WO-A-1017	FM-01T-2634	1201
P	BAND assembly	WO-A-1009	FM-GP-2648	1201	AX	PIN	WO-A-1006	FM-73889-S7	1201
Q	DRUM	WO-A-1002	FM-GPW-2614	1201	AY	QUADRANT	WO-A-1005	FM-GPW-2630	1201
R	NUT	WO-636575	FM-33786-S2	1201	AZ	PIN	WO-A-1004	FM-73928-S7	1201
S	TUBE assembly	WO-A-5226	FM-GPW-2267	1209C	BA	SCREW	WO-A-1019	FM-355352-S7	1201
T	HOSE	WO-637424	FM-GPW-2078	1209B	BB	CAM	WO-A-1003	FM-GPW-2632	1201
U	BRACKET	WO-A-5227	FM-GPW-2274	1209C	BC	PIN	WO-311003	FM-73904-S7	1201
V	TEE	WO-637432	FM-GP-2074	1209C	BD	LINK	WO-A-1228	FM-GPW-2659	1201
W	TUBE assembly	WO-A-5225	FM-GPW-2268	1209C	BE	SPRING	WO-A-5335	FM-GPW-2635	1201
X	WASHER	WO-5010	FM-34807-S7	1203	BF	CRANK assembly	WO-A-1226	FM-GPW-2656	1201
Y	SCREW	WO-A-903	FM-355578-S	1203	BG	PIN	WO-392468	FM-357553-S18	1201
Z	CYLINDER	WO-A-6111	FM-GPW-2135	1207	BH	BOLT	WO-A-1016	FM-01T-2642	1201
AA	PLATE assembly	WO-A-8898	FM-GP-2013	1203	BI	NUT	WO-A-1018	FM-01T-2805	1201
AB	DRUM	WO-A-472	(Willys only)	1302	BJ	NUT	WO-5790	FM-33795-S7	1201
AC	TUBE assembly	WO-A-5224	FM-GPW-2265	1209C	BK	BOLT	WO-A-183	FM-GPW-2462	1205
AD	TANK	WO-637582	FM-GP-2155	1205	BL	FITTING	WO-640038	FM-358006-S8	1204
AE	CAP assembly	WO-637608	FM-GP-2162	1205	BM	NUT	WO-5939	FM-33802-S2	1205
AF	GASKET	WO-637612	FM-GP-2167	1205	BN	ROD assembly	WO-637599	FM-GP-2143-A	1205
AG	NUT	WO-637924	FM-33846-S2	1203B	BO	BOOT	WO-637602	FM-GP-2180	1205
AH	PIN	WO-637899	FM-91A-2027	1203B	BP	BAR	WO-A-1354	FM-GPW-2138	1205

RA PD 305092-A

SNL G-503

1207-1209C

GROUP 12—BRAKE (Cont'd)

Figure Number	Official Stockage Number	Part Number Willys	Part Number Ford	ITEM	Quantity Reqd. per Unit Assy.	Per 100 Major Items 12 Mos. Field Maintenance	Per 100 Major Items Major Overhaul (5th Ech)	Total First Year Procurement	Estimated Reqmts. per 100 Rebuilds
Col. 1	Col. 2	Col. 3	Col. 4	Col. 5	Col. 6	Col. 7	Col. 8	Col. 9	Col. 10
				1207—WHEEL CYLINDERS (Cont'd)					
12-4		WO-637580	FM-GP-2205	SPRING, front wheel cylinder cup (LO-FC-5992)	2	3	3	8	10
		WO-637545	FM-GP-2204	SPRING, rear wheel cylinder cup (LO-FC-6003)	2	3	3	8	10
				1209B—HOSES					
		WO-637426	FM-GP-2087	GASKET, fitting brake hose (after Willys serial 138841)	2	%38	19	110	60
12-5		WO-A-1373	FM-GPW-2078	HOSE, front brake, assembly (11 in.) (LO-FC-8502)	1	%6	3	20	—
12-5		WO-A-1460	FM-GPW-2079	HOSE, brake, front axle, assembly (6 in.) (LO-FC-8553)	2	%13	6	40	—
12-5		WO-637424	FM-GP-2078	HOSE, rear brake, assembly (15 in.) (LO-FC-5784)	1	%6	3	20	—
				1209C—TUBES AND CLIPS					
12-5		WO-A-5227	FM-GPW-2274	BRACKET, axle brake tube tee (issue until stock exhausted)	1	—	4	—	—
		WO-6428	FM-20366-S	BOLT, hex-hd., 5/16-18NC-2 x 7/8 (brkt. and gear cover to housing)	2	—	—	—	—
		WO-A-1515	FM-GPW-2250	CLAMP, axle brake tube (after Willys serial 106763) (issue until stock exhausted)	3	8	12	—	—
		WO-6273	FM-34141-S2	NUT, sq. S., No. 10 (.190)-32NF-2	3	—	—	—	—
		WO-6383	FM-27145-S2	SCREW, rd-hd., S., No. 10 (.190)-32NF-2	3	—	—	—	—
		WO-A-1378	FM-GPW-2244	CLIP, brake pipe (closed, S., 5/16 in. bolt hole 21/64) (issue until stock exhausted)	2	%8	12	—	—
		WO-52839		CLIP, tube (tube to side rail) (Carr fastener) (GM-127753) (Willys only) (issue until stock exhausted)	2	%20	40	—	—
		WO-637439	FM-GPW-2223	CLIP, tube (tube to side rail) (Ford only)	2	—	—	—	—
				CLIP, tube, 1/4 in. (use under head of rear axle cover bolt) (first 6000 Willys trucks) (Willys only) (issue until stock exhausted)	1	%4	8	—	—
12-5		WO-637427	FM-78-2814-A	CLIP, lock, spring (LO-FC-3052)	6	%48	10	75	—
		WO-A-5449		CLIP, tube, S. (open) 5/16 in. bolt hole, 7/32 in. (Willys only) (issue until stock exhausted), (1 used tube to side rail reinforcement after Willys serial 106763, 1 used tube to frame cross member after Willys serial 106763, 1 used tube to master cylinder) (after 20677 Willys trucks)	3	%4	4	—	—
		WO-52840		BOLT, hex-hd., S., cd-pltd., No. 10 (.190) x 1/2 type "B" thd. (clip to master cylinder)	1	—	—	—	—
		WO-6352		NUT, hex., S., No. 10 (.190)-24NC-2	2	—	—	—	—
		WO-5064		SCREW, rd-hd., S., No. 10 (.190)-24NC-2 x 5/8	2	—	—	—	—
		WO-52221		WASHER, lock, S., No. 10 (.190)	2	%52	108	—	—
		WO-384710	FM-GP-2133	NUT, hex. (3/16 in. inverted flared tube)	16	—	—	—	—
12-5		WO-637432	FM-GP-2074	TEE, axle brake tube (LO-S-FC-5778)	2	%6	3	12	10
		WO-6188	FM-20384-S2	BOLT, hex-hd., S., 1/4-20NC-2 x 1/2 (tee to rear axle tube) (Willys) (Ford)	2 1	— —	— —	— —	— —

SNL G-503

1209C-1301B

12-5	WO-52706	FM-33795-S2	NUT, hex, S., reg, ¼-20NC-2 (Ford only)	1	—	—	—	—	—
12-5	WO-A-1377	FM-34805-S2	WASHER, lock, S., ¼ in.	2	—	—	—	—	—
12-5	WO-A-1501	FM-GPW-2264	TUBE, brake, assembly, 3⁄16 in. (21.81 in. lgth.) (master cylinder to front hose) (issue until stock exhausted)	1	%12	24	—	—	—
12-5	WO-A-1376	FM-GPW-2263	TUBE, brake, assembly, 3⁄16 in. (6.5 in. lgth.) (tee to front brake hose, left)	1	%2	2	4	—	—
12-5	WO-A-1488	FM-GPW-2266	TUBE, brake, assembly, 3⁄16 in. (33.12 in. lgth.) (tee to front brake hose, right)	1	%2	2	8	—	—
12-5	WO-A-683	FM-GPW-2298	TUBE, brake, assembly, 3⁄16 in. (5.72 in. lgth.) (wheel cylinder to axle hose)	1	%13	—	16	—	—
	WO-A-630		TUBE, brake, assembly, 3⁄16 in. (38 1⁄16 in. lgth.) (master cylinder to rear hose) (before Willys serial 106763) (Willys only)	1	—	—	—	—	—
	WO-A-631		TUBE, brake, assembly, 3⁄16 in. (13 25⁄32 in. lgth.) (rear axle tee to rear brake, left) (before Willys serial 106763) (Willys only)	1	—	—	—	—	—
			TUBE, brake, assembly, 3⁄16 in. (309⁄32 in. lgth.) (rear axle tee to rear brake, right) (before Willys serial 106763) (Willys only)	1	—	—	—	—	—
12-5	WO-A-5224	FM-GPW-2265	TUBE, brake, assembly, 3⁄16 in. (62.94 in. lgth.) (master cylinder to rear hose) (after Willys serial 106763)	1	%2	2	4	—	5
12-5	WO-A-5225	FM-GPW-2268	TUBE, brake, assembly, 3⁄16 in. (27.38 in. lgth.) (tee to left rear brake) (after Willys serial 106763)	1	%2	2	4	—	5
12-5	WO-A-5226	FM-GPW-2267	TUBE, brake, assembly, 3⁄16 in. (28.81 in. lgth.) (tee to right rear brake) (after Willys serial 106763) (issue until stock exhausted)	1	%12	24	—	—	—

GROUP 13—WHEELS, HUBS AND DRUMS

1301—WHEEL ASSEMBLY

13-2	WO-A-5470	FM-GPW-1029	BOLT, wheel divided rim (S., ½-20NF-2 x 1 17⁄32) (after Willys serial 120700) (KWH-25695)	40	16	16	50	—	—
13-2	WO-A-5471	FM-GPW-1030	NUT, hex, S., ½-20NF-2 (wheel divided rim bolt) (after Willys serial 120700) (KWH-22779)	40	40	40	150	—	—
13-2	WO-A-5472	FM-GPW-1045	PLATE, instruction, combat wheels (KWH-25696) (after Willys serial 120700)	5	—	—	—	—	—
13-2	WO-A-5468	FM-GPW-1016	RIM, wheel, inner half (after Willys serial 120700) (KWH-25693)	5	3	—	5	—	—
	WO-A-5549	FM-GPW-1024	RIM, wheel, outer half (after Willys serial 120700) (KWH-25917)	5	3	—	5	—	—
	WO-A-465	FM-GP-1015	RIM, wheel, w/DISC, assembly (16 x 4:00) (before Willys serial 120700) (KWH-24562)	5	—	—	—	—	—
	WO-A-1799		RIM, wheel, w/DISC, assembly (Willys only) (16 x 4:50, to be used with 16 x 6:50 tires and on trucks when specified)	5	—	—	—	—	—
	WO-A-5488	FM-GPW-1025-C	RING, lock, bead (6:00 x 16 tires on combat wheels) (after Willys serial 120700) (KWH-25930)	5	3	—	5	—	5
	WO-A-5467	FM-GPW-1015	WHEEL assembly (16 x 4:50) (after Willys serial 120700) (KWH-25692)	5	—	—	—	—	—

1301A—BEARINGS

10-2	WO-52942	FM-GP-1201	CONE and ROLLERS, wheel bearing, bore 1.625 x 1 1⁄16 (TIM-18590) BOW-BT-18590	8	%19	6	29	—	20
10-2	WO-52943	FM-GP-1202	CUP, wheel bearing (2.875 O.D. x ½ thk.) (TIM-18520, BOW, BT-18520)	8	%19	6	29	—	20

1301B—SEALS

10-2	WO-778	FM-GP-3031-A	RETAINER, grease, wheel, end (use with Bendix or Rzeppa joints)	2	—	—	—	—	—
10-2	WO-A-864	FM-GP-1177	RETAINER, grease, wheel, hub assembly (SP-17004)	4	%240	80	396	—	250

149

SNL G-503

1301B

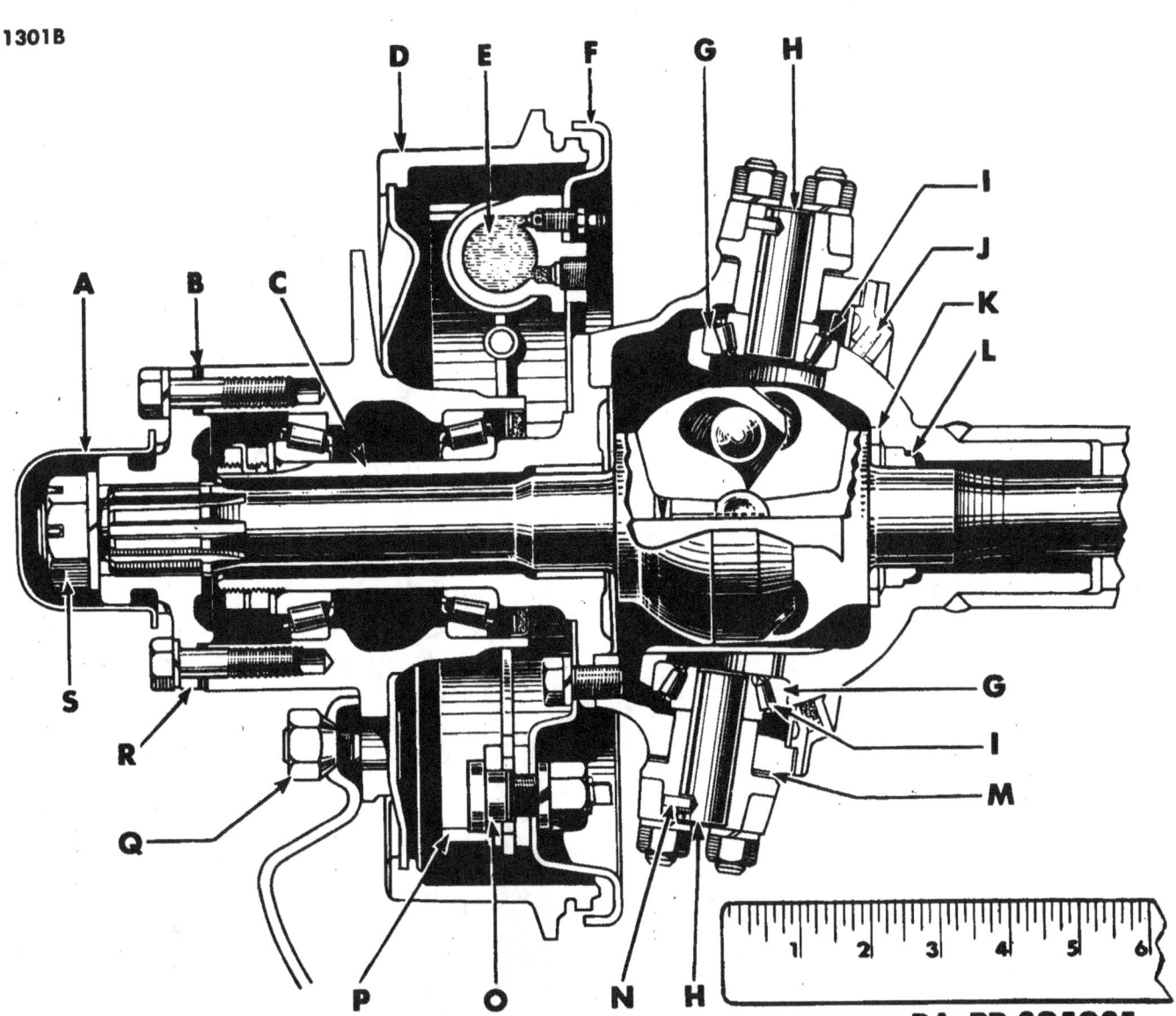

FIGURE 13-1—FRONT WHEEL AND SPINDLE—CROSS-SECTIONAL VIEW

Key	Item	Willys Part No.	Ford Part No.	Gov't Group No.
A	CAP	WO-A-869	FM-GPW-1139	1302
B	SHIM	WO-A-862	FM-GP-3208A	1007
C	SPINDLE	WO-A-851	FM-GP-3105	1006
D	DRUM	WO-A-472	Willys only	1302
E	CYLINDER assembly	WO-A-1484	FM-GPW-2061	1207
F	PLATE	WO-A-8898	FM-GP-2013	1203
G	CUP	WO-52941	FM-GP-3162	1006
H	PIN	WO-A-824	FM-GP-3115	1006
I	CONE	WO-52940	FM-GP-3161	1006
J	SEAL	WO-A-813	FM-GP-1088	1006
K	WASHER	WO-A-780	FM-GP-3374	1007
L	BUSHING	WO-A-6362	Willys only	1007
M	SHIM	WO-A-830	FM-GPW-3117-A	1006
N	PIN	WO-A-825	FM-GP-3122	1006
O	PIN	WO-637899	FM-91A-2030	1203B
P	SHOE assembly	WO-116549	FM-GP-2018	1201
Q	NUT	WO-A-475	FM-GP-1013	1302-A
R	FLANGE	WO-A-868	FM-GP-3204	1007
S	NUT	WO-636569	FM-356126-S	1003

RA PD 305085-A

SNL G-503

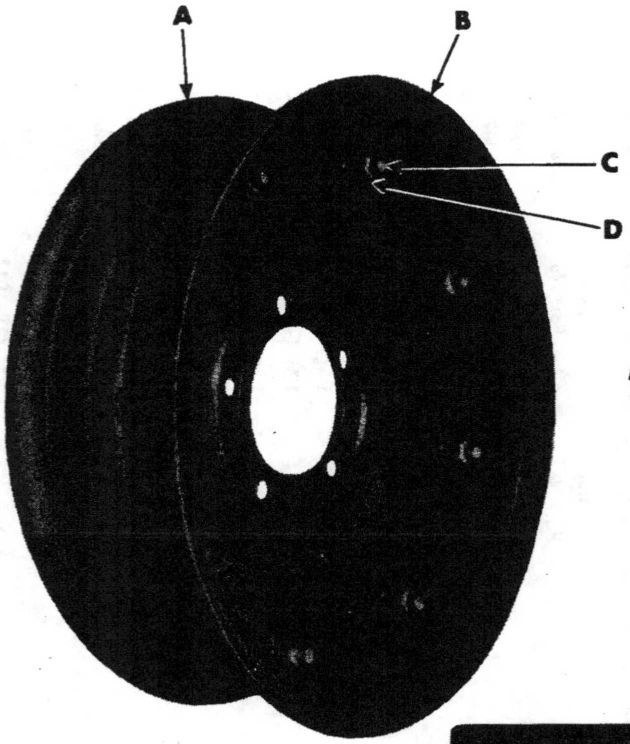

FIGURE 13-2—WHEEL ASSEMBLY

Key	Item	Willys Part No.	Ford Part No.	Gov't Group No.
A	RIM, inner	WO-A-5468	FM-GPW-1016	1301
B	RIM, outer	WO-A-5549	FM-GPW-1024	1301
C	BOLT	WO-A-5470	FM-GPW-1029	1301
D	NUT	WO-A-5471	FM-GPW-1030	1301

RA PD 305159-A

RA PD 305159

GROUP 13—WHEELS, HUBS AND DRUMS (Cont'd)

1301C—RETAINERS

Figure Number	Official Stockage Number	Part Number		ITEM	Quantity Reqd. per Unit Assy.	Per 100 Major Items			Estimated Reqmnts. per 100 Rebuilds
		Willys	Ford			12 Mos. Field Maintenance	Major Overhaul (5th Ech)	Total First Year Procurement	
Col. 1	Col. 2	Col. 3	Col. 4	Col. 5	Col. 6	Col. 7	Col. 8	Col. 9	Col. 10
10-1		WO-636569	FM-356126-S	NUT, axle shaft (S, hex., slotted, ¾-16NF-3) (SP-S-1135) (issue until stock exhausted)	2	% 20	40	—	—
10-2		WO-A-866	FM-GP-4252	NUT, front and rear axle wheel bearing (hex., S., 1⅝-16 (SP-17064)	4	% 13	6	20	20
10-1		WO-5397	FM-72071	PIN, cotter, S., ⅛ x 1½ (axle shaft nut)	2	% 8	—	—	—
10-2		WO-636570	FM-356504-S	WASHER, axle shaft nut (S., ¾ in.) (SP-1056) (issue until stock exhausted)	2	% 8	12	—	—
10-2		WO-A-865	FM-GP-1218	WASHER, wheel bearing (S.)	2	% 6	3	10	10
10-2		WO-A-867	FM-GP-1124	WASHER, lock, wheel bearing nut (S., 1⅝ in.) (SP-17017)	2	% 10	10	30	30

1301C

151

SNL G-503

GROUP 13—WHEELS, HUBS AND DRUMS (Cont'd)

Figure Number	Official Stockage Number	Part Number - Willys	Part Number - Ford	ITEM	Quantity Reqd. per Unit Assy.	Per 100 Major Items - 12 Mos. Field Maintenance	Per 100 Major Items - Major Overhaul (5th Ech)	Per 100 Major Items - Total First Year Procurement	Estimated Reqmts. per 100 Rebuilds
Col. 1	Col. 2	Col. 3	Col. 4	Col. 5	Col. 6	Col. 7	Col. 8	Col. 9	Col. 10
				1302—HUBS AND DRUMS					
10-1		WO-A-869	FM-GP-1139	CAP, wheel hub (SP-17071)	4	9%	6	15	20
11-1		WO-A-472		DRUM, front and rear wheel brakes (KWH-24566) (Willys only)	4	—	—	—	—
		WO-A-1691		HUB, front and rear, wheel, w/BEARING, assembly (KWH-24659) (Willys only)	4	—	—	—	—
		WO-A-1689	FM-GP-1103	HUB, front and rear, left, wheel, w/DRUM, assembly (KWH-25647)	2	2%	2	7	5
		WO-A-1690	FM-GP-1102	HUB, front and rear, right, wheel, w/DRUM, assembly (KWH-25646)	2	2%	2	7	5
				1302A—STUDS AND BOLTS					
11-1		WO-A-473	FM-GP-1108	BOLT, wheel hub, S., L.H. thd., 1/2-20NF-2 x 1 11/32 (KWH-24566)	10	8%	8	30	25
		WO-A-474	FM-GP-1107	BOLT, wheel hub, S., R.H. thd., 1/2-20NF-2 x 1 11/32 (KWH-24568)	10	8%	8	30	—
13-1		WO-A-475	FM-GP-1013	NUT, hex, S., 1/2-20NF-2 L.H. thd. (wheel bolt) (KWH-24576)	10	32%	—	40	—
11-1		WO-A-476	FM-GP-1012	NUT, hex, S., 1/2-20NF-2 R.H. thd. (wheel bolt) (KWH-24575)	10	32%	—	40	—
				GROUP 14—STEERING					
				1401—STEERING CONNECTING ROD (DRAG LINK)					
14-1		WO-A-857	FM-GPW-3171	BEARING, steering bell crank (SP-17212)	2	26	26	60	80
		WO-A-622	FM-GPW-3332-A	COVER, dust (rubber) (CA-S-X-1109)	2	—	—	—	—
		WO-A-8249		CRANK, bell, steering drag link, front (Willys only)	1	2	2	6	5
14-1		WO-640038	FM-358006-S8	FITTING, grease, straight (AD-198) (issue until stock exhausted)	3	92	184	18	20
14-1		WO-A-6791	FM-GPW-18383	KIT, repair, steering connecting rod	as req.	6	6		
				(Includes:					
				2 WO-A-622 FM-GPW-3332-A COVER					
				1 WO-630756 FM-GPW-3323 PLUG, large					
				1 WO-630757 FM-GPW-3328 PLUG, small					
				2 WO-630753 FM-GPW-3326 PLUG, ball seat					
				3 WO-630755 FM-GPW-3320 SEAT, ball					
				2 WO-A-623 FM-GPW-3336 SHIELD					
				2 WO-630754 FM-GPW-3327 SPRING)					
14-1		WO-A-876	FM-356124-S	NUT, steering bell crank shaft (hex., S., 3/4-20NF-2) (SP-S-1106)	1	3	3	12	8
14-1		WO-52527	FM-72062-S	PIN, cotter, S., 1/8 x 1 3/8 (bell crank nut)	1	—	—	—	—
14-1		WO-A-856	FM-GP-3166	PIN, steering bell crank shaft (S., tapered, 3/8 x 1 1/8) (SP-17211) (issue until stock exhausted)	1	4	8	—	—
		WO-630756	FM-GPW-3323	PLUG, adjusting, steering connecting rod ball seat (slotted, large) (CA-S-5070)	1	—	—	—	—
		WO-630757	FM-GPW-3328	PLUG, adjusting, steering connecting rod ball seat (slotted, small) (CA-S-5143)	1	—	—	—	—
14-1		WO-5134	FM-72089-S	PIN, cotter 1/8 x 1 1/4 (ball seat plugs)	2	—	—	—	—

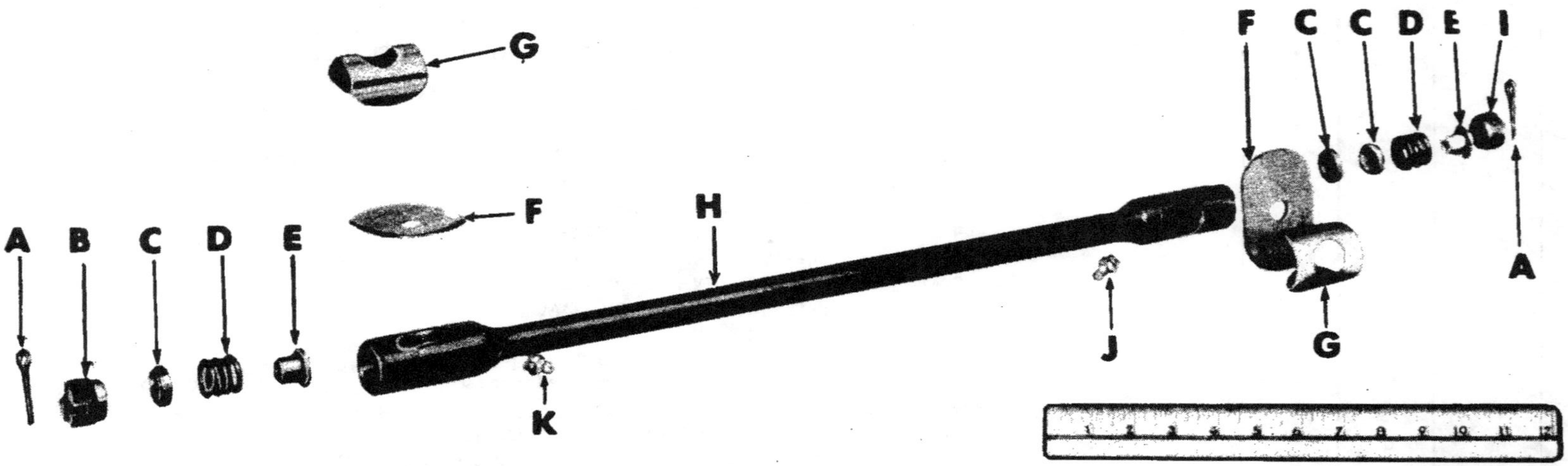

FIGURE 14-1—STEERING CONNECTING ROD

RA PD 305096

Key	Item	Ford Part No.	Willys Part No.	Gov't Group No.
A	PIN	FM-72089-S	WO-5134	1401
B	PLUG	FM-GPW-3323	WO-630756	1401
C	BALL seat	FM-GPW-3320	WO-630755	1401
D	SPRING	FM-GPW-3327	WO-630754	1401
E	PLUG	FM-GPW-3326	WO-630753	1401
F	COVER	FM-GPW-3332-A	WO-A-622	1401
G	SHIELD			
H	ROD assembly			
I	PLUG			
J	FITTING			
K	FITTING			

RA PD 305096-A

	Willys Part No.			Ford Part No.			Gov't Group No.
14-1	WO-630753	2	—	—	FM-GPW-3326	—	—	1401

	Willys Part No.			Ford Part No.			Gov't Group No.
14-1	WO-630753	2	—	—	FM-GPW-3326	—	—	1401
	WO-A-8252	1	—	—	FM-GPW-3304-B	—	—	
	WO-A-8250	1	⅞	3	FM-GPW-3304-B	12	80	
	WO-A-858	2	26		FM-GPW-3167	60		
	WO-630755	3			FM-GPW-3320			
	WO-A-855	1	13	13	FM-GPW-3165	30	40	
	WO-A-861	1	3	3	FM-GPW-3170	9	10	
	WO-A-623	2			FM-GPW-3336			
	WO-630754	2			FM-GPW-3327			
	WO-A-859	1	3	3	FM-GPW-3168	8	10	
	WO-A-860	1	3	3	FM-GPW-3169	9	10	
10-2	WO-A-1706	4	⅞	6	FM-51-3287	15	20	
10-2	WO-A-1707	4	⅞	13	FM-24916-S2	50	40	
10-2	WO-636575	4			FM-34083-S2			
	WO-5010	4			FM-34807-S2			
	WO-A-844	4			FM-78-3336			

PLUG, steering connecting rod ball seat spring (CA-S-5144)
ROD, connecting, steering (CA-D-822)
ROD, connecting, steering, assembly
SEAL, steering bell crank shaft bearing (SP-17213)
SEAT, ball steering connecting rod (CA-S-5069)
SHAFT, steering bell crank
SHIELD, dust, steering bell crank bearing (SP-17205)
SHIELD, dust, steering connecting rod cover (CA-S-X-1110)
SPRING, steering connecting rod ball seat (CA-S-5145)
WASHER, steering bell crank shaft, upper
WASHER, steering bell crank shaft, lower

1402—TIE ROD

CLAMP, steering tie rod socket (SP-13439) (TP-16MN8)
BOLT, hex-hd., S., ⅜-24NF x 1⅛ (SP-6-73-326) (TP-6MN27)
NUT, hex, S., ⅜-24NF-2 (SP-6-73-326) (issue until stock exhausted)
WASHER, lock, S., ⅜ in. (socket clamp screw)
COVER, dust, steering tie rod socket (SP-16067) (TP-14DM-43)

GROUP 14—STEERING (Cont'd)

Figure Number	Official Stockage Number	Part Number Willys	Part Number Ford	ITEM	Quantity Req'd. per Unit Assy.	Per 100 Major Items 12 Mos. Field Maintenance	Per 100 Major Items Major Overhaul (5th Ech)	Total First Year Procurement	Estimated Reqmts. per 100 Rebuilds
Col. 1	Col. 2	Col. 3	Col. 4	Col. 5	Col. 6	Col. 7	Col. 8	Col. 9	Col. 10
				1402—TIE ROD (Cont'd)					
10-2	WO-A-847	FM-GP-3292	END, steering knuckle tie rod, left, assembly (SP-17047-X) (TP-14SV90-A-7)	2	6%	6	18	20
10-1	WO-A-838	FM-GP-3291	END, steering knuckle tie rod, right, assembly (SP-17046-X) (TP-14SV89-A-7)	2	6%	6	18	20
10-2	WO-10558	FM-351059-S7	NUT, hex, S. (dld. f/c-pin) ½-20NP (steering tie rod stud nut) (SP-S-1107)	4	3%	3	25	10
10-2	WO-5152	FM-72025-S	PIN, cotter, 3⁄32 x ⅞ (steering tie rod stud nut)	4				
	WO-A-1704	FM-GPW-3280	ROD, tie, steering, right, assembly (SP-17308-X) (issue until stock exhausted)	1	8%	12		
	WO-A-1708	FM-GPW-3279	ROD, tie, steering, left, assembly (SP-17309-X) (issue until stock exhausted)	1	8%	12		
	WO-A-6305	FM-78-3332-A	SEAL, dust, spindle connecting rod (Ford only)	1				
				SPRING, steering tie rod socket stud (coil, S., cd-pltd.) (SP-17347) (TP-14-DS-20) (Willys only) (issue until stock exhausted)	4	40%	80		
10-1	WO-A-1709	FM-GPW-3282	TUBE, steering left tie rod (14¾ in. lgth.) (SP-17295-2)	1	3%	3	10	10
10-2	WO-A-1705	FM-GPW-3281	TUBE, steering right tie rod (21.80 in. lgth.) (SP-17295-1)	1	3%	3	10	10
				1403A—GEAR ASSEMBLY					
14-2	WO-A-1116	FM-GPW-3590	ARM, steering gear	1	2%	2	5	5
14-2	WO-639104	FM-GPW-3571	BALLS, steel, steering gear cam (RG-400013)	22		299	400	937
14-2	WO-639190	FM-GPW-3517	BEARING, steering column, wheel end, assembly (RG-063996)	1	10	10	60	
14-2	WO-639090	FM-GPW-3587	BUSHING, steering gear housing, inner (RG-063011)	1		13	16	40
14-2	WO-639091	FM-GPW-3576	BUSHING, steering gear housing, outer (RG-063012)	1		1	16	40
14-2	WO-A-1199	FM-GPW-3509	COLUMN, steering, w/BEARING, assembly	1		6	7	20
14-2	WO-639116	FM-GPW-3583	COVER, steering gear housing side, assembly (RG-502540)	1				
	WO-639120	FM-20386-S7	BOLT, hex-hd., s-fin., alloy-S., 5⁄16-18NC-2 x 1 (cover to housing) (BANX1BC) (issue until stock exhausted)	4	12	16		
	WO-639121	FM-356303-S	WASHER, side, cover (cop., 21⁄64 I.D. x 9⁄16 O.D. x .0375 thk.) (issue until stock exhausted)	2	12	16		
	WO-52045	FM-34846-S2	WASHER, lock, hv., S., 5⁄16 in. (19⁄32 O.D. x 11⁄32 I.D. x 3⁄64 thk.) (BECX3H)	2		3	4	10
14-2	WO-A-1760	FM-GPW-3568	COVER, steering gear housing, upper (RG-T-126000)	1				
	WO-637107	FM-355426-S	BOLT, hex-hd., s-fin., alloy-S., 5⁄16-18NC-2 x ¾ (cover to housing) (BANX1BA) (issue until stock exhausted)	3		150		
	WO-52045	FM-34846-S2	WASHER, lock, hv., S., 5⁄16 in. (19⁄32 O.D. x 11⁄32 I.D. x 3⁄64 thk.) (BECX3H)	3				
14-2	WO-390510		CUP, steering gear cam oil hole (Willys only)	1		3	6	10
14-2	WO-639102	FM-GPW-3552	CUP, steering gear cam steel balls (RG-400025)	2		26	30	80
14-2	WO-639119	FM-GPW-3581	GASKET, steering gear side housing cover (RG-T-129001)	1	6	13	20	
14-2	WO-A-1239	FM-GPW-3504	GEAR, steering, w/o WHEEL, assembly (RG-T-13086)	1	3	3	9	
14-2	WO-A-740	FM-GPW-3548	HOUSING, steering gear, assembly (RG-503284) (issue until stock exhausted)	1		4		
	WO-51371	FM-24555-S	BOLT, hex-hd., S., 7⁄16-20NF-2 x 3 (gear to frame) (GM-100044)	2				
	WO-51612	FM-355433-S2	BOLT, hex-hd., S., 7⁄16-20NF-2 x 3¾ (GM-113844)	1				
	WO-5939	FM-33802-S2	NUT, hex., S., 7⁄16-20NF-2	3				

SNL G-503

Key	Item	Willys Part No.	Ford Part No.	Gov't Group No.
S	LEVER shaft assembly	WO-A-745	FM-GPW-3575	1403A
T	GASKET	WO-639119	FM-GPW-3581	1403A
U	SCREW	WO-639118	FM-GPW-3577	1403A
V	NUT	WO-52925	FM-33927-S7	1403A
W	COVER	WO-639116	FM-GPW-3583	1403A
X	HOUSING assembly	WO-A-740	FM-GPW-3548	1403A
Y	PLUG	WO-51091	FM-74121-S	1403A
Z	BUSHING	WO-639090	FM-GPW-3587	1403A
AA	BUSHING	WO-639091	FM-GPW-3576	1403A
AB	ARM	WO-A-1116	FM-GPW-3590	1403A
AC	WASHER	WO-5038	FM-34811-S2	1403A
AD	NUT	WO-639115	FM-356077-S8	1403A
AE	OIL seal	WO-639095	FM-GPW-3591-A	1403A
AF	BALLS	WO-639104	FM-GPW-3571	1403A
AG	COVER	WO-A-1760	FM-GPW-3568	1403A
AH	CONTACT assembly	WO-A-747	FM-GPW-3652	0609B
AI	BRUSH assembly	WO-A-302	FM-GPW-13836	0609B

RA PD 305097-A

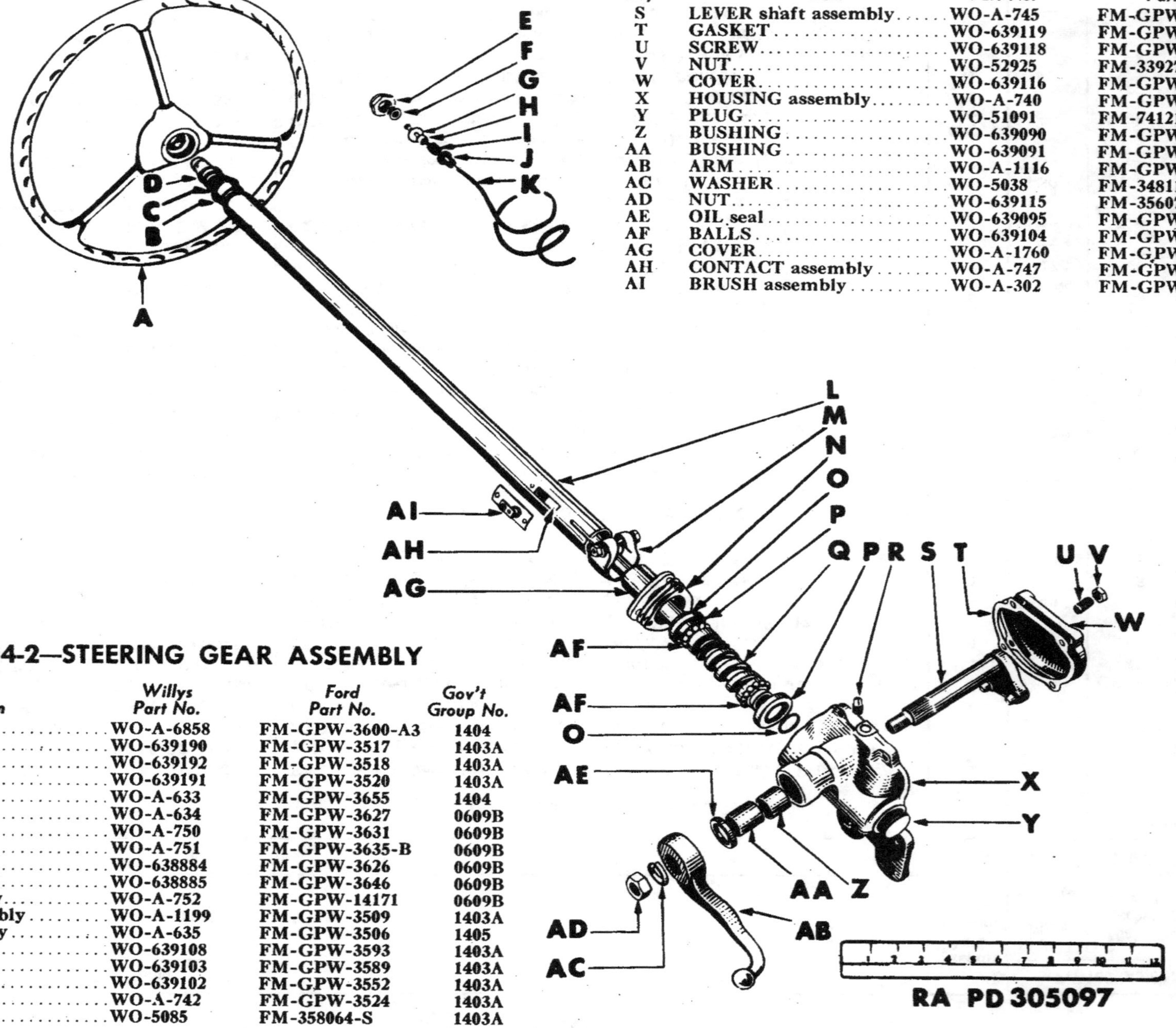

FIGURE 14-2—STEERING GEAR ASSEMBLY

Key	Item	Willys Part No.	Ford Part No.	Gov't Group No.
A	WHEEL	WO-A-6858	FM-GPW-3600-A3	1404
B	BEARING	WO-639190	FM-GPW-3517	1403A
C	SEAT	WO-639192	FM-GPW-3518	1403A
D	SPRING	WO-639191	FM-GPW-3520	1403A
E	NUT	WO-A-633	FM-GPW-3655	1404
F	BUTTON	WO-A-634	FM-GPW-3627	0609B
G	WASHER	WO-A-750	FM-GPW-3631	0609B
H	FERRULE	WO-A-751	FM-GPW-3635-B	0609B
I	SPRING	WO-638884	FM-GPW-3626	0609B
J	CUP	WO-638885	FM-GPW-3646	0609B
K	CABLE assembly	WO-A-752	FM-GPW-14171	0609B
L	COLUMN assembly	WO-A-1199	FM-GPW-3509	1403A
M	CLAMP assembly	WO-A-635	FM-GPW-3506	1405
N	SHIMS	WO-639108	FM-GPW-3593	1403A
O	RING	WO-639103	FM-GPW-3589	1403A
P	CUP	WO-639102	FM-GPW-3552	1403A
Q	TUBE assembly	WO-A-742	FM-GPW-3524	1403A
R	PLUG	WO-5085	FM-358064-S	1403A

GROUP 14—STEERING (Cont'd)

Figure Number	Official Stockage Number	Part Number (Willys)	Part Number (Ford)	ITEM	Quantity Reqd. per Unit Assy.	Per 100 Major Items - 12 Mos. Field Maintenance	Per 100 Major Items - Major Overhaul (5th Ech)	Total First Year Procurement	Estimated Reqmts. per 100 Rebuilds
Col. 1	Col. 2	Col. 3	Col. 4	Col. 5	Col. 6	Col. 7	Col. 8	Col. 9	Col. 10
				1403A—GEAR ASSEMBLY (Cont'd)					
14-2		WO-638381	FM-356439-S	WASHER, S., ½ in. I.D. x 1¼ O.D. (use over slotted hole)	1	10	10	24	—
14-2		WO-52874	FM-34921-S	WASHER, lock, S., shakeproof, ⁷⁄₁₆ in.	3	—	—	—	20
14-2		WO-A-745	FM-GPW-3575	LEVERSHAFT, steering gear, assembly (RG-7698-4⅝)	1	3	6	10	10
14-2		WO-639115	FM-356077-S8	NUT, steering gear, levershaft (RG-025060)	1	—	3	10	—
14-2		WO-52925	FM-33927-S7	NUT, hex., S., ⁷⁄₁₆-20 (steering gear adjusting screw jam nut)	1	—	—	—	40
14-2		WO-639095	FM-GPW-3591-A	OIL SEAL unit, steering gear housing (RG-032075)	1	3	13	20	—
14-2		WO-51091	FM-74121-S	PLUG, expansion, steering gear housing (1¼ in.)	1	—	—	—	—
14-2		WO-5085	FM-358064-S	PLUG, pipe ⅛ in. std. (filler)	1	—	—	—	93
14-2		WO-639103	FM-GPW-3589	RING, retaining, steering gear cam balls (RG-401100)	2	—	30	30	10
14-2		WO-639118	FM-GPW-3577	SCREW, adjusting, steering gear (RG-021116)	1	3	3	10	20
14-2		WO-639192	FM-GPW-3518	SEAT, steering column bearing spring (RG-028093)	1	—	6	6	—
14-2		WO-638918	FM-GPW-3563	SHIM, steering gear mounting	as req.	10	10	24	—
14-2		WO-639108	FM-GPW-3593	SHIM, steering gear housing upper cover, S. (.002 thk.) (RG-033046)	as req.	—	—	—	—
14-2		WO-639109	FM-GPW-3594	SHIM, steering gear housing upper cover, S. (.003 thk.) (RG-033047)	as req.	—	—	—	20
14-2		WO-639110	FM-GPW-3595	SHIM, steering gear housing upper cover, S. (.010 thk.) (RG-033048)	as req.	—	—	—	—
14-2		WO-A-6760	FM-GPW-18374	SHIM SET, steering gear housing upper cover. (Includes: 2 WO-639108 FM-GPW-3593 SHIM, .002; 2 WO-639109 FM-GPW-3594 SHIM, .003; 2 WO-639110 FM-GPW-3595 SHIM, .010)	1	—	6	5	—
14-2			FM-GPW-3521	SNAP RING, steering column (Ford only)	1	—	—	—	—
14-2		WO-A-1422	FM-GPW-5116	SPACER, steering gear to frame bolt	1	—	14	20	20
14-2		WO-639191	FM-GPW-3520	SPRING, steering column bearing (RG-401090)	1	—	6	6	20
14-2		WO-A-742	FM-GPW-3524	TUBE, steering gear cam and wheel, assembly with nut	1	—	6	8	80
14-2		WO-639121	FM-356308	WASHER, steering gear housing side cover, (plain, S.,) (RG-029021)	—	13	26	50	—
14-2		WO-5038	FM-34811-S2	WASHER, lock, S., ⅝ in. (lever shaft nut)	1	—	—	—	—
				1404—WHEEL ASSEMBLY					
14-2		WO-A-6858	FM-GPW-3600-A3	WHEEL, steering, assembly	1	% 2	2	7	—
14-2		WO-A-633	FM-GPW-3655	NUT, hex., S. (RG-026988) (also horn button nut) (issue until stock exhausted)	1	% 4	8	—	—
				1405—BRACKETS					
		WO-A-1277	FM-GPW-3682-A	BUSHING, steering column support	1	—	—	—	—
		WO-A-635	FM-GPW-3506	CLAMP, steering column, assembly (RG-502282) (issue until stock exhausted)	1	% 4	4	—	—
		WO-51308	FM-21492-S2	BOLT, hex-hd, s-fin, alloy-S., ⁵⁄₁₆-24NF-2 x 2 (BAOX1BG)	1	—	—	—	—

SNL G-503

1405-1500

WO-5910	FM-33798-S2	NUT, hex, reg, s-fin, alloy-S., 5/16-24NF-2 (BBBX1BA)	1		
WO-51833	FM-34806-S2	WASHER, lock, light, S., 5/16 in. (19/32 O.D. x 11/32 I.D. x 1/16 thk.) (BECX1H)	1		
WO-A-1276	FM-GPW-3511	CLAMP, steering column support (issue until stock exhausted)	1	5/8	4
WO-51396	FM-24347-S2	BOLT, hex-hd., s-fin, alloy-S., 5/16-24NF-2 x 3/4 (BAOX1BA) (clamp to brkt. to instrument board)	2		
WO-5910	FM-33798-S2	NUT, hex, reg, s-fin, alloy-S., 5/16-24NF-2 (BBB1XBA)	2		
WO-51833	FM-34806-S2	WASHER, lock, light, S., 5/16 (19/32 O.D. x 11/32 I.D. x 1/16 thk.) (BECX1H)	2		
WO-638918	FM-GPW-3563	SHIM, mounting, steering column	as req.	10	10
WO-A-2859	FM-GPW-3658	SUPPORT, steering column (issue until stock exhausted)	1	4	4

GROUP 15—FRAME AND BRACKETS

1500—FRAME AND BRACKETS

WO-A-1152		BRACE, machine gun mounting plate	4		
WO-A-185		BRACKET, brake pedal retracting spring	1		
WO-5267	FM-GPW-5113	RIVET, rd-hd., S., 5/16 x 3/4 (bracket to cross member)	1		
WO-A-415	FM-60416-S	BRACKET, front cross member, intermediate	2		
WO-52832	FM-GPW-5106	RIVET, rd-hd., S., 3/8 x 7/8 (bracket to frame)	6		
WO-A-418	FM-60446-S	BRACKET, engine support, front left (lower half) (Willys only)	1		
		BRACKET, engine support, front, left (upper half) (Willys only)	1		
WO-A-420		BRACKET, engine support, front left (Ford only)	1		
WO-A-421	FM-GPW-6030	BRACKET, engine support, front, right (lower half) (Willys only)	1		
WO-A-419		BRACKET, engine support, front, right (upper half) (Willys only)	1		
	FM-GPW-6029	BRACKET, engine support, front right (Ford only)	1		
WO-50769	FM-60472-S	RIVET, rd-hd., S., 3/8 x 3/4 (brackets to frame)	9		
WO-A-544	FM-GPW-5337	BRACKET, front and rear spring shackle, assembly	4	3	3
WO-5215	FM-352126-S	RIVET, huck, S., 3/8 x 7/8 (front spring shackle bracket to frame) (Willys only)	8		
	FM-352127-S	RIVET, rd-hd., S., 3/8 x 1 3/8 (rear spring shackle bracket to frame) (Ford only)	4		
WO-A-500	FM-GPW-5341	BRACKET, front and rear spring pivot	4		
WO-5215		RIVET, rd-hd., S., 3/8 x 1 (bracket to frame) (Willys only)	8		
WO-A-1341	FM-GPW-5095	BRACKET, master cylinder, assembly	1		
WO-A-1283	FM-357008-S	RIVET, fl-hd., S., 1/4 x 1 1/16 (bracket to frame) (Ford only)	8		
	FM-GPW-2073	BRACKET, mounting, front brake hose (frame end)	2		
WO-5249	FM-357008-S	RIVET, fl-hd., S., 1/4 x 1 1/16 (Ford only)	2		
WO-638809		BRACKET, mounting, rear brake hose (before 6,800 Willys trucks) (Willys only)	2		
WO-A-1201	FM-GPW-5057	BRACKET, radiator	1		
WO-5267	FM-357074-S	RIVET, huck, S., 5/16 x 5/8 (bracket to frame) (Ford only)	2		
WO-A-1431	FM-352103-S	RIVET, rd-hd., S., 5/16 x 13/16 (bracket to frame) (Ford only)	2		
WO-A-1202	FM-60416-S	RIVET, rd-hd., S., 5/16 x 3/4 (bracket to frame) (Willys only)	4		
	FM-GPW-5072	BRACKET, support, body, rear (Willys only)	2		
		BRACKET, support, radiator, left (Willys only)	1		
WO-A-508		BRACKET, support, radiator (Ford only)	2		
		BRACKET, support, radiator, right	1		
WO-A-1514	FM-60093-S	RIVET, flat head, S., 5/16 x 1 1/16 (Ford only)	4		
		BRACKET, support, radiator guard (Willys only)	1		

15-1

15-1

15-1

15-1

157

SNL G-503

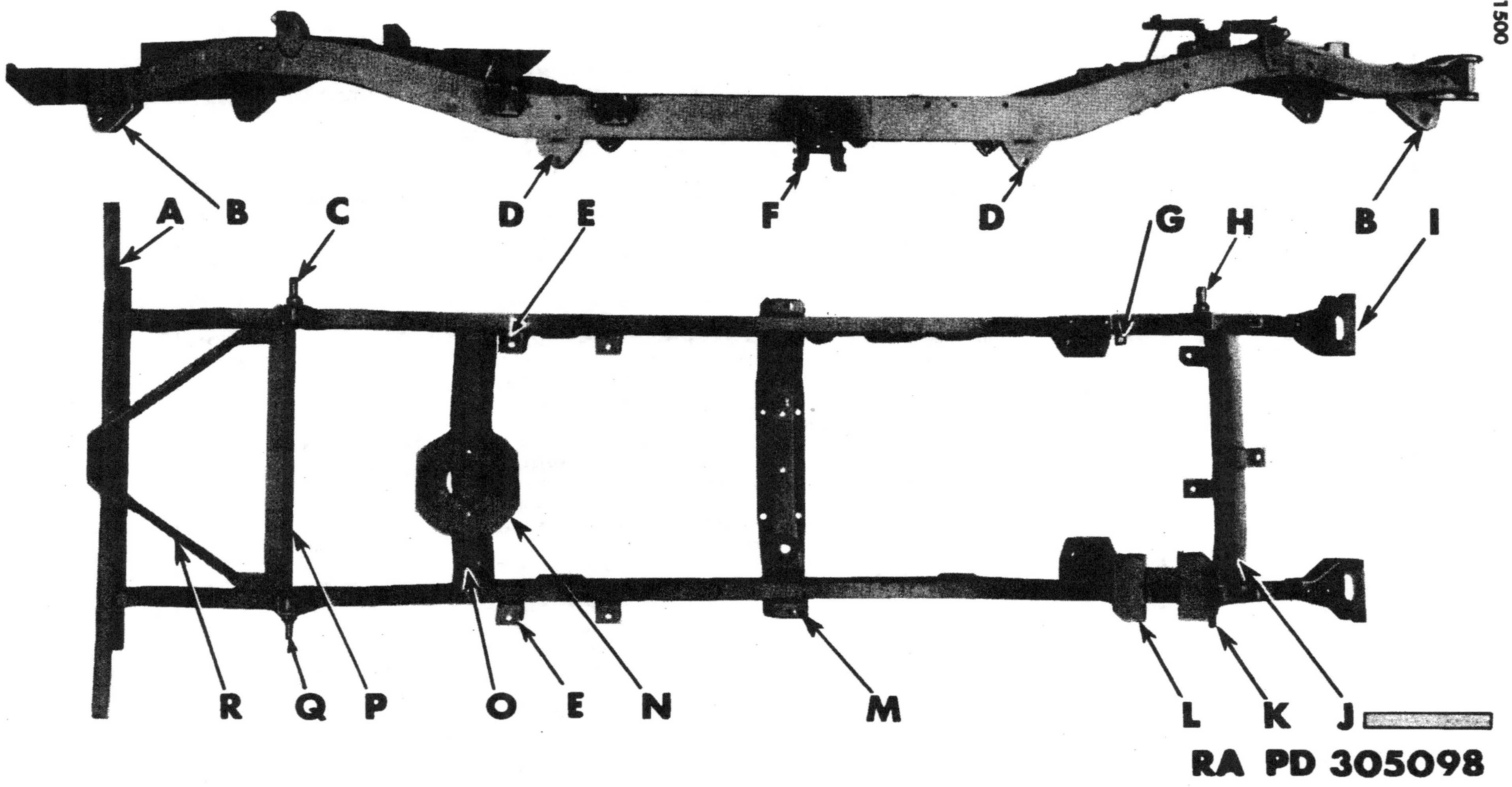

1500

FIGURE 15-1—FRAME ASSEMBLY

Key	Item	Willys Part No.	Ford Part No.	Gov't Group No.
A	MEMBER	WO-A-547	FM-GPW-5035	1500
B	BRACKET assembly	WO-A-544	FM-GPW-5337	1500
C	BRACKET w/SHAFT	WO-A-484	(Willys only)	1500
D	BRACKET	WO-A-500	FM-GPW-5341	1500
E	OUTRIGGER	WO-A-637	FM-GPW-5077	1500
F	BRACKET	WO-A-415	FM-GPW-5106	1500
G	BRACKET	WO-A-1283	FM-GPW-2073	1500
H	BRACKET	WO-A-1204	See GPW-18075	1500
I	GUSSET	WO-A-1127	FM-GPW-17755	1500
J	TUBE	WO-A-1203	FM-GPW-5019	1500
K	BRACKET	WO-A-1205	(Willys Only)	1500
L	SUPPORT assembly	WO-A-1138	(Willys only)	1500
M	MEMBER	WO-A-5127	FM-GPW-5025	1500
N	PLATE	WO-A-1151	FM-GPW-5125	1500
O	MEMBER	WO-A-1150	FM-GPW-5028	1500
P	MEMBER	WO-A-1155	(Willys only)	1500
Q	BRACKET	WO-A-485	See GPW-18040	1500

RA PD 305098-A

SNL G-503

158

GROUP 15—FRAME AND BRACKETS (Cont'd)

Figure Number	Official Stockage Number	Part Number Willys	Part Number Ford	ITEM	Quantity Reqd. per Unit Assy.	12 Mos. Field Maintenance	Major Overhaul (5th Ech)	Total FirstYear Procurement	Estimated Reqmts. per 100 Rebuilds
Col. 1	Col. 2	Col. 3	Col. 4	Col. 5	Col. 6	Col. 7	Col. 8	Col. 9	Col. 10
				1500—FRAME AND BRACKETS (Cont'd)					
15-1			FM-99A-5090	BRACKET, support, radiator grille guard (Ford only)	1	—	—	—	—
15-1		WO-A-484		BRACKET, rear shock absorber, left, w/SHAFT, assembly (Willys only)	1	—	—	—	—
15-1		WO-A-1204		BRACKET, front shock absorber, left, w/SHAFT, assembly (frame end) (Willys only)	1				
			FM-GPW-18075	BRACKET, front shock absorber, w/SHAFT, assembly (Ford only)	2	—	—	—	—
			FM-357074-S	RIVET, huck, S, 5/16 x 5/16	2	—	—	—	—
15-1		WO-A-485		BRACKET, rear shock absorber, right, w/SHAFT, assembly (Willys only)	1	—	—	—	—
15-1		WO-A-1205		BRACKET, front shock absorber, right, w/SHAFT, assembly (frame end) (Willys only)	1				
15-1			FM-GPW-18040	BRACKET, rear shock absorber, w/SHAFT, assembly (Ford only)	2	—	—	—	—
		WO-A-6740	FM-GPW-5084	RIVET, rd.-hd., S., 5/16 x 7/8	2	—	—	—	—
		WO-A-1142	FM-352101-S	CUP, retaining, transfer case insulator	1	—	—	—	—
				RIVET, plain, S., pointed, 5/16 x 3/4 (Ford only)	1				
			FM-GPW-5005	FRAME assembly (Willys only) (Includes BRACES, BRACKETS, CUPS, REINFORCEMENTS, RETAINERS, TUBES, ATTACHING PARTS)					
				FRAME assembly (Ford only) (Includes BRACES, BRACKETS, CUPS, REINFORCEMENTS, RETAINERS, TUBES, ATTACHING PARTS)	1	—	—	—	—
		WO-A-5415		GUARD, exhaust pipe (after Willys serial 120700) (Willys only)	1				
		WO-52945		BOLT, carriage, S., 3/8-16 NC-2 x 7/8 (guard to skid plate) (Willys only)	2	—	—	—	—
		WO-5919		BOLT, hex-hd., S., 3/8-24NF-2 x 1 (guard to frame) (Willys only)	1	—	—	—	—
		WO-5544		NUT, hex, S., 3/8-16NC-2 (Willys only)	2	—	—	—	—
		WO-52101		WASHER, S., 3/8 in., S.A.E. plain (13/16 O.D. x 13/32 I.D. x 1/16 thk.) (Willys only)	3				
		WO-303922		WASHER, S., 3/8 in., plain (1 1/2 O.D. x 1/16 I.D.) (Willys only)	1	—	—	—	—
		WO-5010		WASHER, lock, S., 3/8 in. (1/8 thk. x 3/32 I.D.) (Willys only)	3	—	—	—	—
		WO-A-1129	FM-GPW-17753	GUSSET, front bumper, lower left	1				
15-1		WO-A-1130	FM-GPW-17752	GUSSET, front bumper, lower right	1				
		WO-A-1127	FM-GPW-17755	GUSSET, front bumper, upper left	1				
		WO-A-1128	FM-GPW-17754	GUSSET, front bumper, upper right	1				
15-1		WO-A-1150	FM-GPW-5028	MEMBER, cross, frame, center	1				
		WO-52832	FM-357100-S	RIVET, rd.-hd., S., 3/8 x 7/8 (member to side rail)	2				
15-1			FM-GPW-5019	MEMBER, cross, frame, front (Ford only)	1				
		WO-A-5127	FM-GPW-5025	MEMBER, cross, frame, intermediate, front, assembly	1	—	—	—	—
			FM-20389-S	BOLT, hex-hd., s-fin., alloy-S., 3/8-24NF-2 x 1 (member to brace) (Ford only)	6				
		WO-50151		BOLT, hex-hd., s-fin., alloy-S., 3/8-24NF-2 x 7/8 (member to brace) (Willys only)	6	—	—	—	—

SNL G-503

GROUP 15—FRAME AND BRACKETS (Cont'd)

Figure Number	Official Stockage Number	Part Number Willys	Part Number Ford	ITEM	Quantity Reqd. per Unit Assy.	Per 100 Major Items 12 Mos. Field Maintenance	Per 100 Major Items Major Overhaul (5th Ech)	Total First Year Procurement	Estimated Reqmts. per 100 Rebuilds
Col. 1	Col. 2	Col. 3	Col. 4	Col. 5	Col. 6	Col. 7	Col. 8	Col. 9	Col. 10
				1500—FRAME AND BRACKETS (Cont'd)					
15-1		WO-5901	FM-33800-S2	NUT, hex, S., 3/8-24NF-2	6	—	—	—	—
15-1		WO-52101	FM-34747-S7	WASHER, S., S.A.E., plain (13/16 O.D. x 13/32 I.D. x 1/16 thk.)	2	—	—	—	—
15-1		WO-5010	FM-34807-S2	WASHER, lock, S., 3/8 (3/32 I.D.)	6	—	—	—	—
15-1		WO-547	FM-GPW-5035	MEMBER, cross, frame, rear	1	—	—	—	—
15-1		WO-5326		RIVET, fl-hd., S., 3/8 x 7/8 (member to side rail) (Willys only)	2	—	—	—	—
15-1		WO-50769		RIVET, rd-hd., S., 3/8 x 3/4 (member to frame) (Willys only)	4	—	—	—	—
15-1		WO-50769	FM-357100-S	RIVET, rd-hd., S., 3/8 x 15/16 (member to frame and side rail) (Ford only)	8	—	—	—	—
15-1		WO-A-1153		MEMBER, cross, frame, intermediate rear (Willys only)	1	—	—	—	—
15-1		WO-52832		RIVET, rd-hd., S., 3/8 x 7/8 (member to frame) (Willys only)	2	—	—	—	—
15-1		WO-A-668		NUT, clinch, S., 1/4-28NF-2 (MSP-F-2005) (Willys only)	2	—	—	—	—
15-1		WO-A-549		NUT, clinch, S., 3/8-24NF-2 (dash to frame brace fender brace to frame) (MSP-F-2009) (Willys only)	11	—	—	—	—
15-1		WO-A-548		NUT, clinch, S., 5/16-24NF-2 (splasher apron to frame) (MSP-F-2007) (Willys only)	5	—	—	—	—
15-1		WO-A-637	FM-GPW-5077	OUTRIGGER, body rear	4	—	—	—	—
15-1			FM-357100-S	RIVET, rd-hd., S., 3/8 x 5/6 (outrigger to frame side) (Ford only)	4	—	—	—	—
15-1		WO-50769		RIVET, rd-hd., S., 3/8 x 3/4 (outrigger to frame side) (Willys only)	8	—	—	—	—
15-1		WO-A-1151	FM-GPW-5125	PLATE, mounting, machine gun	1	—	—	—	—
15-1		WO-A-534	FM-GPW-5097	PLATE, reinforcement, frame, rear	1	—	—	—	—
15-1		WO-52906		RIVET, fl-hd., S., 3/8 x 3/4 (plate to frame) (Willys only)	1	—	—	—	—
15-1		WO-5010		WASHER, lock, S., 3/8 (3/32 I.D.) (Willys only)	4	—	—	—	—
15-1		WO-A-493		RETAINER, pedal shaft, assembly (Willys only)	1	—	—	—	—
15-1		WO-A-416		SPACER, frame cross member, intermediate front bracket (Willys only)	2	—	—	—	—
15-1		WO-A-1120	FM-GPW-17759	SPACER, front bumper gusset	2	—	—	—	—
15-1		WO-A-1422	FM-GPW-5116	SPACER, steering gear to frame bolt	1	—	—	—	—
15-1		WO-A-1138		SUPPORT, battery, assembly (before Willys serial 120700) (Willys only)	1	—	—	—	—
15-1		WO-A-5181		SUPPORT, battery, assembly (after Willys serial 120700) (Willys only)	1	—	—	—	—
15-1		WO-A-50769		RIVET, rd-hd., S., 3/8 x 3/4 (support to frame) (Willys only)	1	—	—	—	—
15-1		WO-1203	FM-GPW-5019	TUBE, cross, frame, front (Willys only)	1	—	—	—	—
				1502—PINTLE HOOK					
15-2		WO-A-6393	FM-GPW-5186	BOLT, eye, safety chain (after Willys serial 158372)	2	%	8	—	—
15-2		WO-6163	FM-33845-S2	NUT, hex, S., 1/2-13NC-2 (bolt to frame)	2	8	12	—	—
15-2		WO-5009	FM-34809-S	WASHER, lock, S., 1/2 in. (bolt to frame)	2	%	8	—	—
15-2		WO-A-593	FM-GPW-5182	HOOK, pintle (HLH-T-60-B)	1	—	—	—	—
15-2		WO-6923	FM-24411-S	BOLT, hex-hd., S., 7/16-20NF-2 x 1 1/8 (hook to frame)					
				Before serial 158372	4	—	—	—	—
				After serial 158372	2	—	—	—	—

SNL G-503

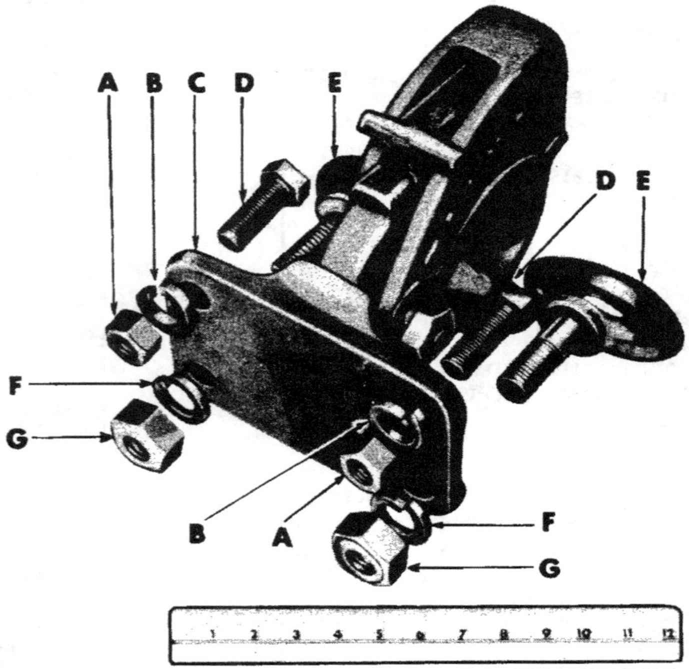

FIGURE 15-2—PINTLE HOOK ASSEMBLY

Key	Item	Willys Part No.	Ford Part No.	Gov't Group No.
A	NUT	WO-5939	FM-33802-S	1502
B	WASHER	WO-52349	FM-34808-S	1502
C	HOOK	WO-A-593	FM-GPW-5182	1502
D	BOLT	WO-6923	FM-24411-S	1502
E	EYE BOLT	WO-A-6393	FM-GPW-5186	1502
F	WASHER	WO-5009	FM-34809-S	1502
G	NUT	WO-6163	FM-N.P.N.	1502

RA PD 305099-A

15-2	WO-5939	FM-33802-S		NUT, hex., S., 7/16-20NF-2	
				Before serial 158372	4
				After serial 158372	2
15-2	WO-52349	FM-34808-S		WASHER, lock, S., external flat, 7/16 in.	
				Before serial 158372	4
				After serial 158372	2
	WO-A-552			REINFORCEMENT, pintle hook frame (Willys only)	1

1505—SPARE WHEEL

	WO-A-11701			BRACKET, support, spare tire, assembly	1
	WO-6299			BOLT, hex-hd., S., 3/8-16NC-2 x 7/8	2
	WO-5544			NUT, hex., S., 3/8-16NC-2	2
	WO-52101			WASHER, S., 3/8 in.	2
	WO-5010			WASHER, lock, S., 3/8 in.	2
	WO-A-2359	FM-GPW-1433		BRACKET, support, spare wheel, assembly	1
	WO-51485	FM-20326-S7		BOLT, hex-hd., s-fin., alloy-S., 5/16-18NC-2 x 5/8 (brkt. to rear panel)	2
	WO-6660	FM-20406-S		BOLT, hex-hd., s-fin., alloy-S., 5/16-18NC-2 x 1¼ (brkt. to rear panel)	2
	WO-6167	FM-33797-S2		NUT, hex., s-fin., alloy-S., 5/16-18NC-2	4
	WO-51840	FM-34806-S2		WASHER, lock, S., 5/16 in.	4
	WO-A-476	FM-GP-1012		NUT, hex., S., ½-20NF-2 (spare wheel to carrier stud)	2
	WO-A-2823	FM-GPW-1144600		STRAINER, spare tire carrier, upper, assembly	1
	WO-A-2820	FM-GPW-1144598		STRAINER, spare tire carrier, w/FILLER, lower, assembly	1

1601-1601A

GROUP 16—SPRINGS AND SHOCK ABSORBERS

Figure Number	Official Stockage Number	Part Number (Willys)	Part Number (Ford)	ITEM	Quantity Reqd. per Unit Assy.	12 Mos. Field Maintenance	Major Overhaul (5th Ech)	Total First Year Procurement	Estimated Reqmts. per 100 Rebuilds
Col. 1	Col. 2	Col. 3	Col. 4	Col. 5	Col. 6	Col. 7	Col. 8	Col. 9	Col. 10
				1601—FRONT SPRING					
		WO-359039	FM-GPW-5781	BUSHING, front spring (.753 O.D. x .565 I.D. x 1.68 long) (spring eye)	2	—	10	12	—
		WO-116609	FM-GPW-5345-A	BOLT, front spring, center	2	48	80	150	—
		WO-116460	FM-GPW-5330-B	CLIP, front spring leaf (large for six leaves)	4	80	80	180	—
		WO-116458	FM-GPW-5330-A	CLIP, front spring leaf (small for three leaves)	4	80	80	180	—
		WO-A-612-1	FM-GPW-5313-B	LEAF, front spring, L., No. 1 w/BUSHING (36¼ in. between eye centers)	1	10	10	23	—
		WO-A-613-1	FM-GPW-5313-A	LEAF, front spring, R., No. 1 w/BUSHING (36¼ in. between eye centers)	1	10	10	23	—
		WO-A-612-2	FM-GPW-5315-B	LEAF, front spring, L., No. 2 (36¼ in. long)	1	10	10	23	—
		WO-A-613-2	FM-GPW-5315-A	LEAF, front spring, R., No. 2 (36¼ in. long)	1	10	10	23	—
		WO-A-612-3	FM-GPW-5316-B	LEAF, front spring, L., No. 3 (30 in. long) (issue until stock exhausted)	1	8	12	—	—
		WO-A-613-3	FM-GPW-5316-A	LEAF, front spring, R., No. 3 (30 in. long) (issue until stock exhausted)	1	8	12	—	—
		WO-A-612-4	FM-GPW-5317-B	LEAF, front spring, L., No. 4 (26 in. long)	1	—	—	—	—
		WO-A-613-4	FM-GPW-5317-A	LEAF, front spring, R., No. 4 (26 in. long)	1	—	—	—	—
		WO-A-612-5	FM-GPW-5318-B	LEAF, front spring, L., No. 5 (23 in. long)	1	—	—	—	—
		WO-A-613-5	FM-GPW-5318-A	LEAF, front spring, R., No. 5 (23 in. long)	1	—	—	—	—
		WO-A-612-6	FM-GPW-5319-B	LEAF, front spring, L., No. 6 (18 in. long)	1	—	—	—	—
		WO-A-613-6	FM-GPW-5319-A	LEAF, front spring, R., No. 6 (18 in. long)	1	—	—	—	—
		WO-A-612-7	FM-GPW-5320-B	LEAF, front spring, L., No. 7 (14 in. long)	1	—	—	—	—
		WO-A-613-7	FM-GPW-5320-A	LEAF, front spring, R., No. 7 (14 in. long)	1	—	—	—	—
		WO-A-612-8	FM-GPW-5321-B	LEAF, front spring, L., No. 8 (10 in. long)	1	—	—	—	—
		WO-A-613-8	FM-GPW-5321-A	LEAF, front spring, R., No. 8 (10 in. long)	1	—	—	—	—
		WO-5910	FM-33798-S	NUT, hex, s-fin., alloy-S., ⁵⁄₁₆-24NF-2 (spring center bolt)	2	—	—	—	—
16-1		WO-A-612	FM-GPW-5311	SPRING, front, left, assembly	1	10 %	5	19	—
		WO-A-613	FM-GPW-5310	SPRING, front, right, assembly	1	10 %	5	19	—
				1601A—REAR SPRINGS					
		WO-359039	FM-GPW-5781	BUSHING, rear spring (.753 O.D. x .565 x 1.68 lgth.) (spring eye) (issue until stock exhausted)	2	160	320	—	—
		WO-116610	FM-GPW-5345-B	BOLT, rear spring center, ⁵⁄₁₆-24NF-2 x 2³⁄₈ (issue until stock exhausted)	2	80	160	40	—
		WO-116589	FM-GPW-5724-B	CLIP, rear spring leaf, large (seven leaves)	4	19	19	40	—
		WO-116459	FM-GPW-5724-A	CLIP, rear spring leaf, small (3 leaves)	4	19	19	40	—
		WO-A-614-1	FM-GPW-5563	LEAF, rear spring, No. 1, w/BUSHING (42 in. between eye centers)	2	3	3	9	—
		WO-A-614-2	FM-GPW-5565	LEAF, rear spring, No. 2 (42 in. long)	2	3	3	9	—
		WO-A-614-3	FM-GPW-5566	LEAF, rear spring, No. 3 (36 in. long) (issue until stock exhausted)	2	72	148	—	—
		WO-A-614-4	FM-GPW-5567	LEAF, rear spring, No. 4 (32 in. long)	2	—	—	—	—
		WO-A-614-5	FM-GPW-5568	LEAF, rear spring, No. 5 (28 in. long)	2	—	—	—	—
		WO-A-614-6	FM-GPW-5569	LEAF, rear spring, No. 6 (23 in. long)	2	—	—	—	—
		WO-A-614-7	FM-GPW-5570	LEAF, rear spring, No. 7 (19 in. long)	2	—	—	—	—

SNL G-503

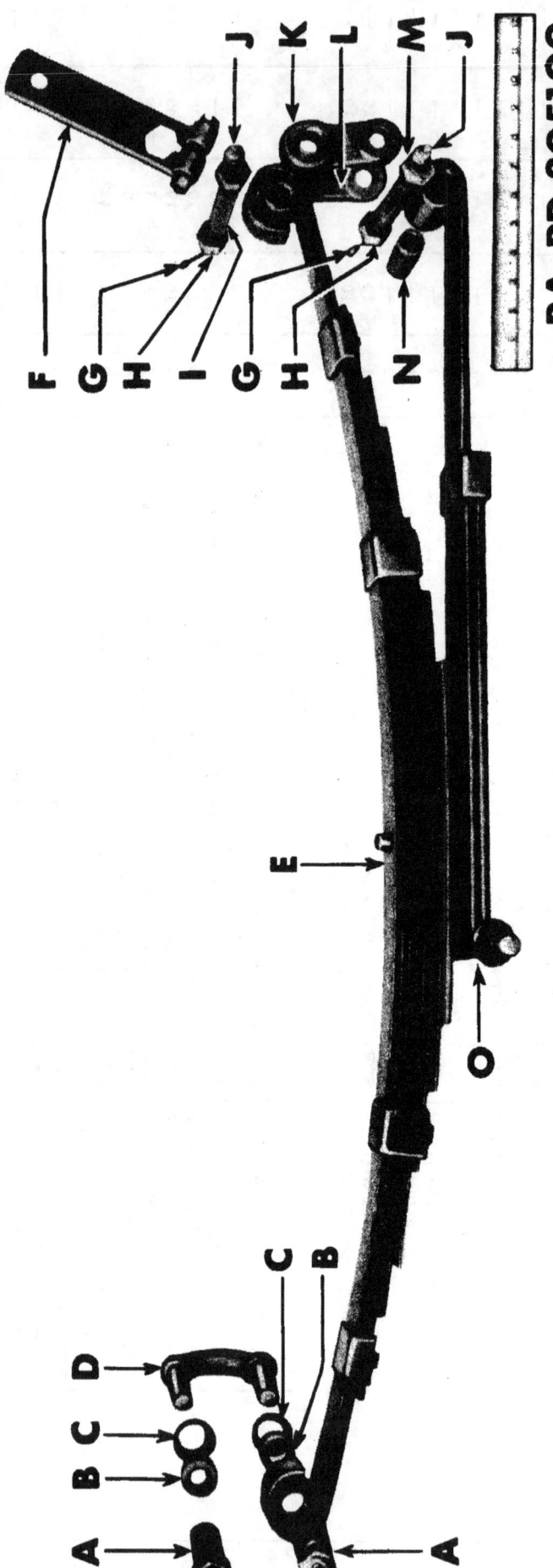

FIGURE 16-1—FRONT SPRING AND TORQUE REACTION SPRING

Key	Item	Willys Part No.	Ford Part No.	Gov't Group No.
A	BUSHING	WO-A-8255	Willys only	1602
B	GREASE seal	WO-A-515	FM-GPW-5481	1602
C	RETAINER	WO-A-1252	FM-GPW-5482	1602
D	U BOLT	WO-A-513	FM-GPW-5778	1602
E	SPRING assembly	WO-A-612	FM-GPW-5311	1601
F	BOLT lock assembly	WO-A-6326	FM-GPW-5601	1602
G	PIN	WO-5021	FM-72035-S	1602
H	NUT	WO-6436	FM-34033-S18	1602
I	BOLT	WO-384228	FM-GPW-5468	1602
J	FITTING	WO-640038	FM-358006-S8	1602
K	SPRING SHACKLE and BUSHING assembly	WO-A-6069	FM-GPW-5605	1602
L	SHACKLE and BUSHING assembly	WO-A-6068	FM-GPW-5602	1602
M	BOLT	WO-A-6067	FM-GPW-5610	1601
N	BUSHING	WO-A-6067	FM-GPW-5610	1601
O	SPRING assembly	WO-A-6066	FM-GPW-5588-A	1601

RA PD 305100-A

GROUP 16—SPRINGS AND SHOCK ABSORBERS (Cont'd)

Figure Number	Official Stockage Number	Part Number Willys	Part Number Ford	ITEM	Quantity Reqd. per Unit Assy.	Per 100 Major Items 12 Mos. Field Maintenance	Per 100 Major Items Major Overhaul (5th Ech)	Total First Year Procurement	Estimated Reqmts. per 100 Rebuilds
Col. 1	Col. 2	Col. 3	Col. 4	Col. 5	Col. 6	Col. 7	Col. 8	Col. 9	Col. 10
				1601A—REAR SPRINGS (Cont'd)					
		WO-A-614-8	FM-GPW-5571	LEAF, rear spring, No. 8 (15 in. long)	2	—	—	—	—
		WO-A-614-9	FM-GPW-5572	LEAF, rear spring, No. 9 (11 in. long)	2	—	—	—	—
		WO-5910	FM-33798-S2	NUT, hex, S, 5/16-24NF-2 (spring center bolt)	2	—	—	—	—
		WO-A-614	FM-GPW-5560	SPRING, rear, assembly	2	%3	3	9	—
				1601B—SPRING BUMPERS					
		WO-A-481	FM-GPW-5783-A	BUMPER, front and rear axle (issue until stock exhausted)	4	8	12	—	—
		WO-51396	FM-24347-S2	BOLT, hex-hd., s-fin., alloy-S., 5/16-24NF-2 x 3/4	8	—	—	—	—
		WO-5910	FM-33798-S	NUT, hex, S, 5/16-24NF-2	4	—	—	—	—
		WO-51833	FM-34806-S2	WASHER, lock, S, 5/16 (19/32 O.D. x 11/32 I.D. x 1/16 thk.)	8	—	—	—	—
				1601C—TORQUE REACTION SPRING					
		WO-A-6067	FM-GPW-5610	BUSHING, torque reaction spring (after Willys serial 146774)	1	5	5	12	—
		WO-A-6169	FM-GPW-5611	CLIP, torque reaction spring (after Willys serial 146774) (issue until stock exhausted)	1	4	4	—	—
		WO-A-6861	FM-GPW-5587	LEAF, No. 1, torque reaction spring (after Willys serial 146774) (issue until stock exhausted)	1	16	24	—	—
		WO-A-6168	FM-GPW-5590-B	LEAF, No. 2, torque reaction spring (after Willys serial 146774) (issue until stock exhausted)	1	16	24	—	—
		WO-A-6066	FM-GPW-5588-A	SPRING, torque reaction, assembly (after Willys serial 146774)	1	2	2	8	—
				1602—SHACKLES AND SPRING ATTACHING PARTS					
16-1		WO-384228	FM-GPW-5468	BOLT, pivot, spring shackle (front for rear spring, rear for front spring)	3	%10	10	28	—
16-1		WO-6436	FM-34033-S18	NUT, hex, castellated, S., 9/16-18NF-2	3	—	—	—	—
16-1		WO-5021	FM-72035-S	PIN, cotter 3/32 x 1	3	—	—	—	—
16-1		WO-A-6075	FM-GPW-5608	BOLT, torque reaction spring shackle (frt. spring rear bolt, left) (after serial 146774)	1	%3	3	8	—
		WO-A-6074	FM-GPW-5609	BOLT, torque reaction spring shackle (shackle to torque spring) (after Willys serial 146774)	1	%3	3	8	—
		WO-6436	FM-34033-S18	NUT, hex, castellated, S., 9/16-18NF-2	1	—	—	—	—
		WO-5021	FM-72035-S	PIN, cotter 3/32 x 1	1	—	—	—	—
16-1		WO-A-513	FM-GPW-5778	BOLT, "U", spring shackle, left hand thd., S. (front end left front spring, rear end right rear spring)	2	%10	3	15	—
		WO-A-514	FM-GPW-5779	BOLT, "U", spring shackle, right hand thd., S. (front end right front spring, rear end left rear spring)	2	%10	3	15	—

1603—SHOCK ABSORBERS AND MOUNTINGS

Fig.	Part No.	Ref. No.	Description						
	WO-A-6072	FM-GPW-5604	BUSHING, spring shackle, inner (after Willys serial 146774)	1	5	5	—	12	—
	WO-A-8256		BUSHING, spring shackle, assembly, right thd. (frt. end of frt. left spring, rear end right rear spring) (Willys only) (issue until stock exhausted)						
16-1	WO-A-8255		BUSHING, spring shackle, assembly, left thd.(frt. end of frt. right spring, rear end rear left spring) (for all frame ends) (Willys only) (issue until stock exhausted)	2	52	108	—	—	—
	WO-A-6073	FM-GPW-5607	BUSHING, spring shackle, outer (after Willys serial 146774)	6	132	268	—	12	—
	WO-A-575	FM-GPW-5705-B	CLIP, front spring, left (axle to spring) (after Willys serial 146744)	1	5	5	—	12	—
	WO-A-1097		CLIP, front spring, right (issue until stock is exhausted, then use WO-A-6511 for maintenance)	2	10	10	—	20	—
	WO-A-6511	FM-GPW-5455	CLIP, front spring, right (axle to spring) (was WO-A-1097)	1	%20	49	—	—	—
	WO-A-574	FM-GPW-5453	CLIP, front spring, right (axle to spring)	1	%2	—	—	—	—
	WO-339372	FM-GPW-5456	NUT, hex, S., special, 7⁄16-20NF-2 (front spring clip, right)	4	%56	2	—	8	—
	WO-638539	FM-33881-S2	NUT, hex, S., 7⁄16-20NF-2 (front spring clip, left) (issue until stock exhausted)	8	%108	56	—	140	—
	WO-5938	FM-34848-S2	WASHER, lock, S., 7⁄16 in.	4	—	216	—	—	—
16-1	WO-638500	FM-353023-S7	FITTING, grease, 90° (SW-1911) (issue until stock exhausted)	4	%16	32	—	—	—
	WO-640038	FM-358006-S8	FITTING, grease, straight (AD-1720) (was WO-A-392909) (issue until stock exhausted)						
16-1	WO-A-6326	FM-GPW-5601	LOCK, spring shackle bolt, assembly (after serial 170307)	2	%92	184	—	—	—
	WO-51308	FM-21492-S2	BOLT, hex-hd. S., 5⁄16-24NF x 2	1	—	—	—	—	—
	WO-51612	FM-350744-S2	BOLT, hex-hd, S., 7⁄16-20NF x 3 3⁄4 (lock to frame) (after Willys serial 170307)	1	—	—	—	—	—
	WO-5910	FM-33798-S	NUT, hex., S., 5⁄16-24 (after Willys serial 170307)	1	—	—	—	—	—
	WO-52510	FM-34941-S	WASHER, lock, S., 5⁄16 in. (after Willys serial 170307)	1	—	—	—	—	—
	WO-A-571	FM-GPW-5460	PLATE, spring clip, w/SHAFT, assembly, rear left (issue until stock exhausted)	2	4	4	—	—	—
	WO-A-572	FM-GPW-5459	PLATE, spring clip, front, w/SHAFT, assembly, left (issue until stock exhausted)						
	WO-A-568	FM-GPW-5458	PLATE, spring clip, front, w/SHAFT, assembly, right (issue until stock exhausted)	2	4	4	—	—	—
16-1	WO-A-1252	FM-GPW-5482	RETAINER, spring shackle grease seal	1	4	—	—	—	—
16-1	WO-A-515	FM-GPW-5481	SEAL, grease, spring shackle	8	%6	3	—	20	—
16-1	WO-A-6068	FM-GPW-5602	SHACKLE, spring, inner, w/BUSHING (after Willys serial 146774)	8	%38	13	—	60	—
16-1	WO-A-6069	FM-GPW-5605	SHACKLE, spring, outer, w/BUSHING (after Willys serial 146774)	1	%2	2	—	6	—
				1	%2	2	—	6	—

1603—SHOCK ABSORBERS AND MOUNTINGS

Fig.	Part No.	Ref. No.	Description						
	WO-A-6902	FM-GPW-18045	ABSORBER, shock, front, assembly (MAE-11465) (after Willys serial 197066) (Willys only)	2	%10	3	—	16	—
			ABSORBER, shock, front, assembly (Gabriel) (Ford only)	2	%19	6	—	30	—
	WO-A-6903		ABSORBER, shock, rear, assembly (MAE-11466) (after Willys serial 197066) (Willys only)	2	%5	2	—	10	—
		FM-GPW-18080	ABSORBER, shock, rear, assembly (Gabriel) (Ford only)	2	%10	3	—	16	—
16-2	WO-116640		BASE, front, shock absorber, assembly (MAE-12626) (Willys only)	2	—	—	—	—	—
	WO-116641		BASE, rear, shock absorber, assembly (MAE-12627) (Willys only)	2	—	—	—	—	—
	WO-637936	FM-GPW-18060	BUSHING, shock absorber mounting pin, rubber (HP-779)	16	14	14	—	50	—
16-2	WO-116631		BUSHING, shock absorber rebound spring (MAE-12448) (Willys only)	4	—	—	—	—	—
16-2	WO-638343		DISC, shock absorber (MAE-108496) (Willys only)	4	—	—	—	—	—
16-2	WO-637810		GASKET, shock absorber bushing (MAE-10875) (Willys only)	4	—	—	—	—	—
16-2	WO-116637		GUIDE, shock absorber piston rod, w/SEAL, assembly (MAE-12507) (Willys only)	4	—	—	—	—	—
16-2	WO-116642		HEAD, front, shock absorber, assembly (MAE-12628) (Willys only)	2	—	—	—	—	—
	WO-116643		HEAD, rear, shock absorber, assembly (MAE-12629) (Willys only)	2	—	—	—	—	—

GROUP 16—SPRINGS AND SHOCK ABSORBERS (Cont'd)

1603—SHOCK ABSORBERS AND MOUNTINGS (Cont'd)

Figure Number	Official Stockage Number	Part Number Willys	Part Number Ford	ITEM	Quantity Reqd. per Unit Assy.	Per 100 Major Items 12 Mos. Field Maintenance	Per 100 Major Items Major Overhaul (5th Ech)	Total First Year Procurement	Estimated Reqmts. per 100 Rebuilds
Col. 1	Col. 2	Col. 3	Col. 4	Col. 5	Col. 6	Col. 7	Col. 8	Col. 9	Col. 10
		WO-A-8810		KIT, repair, rear shock absorber (Willys only)	as req.	6	7	20	—
				(Includes:					
				1 WO-638343 DISC					
				1 WO-637810 GASKET					
				4 WO-637936 GROMMET					
				1 WO-116637 GUIDE w/SEAL					
				1 WO-116632 NUT					
				1 WO-116630 PISTON					
				1 WO-116644 PLATE, adjusting					
				1 WO-116626 PLATE, valve backing					
				1 WO-116625 SEAT					
				1 WO-116636 SLEEVE					
				1 WO-116627 SPACER, metering					
				1 WO-116631 SPACER, piston nut					
				1 WO-637803 SPRING, intake valve					
				1 WO-116633 SPRING, valve					
				1 WO-116639 TUBE					
				1 WO-116624 VALVE, compression assembly					
				1 WO-637804 VALVE, intake					
				1 WO-116629 WASHER, piston support					
				1 WO-116634 WASHER, spring seat)					
		WO-A-8809		KIT, repair, front shock absorber (Willys only)	as req.	13	14	40	—
				(Includes:					
				1 WO-638343 DISC					
				1 WO-637810 GASKET					
				4 WO-637936 GROMMET					
				1 WO-116637 GUIDE w/SEAL					
				1 WO-116632 NUT					
				1 WO-116630 PISTON					
				1 WO-116644 PLATE, adjusting					
				1 WO-116626 PLATE, valve					
				1 WO-116625 SEAT					
				1 WO-116635 SLEEVE					
				1 WO-116628 SPACER, metering					
				1 WO-116631 SPACER, piston nut					
				1 WO-637803 SPRING, intake valve					
				1 WO-116633 SPRING, valve					
				1 WO-116638 TUBE					

SNL G-503

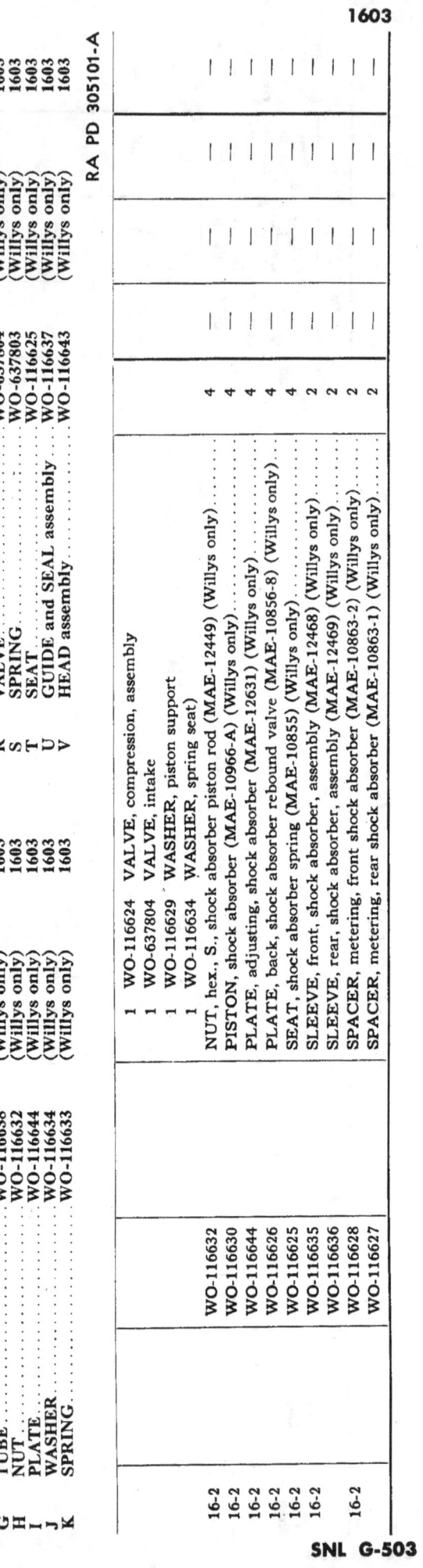

FIGURE 16-2—SHOCK ABSORBER ASSEMBLY

Key	Item	Willys Part No.	Ford Part No.	Gov't Group No.
A	WRENCH	WO-A-7778	(Willys only)	2301
B	THIMBLE	WO-A-7779	(Willys only)	2301
C	GASKET	WO-637810	(Willys only)	1603
D	SLEEVE assembly	WO-116635	(Willys only)	1603
E	BASE assembly	WO-116640	(Willys only)	1603
F	VALVE assembly	WO-116624	(Willys only)	1603
G	TUBE	WO-116638	(Willys only)	1603
H	NUT	WO-116632	(Willys only)	1603
I	PLATE	WO-116644	(Willys only)	1603
J	WASHER	WO-116634	(Willys only)	1603
K	SPRING	WO-116633	(Willys only)	1603
L	BUSHING	WO-116631	(Willys only)	1603
M	WASHER	WO-116629	(Willys only)	1603
N	PLATE	WO-116626	(Willys only)	1603
O	DISC	WO-638343	(Willys only)	1603
P	SPACER	WO-116628	(Willys only)	1603
Q	PISTON	WO-116630	(Willys only)	1603
R	VALVE	WO-637804	(Willys only)	1603
S	SPRING	WO-637803	(Willys only)	1603
T	SEAT	WO-116625	(Willys only)	1603
U	GUIDE and SEAL assembly	WO-116637	(Willys only)	1603
V	HEAD assembly	WO-116643	(Willys only)	1603

1	WO-116624	VALVE, compression, assembly
1	WO-637804	VALVE, intake
1	WO-116629	WASHER, piston support
1	WO-116634	WASHER, spring seat
4		NUT, hex., S., shock absorber piston rod (MAE-12449) (Willys only)
4		PISTON, shock absorber (MAE-10966-A) (Willys only)
4		PLATE, adjusting, shock absorber (MAE-12631) (Willys only)
4		PLATE, back, shock absorber rebound valve (MAE-10856-8) (Willys only)
4		SEAT, shock absorber spring (MAE-10855) (Willys only)
2		SLEEVE, front, shock absorber, assembly (MAE-12468) (Willys only)
2		SLEEVE, rear, shock absorber, assembly (MAE-12469) (Willys only)
2		SPACER, metering, front shock absorber (MAE-10863-2) (Willys only)
2		SPACER, metering, rear shock absorber (MAE-10863-1) (Willys only)

16-2	WO-116632			—	1603
16-2	WO-116630			—	1603
16-2	WO-116644			—	1603
16-2	WO-116626			—	1603
16-2	WO-116625			—	1603
16-2	WO-116635			—	1603
16-2	WO-116636			—	1603
16-2	WO-116628			—	1603
	WO-116627			—	1603

GROUP 16—SPRINGS AND SHOCK ABSORBERS (Cont'd)

Figure Number	Official Stockage Number	Part Number Willys	Part Number Ford	ITEM	Quantity Reqd. per Unit Assy.	Per 100 Major Items 12 Mos. Field Maintenance	Per 100 Major Items Major Overhaul (5th Ech)	Total First Year Procurement	Estimated Reqmts. per 100 Rebuilds
Col. 1	Col. 2	Col. 3	Col. 4	Col. 5	Col. 6	Col. 7	Col. 8	Col. 9	Col. 10
				1603—SHOCK ABSORBERS AND MOUNTINGS (Cont'd)					
16-2		WO-637803		SPRING, shock absorber piston intake valve (MAE-10639-B) (Willys only)	4	—	—	—	—
16-2		WO-116633		SPRING, shock absorber rebound valve (MAE-12463) (Willys only)	4	—	—	—	—
16-2		WO-116638		TUBE, pressure, front shock absorber (MAE-12620) (Willys only)	2	—	—	—	—
16-2		WO-116639		TUBE, pressure, rear shock absorber (MAE-12621) (Willys only)	2	—	—	—	—
16-2		WO-116624		VALVE, compression, shock absorber, assembly (MAE-83-C1) (Willys only)	4	—	—	—	—
16-2		WO-637804		VALVE, intake, shock absorber piston (MAE-10640-B) (Willys only)	4	—	—	—	—
16-2		WO-116634		WASHER, shock absorber adjusting plate (MAE-12464) (Willys only)	4	—	—	—	—
		WO-A-227	FM-356525	WASHER, shock absorber mounting pin (S., 1¼ O.D. x $^{41}/_{64}$ I.D. x $^{5}/_{32}$ thk.)	8	29	10	60	—
		WO-52946	FM-72037-S	PIN, cotter, $^{3}/_{16}$ x 1	8	—	—	—	—
16-2		WO-116629		WASHER, shock absorber piston support (MAE-10906-A) (Willys only)	4	—	—	—	—

GROUP 17—HOOD, FENDERS, APRONS

1701—FENDERS

		WO-A-2662	FM-GPW-16132-	ANTI-SQUEAK, front fender to cowl side panel, lower	2	—	—	—	—
		WO-A-3158	FM-GPW-16159	ANTI-SQUEAK, front fender to cowl side panel, upper (issue until stock exhausted)	2	—	—	—	—
		WO-A-3179		BOLT, welding, S., $^{5}/_{16}$-18NC x $^{5}/_{8}$ (battery holder to fender)	1	% 4	8	—	—
		WO-A-2843	FM-GPW-16095	BRACE, left front fender	1	—	—	—	—
		WO-A-2844	FM-GPW-16094	BRACE, right front fender	1	—	—	—	—
		WO-52600	FM-24449-S2	BOLT, hex-hd., S., $^{3}/_{8}$-24NF x 1$^{3}/_{8}$ (brace to chassis)	4	—	—	—	—
		WO-52819	FM-34707-S2	WASHER, S. (plain) $^{3}/_{8}$ in.	4	—	—	—	—
		WO-51304	FM-34807-S2	WASHER, lock, S., $^{3}/_{8}$ in.	4	—	—	—	—
		WO-A-2942	FM-GPW-16006	FENDER, front left, w/SPLASHER, assembly	1	—	—	—	—
		WO-A-2943	FN-GPW-16005	FENDER, front right, w/SPLASHER, assembly	1	—	—	—	—
		WO-51523	FM-20046-S2	BOLT, hex-hd., S., $^{5}/_{16}$-18NC x $^{3}/_{4}$ (fender to cowl)	8	—	—	—	—
		WO-6167	FM-34130-S2	NUT, hex., S., $^{5}/_{16}$-18NC	2	—	—	—	—
		WO-A-4116		WASHER, S. (plain) $^{5}/_{16}$ in. (was WO-71633) (issue until stock exhausted)	8	% 28	52	—	—
		WO-53029		WASHER, lock, S., external, shakeproof, $^{5}/_{16}$ in.	8	—	—	—	—
		WO-53025		WASHER, lock, S., cd-pltd, internal, external, $^{5}/_{16}$ in. (was WO-A-51833)	10	—	—	—	—
		WO-A-2891	FM-GPW-16105	FILLER, front fender brace, wood	2	—	—	—	—

1702—SPLASH APRON

		WO-A-3159	FM-GPW-16133	ANTI-SQUEAK, apron to frame	3	—	—	—	—
		WO-A-2841	FM-GPW-16083	APRON, splash, left front fender, assembly	1	—	—	—	—
		WO-A-2842	FM-GPW-16082	APRON, splash, right front fender, assembly	1	—	—	—	—

1704—HOOD

WO-50163	FM-20349-S2	BOLT, hex-hd., S., 3/8-24NF-2 x 3/4 (apron to fender)	4
WO-51396	FM-20347-S2	BOLT, hex-hd., S., 5/16-24NF-2 x 3/4 (apron to top of frame)	1
WO-50921	FM-351370-S2	WASHER, S. (plain) 3/8 in. (was WO-5455)	4
WO-71633		WASHER, S. (plain) 5/16 in.	1
WO-53026	FM-34806-S7-8	WASHER, lock, S., cd-pltd., internal, external, 3/8 in. (was WO-5010)	4
WO-51840	FM-GPW-16128	WASHER, lock, S., 5/16 in.	1
WO-A-3096		BRACE, mounting, voltage regulator	2
WO-A-2375	FM-350976-S2	NUT, clinch, 5/16-24NF-2 (regulator to apron)	4
WO-53135		WASHER, lock, S., cd-pltd., external, 3/8 in.	4
WO-A-3059		BRACKET, hood catch	2
WO-51514	FM-GPW-16684	BOLT, hex-hd., S., 1/4-20NC-2 x 5/8 (brkt. to hood)	4
WO-5790	FM-20308-S7-8	NUT, hex, S., 1/4-20NC-2	4
WO-5121	FM-33795-S2	WASHER, S., plain, 1/4 in.	4
WO-52031	FM-34705-S2	WASHER, lock, S., 1/4 in.	4
WO-A-4683	FM-34805-S2	BUMPER, hood top windshield, assembly (wood)	2
WO-A-4679	FM-GTB-16848-A	NUT, S, prong type, sleeve, No. 10 (.190)-32NF-2 (Willys only)	4
WO-6627		SCREW, rd-hd., S., No. 10 (.190)-32NF-2 x 7/8 (bumper to hood) (Willys only)	4
WO-52221		WASHER, lock, No. 10 (.190) (Willys only)	4
WO-51514	FM-34803-S2	CATCH, hood, assembly	2
WO-5790	FM-GPW-16892	BOLT, hex-hd., s-fin., alloy-S., 1/4-20NC-2 x 5/8 (catch to fender)	4
WO-5121	FM-20308-S7-8	NUT, reg. hex, S., 1/4-20NC-2	4
WO-52031	FM-33795-S2	WASHER, S., plain, 1/4 in.	4
WO-A-2188	FM-34706-S2	WASHER, lock, S., 1/4 in.	4
WO-52168	FM-34805-S2	HINGE, hood, assembly	1
WO-52167	FM-GPW-16802	BOLT, hex-hd., S., 1/4-20NC-2 x 5/8 (hinge to cowl top)	5
WO-52702	FM-20324-S2	BOLT, hex-hd., S., 1/4-20NC-2 x 3/4 (hinge to cowl top)	2
WO-53204	FM-24344	WASHER, S., 1/4 in.	5
WO-52031	FM-34705-S2	WASHER, lock, S., cd-pltd., internal, external, 1/4 in. (Willys only)	6
WO-A-3225	FM-34805-S2	WASHER, lock, S., rust proof, 1/4 in.	3
WO-A-2836	FM-GPW-16610	HOOD assembly (after Willys serial 103545)	1
WO-A-4680	FM-GTB-16847	HOOD assembly (before Willys serial 103545)	1
WO-53103	FM-32475-S2	WEBBING, hood top windshield bumper	2
		SCREW, wood, ov-hd., S., No. 6 (.138) x 1 (webbing to bumper)	4

GROUP 18—BODY

1800—BODY

WO-A-12237	FM-GPW-1000000	BODY assembly	1
		(Includes:	
		BOWS, top	
		BRACES	
		BRACKETS	
		CLAMPS	
		COMPARTMENT, glove	

GROUP 18—BODY (Cont'd)

Figure Number	Official Stockage Number	Part Number Willys	Part Number Ford	ITEM	Quantity Reqd. per Unit Assy.	Per 100 Major Items - 12 Mos. Field Maintenance	Per 100 Major Items - Major Overhaul (5th Ech)	Per 100 Major Items - Total First Year Procurement	Estimated Reqmts. per 100 Rebuilds
Col. 1	Col. 2	Col. 3	Col. 4	Col. 5	Col. 6	Col. 7	Col. 8	Col. 9	Col. 10
				1800—BODY (Cont'd)					
				COMPARTMENT, tool					
				HANDLES					
				LIGHTS, tail					
				MIRROR, rear view, outside					
				SEALS					
				SEATS					
				ATTACHING PARTS)					
		WO-A-12237	FM-GPW-1100001	BODY assembly	1	—	—	—	—
				(Includes:					
				FLOORS					
				PANELS)					
				1800A—BODY HANDLES					
		WO-A-2389	FM-GPW-1129672	HANDLE, body, outside, rear corner	2	—	—	—	—
		WO-A-2390	FM-GPW-1129670	HANDLE, body, outside front	2	—	—	—	—
		WO-51485	FM-20326-S2	BOLT, hex-hd., S., 5/16-18NC-2 x 5/8 (handles to body)	16	—	—	—	—
		WO-52350	FM-3379752	NUT, hex., S., 5/16-18NC-2	16	—	—	—	—
		WO-51840	FM-34806-S2	WASHER, lock, S., 5/16 in.	16	—	—	—	—
		WO-A-2168	FM-GPW-1154810-A	LOOP, footman (on body side panel)	5	—	—	—	—
		WO-6352	FM-GPW-355836-S7	NUT, hex., fl.hd., S., No. 10 (.190)-24NC-2 (loop to sides)	10	—	—	—	—
		WO-6290	FM-GPW-355162-S2	SCREW, fl.hd., S., No. 10 (.190)-24NC-2 x 1/2	10	—	—	—	—
		WO-52221	FM-GPW-34803-S7	WASHER, lock, S., No. 10 (.190)	10	—	—	—	—
				1801—BODY MOUNTING PARTS					
		WO-A-4415		LINER, body sill, thick	2	—	—	—	—
		WO-A-4414		LINER, body sill, thin	as req.				
				1802—BODY PILLARS					
		WO-A-2815	FM-GPW-1102511	PILLAR, cowl, left	1	—	—	—	—
		WO-A-2816	FM-GPW-1102510	PILLAR, cowl, right	1	—	—	—	—
		WO-52863		BOLT, hex-hd., S., 3/8-16NC-2 x 1 5/8 (pillar to floor)	2	—	—	—	—
		WO-5544	FM-33799-S	NUT, hex., S., 3/8-16NC-2	2	—	—	—	—

SNL G-503

1802-1802A

1802A—BODY PANELS

WO-5455	FM-34747-S	WASHER, I., plain, 3/8 in.	4	
WO-5010	FM-34807-S7	WASHER, lock, S., 3/8 in. (13/16 O.D. x 13/32 I.D. x 1/16 thd.)	2	
WO-A-2138	FM-GPW-1102396	BOLT, eye, safety strap (S., 3/8-16NC-2)	2	
WO-5544	FM-33799-S2	NUT, hex, S., 3/8-16NC-2 (bolt to body)	2	
WO-51304	FM-34807-S2	WASHER, lock, S., 3/8 in.	2	
WO-A-2853	FM-GPW-1140449	BRACKET, engine crank retaining clamp, assembly	2	
WO-A-2940	FM-GPW-1140441	BRACKET, retaining, engine crank	1	
WO-A-2950	FM-GPW-1110895	FILLER, rear floor pan cross sill (wood)	1	
WO-A-2311		GUSSET, front floor toe board to frame, left assembly	1	
WO-A-2312		GUSSET, front floor toe board to frame, right assembly	1	
WO-50163	FM-355498-S7	BOLT, hex-hd., S., 3/8-24NF-2 x 3/4	2	
WO-50921		WASHER, S., plain, pkzd, 3/8 in. (was WO-5455)	2	
WO-53135		WASHER, lock, S., external, 3/8 in.	2	
WO-53026		WASHER, lock, internal, external, 3/8 in. (GM-178551)	2	
WO-A-2933	FM-355909-S2	NUT, clinch, special, S. (1/4-20NC-2) (for 1/4 in. pipe tap, std.) (for 1/4 in. drain plug in floor)	2	
WO-A-2278		NUT, clinch, S. (1/4-20NC-2) (transmission cover)	7	
WO-A-2375	FM-350976-S2	NUT, clinch, S. (5/16-24NF-2) (transmission cover and accelerator pedal)	1	
WO-A-2263	FM-351015-S7	NUT, clinch, S. (3/8-16NC-2) (front floor rear cross sill assembly)	1	
WO-662010	FM-353202-S	NUT, clinch, S., 5/16-18NC-2 (cowl side panel for front fender)	6	
WO-A-2466	FM-33896-S2	NUT, wing, S., 5/16-18NC-2 (crank retaining brkt.)	1	
WO-A-12003	FM-GPW-1111140	PAN, front floor	1	
WO-51545	FM-20514-S2	BOLT, hex-hd., S., 3/8-16NC x 1 3/4 (thru left front cross sill)	2	
WO-51954		BOLT, hex-hd., S., 3/8-16NC x 2 (thru front cross sills to frame brkt.)	2	
WO-5544	FM-33799-S	NUT, hex, S., 3/8-16NC-2	4	
WO-52893	FM-356016-S7	NUT, hex, S., stamped 3/8-16NC-2	4	
WO-50921		WASHER, S., plain, 3/8 (was WO-5455)	8	
WO-53135		WASHER, lock, S., external, 3/8 in.	4	
WO-53026		WASHER, lock, S., internal, external, 3/8 in.	4	
WO-A-12017	FM-GPW-1111218	PAN, rear floor	1	
WO-51405	FM-20514-S2	BOLT, hex-hd., S., 3/8-16NC x 1 (to frame, left side)	1	
WO-51954		BOLT, hex-hd., S., 3/8-16NC-2 x 2 (to frame, right side)	1	
WO-52893	FM-356016-S7	NUT, hex, S., 3/8-16NC-2 (stamped)	1	
WO-5544	FM-33799-S	NUT, hex, S., 3/8-16NC-2	1	
WO-50921		WASHER, S. Pkzd, plain, 3/8 in. (was WO-5455)	2	
WO-53135		WASHER, lock, S., external, 3/8 in.	3	
WO-53026		WASHER, lock, S., cd-pltd. internal, external, 3/8 in.	3	
WO-A-2756	FM-GPW-1127847	PANEL, body, left side	1	
WO-A-2757	FM-GPW-1127846	PANEL, body, right side	1	
WO-A-2758	FM-GPW-1140324	PANEL, body, rear	1	
WO-52857		BOLT, hex-hd., S., 5/16-18NC-2 x 1 5/8	2	
WO-52845		NUT, hex, S., 5/16-18NC-2 (stamped)	2	
WO-6167		NUT, hex, S., 5/16-18NC-2	2	
WO-A-4116		WASHER, S. Pkzd, plain, 5/16 in.	4	
WO-53025		WASHER, lock, S., internal, external, cd-pltd, 5/16 in.	4	
WO-A-12011	FM-GPW-1102039	PANEL, cowl, left side	1	
WO-A-12010	FM-GPW-1102038	PANEL, cowl, right side	1	
WO-A-12105	FM-GPW-1102015	PANEL, cowl, top	1	

171

SNL G-503

1802A-1803

GROUP 18—BODY (Cont'd)

Part Number				Col. 6	Per 100 Major Items			Col. 10
Official Stockage Number	Willys	Ford	ITEM	Quantity Reqd. per Unit Assy.	12 Mos. Field Maintenance	Major Overhaul (5th Ech)	Total First Year Procurement	Estimated Reqmts. per 100 Rebuilds
Col. 2	Col. 3	Col. 4	Col. 5	Col. 6	Col. 7	Col. 8	Col. 9	Col. 10
			1802A—BODY PANELS (Cont'd)					
	WO-A-12025		PANEL, dash, assembly		—	—	—	—
	WO-A-12023	FM-GPW-1127851	PANEL, rear left wheel house, inner		—	—	—	—
	WO-A-12024	FM-GPW-1127850	PANEL, rear right wheel house, inner		—	—	—	—
	WO-A-2771	FM-GPW-1127849	PANEL, left wheel house, top	1	—	—	—	—
	WO-A-2772	FM-GPW-1127848	PANEL, right wheel house, top	1	—	—	—	—
	WO-A-2939	FM-GPW-1103446-A	PLATE, tapping, axe clamp	1	—	—	—	—
	WO-A-2879	FM-GPW-1111238	PLATE, tapping, driver seat, front floor	1	—	—	—	—
	WO-A-2190	FM-GPW-1102130	PLATE, tapping, hood hinge (on cowl)	1	—	—	—	—
	WO-A-2940	FM-GPW-1128242	PLATE, tapping, shovel clamp	2	—	—	—	—
	WO-A-5120	FM-358019-S	PLUG, body drain hole cover (pipe, slotted, $\frac{1}{4}$ in.)	2	—	—	—	—
	WO-5247		RIVET, rd-hd, S, $\frac{1}{4}$ x $\frac{3}{8}$	8	—	—	—	—
	WO-A-3144	FM-GPW-1162606	REST, foot, rear floor assembly	2	—	—	—	—
	WO-51485	FM-20326-S2	BOLT, hex-hd., S., $\frac{5}{16}$-18NC x $\frac{5}{8}$ (to wheel house panel)	3	—	—	—	—
	WO-51523	FM-20346-S2	BOLT, hex-hd., S., $\frac{5}{16}$-18NC x $\frac{3}{4}$ (to floor pan)	1	—	—	—	—
	WO-6167	FM-33797-S	NUT, hex, S., $\frac{5}{16}$-18NC-2	4	—	—	—	—
	WO-52274	FM-34746-S2	WASHER, S., plain, $\frac{5}{16}$ in.	4	—	—	—	—
	WO-51840	FM-34806-S2	WASHER, lock, S., pltd., $\frac{5}{16}$ in.	4	—	—	—	—
	WO-A-2470	FM-GPW-1140447	RETAINER, engine crank	2	—	—	—	—
	WO-A-3222	FM-GPW-1111155	RISER, front to rear floor pan, assembly	1	—	—	—	—
	WO-51406		BOLT, hex-hd., S., $\frac{3}{8}$-16NC-2 x $1\frac{1}{4}$	2	—	—	—	—
	WO-52893	FM-356016-S7	NUT, hex, S., $\frac{3}{8}$-16NC-2 stamped	2	—	—	—	—
	WO-5544	FM-33799-S	NUT, hex, S., $\frac{3}{8}$-16NC-2	2	—	—	—	—
	WO-50921		WASHER, S., Pkzd, plain $\frac{3}{8}$ in. (was WO-5455)	4	—	—	—	—
	WO-53026		WASHER, lock, S., cd-pltd, internal, external, $\frac{3}{8}$ in.	4	—	—	—	—
	WO-53135		WASHER, lock, S., shakeproof, external, $\frac{3}{8}$ in.	4	—	—	—	—
	WO-52236		SCREW, fl-hd., S., No. 10 (.190)-24-NC-2 x $\frac{3}{8}$ (footman loop hole plug cap)	4	—	—	—	—
	WO-A-2968	FM-GPW-1128244	SHEATH, axe	1	—	—	—	—
	WO-A-2837		SILL, cross, front floor pan, left assembly (Willys only)	1	—	—	—	—
	WO-A-2838	FM-GPW-1110610	SILL, cross, front floor pan, right assembly	1	—	—	—	—
	WO-A-2935	FM-GPW-1110625	SILL, cross, front floor pan, rear assembly	1	—	—	—	—
	WO-A-2948	FM-GPW-1110750	SILL, cross rear floor pan	1	—	—	—	—
	WO-A-2325	FM-GPW-1128286	STRAINER, body side panel to wheel house	1	—	—	—	—
	WO-A-3120	FM-GPW-1144620	SUPPORT, rear body to frame, inner	2	—	—	—	—
	WO-A-2336	FM-GPW-1144606	SUPPORT, rear body to frame, outer	2	—	—	—	—
			1803—FLOOR MATS AND SEALS					
	WO-A-2990	FM-GPW-1112115	COVER, inspection, floor, brake cylinder	1				—

SNL G-503

			Qty	
WO-52170	FM-24308-S2		5	BOLT, hex-hd., S., ¼-20NC x ½
WO-52706	FM-34805-S2		5	WASHER, lock, S., ¼ in.
WO-A-2759	FM-GPW-1111331		1	COVER, rear floor pan, left, shock absorber assembly
WO-A-2760	FM-GPW-1111330		1	COVER, rear floor pan, right, shock absorber assembly
WO-A-4592	FM-GPW-1111162		1	COVER, trailer electric plug housing (after Willys serial 200679)
WO-A-2982	FM-GPW-1112110		1	COVER, floor, transmission
WO-52170	FM-24308-S2		7	BOLT, hex-hd., S., ¼-20NC x ½ (to floor pan)
WO-52706	FM-34805-S2		7	WASHER, lock, S., ¼ in.
WO-A-3784	FM-GPW-1101735-A		1	GROMMET, floor, transfer shift lever (rubber)
WO-12055			1	PLATE, floor seal, dimmer switch (Willys only)
WO-A-2919	FM-GPW-1112119		1	RING, floor gearshift lever housing cover
WO-52167	FM-20344-S2		1	BOLT, hex-hd., S., ¼-20NC-2 x ¾
WO-52142	FM-32866-S2		3	SCREW, binding hd., S., No. 10 (.190) x ⅝
WO-A-2917	FM-GPW-1112158		1	RING, floor seal, steering column
WO-52809	FM-32924-S7-8		4	SCREW, binding hd., S., No. 10 (.190) x ½
WO-A-2918	FM-GPW-1112117		1	RING, floor transmission shift lever grommet
WO-52167	FM-20344-S2		2	BOLT, hex-hd., S., ¼-20NC-2 x ¾
WO-52142	FM-32866-S7		3	SCREW, binding hd., S., No. 10 (.190) x ½
WO-A-3782	FM-GPW-1111520-A		1	SEAL, floor, accelerator pedal rod
WO-A-2932	FM-GPW-1111137-A		1	SEAL, floor brake cylinder inspection cover
WO-A-3116	FM-GPW-1111299-A		1	SEAL, floor, fuel line hole floor
WO-A-3207	FM-GPW-9082-A		1	SEAL, fuel tank stone guard, front
WO-A-3208	FM-GPW-9085-A		1	SEAL, fuel tank stone guard, left
WO-A-3209	FM-GPW-9083-A		1	SEAL, fuel tank stone guard, rear
WO-A-3206	FM-GPW-9071		1	SEAL, fuel tank stone guard, right
WO-A-3783	FM-GPW-M11286-A		1	SEAL, floor gearshift lever housing cover (rubber)
WO-A-3943	FM-GPW-1102330-A		1	SEAL, glove compartment door (after serial 120697)
WO-A-12054			1	SEAL, floor, headlight foot switch (Willys only)
WO-A-3173	FM-GPW-8155		1	SEAL, radiator side air deflector, left (before Willys serial 125809)
WO-A-3575			1	SEAL, radiator side air deflector to radiator (after Willys serial 125809) (Willys only)
WO-A-3182			1	SEAL, radiator side air deflector to radiator, right (before Willys serial 125809) (Willys only)
WO-A-2979	FM-GPW-8222		1	SEAL, radiator top air deflector (before Willys serial 125809)
WO-A-3574	FM-GPW-8166		1	SEAL, radiator top air deflector to radiator (after Willys serial 125809)
WO-A-2992	FM-GPW-11514-A		1	SEAL, floor starter switch
WO-A-2931	FM-GPW-1101698-A		1	SEAL, floor steering column
WO-A-2994	FM-GPW-1112116-A		1	SEAL, floor, transmission cover

1804—SEATS

WO-A-2983	FM-355569-S2		2	BOLT, anchor, safety strap (⅜-16NC-2)

GROUP 18—BODY (Cont'd)

Figure Number	Official Stockage Number	Part Number		ITEM	Quantity Reqd. per Unit Assy.	Per 100 Major Items			Total First Year Procurement	Estimated Reqmts. per 100 Rebuilds
		Willys	Ford			12 Mos. Field Maintenance	Major Overhaul (5th Ech)			
Col. 1	Col. 2	Col. 3	Col. 4	Col. 5	Col. 6	Col. 7	Col. 8		Col. 9	Col. 10

1804—SEATS (Cont'd)

		Willys	Ford	ITEM	Qty
		WO-6169	FM-33799-S2	NUT, hex, S., 3/8-16NC-2	2
		WO-51304	FM-34807-S2	WASHER, lock, S., 3/8 in.	2
		WO-A-2453	FM-GPW-1162552	BRACKET, front seat pivot (passenger)	2
		WO-51561	FM-60375-S	RIVET, rd-hd, S., 3/8 x 9/16	2
		WO-A-2515	FM-GPW-1160314	BRACKET, front seat, rear stop	2
		WO-A-2833	FM-GPW-1162414	BRACKET, pivot, rear seat frame	2
		WO-A-2787	FM-GPW-1162402	BRACKET, retaining, rear seat back frame	4
		WO-52217	FM-33795-S2	NUT, hex, S., 1/4-20NC-2 (brkt. to body)	4
		WO-52702	FM-34745-S	WASHER, S. (plain) 1/4 in.	4
		WO-52706	FM-34805-S2	WASHER, lock, S., 1/4 in.	4
		WO-A-2886	FM-GPW-1160327-B	FRAME, front seat, assembly (driver)	1
		WO-5922	FM-24407-S	BOLT, hex-hd., S., 5/16-24NF-2 x 1 1/8 (frame to wheel house top panel)	1
		WO-51523	FM-20346-S2	BOLT, hex-hd., S., 5/16-18NC-2 x 3/4 (frame to floor)	2
		WO-6167	FM-33797-S	NUT, hex, S., 5/16-18NC-2 (frame to floor)	3
		WO-5437	FM-34706-S2	WASHER, S. (plain) 5/16 in.	1
		WO-71633	FM-351370-S2	WASHER, S. (special) 5/16 in. (frame to floor)	2
		WO-51840	FM-34806-S2	WASHER, lock, S., 5/16 in.	1
		WO-A-2925	FM-GPW-1160326	FRAME, front seat, back assembly (passenger)	1
		WO-51523	FM-20346-S2	BOLT, hex-hd., S., 5/16-18NC-2 x 3/4 (to floor pan)	2
		WO-6167	FM-33797-S	NUT, hex, S., 5/16-18NC-2	2
		WO-71633	FM-351370-S2	WASHER, S. (special) 5/16 in.	2
		WO-51840	FM-34806-S7-8	WASHER, lock, S., 5/16 in.	1
		WO-A-2782	FM-GPW-1161326	FRAME, rear seat, assembly	1
		WO-A-2783	FM-GPW-1161300	FRAME, rear seat back, assembly	1
		WO-A-2607	FM-355965-S2	NUT, wing, S., 5/16-18NC-2 x 1 (on rear seat for tire pump) (after Willys serial 193040)	1
		WO-A-2466	FM-33896-S2	NUT, tee, S., 5/16-24NF-2 (in drivers seat frame)	1
		WO-A-2910	FM-GPW-1162181	PANEL, front seat body side crash pad, left assembly	1
		WO-A-2911	FM-GPW-1162180	PANEL, front seat body side crash pad, right assembly	1
		WO-654934	FM-355900-S2	NUT, tee, S., 1/4-20NC-2 (on crash pads)	8
		WO-51539	FM-36868-S2	SCREW, rd-hd., S., 1/4-20NC-2 x 3/4 (pads to side panels)	8
		WO-52706	FM-34805-S2	WASHER, lock, S., reg., 1/4 in.	8
		WO-A-2945	FM-GPW-1162410	RETAINER, rear seat pivot tube	2
		WO-52170	FM-20308-S2	BOLT, hex-hd., S., 1/4-20NC-2 x 1/2	2
		WO-52217	FM-33799-S2	NUT, hex, S., 1/4-20NC-2	2
		WO-52706	FM-34805-S2	WASHER, lock, S., reg., 1/4 in.	2
		WO-A-4518	FM-GPW-17055	RETAINER, tire pump handle (after Willys serial 193040)	1

SNL G-503

WO-A-4516	FM-GPW-17079	RETAINER, tire pump foot (after Willys serial 193040)	1		
WO-A-2830	FM-GPW-1161302	SPRING, retaining, rear seat frame	2		
WO-52170	FM-20308-S2	BOLT, hex-hd, S., ¼-20NC-2 x ½ (spring to wheel house panel)	4		
WO-52217	FM-33795-S2	NUT, hex, S., ¼-20NC-2	4		
WO-52031	FM-34805-S2	WASHER, lock, S., ¼ in.	4		
WO-A-2883	FM-GPW-1131414	STRAP, safety, assembly	2		
WO-A-3029		SUPPORT, rear seat frame to wheel house (Willys only)	2		

1804A—CUSHIONS

WO-A-2986	FM-GPW-1162900-B	CUSHION, front seat, assembly (driver and passenger)	2		
WO-52651	FM-355836-S7	NUT, hex., S., No. 10 (.190)-24NC-2	10		
WO-50814		SCREW, fl-hd., S., No. 10 (.190)-24NC-2 x ½	10		
WO-52221	FM-34803-S2	WASHER, lock, S., No. 10 (.190)	10		
WO-A-2788	FM-GPW-1160780	FRAME, rear seat cushion, assembly	1		
WO-A-3114	FM-GPW-1166401	PAD, crash, front seat body side, left, assembly	1		
WO-A-3115	FM-GPW-1166400	PAD, crash, front seat body side, right, assembly	1		
WO-A-3108	FM-GPW-1160026	SEAT, trimmed, rear, assembly	1		
WO-A-3107	FM-GPW-1160016-B	SEAT, trimmed, front, assembly	1		
WO-A-3980		SPRING, front seat back assembly	2		
WO-A-3984		SPRING, front seat body side crash pad, left, assembly	1		
WO-A-3985		SPRING, front seat body side crash pad, right, assembly	1		
WO-A-3981	FM-GPW-1163206	SPRING, front seat cushion, assembly (driver and passenger)	2		
WO-A-3982	FM-GPW-1166800-B	SPRING, rear seat back, assembly	1		
WO-A-3983	FM-GPW-1163846-B	SPRING, rear seat cushion, assembly	1		
WO-A-11729		TRIM, front seat back cover, assembly	2		
WO-A-11730		TRIM, front seat cushion cover, assembly	2		
WO-A-11731		TRIM, rear seat back cover, assembly	1		
WO-A-11732		TRIM, rear seat cushion cover, assembly	1		
WO-52990	FM-32989-S2	SCREW, ctsk. ov-hd., S., No. 10 (.190) x ½	32		
WO-52968	FM-34886-S	WASHER, ctsk., S., No. 10 (.190)	42		

1809—INSTRUMENT PANEL AND MOUNTING PARTS

WO-A-3155	FM-GPW-1104364	BRACE, instrument panel, circuit breaker, w/BRACKET, assembly	1		
WO-5247	FM-60332-S2	RIVET, rd-hd., S., ¼ x ⅜ (brace to dash)	2		
WO-A-1765	FM-GPW-1101670	PAD, dash panel	1		
WO-A-3005	FM-GPW-1101610	PANEL, dash, assembly	1		
WO-A-3578	FM-GPW-1104320	PANEL, instrument assembly	1		

1809A—COMPARTMENT AND PARTS

WO-A-3536	FM-GPW-1106085	BOLT, glove compartment lock (after 20,700 Willys trucks up to 37,914 Willys trucks)	1		
WO-A-3531	FM-GPW-1106084-A	CASE, glove compartment lock, assembly (after 20,700 Willys trucks up to 37,914 Willys trucks)	1		

1809A

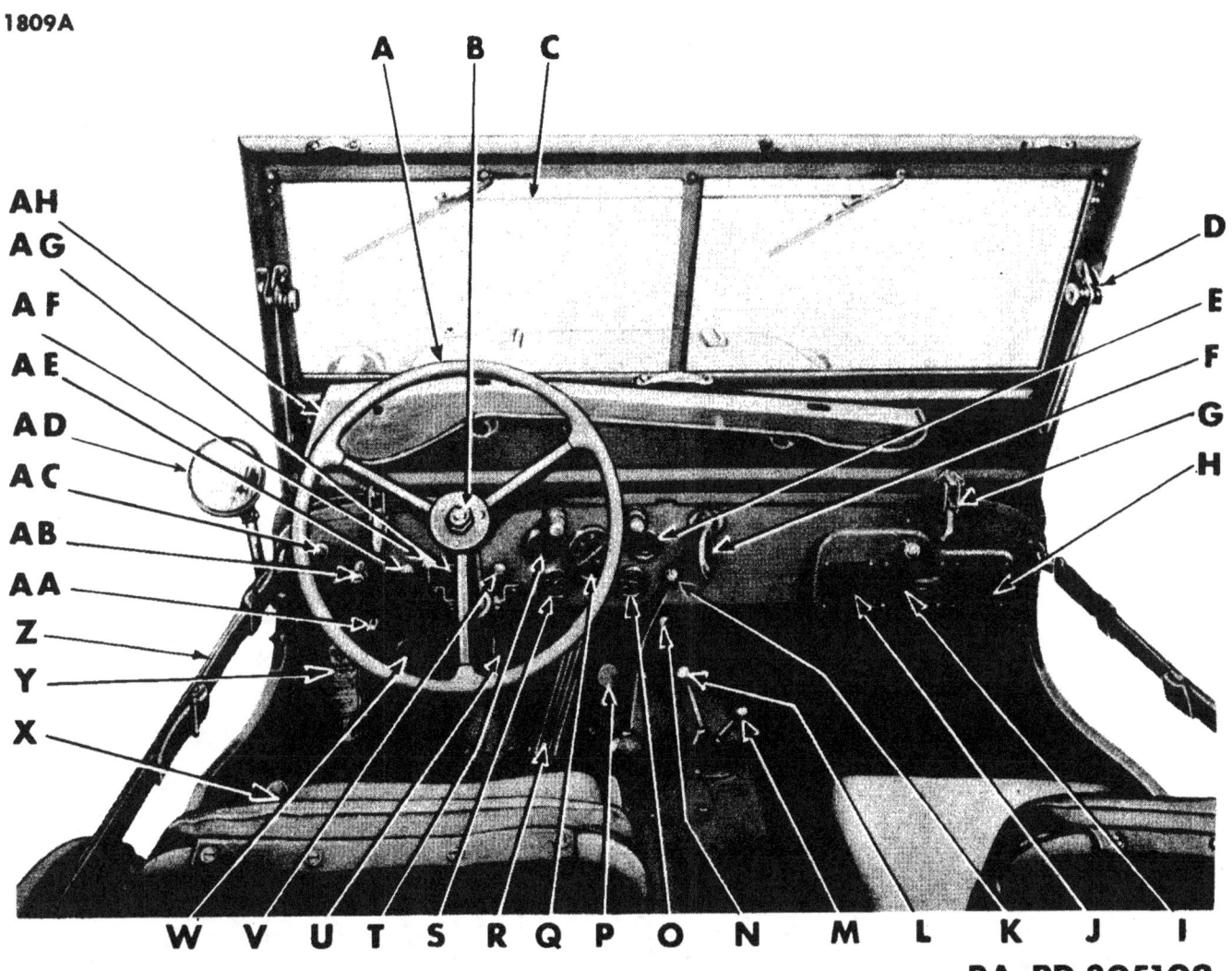

FIGURE 18-1—FRONT COMPARTMENT

Key	Item	Willys Part No.	Ford Part No.	Gov't Group No.
A	WHEEL	WO-A-6858	FM-GPW-3600-A-3	1404
B	BUTTON	WO-A-634	FM-GPW-3627	0609B
C	WIPER	WO-A-11433	FM-GPW-17500	1811D
D	ARM	WO-A-2235	FM-GP-1103302	1811
E	AMMETER	WO-A-8186	FM-GPW-10850	0605
F	HANDLE	WO-A-1242	FM-GPW-2780	1201
G	CLAMP assembly	WO-A-2227	FM-GP-1103482-A	1811
H	PLATE	WO-A-1330	FM-GPW-1101621-A	2203
I	PLATE	WO-A-11757	FM-GPW-4101629-A	2203
J	PLATE	WO-A-1331	FM-GPW-1101627-A	2203
K	LEVER	WO-A-1380	FM-GPW-7210-A	0706A
L	LEVER	WO-A-1506	FM-GPW-7710	0805B
M	LEVER	WO-A-1505	FM-GPW-7793	0805B
N	SWITCH	WO-A-7225	FM-9N-11450-A	0606
O	INDICATOR	WO-A-8188	FM-GPW-10883	0605
P	REST	WO-A-1225	FM-GPW-9716	0303
Q	SPEEDOMETER	WO-A-8180	FM-GPW-17255-A	2204
R	PEDAL	WO-A-6851	FM-GPW-9735-B	0303
S	GAUGE	WO-A-8190	FM-GPW-9273	0605
T	GAUGE	WO-A-8184	FM-GPW-9280	0605
U	PEDAL	WO-A-8253	FM-GPW-2452-B	1204
V	SWITCH	WO-A-1333	FM-GPW-13740	0606
W	PEDAL	WO-A-405	FM-GPW-7250	0204
X	TANK	WO-A-6618	FM-GPW-9002-B	0300
Y	EXTINGUISHER	WO-A-616	FM-GPW-17100	2402
Z	STRAP	WO-A-2883	FM-GPW-1131414	1804
AA	SWITCH	WO-A-12056	See GPW-13532	0606
AB	SWITCH	WO-A-1332	FM-GPW-11649	0606
AC	SWITCH	WO-A-6149	FM-GTB-13739	0606
AD	MIRROR	WO-A-2934	FM-21CS-17682-B	2202
AE	CHOKE	WO-A-1307	FM-GPW-97303	0301B
AF	SWITCH	WO-A-6811	FM-GPW-3686-B	0604C
AG	THROTTLE	WO-A-1308	FM-GPW-9778	0303A
AH	HOLDER	WO-A-11319	FM-GPW-1153100	1820

SNL G-503

GROUP 18—BODY (Cont'd)

Figure Number	Official Stockage Number	Part Number		ITEM	Quantity Reqd. per Unit Assy.	Per 100 Major Items			Estimated Reqmts. per 100 Rebuilds
		Willys	Ford			12 Mos. Field Maintenance	Major Overhaul (5th Ech)	Total First Year Procurement	
Col. 1	Col. 2	Col. 3	Col. 4	Col. 5	Col. 6	Col. 7	Col. 8	Col. 9	Col. 10
				1809A—COMPARTMENT AND PARTS (Cont'd)					
		WO-A-3652	FM-GPW-1106087-A	CUP, retaining, glove compartment door lock (after 20,700 Willys trucks up to 37,914 Willys trucks)		—	—	—	—
		WO-53020	FM-37621-S7	SCREW, S, mach., fil.-hd., No. 6 (.138)-32NC-2 x 5/16 cup to lock) (after 37,914 Willys trucks)	1	—	—	—	—
		WO-52614	FM-34801-S7	WASHER, lock, S., No. 6 (.138)	2	—	—	—	—
		WO-A-3532	FM-GPW-1106068	CYLINDER, glove compartment lock, assembly (after Willys trucks up to 37,914 Willys trucks)	2	—	—	—	—
		WO-A-3434	FM-GPW-1106024-B	DOOR, glove compartment, w/HINGE, assembly (after Willys serial 120698, up to serial Willys 137909)	1	—	—	—	—
		WO-A-3835		DOOR, glove compartment, w/HINGE, assembly (after Willys serial 137909)	1	—	—	—	—
		WO-A-3436	FM-GPW-1106050	HINGE, glove compartment door, assembly (after Willys serial 120698)	1	—	—	—	—
		WO-6352	FM-355836-S7	NUT, hex, S., No. 10 (.190)-24NC-2	3	—	—	—	—
		WO-5182	FM-355158-S2	SCREW, rd-hd., S., No. 10 (.190)-24NC-2 x 3/8 (hinge to panel)	3	—	—	—	—
		WO-52221	FM-34803-S2	WASHER, lock, S., No. 10 (.190)	3	—	—	—	—
		WO-A-3823	FM-GPW-1106084-B	LOCK, glove compartment door, assembly (after 37,914 Willys trucks)	1	—	—	—	—
		WO-A-3537	FM-356123-S7	NUT, glove compartment door lock (hex., S., pkzd., thin, 3/4-28) (after 20,700 Willys trucks up to 37,914 Willys trucks)	1	—	—	—	—
		WO-A-3538		PLATE, tapping, glove compartment hinge (after Willys serial 120,697)	1	—	—	—	—
		WO-A-3943	FM-GPW-1102330-A	SEAL, glove compartment door (after Willys serial 156083)	1	—	—	—	—
		WO-A-3818	FM-GPW-1106064	STRIKER, glove compartment door lock (after Willys serial 137909)	1	—	—	—	—
		WO-5182	FM-355158-S2	SCREW, hr-hd., S., No. 10 (.190)-24NC-2 x 3/8	1	—	—	—	—
		WO-52220		WASHER, S. (plain) No. 10 (.190)	1	—	—	—	—
		WO-52221	FM-34803-S2	WASHER, lock, S., No. 10 (.190)	1	—	—	—	—
				1811—WINDSHIELD ASSEMBLY					
18-1		WO-A-2235	FM-GP-1103302	ARM, adjusting, windshield	1	—	—	—	—
		WO-A-2485	FM-24454-S2	BOLT, windshield clamp bracket (hex. hd., S., 1/4-20NC-2 x 1 1/16) (AN X-1754)	2				
		WO-A-2798	FM-GPW-1103050	BRACKET, windshield mounting clamp catch	4				
		WO-52963	FM-36046-S2	SCREW, S., fl-hd., 1/4-20NC-2 x 3/4	2				
		WO-A-2213	FM-GPW-1103162	BRACKET, windshield pivot, assembly	4				
		WO-51485	FM-355449-S2	BOLT, hex-hd., S., 5/16-18NC-2 x 5/8 (to cowl)	2				
		WO-6167	FM-33797-S2	NUT, hex, S., 5/16-18NC-2	4				
		WO-51840	FM-34806-S2	WASHER, lock, S., 1/4 in.	4				
		WO-A-2232	FM-GPW-1103304	BRACKET, windshield swing arm, w/STUD (5/16-18NC-2)	2				

GROUP 18—BODY (Cont'd)

Figure Number	Official Stockage Number	Part Number: Willys	Part Number: Ford	ITEM	Quantity Reqd. per Unit Assy.	Per 100 Major Items: 12 Mos. Field Maintenance	Per 100 Major Items: Major Overhaul (5th Ech)	Per 100 Major Items: Total FirstYear Procurement	Estimated Reqmts. per 100 Rebuilds
Col. 1	Col. 2	Col. 3	Col. 4	Col. 5	Col. 6	Col. 7	Col. 8	Col. 9	Col. 10
				1811—WINDSHIELD ASSEMBLY (Cont'd)					
		WO-A-4300	FM-GP-1103310	BRACKET, windshield adjusting arm, w/BUSHING, assembly	2	—	—	—	—
		WO-A-2989		CATCH, hold down, windshield, assembly (before Willys serial 103545) (Willys only)	2	—	—	—	—
		WO-A-3197	FM-GPW-1103027	CATCH, hold down, windshield, assembly (after Willys serial 103545)	2	% 6	8	—	—
		WO-51514	FM-20324-S2	BOLT, hex-hd., S., ¼-20NC-2 x ⅝ (catch to hood top panel)	4	—	—	—	—
		WO-5790	FM-33795-S2	NUT, hex., S., ¼-20NC-2	4	—	—	—	—
		WO-5121	FM-34705-S2	WASHER, S. (plain) ¼ in.	4	—	—	—	—
		WO-52031	FM-34805-S2	WASHER, lock, S., ¼ in.	4	—	—	—	—
		WO-A-2791	FM-GPW-1103488	CATCH, windshield clamp (to instrument board)	2	—	—	—	—
		WO-A-2501	FM-GPW-1151289	CHAIN, windshield thumb screw	2	—	—	—	—
18-1		WO-A-2227	FM-GP-1103482-A	CLAMP, windshield, assembly	2	—	—	—	—
		WO-A-4120	FM-352692-S15	FASTENER, windshield curtain (top to windshield) (CNC-519)	10	—	—	—	—
		WO-A-4303	FM-GP-1103350	HANDLE, windshield frame	1	—	—	—	—
		WO-A-2238	FM-B-45482-C	KNOB, windshield adjusting arm (5/16-18NC-2)	2	—	—	—	—
		WO-A-2234	FM-GP-1103334	LOOP, hold down, windshield	2	—	—	—	—
		WO-A-2481	FM-34176-S2	NUT, windshield frame screw (hex., br., No. 10 (.190)-32NF-2 std. acorn) (AN X-1751)	3	—	—	—	—
		WO-A-2301		PANEL, windshield, outer (before Willys serial 103544)	1	—	—	—	—
		WO-A-3190		PANEL, windshield, outer (after Willys serial 103544)	1	—	—	—	—
		WO-A-2939	FM-GPW-1103446-A	PLATE, tapping, windshield clamp catch	2	—	—	—	—
		WO-A-3203	FM-GPW-1103028	RETAINER, windshield hold down catch (BA B-194) (after Willys serial 103545)	2	—	—	—	—
		WO-A-2386	FM-GPW-16892	RETAINER, windshield hold down catch (before Willys serial 103545)	2	—	—	—	—
		WO-51514	FM-20308-S7-8	BOLT, hex-hd., S., ¼-20NC-2 x ⅝ (retainer to hood)	4	—	—	—	—
		WO-5790	FM-33795-S	NUT, hex., S., ¼-20NC-2	4	—	—	—	—
		WO-5121	FM-34706-S2	WASHER, S., plain ¼ in.	4	—	—	—	—
		WO-52031	FM-34805-S2	WASHER, lock, S., ¼ in.	4	—	—	—	—
		WO-A-3189	FM-GPW-1151297	RING, windshield thumb screw chain	4	—	—	—	—
		WO-A-2239	FM-81W-811189	SCREW, pivot, windshield adjusting arm (rd-hd., S., ¼-28NF-2 x .47) (AN X-1747)	2	% 6	6	25	—
		WO-A-2487	FM-33249-S2	SCREW, windshield bracket (ov. ctsk-hd., S., No. 10 (.190) x ½, type "Z") (AN X-1755)	4	—	—	—	—
		WO-A-2480	FM-38259-S2	SCREW, windshield frame (special, rd-hd., S., No. 10 (.190)-32NF-2 x ⅞) (AN X-1750)	4	% 6	6	50	—
		WO-A-2482	FM-38192-S2	SCREW, windshield frame, corner (rd-hd., I., No. 8 (.164)-32NC x 7/16) (AN X-598)	4	% 6	6	50	—

1811-1811D

Part No.	Part No. 2	Description	Qty				
WO-A-2214	FM-355449-S2	SCREW, windshield pivot, thumb (Mail I, 3/8-24)	2	3/8 10	10	20	—
WO-A-2483	FM-99W-8103311	SPACER, windshield adjusting arm (S., 21/32 O.D. x 29/64 I.D.) (AN S-384)	2	3/8 6	6	25	—
WO-A-2490	FM-356286-S	WASHER, windshield adjusting arm (.688 O.D. x .349 I.D. x .067 thk.) (AN X-1637)	6	3/8 6	6	30	—
WO-A-4260	FM-356361-S	WASHER, spring, windshield adjusting arm (S., .56 O.D. x .34 I.D.) (AN S-1044)	4	3/8 10	10	40	—
WO-A-2220	FM-356373-S2	WASHER, windshield pivot spring (S., 1 in. O.D. x 21/32 I.D. x .015 thk., 3/32 crown)	4	3/8 14	14	25	—
WO-A-2486	FM-34805-S2	WASHER, lock, windshield clamp bracket screw (S., 1/4, .475 O.D. x .269 I.D. x .063 thk.) (AN X-1700)	4	—	—	—	—
WO-52960	FM-34803-S7	WASHER, lock, S., pkzd, No. 10 (.190)	3	—	—	—	—
WO-52961	FM-34802-S	WASHER, lock, S., pkzd, No. 8 (.164)	4	—	—	—	—
WO-A-2796		WINDSHIELD, assembly (before Willys serial 103545)	1	—	—	—	—
		(Includes:					
		ARMS					
		BRACKETS					
		CLAMPS					
		GLASS					
		ATTACHING PARTS)					
WO-A-3210	FM-GPW-1103010	WINDSHIELD assembly (after Willys serial 103545)	1	—	—	7	—
		(Includes:					
		ARMS					
		BRACKETS					
		CLAMPS					
		GLASS					
		ATTACHING PARTS)					

1811A—WINDSHIELD GLASS

| WO-A-2226 | FM-GP-1103014 | FRAME, windshield, w/GLASS, assembly | 1 | 2 | 2 | — | — |
| WO-A-2478 | FM-GP-1103100 | GLASS, windshield | 2 | — | — | — | — |

1811B—WINDSHIELD FRAME MOULDING AND SEALS

WO-A-2246	FM-GPW-1103030	FRAME, windshield, lower, assembly	1	—	—	—	—
WO-A-2255	FM-GP-1103020	FRAME, windshield, upper, assembly	1	—	—	—	—
WO-A-2776	FM-GPW-1103268	FRAME, windshield, tubular (before Willys serial 103545) (Willys only)	1	—	—	—	—
WO-A-3204	FM-GP-1103106	FRAME, windshield, tubular (after Willys serial 103545)	1	—	—	—	—
WO-A-2479	FM-GP-1103110	PADDING, windshield glass (Everseal, 1 1/8 in. x 73 in.)	2	10	10	24	—
WO-A-2250	FM-GPW-1103080-A	WEATHERSTRIP, windshield	1	—	—	—	—
WO-A-2476		WEATHERSTRIP, windshield frame to cowl	1	—	—	—	—
WO-A-2489	FM-33184-S2	SCREW, windshield to cowl weatherstrip (S., binding hd., sheet metal type "Z" No. 10 (.190) x 1/2)	15	—	—	—	—

1811D—WINDSHIELD WIPER ARMS AND BLADES

WO-A-2512	FM-GP-17535	ARM, windshield wiper, w/BLADE assembly (includes RIVET)	2	—	—	—	—
WO-A-11519		ARM, windshield wiper, w/CLIP, assembly	2	—	—	—	—
WO-A-4687		BLADE, windshield wiper	2	—	—	—	—

GROUP 18—BODY (Cont'd)

1811D-1817A

Figure Number	Official Stockage Number	Part Number		ITEM	Quantity Reqd. per Unit Assy.	Per 100 Major Items		Total First Year Procurement	Estimated Reqmts. per 100 Rebuilds
		Willys	Ford			12 Mos. Field Maintenance	Major Overhaul (5th Ech)		
Col. 1	Col. 2	Col. 3	Col. 4	Col. 5	Col. 6	Col. 7	Col. 8	Col. 9	Col. 10
				1811D—WINDSHIELD WIPER ARMS AND BLADES (Cont'd)					
		WO-A-2513	FM-GP-17531	HANDLE, wiper, inner w/KNOB, assembly	2	—	—	—	—
		WO-A-11432	FM-GPW-17534	HANDLE, windshield wiper, w/tie ROD, assembly (TRI 80540)	1	% 10	10	50	—
		WO-A-2588	FM 34054-S2	NUT, hex, S., No. 12 (.216)-24NF-2	2	% 2	2	8	—
		WO-53243	FM 63848-S	RIVET, S., tubular, ov-hd., type "A" (blade to arm)	2	—	—	—	—
		WO-A-2584	FM-351303-S2	WASHER, windshield wiper handle to frame (S., 1½ O.D. x ⁹⁄₃₂ I.D. x .032 thk.)	2	% 10	10	100	—
		WO-A-2587	FM-34824-S2	WASHER, lock, S., No. 12 (.216) (.266 I.D. x ½₂ thk.) (TRI 2328)	2	—	—	—	—
18-1		WO-A-11433	FM-GPW-17500	WIPER SET, tandem, windshield	1	—	—	—	—
				(Includes:					
				ARM w/BLADE, assembly					
				HANDLE w/ROD, assembly					
				NUT					
				WASHERS)					
				1817—TOOL BOX					
		WO-A-3198	FM-GPW-1146152	BRACKET, tool compartment lock, assembly	1	—	—	—	—
		WO-A-2774		FRAME, tool compartment, left	1	—	—	—	—
		WO-A-2775		FRAME, tool compartment, right	1	—	—	—	—
		WO-A-2811		HINGE, tool compartment lid, assembly (before Willys serial 118599)	4	—	—	—	—
		WO-A-3226	FM-GPW-1146132	HINGE, tool compartment lid, assembly (after Willys serial 118599)	4	—	—	—	—
		WO-A-2810		LID, tool compartment, w/HINGE, assembly (before Willys serial 118599)	2	—	—	—	—
		WO-A-3227	FM-GPW-1146100	LID, tool compartment, w/HINGE, assembly (after Willys serial 118599)	2	—	—	—	—
		WO-6352	FM-355836-S7-8	NUT, hex, S., No. 10 (.190)-24NC-2	6	—	—	—	—
		WO-5182		SCREW, rd-hd., S., No. 10 (.190)-24NC-2 x ⅜	6	—	—	—	—
		WO-52221	FM-34803-S2	WASHER, lock, S., No. 10 (.190)	6	—	—	—	—
		WO-A-2895	FM-GPW-1143501	LOCK, tool compartment lid, assembly (TA OP-528)	2	—	—	—	—
		WO-6352	FM-355836-S7-8	NUT, hex, S., No. 10 (.190)-24NC-2	4	—	—	—	—
		WO-52236		SCREW, rd-hd., S., No. 10 (.190)-24NC-2 x ⅜	4	—	—	—	—
		WO-52221	FM-34803-S2	WASHER, lock, S., No. 10 (.190)	4	—	—	—	—
		WO-A-3933		SEAL, tool compartment lid (short)	2	—	—	—	—
		WO-A-3934		SEAL, tool compartment lid (long)	2	—	—	—	—
		WO-A-2832	FM-GPW-1146126	STRIKER, tool compartment lid lock	2	—	—	—	—
				1817A—SPARE GASOLINE CARRIER					
		WO-A-4123	FM-GPW-1140330	BRACKET, spare gas line can assembly	1	—	—	—	—
		WO-53031	FM-20348-S2	BOLT, hex-hd., S., ⅜-16NC-2 x ¾ (brkt. to rear panel) –GM-122122)	4	—	—	—	—
		WO-53033	FM-33799-S2	NUT, hex., S., ⅜-16NC-2 (GM-123328)	4	—	—	—	—

SNL G-503

1817A-2103

WO-52046	FM-34807-S2	WASHER, lock, S., 3/8 in.	4	
WO-A-4127		STRAP, spare gasoline can, w/BUCKLE, assembly	1	
WO-53056		RIVET, tubular 9/64 x 11/32	8	
WO-53057		WASHER, S., 7/16 O.D. x 5/32 I.D.	8	
WO-A-4128		STRAP, spare gasoline can, w/TIP, assembly	1	

1820—GUN HOLDERS

WO-A-11581		BRACKET, support, rifle holder to windshield, left, assembly	1	
WO-A-11584		BRACKET, support, rifle holder to windshield, right, assembly	1	
WO-A-11945	FM-GPW-1153124	BUMPER, rifle holder retainer clamp, assembly (BUE-505285)	1	
	FM-GPW-1153106	CAM, rifle holder retainer handle (Ford only)	1	
	FM-62323-S2	RIVET, ov-hd., tubular, 7/64 x 13/32 (Ford only)	1	
WO-A-11942	FM-GPW-1153104	HANDLE, rifle holder retainer, w/CAM, assembly (BUE-505284)	1	
WO-A-11319	FM-GPW-1153100	HOLDER, rifle, assembly (BUE-505275)	1	
WO-53266	FM-23908-S2	BOLT, step, S., 5/16-18NC-2 x 3/4 (holder to windshield)	2	
WO-53037	FM-33797-S2	NUT, reg., hex., S., 5/16-18NC-2 (holder to windshield bolt)	2	
WO-53032		WASHER, flat, S., 3/8 in. (holder to windshield bolt) (Willys only)	2	
	FM-34707-S2	WASHER, flat, S., 5/16 in. (holder to windshield bolt) (Ford only)	2	
WO-53047	FM-34806-S2	WASHER, lock, S., 5/16 in. (holder to windshield bolt)	2	
WO-A-11944	FM-GPW-1153110	PLATE, rifle holder retainer clamp handle (BUE-505288)	1	
	FM-36801-S2	SCREW, rd-hd., No. 12 (.216)-24NC-2 x 3/8 (Ford only)	4	
	FM-34804-S2	WASHER, lock, S., No. 12 (.216) (Ford only)	4	
WO-A-11946	FM-GPW-1153123	SPRING, retaining, rifle holder, assembly (BUE-505282)	1	
WO-A-11940	FM-GPW-1153107	SPRING, rifle holder retainer handle cam (BUE-505287)	1	

GROUP 21—BUMPERS AND GUARDS

2101—BUMPERS

WO-A-1117	FM-GPW-17750	BAR, bumper, front	1	
WO-51823	FM-355456-S	BOLT, hex-hd., S., 5/16-24NF-2 x 4 1/2 (bar to frame)	4	
WO-5910	FM-33798-S2	NUT, hex., S., 5/16-24NF-2	4	
WO-51833	FM-34806-S2	WASHER, lock, S., 5/16 in.	4	
WO-A-1157	FM-GPW-17775	BUMPERETTE, rear	2	
WO-50163	FM-24349-S2	BOLT, hex-hd., S., 3/8-24NF-2 x 3/4 (bumperette to frame)	8	
WO-5901	FM-33800-S	NUT, hex., S., 3/8-24NF-2	8	
WO-5010	FM-34807-S	WASHER, lock, S., 3/8 in.	8	
WO-A-1147	FM-GPW-17751	FILLER, front bumper bar, wood	1	

2103—RADIATOR GUARDS

WO-A-2858		FRAME, radiator guard, assembly (before Willys serial 125809 only)	1	
WO-6299		BOLT, hex-hd., S., 3/8-16NC-2 x 7/8 (frame to chassis) (Willys only)	3	
WO-5544		NUT, hex., S., 3/8-16NC-2 (Willys only)	3	
WO-5455		WASHER, plain S., 3/8 in. (Willys only)	3	
WO-51304		WASHER, lock, S., 3/8 in. (Willys only)	3	
WO-A-3615	FM-GPW-8307	GUARD, radiator, assembly (after Willys serial 125809)	1	
WO-51485	FM-20326-S	BOLT, hex-hd., S., 5/16-18NC-2 x 5/8 (guard to fender)	6	

21-1

181

SNL G-503

FIGURE 21-1—RADIATOR GUARD

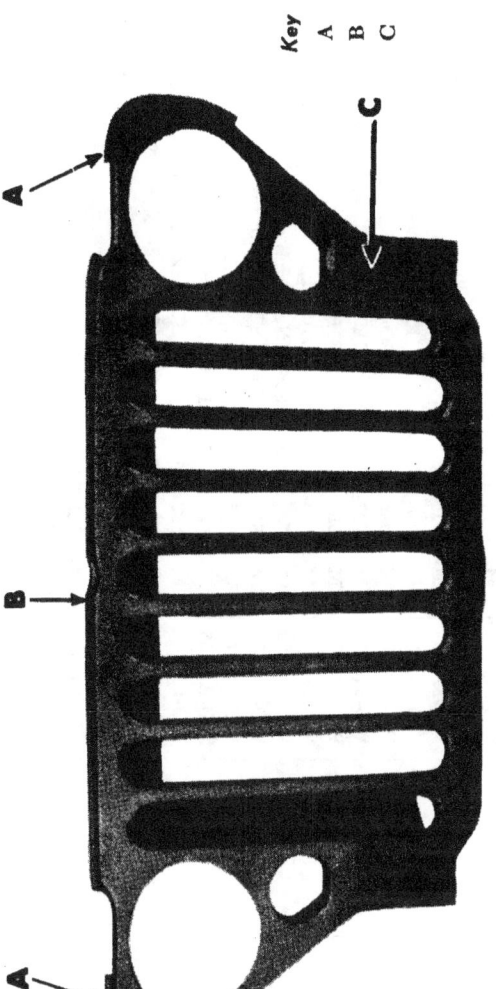

Key	Item	Willys Part No.	Ford Part No.	Gov't Group No.
A	LINER	WO-A-3095	FM-GPW-8384	2103
B	LINER	WO-A-3094	FM-GPW-8349	2103
C	GUARD	WO-A-3615	FM-GPW-8307	213

RA PD 305158-A

GROUP 18—BODY (Cont'd)

Figure Number	Official Stockage Number	Part Number Willys	Part Number Ford	ITEM	Quantity Reqd. per Unit Assy.	12 Mos. Field Maintenance	Major Overhaul (5th Ech)	Total First Year Procurement	Estimated Reqmts. per 100 Rebuilds
Col. 1	Col. 2	Col. 3	Col. 4	Col. 5	Col. 6	Col. 7	Col. 8	Col. 9	Col. 10
				2103—RADIATOR GUARDS (Cont'd)					
		WO-6167	FM-33797-S	NUT, hex, S, 5/16-18NC-2 (before Willys serial 125809)	6	—	—	—	—
		WO-A-3550	FM-34130-S2	NUT, sp, S., 5/16-18NC-2 (after Willys serial 125809)	6	—	—	—	—
		WO-5437	FM-34706-S7	WASHER, S., (plain) 5/16 in.	6	—	—	—	—
		WO-51840	FM-34806-S8	WASHER, lock, S., 5/16 in.	6	—	—	—	—
21-1		WO-A-3094	FM-GPW-8348	LINER, radiator guard hood, center	1	—	—	—	—
21-1		WO-A-3095	FM-GPW-8349	LINER, radiator guard hood, side	2	—	—	—	—
		WO-51182	FM-62627-S3	RIVET, split, 9/64 x 3/8 (5 used center, 6 used sides—Willys) (4 used center, 7 used sides—Ford) (Willys)	11	—	—	—	—
		WO-A-3549		RETAINER, radiator guard to fender nut (after Willys serial 125809) (Willys only)	6	—	—	—	—

2103A—LAMP GUARDS

WO-A-4118	FM-GPW-13176	GUARD, blackout driving lamp, w/SUPPORT, assembly (after Willys serial 163750)	1
WO-53050	FM-20387-S7	BOLT, hex-hd., S., 5/16-24NF-2 x 1 (GP-112657) (after Willys serial 163750)	1
WO-53048	FM-24347-S7	BOLT, hex-hd., S., 5/16-24NF-2 x 3/4 (GM-118772)	2
WO-50802	FM-33798-S7	NUT, hex, S., 5/16-24NF-2 (GM-115729)	3
WO-A-4116		WASHER, S., cd-pltd., 13/16 O.D. x 11/32 I.D. (Willys only)	1
WO-53047	FM-34806-S2	WASHER, lock, S., 5/16 in.	1
WO-53025	FM-356309-S7	WASHER, lock, S., shakeproof, 5/16 in.	4

GROUP 22—MISCELLANEOUS BODY, CHASSIS AND ACCESSORY

2201—TARPAULINS

WO-A-3070	FM-GPW-1102980	COVER, headlight	2
WO-A-3073		COVER, windshield, assembly (before Willys serial 103545)	1
WO-A-3211	FM-GPW-1103214	COVER, windshield, assembly (after Willys serial 103545)	1
WO-A-2909		PAULIN, deck top, assembly (before Willys serial 103545) (Willys only)	1
WO-A-3216	FM-GPW-1152700	PAULIN, deck top, assembly (after Willys serial 103545)	1

2201A—BOWS

WO-A-2898	FM-GPW-1151272	BOW, top front, assembly	1
WO-A-2897	FM-GPW-1151266	BOW, top, assembly (includes bow, top front, assembly)	1
WO-5216	FM-60470-S	RIVET, rd-hd., 5/16 x 1 (bow to pivot)	2
WO-A-2744	FM-GPW-1129069	BRACKET, left hand rail, front, assembly	1
WO-A-2745	FM-GPW-1129068	BRACKET, right hand rail, front, assembly	1
WO-51391	FM-24310-S	BOLT, hex-hd., S., 5/16-18NC-2 x 1/2 (bracket to body)	2
WO-52350	FM-33797-S2	NUT, hex, S., 5/16-18NC-2	2
WO-51840	FM-34806-S2	WASHER, lock, S., 5/16 in.	2
WO-A-2754	FM-GPW-1153030	BRACKET, top bow pivot	2
WO-51495	FM-20326-S2	BOLT, hex-hd., S., 5/16-18NC-2 x 5/8 (bracket to body)	2
WO-52350	FM-33797-S2	NUT, hex, S., 5/16-18NC-2	6
WO-52987	FM-24650-S2	SCREW, fl-hd., S., 5/16-18NC-2 x 5/8	4
WO-51840	FM-34806-S2	WASHER, lock, S., 5/16 in.	6
WO-A-2501	FM-GPW-1151289	CHAIN, thumb screw	2
WO-3113	FM-348148-S	EYELET, curtain fastener (CND-0)	16
WO-3112	FM-95633-S	FASTENER, single post type, curtain (rivet, tubular) (CNC-12)	11
WO-A-2168	FM-GPW-1154810	LOOP, footman	10
WO-6352	FM-355836-S7	NUT, hex., S., No. 10 (.190)-24NC-2	20
WO-6290	FM-355162-S2	SCREW, fl-hd., S., No. 10 (.190)-24NC-2 x 1/2 (loop to body)	20
WO-52221	FM-34803-S7	WASHER, lock, S., No. 10 (.190)	20
WO-A-2901	FM-GPW-1151270	PIVOT, top bow, assembly	2
WO-51389	FM-20328-S2	BOLT, hex-hd., S., 3/8-16NC-2 x 5/8 (pivot to bracket)	2
WO-5455	FM-34747-S	WASHER, S. (plain) 3/8 in.	2
WO-51304	FM-34807-S2	WASHER, lock, S., 3/8 in.	2

GROUP 22—MISCELLANEOUS BODY, CHASSIS AND ACCESSORY (Cont'd)

Figure Number	Official Stockage Number	Part Number		Item	Quantity Reqd. per Unit Assy.	Per 100 Major Items			Estimated Reqmts. per 100 Rebuilds
		Willys	Ford			12 Mos. Field Maintenance	Major Overhaul (5th Ech)	Total First Year Procurement	
Col. 1	Col. 2	Col. 3	Col. 4	Col. 5	Col. 6	Col. 7	Col. 8	Col. 9	Col. 10
				2201A—BOWS (Cont'd)					
		WO-A-2900	FM-GPW-1153142	PIVOT, top bow front	2	—	—	—	—
		WO-A-3111	FM-353104-S	PLATE, back, clincher (CNC-58)	11	—	—	—	—
		WO-A-3189	FM-GPW-1151297	RING, thumb screw chain	4	—	—	—	—
		WO-A-2214	FM-GPW-1150498	SCREW, thumb, top bow, assembly (3/8-24NF)	2	—	—	—	—
		WO-A-3051	FM-95705-S	SOCKET, curtain fastener (CNC-515)	14	—	—	—	—
		WO-A-3110	FM-GPW-1152720	STRAP, top lock, assembly	2	—	—	—	—
		WO-A-3141	FM-GPW-1152950	STRAP, top roll, assembly	2	—	—	—	—
		WO-A-3109	FM-GPW-1152730	STRAP, hold down, top, rear, assembly	6	—	—	—	—
		WO-A-2639	FM-356522-S	WASHER, anti-rattle, top bow pivot	2	—	—	—	—
		WO-A-3054	FM-95627-S	WASHER, curtain eyelet fastener (CNC-98)	32	—	—	—	—
		WO-A-2902	FM-351370-S2	WASHER, top bow, front pivot (S., 1 in. O.D. x 21/64 I.D.)	2	—	—	—	—
				2201F—SIDE CURTAINS					
		WO-A-2998	FM-GPW-1120041	CURTAIN, left, side, assembly	1	—	—	—	—
		WO-A-2999	FM-GPW-1120040	CURTAIN, right, side, assembly	1	—	—	—	—
		WO-A-3052	FM-95626-S	FASTENER, side curtain (male, flush type) (CNC-50)	14	—	—	—	—
		WO-A-3053	FM-95628-S	WASHER, side curtain fastener (CNC 60)	14	—	—	—	—
				2202—REAR VIEW MIRRORS					
18-1		WO-A-2934	FM-21CS-17682-B	MIRROR, rear view, assembly	1	%7	2	15	—
		WO-6989	FM-34054-S2	NUT, hex, S., No. 12 (.216) 24NC-2	6	—	—	—	—
		WO-6988	FM-36845-S2	SCREW, rd-hd, S., No. 12 (.216)-24NC-2 x 5/8	6	—	—	—	—
		WO-51969	FM-34804-S2	WASHER, lock, S., No. 12 (.216)	6	—	—	—	—
		WO-A-11850		MIRROR, rear view with washer and nut, assembly	1	%56	6	73	—
		WO-A-11862		NUT, mirror to tube, (S., hex., 1/4-28NF)	1	—	—	—	—
		WO-A-11861	FM-34804-S2	WASHER, lock, S., 1/4 in.	1	—	—	—	—
				2203—IDENTIFICATION AND CAUTION PLATES					
18-1		WO-A-1330	FM-GPW-1101621-A	PLATE, caution	1	—	—	—	—
18-1		WO-A-11757	FM-GPW-1101629-A	PLATE, name	1	—	—	—	—
18-1		WO-A-1331	FM-GPW-1101627-A	PLATE, shifting	1	—	—	—	—
		WO-53013		NUT, hex, S., stamped, No. 6 (.138)-32NC-2 (Willys only)	12	—	—	—	—

			FM-357016-S19	RIVET, ov-hd., fr., 9/64 x 3/16 (Ford only)	12	—	—	—	—
		WO-53014		SCREW, rd-hd., S., No. 6 (.138)-32NC x 5/16 in. (Willys only)	12	—	—	—	—

2203A—REFLECTORS

		WO-A-1306	FM-GPW-13380-A	REFLECTOR, reflex, assembly (CB-3121441)	4	% 19	10	32	—
		WO-5914	FM-33796-S	NUT, hex., S., reg., 1/4-28NF-2	8	—	—	—	—
		WO-52834	FM-36825-S	SCREW, rd-hd., S., 1/4-28NF-2 x 1/2	8	—	—	—	—
		WO-52706	FM-34805-S	WASHER, lock, light, S., 1/4 in.	8	—	—	—	—

2204—SPEEDOMETER AND PARTS

		WO-A-8242		BRACKET, mounting, speedometer (AL-SPK-66) (AL speedometer only) (Willys only)	1	—	—	—	—
		WO-A-8126		BRACKET, mounting, speedometer (KS 40333) (KS speedometer only) (Willys only)	1	—	—	—	—
		WO-345961	FM-GPW-13434-A2	BUSHING, rubber	1	—	—	—	—
		WO-A-5449	FM-GPW-14561	CLIP, mounting, closed type, S., 5/16 in. (1 used speedometer cable to frame, 1 used speedometer cable to dash with starter to starter switch cable)	2	—	—	—	—
		WO-6352	FM-355836-S7-8	NUT, hex., S., No. 10 (.190)-24NC-2	2	—	—	—	—
		WO-5064	FM-355130-S7	SCREW, rd-hd., S., No. 10 (.190)-24NC-2 x 5/8 in.	2	—	—	—	—
		WO-52221	FM-34803-S2	WASHER, lock, S., 1/4 in.	2	—	—	—	—
		WO-A-1344	FM-GPW-17262	SHAFT, flexible, speedometer drive, assembly (53 3/32 in. long) (AL SPS-1105-X)	1	% 2	29	10	50
18-1		WO-A-8180	FM-GPW-17255-A	SPEEDOMETER, luminous dial, assembly (AL SPK 4003) (KS 40355)	1	% 6	3	12	—
		WO-A-8127		NUT, wing, S., No. 10 (.190)-32NF-2 (AL 8X-72B) (Auto-Lite speedometer to brkt. only) (Willys only)	2	—	—	—	—
		WO-A-8125		NUT, wing, S., No. 10 (.190)-32NF (KS 40350) (King Seeley Speedometer to brkt. only) (Willys only)	2	—	—	—	—
		WO-A-8129		WASHER, lock, S. (special) No. 10 (.190) (AL 12X-196) (Willys only)	2	—	—	—	—
		WO-53194		WASHER, lock, S., No. 10 (.190) (KS-5131) (Willys only)	2	—	—	—	—
		WO-A-1343	FM-GPW-17261	TUBE, speedometer drive, assembly (52 in. long) (AL SPS-L100R)	1	% 13	6	24	—
		WO-A-1267	FM-GPW-17260	TUBE, speedometer drive, w/SHAFT, assembly (AL SPS-2099M)	1	—	—	—	—

GROUP 23—GENERAL USE STANDARDIZED PARTS

2301—TOOLS

23-1		WO-A-11765	FM-GPW-17126	ADAPTER, grease gun, push type (req'd. for universal joints) (SW-6334)	1	3	3	8	—
23-1	41-B-15	WO-A-372	FM-GPW-17005	BAG, tool	1	—	—	—	—
		WO-A-313	FM-GPW-17037	BRACKET, oil can (BA-SK-1670)	1	—	—	—	—
		WO-5790	FM-33795-S2	NUT, hex., S., 1/4-20NC-2	2	—	—	—	—
		WO-51492	FM-26483-S7	SCREW, rd-hd., S., 1/4-20NC x 1/2 (brkt. to body)	2	—	—	—	—
		WO-52706	FM-34805-S2	WASHER, lock, S., 1/4 in.	2	—	—	—	—
23-1		WO-A-239	FM-GP-17036	CRANK, starting	1	—	—	—	—
23-1	8-G-615	WO-A-6855	FM-GPW-18325	GAUGE, tire pressure	1	—	—	—	—
	41-G-1344-40	WO-A-12058		GUN, lubricating, hand type, 16 oz.	1	—	—	—	—

GROUP 23—GENERAL USE STANDARDIZED PARTS (Cont'd)

2301—TOOLS (Cont'd)

Figure Number	Official Stockage Number	Part Number Willys	Part Number Ford	ITEM	Quantity Reqd. per Unit Assy.	12 Mos. Field Maintenance	Major Overhaul (5th Ech)	Total First Year Procurement	Estimated Reqmts. per 100 Rebuilds
Col. 1	Col. 2	Col. 3	Col. 4	Col. 5	Col. 6	Col. 7	Col. 8	Col. 9	Col. 10
23-1		WO-A-213	FM-GP-17125	GUN, lubricating, hand type, 9 oz. capacity (issue until stock is exhausted, then issue 41-G-1344-40)		—	—	—	—
23-1	41-H-523	WO-A-373	FM-GP-17042	HAMMER, machinist, ball peen, 16 oz.	1	3	3	8	—
23-1		WO-306715	FM-GPW-17011-A	HANDLE, spark plug wrench	1	3	3	8	—
23-1	41-J-66	WO-A-1240	FM-GPW-17080	JACK, screw, w/handle (1½ ton)	1	—	—	—	—
23-1	13-O-1530	WO-A-379	FM-GP-17038	OILER, straight spout, spring bottom (capacity ½ pt.)	1	3	3	8	—
23-1	41-P-1650	WO-A-374	FM-GP-17028	PLIERS, combination, slip joint, 6 in.	1	3	3	8	—
23-1		WO-A-1339	FM-GPW-17090	PULLER, wheel hub	1	—	—	—	—
23-1	8-P-5000	WO-A-7511	FM-GPW-17052	PUMP, tire, w/air CHUCK	1	—	—	—	—
23-1	41-S-1076	WO-A-375	FM-GP-17026	SCREW DRIVER, common, heavy duty, integral handle, 6 in. (TGBX1A)	1	3	3	9	—
		WO-A-378		WRENCH SET, engineers	1				
				(Consists of:					
				1 41-W-991 WRENCH, engineers, ⅜ and ⁷⁄₁₆					
				1 41-W-1003 WRENCH, engineers, ½ and ¹⁹⁄₃₂					
				1 41-W-1005-5 WRENCH, engineers, ⁹⁄₁₆ and ¹¹⁄₁₆					
				1 41-W-1008-10 WRENCH, engineers, ⅝ and ²⁵⁄₃₂					
				1 41-W-1012-5 WRENCH, engineers, ¾ and ⅞					
		WO-A-371		TOOL, SET	1	—	—	—	—
				(Consists of:					
				BAG					
				HAMMER					
				PLIERS					
				SCREW DRIVER					
				WRENCH, adjustable					
				WRENCH, bleeder screw					
				WRENCH, crescent					
				WRENCH, set)					
		WO-A-1162	FM-GPW-17003	TOOL, SET	1	—	—	—	—
				(Consists of:					
				ADAPTER, grease gun					
				GAUGE, tire pressure					
				GUN, grease					
				HAMMER					
				HANDLE, spark plug wrench					
				JACK					
				KIT, tool					
				OILER (oil can)					
				PULLER, wheel hub					

SNL G-503

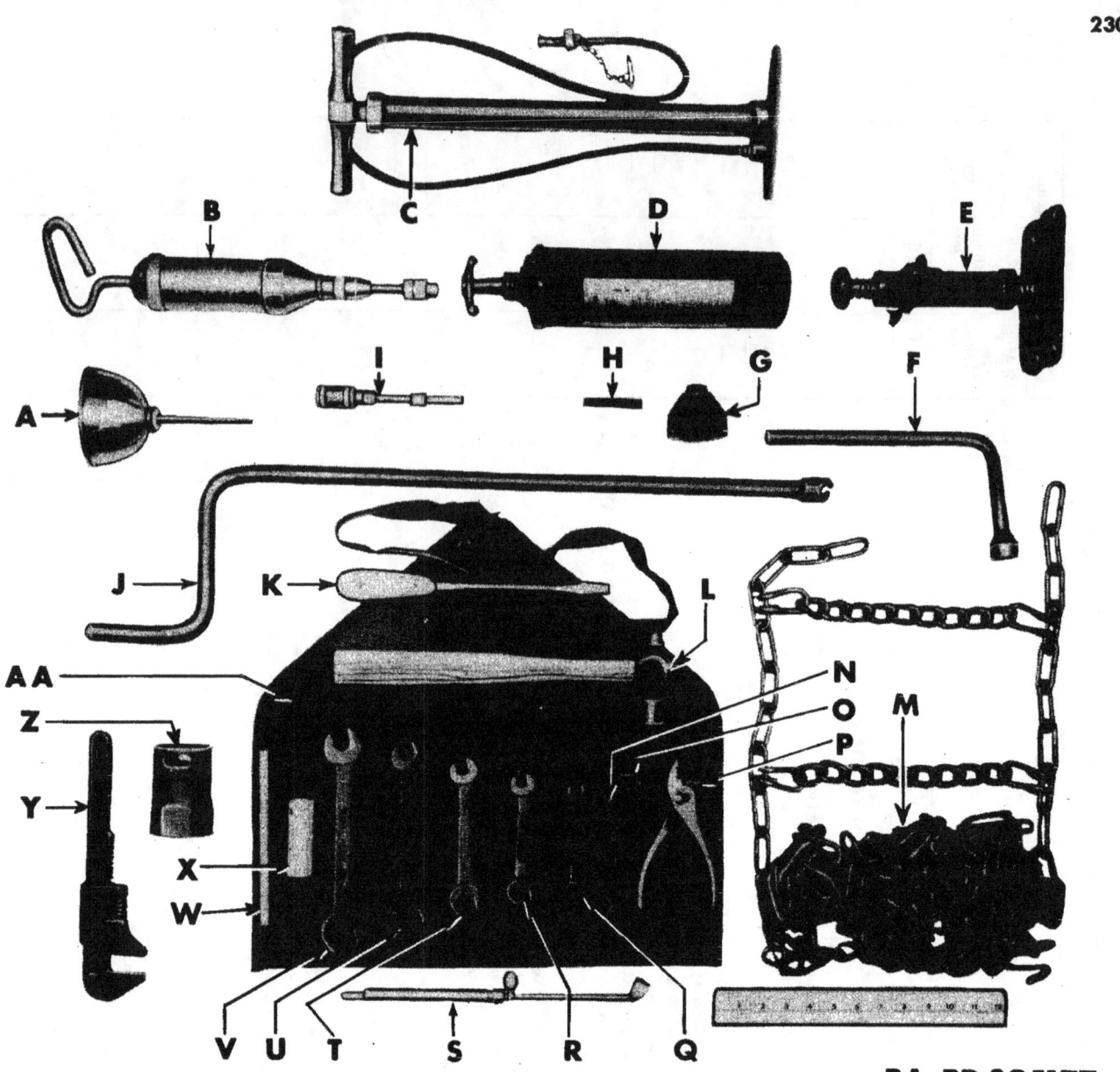

FIGURE 23-1—TOOLS

Key	Item	Willys Part No.	Ford Part No.	Gov't Group No.
A	OILER	WO-A-379	FM-GP-17038	2301
B	GUN, grease	WO-A-213	FM-GP-17125	2301
C	PUMP, tire	WO-A-7511	FM-GPW-17025	2301
D	EXTINGUISHER, fire	WO-A-616	FM-GPW-17100	2402
E	JACK	WO-A-1240	FM-GPW-17080	2301
F	WRENCH, wheel	WO-A-348	FM-GPW-17035	2301
G	PULLER, wheel	WO-A-1339	FM-GPW-17090	2301
H	WRENCH		FM-GP-17062	2301
I	ADAPTER	WO-A-11765	FM-GPW-17126	2301
J	CRANK	WO-A-239	FM-GPW-17036	2301
K	SCREWDRIVER	WO-A-375	FM-GP-17020	2301
L	HAMMER	WO-A-373	FM-GP-17042	2301
M	CHAINS, tire	WO-A-7687	FM-GPW-18136-B	2301B
N	WRENCH	WO-A-5130	FM-GPW-17030	2301
O	WRENCH	WO-A-1492	FM-GPW-17091	2301
P	PLIERS	WO-A-734	FM-GP-17028	2301
Q	WRENCH	WO-A-596	FM-GP-17043	2301
R	WRENCH	WO-A-597	FM-GP-17044	2301
S	GAUGE, tire	WO-A-6855	FM-GPW-18325	2301
T	WRENCH	WO-A-598	FM-GP-17045	2301
U	WRENCH	WO-A-599	FM-GP-17046	2301
V	WRENCH	WO-A-600	FM-GP-17047	2301
W	HANDLE	WO-306715	FM-GPW-17011-A	2301
X	WRENCH, spark plug	WO-637635	FM-GPW-17017-A	2301
Y	WRENCH	WO-A-377	FM-GP-17021	2301
Z	WRENCH	WO-A-692	FM-17033	2301
AA	BAG, tool	WO-A-372	FM-GPW-17005	2301

SNL G-503

2301-2301A

GROUP 23—GENERAL USE STANDARDIZED PARTS (Cont'd)

Figure Number	Official Stockage Number	Part Number Willys	Part Number Ford	ITEM	Quantity Reqd. per Unit Assy.	Per 100 Major Items 12 Mos. Field Maintenance	Per 100 Major Items Major Overhaul (5th Ech)	Total First Year Procurement	Estimated Reqmts. per 100 Rebuilds
Col. 1	Col. 2	Col. 3	Col. 4	Col. 5	Col. 6	Col. 7	Col. 8	Col. 9	Col. 10
				2301—TOOLS (Cont'd)					
				WRENCH, 17⁄64					
				WRENCH, drain plug					
				WRENCH, socket (transmission)					
				WRENCH, socket					
				WRENCH, spark plug					
23-1		WO-A-5130	FM-GPW-17030	WRENCH, hydraulic brake bleeder screw	1				
23-1	41-W-449	WO-A-377	FM-GP-17021	WRENCH, screw, adjustable, auto type, 11 in. (2¾ in. capacity)	1				
23-1		WO-A-1100	FM-GPW-17062	WRENCH, drain plug	1	3	3	8	
23-1	41-W-596	WO-A-596	FM-GP-17043	WRENCH, engineers double hd., alloy S., ⅜ and 7⁄16	1				
23-1	41-W-991	WO-A-597	FM-GP-17044	WRENCH, engineers double hd., alloy S., ½ and 19⁄32	1				
23-1	41-W-1003	WO-A-598	FM-GP-17045	WRENCH, engineers double hd., alloy S., 9⁄16 and 11⁄16	1				
23-1	41-W-1005-5	WO-A-599	FM-GP-17046	WRENCH, engineers double hd., alloy S., ⅝ and 25⁄32	1				
23-1	41-W-1008-10	WO-A-600	FM-GP-17047	WRENCH, engineers double hd., alloy S., ¾ and ⅞	1				
23-1	41-W-1012-5	WO-A-1492	FM-GPW-17091	WRENCH, fluted socket head, (use on transmission) (WG-T-84J-100)	1	5	6	5	
23-1		WO-A-348	FM-GPW-17035	WRENCH, socket (KHW 24832)	1	3	3	10	
23-1		WO-637635	FM-GPW-17017-A	WRENCH, socket, spark plug	1	3	3	10	
16-2		WO-A-7778		WRENCH, special (shock absorbers) (MAE T-317) (Willys only)	1		3	4	4
23-1		WO-A-692		WRENCH, wheel hub	1	3	3	8	
16-2		WO-A-7779	FM-GP-17033	THIMBLE (shock absorber piston rod guide oil seal protector) (MAE T-347) (Willys only)	1				
				2301A—PIONEERING EQUIPMENT					
		WO-A-3082	FM-GPW-1128258	BRACKET, shovel	1				
		WO-52987	FM-24650-S2	NUT, hex., S., 5⁄16-18NC-2 (brkt. to side)	2				
		WO-52350	FM-33797-S2	SCREW, fl-hd, S., 5⁄16-18NC-2 x ⅝	2				
		WO-5455	FM-34747-S	WASHER, S., plain, 5⁄16 in.	2				
		WO-51840	FM-34806-S2	WASHER, lock, S., 5⁄16 in.	2				
		WO-A-2601		BUCKLE (for axe, shovel, and top straps)	12				
		WO-A-2995	FM-GPW-1128256	CLAMP, axe, front	1				
		WO-52350	FM-33797-S2	NUT, hex., S., 5⁄16-28NF-2	4				
		WO-52274	FM-34706-S2	WASHER, S., plain, 5⁄16 in.	2				
		WO-51840	FM-34806-S7-8	WASHER, lock, S., 5⁄16 in.	4				
		WO-A-2984	FM-GPW-1128254	CLAMP, axe rear, assembly	1				
		WO-A-3139	FM-GPW-1128267	STRAP, axe, assembly	1				
		WO-A-3137	FM-GPW-1128237	STRAP, shovel, front, assembly	1				
		WO-A-3135	FM-GPW-1128247	STRAP, shovel, rear, assembly	1				

SNL G-503

2301B—TIRE CHAINS

WO-A-7687	FM-GPW-18136-B	CHAIN, tire, heavy duty, type "TS" 600 x 16	2 pr.	—	—	—	—
WO-A-1798		CHAIN, tire, heavy duty, type "D" 6:50 x 16 (Willys only)	2 pr.	—	—	—	—

2302—BRAKE FLUID

WO-A-8279		FLUID, brake (Lo 21-11)	12 oz.	% 10	10	23	—

2304B—BOLTS

WO-52840		BOLT, hex-hd., S., pltd., No. 10 (.190) x ½, type "B" thd.	—	—	—	—	—
WO-51738	FM-20300-S7	BOLT, hex-hd., S., ¼-20NC-2 x ⅜	—	19	19	100	100
WO-52170	FM-20308-S2	BOLT, hex-hd., S., pkzd., ¼-20NF-2 x ½	—	39	39	100	100
WO-51763	FM-20308-S2	BOLT, hex-hd., S., ¼-20NC-2 x ½	—	20	40	100	100
WO-51732		BOLT, hex-hd., S., ¼-28NF-2 x ½ (issue until stock exhausted)	—	10	10	100	100
WO-51514	FM-20324-S	BOLT, hex-hd., S., ¼-20NC-2 x ⅝	—	—	—	—	—
WO-52168	FM-20324-S2	BOLT, hex-hd., S., pkzd., ¼-20NC-2 x ⅝	—	—	—	—	—
WO-52167	FM-20344-S2	BOLT, hex-hd., S., pkzd., ¼-20NC-2 x ⅜	—	—	—	—	—
WO-6188	FM-20384-S2	BOLT, hex-hd., S., ¼-28NF-2 x ⅞	—	20	40	—	—
WO-52132		BOLT, hex-hd., S., ¼-20NC-2 x ¾	—	4	8	—	—
WO-51732		BOLT, hex-hd., S., ¼-28NF-2 x ⅞ (issue until stock exhausted)	—	—	—	—	—
WO-51798		BOLT, hex-hd., S., ¼-28NF-2 x 3 (issue until stock exhausted)	—	—	—	—	—
WO-51485	FM-20326-S7	BOLT, hex-hd., S., 5/16-18NC-2 x ⅝	—	170	73	300	300
WO-52132	FM-20327-S7	BOLT, hex-hd., S., 5/16-24NF-2 x ⅝	—	36	36	100	100
WO-51523	FM-20346-S2	BOLT, hex-hd., S., 5/16-18NF-2 x ¾	—	98	42	200	200
WO-51396	FM-24347-S2	BOLT, hex-hd., S., 5/16-24NF-2 x ¾	—	10	10	100	100
WO-6428	FM-20366-S	BOLT, hex-hd., S., 5/16-18NC-2 x ⅞	—	46	44	100	100
WO-50929	FM-24367-S2	BOLT, hex-hd., S., 5/16-24NF-2 x ⅞	—	70	10	200	200
WO-51486	FM-24386-S2	BOLT, hex-hd., S., 5/16-18NC-2 x 1	—	42	58	200	200
WO-5934	FM-34387-S7	BOLT, hex-hd., S., 5/16-24NF-2 x 1	—	88	10	100	100
WO-53050	FM-20387-S7	BOLT, hex-hd., S., pkzd., 5/16-24NF-2 x 1 (GM-11267)	—	10	10	100	100
WO-53131	FM-20407-S7	BOLT, hex-hd., S., cd-pltd., 5/16-24NF-2 x 1⅛ (GM-123499)	—	90	10	100	100
WO-6157	FM-24426-S2	BOLT, hex-hd., S., 5/16-18NC-2 x 1¼	—	29	10	50	50
WO-6660	FM-20406-S	BOLT, hex-hd., S., 5/16-18NC-2 x 1¼ (issue until stock exhausted)	—	8	16	—	—
WO-5922	FM-24427-S7	BOLT, hex-hd., S., 5/16-18NF-2 x 1⅛	—	10	—	50	50
WO-52857		BOLT, hex-hd., S., 5/16-18NC-2 x 1⅝	—	—	—	—	—
WO-50992	FM-24505-S7	BOLT, hex-hd., S., 5/16-24NF-2 x 1¾	—	10	10	50	50
WO-51308	FM-21492-S2	BOLT, hex-hd., S., 5/16-24NF-2 x 2	—	10	10	50	50
WO-51858	FM-355442-S	BOLT, hex-hd., S., 5/16-18NC-2 x 2½	—	—	—	—	—
WO-51798	FM-355398-S	BOLT, hex-hd., S., 5/16-18NC-2 x 3	—	—	—	—	—
WO-51823		BOLT, hex-hd., S., 5/16-24NF-2 x 4 (issue until stock exhausted)	—	20	40	—	—
WO-52189	FM-24328-S2	BOLT, hex-hd., S., ⅜-16NC-2 x ⅝	—	10	13	100	100
WO-6412	FM-24348-S	BOLT, hex-hd., S., ⅜-16NC-2 x ¾	—	10	10	100	100
WO-52543		BOLT, hex-hd., S., cd-pltd., ⅜-16NC-2 x ¾	—	90	—	—	—
WO-53031	FM-20348-S2	BOLT, hex-hd., S., rust proofed ⅜-16NC-2 x ¾	—	—	—	—	—
WO-6299		BOLT, hex-hd., S., ⅜-16NC-2 x ⅞	—	—	—	—	—
WO-51405	FM-24388-S	BOLT, hex-hd., S., ⅜-16NC-2 x 1	—	20	20	100	100
WO-52911	FM-24408-S	BOLT, hex-hd., S., ⅜-16NC-2 x 1⅛	—	31	12	100	100

2304B-2304F

GROUP 23—GENERAL USE STANDARDIZED PARTS (Cont'd)

Figure Number	Official Stockage Number	Part Number — Willys	Part Number — Ford	ITEM	Quantity Reqd. per Unit Assy.	Per 100 Major Items — 12 Mos. Field Maintenance	Per 100 Major Items — Major Overhaul (5th Ech)	Total First Year Procurement	Estimated Reqmts. per 100 Rebuilds
Col. 1	Col. 2	Col. 3	Col. 4	Col. 5	Col. 6	Col. 7	Col. 8	Col. 9	Col. 10
				2304B—BOLTS (Cont'd)					
		WO-51406	FM-24428	BOLT, hex-hd, S., ⅜-16NC-2 x 1¼	—	10	10	100	100
		WO-52863		BOLT, hex-hd, S., ⅜-16NC-2 x 1⅝	—	—	—	—	—
		WO-51545		BOLT, hex-hd, S., ⅜-16NC-2 x 1¾	—	—	—	—	—
		WO-51954		BOLT, hex-hd, S., ⅜-16NC-2 x 2	—	—	—	—	—
		WO-6486	FM-24534-S	BOLT, hex-hd, S., ⅜-16NC-2 x 2¼	—	10	10	100	100
		WO-6184	FM-24430-S	BOLT, hex-hd, S., ⁷⁄₁₆-14NC-2 x 1¼	—	—	—	—	—
		WO-52945	FM-23395-S	BOLT, carriage, sq-shank, S., ⅜-16NC-2 x ⅞	—	—	—	—	—
		WO-52983	FM-23393-S2	BOLT, carriage, sq-shank, S., ⁵⁄₁₆-18NC-2 x ⅞	—	8	16	—	—
		WO-53372	FM-23498-S2	BOLT, carriage, sq-shank, S., ⁵⁄₁₆-18NC-2 x 1¾ (issue until stock exhausted)	—	—	—	—	—
				2304C—CHAINS					
		WO-638457	FM-GPW-6260	CHAIN, drive, camshaft, ½ x 1 wide	—	—	—	—	—
				2304D—HOSE CLAMP					
		WO-53108	FM-GPW-6772	CLAMP, hose, S., ¾ in. dia.	—	—	—	—	—
		WO-52226	FM-60-8287	CLAMP, hose, S., 1¹³⁄₁₆ in. I.D. (issue until stock exhausted)	—	100	200	—	—
		WO-635097	FM-GPW-9653	CLAMP, hose, S., 2⁹⁄₃₂ in., I.D.	—	—	—	—	—
				2304E—ELBOWS					
		WO-384549	FM-GPW-9268	ELBOW, inverted flared tube, br.	—	—	—	—	—
		WO-384569	FM-9N-18686	ELBOW, inverted flared tube, ¼ br.	—	—	—	—	—
		WO-A-7180	FM-GPW-14521	ELBOW, Mall-1, z-pltd., ½-14 pipe thd., ¹⁵⁄₁₆-18 U.S.S. thd.	—	—	—	—	—
		WO-A-6885	FM-GPW-6722	ELBOW, pipe ¼ in.	—	—	—	—	—
				2304F—FITTINGS					
		WO-640038	FM-350006-S8	FITTING, grease (AD-1980)	—	23	11	40	40
		WO-392909	FM-353027-A-S7	FITTING, grease, straight, ⅛ inch (AD-1720)	—	32	32	80	80
		WO-638500	FM-353023-S7	FITTING, grease, 90 degrees	—	10	10	40	40
		WO-638792	FM-353043-S7	FITTING, grease, straight, ¼-28NF-2	—	29	10	40	40
		WO-638224	FM-353035-S7	FITTING, grease, 45 degrees	—	5	5	40	40

SNL G-503

2304G—NUTS

Stock No.	Part No.	Description					
WO-53013	FM-34052-S2	NUT, hex, stamped, S, cd-pltd. No. 6 (138)-32NC-2 (GM-174884)	—	—	—	—	—
WO-6214	FM-34052-S2	NUT, hex, reg, S, No. 8 (164)-32NC-2	—	—	—	—	—
WO-52182	FM-355836-S7	NUT, hex, stamped, br, No. 8 (164)-32NC-2	—	—	—	—	—
WO-6352	FM-355836-S7	NUT, hex, reg, S, No. 10 (190)-24NC-2	—	45	45	200	200
WO-52651		NUT, hex, reg, S, cd-pltd. No. 10 (190)-24NC-2	—	—	—	—	—
WO-5848		NUT, hex, reg, br, No. 10 (190)-32NF-2 (AL-2235)	—	10	10	100	100
WO-6536		NUT, hex, reg, S, cd-pltd, No. 10 (190)-32NF-2	—	—	—	—	—
WO-53061	FM-34141-S2	NUT, sq, reg, S, No. 10 (190)-32NF-2	—	54	10	200	200
WO-6273	FM-34054-S2	NUT, hex, reg, S, No. 12 (214)-24NC-2 (issue until stock exhausted)	—	4	8	—	—
WO-6989	FM-33795-S2	NUT, hex, reg, S, pkzd, 1/4-20NC-2	—	—	—	—	—
WO-52217	FM-33795-S2	NUT, hex, reg, S, 1/4-20NC-2 (issue until stock exhausted)	—	132	268	200	200
WO-5790	FM-33823-S	NUT, hex, br, 1/4-20NC-2 (GM-114379)	—	10	10	—	—
WO-53132	FM-34129-S2	NUT, hex, S, pkzd, 1/4-20NC-2	—	—	—	—	—
WO-52165	FM-33795-S2	NUT, hex, S, pkzd, 1/4-20NC-2	—	38	24	100	100
WO-52217	FM-51802	NUT, sq, S, 1/4-20NC-2 (issue until stock exhausted)	—	16	30	—	—
WO-5914	FM-33796-S2	NUT, hex, S, 1/4-28NF-2	—	49	19	200	200
WO-52804	FM-33797-S2	NUT, hex, S, thin, 1/4-28NF-2	—	—	—	—	—
WO-6167		NUT, hex, S, 5/16-18NC-2	—	78	19	200	200
WO-52845	FM-33797-S2	NUT, hex, spring, S, type B, 5/16-18NC-2	—	—	—	—	—
WO-52350		NUT, hex, thin, S, rust proof 5/16-18NC-2	—	—	—	—	—
WO-52891		NUT, hex, spring, S, stamped 5/16-24NF-2 (GM-147510)	—	—	—	—	—
WO-53285	FM-33798-S7	NUT, hex, S, cd-pltd, 5/16-24NF-2	—	—	—	—	—
WO-50802	FM-33909-S7	NUT, hex, S, pkzd, 5/16-24NF-2	—	265	128	500	500
WO-52954	FM-33798-S7	NUT, hex, thin, S, 5/16-24NF-2 (GM-114493)	—	5	5	200	200
WO-5910	FM-33799-S2	NUT, hex, S, 5/16-24NF-2	—	20	40	—	—
WO-53033		NUT, hex, S, rust proof, 3/8-16NC-2 (GM-123228)	—	—	—	—	—
WO-6196		NUT, hex, half hex, S, 3/8-16NC-2 (issue until stock exhausted)	—	—	—	—	—
WO-5544		NUT, hex, S, 3/8-16NC-2	—	—	—	—	—
WO-52893	FM-356016-S7	NUT, hex, stamped, type B, S, 3/8-16NC-2 (GM-147511)	—	391	365	800	800
WO-53289	FM-33800-S7	NUT, hex, Seez-proof, S, 3/8-16NC-2	—	—	—	—	—
WO-5901	FM-33800-S2	NUT, hex, S, 3/8-24NF-2 (CAR-105A-13)	—	—	—	—	—
WO-50922		NUT, hex, S, pkzd, 3/8-24NF-2	—	—	—	—	—
WO-52542		NUT, hex, S, cd-pltd, 3/8-24NF-2	—	—	—	—	—
WO-52909	FM-33800-S7	NUT, hex, stamped, S, cd-pltd, 3/8-24NF-2 (GM-107322)	—	8	16	200	200
WO-53288	FM-33802-S2	NUT, hex, Seez-proof, S, 3/8-24NF-2	—	—	—	—	—
WO-53287		NUT, hex, Seez-proof, S, 3/8-24NF-2	—	—	—	—	—
WO-5336		NUT, hex, S, 7/16-14NC-2 (issue until stock exhausted)	—	20	20	200	200
WO-5939	FM-33927-S7	NUT, hex, S, 7/16-20NF-2	—	16	16	200	200
WO-52925	FM-356028-S	NUT, hex, thin, S, 7/16-20NF-2	—	—	—	—	—
WO-52825	FM-33845-S2	NUT, hex, stamped, S, 7/16-20NF-2 (GM-107322)	—	10	10	40	40
WO-6163	FM-33846-S2	NUT, hex, stamped, S, 1/2-13NC-2	—	68	20	100	100
WO-5916	FM-33832-S2	NUT, hex, stamped, S, 1/2-20NF-2	—	—	—	—	—
WO-52804	FM-351059	NUT, hex, thin, S, 1/2-20NF-2	—	16	6	50	50
WO-10558	FM-34033-S18	NUT, hex, S, dld. f/c-pin, 1/2-20NF-2 (issue until stock exhausted)	—	29	32	100	100
WO-6436		NUT, hex, slotted, S, 9/16-18NF-2	—	29	29		

191 SNL G-503

GROUP 23—GENERAL USE STANDARDIZED PARTS (Cont'd)

Figure Number	Official Stockage Number	Part Number: Willys	Part Number: Ford	ITEM	Quantity Reqd. per Unit Assy.	Per 100 Major Items: 12 Mos. Field Maintenance	Per 100 Major Items: Major Overhaul (5th Ech)	Per 100 Major Items: Total First Year Procurement	Estimated Reqmts. per 100 Rebuilds
Col. 1	Col. 2	Col. 3	Col. 4	Col. 5	Col. 6	Col. 7	Col. 8	Col. 9	Col. 10
				2304Y—PLUGS					
		WO-51091		PLUG, expansion, 1¼ in. (issue until stock exhausted)	—	68	132	—	—
		WO-51460		PLUG, expansion, 1¾ in. (issue until stock exhausted)	—	68	132	—	—
		WO-52525	FM-353052-S	PLUG, pipe, slotted, ⅛ in.	—	3	3	40	40
		WO-5085	FM-358064-S	PLUG, pipe, ⅛ in.	—	3	32	40	40
		WO-5138	FM-353055-S7	PLUG, pipe, ¼ in.	—	26	12	40	40
		WO-376373	FM-358063-S	PLUG, pipe, S., ck, ⅜ in.	—	—	16	20	20
		WO-5140	FM-353064-S	PLUG, pipe, ½ in.	—	14	5	20	20
		WO-636538	FM-353051-S	PLUG, pipe, sq-hd., ¾ x 5/16	—	6	6	40	40
		WO-636577	FM-358048-S	PLUG, pipe, special, ¾ in.	—	6	6	40	40
				2304J—RIVETS					
		WO-635860	FM-62216-S	RIVET, fl-hd., S., 3/16 x 7/16	—	—	—	—	—
		WO-51977	FM-62218-S	RIVET, fl-hd., S., ¼ x 7/16	—	—	—	—	—
		WO-52207	FM-60371-S	RIVET, fl-hd., S., ¼ x 9/16 (issue until stock exhausted)	—	4	8	—	—
		WO-52906		RIVET, fl-hd., S., ⅜ x ¾	—	—	—	—	—
		WO-5326		RIVET, fl-hd., S., ⅜ x ⅞	—	—	—	—	—
		WO-5247		RIVET, rd-hd., S., ¼ x ⅜	—	—	—	—	—
		WO-5249		RIVET, rd-hd., S., ¼ x ⅝	—	—	—	—	—
		WO-5267	FM-60416-S	RIVET, rd-hd., S., 5/16 x ¾	—	—	—	—	—
		WO-5216	FM-60470-S	RIVET, rd-hd., S., 5/16 x 1	—	—	—	—	—
		WO-51561	FM-60375-S	RIVET, rd-hd., S., ⅜ x 9/16	—	—	—	—	—
		WO-50769		RIVET, rd-hd., S., ⅜ x ¾	—	—	—	—	—
		WO-52832	FM-357100-S	RIVET, rd-hd., S., ⅜ x ⅞	—	—	—	—	—
		WO-5215		RIVET, rd-hd., S., ⅜ x 1	—	—	—	—	—
		WO-51182	FM-62627-S3	RIVET, split, tubular, S., 9/64 x 9/16	—	—	—	—	—
		WO-53056		RIVET, ov-hd., tubular, br., 9/64 x 11/32	—	—	—	—	—
		WO-53243	FM-63848-S	RIVET, ov-hd., tubular, br., ⅛ x 15/32	—	—	—	—	—
		WO-51182	FM-62627-S3	RIVET, split tubular, S., 9/64 x 9/16	—	—	—	—	—
		WO-53056		RIVET, ov-hd., tubular, br., 9/64 x 11/32	—	—	—	—	—
		WO-53243	FM-63848-S	RIVET, ov-hd., tubular, br., ⅛ x 15/32	—	—	—	—	—
				2304K—SCREWS					
		WO-52990	FM-32989-S2	SCREW, ctsk., ov-hd., S., No. 10 (.190) x ½	—	—	—	—	—
		WO-50814		SCREW, fl-hd., S., No. 10 (.190)-24NC-2	—	—	—	—	—
		WO-6290	FM-355162-S2	SCREW, fl-hd., S., No. 10 (.190)-24NC-2 x ½	—	—	—	—	—

2304K-2304L

Stock No.	Part No.	Description					
WO-52987	FM-24650-S2	SCREW, fl-hd., S., 5/16-18NC-2 x 5/8	—	—	—	—	—
WO-52350	FM-33797-S2	SCREW, fl-hd., S., 5/16-18NC-2 x 5/8	—	—	—	—	—
WO-52963	FM-36046-S2	SCREW, fl-hd., S., 1/4-20NC-2 x 3/4	—	—	—	—	—
WO-52809	FM-32924-S8	SCREW, binding head, S., No. 10 (.190) x 1/2	—	—	—	—	—
WO-52142	FM-32866-S7	SCREW, binding head, S., No. 10 (.190) x 1/2	—	—	—	—	—
WO-51819	FM-31079	SCREW, ov-hd., S., No. 12 (.216)-24NC-2 x 5/8	—	—	—	—	—
WO-53014		SCREW, rd-hd., machine, S., No. 6 (.138)-32NC-2 x 5/16	—	75	54	200	200
WO-5253	FM-26457-S7	SCREW, rd-hd., machine, S., No. 8 (.164)-32NC-2 x 1/4	—	16	—	100	100
WO-51040	FM-355095-S7	SCREW, rd-hd., machine, S., No. 8 (.164)-32NC-2 x 1/4	—	—	—	—	—
WO-5556	FM-26480-S2	SCREW, rd-hd., machine, S., No. 8 (.164)-32NC-2 x 3/8	—	—	—	—	—
WO-6287	FM-355158-S2	SCREW, rd-hd., machine, S., No. 8 (.164)-32NC-2 x 1/2	—	—	—	—	—
WO-5182		SCREW, rd-hd., machine, S., No. 10 (.190)-24NC-2 x 3/8	—	—	—	—	—
WO-52236		SCREW, rd-hd., machine, S., No. 10 (.190)-24NC-2 x 3/8	—	—	—	—	—
WO-5113		SCREW, rd-hd., machine, S., No. 10 (.190)-24NC-2 x 1/2	—	10	10	200	200
WO-5064	FM-26496-S	SCREW, rd-hd., machine, S., No. 10 (.190)-24NC-2 x 5/8	—	10	10	100	100
WO-5780	FM-355132-S	SCREW, rd-hd., machine, S., No. 10 (.190)-24NC-2 x 5/8	—	16	16	200	200
WO-5272	FM-27698-S	SCREW, rd-hd., machine, S., No. 10 (.190)-24NC-2 x 3/4	—	19	19	200	200
WO-52994		SCREW, rd-hd., machine, S., No. 10 (.190)-32NF-2 x 3/8	—	—	—	—	—
WO-52060		SCREW, rd-hd., machine, S., No. 10 (.190)-32NF-2 x 7/16 (issue until stock exhausted)	—	—	—	—	—
WO-6627		SCREW, rd-hd., machine, S., No. 10 (.190)-32NF-2 x 7/8	—	—	—	—	—
WO-6383	FM-27145-S2	SCREW, rd-hd., machine, S., No. 10 (.190)-32NF-2 x 1	—	32	—	100	100
WO-53267	FM-26498-S2	SCREW, rd-hd., machine, S., No. 12 (.216)-24NC-2	—	—	—	—	—
WO-6988	FM-36845-S2	SCREW, rd-hd., machine, S., No. 12 (.126)-24NC-2 x 5/8	—	3	—	100	100
WO-52036	FM-26147-S7	SCREW, rd-hd., machine, S., 1/4-20NC-2 x 3/8	—	—	—	—	—
WO-51492	FM-26483-S	SCREW, rd-hd., machine, S., 1/4-20NC-2 x 1/2 (GM-113955)	—	48	10	200	200
WO-51539	FM-36868-S2	SCREW, rd-hd., machine, S., 1/4-20NC-2 x 3/4	—	—	—	—	—
WO-52060		SCREW, rd-hd., machine, S., 7/16-32NF	—	—	—	—	—
WO-53020	FM-37621-S7	SCREW, fl-hd., machine, S., No. 6 (.138)-32NC-2 x 5/16	—	—	—	—	—
WO-113439	FM-31628-S7	SCREW, fil-hd., machine, S., No. 10 (.190)-32NF-2 x 1/2	—	—	—	—	—
WO-52993	FM-32852-S	SCREW, rd-hd., sheet metal, S., No. 10 (.190) x 3/8	—	—	—	—	—
WO-51904	FM-92047-S	SCREW, rd-hd., machine, S., br-pltd., No. 6 (.138) x 5/16	—	—	—	—	—
WO-52131	FM-26457-S7	SCREW, rd-hd., machine, S., cd-pltd., No. 8 (.164)-32NC-2	—	—	—	—	—
WO-52781		SCREW, rd-hd., machine, S., cd-pltd., No. 8 (.164)-32NC-2 x 3/8	—	—	—	—	—
WO-51537	FM-355162-S	SCREW, fl-hd., S., cd-pltd, No. 10 (.190)-24NC-2 x 1/2	—	—	—	—	—
WO-53252		SCREW, rd-hd., machine, S., cd-pltd., No. 10 (.190)-32NF-2 x 3/8	—	—	—	—	—
WO-53220		SCREW, rd-hd., machine, S., cd-pltd., No. 10 (.190)-32NF-2 x 3/8	—	—	—	—	—
WO-53103		SCREW, wood, ov-hd., S., No. 6 (.138) x 1	—	—	—	—	—
WO-5580	FM-31866-S	SCREW, wood, rd-hd., S., No. 10 (.190) x 5/8	—	—	—	—	—

2304L—WASHERS

Stock No.	Part No.	Description					
WO-51532	FM-34802-S2	WASHER, plain, S., No. 8 (.164)	—	19	10	200	200
WO-52702	FM-34745-S2	WASHER, plain, S., 1/4 in.	—	20	10	200	200
WO-5437	FM-34706-S2	WASHER, plain, S., 5/16 in.	—	159	39	200	200
WO-53274	FM-34746-S2	WASHER, plain, S., 5/16 in.	—	130	—	100	100
WO-52101	FM-34747-S2	WASHER, plain, S., 3/8 in.	—	46	30	200	200
WO-5455	FM-34707-S7	WASHER, plain, S., 3/8 in.	—	30	10	200	200
WO-52835		WASHER, plain, S., 7/16 in.	—	8	12	—	—

SNL G-503

GROUP 23—GENERAL USE STANDARDIZED PARTS (Cont'd)

Figure Number	Official Stockage Number	Part Number: Willys	Part Number: Ford	ITEM	Quantity Reqd. per Unit Assy.	Per 100 Major Items: 12 Mos. Field Maintenance	Per 100 Major Items: Major Overhaul (5th Ech)	Per 100 Major Items: Total First Year Procurement	Estimated Reqmts. per 100 Rebuilds
Col. 1	Col. 2	Col. 3	Col. 4	Col. 5	Col. 6	Col. 7	Col. 8	Col. 9	Col. 10
				2304L—WASHERS (Cont'd)					
		WO-52768	FM-356305	WASHER, plain, S., $^{21}/_{64}$ in.	—	20	10	200	200
		WO-636570	FM-356504-S	WASHER, plain, S., 1½ O.D. x $^{49}/_{64}$ I.D. x ⅛ thk.	—	4	12	40	40
		WO-52705	FM-34802-S2	WASHER, lock, regular, S., No. 8 (.160)	—	36	20	200	200
		WO-113440	FM-34803-S7	WASHER, lock, regular, S., cd-pltd., No. 10 (.190)	—	126	87	300	300
		WO-51969	FM-34804-S2	WASHER, lock, regular, S., No. 12 (.214)	—	19	—	100	100
		WO-52706	FM-34805-S2	WASHER, lock, regular, S., ¼ in.	—	130	79	300	300
		WO-51833	FM-34806-S2	WASHER, lock, regular, S., $^{5}/_{16}$ in.	—	766	224	1200	1200
		WO-5051		WASHER, lock, regular, S., $^{5}/_{16}$ in.	—	28	52	—	—
		WO-5010	FM-34807-S7	WASHER, lock, regular, S., ⅜ in.	—	571	421	1200	1200
		WO-5059	FM-34808-S2	WASHER, lock, regular, S., $^{7}/_{16}$ in.	—	6	—	100	100
		WO-5938	FM-34838-S	WASHER, lock, regular, S., $^{7}/_{16}$ in. heavy duty	—	56	56	200	200
		WO-5009	FM-34809-S	WASHER, lock, regular, S., ½ in.	—	81	231	400	400
		WO-5038	FM-34811-S2	WASHER, lock, regular, S., ⅝ in.	—	10	10	200	200
		WO-52754	FM-43902-S	WASHER, lock, internal tooth, S., No. 8 (.164)	—	5	5	200	200
		WO-52483	FM-34905-S7	WASHER, lock, internal tooth, S., No. 14	—	19	19	200	200
		WO-52428	FM-34906-S	WASHER, lock, internal tooth, S., $^{5}/_{16}$ in.	—	32	48	200	200
		WO-52332	FM-34907-S7	WASHER, lock, internal tooth, S., ⅜ in.	—	78	59	200	200
		WO-52874	FM-34921-S	WASHER, lock, internal tooth, S., $^{7}/_{16}$ in.	—	26	26	200	200
		WO-636528	FM-34922-S	WASHER, lock, internal tooth, S., ½ in.	—	102	—	100	100
		WO-52510	FM-34941-S7	WASHER, lock, external tooth, S., $^{5}/_{16}$ in.	—	80	80	200	200
				2304M—WIRE LOCKING					
		WO-637598	FM-GP-2174	WIRE, lock piston stop (LO-S-FC-2927)	—	—	—	—	—
		WO-636298	FM-GPW-8576	WIRE, retaining, water pump brg.	—	—	—	—	—
				2305—MISCELLANEOUS COTTER PINS					
		WO-5067	FM-72083-S	PIN, cotter, S., $^{1}/_{16}$ x ½	—	—	—	—	—
		WO-5354	FM-72004-S	PIN, cotter, S., $^{3}/_{32}$ x ½	—	—	—	—	—
		WO-52967	FM-72016-S	PIN, cotter, S., $^{3}/_{32}$ x ⅝	—	—	—	—	—
		WO-5020	FM-72-16-S	PIN, cotter, S., $^{3}/_{32}$ x ¾	—	—	—	—	—
		WO-5152	FM-72025-S	PIN, cotter, S., $^{3}/_{32}$ x ⅞	—	—	—	—	—
		WO-5021	FM-72035-S	PIN, cotter, S., $^{3}/_{32}$ x 1	—	—	—	—	—
		WO-5397	FM-72071	PIN, cotter, S., ⅛ x ½	—	—	—	—	—
		WO-5108	FM-72053-S	PIN, cotter, S., ⅛ x 1¼	—	—	—	—	—
		WO-5257	FM-72062-S	PIN, cotter, S., ⅛ x 1⅜	—	—	—	—	—

2305-2310

2305A—WOODRUFF KEYS

WO-52527	FM-72062-S	PIN, cotter, S., 1/8 x 1 3/8	—	—	—	—	—
WO-5134	FM-72089-S	PIN, cotter, S., 1/8 x 1 1/4	—	—	—	—	—
WO-52946	FM-72037-S	PIN, cotter, S., 3/16 x 1	—	—	—	—	—
WO-A-1641	FM-74144-S	KEY, Woodruff, S., No. 5 (AL-X-260)	—	—	—	—	—
WO-5017	FM-74175-S7	KEY, Woodruff, S., No. 6 (AL-X-261)	—	8	16	—	—
WO-5036	FM-74178-S	KEY, Woodruff, S., No. 9	—	8	16	—	—
WO-50917	FM-74182-S	KEY, Woodruff, S., No. 13	—	8	16	—	—

2310—WINTERIZATION FIELD KIT

WO-A-11848		CRATE, winterization field kit	1
WO-A-11847		KIT, field, winterization (Willys only)	as req.
		(Consists of:	
WO-A-11792		KIT, field, battery plug-in receptacle	as req.
		(Consists of:	
WO-A-12135		CABLE, positive, assembly	1
WO-52600		BOLT, hex-hd., s-fin, alloy-S., 3/8-16NC-2 x 1 3/8 (cable to cranking motor attaching bolt)	1
WO-53026		WASHER, lock, S., internal, external tooth, 3/8 in. (cable to cranking motor attaching bolt)	1
WO-A-11787		CABLE, receptacle to ground, assembly	2
WO-78923		CLIP, wire, 5/8 in., (cable to gusset)	1
WO-6352		NUT, hex, reg., S., No. 10 (.190)-24NC-2 (clip to gusset screw)	2
WO-52548		SCREW, rd-hd., machine, S., No. 10 (.190)-24NC-2 x 5/16 (clip to gusset)	1
WO-5168		WASHER, lock, reg., S., No. (.190) (clip to gusset screw)	1
WO-A-11851		DRAWING, installation, battery plug-in receptacle field kit	1
WO-A-11772		RECEPTACLE, battery plug-in, 2 pole type, assembly	1
WO-53199		BOLT, hex-hd., s-fin, alloy-S., 1/4-20NC-2 x 1 (receptacle to fender)	4
WO-52217		NUT, hex, reg., S., 1/4-20NC-2 (receptacle to fender bolt)	4
WO-53084		WASHER, lock, reg., S., 1/4 in. (receptacle to fender bolt)	4
WO-A-11773		REINFORCEMENT, splash apron	1
WO-52170		BOLT, hex-hd., s-fin, alloy-S., 1/4-20NC-2 x 1/2 (reinforcement to fender)	4
WO-52217		NUT, hex, reg., S., 1/4-20NC-2 (reinforcement to fender bolt)	4
WO-53178		WASHER, plain, SAE, std., 1/4 in. (reinforcement to fender bolt)	4
WO-53084		WASHER, lock, reg., S., 1/4 in. (reinforcement to fender bolt)	4
WO-A-11815		KIT, field, blanket	as req.
		(Consists of:	
WO-A-11816		BLANKET, brush guard, assembly	1
WO-A-12038		BLANKET, hood, assembly	1
WO-A-11828		BLANKET, under fender, left, assembly	1
WO-A-11829		BLANKET, under fender, right, assembly	1
WO-A-11821		BLANKET, under motor, assembly	1
WO-A-11832		CARTON, blanket field kit	1

SNL G-503

2310

FIGURE 23-2—COLD STARTING BLANKET FIELD KIT

Key	Item	Willys Part No.	Ford Part No.	Gov't Group No.
A	BLANKET	WO-A-12038	(Willys only)	2310
B	BLANKET	WO-A-11816	(Willys only)	2310

FIGURE 23-3 — COLD STARTING HEATER FIELD KIT

Key	Item	Willys Part No.	Ford Part No.	Gov't Group No.
A	STACK	WO-A-8214	(Willys only)	2310
B	CLAMP	WO-A-8208	(Willys only)	2310
C	ELBOW	WO-52805	(Willys only)	2310
D	STOVE	WO-A-7156	(Willys only)	2310
E	HOSE	WO-A-7036	(Willys only)	2310
F	COCK	WO-A-7000	(Willys only)	2310
G	LINE	WO-A-1367	FM-GPW-9282	0304
H	LINE	See WO-A-7156	(Willys only)	2310
I	LINE	WO-A-8214	(Willys only)	2310
J	COCK	WO-A-7058	(Willys only)	2310
K	TANK	WO-A-8097	(Willys only)	2310

RA PD 305173-A

SNL G-503

GROUP 23—GENERAL USE STANDARDIZED PARTS (Cont'd)

Figure Number	Official Stockage Number	Part Number Willys	Part Number Ford	ITEM	Quantity Reqd. per Unit Assy.	12 Mos. Field Maintenance	Major Overhaul (5th Ech)	Total First Year Procurement	Estimated Reqmts. per 100 Rebuilds
Col. 1	Col. 2	Col. 3	Col. 4	Col. 5	Col. 6	Col. 7	Col. 8	Col. 9	Col. 10
				2310—WINTERIZING FIELD KIT (Cont'd)					
		WO-A-1173		CLIP, retaining spring (S., 2%2 in., lgth., under motor blanket to flywheel housing)	1	—	—	—	—
		WO-A-11833-1		DRAWING, installation, under motor blanket	1	—	—	—	—
		WO-A-11833-2		DRAWING, installation, radiator and brush guard	1	—	—	—	—
		WO-A-2924		FASTENER, curtain, P.K. type, single stud	8	—	—	—	—
		WO-A-3625		FASTENER, curtain, single stud, No. 10 (.190)-32NF-2	21	—	—	—	—
		WO-52216		NUT, hex, reg., S., No. 10 (.190)-32NF-2	21	—	—	—	—
		WO-52221		WASHER, lock, S., light, No. 10 (.190)	21	—	—	—	—
		WO-A-11950		SPRING, retaining, under motor blanket	4	—	—	—	—
		WO-A-11826		SPRING, retaining, under motor blanket, assembly	12	—	—	—	—
		WO-A-11793		KIT, field, body enclosure	as req.				
				(Consists of:					
		WO-A-11814		CARTON, body enclosure field kit	1	—	—	—	—
		WO-A-11794		CURTAIN, side, left, assembly	1	—	—	—	—
		WO-A-11795		CURTAIN, side, right, assembly	1	—	—	—	—
		WO-A-12200		INSTRUCTIONS, body enclosure field kit	1	—	—	—	—
		WO-A-11813		TOP assembly	1	—	—	—	—
		WO-A-7154		KIT, field, cold starting primer installation	as req.				
				(Consists of:					
		WO-A-7108		CARTON, cold starting primer installation field kit	1	—	—	—	—
		WO-A-6950		CLIP, S., 5⁄16 in.	1	—	—	—	—
		WO-A-7858		EQUIPMENT, cold starting primer (furnished by Ordnance Dept. for packaging by W. O. Co.)	1	—	—	—	—
				(Consists of:					
				1 WO-A-6879 ELBOW, primer (DV-BU-710)					
				1 WO-A-6878 ELBOW, primer, assembly (DV-PR-20-700)					
				1 WO-A-6880 NUT, primer tube, w/SLEEVE, assembly (DV-BU-010)					
				1 WO-A-6876 PRIMER assembly (DV-PR-20)					
				1 WO-A-6877 TEE, primer, assembly (DV-PR-1100))					
		WO-A-6954		TUBING, intake manifold tee to elbow (S., 1⁄8 in., 7 7⁄16 in., lgth.)	1	—	—	—	—
		WO-A-6955		TUBING, intake manifold tee to primer (S., 1⁄8 in., 7 3⁄16 in., lgth.)	1	—	—	—	—
		WO-A-6956		TUBING, primer to gas strainer (S., 3⁄16 in., 30 3⁄16 in., lgth.))	1	—	—	—	—
		WO-A-7155		KIT, field crankcase ventilation	as req.				
				(Consists of:					
		WO-A-6919		BODY, crankcase ventilator assembly	1	—	—	—	—

WO-A-7186		CARTON, crankcase ventilation field kit	1		
WO-53108		CLAMP, hose, ¾ in.	2		
WO-A-6918		CONNECTION, hose, flexible	1		
WO-384549		ELBOW, inverted flared tube, 5⁄16 in., ⅛ in. pipe thread	1		
WO-A-6885		ELBOW, inverted flared tube, 5⁄16 in., ¼ in. pipe thread	1		
WO-634811		GASKET, intake to exhaust manifold	1		
WO-630299		GASKET, ventilator to valve spring cover	1		
WO-A-6525		INDICATOR, oil filler cap and level, assembly	1		
WO-A-6969		INSTALLATION, crankcase ventilator	1		
WO-A-1166		MANIFOLD, intake, assembly	1		
WO-A-9008		PLUG, driving, oil filler tube (wood)	1		
WO-A-6922		TUBE, crankcase ventilator valve, assembly	1		
WO-A-6915		TUBE, oil, filler, assembly	1		
WO-A-6911		TUBE, air cleaner, w/BRACKET, assembly	1		
WO-A-6895		VALVE, crankcase ventilator, assembly	1		
WO-A-11846		KIT, field heater	as req.		
		(Consists of:			
WO-A-11844		CARTON, heater field kit	1		
WO-53308		CLAMP, heater tube coupling hose	8		
WO-A-11865		DRAWING, installation, heater field kit	1		
WO-A-8558		GASKET, cylinder head	1		
WO-A-11839		HEATER assembly	1		
WO-53031		BOLT, hex-hd., s-fin., alloy-S., ⅜-16NC-2 x ¾ (heater to dash)	1		
WO-53033		NUT, hex, reg., S., ⅜-16NC-2	2		
WO-50921		WASHER, S., plain, SAE, std., ⅜ in.	3		
WO-52046		WASHER, lock, reg, S., ⅜ in.	2		
WO-A-11838		HOSE, coupling, heater tube	4		
WO-A-11845		SWITCH, heater, assembly	1		
WO-A-11837		TUBE, inlet, heater	1		
WO-A-11836		TUBE, outlet, heater	1		
WO-A-11840		VALVE, shut off, heater, assembly	2		
WO-53114		CLAMP, hose, ⅞ in.	8		
WO-A-8210		CLAMP, stove stack	1		
WO-6159		BOLT, hex-hd., s-fin., alloy-S., ¼-20NC-2 x 1 (clamp to bracket and stack)	1		
WO-5790	23-3	NUT, hex., reg., S., ¼-20NC-2 (clamp to bracket and stack bolt)	1		
WO-52706		WASHER, lock, SAE, std., ¼ in. (clamp to bracket and stack bolt)	1		
WO-A-7058		COCK, shut off (WH-775)	1		
WO-387249		CONNECTOR, inverted flared tube, 5⁄16 in.	1		
WO-A-7043		CONNECTOR, tee-block (DV-X-697)	1		
WO-A-7001		ELBOW, street, std., 45°, ¼ in.	1		
WO-52805	23-3	ELBOW, street, std., 90°, ¼ in.	1		
WO-A-6999		FITTING, hose, special, ¼ in. pipe to ½ in. hose	1		
WO-639650		GASKET, water outlet elbow	1		
WO-662420		GROMMET, 3⁄16 in., (rubber, split)	1		
WO-A-5080	23-3	WIRE, heater switch to ignition switch, assembly	1		
WO-A-7156		STOVE, winterization field kit	as req.		
		(Consists of:			
WO-A-8206		BRACKET, mounting, stove to frame	1		
WO-53242		BOLT, carriage, S., ⅜-16NC-2 x 2½ (bracket to frame)	1		

2310

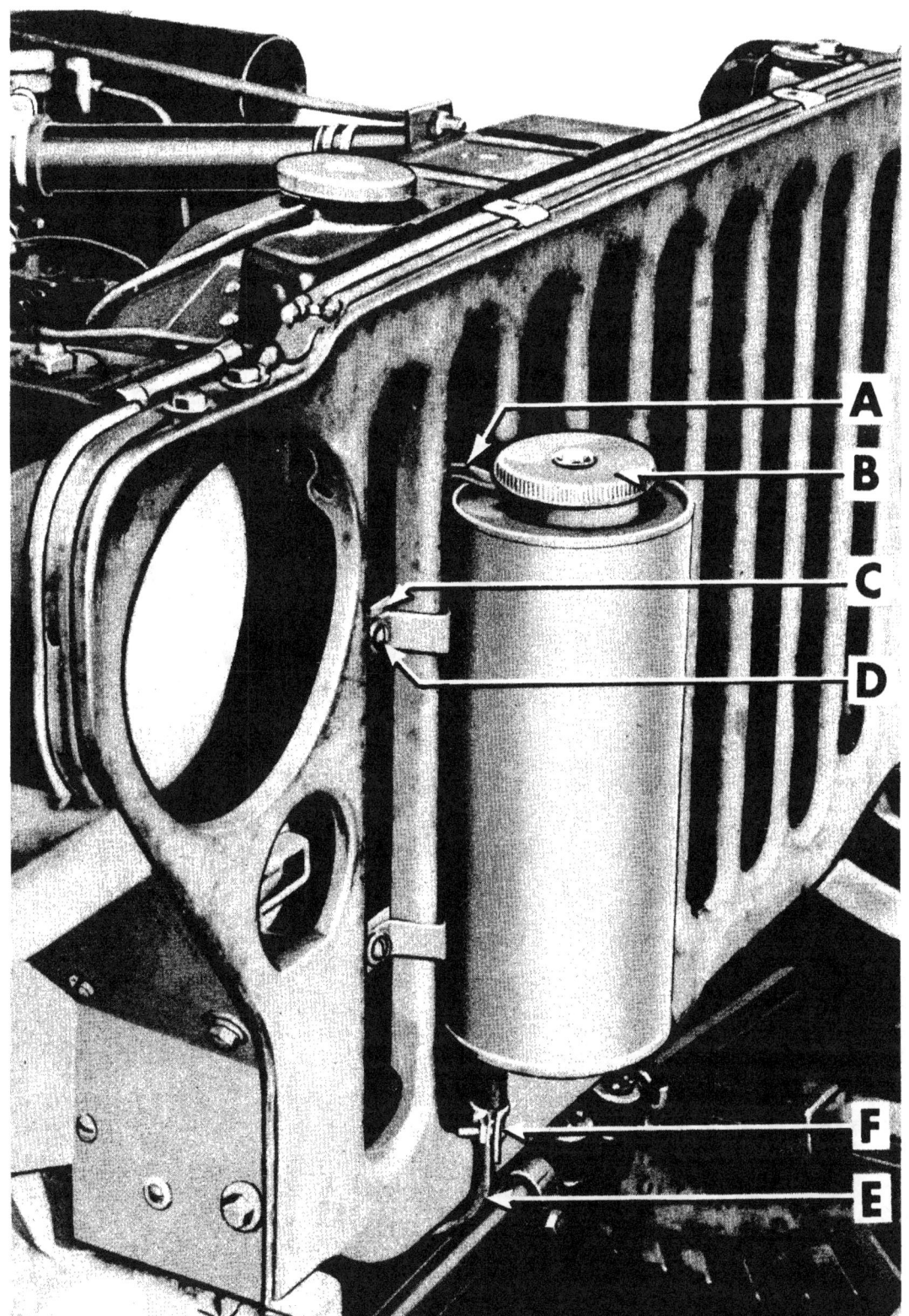

RA PD 305172

FIGURE 23-4—DESERT COOLING KIT

Key	Item	Willys Part No.	Ford Part No.	Gov't Group No.
A	TUBE	WO-A-6945		2309
B	TANK	WO-A-6933		2309
C	STRAP	WO-A-6944		2309
D	SCREW	WO-52468		2309
E	HOSE	WO-A-6946		2309
F	CLAMP	WO-A-6947		2309

RA PD 305172-A

SNL G-503

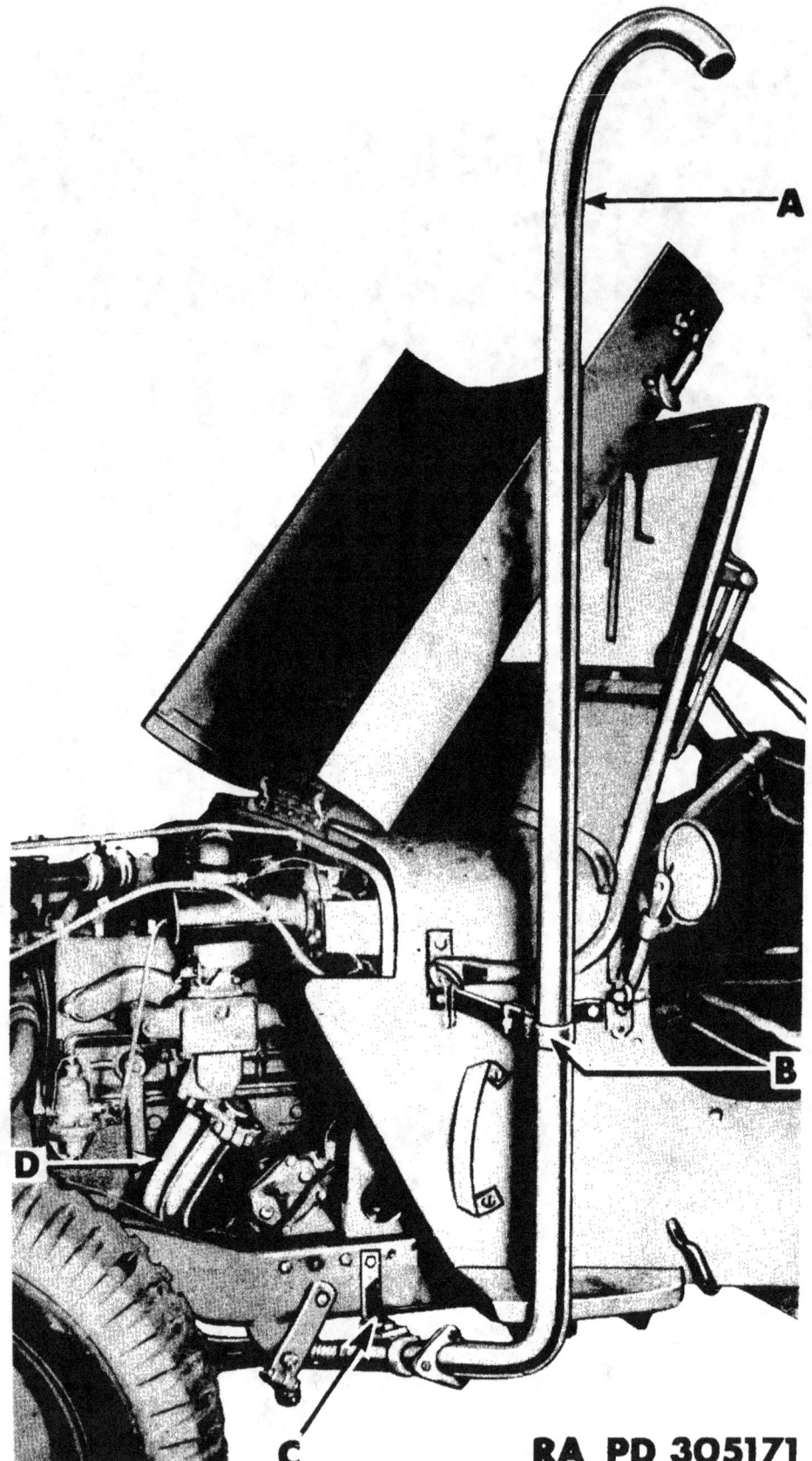

FIGURE 23-5—DEEP WATER EXHAUST

Key	Item	Willys Part No.	Ford Part No.	Gov't Group No.
A	EXTENSION	WO-A-7816		2309
B	BRACKET	WO-A-8196		2309
C	BRACKET	WO-A-8197		2309
D	PIPE	WO-A-7814		2309

2310-2310A

GROUP 23—GENERAL USE STANDARDIZED PARTS (Cont'd)

Figure Number	Official Stockage Number	Part Number Willys	Part Number Ford	ITEM	Quantity Reqd. per Unit Assy.	Per 100 Major Items 12 Mos. Field Maintenance	Per 100 Major Items Major Overhaul (5th Ech)	Total First Year Procurement	Estimated Reqmts. per 100 Rebuilds
Col. 1	Col. 2	Col. 3	Col. 4	Col. 5	Col. 6	Col. 7	Col. 8	Col. 9	Col. 10
				2310—WINTERIZING FIELD KIT (Cont'd)					
		WO-50051		BOLT, hex-hd., s-fin, alloy-S, 3/8-24NF-2 x 2½ (brkt. to frame)	1	—	—	—	—
		WO-5544		NUT, hex, reg., S., 3/8-16NC-2 (bracket to frame bolt)	1	—	—	—	—
		WO-5901		NUT, hex, reg., S., 3/8-24NF-2 (bracket to frame bolt)	1	—	—	—	—
		WO-5010		WASHER, lock, SAE, std., 3/8 in. (bracket to frame bolt)	2	—	—	—	—
23-3		WO-A-8208		BRACKET, stove stack, front	1	—	—	—	—
		WO-A-8209		BRACKET, stove stack, rear	1	—	—	—	—
		WO-A-7059		BUSHING, reducing, pipe 3/8 x 1/4	1	—	—	—	—
		WO-A-7137		CAP, fuel tank	1	—	—	—	—
		WO-A-7187		CARTON, perfection stove field kit	1	—	—	—	—
		WO-A-8212		HEATER, Superfex engine (furnished by Ordnance Dept. for packaging by W. O. Co.)	1	—	—	—	—
		WO-6412		BOLT, hex-hd., s-fin, alloy-S., 3/8-16NC-2 x 3/4 (to bracket)	4	—	—	—	—
		WO-5010		WASHER, lock, reg., S., 3/8 in.	4	—	—	—	—
23-3		WO-A-7036		HOSE, stove, long	2	—	—	—	—
		WO-A-7037		HOSE, stove, short	2	—	—	—	—
		WO-A-8213		INSTRUCTION SET, cold starting stove	1	—	—	—	—
23-3		WO-A-8214		LINE, fuel, tank to stove, 5/16 x 16 3/8	1	—	—	—	—
		WO-A-7040		NIPPLE, hose, 1/4 in. pipe	3	—	—	—	—
		WO-A-7581		PLATE, instruction, cold starting stove	1	—	—	—	—
		WO-53013		NUT, stamped, cd-pltd, S., type C, No. 6 (.138)-3NC-2	4	—	—	—	—
		WO-53014		SCREW, rd-hd, machine, cd-pltd., S., No. 6 (.138)-32NC-2 x 5/16 (plate to windshield)	4	—	—	—	—
23-3		WO-A-8219		SPACER, (steel tubing 5/8 O.D. x 1 3/8 lgth. x 3/32 thk.)	1	—	—	—	—
23-3		WO-A-8211		STACK, Perfection stove	1	—	—	—	—
		WO-A-8097		TANK, fuel	1	—	—	—	—
		WO-52132		BOLT, hex-hd., s-fin, alloy-S., 5/16-24NF-2 x 5/8 (fuel tank to cowl)	3	—	—	—	—
		WO-5910		NUT, hex, reg., S., 5/16-24NF-2 (tank to cowl bolt)	3	—	—	—	—
		WO-5437		WASHER, plain, S., 5/16 in. (tank to cowl bolt)	3	—	—	—	—
		WO-51833		WASHER, lock, reg., S., 5/16 in. (tank to cowl bolt)	3	—	—	—	—
23-3		WO-A-7000		TEE, 1/4 in., std.	1	—	—	—	—
		WO-A-8224		THERMOSTAT, 180°	1	—	—	—	—
		WO-A-8207		TUBE, stove	2	—	—	—	—
				2310A—DESERT COOLING KIT					
		WO-A-6940		KIT, desert cooling	as req.	—	—	—	—
				(Composed of:					
		WO-A-6942		1 CAP, radiator filler, assembly	—	—	—	—	—

SNL G-503

WO-A-6949	1																		1 CARTON, desert cooling kit
WO-A-6947	1	1	1	1	1	1	1	1	1	1	1	1	1	1	1	1	1	1	4 CLAMP, hose, ⅝ in.
WO-A-6941	1																		1 CORE, radiator, and SHROUD, assembly
WO-A-6943	1																		1 DRAWING, installation, desert cooling kit
WO-A-6946	1	1																	2 HOSE, radiator surge tank tube
WO-A-6944																			2 STRAP, mounting radiator surge tank
																			4 WO-52161 NUT, hex, S, ¼-28NF-2
																			4 WO-52468 SCREW, rd-hd., machine, S., ¼-28NF-2 x ⅝ (strap to guard)
																			4 WO-53058 WASHER, lock, S., ¼ in.
WO-A-6933	1																		1 TANK, surge, radiator
WO-A-6945	1																		1 TUBE, radiator surge tank
																			2310E—DEEP WATER EXHAUST KIT
WO-A-7726	as req.																		KIT, deep water exhaust pipe
																			(Composed of:
WO-A-8204	1																		1 BAG (for standard parts)
WO-A-8197	1																		1 BRACKET, support, lower, assembly
																			(Includes:
																			1 WO-A-7986 BRACKET, support, lower
																			1 WO-A-658 INSULATOR, muffler support
																			2 WO-638058 PLATE, muffler support insulator
																			1 WO-50929 BOLT, hex-hd., s-fin., alloy-S., ⁵⁄₁₆-24NF-2 x 1 (plate and insulator to strap)
																			1 WO-5934 BOLT, hex-hd., S., ⁵⁄₁₆-24NF-2 x 1 (plate and insulator to bracket)
																			2 WO-5910 NUT, hex, reg., S., ⁵⁄₁₆-24NF-2
																			2 WO-51833 WASHER, lock, reg., S., ⁵⁄₁₆ in.
																			1 WO-A-657 STRAP, support, muffler)
WO-A-8196	1																		1 BRACKET, support, upper, assembly
																			(Includes:
																			1 WO-A-7988 BRACKET, support, upper
																			1 WO-A-7987 CLAMP, pipe
																			1 WO-50922 BOLT, hex-hd., s-fin., alloy-S., ⁵⁄₁₆-24NF-2 x ⅞ (clamp to plate and insulator)
																			1 WO-5910 NUT, hex, reg., S., ⁵⁄₁₆-24NF-2
																			1 WO-51833 WASHER, lock, reg., S., ⁵⁄₁₆ in.)
WO-A-8203	1																		1 CARTON (accommodates 5 kits)
WO-A-7816	1																		1 EXTENSION, exhaust pipe
																			3 WO-33050 BOLT, hex-hd., s-fin., alloy-S., ⁵⁄₁₆-24NF-2 x 1 (extension to pipe)
																			4 WO-50802 NUT, hex, S., ⁵⁄₁₆-24NF-2
																			4 WO-83047 WASHER, lock, S., ⁵⁄₁₆ in.
WO-634814	1																		1 GASKET, exhaust pipe
WO-A-7967	1																		1 GASKET, exhaust pipe flange
WO-A-7813	1																		1 INSTRUCTIONS, deep water exhaust pipe kit
WO-A-7814	1																		1 PIPE, exhaust, assembly

GROUP 23—GENERAL USE STANDARDIZED PARTS (Cont'd)

Figure Number	Official Stockage Number	Part Number		ITEM	Quantity Reqd. per Unit Assy.	Per 100 Major Items			Estimated Reqmts. per 100 Rebuilds
		Willys	Ford			12 Mos. Field Maintenance	Major Overhaul (5th Ech)	Total First Year Procurement	
Col. 1	Col. 2	Col. 3	Col. 4	Col. 5	Col. 6	Col. 7	Col. 8	Col. 9	Col. 10
				2310E—DEEP WATER EXHAUST KIT (Cont'd)					
				(Includes:					
		WO-A-693	FM-GPW-17097	1 BOLT, hex-hd., S., ⅜-16NC-2 x 2¼ (flange bolt)	3	—	—	—	—
		WO-6989	FM-34054-S	1 CLAMP	6	—	—	—	—
		WO-53267	FM-26498-S2	1 BOLT, hex-hd., S., 5⁄16-24NF-2 x 1⅛ (clamp to pipe)	6	—	—	—	—
		WO-51969	FM-34804-S2	1 NUT, hex, S., reg., 5⁄16-24NF-2	6	—	—	—	—
		WO-A-616	FM-GPW-17100	1 WASHER, plain, SAE. std., 5⁄16 in.	1	—	—	—	—
		WO-6214	FM-34052-S2	1 WASHER, lock, reg., S., 5⁄16 in.	6	—	—	—	—
		WO-6287	FM-26480-S2		6	—	—	—	—
		WO-52705	FM-34802-S2		6	—	—	—	—
				GROUP 24—FIRE EXTINGUISHER					
				2402—PORTABLE SYSTEMS AND PARTS					
23-1		WO-A-8114		BRACKET, mounting, fire extinguisher	1	—	—	—	—
		WO-51514		NUT, hex, S., No. 12 (.216)-24NC-2	4				
				SCREW, rd-hd., S., No. 12 (.216)-24NC-2 (brkt. to body)	4				
		WO-5790		WASHER, lock, S., No. 12 (.216)	4				
				EXTINGUISHER, fire, w/HOLDER (FYR-D-10-A)	1				
				NUT, hex, S., No. 8 (.164)-32NC-2					
				SCREW, rd-hd., S., No. 8 (.164)-32NC-2 x ½ (holder to brkt.)					
				WASHER, lock, S., No. 8 (.164)					
				GROUP 26—RADIO					
				2601—RADIO INSTALLATION PARTS					
			FM-GPA-18861-B	BOX, radio terminal, assembly	1	—	—	—	—
				BOLT, hex-hd., s-fin., alloy-S., ¼-28NF-2 x ½ (box to body panel) (Willys only)	4				
			FM-33795-S7	NUT, hex, reg., S., ¼-28NF-2 (Willys only)	4				
			FM-36813-S7	NUT, hex, reg., S., ¼-20NC-2 (Ford only)	4				
		WO-A-7715		SCREW, rd-hd., machine, ¼-20NC-2 x 7⁄16 (Ford only)	4				
		WO-5121		WASHER, S., flat, ¼ in. (Willys only)	4				
		WO-53024	FM-351274-S7	WASHER, S., plain, ¼ in. (Willys only)	4				
		WO-A-7640		WASHER, lock, internal, external, cd-pltd., S., ¼ in. (GM-174916)	4	—	—	—	—
06-5		WO-52543	FM-GPW-14513	CABLE, ground, w/TERMINALS, assembly	1	—	—	—	—
				BOLT, hex-hd., S., cd-pltd., ⅜-24NF-2 x ¾ (Cable to frame) (Willys only)	1	—	—	—	—

SNL G-503

WO-52542	FM-20346-S7	BOLT, hex-hd., S., cd-pltd., 5/16-18NC-2 x 3/4 (cable to frame) (Ford only)	1	—	—	—	—
	33797-S7	NUT, hex, S., cd-pltd., 3/8-24NF-2 (Willys only)	1	—	—	—	—
	356309-S7	NUT, hex, S., cd-pltd., 5/16-18NC-2 (Ford only)	1	—	—	—	—
WO-53135		WASHER, lock, S., cd-pltd., internal, external tooth, 5/16 in. (Ford only)	3	—	—	—	—
WO-53036		WASHER, lock, S., cd-pltd., internal, external tooth, 3/8 in. (Willys only)	1	—	—	—	—
WO-A-8113	FM-GPW-14480-B	WASHER, lock, S., pkzd, internal, external tooth, 3/8 in. (Willys only)	2	—	—	—	—
WO-A-6809	FM-GPW-14464	CABLE, starting switch to radio box, assembly	1	—	—	—	—
		CLIP, closed, S., 1/2 in., bolt hole 13/64 in. (Willys only)	3	—	—	—	—
		(1 used cable to speedometer clip bolt; 2 used cable to frame side member)					
WO-A-5450		CLIP, closed, S., 3/8 in. (Ford only)	3	—	—	—	—
		(1 used cable to speedometer clip bolt; 2 used cable to frame side member)					
WO-6352	FM-GPW-14481-B	CLIP, closed, S., 7/16 in. (cable to frame side member) (Willys only)	1	—	—	—	—
WO-5113	FM-GPW-14521	NUT, hex, S., No. 10 (.190)-24NC-2 (Willys only)	1	—	—	—	—
WO-52221	FM-GPW-14532	SCREW, rd-hd., machine, S., No. 10 (.190)-24NC-2 x 1/2 (Willys only)	1	—	—	—	—
	FM-GPW-18841	WASHER, lock, sherardized, No. 10 (.190) (Willys only)	1	—	—	—	—
WO-A-8116		CONDUIT, 1/2 in., (thin wall)	1	—	—	—	—
WO-A-7180	FM-GPW-14521	ELBOW, mall-I, z-pltd., 1/2-14 pipe thd., 15/16-18 U.S.S. thd. (AE-9212)	2	—	—	—	—
WO-A-7182	FM-GPW-14532	NUT, hex, thin, S., 1/2 in. pipe thd. (TB-141)	2	—	—	—	—
WO-A-7600	FM-GPW-18841	FILTER, radio terminal box (.5 mfd.) (DT-CHC-201) (SPR-JX-131) (MLL-A-205244) (SOL-EV-128)	1	% 2	2	6	6
WO-A-7181	FM-GPW-14536	GLAND, conduit	2	—	—	—	—
WO-A-7636	FM-GPW-14531	GLAND, radio cable conduit nipple	1	—	—	—	—
WO-7718		LOOM, split, starting switch cable, 7/16 x 1 1/2 in. NIPPLE, radio cable conduit, S., z-pltd., internal thds. 1/2-14 pipe, external thds. 13/32-18 (AE-9821) (Willys only)	3	—	—	—	—
WO-53054	FM-34914-S7	WASHER, flat, S., cd-pltd., internal tooth 7/8 in. (GM-138566)	2	—	—	—	—
WO-A-7947	FM-GPW-14537	RING, sealing, gland to conduit	2	—	—	—	—
WO-A-7638	FM-GPW-14534	SEAL, gland to radio cable conduit	1	—	—	—	—
WC-A-5038	FM-GPW-18876	STRAP, bond, radio box negative terminal to floor ground connection	1	—	—	—	—
WO-53131	FM-20407-S7	BOLT, hex-hd., S., cd-pltd., 5/16-24NF-2 x 1 1/8 (strap to floor) (GM-213499)	1	—	—	—	—
WO-A-1532	FM-33798-S7	NUT, hex, S., cd-pltd., 5/16-24NF-2	2	—	—	—	—
WO-52725	FM-34706-S7	WASHER, S., flat, cd-pltd., 21/64 I.D. x 9/16 O.D. x 1/32 thk. (GM-138485)	1	—	—	—	—
WO-53025	FM-356309-S7	WASHER, lock, S., internal, external tooth cd-pltd., 5/16 in. (GM-178532) (Willys)	3	% 10	10	100	—
WO-A-7466	FM-91BS-14463	TERMINAL, S., 21/64 in. hole (Willys)	4	—	—	—	—
	FM-GPW-18877-B	TERMINAL, S., 21/64 in. hole (Ford only)	2	—	—	—	—
WO-A-7645	FM-GPW-14320	TERMINAL, S., 13/32 in. hole	1	—	—	—	—
WO-371400	FM-11A-14452	TERMINAL, S., 11/32 in. hole, (battery cable)	1	—	—	—	—
WO-53132	FM-33823-S	NUT, hex, br., 1/4-20NC-2 (negative terminal post (GM-114379)	3	—	—	—	—
WO-52667	FM-34805-S7	WASHER, lock, light, S., cd-pltd., 1/4 in. (GM-174916)	3	—	—	—	—

2603—RADIO SUPPRESSION SYSTEM PARTS
(Before Willys serial 289,001)

WO-A-5919		BOLT, hex-hd., cd-pltd., S., 3/8-24NF-2 x 1 (bond No. 10 to engine plate) (GM 100026)	1

GROUP 26—RADIO (Cont'd)

2603

Figure Number	Official Stockage Number	Part Number - Willys	Part Number - Ford	ITEM	Quantity Reqd. per Unit Assy.	Per 100 Major Items - 12 Mos. Field Maintenance	Per 100 Major Items - Major Overhaul (5th Ech)	Total First Year Procurement	Estimated Reqmts. per 100 Rebuilds
Col. 1	Col. 2	Col. 3	Col. 4	Col. 5	Col. 6	Col. 7	Col. 8	Col. 9	Col. 10
				2603—RADIO SUPPRESSION SYSTEM PARTS (Cont'd) (Before Willys Serial 289,001)					
		WO-A-1648	FM-20311-S7	BOLT, hex-hd., cd-pltd., S., 5/16-24NF-3 x 1/2............ (1 used Bond No. 1 to hood) 1 used Bond No. 2 to hood)	2	% 10	—	50	50
		WO-A-1694	FM-GPW-14561	CLIP, closed, S., tinned, 1/4, 7/32 bolt hole (1 used Bond No. 4 to speedometer 4 used Bond No. 5 to gasoline (Willys) 1 used Bond No. 23 (Willys) 1 used blackout driving light cable (Ford) 1 used head light wiring harness (Ford))	6	—	—	—	—
		WO-A-1693	FM-GPW-146211	CLIP, closed, S., tinned, 3/16, 13/64 bolt hole (Willys)....... (1 used Bond No. 4 to heat indicator cable 1 used Bond No. 6 to choke control (Ford) 1 used Bond No. 6 to throttle control 1 used Bond No. 6 to oil gauge line)	2	—	—	—	—
		WO-A-5980	FM-GPW-18960-B	FILTER GROUP, w/COVER, assembly (SOL type Ev. 103, DTC 1107DE, SPR type JX-17, MLL 134739) (after Willys serial 137916).......	4	—	—	—	—
		WO-A-1517	FM-GPW-18936-A	FILTER GROUP, w/COVER, assembly (before Willys serial 137916) (ER type L-7)........	1	% 3	3	9	9
		WO-A-1287	FM-GPW-18935-A	FILTER-UNIT, generator regulator (mounts on regulator) (issue until stock exhausted)............	1	% 4	1	—	—
		WO-A-5337	FM-GPW-112925	FILTER-UNIT, generator to ground to armature circuit (after Willys serial	1	% 5	5	12	12
		WO-662276	FM-GPW-13437-A2	GROMMET, 5/8 in. (harness to filter group cover) (issue until stock exhausted).	1	% 4	8	—	—
		WO-51396	FM-24347-S2	BOLT, hex-hd., s-fin., alloy-S., 5/16-24NF-2 x 3/4 (filter to dash).........	4	—	—	—	—
		WO-5910	FM-33798-S7	NUT, hex-hd., S., 5/16-24NF-2.................	4	—	—	—	—
		WO-51833	FM-34806-S2	WASHER, lock, S., 5/16 in...........	4	—	—	—	—
		WO-A-1550	FM-351025-S7	NUT, hex, S., cd-pltd., 7/16-20NF-2 (Bond No. 3 to cylinder head stud).......	2	% 10	10	100	—
		WO-A-1532	FM-33798-S7	NUT, hex., S., cd-pltd., 5/16-24NF-2................ (1 used Bond No. 3 and 4 to dash stud 1 used Bond No. 6 to dash stud 2 used Bond No. 13 to frame front cross tube stud 2 used Bond No. 14 to frame front cross tube stud 2 used Bond No. 15 to frame front cross member stud 2 used Bond No. 16 to body floor pan stud)	10	—	—	—	—

SNL G-503

WO-A-1546	FM-34084-S	NUT, hex., S., cd-pltd., 7/16-20NF (issue until stock exhausted) (2 used Bond No. 13 radiator to support brkt. and bond strap to stud, right	4			52	
WO-A-1701	FM-355835-S	NUT, hex., S., cd-pltd., No. 10 (.190)-24NC-2 (for SCREW WO-A-1700)	8				
WO-A-1700	FM-355130-S7	SCREW, rd-hd., S., cd-pltd., No. 10 (.190)-24NC-2 × ½ (5 used clip to bond strap)	8				100
WO-53220		3 used head lamp harness clip to fender splasher	6				
WO-53252		SCREW, rd-hd., S., cd-pltd., No. 10 (.190)-32NF-2 × 3/8, terminal screw (for use on SOL, SPR and MLL filter group only)	6				
WO-A-5038	FM-GPW-18876	SCREW, rd-hd., S., cd-pltd., w/WASHER, No. 10 (.190)-32NF-2 × 3/8 (for use with DTC filter group only)					
		STRAP, bond No. 1 (cowl to hood, right, bond No. 2, cowl to hood, left) (issue until stock exhausted)	2	% 16		32	
WO-A-5033	FM-GPW-18858	STRAP, bond No. 3 (cylinder head to dash)	1	% 3			6
WO-A-5035	FM-GPW-18849	STRAP, bond No. 4 (hand brake cable, speedometer cable, heat indicator cable to dash)	1	% 3		16	
WO-A-5039	FM-GPW-18853	STRAP, bond No. 5 (gas line to dash) (issue until stock exhausted)	1	% 8			6
WO-A-5040	FM-GPW-18850	STRAP, bond No. 6 (choke and throttle controls and oil gauge to dash)	1	% 3		16	
WO-A-5037	FM-GPW-18857	STRAP, bond No. 7 (generator mounting bolt to starter motor brkt.) (issue until stock exhausted)					
WO-A-5041	FM-GPW-18846	STRAP, bond No. 8 (generator voltage regulator ground) (issue until stock exhausted)	1	% 8		16	
WO-A-5034	FM-GPW-18873	STRAP, bond No. 9 (ignition coil to cylinder block) (issue until stock exhausted)	1	% 12		16	
WO-A-1098	FM-GPW-14303	STRAP, bond No. 10 (engine front plate to frame right. This is regular engine ground strap) (before serial 288835) (issue until stock exhausted)	1	% 8		16	
WO-A-5036	FM-GPW-18874	STRAP, bond No. 11 (engine front plate to frame left) (issue until stock exhausted)	1			4	
WO-A-5027	FM-GPW-18874	STRAP, bond No. 12 (exhaust pipe ground connection to frame) (issue until stock exhausted)	1	% 8		16	
WO-A-5032	FM-GPW-18859	STRAP, bond No. 13 (radiator stud, right, to frame, and STRAP, bond No. 14 (radiator stud, left, to frame) (issue until stock exhausted)	1	% 8		16	
WO-A-5031	FM-GPW-18872	STRAP, bond No. 15 (rear engine support to frame cross member stud) (issue until stock exhausted)	2	% 16		32	
WO-A-5030	FM-GPW-18871	STRAP, bond No. 16 (transfer case to body floor stud)	1	% 8		16	
WO-A-1699	FM-GPW-18852	STRAP, bond No. 21 (hood to brush guard, right, and STRAP, bond No. 22 (hood to brush guard, left) (issue until stock exhausted)	1	% 3			6
WO-A-6320	FM-GPW-18812-B	SUPPRESSOR, ignition	2	% 16		32	
WO-A-1680		WASHER, S., cd-pltd. (plain) 3/8 in. (used with STRAPS No. 10, 11, 12) (issue until stock exhausted)	4	% 32		8	48
WO-5455		WASHER, S., cd-pltd. (plain) 3/8 in. (used with STRAPS No. 10, 11, 12)	5	% 32		68	
WO-A-1702	FM-34703-S7	WASHER, S., cd-pltd., No. 10 (.190) (harness clip to fender splasher) (issue until stock exhausted)	3				
WO-A-1680	FM-34707-S7	WASHER, S., cd-pltd., 3/8 in. (issue until stock exhausted)	5				
WO-A-1547	FM-34708-S	WASHER, S. (plain) (used with STRAPS No. 13 and 14) (issue until stock exhausted)	2	% 32		68	
WO-51833	FM-34806-S2	WASHER, lock, S., cd-pltd., 5/16 in.	2	% 16		24	
WO-52859		WASHER, lock, S., cd-pltd., internal tooth No. 10 (.190) use with terminal screw on SOL, SPR and MLL filter group units only)	4	%			

GROUP 26—RADIO (Cont'd)

2603

Figure Number	Official Stockage Number	Part Number Willys	Part Number Ford	ITEM	Quantity Reqd. per Unit Assy.	12 Mos. Field Maintenance	Major Overhaul (5th Ech)	Total First Year Procurement	Estimated Reqmts. per 100 Rebuilds
Col. 1	Col. 2	Col. 3	Col. 4	Col. 5	Col. 6	Col. 7	Col. 8	Col. 9	Col. 10
				2603—RADIO SUPPRESSION (After Willys Serial 288,835)					
		WO-A-1694	FM-GPW-14561	CLIP, closed, S., pltd, ¼ in., bolt hole 7/32 in. (bond No. 4)	1	---	---	---	---
		WO-A-1693	FM-GPW-14621	CLIP, closed, S., pltd, 3/16 in., bolt hole 13/64 in., (bond No. 4)	1	---	---	---	---
		WO-A-1701	FM-355836-S7	NUT, hex, S., No. 10 (.190)-24NC-2 (bond No. 4)	2	---	---	---	---
		WO-A-1700	FM-355130-S7	SCREW, rd-hd., S., pltd, No. 10 (.190)-24NC-2 x ½	2	---	---	---	---
		WO-A-1702	FM-34703-S7	WASHER, plain, S., pltd, No. 10 (.190) (¼ I.D. x 9/16 O.D.)	2	---	---	---	---
		WO-52221	FM-34803-S	WASHER, lock, S., std., No. 10 (.190)	1	---	---	---	---
		WO-A-7848	FM-GPW-18938-C	FILTER, 0.1 MFD. (mounted on coil stud)	1	---	---	---	---
		WO-A-8884	FM-GPW-18937	FILTER, 0.01 MFD. (mounted to ignition switch)	1	---	---	---	---
		WO-6536	FM-34079-S7	NUT, hex, S., No. 10 (.190)-32NF-2	1	---	---	---	---
		WO-52250	FM-27068-S7	SCREW, rd-hd., S., No. 10 (.190)-32NF-2 x 3/8	1	---	---	---	---
		WO-53023	FM-356256-S7	WASHER, lock, S., cd-pltd., internal, external teeth, No. 10 (.190)	2	---	---	---	---
		WO-A-7849	FM-GPA-18841	FILTER, 0.5 MFD. (mounted on starter motor)	1	---	---	---	---
		WO-A-8010	FM-GPW-18938-B-1-2	FILTER, 0.25 MFD. (voltage regulator) (mounted to regulator)	1	2	2	---	---
		WO-A-8883	FM-GPW-18937-B	FILTER UNIT, 0.01 MFD., generator regulator field (mounted to regulator)	1	2	2	6	6
		WO-A-5337	FM-GPW-18935-A-1-2-3	FILTER UNIT, generator ground to armature circuit	1	---	---	---	---
		WO-52164	FM-33802-S7	NUT, hex, S., pltd., 7/16-20NF-2 (radiator to support brkt.)	2	---	---	---	---
		WO-52815	FM-356028-S7	NUT, hex, S., stamped, 7/16-20NF-2 (radiator to support brkt.)	2	---	---	---	---
		WO-A-11766	FM-GPW-18876-B	STRAP, bond No. 1 and 2, hood to cowl	1	---	---	---	---
		WO-A-5033		STRAP, bond No. 3, cylinder head to dash	1	---	---	---	---
		WO-A-5035	FM-GPW-18449	STRAP, bond No. 4, hand brake, heat indicator and speedometer cables to dash	1	---	---	---	---
		WO-A-7826	FM-GPW-18840	STRAP, bond No. 7, generator mounting bolt to starter motor bracket	1	---	---	---	---
		WO-A-5036	FM-GPW-18874	STRAP, bond No. 11, engine front plate to frame, left	1	---	---	---	---
		WO-A-1702	FM-34703-S7	WASHER, plain, S., pltd, No. 10 (.190) (clip to heat indicator and speedometer cable strap)	2	---	---	---	---
		WO-52221	FM-34803-S	WASHER, lock, S., std, No. 10 (.190) (clip to heat indicator and speedometer cable strap)	2	---	---	---	---
		WO-53303	FM-34943-S7	WASHER, lock, S., cd-pltd., external, 7/16 in. (radiator to support bracket)	2	---	---	---	---
		WO-53027	FM-356314-S7	WASHER, lock, S., cd-pltd., internal, external, 7/16 in. (radiator to support bracket)	2	---	---	---	---

SNL G-503

MAINTENANCE TOOLS FOR ORDNANCE PERSONNEL

The tools required for Ordnance organizations for the maintenance of the Truck, ¼-Ton, 4 x 4, Command Reconnaissance, are listed in SNL G-27, Volume 1, "Tools, Maintenance, for Repair of Automotive Vehicles," and are issued as required.

MAINTENANCE PARTS FOR ARMAMENT SIGHTING AND FIRE CONTROL EQUIPMENT

For Maintenance Parts see ADDENDUM for the following SNL's:

GUN, Machine, Cal. .50, Browning, M2, Heavy Barrel, Flexible	SNL A-39
or	
GUN, Machine, Cal. .30, Browning, M1919A4, Flexible	SNL A-6
or	
RIFLE, Automatic, Cal. .30, Browning, M1918A2	SNL A-4
MOUNT, Machine Gun, Cal. .30, M48	SNL A-55, Sec. 32
or	
MOUNT, Truck, Pedestal, M31	SNL A-55, Sec. 18

SECTION 3

SECTION 3

VEHICULAR SPARE PARTS AND EQUIPMENT LIST

TABLE OF CONTENTS

Vehicular Spare Parts

Vehicular Accessories

Armament

Armament Spare Parts

Armament Accessories

Major Items and Reference Notes

SNL G-503

VEHICULAR SPARE PARTS AND EQUIPMENT LIST

Figure Number	Official Stockage Number	Ordnance Drawing Number	Mfg.'s Number	Item	Quan. Per Vehicle	Note Symbol
Col. 1	Col. 2	Col. 3	Col. 4	Col. 5	Col. 6	Col. 7
				VEHICULAR SPARE PARTS		
1-A			FM-GP-17008 / WO-A7686	BAG, spare parts (Fed. St. No. 8-B-11)	1	
			FM-GPW-8620 / WO-A1495	BELT, fan, generator and water pump (DAY-No. V-5) (Fed. St. No. 33-B-76)	1	
1-J			FM-GPW-1720 / WO-A5986	CAPS, tire valve (screwdriver type) (boxed five) (Fed. St. No. 8-C-650)	1	
1-H			FM-B1724 / WO-A5491	CORES, tire valve (boxed five) (Fed. St. No. 8-C-6750)	1	
1-F			FM-B13466 / WO-51804	LAMP, elect., incand., 6-8-v., sgle.-tung-fil., No. 63, 3 cp., (D C cand., bay. base, G 6 bulb, C-2R fil.) (Fed. St. No. 17-L-5215)	1	
1-D			FM-GPW-13494A / WO-A1078	LAMP UNIT, blackout stop, sealed, one opening, 6-8-v., 3 cp., assembly (CB-9234) (GL-5933121) (Fed. St. No. 8-L-421)	1	
1-B		C84908K / C84923K / C91706C	FM-GPW-13491A / WO-A1075	LAMP-UNIT, blackout tail, sealed, four opening, 6-8-v., 3 cp., assembly (CB-9225) (GL-5933078) (Fed. St. No. 8-L-415)	1	
1-C		C84908J	FM-GPW-13485A / WO-A1074	LAMP-UNIT, service tail and stop, sealed, 6-8-v., 21-3 cp., assembly (CB-9218) (GL-5933104) (KD-8039) (Fed. St. No. 8-L-419)	1	
1-L	42-P-5347		FM-GPW-18318 / WO-A7683	PIN, cotter, split, S., (type B) (assorted in small box)	1	
				(Consists of:		
				20 PIN, cotter, split, S., (type B), 1/16 x 1½ (Fed. St. No. 42-P-5470)		
				25 PIN, cotter, split, S., (type B), 3/32 x 2 (Fed. St. No. 42-P-5580)		
				20 PIN, cotter, split, S., (type B), 1/8 x 2 (Fed. St. No. 42-P-5690)		
				5 PIN, cotter, split, S., (type B), 5/32 x 2½		
				5 PIN, cotter, split, S., (type B), 3/16 x 2½ (Fed. St. No. 42-P-5890))		
1-G		BFAX1BS / BFAX1CT / BFAX1DT / BFAX2AD	FM-GPW-12405 / WO-A538	PLUG, spark, w/GASKET (Fed. St. No. 17-P-5365)	1	
			FM-GPW-1509	TIRE, 6.00 x 16, mud and snow (Fed. St. No. 8-T-5964)	1	
			FM-GPW-1655	TUBE, 6.00 x 16, heavy duty	1	
			FM-GPW-1015	WHEEL, 16 x 400, assembly	1	
				(Consists of:		
			FM-GPW-1029	8 BOLT, divided rim		
			FM-GPW-1030	8 NUT		
			FM-GPW-1016	1 WHEEL, combat, inner half		
			FM-GPW-1027	1 WHEEL, combat, outer half)		
				VEHICULAR ACCESSORIES		
2-AD			FM-GP-17126 / WO-A11765	ADAPTER, lubricating gun	1	M-3 W
2-J	41-B-15		FM-GPW-17005 / WO-A372	BAG, tool	1	M-8 W
				BOOK, Standard Nomenclature List No. G503	Note 1	

SNL G-503

2-D	8-C-2358	TM-9-803	FM-GPW-18136B	BOOK, Technical Manual, 9-803	Note 1	W
			WO-A7687	CHAINS, tire, single pneumatic, 6.00 x 16 (type TS) (4 link spacing)	4	W
			FM-GPW-1102980	COVER, head light	2	W
			WO-A3070	COVER, windshield	1	W
			FM-GPW-1103214	CRANK, starting, engine	1	W
2-B			WO-3211	CURTAIN, side, left	1	
			FM-GPW-17036	CURTAIN, side, right	1	
			WO-A289			
			FM-GPW-1120041			
			WO-A2998			
			FM-GPW-1120040			
			WO-A2999			
2-P	58-E-202		FM-GPW-17101A	EXTINGUISHER, fire, carbon tetrachloride, 1 qt. size, w/BRACKET, assembly	1	K-2 W
			WO-A8429			
2-Y	8-G-615		FM-GPW-18325	GAGE, tire pressure (general service type)	1	J-9 W
			WO-A6855	GUIDE, lubr., War Dept. No. 501	Note 1	
2-X	41-G-1344-40		FM-GP-17125	GUN, lubricating, hand type	1	M-3 W
			WO-A12058			
2-E	41-H-523		FM-GP-17042	HAMMER, machinist, ball peen, 16 oz.	1	J-6 W
			WO-A373			
2-V			FM-GPW-17011A	HANDLE; spark plug wrench	1	J-4 W
			WO-306715			
2-H	41-J-66		FM-GPW-17080	JACK, screw, w/HANDLE, assembly (1½ Ton)	1	M-3 W
			WO-A1240	KEY, padlock (code H-700)	2	H-8 W
				Note 1. One per vehicle. Additional issue by the Adjutant General is as indicated in FM-21-6.		
				GUIDE, lubr., War Dept. No. 501	Note 1	
2-G	13-O-1530		FM-GP-17038	OILER, S, straight spout, spring bottom (capacity ½ pt.)	1	M-3 W
			WO-A379	PADLOCK, 1¾ in. (type 1-C)	2	H-8 W
2-C	42-L-14280		FM-GP-17028	PLIERS, combination, slip joint, 6 in.	1	J-2 W
	41-P-1650		WO-A374			
2-AB	41-P-2962-700		FM-GPW-17090	PULLER, wheel hub	1	W
			WO-A1339			
2-N	8-P-5000		FM-GPW-17052	PUMP, tire, w/air CHUCK	1	M-3 W
			WO-A7511			
2-F	41-S-1076	TGBX1A	FM-GPW-17020	SCREWDRIVER, common heavy duty, integral handle, 6 in.	1	J-4 W
			WO-A375			
1-K	17-T-805		FM-GPW-17058	TAPE, friction, black, grade A, ¾ in. wide (8 oz. roll)	1	H-9 W
			WO-A7684	TARPAULIN, deck top, assembly	1	
			FM-GPW-1152700			
			WO-A2909			
1-E	22-W-650		FM-GPW-17060	WIRE, iron spool	1	W
			WO-A7685			
2-A	41-W-448		FM-GP-17021	WRENCH, screw, adjustable, auto type, 11 in. (2¾ in. cap.)	1	J-4 W
			WO-A377			
2-AA			FM-GP-17062	WRENCH, drain plug	1	J-4 W
			WO-A1100			

SNL G-503

VEHICULAR SPARE PARTS AND EQUIPMENT LIST—Cont'd.

Figure Number	Official Stockage Number	Ordnance Drawing Number	Mfg.'s Number	Item	Quan. Per Vehicle	Note Symbol
Col. 1	Col. 2	Col. 3	Col. 4	Col. 5	Col. 6	Col. 7
				VEHICULAR ACCESSORIES—Cont'd.		
2-Q	41-W-991		FM-GP-17043 / WO-A596	WRENCH, engrs., dble-hd., alloy-S., 3/8 x 7/16	1	J-4 W
2-R	41-W-1003		FM-GP-17044 / WO-A596	WRENCH, engrs., dble-hd., alloy-S., 1/2 x 19/32	1	J-4 W
2-S	41-W-1005-5		FM-GP-17045 / WO-A598	WRENCH, engrs., dble-hd., alloy-S., 9/16 x 11/16	1	J-4 W
2-T	41-W-1008-10		FM-GP-17046 / WO-A599	WRENCH, engrs., dble-hd., alloy-S., 5/8 x 25/32	1	J-4 W
2-U	41-W-1012-5		FM-GP-17047 / WO-A600	WRENCH, engrs., dble-hd., alloy-S., 3/4 x 7/8	1	J-4 W
2-K	41-W-2459-500		FM-GPW-17091 / WO-A1492	WRENCH, fluted, socket hd., screw (WG-T84J-100)	1	J-4 W
2-L			FM-GPW-17030 / WO-A5130	WRENCH, hydraulic brake, bleeder screw (17/64)	1	J-4 W
2-W	41-W-3335-50		FM-GPW-17017A / WO-637635	WRENCH, spark plug, socket	1	J-4 W
2-AC	41-W-3825-200		FM-GP-17033 / WO-A692	WRENCH, wheel bearing nut	1	J-4 W
2-Z	41-W-3837-55		FM-GP-17035 / WO-A348	WRENCH, wheel nut, socket (Kelsey-Hayes 24832)	1	W
				ARMAMENT		

The armament and mounts listed hereunder are installed in the vehicle prior to issue of the vehicle to using troops. Caliber and model of gun, type of mount and basis of issue dependent upon pertinent T/O & E. Issue only spare parts and accessories for that gun and mount which is authorized.

		51-84		GUN, machine, cal. .30, Browning, M2, M1919A4, flexible (for list of all parts see SNL A-6) or....		A-1 W
		51-70		GUN, machine, cal. .50, Browning, M2, HB, flexible (for list of all parts see SNL A-39) (Note 1) or....		A-1 W
		51-102		RIFLE, automatic, cal. .30, Browning, M1918A2 (for list of all parts see SNL A-4)....		A-1 W
		E6266		MOUNT, machine gun, cal. .30, M48 (for list of all parts see Section 32, SNL A-55) or....		
		D47980		MOUNT, truck, pedestal, M31 (for list of all parts see Section 18, SNL A-55)....		A-1 W

Note 1—BMG, cal. .50, M2, HB not used when Mount, machine gun, cal. .30, M48 is authorized.

ARMAMENT SPARE PARTS; FOR:
GUN, MACHINE, CAL. .30, BROWNING, M2, M1919A4, FLEXIBLE

	A005-01-00010	C64142		ACCELERATOR	1	A-19
	A006-01-00020	A170491		BAND, lock, front barrel, bearing	1	A-6
	A006-01-00050	D35233		BARREL	1	A-6
	A005-01-00180	B147299		BOLT, assembly	1	A-19
	A005-01-00210	A157375		BUSHING, belt feed lever pivot	1	A-19

SNL G-503

A006-01-00340	C9801	COVER, assembly	1	A-6
A005-01-00490	C64139	EXTENSION, barrel, assembly	1	A-19
A005-01-00541	C121076	EXTRACTOR, assembly	1	A-19
A005-01-00570	C9182	FRAME, lock, assembly	1	A-19
A005-01-00800	B131317	LEVER, cocking	2	A-19
A005-01-00820	B17503	LEVER, feed, belt	2	A-19
A005-01-00830	B147214	LOCK, breech	1	A-19
A005-01-00840	A196284	NUT, belt feed lever pivot	2	A-19
A005-01-00900	C8461	PAWL, feed, belt	1	A-19
A005-01-00910	B147216	PAWL, holding, belt	1	A-19
A005-01-00930	B131253	PIN, accelerator, assembly	3	A-19
A005-01-00960	B131255	PIN, belt feed pawl, assembly	1	A-19
A005-01-00970	B147217	PIN, belt, holding pawl, split	1	A-19
A005-01-01010	A20567	PIN, cocking lever	2	A-19
A005-01-01090	C9186	PIN, firing, assembly	2	A-19
A005-01-01170	A20503	PIN, trigger	2	A-19
A005-01-01190	A157434	PIVOT, belt feed lever	1	A-19
A005-01-05200	B131251	PLUNGER, barrel, assembly	1	A-19
A005-01-01780	B147222	ROD, driving spring, assembly	1	A-19
A005-88-02096	A196283	SCREW, belt feed lever pivot	1	A-19
A005-01-01950	C64137	SEAR	1	A-19
A005-01-02040	B131262	SLIDE, feed, belt, assembly	1	A-6
A006-01-01240	A135057	SPRING, barrel plunger	1	A-19
A005-01-02100	B147224	SPRING, belt feed pawl	1	A-19
A005-01-02140	B147225	SPRING, belt feed holding pawl	1	A-19
A005-01-02160	B17513	SPRING, cover extractor	1	A-6
A006-01-01275	B212654	SPRING, driving	1	A-19
A005-01-02240	B147230	SPRING, locking barrel	2	A-19
A005-01-02300	B131265	SPRING, sear, assembly	2	A-19
A005-01-02320	B147231	SPRING, trigger pin	1	A-19
A005-01-02470	C8476	TRIGGER	2	A-19
H001-15-19004	BEAX1D	WASHER, lock, internal teeth, reg, S, No. 5 (0.125)	1	H-1
		GUN, MACHINE, CAL. .50, BROWNING, M2, HB, FLEXIBLE		
A037-01-00010	B8914	ARM, belt, feed pawl	1	A-19
A039-01-00008	D28253A	BARREL (spare)	1	A-39
A019-01-00731	A152835	DISK, buffer	1	A-19
A037-01-00500	B8976	EXTENSION, firing pin, assembly	1	A-19
A037-01-00520	B8959	EXTRACTOR, assembly	1	A-39
	B7918A	LEVER, cocking	1	A-19
A037-01-00940	B8961	PAWL, feed, belt, assembly	1	A-19
A037-01-01010	B8962	PIN, belt feed pawl, assembly	1	H-1
H001-08-11027	BFAX1BE	PIN, cotter, split, S, (type B) 1/16 x 3/4	1	H-1
H001-08-11039	BFAX1CE	PIN, cotter, split, S, (type B) 3/32 x 3/4	1	H-1
	BFAX1DD	PIN, cotter, split, S, (type B) 1/8 x 5/8	1	A-19
A002-01-01100	B17171	PIN, firing	1	A-19
A037-01-01310	A13515	PLUNGER, belt feed lever	1	A-19
A037-01-01470	C64305	ROD, driving spring, w/SPRING, assembly	1	A-19
A037-01-01703	B261110	SLIDE, feed belt, assembly	1	A-19

215

SNL G-503

VEHICULAR SPARE PARTS AND EQUIPMENT LIST (Cont'd.)

Figure Number	Official Stockage Number	Ordnance Drawing Number	Mfg.'s Number	Item	Quan. Per Vehicle	Note Symbol
Col. 1	Col. 2	Col. 3	Col. 4	Col. 5	Col. 6	Col. 7
				ARMAMENT SPARE PARTS; FOR:		
				GUN, MACHINE, CAL. .50, BROWNING, M2, HB, FLEXIBLE (Cont'd)		
	A037-01-01780	A351220		SLIDE, sear	1	A-39
	A037-01-01780	A13516		SPRING, belt feed lever plunger	1	A-19
	A037-01-01775	A9351		SPRING, belt feed pawl	1	A-19
	A036-01-00145	A153146		SPRING, belt holding pawl	1	A-19
	A037-01-01805	B7941		SPRING, cover extractor	1	A-19
	A037-01-01820	B8908		SPRING, locking barrel	1	A-19
	A002-01-02120	A9524		SPRING, sear	1	A-19
	A020-01-00890	A13424		STUD, bolt	1	A-19
				RIFLE, AUTOMATIC, CAL. .30, BROWNING, M1918A2		
	A004-01-00310	B19636		CONNECTOR	1	A-4
	A004-01-00370	C9090		EXTRACTOR	1	A-4
	A004-02-00650	C64076		MAGAZINE, assembly (cap. 20 rds.)	Note 1	A-4
	A004-01-00830	C64074		PIN, retaining, gas cylinder tube, assembly	1	A-4
	A004-01-00850	B19680		PIN, retaining, trigger guard, assembly	1	A-4
	A004-01-00890	A22238		PIN, trigger	2	A-4
				Note 1. Magazines are issued in a quantity sufficient for ammunition authorized.		
	A004-01-01331	B147490		SPRING, change and stop, lever, assembly	1	A-4
	A004-01-01330	B19697		SPRING, change lever stop, assembly	1	A-4
	A004-01-01370	A22202		SPRING, extractor	1	A-4
	A004-01-01410	B147134		SPRING, magazine catch	1	A-4
	A004-01-01450	B147131		SPRING, recoil	1	A-4
	A004-01-01460	B19662		SPRING, sear	1	A-4
				MOUNT, MACHINE GUN, CAL. .30, M48		
				Quantities are per twenty-five mounts.		
		A188031		BUSHING, automatic, rifle, locking pin	2	A-19
		A236999		CHAIN, automatic rifle locking pin. assembly	1	A-55
				(Composed of:		
		SDAX6B		2 CHAIN, welded, machs., short twist link, 0.105 in. (5 links)		H-8
		SCAX1D		1 HOOK, "S", steel, 0.105 x 1⁵⁄₁₆ reach		H-2
	A005-02-01290	A142460		1 SWIVEL, chain)		A-19
	A051-02-00050	A188032		CHAIN, bushing, assembly	1	A-19
				(Composed of:		
		SDAX6B		2 CHAIN, welded, machs., short twist link, 0.105 in. (5 links)		H-8
		SCAX1B		2 HOOK, "S", steel, 0.120 x 1⁷⁄₁₆ reach		H-2
	A005-02-01290	A142460		1 SWIVEL, chain)		A-19

SNL G-503

	A336250	LOCK, pintle, assembly	1	A-55
		(Composed of:		
	A336242	1 LOCK, pintle		A-55
	A336237	1 NUT, hand, pintle lock		A-55
H001-07-25660	BBSX4AC	1 RETAINER, pintle lock, brazed assembly)		A-55
	A188028	NUT, safety, S. (Elastic Stop type), 3/8-24NF-3 ...... (Note 1)		H-1
		PIN, locking, automatic rifle, assembly .... (Note 1)	2	A-19
		(Composed of:		
	CCAX1B	2 BALL, chr-alloy-S, grade 2, 3/16 in.		H-2
	A188027	1 BODY, automatic rifle locking pin		A-55
A055-02-09302	A230490	1 HANDLE, locking pin ((Note 2)		A-55
M005-02-56170	FAAX1B	1 SPRING, compression, S, 0.024 diam. stock, 0.185 O.D., 8 coils		M-5
	A188030	PIN, locking, machine gun, assembly .... (Note 1)	2	A-19
		(Composed of:		
	CCAX1B	2 BALL, chr-alloy-S, grade 2, 3/16 in.		H-2
	A188029	1 BODY, machine gun locking pin		A-55
A055-02-09302	A230490	1 HANDLE, locking pin (Note 2)		A-55
M005-02-56170	FAAX1B	1 SPRING, compression, S, 0.024 diam. stock, 0.185 O.D., 8 coils		M-5
		Note 1—Pin assemblies of early manufacture had body of the same piece mark with integral handle and did not require HANDLE A230490.		
		Note 2—Identical with machine gun locking pin HANDLE A230490, and so carried and reviewed in SNL A-55, Sec. 19.		
H001-15-30009	BCBX1CA	SCREW, cap, hex-hd., S., 3/8-24NF-2 x 3/4	2	H-1
	BEBX1K	WASHER, plain, S, SAE std., 3/8 in.	2	H-1
		MOUNT, TRUCK, PEDESTAL, M31		
		Quantities are per twenty-five mounts.		
	A188028	PIN, locking, automatic rifle, assembly .... (Note 1)	1	A-55
		(Composed of:		
	CCAX1B	2 BALL, chr-alloy-S, grade 2, 3/16 in.		H-2
H002-01-00110	A188027	1 BODY, automatic rifle locking pin		A-55
A055-02-09302	A230490	1 HANDLE, locking pin (Note 2)		A-55
M005-02-56170	FAAX1B	1 SPRING, compression, S. 0.024 diam. stock, 0.185 O.D., 8 coils		M-5
	A188030	PIN, locking, machine gun, assembly .... (Note 1)	1	A-55
		(Composed of:		
	CCAX1B	2 BALL, chr-alloy-S, grade 2, 3/16 in.		H-2
H002-01-00110	A188029	1 BODY, machine gun locking pin		A-55
A055-02-09302	A230490	1 HANDLE, locking pin (Note 2)		A-55
M005-02-56170	FAAX1B	1 SPRING, compression, S, 0.024 diam. stock 0.185 O.D., 8 coils		M-5
	A176083	PIN, locking, traveling, lock, assembly	1	A-55
		(Composed of:		
	CCAX1B	2 BALL, chr-alloy-S, grade 2, 3/16 in.		H-2
H002-01-00110	A176082	1 BODY, traveling lock, locking pin		A-55
A055-02-09302	A230490	1 HANDLE, locking pin (Note 2)		A-55
M005-02-56170	FAAX1B	1 SPRING, compression, S., 0.024 diam. stock, 0.185 O.D., 8 coils		M-5
		Note 1—Pin assemblies of early manufacture had body of the same piece mark with integral handle and did not require HANDLE A230490.		
		Note 2—Identical with machine gun locking pin HANDLE A230490, and so carried and reviewed in SNL A55, Sec. 19.		

VEHICULAR SPARE PARTS AND EQUIPMENT LIST (Cont'd.)

Figure Number	Official Stockage Number	Ordnance Drawing Number	Mfg.'s Number	Item	Quan. Per Vehicle	Note Symbol
Col. 1	Col. 2	Col. 3	Col. 4	Col. 5	Col. 6	Col. 7
				ARMAMENT ACCESSORIES; FOR:		
				GUN, MACHINE, CAL. .30, BROWNING, M1919A4, FLEXIBLE		
	A005-06-00020	C3951		BELT, ammunition, cal. .30, M1917 (250 rd.)	5	T-5 W
		FM 23-50		BOOK, Field Manual 23-50, Browning Machine Gun Caliber .30, HB, M1919A4 (mounted in combat vehicles)	Note 1	
				Note 1—One per packing container. Additional issue by the Adjutant General is as indicated in FM 21-6.		
	A005-06-00081	D44070		BOX, ammunition, cal. .30, M1 (empty)	5	T-5
	M-003-01-01930	B108828		BRUSH, chamber, cleaning, M6	1	M-3
	M003-01-02020	C4035		BRUSH, cleaning, cal. .30, M2	3	M-3
	M003-01-02965	B147310		CAN, tubular (¾ in. diam. x 2½ in. w/screw top)	1	M-3
	M003-01-03160	C6573		CASE, cleaning rod, cal. .30, M1	1	M-3 W
	A019-01-00300	C59656		CASE, spare bolt, M2	3	A-19 W
	M008-01-00570	D28243		CHEST, steel, M5 (STAX1AB modified) (Note 1)	1	M-8 W
	A019-01-00640	D30674		COVER, spare barrel, M9	1	A-19 W
	M003-01-04590	C59696		ENVELOPE, spare parts, M1	1	A-19
	A019-01-00820	C3854		EXTRACTOR, ruptured cartridge, Mk. IV	1	A-19 W
	A019-01-01240	D1262		MACHINE, browning belt filling M1918	Note 2	A-19 W
	M003-01-04570	C59737		OILER, oval, 3 oz., w/cap and chain (Note 3)	1	M-3
	M003-01-10050	C59736		OILER, rectangular, 12 oz., w/cap and chain (Note 3)	1	M-3
	M003-01-12020	B147001		REFLECTOR, barrel, cal. .30	1	M-3 W
	M003-01-12690	D8237		ROD, cleaning, jointed, cal. .30, M1	1	M-3 W
	M008-01-00990	D7349		ROLL, spare parts, M13	1	M-8 W
				Note 1—Issue 49-I-82 CHEST, accessory and spare parts, M1917, until CHEST, steel, M5 is obtainable.		
				Note 2—One per four guns or major fraction thereof.		
				Note 3—To be issued until supply is exhausted.		
	M008-01-01110	D7389		ROLL, tool, M12	1	M-8 W
	J004-01-04360	TGAX1A		SCREWDRIVER, comm., normal duty, 3 in. blade	1	J-4 W
	A006-03-00310	C68334		WRENCH, combination, M6	1	A-19 W
	A006-03-00320	B147277		WRENCH, socket, front barrel bearing plug	1	A-6 W
				GUN, MACHINE, CAL. .50, BROWNING, M2, HB, FLEXIBLE		
	38-B-992-27	FM-23-65		BOOK, Field Manual 23-65, Browning Machine Gun, Caliber .50, HB, M2 (mounted on combat vehicles)	Note 1	A-19
		C4037		BRUSH, cleaning, cal. .50, M4	4	A-19 W
	A019-01-00540	C64274		CASE, cleaning rod, cal. .50, M15	1	A-19 W
	A039-88-00260	C61331		CHUTE, metallic belt link, M1	1	A-19 W
		D33912		COVER, spare barrel, M13 (for 45 in. barrel)	1	A-19
		C59696		ENVELOPE, spare parts M1	2	A-19 W
		C64392		EXTRACTOR, ruptured cartridge, cal. .50, M5	1	A-19 W
		D35441		ROD, cleaning, jointed, cal. .50, M7	1	A-19 W
	A019-01-02710	D28242		WRENCH, combination, cal. .50, M2	1	A-19 W

SNL G-503

RIFLE, AUTOMATIC, CAL. .30, BROWNING, M1918A2

A004-02-00051	FM 23-15	BOOK, Field Manual 23-15, Browning, Automatic Rifle, Cal. .30, M1918A2	Note 1	
38-B-992	D28362	BRUSH, chamber, cleaning M1	1	A-4 W
B003-02-00030	C4035	BRUSH, cleaning, cal. .30, M2	2	M-3
	C64173	BRUSH and THONG, cal. .30, complete	1	M-3
M003-01-15411	C64174	(Consisting of: 1—BRUSH, thong, cal. .30		
	C64175	1—THONG)		
		Note 1—One per packing container. Additional issue by the Adjutant General is as indicated in FM 21-6.		
24-C-534	C6573	CASE, cleaning rod, cal. .30, M1 (for carrying Rod D8237)	1	M-3 W
A004-02-00151	C64177	COVER, front sight	1	A-4 W
M003-01-04570	15-18-102B	ENVELOPE, fabric, one-button, 3 x 3⅛ in.	1	M-3
M003-01-04580	15-18-102A	ENVELOPE, fabric, two-button, 3 x 4⅞ in.	1	M-3
41-E-555	C7912	EXTRACTOR, ruptured cartridge, Mk. II	1	M-3 W
A004-02-00190	C7913	FILLER, magazine	2	A-4 W
13-0-1280	C59737	OILER, oval, 3 oz, w/cap and chain	1	M-3
41-R-2330-975	B147001	REFLECTOR, barrel, cal. .30 ..... (Note 1)	1	M-3 W
41-R-2564	D8398	ROD, cleaning, cal. .30, M2A1	1	M-3
41-A-2567	D8237	ROD, cleaning, cal. .30, jointed, M1	1	M-3 W
B003-02-00151	D44058	SLING, gun, M1 (web) or	1	M-3 W
A004-02-00250	20-18-25	SLING, gun, M1907, modified		M3- W
41-T-3081-110	C64144	TOOL, cleaning, gas cylinder	1	A-4 W
41-T-3085-250	C64145	TOOL, combination	1	A-4 W

MOUNT, MACHINE GUN, CAL. .30, AMMUNITION BOX, M1

	E6288	ADAPTER, machine gun, cal. .30, ammunition box, M1	1	A-55 W
		Note 1—To be issued until supply is exhausted.		

MOUNT, TRUCK, PEDESTAL, M31

	C38571	TRAY, ammunition (for cal. .30 machine gun)	1	W A-55
	D38607	TRAY, ammunition (for cal. .50 machine gun)	1	W A-55

MAJOR ITEM

Note	Symbol	Class	Division	
ϕ	W			TRUCK, ¼-Ton, 4 x 4, Command Reconnaissance

NOTES

Items listed in Section 3 of this SNL will be stored under SNL G-503 unless the note symbol indicates otherwise.
The major item indicated by (ϕ) is stored, issued, and reviewed under "Major Items," SNL G-1.
(a) Items indicated by (W) are non-expendable parts and the issue of such items must be in accordance with par. 3b (2), AR 35-6540.
(b) Items not marked by (W) are expendable in accordance with AR 35-6620.

219

SNL G-503

FIGURE 1—VEHICULAR SPARE PARTS

A	BELT	33-B-76
B	LAMP-UNIT	8-L-415
C	LAMP-UNIT	8-L-419
D	LAMP-UNIT	8-L-421
E	WIRE	22-W-650
F	LAMP	17-L-5215
G	PLUG	17-P-5365
H	CORE	8-C-6750
J	CAP	8-C-650
K	TAPE	17-T-805
L	PIN, COTTER, assorted	42-P-5347

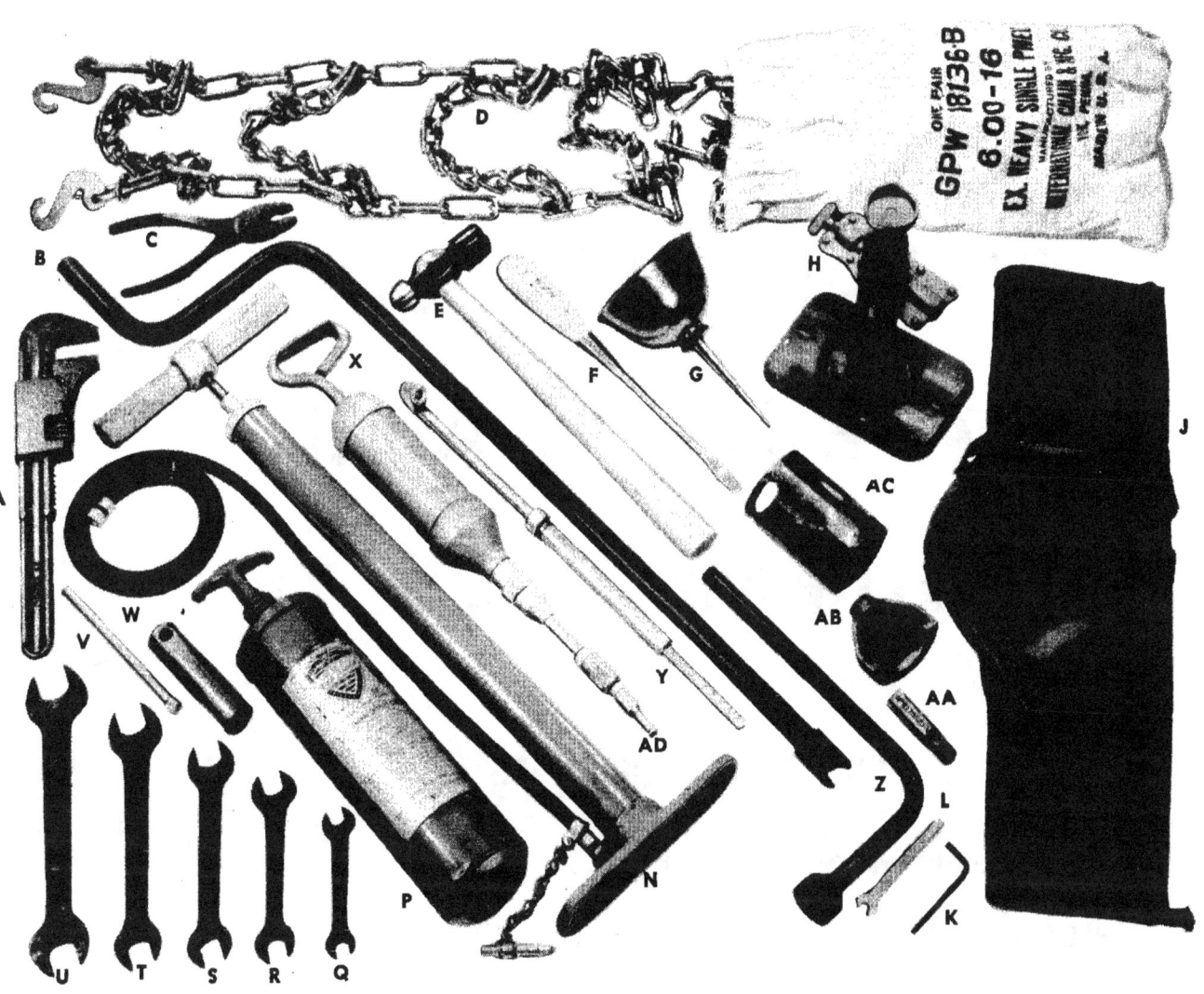

A	WRENCH	41-W-448	K	WRENCH	41-W-2459-500	V	HANDLE	FM-GPW-17011A
B	CRANK	FM-GPW-17036	L	WRENCH	FM-GPW-17030	W	WRENCH	41-W-3335-50
C	PLIERS	41-P-1650	N	PUMP	8-P-5000	X	GUN	41-G-1344
D	CHAINS	8-C-2538	P	EXTINGUISHER	58-E-202	Y	GAGE	8-G-615
E	HAMMER	41-H-523	Q	WRENCH	FM-GP-17043	Z	WRENCH	FM-GP-17035
F	SCREWDRIVER	41-S-1076	R	WRENCH	FM-GP-17044	AA	WRENCH	FM-GP-17062
G	OILER	13-O-1530	S	WRENCH	FM-GP-17045	AB	PULLER	41-P-2962-7
H	JACK	41-J-66	T	WRENCH	FM-GP-17046	AC	WRENCH	41-W-3825-200
J	BAG	41-B-15	U	WRENCH	FM-GP-17047	AD	ADAPTER	FM-GP-17126

RA PD 322889

FIGURE 2—VEHICULAR ACCESSORIES

SECTION 4

ORGANIZATIONAL SPARE PARTS AND EQUIPMENT LIST

TABLE OF CONTENTS

Organizational Spare Parts

Organizational Accessories (Special Tools)

Fire Control Equipment

Applicable Cleaning and Preserving Materials

Articles for Instructional Purposes

Ammunition

Reference Notes

ORGANIZATIONAL SPARE PARTS

Official Stockage Number	Ordnance Drawing Number	Mfr.'s Number		Item	Parts Allowances For:							
		Willys	Ford		Company			Regiment or Separate Battalion				
					1-9 Vehicles Set No. 1	10-25 Vehicles Set No. 2	26 Up Vehicles Set No. 3	1-9 Vehicles Set No. 4	10-35 Vehicles Set No. 5	36-75 Vehicles Set No. 6	76-150 Vehicles Set No. 7	151 Up Vehicles Set No. 8
Col. 1	Col. 2	Col. 3	Col. 4	Col. 5	Col. 6	Col. 7	Col. 8	Col. 9	Col. 10	Col. 11	Col. 12	Col. 13
				GROUP 01—ENGINE								
		WO-A-1236	FM-GPW-18662-A	ELEMENT, oil filter (PU-26637)	2	3	5	2	4	8	15	30
		WO-630299	FM-GPW-6649	GASKET, crankcase ventilator to valve spring cover	0	0	0	1	1	1	2	4
		WO-8558	FM-GPW-6051-B	GASKET, cylinder head	1	1	2	1	2	3	6	12
		WO-A-1235	FM-GPW-18688-A	GASKET, oil filter cover (PU-25802)	1	1	1	1	1	2	4	8
		WO-314338	FM-GPW-6734	GASKET, oil pan drain plug	1	2	3	1	3	5	10	20
		WO-A-1538	FM-GPW-18512	GASKET SET, oil pan	1	1	1	1	1	2	4	8
				(Consists of:								
				1 WO-630398 FM-GPW-6627 GASKET, oil float support								
				1 WO-639980 FM-GPW-6710 GASKET, oil pan								
				1 WO-314338 FM-GPW-6734 GASKET, oil pan drain plug)								
		WO-A-7835	FM-GPW-18323	GASKET SET, manifold	0	0	0	1	1	1	2	4
				(Consists of:								
				1 WO-638640 FM-GPW-9448 GASKET, intake and exhaust manifold								
				1 WO-634811 FM-GPW-9435 GASKET, intake to exhaust manifold								
				1 WO-634814 FM-GPW-9450 GASKET, exhaust pipe flange)								
		WO-51875	FM-GPW-6555	GASKET, valve cover bolt	0	1	1	1	1	2	3	6
		WO-630305	FM-GPW-6521	GASKET, valve spring cover	0	0	0	1	1	1	2	4
		WO-639650	FM-GPW-8255	GASKET, water outlet elbow	0	0	0	0	1	1	1	2
		WO-639979	FM-GPW-6727	PLUG, drain, oil pan (⅞ in.)	0	0	0	1	1	1	2	4
		WO-A-1197	FM-GPW-18667	TUBE, oil filter, inlet, assembly	1	1	1	1	1	2	4	8
		WO-A-1198	FM-GPW-18666	TUBE, oil filter, outlet, assembly	1	1	1	1	1	2	4	8
				GROUP 02—CLUTCH								
		WO-630593	FM-GPW-7523	SPRING, retracting, clutch and brake pedal	0	0	0	0	1	1	1	2

SNL G-503

		GROUP 03—FUEL SYSTEM							
WO-A-6333	FM-GPW-9030-B	CAP, fuel tank, w/safety CHAIN, assembly	0	0	0	0	1	1	2
WO-A-1223	FM-GPW-9510	CARBURETOR, assembly (CAR-539S)	0	0	0	0	1	1	2
WO-A-1325	FM-GPW-9288	CONNECTION, flexible (in fuel line) (FO-HA-8031)	0	0	0	1	1	2	4
WO-A-6357	FM-GPW-9445	GASKET, insulator, carburetor, and intake manifold DIFFUSER, assembly	0	0	0	1	1	1	2
WO-115656	FM-GPW-9364	GASKET, fuel pump bowl (AC-1523096)	0	1	1	1	2	3	6
WO-638737	FM-GPW-9417	GASKET, fuel pump to block (AC-838263)	0	0	1	1	1	2	4
WO-A-6883	FM-GPW-18337	GASKET SET, fuel strainer	1	1	1	1	2	4	8
		(Consists of:							
		1 WO-A-1257 FM-GPW-9184 GASKET, cover cap screw							
		1 WO-A-1259 FM-GPW-9160 GASKET, strainer bowl							
		1 WO-A-1260 FM-GPW-9186 GASKET, strainer unit)							
WO-A-1292	FM-GPW-9276	GAUGE fuel tank unit, assembly (AL-9979-A)	0	0	0	1	1	1	2
WO-A-1311	FM-GPW-9652	HOSE, flexible, air cleaner tube to cleaner	0	0	0	1	1	1	2
WO-A-8323	FM-GPW-9350-B	PUMP, fuel, assembly (AC-1538312)	0	0	0	1	1	1	2
WO-633011	FM-GPW-9799	SPRING, retracting, accelerator	0	0	0	1	1	1	2
		GROUP 04—EXHAUST							
WO-A-1146	FM-GPW-5230-A	MUFFLER, assembly (rd. type)	0	0	1	1	2	3	6
WO-A-1296	FM-GPW-5246	PIPE, exhaust, assembly	1	1	1	1	2	4	8
		GROUP 05—COOLING							
WO-A-1495	FM-GPW-8620	BELT, drive, fan and generator	1	1	1	1	2	4	8
WO-A-1215	FM-GPW-8100	CAP, radiator filler, assembly (AC-846709) (STN-6455-A)	0	0	0	1	1	1	2
WO-52226	FM-60-8287	CLAMP, hose, radiator, 1¹³⁄₁₆ in.	1	2	3	2	4	8	16
WO-A-1216	FM-GPW-8578	GASKET, radiator cap (AC-846732)	0	0	0	1	1	2	4
WO-637053	FM-GPW-8543	GASKET, water pump to cylinder block	0	0	0	1	1	2	4
WO-A-592	FM-GPW-8284	HOSE, radiator, outlet	1	2	3	3	5	10	20
WO-639992	FM-GPW-8501	PUMP, water, assembly	0	0	0	1	1	1	2
WO-637646	FM-GPW-8575	THERMOSTAT, assembly (HR-3108628)	0	0	0	1	1	1	2
		GROUP 06—ELECTRICAL							
WO-A-1238	FM-11AS-10655	BATTERY, wet charged, 6 volt, 116 ampere assembly	1	1	1	1	2	4	8
WO-A-1655	FM-GPW-12106	CAP, distributor, assembly (AL-IG-1324)	0	0	0	1	1	1	2
WO-A-7792	FM-GPW-12000-B	COIL, ignition, w/BRACKET and WASHER, assembly (AL-4070-L)	0	0	0	1	1	1	2
WO-A-1631	FM-GPW-12300	CONDENSER, distributor, assembly (AL-IGW-3139)	0	0	1	1	1	2	4
WO-A-1244	FM-GPW-12100	DISTRIBUTOR, assembly (AL-IGC-4705)	0	0	0	1	1	1	2
WO-637863	FM-01A-12410	GASKET, spark plug (VC-2066C-C1)	0	0	1	1	1	2	4
WO-A-5992	FM-GPW-10000A	GENERATOR, assembly (AL-GEG-5002-D)	0	0	0	1	1	1	2

ORGANIZATIONAL SPARE PARTS (Cont'd)

Official Stockage Number	Ordnance Drawing Number	Mfr.'s Number (Willys)	Mfr.'s Number (Ford)	Item	Parts Allowances For: Company 1-9 Vehicles Set No. 1	Company 10-25 Vehicles Set No. 2	Company 26 Up Vehicles Set No. 3	Regiment or Separate Battalion 1-9 Vehicles Set No. 4	10-35 Vehicles Set No. 5	36-75 Vehicles Set No. 6	76-150 Vehicles Set No. 7	151 Up Vehicles Set No. 8
Col. 1	Col. 2	Col. 3	Col. 4	Col. 5	Col. 6	Col. 7	Col. 8	Col. 9	Col. 10	Col. 11	Col. 12	Col. 13
				GROUP 06—ELECTRICAL (Cont'd)								
		WO-A-6145	FM-GPW-13152	LAMP-UNIT, blackout driving, headlight sealed, 6-V8., assembly (CB-11267)	0	0	0	0				2
C84908K C84934K (C91706C)		WO-A-1075	FM-GPW-13491-A	LAMP-UNIT, blackout tail, sealed, 6-8V., assembly (lower left and right) (CB-9225) (GL-9533078)	0	0	0	0	1	1	1	2
		WO-A-1033	FM-GPW-13007	LAMP-UNIT, head light, sealed, 6-V8., (Seelight unit) (CB-8494)	0	0	0	1	1	1	2	4
C84908J		WO-A-1074	FM-GPW-13494	LAMP-UNIT, service tail and stop, sealed, 6-8V., 21-3 cp, assembly (upper, left) (CB-9218) (GL-5933104)	0	0	0	0	1	1	1	2
C84934J		WO-A-1078	FM-GPW-13485-A	LAMP-UNIT, blackout stop, sealed, 6-8V., assembly (upper, right) (GL-5933121)	0	0	0	0	1	1	1	2
DLAX1F		WO-51804	FM-B-13466	LAMP, elec., incand., min., 6-8 volt, sgle-tung-fil., No. 63, 3 cp. (blackout headlight) (MZ-63)	0	1	1	1	1	2	3	6
		WO-A-1245	FM-GPW-11001-A	MOTOR, cranking, assembly (AL-MZ-4113)	0	0	0	0	1	1	1	2
		WO-A-538	FM-GPW-12405	PLUG, spark, w/GASKET, assembly (CP-type-AN-7)	2	4	8	4	6	12	24	48
		WO-A-1687	FM-GPW-18354	CONTACT SET, contact, distributor (AL-IGP-3028FS)	1	1	2	1	2	3	6	12
		WO-A-1409	FM-GPW-10505	REGULATOR, generator, assembly	0	0	0	0	1	1	1	2
		WO-A-1658	FM-GPW-12200	ROTOR, distributor (AL-IG-1657-R)	0	0	0	1	1	1	2	4
		WO-A-7225	FM-9N-11450-A	SWITCH, starting, assembly (AL-SW-4015)	0	0	0	1	1	1	2	4
		WO-A-638979	FM-GPW-13532	SWITCH, foot, headlight, assembly (CL-9634)	0	0	0	0	1	1	1	2
		WO-A-1271	FM-11A-13480	SWITCH, stop light, assembly	0	0	0	0	1	1	1	2
				GROUP 11—REAR AXLE								
		WO-A-904	FM-GP-4032	GASKET, axle shaft (SP-17146)	1	1	1	1	1	2	4	8
				GROUP 12—BRAKE								
		WO-A-1484	FM-GP-2061	CYLINDER, front wheel brake, assembly (LO-FD-7379)	0	0	0	0	1	1	1	2
		WO-A-556	FM-GP-2140	CYLINDER, master brake, assembly (LO-FE-1444)	0	0	0	0	1	1	1	2
		WO-A-6110	FM-GPW-2261	CYLINDER, rear wheel, 3/4 in. diam., assembly (LO-FD-7568-A)	0	0	0	1	1	1	1	2
		WO-A-1460	FM-GPW-2079	HOSE, brake, front axle, assembly (LO-FC-8553)	0	0	0	1	1	1	2	4

WO-A-1373	FM-GPW-2078	HOSE, brake, front, assembly (11 in.) (LO-FC-8502)	0	0	0	1	1	1	2
WO-637424	FM-GP-2078	HOSE, brake, rear, assembly (15 in.) (LO-FC-5784)	0	0	0	1	1	1	2
WO-116549	FM-GP-2018	SHOE, brake, forward, w/LINING, assembly (BX-1141-S-1)	0	0	0	1	1	1	2
WO-116550	FM-GP-2019	SHOE, brake, reverse, w/LINING, assembly (BX-1141-S-2)	0	0	0	1	1	1	2
WO-637905	FM-GP-2035	SPRING, return, brake shoe (BX-41545)	0	0	1	1	1	2	4
		GROUP 13—WHEELS, HUBS & DRUM							
WO-52942	FM-GP-1201	CONE and ROLLERS, tapered wheel bearing, bore, 1.625 x $^{11}/_{16}$ (front and rear wheels, inner and outer) (TIM-18590) (BOW-BT-18590)	0	0	0	1	1	1	2
WO-52943	FM-GP-1202	CUP, tapered roller bearing, 2.875 x ½ (front and rear wheels, inner and outer) (TIM-18520) (BOW-BT-18520)	0	0	0	1	1	1	2
WO-A-864	FM-GP-1177	OIL SEAL, wheel hub (SP-17004)	1	3	5	2	4	8	15	30
		GROUP 16—SPRING & SHOCK ABSORBER							
WO-A-6902	FM-GPW-5311	ABSORBER, shock, front, assembly ...(Willys only)	0	0	0	0	1	1	2
WO-A-612		SPRING, front, left, assembly	0	0	0	0	1	1	2
WO-A-613		SPRING, front, right, assembly	0	0	0	0	1	1	2
WO-A-2512	FM-GP-17535	ARM, windshield wiper, w/BLADE, assembly	1	1	1	1	1	2	4	8
WO-A-1344	FM-GPW-17262	SHAFT, flexible, speedometer drive (53$^{3}/_{32}$ in.) (AL-SPS-1105X)	0	0	1	1	1	2	4
		GROUP 23—GENERAL USE & STANDARDIZED							
WO-A-7681	FM-GPW-18322	CAP SET, tire valve	1	1	1	1	2	4	8
WO-A-7682	FM-GPW-18320	CORE SET, tire valve	1	1	1	1	2	4	8
WO-638500		FITTING, grease, 90°	0	0	1	1	1	2	4
WO-638224		FITTING, grease, 45°	0	0	1	1	1	2	4
WO-640038	FM-350006-S8	FITTING, grease, straight	0	0	1	1	1	2	4
WO-392909	FM-353027-AS7	FITTING, grease, straight, ⅛ in. (AD-1720)	0	1	1	1	2	3	6
WO-638792	FM-353043-AS7	FITTING, grease, straight, ¼-28NF (AD-1641)	0	0	1	1	1	2	4

NOTE: Where the same part is common to two or more vehicles assigned to the same unit, the part should be issued on basis of the total number of vehicles to which the part applies.

ORGANIZATIONAL SPARE PARTS AND EQUIPMENT LIST

ORGANIZATIONAL ACCESSORIES

The issue of the following organizational tool kits and sets, the component, items of which are listed here-under in columnar form, are as follows:

TOOL KIT, MECHANICS': (Col. 6). 1 per Motor Mechanic (014) in Truck, ¼-Ton, 4 x 4, Command Reconnaissance , equipped unit.
TOOL-SET, COMPANY, (Col. 7). 1 per Truck, ¼-Ton, 4 x 4, Command Reconnaissance , company or troop equipped unit.
TOOL-SET, BATTALION CREW: (Col. 8). As authorized in applicable T/BA, T/A or T/O&E.
TOOL-SET, REGIMENTAL MAINTENANCE PLATOON: (Col. 9). As authorized in applicable T/BA, T/A, or T/O&E.

Figure Number	Official Stockage Number	Ordnance Drawing Number	Mfr.'s Number	Item	Tool Kit, Mechanics	Tool-Set, Company	Tool-Set. Battalion Crew	Tool Set, Regimental Maintenance Platoon	Note Symbol
Col. 1	Col. 2	Col. 3	Col. 4	Col. 5	Col. 6	Col. 7	Col. 8	Col. 9	Col. 10
	41-C-2554-400		MAS-1148	COMPRESSOR, shock absorber rubber grommet				1	W
	41-W-3575	B7076535	KM-J-4056	WRENCH, tappet, double end 1½₂ and 17⁄₃₂ in.				2	W

Figure Number	Official Stockage Number	Ordnance Drawing Number	Mfg.'s Number	Item	Quan. Per Vehicle	Note Symbol
Col. 1	Col. 2	Col. 3	Col. 4	Col. 5	Col. 6	Col. 7
				FIRE CONTROL EQUIPMENT		
				Issue Firing Tables pertinent to Gun authorized.		
				TABLE, firing, 0.30-A-4	Note 1	F-69
				TABLE, firing, 0.30-C-4	Note 1	F-69
				TABLE, firing, 0.30-J-1	Note 1	F-69
				TABLE, firing, 0.50-F-3 (Note 2)	Note 1	F-69
				TABLE, firing, 0.50-H-1	Note 1	F-69
				Note 1—One per gun. Additional issue as authorized in T/o&E.		
				Note 2—Under development.		
				APPLICABLE CLEANING AND PRESERVING MATERIALS		
				The following items extracted from SNL K-1 are requisitioned as required. See TM 9-850 for usage. (To be included when data become available).		
				ARTICLES FOR INSTRUCTIONAL PURPOSES; FOR:		
				The items listed hereunder are for training purposes only and will not be taken into the field. Upon permanent change of station involving movement into another Service Command or upon the departure for the theater of operations units will turn in all equipment held to the commanding officer of the station from which it departs. The receiving officer will make a report to the Commander of the Service Command without delay, showing number, type and conditions of items received. Issue only articles pertinent to Gun authorized.		

SNL G-503

		GUN, MACHINE, CAL. .30, BROWNING, M2, M1919A4, FLEXIBLE		
	B6252	CARTRIDGE, dummy, cal. .30, M1906 (corrugated)	25	T-1
		CHART, instruction, ammunition (chart No. 75) .... (Note 1)	Note 2	A-6
		CHART, instruction, BMG, cal. .30	Note 3	A-6
		(Consisting of:		
		1 PLATE, No. 1-BMG, cal. .30, M1919A4 (ORD. 10987)		
		1 PLATE, No. 3-BMG, cal. .30, M1917 (ORD-10893)		
		1 PLATE, No. 5, BMG, cal. .30, M1917 (ORD-10895))		
		TRAINER, machine gun, cal. .22, M4	Note 4	A-48 W
		Accessories:		
A048-02-00010	B147548	ADAPTER, belt	500	A-48
A037-03-00005	D34338	BAG, metallic, belt link	1	A-19
	C64179	BRUSH, cleaning, cal. .22, M3	1	M-3 W
		Note 1—Under development.		
		Note 2—One per organization; to schools as required.		
		Note 3—One per organization of fifty men or major fraction thereof equipped with machine guns; to schools as required.		
		Note 4—One per three vehicles equipped with machine gun.		
	D28201	CHEST, accessory and spare parts (STAX1AD modified)	1	M-8 W
	C64326	CHUTE, cartridge holder	1	A-19
A048-02-00020	A152916	CLIP, belt adapter	250	A-48
A048-02-00030	A147551	HOLDER, cartridge	500	A-48
A048-02-00051	C3837	ROD, cleaning, cal. .22, M1	1	M-3
A048-02-00060	A152919	TOOL, cartridge ejecting	1	A-19
		GUN, MACHINE, CAL. .50, BROWNING, M2, HB, FLEXIBLE		
	B147421	ATTACHMENT, firing, blank ammunition	1	A-39
	C56579	CARTRIDGE, instruction, cal. .50, M2	25	T-1
		CHART, instruction, Browning Machine Gun, cal. .50, M2, Fixed and Flexible Set	Note 1	A-39
		(Consists of:		
		Plates, 5, 6, 7, 8, 15, 16, 17, 18 and 19)		
		RIFLE, AUTOMATIC, CAL. .30, BROWNING, M1918A2		
	B6252	CARTRIDGE, dummy, cal. .30, M1906 (corrugated)	10	T-1
A004-02-00130		CHART, instruction, Browning automatic rifle, cal. .30, M1918 and M1918A2 ...	Note 2	A-4
		AMMUNITION		
		Ammunition for use with these weapons is shown in SNL T-1.		
		Note 1—One set per organization equipped with BMG, cal. .50, M2; to schools as required.		
		Note 2—One per organization; to schools as required.		

NOTES

Items listed in Section 4 of this SNL will be stored under SNL G-503 unless the note symbol indicates otherwise.
   (a) Items indicated by (W) are non-expendable parts and the issue of such items must be in accordance with par. 3b (2), AR 35-6540.
   (b) Items not marked by (W) are expendable in accordance with AR 35-6620.

W

SECTION 5

ORDNANCE MAINTENANCE UNIT STOCKAGE LIST

ORDNANCE MAINTENANCE UNIT STOCKAGE LIST

BASED ON 30 DAYS MAXIMUM ON HAND AND STOCKAGE

Figure Number	Official Stockage Number	Part Number		ITEM	Parts Allowances For:				
		Willys	Ford		1-9 Vehicles	10-35 Vehicles	36-75 Vehicles	76-150 Vehicles	151-300 Vehicles
Col. 1	Col. 2	Col. 3	Col. 4	Col. 5	Col. 6	Col. 7	Col. 8	Col. 9	Col. 10

GROUP 01—ENGINE

		Willys	Ford	ITEM	1-9	10-35	36-75	76-150	151-300
		WO-A-7233	FM-GPW-18330A	BEARING SET, Connecting Rod (Standard)	3	4	5	7	9
		WO-640070		BOLT, Engine Connecting Rod Bearing Cap	1	1	2	2	3
		WO-638635	FM-GPW-6065	BOLT, Cylinder Head	1	1	2	2	3
		WO-A-1198	FM-GPW-18666	CONNECTION, Flexible (For Oil Pressure Gauge Tube)	1	1	2	2	3
		WO-A-1236	FM-GPW-18662A	ELEMENT, Engine Oil Filter w/GASKET	14	18	24	32	43
		WO-A-5497	FM-GPW-6005	ENGINE, Assembly	0	1	1	1	2
		WO-A-1230	FM-GPW-18660A	FILTER, Oil, Engine, Assembly	0	1	1	1	2
		WO-A-8558	FM-GPW-6051B	GASKET, Engine, Cylinder Head	5	7	9	12	16
		WO-A-1235	FM-GPW-18688A	GASKET, Engine Oil Filter Cover	5	6	8	10	13
		WO-A-1233	FM-GPW-18675A	GASKET, Engine Oil Filter Cover Bolt	5	6	8	10	13
		WO-314338	FM-GPW-6734	GASKET, Engine Oil Pan Drain Plug	9	12	16	21	28
		WO-A-1538	FM-GPW-18512	GASKET SET, Oil Pan	4	5	6	8	11
		WO-630394	FM-GPW-6630	GASKET, Engine Oil Pump to Cylinder Block	0	1	1	1	2
		WO-51875	FM-GPW-6555	GASKET, Engine Valve Cover Screw	2	3	4	5	7
		WO-630305	FM-GPW-6521	GASKET, Engine Valve Spring Cover	2	3	4	5	7
		WO-630299	FM-GPW-6648	GASKET, Engine Crankcase Ventilator to Valve Spring Cover	2	2	3	4	5
		WO-A-1536	FM-GPW-18390	GASKET SET, Engine	2	2	3	4	5
		WO-A-7835	FM-GPW-18323	GASKET SET, Engine Manifold	2	2	3	4	5
		WO-A-6750	FM-GPW-18380	GASKET SET, Engine Oil Pump	2	2	3	4	5
		WO-A-1537	FM-GPW-18387	GASKET SET, Engine Valve Job	4	5	6	8	11
		WO-637045	FM-GPW-6511B	GUIDE, Engine Intake Valve Stem	1	1	2	2	3
		WO-375811	FM-GPW-6510B	GUIDE, Engine Exhaust Valve Stem	1	1	2	2	3
		WO-A-1534	FM-GPW-6050	HEAD, Engine Cylinder	0	1	1	1	2
		WO-A-5168	FM-GPW-6766B	INDICATOR and Breather CAP, Engine Oil Filler, Assembly	0	1	1	1	2
		WO-A-7498	FM-GPW-6038A	INSULATOR, Engine Support Front, Assembly	1	1	2	2	3
		WO-A-6156	FM-GPW-6040B	INSULATOR, Engine Support Rear	0	1	1	1	2
		WO-375994	FM-GPW-6546	LOCK, Engine Valve Spring Retainer, Lower	11	14	18	24	32
		WO-A-912	FM-GPW-9428	MANIFOLD, Exhaust, Assembly	0	1	1	1	2
		WO-637237	FM-GPW-6702	PACKING, Engine Crankshaft, Rear End	1	1	2	2	3
		WO-637098	FM-GPW-6700	PACKING, Engine Crankshaft, Front End	0	1	1	1	2
		WO-A-5105	FM-GPW-6770	PAN, Engine Oil, Assembly	0	1	1	1	2
		WO-639979	FM-GPW-6727	PLUG, Drain, Engine Oil Pan	1	2	2	3	4
		WO-637636	FM-GPW-6600	PUMP, Engine Oil, Assembly	0	1	1	1	2
		WO-637044	FM-GPW-6514	RETAINER, Engine Valve Spring, Lower	2	2	3	4	5
		WO-638636	FM-GPW-6513	SPRING, Engine Valve	3	4	5	7	9
		WO-349368	FM-GPW-6066	STUD, Engine Cylinder Head	1	1	2	2	3
		WO-A-1548	FM-GPW-6067	STUD, Engine Cylinder Head	0	1	1	1	2

SNL G-503

WO-A-1197	FM-GPW-18667	TUBE, Engine Oil Filter Inlet, Assembly	4	5	7	9	12
WO-A-1198	FM-GPW-18666	TUBE, Engine Oil Filter, Outlet	4	5	7	9	12
WO-637183	FM-GPW-6505	VALVE, Exhaust, Engine	2	3	4	5	7
WO-637182	FM-GPW-6507	VALVE, Intake, Engine	1	1	2	2	3

GROUP 02—CLUTCH

WO-635529	FM-GPW-7580B	BEARING, Engine Clutch Release	1	1	2	2	3
WO-639578	FM-GPW-7600	BUSHING, Engine Clutch Pilot	0	1	1	1	2
WO-A-5102	FM-GPW-7530	CABLE, Engine Clutch Control Lever	1	1	1	1	2
WO-A-6751	FM-GPW-18358	FACING SET, Engine Clutch	1	1	2	2	3
WO-630112	FM-GPW-7515	FORK, Engine Clutch Release Shaft	0	1	1	1	2
WO-639654	FM-GPW-7561	CARRIER, Engine Clutch Release Bearing	0	1	1	1	2
WO-638992	FM-GPW-7563	PLATE, Pressure, Engine Clutch, Assembly	1	1	2	2	3
WO-A-7833	FM-GPW-18359B	KIT, Repair, Engine Clutch Cover	0	1	1	1	2
WO-636755	FM-GPW-7550	PLATE, Driven, Engine Clutch Assembly	1	1	2	2	3
WO-638152	FM-GPW-7566	PLATE, Pressure, Engine Clutch	0	1	1	1	2
WO-630593	FM-GPW-7523	SPRING, Retracting, Engine Clutch Pedal	1	1	2	2	3
WO-630117	FM-GPW-7562	SPRING, Engine Clutch Release Bearing Carrier	1	1	2	2	3

GROUP 03—FUEL SYSTEM

WO-A-6333	FM-GPW-9030	CAP, Fuel Tank, w/Safety CHAIN, Assembly	1	1	2	2	3
WO-A-1223	FM-GPW-9510	CARBURETOR, Assembly	0	1	1	1	2
WO-A-1325	FM-GPW-9288	CONNECTION, Flexible (in fuel line)	1	2	2	3	4
WO-A-5630	FM-GPW-9617	ELEMENT, Filter, Air Cleaner, Assembly	0	1	1	1	2
WO-A-6357	FM-GPW-9445	GASKET, Insulator, Carburetor and Intake Manifold DIFFUSER, Assembly	0	1	1	1	2
WO-638737	FM-GPW-9417	GASKET, Fuel Pump to Block	2	3	4	5	7
WO-115656	FM-GPW-9364	GASKET, Fuel Pump Bowl	3	4	5	6	8
WO-A-1293	FM-GPW-9276	GASKET, Fuel Tank Gauge	1	1	2	2	3
WO-A-1292	FM-GPW-9275	GAUGE, Fuel Tank Unit, Assembly	0	1	1	1	2
WO-A-6837	FM-GPW-18352	GASKET SET, Carburetor	2	2	3	4	5
WO-A-6883	FM-GPW-18337	GASKET SET, Fuel Strainer	5	6	8	10	13
WO-A-1311	FM-GPW-9652	HOSE, Flexible, Air Cleaner Tube to Carburetor	1	1	2	2	3
WO-A-6840	FM-GPW-18357B	KIT, Repair, Carburetor	1	2	2	3	4
WO-A-7834	FM-GPW-18373D	KIT, Repair, Fuel Pump	2	3	4	5	7
WO-7517	FM-GPW-9735B	KNOB, Throttle, Assembly	0	1	1	1	2
WO-A-6851	FM-GPW-9350	PEDAL, Accelerator, Assembly	0	1	1	1	2
WO-A-8323	FM-GPW-9799	PUMP, Fuel Assembly	0	1	1	1	2
WO-633011	FM-GPW-9140	SPRING, Retracting, Accelerator Pedal	1	2	2	3	4
WO-A-1261	FM-GPW-9289	STRAINER, Fuel	0	1	1	1	2
WO-A-1368		TUBE, Flexible Connection to Fuel Pump, Assembly	0	1	1	1	2

GROUP 04—EXHAUST

WO-634814	FM-GPW-9450	GASKET, Exhaust Pipe Flange	4	5	6	8	11
WO-A-658	FM-GPW-5283	INSULATOR, Muffler Support	0	1	1	1	2
WO-A-6118	FM-GPW-5230B	MUFFLER, Assembly (oval type)	4	5	6	8	11
WO-A-10198	FM-GPW-5246B	PIPE, Muffler, Assembly	4	5	6	8	11

ORDNANCE MAINTENANCE UNIT STOCKAGE LIST—Cont'd

Figure Number	Official Stockage Number	Part Number		ITEM	Parts Allowances For:				
		Willys	Ford		1-9 Vehicles	10-35 Vehicles	36-75 Vehicles	76-150 Vehicles	151-300 Vehicles
Col. 1	Col. 2	Col. 3	Col. 4	Col. 5	Col. 6	Col. 7	Col. 8	Col. 9	Col. 10
				GROUP 05—COOLING					
		WO-A-1495	FM-GPW-8620A1	BELT, Drive, Fan and Generator	5	6	8	10	13
		WO-A-1215	FM-GPW-8100A	CAP, Radiator	1	1	2	2	3
		WO-52226	FM-60-8287	CLAMP, Hose, 1 13/16 in. Dia.	7	9	12	16	21
		WO-A-1214	FM-GPW-8005	CORE, Radiator, w/TANK and SHROUD, Assembly	0	1	1	1	2
		WO-A-447	FM-GPW-8600	FAN, Assembly	0	1	1	1	2
		WO-A-1216	FM-K-7129-B	GASKET, Radiator Filler Cap	2	3	4	5	7
		WO-639650	FM-GPW-8255	GASKET, Engine Cylinder Head Water Outlet Elbow	1	1	2	2	3
		WO-637053	FM-GPW-8543	GASKET, Water Pump to Cylinder Block	2	2	3	4	5
		WO-A-592	FM-GPW-8284	HOSE, Radiator Water Outlet, Lower	9	12	16	21	28
		WO-A-6839	FM-GPW-18515B	KIT, Repair, Water Pump	2	2	3	4	5
		WO-636299	FM-GPW-8509A	PULLEY, Fan and Water Pump	0	1	1	1	2
		WO-639992	FM-GPW-8501	PUMP, Water, Assembly	0	1	1	1	2
		WO-637646	FM-GPW-8575	THERMOSTAT, Assembly	1	1	2	2	3
				GROUP 06—ELECTRICAL					
		WO-A-8186	FM-GPW-10850	AMMETER	0	1	1	1	2
		WO-A-1637	FM-GPW-10005	ARMATURE, Generator, Assembly	0	1	1	1	2
		WO-A-1568	FM-GPW-11005	ARMATURE, Cranking Motor, Assembly	0	1	1	1	2
		WO-109452	FM-GPW-11077	BAND, Head, Cranking Motor, Assembly	0	1	1	1	2
		WO-A-1767	FM-11AS-10658	BATTERY (Dry-Charged)	4	5	6	8	11
		WO-A-6299	FM-GPW-10094	BEARING, Ball, Generator (ND. 77503)	0	1	1	1	2
		WO-A-1651	FM-GPW-18274	BRUSH SET, Generator	1	1	2	2	3
		WO-A-1573	FM-GPW-11350	DRIVE, Bendix, Assembly	0	1	1	1	2
		WO-335912	FM-350343-S-16	BOLT, Battery Terminal Clamp w/NUT, Lead Plated, Assembly	2	3	4	5	7
		WO-109431	FM-GPW-11055	BRUSH, Insulated, Cranking Motor (AL MZ-12)	1	1	2	2	3
		WO-A-1452	FM-GPW-14300	CABLE, Battery to Starting Switch, Positive	1	1	1	1	2
		WO-A-1655	FM-GPW-12106	CAP, Distributor, Assembly	1	1	2	2	3
		WO-A-1733	FM-GPW-12250A	BREAKER, Circuit, 5 Amps	0	1	1	1	2
		WO-A-1734	FM-GPW-12250B	BREAKER, Circuit, 15 Amps	0	1	1	1	2
		WO-A-1349	FM-GPW-12250C	BREAKER, Circuit, 30 Amps	0	1	1	1	2
		WO-A-7792	FM-GPW-12000B	COIL, Ignition, w/BRACKET and WASHER, Assembly	0	1	1	2	3
		WO-A-1631	FM-GPW-12300	CONDENSER, Distributor, Assembly	2	2	3	4	5
		WO-635981	FM-GPW-14487B	CONNECTOR (2 Wire)	0	1	1	1	2
		WO-A-1687	FM-GPW-18354	CONTACT SET, Distributor	5	7	9	12	16
		WO-A-1244	FM-GPW-12100	DISTRIBUTOR, Assembly	0	1	1	1	2
		WO-A-11760	FM-GPW-18276	ELECTROLYTE, Battery, w/CONTAINER, 160 oz., Assembly	4	5	6	8	11
		WO-A-5980	FM-GPW-18960B	FILTER, and BRACKET, Radio Assembly	0	1	1	1	2
		WO-A-5337	FM-GPW-18935A	FILTER UNIT, Generator Ground	0	1	1	1	2
		WO-637863	FM-01A-12410	GASKET, Spark Plug	2	2	3	4	5

WO-A-8184	FM-GPW-9280	GAUGE, Fuel, Dash Unit	0	1	1	1	2
WO-A-8190	FM-GPW-9273	GAUGE, Oil Pressure, Dash Unit	0	1	1	1	2
WO-A-5992	FM-GPW-10000A	GENERATOR, Assembly	0	1	1	1	2
WO-A-8188	FM-GPW-10883	INDICATOR, Heat	0	1	1	1	2
WO-A-7841	FM-GPW-18319	KIT, Repair, Cranking Motor Field Coil	0	1	1	1	2
WO-A-7843	FM-GPW-18343	KIT, Repair, Distributor	0	1	1	1	2
WO-A-9055	FM-GPW-18363C	KIT, Repair, Generator	0	1	1	1	2
WO-A-7794	FM-GPW-18298	KIT, Repair, Generator Cutout Relay	0	1	1	1	2
WO-A-7805	FM-GPW-18299	KIT, Repair, Generator Current Regulator	0	1	1	1	2
WO-A-7810	FM-GPW-18311	KIT, Repair, Generator Voltage Regulator	0	1	1	1	2
WO-A-7844	FM-GPW-18364B	KIT, Repair, Ignition Wiring	1	1	2	2	3
WO-52837	FM-48-15021	LAMP, Electric 1 CP 6 Volt, Instrument Light	2	3	4	5	7
WO-51804	FM-B-13466	LAMP, Electric 3 CP 6 Volt, Blackout Marker Light	2	3	4	5	7
WO-A-1304	FM-GPW-13006	LIGHT, Head, L. H. Assembly	0	1	1	1	2
WO-A-1305	FM-GPW-13005	LIGHT, Head, R. H. Assembly	0	1	1	1	2
WO-A-6145	FM-GPW-13152	LAMP UNIT, Blackout Driving Headlight, Sealed, Assembly	1	1	2	2	2
WO-A-1075	FM-GP-13491A	LAMP UNIT, Blackout Tail, Sealed, Assembly	1	1	2	2	2
WO-A-1078	FM-GP-13485A	LAMP UNIT, Blackout Stop, Sealed, Assembly	1	2	2	2	3
WO-A-1033	FM-GPW-13007	LAMP UNIT, Head Light, Sealed, Assembly	2	2	3	4	5
WO-A-1074	FM-GP-13494A	LAMP UNIT, Service Tail and Stop, Sealed, Assembly	1	1	2	2	3
WO-A-6142	FM-GPW-13150	LIGHT, Driving, Blackout, Assembly	0	1	1	1	2
WO-A-1245	FM-GPW-11001	MOTOR, Cranking, Assembly	0	1	1	1	2
WO-A-1566	FM-GPW-11049	PLATE, Cranking Motor, Commutator End, Assembly	0	1	1	1	2
WO-A-538	FM-GPW-12405	PLUG, Spark, w/Gasket Assembly	21	28	37	49	65
WO-A-1639	FM-GPW-10505	PULLEY, Drive, Generator, w/SPACER, assembly	0	1	1	1	2
WO-A-1409	FM-GPW-12200	REGULATOR, Generator	0	1	1	1	2
WO-A-1658	FM-GPW-14301	ROTOR	2	2	3	4	5
WO-635883	FM-GPW-18812B	STRAP, Battery Ground	0	1	1	1	2
WO-A-6320	FM-GPW-11649	SUPPRESSOR, Spark Plug	2	2	3	4	5
WO-A-1332	FM-GTB-13739	SWITCH, Lighting, Assembly	0	1	1	1	2
WO-A-6149	FM-9N-11450A	SWITCH, Blackout Driving Light	0	1	1	1	2
WO-A-7225	FM-GPW-13532	SWITCH, Starting	1	2	2	3	4
WO-638979	FM-GPW-13532	SWITCH, Dimmer, Foot, Headlight, Assembly	1	1	2	2	3
WO-A-6811	FM-GPW-3686B	SWITCH, Keyless, Ignition, Assembly	0	1	1	1	2
WO-A-1271	FM-11A-13480	SWITCH, Stop Light, Assembly	0	1	1	1	2
WO-371400	FM-11A-14452	TERMINAL, Battery Cable, $\frac{11}{32}$	0	1	1	1	2
WO-372668	FM-19B-14451	TERMINAL, Battery, Positive Cable Clamp	0	1	1	1	2

GROUP 07—TRANSMISSION

WO-637495	FM-GPW-7051B	GASKET, Transmission Case to Flywheel Housing	1	2	2	3	4
WO-A-7832	FM-GPW-18356B	GASKET SET, Transmission	2	2	3	4	5
WO-A-7837	FM-GPW-18314	KIT, Repair, Speedometer Drive Gear	0	1	1	1	2
WO-640018	FM-GPW-7052	OIL, Seal Transmission Front Bearing Retainer	0	1	1	1	2
WO-A-7596	FM-GPW-7008	TRANSMISSION, Assembly	0	1	1	1	2

GROUP 08—TRANSFER CASE

WO-A-1195	FM-GPW-7700	CASE, Transfer Assembly	0	1	1	1	2
WO-51575	FM-GP-7723	CONE and ROLLERS, Bearing, 1.3125 Diam. x .740 (Transfer Case and Output Clutch Shaft) (Bow-BT-14131) TIM-14131	0	1	1	1	2

ORDNANCE MAINTENANCE UNIT STOCKAGE LIST—Cont'd

Figure Number	Official Stockage Number	Part Number - Willys	Part Number - Ford	ITEM	1-9 Vehicles	10-35 Vehicles	36-75 Vehicles	76-150 Vehicles	151-300 Vehicles
Col. 1	Col. 2	Col. 3	Col. 4	Col. 5	Col. 6	Col. 7	Col. 8	Col. 9	Col. 10
				GROUP 08—TRANSFER CASE—Cont'd					
		WO-52883	FM-01Y-1202	CUP, Bearing, Transfer Case Output Shaft and Output Clutch Shaft (BOW-BT-14276) (TIM 14276)					
		WO-A-1106	FM-GP-7729	FLANGE AND SEALER, Transfer Case Front U Joint, Assembly	0	1	1	1	2
		WO-A-1509	FM-GPW-7707	GASKET, Transfer Case Cover Rear	0	1	1	1	2
		WO-A-7445	FM-GPW-18317	OIL SEAL SET, Transfer Case	1	1	2	2	3
				GROUP 09—PROPELLER SHAFT, UNIVERSAL JOINTS					
		WO-A-942	FM-GP-7077	CAP, Dust, Universal Joint Yoke Sleeve	0	1	1	1	2
		WO-A-1433	FM-GPW-18397	KIT, Repair, Universal Joint Journal	6	8	11	14	19
		WO-A-1326	FM-GPW-3365	SHAFT, Propeller, Front, Assembly	0	1	1	1	2
		WO-A-1327	FM-GPW-4602	SHAFT, Propeller, Rear, Assembly	0	1	1	1	2
		WO-A-490	FM-02Y-4529	U BOLT, Universal Joint	1	1	2	2	3
		WO-A-943	FM-GP-7097	WASHER, Universal Joint Yoke Sleeve	1	1	2	2	3
		WO-A-935	FM-GP-4841	YOKE, Universal Joint Sleeve	0	1	1	1	2
				GROUP 10—FRONT AXLE					
		WO-A-8249	FM-GPW-3131	BELL CRANK, Drag Link, Front	0	1	1	1	2
		WO-A-1712	FM-GPW-3113	ARM, Steering Knuckle, Upper L. H. Assembly	0	1	1	1	2
		WO-A-1710	FM-GPW-3112	ARM, Steering Knuckle, Upper R. H. Assembly	0	1	1	1	2
		WO-A-6029		AXLE, Front, Assembly	0	1	1	1	2
		WO-52940	FM-GP-3161	CONE and ROLLERS, Front Axle Spindle Bearing (Timken 11590)	0	1	1	1	2
		WO-52940	FM-GP-3161	CONE and ROLLERS, King Pin Bearing Steering Pivot Arm	0	1	1	1	2
		WO-52941	FM-GP-3162	CUP, Front Axle Spindle Bearing (TIM-11320)	0	1	1	1	2
		WO-52941	FM-GP-3162	CUP, King Pin Bearing (Steering Pivot Arm)	0	1	1	1	2
		WO-A-782	FM-GP-4035	GASKET, Front Axle Housing Cover	2	2	3	4	5
		WO-A-7830	FM-GPW-18365B	GASKET SET, Front Axle	1	1	2	2	3
		WO-A-820	FM-GP-1092	GASKET, Steering Knuckle Oil Seal, Assembly (Spicer 1704)	7	9	12	16	21
		WO-A-6816	FM-GPW-18384	KIT, Repair, Drive Gear and Pinion	0	1	1	1	2
		WO-A-6743	FM-GPW-18389	KIT, Repair, Front Axle Differential Gear	0	1	1	1	2
		WO-A-6882	FM-GPW-18338	SHIM SET, King Pin Bearing	0	1	1	1	2
		WO-A-6881	FM-GPW-18336	SHIM SET, Universal Joint Adjusting	0	1	1	1	2
		WO-A-812	FM-GP-3149A	KNUCKLE, Steering L. H.	0	1	1	1	2
		WO-A-811	FM-GP-3148A	KNUCKLE, Steering R. H.	0	1	1	1	2
		WO-A-A-819	FM-GP-3135	OIL SEAL, Felt (Half) (Spicer 17019)	8	11	14	18	25
		WO-A-818	FM-GP-3139	OIL SEAL, Felt Pressure Strip (Spicer 16983)	8	11	14	18	25
		WO-A-813	FM-GP-1088	OIL SEAL, Steering Knuckle	1	1	2	2	3
		WO-A-779	FM-GP-3034	OIL SEAL, Carrier, End	1	2	2	3	4
		WO-A-778	FM-GP-3031	RETAINER, Grease, Wheel, End	1	2	2	3	4

WO-A-814	FM-GP-1089	RETAINER, Pivot Oil Seal, Assembly	1	2	2	3	4
WO-A-1728	FM-GPW-3227A	SHAFT, Axle, Left, w/Universal JOINT, Assembly	0	1	1	1	2
WO-A-809	FM-GPW-3206A	SHAFT, Axle, Right, w/Universal JOINT, Assembly	0	1	1	1	2
**WO-A-8836		SHAFT, Axle Front Left, Complete w/Universal JOINT, Assembly (Sp-17128-4X)	0			1	2
**WO-A-8637		SHAFT, Axle Front Right, Complete w/Universal JOINT, Assembly (Sp-17128-3X)	0			1	2
WO-A-855	FM-GPW-3165	SHAFT, Steering Bell Crank	1	1	1	2	3
WO-A-851	FM-GPW-3105	SPINDLE, Front Wheel Bearing, Assembly	0	1	1	1	2
WO-A-5504	FM-GPW-3325	STUD, Steering Arm Dowel	1	2	2	3	4
WO-A-859	FM-GPW-3168	WASHER, Steering Arm, Lower	0	1	1	1	2
WO-A-860	FM-GPW-3169	WASHER, Steering Arm, Upper	0	1	1	1	2
WO-A-851	FM-GPW-3105	SPINDLE, Wheel Bearing, w/BUSHING, Assembly (Spicer 17202-X)	0	1	1	1	2

GROUP 11—REAR AXLE

WO-A-788	FM-GP-4035	DIFFERENTIAL, Assembly (Spicer 16968-X) Front and Rear	0	1	1	1	2
WO-A-782	FM-GP-4032	GASKET, Axle Housing Cover	2	2	3	4	5
WO-A-904		GASKET, Rear Axle Shaft	7	9	12	16	21
WO-A-7831	FM-GPW-18366	GASKET, Set, Rear Axle	1	1	2	2	3
WO-A-6743	FM-GPW-18389	KIT, Repair, Rear Axle Differential Gear	0	1	1	1	3
WO-A-779	FM-GP-3034	OIL SEAL, Axle Inboard (Spicer 17036)	1	2	2	3	4
WO-639265	FM-GP-4676	RETAINER, Differential Pinion	1	2	2	2	3
WO-A-902	FM-GPW-4235	SHAFT, Rear Axle L. H.	0	1	1	1	2
WO-A-901	FM-GPW-4234	SHAFT, Rear Axle R. H.	0	1	1	1	2
WO-A-6745	FM-GPW-18386	SHIM SET, Rear, Axle Drive Pinion	0	1	1	1	2
WO-A-6744	FM-GPW-18388	SHIM SET, Rear Axle Differential Bearing	0	1	1	1	2

GROUP 12—BRAKES

WO-A-1009	FM-GP-2648	BAND, Hand Brake Assembly	0	1	1	1	2
WO-116549	FM-GP-2018	BRAKE, Shoe and Lining Assembly Forward	0	1	1	1	2
WO-116550	FM-GP-2019	BRAKE, Shoe and Lining Assembly Rear	0	1	1	1	2
WO-A-1241	FM-GPW-2853	CABLE, Hand Brake Ratchet, w/TUBE, Assembly	0	1	1	1	2
WO-A-1484	FM-GPW-2061	CYLINDER, Front Wheel Brake	0	1	1	1	2
WO-A-556	FM-GP-2140	CYLINDER, Master, Assembly	0	1	1	1	2
WO-637787	FM-GP-2261	CYLINDER, Rear Brake Lo-FD-4665	0	1	1	1	2
WO-A-6110	FM-GPW-2261	CYLINDER, Rear Wheel Brake Assembly ¾ in. Dia. (Lo-FD-7568) (After Serial No. 134356)	0	1	1	1	2
WO-A-754	FM-GP-2038	ECCENTRIC, Brake Shoe	1	1	2	2	3
WO-637426	FM-GP-2087	GASKET, Fitting, Brake Hose (Lo-FC-5795)	4	5	7	9	12
WO-637612	FM-GP-2167	GASKET, Master Cylinder Cap (Lo-FC-6019)	1	1	2	2	3
WO-A-1460	FM-GPW-2079	HOSE, Front Brake, 6 in. Long	1	2	2	3	4
WO-A-1373	FM-GPW-2078	HOSE, Front Brake, 11 in. Long	1	1	2	2	3
WO-637424	FM-GP-2078	HOSE, Rear Brake, 15 in. Long	1	1	2	2	3
WO-116600	FM-GPW-18367	LINING SET, Brake Shoe w/RIVETS	1	1	2	2	3
WO-A-6759	FM-GPW-18377	LINING SET, Hand Brake w/RIVETS	0	1	1	1	2
WO-A-7838	FM-GPW-18370B	KIT, Repair, Master Cylinder	1	1	1	1	2
WO-115962	FM-GPW-18371	KIT, Repair, Front Wheel Cylinder	1	1	2	2	3

**Willys only

ORDNANCE MAINTENANCE UNIT STOCKAGE LIST—Cont'd

Figure Number	Official Stockage Number	Part Number		ITEM	Parts Allowances For:					
		Willys	Ford		1-9 Vehicles	10-35 Vehicles	36-75 Vehicles	76-150 Vehicles	151-300 Vehicles	
Col. 1	Col. 2	Col. 3	Col. 4	Col. 5	Col. 6	Col. 7	Col. 8	Col. 9	Col. 10	

GROUP 12—BRAKES—Cont'd

		Willys	Ford	Item	Col. 6	Col. 7	Col. 8	Col. 9	Col. 10
		WO-A-6133	FM-GPW-18368	KIT, Repair, Rear Wheel Cylinder	1	1	2	2	3
		WO-637899	FM-91A2027	PIN, Brake, Anchor	0	1	1	1	2
		WO-637540	FM-GP-2208	SCREW, Brake Wheel Cylinder Bleeder	1	1	2	2	3
		WO-630593	FM-GPW-7523	SPRING, Brake Pedal Return	1	1	2	2	3
		WO-637905	FM-GP-2035	SPRING, Return, Brake Shoe	2	2	3	4	5
		WO-A-5335	FM-GPW-2635	SPRING, Retracting, Hand Brake Lever	0	1	1	1	2
		WO-A-1017	FM-01T-2634	SPRING, Releasing Hand Brake	0	1	1	1	2
		WO-A-1488	FM-GPW-2298	TUBE, Brake, Assembly	0	1	1	1	2

GROUP 13—WHEELS, HUBS AND DRUMS

		Willys	Ford	Item	Col. 6	Col. 7	Col. 8	Col. 9	Col. 10
		WO-A-5470	FM-GP-1029	BOLT, Wheel, Divided Rim	2	2	3	4	5
		WO-A-473	FM-GP-1108	BOLT, Wheel Hub, L. H.	1	1	2	2	3
		WO-A-474	FM-GP-1107	BOLT, Wheel Hub, R. H.	1	1	2	2	3
		WO-A-760	FM-GP-1110	BOLT, Wheel Hub to Flange, Front and Rear	2	2	3	4	5
		WO-A-869	FM-GP-1139	CAP, Wheel Hub	0	1	1	1	2
		WO-52942	FM-GP-1201	CONE and ROLLERS, Front and Rear Wheels Inner and Outer Bearing	1	1	2	2	3
		WO-52943	FM-GP-1202	CUP, Front and Rear Wheels Inner and Outer Bearing	1	1	2	2	3
		WO-A-1690	FM-GP-1103	HUB, Front and Rear, Left, w/DRUM, Assembly L. H.	0	1	1	1	2
		WO-A-1689	FM-GP-1102	HUB, Front and Rear, Right, w/DRUM, Assembly R. H.	0	1	1	1	2
		WO-A-866	FM-GP-4252	NUT, Front and Rear Axle Wheel Bearing	1	1	2	2	3
		WO-A-5471	FM-GPW-1030	NUT, Hex, S., 1/2-20NF-2, Right Hand Thread (Wheel Hub Bolt)	1	2	2	3	4
		WO-A-475	FM-GP-1013	NUT, Wheel Bolt, L. H.	1	2	2	3	4
		WO-A-476	FM-GP-1012	NUT, Wheel Bolt, R. H.	1	2	2	3	4
		WO-A-864	FM-GP-1177	RETAINER, Grease, Wheel Hub, Assembly	15	20	26	32	43
		WO-A-865	FM-GP-1218	WASHER, Wheel Bearing	0	1	1	1	2
		WO-A-867	FM-GP-1124	WASHER, Lock, Wheel Bearing Nut	1	1	2	2	3

GROUP 14—STEERING

		Willys	Ford	Item	Col. 6	Col. 7	Col. 8	Col. 9	Col. 10
		WO-A-857	FM-GPW-3171	BEARING, Steering Bell Crank	0	1	1	1	2
		WO-A-847	FM-GPW-3292	END, Steering Knuckle Tie Rod L. H., Assembly	0	1	1	1	2
		WO-A-838	FM-GP-3291	END, Steering Knuckle Tie Rod R. H., Assembly	0	1	1	1	2
		WO-A-6791	FM-GPW-18383	KIT, Repair, Steering Connecting Rod	0	1	1	1	2
		WO-A-8250	FM-GPW-3304	ROD, Connecting	0	1	1	1	2
		WO-A-1709	FM-GPW-3282	ROD, Tie Steering L. H., Assembly	0	1	1	1	2
		WO-A-1705	FM-GPW-3281	ROD, Tie Steering R. H., Assembly	0	1	1	1	2
		WO-A-861	FM-GPW-3170	SHIELD, Dust, Steering, Bell Crank Bearing (SP-17205)	0	1	1	1	2
		WO-A-1239	FM-GPW-2504	STEERING, Gear, w/WHEEL Assembly	0	1	1	1	2
		WO-A-6858	FM-GPW-3600A	WHEEL, Steering, Assembly	0	1	1	1	2

GROUP 16—SPRING

Part No.	Ref.	Description					
*	FM-GPW-18045	ABSORBER, Shock, Front, Assembly	1	1	2	2	3
*	FM-GPW-18080	ABSORBER, Shock, Rear, Assembly	1	1	2	2	3
**WO-A-6902		ABSORBER, Shock, Front	0	1	1	1	2
**WO-A-6903		ABSORBER, Shock, Rear	0	1	1	1	2
WO-116609	FM-GPW-5345A	BOLT, Front Spring, Center	5	7	9	12	16
WO-116610	FM-GPW-5345B	BOLT, Rear Spring, Center	2	3	4	5	7
WO-384228	FM-GPW-5468	BOLT, Pivot Spring Shackle	1	1	2	2	3
WO-A-513	FM-GPW-5778	BOLT U, Shackle, L. H.	0	1	1	1	2
WO-A-514	FM-GPW-5779	BOLT U, Shackle, R. H.	0	1	1	1	2
WO-637936	FM-GPW-18060	BUSHING, Shock Absorber Mounting Pin	2	2	3	4	5
WO-A-8255	FM-GPW-5463	BUSHING, Spring Shackle, L. H. Thread	1	1	2	2	3
WO-8256	FM-GPW-5464	BUSHING, Spring Shackle, R. H. Thread	2	2	3	4	5
WO-116459	FM-GPW-5724A	CLAMP, Rear Spring	1	2	2	3	4
WO-A-1252	FM-GPW-5482	RETAINER, Spring Shackle Grease Seal	1	1	2	2	3
WO-A-515	FM-GPW-5481	SEAL, Grease, Spring Shackle	2	3	4	5	7
WO-A-613	FM-GPW-5310	SPRING, Front Right	1	1	2	2	3
WO-A-612	FM-GPW-5311	SPRING, Left Front	1	1	2	2	3
WO-A-614	FM-GPW-5560	SPRING, Rear	0	1	1	1	2

GROUP 18—BODY AND HULL

Part No.	Ref.	Description					
WO-A-2512	FM-GP-17535	ARM, Windshield Wiper w/BLADE	4	5	6	8	11
WO-A-3197	FM-GP-1103027	CATCH, Hold Down w/SHIELD	0	1	1	1	2
WO-A-2478	FM-GP-1103100	GLASS, w/Shield, Assembly	1	1	2	2	3

GROUP 22—MISCELLANEOUS

Part No.	Ref.	Description					
WO-A-7791	FM-GPW-1509B	CASING, Pneumatic Tire 600 x 16—6 Ply	6	8	11	15	20
WO-A-1343	FM-GPW-17261	CASING, Speedometer Shaft	1	1	2	2	3
WO-A-2934	FM-21CS-17682-B	MIRROR, Rear View, Outside	0	1	1	1	2
WO-A-1344	FM-GPW-17262	SHAFT, Speedometer	2	2	3	4	5
WO-A-8180	FM-GPW-7255A	SPEEDOMETER, Assembly	0	1	1	1	2
WO-A-457	FM-GPW-1655	TUBE, Pneumatic Inner 600 x 16 (Heavy Duty)	6	8	11	15	20

GROUP 23—GENERAL USE STANDARDIZED PARTS

Part No.	Ref.	Description					
WO-A-7682	FM-GPW-18320	CORE, Kit Tire Valve	4	5	6	8	11
WO-638224	FM-353035-S7	FITTING, Grease 45°	1	2	2	3	4
WO-638500	FM-353023-S7	FITTING, Grease 90°	1	2	2	3	4
WO-640038	FM-350006-S8	FITTING, Grease (AD 1980)	1	2	2	3	4
WO-392909	FM-353027-AS7	FITTING, Grease Straight (AD 1720)	3	4	5	6	8
WO-638792	FM-353043-S7	FITTING, Grease Straight 1/4-28NF (AD 1641)	1	2	2	3	4

*Ford only.
**Willys only

SECTION 6

DEPOT STOCKAGE LIST

To determine the quantities of parts required by the various depots, the following procedures will be used:

1. FIELD DEPOT—The figures shown in the 12 Months Field Maintenance Column, No. 7, in Section 2, Maintenance Parts Procurement List, will be used as the basis for computing normal requirements. For example, 1 month of supply for 13 major items is the nearest whole equivalent of 1/12 x 13 items/100 items x the quantities listed in the 12 Months Field Maintenance Column.

2. BASE DEPOT—The figures shown in the Total First Year Procurement Column, No. 9, in Section 2, will be used as the basis for computing normal requirements. For example, 1 month of supply for 17 major items is the nearest whole equivalent of 1/12 x 17 items/100 items x the quantities listed in the Total First Year Procurement Column.

3. MOBILE DEPOT—The figures shown in the Ordnance Maintenance Unit Stockage List will be used as a basis for computing normal requirements. Using the proper column for the number of major items to be serviced and taking into consideration the time intervals being employed as the basis for computation, the quantities of parts required may be determined.

SECTION 7

GEOGRAPHICAL OR SEASONAL MAINTENANCE PARTS LISTS

Information to be included
when data becomes available

SECTION 8

INDEXES

	Page
Illustration Index	246
Alphabetical Index	247
Manufacturers' Numerical Index	260

FIGURE INDEX

Figure No.	Title	Page
A	¼-Ton, 4 x 4 Command Reconnaissance Truck, ¾ Front View...	11
B	¼-Ton, 4 x 4 Command Reconnaissance Truck, ¾ Rear View...	12
C	¼-Ton, 4 x 4 Command Reconnaissance Truck, Front View...	13
D	¼-Ton, 4 x 4 Command Reconnaissance Truck, Rear View...	14
E	¼-Ton, 4 x 4 Command Reconnaissance Truck, Top View...	15
F	Power Plant Assembly, Right Front View...	16
G	Power Plant Assembly, Left Front View...	17
H	Engine, Cross Sectional End View...	18
01-1	Engine, Side View...	21
01-2	Cylinder Block Assembly...	22
01-3	Cylinder Head...	24
01-4	Crankshaft and Piston Assembly...	25
01-5	Camshaft, Chain, Sprockets, Valves...	30
01-6	Valve Cover Plate with Crankcase Ventilator...	32
01-7	Oil Pump Assembly...	34
01-8	Oil Pan Assembly...	36
01-9	Oil Filter Assembly...	38
01-10	Oil Filler Tube and Level Indicator...	39
01-11	Manifold Assembly...	42
01-12	Timing Chain Cover, Cylinder Block and Flywheel Housing...	44
01-13	Crankshaft Chain, and Sprockets...	46
02-1	Clutch Assembly—Sectional View...	48
02-2	Clutch Assembly...	49
02-3	Clutch Controls...	52
03-1	Fuel Tank Assembly...	53
03-2	Carburetor...	56
03-3	Oil Bath Air Cleaner...	62
03-4	Fuel Pump Assembly...	64
03-5	Accelerator Throttle and Choke Controls...	67
03-6	Fuel Filter Assembly...	69
04-1	Exhaust System...	71
05-1	Radiator...	74
05-2	Water Pump Assembly...	75
06-1	Generator Assembly...	78
06-2	Voltage Regulator...	84-85
06-3	Cranking Motor Assembly...	89
06-4	Distributor Assembly...	92
06-5	Wiring Installation...	100
06-6	Trailer Electric Coupling Socket...	101
06-7	Headlight Assembly...	104
06-8	Blackout Driving Light Assembly...	104
06-9	Marker Light...	105
06-10	Tail and Stop Light Assembly...	105
07-1	Transmission Assembly...	110
07-2	Transmission Assembly, Cross Sectional View...	113
08-1	Transfer Case Assembly...	116
08-2	Transfer Case, Cross Sectional View...	119
09-1	Rear Propeller Shaft...	122
10-1	Front Axle...	124
10-2	Front Axle, Steering Knuckle and Wheel Bearing...	130
10-3	Differential Assembly, Cross Sectional View...	126
11-1	Rear Axle Assembly...	134
12-1	Hand Brake Assembly...	138
12-2	Brake Assembly...	141
12-3	Brake Master Cylinder...	143
12-4	Wheel Brake Cylinder Assembly...	143
12-5	Brake System...	146
13-1	Front Wheel and Spindle, Cross Sectional View...	150
13-2	Wheel Assembly...	151
14-1	Steering Connecting Rod...	153
14-2	Steering Gear Assembly...	155
15-1	Frame Assembly...	158
15-2	Pintle Hook Assembly...	161
16-1	Front Spring and Torque Reaction Spring...	163
16-2	Shock Absorber Assembly...	167
18-1	Front Compartment...	176
21-1	Radiator Guard...	182
23-1	Tools...	187
23-2	Cold Starting Blanket Field Kit...	196
23-3	Cold Starting Heater Field Kit...	197
23-4	Desert Cooling Kit...	200
23-5	Deep Water Exhaust Kit...	201
1	Vehicular Spare Parts...	220
2	Vehicular Accessories...	221

ALPHABETICAL INDEX

Description	Gov't Group No.
ABSORBER, shock, front, assembly (Gabriel)	1603
ABSORBER, shock, front, assembly	1603
ABSORBER, shock, rear, assembly (Gabriel)	1603
ABSORBER, shock, rear, assembly	1603
ADAPTER, grease gun push type	2301
ADAPTER, instrument lamp	0605C
AMMETER	0605
ANTI-SQUEAK, front fender to cowl side panel, lower	1701
ANTI-SQUEAK, front fender to cowl side panel, upper	1701
ANTI-SQUEAK, gasoline tank	0300
ANTI-SQUEAK, gasoline tank hold-down strap	0300
ANTI-SQUEAK, gasoline tank hold-down strap, inner	0300
ANTI-SQUEAK, apron to frame	1702
APRON, splash, front fender, left, assembly	1702
APRON, splash, front fender, right, assembly	1702
ARM, adjusting, windshield	1811
ARM, advance control, assembly	0603
ARM, breaker, assembly	0603
ARM, fuel pump, rocker, assembly	0302
ARM, generator brush	0601
ARM, knuckle steering, upper left	1006
ARM, knuckle steering, upper right	1006
ARM, pump, w/COLLAR assembly	0301
ARM, steering gear	1403A
ARM, throttle shaft, w/SCREW	0301
ARM, windshield wiper, w/BLADE, assembly	1811D
ARM, windshield wiper, w/CLIP, assembly	1811D
ARMATURE, cranking motor, assembly	0602
ARMATURE, current regulator coil, assembly	0601B
ARMATURE, cut-out relay coil, assembly	0601B
ARMATURE, generator, assembly	0601
ARMATURE, voltage regulator coil, assembly	0601B
AXLE, front, assembly (Weiss joints)	1000
AXLE, front, assembly (Rzeppa joints)	1000
AXLE, front, assembly (Spicer joints)	1000
AXLE, front, assembly (Tracta joints)	1000
AXLE, rear, assembly	1100
AXLE, front, w/BRAKES, DRUMS and HUBS, assembly	1000
AXLE, front, w/BRAKES, DRUMS, and HUBS, assembly (Ford only)	1000
AXLE, front, w/BRAKES, DRUMS, and HUBS, assembly (Bendix joints)	1000
AXLE, front, w/BRAKES, DRUMS, HUBS, assembly (Rzeppa joints	1000
AXLE, front, w/BRAKES, DRUMS, HUBS, assembly (Tracta joints)	1000
AXLE, rear, w/BRAKES, DRUMS, and HUBS, assembly	1100
BAFFLE, crankcase ventilator	0107N
BAG, tool	2301
BAIL, gas strainer, assembly	0302
BALL, fulcrum, gearshift lever	0706A
BALL, gearshift lever (rubber)	0706A
BALL, poppet, shift rail	0706B
BALL, poppet, shift rod	0804
BALL, shift lever handle	0805B
BALL, steel (for cam)	1403A
BALL, universal joint (with Rzeppa joints)	1007
BAND, generator head	0601
BAND, hand brake, assembly	1201
BAND, head, cranking motor, assembly	0602
BAR, bumper, front	2101
BAR, tie, master cylinder	1205
BASE, carburetor, assembly	0301C
BASE, distributor, assembly	0603
BASE, front, shock absorber, assembly	1603
BASE, rear, shock absorber, assembly	1603
BATTERY (dry, charged)	0610
BATTERY (wet, charged)	0610
BATTERY, (wet dump, unfilled and uncharged)	0610
BEARING, absorbent bronze, distributor	0603
BEARING, annular, single row (main shaft rear end)	0704
BEARING, annular, single rod, annular, single oil shield, snap ring groove	0703
BEARING, annular, single rod, bore (output clutch shaft)	0803
BEARING, ball, generator	0601
BEARING, bell crank	1401
BEARING, clutch release	0203
BEARING, connecting rod	0104A
BEARING, connecting rod (.010" U.S.)	0104A
BEARING, connecting rod, (.020" U.S.)	0104A
BEARING, crankshaft, front lower	0102A
BEARING, crankshaft, front lower (010" U.S.)	0102A
BEARING, crankshaft, front lower (.020" U.S.)	0102A
BEARING, crankshaft, front upper	0102A
BEARING, crankshaft, front upper (.010" U.S.)	0102A
BEARING, crankshaft, front upper (.020" U.S.)	0102A
BEARING, crankshaft, center lower	0102A
BEARING, crankshaft, center lower (.010" U.S.)	0102A
BEARING, crankshaft, center lower (.020" U.S.)	0102A
BEARING, crankshaft, center upper	0102A
BEARING, crankshaft, center upper (.010" U.S.)	0102A
BEARING, crankshaft, center upper (.020" U.S.)	0102A
BEARING, crankshaft, rear lower	0102A
BEARING, crankshaft, rear lower (.010" U.S.)	0102A
BEARING, crankshaft, rear lower (.020" U.S.)	0102A
BEARING, crankshaft, rear upper	0102A
BEARING, crankshaft, rear upper (.010" U.S.)	0102A
BEARING, crankshaft, rear upper (.020" U.S.)	0102A
BEARING, heat control valve shaft	0108C
BEARING, intermediate, cranking motor	0602
BEARING, intermediate, assembly	0602
BEARING, intermediate gear, transfer case	0804
BEARING, roller, transmission	0703
BEARING, steering column, wheel end, assembly	1403A
BEARING, water pump w/SHAFT, assembly double row, annular	0503
BELT, drive, fan and generator	0503B
BELT, drive, fan and generator (Goodrich Neoprene)	0503B
BELLOWS, water pump seal	0503
BEARING SET, repair, crankshaft std.	0102A
BEARING SET, crankshaft (.010" U.S.)	0102A
BEARING SET, crankshaft (.020" U.S.)	0102A
BEARING SET, connecting rod, std.	0104A
BEARING SET, connecting rod, (.010" U.S.)	0104A
BEARING SET, connecting rod bearing (.020" U.S.)	0104A
BLADE, windshield wiper	1811D
BLOCK, adjusting, throttle rod	0303
BLOCK, cylinder, w/BEARING, assembly	0101
BLOCK, cylinder, w/BEARING, and piston, assembly	0101
BLOCK, junction, 6 post (body wiring harness)	0606E
BLOCK, junction, 2 post (headlight wires)	0606E
BODY, air cleaner, assembly	0301C
BODY, assembly	1800
BODY, assembly (includes FLOORS, PANELS)	1800
BODY, crankcase ventilator, assembly	0107N
BODY, fuel pump priming w/LEVER, assembly	0302
BODY, oil pump	0107
BODY, water pump seal	0503

ALPHABETICAL INDEX (Cont'd)

Description	Gov't Group No.
BOOT, master cylinder	1205
BOOT, wheel cylinder, front and rear	1207
BOOT, wheel cylinder, rear	1207
BOW, top, assembly	2201A
BOW, top front, assembly	2201A
BOWL, fuel pump	0302
BOWL, strainer, w/STUD, center, assembly	0306
BOX, radio terminal, assembly	2601
BRACE, front fender	1701
BRACE, front fender	1701
BRACE, generator w/HANDLE, assembly	0601A
BRACE, generator, short	0601A
BRACE, instrument panel circuit breaker, w/BRACKET, assembly	1809
BRACE, machine gun mounting plate	1500
BRACE, voltage regulator mounting	1702
BRACKET, accelerator cross shaft	0308
BRACKET, axle brake tube tee	1101
BRACKET, axle brake tube tee	1209C
BRACKET, choke lever, assembly	0301
BRACKET, clutch control frame	0204A
BRACKET, cross member intermediate, front	1500
BRACKET, adjusting, current regulator coil, assembly	0601B
BRACKET, engine crank retaining clamp, assembly	1802A
BRACKET, engine support, front, left	1500
BRACKET, engine support, front, left (lower half)	1500
BRACKET, engine support, front, right (lower half)	1500
BRACKET, engine support, front, left (upper half)	1500
BRACKET, engine support, front, right (upper half)	1500
BRACKET, engine mounting, rear	0110
BRACKET, front brake hose, left	1001
BRACKET, front brake hose, right	1001
BRACKET, pivot, front and rear spring	1500
BRACKET, front and rear spring shackle, assembly	1500
BRACKET, front seat pivot	1804
BRACKET, front shock absorber, w/SHAFT, right, assembly (frame end)	1500
BRACKET, front shock absorber, w/SHAFT, left, assembly (frame end)	1500
BRACKET, gas tank hold-down	0300
BRACKET, gas tank hold-down, front inner	0300
BRACKET, gas tank hold-down, rear inner	0300
BRACKET, gas tank hold-down, w/SUPPORT, rear inside, assembly	0300
BRACKET, hand brake ratchet tube, w/SPRING, GUIDE, and SCREW, assembly	1201
BRACKET, hand rail front left, assembly	2201A
BRACKET, hand rail front right, assembly	2201A
BRACKET, hood catch	1704
BRACKET, horn	0609
BRACKET, ignition coil	0604
BRACKET, master cylinder, assembly	1500
BRACKET, mounting, speedometer	2204
BRACKET, mounting, speedometer	2204
BRACKET, mounting, ammeter	0605
BRACKET, mounting, windshield, clamp catch	1811
BRACKET, mounting, fire extinguisher	2402
BRACKET, mounting, front brake hose (frame end)	1500
BRACKET, mounting, fuel gauge	0605
BRACKET, mounting, resistor	0601B
BRACKET, mounting, rear brake hose	1500
BRACKET, oil filter, assembly	0107F
BRACKET, pivot, rear seat frame	1804
BRACKET, oil can	2301
BRACKET, oil filler tube support	0107G

Description	Gov't Group No.
BRACKET, radiator	1500
BRACKET, radiator, left	1500
BRACKET, rear stop, front seat	1804
BRACKET, retaining, engine crank	1802A
BRACKET, retaining, rear seat back frame	1804
BRACKET, rear shock absorber w/SHAFT, left, assembly	1500
BRACKET, rear shock absorber, w/SHAFT, right, assembly	1500
BRACKET, shovel	2301A
BRACKET, spare gas can, assembly	1817A
BRACKET, stationary, cut-out relay w/CONTACT	0601B
BRACKET, support, air cleaner, right, assembly	0301C
BRACKET, support, air cleaner, left, assembly	0301C
BRACKET, support, body rear	1500
BRACKET, support, radiator guard	1500
BRACKET, support, radiator, right	1500
BRACKET, support, spare tire, assembly	1505
BRACKET, support, spare wheel, assembly	1505
BRACKET, tool compartment lock, assembly	1817
BRACKET, top bow pivot	2201A
BRACKET, windshield adjusting arm, w/BUSHING, assembly	1811
BRACKET, windshield pivot, assembly	1811
BRACKET, windshield, swing arm, w/STUD	1811
BRAKE, hand, assembly	1201
BRAKE, front left, assembly	1200
BRAKE, front right, assembly	1200
BRAKE, rear right, assembly	1200
BRAKE, rear right, assembly	1200
BRAKE, rear left, assembly	1200
BRAKE, rear left, assembly (¾ in.)	1200
BRAKE, rear right, assembly (⅞ in.)	1200
BREAKER, circuit	0606D
BREAKER, circuit (between ammeter and horn)	0606D
BREAKER, thermal circuit	0606D
BREATHER, transfer case, assembly	0801
BRUSH, contact, horn wire	0609B
BRUSH, cranking motor insulated	0602
BRUSH, grounded, assembly	0602
BRUSH, main	0601
BRUSH SET, generator	0601
BRUSH SET, service, cranking motor	0602
BUCKLE	2301A
BUMPER, axle, front and rear	1601B
BUMPER, hood top windshield, assembly	1704
BUMPERETTE, rear	2101
BUSHING, axle shaft	1007
BUSHING, camshaft (front)	0106A
BUSHING, countershaft gear	0704C
BUSHING, flywheel	0109
BUSHING, cranking motor	0602
BUSHING, front spring	1601
BUSHING, generator brace locking	0601A
BUSHING, steering gear, housing, inner	1403A
BUSHING, steering gear, housing, outer	1403A
BUSHING, insulating, generator	0601
BUSHING, insulating, terminal post	0602
BUSHING, main shaft second speed gear	0704
BUSHING, output clutch shaft, pilot	0803
BUSHING, reducing, pipe	0306B
BUSHING	0301C
BUSHING, rear spring	1601A

SNL G-503

ALPHABETICAL INDEX (Cont'd)

Description	Gov't Group No.
BUSHING, reverse idler gear	0704C
BUSHING, rubber	2204
BUSHING, rubber (controls through dash)	0301B
BUSHING, shock absorber mounting pin, rubber	1603
BUSHING, shock absorber spring seat	1603
BUSHING, speedometer drive gear	0810
BUSHING, spring shackle, assembly	1602
BUSHING, spring shackle, assembly right thd	1602
BUSHING, spring shackle, inner	1602
BUSHING, spring shackle, outer	1602
BUSHING, starting motor	0602
BUSHING, steering column support	1405
BUSHING, torque reaction spring	1601C
BUSHING, wheel bearing spindle	1006
BUTTON, horn	0609B
CABLE, blackout driving lamp connector to switch	0606B
CABLE, blackout driving light, assembly	0607D
CABLE, blackout tail lamp to connector, left	0606B
CABLE, clutch control lever	0203
CABLE, coil primary	0604A
CABLE, coil, secondary	0604A
CABLE, engine stay, assembly	0110
CABLE, filter (reg. batt.)	0606B
CABLE, filter	0606B
CABLE, gasoline gauge to circuit breaker	0605B
CABLE, gasoline gauge	0605B
CABLE, ground, head light	0606B
CABLE, ground, trailer coupling socket to body	0606F
CABLE, ground, w/TERMINALS, assembly	2601
CABLE, hand brake ratchet, w/TUBE, assembly	1201
CABLE, horn assembly	0609B
CABLE, horn to junction block	0609B
CABLE, ignition switch to ammeter to blackout switch	0605B
CABLE, positive, battery to starting switch	0610B
CABLE, spark plug No. 1	0604A
CABLE, spark plug No. 2	0604A
CABLE, spark plug No. 3	0604A
CABLE, spark plug No. 4	0604A
CABLE, starter switch to starter motor	0610B
CABLE, starting switch to radio body, assembly	2601
CABLE, voltage regulator to generator	0606B
CABLE, voltage regulator to junction block	0606B
CAGE, universal joint	1007
CAM, anchor pin	1203B
CAM, hand brake	1201
CAMSHAFT	0106A
CAN, oil	2301
CAP, distributor, assembly	0603
CAP, dust universal joint yoke sleeve	0902
CAP, filler master cylinder	1205
CAP, gas tank well drain	0300
CAP, gasoline tank, w/safety CHAIN	0300
CAP, gasoline tank, w/safety CHAIN and BAR	0300
CAP, gearshift lever	0706A
CAP, hub	1302
CAP, insulator, spark plug	0604B
CAP, king pin bearing, lower	1006
CAP, output clutch, shaft bearing front	0801
CAP, output shaft bearing, w/BUSHING, rear	0801
CAP, radiator filler, assembly	0501
CARBURETOR, assembly	0301
CARRIER, clutch release bearing	0203
CASE, differential	1002
CASE, differential	1103
CASE, glove compartment lock, assembly	1809A
CASE, transfer	0801
CASE, transfer, assembly	0800
CASE, transmission, assembly	0701
CATCH, hold-down, windshield, assembly	1704
CATCH, hood, assembly	1701
CATCH, windshield clamp	1811
CHAIN, thumb screw	1811
CHAIN, thumb screw	2201A
CHAIN, tire, heavy duty, type "D"	2301B
CHAIN, tire, heavy duty, type "TS"	2301B
CHECK, intake ball, assembly	0301
CLAMP, axle brake tube	1101
CLAMP, axle brake tube	1209C
CLAMP, axle, front	2301A
CLAMP, axle rear, assembly	2301A
CLAMP, carburetor air horn, w/BOLT and NUT	0301C
CLAMP, exhaust pipe extension	0402
CLAMP, exhaust pipe to muffler	0402
CLAMP, front axle brake tube	1001
CLAMP, fuel pump valve	0302
CLAMP, gasoline tank hold-down, assembly	0300
CLAMP, hand brake cable tube	1201
CLAMP, hand brake cable tube (at transfer case bearing cap)	1201
CLAMP, hose	0505
CLAMP, oil filter, assembly	0107F
CLAMP, socket	1402
CLAMP, steering column, assembly	1405
CLAMP, steering column support	1405
CLAMP, support, muffler	0401
CLAMP, tail pipe	0402A
CLAMP, tube, assembly	0301
CLAMP, windshield, assembly	1811
CLEANER, air, assembly	0301C
CLIP	0606B
CLIP, Bond No. 4 to speedometer tube	0603
CLIP, Bond No. 4 to heat indicator cable	2603
CLIP, brake pipe	1209C
CLIP, breaker arm spring	0603
CLIP	2601
CLIP, 1 used cable to speedometer clip bolt	2601
CLIP	1201
CLIP, 1 used cable to speedometer clip bolt	2601
CLIP, front spring, left	1000
CLIP, front spring, left	1602
CLIP, front spring, right	1000
CLIP, front spring, right (was WO-A-1097)	1602
CLIP, lock, spring	1209C
CLIP, mounting	2204
CLIP, rear spring	1100
CLIP, retracting spring	0303
CLIP, retracting spring (on manifold)	0303
CLIP, spring leaf	1601
CLIP, spring leaf	1601A
CLIP, spring leaf (large for six leaves)	1601
CLIP, spring leaf, large	1601A
CLIP, support, hand brake cable	1201
CLIP, torque reaction	1601C
CLIP, tube	1209C
CLIP, tube (1 used tube to side rail enforcement)	1209C
CLIP	0304
CLIP	0107J
CLIP (tube to front engine plate, under generator brace bolt)	0304
COCK, drain, cylinder block	0101

249

SNL G-503

ALPHABETICAL INDEX (Cont'd)

Description	Gov't Group No.
COCK, drain, radiator	0501
COIL, current regulator, assembly	0601B
COIL, cut-out relay, assembly	0601B
COIL, field, generator, left, assembly	0601
COIL, field, generator, left and right, assembly	0601
COIL, field, right, assembly	0601
COIL, ignition, w/BRACKET and WASHER, assembly	0604
COIL, lower left, cranking motor	0602
COIL, lower right, cranking motor	0602
COIL, upper left, cranking motor	0602
COIL, upper right, cranking motor	0602
COIL, voltage regulator, assembly	0601B
COLLAR, drive shaft	0603
COLUMN, steering, w/BEARING, assembly	1403A
CONDENSER, assembly	0603
CONDUIT, 1	2601
CONE, and ROLLERS bearing drive pinion, inner	1104
CONE, and ROLLERS bearing, drive pinion, outer	1104
CONE, and ROLLER, differential bearing	1003
CONE, and ROLLERS, differential bearing	1103
CONE, and ROLLER, drive pinion, bearing	1003
CONE and ROLLER, drive pinion bearing, outer	1003
CONE and ROLLERS, king pin bearing	1006
CONE and ROLLERS, wheel bearing	1301A
CONE and ROLLERS, output shaft bearing	0803
CONNECTION, accelerator treadle	0303
CONNECTOR (2 wire)	0606B
CONNECTOR (3 wire)	0606B
CONNECTOR, field coil	0602
CONNECTOR	0107
CONNECTION, flexible	0304
CONNECTOR	0301A
CONNECTOR	0306B
CONNECTOR, series coil	0601B
CONTACT, breaker	0603
CONTACT SET	0603
CONTAINER, steering knuckle, oil seal, assembly	1006
CONTROL, choke or throttle, assembly	0301B
CONTROL, lockout, assembly	0606
CONTROL, throttle, assembly	0303
CORE, radiator, w/TANK and SHROUD, assembly	0501
COUNTERSHAFT	0704C
COVER	0306B
COVER, trailer coupling socket assembly	0606F
COVER, axle housing	1101
COVER, bowl, w/PIN, assembly	0301
COVER, chain, assembly	0106D
COVER, clutch	0202
COVER, commutator end cap	0601
COVER, dust	1401
COVER, duel pump top	0302
COVER, gearshift lever housing	0706A
COVER, headlight	2201
COVER, housing	1001
COVER, housing side, assembly	1403A
COVER, housing, upper	1403A
COVER, inspection, brake cylinder	1803
COVER, oil filter, assembly	0107E
COVER, oil pump, assembly	0107
COVER, rear floor pan shock absorber, left, assembly	1803
COVER, rear floor pan, shock absorber, right assembly	1803
COVER, steering column oil hole	1403A
COVER, terminal	0606F
COVER, terminal slot	0603
COVER, tie rod socket dust	1402
COVER, timing hole	0109C
COVER, trailer electric plug housing	1803
COVER, transfer case bottom	0801
COVER, transfer case	0801
COVER, transmission	1803
COVER, valve spring	0105C
COVER, voltage regulator, assembly	0601B
COVER, windshield, assembly	2201
CRANK BELL, drag link front	1006
CRANK, drag link bell front	1006
CRANK, hand brake cable, assembly	1201
CRANKSHAFT	0102
CUP	1403A
CUP, bearing	1103
CUP, differential bearing	1003
CUP, drive pinion bearing, inner	1003
CUP, drive pinion bearing, inner	1104
CUP, drive pinion bearing, outer	1003
CUP, drive pinion bearing, outer	1104
CUP, horn button spring	0609B
CUP, king pin bearing	1006
CUP, master cylinder check valve	1205
CUP, master cylinder primary, rubber	1205
CUP, master cylinder secondary, rubber	1205
CUP, oil retaining, transmission	0701
CUP, oil cleaner	0301C
CUP, output shaft brg.	0803
CUP, pressure spring	0202
CUP, retaining, door lock	1809A
CUP, retaining, transfer case insulator	1500
CUP, well bearing cone	1301A
CUP, wheel cylinder, front	1207
CUP, wheel cylinder, rear	1207
CURTAIN, side, left, assembly	2201F
CURTAIN, side, right, assembly	2201F
CUSHION, front seat	1804A
CYLINDER, compartment lock, assembly	1809A
CYLINDER, master cylinder, assembly	1205
CYLINDER, front wheel brake	1207
CYLINDER, front wheel brake, assembly	1207
CYLINDER, rear wheel brake	1207
CYLINDER, rear wheel brake, assembly	1207
DEFLECTOR, radiator side, air, w/SEAL, left, assembly	0501H
DEFLECTOR, radiator side air, w/SEAL, right, assembly	0501H
DIAPHRAGM, fuel pump, w/PULL ROD, assembly	0302
DIFFERENTIAL, assembly	1103
DIFFERENTIAL, front, assembly	1002
DISC, metering rod	0301
DISC, oil pump rotor	0107
DISC, shock absorber, rebound spring	1603
DISTRIBUTOR, assembly	0603
DOOR, assembly	0607
DOOR, blackout driving light, assembly	0607D
DOOR, glove compartment, w/HINGE, assembly	1809A
DOOR, marker light, assembly	0607D
DOOR, tail and stop light, left, assembly	0608
DOOR, tail and stop lght, right, assembly	0608
DOWEL, crankshaft bearing	0102
DOWEL, flywheel to crankshaft	0109
DRIVE, Bendix, assembly	0602
DRIVER, screw, common, heavy duty	2301
DRUM, front and rear brakes	1302
DRUM, hand brake	1201
ECCENTRIC, brake shoe	1203A

SNL G-503

ALPHABETICAL INDEX (Cont'd)

Description	Gov't Group No.
ELBOW PIPE	0107N
ELBOW, water outlet	0101A
ELEMENT, w/BOLT	0301C
ELEMENT, filter, assembly	0301C
ELEMENT, oil filter, assembly	0107E
END, yoke adjusting	0203
ENGINE, assembly	0100
EXTENSION, gasoline tank filler tube, assembly	0300
EXTINGUISHER, fire	2402
EYELET, fastener	2201A
FACING, clutch (front)	0201A
FACING, clutch (rear)	0201A
FAN, assembly	0503A
FASTENER, side curtain	2201F
FASTENER, single post type	2201A
FASTENER, windshield curtain	1811
FENDER, w/SPLASHER, front left, assembly	1701
FENDER, w/SPLASHER, front right, assembly	1701
FERRULE, insulating	0609B
FILLER, front bumper bar, wood	2101
FILLER, front fender brace, wood	1701
FILLER, rear floor pan cross sill	1802A
FILTER, oil, assembly	0107E
FILTER, radio terminal box	2601
FILTER GROUP, w/COVER, assembly	2603
FILTER-UNIT, generator to ground to armature circuit	2603
FILTER-UNIT, generator regulator	2603
FITTING, grease	1602
FITTING, grease, straight	0204
FITTING, grease, straight	1401
FITTING, grease, straight	1602
FITTING, grease, straight, bell crank	1006
FITTING, master cylinder outlet	1205
FLANGE, body, assembly	0301
FLANGE, companion, rear	0803
FLANGE, drive, axle shaft	1007
FLANGE, exhaust pipe	0402
FLOAT, oil, assembly	0107A-1
FLOAT, w/LEVER, assembly	0301
FLUID, brake	2302
FLYWHEEL	0109
FORK, front wheel drive shaft	0805A
FORK, shift, high and intermediate	0706C
FORK, shift, low and reverse	0706C
FORK, underdrive and direct shift	0805A
FRAME, assembly	1500
FRAME, front seat, assembly	1804
FRAME, back front seat, w/PAN and PANEL, assembly	1804
FRAME, hold-down, batter	0610A
FRAME, radiator guard, assembly	2103
FRAME, rear seat, assembly	1804
FRAME, rear seat back, assembly	1804
FRAME, rear seat cushion, assembly	1804A
FRAME, tool compartment, left	1817
FRAME, tool compartment, right	1817
FRAME, windshield, lower, assembly	1811B
FRAME, windshield, tubular	1811B
FRAME, windshield, upper, assembly	1811B
FRAME, windshield w/GLASS, assembly	1811A
FULCRUM, clutch control lever	0203
GASKET, axle shaft	1102
GASKET, body flange	0301
GASKET, body, upper	0301C
GASKET, bowl cover	0301
GASKET, chain cover	0106D

Description	Gov't Group No.
GASKET, cover	0601B
GASKET, crankshaft oil slinger	0102
GASKET, cylinder block, front	0101A
GASKET, cover cap screw	0306B
GASKET, drive pinion oil seal	1003
GASKET, gasoline tank cap	0300
GASKET, cylinder head	
GASKET, exhaust pipe flange	0402
GASKET, fitting brake hose	1209B
GASKET, fuel pump to block	0302A
GASKET, fuel pump bowl	0302
GASKET, fuel pump valve	0302
GASKET, gasoline tank gauge	0300A
GASKET, gearshift lever housing	0706A
GASKET, generator, commutator end, cap cover	0601
GASKET, housing cover	1001
GASKET, housing cover	1101
GASKET, insulator, carburetor, and intake manifold deffuse, assembly	0301A
GASKET, intake and exhaust manifold	0108
GASKET, intake to exhaust manifold	0108
GASKET, marker light door	0607D
GASKET, master cylinder filler cap	1205
GASKET, master cylinder outlet fitting	1205
GASKET, master cylinder outlet fitting bolt	1205
GASKET, metering rod jet and plug	0301
GASKET, needle seat and plug	0301
GASKET, nozzle	0301
GASKET, oil cup, lower	0301C
GASKET, oil filler cap	0107H
GASKET, oil filler cover	0107E
GASKET, oil filter cover bolt	0107E
GASKET, oil float support	0107A-1
GASKET, oil pan	0107A
GASKET, oil pan drain plug	0107A
GASKET, oil pump cover	0107
GASKET, oil pump cover (vellumoid)	0107
GASKET, oil pump cover to body	0107
GASKET, oil pump to cylinder block	0107
GASKET, oil pump shaft	0107
GASKET, steering knuckle, oil seal, assembly	1006
GASKET, oil relief spring retainer	0107
GASKET, output clutch shaft bearing cap	0801
GASKET, output clutch shaft bearing cap oil seal	0803
GASKET, pinion oil seal, leather	1104
GASKET, pump to cylinder block	0503
GASKET, radiator filler cap	0501
GASKET, rear cover	0801
GASKET, shock absorber bushing	1603
GASKET, side cover	1403A
GASKET, spark plug	0604B
GASKET, strainer bowl	0306B
GASKET, strainer plug	0301
GASKET, strainer unit	0306B
GASKET, terminal	0601B
GASKET, transfer case bottom cover	0801
GASKET, transfer case to transmission	0801
GASKET, transmission case to flywheel housing	0700
GASKET, universal joint trunnion	0902
GASKET, valve spring cover	0105C
GASKET, valve spring cover bolt	0105C
GASKET, valve spring cover to ventilator	0105C
GASKET, ventilator to valve spring cover	0107N
GASKET, water outlet elbow	0101A
GASKET, water pump to cylinder block	0503

ALPHABETICAL INDEX (Cont'd)

Description	Gov't Group No.
GASKET SET, engine overhaul	0101
GASKET SET, front axle	1001
GASKET SET, fuel strainer	0306B
GASKET SET, oil pan	0107A
GASKET SET, overhaul, transfer case	0800
GASKET SET, rear axle	1100
GASKET SET, service, carburetor	0301
GASKET SET, service, manifold	0108
GASKET SET, service, oil pump	0107
GASKET SET, transfer case overhaul	0800
GASKET SET, transmission	0700
GASKET SET, valve job	0101
GAUGE, fuel	0605
GAUGE, gasoline, assembly	0300A
GAUGE, oil pressure, assembly	0604C
GAUGE, pressure, tire	2301
GEAR, clutch, output shaft	0802
GEAR, drive	0802
GEAR, drive, differential, w/PINION	1003
GEAR, drive, differential, w/PINION	1003
GEAR, drive, speedometer	0810
GEAR, driven speedometer	0810
GEAR, flywheel ring	0109B
GEAR, intermediate	0804
GEAR, main drive	0703
GEAR, main drive, w/BEARINGS, assembly	0703
GEAR, main shift, second speed, w/BUSHING, assembly	0704B
GEAR, oil pump, driven	0107
GEAR, output shaft	0803
GEAR, reverse idler, w/BUSHING, assembly	0704B
GEAR, side	1103
GEAR, side, differential	1003
GEAR, sliding, main shift low and reverse	0704B
GEAR, sliding, output shaft	0803
GEAR, steering, w/o wheel, assembly	1403A
GEARS, cluster, countershaft, w/BUSHING and SPACER, assembly	0704B
GENERATOR, assembly	0601
GLAND, conduit	2601
GLAND, radio cable conduit nipple	2601
GLASS, windshield	1811A
GROMMET	2603
GROMMET, heat indicator through dash	0605
GROMMET, pedal shaft	0204
GROMMET	0107J
GROMMET, transfer shift lever	0805B
GROMMET, transfer shift lever	1803
GROMMET	0606B
GROMMET, (wiring harness through dash) (trailer coupling socket cable through floor pan)	0606B
GUARD, blackout lamp, w/SUPPORT, assembly	2013A
GUARD, exhaust pipe	1500
GUARD, front wheel brake hose, assembly	1001
GUARD, oil, flat	0601
GUARD, radiator, assembly	2103
GUIDE, adjusting, generator brace	0601A
GUIDE, axle shaft	1102
GUIDE, valve stem	0105
GUIDE, front wheel cylinder, piston and shoe, assembly	1207
GUIDE, intake valve stem	0105
GUIDE, rear wheel cylinder piston and shoe, assembly	1207
GUIDE, shock absorber piston rod, w/SEAL, assembly	1603
GUN, grease, hydraulic hand type	2301
GUSSET, front bumper, lower left	1500
GUSSET, front bumper, lower right	1500
GUSSET, front bumper, upper left	1500
GUSSET, front bumper, upper right	1500
GUSSET, front floor toe board to frame, left, assembly	1802A
GUSSET, front floor toe board to frame, right assembly	1802A
HAMMER, ball pein	2301
HANDLE, body outside, front	1800A
HANDLE, body outside, rear corner	1800A
HANDLE, hand brake	1201
HANDLE, hand brake, w/TUBE and CABLE, assembly	1201
HANDLE, ignition switch	0604C
HANDLE, spark plug wrench	2301
HANDLE, windshield frame	1811
HANDLE, wiper inner w/KNOB, assembly	1811D
HANDLE, wiper tie w/ROD, assembly	1811D
HARNESS, generator to voltage regulator and filter	0606B
HARNESS, wiring, body, left side, short	0606B
HARNESS, wiring, body, long	0606B
HARNESS, wiring, body, right side	0606B
HARNESS, wiring, chassis, left side	0606B
HARNESS, wiring, filter	0605B
HARNESS, wiring, generator to voltage regulator	0606B
HARNESS, wiring, headlight	0606B
HEAD, commutator end, assembly	0601
HEAD, cylinder	0101A
HEAD, drive end, generator	0601
HEAD, drive end, assembly	0601
HEAD, driving, cranking motor	0602
HEAD, shock absorber, front, assembly	1603
HEAD, shock absorber, rear, assembly	1603
HINGE, accelerator treadle	0303
HINGE, compartment door, assembly	1809A
HINGE, head light support	0607
HINGE, hood, assembly	1704
HINGE, tool compartment lid, assembly	1817
HOLDER, brush, cranking motor	0602
HOOD assembly	1704
HOOK, pintle	1502
HORN, air, assembly	0301
HORN assembly	0609
HORN, air cleaner	0301C
HOSE, air cleaner tube to oil filler tube	0301C
HOSE, brake, front, assembly	1209B
HOSE, brake front axle, assembly	1209B
HOSE, rear brake, assembly	1209B
HOSE, connecting	0301C
HOSE, radiator, upper	0505
HOSE, radiator water outlet, upper	0505
HOSE, water outlet, radiator, lower	0505
HOSE, water outlet, radiator, upper	0505
HOUSING assembly	1403A
HOUSING, blackout driving light, assembly	0607D
HOUSING, flywheel	0109C
HOUSING, gear shift lever, assembly	0706A
HOUSING, headlight, assembly	0607
HOUSING, market light w/WIRE, left, assembly	0607D
HOUSING, market light w/WIRE, right, assembly	0607D
HOUSING, pinion, assembly	0602
HOUSING SUB assembly	0608
HOUSING, w/TUBE, assembly	1101
HOUSING, w/TUBE, assembly	1001
HUB, w/BEARING, front and rear, assembly	1302
HUB, w/DRUM, front and rear, left, assembly	1302
HUB, w/DRUM, front and rear, right, assembly	1302
HUB, intermediate and high clutch	0704B
IMPELLER, water pump	0503

SNL G-503

ALPHABETICAL INDEX (Cont'd)

Description	Gov't Group No.
INDICATOR, and breather CAP, oil filler, assembly	0107H
INDICATOR, heat, assembly	0605
INSULATION, cut-out relay stationary contact bracket	0601B
INSULATION, field connection	0602
INSULATION, terminal	0601B
INSULATOR	0301
INSULATOR, armature terminal post, inner	0601
INSULATOR, armature terminal post, top	0601
INSULATOR, engine support, front, assembly	0110
INSULATOR, engine support, rear	0110
INSULATOR, field connection	0601
INSULATOR, field terminal post, inner	0601
INSULATOR, generator support	0601A
INSULATOR, ignition cable	0604A
INSULATOR, muffler support	0401
INSULATOR, transfer case support	0809
INTERLOCK, shift rod	0805A
JACK	2301
JET, idle well	0301
JET, low speed, assembly	0301
JET, metering rod, w/GASKET, assembly	0301
JET, pump	0301
JOINT, universal axle shaft, assembly	1007
KEY, heat control lever	0108C
KEY, ignition switch	0604C
KEY, Woodruff, No. 9	0102
KIT, clutch cover repair	0202
KIT, clutch disc facing repair	0201A
KIT, repair, axle shaft and universal joint, left	1007
KIT, repair, brake master cylinder	1205
KIT, repair, brake shoe lining and rivets	1202
KIT, repair, cranking motor	0602
KIT, repair, cranking motor Bendix drive	0602
KIT, repair, cranking motor field coil	0602
KIT, repair crankshaft and dowel bolt	0102
KIT, repair, current, regulator	0601B
KIT, repair distributor	0603
KIT, repair, drive gear and pinion	1003
KIT, repair, drive gear and pinion	1103
KIT, repair, field coil	0601
KIT, repair, flywheel and dowel bolt	0109
KIT, repair, front axle differential gear	1003
KIT, repair, front wheel cylinder	1207
KIT, repair, generator	0601
KIT, repair, generator cut-out relay	0601B
KIT, repair hand brake lining	1201
KIT, repair, ignition wiring	0604A
KIT, repair oil pump	0107
KIT, repair, rear axle differential gear	1103
KIT, repair, rear wheel cylinder	1207
KIT, repair, shock absorber front	1603
KIT, repair, shock absorber rear	1603
KIT, repair, speedometer drive gear	0810
KIT, repair, steering connecting rod	1401
KIT, repair, universal joint journal	0902
KIT, repair, voltage regulator	0601B
KIT, repair, water pump	0503
KIT, repair, wiring harness	0606B
KIT, service, horn button	0609B
KNOB, blackout driving light 2/set SCREW, assembly	0606
KNOB, instrument panel light switch w/screw SET, assembly	0606
KNOB, lighting switch w/set SCREW, assembly	
KNOB, windshield adjusting arm	1811
KNOB, PLUNGER, choke and WIRE, assembly	0301B
KNOB, PLUNGER, choke or throttle and WIRE, assembly	0301B
KNUCKLE, steering, left	1006
KNUCKLE, steering, right	1006
LAMP, electric incandescent	0607B
LAMP UNIT, blackout driving light sealed	0607A
LAMP UNIT, blackout stop, sealed	0607A
LAMP UNIT, blackout tail, sealed	0607A
LAMP UNIT, head light, sealed	0607A
LAMP UNIT, service, tail and stop, sealed	0607A
LEAD, armature terminal, assembly	0601
LEAD, voltage regulator armature to base	0601B
LEAF, front spring, L., No. 1 w/BUSHING	1601
LEAF, front spring, L., No. 2	1601
LEAF, front spring, L., No. 3	1601
LEAF, front spring, L., No. 4	1601
LEAF, front spring, L., No. 5	1601
LEAF, front spring, L., No. 6	1601
LEAF, front spring, L., No. 7	1601
LEAF, front spring, L., No. 8	1601
LEAF, front spring, R., No. 1	1601
LEAF, front spring, R., No. 2	1601
LEAF, front spring, R., No. 3	1601
LEAF, front spring, R., No. 4	1601
LEAF, front spring, R., No. 5	1601
LEAF, front spring, R., No. 6	1601
LEAF, front spring, R., No. 7	1601
LEAF, front spring, R., No. 8	1601
LEAF, rear spring No. 1, w/BUSHING	1601A
LEAF, rear spring, No. 2	1601A
LEAF, rear spring, No. 3	1601A
LEAF, rear spring, No. 4	1601A
LEAF, rear spring, No. 5	1601A
LEAF, rear spring, No. 6	1601A
LEAF, rear spring, No. 7	1601A
LEAF, rear spring, No. 8	1601A
LEAF, rear spring, No. 9	1601A
LEAF, No. 1, torque reaction spring	1601C
LEAF, No. 2, torque reaction spring	1601C
LEVER, accelerator cross shaft, w/BRACKET, assembly	0303
LEVER, choke, w/BRACKET, assembly	0301
LEVER, choke control, w/SHAFT, assembly	0301
LEVER, clutch control	0203
LEVER, clutch control, w/TUBE, assembly	0204A
LEVER, counterweight, heat control valve	0108C
LEVER, gearshift, assembly	0706A
LEVER, operating, pump, assembly	0301
LEVER, release	0202
LEVER, shift front wheel drive	0805B
LEVER, shift underdrive and direct	0805B
LEVERSHAFT, assembly	1403A
LID, tool compartment, w/HINGE, assembly	1817
LIGHT, driving blackout, assembly	0607D
LIGHT, marker left, assembly	0607D
LIGHT, marker right, assembly	0607D
LIGHT, service head, left, assembly	0607
LIGHT, service head, right, assembly	0607
LIGHT, tail and stop, left, assembly	0608
LIGHT, tail and stop, right, assembly	0608
LINER, body sill, thick	1801
LINER, body sill, thin	1801
LINER, radiator guard hood, center	2103
LINER, radiator guard hood, side	2103
LINING, brake shoe, forward, ground	1202
LINING, brake shoe, reverse, ground	1202

ALPHABETICAL INDEX (Cont'd)

Description	Gov't Group No.
LINING, hand brake	1201
LINK, choke	0301
LINK, connecting, accelerator	0303
LINK, connector, pump	0301
LINK, fuel pump rocker arm	0302
LINK, hand brake cable	1201
LOCK, compartment door, assembly	1809A
LOCK, spring shackle bolt, assembly	1602
LOCK, tool compartment lid, assembly	1817
LOCK, valve spring retainer (lower)	0105
LOOM, split, starting switch cable	2601
LOOP, footman	2201A
LOOP, footman	1800A
LOOP, footman	1802A
LOOP, hold-down, windshield	1811
MANIFOLD, exhaust, assembly	0108B
MANIFOLD, intake, assembly	0108A
MANIFOLD, intake and exhaust, assembly	0108
MEMBER, cross, center	1500
MEMBER, cross, intermediate front, assembly	1500
MEMBER, cross, intermediate rear	1500
MEMBER, cross, rear	1500
MIRROR, rear view, assembly	2202
MIRROR, rear view with washer and nut assembly	2202
MOTOR, cranking, assembly	0602
MUFFLER, assembly	0401
MUFFLER assembly	0401
NECK, gas tank well drain hole	1802A
NIPPLE, rubber marker light	0607D
NOZZLE	0301
OILER, distributor shaft	0603
OILER, press in type	0601
OIL SEAL, carrier end	1006
OIL SEAL, drive pinion	1003
OIL SEAL, felt steering knuckle	1006
OIL SEAL, hub, assembly	1007
OIL SEAL, hub, assembly	1102
OIL SEAL, inboard	1102
OIL SEAL, output shaft	0803
OIL SEAL, pinion, leather	1104
OIL SEAL, shift rod	0805B
OIL SEAL, steering knuckle, assembly	1006
OIL SEAL, transmission front bearing retainer	0703
OIL SEAL, unit	1403A
OIL SEAL SET, transfer case	0800
OUTLET, air cleaner	0301C
OUTRIGGER, body rear	1500
PACKING, crankshaft bearing cap, rear	0102A
PACKING, crankshaft, front end	0102A
PACKING, crankshaft, rear end	0102A
PACKING, gearshift lever housing	0706A
PAD, brake pedal, assembly	1204
PAD, clutch control tube	0204A
PAD, clutch pedal, assembly	0204
PAD, dash	1809
PAD, front seat body side crash, right, assembly	1804A
PAD, front seat body side crash, left, assembly	1804A
PAD, mounting, blackout light	0607D
PAD, pedal shank draft	0204
PADDING, windshield glass	1811B
PAN, front floor	1802A
PAN, oil, assembly	0107A
PAN, rear floor	1802A
PANEL, body, rear	1802A
PANEL, body side, left	1802A
PANEL, body side, right	1802A
PANEL, cowl side, left	1802A
PANEL, cowl side, right	1802A
PANEL, cowl top	1802A
PANEL, dash, assembly	1802A
PANEL, dash, assembly	1809
PANEL, front seat body side crash pad, left, assembly	1804
PANEL, front seat body side crash pad, right, assembly	1804
PANEL, instrument, assembly	1809
PANEL, rear wheel house, inner left	1802A
PANEL, rear wheel house, inner right	1802A
PANEL, wheel house, top left	1802A
PANEL, wheel house, top right	1802A
PANEL, windshield, outer	1811
PEDAL, brake, assembly	1204
PEDAL, clutch	0204
PIECE, pole	0601
PIECE, pole	0602
PILLAR, cowl, left	1802
PILLAR, cowl, right	1802
PILOT, universal joint	1007
PIN, accelerator treadle hinge	0303
PIN, anchor	1203B
PIN, bell crank shaft lock	1401
PIN, clevis	1201
PIN, clevis	0607
PIN, clevis	0203
PIN, countershaft washer	0704C
PIN, dowel commutator end plate, assembly	0601
PIN, drain tube to rear bearing cap	0102
PIN, float lever	0301
PIN, fuel pump rocker arm	0302
PIN, gearshift lever fulcrum	0706A
PIN, hand brake cam	1201
PIN, hand brake support quadrant	1201
PIN, intake needle	0301
PIN, king	1006
PIN, locating, dowel, intermediate bearing	0602
PIN, lock, differential pinion shaft	1003
PIN, lock, king pin	1006
PIN, lock, pinion shaft	1103
PIN, metering rod	0301
PIN, oil pump driven gear straight	0107
PIN, piston	0103B
PIN, pivot, shift lever	0805B
PIN, release lever pivot	0202
PIN, shift fork guide	0706C
PIN, shifter plate fulcrum	0706C
PIN, starting crank	0102
PIN, universal joint pilot	1007
PIN, distributor cap hinge	0603
PINION, differential	1003
PINION, differential	1103
PINION, oil pump	0107
PIPE, drain	0102A
PIPE, exhaust, assembly	0402
PISTON, assembly (std.)	0103
PISTON, assembly	0103
PISTON, master cylinder, assembly	1205
PISTON, shock absorber	1603
PISTON w/PIN, assembly	0103
PIVOT, top bow, assembly	2201A
PIVOT, top bow front	2201A
PLANT, power	0100
PLANT, adjusting, shock absorber	1603

SNL G-503

ALPHABETICAL INDEX (Cont'd)

Description	Gov't Group No.
PLATE, anchor pin	1203B
PLATE, back, clincher	2201A
PLATE, back, shock absorber rebound valve	1603
PLATE, backing, front and rear, brake, assembly	1203
PLATE, breaker, assembly	0603
PLATE, breaker part, assembly	0603
PLATE, cam and stop, 4 cyl., left hand	0603
PLATE, caution	2203
PLATE, clutch driven, w/HUB, assembly	0201B
PLATE, clutch pressure, assembly	0201B
PLATE, commutator end, assembly	0601
PLATE, commutator end, assembly	0602
PLATE, dimmer switch floor seal	1802A
PLATE, dimmer switch floor seal	1803
PLATE, drive end, assembly	0601
PLATE, engine, front, assembly	0110
PLATE, engine, rear, assembly	0110
PLATE, guide, gearshift lever	0706A
PLATE, instruction	1301
PLATE, lock, countershaft and idler shaft	0704C
PLATE, lock, intermediate	0804
PLATE, machine gun mounting	1500
PLATE, muffler support insulator	0401
PLATE, name	2203
PLATE, pressure	0202
PLATE, reinforcement rear	1500
PLATE, shifter, assembly	0706C
PLATE, shifting	2203
PLATE, skid, under frame	1500
PLATE, spring clip w/SHAFT, left, assembly	1100
PLATE, spring clip, w/SHAFT, front, assembly, left	1000
PLATE, spring clip, w/SHAFT, front, assembly, left	1602
PLATE, spring clip, w/SHAFT, front, assembly, right	1000
PLATE, spring clip, w/SHAFT, front, assembly, right	1602
PLATE, spring, w/SHAFT, assembly	1602
PLATE, spring clip, w/SHAFT, right, assembly	1100
PLATE, stop, master cylinder piston	1205
PLATE, synchronizer, shifting	0704B
PLATE, tapping, axe clamp	1802A
PLATE, tapping, compartment hinge	1809A
PLATE, tapping, driver seat, front floor	1802A
PLATE, tapping, hood hinge	1802A
PLATE, tapping, shovel clamp	1802A
PLATE, tapping, windshield clamp catch	1811
PLIERS	2301
PLUG, adjusting, ball seat	1401
PLUG, adjusting, ball seat	1401
PLUG, ball seat spring	1401
PLUG, discharge disc check, assembly	0301
PLUG, drain	0107E
PLUG, drain	1001
PLUG, drain, housing	1101
PLUG, drain, strainer	0306B
PLUG, filler, housing	1101
PLUG, filler, housing cover	1001
PLUG, idle port rivet	0301
PLUG, low speed jet and idle well passage, w/GASKET	0301
PLUG, nozzle retainer	0301
PLUG, oil pan drain	0107A
PLUG, pipe	0101
PLUG	0801
PLUG, pipe, slotted	1802A
PLUG, rivet	0301
PLUG, spark, w/GASKET, assembly	0604B
PLUG, strainer passage, w/GASKET, assembly	0301
PLUG, vent	1001
PLUG, shaft rod poppet	0805
PLUNGER, camshaft thrust	0106
PLUNGER, contact spring	0603
PLUNGER, gearshift interlock	0706B
PLUNGER, oil relief	0107
PLUNGERk pump, w/ROD, assembly	0301
POST, terminal	0601
POST, field terminal	0601
POST, terminal	0602
PROTECTOR, gasoline gauge	0300A
PULIER, wheel hub	2301
PULLEY, drive, fan	0102
PULLEY, drive, generator	0601
PULLEY, fan and water pump	0503
PUMP, fuel, assembly	0302
PUMP, oil, assembly	0107
PUMP, tire w/CHAIN	2301
PUMP, water, assembly	0503
QUADRANT, support, hand brake	1201
RACE, universal joint, inner	1007
RAIL, shift, high and intermediate	0706B
RAIL, shift, low and reverse	0706B
REFLECTOR, reflex, assembly	2203A
REGULATOR, generator	0601B
REINFORCEMENT, axle bumper frame, rear left	1500
REINFORCEMENT, axle bumper frame, rear right	1500
REINFORCEMENT, frame side member, left, assembly	1500
REINFORCEMENT, frame side member, right, assembly	1500
REINFORCEMENT, front bumper, left	1500
REINFORCEMENT, front bumper, right	1500
REINFORCEMENT, front engine support bracket, right	1500
REINFORCEMENT, rear	1500
REINFORCEMENT, rear spring front bracket to frame, left	1500
REINFORCEMENT, rear spring front bracket to frame, RIGHT	1500
REINFORCEMENT, pedal shaft	1500
REINFORCEMENT, pintle hook frame	1502
REINFORCEMENT, wheel house to rear panel, left	1802A
REINFORCEMENT, wheel house to rear panel, right	1802A
REPAIR KIT, carburetor	0301
REPAIR KIT, fuel pump	0302
REPAIR KIT, fuel pump	0302
REPAIR KIT, throttle shaft and lever	0301
RESISTANCE, carbon	0601B
RESISTANCE, carbon	0601B
REST, foot, accelerator	0303
REST, rear floor foot, assembly	1802A
RETAINER, axle shaft	1007
RETAINER, clutch control tube spring	0204A
RETAINER, crankshaft packing	0102
RETAINER, drive end bearing	0601
RETAINER, dust shield	0606F
RETAINER, engine crank	1802A
RETAINER, hub grease, assembly	1301B
RETAINER GREASE, wheel end	1301B
RETAINER	2103
RETAINER, main drive gear bearing	0703
RETAINER, master cylinder check valve cup	1205
RETAINER, mounting bolt	0607
RETAINER, oil relief spring	0107
RETAINER, pedal shaft, assembly	1500
RETAINER, rear seat pivot tube	1804

ALPHABETICAL INDEX (Cont'd)

Description	Gov't Group No.
RETAINER, spring	0301
RETAINER, spring shackle grease seal	1602
RETAINER, thermostat	0502
RETAINER, tire pump foot	1804
RETAINER, tire pump, handle	1804
RETAINER, trunnion gasket	0902
RETAINER, valve spring	0105
RETAINER, windshield hold-down catch	1811
RETAINER, windshield hold-down catch	1811
RIM, w/DISC, assembly (16 x 4:00)	1301
RIM, w/DISC, assembly (16 x 4:50)	1301
RIM, wheel, inner half	1301
RIM, wheel, outer half	1301
RING, contact, horn, assembly	0609B
RING, gearshift lever housing cover	1803
RING, lock bead	1301
RING, retaining, ball	1403A
RING, retaining, chain cover packing	0106D
RING, sealing, gland to conduit	2601
RING, steering column floor seal	1803
RING, synchronizer, blocking	0704B
RING, take-up, Bendix	0602
RING, thumb screw chain	1811
RING, thumb screw chain	2201A
RING, transmission shift lever grommet	1803
RING, unit mounting, blackout driving light	0607D
RING, piston	0103A
RING, piston	0103A
RING, piston std.	0103A
RING, piston .030"	0103A
RING, piston .020°	0103
RING, piston std., conventional type	0103A
RING, piston std., conventional re-ring type	0103A
RING, piston, .020" O.S., compression, re-ring type	0103A
RING, piston .020" O.S., conventional type	0103A
RING, piston .030" O.S., compression type	0103A
RING, piston .030" O.S., compression re-ring type	0103A
RING, piston .030" O.S., conventional type	0103A
RING SET, piston, std., conventional type	0103A
RING SET, piston, .030" O.S., conventional type	0103A
RING SET, piston, .020" O.S., conventional type	0103A
RING SET, piston, std., re-ring	0103A
RING SET, piston, .020" O.S. re-ring	0103A
RING SET, piston, .030" O.S. re-ring	0103A
RISER, front to rear floor pan, assembly	1802A
ROD, brace, radiator	0501
ROD, clutch release pedal	0204A
ROD, connecting, assembly 1 and 3	0104
ROD, connecting, assembly 2 and 4	0104
ROD, connecting, steering	1401
ROD, connecting, steering, assembly	1401
ROD, connecting, w/BEARING, assembly 1 and 3	0104
PLUG, pipe	0300
PLUG, pump jet and nozzle passage, w/GASKET	0301
ROD, connecting, w/BEARING, assembly 2 and 3	0104
ROD, metering	0301
ROD, push, master cylinder, assembly	1205
ROD, shift, front wheel drive	0805
ROD, shift, underdrive and direct	0805
ROD, throttle	0303
ROD, throttle connector	0301
ROD, tie, left, assembly	1402
ROD, tie, right, assembly	1402
ROTOR	0603
SCREEN, fuel pump filtering, assembly	0302
SEAL, accelerator treadle rod floor	0303
SEAL, accelerator treadle rod floor	1803
SEAL, bell crank shaft bearing	1401
SEAL, brake cylinder inspection cover	1803
SEAL, clutch control tube dust	0204A
SEAL, compartment door	1809A
SEAL, dust, spindle connecting rod	1402
SEAL, fuel pump pull rod	0302
SEAL, gas line hole floor	1803
SEAL, gas tank stone guard, front	1803
SEAL, gas tank stone guard, left	1803
SEAL, gas tank stone guard, rear	1803
SEAL, gas tank stone guard, right	1803
SEAL, gearshift lever housing cover	1803
SEAL, gland to radio cable conduit	2601
SEAL, glove compartment door	1803
SEAL, headlight foot switch	1803
SEAL, lead and wire	0601B
SEAL, spark plug ignition wiring, distributor end	0603
SEAL, spring shackle grease	1602
SEAL, steering column floor	1803
SEAL, radiator side air deflector, left	1803
SEAL, radiator side air deflector to radiator, right	1803
SEAL, radiator side air deflector to radiator	1803
SEAL, radiator top air deflector to radiator	1803
SEAL, radiator top air deflector	1803
SEAL, starter switch floor	1803
SEAL, transmission cover	1803
SEAL, weather, spark plug cable, distributor end	0604A
SEAL, tool compartment lid	1817
SEAL, tool compartment lid	1817
SEAL, water pump assembly	0503
SEAT, ball	1401
SEAT, gas strainer bowl	0302
SEAT, master cylinder valve	1205
SEAT, shock absorber spring	1603
SEAT, steering column bearing spring	1403A
SEAT, trimmed, rear, assembly	1804A
SERVICE KIT, water pump	0503
SET, drive gear and pinion	1103
SET, engineers double end wrench	2301
SET, piston ring and spring 2nd groove, re-ring (.030" to .039" O.S.)	0103A
SET, piston ring and spring 2nd groove, re-ring (.020" to .029" O.S.)	0103A
SET, piston ring and spring, bottom groove, re-ring (std. to .009" O.S.)	0103A
SET, piston ring and spring, bottom groove, re-ring (.010" to .019" O.S.)	0103A
SET, ring and spring, piston bottom groove, re-ring (.020" to .029" O.S.)	0103A
SET, piston ring and spring, 2nd groove re-ring (std., to .009" O.S.)	
SET, ring and spring, piston bottom groove, re-ring (.030" to .039" O.S.)	0105A
SET, tandem windshield wiper	1811D
SHACKLE, spring, w/BUSHING, outer	1602
SHACKLE, spring, w/BUSHING, inner	1602
SHAFT, axle, inner left	1007
SHAFT, axle, left	1007
SHAFT, axle, left w/universal JOINT, assembly	1007
SHAFT, axle, outer	1007
SHAFT, axle, inner ring	1007
SHAFT, axle, left w/universal JOINT, assembly	1007
SHAFT, axle, left universal JOINT, assembly	1007

ALPHABETICAL INDEX (Cont'd)

Description	Gov't Group No.
SHAFT, axle, outer, w/universal JOINT, assembly	1007
SHAFT, axle, right, w/universal JOINT, assembly	1007
SHAFT, axle, right	1007
SHAFT, axle, right w/universal JOINT, assembly	1007
SHAFT, bell crank	1401
SHAFT, Bendix, drive w/PINION, assembly	0602
SHAFT, brake pedal, assembly	1204
SHAFT, cross, accelerator w/LEVER	0303
SHAFT, differential pinion	1003
SHAFT, drive, assembly	0603
SHAFT, drive, distributor, w/GOVERNOR, assembly	0603
SHAFT, heat control valve	0108C
SHAFT, intermediate	0804
SHAFT, main	0704
SHAFT, main, w/GEARS, assembly	0704
SHAFT, oil pump, assembly	0107
SHAFT, output, w/BUSHING, assembly	0803
SHAFT, output, w/NUT and WASHER, assembly	0803
SHAFT, output clutch	0803
SHAFT, output clutch w/NUT and WASHER, assembly	0803
SHAFT, pedal assembly	0204
SHAFT, pinion	1103
SHAFT, propeller, front, assembly	0901
SHAFT, propeller, rear, assembly	0901
SHAFT, propeller, rear, assembly	0902
SHAFT, rear axle, left	1102
SHAFT, rear axle, right	1102
SHAFT, reverse idler gear	0704C
SHAFT, right, axle, w/universal JOINT, assembly	1007
SHAFT, speedometer, assembly	2204
SHAFT, throttle, w/INNER. assembly	0301
SHEATH, axe	1802A
SHIELD, air cleaner, assembly	0301C
SHIELD, dust	0606F
SHIELD, dust	0803
SHIELD, dust, bell crank bearing dust	1401
SHIELD, dust cover	1401
SHIELD, dust, front universal joint end yoke	0902
SHIELD, dust universal joint end yoke	1003
SHIELD, dust, universal joint end yoke	1104
SHIELD, fan pulley	0503A
SHIELD, front seat to rear floor gas tank	0300
SHIELD, instrument, lamp assembly	0605C
SHIELD, master cylinder, assembly	1205
SHIM, adjusting (.003 thk.)	1103
SHIM, adjusting (.005 thk.)	1103
SHIM, adjusting (.010 thk.)	1103
SHIM, adjusting (.015 thk.)	1103
SHIM, adjusting (.030 thk.)	1103
SHIM, adjusting, bearing (small .003)	1104
SHIM, adjusting, bearing (large, .005)	1104
SHIM, adjusting, bearing (large, .003)	1104
SHIM, adjusting, bearing (large, .010)	1104
SHIM, adjusting, bearing (small, .010)	1104
SHIM, adjusting, bearing (small, .030)	1104
SHIM, adjusting, bearing (small, .005)	1104
SHIM, adjusting, differential bearing (.003 thk.)	1003
SHIM, adjusting, differential bearing (.005 thk.)	1003
SHIM, adjusting, differential bearing (.015 thk.)	1003
SHIM, adjusting, differential bearing (.010 thk.)	1003
SHIM, adjusting, differential bearing (.030 thk.)	1003
SHIM, adjusting, drive pinion, bearing (.003 thk.)	1003
SHIM, adjusting, drive pinion bearing (.005 thk.)	1003
SHIM, adjusting, drive pinion, bearing small (.010 thk.)	1003
SHIM, adjusting, drive pinion, bearing (.030 thk.)	1003
SHIM, adjusting, drive pinion, large (.003 thk.)	1003
SHIM, adjusting, drive pinion, large (.005 thk.)	1003
SHIM, adjusting, drive pinion, large (.010 thk.)	1003
SHIM, adjusting, king pin bearing (.003 thk.)	1006
SHIM, adjusting, king pin bearing (.005 thk.)	1006
SHIM, adjusting, king pin bearing (.010 thk.)	1006
SHIM, adjusting, king pin bearing (.030 thk.)	1006
SHIM, adjusting, universal joint (.010 thk.)	1007
SHIM, adjusting, universal joint (.030 thk.)	1007
SHIM, crankshaft (.002 thk.)	0102
SHIM, engine support insulator	0110
SHIM, mounting	1405
SHIM, oil relief spring	0107
SHIM, output shaft rear bearing, cap (.003 thk.)	0801
SHIM, output shaft rear bearing cap (.010 thk.)	0801
SHIM, output shaft rear bearing cap (.031 thk.)	0801
SHIM, radiator	0504
SHIM, radiator to support bracket	0501
SHIM, steering gear mounting	1403A
SHIM, upper cover, S. (.010 thk.)	1403A
SHIM, upper cover, S. (.003 thk.)	1403A
SHIM, upper cover, S. (.002 thk.)	1403A
SHIM SET, king pin bearing	1006
SHIM SET, universal joint adjusting	1007
SHIM SET, output shaft	0801
SHIM SET, rear axle drive pinion	1104
SHIM SET, axle differential bearing	1103
SHIM SET, steering gear housing upper cover	1403A
SHOE, brake, forward w/LINING, assembly	1202
SHOE, brake, reverse w/LINING, assembly	1202
SILL, cross, front floor pan, left, assembly	1802A
SILL, cross, front floor pan, rear, assembly	1802A
SILL, cross, front floor pan, rear, assembly	1802A
SILL, cross, front floor pan, right, assembly	1802A
SILL, cross, rear floor pan	1802A
SLEEVE, compression	0602
SLEEVE, second and direct speed clutch	0704B
SLEEVE, shock absorber, front, assembly	1603
SLEEVE, shock absorber, rear, assembly	1603
SLEEVE, speedometer driven gear	0810
SLINGER, drive pinion bearing	1104
SLINGER, oil, crankshaft	0102
SLINGER, oil, drive pinion bearing	1003
SNAP RING, axle shaft retainer	1007
SNAP RING, high and intermediate clutch hub	0704B
SNAP RING, main drive gear bearing	0703
SNAP RING, main shaft	0703
SNAP RING, main shaft	0704
SNAP RING, main shaft, pilot roller bearing	0703
SNAP RING, output clutch shaft bearing	0803
SNAP RING, output shaft	0802
SNAP RING, retaining, distributor cam	0603
SNAP RING, steering column	1403A
SNAP RING, trunnion bearing	0902
SNUBBER, transfer case support insulator	0809
SOCKET, coupling, electric trailer, assembly	0606F
SOCKET, coupling electric trailer, assembly (includes COVER, INTERNAL PARTS, ATTACHING PARTS)	0606F
SOCKET, curtain fastener	2201A
SOCKET, tie rod left, assembly	1402
SOCKET, tie rod right, assembly	1402
SOCKET, instrument lamp, w/CABLE, assembly	0605C
SPACER, cam	0603
SPACER, crankshaft sprocket	0106C

ALPHABETICAL INDEX (Cont'd)

Description	Gov't Group No.
SPACER, countershaft bushing	0704C
SPACER, cross member intermediate front bracket	1500
SPACER, drive pinion bearing	1003
SPACER, drive pinion bearing	1104
SPACER, front bumper gusset	1500
SPACER, main shaft bearing	0704
SPACER, metering, shock absorber, front	1603
SPACER, shaft	0601
SPACER, shock absorber, metering, rear	1603
SPACER, steering gear to frame bolt	1500
SPACER, thrust, bearing	0602
SPACER, windshield adjusting arm	1811
SPARE PARTS SET	2309
SPEEDOMETER, luminous dial, assembly	2204
SPINDLE, wheel bearing, w/BUSHING, assembly	1006
SPRING, accelerator	0303
SPRING, accelerator cross shaft	0303
SPRING, accelerator treadle	0303
SPRING, anchor clip bolt	1201
SPRING, anti-rattle, distributor	0603
SPRING, ball seat	1401
SPRING, breaker arm	0603
SPRING, brush	0601
SPRING, brush	0602
SPRING, camshaft thrust plunger	0106
SPRING, choke pull-back	0301
SPRING, clutch control tube	0204A
SPRING, clutch release bearing carrier	0203
SPRING, connector link	0301
SPRING, connector rod	0301
SPRING, contact distributor	0603
SPRING, cut-out relay	0601B
SPRING, distributor cap	0603
SPRING, drive	0602
SPRING, friction, distributor shaft	0603
SPRING, front, L., assembly	1601
SPRING, front, R., assembly	1601
SPRING, front seat back, assembly	1804A
SPRING, front seat body side crash pad, left assembly	1804A
SPRING, front seat body side crash pad, right assembly	1804A
SPRING, front seat, cushion, assembly	1804A
SPRING, front wheel, cylinder cup	1207
SPRING, fuel pump diaphragm	0302
SPRING, fuel pump rocker arm	0302
SPRING, generator brace	0601A
SPRING, heat control valve	0108C
SPRING, horn button	0609B
SPRING, idle adjustment screw	0301
SPRING, intake needle	0301
SPRING, metering rod	0301
SPRING, needle, w/SEAT, assembly	0301
SPRING, oil filter cover bolt	0107E
SPRING, oil relief plunger	0107
SPRING, pedal retracting	0204
SPRING, pedal shank draft pad	0204
SPRING, pin	0301
SPRING, plunger	0301
SPRING, pressure	0202
SPRING, pressure plate return	0202
SPRING, pump	0301
SPRING, pump arm	0301
SPRING, ratchet tube	1201
SPRING, rear, assembly	1601A
SPRING, rear seat back assembly	1804A
SPRING, rear seat cushion assembly	1804A
SPRING, rear wheel, cylinder cup	1207
SPRING, releasing, hand brake	1201
SPRING, retaining, rear seat frame	1804
SPRING, retaining, water pump bearing	0503
SPRING, return, brake shoe	1203A
SPRING, return, master cylinder piston, w/RETAINER assembly	1205
SPRING, retracting, hand brake	1201
SPRING, shifter plate	0706C
SPRING, shift lever	0805B
SPRING, shift rail, poppet ball	0706B
SPRING, shift rod poppet	0805
SPRING, shock absorber, piston intake valve	1603
SPRING, shock absorber, rebound valve	1603
SPRING, steering column bearing	1403A
SPRING, strainer unit	0306B
SPRING, support, gearshift lever	0706A
SPRING, synchronizer	0704B
SPRING, tie rod socket stud	1402
SPRING, toggle	0301C
SPRING, torque reaction assembly	1601C
SPRING, valve	0105
SPRING, voltage regulator and current regulator coil	0601B
SPRING, water pump seal	0503
SPRING SET, governor weight	0603
SPROCKET, camshaft	0106C
SPROCKET, crankshaft	0102
STRAINER, body side panel to wheel house	1802A
STRAINER, gasoline assembly	0306B
STRAINER, pump check	0301
STRAINER, spare tire carrier, upper, assembly	1505
STRAINER, spare tire carrier, w/FILLER, lower, assembly	1505
STRAINER UNIT assembly	0306B
STRAP, axe assembly	2301A
STRAP, battery to front fender	0610A
STRAP, battery ground	0601B
STRAP, battery ground	0610B
STRAP, bond No. 1	2603
STRAP, bond No. 3	2603
STRAP, bond No. 4	2603
STRAP, bond No. 5	2603
STRAP, bond No. 6	2603
STRAP, bond No. 7	2603
STRAP, bond No. 8	2603
STRAP, bond No. 9	2603
STRAP, bond No. 10	2603
STRAP, bond No. 11	2603
STRAP, bond No. 12	2603
STRAP, bond No. 13	2603
STRAP, bond No. 15	2603
STRAP, bond No. 16	2603
STRAP, bond No. 21 and STRAP, bond No. 22	2603
STRAP, bond, radio box negative terminal to floor ground connection	2601
STRAP, hold-down, gasoline tank inner, assembly	0300
STRAP, hold-down, gasoline tank outer, assembly	0300
STRAP, hold-down, top, rear assembly	2201A
STRAP, lock, differential drive gear screw	1003
STRAP, lock, drive gear screw	1103
STRAP, muffler support	0401
STRAP, safety assembly	1804
STRAP, shovel front assembly	2301A
STRAP, shovel rear assembly	2301A
STRAP, spare gas can, w/BUCKLE, assembly	1817A
STRAP, spare gas can, w/TIP, assembly	1817A
STRAP, top lock, assembly	2201A

ALPHABETICAL INDEX (Cont'd)

Description	Gov't Group No.
STRAP, top roll assembly	2201A
STRIKER, compartment door lock	1809A
STRIKER, tool compartment lid lock	1817
STRIP, pressure, felt, steering knuckle, oil seal	1006
STOP, heat control valve spring	0108C
STOP, throttle wire	0303A
STOP, throttle wire, w/SCREW, assembly	0303A
STUD	0108
STUD	0108
STUD	0107
STUD, camshaft thrust plunger	0106D
STUD, carburetor to manifold	0301
STUD, carburetor to manifold	0301A
STUD, clutch control ball	0204A
STUD, clutch control ball w/BRACKET, assembly	0204A
STUD, coil to cylinder block	0101
STUD, cross shaft bracket	0303
STUD, cylinder head holes 5, 7, 9, 15	0101A
STUD, cylinder head hole 10	0101
STUD, cylinder head hole 12	0101
STUD, exhaust pipe to manifold	0108B
STUD, front plate and chain cover to cylinder block	0106D
STUD, inlet and exhaust manifold	0108
STUD, oil pump to cylinder block	0107
STUD, steering arm	1006
STUD, steering arm dowel	1006
SUPPRESSOR, ignition	2603
SUPPORT, battery assembly	1500
SUPPORT, cranking motor	0602
SUPPORT, generator, assembly	0601A
SUPPORT, hand brake ratchet tube bracket	1201
SUPPORT, headlight, left assembly	0607
SUPPORT, headlight, right assembly	0607
SUPPORT, oil float	0107A-1
SUPPORT, rear body to frame, inner	1802A
SUPPORT, rear body to frame, outer	1802A
SUPPORT, rear seat frame to wheel house	1804
SUPPORT, steering column	1405
SWITCH, blackout driving light, assembly	0606
SWITCH, blackout driving light, assembly	0606
SWITCH, dimmer, foot, headlight, assembly	0606
SWITCH, dimmer, headlight foot, assembly	0606
SWITCH, keyless, ignition assembly	0604C
SWITCH, lighting	0606
SWITCH, lighting, assembly	0606
SWITCH, panel light	0606
SWITCH, starting	0606
SYNCHRONIZER UNIT, blocking RING, assembly	0704B
TANK, gasoline, assembly	0300
TANK, gasoline, w/EXTENSION and CAP, assembly	0300
TANK, supply, master cylinder	1205
TAPPET, valve	0105A
TAPPET, valve, .004 oversize	0105A
TEE, axle brake tube	1209C
TERMINAL	2601
TERMINAL	2601
TERMINAL, armature lead	0601
TERMINAL, battery cable	2601
TERMINAL, clamp, battery positive cable	0610B
TERMINAL, coil secondary cable	0604A
TERMINAL, field coil	0602
TERMINAL, field	0601
TERMINAL, high tension, spark plug cables	0604A
TERMINAL, spark plug cables and coil secondary cable	0604A
THERMOSTAT assembly	0502
THIMBLE	2301
TIP, coil primary cable	0604A
TOP, assembly	2201
TOOL SET	2301
TRANSMISSION, w/gearshift LEVER, assembly	0700
TREADLE, accelerator, assembly	0303
TREADLE, accelerator, w/HINGE and LINK, assembly	0303
TRIM, front seat back cover, assembly	1804A
TRIM, front seat cushion cover, assembly	1804A
TRIM, rear seat back cover, assembly	1804A
TRIM, rear seat cushion cover, assembly	1804A
TUBE, air cleaner, w/BRACKET, assembly	0301C
TUBE, brake assembly	1209C
TUBE, connecting, water inlet	0505
TUBE, connecting, water outlet	0505
TUBE, cam and wheel assembly	1403A
TUBE, crankcase ventilator valve assembly	0107N
TUBE, cross, front	1500
TUBE, front, propeller shaft assembly	0901
TUBE, gasoline, assembly	0304
TUBE, oil filler assembly	0107G
TUBE, oil filler	0107G
TUBE, oil filter inlet, assembly	0107E
TUBE, oil filter outlet, assembly	0107E
TUBE, pump to carburetor assembly	0304
TUBE, rear propeller shaft, assembly	0902
TUBE, shock absorber, pressure, front	1603
TUBE, shock absorber, pressure, rear	1603
TUBE, speedometer assembly	2204
TUBE, speedometer, w/SHAFT assembly	2204
TUBE, tie rod	1402
TUBE, tie rod, right	1402
VALVE, check, master cylinder, assembly	1205
VALVE, choker, assembly	0301
VALVE, crankcase ventilator assembly	0107N
VALVE, exhaust	0105
VALVE, fuel pump, assembly	0302
VALVE, heat control	0108C
VALVE, intake	0105
VALVE, shock absorber, compression, assembly	1603
VALVE, shock absorber, piston intake	1603
VALVE, throttle	0301
VENTILATOR, crankcase, assembly	0107N
WEATHERSTRIP, windshield	1811B
WEATHERSTRIP, windshield frame to cowl	1811B
WEBBING, hood top windshield bumper	1704
WEIGHT assembly	0603
WELL, gas tank, assembly	0300
WICK, oiling, felt	0603
WHEEL assembly	1301
WHEEL, steering, assembly	1404
WINDSHIELD, assembly	1811
WIRE, ground, blackout driving light, assembly	0607D
WIRE, left headlight, assembly	0607
WIRE, right headlight, assembly	0607
WRENCH, adjustable, auto type	2301
WRENCH, brake bleeder screw	2301
WRENCH, crescent	2301
WRENCH, drain plug	2301
WRENCH, engineer's double end, 15° angle	2301
WRENCH, fluted socket head	2301
WRENCH, socket	2301
WRENCH, socket, spark plug	2301
WRENCH, special	2301
WRENCH, wheel hub	2301
YOKE, end, universal joint, assembly	1003

VENDORS' NUMERICAL INDEX

Vendor Number	Willys Part Number	Ford Part Number	Gov't Group No.
AETNA BALL BEARING MFG. CO. (AB)			
AB-A-935-1	WO-635529	FM-GPW-7580	0203
A.C. SPARK PLUG DIV. (AC)			
AC-119922-21	WO-A-1265	FM-GPW-9154	0306B
AC-127951	WO-A-1264	FM-GPW-9185	0306B
AC-132629	WO-51546	FM-26466-S7	0302
AC-846709	WO-A-1215	FM-GPW-8100-A	0501
AC-846732	WO-A-1216	FM-GPW-8578	0501
AC-850024	WO-A-1275	FM-GPW-9035-B	0300
AC-853558	WO-A-1259	FM-GPW-9160	0306B
AC-853562	WO-A-1257	FM-GPW-9184	0306B
AC-853572	WO-A-1260	FM-GPW-9186	0306B
AC-854005	WO-113460	FM-GPW-9388	0302
AC-855064	WO-113440	FM-34803-S7	0302
AC-855493	WO-113439	FM-31628-S7	0302
AC-855763	WO-113461	FM-GPW-9373	0302
AC-1504117	WO-A-1263	FM-GPW-9162	0306B
AC-1504118	WO-A-1262	FM-GPW-9182	0306B
AC-1504212	WO-A-1258	FM-GPW-9149	0306B
AC-1521288	WO-A-1047	FM-GPW-9377	0302
AC-1521578	WO-A-1046	FM-GPW-9378	0302
AC-1521708	WO-115880	FM-INC-9381	0302
AC-1521880	WO-115870	FM-GPW-19469	0302
AC-1521953	WO-115652	FM-GPW-9363	0302
AC-1521956	WO-115653	FM-11A-9361	0302
AC-1521960	WO-115641	FM-GPW-9399	0302
AC-1521985	WO-115869	FM-GPW-9468	0302
AC-1522046	WO-115643	FM-GPW-9380	0302
AC-1523068	WO-116694	....	0302
AC-1523084	WO-115650	FM-GPW-9354	0302
AC-1523096	WO-115656	FM-GPW-9364	0302
AC-1523099	WO-115654	FM-GPW-9365	0302
AC-1523106	WO-115651	FM-11A-9352	0302
AC-1523231	WO-115657	FM-GPW-9387	0302
AC-1537065	WO-A-1494	FM-GPW-9355	0302
AC-1537812	WO-A-1045	FM-GPW-9386	0302
AC-1538205	WO-116695	FM-GPW-9398	0302
AC-1538312	WO-A-8323	....	0302
AC-1542508	WO-A-1313	....	0301C
AC-1542509	WO-A-1314	....	0301C
AC-1542510	WO-A-1315	....	0301C
AC-1595235	WO-A-1255	FM-GPW-9155	0306B
AC-1595823	WO-A-1261	FM-GPW-9140	0306B
ALEMITE DIV. (AD)			
AD-1636	WO-368224	FM-353035-S7	0805B
AD-1641	WO-638792	FM-353043-A-S7	0204
AD-1980	WO-640038	FM-358006-S8	1204
ELECTRIC AUTO-LITE (AL)			
AL-GG-6	WO-106313	FM-GPW-10098	0601
AL-DH-7	WO-A-1624	FM-GPW-10116	0601
AL-MZ-12	WO-109431	FM-GPW-11055	0602
AL-MU-14	WO-A-1554	FM-GPW-11102	0602
AL-MZ-16	WO-109442	FM-GPW-11061	0602
AL-MZ-19	WO-109445	FM-B-11059	0602
AL-MN-21	WO-A-1609	FM-GPW-10218	0601
AL-DA-22	WO-A-1621	FM-356208-S7	0601
AL-DK-23	WO-A-1590	FM-GPW-10120	0601
AL-GCT-25	WO-A-1598	FM-GPW-10104	0601
AL-GCY-25	WO-A-1599	FM-GPW-10206-A	0601
AL-GBJ-25	WO-A-1625	FM-GPW-10119	0601
AL-GC-26	WO-A-1597	FM-GPW-10208-A	0601
AL-GEB-27	WO-A-1602	FM-GPW-10211-A	0601
AL-MU-28	WO-109433	FM-GPW-11103	0602
AL-IGH-28	WO-A-1671	FM-GPW-12133	0603
AL-GEB-29	WO-A-1600	FM-GPW-10041	0601
AL-M-29	WO-A-1556	FM-GPW-11120	0602
AL-MZ-30-A	WO-A-1557	FM-GPW-11089	0602
AL-IGP-30	WO-A-1587	FM-GPW-12169	0603
AL-GEW-31	WO-A-1638	FM-34709-S7	0601
AL-GAU-31	WO-A-1644	FM-78-10212-A	0601
AL-MU-31	WO-109436	FM-GPW-11107	0602
AL-MAB-31	WO-A-1553	FM-GPW-11094	0602
AL-GBJ-32-A	WO-A-1626	FM-GPW-10118	0601
AL-GAA-32	WO-A-1591	FM-GPW-10202-A	0601
AL-MZ-32	WO-A-1558	FM-GPW-11090	0602
AL-GBW-34	WO-A-1593	FM-GPW-10206-A	0601
AL-MU-37	WO-A-1555	FM-34706-S	0602
AL-MZ-38-E	WO-A-1559	FM-355385-S	0602
AL-GBY-38-A	WO-A-1596	FM-355486-S7	0601
AL-DA-39	WO-A-1622	FM-GPW-10124	0601
AL-MU-39	WO-109437	FM-GPW-11095	0602
AL-GEB-44	WO-A-1601	FM-GPW-10100	0601
AL-GAL-44	WO-A-1592	FM-OIA-10193	0601
AL-GBA-46	WO-A-1573	FM-GPW-11350	0602
AL-MZ-51	WO-A-1569	FM-GPW-11053	0602
AL-MZ-52	WO-A-1584	FM-355164-S2	0602
AL-GCE-53	WO-A-1628	FM-GPW-10057	0601
AL-GCE-54	WO-A-1629	FM-GPW-10105	0601
AL-MU-54	WO-109455	FM-GPW-11036-A	0602
AL-GEB-58	WO-A-1605	FM-GPW-10211-B	0601
AL-DA-60	WO-A-1623	FM-355260-S7	0601
AL-GBW-66	WO-A-1594	FM-GPW-10106	0601
AL-SPK-66	WO-A-8242	....	2204
AL-GBW-67	WO-A-1595	FM-GPW-10202-C	0601
AL-IGT-69	WO-A-1684	FM-GPW-12191	0603
AL-8X-72-B	WO-A-8127	....	2204
AL-MG-77-A	WO-A-1583	FM-GPW-11395	0602
AL-GT-78	WO-A-1646	FM-GPW-10212	0601
AL-MAB-88	WO-A-1586	FM-72798-S7-8	0602
AL-IG-90	WO-106740	FM-GPW-12193	0603
AL-IG-94	WO-A-1652	FM-GPW-12006	0604A
AL-IGS-99	WO-A-1672	FM-GPW-12120	0603
AL-IGS-104	WO-A-1673	FM-GPW-12182	0603
AL-EB-108	WO-A-1574	FM-B-11379	0602
AL-IGC-117	WO-A-1660	FM-GPW-12174	0603
AL-GCE-125-A	WO-A-6301	FM-GPW-10139	0601
AL-8X-140	WO-A-1610	FM-34051-S7-8	0601
AL-X-156	WO-A-1640	FM-34019-S7	0601
AL-8X-177	WO-A-1611	FM-355883-S	0601
AL-12X-193	WO-A-1612	FM-356263-S7	0601
AL-12X-194	WO-A-1613	FM-34801-S7	0601
AL-X-195	WO-51532	FM-34802-S	0601
AL-X-196	WO-5168	FM-78-10212-A	0601
AL-12X-196	WO-A-1614	FM-34803-S7	0601
AL-IGB-199	WO-A-1659	FM-GPW-12195	0603
AL-X-199	WO-5045	FM-34805-S7-8	0601
AL-IGB-202-S	WO-A-1677	FM-GPW-12084	0603
AL-12X-203	WO-A-1619	FM-34806-S2	0601
AL-XS-213	WO-A-1685	FM-72867-S	0603
AL-TS-215	WO-A-5433	FM-11AS-10657	0610
AL-TSR-215	WO-A-1767	FM-11-AS-10658	0610
AL-X-260	WO-A-1641	FM-74144-S	0601
AL-X-261	WO-5017	FM-74175-S	0602
AL-X-295	WO-51248	FM-GPW-10094	0601
AL-8X-305	WO-A-1632	FM-26457-S7	0603
AL-8X-311	WO-A-1647	FM-36800-S7-8	0601
AL-8X-323	WO-A-1633	FM-36787-S7	0603

SNL G-503

VENDOR'S NUMERICAL INDEX (Cont'd)

Vendor Number	Willys Part Number	Ford Part Number	Gov't Group No.
ELECTRIC AUTO-LITE (AL) Cont'd			
AL-8X-349	WO-A-1615	FM-34703-S7-8	0601
AL-8X-361	WO-A-1616	FM-34705-S	0601
AL-X-404	WO-A-1642	FM-72034-S	0601
AL-SP-484-A	WO-A-1639	FM-GPW-10130	0601
AL-X-489	WO-302347	FM-B-10141	0601
AL-X-490	WO-107128	FM-B-10141	0603
AL-IG-514	WO-A-1656	FM-GPW-12011	0603
AL-IG-515	WO-A-1657	FM-GPW-12012	0603
AL-X-532	WO-A-1567	FM-GPW-11069	0602
AL-12X-544	WO-A-1635	FM-34803-S7	0601
AL-X-544	WO-A-1571	FM-34803-S7	0602
AL-IG-579-A	WO-A-1681	FM-GPW-12082	0603
AL-IG-676	WO-A-1663	FM-GPW-12217	0603
AL-IG-680	WO-A-1653	FM-GPW-12177	0603
AL-IG-694	WO-A-1682	FM-GPW-12144	0603
AL-X-714	WO-A-1588	FM-36954-S7-8	0602
AL-8X-715	WO-A-1650	FM-27161-S7-8	0601
AL-8X-794	WO-A-1589	FM-34141-S7-8	0601
AL-IG-816-C	WO-A-1654	FM-GPW-12267	0603
AL-X-847	WO-A-1603	....	0601
AL-8X-870	WO-A-1636	FM-31588-S7-8	0601
AL-8X-872	WO-A-1686	FM-31583-S	0603
AL-8X-884	WO-A-1668	FM-31027-S8-7	0603
AL-X-902	WO-A-1572	FM-31596-S	0602
AL-X-959	WO-A-1605	FM-GPW-10211-B	0601
AL-GEB-1005	WO-A-1604	FM-GPW-10175	0601
AL-GEB-1007-A	WO-A-1606	FM-GPW-10192	0601
AL-MZ-1007	WO-A-1560	FM-GPW-11083	0602
AL-GEB-1008-B	WO-A-1607	FM-GPW-10191	0601
AL-MZ-1008	WO-109428	FM-GPW-11084	0602
AL-MZ-1009	WO-109427	FM-GPW-11082	0602
AL-MZ-1010	WO-A-1563	FM-GPW-11085	0602
AL-GCE-1012	WO-A-1630	FM-GPW-10069	0601
AL-X-1012	WO-A-1669	FM-34801-S	0603
AL-12X-1014	WO-A-5288	FM-34806-S7	0601
AL-X-1014	WO-5051	FM-34806-S2	0602
AL-GCE-24	WO-A-1649	FM-GPW-10142	0601
AL-MZ-1024	WO-109452	FM-GPW-11077	0602
AL-MZ-1034	WO-109446	FM-GPW-11056	0602
AL-PS-1079-A	WO-A-1585	FM-GPW-11131	0602
AL-IGS-1080	WO-A-1647	FM-36800-S7-8	0601
AL-SPS-1100-R	WO-A-1343	FM-GPW-17261	2204
AL-SPS-1105-X	WO-A-1344	FM-GPW-17262	2204
AL-GCE-1125-A	WO-A-6300	FM-GPW-10138	0601
AL-IGC-1132-LB	WO-A-1661	FM-GPW-12176	0603
AL-IGS-1134-L	WO-A-1678	FM-GPW-12178	0603
AL-IGC-1148	WO-A-1664	FM-GPW-12010	0603
AL-IGC-1149	WO-A-1654	FM-GPW-12267	0603
AL-IG-1324	WO-A-1655	FM-GPW-12106	0603
AL-8X-1368	WO-A-6297	FM-37789-S7	0601
AL-5X-1376	WO-A-1565	FM-355944-S5	0602
AL-8X-1377	WO-A-1617	FM-350853-S7	0601
AL-8X-1420	WO-A-1618	FM-36009S7	0601
AL-X-1448	WO-A-1683	FM-GPW-12145	0603
AL-8X-1546	WO-A-1670	FM-31026-S7	0603
AL-X-1655	WO-A-6299	FM-B-10094	0601
AL-IG-1657	WO-A-1658	FM-GPW-12200	0604
AL-IG-1798-D	WO-A-1526	FM-GPW-12030	0604
AL-MZ-2012-S	WO-A-1552	FM-GPW-18535	0602
AL-GCG-2017-S	WO-A-1651	FM-GPW-18274	0601
AL-MAB-2040-A	WO-A-1582	FM-GPW-11394	0602
AL-MZ-2089	WO-A-1568	FM-GPW-11005	0602
AL-SPS-2099-M	WO-A-1267	FM-GPW-17260	2204
AL-GCE-2118-A	WO-A-6298	FM-GPW-10050	0601
AL-GCE-2118	WO-A-1627	FM-GPW-10050	0601
AL-GEG-2120-F	WO-A-1637	FM-GPW-10005	0601
AL-IGS-2134-L	WO-A-1675	FM-GPW-12175	0603
AL-IGS-2135	WO-A-1679	FM-GPW-12139	0603
AL-IGC-2148-C	WO-A-1662	FM-GPW-12151	0603
AL-MZ-2156	WO-A-1566	FM-GPW-11949	0601
AL-2229	WO-A-8129	....	0605
AL-2235	WO-6536	FM-34079-S7	0605
AL-IG-2456	WO-A-1676	FM-GPW-12188	0603
AL-SW-4015	WO-A-7225	FM-9N-11450-A	0605
AL-IGP-3028	WO-A-1570	FM-GPW-12162	0603
AL-IGP-3028-S	WO-A-1687	FM-GPW-18354	0603
AL-IGW-3139	WO-A-1631	FM-GPW-12300	0604
AL-SPK-4003	WO-A-8180	FM-GPW-17255-A	2200
AL-MZ-4113	WO-A-1245	FM-GPW-11001-A	0602
AL-VRY-4203-A	WO-A-1409	FM-GPW-10505	0601B
AL-EBA-4611	WO-A-1581	FM-GPW-11354	0602
AL-IGC-4705	WO-A-1244	FM-GPW-12100	0603
AL-GED-5002-D	WO-A-5992	FM-GPW-10000-A	0601
AL-E-7819-S	WO-A-1575	FM-B-11357-A	0602
AL-8170	WO-5848	FM-34079-S8	0605
AL-EB-8503	WO-A-1576	FM-B-11381	0601
AL-EB-8505	WO-A-1577	FM-GPW-11375	0602
AL-RB-8506	WO-A-1578	FM-GPW-11377	0602
AL-EB-8507	WO-A-1579	FM-GPW-11382	0602
AL-EB-8734	WO-A-1580	FM-B-11371	0602
AL-9288-A	WO-A-8130	....	0605
AL-9979-A	WO-A-1292	FM-GPW-9275	0300
AL-10310-A	WO-A-8190	FM-GPW-9273	0605
AL-10311-A	WO-A-8186	FM-GPW-10850	0605
AL-10312-A	WO-A-8188	FM-GPW-10883	0605
AL-10063	WO-A-8132	....	0605
AL-21608	WO-A-8131	....	0605
AL-IG-40700	WO-A-7792	FM-GPW-12000-B	0604
AINSWORTH MFG. CO. (AN)			
AN-S-384	WO-A-2483	....	1811
AN-X-598	WO-A-2482	FM-38192-S2	1811
AN-S-1044	WO-A-4260	....	1811
AN-X-1637	WO-A-2490	FM-356286-S	1811
AN-X-1700	WO-A-2486	FM-34805-S2	1811
AN-X-1747	WO-A-2239	FM-81W-811189	1811
AN-X-1750	WO-A-2480	FM-38259-S2	1811
AN-X-1751	WO-A-2481	FM-34176-S2	1811
AN-X-1755	WO-A-2487	FM-33249-S2	1811
ATWOOD VACUUM MFG. CO. (AVM)			
AVM-TP-28-7-1	WO-638992	FM-GPW-7563	0201B
AVM-TP-283	WO-638157	FM-GPW-7567	0202
AVM-TP-287	WO-638159	FM-GPW-7564	0202
AVM-TP-2811	WO-638151	FM-GPW-7580	0202
AVM-TP-2817	WO-638153	FM-GPW-7590	0202
AVM-TP-2818	WO-638154	FM-24325-S	0202
AVM-TP-2819	WO-638155	FM-33921-S7	0202
AVM-TP-2820	WO-638158	FM-GPW-7591	0202
AVM-TP-2827	WO-638305	FM-34745-S2	0202
AVM-TP-2831	WO-638993	FM-GPW-7572	0202
AVM-TP-2851	WO-638152	FM-GPW-7566	0202
BASSICK CO. (THE) (BA)			
BA-B-194	WO-A-3203	FM-GPW-1103028	1811
BA-SK-1670	WO-A-313	FM-GP-17037	2301

VENDOR'S NUMERICAL INDEX (Cont'd)

Vendor Number	Willys Part Number	Ford Part Number	Gov't Group No.
BORG AND BECK (B-B)			
B-B-2940	WO-371567	FM-GPW-7549	0201A
B-B-4324	WO-636778	FM-GPW-7577	0201A
B-B-11123	WO-636755	FM-GPW-7550	0201B
BOWER ROLLER BEARING CO. (BOW)			
BOW-BT-14131	WO-51575	FM-GP-7175	0803
BOW-BT-14276	WO-52883	FM-O2Y-1202	0803
BOW-BT-18250	WO-52942	FM-GP-1201	1301A
BOW-BT-18590	WO-52943	FM-GP-1202	1301A
BUDD MFG. CO., EDW. G. (BUE)			
BUE-5052751	WO-A-11319	FM-GPW-1153100	1820
BENDIX PRODUCTS CORP. (BX)			
BX-40-S-33	WO-5010	FM-34807-S2	1500
BX-62-S-176	WO-51738	FM-20300-S7	1207
BX-76-S-25	WO-52483	FM-34905-S2	1207
BX-179-S-6	WO-374586	FM-351915-S	0201A
BX-1067-S-3	WO-116551	FM-GP-2021	1202
BX-1067-S-4	WO-116552	FM-GP-2022	1202
BX-1141-S-1	WO-116549	FM-GP-2018	1202
BX-1141-S-2	WO-116550	FM-GP-2019	1202
BX-39953	WO-637899	FM-91A-2027	1203B
BX-39956	WO-637901	FM-91A-2030	1203B
BX-41545	WO-637905	FM-GP-2035	1203A
BX-41665	WO-637923	FM-351466-S24	1203B
BX-41708	WO-637924	FM-33846-S2	1203B
BX-41876	WO-637900	FM-GP-2028	1203B
BX-41887	WO-637787	FM-GP-2261	1207
BX-44517	WO-116600	FM-GPW-18367	1202
BX-45322	WO-A-927	FM-GPW-2211	1200
BX-45323	WO-A-924	FM-GP-7718-A	0804
BX-45771	WO-A-751	FM-GPW-3635	0609B
BX-45908	WO-A-1484	FM-GPW-2061	1207
BX-46491	WO-A-6110	....	1207
BX-46752	WO-A-755	FM-33800-S7	1203A
BX-47047	WO-A-8894	FM-GPW-2011	1200
BX-47048	WO-A-8895	FM-GPW-2010	1200
BX-47050	WO-A-8896	FM-GPW-2211-B	1200
BX-47051	WO-A-8897	FM-GPW-2210-B	1200
BX-47054	WO-A-8898	FM-GP-2013	1203
COLUMBUS AUTO PARTS (CA)			
CA-SX-1109	WO-A-622	FM-GPW-3332-A	1401
CA-SX-1110	WO-A-623	FM-GPW-3336	1401
CA-S-5069	WO-630755	FM-GPW-3320	1401
CA-S-5070	WO-630756	FM-GPW-3323	1401
CA-S-5143	WO-630757	FM-GPW-3328	1401
CA-S-5144	WO-630753	FM-GPW-3326	1401
CA-S-5145	WO-630754	FM-GPW-3327	1401
CA-D-8822	WO-A-8252	FM-GPW-3305-B	1401
CARTER CARBURETOR (CAR)			
CAR-1-407	WO-116584	FM-GPW-9518	0301
CAR-2-89	WO-116154	FM-GPW-9585	0301
CAR-3-465-S	WO-116585	FM-GPW-9581	0301
CAR-3-4664	WO-A-5501	....	0301
CAR-6-312-S	WO-116544	FM-GPW-9520	0301
CAR-7-116S	WO-116157	FM-GPW-9549	0301
CAR-11-180S	WO-116539	FM-GPW-9533	0301
CAR-11B-35	WO-116159	FM-GPW-9522	0301
CAR-11B-79	WO-116160	FM-GPW-9523	0301
CAR-11B-105	WO-116161	FM-GPW-9562	0301
CAR-11B-108	WO-116162	FM-GPW-9579	0301
CAR-11B-125S	WO-116163	FM-GPW-9696	0301
CAR-11B-127S	WO-116164	FM-GPW-9928	0301
CAR-11B-129S	WO-116165	FM-GPW-9543	0301
CAR-12-255	WO-116166	FM-GPW-9922	0301
CAR-14-246S	WO-116545	FM-GPW-9546	0301
CAR-20-22	WO-116168	FM-GPW-9569	0301
CAR-20-26	WO-116169	FM-GPW-9608	0301
CAR-20-61	WO-116170	FM-GPW-9574	0301
CAR-20-72	WO-116171	FM-GPW-9926	0301
CAR-21-74S	WO-116172	FM-GPW-9550	0301
CAR-24-23	WO-116173	FM-GPW-9558	0301
CAR-25-93S	WO-116174	FM-GPW-9567	0301
CAR-30-20	WO-116175	FM-GPW-9575	0301
CAR-30A-39	WO-116176	FM-GPW-9541	0301
CAR-39-10	WO-116216	FM-GPW-9586	0301
CAR-39-11	WO-116217	FM-GPW-9588	0301
CAR-43-67	WO-116179	FM-GPW-9544	0301
CAR-48-84	WO-116180	FM-GPW-9940	0301
CAR-53A-168S	WO-116181	FM-GPW-9528	0301
CAR-53A-251S	WO-116537	FM-GPW-9529	0301
CAR-61-57	WO-116183	FM-GPW-9578	0301
CAR-61-119	WO-116184	FM-GPW-9624	0301
CAR-61-128	WO-116185	FM-GPW-9615	0301
CAR-61-143	WO-116186	FM-GPW-9650	0301
CAR-61-169	WO-116187	FM-GPW-9570	0301
CAR-61-171	WO-116188	FM-GPW-9636	0301
CAR-61-190	WO-116189	FM-GPW-9587	0301
CAR-61-207	WO-116191	FM-GPW-9935	0301
CAR-61-272	WO-116538	FM-GPW-9907	0301
CAR-62-108S	WO-116543	FM-GPW-6333-D	0301
CAR-62-131S	WO-116548	....	0301
CAR-62-134	WO-116587	FM-GPW-9595	0301
CAR-62-135S	WO-116586	FM-GPW-9526	0301
CAR-63-35	WO-116194	FM-GPW-9614	0301
CAR-64-62S	WO-116195	FM-GPW-9631	0301
CAR-75-547	WO-116540	FM-GPW-9906	0301
CAR-86-10	WO-6330	FM-34802-S2	0301
CAR-86-11	WO-5045	FM-34805-S2	0301
CAR-86-15	WO-5010	FM-34807-S2	0301
CAR-101-10	WO-116211	FM-31032-S7	0301
CAR-101-28	WO-52290	FM-31588-S7	0301
CAR-101-82	WO-116213	FM-31061-S8	0301
CAR-101-120	WO-116651	FM-GPW-9610	0301
CAR-101-122	WO-116215	FM-31662-S	0301
CAR-101-142S	WO-116384	FM-355067-S7	0301
CAR-101-150S	WO-116385	FM-355200-S7	0301
CAR-105-11	WO-116218	....	0301
CAR-105-13	WO-116588	FM-355132-S	0301
CAR-114-21S	WO-116197	FM-GPW-9583	0301
CAR-115-59	WO-116198	FM-GPW-9531	0301
CAR-117-58	WO-116199	FM-GPW-9527	0301
CAR-117-106	WO-116542	FM-GPW-9598	0301
CAR-120-151S	WO-116541	FM-GPW-9914	0301
CAR-121-56	WO-116202	FM-GPW-9516	0301
CAR-121-73	WO-116203	FM-GPW-9519	0301
CAR-122-47S	WO-116204	FM-GPW-119594	0301
CAR-122-64S	WO-116205	FM-GPW-9576	0301
CAR-129-15	WO-116206	FM-GPW-9905	0301
CAR-136-39	WO-116207	FM-34711-S	0301
CAR-146-95S	WO-116208	FM-GPW-9515	0301
CAR-150-97	WO-116209	FM-GPW-9930	0301
CAR-150-98	WO-116177	FM-GPW-9566	0301
CAR-150A-8	WO-116219	FM-355858-S	0301
CAR-150A-13	WO-50922	FM-33800-S2	0301
CAR-150A-19	WO-52615	FM-34051-S7	0301

SNL G-503

VENDOR'S NUMERICAL INDEX (Cont'd)

Vendor Number	Willys Part Number	Ford Part Number	Gov't Group No.
CARTER CARBURETOR (CAR) Cont'd			
CAR-183-19	WO-116210	FM-GPW-9554	0301
CAR-595	WO-A-1223	FM-GPW-9510	0301
CHICAGO SCREW CO. (CCG)			
CCG-W-41-25	WO-637427	FM-78-2814A	1209C
CORCORAN BROWN (CB)			
CB-CB-300	WO-52221	FM-34803-S	0607
CB-CB-4057	WO-A-1032	FM-37206-S	0607
CB-CB-5099	WO-A-1037	FM-38095-S	0607
CB-CB-6012	WO-A-1361	FM-GPW-13022	0607
CB-CB-7783	WO-53071	FM-40631-S7	0607D
CB-CB-7784	WO-A-6313	FM-GPW-13180	0607D
CB-7861	WO-A-1072	FM-28378-S2	0607D
CB-CB-8010	WO-A-1031	FM-GPW-13022	0607
CB-CB-8494	WO-A-1033	FM-GPW-13007	0607A
CB-CB-8523	WO-A-1036	FM-GPW-13048	0607
CB-9204	WO-A-1070	FM-GP-13210-B	0607D
CB-CB-9212	WO-A-1073	FM-GPW-13408-B	0608
CB-CB-9218	WO-A-1074	FM-GPW-13494	0607A
CB-CB-9225	WO-A-1075	FM-GPW-13491-A	0607A
CB-CB-9231	WO-A-1076	FM-GPW-13448-B	0608
CB-CB-9232	WO-A-1079	FM-GPW-13349-A	0608
CB-CB-9233	WO-A-1077	FM-36931-S	0608
CB-CB-9234	WO-A-1078	FM-GPW-13485	0607A
CB-B-9276	WO-A-5806	....	0607D
CB-CB-9281	WO-A-1071	FM-GP-13209	0607D
CB-CB-10436	WO-A-5586	FM-GPW-13012	0607
CB-CB-10461	WO-A-1439	FM-GPW-13217	0607D
CB-CB-10462	WO-A-1440	FM-GPW-13216-B	0607D
CB-CB-11193	WO-A-6144	FM-GPW-13162	0607D
CB-CB-11199	WO-A-6783	FM-GPW-13166	0607D
CB-CB-11200	WO-53070	....	0607D
CB-CB-11261	WO-A-6145	FM-GPW-13153	0607A
CB-CB-11263	WO-A-6148	FM-131045-S2	0607D
CB-11495	WO-A-6142	FM-GPW-13150	0607D
CB-CB-11500	WO-A-6143	FM-GWP-13170	0607D
CB-CB-11503	WO-A-6146	FM-GPW-13175	0607D
CB-60142	WO-A-1064	FM-GPW-13405-B	0608
CB-60242	WO-A-1065	FM-GPW-13040-B	0608
CB-CB-215142	WO-A-1304	FM-GPW-13006	0607
CB-CB-215242	WO-A-1305	FM-GPW-13005	0607
CB-415342	WO-A-1346	....	0606
CB-415442	WO-A-1347	....	0606
CLUM MFG. CO. (CL)			
CL-9654	WO-638979	FM-GPW-13532	0606
COPELAND GIBSON PRODUCTS CORP. (CN)			
CN-310-0	WO-637237	FM-GPW-6702	0102A
CINCH MFG. CO. (CNC)			
CNC-0	WO-A-3113	FM-348148-S	2201A
CNC-12	WO-A-1312	FM-GPW-13802	0609
CNC-50	WO-A-3052	FM-95626-S	2201F
CNC-58	WO-A-1311	FM-GPW-9652	0301C
CNC-60	WO-A-3053	FM-95628-S	2201F
CNC-98	WO-A-3054	FM-95627-S	2201A
CNC-515	WO-A-3051	FM-95705-S	2201A
CNC-519	WO-A-4120	FM-352699-S15	1811
CHAMPION SPARK PLUG CO. (CP)			
CP-QM2	WO-A-538	FM-GPW-12405	0604B

Vendor Number	Willys Part Number	Ford Part Number	Gov't Group No.
DOUGLAS MFG. CO., H. A. (DM)			
DM-D-398	WO-A-6149	FM-GP-13739	0606
DM-5943	WO-A-1347	....	0606
DM-5944	WO-A-1348	....	0606
DM-5969	WO-A-1345	FM-34907-S7-8	0606
DM-5970	WO-A-1332	FM-GPW-11649	0606
DM-5995	WO-A-1333	FM-GPW-13740	0606
DM-6000	WO-A-1351	....	0606
DM-5999	WO-A-1352	....	0606
DM-29392	WO-A-1748	FM-GPW-13713	0605C
DM-5033	WO-52131	FM-26457-S7	0604C
DM-51104	WO-A-1350	FM-356229-S	0606
DM-52947	WO-52332	....	0606
DM-53097-A	WO-A-1349	FM-GPW-12250-C	0606
DM-53170-A	WO-A-6813	....	0604C
DM-53175	WO-A-1346	....	0606
DM-53414	WO-A-1353	....	0606
DEUTSCHMAN, TOBE, CORP. (DTC.)			
DTC-1107G	WO-A-1517	....	2603
DTC-1126	WO-A-1287	FM-GPW-18936-A	2603
EATON MFG. CO. (EAT)			
EAT-LE-1212	WO-A-1126	FM-9N-8115	0501
ERIE RESISTOR CORP. (ER)			
ER-L7	WO-A-6320	....	2603
FEDERAL BEARINGS CO. (FB)			
FB-1207MGF	WO-636885	FM-GPW-7025	0703

Ford Part Number	Willys Part Number	Gov't Group No
FORD MOTOR CO. (FM)		
FM-GP-1012	WO-A-476	1302A
FM-GP-1013	WO-A-475	1302A
FM-GP-1015	WO-A-465	1301
FM-GPW-1015	WO-A-5467	1301
FM-GPW-1016	WO-A-5468	1301
FM-GPW-1024	WO-A-5549	1301
FM-GPW-1025-C	WO-A-5488	1301
FM-GPW-1029	WO-A-5470	1301
FM-GPW-1030	WO-A-5471	1301
FM-GPW-1045	WO-A-5472	1301
FM-GP-1088	WO-A-813	1006
FM-GP-1089	WO-A-814	1006
FM-GP-1092	WO-A-820	1006
FM-GP-1102	WO-A-1690	1302
FM-GP-1103	WO-A-1689	1302
FM-GP-1107	WO-A-474	1302A
FM-GP-1108	WO-A-473	1302A
FM-GP-1110	WO-A-760	1007
FM-GP-1124	WO-A-867	1102
FM-GP-1139	WO-A-869	1302
FM-GP-1177	WO-A-864	1301B
FM-GP-1201	WO-52942	1301A
FM-O1Y-1202	WO-52883	0803
FM-GP-1202	WO-52943	1301A
FM-GP-1218	WO-A-865	1102
FM-GPW-1418	WO-A-11701	1505
FM-GPW-1433	WO-A-2359	1505
FM-GPW-2010	WO-A-8895	1200
FM-GPW-2011	WO-A-8894	1200
FM-GP-2013	WO-A-8898	1203
FM-GP-2018	WO-116549	1202
FM-GP-2019	WO-116550	1202

VENDOR'S NUMERICAL INDEX (Cont'd)

Ford Part Number	Willys Part Number	Gov't Group No.
FORD MOTOR CO. (FM) Cont'd		
FM-GP-2021	WO-116551	1202
FM-GP-2022	WO-116552	1202
FM-91A-2027	WO-637899	1203B
FM-GP-2028	WO-637900	1203B
FM-91A-2030	WO-637901	1203B
FM-GP-2035	WO-637905	1203A
FM-GP-2038	WO-A-754	1203A
FM-GPW-2061	WO-A-1484	1207
FM-GPW-2063	WO-A-1502	1207
FM-GPW-2073	WO-A-1283	1500
FM-GP-2074	WO-637432	1209C
FM-GP-2076	WO-A-557	1205
FM-GP-2077	WO-637605	1205
FM-GP-2078	WO-637424	1209B
FM-GPW-2078	WO-A-1373	1209B
FM-GPW-2079	WO-A-1460	1209B
FM-GPW-2082	WO-A-1487	1001
FM-GPW-2084	WO-A-1486	1001
FM-GP-2087	WO-637426	1209B
FM-GPW-2096	WO-A-1457	1001
FM-GP-2133	WO-384710	1209C
FM-GPW-2135	WO-A-6111	1207
FM-GPW-2138	WO-A-1354	1205
FM-GP-2140	WO-A-556	1205
FM-GP-2143-A	WO-637599	1205
FM-GP-2145	WO-637587	1205
FM-91A-2151	WO-637606	1205
FM-91A-2152	WO-637604	1205
FM-GP-2155	WO-637582	1205
FM-GP-2160	WO-637583	1205
FM-GP-2162	WO-637608	1205
FM-GP-2167	WO-637612	1205
FM-GP-2169	WO-637591	1205
FM-GP-2170	WO-637595	1205
FM-GP-2173	WO-637590	1205
FM-GP-2174	WO-637598	1205
FM-GP-2175	WO-637584	1205
FM-GP-2176	WO-637585	1205
FM-GP-2180	WO-637602	1205
FM-GP-2183	WO-637586	1205
FM-GP-2188	WO-637597	1205
FM-GP-2192	WO-637789	1207
FM-GP-2194	WO-637577	1207
FM-GP-2196	WO-637541	1207
FM-GPW-2196	WO-A-6113	1207
FM-91A-2201	WO-637591	1205
FM-WP-2201	WO-635544	1207
FM-GPW-2201	WO-A-6116	1207
FM-GP-2204	WO-637545	1207
FM-GP-2205	WO-637580	1207
FM-GPW-2206	WO-A-6117	1207
FM-GP-2206-A	WO-637546	1207
FM-GP-2208	WO-637540	1207
FM-GPW-2210	WO-A-928	1200
FM-GPW-2210-B	WO-A-8897	1200
FM-GPW-2211	WO-A-927	1200
FM-GPW-2211-B	WO-A-8896	1200
FM-GPW-2223	....	1209C
FM-GPW-2244	WO-A-1378	1209C
FM-GPW-2250	WO-A-1515	1209C
FM-GP-2261	VENDOR-637787	1207
FM-GPW-2261	WO-A-6110	1207
FM-GPW-2263	WO-A-1501	1209C
FM-GPW-2264	WO-A-1377	1209C
FM-GPW-2265	WO-A-5224	1209C
FM-GPW-2266	WO-A-1376	1209C
FM-GPW-2267	WO-A-5226	1209C
FM-GPW-2268	WO-A-5225	1209C
FM-GPW-2270	WO-A-1795	1201
FM-GPW-2270-B	WO-A-5393	1201
FM-GPW-2272	WO-A-1735	1201
FM-GPW-2274	WO-A-5227	1209C
FM-GPW-2279	WO-638780	1201
FM-GPW-2281	WO-A-5449	0606B
FM-GPW-2298	WO-A-1488	1209C
FM-GPW-2452-B	WO-A-8253	1204
FM-GPW-2454	WO-A-1359	1204
FM-GPW-2473	WO-A-495	0204
FM-GPW-2476-A	....	0204
FM-GPW-2598	WO-A-1008	1201
FM-GPW-2614	WO-A-1002	1201
FM-O1T-2616	WO-A-1020	1201
FM-GP-2620	WO-A-1014	1201
FM-GPW-2630	WO-A-1005	1201
FM-GPW-2632	WO-A-1003	1201
FM-O1T-2634	WO-A-1017	1201
FM-GPW-2635	WO-A-5335	1201
FM-O1T-2640	WO-A-1021	1201
FM-O1T-2642	WO-A-1016	1201
FM-GP-2648	WO-A-1009	1201
FM-GPW-2656	WO-A-1226	1201
FM-GPW-2659	WO-A-1228	1201
FM-GPW-2780	WO-A-1242	1201
FM-GPW-2782	WO-639244	0703
FM-O1T-2805	WO-A-1018	1201
FM-78-2814-A	WO-637427	1209C
FM-GPW-2848	WO-639010	1201
FM-GPW-2852	WO-A-2892	1201
FM-GPW-2853	WO-A-1241	1201
FM-GPW-3001	....	1000
FM-GP-3016-A	WO-A-1727	1007
FM-GP-3017-A	WO-A-1729	1007
FM-GP-3031-A	WO-A-778	1301B
FM-GP-3034	WO-A-779	1006
FM-GPW-3074	....	1001
FM-GP-3105	WO-A-851	1006
FM-GPW-3112	WO-A-1710	1006
FM-GPW-3113	WO-A-1712	1006
FM-GP-3115	WO-A-824	1006
FM-GP-3117-A	WO-A-830	1006
FM-GP-3117-B	WO-A-831	1006
FM-GP-3117-C	WO-A-832	1006
FM-GP-3117-D	WO-A-833	1006
FM-GP-3122	WO-A-825	1006
FM-GPW-3131-B	WO-A-8249	1006
FM-GP-3135	WO-A-819	1006
FM-GP-3139	WO-A-818	1006
FM-GP-3140	WO-A-828	1006
FM-GP-3148-A	WO-A-811	1006
FM-GP-3149-A	WO-A-812	1006
FM-GP-3161	WO-52940	1006
FM-GP-3162	WO-52941	1006
FM-GPW-3165	WO-A-855	1401
FM-GP-3166	WO-A-856	1401
FM-GPW-3167	WO-A-858	1401
FM-GPW-3168	WO-A-859	1401
FM-GPW-3169	WO-A-860	1401

SNL G-503

VENDOR'S NUMERICAL INDEX (Cont'd)

Ford Part Number	Willys Part Number	Gov't Group No.
FORD MOTOR CO. (FM) Cont'd		
FM-GPW-3170	WO-A-861	1401
FM-GPW-3171	WO-A-857	1006
FM-GP-3200-A	WO-A-1716	1007
FM-GP-3204	WO-A-868	1007
FM-GPW-3206-A1	WO-A-1715	1007
FM-GPW-3207-A1	WO-A-1728	1007
FM-GP-3208-A	WO-A-862	1007
FM-GP-3208-B	WO-A-863	1007
FM-GP-3215-A	WO-A-1719	1007
FM-GP-3216	WO-A-1726	1007
FM-GP-3217	WO-A-1724	1007
FM-GP-3218	WO-A-1723	1007
FM-GP-3219	WO-A-1722	1007
FM-GP-3221-A	WO-A-1720	1007
FM-GPW-3279	WO-A-1708	1402
FM-GPW-3280	WO-A-1704	1402
FM-GPW-3281	....	1402
FM-GPW-3282	WO-A-1709	1402
FM-51-3287	WO-A-1706	1402
FM-GP-3291	WO-A-838	1402
FM-GP-3292	WO-A-847	1402
FM-GPW-3304-B	WO-A-8250	1401
FM-GPW-3305-B	WO-A-8252	1401
FM-GPW-3320	WO-630755	1401
FM-GPW-3323	WO-630756	1401
FM-GPW-3325	WO-A-5504	1006
FM-GPW-3326	WO-630753	1401
FM-GPW-3327	WO-630754	1401
FM-GPW-3328	WO-630757	1401
FM-78-3332-A	....	1402
FM-GPW-3332-A	WO-A-622	1401
FM-78-3336	WO-A-844	1402
FM-GPW-3336	WO-A-623	1401
FM-GPW-3365	WO-A-1326	0901
FM-GPW-3370	WO-A-1428	0901
FM-GP-3374	WO-A-780	1007
FM-GPW-3504	WO-A-1239	1403A
FM-GPW-3506	WO-A-635	1405
FM-GPW-3509	WO-A-1199	1403A
FM-GPW-3511	WO-A-1276	1405
FM-GPW-3517	WO-639190	1403A
FM-GPW-3518	WO-639192	1403A
FM-GPW-3520	WO-639191	1403A
FM-GPW-3524	WO-A-742	1403A
FM-GPW-3548	WO-A-740	1403A
FM-GPW-3552	WO-639102	1403A
FM-GPW-3563	WO-638918	1405
FM-GPW-3568	WO-A-1760	1403A
FM-GPW-3571	WO-639104	1403A
FM-GPW-3575	WO-A-745	1403A
FM-GPW-3576	WO-639091	1403A
FM-GPW-3577	WO-639118	1403A
FM-GPW-3581	WO-639119	1403A
FM-GPW-3583	WO-639116	1403A
FM-GPW-3587	WO-639090	1403A
FM-GPW-3589	WO-639103	1403A
FM-GPW-3590	WO-A-1116	1403A
FM-GPW-3591-A	WO-639095	1403A
FM-GPW-3593	WO-639108	1403A
FM-GPW-3594	WO-639109	1403A
FM-GPW-3595	WO-639110	1403A
FM-GPW-3600-A3	WO-A-6858	1404
FM-GPW-3626	WO-638884	0609B
FM-GPW-3627	WO-A-634	0609B
FM-GPW-3631	WO-A-750	0609B
FM-GPW-3635-B	WO-A-751	0609B
FM-GPW-3646	WO-638885	0609B
FM-GPW-3652	WO-A-747	0609B
FM-GPW-3655	WO-A-633	0609B
FM-GPW-3658	WO-A-2859	1405
FM-GPW-3682-A	WO-A-1277	1405
FM-GPW-3685	WO-A-2518	0604C
FM-GPW-3685-B	WO-A-6814	0604C
FM-GPW-3686-B	WO-A-6811	0604C
FM-GPW-4001	WO-A-5500	1100
FM-GPW-4004	WO-A-888	1101
FM-GP-4016	WO-A-781	1001
FM-GPW-4022	WO-A-870	1001
FM-GP-4032	WO-A-904	1102
FM-GP-4035	WO-A-782	1001
FM-GP-4206	WO-A-793	1103
FM-GPW-4209	....	1103
FM-GP-4211	WO-A-798	1103
FM-GPW-4212	WO-A-5566	1103
FM-GPW-4215	WO-A-796	1103
FM-GP-4221	WO-52880	1003
FM-GP-4222	WO-52881	1003
FM-GP-4228	WO-A-795	1003
FM-GP-4229-A	WO-A-784	1103
FM-GP-4229-B	WO-A-785	1003
FM-GP-4229-C	WO-A-786	1003
FM-GP-4229-D	WO-A-787	1003
FM-GP-4229-E	....	1103
FM-GP-4230	WO-A-797	1003
FM-GPW-4234	WO-A-901	1102
FM-GP-4235	WO-A-902	1102
FM-GPW-4236	WO-A-794	1103
FM-GP-4241	WO-636360	1103
FM-GP-4252	WO-A-866	1301C
FM-GPW-4259	WO-A-6439	1102
FM-GP-4281	WO-A-792	1003
FM-O1Y-4529	WO-A-490	0902
FM-GPW-4602	WO-A-1327	0901
FM-GPW-4605	WO-A-1429	0902
FM-86H-4616	WO-52877	1104
FM-GP-4619	WO-636566	1104
FM-86H-4621	WO-52876	1104
FM-GP-4628	WO-52879	1003
FM-GP-4630	WO-52878	1104
FM-GP-4659-A	WO-A-803	1003
FM-GP-4659-B	WO-A-804	1003
FM-GP-4659-C	WO-A-805	1003
FM-GP-4659-D	WO-A-806	1003
FM-GP-4660-A	WO-A-800	1104
FM-GP-4660-B	WO-A-801	1003
FM-GP-4660-C	WO-A-802	1003
FM-GP-4661	WO-636565	1104
FM-GP-4666	WO-636568	1104
FM-GP-4668	WO-A-799	1104
FM-GP-4676-A	WO-639265	1003
FM-GP-4841	WO-A-935	0902
FM-GP-4842	WO-A-1445	1104
FM-GP-4863	WO-A-1105	0803
FM-GP-4866	WO-A-950	0902
FM-GPW-5005	WO-A-1142	1500
FM-GPW-5019	....	1500
FM-GPW-5025	WO-A-5127	1500
FM-GPW-5028	WO-A-1150	1500
FM-GPW-5035	WO-A-547	1500

VENDOR'S NUMERICAL INDEX (Cont'd)

Ford Part Number	Willys Part Number	Gov't Group No.
FORD MOTOR CO. (FM) Cont'd		
FM-GPW-5057	WO-A-1201	1500
FM-GPW-5072	....	1500
FM-GPW-5084	WO-A-6740	1500
FM-GPW-5095	WO-A-1341	1500
FM-GPW-5097	WO-A-534	1500
FM-GPW-5106	WO-A-415	1500
FM-GPW-5113	WO-A-185	1500
FM-GPW-5116	WO-A-1422	1500
FM-GPW-5125	WO-A-1151	1500
FM-GPW-5165	WO-A-1291	0610A
FM-GPW-5168	WO-A-1757	0610A
FM-GPW-5175	WO-A-1164	0610A
FM-GPW-5182	WO-A-593	1502
FM-GPW-5186	WO-A-6393	1502
FM-GPW-5196	WO-A-522	1502
FM-GPW-5230-A	WO-A-1146	0401
FM-GPW-5230-B	WO-A-6118	0401
FM-GPW-5246	WO-A-1296	0402
FM-GPW-5251	WO-A-1300	0402
FM-GPW-5264-A	WO-A-655	0401
FM-GPW-5264-B	WO-A-5753	0401
FM-GPW-5270	WO-636004	0402
FM-GPW-5274	WO-638058	0401
FM-GPW-5283	WO-A-658	0401
FM-GPW-5291-B	WO-A-1253	1500
FM-GPW-5298	WO-A-6119	0402A
FM-GPW-5310	WO-A-613	1601
FM-GPW-5311	WO-A-612	1601
FM-GPW-5313-A	WO-A-613-1	1601
FM-GPW-5313-B	WO-A-612-1	1601
FM-GPW-5315-A	WO-A-613-2	1601
FM-GPW-5315-B	WO-A-612-2	1601
FM-GPW-5316-A	WO-A-613-3	1601
FM-GPW-5316-B	WO-A-612-3	1601
FM-GPW-5317-A	WO-A-613-4	1601
FM-GPW-5317-B	WO-A-612-4	1601
FM-GPW-5318-A	WO-A-613-5	1601
FM-GPW-5318-B	WO-A-612-5	1601
FM-GPW-5319-A	WO-A-613-6	1601
FM-GPW-5319-B	WO-A-612-6	1601
FM-GPW-5320-A	WO-A-613-7	1601
FM-GPW-5320-B	WO-A-612-7	1601
FM-GPW-5321-A	WO-A-613-8	1601
FM-GPW-5321-B	WO-A-612-8	1601
FM-GPW-5330-A	WO-116458	1601
FM-GPW-5330-B	WO-116460	1601
FM-GPW-5337	WO-A-544	1500
FM-GPW-5341	WO-A-500	1500
FM-GPW-5345-A	WO-116609	1601
FM-GPW-5345-B	WO-116610	1601A
FM-GPW-5453	WO-A-574	1000
FM-GPW-5455	WO-A-6511	1000
FM-GPW-5456	WO-339372	1100
FM-GPW-5458	WO-A-568	1000
FM-GPW-5459	WO-A-572	1000
FM-GPW-5460	WO-A-571	1100
FM-GPW-5463	....	1602
FM-GPW-5464	....	1602
FM-GPW-5468	WO-384228	1602
FM-GPW-5481	WO-A-515	1602
FM-GPW-5482	WO-A-1252	1602
FM-GPW-5560	WO-A-614	1601A
FM-GPW-5563	WO-A-614-1	1601A
FM-GPW-5565	WO-A-614-2	1601A
FM-GPW-5566	WO-A-614-3	1601A
FM-GPW-5567	WO-A-614-4	1601A
FM-GPW-5568	WO-A-614-5	1601A
FM-GPW-5569	WO-A-614-6	1601A
FM-GPW-5570	WO-A-614-7	1601A
FM-GPW-5571	WO-A-614-8	1601A
FM-GPW-5572	WO-A-614-9	1601A
FM-GPW-5601	WO-A-6326	1601C
FM-GPW-5602	WO-A-6068	1601C
FM-GPW-5605	WO-A-6069	1601C
FM-GPW-5607	WO-A-6073	1601C
FM-GPW-5608	WO-A-6075	1601C
FM-GPW-5609	WO-A-6074	1601C
FM-GPW-5705-B	WO-A-575	1000
FM-GPW-5724-A	WO-116459	1601A
FM-GPW-5724-B	WO-116589	1601A
FM-GPW-5778-A	WO-A-513	1602
FM-GPW-5779	WO-A-514	1602
FM-GPW-5781	WO-359039	1601A
FM-GPW-5783-A	WO-A-481	1601B
FM-GPW-6002	WO-A-5338	0100
FM-GPW-6005	WO-A-5497	0100
FM-GPW-6009	WO-A-5793	0101
FM-GPW-6010	WO-A-1272	0101
FM-GPW-6016	WO-A-1190	0106
FM-GPW-6020	WO-630359	0101A
FM-GPW-6031-A3	WO-A-1463	0101
FM-74-6038	WO-634758	0809
FM-GPW-6038-A	WO-A-7498	0110
FM-GPW-6040-B	WO-A-6156	0110
FM-GPW-6043	WO-A-146	0110
FM-GPW-6044	WO-A-5125	0110
FM-GPW-6050	WO-A-1534	0101A
FM-GPW-6051-B	WO-A-8558	0101A
FM-GPW-6065	WO-638635	0101A
FM-GPW-6066	WO-349368	0101A
FM-GPW-6067	WO-A-1548	0101
FM-GPW-6105-A	WO-637041	0103
FM-GPW-6105-C	WO-116019	0103
FM-GPW-6105-D	WO-116020	0103
FM-GPW-6135-A	WO-636961	0103B
FM-GPW-6149-A	WO-116110	0103A
FM-GPW-6149-C	WO-116112	0103A
FM-GPW-6149-D	WO-116113	0103A
FM-GPW-6149-E	WO-A-6794	0103A
FM-GPW-6149-G	WO-A-6796	0103A
FM-GPW-6149-H	WO-A-6797	0103A
FM-GPW-6150-A	WO-639864	0103A
FM-GPW-6150-C	WO-116502	0103A
FM-GPW-6150-D	WO-116503	0103A
FM-GPW-6152-A	WO-116562	0103A
FM-GPW-6152-C	WO-116564	0103A
FM-GPW-6152-D	WO-116565	0103A
FM-GPW-6155-A	WO-637042	0103A
FM-GPW-6155-G	WO-116023	0103A
FM-GPW-6155-H	WO-116024	0103A
FM-GPW-6156-F	WO-116616	0103A
FM-GPW-6156-H	WO-116116	0103A
FM-GPW-6156-J	WO-116117	0103A
FM-GPW-6159-A	WO-116566	0103A
FM-GPW-6159-C	WO-116568	0103A
FM-GPW-6159-D	WO-116569	0103A
FM-GPW-6200	WO-640071	0104
FM-GPW-6201	WO-640072	0104
FM-GPW-6211-A	WO-639862	0104A

SNL G-503

VENDOR'S NUMERICAL INDEX (Cont'd)

Ford Part Number	Willys Part Number	Gov't Group No.
FORD MOTOR CO. (FM) Cont'd		
FM-GPW-6211-B	WO-116534	0104A
FM-GPW-6211-C	WO-116535	0104A
FM-GPW-6243	WO-375907	0106
FM-GPW-6244	WO-375908	0106
FM-GPW-6245	WO-375900	0106
FM-GPW-6250	WO-637065	0106A
FM-GPW-6256	WO-638458	0106C
FM-GPW-6260	WO-638457	0106C
FM-GPW-6262-A	WO-639051	0106A
FM-GPW-6269	WO-315932	0601C
FM-GPW-6285	WO-630364	0106D
FM-GPW-6286	WO-375917	0106D
FM-GPW-6287	WO-375920	0102
FM-GPW-6288	WO-630365	0106D
FM-GPW-6306	WO-638459	0102
FM-GPW-6308	WO-634796	0102
FM-GPW-6310	WO-375877	0102
FM-GPW-6312	WO-638113	0102
FM-GPW-6319	WO-387633	0102
FM-GPW-6326	WO-630294	0102A
FM-GPW-6331-A	WO-638732	0102A
FM-GPW-6331-B	WO-639239	0102A
FM-GPW-6331-C	WO-116530	0102A
FM-GPW-6333-A	WO-637007	0102A
FM-GPW-6333-B	WO-637724	0102A
FM-GPW-6333-C	WO-116522	0102A
FM-GPW-6337-A	WO-638733	0102A
FM-GPW-6337-B	WO-639240	0102A
FM-GPW-6337-C	WO-116532	0102A
FM-GPW-6338-A	WO-637008	0102A
FM-GPW-6338-B	WO-637725	0102A
FM-GPW-6338-C	WO-116524	0102A
FM-GPW-6339-A	WO-638730	0102A
FM-GPW-6339-B	WO-639237	0102A
FM-GPW-6339-C	WO-116526	0102A
FM-GPW-6341-A	WO-638731	0102A
FM-GPW-6341-B	WO-639238	0102A
FM-GPW-6341-C	WO-116528	0102A
FM-GPW-6342-A	WO-630727	0106C
FM-GPW-6342-B	WO-630262	0102
FM-GPW-6345	WO-381519	0102
FM-GPW-6353	WO-334103	0102
FM-GPW-6369	WO-635377	0102
FM-GPW-6384	WO-635394	0109B
FM-GPW-6387	WO-632156	0109
FM-GPW-6290	WO-116295	0109
FM-GPW-6392	WO-A-439	0109C
FM-GPW-6500-A	WO-637047	0105A
FM-GPW-6500-B	WO-115948	0105A
FM-GPW-6510-B	WO-375811	0105
FM-GPW-6511-B	WO-637045	0105
FM-GPW-6513	WO-638636	0105
FM-GPW-6514	WO-637044	0105
FM-GPW-6520	WO-630303	0105C
FM-GPW-6521	WO-630305	0105C
FM-GPW-6546	WO-375994	0105
FM-GPW-6549-B	WO-640020	0105C
FM-GPW-6555	WO-51875	0105C
FM-GPW-6600	WO-637636	0107
FM-GPW-6604	WO-630384	0107
FM-GPW-6608	WO-636599	0107
FM-GPW-6610	WO-637425	0107
FM-GPW-6614	VENDOR-343306	0107
FM-GPW-6615	WO-630396	0107A-1
FM-GPW-6617	WO-630397	0107A-1
FM-GPW-6619	WO-630392	0107
FM-GPW-6625	WO-375927	0107
FM-GPW-6627	WO-630398	0107A-1
FM-GPW-6628	WO-630389	0107
FM-GPW-6630	WO-630394	0107
FM-GPW-6642	WO-634813	0107
FM-GPW-6644	WO-630390	0107
FM-GPW-6648	WO-630299	0105C
FM-GPW-6654	WO-356155	0107
FM-GPW-6659	WO-639870	0107
FM-GPW-6663	WO-630518	0107
FM-GPW-6664	WO-630387	0107
FM-GPW-6673	WO-636600	0107
FM-GPW-6675	WO-A-7238	0107A
FM-GPW-6684	WO-330964	0107
FM-GPW-6700	WO-637098	0102A
FM-GPW-6701	WO-637790	0102A
FM-GPW-6702	WO-637237	0102A
FM-GPW-6718	WO-A-6885	0107N
FM-GPW-6722	WO-A-6885	0107N
FM-GPW-6727	WO-639979	0107A
FM-GPW-6734	WO-314338	0107A
FM-GPW-6756	WO-A-6922	0107N
FM-GPW-6758	WO-A-1061	0107N
FM-GPW-6758-B	WO-A-6919	0107N
FM-GPW-6762	WO-630298	0107N
FM-GPW-6763-A	WO-639555	0107G
FM-GPW-6763-B	WO-A-5165	0107G
FM-GPW-6763-C	WO-A-6915	0107G
FM-GPW-6766-A	WO-639556	0107H
FM-GPW-6766-B	WO-A-5168	0107H
FM-GPW-6766-C	WO-A-6525	0107H
FM-GPW-6769	WO-A-6895	0107N
FM-GPW-6770	WO-A-5105	0107G
FM-GPW-6771	WO-A-6918	0301C
FM-GPW-6772	WO-53108	0301C
FM-GPW--6789	WO-A-7280	0107H
FM-GPW-7000	WO-A-1145	0700
FM-GPW-7005	WO-A-1148	0701
FM-GPW-7008	WO-A-7596	0700
FM-GPW-7015	WO-A-5553	0703
FM-GPW-7017	WO-A-5554	0703
FM-GPW-7023	WO-375217	0109C
FM-GPW-7025	WO-636885	0703
FM-GPW-7050	WO-640017	0703
FM-GPW-7051-B	WO-637495	0700
FM-GPW-7052	WO-640018	0703
FM-GPW-7056	WO-637503	0701
FM-GPW-7059	WO-637835	0704B
FM-GPW-7060	WO-A-6317	0704
FM-GPW-7061	WO-A-519	0704
FM-GPW-7062	WO-A-738	0704
FM-GPW-7063	WO-639423	0703
FM-GPW-7064	WO-635844	0703
FM-GP-7065	WO-A-916	0704
FM-B-7070	WO-635846	0703
FM-GP-7077	WO-A-942	0902
FM-O1T-7078-A	WO-A-941	0902
FM-GPW-7080	WO-A-410	0704
FM-O1Y-7083	WO-A-940	0902
FM-O1Y-7096	WO-A-945	0902
FM-GP-7097	WO-A-943	0902
FM-GPW-7100	WO-636879	0704B
FM-GPW-7102	WO-638798	0704B

VENDOR'S NUMERICAL INDEX (Cont'd)

Ford Part Number	Willys Part Number	Gov't Group No.
FORD MOTOR CO. (FM) Cont'd		
FM-GPW-7104-A2	WO-640006	0704
FM-GPW-7105	WO-A-6319	0704B
FM-GPW-7106	WO-637833	0704B
FM-GPW-7107	WO-637834	0704B
FM-GPW-7109	WO-637831	0704B
FM-GPW-7111	WO-638948	0704C
FM-GPW-7113	WO-A-739	0704B
FM-GPW-7115	WO-A-880	0704C
FM-GPW-7116	WO-637832	0704B
FM-GPW-7119	WO-635812	0704C
FM-GPW-7120	WO-639422	0703
FM-GPW-7121	WO-A-878	0704C
FM-GPW-7124	WO-A-6318	0704B
FM-GPW-7126	WO-A-879	0704C
FM-GPW-7129-A	WO-635811	0704C
FM-K-7129-B	WO-A-1216	0501
FM-GPW-7140	WO-638952	0704C
FM-GPW-7141	WO-636882	0704B
FM-GPW-7143	WO-635804	0704C
FM-GPW-7155	WO-638949	0704C
FM-GPW-7206	WO-635836	0706C
FM-GPW-7207	WO-635862	0706A
FM-GPW-7208	WO-635839	0706C
FM-GPW-7210-A	WO-A-1380	0706A
FM-GPW-7211-B	WO-A-7260	0706C
FM-GP-7213	WO-A-971	0805B
FM-GP-7213	WO-A-971	0706A
FM-GPW-7214	WO-635859	0706A
FM-GPW-7217	WO-A-1381	0706A
FM-BB-7220	WO-A-1379	0706A
FM-GPW-7221	WO-A-1382	0706A
FM-GPW-7223	WO-635861	0706A
FM-GPW-7227	WO-392328	0706A
FM-GPW-7230	WO-636196	0706C
FM-GPW-7231	WO-636197	0706C
FM-GPW-7233	WO-A-1385	0706B
FM-GPW-7234	WO-635837	0706A
FM-GPW-7240	WO-A-1156	0706B
FM-GPW-7241	WO-A-1155	0706B
FM-GPW-7245	WO-636200	0706B
FM-GPW-7503	WO-A-1355	0204A
FM-GPW-7507	WO-A-179	0204A
FM-GPW-7508	WO-A-180	0204A
FM-GPW-7512	WO-A-887	0204A
FM-GPW-7514	WO-A-181	0204A
FM-GPW-7515	WO-630112	0203
FM-GPW-7516	WO-630068	0204A
FM-GPW-7517	WO-A-177	0204A
FM-GPW-7518	WO-630103	0109C
FM-GPW-7520	WO-A-405	0204
FM-GPW-7521	WO-A-499	0204A
FM-GPW-7525	WO-A-1360	0204
FM-GPW-7530	WO-A-5102	0203
FM-GPW-7532	WO-632177	0203
FM-GPW-7539	WO-A-176	0204A
FM-GPW-7545	WO-A-178	0204A
FM-GPW-7549	WO-371567	0201A
FM-GPW-7550	WO-636755	0201B
FM-GPW-7561	WO-639654	0203
FM-GPW-7562	WO-630117	0203
FM-GPW-7563	WO-638992	0201B
FM-GPW-7565	WO-638159	0202
FM-GPW-7566	WO-638152	0202
FM-GPW-7567	WO-638157	0202
FM-GPW-7570	WO-638151	0202
FM-GPW-7572	WO-638993	0202
FM-GPW-7577	WO-636778	0201A
FM-74-7580-B	WO-635529	0203
FM-GPW-7590	WO-638153	0202
FM-GPW-7591	WO-638158	0202
FM-GPW-7600	WO-639578	0109
FM-GPW-7700	WO-A-1195	0800
FM-GPW-7705	WO-A-1503	0801
FM-GPW-7706	WO-A-1508	0801
FM-GPW-7707	WO-A-1509	0801
FM-GP-7708	WO-A-953	0801
FM-GP-7709	WO-A-954	0801
FM-GPW-7710	WO-A-1506	0805B
FM-GP-7711	WO-A-960	0805A
FM-GP-7712	WO-A-959	0805A
FM-GP-7718-A	WO-A-924	0804
FM-GP-7719	WO-A-1007	0803
FM-GP-7722	WO-A-1510	0802
FM-GP-7723	WO-51575	0803
FM-GP-7729	WO-A-1106	0803
FM-GP-7742	WO-A-999	0804
FM-GP-7743	WO-A-998	0804
FM-GP-7744	WO-A-1000	0804
FM-GP-7746	WO-A-1134	0803
FM-GP-7754	WO-A-934	0801
FM-GPW-7756	WO-A-1435	0801
FM-GPW-7761	WO-A-975	0803
FM-GP-7762	WO-A-992	0802
FM-GP-7763	WO-A-1764	0803
FM-GP-7765	WO-A-988	0803
FM-GP-7766	WO-A-989	0803
FM-GP-7767	WO-A-1001	0804
FM-GPW-7768	WO-A-1507	0801
FM-GP-7770-A	WO-A-958	0803
FM-GP-7771	WO-A-990	0803
FM-GPW-7773	WO-A-957	0801
FM-GPW-7774	WO-A-956	0801
FM-GPW-7776	WO-A-1111	0803
FM-GP-7777-A	WO-A-987	0803
FM-GPW-7781-A	WO-634759	0809
FM-GP-7782-A	WO-A-982	0801
FM-GP-7782-B	WO-A-983	0801
FM-GP-7782-C	WO-A-984	0801
FM-GP-7783	WO-A-976	0803
FM-GP-7784	WO-A-991	0802
FM-GPW-7786	WO-A-1504	0805
FM-GP-7787	WO-A-962	0805
FM-GP-7788	WO-A-966	0805
FM-GP-7789	WO-A-965	0805A
FM-GPW-7793	WO-A-1505	0805B
FM-GP-7796	WO-A-972	0805B
FM-GP-7798-A	WO-A-974	0805B
FM-GP-7799	WO-A-970	0805B
FM-GPW-8005	WO-A-1214	0501
FM-GPW-8100-A	WO-A-1215	0501
FM-GPW-8102	WO-A-3175	0501H
FM-GPW-8103	WO-A-3176	0501H
FM-9N-8115	WO-A-1126	0501
FM-GPW-8125-A	WO-A-4413	0504
FM-GPW-8133	WO-A-1217	0501
FM-GPW-8155	WO-A-3173	1803
FM-GPW-8162	WO-A-2977	0501
FM-GPW-8166	WO-A-3574	1803
FM-GPW-8222	WO-A-2979	1803

VENDOR'S NUMERICAL INDEX (Cont'd)

Ford Part Number	Willys Part Number	Gov't Group No.
FORD MOTOR CO. (FM) Cont'd		
FM-GPW-8240	WO-A-1124	0503A
FM-GPW-8250	WO-A-1192	0101A
FM-GPW-8255	WO-639650	0101A
FM-1GT-8260	WO-630512	0505
FM-GPW-8264	WO-A-592	0505
FM-GPW-8269	WO-636109	0505
FM-GPW-8285	WO-A-6373	0505
FM-60-8287	WO-52226	0505
FM-GPW-8290	WO-A-6374	0505
FM-GPW-8307	WO-A-3615	2103
FM-GPW-8348	WO-A-3094	2103
FM-GPW-8349	WO-A-3095	2103
FM-GPW-8501	WO-639992	0503
FM-GPW-8505	WO-637052	0503
FM-GPW-8509-A	WO-636299	0503A
FM-GPW-8512	WO-639993	0503
FM-GPW-8524-A2	WO-640031	0503
FM-GPW-8530	WO-636297	0503
FM-GPW-8543	WO-637053	0503
FM-GPW-8548-B	WO-640032	0503
FM-GPW-8557-A	WO-640034	0503
FM-GPW-8572-B	WO-640033	0503
FM-GPW-8575	WO-637646	0502
FM-GPW-8576	WO-636298	0503
FM-GPW-8578	WO-639651	0502
FM-GPW-8600	WO-A-447	0503A
FM-GPW-8620	WO-A-1495	0503B
FM-GPW-9001	WO-A-6897	0300
FM-GPW-9002-A	WO-A-1221	0300
FM-GPW-9002-B	WO-A-6618	0300
FM-GPW-9019	WO-A-3195	0300
FM-GPW-9030-A	WO-A-1254	0300
FM-GPW-9034-A	WO-A-6424	0300
FM-GPW-9035-B	WO-A-1275	0300
FM-GPW-9051	WO-A-2970	0300
FM-GPW-9062	WO-A-2952	0300
FM-GPW-9063	WO-A-2954	0300
FM-GPW-9065	WO-A-2953	0300
FM-GPW-9069	WO-A-1738	0300
FM-GPW-9071	WO-A-1741	0300
FM-GPW-9071	WO-A-3206	1803
FM-GPW-9074	WO-A-1476	0300
FM-GPW-9075	WO-A-1740	0300
FM-GPW-9078	WO-A-1480	0300
FM-GPW-9079	WO-A-1739	0300
FM-GPW-9082-A	WO-A-3207	1803
FM-GPW-9083-A	WO-A-3209	1803
FM-GPW-9085-A	WO-A-3208	1803
FM-GPW-9140-A	WO-A-1261	0306B
FM-GPW-9149	WO-A-1258	0306B
FM-GPW-9154	WO-A-1265	0306B
FM-GPW-9155-A	WO-A-1255	0306B
FM-GPW-9160	WO-A-1259	0306B
FM-GPW-9162	WO-A-1263	0306B
FM-GPW-9182	WO-A-1262	0306B
FM-GPW-9183	WO-A-1256	0306B
FM-GPW-9184	WO-A-1257	0306B
FM-GPW-9185	WO-A-1265	0306B
FM-GPW-9186	WO-A-1260	0306B
FM-GPW-9211	WO-A-1763	0300A
FM-GPW-9237	WO-A-1366	0304
FM-GPW-9267-A	WO-387249	0306B
FM-GPW-9268	WO-384549	0306B
FM-GPW-9273	WO-A-8184	0605
FM-GPW-9273	WO-A-8190	0605
FM-GPW-9275	WO-A-1292	0300A
FM-GPW-9276	WO-A-1293	0300A
FM-GPW-9282	WO-A-1367	0304
FM-GPW-9288	WO-A-1325	0304
FM-GPW-9289	WO-A-1368	0304
FM-GPW-9295	WO-A-5450	0304
FM-GPW-9315	WO-A-8131	0605
FM-GPW-9316	WO-A-1450	0107J
FM-GPW-9319-A	WO-662420	0605
FM-GPW-9323	WO-A-1456	0107J
FM-11A-9352	WO-115651	0302
FM-GPW-9354	WO-115650	0302
FM-GPW-9355	WO-A-1494	0302
FM-11A-9361	WO-115653	0302
FM-GPW-9363	WO-115652	0302
FM-GPW-9364	WO-115656	0302
FM-GPW-9365	WC-115654	0302
FM-GPW-9369	WO-A-1369	0304
FM-GPW-9373	WO-113461	0302
FM-GPW-9377	WO-A-1047	0302
FM-GPW-9378	WO-A-1046	0302
FM-GPW-9380	WO-115643	0302
FM-1NC-9381	WO-115880	0302
FM-GPW-9386	WO-A-1045	0302
FM-GPW-9387	WO-115657	0302
FM-GPW-9388	WO-113460	0302
FM-GPW-9396	WO-116694	0302
FM-GPW-9398	WO-113460	0302
FM-GPW-9399	WO-115641	0302
FM-GPW-9410-B	WO-A-1165	0108
FM-GPW-9424-B	WO-A-1166	0108A
FM-GPW-9428	WO-A-912	0108B
FM-GPW-9435	WO-634811	0108
FM-GPW-9443	WO-344732	0108
FM-GPW-9448	WO-638640	0108
FM-GPW-9450	WO-634814	0402
FM-GPW-9456	WO-637206	0108C
FM-GPW-9458	WO-637210	0108C
FM-GPW-9460	WO-636439	0108C
FM-GPW-9462	WO-636438	0108C
FM-GPW-9463	WO-639743	0108C
FM-GPW-9465	WO-637211	0108C
FM-GPW-9467-A	WO-637208	0108C
FM-GPW-9468	WO-115869	0302
FM-GPW-9469	WO-115870	0302
FM-GPW-9484	WO-637209	0108C
FM-GPW-9510	WO-A-1223	0301
FM-GPW-9515	WO-116208	0301
FM-GPW-9516	WO-116202	0301
FM-GPW-9518	WO-116584	0301
FM-GPW-9519	WO-116203	0301
FM-GPW-9520	WO-116544	0301
FM-GPW-9522	WO-116159	0301
FM-GPW-9523	WO-116160	0301
FM-GPW-9526	WO-116586	0301
FM-GPW-9527	WO-116199	0301
FM-GPW-9528	WO-116181	0301
FM-GPW-9529	WO-116537	0301
FM-GPW-9531	WO-116198	0301
FM-GPW-9533	WO-116539	0301
FM-GPW-9541	WO-116176	0301
FM-GPW-9543	WO-116165	0301
FM-GPW-9544	WO-116179	0301
FM-GPW-9546	WO-116545	0301

VENDOR'S NUMERICAL INDEX (Cont'd)

Ford Part Number	Willys Part Number	Gov't Group No.
FORD MOTOR CO. (FM) Cont'd		
FM-GPW-9549	WO-116157	0301
FM-GPW-9550	WO-116172	0301
FM-GPW-9554	WO-116210	0301
FM-GPW-9558	WO-116173	0301
FM-GPW-9562	WO-116161	0301
FM-GPW-9566	WO-116177	0301
FM-GPW-9567	WO-116174	0301
FM-GPW-9569	WO-116168	0301
FM-GPW-9570	WO-116187	0301
FM-GPW-9574	WO-116170	0301
FM-GPW-9575	WO-116175	0301
FM-GPW-9576	WO-116205	0301
FM-GPW-9578	WO-116183	0301
FM-GPW-9579	WO-116162	0301
FM-GPW-9581	WO-116585	0301
FM-GPW-9583	WO-116197	0301
FM-GPW-9585	WO-116154	0301
FM-GPW-9586	WO-116216	0301
FM-GPW-9587	WO-116189	0301
FM-GPW-9588	WO-116217	0301
FM-GPW-9594	WO-116204	0301
FM-GPW-9595	WO-116587	0301
FM-GPW-9598	WO-116542	0301
FM-GPW-9599	WO-116178	0301
FM-GPW-9608	WO-116169	0301
FM-GPW-9609	WO-A-5629	0301C
FM-GPW-9610	WO-116651	0301
FM-GPW-9612	WO-A-7191	0301C
FM-GPW-9614	WO-116194	0301
FM-GPW-9615	WO-116185	0301
FM-GPW-9617	WO-A-5630	0301C
FM-GPW-9621	WO-A-5632	0301C
FM-GPW-9623	WO-A-5633	0301C
FM-GPW-9624	WO-116184	0301
FM-GPW-9628	WO-A-281	0301C
FM-GPW-9631	WO-116195	0301
FM-GPW-9632	WO-A-463	0301C
FM-GPW-9636	WO-116188	0301
FM-GPW-9637-A	WO-A-1290	0301C
FM-GPW-9637-B	WO-A-6911	0301C
FM-GPW-9647	WO-A-642	0301C
FM-GPW-9650	WO-116186	0301
FM-GPW-9652	WO-A-1311	0301C
FM-GPW-9653	WO-635097	0301C
FM-GPW-9656	WO-A-1278	0301C
FM-GPW-9657	WO-A-1279	0301C
FM-GPW-9658	WO-A-5631	0301C
FM-GPW-9686-A	WO-A-1451	0301C
FM-GPW-9696	WO-116163	0301
FM-GPW-9711	WO-650482	0303
FM-GPW-9716	WO-A-1225	0303
FM-GPW-9719	WO-A-6710	0303
FM-GPW-9726	WO-A-1910	0303
FM-GPW-9727	WO-A-1174	0303
FM-GPW-9728	WO-639607	0303
FM-GPW-9731-B	WO-651298	0303
FM-GPW-9732-A	WO-A-1084	0303
FM-GPW-9735-B	WO-A-6851	0303
FM-GPW-9737	WO-632174	0303
FM-GPW-9739	WO-A-1243	0303
FM-GPW-9742	WO-A-1175	0303
FM-GPW-9745	WO-639610	0303
FM-GPW-9751	WO-A-1173	0303
FM-GPW-9752	WO-633013	0303
FM-GPW-9775-B	WO-A-7517	0303
FM-GPW-9778	WO-A-7518	0303
FM-GPW-9795	WO-650483	0303
FM-GPW-9799	WO-633011	0303
FM-GPW-9905	WO-116206	0301
FM-GPW-9906	WO-116540	0301
FM-GPW-9907	WO-116538	0301
FM-GPW-9914	WO-116541	0301
FM-GPW-9922	WO-116166	0301
FM-GPW-9926	WO-116171	0301
FM-GPW-9928	WO-116164	0301
FM-GPW-9930	WO-116209	0301
FM-GPW-9935	WO-116191	0301
FM-GPW-9940	WO-116180	0301
FM-GPW-10000-A	WO-A-5992	0601
FM-GPW-10005	WO-A-1637	0601
FM-GPW-10041	WO-A-1600	0601
FM-GPW-10050	WO-A-6298	0601
FM-GPW-10057	WO-A-1628	0601
FM-GPW-10069	WO-A-1630	0601
FM-B-10094	WO-A-6299	0601
FM-GPW-10094	WO-51248	0601
FM-GPW-10098	WO-106313	0601
FM-GPW-10100	WO-A-1601	0601
FM-GPW-10104	WO-A-1598	0601
FM-GPW-10105	WO-A-1629	0601
FM-GPW-10116	WO-A-1624	0601
FM-GPW-10118	WO-A-1626	0601
FM-GPW-10119	WO-A-1625	0601
FM-GPW-10120	WO-A-1590	0601
FM-GPW-10124	WO-A-1622	0601
FM-GPW-10130	WO-A-1639	0601
FM-GPW-10134	WO-A-1638	0601
FM-GPW-10138	WO-A-6300	0601
FM-GPW-10139	WO-A-6301	0601
FM-B-10141	WO-107128	0603
FM-GPW-10142	WO-A-1649	0601
FM-GPW-10143	WO-A-1399	0601A
FM-GPW-10153-A	WO-A-1491	0601A
FM-GPW-10155	WO-A-1469	0601A
FM-GPW-10162	WO-A-1400	0601A
FM-GPW-10166	WO-A-1392	0601A
FM-GPW-10175	WO-A-1604	0601
FM-GPW-10176	WO-A-1468	0601A
FM-GPW-10177	WO-A-1470	0601A
FM-GPW-10178-A	WO-A-1395	0601A
FM-GPW-10191	WO-A-1607	0601
FM-GPW-10192	WO-A-1606	0601
FM-O1A-10193	WO-A-1592	0601
FM-GPW-10202-A	WO-A-1591	0601
FM-GPW-10202-C	WO-A-1594	0601
FM-GPW-10206-A	WO-A-1599	0601
FM-GPW-10208-A	WO-A-1597	0601
FM-GPW-10211-A	WO-A-1602	0601
FM-GPW-10211-B	WO-A-1605	0601
FM-78-10212-A	WO-A-1644	0601
FM-GPW-10212	WO-A-1646	0601
FM-GPW-10218	WO-A-1609	0601
FM-GPW-10505	WO-A-1409	0601B
FM-11AS-10655	WO-A-1238	0610
FM-11AS-10657	WO-A-5433	0610
FM-11AS-10658	WO-A-1767	0610
FM-GPW-10850	WO-A-8186	0605
FM-GPW-10883	WO-A-8188	0605
FM-GPW-11001-A	WO-A-1245	0602

SNL G-503

270

VENDOR'S NUMERICAL INDEX (Cont'd)

Ford Part Number	Willys Part Number	Gov't Group No.
FORD MOTOR CO. (FM) Cont'd		
FM-GPW-11005	WO-A-1568	0602
FM-GPW-11036-B	WO-109455	0602
FM-GPW-11049	WO-A-1566	0602
FM-GPW-11053	WO-A-1569	0602
FM-GPW-11055	WO-109431	0602
FM-GPW-11056	WO-109446	0602
FM-B-11059	WO-109445	0602
FM-GPW-11061	WO-109442	0602
FM-GPW-11069	WO-A-1567	0602
FM-GPW-11077	WO-109452	0602
FM-GPW-11082	WO-109427	0602
FM-GPW-11083	WO-A-1560	0602
FM-GPW-11084	WO-109428	0602
FM-GPW-11085	WO-A-1563	0602
FM-GPW-11089	WO-A-1557	0602
FM-GPW-11090	WO-A-1558	0602
FM-GPW-11094	WO-A-1553	0602
FM-GPW-11095	WO-109437	0602
FM-GPW-11102	WO-A-1554	0602
FM-GPW-11103	WO-109433	0602
FM-GPW-11107	WO-109436	0602
FM-GPW-11120	WO-A-1556	0602
FM-GPW-11130	WO-A-1582	0602
FM-GPW-11131	WO-A-1585	0602
FM-GPW-11134	WO-639734	0602
FM-GPW-11135	WO-A-1583	0602
FM-GPW-11140	WO-638646	0602
FM-GPW-11350	WO-A-1573	0602
FM-GPW-11354	WO-A-1581	0602
FM-B-11357-A	WO-A-1575	0602
FM-B-11371	WO-A-1580	0602
FM-GPW-11375	WO-A-1577	0602
FM-GPW-11377	WO-A-1578	0602
FM-B-11379	WO-A-1574	0602
FM-B-11381	WO-A-1576	0602
FM-GPW-11382	WO-A-1579	0602
FM-9N-11450-A	WO-A-7225	0606
FM-GPW-11474	WO-372438	0303A
FM-GPW-11514-A	WO-A-2992	1803
FM-GPW-11649	WO-A-1332	0606
FM-GPW-12000	WO-A-1424	0604
FM-GPW-12000-B	WO-A-7792	0604
FM-GPW-12006	WO-A-1652	0604A
FM-GPW-12010	WO-A-1664	0603
FM-GPW-12011	WO-A-1656	0603
FM-GPW-12012	WO-A-1657	0603
FM-GPW-12030	WO-A-1536	0604
FM-GPW-12064	WO-631105	0604
FM-GPW-12082	WO-A-1681	0603
FM-GPW-12083	WO-637615	0603
FM-GPW-12084	WO-A-1677	0603
FM-GPW-12100	WO-A-1244	0603
FM-GPW-12106	WO-A-1655	0603
FM-9N-12113-A2	WO-327257	0604A
FM-GPW-12120	WO-A-1672	0603
FM-GPW-12133	WO-A-1671	0603
FM-GPW-12139	WO-A-1679	0603
FM-GPW-12144	WO-A-1682	0603
FM-GPW-12145	WO-A-1683	0603
FM-GPW-12151	WO-A-1662	0603
FM-GPW-12155	WO-A-1674	0603
FM-GPW-12162	WO-A-1570	0603
FM-GPW-12169	WO-A-1587	0603
FM-GPW-12172	WO-A-1564	0603
FM-GPW-12174	WO-A-1660	0603
FM-GPW-12175	WO-A-1675	0603
FM-GPW-12176	WO-A-1661	0603
FM-GPW-12177	WO-A-1653	0603
FM-GPW-12178	WO-A-1678	0603
FM-GPW-12182	WO-A-1673	0603
FM-GPW-12188	WO-A-1676	0603
FM-GPW-12191	WO-A-1684	0603
FM-GPW-12193	WO-106740	0603
FM-GPW-12195	WO-A-1659	0603
FM-GPW-12200	WO-A-1658	0603
FM-GPW-12217	WO-A-1663	0603
FM-GPW-12250-A	WO-A-1733	0606D
FM-GPW-12250-B	WO-A-1734	0606D
FM-GPW-12250-C	WO-A-1349	0606D
FM-GPW-12267	WO-A-1654	0603
FM-GPW-12283	WO-A-1416	0604A
FM-GPW-12284	WO-A-1414	0604A
FM-GPW-12286	WO-A-1418	0604A
FM-GPW-12287	WO-A-1412	0604A
FM-GPW-12298-B	WO-A-1420	0604A
FM-GPW-12300	WO-A-1631	0603
FM-GPW-12405	WO-A-538	0604B
FM-01A-12410	WO-637863	0604B
FM-11A-12425	WO-A-1096	0604B
FM-GPW-13005-A	WO-A-1305	0607
FM-GPW-13006-A	WO-A-1304	0607
FM-GPW-13007	WO-A-1033	0607A
FM-GPW-13012	WO-A-5586	0607
FM-GPW-13015	WO-A-1031	0607
FM-GPW-13020	WO-A-2873	0607
FM-GPW-13022	WO-A-3161	0607
FM-GPW-13032	WO-A-2878	0607
FM-GPW-13043	WO-A-1036	0607
FM-GPW-13070	WO-A-2871	0607
FM-GPW-13071	WO-A-2870	0607
FM-GPW-13075	WO-A-1363	0607
FM-GPW-13076	WO-A-1362	0607
FM-GPW-13150	WO-A-6142	0607D
FM-GPW-13152	WO-A-6145	0607D
FM-GPW-13162	WO-A-6144	0607D
FM-GPW-13166	WO-A-6783	0607D
FM-GPW-13174	WO-A-6147	0607D
FM-GPW-13175	WO-A-6146	0607D
FM-GPW-13176	WO-A-4118	2103A
FM-GPW-13180	WO-A-6313	0607D
FM-GPW-13181	WO-A-6153	0606B
FM-GPW-13200	WO-A-1437	0607D
FM-GPW-13201	WO-A-1436	0607D
FM-GPW-13209	WO-A-1071	0607D
FM-GP-13210-B	WO-A-1070	0607D
FM-GPW-13216-C	WO-A-1440	0607D
FM-GPW-13217	WO-A-1439	0607D
FM-GPW-13351	WO-A-2993	1802A
FM-GP-13404-B	WO-A-1065	0608
FM-GP-13405-B	WO-A-1064	0608
FM-GPW-13408-B	WO-A-1073	0608
FM-GPW-13410	WO-A-719	0606B
FM-GPW-13434-A	WO-345961	0606B
FM-GPW-13434-A2	WO-345961	0606B
FM-GPW-13437-A2	WO-662276	0606B
FM-GP-13448-B	WO-A-1076	0608
FM-GP-13449-A	WO-A-1079	0608
FM-B-13466	WO-51804	0607B
FM-GP-13485-A	WO-A-1078	0607A

271

SNL G-503

VENDOR'S NUMERICAL INDEX (Cont'd)

Ford Part Number	Willys Part Number	Gov't Group No.
FORD MOTOR CO. (FM) Cont'd		
FM-GP-13491-A	WO-A-1075	0607A
FM-GPW-13494-A	WO-A-1074	0607A
FM-GPW-13532	WO-638979	0606
FM-GPW-13549-A2	WO-A-2991	0606B
FM-GPW-13704	WO-A-1334	0605C
FM-GPW-13710	WO-A-1411	0605C
FM-GPW-13713	WO-A-1748	0605C
FM-GTB-13739	WO-A-6149	0606
FM-GPW-13740	WO-A-1333	0606
FM-GPW-13802	WO-A-1312	0609
FM-GPW-13831	WO-A-1389	0609
FM-GPW-13836	WO-A-302	0609B
FM-GPW-14171	WO-A-752	0609B
FM-GPW-14300	WO-A-1452	0610B
FM-GPW-14301	WO-635883	0601B
FM-GPW-14303	WO-A-1098	2603
FM-GPW-14305	WO-A-5074	0606B
FM-GPW-14305-C	WO-A-7824	0606B
FM-GPW-14320	WO-A-7645	2601
FM-GPW-14321	WO-A-5083	0604A
FM-09B-14362	WO-A-6356	0606F
FM-GPW-14401-D	WO-A-5048	0606B
FM-GPW-14406	WO-A-5070	0605B
FM-GPW-14409	WO-A-5081	0609B
FM-GPW-14416	WO-A-5080	0605B
FM-GPW-14425	WO-A-1665	0606B
FM-GPW-14425-B	WO-A-7825	0606B
FM-GPW-14431	WO-A-1454	0610B
FM-GPW-14432	WO-A-5981	0605B
FM-GPW-14432-B	WO-A-7823	0606B
FM-GPW-14436	WO-A-1731	0606B
FM-GPW-14446	WO-A-5061	0606B
FM-GPW-14448-C	WO-639599	0606E
FM-GPW-14448-D	WO-A-1490	0606E
FM-11A-14452	WO-371400	0610A
FM-GPW-14456	WO-A-5079	0606B
FM-GPW-14457	WO-A-5078	0606B
FM-GPW-14458	WO-A-5072	0605B
FM-GPW-14459	WO-A-5073	0606B
FM-B-14463	WO-307556	0604A
FM-91-BS-14463	WO-A-7466	2601
FM-GPW-14464	WO-78318	0606B
FM-GPW-14465	WO-A-5082	0606B
FM-B-14466	WO-314369	0604A
FM-GPW-14480-B	WO-A-8113	2601
FM-GPW-14481-B	WO-A-8116	2601
FM-GPW-14487-A	WO-635985	0606B
FM-GPW-14487-B	WO-635981	0606B
FM-GPW-14513	WO-A-7640	2601
FM-GPW-14521	WO-A-7180	2601
FM-GPW-14531	WO-A-7636	2601
FM-GPW-14532	WO-A-7182	2601
FM-GPW-14534	WO-A-7638	2601
FM-GPW-14535	WO-A-7637	2601
FM-GPW-14536	WO-A-7181	2601
FM-GPW-14537	WO-A-7947	2601
FM-GPW-14561	WO-A-1694	2603
FM-GPW-14561	WO-A-5598	0606B
FM-GPW-14566	WO-78932	0606B
FM-GPW-14585	WO-A-1289	0304
FM-GPW-14589	WO-A-1289	0107J
FM-48-15021	WO-52837	0607B
FM-GPW-16005	VENDOR-A-2942	1701
FM-GPW-16006	WO-52031	0300
FM-GPW-16082	WO-A-2842	1702
FM-GPW-16083	WO-A-2841	1702
FM-GPW-16094	WO-A-2844	1701
FM-GPW-16095	WO-A-2843	1701
FM-GPW-16105	WO-A-2891	1701
FM-GPW-16128	WO-A-3096	1702
FM-GPW-16132	WO-A-2662	1701
FM-GPW-16133	WO-A-3159	1702
FM-GPW-16159	WO-A-3158	1701
FM-GPW-16610	WO-A-3225	1704
FM-GPW-16684	WO-A-3059	1704
FM-GPW-16802	WO-A-2188	1704
FM-GTB-16847	WO-A-4680	1704
FM-GTB-16848-A	WO-A-4683	1704
FM-9N-16868	WO-384569	0107J
FM-GPW-16892	WO-A-2896	1701
FM-GPW-17003-B	WO-A-1162	2301
FM-GPW-17005	WO-A-372	2301
FM-GPW-17011-A	WO-306715	2301
FM-GPW-17017-A	WO-637635	2301
FM-GP-17020	WO-A-375	2301
FM-GP-17021	WO-A-377	2301
FM-GP-17023	WO-A-376	2301
FM-GP-17028	WO-A-574	2301
FM-GPW-17030	WO-A-5130	2301
FM-GP-17033	WO-A-692	2301
FM-GPW-17035	WO-A-348	2301
FM-GPW-17036	WO-A-289	2301
FM-GP-17037	WO-A-313	2301
FM-GP-17038	WO-A-379	2301
FM-GP-17042	WO-A-373	2301
FM-GP-17043	WO-A-596	2301
FM-GP-17044	WO-A-597	2301
FM-GP-17045	WO-A-598	2301
FM-GP-17046	WO-A-599	2301
FM-GP-17047	WO-A-600	2301
FM-GPW-17052	WO-A-7511	2301
FM-GPW-17055	WO-A-4518	1804
FM-GPW-17062	WO-A-1100	2301
FM-GPW-17079	WO-A-4516	1804
FM-GPW-17080	WO-A-1240	2301
FM-GPW-17090	WO-A-1339	2301
FM-GPW-17091	WO-A-1492	2301
FM-GPW-17097	WO-A-693	2402
FM-GPW-17100	WO-A-616	2402
FM-GP-17125	WO-A-213	2301
FM-GPW-17126-B	WO-A-11765	2301
FM-GPW-17255-A	WO-A-8180	2204
FM-GPW-17260	WO-A-1267	2204
FM-GPW-17261	WO-A-1343	2204
FM-GPW-17262	WO-A-1344	2204
FM-GPW-17271	WO-A-1512	0810
FM-GP-17277	WO-A-985	0810
FM-GP-17285	WO-A-1511	0810
FM-GP-17333	WO-636396	0810
FM-GPW-17500	WO-A-11433	1811D
FM-GP-17535	WO-A-2512	1811D
FM-GP-17541	WO-A-2513	1811D
FM-21CS-17682-B	WO-A-2934	2202
FM-GPW-17750	WO-A-1117	2101
FM-GPW-17751	WO-A-1147	2101
FM-GPW-17752	WO-A-1130	1500
FM-GPW-17753	WO-A-1129	1500
FM-GPW-17754	WO-A-1128	1500
FM-GPW-17755	WO-A-1127	1500

SNL G-503

VENDOR'S NUMERICAL INDEX (Cont'd)

Ford Part Number	Willys Part Number	Gov't Group No.
FORD MOTOR CO. (FM) Cont'd		
FM-GPW-17759	WO-A-1120	1500
FM-GPW-17775	WO-A-1157	2101
FM-GPW-1151266	WO-A-2897	2201A
FM-GPW-1151270	WO-A-2901	2201A
FM-GPW-1151272	WO-A-2898	2201A
FM-GPW-1151289	WO-A-2501	1811
FM-GPW-1151297	WO-A-3189	1811
FM-GPW-1152700	WO-A-3216	2201
FM-GPW-1152720	WO-A-3110	2201A
FM-GPW-1152730	WO-A-3109	2201A
FM-GPW-1152950	WO-A-3141	2201A
FM-GPW-1153030	WO-A-2754	2201A
FM-GPW-1153142	WO-A-2900	1804
FM-GPW-1154810-A	WO-A-2168	1802A
FM-GPW-146211	WO-A-1693	2603
FM-GPW-18040	....	1500
FM-GPW-18045	....	1603
FM-GPW-18060	WO-637936	1603
FM-GPW-18075	....	1500
FM-GPW-18080	....	1500
FM-GPW-18136-B	WO-A-7687	2301B
FM-11YS-18142-B	WO-A-6019	0606F
FM-11YS-18143-B	WO-A-6589	0606F
FM-11YS-18144-B	WO-A-6588	0606F
FM-11YS-18149-B	WO-A-6587	0606F
FM-11YS-18151-B	WO-A-6586	0606F
FM-GPW-18205-B	WO-A-5621	0301C
FM-GPW-18274	WO-A-1650	0601
FM-GPW-18288	WO-A-7568	0102
FM-GPW-18289	WO-A-7503	0109
FM-GPW-18308	WO-A-8279	2302
FM-GPW-18314	WO-A-7837	0810
FM-GPW-18317	WO-A-7445	0800
FM-GPW-18323	WO-A-7835	0108
FM-GPW-18325	WO-A-6855	2301
FM-GPW-18330-A	WO-A-7233	0104A
FM-GPW-18330-B	WO-A-7234	0104A
FM-GPW-18330-C	WO-7235	0104A
FM-GPW-18336	WO-A-6881	1007
FM-GPW-18337	WO-6883	0306B
FM-GPW-18338	WO-A-6882	1006
FM-GPW-18342	WO-A-7840	0601
FM-GPW-18343	WO-A-7843	0603
FM-GPW-18347	WO-A-6798	0102A
FM-GPW-18348	WO-A-6746	0102A
FM-GPW-18349	WO-A-6747	0102A
FM-GPW-18352	WO-A-6837	0301
FM-GPW-18353	WO-A-7680	2309
FM-GPW-18354	WO-A-1687	0603
FM-GPW-18355-B	WO-A-7443	0800
FM-GPW-18356-B	WO-A-7832	0700
FM-GPW-18357-B	WO-A-6840	0301
FM-GPW-18358	WO-A-6751	0201A
FM-GPW-18359-B	WO-A-7833	0202
FM-GPW-18360	WO-A-6753	0801
FM-GPW-18361-B	WO-A-7845	0606B
FM-GPW-18363-B	WO-A-7844	0604A
FM-GPW-18363-C	WO-A-9055	0601
FM-GPW-18365-B	WO-A-7830	1001
FM-GPW-18366-B	WO-A-7831	1100
FM-GPW-18367	WO-116600	1202
FM-GPW-18368	WO-A-6133	1207
FM-GPW-18370-B	WO-A-7838	1205
FM-GPW-18371	WO-115962	1207
FM-GPW-18372	WO-115963	1207
FM-GPW-18374	WO-A-6760	1403A
FM-GPW-18376-B	WO-A-7836	0602
FM-GPW-18377	WO-A-6759	1201
FM-GPW-18359	WO-A-6749	0107
FM-GPW-18380	WO-A-6750	0107
FM-GPW-18382	WO-A-6742	0609B
FM-GPW-18383	WO-A-6791	1401
FM-GPW-18384	WO-A-6816	1003
FM-GPW-18386	WO-A-6745	1104
FM-GPW-18387	WO-A-1537	0101
FM-GPW-18388	WO-A-6744	1103
FM-GPW-18389	WO-A-6743	1103
FM-GPW-18390	WO-A-1536	0101
FM-21C-18397-B	WO-A-1433	0902
FM-GPW-18512	WO-A-1538	0107A
FM-GPW-18515-B	WO-A-6839	0503
FM-GPW-18535	WO-A-1552	0602
FM-GPW-18660	WO-A-1230	0107E
FM-GPW-18662-A	WO-A-1236	0107E
FM-GPW-18663	WO-A-1247	0107F
FM-GPW-18664	WO-A-1251	0107F
FM-GPW-18666	WO-A-1198	0107E
FM-GPW-18667	WO-A-1197	0107E
FM-GPW-18675-A	WO-A-1233	0107E
FM-9N-18679	WO-387891	0107J
FM-GPW-18685	WO-A-1234	0107E
FM-9N-18686	WO-384569	0107J
FM-GPW-18687	WO-A-1231	0117E
FM-GPW-18688-A	WO-A-1235	0107E
FM-GPW-18691-A	WO-A-1232	0107E
FM-GPW-18812-B	WO-A-6320	2603
FM-GPW-18840	WO-A-7826	2601
FM-GPW-18841	WO-A-7600	2601
FM-GPW-18846	WO-A-5041	0606B
FM-GPW-18849	WO-A-5035	2603
FM-GPW-18850	WO-A-5040	2603
FM-GPW-18852	WO-A-1699	2603
FM-GPW-18853	WO-A-5039	2603
FM-GPW-18854	WO-A-5027	2603
FM-GPW-18857	WO-A-5037	2603
FM-GPW-18858	WO-A-5033	2603
FM-GAA-18861-A	WO-A-8114	2601
FM-GPW-18871	WO-A-5030	2603
FM-GPW-18872	WO-A-5031	2603
FM-GPW-18873	WO-A-5034	2603
FM-GPW-18874	WO-A-5036	2603
FM-GPW-18876	WO-A-5038	2601
FM-GPW-18876-B	WO-A-11766	2601
FM-GPW-18877-B	....	2601
FM-GPW-18935-A	WO-A-5337	2603
FM-GPW-18936-A	WO-A-8883	2603
FM-GPW-18937-A	WO-A-8884	2603
FM-GPW-18938-C	WO-A-7848	2603
FM-GPW-18960-B	WO-A-5980	2603
FM-GPW-1100000	WO-A-3565	1800
FM-GPW-1100001	WO-A-3563	1800
FM-GPW-1101610	WO-A-3005	1809
FM-GPW-1101621-A	WO-A-1330	2203
FM-GPW-1101627-A	WO-A-1331	2203
FM-GPW-1101670	WO-A-3132	1804
FM-GPW-1101698-A	WO-A-2931	1803
FM-GPW-1101735-A	WO-A-3784	0805B
FM-GPW-1102015	WO-A-2747	1802A

VENDOR'S NUMERICAL INDEX (Cont'd)

Ford Part Number	Willys Part Number	Gov't Group No.
FORD MOTOR CO. (FM) Cont'd		
FM-GPW-1102038	WO-A-3007	1802A
FM-GPW-1102039	WO-A-3008	1802A
FM-GPW-1102130	WO-A-2190	1802A
FM-GPW-1102330	WO-A-3943	1803
FM-GPW-1102396	WO-A-2138	1802A
FM-GPW-1102510	WO-A-2816	1802
FM-GPW-1102511	WO-A-2815	1802
FM-GPW-1103010	WO-A-3210	1811
FM-GP-1103014	WO-A-2226	1811A
FM-GP-1103020	WO-A-2255	1811B
FM-GPW-1103027	WO-A-3197	1811
FM-GPW-1103028	WO-A-3203	1811
FM-GPW-1103030	WO-A-2246	1811B
FM-GPW-1103050	WO-A-2798	1811
FM-GPW-1103080-A	WO-A-2476	1811B
FM-GP-1103100	WO-A-2478	1811A
FM-GP-1103106	WO-A-2479	1811B
FM-GP-1103110	WO-A-2250	1811B
FM-GPW-1103162	WO-A-2213	1811
FM-GPW-1103214	WO-A-3211	2201
FM-GPW-1103268	WO-A-3204	1811B
FM-GP-1103302	WO-A-2235	1811
FM-GPW-1103304	WO-A-2232	1811
FM-GP-1103310	WO-A-4300	1811
FM-GP-1103334	WO-A-2234	1811
FM-GP-1103350	WO-A-4303	1811
FM-GPW-1103446-A	WO-A-2939	1802A
FM-GP-1103482-A	WO-A-2227	1811
FM-GPW-1103488	WO-A-2791	1811
FM-GPW-1104320	WO-A-3578	1809
FM-GPW-1104364	WO-A-3155	1809
FM-GPW-1106024-B	WO-A-3434	1809A
FM-GPW-1106050	WO-A-3436	1809A
FM-GPW-1106064	WO-A-3818	1809A
FM-GPW-1106083	WO-A-3532	1809A
FM-GPW-1106084-A	WO-A-3531	1809A
FM-GPW-1106084-B	WO-A-3823	1809A
FM-GPW-1106085	WO-A-3536	1809A
FM-GPW-1106087-A	WO-A-3652	1809A
FM-GPW-1110610	WO-A-2838	1505
FM-GPW-1110625	WO-A-2935	1802A
FM-GPW-1110750	WO-A-2948	1802A
FM-GPW-1110895	WO-A-2950	1802A
FM-GPW-1111137-A	WO-A-2932	1803
FM-GPW-1111140	WO-A-2768	1802A
FM-GPW-1111155	WO-A-3222	1802A
FM-GPW-1111218	WO-A-2773	1802A
FM-GPW-1111238	WO-A-2879	1802A
FM-GPW-1111286-A	WO-A-3783	1803
FM-GPW-1111286-A3	WO-A-3783	1803
FM-GPW-1111299-A	WO-A-3116	1803
FM-GPW-1111322	WO-A-3055	0300
FM-GPW-1111323	WO-A-3056	1802A
FM-GPW-1111324	WO-A-3497	0300
FM-GPW-1111330	WO-A-2760	1803
FM-GPW-1111331	WO-A-2759	1803
FM-GPW-1111520	WO-A-3782	1803
FM-GPW-1112110	WO-A-2982	1803
FM-GPW-1112115	WO-A-2990	1803
FM-GPW-1112116	WO-A-2994	1803
FM-GPW-1112117	WO-A-2918	1803
FM-GPW-1112119	WO-A-2919	1803
FM-GPW-1112158	WO-A-2917	1803
FM-GPW-1160026	WO-A-3108	1804A
FM-GPW-1160314	WO-A-2515	1804
FM-GPW-1150326	WO-A-2925	1804
FM-GPW-1160327-B	WO-A-2886	1804
FM-GPW-1160780	WO-A-2788	1804A
FM-GPW-1161300	WO-A-2783	1804
FM-GPW-1161302	WO-A-2830	1804
FM-GPW-1161326	WO-A-2782	1804
FM-GPW-1162180	WO-A-2911	1804
FM-GPW-1162181	WO-A-2910	1804
FM-GPW-1162400	WO-A-3029	1804
FM-GPW-1162402	WO-A-2787	1804
FM-GPW-1162410	WO-A-2945	1804
FM-GPW-1162414	WO-A-2833	1804
FM-GPW-1162552	WO-A-2453	1804
FM-GPW-1162606	WO-A-3144	1802A
FM-GPW-1162900-B	WO-A-2968	1802A
FM-GPW-1163846-B	WO-A-3983	1804A
FM-GPW-1166400	WO-A-3115	1804A
FM-GPW-1166401	WO-A-3114	1804A
FM-GPW-1166800-B	WO-A-3982	1804A
FM-G8T-8103074-B	WO-A-2239	1811
FM-99W-8103311	WO-A-2483	1811
FM-B-45482-C	WO-A-2238	1811
FM-GPW-1120040	WO-A-2999	2201F
FM-GPW-1120041	WO-A-2998	2201F
FM-GPW-1127846	WO-A-2757	1802A
FM-GPW-1127847	WO-A-2756	1802A
FM-GPW-1127848	WO-A-2772	1802A
FM-GPW-1127849	WO-A-2771	1802A
FM-GPW-1127850	WO-A-2770	1802A
FM-GPW-1127851	WO-A-2769	1802A
FM-GPW-1128237	WO-A-3137	2301A
FM-GPW-1128242	WO-A-2940	1802A
FM-GPW-1128244	WO-A-2986	1804A
FM-GPW-1128247	WO-A-3135	2301A
FM-GPW-1128254	WO-A-2984	2301A
FM-GPW-1128256	WO-A-2995	2301A
FM-GPW-1128258	WO-A-3082	2301A
FM-GPW-1128267	WO-A-3139	2301A
FM-GPW-1128286	WO-A-2325	1802A
FM-GPW-1129068	WO-A-2745	2201A
FM-GPW-1129069	WO-A-2744	2201A
FM-GPW-1129200	WO-A-2946	2301
FM-GPW-1129670	WO-A-2390	1800A
FM-GPW-1129672	WO-A-2389	1800A
FM-GPW-1131414	WO-A-2883	1804
FM-GPW-1140326	WO-A-4607	1802A
FM-GPW-1140327-B	WO-A-4606	1802A
FM-GPW-1140330	WO-A-4123	1817A
FM-GPW-1140334	WO-A-4127	1817A
FM-GPW-1140344	WO-A-4128	1817A
FM-GPW-1140447	WO-A-2470	1802A
FM-GPW-1140449	WO-A-2853	1802A
FM-GPW-1140474	WO-A-2930	0300
FM-GPW-1143501	WO-A-2895	1817
FM-GPW-1144598	WO-A-2820	1505
FM-GPW-1144600	WO-A-2823	1505
FM-GPW-1144620	WO-A-3120	1802A
FM-GPW-1146100	WO-A-3227	1817
FM-GPW-1146132	WO-A-3226	1817
FM-GPW-1146152	WO-A-3198	1817
FM-20027-S2	WO-52132	
FM-20046-S	WO-51523	
FM-20046-S2	WO-51523	
FM-20066-S2	WO-6428	

SNL G-503

VENDOR'S NUMERICAL INDEX (Cont'd)

Ford Part Number	Willys Part Number	Gov't Group No.
FORD MOTOR CO. (FM) Cont'd		
FM-20300-S7	WO-51738	
FM-20308-S2	WO-51514	
FM-20308-S2	WO-51763	
FM-20308-S2	WO-52170	
FM-20308-S7-8	WO-51514	
FM-20309-S7	WO-51732	
FM-20310-S2	WO-51391	
FM-20311-S7	WO-A-1648	
FM-20324-S	WO-51514	
FM-20324-S2	WO-51514	
FM-20324-S2	WO-52168	
FM-20325-S7	WO-5920	
FM-20326-S	WO-51485	
FM-20326-S2	WO-51485	
FM-20326-S7	WO-51485	
FM-20328-S2	WO-52189	
FM-20344-S2	WO-52167	
FM-20344-S2	WO-52168	
FM-20346-S2	WO-51523	
FM-20347-S2	WO-51396	
FM-20348-S2	WO-53031	
FM-20349-S2	WO-50163	
FM-20364-S7	....	
FM-20366-S	WO-6428	
FM-20366-S	WO-639689	
FM-20366-S2	WO-6468	
FM-20367-S7	WO-50929	
FM-20384-S2	WO-6188	
FM-20386-S7	WO-639120	
FM-20387-S2	WO-5934	
FM-20388-S2	WO-51405	
FM-20388-S7	WO-51405	
FM-20389-S	....	
FM-20406-S7	WO-6660	
FM-20411-S4	WO-6923	
FM-20486-S2	....	
FM-20514-S2	VEN-51545	
FM-21492-S2	WO-51308	
FM-23017-S16	WO-323397	
FM-23393-S2	WO-52983	
FM-23395-S	WO-52945	
FM-23498-S2	WO-52372	
FM-24308-S	WO-51763	
FM-24308-S2	WO-52170	
FM-24325-S	WO-638154	
FM-24327-S2	WO-A-872	
FM-24327-S2	WO-52132	
FM-24327-S7	WO-52132	
FM-24328-S2	WO-52189	
FM-24346-S2	WO-A-2943	
FM-24347-S2	WO-51396	
FM-24348-S	WO-6412	
FM-24349-S2	WO-50163	
FM-24366-S2	....	
FM-24367-S2	WO-50929	
FM-24368-S	WO-6299	
FM-24369-S	WO-50151	
FM-24386-S2	WO-51486	
FM-24389-S2	WO-5919	
FM-24407-S	WO-5922	
FM-24408-S	WO-52911	
FM-24409-S7	WO-6606	
FM-24426-S	VENDOR-6157	
FM-24426-S2	WO-6157	
FM-24427-S7	WO-6609	
FM-24428-S	WO-51406	
FM-24429-S7	WO-50878	
FM-24430-S	WO-6184	
FM-24449-S2	WO-52600	
FM-24454-S2	WO-A-2485	
FM-24489-S	WO-52379	
FM-24505-S7	WO-50992	
FM-24555-S	WO-51371	
FM-24622-S2	WO-1725	
FM-24650-S2	WO-52987	
FM-24916-S2	WO-A-1707	
FM-26147-S7	WO-52036	
FM-26147-S7	WO-51040	
FM-26457-S7	WO-1632	
FM-26457-S7	WO-51040	
FM-26457-S7	WO-52131	
FM-26466-S7	WO-51546	
FM-26480-S2	WO-6287	
FM-26483-S	WO-51492	
FM-26483-S2	WO-51492	
FM-26483-S7	WO-51492	
FM-26496-S2	WO-52131	
FM-26498-S2	WO-53267	
FM-27063-S7	WO-1620	
FM-27068-S	WO-52236	
FM-27145-S2	WO-6383	
FM-27161-S7	WO-1650	
FM-27698-S	WO-52994	
FM-28378-S2	WO-A-1072	
FM-31026-S7	WO-A-1670	
FM-31027-S8	WO-A-1668	
FM-31032-S7	WO-116211	
FM-31037-S7	WO-A-5197	
FM-31045-S2	WO-A-6148	
FM-31061-S8	WO-116213	
FM-31079-S	WO-51819	
FM-31583-S7	WO-A-1686	
FM-31588-S	WO-A-1636	
FM-31588-S7	WO-A-1636	
FM-31596-S	WO-A-1572	
FM-31628-S7	WO-113439	
FM-31662-S	WO-116215	
FM-31866-S	WO-5580	
FM-32475-S2	WO-53103	
FM-32852-S	WO-52993	
FM-32866-S2	WO-52142	
FM-32866-S2	WO-52142	
FM-32866-S7	WO-52889	
FM-32924-S2	WO-52809	
FM-32924-S7-8	WO-52809	
FM-32989-S2	WO-52990	
FM-33184-S2	WO-A-2489	
FM-33249-S2	WO-A-2487	
FM-33784-S2	WO-A-491	
FM-33786-S2	WO-636575	
FM-33795-S	WO-5790	
FM-33795-S2	{ WO-5790 / WO-52217 }	
FM-33795-S7	{ WO-5790 / WO-52847 }	
FM-33796-S2	WO-5914	
FM-33797-S	WO-6167	
FM-33797-S2	WO-6167	
FM-33797-S2	WO-52350	

275

SNL G-503

VENDOR'S NUMERICAL INDEX (Cont'd)

Ford Part Number	Willys Part Number	Gov't Group No.	Ford Part Number	Willys Part Number	Gov't Group No
FORD MOTOR CO. (FM) Cont'd			FM-34707-S2	WO-5455	
FM-33798-S	WO-5910		FM-34707-S2	WO-52819	
FM-33798-S2	WO-5910		FM-34707-S7	WO-A-1680	
FM-33798-S7	WO-5910		FM-34707-S7	WO-50921	
FM-33798-S7	WO-111062		FM-34708-S	WO-A-1547	
FM-33799-S	WO-5544		FM-34711-S	WO-116207	
FM-33799-S2	WO-5544		FM-34745-S	WO-52702	
FM-33799-S2	WO-6169		FM-34745-S2	WO-52702	
FM-33799-S2	WO-53033		FM-34745-S2	WO-638305	
FM-33799-S7	WO-5544		FM-34746-S2	WO-A-2490	
FM-33799-S7	WO-53033		FM-34746-S2	WO-52274	
FM-33800-S	WO-5901		FM-34746-S7	WO-A-1533	
FM-33800-S2	WO-5901		FM-34747-S	WO-5455	
FM-33800-S2	WO-50922		FM-34747-S	WO-52101	
FM-33800-S7	WO-A-755		FM-34747-S2	WO-A-1680	
FM-33801-S2	WO-5336		FM-34747-S2	WO-52101	
FM-33802-S	WO-5939		FM-34747-S7	WO-A-1680	
FM-33802-S2	WO-5939		FM-34753-S	WO-A-8087	
FM-33832-S2	WO-52804		FM-34801-S7	WO-A-1613	
FM-33845-S2	WO-6163		FM-34801-S7	WO-A-1669	
FM-33846-S	WO-5916		FM-34801-S7	WO-52614	
FM-33846-S2	WO-5916		FM-34802-S	WO-51532	
FM-33846-S2	WO-637924		FM-34802-S	WO-52961	
FM-33880-S	WO-5901		FM-34802-S2	WO-6330	
FM-33881-S2	WO-638539		FM-34802-S2	WO-52652	
FM-33896-S2	WO-A-2466		FM-34802-S2	WO-52705	
FM-33896-S2	WO-A-300329		FM-34803-S	WO-5168	
FM-33909-S7	WO-52954		FM-34803-S2	WO-52221	
FM-33911-S2	WO-A-873		FM-34803-S2	WO-52960	
FM-33921-S7	WO-638155		FM-34803-S7	WO-A-1571	
FM-33925-S7	WO-5901		FM-34803-S7	WO-5186	
FM-33927-S7	WO-52925		FM-34803-S7	WO-52221	
FM-34032-S7	WO-A-1640		FM-34803-S7	WO-52960	
FM-34033-S18	WO-6436		FM-34803-S7	WO-113440	
FM-34051-S7	WO-A-1610		FM-34804-S	WO-52164	
FM-34051-S7	WO-52615		FM-34804-S2	WO-51969	
FM-34052-S2	WO-6214		FM-34805-S	WO-52706	
FM-34052-S2	WO-52182		FM-34805-S2	WO-A-2486	
FM-34054-S	WO-6989		FM-34805-S2	WO-5045	
FM-34054-S2	WO-A-2586		FM-34805-S2	WO-52031	
FM-34054-S2	WO-6989		FM-34805-S2	WO-52706	
FM-34056-S16	WO-A-9094		FM-34805-S2	WO-53058	
FM-34079-S7	WO-6536		FM-34805-S7	WO-5045	
FM-34079-S7	WO-53061		FM-34805-S7	WO-52706	
FM-34079-S8	WO-5848		FM-34805-S8	WO-52707	
FM-34083-S2	WO-636575		FM-34806-S2	WO-A-1619	
FM-34084-S	WO-A-1546		FM-34806-S2	WO-51833	
FM-34129-S2	WO-52165		FM-34806-S2	WO-51840	
FM-34130-S2	WO-A-3550		FM-34806-S2	WO-52045	
FM-34130-S2	WO-6167		FM-34806-S2	WO-53047	
FM-34141-S2	WO-6273		FM-34806-S7	WO-A-5288	
FM-34141-S7	WO-1589		FM-34806-S7	WO-51833	
FM-34176-S2	WO-2481		FM-34806-S7-8	WO-51840	
FM-34701-S7	WO-A-1667		FM-34807-S	WO-5010	
FM-34702-S2	WO-A-1666		FM-34807-S2	WO-5010	
FM-34703-S7	WO-A-1615		FM-34807-S2	WO-51304	
FM-34703-S7	WO-A-1702		FM-30807-S2	WO-52046	
FM-34705-S	WO-52702		FM-34807-S7	WO-5010	
FM-34705-S2	WO-5121		FM-34807-S7	WO-52046	
FM-34706-S2	WO-A-1555		FM-34808-S	WO-5059	
FM-34706-S2	WO-5121		FM-34808-S	WO-52349	
FM-34706-S2	WO-5437		FM-34808-S2	WO-5039	
FM-34706-S2	WO-52274		FM-34809-S	WO-5009	
FM-34706-S2	VENDOR-52768		FM-34809-S	WO-5009	
FM-34706-S7	WO-5437				

SNL G-503

VENDOR'S NUMERICAL INDEX (Cont'd)

Ford Part Number	Willys Part Number	Gov't Group No.
FORD MOTOR CO. (FM) Cont'd		
FM-34809-S2	WO-5009	
FM-34811-S2	WO-5038	
FM-34824-S2	WO-2587	
FM-34836-S7	WO-5051	
FM-34846-S2	WO-52045	
FM-34848-S2	WO-5938	
FM-34886-S	WO-52968	
FM-34902-S	WO-52754	
FM-34903-S2	WO-53194	
FM-34905-S	WO-352760	
FM-34905-S2	WO-52483	
FM-34906-S	WO-52428	
FM-34907-S2	WO-52332	
FM-34909-S2	WO-52330	
FM-34921-S	WO-52874	
FM-34922-S	WO-636528	
FM-34941-S	WO-52510	
FM-34941-S2	WO-52510	
FM-36002-S2	WO-51541	
FM-36009-S	WO-A-1618	
FM-36009-S7	WO-A-1618	
FM-36046-S2	WO-52963	
FM-36787-S7	WO-A-1633	
FM-36800-S	WO-A-1647	
FM-36845-S2	WO-6988	
FM-36868-S2	WO-51539	
FM-36931-S2	WO-A-1077	
FM-36954-S7	WO-A-1588	
FM-37206-S	WO-A-1032	
FM-36364-S7	WO-A-1572	
FM-37621-S7	WO-53020	
FM-37789-S7	WO-A-6297	
FM-38095-S	WO-A-1037	
FM-38192-S2	WO-2482	
FM-38259-S2	WO-A-2480	
FM-39114-S2	....	
FM-40631-S7	WO-53071	
FM-60093-S	....	
FM-60332-S2	WO-5247	
FM-60371-S	WO-52207	
FM-60275-S	WO-51561	
FM-60416-S	WO-5267	
FM-60446-S	WO-52832	
FM-60466-S	WO-52832	
FM-60470-S	WO-5216	
FM-60472-S	WO-50769	
FM-62216-S	WO-635860	
FM-62218-S	WO-51977	
FM-62627-S3	WO-51182	
FM-63848-S	WO-53243	
FM-64647-S	WO-A-1015	
FM-72003-S	WO-5067	
FM-72004-S	WO-5354	
FM-72016-S	WO-5020	
FM-72016-S	WO-52967	
FM-72017-S	WO-5020	
FM-72025-S	WO-5152	
FM-72034-S	WO-A-1642	
FM-72035-S	WO-5021	
FM-72037-S	WO-52946	
FM-72043-S	WO-5802	
FM-72053-S	WO-5108	
FM-72062-S	WO-52527	
FM-72063-S	WO-52944	
FM-72071-S	WO-5397	
FM-72798-S7-8	WO-A-1586	
FM-72809-S	WO-5134	
FM-72867-S7	WO-A-1685	
FM-73880-S	WO-339043	
FM-73889-S7	WO-A-1006	
FM-73904-S7	WO-311003	
FM-73928-S7	WO-A-1004	
FM-74019-S	WO-650484	
FM-74113-S	WO-51921	
FM-74121-S	WO-51091	
FM-74127-S	WO-15460	
FM-74144-S	WO-A-1641	
FM-74175-S7	WO-5017	
FM-74178-S	WO-5036	
FM-74182-S	WO-50917	
FM-88022-S	WO-384958	
FM-88032-S	WO-332515	
FM-88042-S	WO-300143	
FM-88057-S7	WO-632159	
FM-88082-S	WO-349712	
FM-88141-S	WO-375981	
FM-88350-S	WO-337304	
FM-92047-S	WO-51904	
FM-95626-S	WO-A-3052	
FM-95627-S	WO-A-3054	
FM-95628-S	WO-A-3053	
FM-95633-S	WO-A-1312	
FM-95705-S	WO-A-3051	
FM-348148-S	WO-A-3113	
FM-350343-S16	WO-A-9073	
FM-350744-S2	WO-51612	
FM-350796-S2	WO-A-2375	
FM-350850-SA2	WO-A-4679	
FM-350853-S7	WO-A-1617	
FM-350976-S2	WO-A-2375	
FM-351015-S7	WO-A-2263	
FM-351023-S	WO-53287	
FM-351025-S7	WO-A-1550	
FM-351025-S8	WO-638539	
FM-351027-S	WO-53289	
FM-351193-S	WO-53057	
FM-351274-S7	WO-53024	
FM-351303-S2	WO-A-2584	
FM-351355-S7	WO-5051	
FM-351359-S7	WO-4116	
FM-351359-S7	WO-52274	
FM-351370-S2	WO-2902	
FM-351370-S2	WO-71633	
FM-351838-S2	WO-5455	
FM-351466-S24	637923	
FM-351915-S	WO-374586	
FM-351926-S	WO-374681	
FM-352101-S	....	
FM-352103-S	....	
FM-352104-S	....	
FM-352126-S	WO-5215	
FM-352127-S	....	
FM-352692-S15	WO-A-4120	
FM-353023-S7	WO-638500	
FM-353043-A-S7	WO-638792	
FM-353051-S	WO-636538	
FM-353052-S	WO-52525	
FM-353053-S	WO-A-1104	
FM-353055-S	WO-5138	

VENDOR'S NUMERICAL INDEX (Cont'd)

Ford Part Number	Willys Part Number	Gov't Group No.
FORD MOTOR CO. (FM) Cont'd		
FM-353055-S7	WO-A-5120	
FM-353055-S7	WO-5138	
FM-353064-S	WO-5140	
FM-353075-S	WO-5599	
FM-353081-S	WO-635838	
FM-353104-S	WO-3111	
FM-353202-S	WO-662010	
FM-355035-S7	WO-638224	
FM-355067-S7	WO-116384	
FM-355095-S7	....	
FM-355130-S7	WO-A-1700	
FM-355130-S7	WO-5113	
FM-355131-S	WO-5066	
FM-355132-S	WO-5272	
FM-355132-S	WO-116588	
FM-355151-S	WO-5064	
FM-355158-S2	WO-5182	
FM-355160-S2	WO-5272	
FM-355162-S2	WO-6290	
FM-355162-S2	WO-51537	
FM-355163-S2	....	
FM-355164-S	WO-A-1584	
FM-355165-S2	....	
FM-355200-S7	WO-116385	
FM-355253-S	WO-115905	
FM-355260-S7	WO-A-1623	
FM-355262-S	WO-380197	
FM-355319-S	WO-A-4590	
FM-355351-S	....	
FM-355351-S7	....	
FM-355352-S7	WO-A-1019	
FM-355378-S	WO-A-973	
FM-355398-S2	WO-51798	
FM-355403-S2	....	
FM-355426-S	WO-639107	
FM-355433-S2	WO-51612	
FM-355442-S2	WO-51858	
FM-355444-S	WO-52836	
FM-355449-S2	WO-A-2214	
FM-355451-S	WO-632158	
FM-355452-S	WO-639052	
FM-355455-S	WO-A-1397	
FM-355456-S	WO-51823	
FM-355470-S2	WO-A-2473	
FM-355476-S	WO-630129	
FM-355485-S7	WO-A-1559	
FM-355486-S7	WO-A-1596	
FM-355496-S	WO-633949	
FM-355497-S	WO-632157	
FM-355498-S7	WO-50163	
FM-355499-S	WO-634850	
FM-355511-S	WO-A-871	
FM-355528-S	WO-A-821	
FM-355533-S	WO-A-1137	
FM-355550-S	WO-A-963	
FM-355551-S	WO-A-997	
FM-355552-S	WO-A-877	
FM-355554-S	WO-A-1136	
FM-355569-S2	WO-A-2983	
FM-355578-S	WO-A-903	
FM-355597-S	WO-630101	
FM-355698-S	WO-A-967	
FM-355699-S2	WO-636527	
FM-355741-S	WO-A-147	
FM-355752-S	WO-A-1227	
FM-355835-S	WO-A-1701	
FM-355836-S7	WO-5352	
FM-355836-S7	WO-6352	
FM-355836-S7	WO-52651	
FM-355858-S	WO-116219	
FM-355883-S	WO-A-1611	
FM-355900-S2	WO-654934	
FM-355909-S	WO-A-2933	
FM-355943-S7	WO-52845	
FM-355944-S5	WO-A-1565	
FM-355965-S2	WO-A-2607	
FM-355982-S2	WO-53285	
FM-356016-S7	WO-52893	
FM-356021-S	WO-636962	
FM-356028-S	WO-52825	
FM-356077-S8	WO-639115	
FM-356123-S7	WO-A-3537	
FM-356124-S	WO-A-876	
FM-356125-S	WO-A-980	
FM-356126-S	WO-636569	
FM-356134-S18	WO-A-520	
FM-356200-S7	WO-53026	
FM-356201-S	WO-51662	
FM-356205-S	WO-A-1135	
FM-356208-S7	WO-A-1621	
FM-356229-S	WO-A-1350	
FM-356235-S8	WO-8129	
FM-356263-S7	WO-A-1612	
FM-356264-S7	WO-A-1616	
FM-356299-S	WO-A-1089	
FM-356303-S	WO-639121	
FM-356305-S	WO-52768	
FM-356309-S7	WO-53025	
FM-356361-S	WO-A-4260	
FM-356371-S	WO-A-1401	
FM-356373-S2	WO-2220	
FM-356376-S	WO-A-1125	
FM-356394-S2	WO-52835	
FM-356436-S	WO-A-1396	
FM-356439-S	WO-638381	
FM-356504-S	WO-A-1028	
FM-356504-S	WO-636570	
FM-356519-S	WO-A-1410	
FM-356522-S2	WO-A-2639	
FM-356524-S	WO-634762	
FM-356561-S	WO-A-498	
FM-357008-S	....	
FM-357016-S2	....	
FM-357074-S	....	
FM-357076-S	....	
FM-357100-S	WO-52832	
FM-357155-S	WO-53056	
FM-357202-S	WO-636571	
FM-357417-S	WO-A-524	
FM-357418-S	WO-635840	
FM-357419-S2	WO-2875	
FM-357420-S	WO-337112	
FM-357553-S18	WO-392468	
FM-357574-S	WO-28023	
FM-357699-S2	WO-A-834	
FM-357703-S	WO-A-1714	
FM-358006-S7	WO-640038	
FM-358006-S8	WO-640038	
FM-358019-S	WO-A-5120	

VENDOR'S NUMERICAL INDEX (Cont'd)

	Ford Part Number	Willys Part Number	Gov't Group No.
FORD MOTOR CO. (FM) Cont'd			
	FM-358040-S	WO-A-1237	
	FM-358048-S	WO-636577	
	FM-358063-S	WO-376373	
	FM-358064-S	WO-5085	
	FM-358074-S	WO-A-1721	

Vendor Number	Willys Part Number	Ford Part Number	Gov't Group No
FLEX-O-TUBE CO. (FO)			
FO-HA-8031	WO-A-1325	FM-GPW-9288	0304
FYR-FIGHTER CO. (FYR)			
FYR D-10-A	WO-A-616	FM-GPW-17100	2402
GENERAL MOTORS CORP. (GM)			
GM-100014	WO-5934	FM-2046-S2	0401
GM-100025	WO-50163	FM-20349-S2	0106D
GM-100026	WO-5919	FM-24389-S2	0106D
GM-100027	WO-50878	FM-24449-S7	0402
GM-100044	WO-51371	FM-24555-S	1403A
GM-100121	WO-51523	FM-20346-S2	1802A
GM-100768	WO-6383	FM-27145-S2	0301C
GM-102634	WO-6167	FM-33797-S2	0601A
GM-103024	WO-5914	FM-33796-S2	0301C
GM-103026	WO-5901	FM-33800-S2	0601A
GM-106262	WO-5437	FM-34706-S2	0601B
GM-106279	WO-52132	FM-20027-S2	0301C
GM-106281	WO-5922	FM-24407	0402
GM-106331	WO-52911	FM-24408-S	0101A
GM-106333	WO-6486	FM-24534-S	0402
GM-107322	WO-52909	....	0110
GM-107381	WO-52825	FM-356028-S	0104
GM-110347	WO-638792	FM-353043-S7	0902
GM-112657	WO-53050	FM-20387-S7	2103A
GM-113844	WO-51612	FM-355433-S2	1403A
GM-114493	WO-52945	FM-23395-S	1500
GM-115093	WO-52046	FM-34807-S2	1817A
GM-115729	WO-50802	FM-33798-S7	2103A
GM-118613	WO-52217	FM-33799-S2	1804
GM-118772	WO-53048	FM-24347-S7	2103A
GM-119034	WO-52983	FM-23393-S2	0402
GM-122122	WO-53031	FM-20348-S2	1817A
GM-123228	WO-53033	FM-33799-S2	1817A
GM-123499	WO-53131	FM-20407-S7	2601
GM-127753	WO-52839	....	1209C
GM-128854	WO-6352	FM-355836-S8	0303
GM-136837	WO-53303	....	2601
GM-138485	WO-52725	FM-34706-S7	2601
GM-138489	WO-53135	....	1802A
GM-174916	WO-53024	FM-351274-57	0301C
GM-178378	WO-53023	....	2601
GM-178532	WO-53025	FM-356309-S7	2601
GM-178551	WO-53036	....	2601
GM-263549	WO-A-302	FM-GPW-13836	0609B
OAKES PRODUCTS DIV. (HH)			
HH-613300	WO-A-5621	FM-GPW-18205-A	0301C
HH-613306	WO-A-5631	FM-GPW-9658	0301C
HH-613313	WO-A-5632	FM-GPW-9621	0301C
HH-613314	WO-A-5633	FM-GPW-9623	0301C
HH-613380	WO-A-7191	FM-GPW-9612	0301C
HH-613387	WO-A-5630	FM-GPW-9617	0301C
HH-613455	WO-A-5629	FM-GPW-9609	0301C

Vendor Number	Willys Part Number	Ford Part Number	Gov't Group No.
HOLLAND HITCH CO. (HLH)			
HLH-T-60-B	WO-A-593	FM-GPW-5182	1502
HOOVER BALL AND BEARING CO. (HO)			
HO-88541	WO-636297	FM-GPW-8530	0503
HARRIS PRODUCTS CO. (HP)			
HP-770	WO-637936	FM-GPW-18060	1603
HARRISON RADIATOR CORP. (HR)			
HR-3108628	WO-637646	FM-GPW-8575	0502
HYATT BEARINGS DIV. (HY)			
HY-94322	WO-A-924	FM-GP-7718-A	0804
KELSEY HAYES WHEEL CO. (KHW)			
KHW-24382	WO-A-348	FM-GPW-17035	2301
KHW-24562	WO-A-465	FM-GPW-1015	1301
KHW-24566	WO-A-472	....	1302
KHW-24568	WO-A-473	FM-GP-1108	1302A
KHW-24575	WO-A-475	FM-GP-1013	1302A
KHW-24576	WO-A-474	FM-GP-1107	1302A
KHW-25646	WO-A-1690	FM-GP-1102	1302
KHW-25647	WO-A-1689	FM-GP-1103	1302
KHW-25649	WO-A-1691	....	1302
KHW-25692	WO-A-5467	FM-GPW-1015	1301
KHW-25693	WO-A-5468	FM-GPW-1016	1301
KHW-25695	WO-A-5470	FM-GPW-1029	1301
KHW-25696	WO-A-5472	FM-GPW-1045	1301
KHW-25779	WO-A-5471	FM-GPW-1030	1301
KHW-25917	WO-A-5539	FM-GPW-1024	1301
KHW-25930	WO-A-5488	FM-GPW-1025-C	1301
KING SEELEY CORP. (KS)			
KS-513	WO-53194	FM-34903-S2	0606F
KS-40333	WO-A-8242	....	2204
KS-40350	WO-A-8125	....	2204
KS-40355	WO-A-8180	FM-GPW-17255-A	2204
LINK-BELT CO. (LK)			
LK-S-35116-1	WO-634796	FM-GPW-6308	0102
LK-S-35117-1	WO-638458	FM-GPW-6256	0106C
LK-S-40936	WO-638457	FM-GPW-6260	0106C
WAGNER ELECTRIC CO. (LO)			
LO-21-11	WO-A-8279	....	2302
LO-S-FC-602	WO-637604	FM-91A-2152	1205
LO-S-FC-603	WO-637606	FM-90A-2151	1205
LO-FE-1444	WO-A-556	FM-GP-2140	1205
LO-S-FC-1499	WO-637579	FM-91A-2201	1207
LO-S-FD-2108-E	WO-637590	FM-GP-2173	1205
LO-S-FD-2109-B	WO-637595	FM-GP-2170	1205
LO-S-FC-2917	WO-637584	FM-GP-2175	1205
LO-S-FC-2918-A	WO-637585	FM-GP-2176	1205
LO-S-FC-2919-E	WO-637586	FM-GP-2183	1205
LO-S-FC-2926	WO-637597	FM-GP-2188	1205
LO-S-FC-2927	WO-637598	FM-GP-2174	1205
LO-S-FC-3023	WO-638544	....	1207
LO-FC-3052	WO-637427	FM-78-2814-A	1209C
LO-FC-4158	WO-A-6116	FM-GPW-2201	1207
LO-S-FD-4564	WO-637582	FM-GP-2155	1205
LO-FD-4664	WO-637789	FM-GP-2192	1207
LO-FD-4665	WO-637787	FM-GP-2261	1207
LO-FD-5381	WO-115962	FM-GPW-18371	1207
LO-S-FC-5727-A	WO-A-557	FM-GP-2076	1205

VENDOR'S NUMERICAL INDEX (Cont'd)

Vendor Number	Willys Part Number	Ford Part Number	Gov't Group No.
WAGNER ELECTRIC CO. (LO) Cont'd			
LO-FC-5778	WO-637432	FM-GP-2074	1209C
LO-FC-5784	WO-637424	FM-GP-2078	1209B
LO-S-FC-5992	WO-637580	FM-GP-2205	1207
LO-S-FC-5993	WO-637540	FM-GP-2208	1207
LO-FC-5994	WO-637546	FM-GP-2206	1207
LO-FD-5997	WO-637577	FM-GP-2194	1207
LO-FD-5998	WO-637541	FM-GP-2196	1207
LO-FC-6003	WO-637545	FM-GP-2204	1207
LO-S-FC-6007	WO-637591	FM-GP-2169	1205
LO-S-FC-6009	WO-637587	FM-GPW-2145	1205
LO-S-FC-6010	WO-637583	FM-GP-2160	1205
LO-S-FC-6011	WO-637602	FM-GP-2180	1205
LO-S-FC-6014	WO-637599	FM-GP-2143-A	1205
LO-S-FC-6018-E	WO-637608	FM-GP-2162	1205
LO-S-FC-6019	WO-637612	FM-GP-2167	1205
LO-S-FD-7379	WO-A-1484	GM-GPW-2061	1207
LO-FD-7568-A	WO-A-6110	....	1207
LO-FC-8502	WO-A-1373	FM-GPW-2078	1209B
LO-FD-8547	WO-A-1502	FM-GPW-2063	1502
LO-FC-8553	WO-A-1460	FM-GPW-2079	1209B
LO-FC-8772	WO-A-6111	FM-GP-2135	1207
LO-FC-8779	WO-A-6117	FM-GPW-2206	1207
LO-FC-8782	WO-A-6111	FM-GP-2135	1207
MONROE AUTO EQUIPMENT (MAE)			
MAE-T-317	WO-A-7778	....	2301
MAE-T-347	WO-A-7779	....	2301
MAE-8301	WO-116624	....	1603
MAE-10639-B	WO-637803	....	1603
MAE-10640-B	WO-637804	....	1603
MAE-10855	WO-116625	....	1603
MAE-10856-8	WO-116626	....	1603
MAE-10863-1	WO-116627	....	1603
MAE-10863:2	WO-116628	....	1603
MAE-10875	WO-637810	....	1603
MAE-10906-A	WO-116629	....	1603
MAE-10966	WO-116630	....	1603
MAE-11465	WO-A-6902	....	1603
MAE-11466	WO-A-6903	....	1603
MAE-12448	WO-116631	....	1603
MAE-12449	WO-116632	....	1603
MAE-12463	WO-116633	....	1603
MAE-12464	WO-116634	....	1603
MAE-12468	WO-116635	....	1603
MAE-12469	WO-116636	....	1603
MAE-12507	WO-116637	....	1603
MAE-12620	WO-116638	....	1603
MAE-12621	WO-116639	....	1603
MAE-12626	WO-116640	....	1603
MAE-12627	WO-116641	....	1603
MAE-12628	WO-116642	....	1603
MAE-12629	WO-116643	....	1603
MAE-12631	WO-116644	....	1603
MAE-108496	WO-638343	....	1603
McCORD RADIATOR & MFG. CO. (MDR)			
MDR-AM-504K	WO-637863	FM-O1A-12410	0604B
MALLORY & CO., P. R. (MLL)			
MLL-205244	WO-A-7600	FM-GPW-18841	2601
MARLIN ROCKWELL CORP. (MRC)			
MRC-206	WO-A-1007	FM-GP-7777-A	0803
MRC-307	WO-A-916	FM-GP-7065	0704
MRC-DZ-13567	WO-636297	FM-GPW-8530	0503
MIDLAND STEEL PRODUCTS CO. (MSP)			
MSP-F-2005	WO-A-668	....	1500
MSP-F-2009	WO-A-549	....	1500
MAZDA (GENERAL ELECTRIC) (MZ)			
MZ-51	WO-52837	FM-48-15021	0607B
MZ-1245	WO-51804	FM-B-13466	0607B
NEW DEPARTURE MFG. CO. (ND)			
ND-1203	WO-51248	FM-B-13466	0607B
ND-3206	WO-A-1007	FM-GPW-7777-A	0803
ND-7607	WO-A-916	FM-GP-7065	0704
ND-77503	WO-A-6299	FM-B-10094	0601
ND-885141	WO-636297	FM-GPW-8530	0503
NEW PROCESS GEAR, INC. (NP)			
NP-33873	WO-A-6382	....	1007
NP-38487	WO-A-6361	....	1007
NP-38491	WO-A-6383	....	1007
NP-38492	WO-A-6384	....	1007
NP-38493	WO-A-6362	....	1007
PUROLATOR PRODUCTS (PU)			
PU-25755	WO-A-1232	FM-GPW-18691-A	0107E
PU-25756	WO-A-1233	FM-GPW-18675-A	0107E
PU-25757	WO-A-1234	FM-GPW-18685-A	0107E
PU-25791	WO-A-1231	FM-GPW-18687-A	0107E
PU-25795	WO-A-1237	FM-358040-S	0107E
PU-25802	WO-A-1235	FM-GPW-18688-A	0107E
PU-26637	WO-A-1236	FM-GPW-18866	0107E
PU-27078	WO-A-1230	FM-GPW-18660-A	0107E
PU-27081	WO-A-1251	FM-GPW-18664-A	0107F
RICH MFG. CORP. (RMC)			
RMC-A-365	WO-637183	FM-GPW-6505	0105
RMC-A-366	WO-637182	FM-GPW-6507	0105
ROSS GEAR & TOOL CO. (RG)			
RG-7698-4-5/8	WO-A-745	FM-GPW-3575	1403A
RG-8287-32	WO-A-752	FM-GPW-14171	0609B
RG-T-13086	WO-A-1239	FM-GPW-3504	1403
RG-021116	WO-639118	FM-GPW-3577	1403A
RG-025060	WO-639115	FM-356077-S8	1403A
RG-026083	WO-A-633	FM-GPW-3655	0609A
RG-028093	WO-639992	FM-GPW-8501	0503
RG-029021	WO-639121	FM-356303-S	1403
RG-029046	WO-638885	FM-GPW-3646	0609B
RG-029049	WO-A-750	FM-GPW-3631	0609B
RG-032075	WO-639095	FM-GPW-3591-A	1403A
RG-032087	WO-A-302	FM-GPW-13836	0609B
RG-033046	WO-639108	FM-GPW-3593	1403A
RG-033047	WO-639109	FM-GPW-3594	1403A
RG-033048	WO-639110	FM-GPW-3595	1403A
RG-051035	WO-A-751	FM-GPW-3635	0609B
RG-063011	WO-639090	FM-GPW-3587	1403A
RG-063012	WO-639091	FM-GPW-3576	1403A
RG-063991	WO-A-747	FM-GPW-3652	0609B
RG-063996	WO-639190	FM-GPW-3517	1403A
RG-T-126000	WO-A-1760	FM-GPW-3568	1403A
RG-T-129001	WO-639119	FM-GPW-3581	1403A
RG-400013	WO-639104	FM-GPW-3571	1403A
RG-400025	WO-639102	FM-GPW-3552	1403A
RG-401090	WO-639191	FM-GPW-3520	1403A
RG-401100	WO-639103	FM-GPW-3589	1403A
RG-401107	WO-638884	FM-GPW-3626	0609B
RG-450054	WO-A-634	FM-GPW-3627	0609B

VENDOR'S NUMERICAL INDEX (Cont'd)

Vendor Number	Willys Part Number	Ford Part Number	Gov't Group No.
ROSS GEAR & TOOL CO. (RG) Cont'd			
RG-502282	WO-A-635	FM-GPW-3506	1405
RG-5025401	WO-639116	FM-GPW-3583	1403A
RG-503284	WO-A-740	FM-GPW-3548	1403A
RG-503308	WO-A-742	FM-GPW-3524	1403A
STANT MFG. CO. (SA)			
SA-6455-A	WO-A-6424	FM-GPW-9034-A	0300
SA-6935-A	WO-A-1215	FM-GPW-8100-A	0501
SKF INDUSTRIES, INC. (SKF)			
SKF-6206	WO-A-1007	FM-GP-7777-A	0803
SKF-6307-Z	WO-A-916	FM-GP-7065	0704
SPICER MFG. CORP. (SP)			
SP-LW-1	WO-A-964	....	0805
SP-42-G	WO-5354	FM-72004-S	0204A
SP-S-58	WO-A-784	FM-GP-4229-A	1103
SP-S-59	WO-A-785	FM-GP-4229-B	1003
SP-S-74	WO-A-786	FM-GP-4229-C	1003
SP-S-75	WO-A-787	FM-GP-4229-D	1003
SP-S-112	WO-A-800	FM-GP-4660-A	1003
SP-S-113	WO-A-801	FM-GP-4660-B	1003
SP-S-114	WO-A-802	FM-GP-4660-C	1003
SP-122-D	WO-A-1136	FM-355554-S	0801
SP-124-J	WO-A-980	FM-356125-S	0803
SP-145-W	WO-A-1028	FM-356504-S	0803
SP-153-SP	WO-A-1006	FM-73889-S7-8	1201
SP-S-171	WO-636565	FM-GP-4661	1104
SP-232-SP	WO-A-1004	FM-73928-S7-8	1201
SP-238-R	WO-A-1015	FM-64647-S	1201
SP-363-SP	WO-A-972	FM-GP-7796	0805B
SP-373-J	WO-A-1018	FM-O1T-2805	1201
SP-433-D	WP-A-973	FM-355378-S	0805F
SP-477-W	WO-A-1135	FM-356205-S	0801
SP-529-W	WO-52510	FM-34941-S	0607D
SP-634-D	WO-A-1137	FM-355533-S	0801
SP-690-W	WO-A-1000	FM-GP-7744	0804
SP-710-W	WO-A-990	FM-GP-7771	0803
SP-745-D	WO-A-963	FM-355550-S	0805A
SP-S-780	WO-636577	FM-358048-S	1101
SP-854-D	WP-A-967	FM-355698-S	0805
SP-886-D	WO-A-1019	FM-355352-S7	1201
SP-887-D	WO-A-1020	FM-O1T-2616	1201
SP-S-953	WO-A-780	FM-GP-3374	1007
SP-S-957	WO-A-825	FM-GP-3122	1006
SP-S-962	WO-A-1714	FM-357703-S	1006
SP-S-1056	WO-636570	FM-356504-S	1301C
SP-S-1106	WO-A-876	FM-356124-S	1401
SP-S-1107	WO-10558	FM-351059-S7	1402
SP-S-1135	WO-636569	FM-356126-S	1301C
SP-IS-1310	WO-A-877	FM-355552-S	1203
SP-K2-62-210-212-1718	WO-A-1429	FM-GPW-4605	0902
SP-K2-62-210-212-1907	WO-A-1428	FM-GPW-3370	0901
SP-5815-X	WO-A-1008	FM-GPW-2598	1201
SP-5900-X	WO-A-1009	FM-GPW-2648	1201
SP-5951-X	WO-A-934	FM-GP-7754	0801
SP-5962-X	WO-A-1111	FM-GP-7776	0803
SP-8996-SF	WP-A-1326	FM-GPW-3365	0901
SP-8997-S7	WO-A-1327	FM-GPW-4602	0901
SP-11586	WO-636528	FM-34922-S	1001
SP-13439	WO-A-1706	FM-51-3287	1402
SP-13449	WO-636360	FM-GP-4241	1003
SP-13575	WO-636566	FM-GP-4619	1003
SP-14223	WO-639265	FM-GP-4676A	1104
SP-15099	WO-636568	FM-GP-4666	1104
SP-15367	WO-A-799	FM-GP-4668	1104
SP-15926	WO-A-796	FM-GPW-4215	1103
SP-16067	WO-A-844	FM-78-3336	1402
SP-16075	WO-A-798	FM-GP-4211	1003
SP-16383	WO-A-793	FM-GP-4206	1002
SP-16385	WO-A-794	FM-GPW-4236	1003
SP-16409	WO-A-782	FM-GP-4035	1001
SP-16412-X	WO-A-789	FM-GPW-4209	1003
SP-16866	WO-A-792	FM-GP-4281	1103
SP-16968-X	WO-A-788	....	1103
SP-16976	WO-A-781	FM-GP-4016	1101
SP-16979	WO-A-870	FM-GP-4022	1001
SP-16983	WO-A-818	FM-GP-3139	1006
SP-16991	WO-A-824	FM-GP-3115	1006
SP-17004	WO-A-864	FM-GP-1177	1301B
SP-17015	WO-A-864	FM-GP-1218	1301C
SP-17016	WO-A-866	FM-GP-4252	1007
SP-17017	WO-A-867	FM-GP-1124	1102
SP-17018	WO-A-778	FM-GP-3031-A1	1301B
SP-17019	WO-A-819	FM-GP-3135	1006
SP-17036	WO-A-779	FM-GP-3034	1006
SP-17041	WO-A-820	FM-GP-1092	1006
SP-17046-X	WO-A-838	FM-GP-3289	1402
SP-17047-X	WO-A-847	FM-GP-3290	1402
SP-17048-X	WO-A-828	FM-GP-3140	1006
SP-17071	WO-A-869	FM-GP-1139	1302
SP-17121-X	WO-A-1716	FM-GP-3200-A1	1007
SP-17133-X	WO-A-814	FM-GP-1089	1006
SP-17135-X	WO-A-813	FM-GPW-1088	1006
SP-17146	WO-A-904	FM-GP-4032	1102
SP-17153	WO-A-868	FM-GPW-3204	1007
SP-17202-X	WO-A-851	FM-GP-3105	1006
SP-17205	WO-A-861	FM-GPW-3170	1401
SP-17210	WO-A-855	FM-GPW-3165	1401
SP-17211	WO-A-856	FM-GPW-3166	1401
SP-17213	WO-A-858	FM-GPW-3167	1401
SP-17214	WO-A-859	FM-GPW-3168	1401
SP-17215	WO-A-860	FM-GPW-3169	1401
SP-17216	WO-A-1724	FM-GP-3217	1007
SP-17217	WO-A-1725	FM-24622-S	1007
SP-17218	WO-A-1726	FM-GP-3216	1007
SP-17221	WO-A-811	FM-GP-3148-A2	1006
SP-17222	WO-A-812	FM-GP-3149-A2	1006
SP-17224	WO-A-1723	FM-GP-3218	1007
SP-17226-X	WO-A-888	FM-GPW-4004	1101
SP-17230	WO-A-1719	FM-GP-3215-B	1007
SP-17231	WO-A-1720	FM-GP-3221-A	1007
SP-17232	WO-A-1721	FM-358074-S	1007
SP-17233	WO-A-1722	FM-GP-3219	1007
SP-17301-X	WO-A-1710	FM-GPW-3112	1006
SP-17302-X	WO-A-1712	FM-GPW-3113	1006
SP-17307	WO-A-8249	FM-GPW-3131-13	1006
SP-17308-X	WO-A-1704	FM-GPW-3280	1402
SP-17309-X	WO-A-1708	FM-GPW-3279	1402
SP-17310-X	WO-A-1703	FM-GPW-3074	1001
SP-17347	WO-A-6305	....	1402
SP-17356-X	WO-A-6361	....	1007
SP-17359	WO-A-6382	....	1007
SP-17361	WO-A-6362	....	1007
SP-17377	WO-A-6439	FM-GPW-4259	1102
SP-SKA-34172	WO-A-1765	FM-GPW-1101670	1809
SP-10-B-18	WO-A-1016	FM-O1T-2642	1201
SP-12-B-5	WO-A-1017	FM-O1T-2634	1201
SP-14-B-6	WO-A-1021	FM-O1T-2640	1201

VENDOR'S NUMERICAL INDEX (Cont'd)

Vendor Number	Willys Part Number	Ford Part Number	Gov't Group No.
SPICER MFG. CORP. (SP) Cont'd			
SP-20-B-62	WO-A-1002	FM-GP-2614	1201
SP-3-O-74	WO-A-1005	FM-GPW-2630	1201
SP-3-Q-76	WO-311003	FM-73904-S	1201
SP-4-B-26	WO-A-1016	FM-GP-2620	1201
SP-5-B-15	WO-A-1003	FM-GPW-2632	1201
SP-5-B-16	WO-A-1226	FM-GPW-2656	1201
SP-K-5-21-X	WO-A-1433	FM-21C-18397-B	0902
SP-39-Q-20	WO-A-971	FM-GP-7213	0706A
SP-97-Q-13	WO-A-970	FM-GP-7799	0805B
SP-475-2	WO-A-985	FM-GP-17277	0810
SP-S-638	WO-A-803	FM-GPW-4659-A	1104
SP-S-638-1	WO-A-804	FM-GP-4659-B	1104
SP-S-638-2	WO-A-805	FM-GP-4659-C	1104
SP-S-638-3	WO-A-806	FM-GP-4659-D	1104
SP-2058-1	WO-A-1212	....	1000
SP-2058-2	WO-A-1387	....	1000
SP-16322-2	WO-A-797	FM-GP-4230	1003
SP-16323-2	WO-A-795	FM-GP-4228	1103
SP-16992-1	WO-S-830	FM-GP-3117-A	1006
SP-16992-2	WO-A-831	FM-GP-3117-B	1006
SP-16992-3	WO-A-832	FM-GP-3117-C	1006
SP-16992-4	WO-A-833	FM-GP-3117-D	1006
SP-17120-3-X	WO-A-1715	FM-GPW-3206-A1	1007
SP-17120-4-X	WP-A-1728	FM-GPW-3207-A1	1007
SP-17123-3	WO-A-1727	FM-GPW-3016-A	1007
SP-17122-4	WO-A-1729	FM-GPW-3017-A	1007
SP-17128-2-X	WO-A-810	....	1007
SP-17128-3-X	WO-A-809	....	1007
SP-17144-3	WO-A-901	FM-GPW-4234	1102
SP-17144-4	WO-A-902	FM-GPW-4235	1102
SP-17155-1	WO-A-862	FM-GPW-3208-A	1007
SP-17155-2	WO-A-863	FM-GPW-3208-B	1007
SP-17295-1	WO-A-1705	FM-GPW-3281	1402
SP-17295-2	WO-A-1709	FM-GPW-3282	1402
SP-17360-1	WO-A-6384	....	1007
SP-17360-2	WO-A-6383	FM-27145-S2	1007
SP-K2-1-28	WO-A-1105	FM-GP-4863	0803
SP-K2-4-88-X	WO-A-1106	FM-GP-7729	0803
SP-K2-7-29	WO-A-945	FM-O1Y-7096	0902
SP-K2-14-69	WO-A-942	FM-GP-7077	0902
SP-K2-15-63	WO-A-943	FM-GP-7097	0902
SP-K2-76-17	WO-A-940	FM-O1Y-7083	0902
SP-K2-94-29-V	WO-A-490	FM-O1Y-4529	0902
SP-K2-3-198-X	WO-A-935	FM-GP-4841	0902
SP-K2-4-108-X	WO-A-1445	FM-GP-4842	1104
SP-K3-86-89	WO-A-941	FM-O1T-7078-A	0902
SP-5-74-11	WO-A-491	FM-33784-S2	0902
SP-5-75-19	WO-51833	FM-34806-S7	0902
SP-5-73-310	WO-A-872	FM-24327-S2	1006
SP-6-74-11	WO-636575	FM-33786-S2	1402
SP-6-73-1117	WO-A-821	FM-355526-S	1006
SP-6-73-124	WO-A-760	FM-GP-1110	1007
SP-6-73-218	WO-A-997	FM-355551-S	1201
SP-6-73-414	WO-A-871	FM-355511-S	1003
SP-6-73-513	WO-A-903	FM-355578-S	1203
SP-6-74-101	WO-A-873	FM-33911-S	1006
SP-7-72-39	WO-636571	FM-357202-S	1003
SP-13-463-1	WO-A-974	....	0805B
SP-22-381-19	WO-A-976	FM-GP-7783	0803
SP-31-246-2	WO-A-1001	FM-GP-7767	0804
SP-50-39-2	WO-5140	FM-353064-S	0701
SP-50-39-4	WO-636538	FM-353051-S	1001
SP-50-80-1	WO-5599	FM-353075-S	0804
SP-62-463-2	WO-A-958	....	0803
SP-18-5-1	WO-A-999	FM-GP-7742	0804
SP-18-8-1	WO-A-989	FM-GP-7766	0803
SP-18-8-2	WO-A-988	FM-GP-7765	0803
SP-18-8-7	WO-A-1510	FM-GP-7722	0802
SP-18-12-1	WO-A-965	FM-GP-7789	0805A
SP-18-15-9	WO-A-1503	FM-GPW-7705	0801
SP-18-16-3	WO-A-953	FM-GP-7708	0801
SP-18-19-8	WO-A-956	FM-GPW-7774	0801
SP-18-19-18	WO-A-1507	FM-GPW-7769	0801
SP-18-24-1	WO-A-987	FM-GP-7777-A2	0803
SP-18-66-1	WO-A-959	FM-GP-7712	0805A
SP-18-66-7	WO-A-960	FM-GPW-7711	0805A
SP-18-67-2	WO-A-962	FM-GP-7787	0805
SP-18-67-5	WO-A-1504	FM-GPW-7786	0805
SP-18-155-2	WO-A-954	FM-GP-7709	0801
SP-18-187-1	WO-A-998	FM-GP-7743	0804
SP-18-223-2	WO-A-957	FM-GPW-9773	0801
SP-18-223-3	WO-A-1134	FM-GP-7746	0803
SP-18-267-4	WO-A-1508	FM-GPW-7706	0801
SP-18-324-2	WO-A-1509	FM-GPW-7707	0801
SP-18-352-5	WO-A-1435	FM-GPW-7756	0801
SP-18-362-3	WO-A-975	FM-GP-7761	0803
SP-18-363-4-566-X	WO-A-1764	FM-GP-7763	0803
SP-18-381-2	WO-A-991	FM-GP-7784	0802
SP-18-452-2	WO-A-1511	FM-GP-17285	0810
SP-18-453-3	WO-A-1512	FM-GPW-17271	0810
SP-18-454-4	WO-636396	FM-GP-17333	0810
SP-18-466-1	WO-A-992	FM-GP-7762	0802
SP-25-Q-1055	WO-A-1505	FM-GPW-7793	0805B
SP-25-Q-1056	WO-A-1506	FM-GPW-7710	0805B
SPARKS WITHINGTON CO. (SPW)			
SPW-B-9427	WO-A-1312	FM-GPW-13082	0609
SCHREADER'S SON, A. (SV)			
SV-7188-BT	WO-A-6855	FM-GPW-6722	0107N
STEWART-WARNER CORP. (SW)			
SW-5585	WO-A-213	FM-GP-17125	2301
SCHWARZE ELECTRIC CO. (SZE)			
SZE-61400	WO-A-1312	FM-GPW-13082	0609
TIMKEN ROLLER BEARING CO. (TM)			
TM-02820	WO-52879	FM-GP-4628	1104
TM-02872	WO-52878	FM-GP-4630	1104
TM-11520	WO-52941	FM-GP-3162	1006
TM-11590	WO-52940	FM-GP-3161	1006
TM-14131	WO-51575	FM-O1Y-1202	0803
TM-14276	WO-52883	FM-O1Y-1202	0803
TM-18250	WO-52942	FM-GP-1201	1301A
TM-18590	WO-52943	FM-GP-1202	1301A
TM-24721	WO-52881	FM-GP-4222	1003
TM-24780	WO-52880	FM-GP-4221	1003
TM-31530	WO-52877	FM-86H-4616	1003
TM-31593	WO-52876	FM-86H-4621	1003
TORRINGTON CO. (TR)			
TR-B-1210	WO-A-857	FM-GPW-3171	1006
TRICO PRODUCTS CO. (TRI)			
TRI-80540	WO-A-11432	FM-GPW-17534	1811D

SNL G-503

VENDOR'S NUMERICAL INDEX (Cont'd)

Vendor Number	Willys Part Number	Ford Part Number	Gov't Group No.
AMERICAN CHAIN & CABLE (TRU)			
TRU-SA-2844-1	WO-630068	FM-GPW-7516	0204A
TAYLOR SALES ENGINEERING CO. (TSE)			
TSE-215-B	WO-630396	FM-GPW-6615	0107A
TOMPSON PRODUCTS (TP)			
TP-6MN27	WO-A-1707	FM-24916-S2	1402
TP-14DS20	WO-A-6305	. . . .	1402
TP-14DM-43	WO-A-844	FM-78-3336	1402
TP-14SV79-A-7	WO-A-838	FM-GP-3291	1402
TP-148V90-A-7	WO-A-847	FM-GP-3292	1402
TP-16MN8	WO-A-1706	FM-51-3287	1402
USL BATTERY CORP. (USL)			
USL-461389	WO-635097	FM-GPW-9653	0301C
VICTOR MFG. & GASKET CO. (VG)			
VG-2066-C-C1	WO-637863	FM-O1A-12410	0604B
WILLARD STORAGE BATTERY CO. (WB)			
WB-SW-2-119	WO-A-1238	FM-11AS-10655	0610
WB-SR-2-119	WO-A-1767	FM-11AS-10658	0610
WARNER ELECTRIC BRAKE (WEB)			
WEB-3529	WO-A-6586	FM-11YS-18151B	0606F
WEB-3604	WO-A-6019	FM-11YS-18142-B	0606F
WEB-11935-B	WO-A-6587	. . . .	0606F
WEB-20098	WO-A-6589	FM-11YS-18193-B	0606F
WEB-20099	WO-A-6588	FM-11YS-18198-B	0606F
WEB-110242	WO-A-8088	. . . .	0606F
WEB-110477	WO-53061	. . . .	0606F
WARNER GEAR (WG)			
WG-X-802	WO-639689	FM-20366-S	0706A
WG-X-2136	WO-635838	FM-353081-S	0706B
WG-X-2428-A	WO-A-1019	FM-355352-S7-8	1201
WG-X-3204-ML	WO-636885	FM-GPW-7025	0703
WG-4418-M	WO-636200	FM-GPW-7245	0706B
WG-4496-K	WO-A-1379	FM-BB-7220	0706A
WG-4498-K	WO-392328	FM-GPW-7227	0706A
WG-4682-K	WO-637831	FM-GPW-7109	0704B
WG-4686	WO-637835	FM-GPW-7059	0704B
WG-B-7070	WO-635846	FM-B-7070	0703
WG-C-8-2-½	WO-A-1381	FM-GPW-7217	0706A
WG-T-9-50P	WO-A-520	FM-356134-S18	0802
WG-ASI-T-84-J	WO-A-1145	FM-GPW-7000	0700
WG-AT-84-J-1A	WO-A-1148	FM-GPW-7005	0701
WG-2AT-84-H-2	WO-A-6317	FM-GPW-7060	0704
WG-T-84-H-2	WO-A-519	FM-GPW-7061	0704
WG-T-84-J-2½	WO-A-6319	FM-GPW-7105	0704B
WG-AT-84-F-2½A	WO-A-6318	FM-GPW-7124	0704B
WG-AC-84-J-2A	WO-A-1380	FM-GPW-7210-A	0706A
WG-T-84-C-3	WO-638948	FM-GPW-7111	0704C
WG-T-84-J-6	WO-640017	FM-GPW-7050	0703
WG-T-84-J-8	WO-A-739	FM-GPW-7113	0704B
WG-AT-84-F-10A	WO-636882	FM-GPW-7141	0704B
WG-T-84-F-11-A	WO-638798	FM-GPW-7102	0704B
WG-T-84-F-12-A	WO-636879	FM-GPW-7100	0704B
WG-C-84-B-12-A	WP-A-1382	FM-GPW-7221	0706A
WG-T-84-F-13	WO-637832	FM-GPW-7116	0704B
WG-T-84-F-14	WO-637834	FM-GPW-7107	0704B
WG-T-84-F-15	WO-637833	FM-GPW-7106	0704B
WG-AT-84-J-16-A	WO-A-5553	FM-GPW-7015	0703
WG-T-84-J-16-A	WO-A-5554	FM-GPW-7017	0703
WG-T-84-17	WO-635844	FM-GPW-7064	0703
WG-T-84-C-19	WO-640006	FM-GPW-7104-A2	0704
WG-T-84-J-20-A	WO-A-1155	FM-GPW-7241	0706B
WG-T-84-J-21	WO-A-1156	FM-GPW-7240	0706B
WG-T-84-22	WO-635836	FM-GPW-7206	0706C
WG-T-84-C-23-A	WO-636196	FM-GPW-7230	0706C
WG-T-84-C-24-A	WO-636197	FM-GPW-7231	0706C
WG-T-84-G-25	WO-639423	FM-GPW-7063	0703
WG-AT-84-J-25	WO-A-7260	FM-GPW-7211-B	0706C
WG-T-84-G-26	WO-639422	FM-GPW-7120	0703
WG-T-84-J-28	WO-A-738	FM-GPW-7062	0704
WG-T-84-J-28	WO-A-880	FM-GPW-7115	0704C
WG-T-84-J-29	WO-A-879	FM-GPW-7126	0704C
WG-T-84-29	WO-635811	FM-GPW-7129A	0704C
WG-T-84-B-30-A	WO-635812	FM-GPW-7119	0704C
WG-T-84-30	WO-635840	FM-357418-S	0706C
WG-T-84-31	WO-635839	FM-GPW-7208	0706C
WG-T-84-J-31	WO-A-524	FM-357417-S	0704C
WG-T-84-B-32	WO-635859	FM-GPW-7214	0706A
WG-T-84-G-35	WO-638952	FM-GPW-7140	0704C
WG-T-84-42	WO-635837	FM-GPW-7234	0706A
WG-T-84-C-48	WO-638949	FM-GPW-7155	0704C
WG-T-84-J-50½A	WO-A-1410	FM-356519-S	0802
WG-T-84-J-54	WO-640018	FM-GPW-7052	0703
WG-T-84-85A	WO-635804	FM-GPW-7143	0704C
WG-T-84-J-86A	WO-A-1385	FM-GPW-7233	0706B
WG-T-84-J-100	WO-A-1492	FM-GPW-17091	2301
WG-T-84-115	WO-635861	FM-GPW-7223	0706A
WG-T-84-H-137	WO-A-410	FM-GPW-7080	0704
WG-T-84-J-167	WO-A-878	FM-GPW-7121	0704C
WEATHERHEAD CO. (WH)			
WH-145-A	WO-A-1126	FM-9N-8115	0501
WILCOX RICH (WIL)			
WIL-ST-1265	WO-637047	FM-GPW-6500-A	0105A

Willys Part Number	Ford Part Number	Gov't Group No
WILLYS-OVERLAND MOTORS, INC. (WO)		
WO-A-146	FM-GPW-6043	0110
WO-A-147	FM-355741-S	0809
WO-A-168	FM-34747-S7	0610B
WO-A-176	FM-GPW-7539	0204A
WO-A-177	FM-GPW-7517	0204A
WO-A-178	FM-GPW-7545	0204A
WO-A-179	FM-GPW-7507	0204A
WO-A-180	FM-GPW-7508	0204A
WO-A-181	FM-GPW-7514	0204A
WO-A-183	FM-GPW-2462	1205
WO-A-185	FM-GPW-5113	1500
WO-A-213	FM-GP-17125	2301
WO-A-227	FM-356525	1603
WO-A-281	FM-GPW-9628	0301C
WO-A-302	FM-GPW-13836	0609B
WO-A-313	FM-GP-17037	2301
WO-A-348	FM-GPW-17035	2301
WO-A-371	. . . .	2301
WO-A-372	FM-GPW-17005	2301
WO-A-373	FM-GP-17042	2301
WO-A-374	FM-GP-17028	2301
WO-A-375	FM-GP-17020	2301
WO-A-376	FM-GP-17023	2301
WO-A-377	FM-GP-17021	2301

VENDOR'S NUMERICAL INDEX (Cont'd)

WILLYS-OVERLAND MOTORS, INC. (WO) Cont'd

Willys Part Number	Ford Part Number	Gov't Group No.
WO-A-378	....	2301
WO-A-379	FM-GP-17038	2301
WO-A-405	FM-GPW-7250	0204
WO-A-410	FM-GPW-7080	0704
WO-A-415	FM-GPW-5106	1500
WO-A-416	....	1500
WO-A-417	....	1500
WO-A-418	....	1500
WO-A-419	....	1500
WO-A-420	....	1500
WO-A-421	....	1500
WO-A-439	FM-GPW-6392	1091C
WO-A-445	....	1100
WO-A-447	FM-GPW-8600	0503A
WO-A-463	FM-GPW-9632	0301C
WO-A-465	FM-GPW-1015	1301
WO-A-472	....	1302
WO-A-473	FM-GP-1108	1302A
WO-A-474	FM-GP-1107	1302A
WO-A-475	FM-GP-1013	1302A
WO-A-476	FM-GP-1012	1505
WO-A-481	FM-GPW-5783-A	1601B
WO-A-484	....	1500
WO-A-485	....	1500
WO-A-490	FM-O1Y-4529	0902
WO-A-491	FM-33784-S2	0902
WO-A-493	....	1500
WO-A-495	FM-GPW-2473	1204
WO-A-498	FM-356561-S	0204
WO-A-499	FM-GPW-7521	0204A
WO-A-500	FM-GPW-5341	1500
WO-A-508	....	1500
WO-A-510	....	1500
WO-A-513	FM-GPW-5778	1602
WO-A-514	FM-GPW-5779	1602
WO-A-515	FM-GPW-5481	1602
WO-A-519	FM-GPW-7061	0704
WO-A-520	FM-356134-S18	0802
WO-A-524	FM-357417-S	0704C
WO-A-534	FM-GPW-5097	1500
WO-A-538	FM-GPW-12405	0604B
WO-A-544	FM-GPW-5337	1500
WO-A-547	FM-GPW-5035	1500
WO-A-548	....	1500
WO-A-549	....	1500
WO-A-552	....	1502
WO-A-556	FM-GP-2140	1205
WO-A-557	FM-GP-2076	1205
WO-A-568	FM-GPW-5458	1000
WO-A-571	FM-GPW-5460	1100
WO-A-572	FM-GPW-5459	1000
WO-A-574	FM-GPW-5453	1602
WO-A-575	FM-GPW-5705-B	1602
WO-A-592	FM-GPW-8284	0505
WO-A-593	FM-GPW-5182	1502
WO-A-596	FM-GP-17043	2301
WO-A-597	FM-GP-17044	2301
WO-A-598	FM-GP-17045	2301
WO-A-599	FM-GP-17046	2301
WO-A-600	FM-GP-17047	2301
WO-A-612	FM-GPW-5311	1601
WO-A-612-1	FM-GPW-5313-B	1601
WO-A-612-2	FM-GPW-5315-B	1601
WO-A-612-3	FM-GPW-5316-B	1601
WO-A-612-4	FM-GPW-5317-B	1601
WO-A-612-5	FM-GPW-5318-B	1601
WO-A-612-6	FM-GPW-5319-B	1601
WO-A-612-7	FM-GPW-5320-B	1601
WO-A-612-8	FM-GPW-5321-B	1601
WO-A-613	FM-GPW-5310	1601
WO-A-613-1	FM-GPW-5313-A	1601
WO-A-613-2	FM-GPW-5315-A	1601
WO-A-613-3	FM-GPW-5316-A	1601
WO-A-613-4	FM-GPW-5317-A	1601
WO-A-613-5	FM-GPW-5318-A	1601
WO-A-613-6	FM-GPW-5319-A	1601
WO-A-613-7	FM-GPW-5320-A	1601
WO-A-613-8	FM-GPW-5321-A	1601
WO-A-614	FM-GPW-5560	1601A
WO-A-614-1	FM-GPW-5563	1601A
WO-A-614-2	FM-GPW-5565	1601A
WO-A-614-3	FM-GPW-5566	1601A
WO-A-614-4	FM-GPW-5567	1601A
WO-A-614-5	FM-GPW-5568	1601A
WO-A-614-6	FM-GPW-5569	1601A
WO-A-614-7	FM-GPW-5570	1601A
WO-A-614-8	FM-GPW-5571	1601A
WO-A-614-9	FM-GPW-5572	1601A
WO-A-616	FM-GPW-17100	2402
WO-A-622	FM-GPW-3332-A	1401
WO-A-623	FM-GPW-3336	1401
WO-A-630	....	1209C
WO-A-631	....	1209C
WO-A-633	FM-GPW-3655	0609B
WO-A-634	FM-GPW-3627	0609B
WO-A-635	FM-GPW-3506	1405
WO-A-637	FM-GPW-5077	1500
WO-A-642	FM-GPW-9647	0301C
WO-A-647	FM-GPW-5118	1205
WO-A-655	FM-GPW-5264-A	0401
WO-A-657	....	0401
WO-A-658	FM-GPW-5283	0401
WO-A-668	....	1500
WO-A-683	....	1209C
WO-A-692	FM-GP-17033	2301
WO-A-693	FM-GPW-17097	2402
WO-A-719	FM-GPW-13410	0606B
WO-A-738	FM-GPW-7062	0704
WO-A-739	FM-GPW-7113	0704B
WO-A-740	FM-GPW-3548	1403A
WO-A-742	FM-GPW-3524	1403A
WO-A-745	FM-GPW-3575	1403A
WO-A-747	FM-GPW-3652	0609B
WO-A-750	FM-GPW-3631	0609B
WO-A-751	FM-GPW-3635	0609B
WO-A-752	FM-GPW-14171	0609B
WO-A-754	FM-GP-2038	1203A
WO-A-755	FM-33800-S7	1203A
WO-A-760	FM-GP-1110	1007
WO-A-778	FM-GP-3031-A	1301B
WO-A-779	FM-GP-3034	1006
WO-A-780	FM-GP-3374	1007
WO-A-781	FM-GP-4015	1001
WO-A-782	FM-GF-4035	1101
WO-A-784	FM-GP-4229-A	1003
WO-A-785	FM-GP-4229-B	1003
WO-A-786	FM-GP-4229-C	1003
WO-A-787	FM-GP-4229-D	1003
WO-A-788	....	1002

SNL G-503

VENDOR'S NUMERICAL INDEX (Cont'd)

Willys Part Number	Ford Part Number	Gov't Group No.
WILLYS-OVERLAND MOTORS, INC. (WO) Cont'd		
WO-A-789	FM-GPW-4209	1103
WO-A-792	FM-GP-4281	1003
WO-A-793	FM-GP-4206	1103
WO-A-794	FM-GPW-4236	1003
WO-A-795	FM-GP-4228	1003
WO-A-796	FM-GPW-4215	1003
WO-A-797	FM-GP-4230	1103
WO-A-798	FM-GP-4211	1003
WO-A-799	FM-GP-4668	1104
WO-A-800	FM-GP-4660-A	1104
WO-A-801	FM-GP-4660-B	1003
WO-A-802	FM-GP-4660-C	1003
WO-A-803	FM-GP-4659-A	1003
WO-A-804	FM-GP-4659-B	1003
WO-A-805	FM-GP-4659-C	1003
WO-A-806	FM-GP-4659-D	1003
WO-A-809	....	1007
WO-A-810	....	1007
WO-A-811	FM-GP-3148-A	1006
WO-A-812	FM-GP-3149-A	1006
WO-A-813	FM-GP-1088	1006
WO-A-814	FM-GP-1089	1006
WO-A-818	FM-GP-3139	1006
WO-A-819	FM-GP-3135	1006
WO-A-820	FM-GP-1092	1006
WO-A-821	FM-355526-S	1006
WO-A-824	FM-GP-3115	1006
WO-A-825	FM-GP-3112	1006
WO-A-828	FM-GP-3140	1006
WO-A-830	FM-GP-3117-A	1006
WO-A-831	FM-GP-3117-B	1006
WO-A-832	FM-GP-3117-C	1006
WO-A-833	FM-GP-3117-D	1006
WO-A-838	FM-GP-3291	1402
WO-A-844	FM-78-3336	1402
WO-A-847	FM-GP-3292	1402
WO-A-851	FM-GP-3105	1006
WO-A-853	....	1006
WO-A-855	FM-GPW-3165	1401
WO-A-856	FM-GP-3166	1401
WO-A-857	FM-GPW-3171	1401
WO-A-858	FM-GPW-3167	1401
WO-A-859	FM-GPW-3168	1401
WO-A-860	FM-GPW-3169	1401
WO-A-861	FM-GPW-3170	1401
WO-A-862	FM-GP-3208-A	1007
WO-A-863	FM-GP-3208-B	1007
WO-A-864	FM-GP-1177	1007
WO-A-865	FM-GP-1218	1102
WO-A-866	FM-GP-4252	1007
WO-A-867	FM-GP-1124	1102
WO-A-868	FM-GP-3204	1007
WO-A-869	FM-GP-1139	1302
WO-A-870	FM-GP-4022	1001
WO-A-871	FM-355511-S	1103
WO-A-872	FM-24327-S2	1006
WO-A-873	FM-33911-S2	1006
WO-A-876	FM-33911-S2	1401
WO-A-877	FM-355552-S	1203
WO-A-878	FM-GPW-7121	0704C
WO-A-879	FM-GPW-7126	0704C
WO-A-880	FM-GPW-7115	0704C
WO-A-887	FM-GPW-7512	0204A
WO-A-888	FM-GPW-4004	1101
WO-A-901	FM-GPW-4234	1102
WO-A-902	FM-GPW-4235	1102
WO-A-903	FM-355578-S	1203
WO-A-904	FM-GP-4032	1102
WO-A-912	FM-GPW-9428	0108B
WO-A-916	FM-GP-7065	0704
WO-A-924	FM-GP-7718-A	0804
WO-A-927	FM-GPW-2211	1200
WO-A-928	FM-GPW-2210	1200
WO-A-934	FM-GP-7754	0801
WO-A-940	FM-O1Y-7083	0902
WO-A-941	FM-O1T-7078-A	0902
WO-A-942	FM-GP-7077	0902
WO-A-943	FM-GP-7097	0902
WO-A-945	FM-O1Y-7096	0902
WO-A-953	FM-GPW-7708	0801
WO-A-954	FM-GP-7709	0801
WO-A-956	FM-GPW-7774	0801
WO-A-957	FM-GPW-7773	0801
WO-A-958	....	0803
WO-A-959	FM-GP-7712	0805A
WO-A-960	FM-GP-7711	0805A
WO-A-962	FM-GP-7787	0805
WO-A-963	FM-355550-S	0805A
WP-A-964	....	0805
WO-A-965	FM-GP-7789	0805A
WO-A-966	FM-GP-7788	0805
WO-A-967	FM-355698-S	0805
WO-A-970	FM-GP-7799	0805B
WO-A-971	FM-GPW-7213	0706A
WO-A-972	FM-GP-7796	0805B
WO-A-973	FM-355378-S	0805B
WO-A-974	....	0805B
WO-A-975	FM-GP-7761	0803
WO-A-976	FM-GP-7783	0803
WO-A-980	FM-356125-S	0803
WO-A-982	FM-GP-7782-A	0801
WO-A-983	FM-GP-7782-B	0801
WO-A-984	FM-GP-7782-C	0801
WO-A-985	FM-GP-17277	0810
WO-A-987	FM-GP-7777-A	0803
WO-A-988	FM-GP-7765	0803
WO-A-989	FM-GP-7766	0803
WO-A-990	FM-GP-7771	0803
WO-A-991	FM-GP-7784	0802
WO-A-992	FM-GP-7762	0802
WO-A-997	FM-355551-S	1201
WO-A-998	FM-GP-7743	0804
WO-A-999	FM-GP-7742	0804
WO-A-1000	FM-GP-7744	0804
WO-A-1001	FM-GP-7767	0804
WO-A-1002	FM-GP-2614	1201
WO-A-1003	FM-GPW-2632	1201
WO-A-1004	FM-73928-S7	1201
WO-A-1005	FM-GPW-2630	1201
WO-A-1006	FM-73889-S7	1201
WO-A-1007	GP-7777-A	0803
WO-A-1008	FM-GPW-2598	1201
WO-A-1009	FM-GP-2648	1201
WO-A-1014	FM-GP-2620	1201
WO-A-1015	FM-64647-S	1201
WO-A-1016	FM-O1T-2642	1201
WO-A-1017	FM-O1T-2634	1201
WO-A-1018	FM-O1T-2805	1201
WO-A-1019	FM-355352-S7	1201

VENDOR'S NUMERICAL INDEX (Cont'd)

Willys Part Number	Ford Part Number	Gov't Group No.
WILLYS-OVERLAND MOTORS, INC. (WO) Cont'd		
WO-A-1020	FM-GP-2616	1201
WO-A-1021	FM-O1T-2640	1201
WO-A-1028	FM-356504-S	0803
WO-A-1031	FM-GPW-13022	0607
WO-A-1032	FM-37206-S	0607
WO-A-1033	FM-GPW-13007	0607A
WO-A-1036	FM-GPW-13043	0607
WO-A-1037	FM-38095-S	0607
WO-A-1045	FM-GPW-9386	0302
WO-A-1046	FM-GPW-9378	0302
WO-A-1047	FM-GPW-9377	0302
WO-A-1051	FM-GPW-6098	0110
WO-A-1061	FM-GPW-6758	0107N
WO-A-1064	FM-GPW-13405-B	0608
WO-A-1065	FM-GPW-13404-B	0608
WO-A-1070	FM-GP-13210-B	0607D
WO-A-1071	FM-GP-13209-B2	0607D
WO-A-1072	FM-28378-S2	0607D
WO-A-1073	FM-GPW-13408-B	0608
WO-A-1074	FM-GPW-13494	0607A
WO-A-1075	FM-GPW-13491-A	0607A
WO-A-1076	FM-GPW-13448-B	0608
WO-A-1077	FM-36931-S	0608
WO-A-1078	FM-GPW-13485	0607A
WO-A-1079	FM-GPW-13449-A	0608
WO-A-1082	....	0607D
WO-A-1084	FM-GPW-9732-A	0303
WO-A-1089	FM-356299	0606E
WO-A-1096	FM-11A-12425	0604B
WO-A-1097	....	1602
WO-A-1098	FM-GPW-14303	2603
WO-A-1100	FM-GPW-17062	2301
WO-A-1104	FM-353053	0801
WO-A-1105	FM-GP-4863	0803
WO-A-1111	FM-GP-7776	0803
WO-A-1116	FM-GPW-3590	1403A
WO-A-1117	FM-GPW-17750	2101
WO-A-1120	FM-GPW-17759	1500
WO-A-1122	....	1500
WO-A-1124	FM-GPW-8240	0503A
WO-A-1125	FM-356376-S	0503A
WO-A-1126	FM-9N-8115	0501
WO-A-1127	FM-GPW-17755	1500
WO-A-1128	FM-GPW-17754	1500
WO-A-1129	FM-GPW-17753	1500
WO-A-1130	FM-GPW-17752	1500
WO-A-1134	FM-GP-7746	0803
WO-A-1135	FM-356205-S	0801
WO-A-1136	FM-355554-S	0801
WO-A-1137	FM-355533	0801
WO-A-1138	....	1500
WO-A-1142	....	1500
WO-A-1145	FM-GPW-7000	0700
WO-A-1146	FM-GPW-5230-A	0401
WO-A-1147	FM-GPW-17751	2101
WO-A-1148	FM-GPW-7005	0701
WO-A-1150	FM-GPW-5028	1500
WO-A-1151	FM-GPW-5125	1500
WO-A-1152	....	1500
WO-A-1153	....	1500
WO-A-1154	....	1500
WO-A-1155	FM-GPW-7241	0706B
WO-A-1156	FM-GPW-7240	0706B
WO-A-1157	FM-GPW-17775	2101
WO-A-1159	....	1500
WO-A-1160	....	1500
WO-A-1162	FM-GPW-17003	2301
WO-A-1164	FM-GPW-5175	0610A
WO-A-1165	FM-GPW-9410-B	0108
WO-A-1166	FM-GPW-9424-B	0108A
WO-A-1173	FM-GPW-9751	0303
WO-A-1174	FM-GPW-9727	0303
WO-A-1175	FM-GPW-9742	0303
WO-A-1190	FM-GPW-6016	0106D
WO-A-1192	FM-GPW-8250	0101A
WO-A-1195	FM-GPW-7700	0800
WO-A-1197	FM-GPW-18667	0107E
WO-A-1198	FM-GPW-18666	0107E
WO-A-1199	FM-GPW-3509	1403A
WO-A-1201	FM-GPW-5057	1500
WO-A-1202	....	1500
WO-A-1203	FM-GPW-5019	1500
WO-A-1204	....	1500
WO-A-1205	....	1500
WO-A-1207	....	1500
WO-A-1210	....	1500
WO-A-1212	....	1000
WO-A-1214	FM-GPW-8005	0501
WO-A-1215	FM-GPW-8100-A	0501
WO-A-1216	FM-GPW-8578	0501
WO-A-1217	FM-GPW-8133	0501
WO-A-1221	FM-GPW-9002-A	0300
WO-A-1223	FM-GPW-9510	0301
WO-A-1224	....	0301C
WO-A-1225	FM-GPW-9716	0303
WO-A-1226	FM-GPW-2656	1201
WO-A-1227	FM-355752-S	1201
WO-A-1228	FM-GPW-2659	1201
WO-A-1230	FM-GPW-18660-A	0107E
WO-A-1231	FM-GPW-18687-A	0107E
WO-A-1232	FM-GPW-18691-A	0107E
WO-A-1233	FM-GPW-18675-A	0107E
WO-A-1234	FM-GPW-18685-A	0107E
WO-A-1235	FM-GPW-18688-A	0107E
WO-A-1236	FM-GPW-18662-A	0107E
WO-A-1237	FM-358040-S	0107E
WO-A-1238	FM-11AS-10655	0610
WO-A-1239	FM-GPW-3504	1403A
WO-A-1240	FM-GPW-17080	2301
WO-A-1241	FM-GPW-2853	1201
WO-A-1242	FM-GPW-2780	1201
WO-A-1243	FM-GPW-9738	0303
WO-A-1244	FM-GPW-12100	0603
WO-A-1245	FM-GPW-11001-A	0602
WO-A-1247	FM-GPW-18663	0107F
WO-A-1251	FM-GPW-18664-A	0107F
WO-A-1252	FM-GPW-5482	1602
WO-A-1253	FM-GPW-5291-B	1500
WO-A-1254	FM-GPW-9030-A	0300
WO-A-1255	FM-GPW-9155	0306B
WO-A-1256	FM-GPW-9183	0306B
WO-A-1257	FM-GPW-9184	0306B
WO-A-1258	FM-GPW-9149	0306B
WO-A-1259	FM-GPW-9160	0306B
WO-A-1260	FM-GPW-9186	0306B
WO-A-1261	FM-GPW-9140	0306B
WO-A-1262	FM-GPW-9182	0306B
WO-A-1263	FM-GPW-9162	0306B
WO-A-1264	FM-GPW-9185	0306B

VENDOR'S NUMERICAL INDEX (Cont'd)

Willys Part Number	Ford Part Number	Gov't Group No.	Willys Part Number	Ford Part Number	Gov't Group No.
WILLYS-OVERLAND MOTORS, INC. (WO) Cont'd			WO-A-1377	FM-GPW-2264	1209C
WO-A-1265	FM-GPW-9154	0306B	WO-A-1378	FM-GPW-2244	1209C
WO-A-1267	FM-GPW-17260	2204	WO-A-1379	FM-BB-7220	0706A
WO-A-1272	FM-GPW-6010	0101	WO-A-1380	FM-GPW-7210-A	0706A
WO-A-1275	FM-GPW-9035-B	0300	WO-A-1381	FM-GPW-7217-A	0706A
WO-A-1276	FM-GPW-3511	1405	WO-A-1382	FM-GPW-7221	0706A
WO-A-1277	FM-GPW-3682-A	1405	WO-A-1385	FM-GPW-7233	0706B
WO-A-1278	FM-GPW-9656	0301C	WO-A-1386	FM-GPW-2452-A	1204
WO-A-1279	FM-GPW-9657	0301C	WO-A-1387	....	1000
WO-A-1283	FM-GPW-2073	1500	WO-A-1389	FM-GPW-13831	0609
WO-A-1287	FM-GPW-18936-A	2603	WO-A-1392	FM-GPW-10166	0601A
WO-A-1289	FM-GPW-14589	0606B	WO-A-1395	FM-GPW-10178-A	0601A
WO-A-1290	FM-GPW-9637-A	0301C	WO-A-1396	FM-356436-S	0601A
WO-A-1291	FM-GPW-5165	0610A	WO-A-1397	FM-355455-S	0601A
WO-A-1292	FM-GPW-9275	0300A	WO-A-1399	FM-GPW-10143	0601A
WO-A-1293	FM-GPW-9276	0300A	WO-A-1400	FM-GPW-10162	0601A
WO-A-1296	FM-GPW-5246	0402	WO-A-1401	FM-356371-S7	0601A
WO-A-1300	FM-GPW-5251	0402	WO-A-1409	FM-GPW-10505	0601B
WO-A-1302	FM-GPW-9775	0303	WO-A-1410	FM-356519-S	0802
WO-A-1304	FM-GPW-13006	0607	WO-A-1411	FM-GPW-13710	0605C
WO-A-1305	FM-GPW-13005	0607	WO-A-1412	FM-GPW-12287	0604A
WO-A-1306	FM-GPW-13380-A	2203A	WO-A-1414	FM-GPW-12284	0604A
WO-A-1307	FM-GPW-97303	0301B	WO-A-1416	FM-GPW-12283	0604A
WO-A-1311	FM-GPW-9652	0301C	WO-A-1418	FM-GPW-12286	0604A
WO-A-1312	FM-GPW-13802	0609	WO-A-1420	FM-GPW-12298-B	0604A
WO-A-1313	....	0301C	WO-A-1422	FM-GPW-5116	1500
WO-A-1314	....	0301C	WO-A-1423	....	1500
WO-A-1315	....	0301C	WO-A-1424	FM-GPW-12030	0604
WO-A-1320	FM-GPW-14301	0610B	WO-A-1428	FM-GPW-3370	0901
WO-A-1325	FM-GPW-9288	0304	WO-A-1429	FM-GPW-4605	0902
WO-A-1326	FM-GPW-3335	0901	WO-A-1431	....	1500
WO-A-1327	FM-GPW-4602	0901	WO-A-1433	FM-21C-18397-B	0902
WO-A-1330	FM-GPW-1101621-A	2203	WO-A-1434	FM-GPW-7074	0902
WO-A-1331	FM-GPW-1101627-A	2203	WO-A-1435	FM-GPW-7756	0801
WO-A-1332	FM-GPW-11649	0606	WO-A-1436	FM-GPW-13021	0607D
WO-A-1333	FM-GPW-13740	0606	WO-A-1437	FM-GPW-13200	0607D
WO-A-1334	FM-GPW-13704	0605C	WO-A-1439	FM-GPW-13217	0607D
WO-A-1339	FM-GPW-17090	2301	WO-A-1440	FM-GPW-13216-B	0607D
WO-A-1341	FM-GPW-5095	1500	WO-A-1443	FM-GPW-6375	0109
WO-A-1343	FM-GPW-17261	2204	WO-A-1445	FM-GP-4842	1003
WO-A-1344	FM-GPW-17262	2204	WO-A-1450	FM-GPW-9316	0107J
WO-A-1345	FM-34907-S7-8	0606	WO-A-1451	FM-GPW-9686-A	0301C
WO-A-1346	....	0606	WO-A-1452	FM-GPW-14300	0610B
WO-A-1347	....	0606	WO-A-1454	FM-GPW-14431	0610B
WO-A-1348	....	0606	WO-A-1456	FM-GPW-9323	0107J
WO-A-1349	FM-GPW-12250-C	0606D	WO-A-1457	FM-GPW-2096	1001
WO-A-1350	FM-356229-S	0604C	WO-A-1460	FM-GPW-2079	1209B
WO-A-1351	....	0606	WO-A-1463	FM-GPW-6031	0110
WO-A-1352	....	0606	WO-A-1468	FM-GPW-10176	0601A
WO-A-1353	....	0606	WO-A-1469	FM-GPW-10155	0601A
WO-A-1354	FM-GPW-2138	1205	WO-A-1470	FM-GPW-10177	0601A
WO-A-1355	FM-GPW-7503	0204A	WO-A-1472	FM-GPW-9095	0300
WO-A-1359	FM-GPW-2454	1204	WO-A-1476	FM-GPW-9074	0300
WO-A-1360	FM-GPW-7525	0204	WO-A-1477	FM-GPW-9057	0300
WO-A-1361	FM-GPW-13022	0607	WO-A-1480	FM-GPW-9078	0300
WO-A-1362	FM-GPW-13076	0607	WO-A-1481	FM-GPW-9066	0300
WO-A-1363	FM-GPW-13075	0607	WO-A-1483	FM-24505-S7	0300
WO-A-1366	FM-GPW-9237	0304	WO-A-1484	FM-GPW-2061	1207
WO-A-1367	FM-GPW-9282	0304	WO-A-1486	FM-GPW-2084	1001
WO-A-1368	FM-GPW-9289	0304	WO-A-1487	FM-GPW-2082	1001
WO-A-1369	FM-GPW-9369	0304	WO-A-1488	FM-GPW-2298	1209C
WO-A-1373	FM-GPW-2078	1209B	WO-A-1490	FM-GPW-14448-D	0606E
WO-A-1376	FM-GPW-2266	1209C	WO-A-1491	FM-GPW-10153-A	0601A
			WO-A-1492	FM-GPW-17091	2301
			WO-A-1493	....	0100

VENDOR'S NUMERICAL INDEX (Cont'd)

Willys Part Number	Ford Part Number	Gov't Group No.
WILLYS-OVERLAND MOTORS, INC. (WO) Cont'd		
WO-A-1494	FM-GPW-9355	0302
WO-A-1495	FM-GPW-8260	0503B
WO-A-1501	FM-GPW-2263	1209C
WO-A-1502	FM-GPW-2063	1207
WO-A-1503	FM-GPW-7705	0801
WO-A-1504	FM-GPW-7786	0805
WO-A-1505	FM-GPW-7793	0805B
WO-A-1506	FM-GPW-7710	0805B
WO-A-1507	FM-GPW-7768	0801
WO-A-1508	FM-GPW-7706	0801
WO-A-1509	FM-GPW-7707	0801
WO-A-1510	FM-GP-7722	0802
WO-A-1511	FM-GP-17285	0810
WO-A-1512	FM-GPW-17271	0810
WO-A-1514	....	1500
WO-A-1515	FM-GPW-2250	1101
WO-A-1517	....	2603
WO-A-1526	FM-GPW-12030	0604
WO-A-1532	FM-33798-S7	2601
WO-A-1533	FM-34746-S7	1201
WO-A-1534	FM-GPW-6050	0101A
WO-A-1536	FM-GPW-18390	0101
WO-A-1537	FM-GPW-18387	0101
WO-A-1538	FM-GPW-18512	0107A
WO-A-1542	FM-GPW-18356	0700
WO-A-1543	FM-GPW-18355	0800
WO-A-1545	FM-GPW-18366	1100
WO-A-1546	FM-34084-S	0501
WO-A-1547	FM-34708-S	0501
WO-A-1548	FM-GPW-6067	0101
WO-A-1549	....	0101
WO-A-1550	FM-356025-S	0101A
WO-A-1552	FM-GPW-18535	0602
WO-A-1553	FM-GPW-11094	0602
WO-A-1554	FM-GPW-11102	0602
WO-A-1555	FM-34706-S2	0602
WO-A-1556	FM-GPW-11120	0602
WO-A-1557	FM-GPW-11089	0602
WO-A-1558	FM-GPW-11090	0602
WO-A-1559	FM-355485-S7	0602
WO-A-1560	FM-GPW-11083	0602
WO-A-1563	FM-GPW-11085	0602
WO-A-1564	FM-GPW-12172	0603
WO-A-1565	FM-355944-S5	0602
WO-A-1566	FM-GPW-11049	0602
WO-A-1567	FM-63543-S	0602
WO-A-1568	FM-GPW-11005	0602
WO-A-1569	FM-GPW-11053	0602
WO-A-1570	FM-GPW-12164	0603
WO-A-1571	....	0602
WO-A-1572	FM-31596-S	0602
WO-A-1573	FM-GPW-11350	0602
WO-A-1574	FM-B-11379	0602
WO-A-1575	FM-B-11357-A	0602
WO-A-1576	FM-B-11381	0602
WO-A-1577	FM-GPW-11375	0602
WO-A-1578	FM-GPW-11377	0602
WO-A-1579	FM-GPW-11382	0602
WO-A-1580	FM-B-11371	0602
WO-A-1581	FM-GPW-11354	0602
WO-A-1582	FM-GPW-11394	0602
WO-A-1583	FM-GPW-11395	0602
WO-A-1584	FM-355164-S	0602
WO-A-1585	FM-GPW-11131	0602
WO-A-1586	FM-72798-S7-8	0602
WO-A-1587	FM-GPW-12169	0603
WO-A-1588	FM-36954-S7	0602
WO-A-1589	FM-34141-S2	0602
WO-A-1590	FM-GPW-10120	0601
WO-A-1591	FM-GPW-10202-A	0601
WO-A-1592	FM-01A-10193	0601
WO-A-1593	FM-GPW-10202-A	0601
WO-A-1594	FM-GPW-10202-C	0601
WO-A-1595	FM-GPW-10208-A	0601
WO-A-1596	FM-355486-S7	0601
WO-A-1597	FM-GPW-10208-A	0601
WO-A-1598	FM-GPW-10104	0601
WO-A-1599	FM-GPW-10206-A	0601
WO-A-1600	FM-GPW-10041	0601
WO-A-1601	FM-GPW-10100	0601
WO-A-1602	FM-GPW-10211-A	0601
WO-A-1603	....	0601
WO-A-1604	FM-GPW-10175	0601
WO-A-1605	FM-GPW-10211-B	0601
WO-A-1606	FM-GPW-10192	0601
WO-A-1607	FM-GPW-10191	0601
WO-A-1608	....	0601
WO-A-1609	FM-GPW-10218	0601
WO-A-1610	FM-34051-S7	0601
WO-A-1611	FM-35583-S	0601
WO-A-1612	FM-356263-S7	0601
WO-A-1613	FM-34801-S7	0601
WO-A-1614	FM-34803-S7	0601
WO-A-1615	FM-34703-S7	0601
WO-A-1616	FM-34705-S2	0601
WO-A-1617	FM-350853-S7	0601
WO-A-1618	FM-36009-S	0601
WO-A-1619	FM-34806-S2	0601
WO-A-1620	FM-27063-S7	0601B
WO-A-1621	FM-356208-S7	0601
WO-A-1622	FM-GPW-10124	0601
WO-A-1623	FM-355267-S7	0601
WO-A-1624	FM-GPW-10116	0601
WO-A-1625	FM-GPW-10119	0601
WO-A-1626	FM-GPW-10118	0601
WO-A-1628	FM-GPW-10057	0601
WO-A-1629	FM-GPW-10105	0601
WO-A-1630	FM-GPW-10069	0601
WO-A-1631	FM-GPW-12300	0603
WO-A-1632	FM-26457-S7	0601
WO-A-1633	FM-36787-S7	0603
WO-A-1635	FM-34803-S7	0601
WO-A-1636	FM-31588-S	0601
WO-A-1637	FM-GPW-10005	0601
WO-A-1638	FM-GPW-10134	0601
WO-A-1639	FM-GPW-10130	0601
WO-A-1640	FM-34032-S7	0601
WO-A-1642	FM-72034-S	0601
WO-A-1644	FM-78-10212-A	0601
WO-A-1645	FM-GPW-10138	0601
WO-A-1646	FM-GPW-10212	0601
WO-A-1647	FM-36800-S	0601
WO-A-1648	FM-20311-S7	2603
WO-A-1649	FM-GPW-10142	0601
WO-A-1650	FM-27161-S7	0601
WO-A-1651	FM-GPW-18274	0601
WO-A-1652	FM-GPW-12006	0604A
WO-A-1653	FM-GPW-12177	0603
WO-A-1654	FM-GPW-12267	0603

SNL G-503

VENDOR'S NUMERICAL INDEX (Cont'd)

Willys Part Number	Ford Part Number	Gov't Group No.
WILLYS-OVERLAND MOTORS, INC. (WO) Cont'd		
WO-A-1655	FM-GPW-12106	0603
WO-A-1656	FM-GPW-12011	0603
WO-A-1657	FM-GPW-12012	0603
WO-A-1658	FM-GPW-12200	0603
WO-A-1659	FM-GPW-12195	0603
WO-A-1660	FM-GPW-12174	0603
WO-A-1661	FM-GPW-12176	0603
WO-A-1662	FM-GPW-12151	0603
WO-A-1663	FM-GPW-12217	0603
WO-A-1664	FM-GPW-12010	0603
WO-A-1665	FM-GPW-14425	0606B
WO-A-1666	FM-34702-S2	0603
WO-A-1667	FM-34701-S7	0603
WO-A-1668	FM-31027-S8	0603
WO-A-1669	FM-34801-S7	0603
WO-A-1670	FM-31026-S7	0603
WO-A-1671	FM-GPW-12133	0603
WO-A-1672	FM-GPW-12120	0603
WO-A-1673	FM-GPW-12182	0603
WO-A-1674	FM-GPW-12155	0603
WO-A-1675	FM-GPW-12175	0603
WO-A-1676	FM-GPW-12188	0603
WO-A-1677	FM-GPW-42084	0603
WO-A-1678	FM-GPW-12178	0603
WO-A-1679	FM-GPW-12139	0603
WO-A-1680	FM-34707-S7	2603
WO-A-1681	FM-GPW-12082	0603
WO-A-1682	FM-GPW-12144	0603
WO-A-1683	FM-GPW-12145	0603
WO-A-1684	FM-GPW-12191	0603
WO-A-1685	FM-72867-S7	0603
WO-A-1686	FM-31583-S7	0603
WO-A-1687	FM-GPW-18354	0603
WO-A-1689	FM-GP-1103	1302
WO-A-1690	FM-GP-1102	1302
WO-A-1691	....	1302
WO-A-1693	FM-GPW-14621	2603
WO-A-1694	FM-GPW-14561	2603
WO-A-1699	FM-GPW-18852	2603
WO-A-1700	FM-355130-S7	2603
WO-A-1701	FM-355835-S	2603
WO-A-1702	FM-34703-S7	2603
WO-A-1703	....	1001
WO-A-1704	FM-GPW-3280	1402
WO-A-1705	FM-GPW-3281	1402
WO-A-1706	FM-51-3287	1402
WO-A-1707	FM-24916-S2	1402
WO-A-1708	FM-GPW-3279	1402
WO-A-1709	FM-GPW-3282	1402
WO-A-1710	FM-GPW-3112	1006
WO-A-1712	FM-GPW-3113	1006
WO-A-1714	FM-357703-S	1006
WO-A-1715	FM-GPW-3206	1007
WO-A-1716	FM-GP-3200-A	1007
WO-A-1719	FM-3215-A	1007
WO-A-1720	FM-GP-3221-A	1007
WO-A-1721	FM-358074-S	1007
WO-A-1722	FM-GP-3219	1007
WO-A-1723	FM-GP-3218	1007
WO-A-1724	FM-GP-3217	1007
WO-A-1725	FM-24622-S2	1007
WO-A-1726	FM-GP-3216	1007
WO-A-1727	FM-GP-3016-A	1007
WO-A-1728	FM-GPW-3207-A1	1007
WO-A-1729	FM-GP-3017-A	1007
WO-A-1731	FM-GPW-14436	0606B
WO-A-1733	FM-GPW-12250-A	0606D
WO-A-1734	FM-GPW-12250-B	0606D
WO-A-1735	FM-GPW-2272	1201
WO-A-1738	FM-GPW-9069	0300
WO-A-1739	FM-GPW-9079	0300
WO-A-1740	FM-GPW-9075	0300
WO-A-1741	FM-GPW-9071	0300
WO-A-1746	....	0602
WO-A-1748	FM-GPW-13713	0605C
WO-A-1755	....	1500
WO-A-1756	....	1500
WO-A-1757	FM-GPW-5168	0610A
WO-A-1760	FM-GPW-3568	1403A
WO-A-1763	FM-GPW-9211	0300A
WO-A-1764	FM-GP-7763	0803
WO-A-1765	FM-GPW-1101670	1809
WO-A-1767	FM-11AS-10658	0610
WO-A-1795	....	1201
WO-A-1798	....	2301B
WO-A-1799	....	1301
WO-A-1910	....	0303
WO-A-2138	FM-GPW-1102396	1802A
WO-A-2168	FM-GPW-1154810-A	1800A
WO-A-2188	FM-GPW-16802	1704
WO-A-2190	FM-GPW-1102130	1802A
WO-A-2190	FM-GPW-1102130	1802A
WO-A-2213	FM-GPW-1103162	1811
WO-A-2214	FM-GPW-1150498	1811
WO-A-2220	....	1811
WO-A-2226	FM-GP-1103014	1811A
WO-A-2227	FM-GP-1103482-A	1811
WO-A-2232	FM-GPW-1103304	1811
WO-A-2234	FM-GP-1103334	1811
WO-A-2235	FM-GP-1103302	1811
WO-A-2238	FM-B-45482-C	1811
WO-A-2239	FM-GBT-8103074-B	1811
WO-A-2246	FM-GPW-1103030	1811B
WO-A-2250	FM-GP-1103110	1811B
WO-A-2263	FM-351015-S7	1802A
WO-A-2278	....	1802A
WO-A-2301	....	1811
WO-A-2311	....	1802A
WO-A-2312	....	1802A
WO-A-2325	FM-GPW-1128286	1802A
WO-A-2336	FM-GPW-1144606	1802A
WO-A-2359	FM-GPW-1433	1505
WO-A-2375	FM-350976-S2	1802A
WO-A-2386	FM-GPW-16892	1811
WO-A-2389	FM-GPW-1129672	1800A
WO-A-2390	FM-GPW-1129670	1800A
WO-A-2453	FM-GPW-1162552	1804
WO-A-2466	FM-33896	1802A
WO-A-2470	FM-GPW-1140447	1802A
WO-A-2476	FM-GPW-1103080-A	1811B
WO-A-2478	FM-GPW-1103100	1811A
WO-A-2479	FM-GP-1103106	1811B
WO-A-2480	FM-38259-S2	1811
WO-A-2481	FM-34176-S2	1811
WO-A-2482	FM-38192-S2	1811

VENDOR'S NUMERICAL INDEX (Cont'd)

Willys Part Number	Ford Part Number	Gov't Group No.
WILLYS-OVERLAND MOTORS, INC. (WO) Cont'd		
WO-A-2483	FM-99N-8103311	1811
WO-A-2485	FM-24454-S2	1811
WO-A-2486	FM-34805-S2	1811
WO-A-2487	FM-33249-S2	1811
WO-A-2489	FM-33184-S2	1811B
WO-A-2490	FM-356286-S	1811
WO-A-2501	FM-GPW-1151289	1811
WO-A-2512	FM-GP-17535	1811D
WO-A-2513	FM-GP-17531	1811D
WO-A-2515	FM-GPW-1160314	1804
WO-A-2518	FM-GPW-3685	0604C
WO-A-2584	FM-351303-S2	1811D
WO-A-2587	....	1811D
WO-A-2588	FM-34054-S2	1811D
WO-A-2601	....	2301A
WO-A-2607	FM-355965-S2	1804
WO-A-2639	FM-356522-S2	2201A
WO-A-2662	FM-GPW-16132	1701
WO-A-2744	FM-GPW-1129069	2201A
WO-A-2745	FM-GPW-1129068	2201A
WO-A-2747	FM-GPW-1102015	1802A
WO-A-2754	FM-GPW-1153030	2201A
WO-A-2756	FM-GPW-1127847	1802A
WO-A-2757	FM-GPW-1127846	1802A
WO-A-2758	FM-GPW-1140324	1802A
WO-A-2759	FM-GPW-1111331	1803
WO-A-2760	FM-GPW-1111330	1803
WO-A-2768	FM-GPW-1111140	1802A
WO-A-2769	FM-GPW-1127851	1802A
WO-A-2769	FM-GPW-1127851	1802A
WO-A-2770	FM-GPW-1127850	1802A
WO-A-2771	FM-GPW-1127849	1802A
WO-A-2772	FM-GPW-1127848	1802A
WO-A-2773	FM-GPW-1111218	1802A
WO-A-2774	....	1817
WO-A-2775	....	1817
WO-A-2776	....	1811B
WO-A-2782	FM-GPW-1161326	1804
WO-A-2783	FM-GPW-1161300	1804
WO-A-2787	FM-GPW-1162402	1804
WO-A-2788	FM-GPW-1160780	1804A
WO-A-2791	FM-GPW-1103488	1811
WO-A-2796	....	1811
WO-A-2798	FM-GPW-1103050	1811
WO-A-2810	....	1817
WO-A-2811	....	1817
WO-A-2815	FM-GPW-1102511	1802
WO-A-2816	FM-GPW-1102510	1802
WO-A-2820	FM-GPW-1144598	1505
WO-A-2823	FM-GPW-1144600	1505
WO-A-2830	FM-GPW-1161302	1804
WO-A-2832	FM-GPW-1146126	1817
WO-A-2833	FM-GPW-1162414	1804
WO-A-2836	....	1704
WO-A-2837	....	1802A
WO-A-2838	FM-GPW-1110610	1802A
WO-A-2841	FM-GPW-16083	1702
WO-A-2842	FM-GPW-16082	1702
WO-A-2843	FM-GPW-16095	1701
WO-A-2844	FM-GPW-16094	1701
WO-A-2853	FM-GPW-1140449	1802A
WO-A-2858	....	2103
WO-A-2859	FM-GPW-3658	1405
WO-A-2870	FM-GPW-16006	0607
WO-A-2871	FM-GPW-16005	0607
WO-A-2873	FM-GPW-13020	0607
WO-A-2875	FM-357419-S2	0607
WO-A-2878	FM-GPW-13032	0607
WO-A-2879	FM-GPW-1111238	1802A
WO-A-2883	FM-GPW-1131414	1804
WO-A-2886	FM-GPW-1160327-B	1804
WO-A-2891	FM-GPW-16105	1701
WO-A-2892	FM-GPW-2852	1201
WO-A-2895	FM-GPW-1143501	1817
WO-A-2896	FM-GPW-16892	1701
WO-A-2897	FM-GPW-1151266	2201A
WO-A-2898	FM-GPW-1151272	2201A
WO-A-2900	FM-GPW-1153142	2201A
WO-A-2901	FM-GPW-1151270	2201A
WO-A-2902	FM-351370-S2	2201A
WO-A-2909	....	2201
WO-A-2910	FM-GPW-1162181	1804
WO-A-2911	FM-GPW-1162180	1804
WO-A-2917	FM-GPW-1112158	1803
WO-A-2918	FM-GPW-1112117	1803
WO-A-2919	FM-GPW-1112119	1803
WO-A-2925	FM-GPW-1160326	1804
WO-A-2930	FM-GPW-1140474	0300
WO-A-2931	FM-GPW-1101698-A	1803
WO-A-3932	FM-GPW-1111137-A	1803
WO-A-2933	FM-355909-S2	1802A
WO-A-2934	FM-21CS-17682-B	2202
WO-A-2935	FM-GPW-1110625	1802A
WO-A-2939	FM-GPW-1103446-A	1802A
WO-A-2940	FM-GPW-1128242	1802A
WO-A-2942	FM-GPW-16006	1701
WO-A-2943	FM-GPW-16005	1701
WO-A-2945	FM-GPW-1162410	1804
WO-A-2948	FM-GPW-1110750	1802A
WO-A-2950	FM-GPW-1110895	1802A
WO-A-2952	FM-GPW-9062	0300
WO-A-2953	FM-GPW-9065	0300
WO-A-2954	FM-GPW-9063	0300
WO-A-2968	FM-GPW-1128244	1802A
WO-A-2970	FM-GPW-9051	0300
WO-A-2977	FM-GPW-8162	0501
WO-A-2979	FM-GPW-8222	1803
WO-A-2982	FM-GPW-1112110	1803
WO-A-2983	FM-355569-S2	1804
WO-A-2984	FM-GPW-1128254	2301A
WO-A-2986	FM-GPW-1162900-B	1804A
WO-A-2989	....	1704
WO-A-2990	FM-GPW-1112115	1803
WO-A-2992	FM-GPW-11514-A	1803
WO-A-2993	FM-GPW-13351	1802A
WO-A-2994	FM-GPW-1112116-A	1803
WO-A-2995	FM-GPW-1128256	2301A
WO-A-2998	FM-GPW-1120041	2201F
WO-A-2999	FM-GPW-1120040	2201F
WO-A-3005	FM-GPW-1101610	1809
WO-A-3007	FM-GPW-1102038	1802A
WO-A-3008	FM-GPW-1102039	1802A
WO-A-3029	FM-GPW-1162400	1804

SNL G-503

VENDOR'S NUMERICAL INDEX (Cont'd)

Willys Part Number	Ford Part Number	Gov't Group No.
WILLYS-OVERLAND MOTORS, INC. (WO) Cont'd		
WO-A-3051	FM-95705-S	2201A
WO-A-3052	FM-95626-S	2201F
WO-A-3053	FM-95628-S	2201F
WO-A-3054	FM-95627-S	2201A
WO-A-3055	FM-GPW-1111322	0300
WO-A-3056	FM-GPW-1111323	1802A
WO-A-3059	FM-GPW-16684	1704
WO-A-3070	FM-GPW-1102980	2201
WO-A-3073	....	2201
WO-A-3082	FM-GPW-1128258	2301A
WO-A-3094	FM-GPW-8348	2103
WO-A-3095	FM-GPW-8349	2103
WO-A-3096	FM-GPW-16128	1702
WO-A-3108	FM-GPW-1160026	1804A
WO-A-3109	FM-GPW-1152730	2201A
WO-A-3110	FM-GPW-1152720	2201A
WO-A-3111	FM-353104-S	2201A
WO-A-3112	FM-95633-S	2201A
WO-A-3113	FM-348148-S	2201A
WO-A-3114	FM-GPW-1166401	1804A
WO-A-3115	FM-GPW-1166400	1804A
WO-A-3116	FM-GPW-1111299-A	1803
WO-A-3120	FM-GPW-1144620	1802A
WO-A-3135	FM-GPW-1128247	2301A
WO-A-3137	FM-GPW-1128237	2301A
WO-A-3139	FM-GPW-1128267	2301A
WO-A-3141	FM-GPW-1152950	2201A
WO-A-3144	FM-GPW-1162606	1802A
WO-A-3155	FM-GPW-1104364	1809
WO-A-3158	FM-GPW-16159	1701
WO-A-3159	FM-GPW-16133	1702
WO-A-3173	FM-GPW-8155	1803
WO-A-3175	FM-GPW-8102	0501H
WO-A-3176	FM-GPW-8103	0501H
WO-A-3182	....	1803
WO-A-3189	FM-GPW-1151297	1811
WO-A-3190	....	1811
WO-A-3197	FM-GPW-1103027	1811
WO-A-3198	FM-GPW-1146152	1817
WO-A-3203	FM-GPW-1103028	1811
WO-A-3204	FM-GPW-1103268	1811B
WO-A-3206	FM-GPW-9071	1803
WO-A-3207	FM-GPW-9082-A	1803
WO-A-3208	FM-GPW-9085-A	1803
WO-A-3209	FM-GPW-9083-A	1803
WO-A-3210	FM-GPW-1103010	1811
WO-A-3211	FM-GPW-1103214	2201
WO-A-3216	FM-GPW-1152700	2201
WO-A-3222	FM-GPW-1111155	1802A
WO-A-3225	FM-GPW-16610	1704
WO-A-3226	FM-GPW-1146132	1817
WO-A-3227	FM-GPW-1146100	1817
WO-A-3434	FM-GPW-1106024	1809A
WO-A-3436	FM-GPW-1106050	1809A
WO-A-3497	FM-GPW-1111324	0300
WO-A-3531	FM-GPW-1106084-A	1809A
WO-A-3532	FM-GPW-1106068	1809A
WO-A-3536	FM-GPW-1106085	1809A
WO-A-3537	FM-356123-S7	1809A
WO-A-3538	....	1809A
WO-A-3549	....	2103
WO-A-3550	FM-34130-S2	2103
WO-A-3563	FM-GPW-1100001	1800
WO-A-3565	FM-GPW-1000000	1800
WO-A-3574	FM-GPW-8166	1803
WO-A-3575	....	1803
WO-A-3578	FM-GPW-1104320	1809
WO-A-3615	FM-GPW-8307	2103
WO-A-3652	FM-GPW-1106087-A	1809A
WO-A-3728	....	0610A
WO-A-3782	FM-GPW-1111520-A	1803
WO-A-3783	FM-GPW-1111286-A	1803
WO-A-3784	FM-GPW-1101735-A	0805B
WO-A-3818	FM-GPW-1106064	1809A
WO-A-3823	FM-GPW-1106084-A3	1809A
WO-A-3835	....	1809A
WO-A-3933	....	1817
WO-A-3934	....	1817
WO-A-3940	....	1811D
WO-A-3943	FM-GPW-1102330-A	1809A
WO-A-3980	....	1804A
WO-A-3981	FM-GPW-1163206	1804A
WO-A-3982	FM-GPW-1166800-B	1804A
WO-A-3983	FM-GPW-1163846-B	1804A
WO-A-3984	....	1804A
WO-A-3985	....	1804A
WO-A-4116	....	2103A
WO-A-4118	FM-GPW-13176	2103A
WO-A-4120	FM-352699-S15	1811
WO-A-4123	FM-GPW-1140330	1817A
WO-A-4127	FM-GPW-1140334	1817A
WO-A-4128	FM-GPW-1140344	1817A
WO-A-4260	....	1811
WO-A-4300	FM-GP-1103310	1811
WO-A-4303	FM-GP-1103350	1811
WO-A-4413	FM-GPW-8125-A	0501
WO-A-4414	....	1801
WO-A-4415	....	1801
WO-A-4416	FM-GPW-17079	1804
WO-A-4518	FM-GPW-17055	1804
WO-A-4592	FM-GPW-1111162	1803
WO-A-4606	FM-GPW-1140327-B	1802A
WO-A-4607	FM-GPW-1140326-B	1802A
WO-A-4679	....	1704
WO-A-4680	FM-GTB-16847	1704
WO-A-4683	FM-GTB-16848-A	1704
WO-A-4687	....	1811D
WO-5009	FM-34809-S	....
WO-5010	FM-34807-S2	....
WO-5020	FM-72017-S	....
WO-5021	FM-72035-S	....
WO-5027	FM-GPW-18874	2603
WO-A-5030	FM-GPW-13871	2603
WO-A-5031	FM-GPW-18872	2603
WO-A-5032	FM-GPW-18859	2603
WO-A-5033	FM-GPW-18858	2603
WO-A-5034	FM-GPW-18873	2603

VENDOR'S NUMERICAL INDEX (Cont'd)

Willys Part Number	Ford Part Number	Gov't Group No.
WILLYS-OVERLAND MOTORS, INC. (WO) Cont'd		
WO-A-5035	FM-GPW-18849	2603
WO-A-5036	FM-GPW-18874	2603
WO-5036	FM-74178-S	....
WO-A-5037	FM-GPW-18857	2603
WO-A-5038	FM-GPW-18876	2601
WO-A-5039	FM-GPW-18853	2603
WO-A-5040	FM-GPW-18850	2603
WO-A-5041	FM-GPW-18846	2603
WO-5045	FM-34805-S2	....
WO-A-5048	FM-GPW-14401-C	0606B
WO-5051	FM-34836-S-7	....
WO-A-5061	FM-GPW-14446	0606B
WO-5064	FM-355130-S7	....
WO-5067	FM-72003-S	....
WO-A-5070	FM-GPW-14406	0605B
WO-A-5072	FM-GPW-14458	0605B
WO-A-5073	FM-GPW-14459	0606B
WO-A-5074	FM-GPW-14305	0606B
WO-A-5078	FM-GPW-14457	0606B
WO-A-5079	FM-GPW-14456	0606B
WO-A-5080	FM-GPW-14416	0605B
WO-A-5081	FM-GPW-14409	0609B
WO-A-5082	FM-GPW-14465	0606B
WO-A-5083	FM-GPW-14321	0604A
WO-5085	FM-358064-S	....
WO-A-5102	FM-GPW-7530	0203
WO-A-5105	FM-GPW-6770	0107G
WO-5108	FM-72053	....
WO-5113	FM-355130-S7	....
WO-A-5120	FM-358019-S	1802A
WO-5121	FM-34705-S2	....
WO-A-5125	FM-GPW-6044	0110
WO-A-5127	FM-GPW-5025	1500
WO-A-5130	FM-GPW-17030	2301
WO-5138	FM-353055	....
WO-5140	FM-353064-S	0801
WO-A-5165	FM-GPW-6763-B	0107G
WO-A-5168	FM-GPW-6766-B	0107H
WO-5168	FM-34703-S	....
WO-A-5181	....	....
WO-5182	FM-355158-S2	....
WO-A-5197	FM-31037-S7	0606
WO-5215	....	....
WO-52221	FM-GPW-34803-S7	....
WO-A-5224	FM-GPW-2265	1209C
WO-A-5225	FM-GPW-2268	1209C
WO-A-5226	FM-GPW-2267	1209C
WO-A-5227	FM-GPW-2274	1209C
WO-5247	....	1802A
WO-A-5256	....	0601B
WO-A-5260	....	0601B
WO-A-5262	....	0601B
WO-5267	....	....
WO-5272	FM-355160-S2	....
WO-A-5288	FM-34806-S7	0601
WO-A-5335	FM-GPW-2635	1201
WO-A-5337	FM-GPW-18935-A	2603
WO-A-5338	FM-GPW-6002	0100
WO-5354	FM-72004-S	....
WO-A-5393	FM-GPW-2270-B	1201
WO-5397	FM-72071	....
WO-A-5415	....	1500
WO-A-5433	FM-11AS-10657	0610
WO-5437	FM-34706-S2	....
WO-A-5449	FM-GPW-14561	2204
WO-A-5450	....	2601
WO-5455	FM-34747-S	....
WO-A-5467	FM-GPW-1015	1301
WO-A-5468	FM-GPW-1016	1301
WO-A-5470	FM-GPW-1029	1301
WO-A-5471	FM-GPW-1030	1301
WO-A-5472	FM-GPW-1045	1301
WO-A-5488	FM-GPW-1025-C	1301
WO-A-5497	FM-GPW-6005	0100
WO-A-5498	....	1000
WO-A-5499	....	1000
WO-A-5500	FM-GPW-4001	1100
WO-A-5501	....	0301
WO-A-5504	FM-GPW-3325	1006
WO-5544	FM-33799-S	....
WO-A-5549	FM-GPW-1024	1301
WO-A-5553	FM-GPW-7015	0703
WO-A-5554	FM-GPW-7017	0703
WO-A-5586	FM-GPW-13012	0607
WO-A-5598	FM-GPW-14561	0606B
WO-5599	FM-353075-S	....
WO-A-5621	FM-GPW-18205-A	0301C
WO-A-5629	FM-GPW-9609	0301C
WO-A-5630	FM-GPW-9617	0301C
WO-A-5631	FM-GPW-9658	0301C
WO-A-5632	FM-GPW-9621	0301C
WO-A-5633	FM-GPW-9623	0301C
WO-A-5753	FM-GPW-5264-B	0401
WO-5790	FM-33795-S	....
WO-A-5806	....	0607D
WO-5901	FM-33800-S	....
WO-5910	FM-33798-S2	....
WO-5914	FM-33796-S	....
WO-5916	FM-33846-S	....
WO-5919	....	....
WO-5920	FM-20325-S7	....
WO-5922	FM-2407-S	....
WO-5934	....	....
WO-5938	FM-34838-S	....
WO-5939	FM-33802-S2	....
WO-A-5980	FM-GPW-18960-B	2603
WO-A-5981	FM-GPW-14432	0605B
WO-A-5992	FM-GPW-10000-A	0601
WO-A-6019	FM-11YS-18142-B	0606F
WO-A-6029	....	1000
WO-A-6030	....	1007
WO-A-6031	....	1007
WO-A-6066	FM-GPW-5588-A	1601C
WO-A-6067	FM-GPW-5610	1601C
WO-A-6068	FM-GPW-5602	1601C
WO-A-6069	FM-GPW-5605	1601C
WO-A-6072	FM-HPW-5604	1601C
WO-A-6073	FM-GPW-5607	1601C
WO-A-6074	FM-GPW-5609	1601C
WO-A-6075	FM-GPW-5608	1601C
WO-A-6110	....	1207
WO-A-6111	FM-GP-2135	1207
WO-A-6113	FM-GPW-2196	1207
WO-A-6116	FM-GPW-2201	1207
WO-A-6117	FM-GPW-2206	1207
WO-A-6118	FM-GPW-5230-B	0401
WO-A-6119	FM-GPW-5298	0402A
WO-A-6133	FM-GPW-18368	1207
WO-A-6142	FM-GPW-13150	0607D

SNL G-503

VENDOR'S NUMERICAL INDEX (Cont'd)

Willys Part Number	Ford Part Number	Gov't Group No.
WILLYS-OVERLAND MOTORS, INC. (WO) Cont'd		
WO-A-6143	FM-GPW-13170	0607D
WO-A-6144	FM-GPW-13162	0607D
WO-A-6145	FM-GPW-13153	0607A
WO-A-6146	FM-GPW-13175	0607D
WO-A-6147	FM-GPW-13174	0607D
WO-A-6148	FM-131045-S2	0607D
WO-A-6149	FM-GT-13739	0606
WO-A-6152	....	0606
WO-A-6153	FM-GPW-13181	0606B
WO-A-6154	FM-GPW-14402	0606B
WO-A-6156	FM-GPW-6040-B	0110
WO-6157	FM-24426-S2	....
WO-6167	FM-33797-S2	....
WO-A-6168	FM-GPW-5590-B	1601C
WO-A-6169	FM-GPW-5611	1601C
WO-6184	FM-24430-S	....
WO-6188	FM-20384-S2	....
WO-6273	FM-34141-S2	....
WO-6290	FM-GPW-355162-S2	....
WO-A-6297	FM-37789-S7	0601
WO-A-6298	FM-GPW-10050	0601
WO-6299	....	....
WO-A-6299	FM-B-10094	0601
WO-A-6300	....	0601
WO-A-6301	FM-GPW-10139	0601
WO-A-6305	....	1402
WO-A-6313	FM-GPW-13180	0607D
WO-A-6317	FM-GPW-7060	0704
WO-A-6318	FM-GPW-7124	0704B
WO-A-6319	FM-GPW-7105	0704B
WO-A-6320	FM-GPW-18812	2603
WO-A-6321	....	0604A
WO-A-6326	FM-GPW-5601	1602
WO-A-6333	FM-GPW-9030-B	0300
WO-6352	FM-355836-S7	....
WO-A-6356	FM-O9B-14362	0606F
WO-A-6357	FM-GPW-9445	0301A
WO-A-6359	....	0204
WO-A-6360	....	0204
WO-A-6361	....	1007
WO-A-6362	....	1007
WO-A-6373	FM-GPW-8285	0505
WO-A-6374	FM-GPW-8290	0505
WO-A-6382	....	1007
WO-A-6383	....	1007
WO-6383	FM-27145-S2	....
WO-A-6384	....	1007
WO-A-6393	FM-GPW-5186	1502
WO-6412	FM-24348-S	....
WO-A-6424	FM-GPW-9034-A	0300
WO-6428	FM-20366-S	....
WO-6436	FM-34033-S-18	....
WO-A-6439	FM-GPW-4259	1102
WO-A-6442	....	1000
WO-6470	....	....
WO-A-6472	....	1007
WO-6486	FM-GP-106333	....
WO-A-6511	FM-GPW-5455	1602
WO-A-6525	....	0107H
WO-A-6586	FM-11YS-18151-B	0606F
WO-A-6587	....	0606F
WO-A-6588	FM-11YS-18198-B	0606F
WO-A-6589	FM-AAYS-18193-B	0606F
WO-6606	FM-24409-S7	....
WO-6609	FM-24427-S7	....
WO-A-6618	FM-GPW-9002-B	0300
WO-A-6701	....	0503A
WO-A-6710	....	0303
WO-A-6740	FM-GPW-5084	1500
WO-A-6742	FM-GPW-18382	0609B
WO-A-6743	FM-GPW-18389	1003
WO-A-6744	FM-GPW-18388	1103
WO-A-6745	FM-GPW-18386	1104
WO-A-6746	FM-GPW-18348	0102A
WO-A-6747	FM-GPW-18349	0102A
WO-A-6749	FM-GPW-18379	0107
WO-A-6750	FM-GPW-18380	0107
WO-A-6751	FM-GPW-18358	0201A
WO-A-6752	....	0202
WO-A-6753	FM-GPW-18360	0801
WO-A-6756	FM-GPW-18376	0602
WO-A-6759	FM-GPW-18377	1201
WO-A-6760	FM-GPW-18374	1403A
WO-A-6783	FM-GPW-13166	0607D
WO-A-6791	FM-GPW-18383	1401
WO-A-6793	FM-GPW-6009	0101
WO-A-6794	FM-GPW-6149-E	0103A
WO-A-6796	FM-GPW-6149-G	0103A
WO-A-6797	FM-GPW-6149-H	0103A
WO-A-6798	FM-GPW-18347	0102A
WO-A-6809	....	2601
WO-A-6811	FM-GPW-3686-B	0604C
WO-A-6813	....	0604C
WO-A-6814	FM-GPW-3685-B	0604C
WO-A-6816	FM-GPW-18384	1003
WO-A-6816	FM-GPW-18384	1103
WO-A-6837	FM-GPW-18352	0301
WO-A-6839	FM-GPW-18515	0503
WO-A-6840	FM-GPW-18357-B	0301
WO-A-6851	....	0303
WO-A-6855	FM-GPW-18325	2301
WO-A-6858	FM-GPW-3600-A3	1404
WO-A-6861	FM-GPW-5587	1601C
WO-A-6881	FM-GPW-18336	1007
WP-A-6882	FM-GPW-18388	1006
WO-A-6883	FM-GPW-18337	0306B
WO-A-6895	FM-GPW-6769	0107N
WO-A-6897	FM-GPW-9001	0300
WO-A-6902	....	1603
WO-A-6903	....	1603
WO-A-6911	FM-GPW-9637-B	0301C
WO-A-6915	FM-GPW-6763-C	0107G
WO-A-6918	FM-GPW-6771	0301C
WO-A-6919	FM-GPW-6758-B	0107N
WO-A-6922	FM-GPW-6756	0107N
WO-6923	FM-24411-S	....
WO-6989	FM-34054-S	2402
WO-A-7180	FM-GPW-14521	2601
WO-A-7181	FM-GPW-14536	2601
WO-A-7182	FM-GPW-14532	2601
WO-A-7191	FM-GPW-9612	0301C
WO-A-7225	....	0606
WO-A-7233	FM-GPW-18330-A	0104A
WO-A-7234	FM-GPW-18330-B	0104A
WO-A-7235	FM-GPW-18330-C	0104A
WO-A-7238	....	0107A
Cont-A-7260	FM-GPW-7211-B	0706C
WO-A-7280	FM-GPW-6789	0107H

293

SNL G-503

VENDOR'S NUMERICAL INDEX (Cont'd)

Willys Part Number	Ford Part Number	Gov't Group No.
WILLYS-OVERLAND MOTORS, INC. (WO) Cont'd		
WO-A-7443	FM-GPW-18355-B	0800
WO-A-7445	FM-GPW-18317-B	0800
WO-A-7466	FM-91BS-14463	2601
WO-A-7498	FM-GPW-6038-A	0110
WO-A-7503	FM-GPW-18289	0109
WO-A-7511	FM-GPW-17052	2301
WO-A-7517	FM-GPW-9775-B	0301B
WO-A-7518	FM-GPW-9778	0301B
WO-A-7568	FM-GPW-18288	0102
WO-A-7596	....	0700
WO-A-7600	FM-GPW-18841	2601
WO-A-7636	FM-GPW-14531	2601
WO-A-7637	FM-GPW-14535	2601
WO-A-7638	FM-GPW-14534	2601
WO-A-7640	FM-GPW-14513	2601
WO-A-7645	FM-GPW-14320	2601
WO-A-7680	FM-GPW-18353	2309
WO-A-7687	FM-GPW-18136-B	2301B
WO-A-7715	....	2601
WO-A-7718	....	2601
WO-A-7778	....	2301
WO-A-7779	....	2301
WO-A-7792	FM-GPW-12000-B	0604
WO-A-7794	....	0601B
WO-A-7795	....	0601B
WO-A-7796	....	0601B
WO-A-7797	....	0601B
WO-A-7798	....	0601B
WO-A-7799	....	0601B
WO-A-7800	....	0601B
WO-A-7801	....	0601B
WO-A-7802	....	0601B
WO-A-7803	....	0601B
WO-A-7805	....	0601B
WO-A-7806	....	0601B
WO-A-7807	....	0601B
WO-A-7808	....	0601B
WO-A-7809	....	0601B
WO-A-7810	....	0601B
WO-A-7823	....	0606B
WO-A-7824	....	0606B
WO-A-7830	FM-GPW-18365-B	1001
WO-A-7831	FM-GPW-18366-B	1100
WO-A-7832	FM-GPW-18356-B	0700
WO-A-7833	FM-GPW-18359-B	0202
WO-A-7834	FM-GPW-18373-C	0302
WO-A-7835	FM-GPW-18323	0108
WO-A-7836	FM-GPW-18376-B	0602
WO-A-7837	FM-GPW-18314	0810
WO-A-7838	FM-GPW-18370-B	1205
WO-A-7840	FM-GPW-18342	0601
WO-A-7841	FM-GPW-18319	0602
WO-A-7842	FM-GPW-18329	0602
WO-A-7843	FM-GPW-18343	0603
WO-A-7844	FM-GPW-18363-B	0604A
WO-A-7845	....	0606B
WO-A-7848	FM-GPW-18938-C	2603
WO-A-7895	FM-GPW-18363-B	0601
WO-A-7947	FM-GPW-14537	2601
WO-A-7956	....	0302
WO-A-8087	FM-34753-S	0606F
WO-A-8088	....	0606F
WO-A-8113	FM-GPW-14480-B	2601
WO-A-8114	FM-GPW-18861-B	2601
WO-A-8116	FM-GPW-14481-B	2601
WO-A-8124	FM-GPW-9273	0605
WO-A-8125	....	2204
WO-A-8126	....	2204
WO-A-8127	....	2204
WO-A-8129	....	2204
WO-A-8130	....	0605
WO-A-8132	....	0605
WO-A-8180	FM-GPW-17255-A	2204
WO-A-8186	FM-GPW-10850	0605
WO-A-8188	FM-GPW-10883	0605
WO-A-8190	FM-GPW-9273	0604C
WO-A-8242	....	2204
WO-A-8249	....	1006
WO-A-8250	FM-GPW-3304-B	1401
WO-A-8252	FM-GPW-3305-B	1401
WO-A-8253	FM-GPW-2452-B	1204
WO-A-8255	....	1602
WO-A-8256	....	1602
WO-A-8279	....	2302
WO-A-8322	....	0601B
WO-A-8323	....	0302
WO-A-8498	....	0601B
WO-A-8558	FM-GPW-6051-B	0101A
WO-A-8809	....	1603
WO-A-8810	....	1603
WO-A-8834	....	0303A
WO-A-8835	....	0803
WO-A-8841	FM-GPW-7736	0803
WO-A-8842	....	0601
WO-A-8883	FM-GPW-18958-C	2603
WO-A-8884	FM-GPW-18937	2603
WO-A-8894	FM-GPW-2011	1200
WO-A-8895	FM-GPW-2010	1200
WO-A-8896	FM-GPW-2211-B	1200
WO-A-8897	FM-GPW-2210-B	1200
WO-A-8898	FM-GPW-2013	1203
WO-A-8914	....	0601B
WO-A-8993	....	0601B
WO-A-8997	....	0601B
WO-A-9040	....	0601B
WO-A-9041	....	0601B
WO-A-9042	....	0601B
WO-A-9043	....	0601B
WO-A-9044	....	0601B
WO-A-9046	....	0601B
WO-A-9047	....	0601B
WO-A-9048	....	0601B
WO-A-9049	....	0601B
WO-A-9050	....	0601B
WO-A-9051	....	0601B
WO-A-9052	....	0601B
WO-A-9053	....	0601B
WO-A-9054	....	0601B
WO-A-9055	....	0601
WO-A-9220	....	0606B
WO-A-9490	FM-GPW-8620-A2	0503B
WO-A-9492	FM-GPW-10130	0601
WO-A-11432	FM-GPW-17534	1811D
WO-A-11433	FM-GPW-17500	1811D
WO-A-11519	....	1811D
WO-A-11701	....	1505
WO-A-11729	....	1804A
WO-A-11730	....	1804A
WO-A-11731	....	1804A

SNL G-503

VENDOR'S NUMERICAL INDEX (Cont'd)

Willys Part Number	Ford Part Number	Gov't Group No.
WILLYS-OVERLAND MOTORS, INC. (WO) Cont'd		
WO-A-11732	....	1804A
WO-A-11757	FM-GPW-1101629-A	2203
WO-A-11765	FM-GPW-17126	2301
WO-A-11768	....	0607D
WO-A-11770	....	0607D
WO-A-11850	FM-34804-S2	2202
WO-A-11861	....	2202
WO-A-11862	....	2202
WO-A-12025	....	1802A
WO-A-12054	....	1803
WO-A-12055	....	1803
WO-A-12056	....	0606
WO-A-2940	FM-GPW-1128242	1802A
WO-A-28023	FM-357574-S	
WO-50151	....	
WO-50163	FM-355498-S7	
WO-50769	....	
WO-50878	FM-24449-S7	
WO-50921	....	
WO-50922	FM-33800-S2	
WO-50929	FM-20367-S7	
WO-50992	FM-24505-S7	
WO-51040	FM-26457-S	
WO-51091	FM-74121-S	
WO-51248	FM-GPW-10094	
WO-51304	FM-34807-S2	
WO-51308	FM-21492-S2	
WO-51371	FM-24555-S	
WO-51391	FM-24310-S	
WO-51396	FM-24347-S2	
WO-51405	....	
WO-51406	FM-24428-S	
WO-51485	FM-20326-S2	
WO-51486	FM-24386-S2	
WO-51492	FM-26483-S2	
WO-51514	FM-20324	
WO-51523	FM-20346-S2	
WO-51532	....	
WO-51545	....	
WO-51546	FM-26466-S7	
WO-51575	FM-O1Y-1202	
WO-51612	FM-350744-S2	
WO-51662	FM-356201-S	
WO-51732	FM-20309-S7	
WO-51738	FM-20300-S7	
WO-51763	FM-20308-S2	
WO-51798	FM-355398-S2	
WO-51804	FM-B-13466	
WO-51823	FM-355456-S	
WO-51833	FM-34806-S2	
WO-51840	FM-34806-S2	
WO-51858	FM-355442-S	
WO-51875	FM-GPW-6555	
WO-51921	FM-74113-S	
WO-51954	....	
WO-51969	FM-34804-S2	
WO-52031	FM-34805-S2	
WO-52045	FM-34846-S2	
WO-52046	FM-34807-S2	
WO-52101	....	
WO-52131	FM-26457-S7	
WO-52132	FM-24327-S7	
WO-52142	FM-32866-S2	
WO-52167	FM-20344-S2	
WO-52168	FM-20324-S2	
WO-52170	FM-24308-S2	
WO-52189	FM-24328-S2	
WO-52217	FM-33799-S2	
WO-52221	FM-34803-S2	
WO-52226	FM-60-8287	
WO-52236	FM-27068-S	
WO-52274	FM-34746-S2	
WO-52332	....	
WO-52350	FM-33797-S2	
WO-52379	FM-24489-S	
WO-52424	FM-34806-S7	
WO-52510	FM-34941-S2	
WO-52600	FM-24449-S2	
WO-52615	FM-34051-S7	
WO-52702	FM-34745-S	
WO-52705	....	
WO-52706	FM-34805-S7	
WO-52700	FM-34805-S2	
WO-52706	FM-34805-S2	
WO-52768	FM-34706-S2	
WO-52781	....	
WO-52809	FM-32924-S2	
WO-52832	....	
WO-52836	FM-355444-S	
WO-52837	FM-48-15021	
WO-52839	....	
WO-52857	....	
WO-52863	....	
WO-52876	FM-86H-4621	
WO-52877	FM-86H-4616	
WO-52877	FM-86H-4616	
WO-52878	FM-GP-4630	
WO-52879	FM-GP-4628	
WO-52880	FM-GP-4221	
WO-52881	FM-GP-4222	
WO-52883	FM-O1Y-1202	
WO-52893	FM-356016-S7	
WO-52909	....	
WO-52911	FM-24408-S	
WO-52921	....	
WO-52925	FM-33927-S7	
WO-52940	FM-GP-3161	
WO-52941	FM-GP-3162	
WO-52942	FM-GP-1201	
WO-52943	FM-GP-1202	
WO-52944	FM-72063-S	
WO-52945	....	
WO-52954	FM-33909-S	
WO-53023	FM-351274-S7	
WO-53024	....	
WO-53025	FM-356309-S7	
WO-53026	....	
WO-53031	FM-20348-S2	
WO-53048	FM-24347-S7	
WO-53069	....	
WO-53070	....	
WO-53071	FM-40631-S7	
WO-53135	....	
WO-53194	FM-34903-S2	
WO-53024	....	
WO-53036	....	
WO-53135	....	
WO-53285	FM-33798-S7	
WO-53303	....	

295

SNL G-503

VENDOR'S NUMERICAL INDEX (Cont'd)

Ford Part Number	Willys Part Number	Gov't Group No.	Willys Part Number	Ford Part Number	Gov't Group No.
WILLYS-OVERLAND MOTORS, INC. (WO) Cont'd			WO-116175	FM-GPW-9575	0301
WO-71633	FM-351370-S2	1804	WO-116176	FM-GPW-9541	0301
WO-78932	FM-GPW-14566	0606B	WO-116177	FM-GPW-9566	0301
WO-106313	FM-GPW-10098	0601	WO-116178	FM-GPW-9599	0301
WO-106740	FM-GPW-12193	0603	WO-116179	FM-GPW-9544	0301
WO-107128	FM-B-10141	0603	WO-116180	FM-GPW-9940	0301
WO-109427	....	0602	WO-116181	FM-GPW-9528	0301
WO-109428	....	0602	WO-116183	FM-GPW-9578	0301
WO-109431	FM-GPW-11055	0602	WO-116184	FM-GPW-9624	0301
WO-109433	FM-GPW-11103	0602	WO-116185	FM-GPW-9615	0301
WO-109436	FM-GPW-11107	0602	WO-116186	FM-GPW-9650	0301
WO-109437	FM-GPW-11071	0602	WO-116187	FM-GPW-9570	0301
WO-109442	FM-GPW-11061	0602	WO-116188	FM-GPW-9636	0301
WO-109445	FM-GPW-11036-B	0602	WO-116189	FM-GPW-9587	0301
WO-109446	FM-GPW-11056	0602	WO-116191	FM-GPW-9935	0301
WO-109452	FM-GPW-11077	0602	WO-116194	FM-GPW-9614	0301
WO-109453	....	0603	WO-116195	FM-GPW-9631	0301
WO-111063	FM-33798-S7	0606	WO-116197	FM-GPW-9583	0301
WO-113440	FM-34803-S7	0302	WO-116198	FM-GPW-9531	0301
WO-113460	FM-GPW-9388	0302	WO-116199	FM-GPW-9527	0301
WO-113461	FM-GPW-9373	0302	WO-116202	FM-GPW-9516	0301
WO-115641	FM-GPW-9399	0302	WO-116203	FM-GPW-9519	0301
WO-115643	FM-GPW-9380	0302	WO-116204	FM-GPW-119594	0301
WO-115650	FM-GPW-9354	0302	WO-116205	FM-GPW-9576	0301
WO-115651	FM-11A-9352	0302	WO-116206	FM-GPW-9905	0301
WO-115652	FM-GPW-9363	0302	WO-116207	FM-34711-S	0301
WO-115653	FM-11A-9361	0302	WO-116208	FM-GPW-9515	0301
WO-115654	FM-GPW-9365	0302	WO-116209	FM-GPW-9930	0301
WO-115656	FM-GPW-9364	0302	WO-116210	FM-GPW-9554	0301
WO-115657	FM-GPW-9387	0302	WO-116211	FM-31032-S7	0301
WO-115869	FM-GPW-9468	0302	WO-116213	FM-31061-S8	0301
WO-115870	FM-GPW-19469	0302	WO-116215	FM-31662-S	0301
WO-115880	FM-INC-9381	0302	WO-116216	FM-GPW-9586	0301
WO-115905	FM-355253-S	0301C	WO-116217	FM-GPW-9588	0301
WO-115948	FM-GPW-6552-C	0105A	WO-116218	....	0301
WO-115962	FM-GPW-18371	1207	WO-116219	FM-355858-S	0301
WO-115963	FM-GPW-18372	1207	WO-116295	FM-GPW-6390	0109
WO-116017	....	0103	WO-116384	FM-355067-S7	0301
WO-116018	....	0103	WO-116385	FM-355200-S7	0301
WO-116019	FM-GPW-6105-C	0103	WO-116458	FM-GPW-5330-A	1601
WO-116020	FM-GPW-6105-D	0103	WO-116459	FM-GPW-5724-A	1601A
WO-116023	FM-GPW-6155-G	0130A	WO-116460	FM-GPW-5330-B	1601
WO-116024	FM-GPW-6155-H	0103A	WO-116502	FM-GPW-6150-C	0103A
WO-116110	FM-GPW-6149-A	0103A	WO-116503	FM-GPW-6150-D	0103A
WO-116112	FM-GPW-6149-C	0103A	WO-116522	FM-GPW-6333-C	0102A
WO-116113	FM-GPW-6149-D	0103A	WO-116524	FM-GPW-6338-C	0102A
WO-116116	FM-GPW-6156-H	0103A	WO-116526	FM-GPW-6339-C	0102A
WO-116117	FM-GPW-6156-J	0103A	WO-116528	FM-GPW-6341-C	0102A
WO-116154	FM-GPW-9585	0301	WO-116530	FM-GPW-6331-C	0102A
WO-116157	FM-GPW-9549	0301	WO-116532	FM-GPW-6337-C	0102A
WO-116159	FM-GPW-9522	0301	WO-116534	FM-GPW-6211-B	0104A
WO-116160	FM-GPW-9523	0301	WO-116535	FM-GPW-6211-C	0104A
WO-116161	FM-GPW-9562	0301	WO-116537	FM-GPW-9529	0301
WO-116162	FM-GPW-9579	0301	WO-116538	FM-GPW-9907	0301
WO-116163	FM-GPW-9696	0301	WO-116539	FM-GPW-9553	0301
WO-116164	FM-GPW-9928	0301	WO-116540	FM-GPW-9906	0301
WO-116165	FM-GPW-9543	0301	WO-116541	FM-GPW-9914	0301
WO-116166	FM-GPW-9922	0301	WO-116542	FM-GPW-9598	0301
WO-116168	FM-GPW-9569	0301	WO-116543	FM-GPW-6333-D	0301
WO-116169	FM-GPW-9608	0301	WO-116544	FM-GPW-9520	0301
WO-116170	FM-GPW-9574	0301	WO-116545	FM-GPW-9546	0301
WO-116171	FM-GPW-9926	0301	WO-116548	....	0301
WO-116172	FM-GPW-9550	0301	WO-116549	FM-GP-2018	1202
WO-116173	FM-GPW-9558	0301	WO-116550	FM-GP-2019	1202
WO-116174	FM-GPW-9567	0301	WO-116551	FM-GP-2021	1202

VENDOR'S NUMERICAL INDEX (Cont'd)

Willys Part Number	Ford Part Number	Gov't Group No
WILLYS-OVERLAND MOTORS, INC. (WO) Cont'd		
WO-116552	FM-GP-2022	1202
WO-116558	....	0103
WO-116560	....	0103
WO-116561	....	0103
WO-116562	FM-GPW-6152-A	0103A
WO-116564	FM-GPW-6152-C	0103A
WO-116565	FM-GPW-6152-D	0103A
WO-116566	FM-GPW-6159-A	0103A
WO-115667	FM-GPW-6159-B	0103A
WO-116568	FM-GPW-6159-C	0103A
WO-116569	FM-GPW-6159-D	0103A
WO-116584	FM-GPW-9518	0301
WO-116585	FM-GPW-9581	0301
WO-116586	FM-GPW-9526	0301
WO-116587	FM-GPW-9595	0301
WO-116588	FM-355132-S	0301
WO-116589	FM-GPW-5724-B	1601A
WO-116600	FM-GPW-18367	1202
WO-116609	FM-GPW-5345-A	1601
WO-116610	FM-GPW-5345-B	1601
WO-116616	FM-GPW-6156-F	0103A
WO-116624	....	1603
WO-116625	....	1603
WO-116626	....	1603
WO-116627	....	1603
WO-116628	....	1603
WO-116629	....	1603
WO-116630	....	1603
WO-116631	....	1603
WO-116632	....	1603
WO-116633	....	1603
WO-116634	....	1603
WO-116635	....	1603
WO-116636	....	1603
WO-116637	....	1603
WO-116638	....	1603
WO-116639	....	1603
WO-116640	....	1603
WO-116641	....	1603
WO-116642	....	1603
WO-116643	....	1603
WO-116644	....	1603
WO-116651	FM-GPW-9610	0301
WO-116694	FM-GPW-9396	0302
WO-116695	FM-GPW-9398	0302
WO-52217	FM-33795-S2	0501
WO-300143	FM-88042	0108
WO-300329	FM-33896-S2	0610A
WO-301232	FM-31037-S7	0301
WO-302347	FM-GPW-10141	0601
WO-303922	....	1500
WO-306715	FM-GPW-17011-A	2301
WO-307556	FM-B-14463	0604A
WO-311003	FM-73904-S7	1201
WO-314338	FM-GPW-6734	0107A
WO-314369	FM-B-14466	0604A
WO-315932	FM-GPW-6269	0106C
WO-323397	FM-23017-S16	0610B
WO-323457	FM-34056-S16	0610B
WO-327257	FM-9N-12113-A	0604A
WO-330964	FM-GPW-6684	0107
WO-332515	FM-88032-S	0108B
WO-334103	FM-GPW-6353	0102
WO-335912	FM-350343-S-16	0610B
WO-337112	FM-357420-S	0102
WO-337304	FM-88350-S	0303
WO-339043	FM-73880-S	0203
WO-339372	FM-GPW-5456	1602
WO-343306	FM-GPW-6614	0107
WO-344732	FM-GPW-9443	0108
WO-345961	FM-GPW-13434-A	0606B
WO-349368	FM-GPW-6066	0101A
WO-349712	FM-88082-S	0108
WO-352760	FM-34905-S	0608
WO-356155	FM-GPW-6654	0107
WO-359039	FM-GPW-5781	1601A
WO-371400	FM-11A-14452	2601
WO-371567	FM-GPW-7549	0201A
WO-372438	FM-GPW-11474	0303A
WO-374586	FM-361915-S	0201A
WO-375217	FM-GPW-7023	0109C
WO-375811	FM-GPW-6510-B	0105
WO-375877	FM-GPW-6310	0102
WO-375900	FM-GPW-6245	0106
WO-375907	FM-GPW-6243	0106
WO-375908	FM-GPW-6244	0106
WO-375917	FM-GPW-6286	0106D
WO-375920	FM-GPW-6287	0102
WO-375927	FM-GPW-6625	0107
WO-375981	FM-88141-S	0107
WO-375994	FM-GPW-6546	0105
WO-376373	FM-358063-S	0101
WO-380197	FM-355262	0107
WO-381519	FM-GPW-6345	0102
WO-384228	FM-GPW-5468	1602
WO-384549	FM-GPW-9268	0107N
WO-384710	FM-GPW-2133	1209C
WO-384958	FM-88022-S	0106D
WO-387249	FM-GPW-9267-A	0306B
WO-387633	FM-GPW-6319	0102
WO-387891	FM-9N-18679	0107J
WO-390510	....	1403A
WO-392328	FM-GPW-7227	0706A
WO-392468	FM-357553-S-18	1201
WO-393594	....	0603
WO-630068	FM-GPW-7516	0203
WO-630101	FM-355597	0109C
WO-630103	FM-GPW-7518	0109C
WO-630112	FM-GPW-7515	0203
WO-630117	FM-GPW-7562	0203
WO-630129	FM-355476-S	0201B
WO-630262	FM-GPW-6342-B	0102
WO-630294	FM-GPW-6326	0102A
WO-630298	FM-GPW-6762	0107N
WO-630299	FM-GPW-6648	0105C
WO-630303	FM-GPW-6519	0105C
WO-630305	FM-GPW-6521	0105C
WO-630359	FM-GPW-6020	0101A
WO-630364	FM-GPW-6285	0106D
WO-630365	FM-GPW-6288	0106D
WO-630384	FM-GPW-6604	0107
WO-630387	FM-GPW-6664	0107
WO-630389	FM-GPW-6628	0107
WO-630390	FM-GPW-6644	0107
WO-630392	FM-GPW-6619	0107
WO-630394	FM-GPW-6630	0107
WO-630396	FM-GPW-6615	0107A-1
WO-630397	FM-GPW-6617	0107A-1
WO-630398	FM-GPW-6627	0107A-1

SNL G-503

VENDOR'S NUMERICAL INDEX (Cont'd)

Willys Part Number	Ford Part Number	Gov't Group No.
WILLYS-OVERLAND MOTORS, INC. (WO) Cont'd		
WO-630512	FM-IGT-8260	0505
WO-630518	FM-GPW-6663	0107
WO-630526	FM-GPW-5269	0402
WO-630593	FM-GPW-7523	0204
WO-630727	FM-GPW-6342-A	0106C
WO-630753	FM-GPW-3326	1401
WO-630754	FM-GPW-3327	1401
WO-630755	FM-GPW-3320	1401
WO-630756	FM-GPW-3323	1401
WO-630757	FM-GPW-3328	1401
WO-631105	FM-GPW-12064	0604
WO-632156	FM-GPW-6387	0109
WO-632157	FM-355497-S	0109
WO-632158	FM-355451-S	0105C
WO-632159	FM-88057-S7	0301A
WO-632174	FM-GPW-9737	0303
WO-632177	FM-GPW-7532	0203
WO-633011	FM-GPW-9799	0303
WO-633013	FM-GPW-9752	0303
WO-633949	FM-355496-S	0601A
WO-634758	FM-74-6038	0809
WO-634759	FM-GPW-7781-A	0809
WO-634762	FM-356524-S	0809
WO-634796	FM-GPW-6308	0102
WO-634811	FM-GPW-9435	0108
WO-634813	FM-GPW-6642	0107
WO-634814	FM-GPW-9450	0402
WO-634850	FM-355499-S	0106C
WO-635377	FM-GPW-6369	0102
WO-635394	FM-GPW-6384	0109B
WO-635529	FM-GPW-7580	0203
WO-635681	FM-GPW-7291	1201
WO-635804	FM-GPW-7143	0704C
WO-635811	FM-GPW-7129	0704C
WO-635812	FM-GPW-7119	0704C
WO-635836	FM-GPW-7206	0706C
WO-635837	FM-GPW-7234	0706B
WO-635839	FM-GPW-7208	0706C
WO-635840	FM-357418-S	0706C
WO-635844	FM-GPW-7064	0703
WO-635846	....	0703
WO-635859	FM-GPW-7214	0706A
WO-635861	FM-GPW-7223	0706A
WO-635862	FM-GPW-7267	0706A
WO-635863	FM-GPW-7227	0706A
WO-635868	....	0706A
WO-635883	....	0601B
WO-635886	FM-357689-S	0101
WO-635981	FM-GPW-11487-B	0606B
WO-635985	FM-GPW-11487-A	0606B
WO-636004	FM-GPW-5270	0402
WO-636109	FM-GPW-8269	0505
WO-636196	FM-GPW-7230	0706C
WO-636197	FM-GPW-7231	0706C
WO-636200	FM-GPW-7245	0706B
WO-636297	FM-GPW-8530	0503
WO-636298	FM-GPW-8576	0503
WO-636299	FM-GPW-8509-A	0503A
WO-636360	FM-GP-4241	1003
WO-636396	FM-GP-17333	0810
WO-636438	FM-GPW-9462	0108C
WO-636439	FM-GPW-9460	0108C
WO-636527	FM-355699-S2	1001
WO-636528	FM-34922-S	1001
WO-636538	FM-353051-S	1001
WO-636565	FM-GP-4661	1003
WO-636566	FM-GP-4619	1003
WO-636568	FM-GP-4666	0902
WO-636569	FM-356126-S	1104
WO-636570	FM-356504-S	1003
WO-636571	FM-357202-S	1104
WO-636575	FM-33786-S2	1201
WO-636577	FM-358048-S	1101
WO-636599	FM-GPW-6608	0107
WO-636600	FM-GPW-6673	0107
WO-636755	FM-GPW-7550	0201B
WO-636778	FM-GPW-7577	0201A
WO-636796	FM-355396-S	0107A-1
WO-636879	FM-GPW-7100	0704B
WO-636882	FM-GPW-7141	0704B
WO-636885	FM-GPW-7025	0703
WO-636961	FM-GPW-6135-A	0103B
WO-636962	FM-356021-S	0104
WO-637007	FM-GPW-6333-A	0102A
WO-637008	FM-GPW-6338-A	0102A
WO-637037	....	0103
WO-637041	FM-GPW-6105-A	0103
WO-637042	FM-GPW-6155-A	0103A
WO-637044	FM-GPW-6514	0105
WO-637045	FM-GPW-6511-B	0105
WO-637047	FM-GPW-6500-A	0105A
WO-637052	FM-GPW-8505	0503
WO-637053	FM-GPW-8543	0503
WO-637065	FM-GPW-6250	0106A
WO-637098	FM-GPW-6700	0102A
WO-637107	FM-355426-S	1403A
WO-637182	FM-GPW-6507	0105
WO-637183	FM-GPW-6505	0105
WO-637206	FM-GPW-9456	0108C
WO-637208	FM-GPW-9467-A	0108C
WO-637209	FM-GPW-9484	0108C
WO-637210	FM-GPW-9458	0108C
WO-637211	FM-GPW-9465	0108C
WO-637237	FM-GPW-6702	0102A
WO-637424	FM-GP-2078	1209B
WO-637425	FM-GPW-6610	0107
WO-637426	FM-GP-2087	1209B
WO-637427	FM-78-2814-A	1209B
WO-637432	FM-GP-2074	1209B
WO-637439	....	1209C
WO-637495	FM-GPW-7051-B	0700
WO-637503	FM-GPW-7056	0701
WO-637540	FM-GP-2208	1207
WO-637541	FM-GP-2196	1207
WO-637544	....	1207
WO-637545	FM-GP-2204	1207
WO-637546	FM-GP-2206-A	1207
WO-637577	FM-GP-2194	1207
WO-637579	FM-91A-2201	1207
WO-637580	FM-GP-2205	1207
WO-637582	FM-GP-2155	1205
WO-637583	FM-GP-2160	1205
WO-637584	FM-GP-2175	1205
WO-637585	FM-GP-2176	1205
WO-637586	FM-GP-2183	1205
WO-637587	FM-GP-2145	1205
WO-637590	FM-GP-2173	1205
WO-637591	FM-GP-2169	1205
WO-637595	FM-GP-2170	1205

SNL G-503

VENDOR'S NUMERICAL INDEX (Cont'd)

Willys Part Number	Ford Part Number	Gov't Group No.
WILLYS-OVERLAND MOTORS, INC. (WO) Cont'd		
WO-637597	FM-GP-2188	1205
WO-637599	FM-GP-2143-A	1205
WO-637602	FM-GP-2180	1205
WO-637604	FM-91A-2152	1205
WO-637605	FM-GP-2077	1205
WO-637606	FM-91A-2151	1205
WO-637608	FM-GP-2162	1205
WO-637612	FM-GP-2167	1205
WO-637615	FM-GPW-12083	0603
WO-637635	FM-GPW-17017-A	2301
WO-637636	FM-GPW-6600	0107
WO-637646	FM-GPW-8575	0502
WO-637724	FM-GPW-6333-B	0102A
WO-637725	FM-GPW-6338-B	0102A
WO-637787	FM-GP-2261	1207
WO-637789	FM-GP-2192	1207
WO-637790	FM-GPW-6701	0102A
WO-637803	....	1603
WO-637804	....	1603
WO-637810	....	1603
WO-637831	FM-GPW-7109	0704B
WO-637832	FM-GPW-7116	0704B
WO-637833	FM-GPW-7106	0704B
WO-637834	FM-GPW-7107	0704B
WO-637835	FM-GPW-7059	0704B
WO-637863	FM-O1A-12410	0604B
WO-637899	FM-91A-2027	1203B
WO-637900	FM-GP-2028	1203B
WO-637901	FM-91A-2030	1203B
WO-637905	FM-GP-2035	1203A
WO-637923	FM-351466-S-24	1203B
WO-637924	FM-33846-S2	1203B
WO-637936	FM-GPW-18060	1603
WO-638058	FM-GPW-5274	0401
WO-638113	FM-GPW-6312	0102
WO-638121	FM-GPW-6303-A	0102
WO-638151	FM-GPW-7580	0202
WO-638152	FM-GPW-7566	0202
WO-638153	FM-GPW-7590	0202
WO-638154	FM-24325-S	0202
WO-638155	FM-33921-S7	0202
WO-638157	FM-GPW-7567	0202
WO-638158	FM-GPW-7591	0202
WO-638159	FM-GPW-7564	0202
WO-638305	FM-34745-S2	0202
WO-638343	....	1603
WO-638381	FM-356439-S	1403A
WO-638458	FM-GPW-6256	0106A
WO-638459	FM-GPW-6306	0102
WO-638500	FM-353023-S7	1602
WO-638539	FM-33881-S2	1602
WO-638635	FM-GPW-6065	0101A
WO-638636	FM-GPW-6513	0105
WO-638640	FM-GPW-9448	0108
WO-638646	FM-GPW-11140	0602
WO-638730	FM-GPW-6339-A	0102A
WO-638731	FM-GPW-6341-A	0102A
WO-638732	FM-GPW-6331-A	0102A
WO-638733	FM-GPW-6337-A	0102A
WO-638737	FM-GPW-9417	0302A
WO-638780	FM-GPW-2279	1201
WO-638792	FM-353043-A-57	0204
WO-638798	FM-GPW-7102	0704B
WO-638809	....	1500
WO-638884	FM-GPW-3626	0609B
WO-638885	FM-GPW-3646	0609B
WO-638918	FM-GPW-3563	1403A
WO-638918	FM-GPW-3563	1405
WO-638948	FM-GPW-7111	0704C
WO-638949	FM-GPW-7155	0704C
WO-638952	FM-GPW-7140	0704C
WO-638979	FM-GPW-13532	0606
WO-638992	FM-GPW-7563	1201
WO-638993	FM-GPW-7572	0202
WO-639010	FM-GPW-28482	1201
WO-639051	FM-GPW-6262-A	0105A
WO-639052	FM-355452-S	0105C
WO-639090	FM-GPW-3587	1403A
WO-639091	FM-GPW-3576	1403A
WO-639095	FM-GPW-3591-A	1403A
WO-639102	FM-GPW-3552	1403A
WO-639103	FM-GPW-3589	1403A
WO-639104	FM-GPW-3571	1403A
WO-639108	FM-GPW-3593	1403A
WO-639109	FM-GPW-3594	1403A
WO-639110	FM-GPW-3595	1403A
WO-639116	FM-GPW-3583	1403A
WO-639118	FM-GPW-3577	1403A
WO-639119	FM-GPW-3581	1403A
WO-639120	FM-20386-S7	1403A
WO-639121	FM-356303-S	1403A
WO-639190	FM-GPW-3517	1403A
WO-639191	FM-GPW-3520	1403A
WO-639192	FM-GPW-3518	1403A
WO-639237	FM-GPW-6339-B	0102A
WO-639238	FM-GPW-6341-B	0102A
WO-639239	FM-GPW-6331-B	0102A
WO-639240	FM-GPW-6337-B	0102A
WO-639244	FM-GPW-2782	1201
WO-639265	FM-GP-4676-A	1003
WO-639422	FM-GPW-7120	0703
WO-639423	FM-GPW-7063	0703
WO-639555	FM-GPW-6763-A	0107G
WO-639556	FM-GPW-6766-A	0107H
WO-639578	FM-GPW-7600	0109
WO-639599	FM-GPW-14448-C	0606E
WO-639607	FM-GPW-9728	0303
WO-639610	FM-GPW-9745	0303
WO-639650	FM-GPW-8255	0101A
WO-639651	FM-GPW-8578	0502
WO-639654	FM-GPW-7561	0203
WO-639689	FM-20366-S	0706A
WO-639734	FM-GPW-11134	0601B
WO-639743	FM-GPW-9463	0108C
WO-639862	FM-GPW-6211-A	0104A
WO-639864	FM-GPW-6150-A	0103A
WO-639870	FM-GPW-6659	0107
WO-639979	FM-GPW-6727	0107A
WO-639980	FM-GPW-6710	0107A
WO-639992	FM-GPW-8501	0503
WO-639993	FM-GPW-8512	0503
WO-640006	FM-GPW-7104-A2	0704
WO-640017	FM-GPW-7050	0703
WO-640018	FM-GPW-7052	0703
WO-640020	FM-GPW-6549-B	0105A
WO-640031	FM-GAA-8524-A2	0503
WO-640032	FM-GPW-8549	0503
WO-640033	FM-GPW-8572-B	0503
WO-640034	FM-GPW-8557-A	0503

VENDOR'S NUMERICAL INDEX (Cont'd)

Willys Part Number	Ford Part Number	Gov't Group No.
WILLYS-OVERLAND MOTORS, INC. (WO) Cont'd		
WO-640038	FM-358006-S8	1602
WO-640066	....	0104
WO-640067	....	0104
WO-640070	....	0104
WO-640071	FM-GPW-6200	0104
WO-640072	FM-GPW-6201	0104
WO-635838	FM-353081-S	0706B
WO-650482	FM-GPW-9711	0303
WO-650483	FM-GPW-9795	0303

Willys Part Number	Ford Part Number	Gov't Group No.
WO-650484	FM-74019-S	0303
WO-650684	....	0204
WO-651298	FM-GPW-9731-B	0303
WO-662010	FM-353202-S	1802A
WO-662276	FM-GPW-13437-A	0606B
WO-662420	FM-GPW-9319-A	0605
YALE & TOWNE MFG. CO. (YA)		
YA-OP-528 WO-A-2895	FM-GPW-1143501	1817

INDEX
Section I Clothing Para 1-2

Section II Equipment Para 3-5

Section III Vehicles Para 6-14
Motorcycles, Passenger Cars, 1/4ton Reconnaissance,
1/2ton Command Reconnaissance, 1/2ton Carry All, Pick Up, 1/2ton Weapons Carrier,
Panel Truck, 3/4ton Command Reconnaissance,
3/4ton Carry All, 3/4ton Weapons Carrier, $1^1/_2$ton Panel Truck, $1^1/_2$ton Open Cargo,
$1^1/_2$ton (Enclosed) Cargo Truck, $1^1/_2$ton Tractor Truck, $1^1/_2$-3ton Van,
$2^1/_2$ton Open Cab 6x6, $2^1/_2$ton (Enclosed) 6x6, Open Tanker, (Enclosed) Tanker,
Searchlight Truck, 4ton Wrecker, 4ton Cargo 6x6, 4 -5ton Tractor Truck,
Pontoon Tractor 5-6ton, 6ton Cargo 6x6, Semi 6T, Trailer 1-T, T-26 Trailer,
Scout car, Half Track, Light Tank, Medium Tank,
Letters & Numbers

Section IV Baggage & Property Para 6-14

Section V Miscellaneous Para 18-20

CHANGES

Change #1 - August 24, 1942   Para 4 (e) "Organization"

Change #2 - October 15, 1942 Para 4 (b) "Organization" Para 20 "New types of Vehicles"

Change #3 - November 12,1942   Para 4 (e) "Organization

Change #4 - December 11, 1942   Para 5 (a, d, e) "Individual"

Change #5 - December 24, 1942   Para 5 (f) "Individual"

Change #6 - March 4, 1943   Para 5 (f) "Individual"

Change #7 - June 12, 1943   Para 10 (b)"Unit markings"

Change #8 - 8 October, 1943   Para 10 (a) "Unit markings"

Change #9 - 27 January, 1944   Para 10 (a) "Unit markings"
Para 12 "Special markings"
Para 14 (b) "Armored Vehicles"
M8, M3A1, M3A3, M5, M5A1, T9E1, M4, M6, M8, M7, M10 & M10A1, M12, M30(T14)
1/4ton Trailer, 6ton Van, 5ton refrig, 6ton Repair, 6ton Laundry, 6ton Records,
GPA, $2^1/_2$ ton 6x6 Amphibian.

Change 10 - 25 March, 1944
Para 10 (a) "Unit markings"

*AR 850–5
1–4

ARMY REGULATIONS
No. 850–5

WAR DEPARTMENT,
WASHINGTON, August 5, 1942.

MISCELLANEOUS

MARKING OF CLOTHING, EQUIPMENT, VEHICLES, AND PROPERTY

	Paragraphs
SECTION I. Clothing	1–2
II. Equipment	3–5
III. Vehicles	6–14
IV. Baggage and property	15–17
V. Miscellaneous	18–20

SECTION I
CLOTHING

	Paragraph
Marks required; where placed	1
To be indelible	2

1. Marks required; where placed.—Clothing will be marked on the inside of each garment with the initial letter of the enlisted man's last name, followed by a dash and the last four figures of the enlisted man's Army serial number, for example, W–6046.

2. To be indelible.—The marking required by paragraph 1 will be done with indelible ink.

SECTION II
EQUIPMENT

	Paragraph
General	3
Organization	4
Individual	5

3. General.—*a. Application.*—The provisions of *b*, *c*, and *d* below apply to both organization equipment (par. 4) and individual equipment (par. 5).

b. Letters "U.S."—The letters "U.S." will invariably be placed on all articles of equipment, except the identification tag, by the procuring agency, but will not be so placed as to occupy space required by these regulations for other marking.

c. Articles with wooden handles.—Articles of equipment which have wooden handles, such as shovels, will be marked on the handle. Leather-stamping dies may be used, if available; if not, the marking will be done by improvised methods, such as branding or painting.

d. Equipment not specifically provided for.—The marking of equipment not specifically provided for in these regulations will conform as far as practicable to the marking herein prescribed for similar equipment.

4. Organization.—*a. Not to be marked until placed in use.*—No organization equipment will be marked, except with the letters "U.S." (par. 3*b*), prior to being placed in actual use.

---

*This pamphlet supersedes AR 850–5, September 25, 1936, including C2, April 22, 1942; section II, Circular No. 239, War Department, 1941; and section III, Circular No. 174, War Department, 1942.

b. Marking required.

(1) In regiments, separate battalions, or independent administrative units similar thereto, articles of organization equipment will be marked with the letter or number of the company. Numbers will be separated by a dash, but a number and a letter will not be separated; for example, 2–49 and 12A.

(2) In independent administrative units smaller than a battalion, the organization equipment will be marked with the number of the unit, and letters suitably representing the abbreviated designation thereof.

c. How marked.

(1) Except as authorized by paragraph 3c, the marking required by *b* above will be done with paint and by means of a stencil whenever possible.

(2) Letters and numerals will be of a size suitable to the space available, but will not ordinarily exceed 1½ inches in height, nor will they be so small as not to be readily distinguishable.

d. Where marked.— Marking should be so placed as to be readily visible, but if it can be avoided, not so exposed as to be liable to eradication.

5. **Individual.**—*a. Articles not to be marked.*—Articles of individual equipment which will not be marked as prescribed in *b* below are—

(1) Metal articles which, when issued, have a serial number stamped thereon, for example, rifle and pistol.
(2) Batons.
(3) Brassards, except Red Cross.
(4) Bugles.
(5) Canteens.
(6) Covers of cans, meat. (See *d*(3) below.)
(7) Goggles.
(8) Masks, gas.
(9) Packets, first-aid.
(10) Stands, music.
(11) Tags, identification. (See AR 600–40.)
(12) Tapes, steel.
(13) Trumpets.
(14) Whistles.

b. Marking required.—Subject to the exceptions enumerated in *a* above and (1) and (2) below, all articles of individual equipment will be marked with the last four figures of the enlisted man's Army serial number (for example, 6046) without addition of company letters or numbers.

(1) Articles which, though issued to an individual, are intended for the use of the squad or section as a whole and which, upon transfer of the individual, will be left in the squad or section, for example, grenade carriers. Such articles will be marked as required in paragraph 4*b*, adding the number of the squad or section.

(2) Articles issued to a company and intended for the use of the company as a whole. Such articles will be marked as required in paragraph 4*b*.

c. When to use dash.— In the cases of numbered companies, the number of the company will be separated by a dash from the squad or section's company number.

MARKING OF CLOTHING, EQUIPMENT, VEHICLES, PROPERTY

AR 850–5

d. How marked.
  (1) *General.*
    (a) As prescribed in (2) and (3) below, by ½-inch stencil for textile equipment, and by marking outfit, metal, and marking outfit, leather, respectively, for metal and leather equipment.
    (b) By the inside of a flap is meant the inner side when the flap is closed.
    (c) In case subsequent markings are necessary, the original markings will be eradicated, by erasure and scouring, if possible. Where eradication is not possible the original markings will be nullified as follows: Markings on textile equipment, by two indelible horizontal lines through each line of the original markings, preferably of the same color as the original. Markings on leather and metal equipment, by stamping over the original marking with the "dash" die.
    (d) Subsequent markings, where they cannot be placed over eradicated original markings legibly, will be placed so as to follow immediately the nullified original marking, that is, adjacent and to the right or below, as the article best facilitates. In the case of ammunition pocket flaps the subsequent marking to be placed on adjacent flap to the right.
    (e) Spare articles of individual equipment in an organization will not be marked until issued to individuals. Such articles will be used as far as practicable for transfer to other organizations when accompanying individuals.
  (2) *Stenciling of textile equipment; where placed.*

Article	Location of stenciling
Bag, feed	Bottom to be ½ inch above center of the "U. S."
Bag, grain	Top to be 4 inches below top tie string.
Belt, cartridge	Inside of flap of first right rifle ammunition pocket.
Belt, magazine	Inside of flap of first automatic or machine rifle ammunition pocket.
Belt, pistol or revolver	Inside of loop next to end hook.
Blanket	On lower left-hand corner, 1½ inches from edges, of blankets marked with the letters "U. S.," and similarly placed on any corner of those without the letters "U. S."
Blanket, saddle	On outside corner, when folded; center to be 1½ inches from edges of blanket.
Carrier, gas mask	Inside of flap near top seam to right of left eyelet reinforcing seam.
Carrier, grenade, 4-inch pocket.	Inside of flap of upper right-hand pocket.
Carrier, ax, intrenching	Inside of flap as near left as practicable.
Carrier, pack	Outside between first and second buttonholes for coupling strap in one or two lines, beginning on the same side as the "U" of the "U. S." Top of lettering to be parallel to and ¼ inch from top edge of pack carrier.

MISCELLANEOUS

Article	Location of stenciling
Carrier, pick mattock, intrenching.	Inside body near lower end.
Carrier, shovel, intrenching.	Inside near lower edge opposite face of shovel.
Case, carrying, automatic rifle.	Inside of flap near lower edge.
Cover, canteen	Bottom.
Cover, horse	Midway between buckle chapes on near side of cover.
Cover, rifle	Inside of flap near lower edge.
Haversack	Inside of flap of meat-can pouch near left-hand edge.
Pocket, magazine, pistol	Inside of flap.
Pouch, first-aid packet	Inside of flap near left edge.
Pouch, medical	Inside of flap near left edge.
Respirator	On flap, as near left edge as possible.
Surcingle	Inside of strap near end.
Suspenders, for belts, cartridge and pistol.	Inside of adjusting strap where stenciling will not touch clothing.
Tent, shelter	On the same side as buttons, 1½ inches below center button of row along the ridge of the tent.

(3) *Stamping of metal or leather articles.*—The marking required by *b* above will be stamped directly upon the articles of equipment enumerated in the following table, by means of metal-stamping dies, as specified opposite each article.

Article	Where marking placed
Bridle	On headstall.
Brush, horse	1 inch to the right of the "U. S."
Can, meat	On center of handle and below the ridge (with the hinge to the left). Lid not to be marked. (See *a*(6) above.)
Cup	On handle, immediately below hinge.
Currycomb	On back, where most convenient.
Fork	Across back of head of handle.
Harness	Where most convenient on each major portion.
Holster, pistol	On inside of flap.
Knife	On handle to the right of the "U. S."
Pouch, music	Inside flap near lower left-hand corner.
Saddle	On near side of pommel, where most convenient.
Saddlebag	On center of clip and strap, near reinforcement.
Scabbard, automatic rifle	On flap, bottom to be ½ inch above strap.
Scabbard, bayonet, M1910.	On outside of leather tip.
Scabbard, bayonet, M1917.	On outside of metal tip.
Scabbard, rifle	Top to be 2 inches below mouth of scabbard.
Scabbard, saber	On side of metal mouthpiece between rings.
Sling, gun	On outer face of long strap.
Spoon	Across back of head of handle.
Strap, saber	On strap.
Strap, sling for automatic rifle.	On outer face of long strap.

e. Identification tag.—See AR 600–40.

MARKING OF CLOTHING, EQUIPMENT, VEHICLES, PROPERTY

Section III
VEHICLES

	Paragraph
Motor	6
Registration markings	7
Automobile plates, other than registration	8
Removal	9
Unit markings	10
Tactical markings	11
Special markings	12
Animal-drawn	13
Tractors and tanks	14

6. Motor.—*a. Definition.*—See AR 850–10.

b. Registration.—See AR 850–10.

c. Inventory.—See AR 850–10.

7. Registration markings.—U. S. registration symbols and numbers.

a. When marked.—Motor vehicles will be marked with the U. S. registration symbols and numbers prescribed in AR 850–10. Original markings will be applied by manufacturers.

b. How marked.—The marking on the vehicle will be with blue-drab lusterless enamel by means of a stencil.

c. Height of letters.—The height of the letters and figures will be 1 inch on motorcycles, 2 inches on trailers and on all other types of motor vehicles. The character and style of marking will be as follows:

U. S. A.
123786

d. Where marked.—The marking required above will be placed on the outside of the vehicle except where otherwise indicated and will be so placed as not to interfere with international and unit markings.

Type of vehicle*	Location of marking
Passenger cars	All letters and figures will be placed on outer surface of the hood, both sides and also on the rear panel.
Ambulances, metropolitan	Same as passenger cars.
Ambulances, field	On the outer surface of the hood, both sides and also the rear doors.
Motorcycles (both solo and those with side cars).	On rear fender only.
Trucks	On the outer surface of the hood, both sides.
Reconnaissance trucks	Same as trucks.
Panel and sedan delivery	Same as ambulances, field.
Trailers	In the center of the back.
Trailer, bomb, M5	U. S. A. on both sides of tool box and registration markings on both sides of front splash guard.
Trailer, tractor crane, T–26	Outer surface forward frame, both sides.
Combat vehicles track, wheel or wheel and track.	To conform as closely as practicable to marking prescribed for other vehicles.

* Special vehicles of the types enumerated will be marked so as to conform as closely as practicable to the marking prescribed for the type to which they belong.

AR 850-5
7-10 MISCELLANEOUS

e. Changing registration number prohibited.—The registration number assigned to a motor vehicle establishes its permanent identity and under no circumstances will this number, when once assigned, be changed or transferred to another vehicle without the authority of the procuring agency.

8. Automobile plates, other than registration.—*a. Material and size.*—See AR 260-10.

b. Design.—See AR 260-10.

c. Paint.—Paint will be used as prescribed by the War Department.

d. Display.—For individuals (general officers), the automobile plates may be displayed in a suitable bracket on the front and rear of the vehicle only. The automobile plates for individuals will be removed or covered when the individual is not in the vehicle.

9. Removal.—All War Department markings on motor vehicles transferred to other parts of the Government or sold at public auctions will be removed or obliterated.

10. Unit markings.—Gasoline solvent paint or paint as prescribed by the War Department will be used.

a. Unit markings.—National symbol.

(1) A white five-pointed star will be the national symbol of all motor vehicles assigned to tactical units. Administrative motor vehicles operating in an active theater of operations will be similarly marked when directed by the theater commander.

(2) The size of the national symbol will be determined for each type of motor vehicle and will be large enough to take advantage of the surface upon which to be painted. See figures 1 to 34.

(3) Whenever requirements for camouflage and concealment outweigh the requirements for recognition, the national symbol may be covered by lusterless olive-drab gasoline solvent paint, camouflage nets, oil and dirt, etc., or will be removed.

b. Unit identification symbols.

(1) *Front and rear markings.*—See figures 1 to 34 for location.

(a) *General.*—Front and rear markings are identical and consist of four groups in consecutive order, from the left to right when facing the vehicle, separated by a dash 1 inch long. Markings may be on a single line or on two lines, but in no case will groups be broken nor order of groupings be changed. If two lines are used, the first and third groups will be on the top row. Composition of groups is indicated below.

(b) *First group.*—The first group will designate the smallest appropriate unit listed below in accordance with the following code:

MARKING OF CLOTHING, EQUIPMENT, VEHICLES, PROPERTY   AR 850-5   10

Unit	Designation
Division (infantry)	Arabic numeral.
Division (armored)	Arabic numeral followed by triangle 3 inches high with ¼-inch stroke.
Division (cavalry)	Arabic numeral followed by letter C.
Corps (army)	Roman numeral.
Corps (cavalry)	Roman numeral followed by letter C.
Corps (armored)	Roman numeral followed by triangle 3 inches high with ¼-inch stroke.
Army	Arabic numeral followed by letter A.
Air force	Arabic numeral followed by a star 3 inches high.
Zone of communications.	ZC.
Army Ground Forces	AGF.
Services of Supply	SOS.
General Headquarters	GHQ.
Zone of interior	ZI.
Reception center	RC.
Replacement training center.	RTC preceded by arm or service symbol.
Training center	TC preceded by arm or service symbol.
Firing center	FC preceded by arm or service symbol.
All others	Nonconflicting letters.

(c) *Second group.*—The second group will designate separate regiment, separate brigades, groups, separate battalions, or separate companies, and similar units by appropriate number of symbol, followed by arm or service in accordance with abbreviations listed below. When indicating headquarters and headquarters companies or special companies of units identified in first group, the second group will consist only of the letter "X." When indicating brigades, the numeral will be underlined.

MISCELLANEOUS

Arm or service	Designation
Airborne	AB
Army Air Forces units	Star 3 inches high.
Antiaircraft	AA
Amphibious	AM
Armored regiment	Triangle 3 inches high with ¼-inch stroke.
Cavalry	C
Chemical Warfare Service	G
Coast Artillery Corps	CA
Corps of Engineers	E
Field Artillery	F
Infantry	I (preceded by dash ½-inch square)
Medical Department	M
Military police	P
Ordnance Department	O (preceded by dash ½-inch square).
Quartermaster Corps	Q
Signal Corps	S
Tank Destroyer	TD
Tank group	TG

(d) *Third group.*—The third group will designate companies and similar organizations by letters in accordance with the following code:

Organization	Designation
Headquarters and headquarters company (or headquarters and headquarters and service company) of lowest unit identification in previous groups.	HQ
Service company of lowest unit identified in service groups.	SV
Headquarters and headquarters company of battalion not previously identified.	Numerical designation of battalion followed by letters HQ.
Service company of battalion not previously identified.	Battalion number followed by letters SV.
Lettered company	Letter designation.
Separate company identified in second group.	X or abbreviation of company.
Antitank	AT
Maintenance	MT
Heavy weapons	HW
Cannon	CN
Reconnaissance	R
Train	TN
Weapons	W
"Name" company (other than headquarters company, headquarters and service company or service company).	Nonconflicting letters assigned for identification purposes, preceded by the battalion number, when necessary.

MARKING OF CLOTHING, EQUIPMENT, VEHICLES, PROPERTY

AR 850-5

(e) *Fourth group.*—The fourth group will designate the serial number of the vehicle in normal order of march within the organization to which it is assigned. Vehicles assigned to any headquarters will be combined for purposes of numbering with those of the appropriate headquarters company or similar organization, and will be given the smaller serial numbers used therein.

(2) *Examples.*

(a)
1—X—HQ—10	10th vehicle, Hq Co, 1st Inf Div.
1—X—1S—10	10th vehicle, 1st Sig Co, 1st Inf Div.
1—X—1R—10	10th vehicle, 1st Rcn Tr, 1st Inf Div.
1—16-I—A—10	10th vehicle, Co A, 16th Inf, 1st Inf Div.
1—F—HQ—10	10th vehicle, Hq Btry, Div Arty, 1st Inf Div.
1—33F—D—10	10th vehicle, Btry D, 33d FA Bn, 1st Inf Div.
1—1E—A—10	10th vehicle, Co A, 1st Engr Bn, 1st Inf Div.
1—1M—A—10	10th vehicle, Co A (Coll), 1st Med Bn, 1st Inf Div.
1—1Q—A—10	10th vehicle, Co A, 1st QM Bn, 1st Inf Div.

(b)
1△—X—SV—10	10th vehicle, Serv. Co, 1st Armd Div.
1△—81R—A—10	10th vehicle, Co A, 81st Rcn Bn, 1st Armd Div.
1△—1△—A—10	10th vehicle, Co A, 1st Armd Regt, 1st Armd Div.
1△—27F—A—10	10th vehicle, Btry A, 27th FA Bn, 1st Armd Div.
1△—16E—A—10	10th vehicle, Co A, 16th Engr Bn, 1st Armd Div.
1△—6-I—A—10	10th vehicle, Co A, 6th Inf, 1st Armd Div.
1△—TN—HQ—10	10th vehicle, Hq Co, Div Tn, 1st Armd Div.
1△—1-O—A—10	10th vehicle, Co A, Maint Bn, 1st Armd Div.
1△—1Q—A—10	10th vehicle, Co A, Sup Bn, 1st Armd Div.

(c)
1C—X—HQ—10	10th vehicle, Hq Tr, 1st Cav. Div.
1C—X—1S—10	10th vehicle, 1st Sig Tr, 1st Cav Div.
1C—91R—A—10	10th vehicle, Tr A, 91st Rcn Sq, 1st Cav Div.
1C—X—27-O—10	10th vehicle, 27th Ord Co, 1st Cav Div.
1C—1C—HQ—10	10th vehicle, Hq Tr, 1st Cav Brig, 1st Cav Div.
1C—12C—HQ—10	10th vehicle, Hq Tr, 12th Cav, 1st Cav Div.

AR 850–5

MISCELLANEOUS

	1C—F—HQ—10	10th vehicle, Hq Btry, Div Arty, 1st Cav Div.
	1C—61F—A—10	10th vehicle, Btry A, 61st FA Bn, 1st Cav Div.
	1C—8E—A—10	10th vehicle, Tr A, 8th Engr Bn, 1st Cav Div.
	1C—1M—A—10	10th vehicle, Tr A, 1st Med Sq, 1st Cav. Div.
	1C—1Q—A—10	10th vehicle, Tr A, 1st QM Sq, 1st Cav Div.
(d)	2A—21Q—A—10	10th vehicle, Co A, 21st QM Regt, Second Army.
	3A—61Q—A—10	10th vehicle, Co A, 61st QM Bn, Third Army.
	★—1Q—SVAV—10	10th vehicle, 1st QM Co, Serv Gp, (Avn) AAF.
	3A—21Q—CAR—10	10th vehicle, 21st QM Car Co, Third Army.
	3A—56Q—COM—10.	10th vehicle, 56th QM Sales Comm Co, Third Army.
	3A—67Q—REF—10	10th vehicle, 67th QM Ref Co, Third Army.
	4A—79Q—DP—10	10th vehicle, 79th QM Depot Co, Fourth Army.
	1A—87Q—RHD—10	10th vehicle, 87th QM Rhd Co, First Army.
	1A—132Q—TRK—10.	10th vehicle, 132d QM Trk Co, First Army.
	★—862Q—MT—10	10th vehicle, 862d QM L Maint Co, AAF.
(e)	GHQ—101M—A—10.	10th vehicle, Co A, 101 Med Regt, GHQ.
	II—40M—A—10	10th vehicle, Co A, 40th Med Bn, II Army Corps.
	3A—1M—VET—10	10th vehicle, 1st Vet Co, Third Army.
	II—701M—SAN—10.	10th vehicle, 701st Sn Co, II Army Corps.
	II—3M—EV—10	10th vehicle, 3d Evac Hosp, II Army Corps.
	3A—53M—GEN—10.	10th vehicle, 53d Gen Hosp, Third Army.
	1A—151M—STA—10.	10th vehicle, 151st Sta Hosp, First Army.
	3A—63M—SUR—10.	10th vehicle, 63d Surg Hosp, Third Army.
	3A—16M—VET—10	10th vehicle, 16th Vet Evac Hosp, Third Army.
	1A—1M—DP—10	10th vehicle, 1st Med Sup Dep, First Army.

MARKING OF CLOTHING, EQUIPMENT, VEHICLES, PROPERTY

	3A—2M—LAB—10	10th vehicle, 2d Med Lab, Third Army.
(f)	1A—101P—A—10	10th vehicle, Co A, 101st MP Bn, First Army.
	ZI—704P—A—10	10th vehicle, Co A, 704th MP Bn, ZI.
	IV—X—44P—10	10th vehicle, 44th MP Co, IV Army Corps.
(g)	IX—18E—A—10	10th vehicle, Co A, 18th Engr, IX Army Corps.
	II—62E—D—10	10th vehicle, Co D, 62d Top Engr, II Army Corps.
(h)	1A—41-O—A—10	10th vehicle, Co A, 41st Ord Bn, First Army.
	I—1-O—HQ—10	10th vehicle, Hq & Hq Det, 1st Ord Bn, I Army Corps.
	3A—X—60-O—10	10th vehicle, 60th Ord Co, Third Army.
	★—714-O—AB—10	10th vehicle, 714 Ord Co, AB, AAF.
(i)	II△—1S—A—10	10th vehicle, Co A, 1st Sig Bn, I Armd Corps.
	III—26S—A—10	10th vehicle, Co A, 26th Sig Bn, III Army Corps.
	★—X—S—10	10th vehicle, Sig Co, AAF.
	III—X—280S—10	10th vehicle, 280th Sig Co, III Army Corps.
	ZI—1S—SV—10	10th vehicle, 1st Sig Serv Co, ZI.
(j)	1A—2G—A—10	10th vehicle, Co A, 2d Cml Bn, First Army.
	3A—3G—MT—10	10th vehicle, 3d Cml Co Maint, Third Army.
	3A—3G—DP—10	10th vehicle, 3d Cml Co, Dep, Third Army.
(k)	II—13F—HQ—10	10th vehicle, Hq Btry, 13th FA Brig, II Army Corps.
	I—17F—A—10	10th vehicle, Btry A, 17th FA, I Army Corps.
	1A—6F—A—10	10th vehicle, Btry A, 6th FA Bn, First Army.
	GHQ—71F—A—10	10th vehicle, Btry A, 71st FA Bn, GHQ.
	I—1FOB—A—10	10th vehicle, Btry A, 1st FA Obsn Bn, I Army Corps.
(l)	3A—33CA—HQ—10	10th vehicle, Hq Btry, 33 CAC Brig, Third Army.
	1A—2CA—A—10	10th vehicle, Btry A, 2d CAC, First Army.
	AA—302CA—A—10	10th vehicle, Btry A, 302 CAC, AA Comd.

AR 850-5
10-11 MISCELLANEOUS

    (*m*)   2A—601TD—A—10. 10th vehicle, Co A, 601TD Bn, Second Army.

           TD—693TD—A—10. 10th vehicle, Co A, 693 TD Bn, TD Comd.

    (*n*)   ★—850E—A—10. 10th vehicle, Co A, Engr Bn Avn, AAF.

    (3) *How marked.*—Unit identification markings will be painted in white letters on olive-drab background.

    (4) *Height of letters and figures.*—Letters and figures will be of a character style and dimensions as designed for use for traffic signs. Where space does not permit this size, letters and figures will conform to space available. See figure 35.

    (5) *Marking not required.*—Rear marking only is required on trailers, except on trailer, bomb, M5 and trailer, tractor, crane (see figs. 29 and 30). Unit marking will be placed on motorcycles when practicable.

    (6) *Removal of unit markings.*—The first two groups of unit markings will be removed from all motor vehicles when leaving home stations for movement to theaters of operations or ports of embarkation and in the theater of operations when directed by the theater commanders.

    (7) Administrative vehicles, motor vehicles assigned to administrative units or functions at posts, camps, and stations or similar activities will substitute abbreviations or words for the first two groups in order to show station and activity represented. Abbreviations will be those commonly used and where practical will conform to those prescribed in these regulations.

11. Tactical markings.—*a.* Division, separate brigades, combat commands, combat teams, similar separate unit commanders, and higher unit commanders may prescribe a system of tactical markings for units of their commands. This may include the naming of individual vehicles.

b. How marked.—Tactical markings will be painted on vehicles using gasoline solvent paints or paints approved by the War Department. Colors may be used. The headquarters directing use of tactical marking will specify the location of tactical marking.

c. Forms.—Tactical markings may consist of stripes, geometrical figures, combinations of geometrical figures, or combinations of geometrical figures and stripes.

d. Size.—Tactical markings will be of such size as to make ground to ground identification of vehicles possible.

e. Concealment.—Tactical markings will be removed when camouflage or concealment outweigh the requirements for recognition.

f. Restrictions.

    (1) The system of tactical markings prescribed will in no way represent the numerical designation or the distinctive insignia of any unit.

    (2) In a combat zone, the system of tactical markings will be changed whenever conditions require.

    (3) In a theater of operations, no written record will be made or published of the system of tactical markings.

12. Special markings.—*a. Markings, recruiting service vehicles.*—Motor vehicles of the Army Recruiting Service will be marked on the sides and rear of the body with dark blue block letters "U. S. ARMY RECRUITING SERVICE" at least 3 inches in height, and with the insignia of The Adjutant General's Department in approved colors in the center of the front door panels and in the middle of the rear below the lettering. On mobile recruiting stations (trailers), the lettering will include the words "MOBILE STATION No. —," and the insignia will be placed in the middle of the sides and rear below the lettering.

b. Red cross.—Both sides of the body of an ambulance will be marked with a Geneva cross, bright red, on a snow-white field, the cross to be 18 inches in the horizontal and vertical dimensions and each limb to represent a 6-inch square. It will be located in the center of the middle or advertising panel of the body. A Geneva cross on a snow-white field will be placed in the center of the top and on the outside. This cross will be of such size that the transverse arm will reach entirely across the top. On the visor, or directly above the windshield, will be marked in suitable size white letters, block style, the word AMBULANCE. On each side of the word AMBULANCE will be placed a small bright-red Geneva cross on a white field, or two small red Geneva crosses may be placed on either side of the windshield in a location that will not interfere with the operator's vision. On the perpendicular center line of each of the rear doors, below the windows, will be placed a 6-inch bright-red Geneva cross on an 8-inch white field. All lettering executed in white will be properly shaded to give depth.

c. Green cross.—Veterinary ambulances only will be marked as prescribed in *b* above for medical ambulances, except that—

    (1) Color of the cross will be green.

    (2) Marking will be placed as prescribed in *b* above, with the exception that the marking in the center of the top and in the rear will be omitted.

d. Caduceus.—A caduceus, the insignia of the Medical Department, in maroon, will be painted on both sides of the body of an ambulance below the lower molding and 7 inches to the rear of the front body as follows: A caduceus 6 inches in height outlined with a narrow white stripe will be painted with its anterior edge 7 inches from the rear edge of the side door. A drawing of the caduceus will be furnished on request by the Medical Department Equipment Laboratory, Carlisle Barracks, Pa. Two inches below this caduceus in 1-inch white letters, block style, in two lines, will be painted the words:

UNITED STATES
ARMY

e. Paints.—All paints used in marking ambulances will be of lusterless synthetic enamel of the proper colors prescribed above.

13. Animal-drawn.—*a. Letters "U. S."*

    (1) All animal-drawn vehicles, including canvas covers therefor, if any, will be marked with the letters "U. S. A."

    (2) *When and how marked.*

        (*a*) The marking required by (1) above will be done by the issuing agency before the vehicle is placed in service.

        (*b*) This marking will be done with lusterless blue paint by means of a stencil.

        (*c*) The marking required by (1) above will be placed on the vehicles on both sides unless otherwise prescribed as follows:

Type of vehicle	Location of marking
Ambulances	On the body, in the center of the space between prolongation of the first and second bows from the front.
Canvas covers	To appear in middle of cover, on both sides, when cover is in place on the vehicle; bottom of marking to be 24 inches above hem of cover.
Escort wagons	In the center of the panel abreast of the driver's seat.
Mountain wagon	In the center of the end of the driver's seat, if practicable; if not, as close thereto as practicable.
Types not enumerated above.	Same as mountain wagon.

b. Side marking.—Animal-drawn, tactical transportation will be marked on both sides in the same manner as prescribed for international and tactical marking of motor vehicles (pars. 10 and 11). The international marking will be placed to the rear of the wagon body and will take advantage of the flat surface presented but will not exceed 24 inches in size.

c. Tailgate marking.—The rear marking prescribed in paragraph 10*b*(5) for unit identification marking of motor vehicles will be painted directly upon the tailgate of the vehicle.

d. Animal-drawn vehicles which are not assigned to tactical units and are for general utility purposes at posts, camps, or stations will be given a serial number which will replace the marking prescribed in *b* and *c* above and in addition may be marked on both sides with the name of the activity concerned.

e. Red cross.
  (1) *Large.*
    (a) The marking prescribed in paragraph 7*d* for motor ambulances will also be placed upon animal-drawn medical ambulances.
    (b) The marking will be placed on the vehicles in three places, as follows:
      1. In the center of the outside of the rear half of the side curtains, on both sides.
      2. In the center of the top, outside.
      3. In the center of the rear curtain, outside.
  (2) *Small.*
    (a) In addition to the marking required by (1) above, medical ambulances will also be marked with a red cross 6 inches high and 6 inches wide, having arms 2 inches wide, centrally placed upon a white field 8 inches square.
    (b) This marking will be placed in the center of the space between the prolongation of the first and second bows from the rear on both sides.

f. Green cross.—Veterinary ambulance only will be marked as prescribed in *e* above for medical ambulances, except that—
  (1) Color of the cross will be green.
  (2) Marking will be placed as prescribed in *e* (1)(b) and (2)(b) above only, the markings in the center of the top and the center of the rear curtain being omitted.

MARKING OF CLOTHING, EQUIPMENT, VEHICLES, PROPERTY 13-14

g. Cuduceus.
  (1) Medical and veterinary ambulances only will be marked as prescribed in paragraph 7d for motor ambulances.
  (2) The marking required by (1) above will be placed on the body in the center of the space between the prolongation of the third and fourth bows from the front, on both sides.

14. **Tractors and tanks.**—*a. Tractors.*—The vehicles should be marked as prescribed in paragraphs 6 to 11, inclusive, for other motor vehicles so far as applicable, otherwise marking should be placed in a visible location as near as possible to that prescribed. In general, the international symbol will be placed as follows: 20-inch star on hood, 10-inch star on both sides of tool boxes, and 10-inch star on gas tank in rear.

b. Tanks and other combat vehicles.—Tanks and other combat vehicles armored or on which armament is mounted should be marked as prescribed in paragraphs 6 to 11 inclusive, for other motor vehicles so far as applicable, otherwise marking should be placed in a visible location as near as possible to that prescribed. See figures 33 and 34.

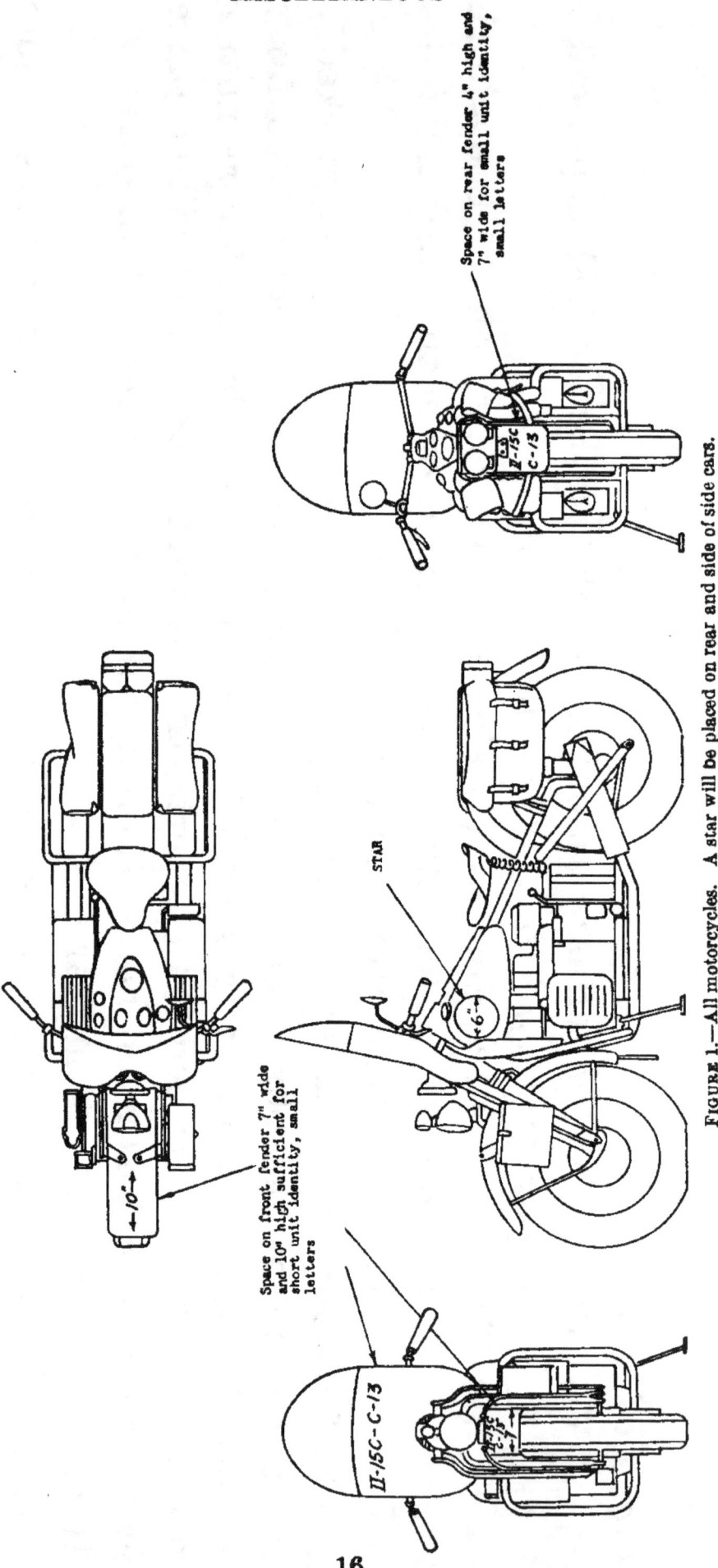

FIGURE 1.—All motorcycles. A star will be placed on rear and side of side cars.

AR 850–5
MARKING OF CLOTHING, EQUIPMENT, VEHICLES, PROPERTY 14

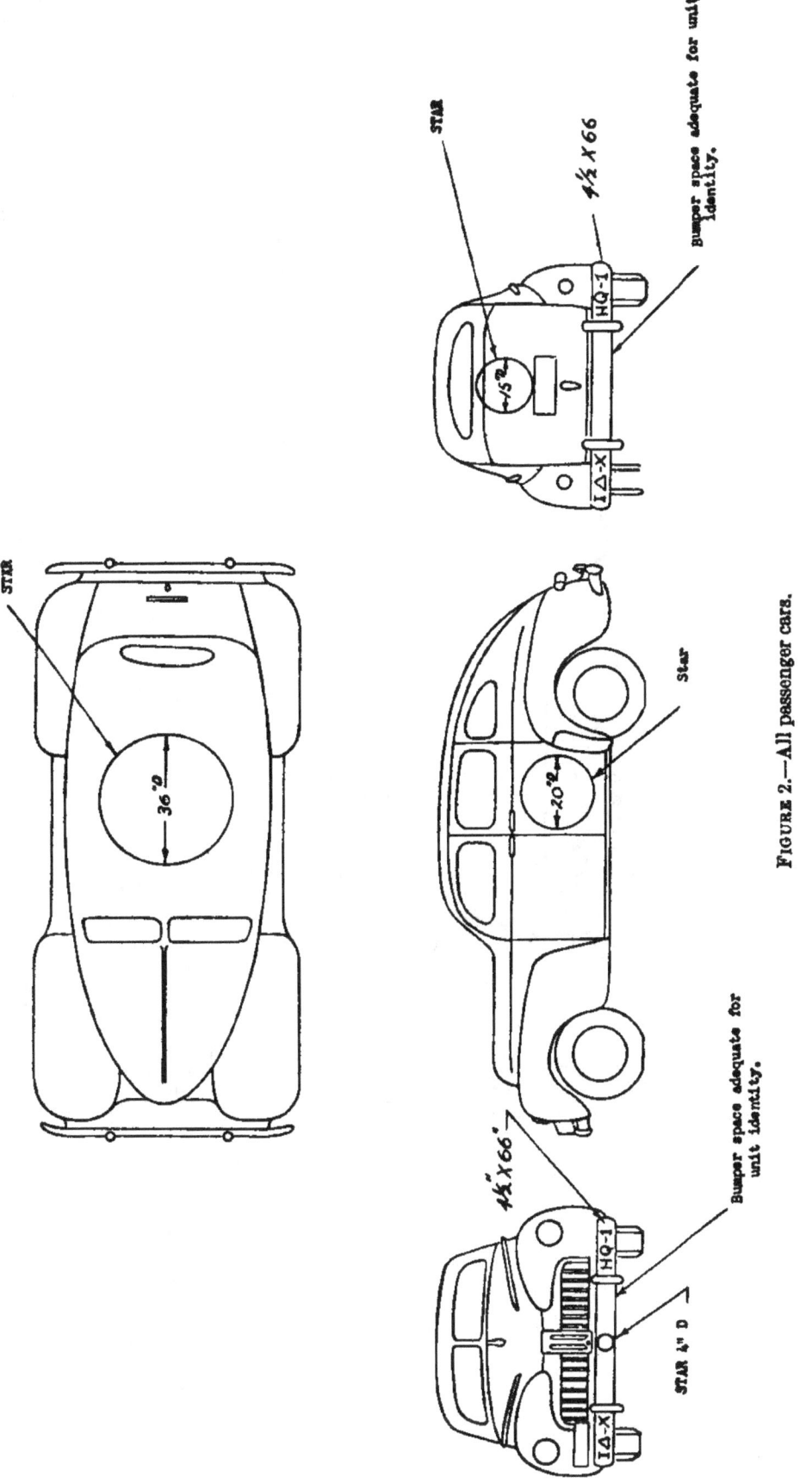

FIGURE 2.—All passenger cars.

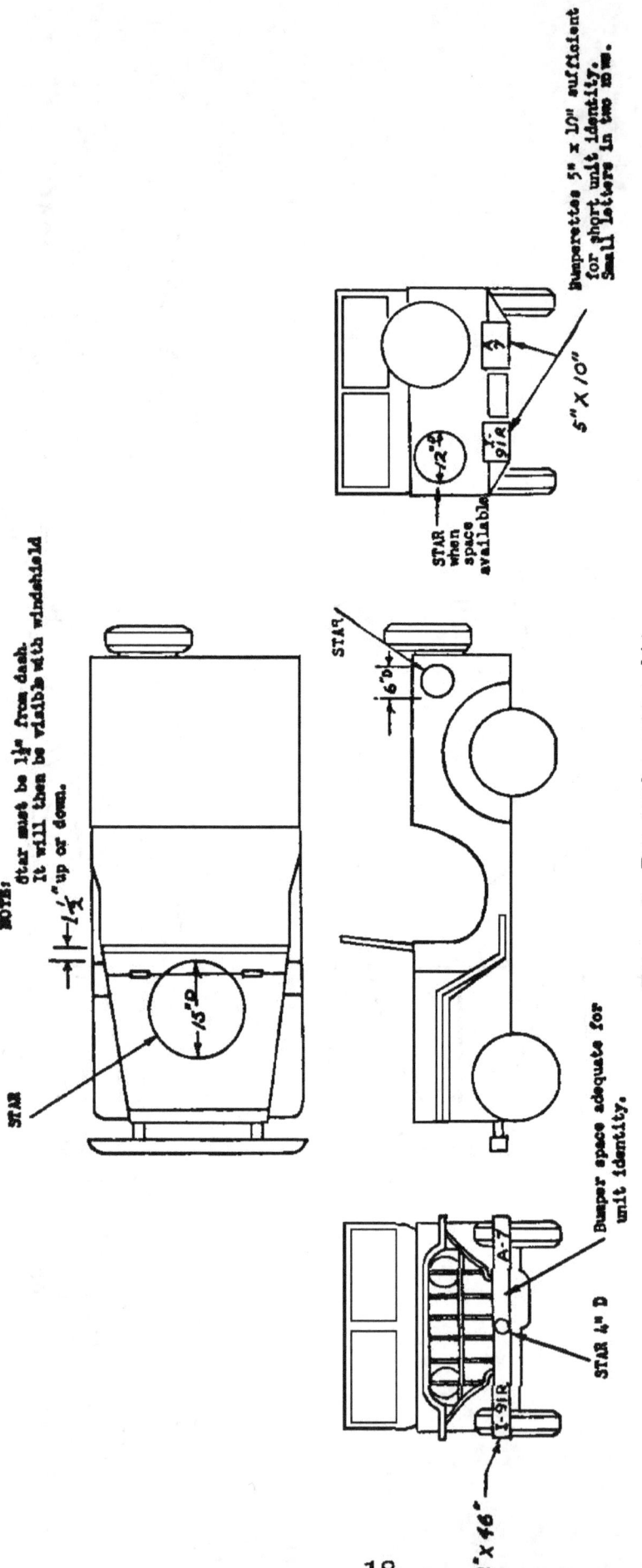

FIGURE 3.—Reconnaissance car, ¼-ton.

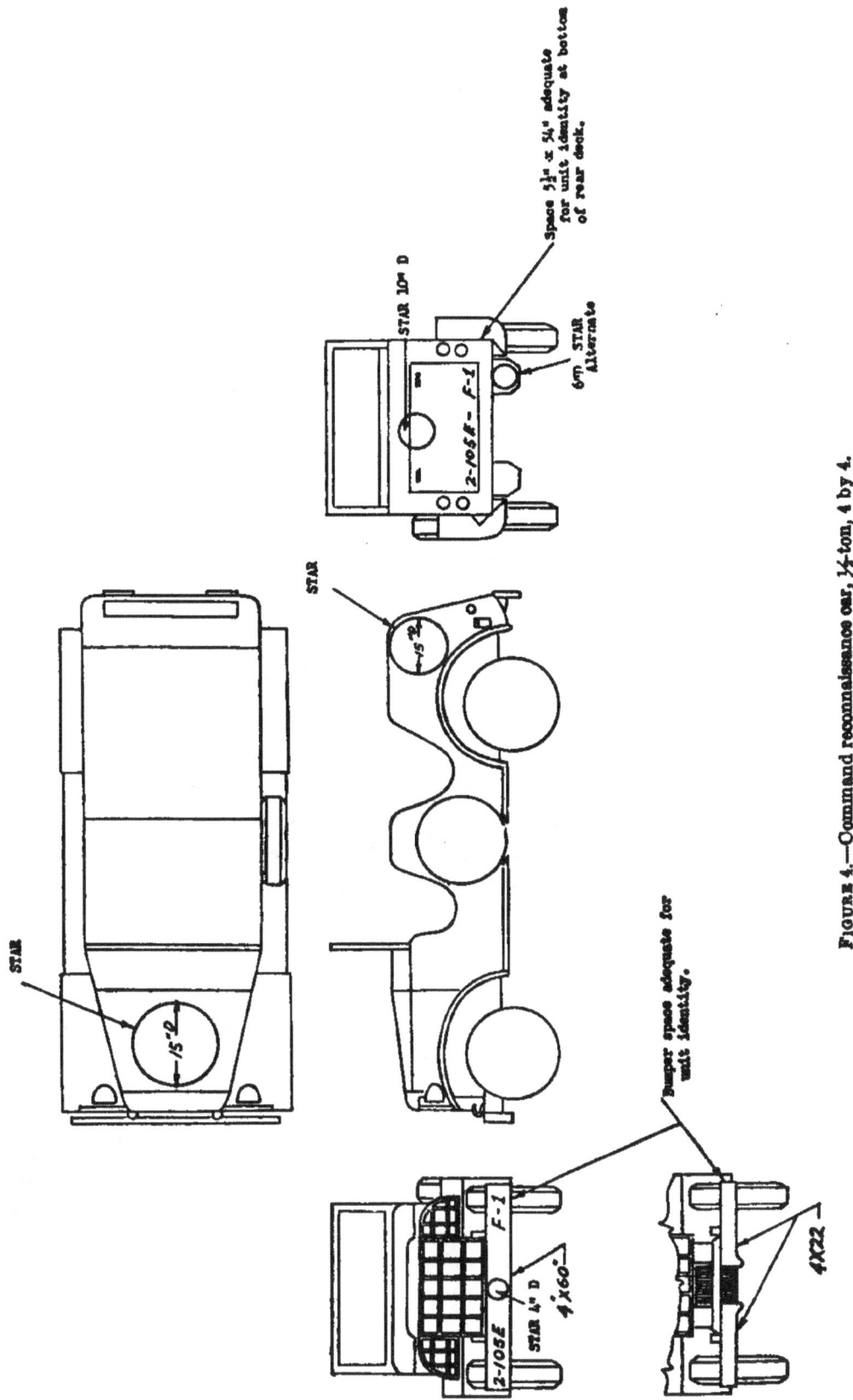

FIGURE 4.—Command reconnaissance car, ½-ton, 4 by 4.

AR 850–5
14 MISCELLANEOUS

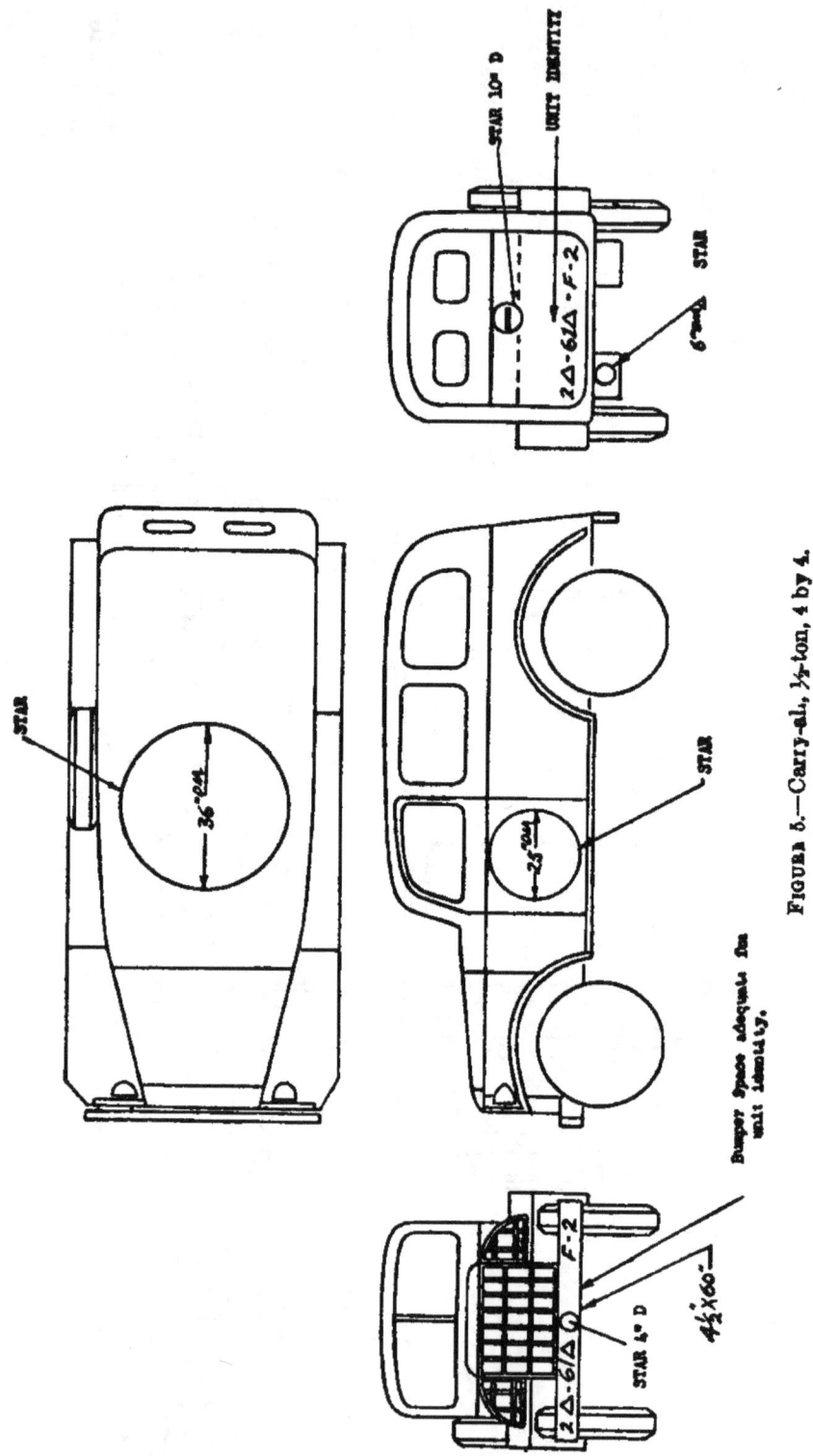

FIGURE 5.—Carry-al, ½-ton, 4 by 4.

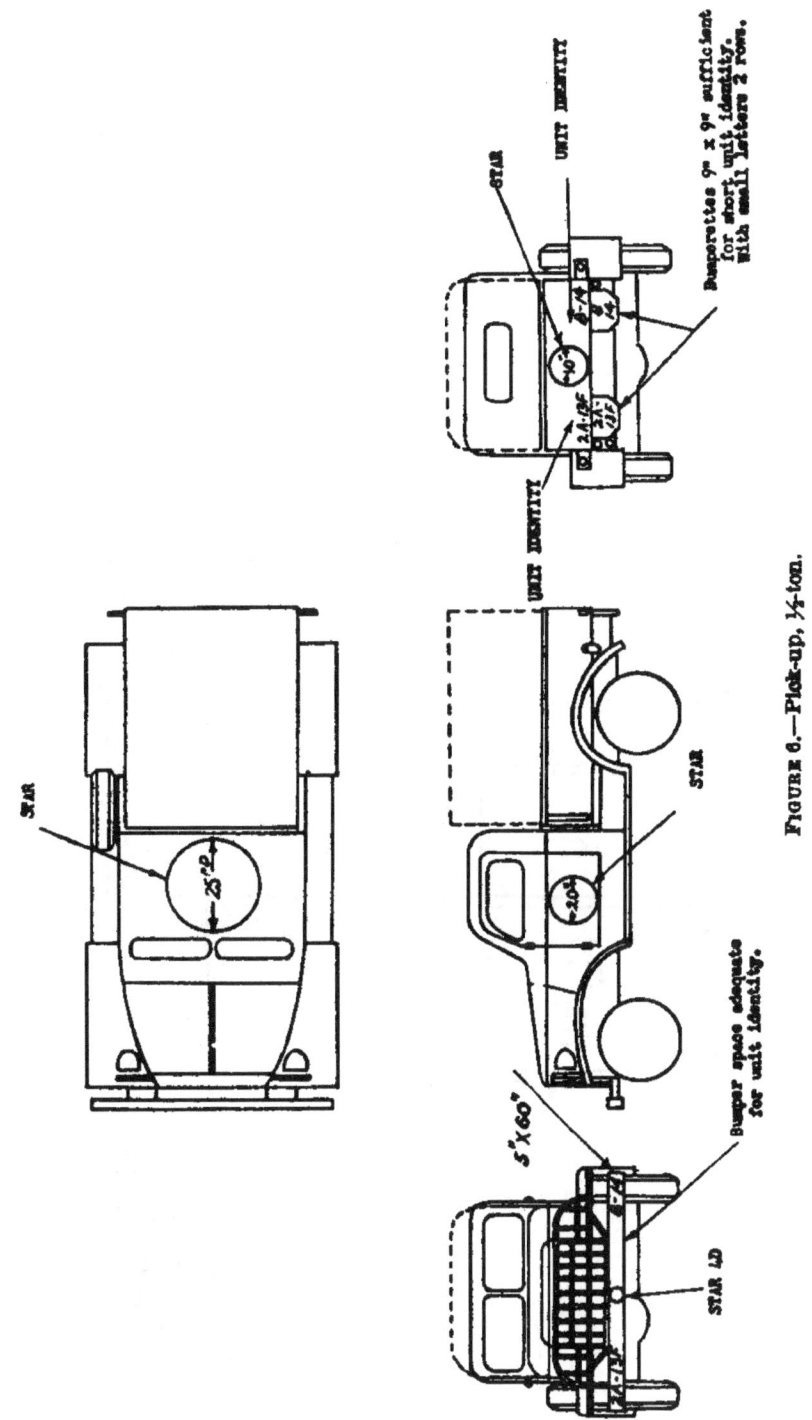

FIGURE 6.—Pick-up, ½-ton.

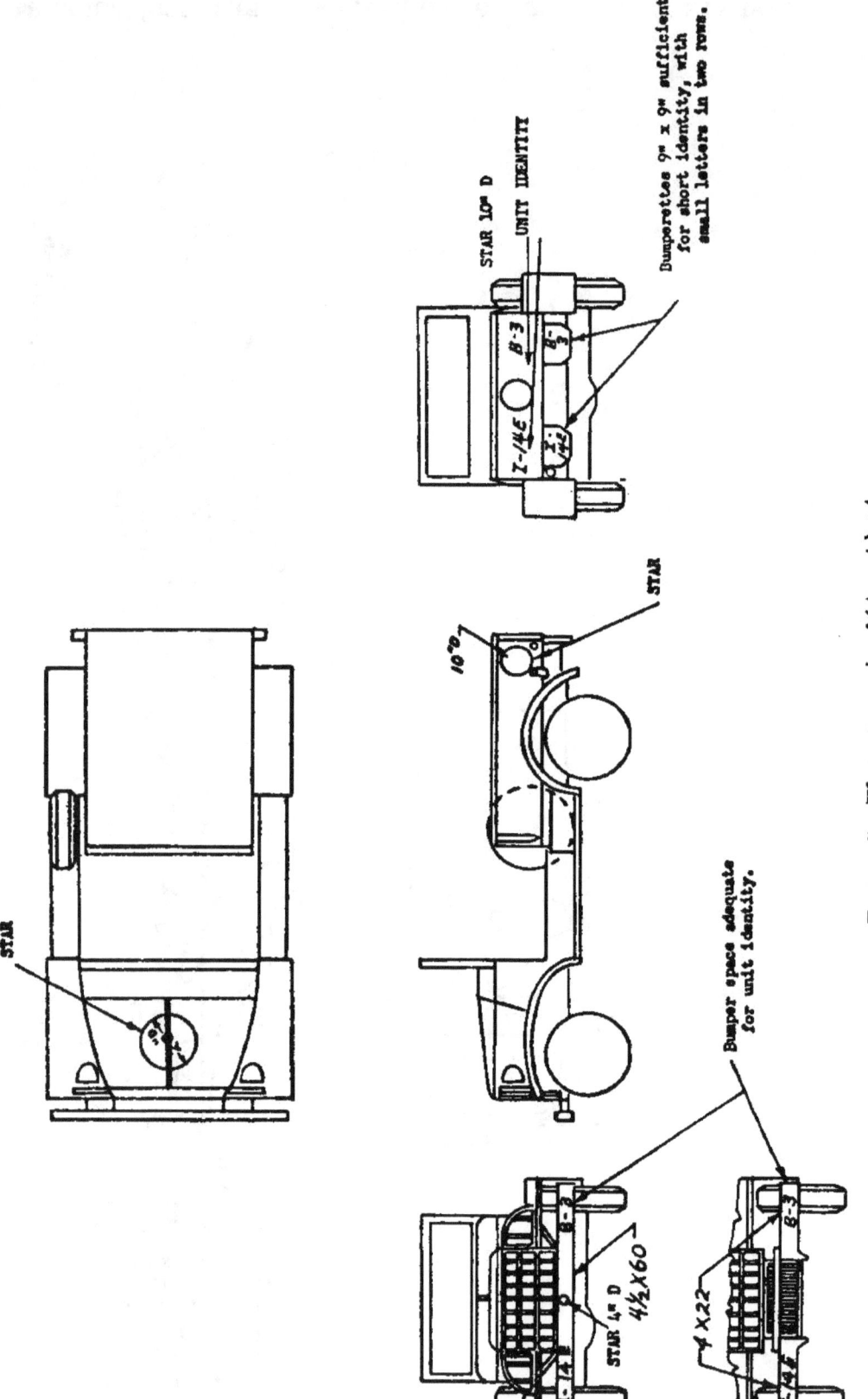

FIGURE 7.—Weapons carrier, ½-ton, 4 by 4.

MARKING OF CLOTHING, EQUIPMENT, VEHICLES, PROPERTY

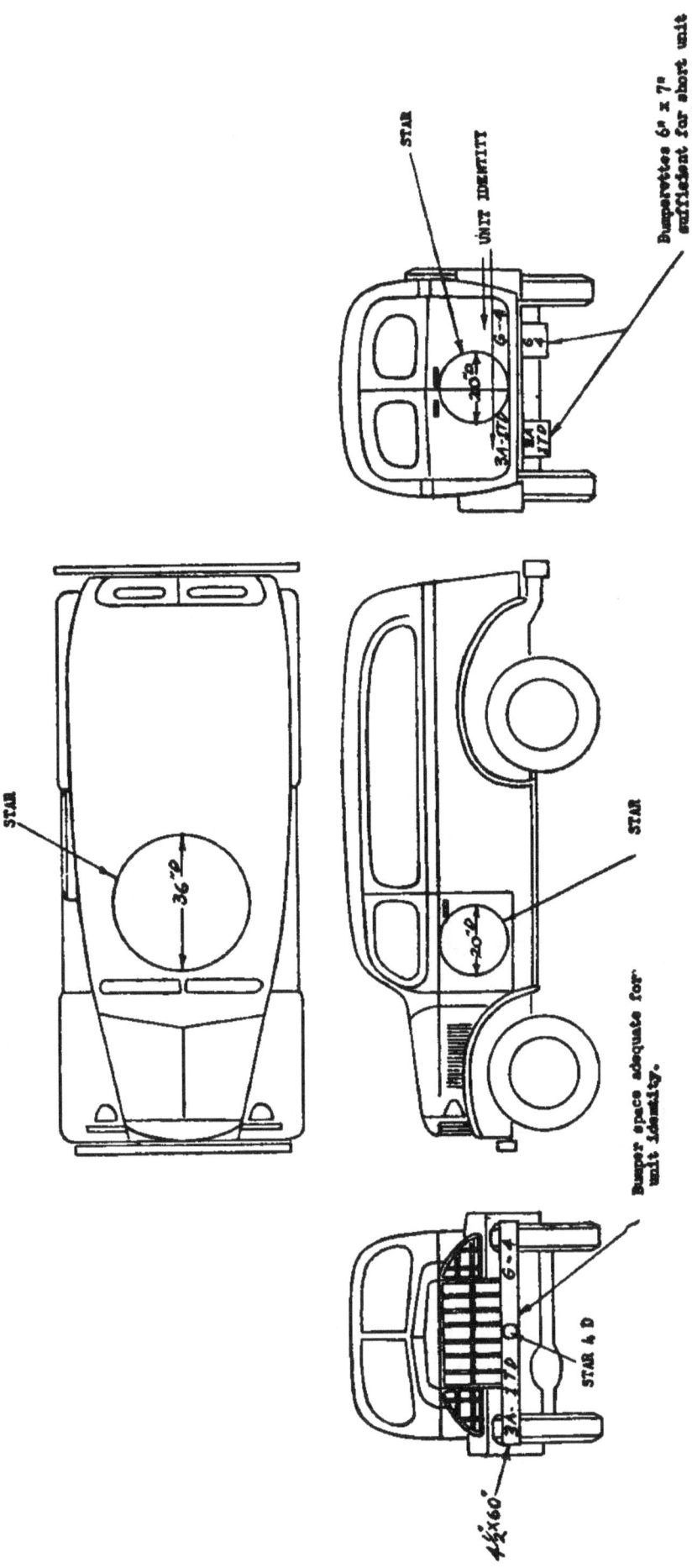

FIGURE 8.—Panel truck, ¼-ton, 4 by 4.

AR 850–5
14   MISCELLANEOUS

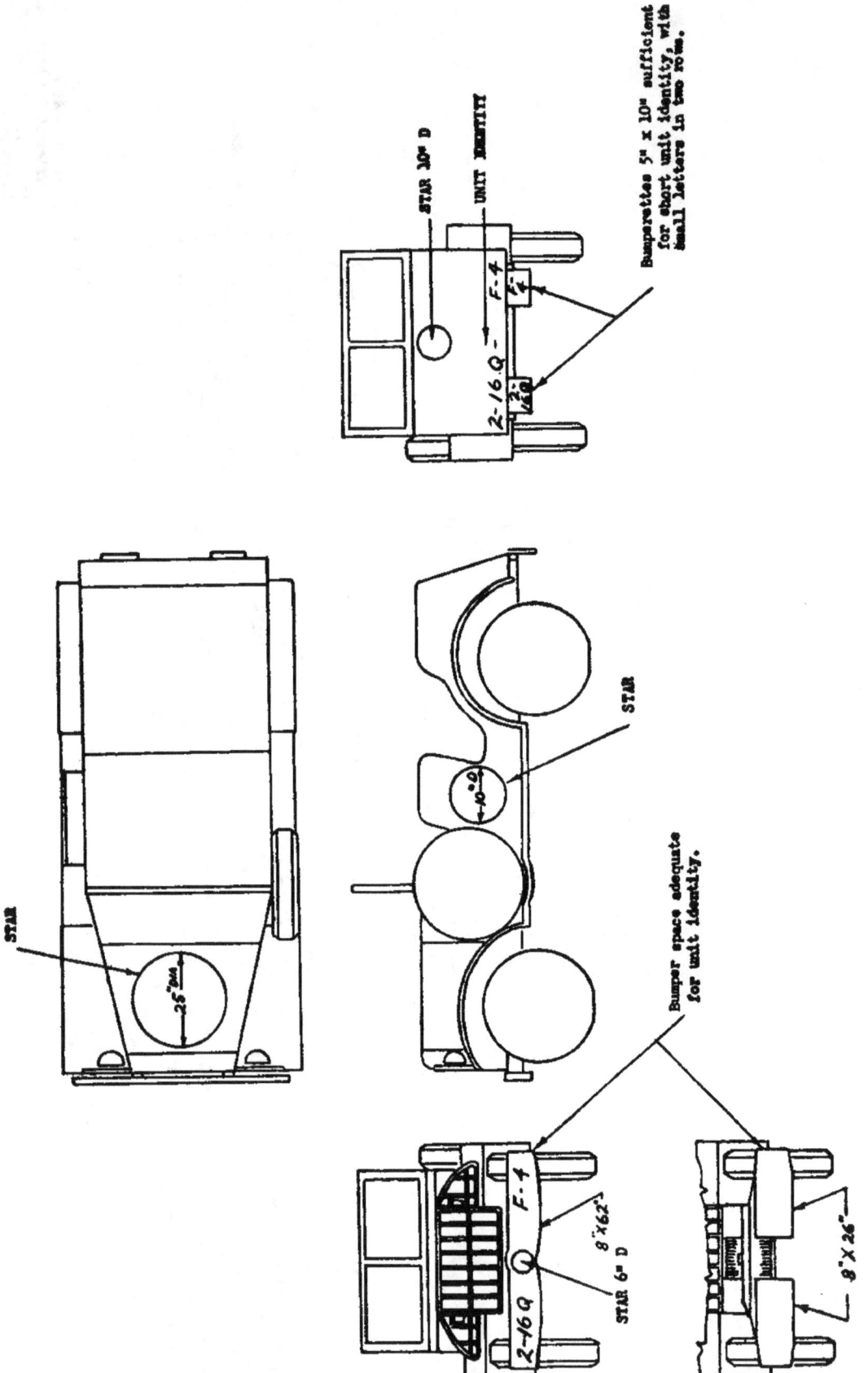

Figure 9.—Command reconnaissance car, ¾-ton, 4 by 4.

MARKING OF CLOTHING, EQUIPMENT, VEHICLES, PROPERTY

AR 850–5

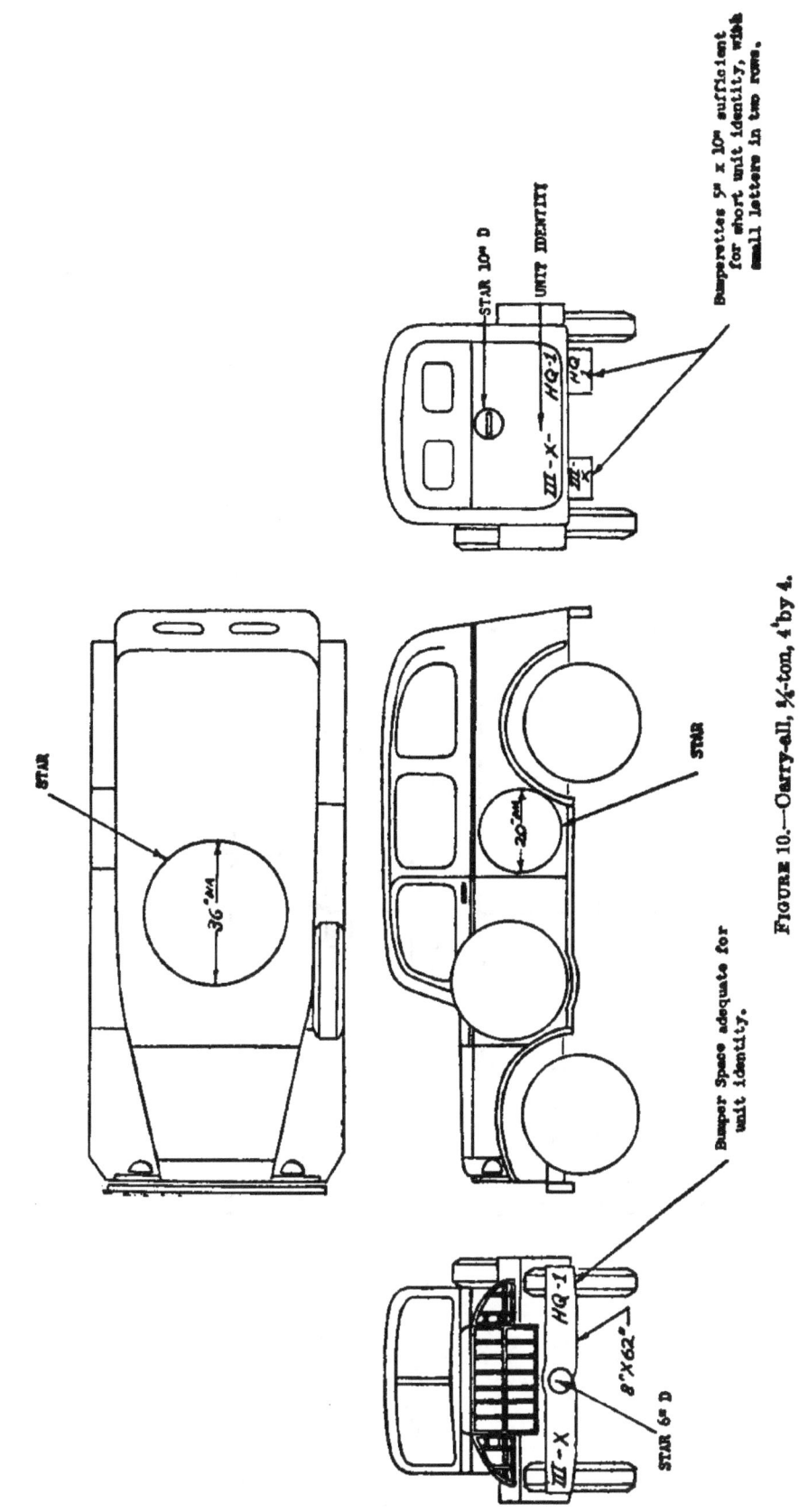

Figure 10.—Carry-all, ¾-ton, 4 by 4.

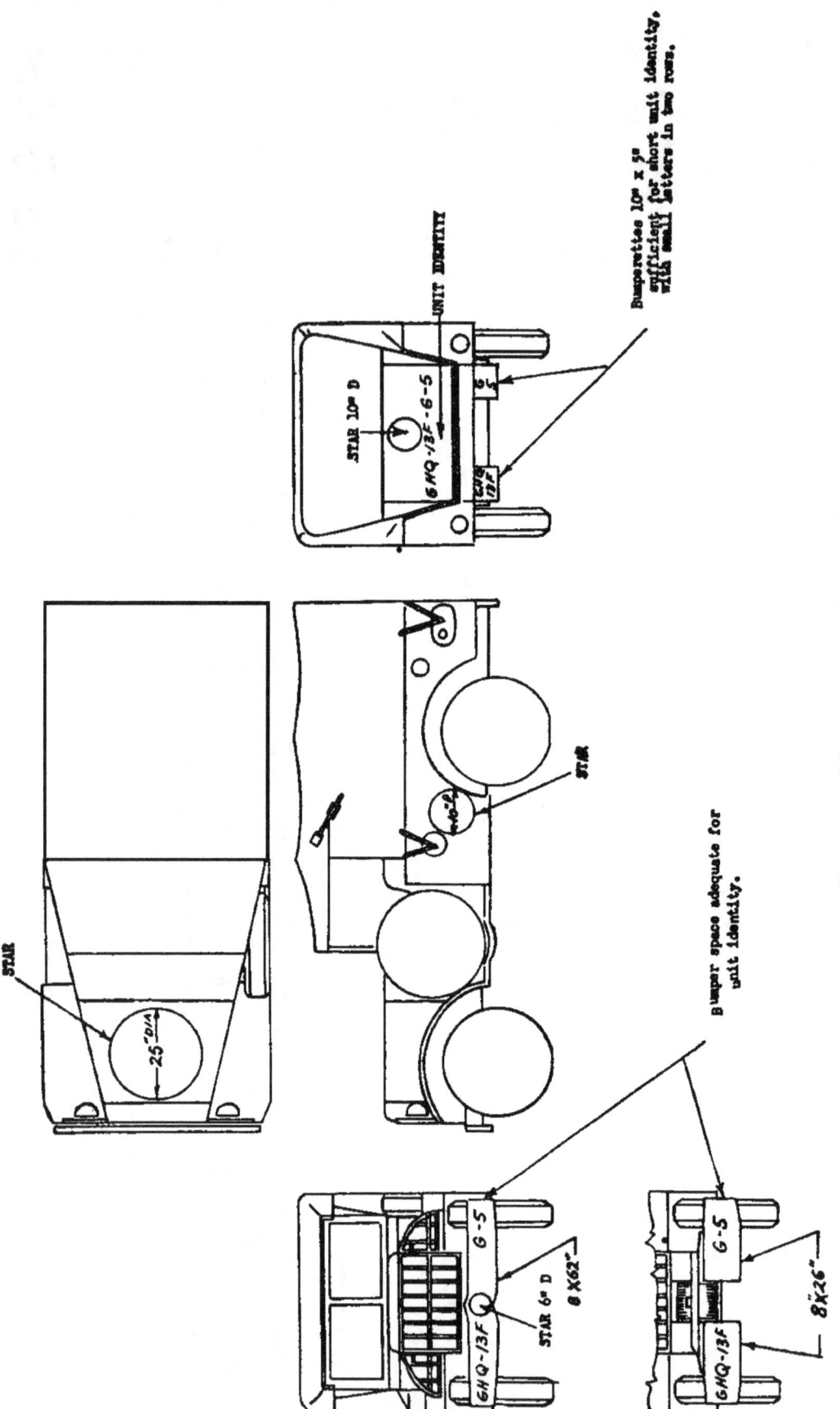

Figure 11.—Weapons carrier, ¾-ton, 4 by 4.

AR 850-5
MARKING OF CLOTHING, EQUIPMENT, VEHICLES, PROPERTY   14

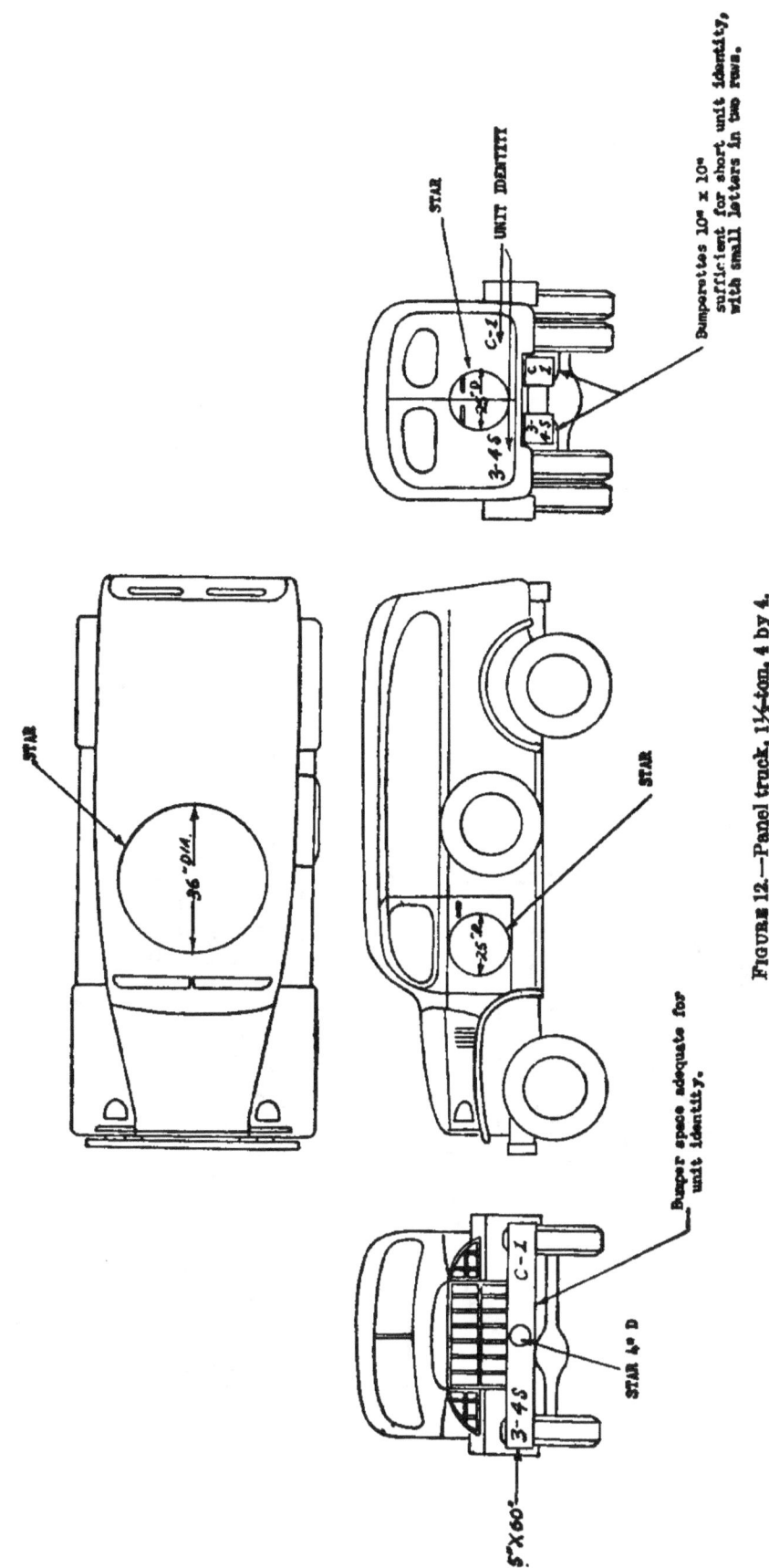

FIGURE 12.—Panel truck, 1¼-ton, 4 by 4.

AR 850-5
14 MISCELLANEOUS

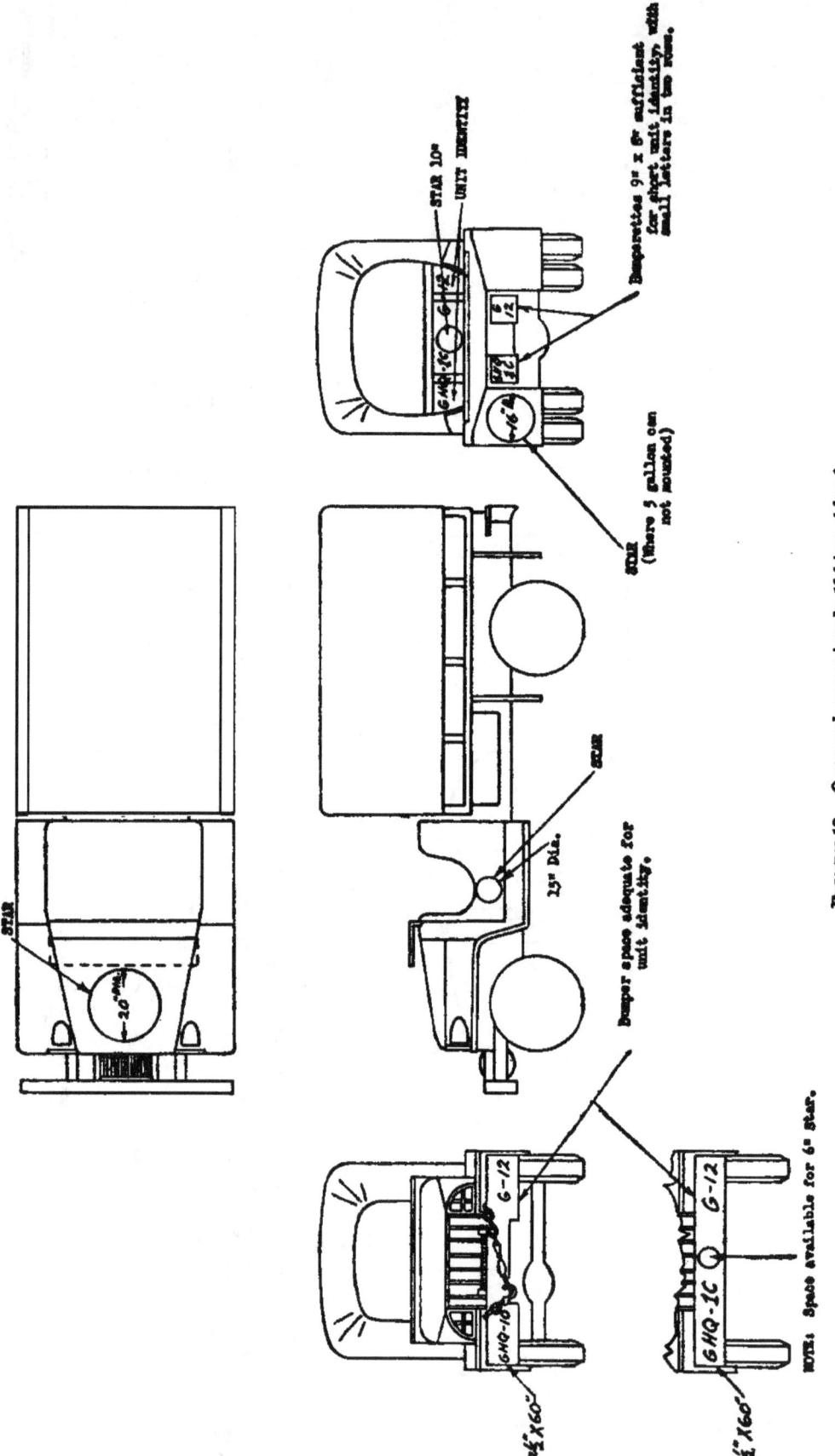

FIGURE 13.—Open cab cargo truck, 1½-ton, 4 by 4.

MARKING OF CLOTHING, EQUIPMENT, VEHICLES, PROPERTY **AR 850–5 14**

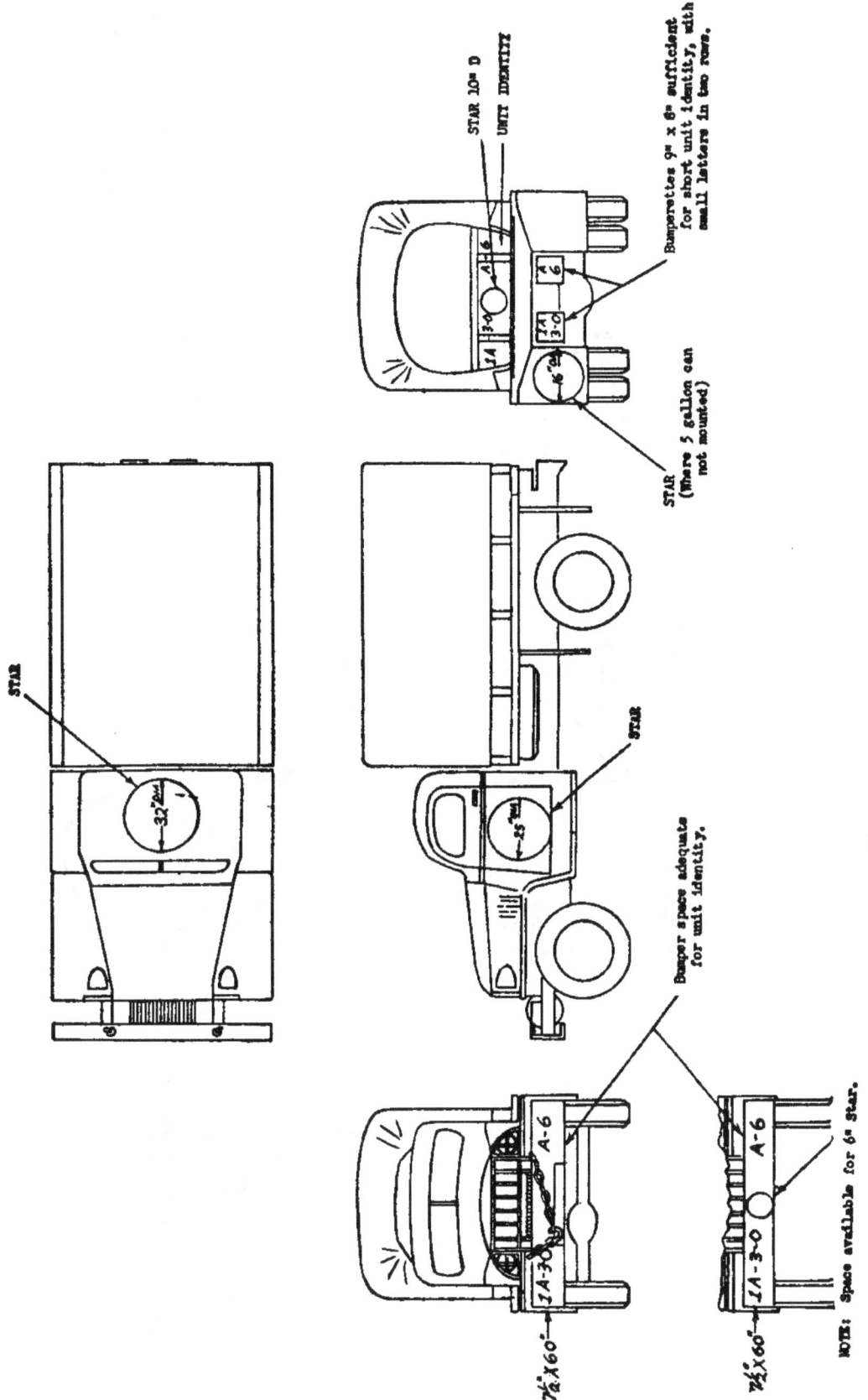

FIGURE 14.—Cargo truck, 1½-ton.

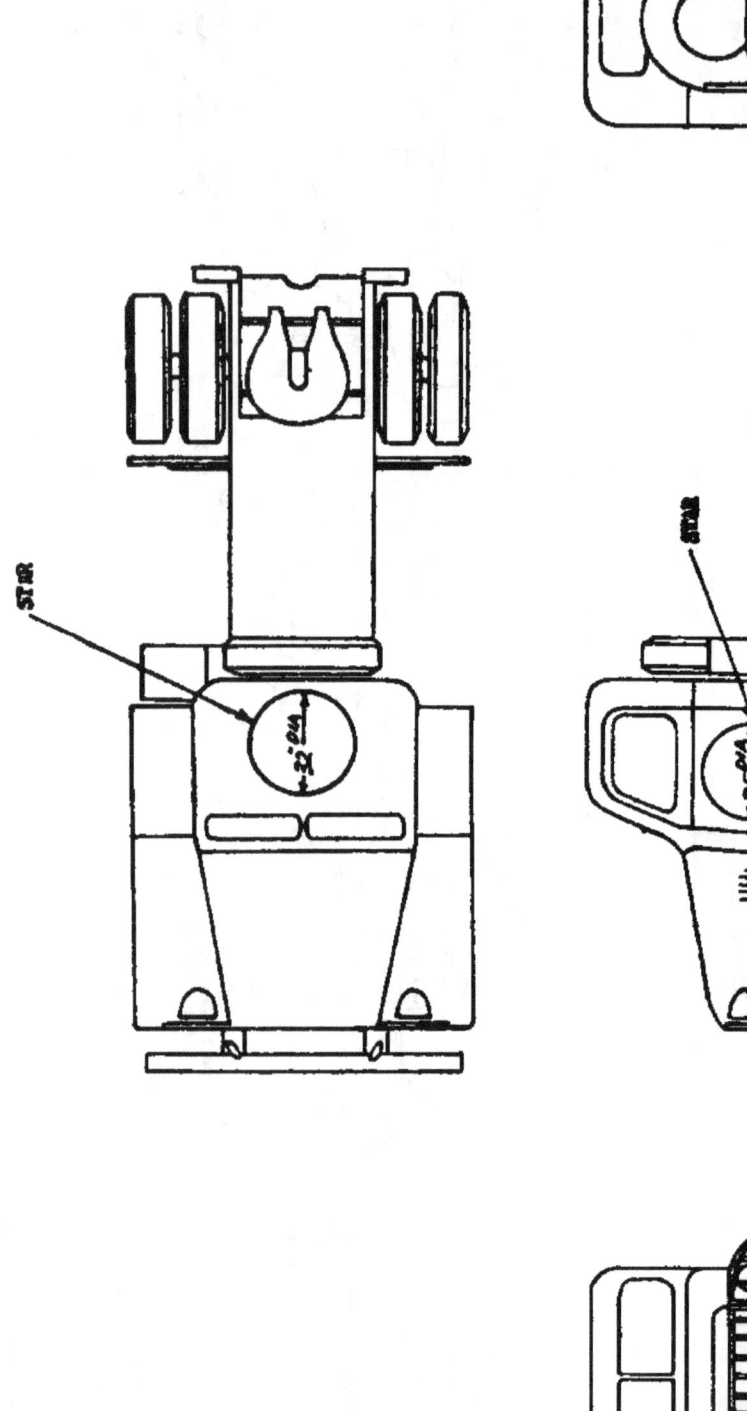
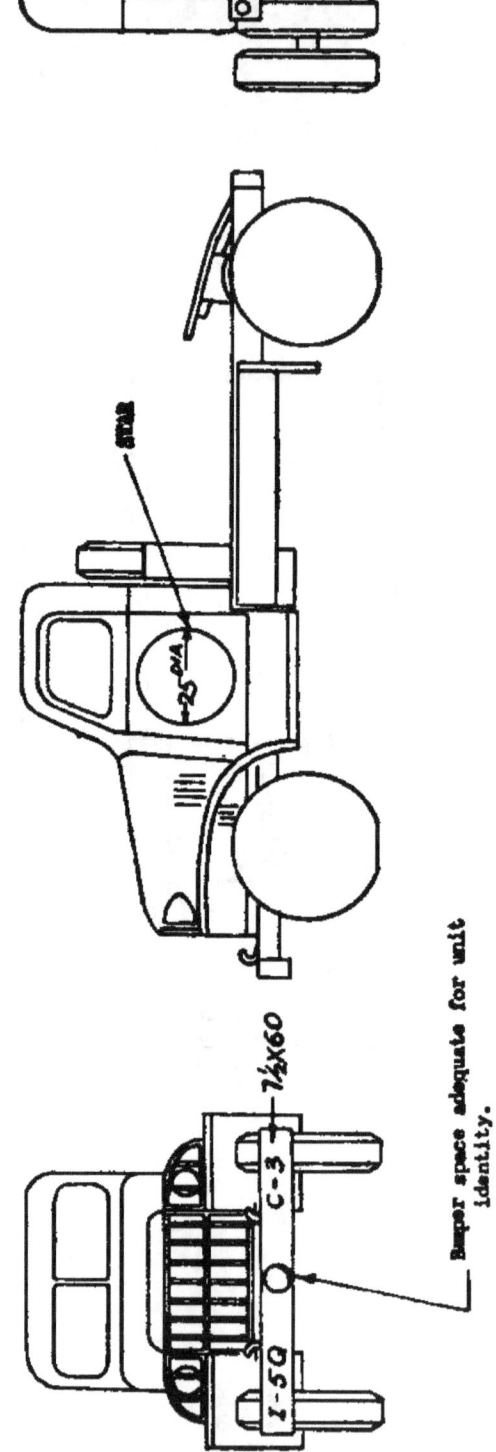

FIGURE 15.—Tractor truck, 1½-ton, 4 by 4.

MARKING OF CLOTHING, EQUIPMENT, VEHICLES, PROPERTY 14

AR 850-5

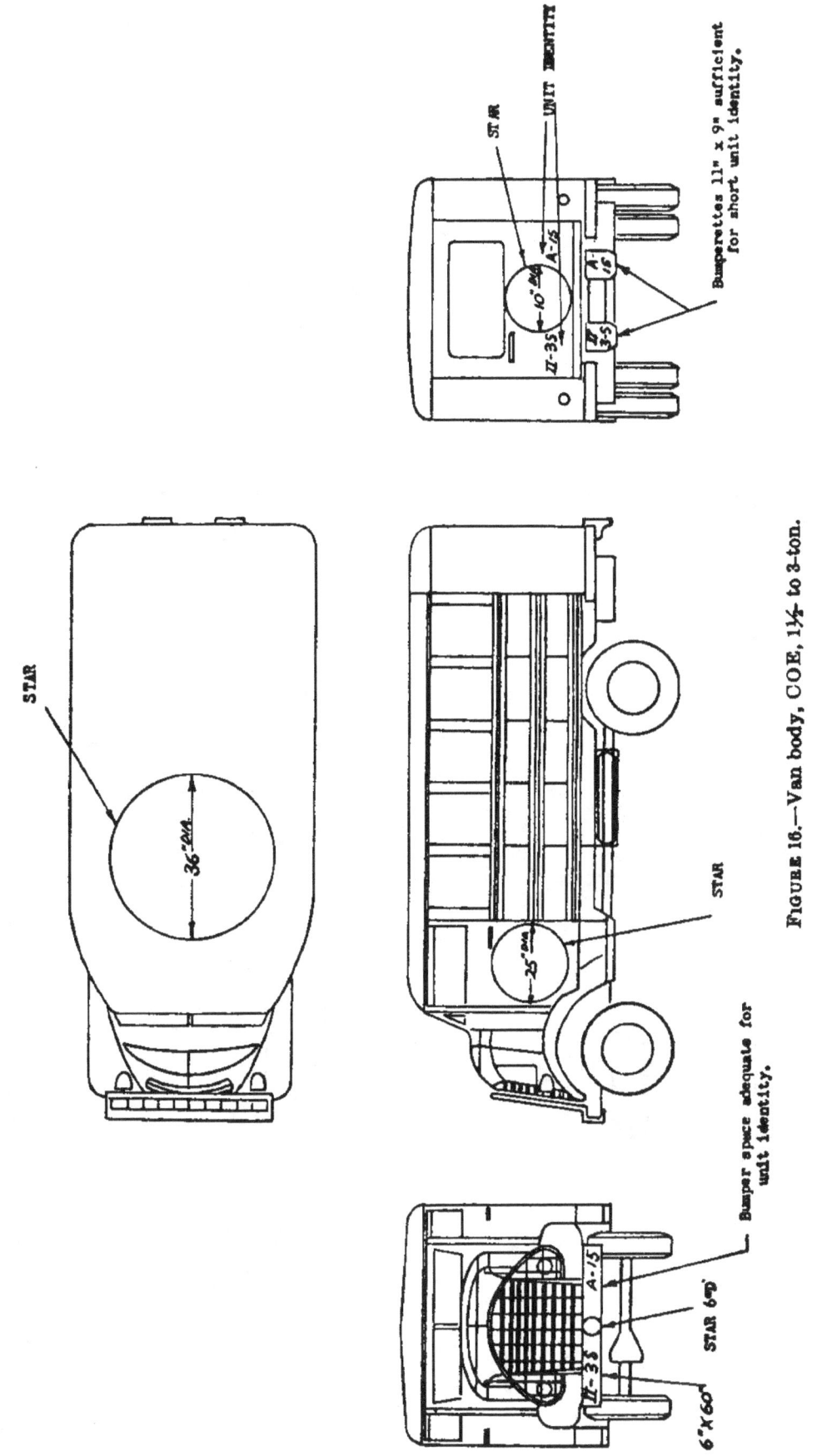

FIGURE 16.—Van body, COE, 1½ to 3-ton.

AR 850-5
14 MISCELLANEOUS

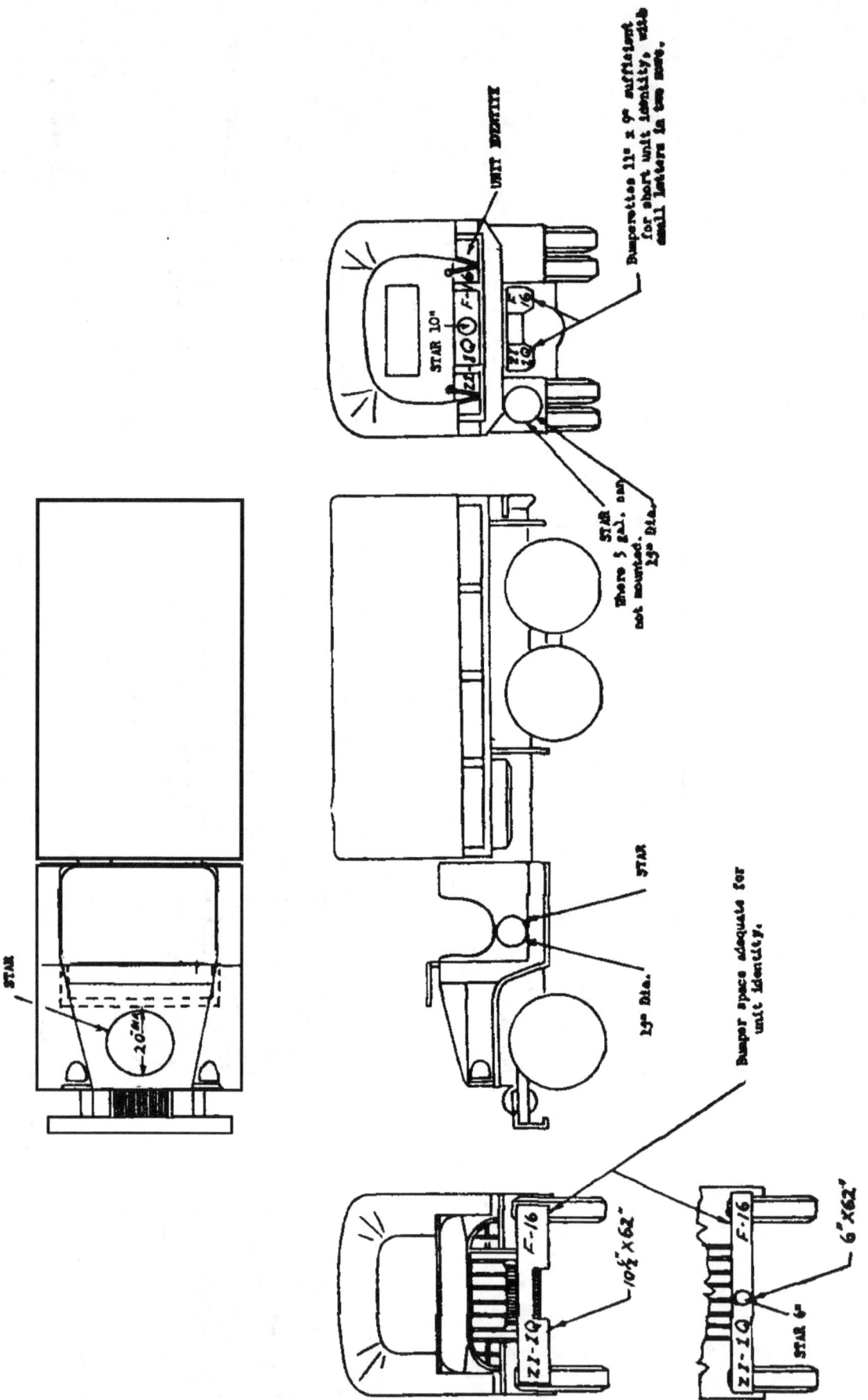

FIGURE 17.—Open cab cargo truck, 2½-ton, 6 by 6.

MARKING OF CLOTHING, EQUIPMENT, VEHICLES, PROPERTY **14**

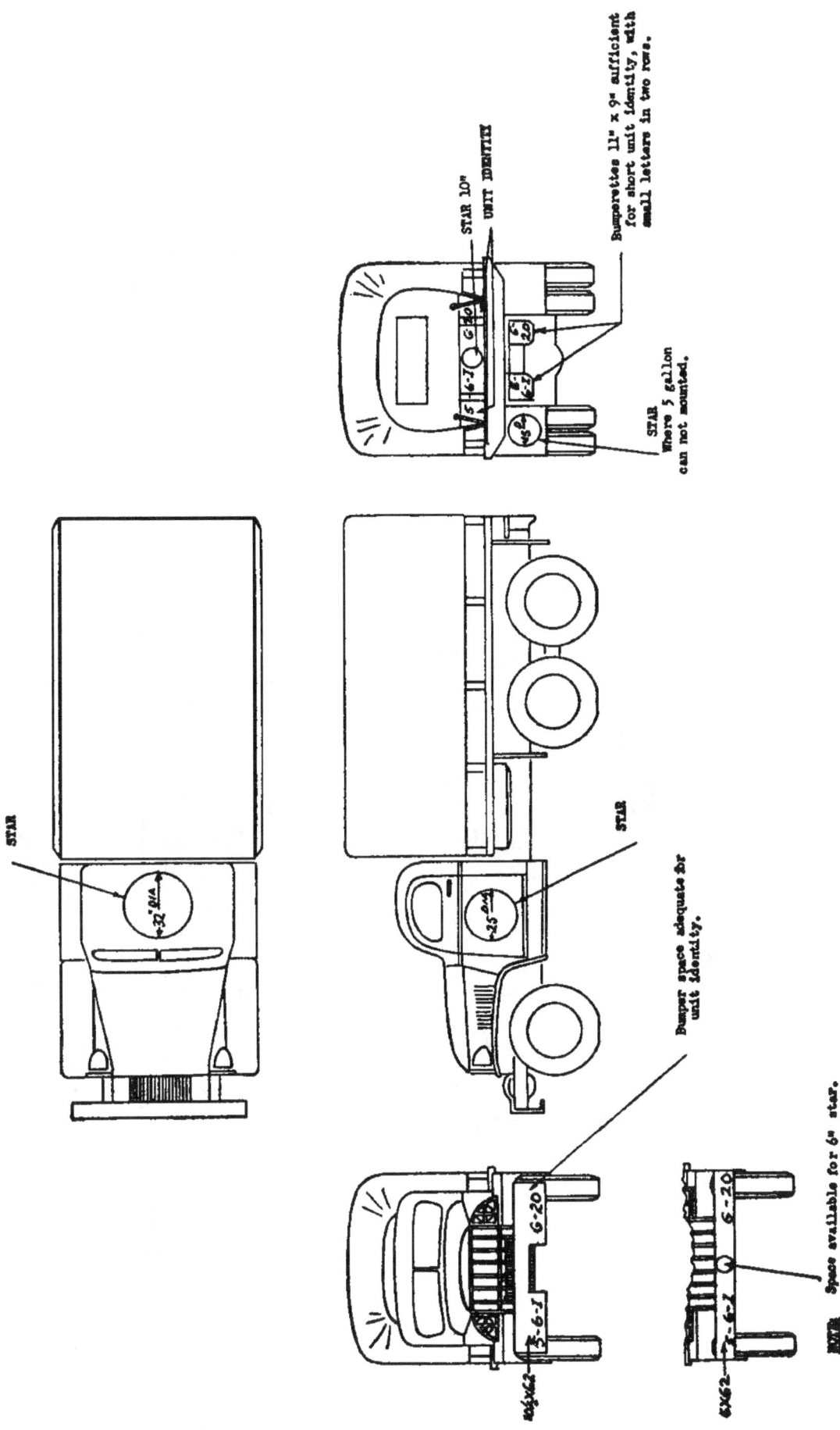

FIGURE 18.—Cargo truck, 2½-ton, 6 by 6.

AR 850-5
MISCELLANEOUS

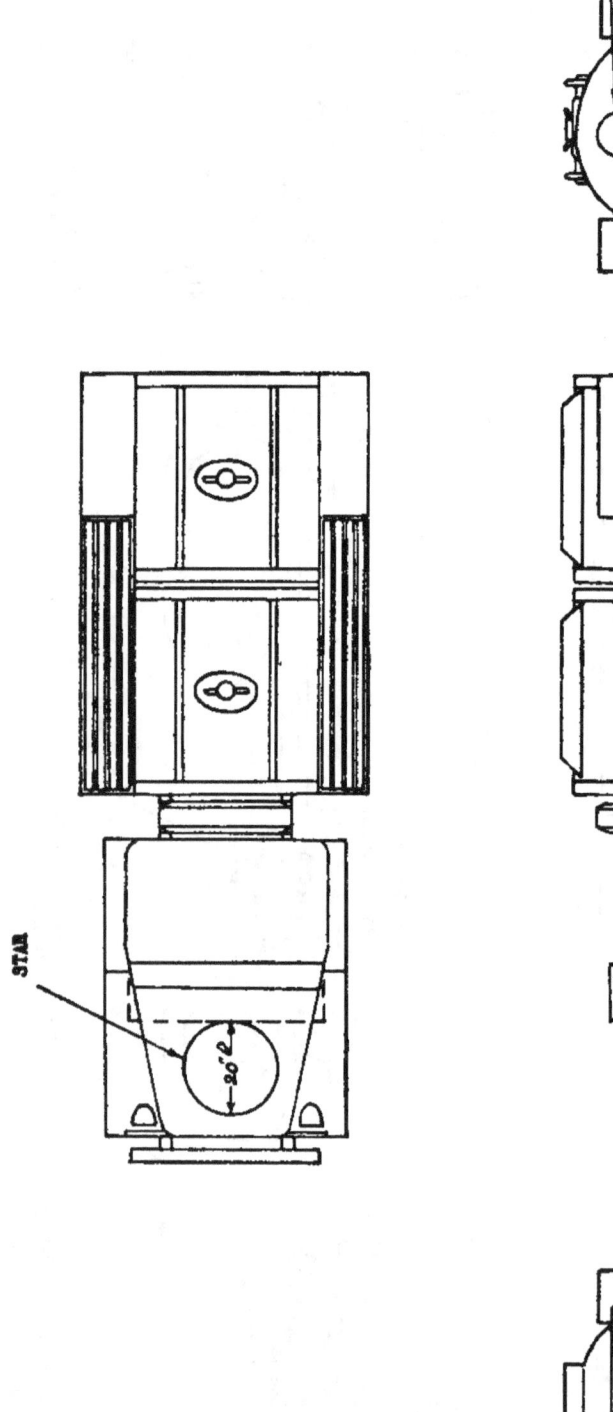
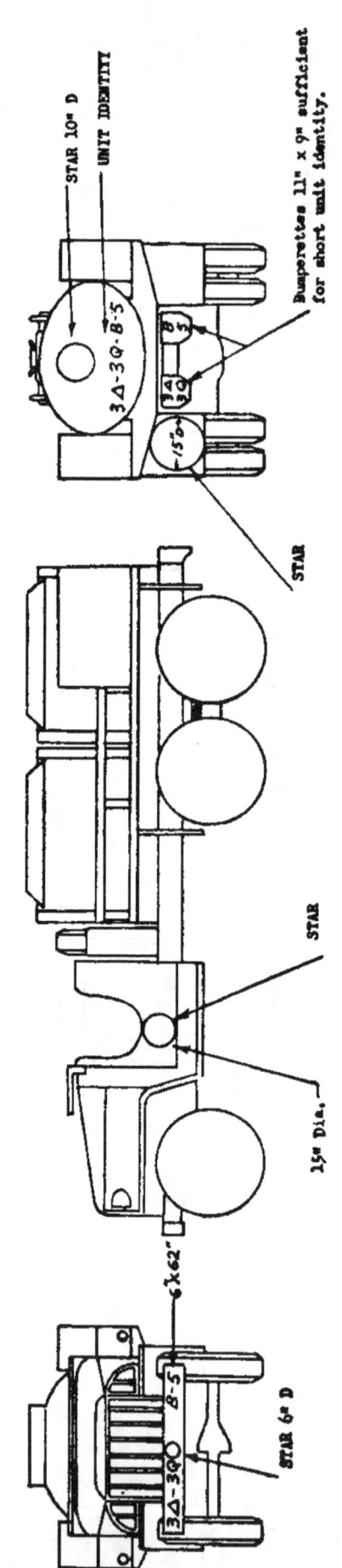

FIGURE 10.—Open cab tanker, 2½-ton.

MARKING OF CLOTHING, EQUIPMENT, VEHICLES, PROPERTY

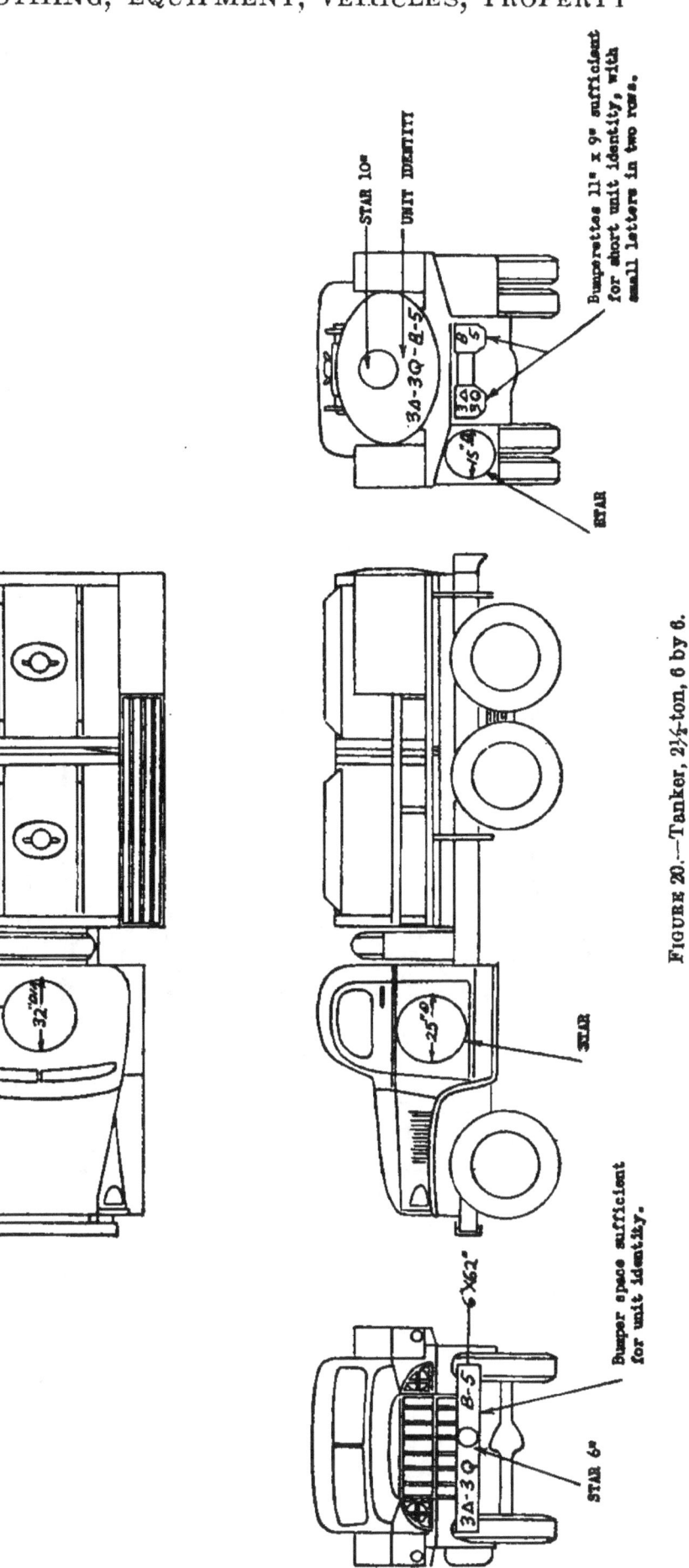

FIGURE 20.—Tanker, 2½-ton, 6 by 6.

AR 850-5
14 MISCELLANEOUS

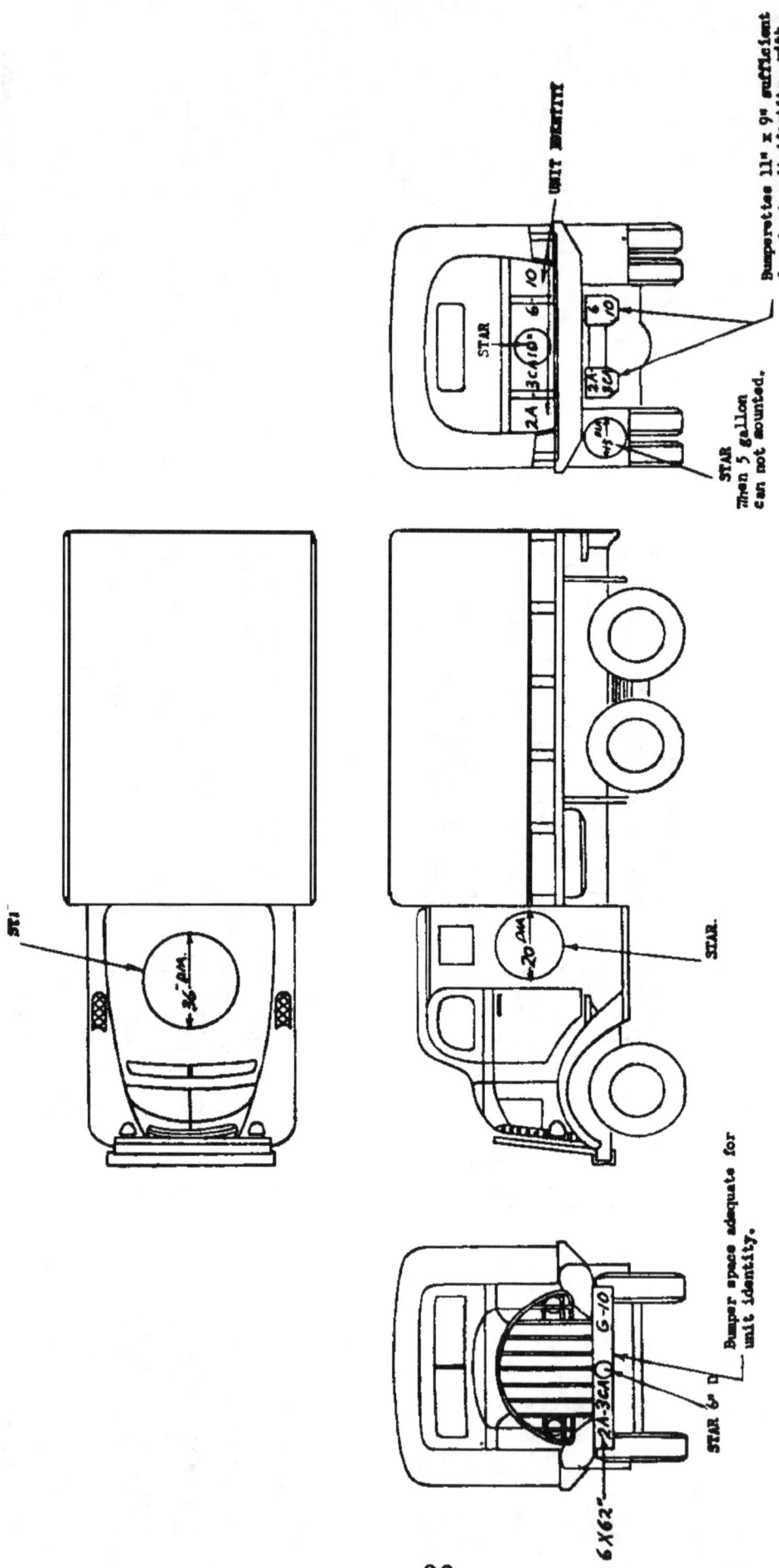

Figure 21.—Searchlight truck, 2½-ton, 6 by 4.

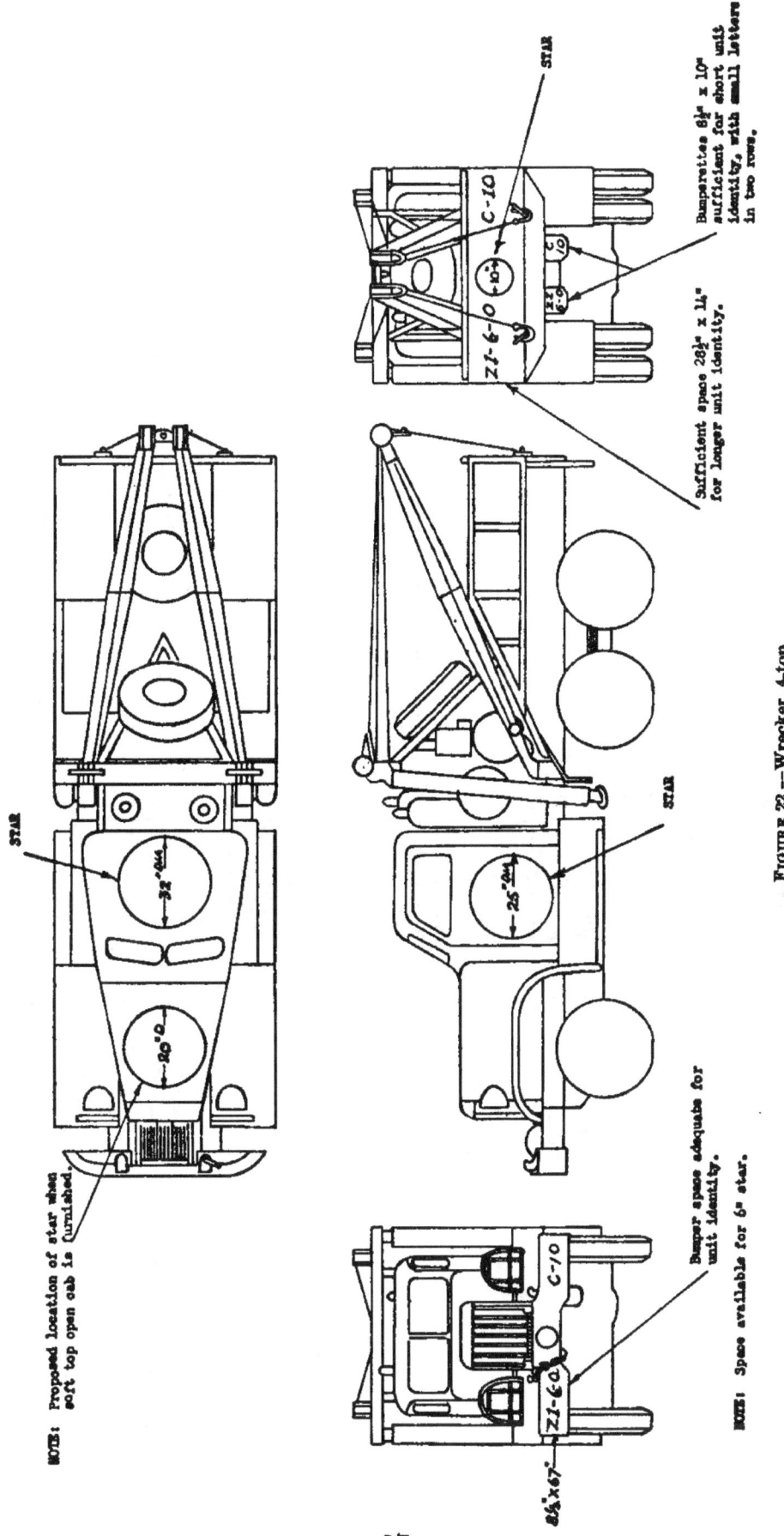

Figure 22.—Wrecker, 4-ton.

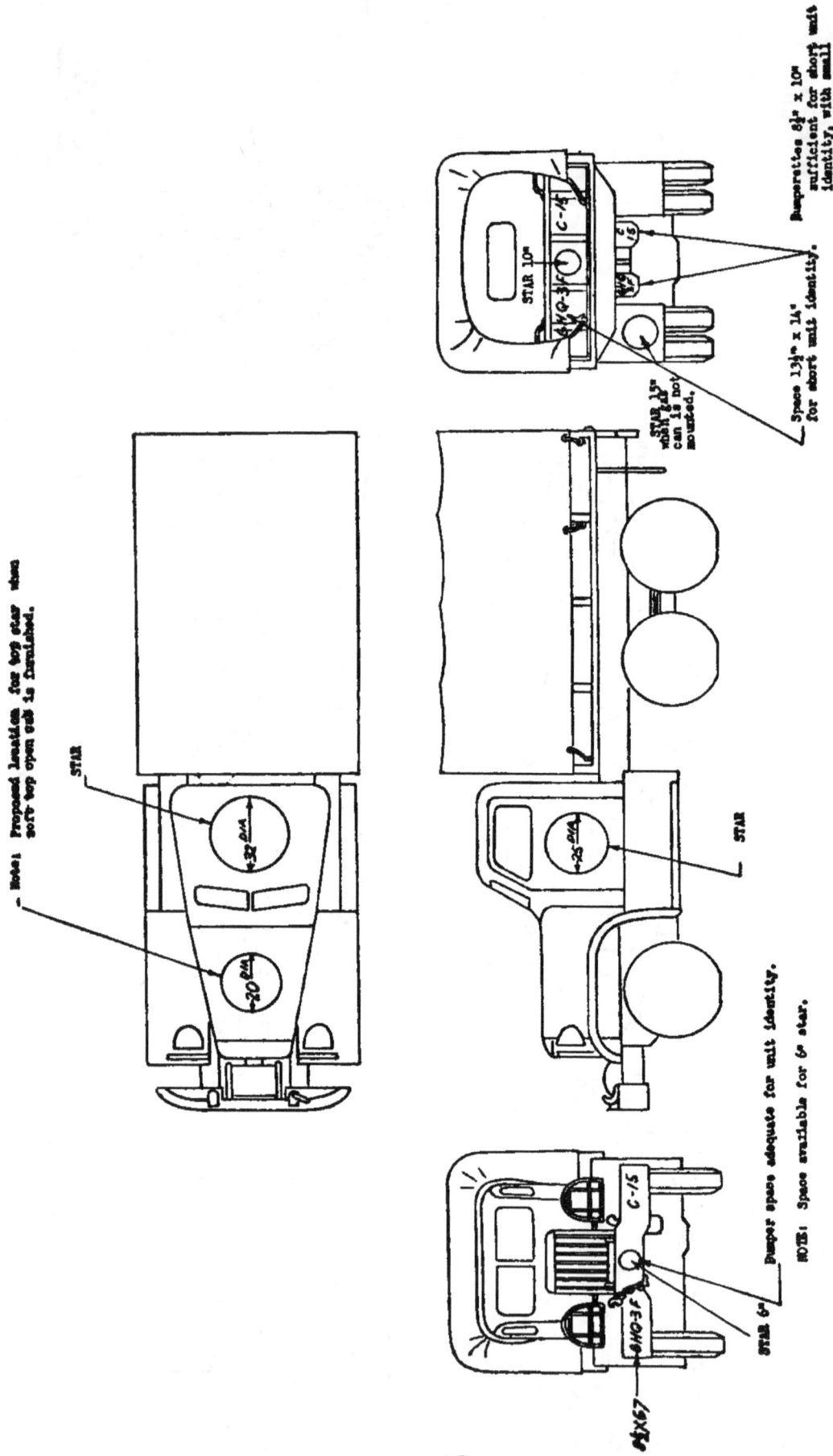

FIGURE 23.—Cargo truck, 4-ton, 6 by 6.

AR 850–14

MARKING OF CLOTHING, EQUIPMENT, VEHICLES, PROPERTY

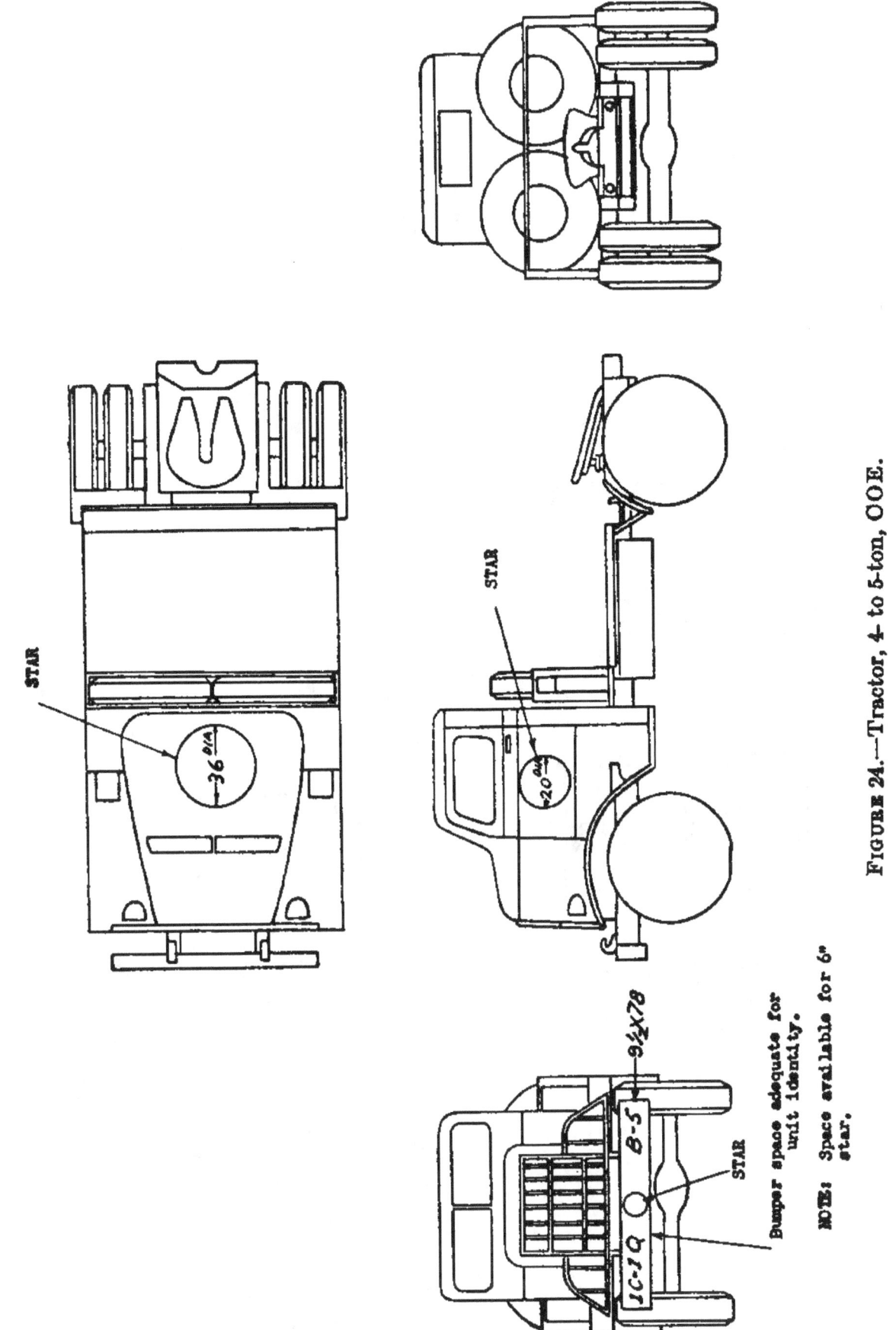

FIGURE 2A.—Tractor, 4 to 5-ton, OOE.

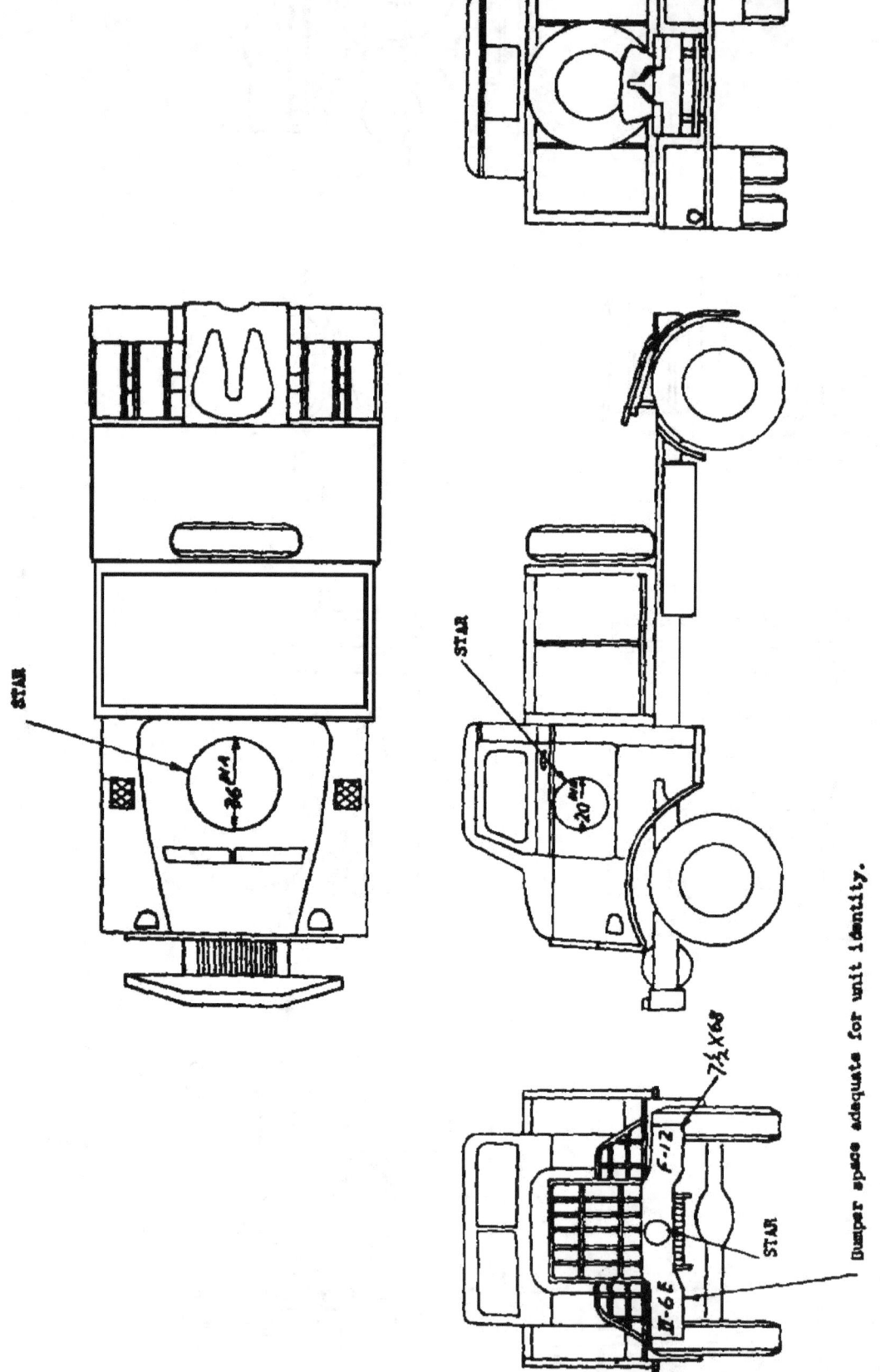

FIGURE 25.—Pontoon tractor, 5- to 6-ton, COE.

MARKING OF CLOTHING, EQUIPMENT, VEHICLES, PROPERTY 14

AR 850-5

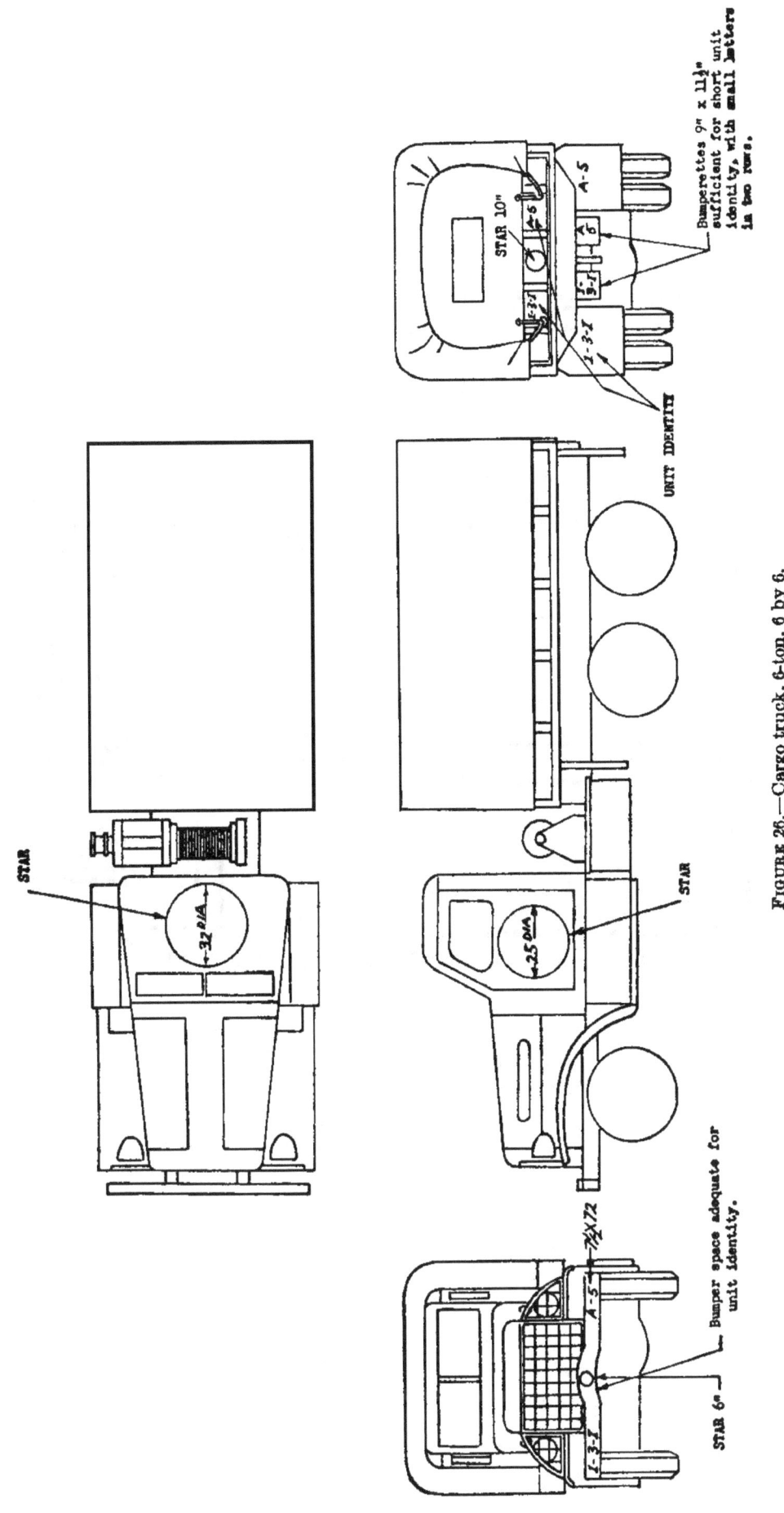

FIGURE 26.—Cargo truck, 6-ton, 6 by 6.

AR 850-5
14

MISCELLANEOUS

FIGURE 27.—Semitrailer, 6–T, combination animal and cargo.

42

MARKING OF CLOTHING, EQUIPMENT, VEHICLES, PROPERTY

AR 850–5
14

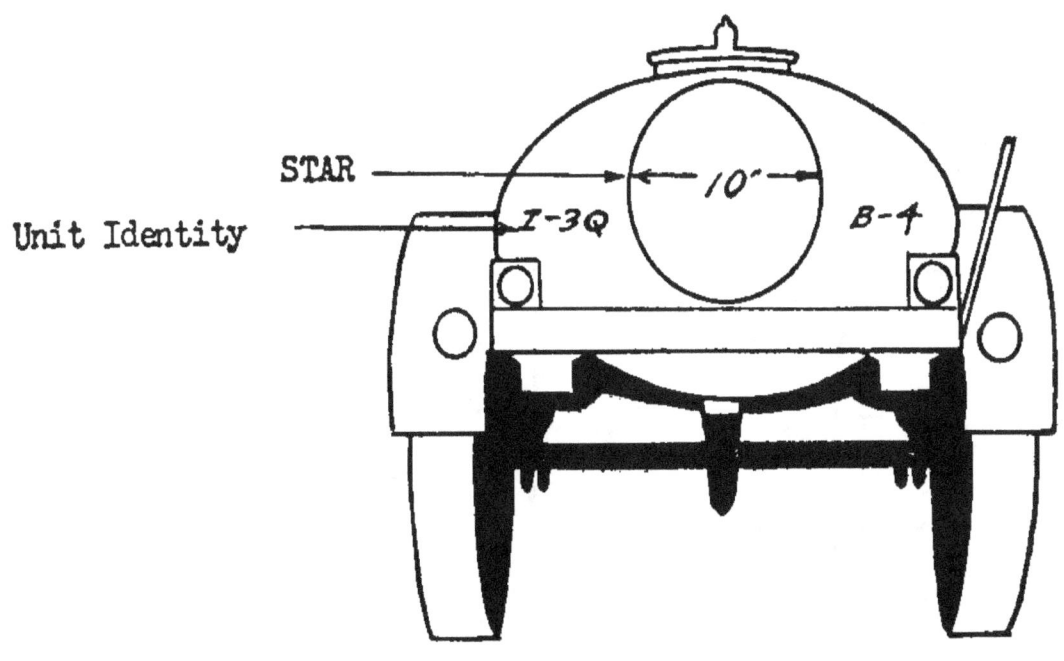

① Trailer, 1-T, 2-wheel, water tank (250-gal.).

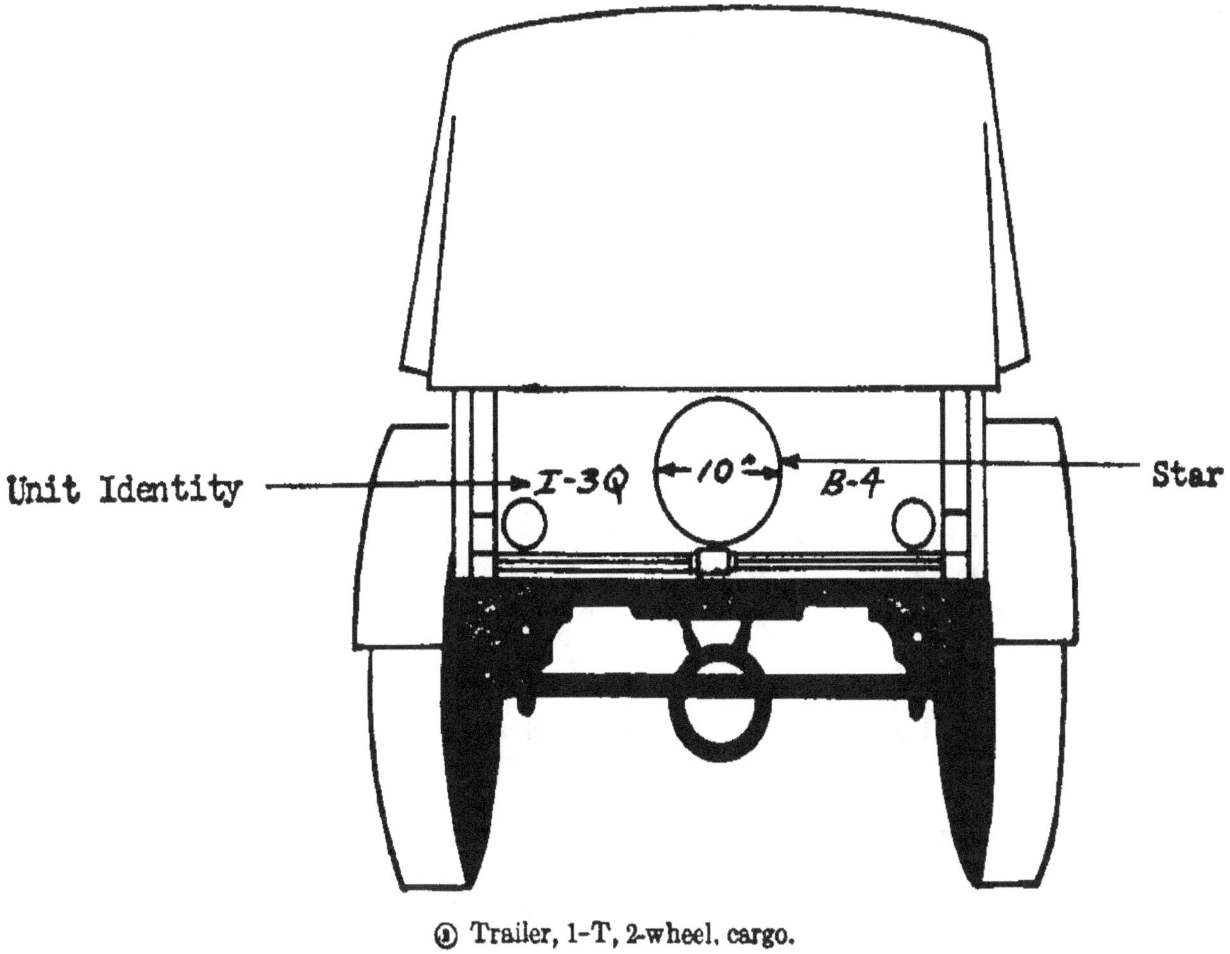

② Trailer, 1-T, 2-wheel, cargo.

FIGURE 28.

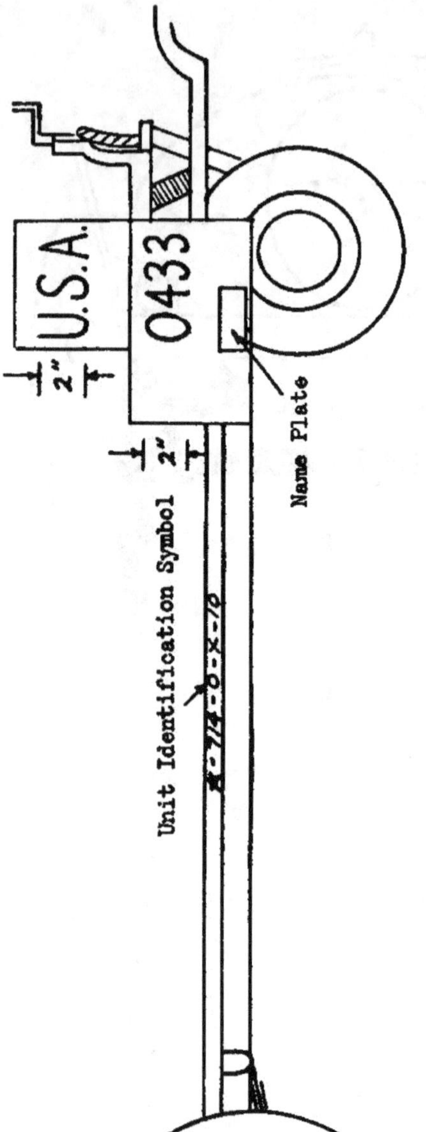

AR 850-5
14   MISCELLANEOUS

FIGURE 29.—Trailer, bomb M6.

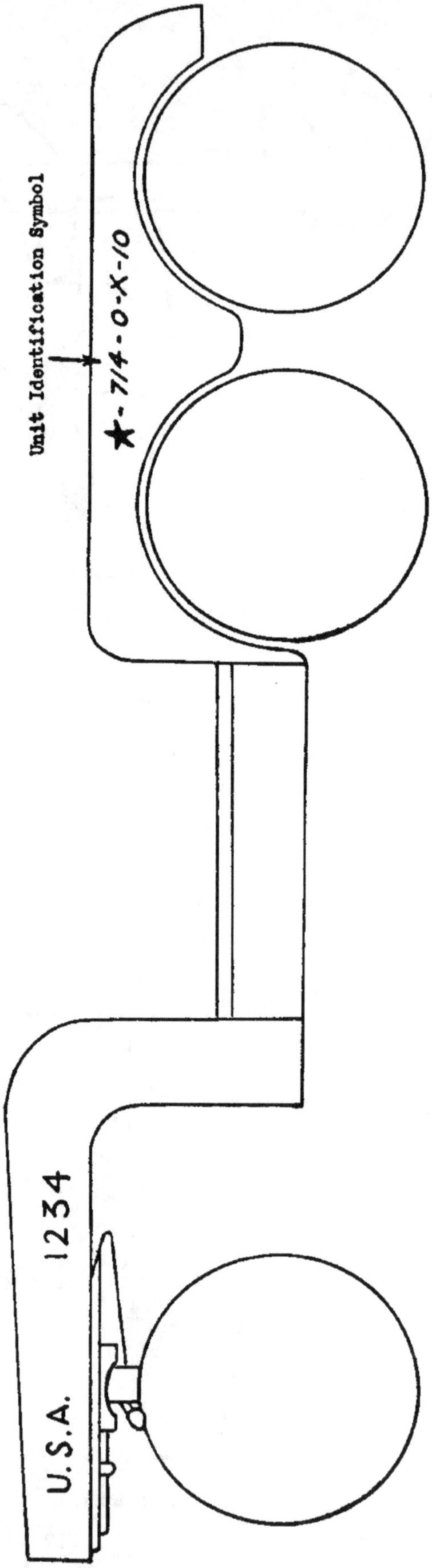

FIGURE 30.—Trailer tractor crane, T-26.

AR 850-5
MARKING OF CLOTHING, EQUIPMENT, VEHICLES, PROPERTY 14

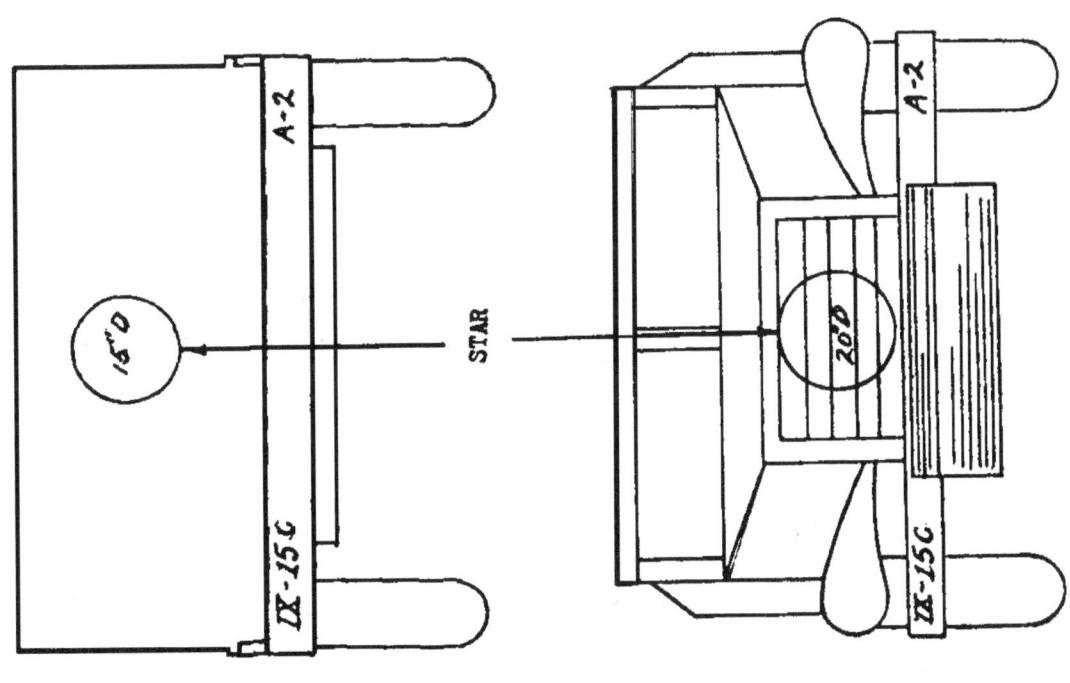
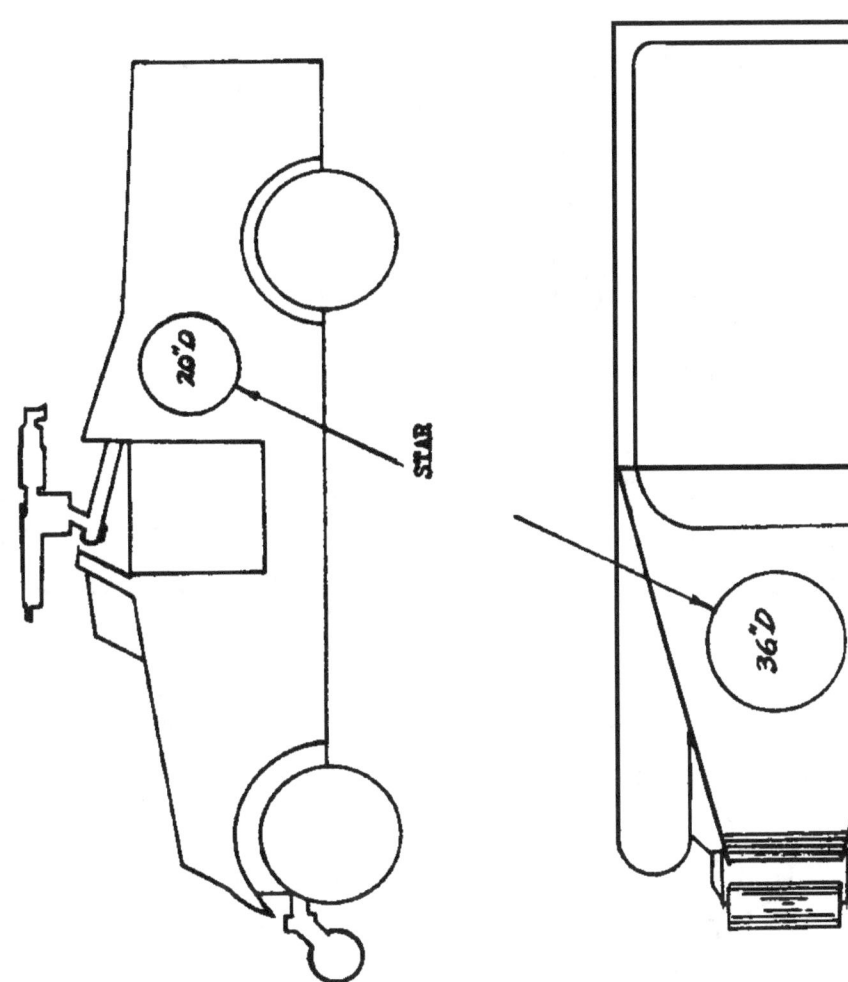

FIGURE 31.—Scout car.

AR 850–5
14 MISCELLANEOUS

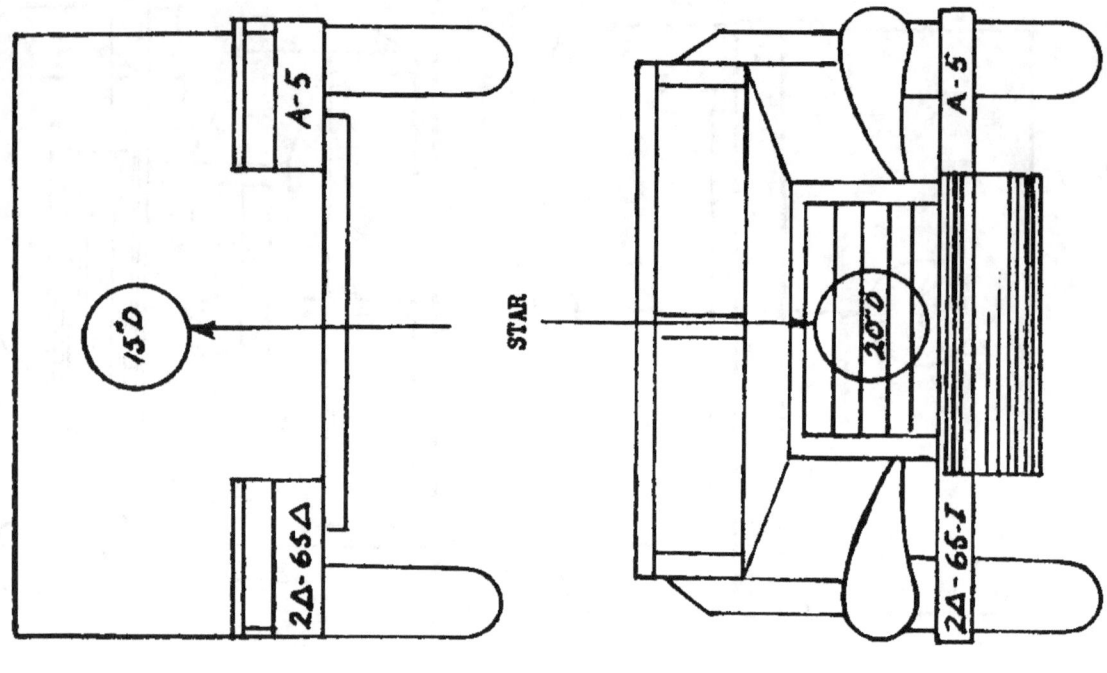

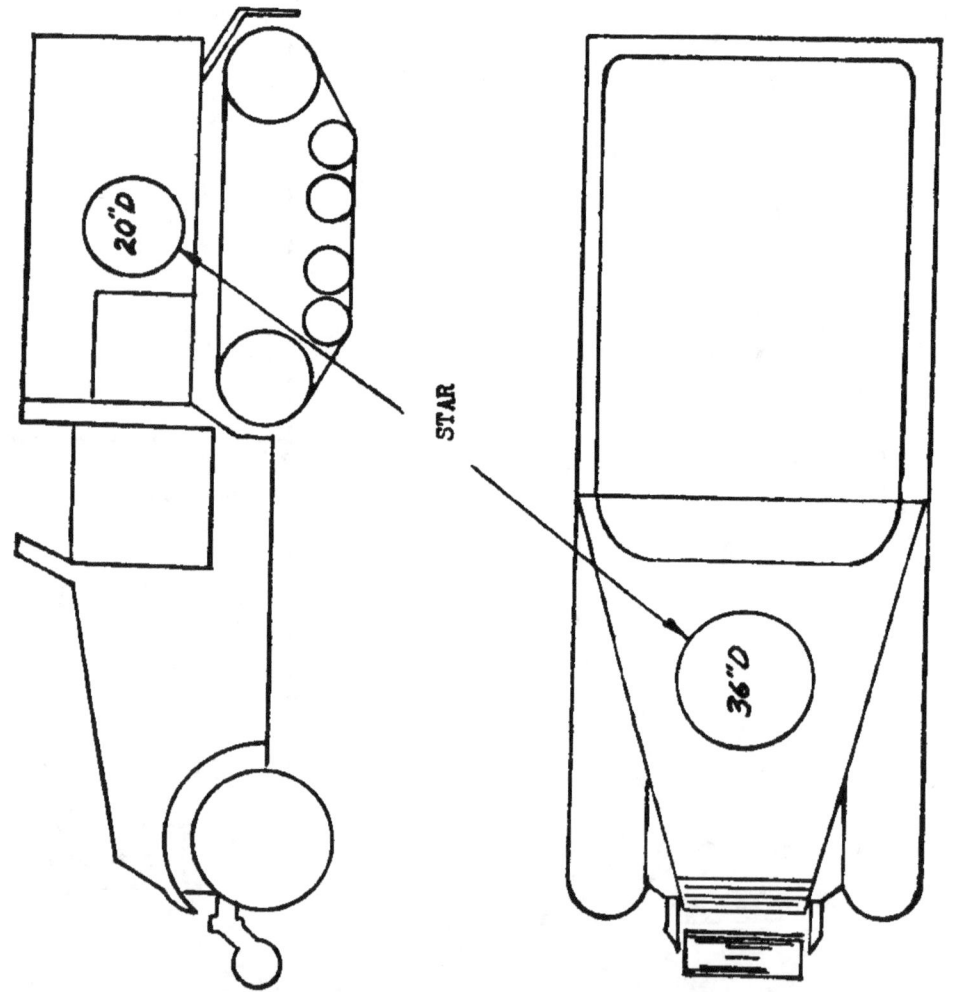

FIGURE 32.—Half track.

46

AR 850–5
MARKING OF CLOTHING, EQUIPMENT, VEHICLES, PROPERTY 14

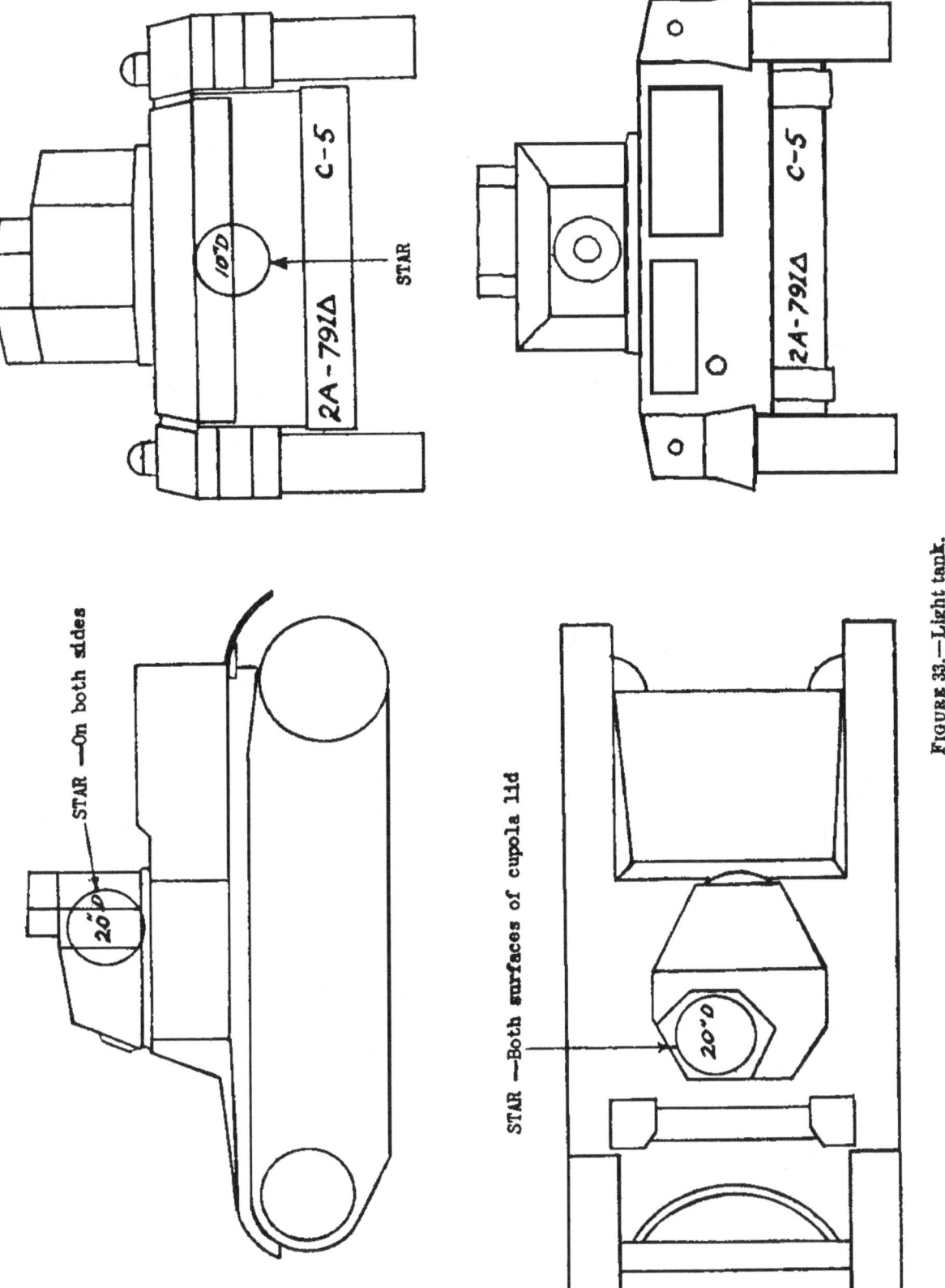

FIGURE 33.—Light tank.

AR 850-5
14   MISCELLANEOUS

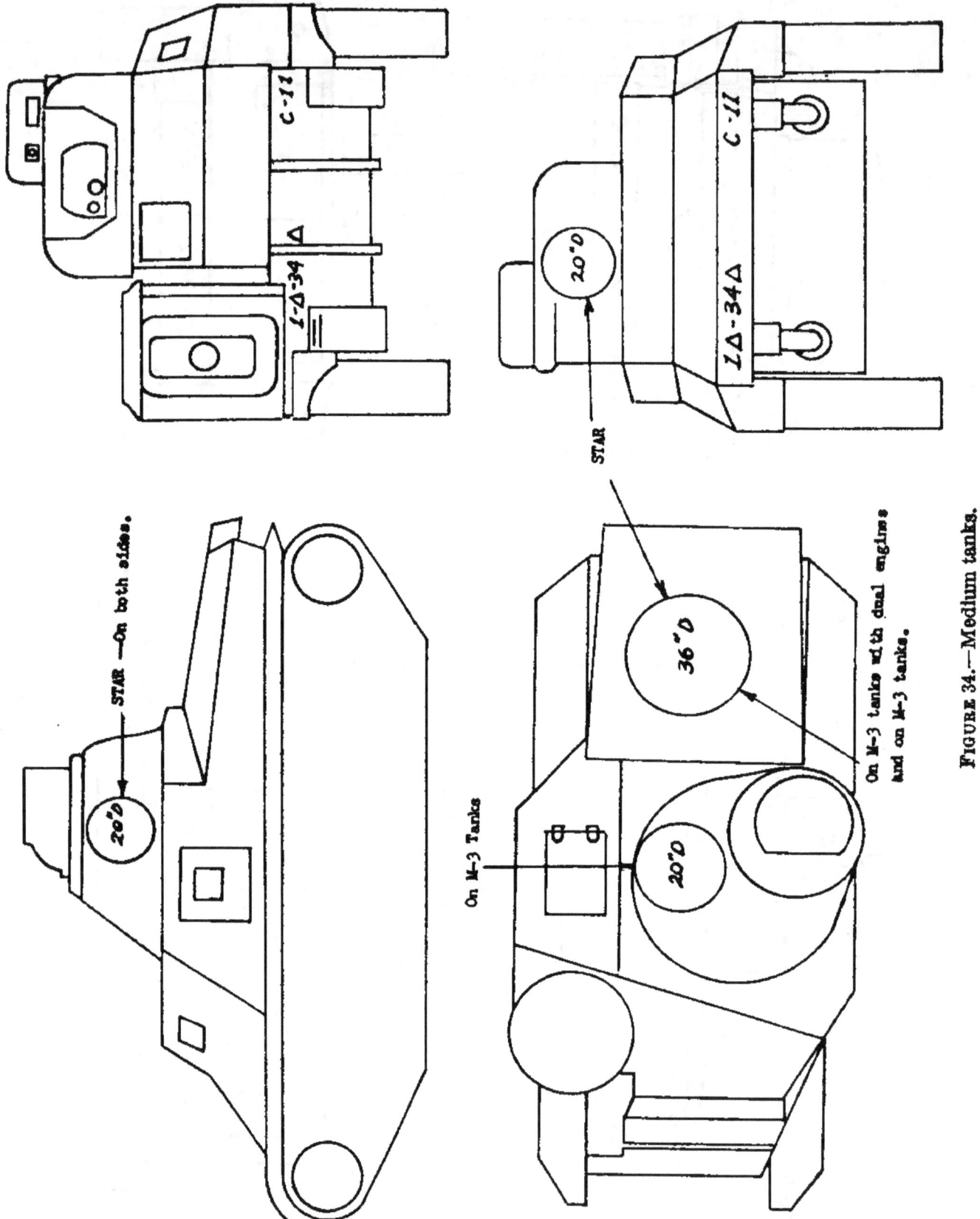

FIGURE 34.—Medium tanks.

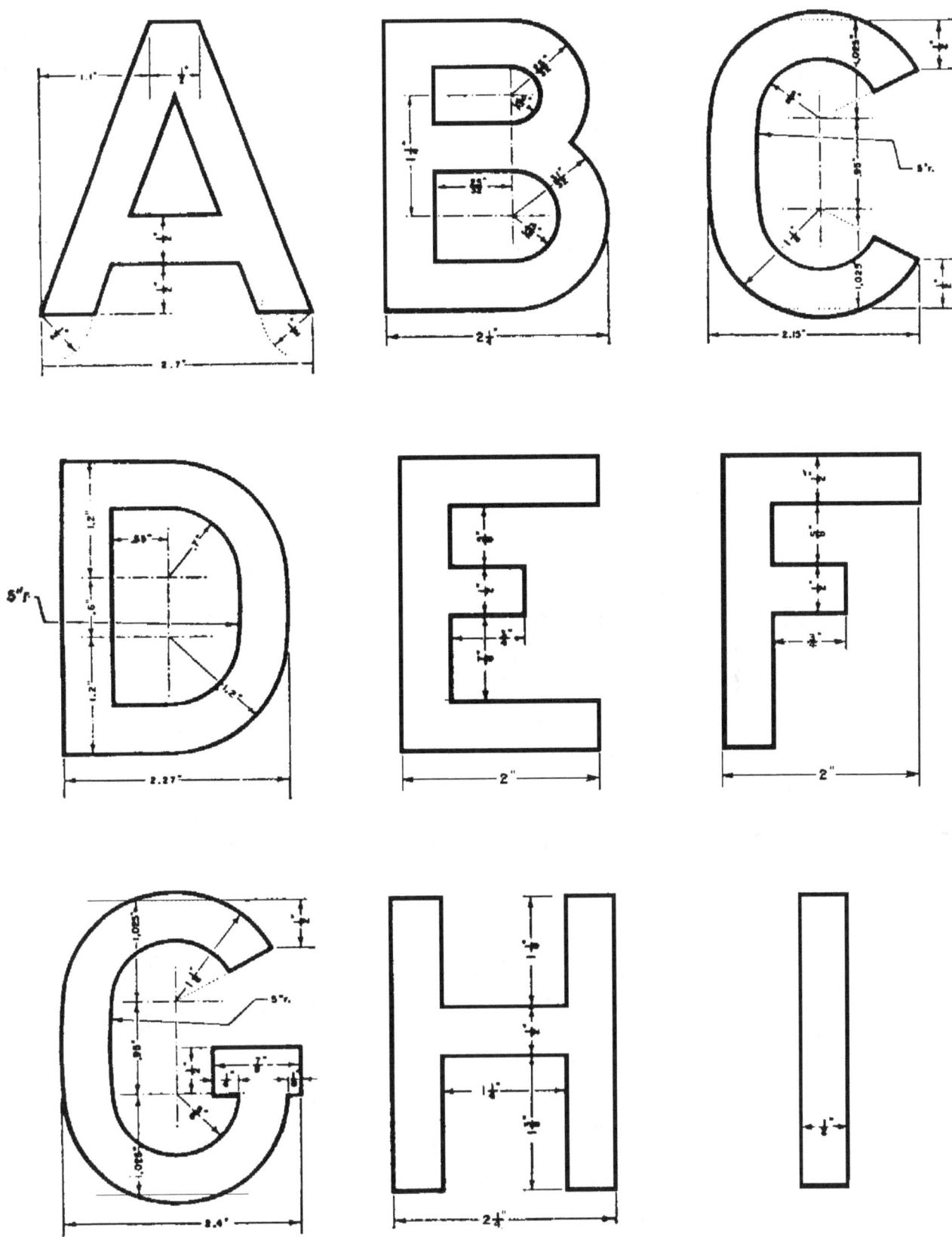

FIGURE 35.—Letters and numerals (3-inch).

FIGURE 35.—Letters and numerals (3-inch)—Continued.

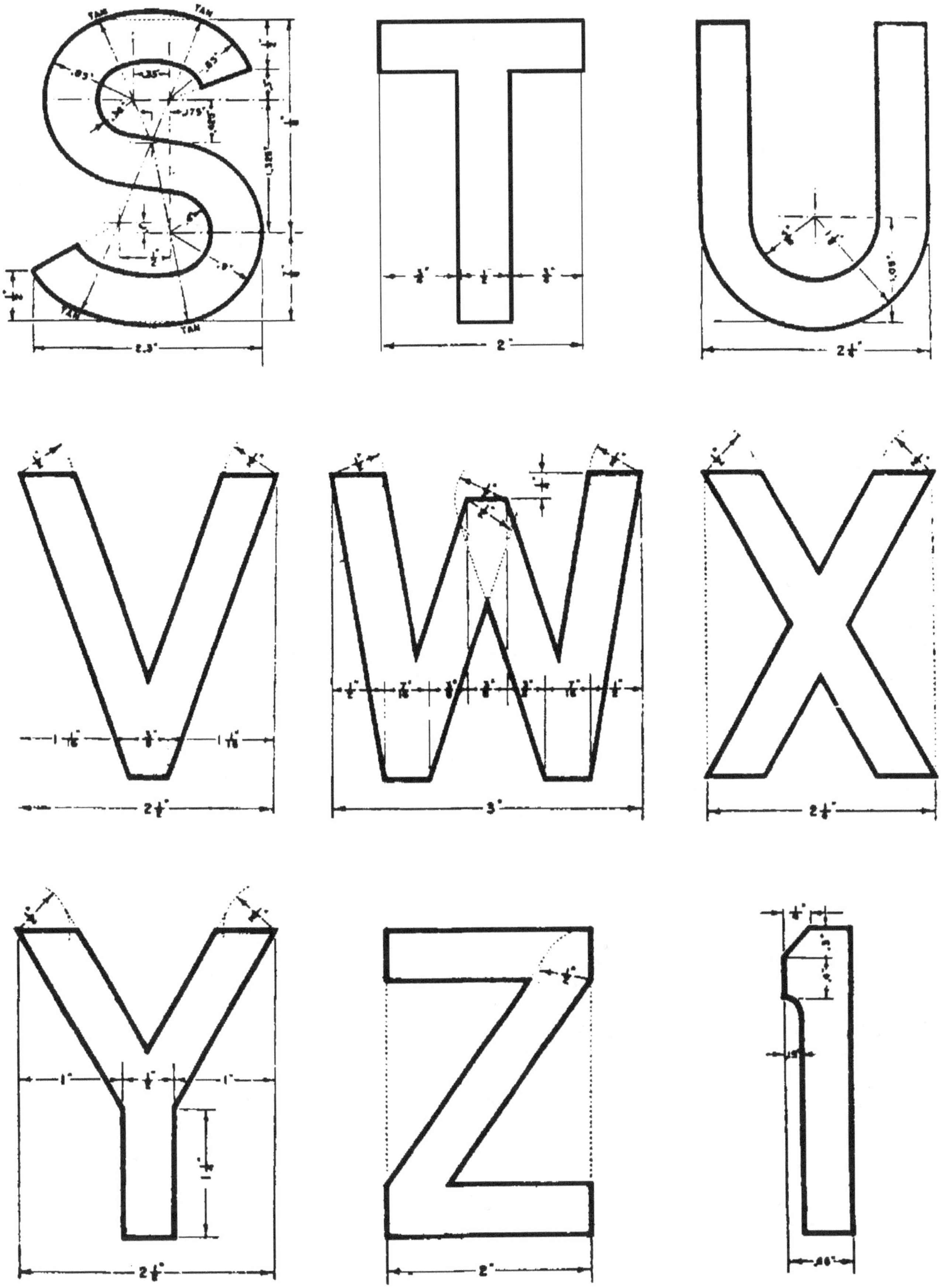

AR 850–5
14
MARKING OF CLOTHING, EQUIPMENT, VEHICLES, PROPERTY

FIGURE 35.—Letters and numerals (3-inch)—Continued.

FIGURE 35.—Letters and numerals (3-inch)—Continued.

MARKING OF CLOTHING, EQUIPMENT, VEHICLES, PROPERTY

Section IV

BAGGAGE AND PROPERTY

	Paragraph
Definitions	15
Primary mark	16
Secondary marking	17

15. Definitions.—*a. Rectangular containers.*—The ends will be considered to be the surfaces at the extremities of the longest dimension, and the sides to be the surfaces, other than the top and bottom, between the ends.

b. Odd shapes.—The above definitions will be applied in principle to odd shapes. When the top and bottom are not the same size, the larger will be considered to be the bottom.

16. Primary mark.—*a.* All Government-owned property will be marked on one side only with the letters "U. S." as a primary mark.

b. The height of these letters will be appropriate to the space available, but not so small as not to be readily visible.

c. This mark will be made with black paint and, when practicable, by means of a stencil, except that metal articles will be marked with a metal stamp.

17. Secondary marking.

a. (1) Separate brigades, regiments, separate battalions, and similar units or independent administrative units similar thereto, composed of troops of one arm or service only, will adopt a system of colored stripes to be painted midway between the top and bottom of article being marked or geometrical figures or combination of geometrical figures of appropriate size and colors to be painted on each end of the article being marked. A combination of stripes and geometrical figures may be used.

(2) The number of stripes and the geometrical figures, combination of geometrical figures and stripes, will in no way represent the numerical designation of the unit or higher unit, or resemble any organizational or higher unit insignia.

b. Division or similar higher unit commanders may prescribe the system of marking described above for units of their command.

c. The record of a system prescribed under provisions of *b* above will be retained as a confidential document but will not be taken into a combat area.

Section V

MISCELLANEOUS

	Paragraph
Wall lockers and refrigerators	18
Miscellaneous movable public property	19
New types of vehicles	20

18. **Wall lockers and refrigerators.**—*a.* Wall lockers and refrigerators will be marked with the post number of the building in which installed.

b. This marking will be done by branding, if of wood; otherwise with paint.

19. **Miscellaneous movable public property.**—*a.* All movable public property for which special marking is not hereinbefore specifically prescribed will, if practicable, be branded or stamped with the letters "U. S." before being used.

b. All such property issued to a unit for its exclusive use will, if practicable, be marked or stamped with the number or letter of the unit in addition to the marking required by *a* above.

20. **New types of vehicles.**—When new vehicles of any type are issued to the Army, instructions pertaining to the location and size of the international symbol and unit identification will be issued by the War Department. It will be the responsibility of the procuring service to recommend these locations and to prepare necessary plates for publishing.

[A. G. 400.161 (7-1-42).]

By order of the Secretary of War:

G. C. MARSHALL,
Chief of Staff.

Official:
J. A. ULIO,
Major General,
The Adjutant General.

AR 850-5
C 1

MISCELLANEOUS

MARKING OF CLOTHING, EQUIPMENT, VEHICLES, AND PROPERTY

Changes | WAR DEPARTMENT,
No. 1 | Washington, August 24, 1942.

AR 850-5, August 5, 1942, is changed as follows:

4. Organization.

* * * * * * *

e. Organizational designs authorized.

(1) The custom of decorating organizational equipment with an individual characteristic design is authorized as a means of increasing the morale of the unit. Squadrons, companies, and similar organizations may select and submit a design, which the members consider representative of their particular unit, for decorating their organizational equipment and distinguishing such equipment from all similar equipment. A proposed design, which may be disclosed by a finished drawing, a rough drawing, or a word description of the idea, will be submitted for approval before it may be reproduced on the equipment.

(2) Definitions:

   (*a*) "Equipment," as used in this section, means operating equipment such as airplanes, tanks, etc.

   (*b*) "Design" or "organizational design" as used in this section refers to the markings applied to organizational equipment, and not to regimental coats of arms, uniform insignia or shoulder sleeve insignia, as covered by AR 600-40.

(3) Organizations desiring special markings will endeavor to develop individual designs. When an organization desires assistance in developing its design, written suggestions may be sent to the office of The Quartermaster General for preparation of a design embodying the ideas submitted. The developed design will be returned to the organization for consideration.

(4) The Commanding Generals, Army Ground Forces, Army Air Forces and Services of Supply will have sole authority to approve organizational designs, as covered by these regulations, for any organization under his command.

(5) Some organizations may choose designs which are covered by copyright. When such a choice is made, release must be obtained from the owner of the copyright to use the particular motif. Where an organization determines that a copyrighted design is its choice, the arm or service concerned will obtain from the owner of the copyright a release for the use of the design by the particular organization before use of the design is made.

AR 850-5
C 1

MISCELLANEOUS

(6) The office of The Quartermaster General is designated as the coordinating agency for all organizational designs. Designs which are acceptable to the various arms and services will be forwarded to the office of The Quartermaster General for coordination prior to approval and obtaining any required release. Where the design submitted is a duplicate of one already in use, the arm or service will be advised of that fact and the design will be returned for preparation of another design. All approved designs will be sent to the office of The Quartermaster General for registration at the time the design is authorized for organizational use.

[A. G. 400.161 (8-18-42).] (C 1, Aug. 24, 1942.)

By order of the Secretary of War:

G. C. MARSHALL,
Chief of Staff.

Official:
J. A. ULIO,
*Major General,
The Adjutant General.*

Distribution:
A; E.

AR 850-5
C 2

MISCELLANEOUS

MARKING OF CLOTHING, EQUIPMENT, VEHICLES, AND PROPERTY

Changes }   WAR DEPARTMENT,
No. 2 }   Washington, October 15, 1942.

AR 850-5, August 5, 1942, is changed as follows:

4. Organization.

*   *   *   *   *   *   *

b. Marking required.

    (1) In regiments, separate battalions, or units similar thereto, articles of organizational equipment will be marked with the letter or letters suitably representing the abbreviated designation of each company or similar unit.

    (2) In independent administrative units smaller than a battalion, the organizational equipment will be marked with letters suitably representing the abbreviated designation thereof.

*   *   *   *   *   *   *

[A. G. 400.161 (10-10-42).]   (C 2, Oct. 15, 1942.)

20. New types of vehicles.—When new vehicles of any type are issued to the Army, instructions pertaining to the location and size of the national symbol and unit identification will be issued by the War Department. It will be the responsibility of the procuring service to recommend these locations, to prepare necessary plates for publishing, and to take necessary action for the processing of changes to these regulations.

[A. G. 400.161 (10-10-42).]   (C 2, Oct. 15, 1942.)

By order of the Secretary of War:

                                                      G. C. MARSHALL,
                                                                Chief of Staff.

Official:
    J. A. ULIO,
        Major General,
            The Adjutant General.

Distribution:
    A; E.

AR 850–5
C 3

MISCELLANEOUS

MARKING OF CLOTHING, EQUIPMENT, VEHICLES, AND PROPERTY

CHANGES } WAR DEPARTMENT,
No. 3 } WASHINGTON, November 12, 1942.

AR 850–5, August 5, 1942, is changed as follows:

4. Organization.

* * * * *

e. *Organizational designs authorized.*—Rescinded.
[A. G. 400.161 (9–17–42).] (C 3, Nov. 12, 1942.)

By order of the Secretary of War:

G. C. MARSHALL,
Chief of Staff.

OFFICIAL:
J. A. ULIO,
Major General,
The Adjutant General.

DISTRIBUTION:
A; E.

U. S. GOVERNMENT PRINTING OFFICE: 1942

AR 850–5
C 4

MISCELLANEOUS

MARKING OF CLOTHING, EQUIPMENT, VEHICLES, AND PROPERTY

CHANGES }
No. 4 }

WAR DEPARTMENT,
WASHINGTON, December 11, 1942.

AR 850–5, August 5, 1942, is changed as follows:

5. **Individual.**—*a. Articles not to be marked.*—Articles of individual equipment which will not be marked as prescribed in *b* below are—

* * * * * * *

(6) Covers of cans, meat.

* * * * * * *

d. How marked.
   (1) *General.*
      (*a*) As prescribed in (2) and (3) below, by ½-inch stencil for textile equipment, and by marking outfit, leather, for leather equipment.

* * * * * * *

   (*c*) In case subsequent markings are necessary, the original markings will be eradicated, by erasure and scouring, if possible. Where eradication is not possible the original markings will be nullified as follows: Markings on textile equipment, by two indelible horizontal lines through each line of the original markings, preferably of the same color as the original. Markings on leather equipment, by stamping over the original marking with the "dash" die.

* * * * * * *

(2) *Stenciling of textile equipment; where placed.*

Article	Location of stenciling
* * *	* * * *
Cover, rifle	Inside of flap near lower edge.
Currycomb	On handle where most convenient.
Haversack	Inside of flap of meat-can pouch near left-hand edge.
* * *	* * * *

(3) *Stamping of leather articles.*—The marking required by *b* above will be stamped directly upon the articles of equipment enumerated in the following table, by means of stamping dies, as specified opposite each article.

AR 850-5
C 4   MISCELLANEOUS

Article	Where marking placed
Bridle	On headstall.
Brush, horse	1 inch to the right of the "U. S."
Harness	Where most convenient on each major portion.
Holster, pistol	On inside of flap.
Pouch, music	Inside flap near lower left-hand corner.
Saddle	On near side of pommel, where most convenient.
Saddlebag	On center of clip and strap, near reinforcement.
Scabbard, automatic rifle	On flap, bottom to be ½ inch above strap.
Scabbard, bayonet, M1910	On outside of leather tip.
Scabbard, bayonet, M1917	Where most convenient on major portion.
Scabbard, rifle	Top to be 2 inches below mouth of scabbard.
Scabbard, saber	Where most convenient on major portion.
Sling, gun	On outer face of long strap.
Strap, saber	On strap.
Strap, sling for automatic rifle	On outer face of long strap.

e. Identification tag.—See AR 600-40.

[A. G. 400.161 (11-19-42).]   (C 4, Dec. 11, 1942.)

BY ORDER OF THE SECRETARY OF WAR:

                                          G. C. MARSHALL,
                                                Chief of Staff.

OFFICIAL:
    J. A. ULIO,
        Major General,
            The Adjutant General.

DISTRIBUTION:
    A; E.

MISCELLANEOUS

MARKING OF CLOTHING, EQUIPMENT, VEHICLES, AND PROPERTY

CHANGES }   WAR DEPARTMENT,
No. 5 }   WASHINGTON, December 24, 1942.

AR 850–5, August 5, 1942, is changed as follows:

5. **Individual.**

f. Barrack bags.—Barrack bags will be stenciled with the enlisted man's full name (first name, middle initial, and last name) and full Army serial number (letters and numbers not smaller than 1 inch nor larger than 2 inches). In addition one barrack bag will be marked with a 3-inch "A" and the second bag (when issued) with a 3-inch "B." At the appropriate time the oversea shipment number and letter (code designation) will be stenciled in letters or numbers not smaller than 1 inch nor larger than 2 inches.

    (1) *How marked.*—The markings on blue denim barrack bags will be made with white lead paste (paint). Khaki barrack bags will be marked with black lead paste (paint). Markings will be made by use of a stencil of the size indicated in (2) below, except for letters A and B.

    (2) *Where marked.*—One barrack bag will be marked with a 3-inch "A" and the second (when issued) with a 3-inch "B." The bottom of each letter "A" or "B" will be 2 inches from the bottom seam. A 5-inch blank space will be left between the top of the letter A or B and the bottom of the Army serial number. The full Army serial number will be stenciled above the 5-inch blank space in numbers not less than 1 inch nor more than 2 inches. A 1-inch space will be left above the Army serial number and the full name (first name, middle initial, and last name) will be stenciled in above the 1-inch space in letters not less than 1 inch nor more than 2 inches. At the appropriate time the oversea shipment number and letter (code designation) will be centered in the 5-inch blank space by means of a stencil, in letters or numbers not smaller than 1 inch nor larger than 2 inches.

    (3) *When marked.*—All barrack bags now in the possession of enlisted personnel will be marked immediately. Following induction, personnel will mark their barrack bags at the time they are issued. Barrack bags will be re-marked as often as it is necessary in order to maintain legibility.

[A. G. 400.161 (12–19–42).]   (C 5, Dec. 24, 1942.)

BY ORDER OF THE SECRETARY OF WAR:

                                               G. C. MARSHALL,
                                                      Chief of Staff.

OFFICIAL:
    J. A. ULIO,
        Major General,
            The Adjutant General.

DISTRIBUTION:
    A; E.

AR 850-5
C 6

MISCELLANEOUS

MARKING OF CLOTHING, EQUIPMENT, VEHICLES, AND PROPERTY

Changes No. 6 }   WAR DEPARTMENT,
                                      WASHINGTON, March 4, 1943.

AR 850-5, August 5, 1942, is changed as follows:

5. Individual.

*   *   *   *   *   *   *

f. Barrack bags.—Barrack bags will be stenciled with the enlisted man's full name (first name, middle initial, and last name) and full Army serial number (letters and numbers not smaller than 1 inch nor larger than 2 inches). In addition one barrack bag will be marked with a 3-inch "A" and the second bag (when issued) with a 3-inch "B." At the appropriate time the oversea shipment number and letter (code designation) will be stenciled in letters or numbers not smaller than 1 inch nor larger than 2 inches.

*   *   *   *   *   *   *

     (3) *When marked.*—All barrack bags now in the possession of enlisted personnel will be marked immediately. Following induction, personnel will mark their barrack bags at first station of assignment after reception centers. Barrack bags will be re-marked as often as it is necessary in order to maintain legibility.

[A. G. 400.161 (2-27-43).]   (C 6, Mar. 4, 1943.)

BY ORDER OF THE SECRETARY OF WAR:

                                                 G. C. MARSHALL,
                                                        Chief of Staff.

OFFICIAL:
     J. A. ULIO,
         Major General,
             The Adjutant General.

DISTRIBUTION:
     A; B.

AR 850-5
C 7

MISCELLANEOUS

MARKING OF CLOTHING, EQUIPMENT, VEHICLES, AND PROPERTY

CHANGES }   WAR DEPARTMENT,
No. 7 }   WASHINGTON, 12 June 1943.

AR 850-5, 5 August 1942, is changed as follows:

10. *Unit markings.*—Gasoline solvent paint or paint as prescribed by the War Department will be used.

*   *   *   *   *   *   *

    b. Unit identification symbols.

        (1) *Front and rear markings.*—See figures 1 to 34 for location.

*   *   *   *   *   *   *

            (d) *Third group.*—The third group will designate companies and similar organizations by letters in accordance with the following code:

Organization	Designation
Headquarters and headquarters company (or headquarters and headquarters and service company) of lowest unit identification in previous groups of markings, except for headquarters and headquarters companies of groups (all types).	HQ
Headquarters and headquarters companies of groups (all types).	GP
Service company of lowest unit indentified in service groups.	SV
* * * * * *	*

*   *   *   *   *   *   *

[A. G. 400.161 (7 Apr 48).]   (C 7, 12 Jun. 43.)

BY ORDER OF THE SECRETARY OF WAR:

                                                                 G. C. MARSHALL,
                                                                              Chief of Staff.

OFFICIAL:
    J. A. ULIO,
        Major General,
            The Adjutant General.

DISTRIBUTION:
    A; E.

AR 850–5
C 8

MISCELLANEOUS

MARKING OF CLOTHING, EQUIPMENT, VEHICLES, AND PROPERTY

Changes No. 8	WAR DEPARTMENT, Washington 25, D. C., 8 October 1943.

AR 850–5, 5 August 1942, is changed as follows:

10. **Unit markings.**—Gasoline solvent paint or paint as prescribed by the War Department will be used.

a. Unit markings.—National symbol.

* * * * * * *

> (8) Whenever requirements for camouflage and concealment outweigh the requirements for recognition, the national symbol may, at the discretion of the theater commander, be covered by lusterless olive-drab gasoline solvent paint or removed. Under similar circumstances, the national symbol may, at the discretion of lower commanders, be covered temporarily with camouflage nets or obscured with oil and dirt or some other readily removable field expedient.

* * * * * * *

[A. G. 400.161 (15 Sep 43).]   (C 8, 8 Oct 43.)

By order of the Secretary of War:

G. C. MARSHALL,
Chief of Staff.

Official:
   J. A. ULIO,
      Major General,
         The Adjutant General.

Distribution:
   A; E.

552721°—43—AGO 54

U. S. GOVERNMENT PRINTING OFFICE, 1943

AR 850-5
C 9

MISCELLANEOUS

MARKING OF CLOTHING, EQUIPMENT, VEHICLES, AND PROPERTY

| Changes | WAR DEPARTMENT, |
| No. 9 | Washington 25, D. C., 27 January 1944. |

AR 850-5, 5 August 1942, is changed as follows:

10. **Unit markings.**—Gasoline solvent paint or paint as prescribed by the War Department will be used.

   a. Unit markings.—National symbol.

   * * * * * * *

   (2) The size of the national symbol will be determined for each type of *motor vehicle and will be large enough to take advantage of the* surface upon which it is to be painted. See figures 1 to 32, inclusive; figure 34; and figures 36 to 56, inclusive.

   * * * * * * *

   b. Unit identification symbols.

   (1) *Front and rear markings.*—See figures 1 to 32, inclusive; figure 34; and figures 36 to 56, inclusive, for location.

   * * * * * * *

   [A. G. 400.161 (10 Nov 43).]   (C 9, 27 Jan 44.)

12. Special markings.

   * * * * * * *

   f. Letter "S."

   (1) The letter "S" will be used on a military vehicle to indicate that the vehicle so marked has been suppressed to eliminate radio interference, caused by the vehicle electrical system, over a frequency range 0.5 to 30 megacycles.

   (2) The letter "S" will be placed after the registration number and will be made as conspicuous as possible by leaving a space between the last numeral and the "S." This space should be equivalent to the space required for a numeral. Design, size, and color of the letter will be the same as that of the digits of the registration number. The character and style of markings of a vehicle suppressed to eliminate radio interference will be as follows:

   <u>U. S. A.</u>
   <u>123726 S</u>

   (3) The letter "S" will be applied originally to a vehicle either by the vehicle manufacturer at the time of assembly or by the authorized persons effecting suppression in the field. If a suppressed vehicle is repainted, the letter "S" will be reapplied in the manner and location described in (2) above.

   Note.—Where any choice exists, radio sets will be installed in suppressed vehicles rather than in nonsuppressed vehicles of the same type.

   [A. G. 400.161 (10 Nov 43).]   (C 9, 27 Jan 44.)

AR 850-5
C 9

MISCELLANEOUS

14. Tractors and armored vehicles.

* * * * * * *

b. Armored vehicles.—Tanks and other combat vehicles armored or on which armament may be mounted should be marked as prescribed in paragraphs 6 to 11, inclusive, so far as applicable, otherwise marking should be placed in a visible location as near as possible to that prescribed. See figure 34 and figures 36 to 48, inclusive.

[A. G. 400.161 (10 Nov 43).] (C 9, 27 Jan 44.)

* * * * * * *

FIGURE 33.—Rescinded.

[A. G. 400.161 (10 Nov 43).] (C 9, 27 Jan 44.)

* * * * * * *

FIGURE 34.—Medium tanks, M3 series.

[A. G. 400.161 (10 Nov 43).] (C 9, 27 Jan 44.)

* * * * * * *

AR 850–5
MARKING OF CLOTHING, EQUIPMENT, ETC. C 9

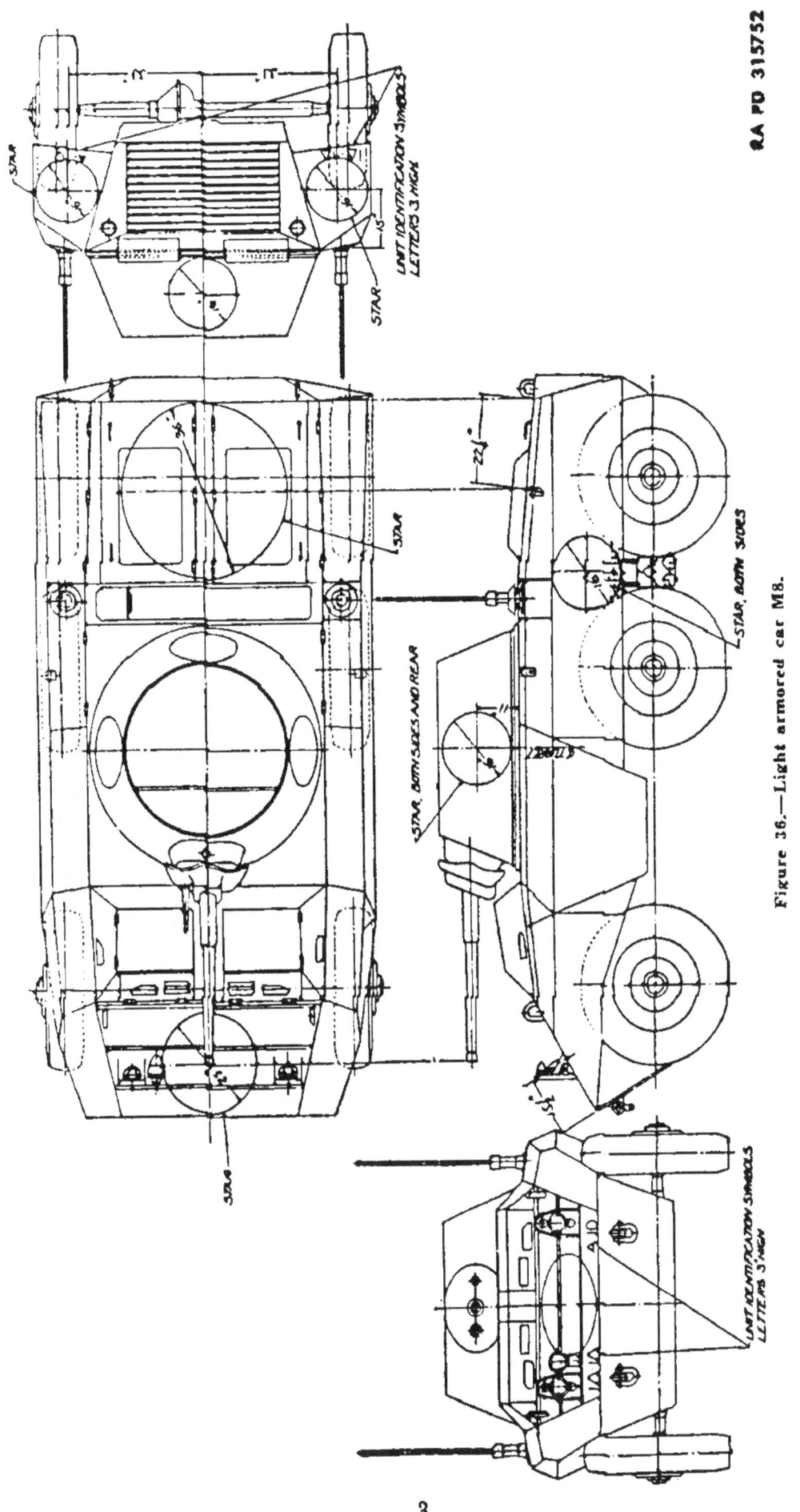

Figure 36.—Light armored car M8.

[A. G. 400.161 (10 Nov 43).] (C 9, 27 Jan 44.)

AR 850-5
C 9   MISCELLANEOUS

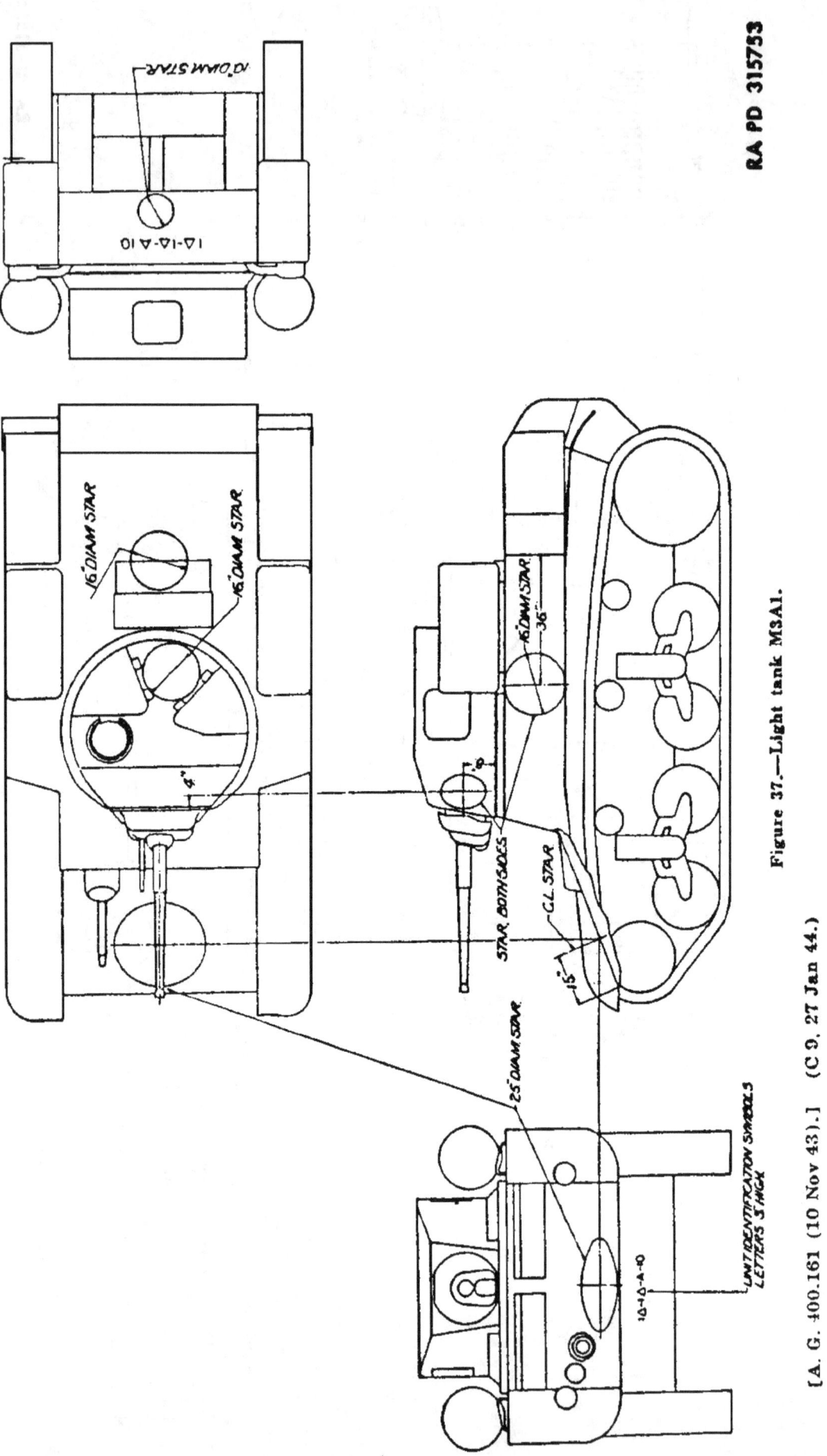

Figure 37.—Light tank M3A1.

[A. G. 400.161 (10 Nov 43).]   (C 9, 27 Jan 44.)

AR 850–5

MARKING OF CLOTHING, EQUIPMENT, ETC. C 9

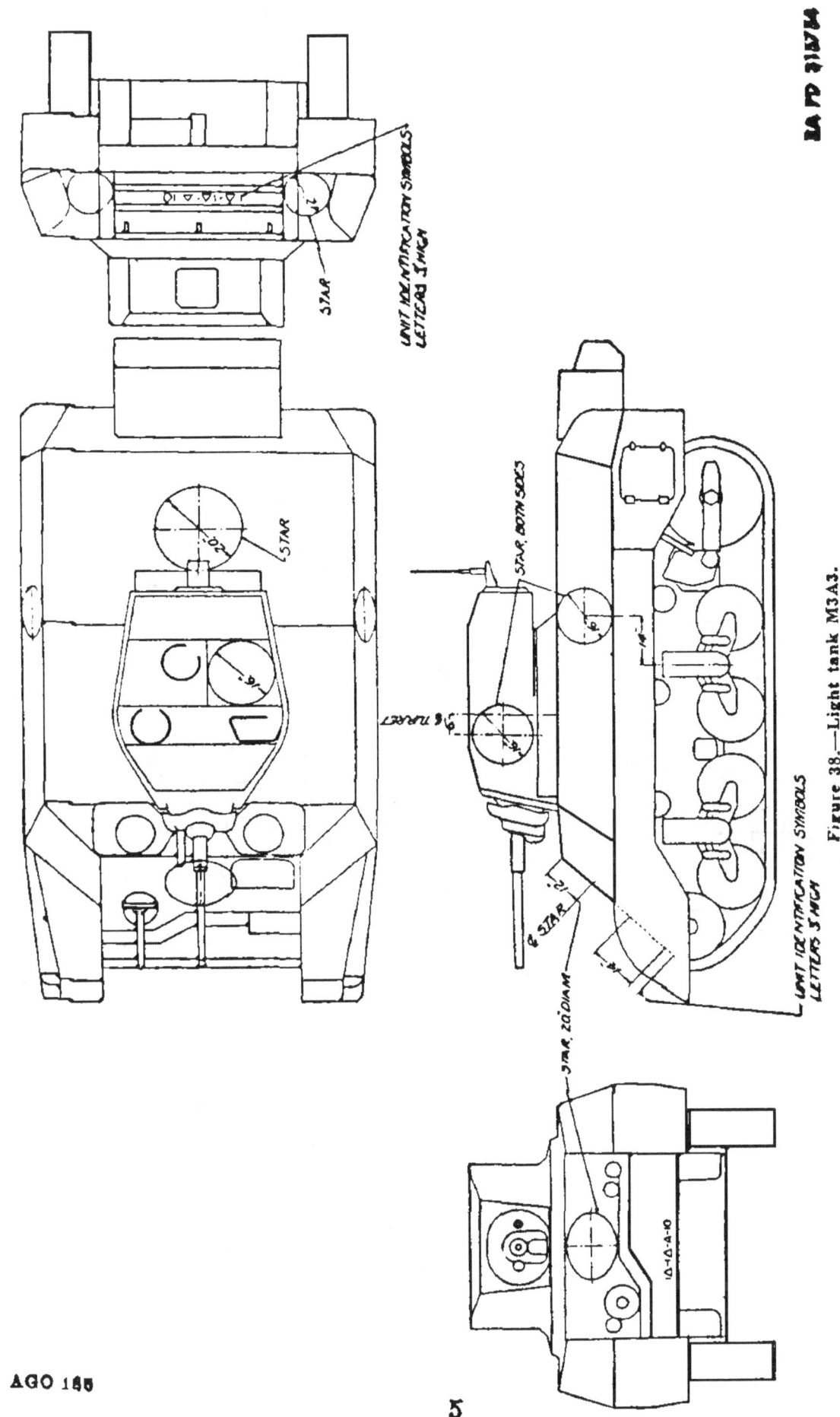

Figure 38.—Light tank M3A3.

[A. G. 400.161 (10 Nov 43).] (C. O. 27 Jan 44.)

AR 850-5
C 9

MISCELLANEOUS

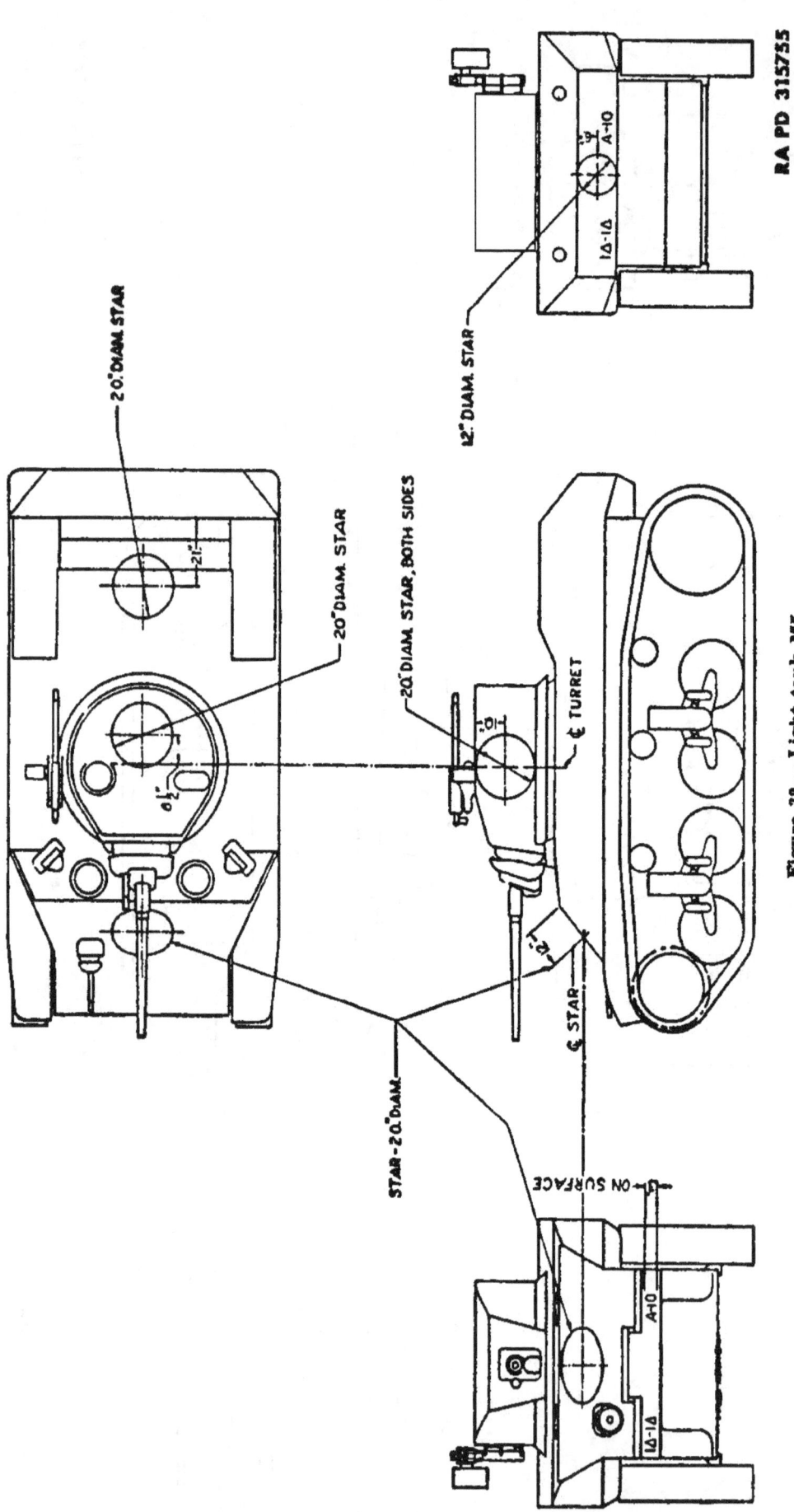

Figure 39.—Light tank M5.

[A. G. 400.161 (10 Nov 43).] (C 9, 27 Jan 44.)

MARKING OF CLOTHING, EQUIPMENT, ETC.   AR 850-5 C 9

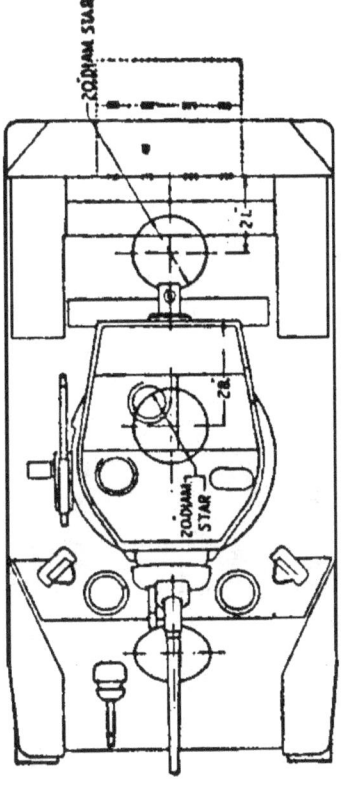
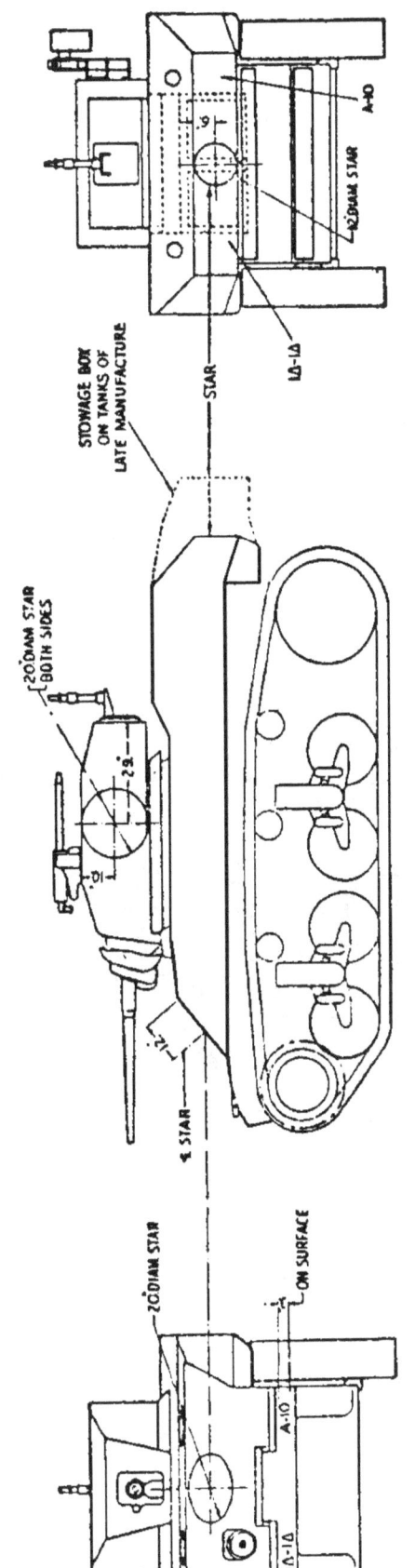

Figure 40.—Light tank M5A1.

[A. G. 400.161 (10 Nov 43).] (C 9, 27 Jan 44.)

AR 850-5
C 9
MISCELLANEOUS

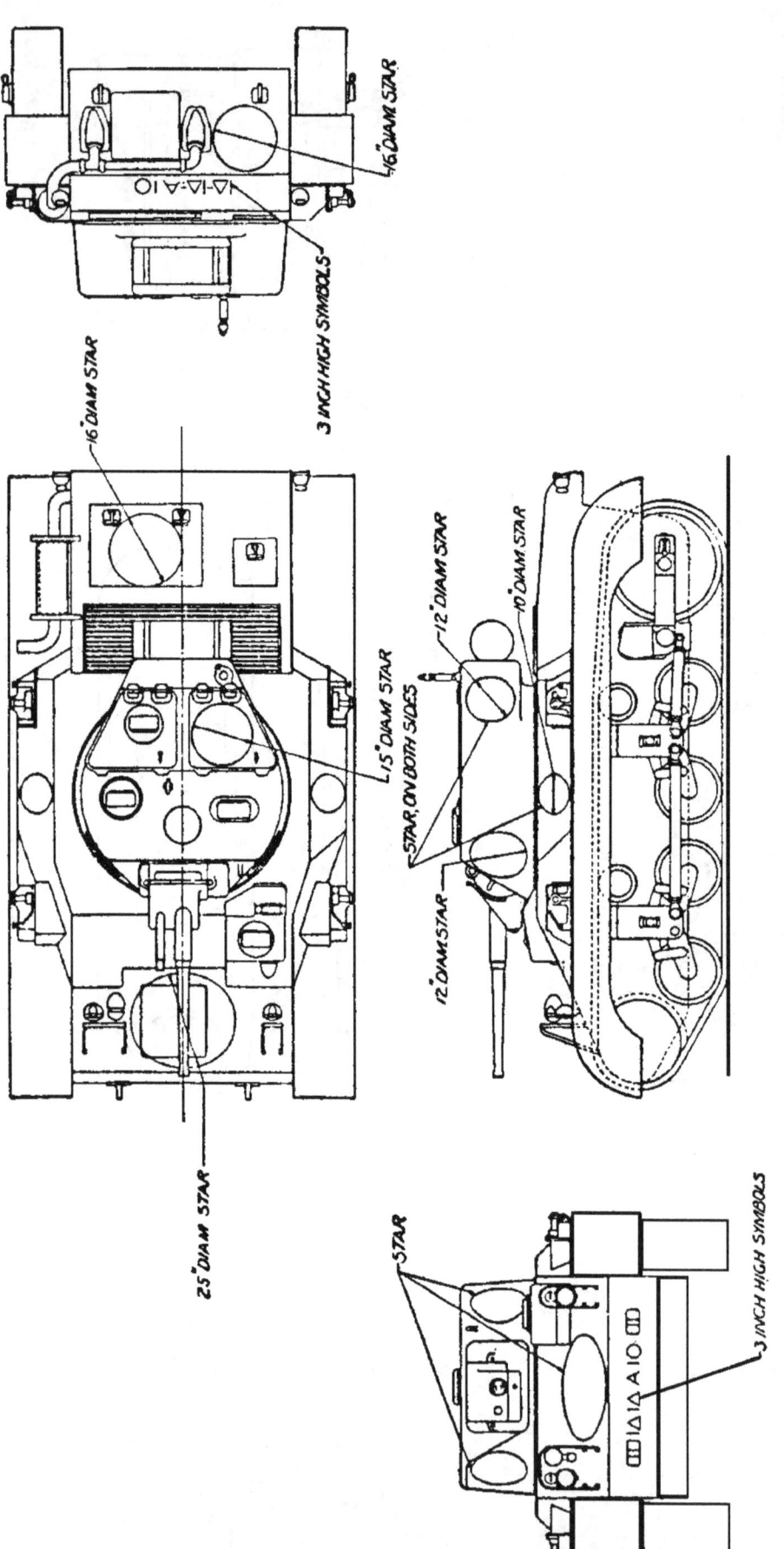

Figure 41.—Light tank T9E1.

[A. G. 400.161 (10 Nov 43).] (C 9, 27 Jan 44.)

MARKING OF CLOTHING, EQUIPMENT, ETC.

AR 850–5
C 9

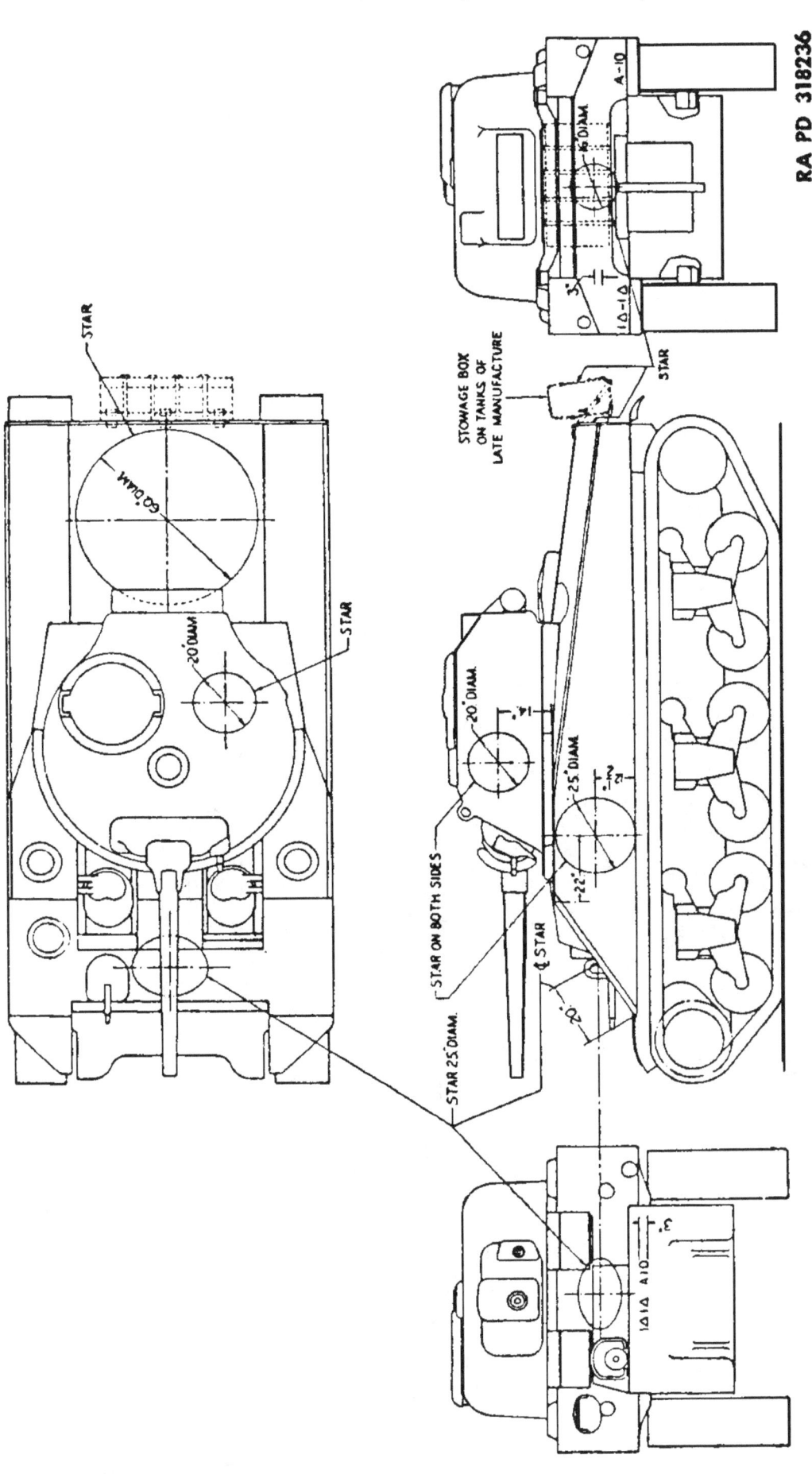

Figure 42.—Medium tank M4 series.

[A. G. 400.161 (10 Nov 43).] (C 9, 27 Jan 44.)

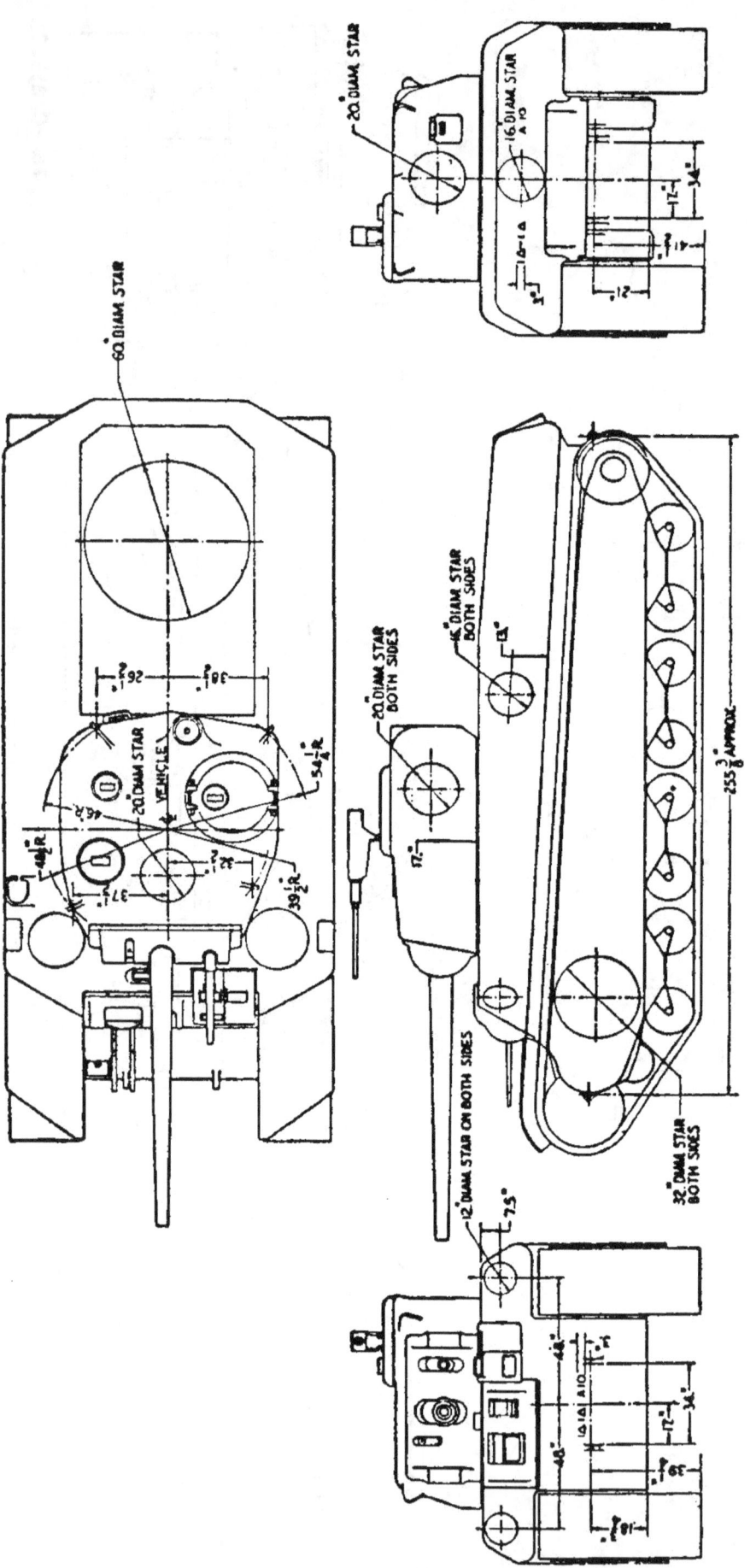

Figure 43.—Heavy tank M6.

MARKING OF CLOTHING, EQUIPMENT, ETC.

AR 850–5
C 9

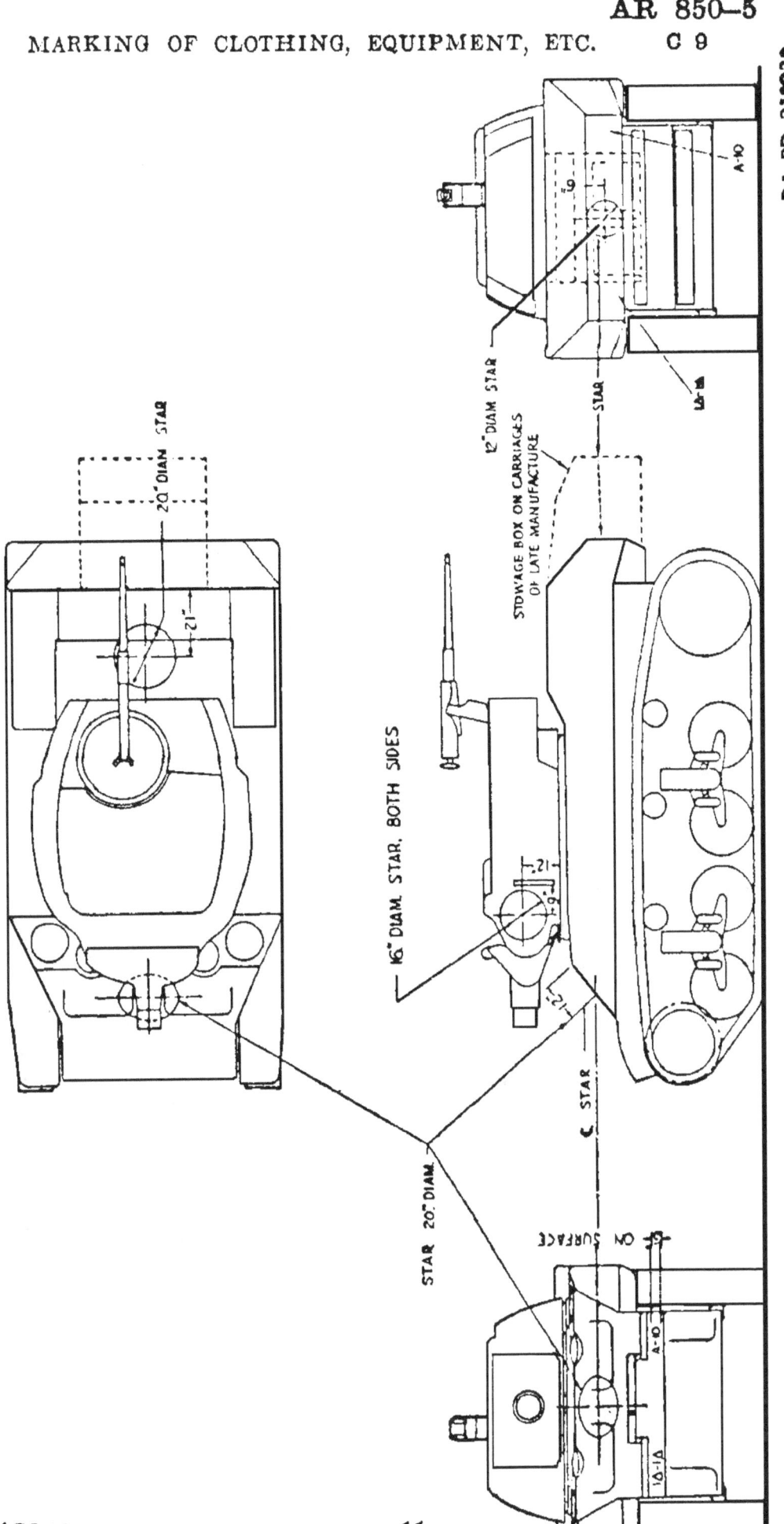

Figure 44.—75-mm howitzer motor carriage M8.

AR 850-5
C 9   MISCELLANEOUS   RA PD 318239

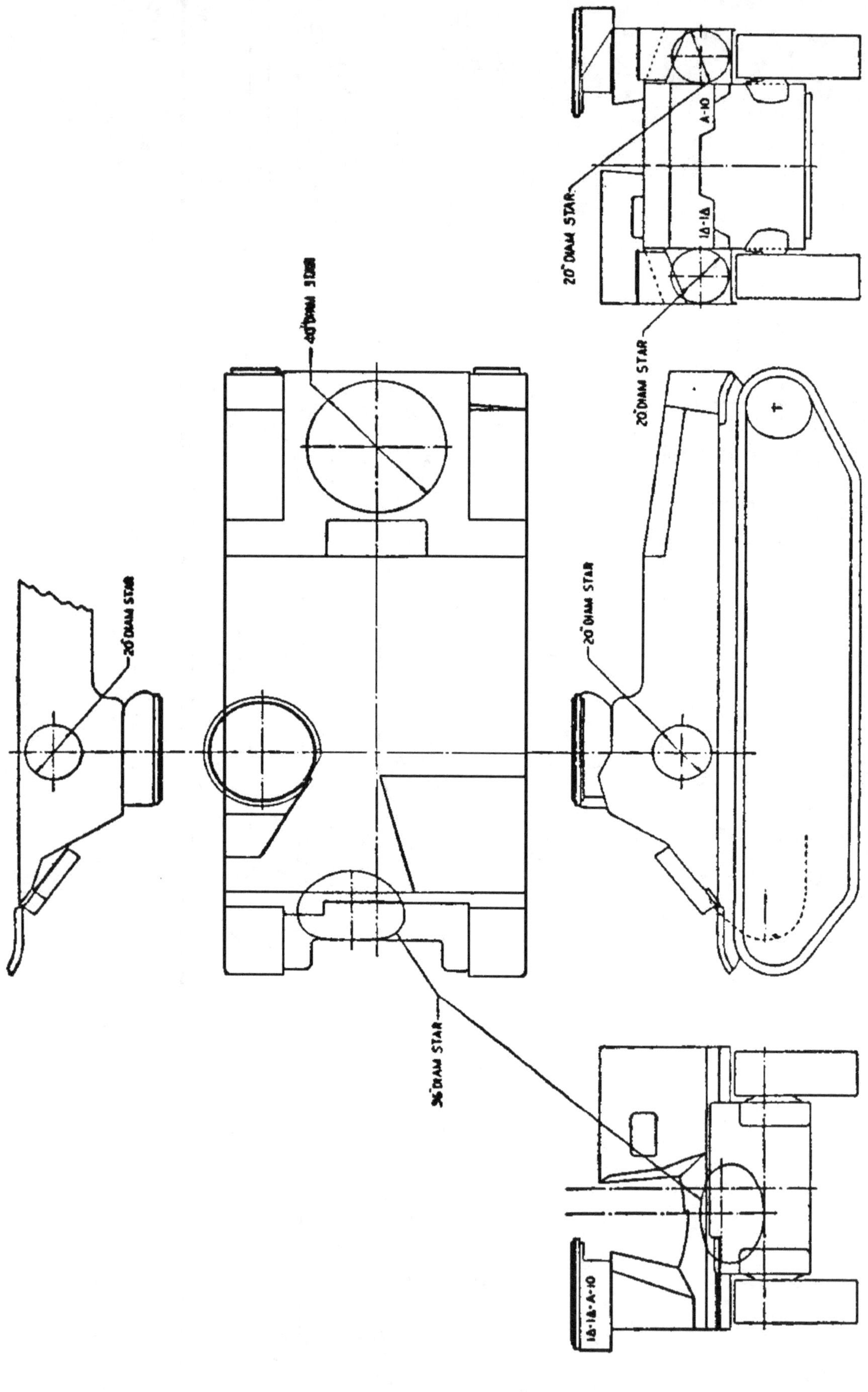

Figure 45.—105-mm howitzer motor carriage M7.

[A. G. 400.161 (10 Nov 43).]   (C 9, 27 Jan 44.)

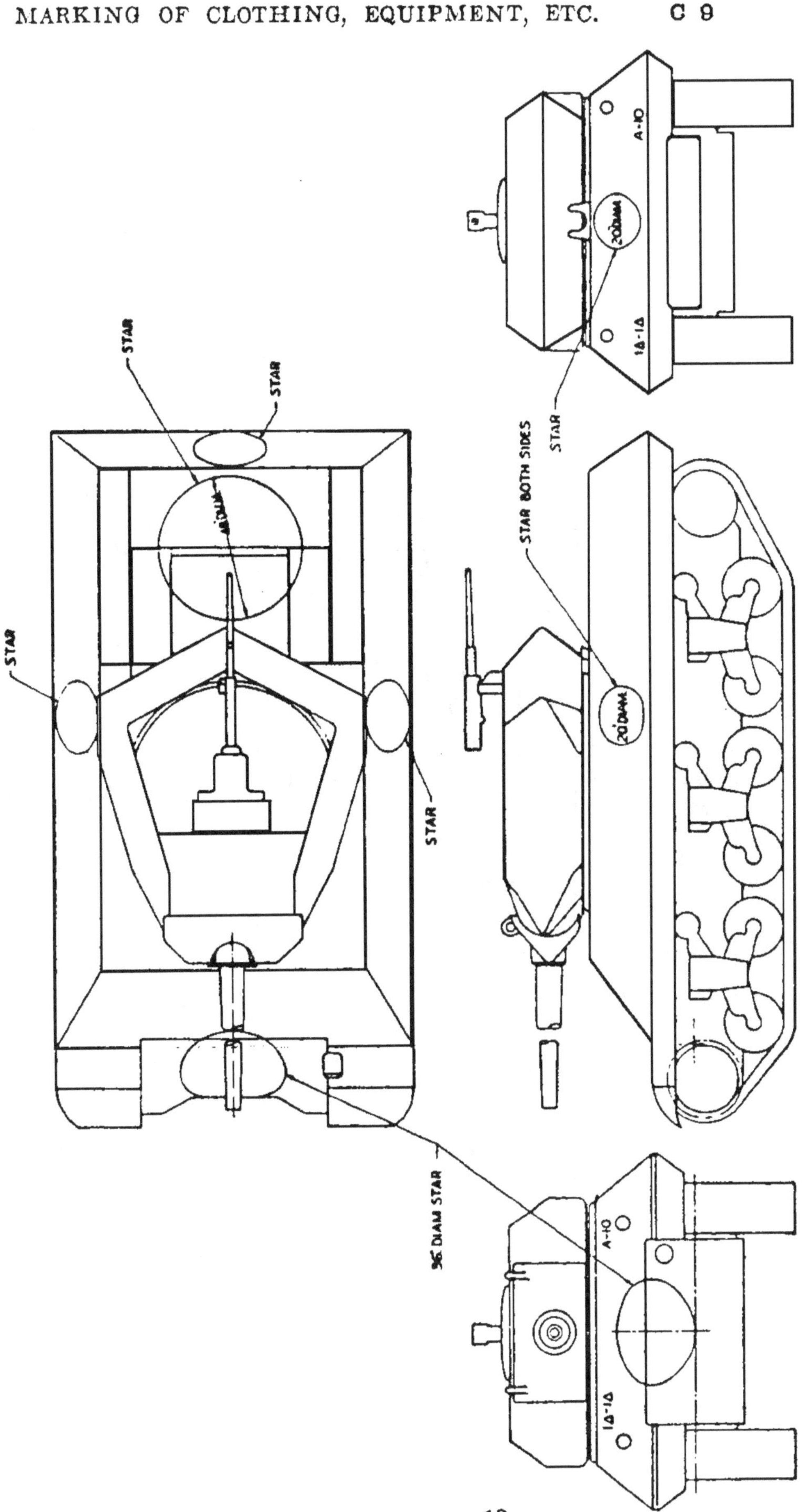

MARKING OF CLOTHING, EQUIPMENT, ETC. AR 850–5
C 9

Figure 46.—3-inch gun motor carriages M10 and M10A1.

[A. G. 400.161 (10 Nov 43).]   (C 9, 27 Jan 44.)

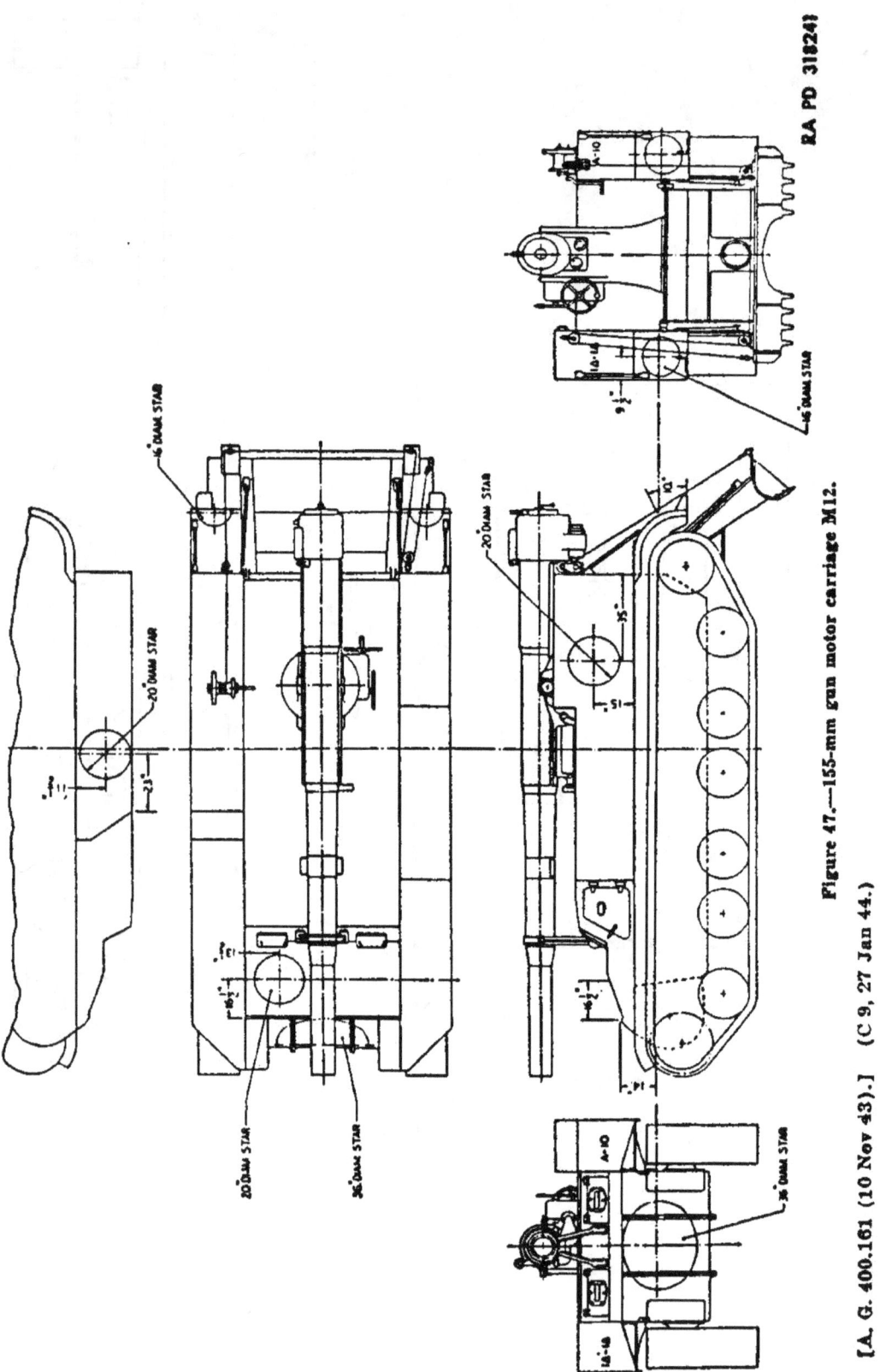

Figure 47.—155-mm gun motor carriage M12.

[A. G. 400.161 (10 Nov 43).] (C 9, 27 Jan 44.)

MARKING OF CLOTHING, EQUIPMENT, ETC. AR 850-5 C 9

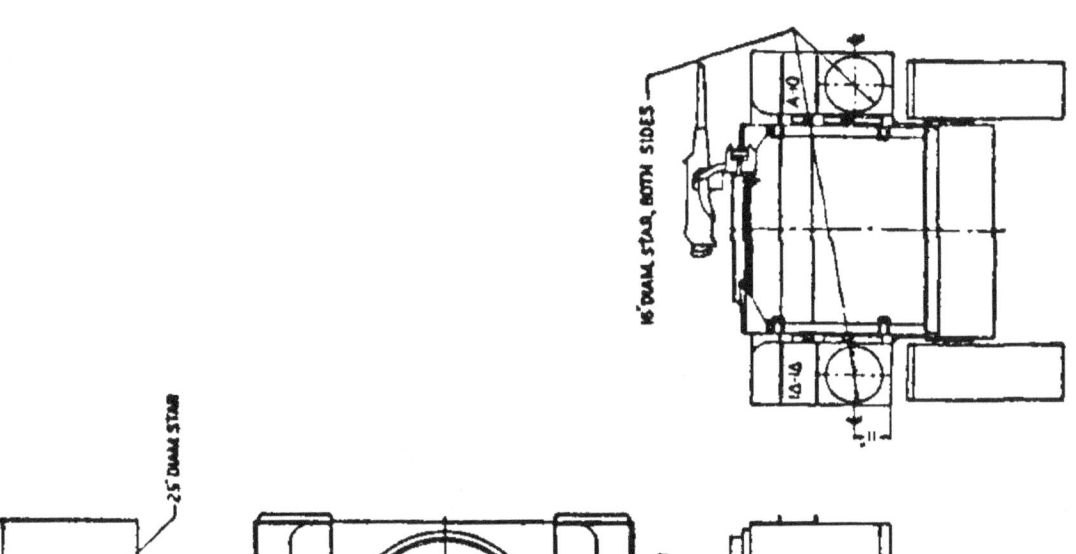
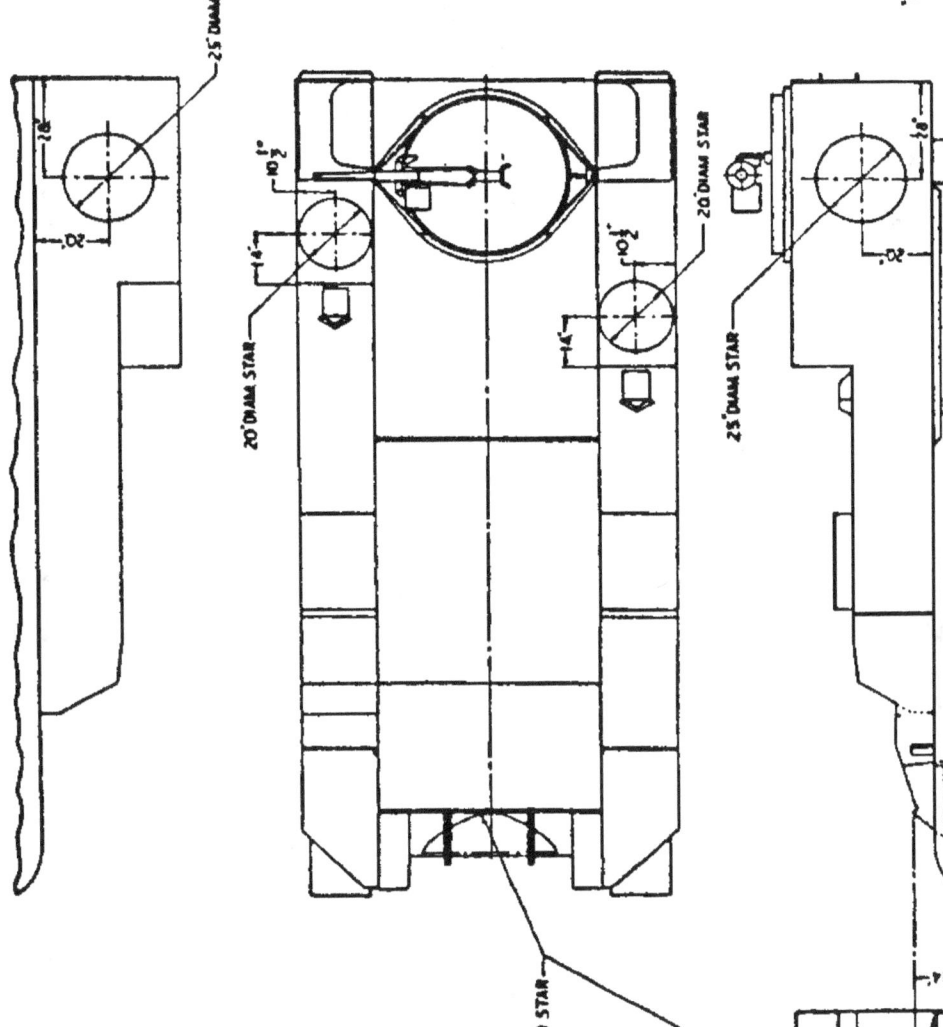

Figure 48.—Cargo carrier M30 (T14).

[A. G. 400.161 (10 Nov 43).] (C 9, 27 Jan 44.)

AR 850–5
C 9

MISCELLANEOUS

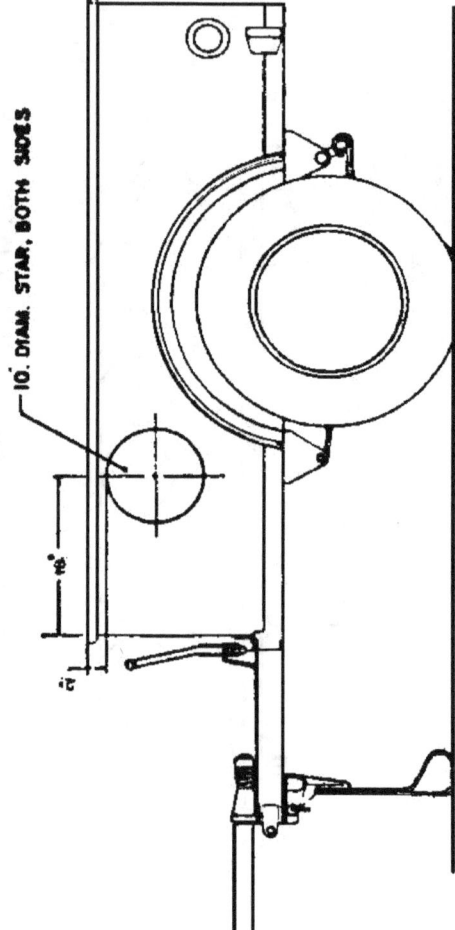

Figure 49.—¼-ton payload, 2-wheel cargo trailer.

[A. G. 400.181 (10 Nov 43).] (C 9, 27 Jan 44.)

AR 850-5

MARKING OF CLOTHING, EQUIPMENT, ETC. C 9

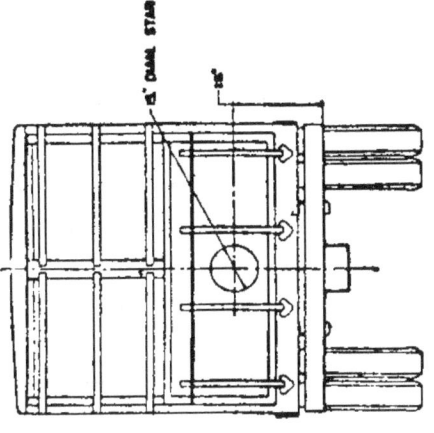
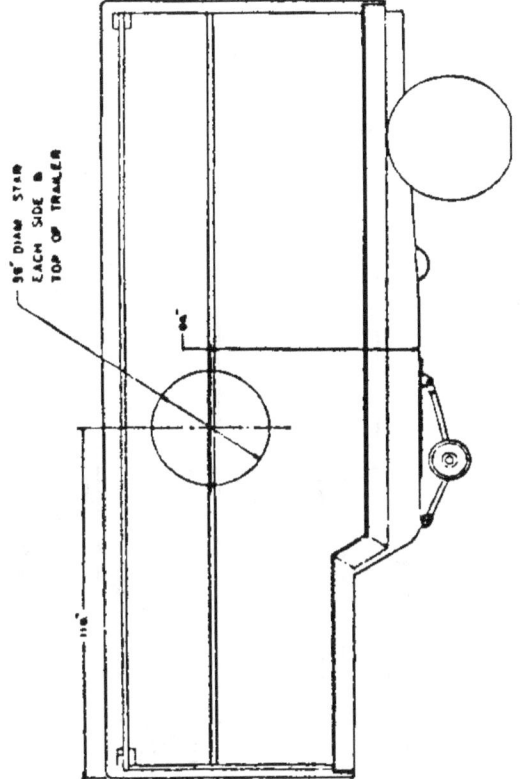
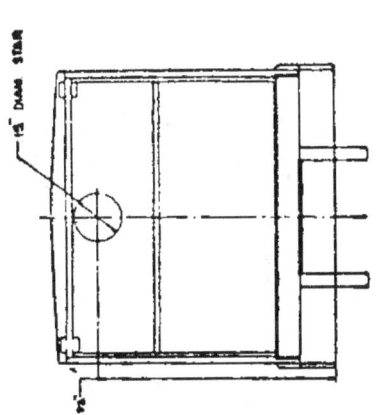

Figure 50.—6-ton payload, 10-ton gross, 2-wheel, van semitrailer.

[A. G. 400.161 (10 Nov 43).] (C 9, 27 Jun 44.)

AR 850-5
C 9
MISCELLANEOUS

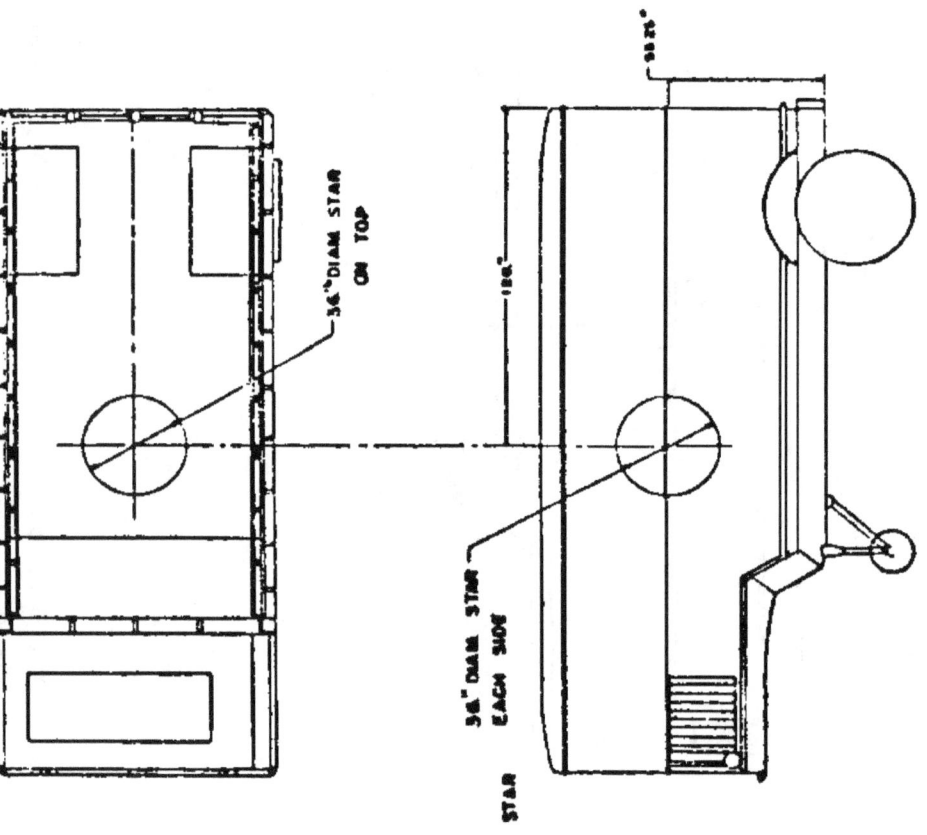

Figure 51.—5-ton payload, 10-ton gross, 2-wheel refrigerator semitrailer.

[A. G. 400.161 (10 Nov 43).]   (C 9, 27 Jun 44.)

AR 850-5
MARKING OF CLOTHING, EQUIPMENT, ETC. C 9

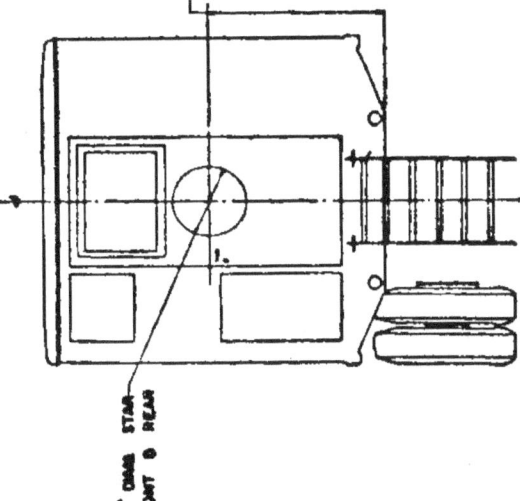

Figure 52.—6-ton payload, 10-ton gross, 2-wheel shoe, clothing and textile repair semitrailer.

[A. G. 400.161 (10 Nov 43).] (C 9, 27 Jan 44.)

AR 850-5
C 9
MISCELLANEOUS

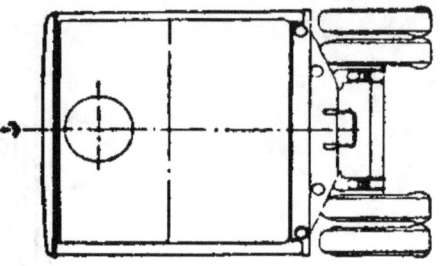

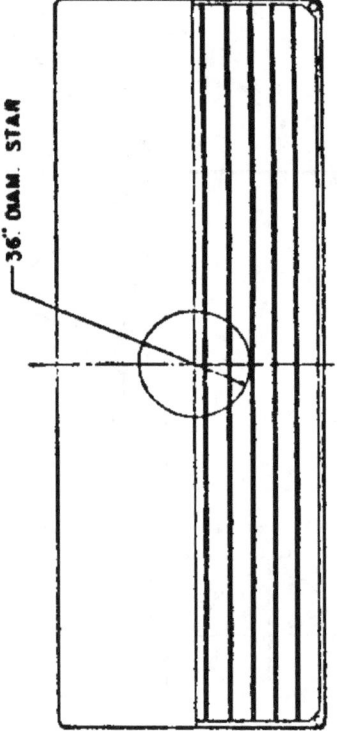

Figure 53.—6-ton payload, 10-ton gross, 2-wheel laundry and sterilizer semitrailer.

MARKING OF CLOTHING, EQUIPMENT, ETC. C 9 AR 850–5

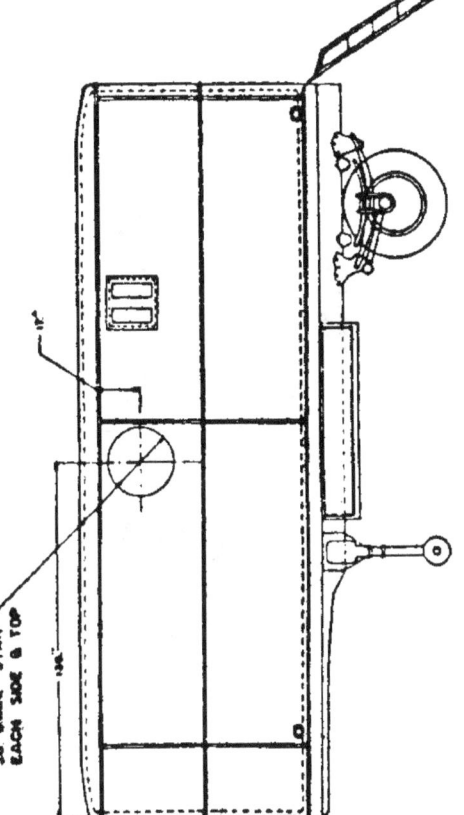

Figure 54.—6-ton payload, 10-ton gross, 2-wheel mobile records semitrailer.

[A. G. 400.161 (10 Nov 43).] (C 9, 27 Jan 44.)

AR 850-5
C 9
MISCELLANEOUS

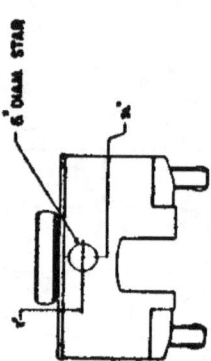

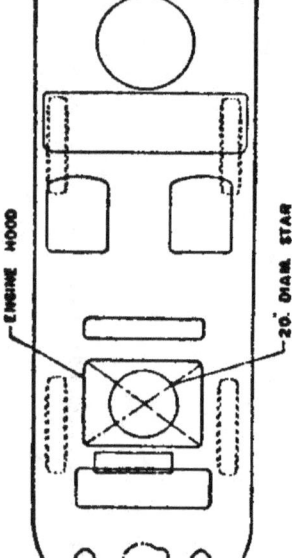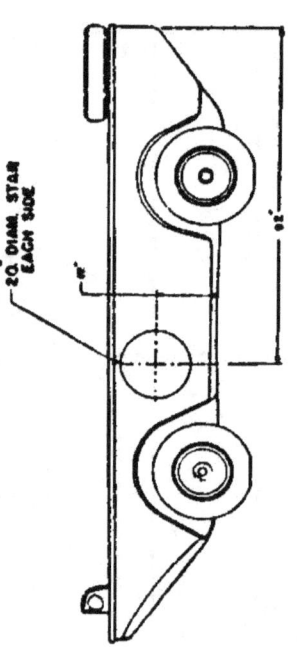

Figure 55.—¼-ton, 4 x 4 amphibian truck.

[A. G. 400.161 (10 Nov 43).] (C 9, 27 Jan 44.)

MARKINGS OF CLOTHING, EQUIPMENT, ETC.

AR 850–5
C 9

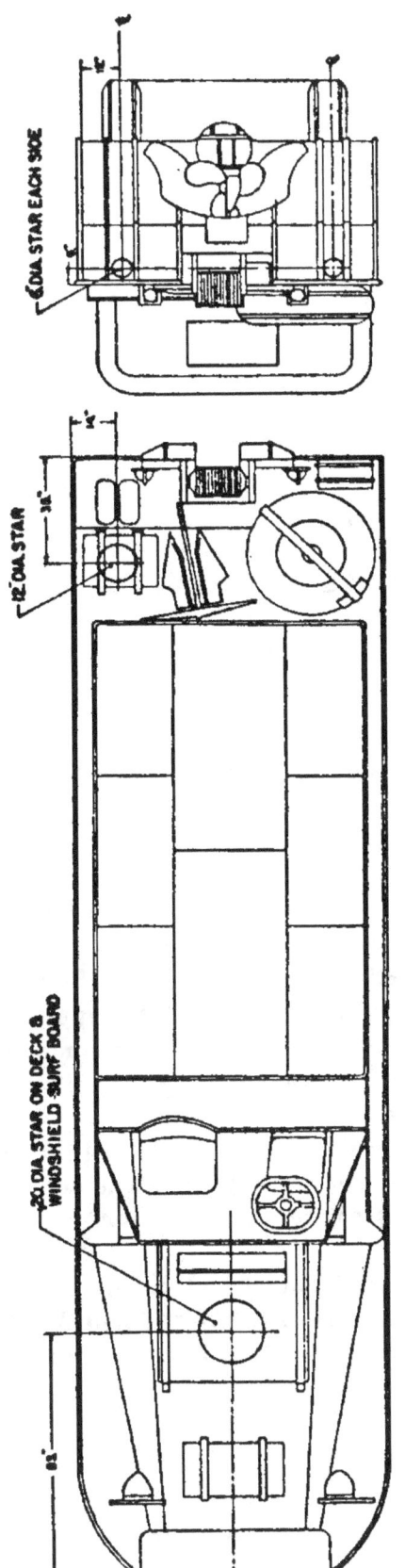
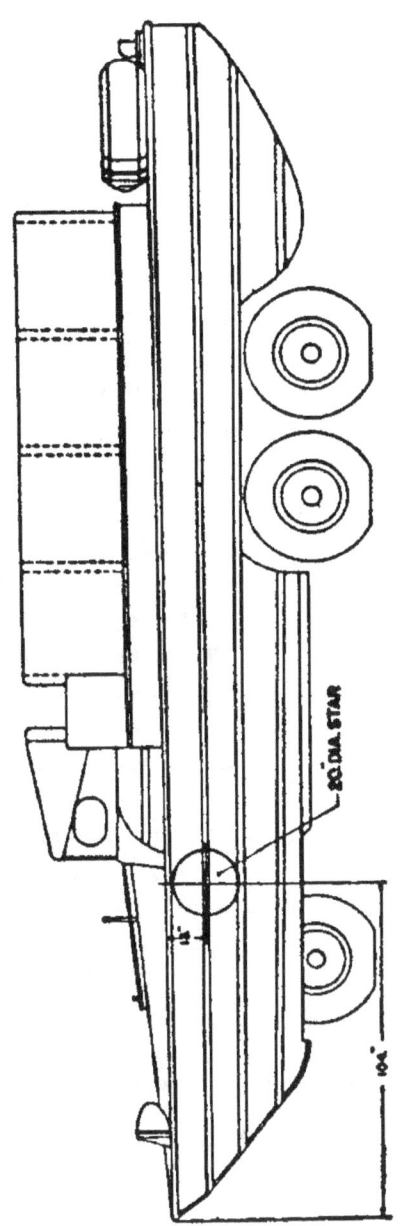

Figure 56.—2½-ton, 6 x 6 amphibian truck.

[A. G. 400.161 (10 Nov 43).]   (C 9, 27 Jan 44.)

AR 850–5
C 9   MISCELLANEOUS

By order of the Secretary of War:

G. C. MARSHALL,
Chief of Staff.

Official:
   J. A. ULIO,
      Major General,
         The Adjutant General.

Distribution:
   A; E.

MISCELLANEOUS   AR 850–5
                    C 10

MARKING OF CLOTHING, EQUIPMENT, VEHICLES, AND PROPERTY

Changes }   WAR DEPARTMENT,
No. 10 }   Washington 25, D. C., 25 March 1944.

AR 850–5, 5 August 1942, is changed as follows:

10. **Unit markings.**—Gasoline solvent paint or paint as prescribed by the War Department will be used.

a. Unit markings.—National symbol.

* * * * * * *

(4) (a) The national symbol for all Army Service Forces vehicles not included in (1) above, except contractor operated vehicles at class IV installations, will be the insignia of the respective service command and such symbol will be applied to all vehicles assigned to installations located within the territorial limits of the service command except as further provided in these regulations.

(b) All vehicles not included in (1) above which are operated by Army Air Forces installations will be identified by such symbol as may be directed by the Commanding General, Army Air Forces.

(c) All vehicles not included in (1) above which are operated by Army Ground Forces installations will be identified by such symbol as may be directed by the Commanding General, Army Ground Forces.

* * * * * * *

[A. G. 400.161 (17 Mar 44).]   (C 10, 25 Mar 44.)

By order of the Secretary of War:

G. C. MARSHALL,
Chief of Staff.

Official:
   J. A. ULIO,
      Major General,
         The Adjutant General.

Distribution:
   A; E.

www.ingramcontent.com/pod-product-compliance
Lightning Source LLC
Chambersburg PA
CBHW081341070526
44578CB00005B/689